Table of the Atomic Masses* (Based on Carbon-12)

	Symbol	Atomic Number	Atomic Mass		Symbol	Atomic Number	Atomic Mass
Actinium	Ac	89	227.0278	Mercury	Hg	80	200.59
Aluminum	Al	13	26.981539	Molybdenum	Mo	42	95.94
Americium	Am	95	[243]**	Neodymium	Nd	60	144.24
Antimony	Sb	51	121.757	Neon	Ne	10	20.1797
Argon	Ar	18	39.948	Neptunium	Np	93	237.0482
Arsenic	As	33	74.92159	Nickel	Ni	28	58.6934
Astatine	At	85	[210]	Niobium	Nb	41	92.90638
Barium	Ba	56	137.327	Nitrogen	N	7	14.00674
Berkelium	Bk	97	[247]	Nobelium	No	102	[259]
Beryllium	Be	4	9.012182	Osmium	Os	76	190.2
Bismuth	Bi	83	208.98037	Oxygen	O	8	15.9994
Boron	B	5	10.811	Palladium	Pd	46	106.42
Bromine	Br	35	79.904	Phosphorus	P	15	30.973762
Cadmium	Cd	48	112.411	Platinum	Pt	78	195.08
Calcium	Ca	20	40.078	Plutonium	Pu	94	[244]
Californium	Cf	98	[251]	Polonium	Po	84	[210]
Carbon	C	6	12.011	Potassium	K	19	39.0983
Cerium	Ce	58	140.115	Praseodymium	Pr	59	140.90765
Cesium	Cs	55	132.90543	Promethium	Pm	61	[145]
Chlorine	Cl	17	35.4527	Protactinium	Pa	91	231.03588
Chromium	Cr	24	51.9961	Radium	Ra	88	226.0254
Cobalt	Co	27	58.93320	Radon	Rn	86	[220]
Copper	Cu	29	63.546	Rhenium	Re	75	186.207
Curium	Cm	96	[247]	Rhodium	Rh	45	102.90550
Dysprosium	Dy	66	162.50	Rubidium	Rb	37	85.4678
Einsteinium	Es	99	[252]	Ruthenium	Ru	44	101.07
Erbium	Er	68	167.26	Samarium	Sm	62	150.36
Europium	Eu	63	151.965	Scandium	Sc	21	44.9559
Fermium	Fm	100	[257]	Selenium	Se	34	78.96
Fluorine	F	9	18.9984032	Silicon	Si	14	28.0855
Francium	Fr	87	[223]	Silver	Ag	47	107.8682
Gadolinum	Gd	64	157.25	Sodium	Na	11	22.989768
Gallium	Ga	31	69.723	Strontium	Sr	38	87.62
Germanium	Ge	32	72.61	Sulfur	S	16	32.066
Gold	Au	79	196.96654	Tantalum	Ta	73	180.9479
Hafnium	Hf	72	178.49	Technetium	Tc	43	[98]
Helium	He	2	4.002602	Tellurium	Te	52	127.60
Holmium	Ho	67	164.93032	Terbium	Tb	65	158.92534
Hydrogen	H	1	1.007944	Thallium	Tl	81	204.3833
Indium	In	49	114.82	Thorium	Th	90	232.0381
Iodine	I	53	126.90447	Thulium	Tm	69	168.93421
Iridium	Ir	77	192.22	Tin	Sn	50	118.710
Iron	Fe	26	55.847	Titanium	Ti	22	47.88
Krypton	Kr	36	83.80	Tungsten	W	74	183.85
Lanthanum	La	57	138.9055	Uranium	U	92	238.0289
Lawrencium	Lr	103	[262]	Vanadium	V	23	50.9415
Lead	Pb	82	207.2	Xenon	Xe	54	131.29
Lithium	Li	3	6.941	Ytterbium	Yb	70	173.04
Lutetium	Lu	71	174.967	Yttrium	Y	39	88.90585
Magnesium	Mg	12	24.3050	Zinc	Zn	30	65.39
Manganese	Mn	25	54.93805	Zirconium	Zr	40	91.224
Mendelevium	Md	101	[256]				

* Atomic Masses given here are 1989 IUPAC values.

** A Value given in brackets denotes the mass number of the longest-lived isotope.

CHEMISTRY

Principles and Reactions

Second Edition

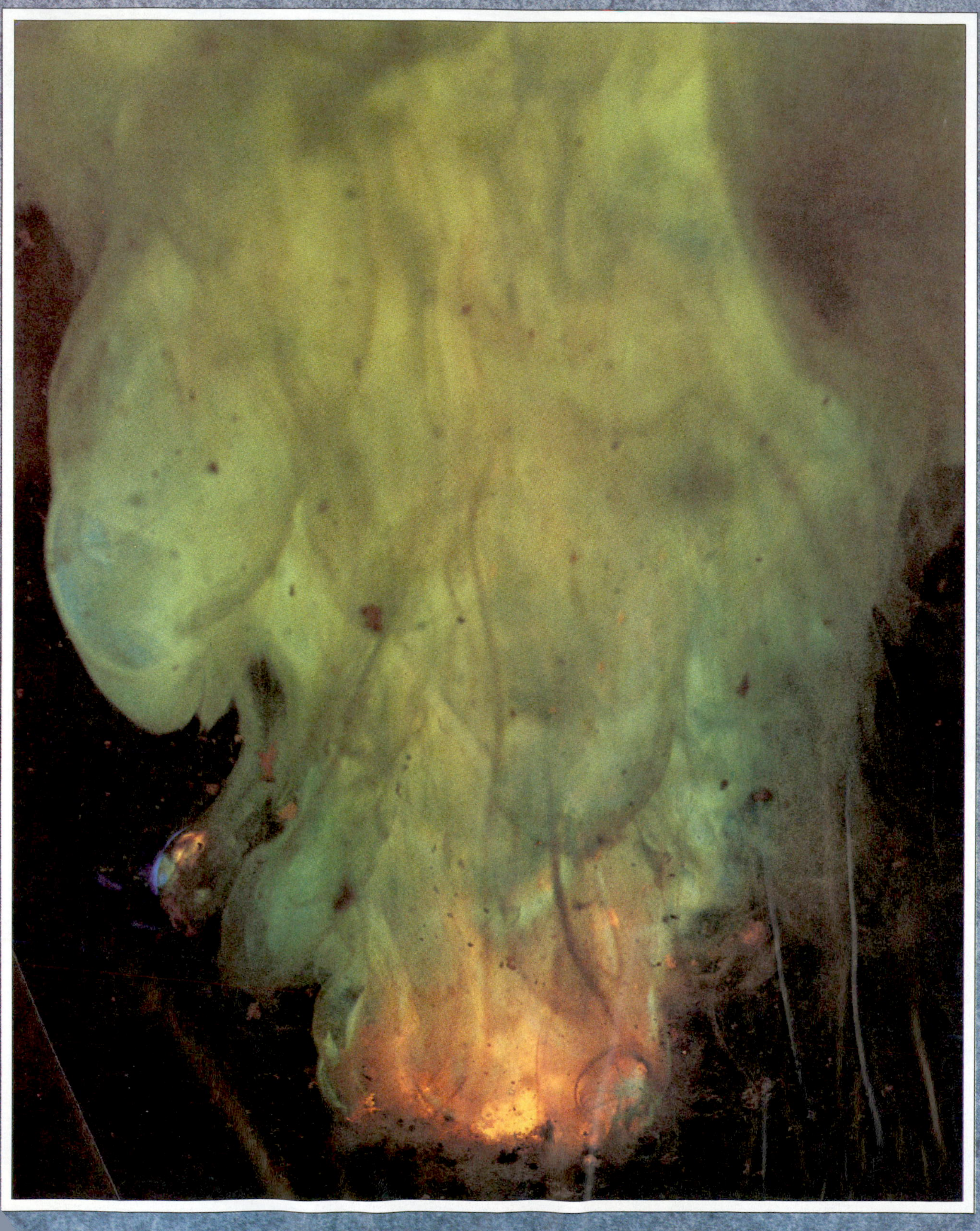

CHEMISTRY

Principles and Reactions

Second Edition

William L. Masterton
University of Connecticut

Cecile N. Hurley
University of Connecticut

Saunders Golden Sunburst Series

Saunders College Publishing

Harcourt Brace College Publishers

Fort Worth Philadelphia
San Diego New York
Orlando Austin
San Antonio Toronto
Montreal London
Sydney Tokyo

Text Typeface: Cheltenham Light
Compositor: York Graphic Services
Publisher: John Vondeling
Developmental Editor: Elizabeth Rosato
Managing Editor: Carol Field
Senior Project Manager: Marc Sherman
Copy Editor: Rebecca Gruliow
Manager of Art and Design: Carol Bleistine
Art & Design Coordinator: Caroline McGowan
Text Designer: Tracy Baldwin
Layout Artist: Dorothy Chattin
Cover Designer: Lawrence R. Didona
Director of EDP: Tim Frelick
Production Manager: Charlene Squibb
Marketing Manager: Marjorie Waldron

Cover Credit: Copper wire heated over a bunsen burner, by Charles D. Winters.

CHEMISTRY: PRINCIPLES & REACTIONS, second edition
0-03-074609-4

Library of Congress Catalog Number: 92-050140

Printed in the United States of America

345 032 98765432

To Loris and Jim

For being both buffer and catalyst

Preface

Our goal in writing the second edition of *Chemistry: Principles and Reactions* was to produce a book of about 700 pages which could reasonably be covered in its entirety in the mainstream year course in general chemistry. We wanted to achieve this without deleting any fundamental topics; beyond that, we were determined to include enough explanatory material and applications to make chemistry intelligible and interesting to the student. Somewhat to our surprise, we achieved our goal of brevity; this book contains 21 chapters with an honest page count of 679. Enthusiastic reviewers tell us that we have indeed written a user-friendly text covering all the basics. Naturally we agree with them, but you, and particularly your students, will be the final judge of how well this book is written.

CHANGES IN THE SECOND EDITION

To reduce the length of the text, we sought to eliminate repetition and duplication wherever possible. For example, the book contains

— one and only one method of balancing redox equations, the half-equation approach introduced in Chapter 4.
— one and only one equilibrium constant for gas-phase reactions (Chapter 12), the thermodynamic constant K, commonly referred to as K_p. This simplifies and clarifies not only the treatment of gaseous equilibrium, but also the discussion of reaction spontaneity (Chapter 16) and electrochemistry (Chapter 17).
— one and only one bonding model for complex ions (Chapter 15). This is the crystal-field model, the basis for all modern treatments of the structure of coordination compounds.

Nearly all of the topics covered in the typical general chemistry text are found in this book, with a couple of notable exceptions.

— **molecular orbitals** A great many people have suggested deleting this material from general chemistry texts; we did it. It is our feeling that MO theory simply doesn't come across to first year students; they don't have the background to appreciate its importance.
— **biochemistry,** traditionally covered in the last chapter of every general chemistry text. Fascinating as this material is, it requires an understanding of organic chemistry that few students have. Our last chapter (Chapter 21) is devoted to organic chemistry, with emphasis on practical applications such as synthetic polymers.

Several topics discussed at some length in our first edition have been condensed in this revision. These include:

— **atmospheric chemistry,** eliminated as a separate chapter. Several relevant topics in this area now appear as end-of-chapter perspectives (the greenhouse effect in Chapter 5, the ozone story in Chapter 11, acid rain in Chapter 12).

— **qualitative analysis,** reduced from one chapter to one section (Section 19.4) which is flexible enough to be covered whenever students start cation analysis in the laboratory.
— **descriptive inorganic chemistry,** used throughout the book to illustrate chemical principles, is covered formally in two chapters (Chapter 19 on metals, Chapter 20 on nonmetals). This book is short enough so that both of these chapters can easily be covered. To help you do that, we've rewritten all the descriptive material to make it easier to teach, more interesting and less encyclopedic.

We believe that the selective pruning just described has produced a text of manageable length that retains all the basic topics of general chemistry. However, opinions will differ on that matter; what seems peripheral to us may be basic to you. Consequently, we have arranged with our publisher to make available to your students, selected topics from our 1000-page first edition. Essentially, this amounts to custom publishing; the basic core package (this text) can be supplemented with material from the publisher. Your local Saunders representative can furnish further details.

ORGANIZATION

As you can see from the Table of Contents on p. xxi, the organization of this text is more or less conventional. A few features are worthy of comment.

— Chapter 4 (Reactions in Aqueous Solutions) shows students how to write, balance and apply net ionic equations for precipitation, acid-base, and redox reactions. It seems reasonable to introduce this fundamental material early; among other things it establishes a background for a meaningful first-semester laboratory. If you prefer, this material can be postponed until Chapter 10, where the physical properties of solutions are covered.
— the two factors that determine reaction feasibility, rate and equilibrium, are presented back-to-back in Chapters 11 and 12. This seems a logical starting point for the second half of the year course.
— a descriptive treatment of coordination chemistry (Chapter 15) is presented somewhat earlier than is customary. This is our subtle way of encouraging you to spend time on an area of inorganic chemistry that students find fascinating albeit challenging.

FEATURES

Returning to the theme of the opening paragraph, let us see how this book is designed to stimulate student interest in chemistry. First, and most important, brief discussions of current research have been integrated into the text; see, for example, the reference to "buckyball" on p. 238. Beyond that, there are:

— **end-of-chapter perspectives,** which apply chemical principles to the world around us. Several of these deal with the environment (Chaps. 5, 11, 12, 16, 18), others with nutrition (Chaps. 8, 19, 20). One that we strongly recommend discusses the production of maple syrup (Chap. 10; note particularly Figure 10.A).

— **chapter opening photographs and quotations,** chosen by CNH to show that chemistry is not as far removed from abstract art and literature as most people think.
— **historical boxes** that discuss the human qualities as well as the scientific accomplishments of some of the giants of chemistry, including G. N. Lewis (p. 169), J. Willard Gibbs (p. 442), and Marie Curie (p. 499).

The many learning aids incorporated into this text are discussed on p. xiii under the heading "To The Student." Three of these are new to this edition.

1. Each in-text Example includes a "Strategy" section outlining the reasoning behind the solution. This helps the student distinguish the "why" from the "how" of problem solving.
2. Each chapter ends with a "Highlights" section which includes key concepts, terms, and equations introduced in the chapter. The section closes with a Summary Problem designed to tie together all the major concepts.
3. Appendix 5 correlates and summarizes material on net ionic equations, spread over several chapters of the text. Typically, students have a great deal of trouble with this topic; we are sure this appendix will help.

ANCILLARY PACKAGE

A complete resource package has been prepared to enhance the student's learning and the professor's teaching of *Chemistry: Principles and Reactions,* 2/e.

Instructor's Manual by William L. Masterton. Included are lecture outlines and quizzes for discussion sections. Each chapter is accompanied by a list of references to appropriate demonstrations, emphasizing Volumes 1–4 of the Shakhashiri handbook and videotapes 1–3 (see below). Worked-out solutions are provided for all the end-of-chapter problems not answered in the text.

Study Guide/Workbook by Cecile N. Hurley. Worked examples and problem-solving techniques help the student understand the principles of general chemistry. Each chapter is outlined for the student with fill-in-the-blanks, and exercises and self-tests allow the students to gauge their mastery of the chapter.

Student Solutions Manual by John E. Bauman (University of Missouri). Detailed solutions to all the problems answered in the text, including the Challenge Problems.

Overhead Transparencies One hundred four-color figures from the text.

Test Bank and Computerized Test Bank Over 1000 multiple choice test bank questions reviewed by the text authors. Available in computerized versions for IBM and Macintosh computers.

Lecture Outline by Ronald O. Ragsdale (University of Utah). Organized to follow class lectures to free students from extensive note taking.

Chemical Principles in the Laboratory, 5/e, by Emil J. Slowinski, Wayne C. Wolsey and William L. Masterton. The industry's best-selling general chemistry laboratory manual includes 43 experiments, each with a pre-lab study assignment. An instructor's manual is also available.

Audiotape Lessons and Workbook, 2/e, by Bassam Shakhashiri, Rodney Schreiner and Phyllis Anderson Meyer (all of the University of Wisconsin–Madison). Enables students to learn chemistry at their own pace. Students listen to instructions on the tape and follow the examples in the workbook.

Shakhashiri Videotapes by Bassam Shakhashiri (University of Wisconsin–Madison). Fifty 3–5 minute classroom experiments.

Videodisc and Barcode Manual contains all the Shakhashiri demonstrations and over 600 images drawn from various Saunders sources. Barcode manual allows easy access.

Periodic Table Videodisc: Reactions of the Elements by Alton J. Banks (Southwest Texas State University). Also published by JCE: Software. A visual compilation of information about chemical elements, their uses and their reactions with air, water, acids and bases. It is particularly useful as a way to demonstrate chemical reactions in a large lecture room.

Saunders Chemical Update Newsletters

ACKNOWLEDGMENTS

In this revision we have been guided by comments and suggestions from students and colleagues. We are indebted to the many reviewers who devoted so much time to this project, earning somewhat less than the minimum wage. These include:

Deborah M. Nycz, *Broward Community College*
Charles A. Wilkie, *Marquette University*
Pushkar Kaul, *Boston College*
Virgil L. Payne, *Florida Atlantic University*
Paul B. Kelter, *The University of Wisconsin–Oshkosh*
John E. Bauman, *University of Missouri–Columbia*
James D. Cherry, *Enrico Fermi High School*
Paul S. Poskozim, *Northeastern Illinois University*
William E. Ohnesorge, *Lehigh University*
T. W. Sottery, *University of Southern Maine*
Wyman K. Grindstaff, *Southwest Missouri State University*
Elizabeth S. Friedman, *Los Angeles Valley College*
Douglas W. Hensley, *Louisiana Tech University*
Leslie Kinsland, *University of Southwestern Louisiana*
Robert Conley, *New Jersey Institute of Technology*
Donald Titus, *Temple University*
Sidney H. Young, *University of South Alabama*
William Van Doorne, *Calvin College*
Jesse Binford, *University of South Florida*
Michael Kenney, *Marquette University*
Ronald Ragsdale, *University of Utah*
Coran Cluff, *Brigham Young University*
Peter Baine, *California State University, Long Beach*
Stanley M. Williamson, *University of California*
Frederick A. Grimm, *The University of Tennessee–Knoxville*

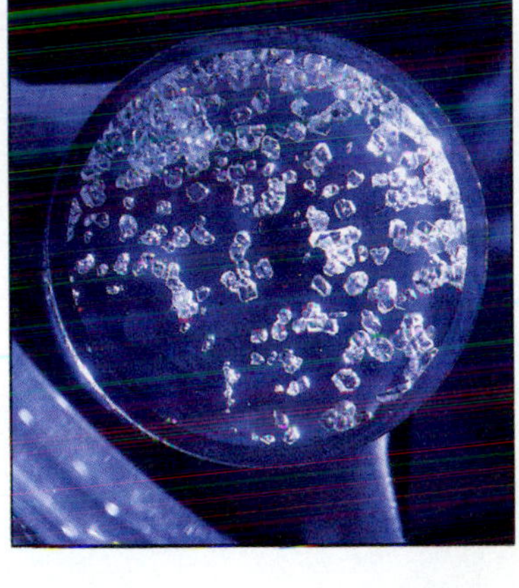

Daniel T. Haworth, *Marquette University*
Grant N. Holder, *Appalachian State University*
David Hilderbrand, *South Dakota State University*
M. Davis, *University of Texas–El Paso*
Richard L. Snow, *Brigham Young University*
Steven D. Gammon, *University of Idaho*

Many people at Saunders made major contributions to this book. **John Vondeling,** publisher, supported our efforts from the start, knowing full well that an innovative text such as this one is a risky investment. Perhaps the smartest move he made was to persuade **Mary Castellion,** a distinguished author in her own right, to help us develop the manuscript. Mary taught us a great deal about writing (really somewhat more than we wanted to know). **Elizabeth Rosato,** our developmental editor, used her knowledge of chemistry to do everything we asked of her, no matter how unreasonable. The greatest compliment we can pay Beth is to thank her for making the task of writing this book a (relatively) pleasant one.

William L. Masterton
Cecile N. Hurley
Storrs, Connecticut
June 1992

To the Student

Over the next several months, you will probably receive a lot of advice from your instructor, teaching assistant, and fellow students about how to study chemistry. We hesitate to add our advice; experience as teachers and parents has taught us that students tend to do surprisingly well without it. We would, however, like to acquaint you with some of the learning tools in this text. They are described and illustrated in the pages that follow.

Chapter Opening Photograph and Quotation

Chosen by CNH to show that chemistry is not as far removed from abstract art and literature as most people think, these illustrate, in a somewhat abstract way, chemical reactions of one type or another. Each photograph is accompanied by a chemical equation and a page reference identifying where in the text the reaction is discussed.

YOAV LEVY/PHOTAKE NYC

$$Cu(s) + 2Ag^{+}(aq) \longrightarrow Cu^{2+}(aq) + 2Ag(s)$$

(See p. 491)

8

Thermochemistry

Some say the world will end
in fire,
Some say in ice.
From what I've tasted of
desire
I hold with those
who favour fire.

ROBERT FROST
Fire and Ice

Example 3.12 Ammonia used to make fertilizers for lawns and gardens is made by reacting nitrogen of the air with hydrogen. The balanced equation for the reaction is

$$3H_2(g) + N_2(g) \longrightarrow 2NH_3(g)$$

Determine

(a) the mass in grams of ammonia, NH_3, formed when 1.34 mol N_2 reacts.
(b) the mass in grams of N_2 required to form 1.00 kg NH_3.
(c) the mass in grams of H_2 required to react with 6.00 g N_2.

To go from moles of A to mass of A, use molar mass

Strategy In each case, you use the mole ratios given by the coefficients of the balanced equation to relate moles of one substance to moles of another. Beyond that, use molar mass to relate moles of one substance to mass in grams of that substance. Before starting, decide upon the path you will follow to go from the quantity given to that required, i.e.,

(a) $n_{N_2} \longrightarrow n_{NH_3} \longrightarrow$ mass of NH_3
(b) mass of $NH_3 \longrightarrow n_{NH_3} \longrightarrow n_{N_2} \longrightarrow$ mass of N_2
(c) mass of $N_2 \longrightarrow n_{N_2} \longrightarrow n_{H_2} \longrightarrow$ mass of H_2

To go from moles of A to moles of B, use the coefficients of the balanced equation

Conversions indicated by colored arrows involve mole ratios given by the coefficients of the balanced equation. The other conversions require only molar masses and are essentially identical to those carried out in Examples 3.4 and 3.5.

Solution

$$\text{(a) mass of } NH_3 = 1.34 \text{ mol } N_2 \times \frac{2 \text{ mol } NH_3}{1 \text{ mol } N_2} \times \frac{17.03 \text{ g } NH_3}{1 \text{ mol } NH_3} = \boxed{45.6 \text{ g } NH_3}$$

$$\text{(b) mass of } N_2 = 1000 \text{ g } NH_3 \times \frac{1 \text{ mol } NH_3}{17.03 \text{ g } NH_3} \times \frac{1 \text{ mol } N_2}{2 \text{ mol } NH_3} \times \frac{28.02 \text{ g } N_2}{1 \text{ mol } N_2} = \boxed{823 \text{ g } N_2}$$

$$\text{(c) mass of } H_2 = 6.00 \text{ g } N_2 \times \frac{1 \text{ mol } N_2}{28.02 \text{ g } N_2} \times \frac{3 \text{ mol } H_2}{1 \text{ mol } N_2} \times \frac{2.016 \text{ g } H_2}{1 \text{ mol } H_2} = \boxed{1.30 \text{ g } H_2}$$

Examples

In a typical chapter, you will find 10 or more examples, each designed to illustrate a particular principle. These have answers, screened in yellow. More important, they contain a strategy statement which describes the reasoning behind the solution. You will find it helpful to get into the habit of working all problems this way. First, spend a few moments deciding how the problem should be solved. Then, and only then, set up the arithmetic to solve it.

Marginal Notes

Sprinkled throughout the text are a number of short notes that have been placed in the margin. Many of these are of the "now, hear this" variety; a few bring you up to date on current research in chemistry, in progress when the book was written. Some, probably fewer than we think, are humorous.

Chapter Highlights

At the end of each chapter, you will find a brief review of the material covered in that chapter. The "Chapter Highlights" include:

— the **key concepts** introduced in the chapter. These are indexed to the corresponding examples and end-of-chapter problems. If you have trouble working a particular problem, it may help to go back and re-read the example that covers the same concept.

— the **key equations** and **key terms** in the chapter. If a particular term is unfamiliar to you, refer to the index at the back of the book. You will find the term defined in a glossary incorporated into the index.

— a **summary problem,** covering all or nearly all of the key concepts in the chapter. You can test your understanding of the chapter by working this problem; you may wish to do this as part of your preparation for examinations. A major advantage of a summary problem is that it ties together many different ideas, showing how they correlate with one another.

CHAPTER HIGHLIGHTS

KEY CONCEPTS

1. *Convert between °F, °C, and K*
   (Example 1.1; Problems 5–8, 58)
2. *Determine the number of significant figures in a measured quantity*
   (Example 1.2; Problems 13, 14)
3. *Determine the number of significant figures in a calculated quantity*
   (Examples 1.3, 1.4; Problems 15–18)
4. *Use conversion factors to change the units of a measured quantity*
   (Examples 1.5, 1.6, 1.8; Problems 11, 12, 21–32, 43, 44, 59, 61)
5. *Relate density to mass and volume*
   (Example 1.7; Problems 35–42, 53–57, 60)

KEY EQUATIONS

$t_{°F} = 1.8t_{°C} + 32°$ $\qquad$ $T_K = t_{°C} + 273.15$

KEY TERMS

boiling point	kilo-	—chemical
centi-	melting point	—extensive
compound	milli-	—intensive
conversion factor	mixture	—physical
density	nano-	significant figure
element	property	solution
joule		

SUMMARY PROBLEM

Potassium dichromate is a reddish-orange compound containing the three elements potassium, chromium, and oxygen. It has a density of 2.68 g/cm^3; its melting point is 398°C. At 20°C, its solubility is 12 g/100 g water; at 100°C, the solubility is 80 g/100 g water.

a. What are the symbols of the three elements present in potassium dichromate?
b. List the physical properties for potassium dichromate given above.
c. What is the volume of a sample of potassium dichromate weighing 32.349 g?
d. Express the density in pounds per cubic foot.
e. Express the melting point of potassium dichromate in °F and K.
f. How much water at 100°C is required to dissolve 88 g of potassium dichromate?
g. When the solution in (f) is cooled to 20°C, how much potassium dichromate remains in solution? How much crystallizes out?

Express all your answers to the correct number of significant figures; use the conversion factor approach throughout.

Answers

a. K, Cr, O
b. density, melting point, solubility
c. 12.1 cm^3 **d.** 167 lb/ft^3 **e.** 748°F; 671 K
f. 1.1×10^2 g **g.** 13 g, 75 g

Questions and Problems

At the end of each chapter is a set of questions/problems. Most of these are classified, that is, grouped by type under a particular heading. The classified problems are in matched pairs. The second member of each pair illustrates the same principle as the first; it is numbered in color and answered in Appendix 4. Your instructor may assign unanswered problems as homework. After these problems have been discussed, you should work the corresponding answered problems to make sure you know what's going on. Each chapter also contains a smaller number of unclassified and "challenge" problems. All of the challenge problems are answered in Appendix 4.

QUESTIONS & PROBLEMS

Symbols, Formulas, and Equations

1. Using information given in this chapter, write the chemical formula for

- **a.** ammonia.
- **b.** water.
- **c.** methane.
- **d.** hydrochloric acid.

2. Using the figures in this chapter, give the color and physical state at room temperature of

a. Cr_2O_3 **b.** CrO_3

3. Using the figures in the chapter, give the color, formula, and physical state of

- **a.** anhydrous copper sulfate.
- **b.** cobalt(II) chloride hexahydrate.

4. Using the figures in the chapter, give the color, formula, and physical state of

- **a.** copper(II) sulfate pentahydrate.
- **b.** cobalt(II) chloride tetrahydrate.

Atomic Theory and Laws

5. Who is the father of atomic theory? State in your own words the law of constant composition.

6. Who first stated the law of conservation of mass? State the law in its modern form.

7. Which of the three laws (if any) listed on p. 27 is illustrated by each of the following statements?

a. Lavoisier found that when mercury(II) oxide, HgO, decomposes, the total mass of mercury and oxygen formed equals the mass of mercury(II) oxide decomposed.

b. Analysis of the calcium carbonate, $CaCO_3$, found in the marble of Carrara, Italy, and in the stalactites of the Carlsbad Caverns of New Mexico gives the same value for the percent calcium.

c. The atom ratio of oxygen to hydrogen is twice as large in one compound as it is in another compound made up of the two elements.

d. Hydrogen occurs as a mixture of two isotopes, one of which is twice as heavy as the other.

8. Which of the three laws (if any) listed on p. 27 is illustrated by each of the following statements?

a. A cold pack has the same mass before and after the seal between the two compartments is broken.

b. It is highly improbable that the formula for carbon monoxide gas found in London, England, is $C_{1.1}O_{2.5}$.

c. The mass of phosphorus, P, combined with one gram of hydrogen, H, in the highly toxic gas phosphine, experiment, 3.56 g of magnesium ribbon is completely consumed in reacting with 7.00 g of oxygen to produce 5.93 g of magnesium oxide; some oxygen remains unreacted. In a second experiment, 2.50 g of magnesium ribbon reacts with 1.10 g of oxygen gas. This time, all the oxygen is consumed; some unreacted magnesium remains and 2.75 g of magnesium oxide is produced. Show that these results are consistent with the law of constant composition.

10. Mercury(II) oxide, a red powder, can be decomposed by heating to produce liquid mercury and oxygen gas. When a sample of this compound is decomposed, 3.87 g of oxygen and 48.43 g of mercury are produced. In a second experiment, 15.68 g of mercury is allowed to react with an excess of oxygen; 16.93 g of red mercury(II) oxide is produced. Show that these results are consistent with the law of constant composition.

Nuclear Symbols and Isotopes

11. Who discovered the electron? Describe the experiment that led to the deduction that electrons are negatively charged particles.

12. Who discovered the nucleus? Describe the experiment that led to this discovery.

13. Studies show that there is an inverse relationship between the selenium content of the blood and the incidence of breast cancer in women. ${}^{80}_{34}Se$ is the most abundant form of naturally occurring selenium. How many protons are there in an Se-80 atom? How many neutrons?

14. The eruption of Mount St. Helens in Washington state produced a considerable amount of a radioactive gas, radon-222. Write the nuclear symbol for this isotope of radon (Rn).

15. a. Do the symbols ${}^{57}_{26}Fe$ and ${}_{26}Fe$ have the same meaning?

b. Do the symbols ${}^{57}_{26}Fe$ and ${}^{57}Fe$ convey the same information?

16. Explain how the two isotopes of copper, Cu-63 and Cu-65, differ from each other. Write nuclear symbols for each isotope.

17. Lithium is an element that is used by physicians in the treatment of some mental disorders. Lithium-7 is one of its isotopes. How many

- **a.** protons are in its nucleus?
- **b.** neutrons are in its nucleus?
- **c.** electrons are in a lithium atom?
- **d.** neutrons, protons, and electrons are in the Li^+ ion formed from this isotope?

18. An isotope of iodine used in thyroid disorders is ${}^{131}_{53}I$. How many

Perspective

At the end of each chapter are essays that apply the chemical principles just learned to the world around us. Several of these essays deal with the environment (Chaps. 5, 11, 12, 16 and 18), others with nutrition (Chaps. 8, 19 and 20). One that we strongly recommend discusses the production of maple syrup (Chap. 10).

PERSPECTIVE

•

The Greenhouse Effect

This chapter has emphasized the common properties of different gases. Many properties, however, differ tremendously from one gas to another. One of these is the ability to absorb infrared radiation (heat). Among the major components of the atmosphere, only carbon dioxide and water vapor show this behavior. They absorb much of the infrared radiation given off by the warm earth. In this way, CO_2 and H_2O act as an insulating blanket to prevent heat from escaping into outer space; this is often referred to as the greenhouse effect.

Of the two gases, water vapor absorbs more infrared radiation than carbon dioxide because its concentration is higher. This property of water vapor accounts for the fact that the temperature drops less on nights when there is a heavy cloud cover. In desert regions, where there is very little water vapor, large variations between day and night temperatures are common.

Although the concentration of water vapor in the atmosphere varies greatly with location, it remains relatively constant over time. In contrast, the concentration of carbon dioxide has increased by more than 20% over the past century, due to human activities. Increased combustion of fossil fuels is mainly responsible. Every gram of fossil fuel burned releases about three grams of carbon dioxide into the atmosphere. Part of this CO_2 is used by plants in photosynthesis or is absorbed by the oceans, but at least half of it remains. Extensive land clearing, which reduces the amount of carbon dioxide consumed by photosynthesis, is also a factor in raising the CO_2 content of the atmosphere. This is one of the adverse effects of the destruction of tropical rain forests for agricultural purposes.

It has been estimated that, unless preventive action is taken, increasing CO_2 levels could raise the earth's temperature by 3°C over the next century. This could raise sea level by as much as 1 m, flooding many coastal areas, including much of the state of Florida. On a more optimistic note, an increase in CO_2 concentration would promote photosynthesis, perhaps increasing the world's food supply.

Recent studies show that average global temperatures have indeed increased over the past century, by about 0.5°C (1°F). Beyond that, the three years 1989, 1988, and 1987 were, in that order, the warmest on record. Nobody knows whether these data reflect increased concentrations of CO_2 and other greenhouse gases (Fig. 5.B) or statistical fluctuations. The general consensus among atmospheric scientists is that carbon dioxide emissions should be reduced to avoid a worst-case scenario. There are several ways to do this:

— raise fuel efficiency standards for automobiles
— impose a surtax on all carbon-containing fuels
— develop "clean" sources of energy, notably solar

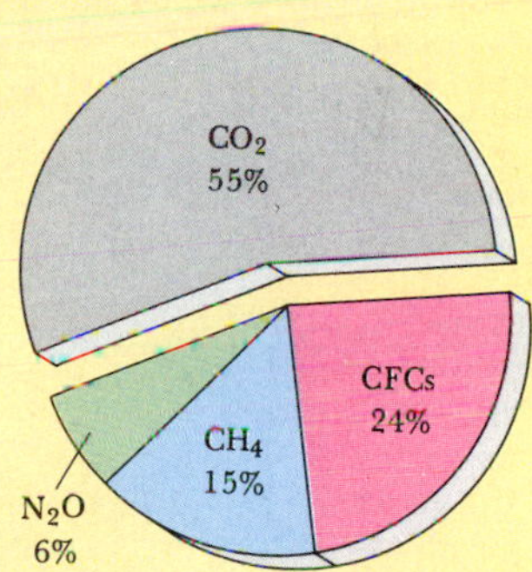

Figure 5.B
Contributions of different gases to global warming (1980–1990). Carbon dioxide is the major factor, but chlorofluorocarbons, such as CF_2Cl_2, make a contribution. These compounds are used in refrigerators, air conditioners, and, until recently, aerosol sprays.

J. Willard Gibbs (yearbook portrait as a graduating senior at Yale, Class of 1858.)

Two theoreticians working in the latter half of the nineteenth century changed the very nature of chemistry by deriving the mathematical laws that govern the behavior of matter undergoing physical or chemical change. One of these was James Clerk Maxwell, whose contributions to kinetic theory were discussed in Chapter 5. The other was J. Willard Gibbs, Professor of Mathematical Physics at Yale from 1871 until his death in 1903.

In 1876 Gibbs published the first portion of a remarkable paper in the *Transactions of the Connecticut Academy of Sciences* entitled "On the Equilibrium of Heterogeneous Substances." When the paper was completed in 1878 (it was 323 pages long), the foundation was laid for the science of chemical thermodynamics. Here, for the first time, the concept of free energy appeared. Included as well were the basic principles of chemical equilibrium (Chap. 12), phase equilibrium (Chap. 9), and the relations governing energy changes in electrical cells (Chap 16).

If Gibbs had never published another paper, this single contribution would have placed him among the greatest theoreticians in the history of science. Generations of experimental scientists have established their reputations by demonstrating in the laboratory the validity of the relationships that Gibbs derived at his desk. Many of these relationships were rediscovered by others; an example is the Gibbs-Helmholtz equation developed in 1882 by Helmholtz, a prestigious German physiologist and physicist who was completely unaware of Gibbs' work.

J. Willard Gibbs is often cited as an example of the "prophet without honor in his own country." His colleagues in New Haven and elsewhere in the United States seem not to have realized the significance of his work until late in his life. During his first 10 years as a professor at Yale he received no salary. In 1920, when he was first proposed for the Hall of Fame of Distinguished Americans at New York University, he received 9 votes out of a possible 100. Not until 1950 was he elected to that body. Even today the name of J. Willard Gibbs is generally unknown among educated Americans outside of those interested in the natural sciences.

Maxwell was perhaps the first to recognize Gibbs' genius

Admittedly, Gibbs himself was largely responsible for the fact that for many years his work did not attract the attention it deserved. He made little effort to publicize it; the *Transactions of the Connecticut Academy of Sciences* was hardly the leading scientific journal of its day. Gibbs was one of those rare individuals who seem to have no inner need for recognition by contemporaries. His satisfaction came from solving a problem in his mind; having done so, he was ready to proceed to other problems. His papers are not easy to read; he seldom cites examples to illustrate his abstract reasoning. Frequently, the implications of the laws that he derives are left for the readers to grasp on their own. One of his colleagues at Yale confessed many years later that none of the members of the Connecticut Academy of Sciences understood his paper on thermodynamics; as he put it, "We knew Gibbs and took his contributions on faith."

Historical Boxes

Several chapters contain boxed material on some of the giants of chemistry, including G. N. Lewis (p. 169), J. Willard Gibbs (p. 442) and Marie Curie (p. 499). The boxes focus on both the human qualities as well as the scientific accomplishments of the historical figure.

Appendices

In addition to Appendix 4, there are two other appendices designed to help you learn chemistry. These are

— Appendix 3, Review of Mathematics, which touches on just about all the mathematical techniques you will use in general chemistry. Exponential notation and logarithms (natural and base 10) are emphasized. There is also a short discussion of how to use your scientific calculator.

— Appendix 5, Net Ionic Equations. This topic is introduced in Chapter 4 and referred to in several later chapters (Chaps. 10, 13, 14, 17, 19, 20). The purpose of Appendix 5 is to tie all this material together. This should make you more proficient in writing, balancing, and interpreting chemical equations.

Contents Overview

Contents

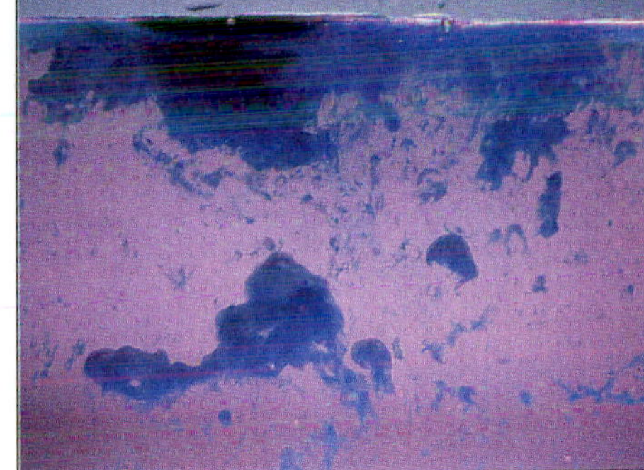

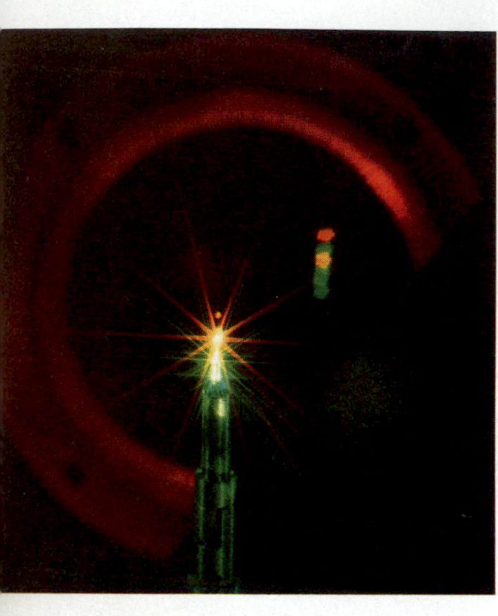

CHARLES D. WINTERS

$$2Zn(s) + O_2(g) \longrightarrow 2ZnO(s)$$

(See p. 85)

1 Matter and Measurement

Inchworm, inchworm, measuring the marigolds,
You and your arithmetic will probably go far.
Inchworm, inchworm, measuring the marigolds,
Seems to me you'd stop and see how beautiful they are.

—NURSERY RHYME

CHAPTER OUTLINE

Almost certainly, this is your first college course in chemistry; perhaps it is your first exposure to chemistry at any level. Unless you are a chemistry major, you may wonder why you are taking this course and what you can expect to gain from it. To address that question, it will be helpful to look at some of the ways in which chemistry contributes to other disciplines.

If you're planning to be an engineer, you can be sure that many of the materials you will work with have been synthesized by chemists. Some of these materials are organic (carbon-containing). They could be familiar plastics like polyethylene (Chapter 21) or the more esoteric plastics used in unbreakable windows and nonflammable clothing. Other materials, including metal alloys, semiconductors, and superconductors, are inorganic in nature.

Perhaps you are a health science major, looking forward to a career in medicine or pharmacy. If so, you will want to become familiar with the properties of aqueous solutions (Chapters 4, 10, 14), which include blood and other body fluids. Chemists have made many life-saving products over the past few decades. These range from drugs used in chemotherapy to new antibiotics used against resistant microorganisms.

Beyond career preparation, an objective of a college education is to stimulate your curiosity about the world around you, or, to abuse a cliché, to make you a "better informed citizen." In this text, we will look at some of the chemistry-related issues facing society today, including

— nuclear power (Chapter 18)
— the greenhouse effect (Chapter 5)
— depletion of the ozone layer (Chapter 11)
— the phenomenon of acid rain (Chapter 12)

We hope that when you complete this course you too will be convinced of the importance of chemistry in today's world. We should, however, caution you upon one point. Although we will talk about many of the applications of chemistry, *our main concern will be with the principles that govern chemical reactions.* Only by mastering these principles will you understand the basis of the applications referred to in the preceding paragraphs.

Chemistry deals with the properties and reactions of substances.

This chapter begins the study of chemical principles by

— considering the different types of matter: pure substances vs. mixtures, elements vs. compounds (Section 1.1)
— looking at the nature of measurements (Section 1.2), the uncertainties associated with them (Section 1.3), and the method used to convert measured quantities from one set of units to another (Section 1.4)
— focusing on certain physical properties, including density and water solubility, used to identify substances (Section 1.5)

1.1 TYPES OF MATTER

Matter is anything that has mass and occupies space. It exists in three phases: solid, liquid, and gas. A solid has a rigid shape and a fixed volume. A liquid has a fixed volume but is not rigid in shape; it takes on the shape of the container. A gas has neither a fixed volume nor a rigid shape; it takes on both the volume and the shape of the container.

Most materials you encounter are mixtures

Matter can be classified into two categories: pure substances and mixtures. Pure substances are either elements or compounds, while mixtures can be either homogeneous or heterogeneous (Fig. 1.1).

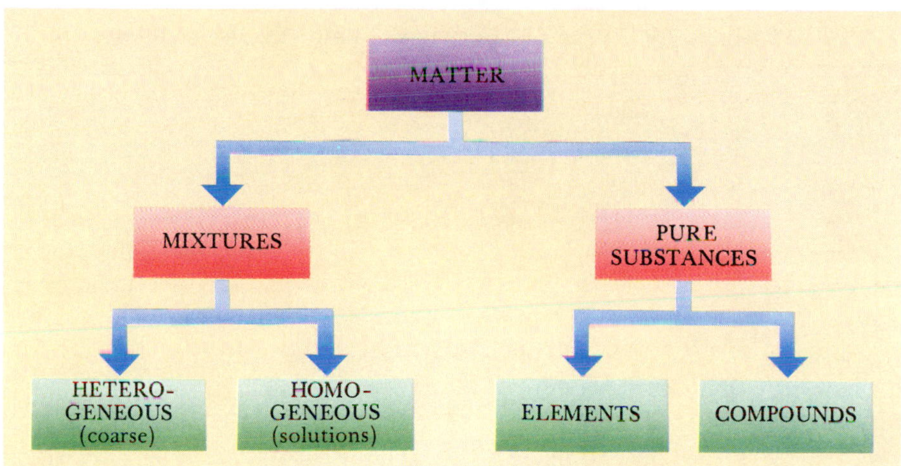

Figure 1.1
Classification of matter.

Elements

An **element** is a type of matter which cannot be broken down into two or more pure substances. There are 109 known elements, of which 91 occur naturally.

Many elements are familiar to all of us. The charcoal used in outdoor grills is nearly pure carbon. Electrical wiring, jewelry, and water pipes are often made from copper, a metallic element. Another such element, aluminum, is used in many household utensils. The shiny liquid in the thermometers you use is still another metallic element, mercury.

Some elements come in and out of fashion, so to speak. Fifty years ago, elemental silicon was a chemical curiosity. Today, ultrapure silicon has become the basis for the multibillion-dollar semiconductor industry. Lead, on the other hand, is an element moving in the other direction. A generation ago it was widely used to make paint pigments, plumbing connections, and gasoline additives. Today, because of the toxicity of lead compounds, all of these applications have been banned in the United States.

In chemistry, an element is identified by its symbol. This consists of one or two letters, usually derived from the name of the element. Thus the symbol for carbon is C; that for aluminum is Al. Sometimes the symbol comes from the Latin name of the element or one of its compounds. The two elements copper and mercury, which were known in ancient times, have the symbols Cu *(cuprum)* and Hg *(hydrargyrum)*. Table 1.1 lists the names and symbols of the elements that we will be most concerned with in this text.

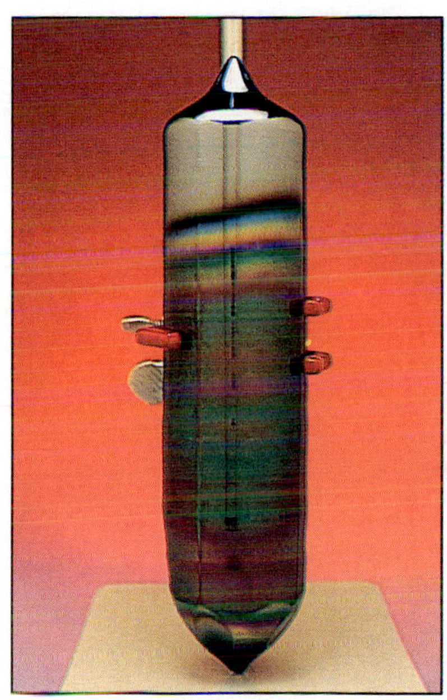

A solid cylinder of nearly pure elemental silicon. The cylinder is sliced into thin wafers for use in the integrated circuits of semiconductors. (Charles D. Winters)

Compounds

A **compound** is a pure substance that contains more than one element. Water is a compound of hydrogen and oxygen. The compounds methane, acetylene, and naphthalene all contain the elements carbon and hydrogen, in different proportions.

The properties of compounds are very different from those of the elements they contain. Ordinary table salt, sodium chloride, is a white, unreactive solid.

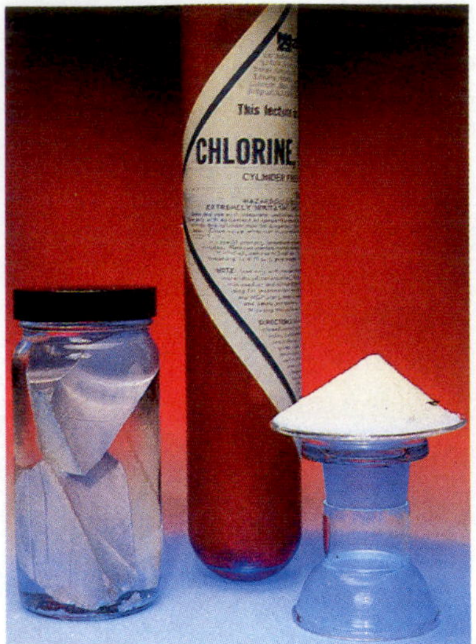

Figure 1.2
Sodium is a shiny, highly reactive metal, ordinarily stored under toluene (bottle) to prevent reaction with air and water. Chlorine is a greenish-yellow gas, shown here in a high-pressure cylinder. Sodium chloride is a white, nontoxic solid (table salt). (Charles D. Winters)

Table 1.1 Names and Symbols of Some of the More Familiar Elements

Aluminum	Al	Chlorine	Cl	Lithium	Li	Rubidium	Rb
Antimony	Sb	Chromium	Cr	Magnesium	Mg	Selenium	Se
Argon	Ar	Cobalt	Co	Manganese	Mn	Silicon	Si
Barium	Ba	Copper	Cu	Mercury	Hg	Silver	Ag
Beryllium	Be	Fluorine	F	Neon	Ne	Sodium	Na
Bismuth	Bi	Gold	Au	Nickel	Ni	Strontium	Sr
Boron	B	Helium	He	Nitrogen	N	Sulfur	S
Bromine	Br	Hydrogen	H	Oxygen	O	Tin	Sn
Cadmium	Cd	Iodine	I	Phosphorus	P	Uranium	U
Calcium	Ca	Iron	Fe	Platinum	Pt	Xenon	Xe
Carbon	C	Krypton	Kr	Plutonium	Pu	Zinc	Zn
Cesium	Cs	Lead	Pb	Potassium	K		

As you can guess from its name, it contains the two elements sodium and chlorine. Sodium (Na) is a shiny, extremely reactive metal. Chlorine (Cl) is a poisonous, greenish-yellow gas. Clearly, when these two elements combine to form sodium chloride, a profound change takes place (Fig. 1.2).

Making a compound from the elements is often easier than the reverse process

Many different methods can be used to resolve compounds into their elements. Sometimes, but not often, heat alone is sufficient. Mercury(II) oxide, a compound of mercury and oxygen, decomposes to its elements when heated to 600°C. Joseph Priestley, an English chemist, discovered oxygen 200 years ago when he carried out this reaction by exposing a sample of mercury(II) oxide to an intense beam of sunlight focused through a powerful lens. Another method of resolving compounds into elements is called electrolysis. This involves passing an electric current through a compound, usually in the liquid state. Through the process of electrolysis, it is possible to separate water into the two elements hydrogen and oxygen.

Mixtures

You can separate this mixture using a magnet

A **mixture** contains two or more pure substances combined in such a way that each substance retains its chemical identity. When you shake iron filings with powdered sulfur, a mixture is formed; the two elements are chemically unchanged. In contrast, when sodium is exposed to chlorine gas, a compound, sodium chloride, is formed; the two elements lose their chemical identity. There are two types of mixtures.

1. Homogeneous or uniform mixtures, in which the composition of the mixture is the same throughout. Another name for a homogeneous mixture is a **solution**. A solution is made up of a solvent, the substance present in largest amount, and one or more solutes. Most commonly, the solvent is a liquid, while solutes may be solids, liquids, or gases. Soda water is a solution of carbon dioxide (solute) in water (solvent). Seawater is a more complex solution in which there are several solid solutes, including sodium chloride; the solvent is water. It is also possible to have solutions in the solid state. Brass (Fig. 1.3) is a solid solution containing the two metals copper (67–90%) and zinc (10–33%).

2. Heterogeneous or nonuniform mixtures are those in which the composition varies throughout. Most rocks fall into this category. In a piece of granite (Fig. 1.3), several components can be distinguished, differing from one another in color.

Many different methods can be used to separate the components of a mixture from one another. A couple of methods that you may have carried out in the laboratory are

— *filtration*, used to separate a heterogeneous solid–liquid mixture. The mixture is passed through a barrier with fine pores such as filter paper.
— *distillation* (Fig. 1.4), used to resolve a homogeneous solid–liquid mixture. The liquid vaporizes, leaving a residue of the solid in the distilling flask. Pure liquid is obtained by condensing the vapor.

A more complex but more versatile separation method is **chromatography**, a technique widely used in teaching, research, and industrial laboratories to separate all kinds of mixtures. This method takes advantage of differences in solubility and/or extent of adsorption on a solid surface. In *gas-liquid chromatography*, a mixture of volatile liquids and gases is introduced into one end of a heated glass tube. As little as one microliter (10^{-6} L) of sample may be used.

Figure 1.3
Mixtures can be *homogeneous*, as with brass, which is a solid solution of copper and zinc. Alternatively, they can be *heterogeneous*, as with granite, which contains discrete regions of different minerals (feldspar, mica, and quartz). (Charles D. Winters)

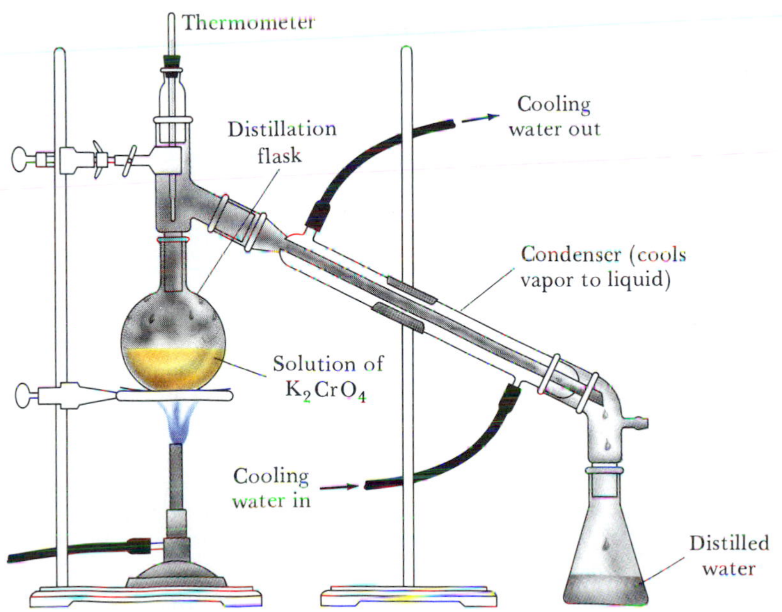

Figure 1.4
The two components of a water solution of potassium chromate can be separated from each other by distillation. Water is vaporized by heating; cooling the vapor causes it to condense to a pure liquid in the flask at the lower right. Since solid potassium chromate is not volatile, it remains in the distillation flask as a yellow residue.

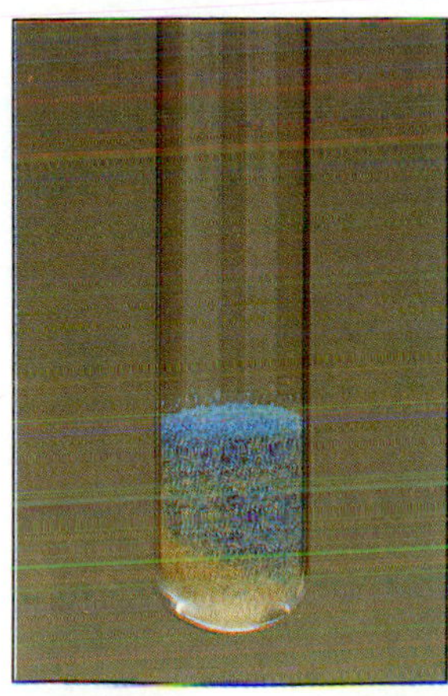

A heterogeneous mixture of copper sulfate crystals (blue) and sand. (Charles D. Winters)

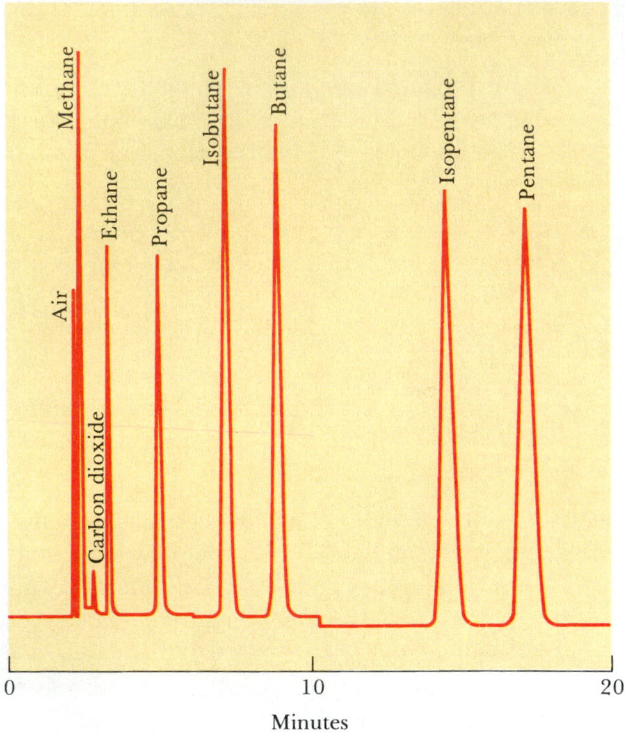

Figure 1.5
The components of natural gas (mostly methane) can be separated by gas-liquid chromatography. With some volatile mixtures, the sample can be as small as 10^{-6} L.

The tube is packed with an inert solid whose surface is coated with a viscous liquid. An unreactive "carrier gas," often helium, is passed through the tube. The components of the sample gradually separate as they vaporize into the helium or condense into the viscous liquid. Usually the more volatile fractions move faster and emerge first; successive fractions activate a detector and recorder. The end result is a plot such as that shown in Figure 1.5.

Gas-liquid chromatography finds many applications outside the chemistry laboratory. Many procedures for drug testing (marijuana, cocaine) depend upon this technique. It also offers a rapid, precise method of determining the percentage of alcohol in a blood sample and hence the degree of intoxication.

1.2 MEASUREMENTS; QUANTITIES AND UNITS

When you measure a quantity such as length or mass you express your result as a number attached to a unit, e.g., "6.02 *inches*," "148 *pounds*." Scientists ordinarily use a set of units, developed in France in the eighteenth century, known as the **metric system**. This, as you probably know, is a decimal-based system in which all the units of a particular quantity are related to each other by powers of ten.

Surprising as it may seem, all the quantities that scientists measure can be expressed in terms of seven base quantities (Table 1.2). Putting it another way, if base units can be established for each of these quantities (length, mass, and

Table 1.2 SI Base Units and Selected Prefixes

Base Quantity	Base Unit	Factor*	Prefix
Length	meter (m)	10^6	mega (M)
Mass	kilogram (kg)	10^3	kilo (k)
Temperature	kelvin (K)	10^{-1}	deci (d)
Time	second (s)	10^{-2}	centi (c)
Amount of substance	mole (mol)	10^{-3}	milli (m)
Electric current	ampere (A)	10^{-6}	micro (μ)
Luminous intensity	candela (cd)	10^{-9}	nano (n)
		10^{-12}	pico (p)

* Exponential notation is discussed in Appendix 3.

You can forget about the candela, at least in this course

so forth), then units for all other quantities (e.g., volume, force, pressure, energy) can be related to them. Recognizing this fact, the General Conference of Weights and Measures in 1960 recommended an International System of Units **(SI)** constructed upon the seven base units listed in Table 1.2. Decimal multiples of these units are shown using the prefixes listed at the right of the table.

To see how this all works out, let us examine three base quantities commonly measured in chemistry: length, mass, and temperature. We will consider how these quantities are measured as well as the units in which they are expressed.

Length

The meter stick found in every general chemistry laboratory reproduces the SI base unit of length, the meter. The meter was originally intended to be 1/40,000,000 of the earth's meridian that passes through Paris. It is now defined in terms of the speed of light, which travels a distance of one meter in 1/299,792,458 second.

Other units of length are expressed in terms of the meter, using the prefixes listed in Table 1.2. You are familiar with the centimeter, the millimeter, and the kilometer:

$$1\ \text{cm} = 10^{-2}\ \text{m} \qquad 1\ \text{mm} = 10^{-3}\ \text{m} \qquad 1\ \text{km} = 10^3\ \text{m}$$

The dimensions of very tiny particles are often expressed in nanometers:

Picometers are also used; 1 nm corresponds to 1000 pm

$$1\ \text{nm} = 10^{-9}\ \text{m}$$

Mass

In the metric system, mass may be expressed in grams, kilograms, or milligrams; the base SI unit is the kilogram.

$$1\ \text{g} = 10^{-3}\ \text{kg} \qquad 1\ \text{mg} = 10^{-6}\ \text{kg}$$

The megagram, more frequently called the **metric ton**, is

$$1\ \text{Mg} = 10^6\ \text{g} = 10^3\ \text{kg}$$

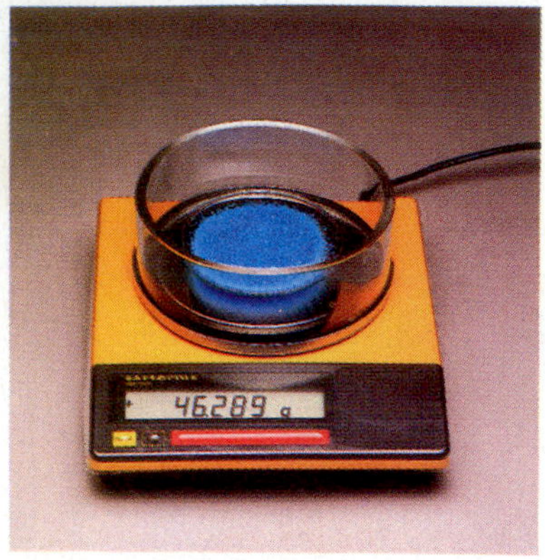

Figure 1.6
Single-pan analytical balance with digital readout; the solid sample and container weigh 46.289 g. (Marna G. Clarke)

Properly speaking, there is a distinction between mass and weight. **Mass** is a measure of the amount of matter in an object; *weight* is a measure of the gravitational force acting on the object. Chemists often use these terms interchangeably; we determine the mass of an object by "weighing" it on a balance (Fig. 1.6).

Temperature

Temperature is the factor that determines the direction of heat flow. When two objects at different temperatures are placed in contact with one another, heat flows from the one at the higher temperature to the one at the lower temperature.

Temperature is measured indirectly, by observing its effect upon the properties of a substance. A mercury-in-glass thermometer takes advantage of the fact that mercury, like other substances, expands as temperature increases. When the temperature rises, the mercury in the thermometer expands up a narrow tube. The total volume of the tube is only about 2% of that of the bulb at the base. In this way, a rather small change in volume is made readily visible.

Thermometers used in chemistry are marked in degrees *Celsius* (centigrade), named after the Swedish astronomer Anders Celsius (1701–1744). On this scale, the freezing point of water is taken to be 0°C. The normal boiling point of water is 100°C. Household thermometers in the United States are commonly marked in *Fahrenheit* degrees. Daniel Fahrenheit (1686–1736) was a German instrument maker who was the first to use the mercury-in-glass thermometer. On this scale, the normal freezing and boiling points of water are taken to be 32° and 212°, respectively (Fig. 1.7). Hence

$$32°\text{F} = 0°\text{C} \qquad 212°\text{F} = 100°\text{C}$$

The general relation between the two scales is

$$t_{°\text{F}} = 1.8t_{°\text{C}} + 32°$$

The SI base unit for temperature is the **kelvin (K)**; note the absence of the degree sign. The kelvin is defined to be 1/273.16 of the difference between the lowest attainable temperature (0 K) and the triple point of water* (0.01°C). The relationship between temperature in K and in °C is:

$$T_{\text{K}} = t_{°\text{C}} + 273.15$$

This scale is named after Lord Kelvin (1824–1907), an English physicist. He showed that it is impossible to reach a temperature lower than 0 K (−273.15°C).

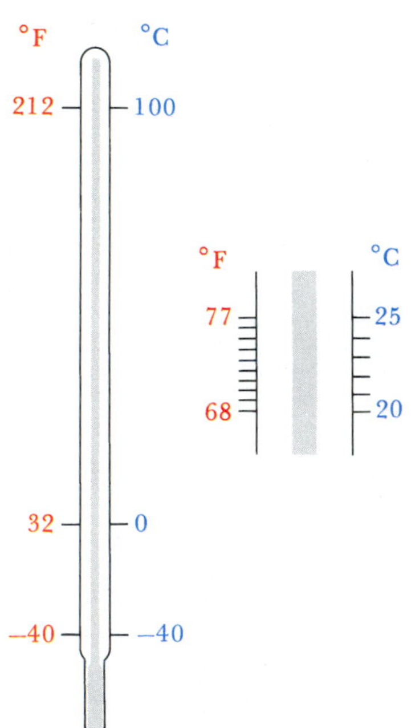

Figure 1.7
On the Celsius scale, the distance between the freezing and boiling points of water is 100°; on the Fahrenheit scale, it is 180°. This means that the Celsius degree is $\frac{9}{5}$ as large as the Fahrenheit degree, as is evident from the magnified section of the thermometer at the right.

Derived Quantities and Units

As pointed out earlier, units for any measured quantity can be derived from the base units listed in Table 1.2. We will consider the units used to express four derived quantities: volume, force, pressure, and energy.

* The triple point of water (Chapter 9) is the one, unique temperature at which ice, liquid water, and water vapor can coexist in contact with one another.

Example 1.1 Express normal body temperature, 98.60°F, in °C and K.

Strategy Use the relations $t_{°F} = 1.8t_{°C} + 32°$ and $T_K = t_{°C} + 273.15$. Solve algebraically for the desired quantity, $t_{°C}$ in the first case, T_K in the second.

Solution

To find $t_{°C}$: $98.60° = 1.8t_{°C} + 32°$

Solving: $$t_{°C} = \frac{98.60° - 32°}{1.8} = 37.00°C$$

To find T_K: $T_K = 37.00 + 273.15 = 310.15\ K$

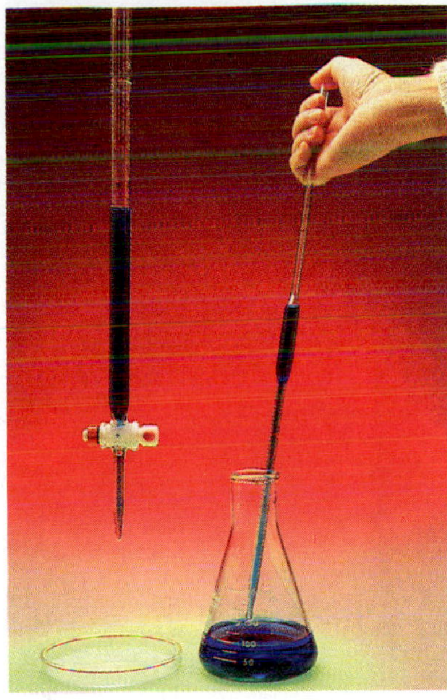

Figure 1.8
A buret *(left)* is used to deliver a variable volume of liquid, accurately measured. A pipet *(right)* is used to deliver a fixed volume (e.g., 25.00 mL) of liquid. (Charles D. Winters)

Volume The SI unit for volume is the *cubic meter*, m^3. This is a very large unit; a cubic meter of water weighs about a ton. We will more often express volumes in

— cubic centimeters $1\ cm^3 = (10^{-2}\ m)^3 = 10^{-6}\ m^3$
— liters (L) $1\ L = 10^{-3}\ m^3 = 10^3\ cm^3$
— milliliters (mL) $1\ mL = 10^{-3}\ L = 10^{-6}\ m^3$

Notice that a milliliter is equal to one cubic centimeter: $1\ mL = 1\ cm^3$.

The device most commonly used to measure volumes in general chemistry is the graduated cylinder. A pipet or buret (Fig. 1.8) is used when greater accuracy is required. A pipet is calibrated to deliver a fixed volume of liquid, e.g., 25.00 mL, when filled to the mark and allowed to drain. Variable volumes can be delivered accurately by a buret, perhaps to ±0.01 mL.

A graduated cylinder is marked to contain a given volume

Force The SI unit of force is the *newton* (N); it is the force required to give a mass of one kilogram an acceleration of one meter per second squared. That is

$$1\ N = 1\ kg \cdot m/s^2$$

The newton, by itself, is not widely used in general chemistry. It is, however, involved in the definitions of the SI units for pressure and energy.

Pressure The SI unit for pressure is the **pascal (Pa)**, which is the pressure exerted by a force of one newton operating on an area of one square meter.

$$1\ Pa = 1\ N/m^2 = 1\ kg/m \cdot s^2$$

A pascal is a very small unit; a film of water 0.1 mm thick exerts a pressure of about 1 Pa on a surface with which it is in contact. The *kilopascal* is used more frequently; atmospheric pressure is of the order of 100 kPa (or 10^5 Pa).

Chemists use a variety of units to express pressure, as you will see in Chapter 5. Perhaps the most common is the **standard atmosphere (atm)**, defined by the relation

$$1\ atm = 1.01325 \times 10^5\ Pa$$

Another pressure unit is the millimeter of mercury (Chap. 5)

Antoine Lavoisier
(1743–1794)

The discussion in Section 1.2 emphasizes the importance of numerical measurements, involving such quantities as mass, temperature, and volume. When a chemist carries out an experiment in the laboratory, he or she almost always makes quantitative measurements. Consider, for example, the following directions for the preparation of aspirin, abstracted from a laboratory manual in organic chemistry.

> Add *2.0 g* of salicylic acid, *5.0 mL* of acetic anhydride, and 5 drops of 85% H_3PO_4 to a 50-mL Erlenmeyer flask. Heat in a water bath at *75°C* for *15 minutes.* Add *cautiously 20 mL* of water and transfer to an ice bath at *0°C.* Scratch the inside of the flask with a stirring rod to initiate crystallization. Separate aspirin from the solid-liquid mixture by filtering through a Buchner funnel *10 cm* in diameter.

Chemistry was not always so quantitative. The following recipe for finding the philosopher's stone was recorded about 300 years ago.

> Take all the mineral salts there are, also all salts of animal and vegetable origin. Add all the metals and minerals, omitting none. Take two parts of the salts and grate in one part of the metals and minerals. Melt this in a crucible, forming a mass that reflects the essence of the world in all its colors. Pulverize this and pour vinegar over it. Pour off the red liquid into English wine bottles, filling them half-full. Seal them with the bladder of an ox (*not* that of a pig). Punch a hole in the top with a coarse needle. Put the bottles in hot sand for three months. Vapor will escape through the hole in the top, leaving a red powder . . . .

This experiment, as you can imagine, is not easy to carry out. You could spend a lifetime collecting *all* the salts, metals, and minerals in the world. So much for the good old days.

Lavoisier was executed because he was a tax collector; chemistry had nothing to do with it

One man more than any other transformed chemistry from an art to a science. Antoine Lavoisier was born in Paris; he died on the guillotine during the French Revolution. Above all else, Lavoisier understood the importance of carefully controlled, quantitative experiments. These were described in his book *Elements of Chemistry.* Published in 1789, it is illustrated with diagrams drawn by his wife.

The results of one of Lavoisier's quantitative experiments are shown in Table 1.A; the data are taken directly from Lavoisier. If you add up the masses of reactants and products (expressed in arbitrary units), you find them to be the same, 510. As Lavoisier put it, "In all of the operations of men and nature, nothing is created. An equal quantity of matter exists before and after the experiment."

Table 1.A Quantitative Experiment on the Fermentation of Wine (Lavoisier)

Reactants	Mass (Relative)	Products	Mass (Relative)
Water	400	carbon dioxide	35.3
Sugar	100	alcohol	57.7
Yeast	10	acetic acid	2.5
		water	409.0
		sugar (unreacted)	4.1
		yeast (unreacted)	1.4

E. I. Du Pont (1772–1834) was a student of Lavoisier

This was the first statement of the law of conservation of mass, which was the cornerstone for the growth of chemistry in the nineteenth century. You can readily see why Lavoisier is called the father of modern chemistry.

Energy The SI unit of energy is the **joule (J)**, which is the work done when a force of one newton acts through a distance of one meter.

$$1\ \text{J} = 1\ \text{N} \cdot \text{m} = 1\ \text{kg} \cdot \text{m}^2/\text{s}^2$$

In the past, chemists used the calorie,* which is the amount of heat required to raise the temperature of one gram of water one degree Celsius. The calorie is now defined by the relation

A burning match gives off about 500 cal or 2000 J of heat

$$1\ \text{cal} = 4.184\ \text{J}$$

A joule is a rather small energy unit. One joule of electrical energy would keep a 10-watt light bulb burning for only a tenth of a second. For that reason, we will often express energy changes in kilojoules.

$$1\ \text{kJ} = 10^3\ \text{J}$$

1.3 UNCERTAINTIES IN MEASUREMENTS; SIGNIFICANT FIGURES

Every measurement carries with it a degree of uncertainty whose magnitude depends upon the nature of the measuring device and the skill with which it is used. Suppose, for example, you measure out 8 mL of liquid using the 100-mL graduated cylinder shown in Figure 1.9. Here the volume is uncertain to perhaps ±1 mL. With such a crude measuring device, you would be lucky to obtain a volume between 7 and 9 mL. To obtain greater precision, you could use a narrow 10-mL cylinder, which has divisions in smaller increments. You might now measure a volume within 0.1 mL of the desired value, in the range of 7.9 to 8.1 mL. Using a buret, the uncertainty could be reduced to ±0.01 mL.

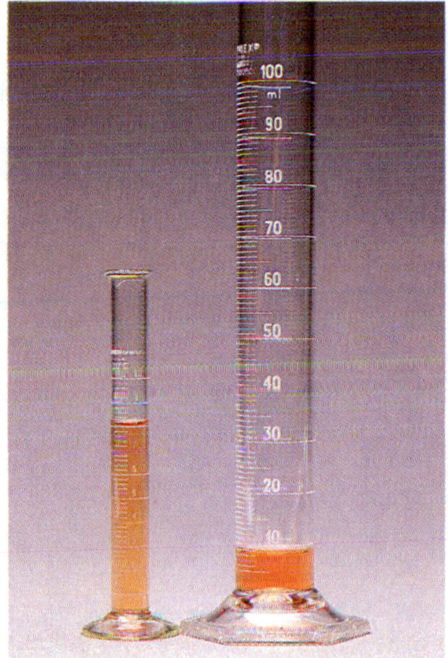

Figure 1.9
The uncertainty associated with a measurement depends upon the nature of the measuring device. To measure out small volumes of liquids precisely, a 10-mL graduated cylinder is much more effective than one with a volume of 100 mL. (Marna G. Clarke)

Anyone making a measurement has a responsibility to indicate the uncertainty associated with it. Such information is vital to anyone who wants to repeat the experiment or judge its precision. The three volume measurements referred to above could be reported as

8 ± 1 mL	(large graduated cylinder)
8.0 ± 0.1 mL	(small graduated cylinder)
8.00 ± 0.01 mL	(buret)

In this text, we will drop the ± notation and simply write

8 mL   8.0 mL   8.00 mL

When we do this, it is understood that there is an ***uncertainty of at least one unit in the last digit,*** that is, 1 mL, 0.1 mL, 0.01 mL, respectively. This method of citing the degree of confidence in a measurement is often described in terms of **significant figures**, the meaningful digits obtained in a measurement. In 8.00 mL, there are three significant figures; each of the three digits has experimental meaning. Similarly, there are two significant figures in 8.0 mL and one significant figure in 8 mL.

There's a big difference between 8 mL and 8.00 mL, maybe as much as half a milliliter

* The "calorie" referred to by nutritionists is actually a kilocalorie (1 kcal = 10^3 cal). On a "2000-calorie" per day diet, you eat food capable of producing 2000 kcal = 2×10^3 kcal = 2×10^6 cal of energy.

Counting Significant Figures

Frequently, you need to know the number of significant figures in a measurement that someone else has reported. To do this, you apply the following common-sense rules.

1. *All nonzero digits are significant.* There are three significant figures in 5.37 cm and four significant figures in 4.293 cm.
2. *Zeros between nonzero digits are significant.* There are three significant figures in 106 g or in 1.02 g.
3. *Zeros beyond the decimal point at the end of a number are significant.* When the volume of a liquid is reported as 8.00 mL, this implies that the two zeros are experimentally meaningful. The quantity 8.00 mL carries the same degree of precision as 8.13 mL or 7.98 g; all of these quantities have three significant figures.
4. *Zeros preceding the first nonzero digit in a number are not significant.* In a mass measurement of 0.002 g, there is only one significant figure—the "2" at the end. The zeros serve only to fix the position of the decimal point. This becomes obvious if the mass is expressed in **exponential** (scientific) **notation** (Appendix 3). In that case, 0.002 g is written as

$$2 \times 10^{-3}\ \text{g}$$

Now, clearly, there is only one significant figure. The uncertainty is $\pm 1 \times 10^{-3}$ g.

Unfortunately, the uncertainty is uncertain

Sometimes the number of significant figures in a reported measurement is ambiguous. Suppose that a piece of metal is reported to weigh "500 g." You cannot be sure how many of these digits are meaningful. Perhaps the metal was weighed to the nearest gram (500 ±1 g). If so, the "5" and the two zeros are significant; there are three significant figures. Then again, the metal might have been weighed only to the nearest 10 g (500 ±10 g). In this case, only the "5" and one zero are known accurately; there are two significant figures. About all you can do in such cases is to wish the person who carried out the weighing had used exponential notation. The mass should have been reported as

$$5.00 \times 10^2\ \text{g} \quad \text{(3 significant figures)}$$

or

$$5.0 \times 10^2\ \text{g} \quad \text{(2 significant figures)}$$

or

$$5 \times 10^2\ \text{g} \quad \text{(1 significant figure)}$$

In general, *any ambiguity concerning the number of significant figures in a measurement can be resolved by using exponential notation.*

Example 1.2 Three different students weigh the same object, using different balances. They report the following masses:

(a) 15.02 g (b) 15.0 g (c) 0.01502 kg

How many significant figures are there in each value?

Strategy Follow the rules for counting significant figures. If you have trouble deciding whether a zero is significant or not, put the number in exponential notation.

Solution

(a) 4.

(b) 3. The zero after the decimal point is significant. It indicates that the object was weighed to the nearest 0.1 g.

(c) 4. The zeros at the left are not significant. They are there only because the mass was expressed in kilograms rather than grams. Note that "15.02 g" and "0.01502 kg" represent the same mass.

Significant Figures in Multiplication and Division

Most measured quantities are not end results in themselves. Instead, they are used to calculate other quantities, often by multiplication or division. The precision of any such derived result is limited by that of the measurements on which it is based. *When measured quantities are multiplied or divided, the number of significant figures in the result is the same as that in the quantity with the smallest number of significant figures.*

This rule is approximate but sufficient for our purposes

Example 1.3 An overseas flight leaves New York in the late afternoon and arrives in London 8.50 hours later. The airline distance from New York to London is about 5.6×10^3 km, depending to some extent upon the flight path followed. What is the average speed of the plane, in kilometers per hour?

Strategy Calculate the average speed by taking the quotient

$$\text{speed} = \frac{\text{distance traveled}}{\text{time elapsed}}$$

Count the number of significant figures in the numerator and in the denominator; the smaller of these two numbers is the number of significant figures in the quotient.

Solution The average speed will appear on your calculator as

$$\frac{5.6 \times 10^3 \text{ km}}{8.50 \text{ h}} = 658.82353 \text{ km/h}$$

It makes no sense to report all these numbers, so don't do it

There are three significant figures in the denominator and two in the numerator. The answer should have two significant figures; round off the average speed to 6.6×10^2 km/h.

The rules for "rounding off" a measurement, which were applied in Example 1.3, are as follows:

1. *If the digits to be discarded are less than --500 . . . , leave the last digit unchanged.* Masses of 23.315 g and 23.487 g both round off to 23 g if only two significant digits are required.

2. *If the digits to be discarded are greater than --500 . . . , add one to the last digit.* Masses of 23.692 g and 23.514 g round off to 24 g.
3. *If, perchance, the digits to be discarded are --500 . . . (or simply --5 by itself), round off so that the last digit is an even number.* Masses of 23.500 g and 24.5 g both round off to 24 g (two significant figures).

This way, you round up as often as you round down

Uncertainties in Addition and Subtraction

When measured quantities are added or subtracted, the uncertainty in the result is found in a quite different way than in multiplication and division. It is determined by counting the number of "decimal places," i.e., the number of digits to the right of the decimal point for each measured quantity. *The sum or difference should be rounded off to the same number of decimal places as there are in the measured quantity with the smallest number of decimal places.*

Suppose you want to find the total mass of a solution made up of 10.21 g of instant coffee, 0.2 g of sugar, and 256 g of water.

	Mass	**Uncertainty**	
Instant coffee	10.21 g	± 0.01 g	2 decimal places
Sugar	0.2 g	± 0.1 g	1 decimal place
Water	256 g	± 1 g	0 decimal places
Total mass	266 g		

The mass is expressed as 266 g; it has an uncertainty of ±1 g, as does the mass of water, the quantity with the greatest uncertainty.

Example 1.4 A beaker containing lead pellets has a mass of 185.36 g. The empty beaker has a mass of 75.681 g. What is the mass of the lead pellets?

Strategy Count the number of decimal places in both measurements. Your answer should have the same number of decimal places as the measurement with the smallest number of decimal places.

Solution The mass of the lead pellets is

185.36 g	2 decimal places
75.681 g	3 decimal places
109.679 g	

The answer should have two decimal places and is 109.68 g.

Exact Numbers

In applying the principles just described, keep in mind one important point. Some numbers involved in calculations are exact rather than approximate because they are defined rather than measured quantities. Exact numbers never limit the precision of any calculated result. To illustrate the situation, consider the equation relating Fahrenheit and Celsius temperatures:

$$t_{°F} = 1.8t_{°C} + 32°$$

The numbers 1.8 and 32 are exact. Hence, they do not limit the number of significant figures in a temperature conversion; that limit is determined only by the precision of the thermometer used to measure temperature.

A different type of exact number arises in certain calculations. Suppose you are asked to determine the amount of heat evolved when *one kilogram* of coal burns. The implication is that since "one" is spelled out, *exactly* one kilogram of coal burns. The uncertainty in the answer should be independent of the amount of coal.

"One" kilogram means 1.000000 . . . kg, *not* 1 kg

1.4 CONVERSION OF UNITS

It is often necessary to convert measurements expressed in one unit (e.g., grams) to another unit (milligrams or kilograms). To do this, we follow what is known as a **conversion factor** approach. For example, to convert a volume of 536 cm^3 to liters, the relation

$$1 \text{ L} = 1000 \text{ cm}^3$$

is used. Dividing both sides of this equation by 1000 cm^3 gives a quotient equal to one:

$$\frac{1 \text{ L}}{1000 \text{ cm}^3} = \frac{1000 \text{ cm}^3}{1000 \text{ cm}^3} = 1$$

The quotient 1 L/1000 cm^3, which is called a *conversion factor,* is multiplied by 536 cm^3. Since the conversion factor equals one, this does not change the value of the volume. However, it does accomplish the desired conversion of units. The cm^3 in the numerator and denominator cancel to give the desired unit—liters.

$$536 \cancel{\text{cm}^3} \times \frac{1 \text{ L}}{1000 \cancel{\text{cm}^3}} = 0.536 \text{ L}$$

The relation 1 L = 1000 cm^3 can be used equally well to convert a volume in liters, say, 1.28 L, to cubic centimeters. In this case, the necessary conversion factor is obtained by dividing both sides of the equation by 1 L:

$$\frac{1000 \text{ cm}^3}{1 \text{ L}} = \frac{1 \text{ L}}{1 \text{ L}} = 1$$

Multiplying 1.28 L by the quotient 1000 cm^3/1 L converts the volume from liters to cubic centimeters:

$$1.28 \cancel{\text{L}} \times \frac{1000 \text{ cm}^3}{1 \cancel{\text{L}}} = 1280 \text{ cm}^3 = 1.28 \times 10^3 \text{ cm}^3$$

Notice that a single relation (1 L = 1000 cm^3) gives two conversion factors:

$$\frac{1 \text{ L}}{1000 \text{ cm}^3} \quad \text{and} \quad \frac{1000 \text{ cm}^3}{1 \text{ L}}$$

To go from cubic centimeters to liters, use the ratio 1 L/1000 cm^3; to go from liters to cubic centimeters, use the ratio 1000 cm^3/1 L. In general, when you make a conversion, *choose the factor that cancels out the initial unit.*

initial quantity × conversion factor(s) = desired quantity

Table 1.3 Relations Between Length, Volume, and Mass Units

Metric		English		Metric-English	
Length					
1 km	$= 10^3$ m	1 ft	= 12 in	1 in	= 2.54 cm*
1 cm	$= 10^{-2}$ m	1 yd	= 3 ft	1 m	= 39.37 in
1 mm	$= 10^{-3}$ m	1 mile	= 5280 ft	1 mile	= 1.609 km
1 nm	$= 10^{-9}$ m = 10 Å				
Volume					
1 m^3	$= 10^6$ cm^3 $= 10^3$ L	1 gallon	= 4 qt = 8 pt	1 ft^3	= 28.32 L
1 cm^3	= 1 mL $= 10^{-3}$ L	1 qt (U.S. liq.)	= 57.75 in^3	1 L	= 1.057 qt (U.S. liq.)
Mass					
1 kg	$= 10^3$ g	1 lb	= 16 oz	1 lb	= 453.6 g
1 mg	$= 10^{-3}$ g	1 short ton	= 2000 lb	1 g	= 0.03527 oz
1 metric ton	$= 10^3$ kg			1 metric ton	= 1.102 short ton

* This conversion factor is exact; the inch is defined to be exactly 2.54 cm. The other factors listed in this column are approximate, quoted to four significant figures. Additional digits are available if needed for very accurate calculations. For example, the pound is defined to be 453.59237 g.

Conversions between English and metric units can be made using Table 1.3.

Later (Chapters 3, 4), we'll use conversion factors with chemical units

Example 1.5 According to a highway sign, the distance from St. Louis to Chicago is 295 miles. Express this distance in kilometers.

Strategy Use Table 1.3 to find a relation between miles and kilometers. Write the conversion factor in such a way that miles cancel out and are replaced by kilometers.

Solution From Table 1.3, the required relation is

$$1 \text{ mile} = 1.609 \text{ km}$$

Since the initial unit, miles, is in the numerator, the conversion factor must have miles in the denominator, i.e., 1.609 km/1 mile:

$$295 \text{ miles} \times \frac{1.609 \text{ km}}{1 \text{ mile}} = 475 \text{ km}$$

Frequently, it is necessary to carry out more than one conversion to work a problem. This can be done by setting up successive conversion factors (Example 1.6).

Example 1.6 A certain U.S. car has a fuel efficiency rating of 36.2 miles per gallon. Convert this to kilometers per liter.

Strategy Use Table 1.3 to find a relation between kilometers and miles and between gallons and liters. Sometimes there is no direct relation shown in the table; in that case, use two relations to accomplish the desired conversion. Set up the arithmetic in a single expression, writing conversion factors in such a way that, after cancellation, only the desired unit remains.

Solution The relations to be used are

$$1 \text{ mile} = 1.609 \text{ km}$$

$$1 \text{ gallon} = 4 \text{ quarts} \qquad \text{(This is an exact relation.)}$$

$$1 \text{ L} = 1.057 \text{ quart}$$

Relations within a given system, either English or metric, are usually exact

Using these relations as conversion factors gives

$$36.2 \frac{\cancel{\text{miles}}}{\cancel{\text{gallon}}} \times \frac{1.609 \text{ km}}{1 \cancel{\text{mile}}} \times \frac{1 \cancel{\text{gallon}}}{4 \cancel{\text{qt}}} \times \frac{1.057 \cancel{\text{qt}}}{1 \text{ L}} = 15.4 \text{ km/L}$$

Notice that three conversion factors are required. First miles are converted to kilometers to obtain the fuel efficiency in kilometers per gallon. Then gallons are converted to quarts, and, finally, quarts to liters.

The conversion factor approach shown in Examples 1.5 and 1.6 will be used throughout this text. If this is your first contact with it, it may seem awkward or artificial. You will find, however, that it is the best way to solve a wide variety of problems in chemistry. It is particularly useful when multiple conversions are required (Example 1.6).

Conversion factors are certainly the way to go here

1.5 PROPERTIES OF SUBSTANCES

Every pure substance has its own unique set of properties that serve to distinguish it from all other substances. A chemist most often identifies an unknown substance by measuring its properties and comparing them to the properties recorded in the chemical literature for known substances.

The properties used to identify a substance must be *intensive;* that is, they must be independent of amount. The fact that a sample weighs 4.02 g or has a volume of 229 mL tells us nothing about its identity; mass and volume are *extensive* properties; that is, they depend on amount. Beyond that, substances may be identified on the basis of their

— **chemical properties**, observed when the substance takes part in a **chemical reaction**, a change that converts it to a new substance. For example, the fact that mercury(II) oxide decomposes to mercury and oxygen upon heating to 600°C can be used to identify it.

— **physical properties**, observed without changing the chemical identity of a substance. One such property is color; the fact that potassium chromate is yellow serves to distinguish it from a great many other substances.

To measure its chemical properties, a substance must be destroyed

In this section, we consider four physical properties that you may well have occasion to measure in the general chemistry laboratory.

Density

The density of a substance is the ratio of mass to volume:

$$\text{density} = \frac{\text{mass}}{\text{volume}} \qquad d = \frac{m}{V}$$

Note that even though mass and volume are extensive properties, the ratio of mass to volume is intensive. Samples of copper weighing 1.00 g, 10.5 g, 264 g, . . . all have the same density, 8.94 g/mL at 25°C.

Gases typically have much lower densities than liquids or solids

For liquids or gases, density can be found in a straightforward way by measuring, independently, the mass and volume of a sample (Example 1.7). For liquids, density is most often expressed in g/mL; for gases, g/L is more common.

The three layers are made up of three liquids with different densities. Gasoline is the top layer, water is the middle layer, and mercury, the densest of the three, is at the bottom. Cork floats on gasoline, oak wood sinks in gasoline but floats on water, while brass sinks in gasoline and water but floats on mercury. (Charles Steele)

Example 1.7 To determine the density of ethyl alcohol, a student pipets a 5.00 mL sample into an empty flask weighing 15.246 g. He finds that the mass of the flask + ethyl alcohol = 19.171 g. Calculate the density of ethyl alcohol.

Strategy Determine the mass of the alcohol by subtracting the mass of the empty flask from the mass of the flask and the alcohol. The volume is given. Take the quotient of mass/volume as the density.

Solution

mass of ethyl alcohol = 19.171 g − 15.246 g = 3.925 g

volume of ethyl alcohol = 5.00 mL

density = 3.925 g/5.00 mL = 0.785 g/mL

For solids, density is a bit more difficult to determine. A common approach (for insoluble solids) is shown in Figure 1.10. The mass of the solid sample is found in the usual way; its volume is found indirectly.

Melting Point and Boiling Point

The **melting point** is the temperature at which a substance changes from the solid to the liquid state. If the substance is pure, the temperature stays constant during melting. This means that, for a pure substance, the melting point is identical with the freezing point. Ice melts at 0°C; pure water freezes at that same temperature.

The **boiling point** of a liquid is the temperature at which bubbles filled with vapor form within the liquid. For reasons to be discussed in Chapter 9, boiling point depends upon the pressure above the liquid. The normal boiling point is the temperature at which a liquid boils when the pressure above it is one atmosphere. For a pure liquid, temperature remains constant during the boiling process.

If you find that a colorless liquid freezes at 0°C and boils at 100°C (at 1 atm pressure), chances are the liquid is water. To be sure, you might check the density, which should be 0.997 g/mL at 25°C. Melting point and boiling point, like density, are intensive properties. The melting point of ice is 0°C whether you're dealing with a single ice cube or a skating rink.

Solubility

Solids are usually less soluble at low temperatures

The extent to which a substance dissolves in a particular solvent can be expressed in various ways. A common method is to state the number of grams of the substance that dissolves in 100 g of solvent at a given temperature. At 20°C, about 32 g of potassium nitrate dissolves in 100 g of water. At 100°C, the solubility of this solid is considerably greater, about 246 g/100 g of water.

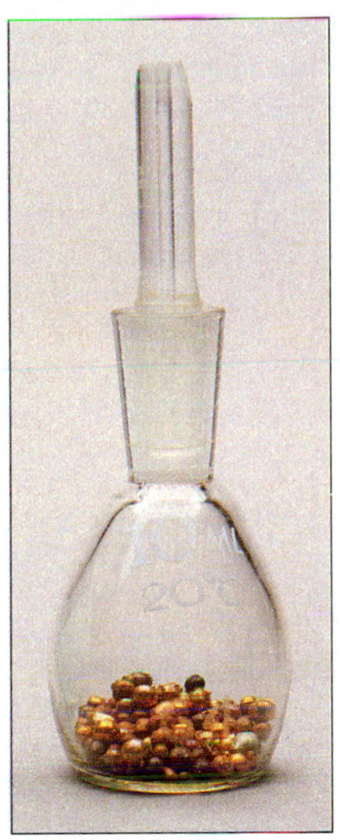

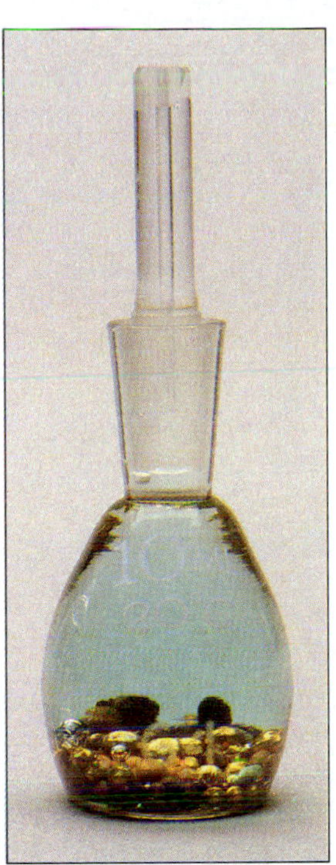

Figure 1.10
To determine the density of a solid, you first find its mass, m. To find its volume, add the solid to a flask of known volume, V, and determine the volume of water, V_m, required to fill the flask. The density of the solid is $m/(V - V_m)$. (Marna G. Clarke)

Example 1.8 Taking the solubility of potassium nitrate, KNO_3, to be 246 g/100 g water at 100°C and 32 g/100 g water at 20°C, calculate:

(a) the mass of water required to dissolve one hundred grams of KNO_3 at 100°C.
(b) The amount of KNO_3 that remains in solution when the mixture in (a) is cooled to 20°C.

Strategy The solubility at a particular temperature gives you a relationship between grams of solute (KNO_3) and grams of solvent (water). This in turn leads to the conversion factor required to calculate the mass of water in (a) or that of KNO_3 in (b). Note that the temperature is 100°C in (a), 20°C in (b).

Solubility can be used as a conversion factor

Solution

(a) Mass water required $= 100 \text{ g } KNO_3 \times \dfrac{100 \text{ g water}}{246 \text{ g } KNO_3} = \boxed{40.7 \text{ g water}}$

(b) Since the solution contains 40.7 g of water,

$$\text{mass } KNO_3 \text{ in solution} = 40.7 \text{ g water} \times \frac{32 \text{ g } KNO_3}{100 \text{ g water}} = \boxed{13 \text{ g } KNO_3}$$

The remaining 87 g of potassium nitrate crystallizes out of solution when the temperature drops from 100°C to 20°C.

PERSPECTIVE

•

Arsenic

An element everyone has heard about but almost no one has ever seen is arsenic, symbol As. It is a gray solid with some metallic properties, melts at 816°C, and has a density of 5.78 g/mL. The element occurs naturally at low concentrations in a variety of sulfide ores; it is commonly obtained as a by-product in the metallurgy of lead or copper.

The principal use of elemental arsenic is in its alloys with lead. The "lead" storage battery contains a trace of arsenic along with 3% antimony. Lead shot, which are formed by allowing drops of molten metal to fall through air, contain from 0.5 to 2.0% arsenic. It was found many years ago that adding arsenic makes the shot more nearly spherical. Nowadays, increasing amounts of arsenic are being used to make gallium arsenide semiconductors.

The "arsenic poison" referred to in true crime dramas is actually the oxide of arsenic, As_2O_3, rather than the element itself. The classic symptoms of acute arsenic poisoning involve various unpleasant gastrointestinal disturbances, severe abdominal pain, and burning of the mouth and throat. The Marsh test for arsenic (Fig. 1.A), in use until about 1960, involves the generation of the deadly poison arsine (AsH_3). This test brought a host of murderers (and a few analytical chemists) to an untimely end.

In the modern forensic chemistry laboratory, arsenic is detected by analysis of hair samples; a single strand of hair is sufficient. The technique most commonly used is neutron activation analysis, described in Chapter 18. If the concentration found is greater than about 0.0003%, poisoning is indicated; normal arsenic levels are much lower than this.

This technique was applied in the early 1960s to a lock of hair taken from Napoleon Bonaparte (1769–1821) on St. Helena. Arsenic levels of up to 50 times normal suggested he may have been a victim of poisoning, perhaps on orders from the French royal family. More recently (1991), U.S. President Zachary Taylor (1785–1850) was exhumed on the unlikely hypothesis that he had been poisoned by Southern symphathizers concerned about his opposition to the extension of slavery. The results indicated normal arsenic levels. Apparently "Old Rough and Ready" died of cholera morbus, brought on by overindulgence in overripe fruit.

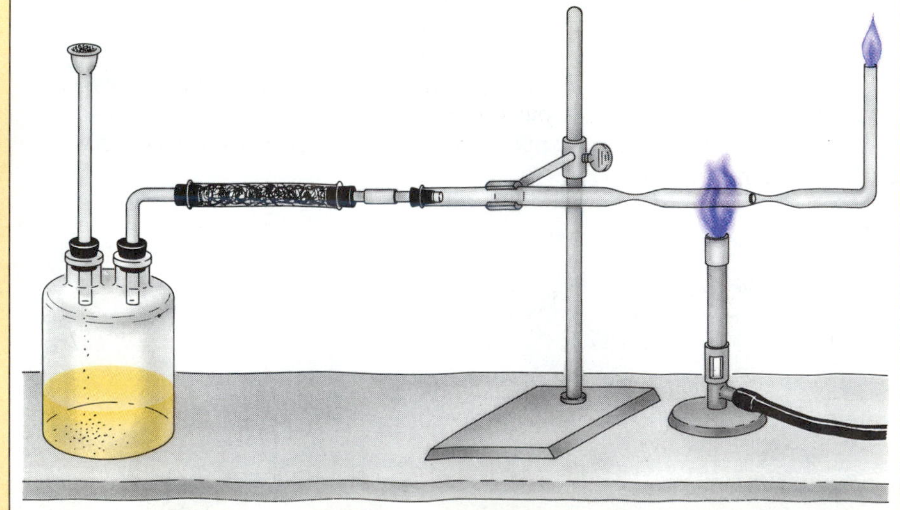

Figure 1.A
The Marsh test for arsenic involves adding zinc to an acidic solution of the unknown. Any arsenic compound present is converted to arsine, AsH_3. Heating arsine strongly decomposes it to the elements; the arsenic deposits as a shiny metallic ring on a glass tube.

CHAPTER HIGHLIGHTS

KEY CONCEPTS

1. *Convert between °F, °C, and K*
   (Example 1.1; Problems 5–8, 58)
2. *Determine the number of significant figures in a measured quantity*
   (Example 1.2; Problems 13, 14)
3. *Determine the number of significant figures in a calculated quantity*
   (Examples 1.3, 1.4; Problems 15–18)
4. *Use conversion factors to change the units of a measured quantity*
   (Examples 1.5, 1.6, 1.8; Problems 11, 12, 21–32, 43, 44, 59, 61)
5. *Relate density to mass and volume*
   (Example 1.7; Problems 35–42, 53–57, 60)

KEY EQUATIONS

$t_{°F} = 1.8t_{°C} + 32°$ $\quad$ $T_K = t_{°C} + 273.15$

KEY TERMS

boiling point	kilo-	—chemical
centi-	melting point	—extensive
compound	milli-	—intensive
conversion factor	mixture	—physical
density	nano-	significant figure
element	property	solution
joule		

SUMMARY PROBLEM

Potassium dichromate is a reddish-orange compound containing the three elements potassium, chromium, and oxygen. It has a density of 2.68 g/cm^3; its melting point is 398°C. At 20°C, its solubility is 12 g/100 g water; at 100°C, the solubility is 80 g/100 g water.

a. What are the symbols of the three elements present in potassium dichromate?
b. List the physical properties for potassium dichromate given above.
c. What is the volume of a sample of potassium dichromate weighing 32.349 g?
d. Express the density in pounds per cubic foot.
e. Express the melting point of potassium dichromate in °F and K.
f. How much water at 100°C is required to dissolve 88 g of potassium dichromate?
g. When the solution in (f) is cooled to 20°C, how much potassium dichromate remains in solution? How much crystallizes out?

Express all your answers to the correct number of significant figures; use the conversion factor approach throughout.

Answers

a. K, Cr, O
b. density, melting point, solubility
c. 12.1 cm^3 **d.** 167 lb/ft^3 **e.** 748°F; 671 K
f. 1.1×10^2 g **g.** 13 g, 75 g

QUESTIONS & PROBLEMS

The questions and problems listed here are typical of those at the end of each chapter. Some involve discussion and most require calculations, writing equations, or other quantitative work. The topic emphasized in each question or problem is indicated in the heading, such as "Symbols and Formulas" or "Significant Figures." Those in the "Unclassified" category may involve more than one concept, including, perhaps, topics from a preceding chapter. "Challenge Problems," listed at the end of the set, require extra skill and/or effort.

The "classified" questions and problems (Problems 1–44 in this set) are arranged in matched pairs, one below the other, and illustrate the same concept. For example, Questions 1 and 2 below are nearly identical in nature; the same is true of Questions 3 and 4, and so on. Problems numbered in color are answered in Appendix 4.

Measurements

1. Classify each of the following as units of mass, volume, length, density, energy, or pressure.

a. mg **b.** mL **c.** cm^3 **d.** mm
e. kg/m^3 **f.** Pa **g.** kJ

2. Classify each of the following as units of mass, volume, length, density, energy, or pressure.

a. nm **b.** kg **c.** J **d.** m^3
e. g/cm^3 **f.** atm **g.** kcal

3. Select the smaller member of each pair.

a. 303 m or 0.300 km
b. 500 kg or 0.0500 g
c. 1.50 cm^3 or 1.50×10^3 nm^3
d. 25.0 g/cm^3 or 2.50×10^{-3} kg/m^3

4. Select the smaller member of each pair.

a. 27.12 g or 27.12 kg
b. 35 cm^3 or 0.035 m^3
c. 2.87 g/L or 2.87 g/cm^3
d. 525 mm or 5.25×10^{-3} km

5. Most laboratory experiments are done at 25°C. Express this in

a. °F **b.** K

6. A child has a temperature of 104°F. What is his temperature in

a. °C **b.** K

7. Carbon dioxide, CO_2, at room temperature (70°F) is a gas. It can be frozen at −69.7°F and 5 atm pressure to solid carbon dioxide, popularly known as dry ice. What is the freezing point of carbon dioxide in °C?

8. Superconductors use liquid nitrogen as a coolant. Liquid nitrogen boils at −195.8°C. What is its boiling point in

a. °F **b.** K

9. Express the following derived units in terms of the base SI units listed in Table 1.2 (e.g., 1 $cm^3 = 10^{-6}$ m^3).

a. milliliter **b.** joule **c.** pascal

10. Follow the directions in Question 9 for the following derived units.

a. liter **b.** newton **c.** kilopascal

11. Convert

a. 0.863 atm to kilopascals **b.** 226 kJ to calories.

12. Convert

a. 2.63×10^4 Pa to atmospheres
b. 363 kcal to joules.

Significant Figures

13. How many significant figures are there in each of the following?

a. 1.92 mm **b.** 0.032100 g
c. 6.022×10^{23} atoms **d.** 460.00 L
e. 0.00036 cm^3 **f.** 2×10^9 nm

14. How many significant figures are there in each of the following?

a. 23.437 m **b.** 0.002017 g **c.** 30.0×10^{20} nm
d. 50.010 L **e.** 2.30790 atm **f.** 350 miles

15. A student prepares a salt solution by dissolving 85.638 g of sodium chloride (NaCl) in enough water to form 237 cm^3 of solution. Calculate the number of grams of salt per cubic centimeter of solution.

16. Calculate the volume of a sodium atom, which has a radius of 0.186 nm. Assume the atom is spherical. The volume of a sphere is given by the expression $V = 4\pi r^3/3$.

17. Calculate the following to the correct number of significant figures.

a. $x = \dfrac{1.27\text{ g}}{5.296\text{ cm}^3}$ **b.** $x = \dfrac{12.235\text{ g}}{1.01\text{ L}}$

c. $x = 12.2\text{ g} + 0.38\text{ g}$ **d.** $x = \dfrac{17.3\text{ g} + 2.785\text{ g}}{30.20\text{ cm}^3}$

18. How many significant figures are there in the values of x obtained from

a. $x = \dfrac{34.0300\text{ g}}{12.09\text{ cm}^3}$
b. $x = 32.647\text{ g} - 32.327\text{ g}$
c. $x = (0.00630\text{ cm})(2.003\text{ cm})(200.0\text{ cm})$
d. $x = \dfrac{236.45\text{ g} - 1.3\text{ g}}{(3.4561\text{ cm})(32.675\text{ cm}^2)}$

19. Round off the following quantities to the indicated number of significant figures.

a. 12.2654 g (4 significant figures)
b. 32.4892 cm (5 significant figures)
c. 75.648 mL (3 significant figures)
d. 126.5 oz (3 significant figures)

20. Round off the following quantities to the indicated number of significant figures.

a. 3.5500 L (2 significant figures)
b. 8394 nm (2 significant figures)
c. 72.4503 g (3 significant figures)
d. 0.39261 kJ (4 significant figures)

Conversion Factors

21. Using Table 1.3, convert 17.5 quarts to

a. liters **b.** cubic meters **c.** cubic feet

22. Using Table 1.3, convert 38.75 ft^2 to

a. square miles **b.** km^2 **c.** m^2

23. The unit of land measure in the metric system is the hectare; in the English system it is the acre. A square exactly 100 m on a side has an area of one hectare; if it is 208.7 ft on a side, the area is one acre. How many acres are there in one hectare?

24. In the United States, cigarettes are smoked at the rate of 2.0×10^4 cigarettes/s. At this rate, how many are smoked in one day?

25. A day on Mars is 8.864×10^4 seconds long and a year is 5.935×10^7 seconds long.

- **a.** How many earth days are there in one Mars day?
- **b.** How many earth days are there in one Mars year?

26. A typical aerobic walker on a track covers a lap (0.25 mile) in about 3.4 minutes. A champion marathon runner takes about five minutes to cover a mile. If an aerobic walker competed in the Boston marathon (42.16 km), how many minutes would you expect him to finish behind the winner?

27. During earlier times in England, land was measured in units such as fardells, nookes, yards, and kides:

2 fardells = 1 nooke 4 nookes = 1 yard
4 yards = 1 kide

Thus,

a. 6.00 kides = ______________ fardells
b. 15 nookes = ______________ kides
= ______________ fardells

28. When the Pharmacopoeia of London was compiled in 1618, the troy system of measure was used to prepare medicine. Among the units used were

20 grains = 1 scruple 3 scruples = 1 drachm
8 drachms = 1 ounce 12 ounces = 1 pound

Thus,

a. 12.05 pounds = ______________ drachms
b. 25.0 drachms = ______________ pounds
= ______________ ounces

29. In Germany, nutritional information is given in kilojoules instead of nutritional calories (1 nutritional calorie = 1 kcal). A packet of Rindfleisch-Suppe (meat soup) has the following information:

250 mL of prepared soup = 235 kJ

A packet of the same soup sold in the United States would have the same nutritional information in kilocalories per cup. There are two cups to a pint. How many nutritional calories would be quoted per cup of prepared soup?

30. White gold is an alloy that typically contains 60.0% by mass of gold; the rest is platinum. If 175 g of gold are available, how many grams of platinum are required to combine with the gold to form this alloy?

31. Because of the rapid rise in the price of copper, the Bureau of the Mint in 1982 changed the composition of the penny. The penny is now an alloy of 97.6% Zn and 2.4% Cu. Pre-1982 pennies were made up of 95% Cu and 5% Zn. If a new penny has a mass of 2.507 g, how many grams of each metal are there in the new penny?

32. One type of antiperspirant uses aluminum chlorohydrate as its active ingredient. This compound is made up of 30.93% Al, 45.86% O, 2.89% H, and 20.32% Cl. How many grams of each element are there in one ounce of this compound?

Physical and Chemical Properties

33. The following data refer to the element carbon. Classify each as a physical or chemical property.

- **a.** It is virtually insoluble in water.
- **b.** It exists in several forms, e.g., diamond, graphite.
- **c.** It is a solid at 25°C and 1 atm.
- **d.** It reacts with oxygen to form carbon dioxide.

34. The following data refer to the element iodine. Classify each as a physical or chemical property.

- **a.** Its density at 20°C and 1 atm is 4.93 g/cm^3.
- **b.** Iodine has a purple color.
- **c.** It reacts with chlorine.
- **d.** Its normal melting point is 113.5°C.

35. The mass and volume of a piece of pure graphite are 0.600 g and 0.270 mL. What is the density of graphite?

36. A sample of mercury has a volume of 0.250 L and weighs 3.40 kg. Calculate the density of mercury in g/cm^3.

37. A piece of metal weighing 20.32 g is added to a flask with a volume of 24.5 cm^3. It is found that 18.52 g of water (d = 1.00 g/cm^3) must be added to the metal to fill the flask. What is the density of the metal?

38. A solid with an irregular shape weighing 13.56 g is added to a flask with a volume of 20.35 cm^3. It is found that 10.35 g of benzene (d = 0.879 g/cm^3) must be added to the metal to fill the flask. What is the density of the metal?

39. A water bed filled with water has the dimensions 8.0 ft × 7.0 ft × 0.75 ft. Taking the density of water to be 1.00 g/cm^3, determine the number of kilograms of water required to fill the water bed.

40. Blood plasma volume for adults is about 3.1 L. Its density is 1.020 g/cm^3. How many pounds of blood plasma are there in your body?

41. Air is 21% oxygen by volume. Oxygen has a density of 1.31 g/L. What is the volume in liters of a room that holds enough air to contain 75 kg of oxygen?

42. A water solution contains 10.0% ethyl alcohol by mass and has a density of 0.983 g/cm^3. What mass in grams of ethyl alcohol is present in 7.50 L of this solution?

43. The solubility of potassium chloride is 37.0 g/100 g water at 30°C. Calculate at 30°C:

- **a.** the mass of potassium chloride that dissolves in 34.5 g of water.
- **b.** the mass of water required to dissolve 34.5 g of potassium chloride.

44. At 25°C and 1 atm, 0.1449 g of carbon dioxide dissolves in 100.0 grams of water. Calculate at 25°C and 1 atm:

- **a.** the mass of carbon dioxide that dissolves in 5.00 mL of water (d = 1.00 g/cm^3).
- **b.** the mass of water required to dissolve 5.00 g of carbon dioxide.

Arsenic

45. What is the symbol for arsenic? Give two physical properties of arsenic.

46. Describe three uses for elemental arsenic.

47. How is arsenic poisoning detected today? What levels of arsenic indicate poisoning?

48. Analysis of a human hair weighing 1.23 mg shows that it contains 1.25×10^{16} atoms of arsenic (6.022×10^{23} As atoms weigh 74.92 g). Determine the mass percent of arsenic in the hair. How likely is it that the person from whom the hair was taken was subjected to arsenic poisoning?

Unclassified

49. How do you distinguish
- **a.** chemical properties from physical properties?
- **b.** a solute from a solution?
- **c.** a compound from a mixture?

50. How do you distinguish
- **a.** an element from a compound?
- **b.** an element from a mixture?
- **c.** a solution from a heterogeneous mixture?
- **d.** distillation from filtration?

51. An intensive property is independent of the mass of the sample. Which of the following properties are intensive?
- **a.** heat required to raise the temperature 1°C
- **b.** boiling point
- **c.** solubility per 100 g of water

52. An extensive property is one that depends on the amount of the sample. Which of the following properties are extensive?

a. volume **b.** density **c.** temperature
d. energy **e.** melting point

53. Diamonds ($d = 3.51$ g/cm^3) are commonly measured in carats (1 carat = 200 mg). What is the volume of a 2.0-carat diamond?

54. A swimming pool that is 4.05 m wide and 6.10 m long has an average depth of 8.0 ft. The water solution in the pool ($d = 1.00$ g/cm^3) contains 0.70% chlorine by mass as a disinfectant. What is the mass of chlorine in the pool?

55. Titanium is used in airplane bodies because it is both strong and light. It has a density of 4.55 g/cm^3. If a cylinder of titanium is 4.75 cm long and has a mass of 104.2 g, calculate the diameter of the cylinder ($V = \pi r^2 l$, where V is the volume of the cylinder, r is its radius, and l is the length.)

56. Take the price of gold to be $310 an ounce. Its density is 19.3 g/cm^3. What is the value of a bar of gold two inches thick, a foot long, and 6.50 inches wide?

57. A pycnometer is a device used to determine density. It weighs 20.455 g empty and 31.486 g when filled with water ($d = 1.000$ g/cm^3). Pieces of an alloy are put into the empty, dry pycnometer. The mass of the alloy and pycnometer is 28.695 g. Water is added to the alloy to exactly fill the pycnometer. The mass of the pycnometer, water, and alloy is 38.689 g. What is the density of the alloy?

Challenge Problems

58. At what point is the temperature in °C exactly twice that in °F?

59. Oil spreads on water to form a film about 120 nm thick (two significant figures). How many square kilometers of ocean will be covered by the slick formed when two barrels of oil are spilled (1 barrel = 31.5 U.S. gallons)?

60. A laboratory experiment requires 0.750 g of aluminum wire ($d = 2.70$ g/cm^3). The diameter of the wire is 0.0179 inches. Determine the length of the wire, in centimeters, to be used for this experiment. The volume of a cylinder is $\pi r^2 l$, where r = radius and l = length.

61. An average human male breathes about 8.50×10^3 L of air per day. The concentration of lead (Pb) in highly polluted urban air is 7.0×10^{-6} g Pb/m^3 of air. Assume that 75% of the lead is present as particles less than 1.0×10^{-6} m in diameter and that 50% of the particles below that size are retained in the lungs. Calculate the mass of lead absorbed in this manner in one year by an average male living in this environment.

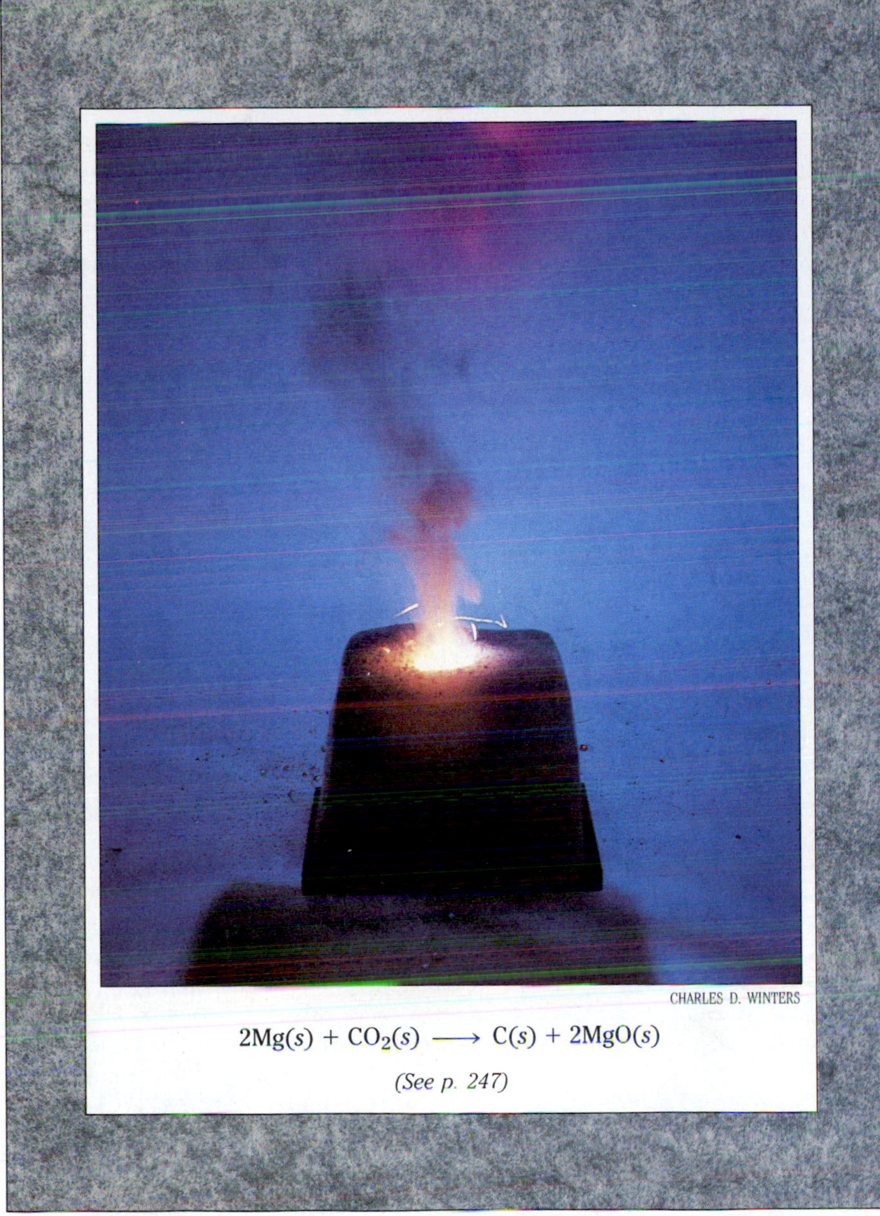

CHARLES D. WINTERS

$$2Mg(s) + CO_2(s) \longrightarrow C(s) + 2MgO(s)$$

(See p. 247)

2 Atoms, Molecules, and Ions

Atom from atom yawns as far
As moon from earth, or star
from star.

—RALPH WALDO EMERSON
Atoms

CHAPTER OUTLINE

To learn chemistry, you must become familiar with the "building blocks" that chemists use to describe the structure of matter. These include

— **atoms** (Section 2.1), which in turn are composed of *electrons, protons,* and *neutrons* (Section 2.2). Each type of atom is characteristic of a particular element. A classification system for elements known as the periodic table is introduced in Section 2.3.

— **molecules**, the building blocks of several elements and a great many compounds. Molecular substances can be identified by their formulas (Section 2.4) or their names (Section 2.6).
— **ions**. An ionic compound is made up of two different kinds of ions of opposite charge. Using relatively simple principles, it is possible to derive the formulas (Section 2.5) and names (Section 2.6) of ionic compounds.

2.1 ATOMS AND THE ATOMIC THEORY

In 1808, an English scientist and schoolteacher, John Dalton, developed the atomic model of matter that underlies modern chemistry. Three of the main postulates of modern atomic theory, all of which Dalton suggested, are stated below and illustrated in Figure 2.1.

Atoms are indeed "tiny"; they have diameters of about 10^{-10} m (0.1 nm)

1. *An element is composed of tiny particles called atoms.* All atoms of a given element show the same chemical properties. Atoms of different elements show different properties.
2. *In an ordinary chemical reaction, no atom of any element disappears* or is changed into an atom of another element.

In the compound ammonia, there are three H atoms for every N atom

3. *Compounds are formed when atoms of two or more elements combine.* In a given compound, the relative numbers of atoms of each kind are definite

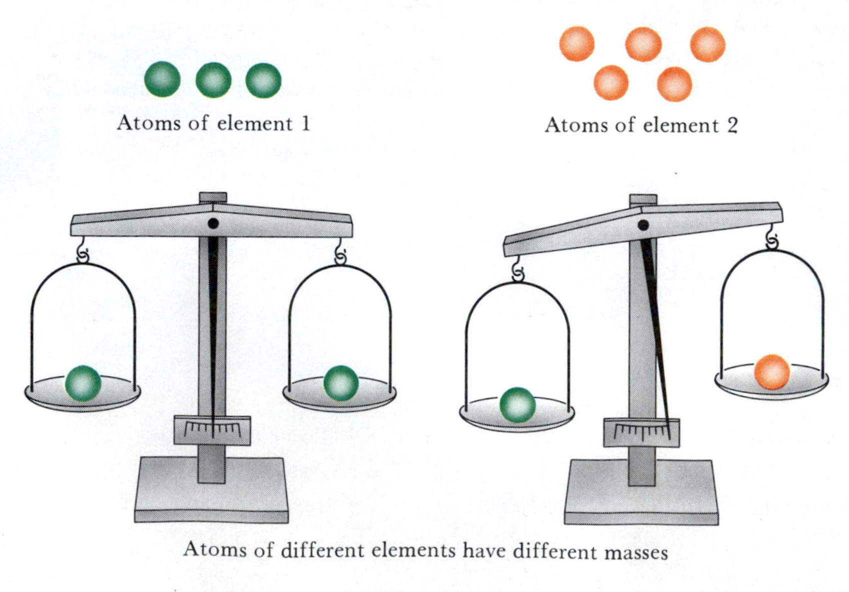

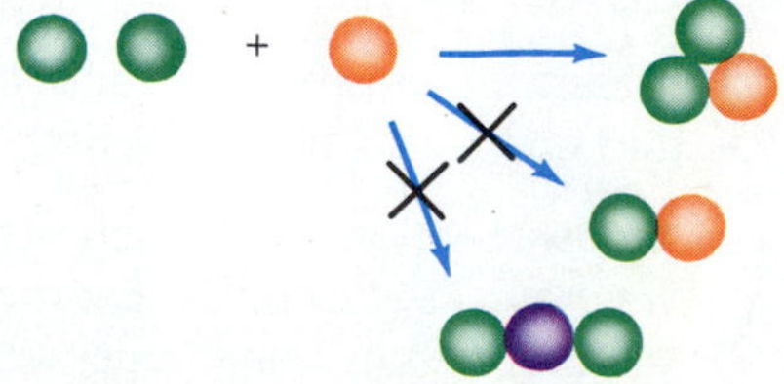

+
No atom disappears or is changed in a chemical reaction

Figure 2.1
Some features of Dalton's theory.

and constant. In general, these relative numbers can be expressed as integers or simple fractions.

On the basis of Dalton's theory, the atom can be defined as the smallest particle of an element that can enter into a chemical reaction.

Dalton was a quiet, unassuming man and a devout Quaker. When presented to King William IV of England, Dalton refused to wear the colorful court robes because of his religion. His friends persuaded him to wear the scarlet robes of Oxford University, from which he had a doctor's degree. Unfortunately, Dalton was colorblind, so he saw himself clothed only in gray. He had little time for anything but science and teaching, never married, and indulged in only one sport: lawn bowling.

THE E.F. SMITH MEMORIAL COLLECTION IN THE HISTORY OF CHEMISTRY, DEPARTMENT OF SPECIAL COLLECTIONS, VAN PELT-DIETRICH LIBRARY, UNIVERSITY OF PENNSYLVANIA

Dalton's atomic theory was successful because it explained three of the basic laws of chemistry:

The **law of conservation of mass**: This law was first stated (Chapter 1) by Lavoisier in 1789. In modern form, it says that *there is no detectable change in mass in an ordinary chemical reaction.* If atoms are "conserved" in a reaction (Postulate 2 above), mass will also be conserved.

The **law of constant composition**: This tells us that *a compound always contains the same elements in the same proportions by mass.* If the atom ratio of the elements in a compound is fixed (Postulate 3), their proportions by mass must also be fixed.

The **law of multiple proportions**: This law, formulated by Dalton himself, was crucial to the establishment of atomic theory. It applies to situations in which two elements form more than one compound. The law states that in these compounds, *the masses of one element that combine with a fixed mass of the second element are in a ratio of small whole numbers.*

The validity of this law depends upon the fact that atoms combine in simple, whole-number ratios (Postulate 3). Its relation to atomic theory is further illustrated in Figure 2.A.

Figure 2.A
Chromium forms two different compounds with oxygen, as shown by their different colors. In the green compound on the left, there are two chromium atoms for every three oxygen atoms (2 Cr:3 O) and 2.167 g of chromium per gram of oxygen. In the red compound on the right, there is one chromium atom for every three oxygen atoms (1 Cr:3 O) and 1.083 g of chromium per gram of oxygen. The ratio of the chromium masses, 2.167:1.083, is that of two small whole numbers, 2.167:1.083 = 2:1, an illustration of the law of multiple proportions. (Marna G. Clarke)

Figure 2.2
J. J. Thomson and Ernest Rutherford (*right*) talking, perhaps, about nuclear physics, more likely about yesterday's cricket match. (AIP Niels Bohr Library, Bainbridge Collection)

2.2 COMPONENTS OF THE ATOM

Like any useful scientific theory, the atomic theory raised more questions than it answered. Scientists wondered whether atoms, tiny as they are, could be broken down into still smaller particles. Nearly 100 years passed before the existence of subatomic particles was confirmed by experiment. Two future Nobel laureates did pioneer work in this area. J.J. Thomson was an English physicist working at the Cavendish Laboratory at Cambridge. Ernest Rutherford, at one time a student of Thomson (Fig. 2.2), was a native of New Zealand. Rutherford carried out his research at McGill University in Montreal and at Manchester and Cambridge in England.

Electrons

The first evidence for the existence of subatomic particles came from studies of the conduction of electricity through gases at low pressures. When the glass tube shown in Figure 2.3 is partially evacuated and connected to a spark coil, an electric current flows through it. Associated with this flow are colored rays of light called *cathode rays*, which are bent by both electric and magnetic fields. From a careful study of this deflection, J.J. Thomson showed in 1897 that the rays consist of a stream of negatively charged particles, which he called **electrons**. Electrons are common to all atoms, carry a unit negative charge (−1), and have a very small mass, roughly 1/2000 of that of the lightest atom.

Every atom contains a definite number of electrons. This number, which runs from 1 to over 100, is characteristic of a neutral atom of a particular element. All atoms of hydrogen contain one electron; all atoms of the element uranium contain 92 electrons. We will have more to say in Chapter 6 about how these electrons are arranged relative to one another. Right now, you need only know that they are found in the outer regions of the atom, where they form what amounts to a cloud of negative charge.

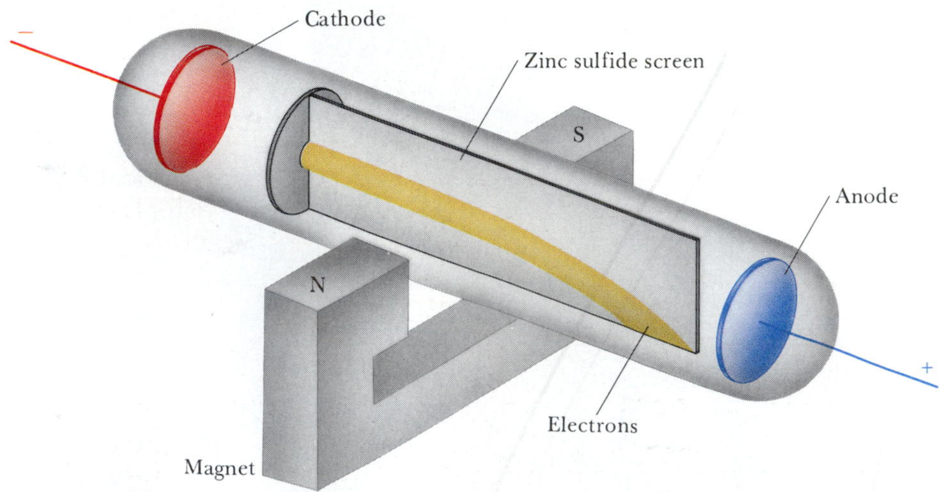

Figure 2.3
Cathode ray tube. The ray, shown here as a yellow beam, is made up of fast-moving electrons. In an electric or magnetic field, the beam is deflected in such a way as to indicate that it carries a negative charge.

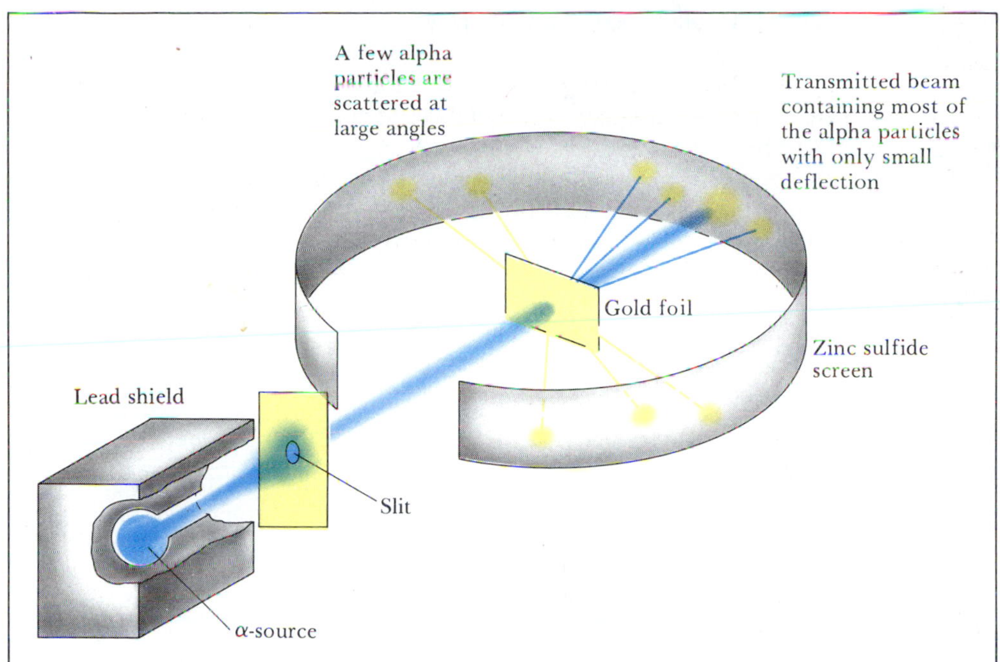

Figure 2.4
Rutherford's scattering experiment. Most of the α-particles are essentially undeflected, but a few are scattered at large angles. In order to cause the large deflections, atoms must contain heavy, positively charged nuclei.

Protons and Neutrons; the Atomic Nucleus

A series of experiments carried out under the direction of Ernest Rutherford in 1911 shaped our ideas about the nature of the atom. He and his students bombarded a piece of thin gold foil (Fig. 2.4) with α-particles (helium atoms minus their electrons). With a fluorescent screen, they observed the extent to which the α-particles were scattered. Most of the particles went through the foil unchanged in direction; a few, however, were reflected back at acute angles. This was a totally unexpected result, inconsistent with the model of the atom in vogue at that time. In Rutherford's words, "It was as though you had fired a 15-inch shell at a piece of tissue paper and it had bounced back and hit you." By a mathematical analysis of the forces involved, Rutherford showed that the scattering was caused by a small, positively charged **nucleus** at the center of the gold atom.

Since the time of Rutherford, scientists have learned a great deal about the properties of atomic nuclei. For our purposes in chemistry, the nucleus of an atom can be considered to consist of two different types of particles (Table 2.1).

Table 2.1 Properties of Subatomic Particles

Particle	Location	Relative Charge	Relative Mass*
Proton	nucleus	+1	1.00728
Neutron	nucleus	0	1.00867
Electron	outside nucleus	−1	0.00055

* These are expressed in atomic mass units (Chap. 3).

The "ordinary" H atom contains one proton, one electron, and no neutrons

1. The **proton**, which has a mass nearly equal to that of an ordinary hydrogen atom. The proton carries a unit positive charge (+1), equal in magnitude to that of the electron (−1).
2. The **neutron**, an uncharged particle with a mass slightly greater than that of a proton.

Atomic Number

The atomic number is always an integer, 1, 2, 3, . . .

All the atoms of a particular element have the same number of protons in the nucleus. This number is a basic property of an element, called its **atomic number** and given the symbol Z:

$$Z = \text{number of protons}$$

In a neutral atom, the number of protons in the nucleus is exactly equal to the number of electrons outside the nucleus. Consider, for example, the elements hydrogen ($Z = 1$) and uranium ($Z = 92$). All hydrogen atoms have one proton in the nucleus; all uranium atoms have 92. In a neutral hydrogen atom there is one electron outside the nucleus; in a uranium atom there are 92.

Mass Numbers; Isotopes

The **mass number** of an atom, given the symbol A, is found by adding up the number of protons and neutrons in the nucleus:

$$A = \text{number of protons} + \text{number of neutrons}$$

All atoms of a given element have the same number of protons, hence the same atomic number. They may, however, differ from one another in mass and hence in mass number. This can happen because, although the number of protons in the nucleus of a given kind of atom is fixed, the number of neutrons is not. It may vary and often does. Consider the element hydrogen ($Z = 1$). There are three different kinds of hydrogen atoms. They all have one proton in the nucleus. A "light" hydrogen atom (the most common type) has no neutrons in the nucleus ($A = 1$). Another type of hydrogen atom (deuterium) has one neutron ($A = 2$). Still a third type (tritium) has two neutrons ($A = 3$).

Atoms that contain the same number of protons but a different number of neutrons are called **isotopes**. The three kinds of hydrogen atoms just described are isotopes of that element. They have masses that are very nearly in the ratio 1:2:3. Among the isotopes of the element uranium are the following:

Isotope	Z	A	Number of Protons	Number of Neutrons
Uranium-235	92	235	92	143
Uranium-238	92	238	92	146

The composition of a nucleus is shown by its **nuclear symbol**. Here, the atomic number appears as a subscript at the lower left of the symbol of the element. The mass number is written as a superscript at the upper left.

No. of neutrons = $A - Z$

$$\begin{array}{l} \text{mass number} \longrightarrow \\ \text{atomic number} \longrightarrow \end{array} {}^{A}_{Z}\text{X} \longleftarrow \text{element symbol}$$

The nuclear symbols for the isotopes of hydrogen and uranium referred to above are

$$^{1}_{1}H,\ ^{2}_{1}H,\ ^{3}_{1}H \qquad ^{235}_{92}U,\ ^{238}_{92}U$$

Quite often, isotopes of an element are distinguished from one another by writing the mass number after the symbol of the element. The isotopes of uranium are often referred to as U-235 and U-238.

Example 2.1

(a) An isotope of cobalt (Co) is used in radiation therapy for certain types of cancer. Write nuclear symbols for three isotopes of cobalt ($Z = 27$) in which there are 29, 31, and 33 neutrons, respectively.
(b) One of the most harmful components of nuclear waste is a radioactive isotope of strontium, $^{90}_{38}Sr$; it can be deposited in your bones, where it replaces calcium. How many protons are there in the nucleus of Sr-90? How many neutrons?

Strategy Remember the definitions of atomic number and mass number and where they appear in the nuclear symbol.

Solution

(a) The mass numbers are

$$27 + 29 = 56 \qquad 27 + 31 = 58 \qquad 27 + 33 = 60$$

Thus the nuclear symbols are $^{56}_{27}Co$, $^{58}_{27}Co$, $^{60}_{27}Co$.

(b) The number of protons is given by the atomic number (left subscript) and is 38. The mass number (left superscript) is 90. The number of neutrons is $90 - 38 = 52$

2.3 INTRODUCTION TO THE PERIODIC TABLE

From a structural point of view, an element is a substance all of whose atoms have the same number of protons, i.e., the same atomic number. The chemical properties of elements depend upon their atomic numbers, which can be read off the **periodic table** (inside front cover of this text). The atomic number is given directly above the symbol of the element. For example, sulfur (S) has an atomic number of 16; lead (Pb) has an atomic number of 82.

The periodic table is useful for a great many purposes. Here, we will take a brief look at some of the characteristics of the table. Later, in Chapter 6, the periodic table will be examined in greater detail.

Periods and Groups

The horizontal rows in the table are referred to as **periods**. The first period consists of the two elements, hydrogen (H) and helium (He). The second period starts with lithium (Li) and ends with neon (Ne).

The vertical columns are known as **groups** or **families**. Historically, a variety of different systems have been used to designate the different groups. Both

Other periodic tables label the groups differently, but the elements are in the same position

Arabic and Roman numerals have been used in combination with the letters A and B. The system used in this text is the one recommended by the International Union of Pure and Applied Chemistry (IUPAC) in 1985. The groups are numbered from 1 to **18**, starting at the left.

Elements falling in Groups 1, 2, **13**, **14**, **15**, **16**, **17** and **18*** are referred to as **main-group elements**. The ten elements in the center of periods 4 through 6 are called **transition elements**; they fall in Groups 3 through 12. The first transition series (period 4) starts with Sc (Group 3) and ends with Zn (Group 12).

Certain main groups are given special names. The elements in Group 1, at the far left of the periodic table, are called *alkali metals;* those in Group 2 are referred to as *alkaline earth metals.* Moving to the right, the elements in Group **17** are called *halogens;* at the far right, the *noble* (unreactive) *gases* constitute Group **18**.

Elements in the same main group show very similar chemical properties. For example,

— lithium (Li), sodium (Na), and potassium (K) in Group 1 all react vigorously with water to produce hydrogen gas.
— helium (He), neon (Ne), and argon (Ar) in Group **18** do not react with any other substances.

On the basis of observations such as these, we can say that *the periodic table is an arrangement of elements, in order of increasing atomic number, in horizontal rows of such a length that elements with similar chemical properties fall directly beneath one another in vertical groups.*

Metals and Nonmetals

The diagonal line or stairway that starts to the left of boron in the periodic table separates metals from nonmetals. The more than 80 elements to the left and below that line, shown in blue in the table, have the properties of **metals**; in particular they have high electrical conductivities. Elements above and to the right of the stairway are **nonmetals** (red); about 20 elements fit in that category.

Along the diagonal line in the periodic table are several elements that are difficult to classify exclusively as metals or nonmetals. They have properties in between those of elements in the two classes. In particular, their electrical conductivities are intermediate between those of metals and nonmetals. The six elements

These elements, particularly Si, are used in semiconductor devices such as transistors and solar cells

B	Si	Ge	As	Sb	Te
boron	silicon	germanium	arsenic	antimony	tellurium

are often called **metalloids**.

2.4 MOLECULES AND IONS

Isolated atoms rarely occur in nature; only the noble gases (He, Ne, Ar, . . .) consist of individual, nonreactive atoms. Atoms tend to combine with one another in various ways to form more complex structural units. Two such units, which serve as building blocks for a great many elements and compounds, are molecules and ions.

* Prior to 1985, Groups **13–18** were commonly numbered 3–8 or 3A–8A in the United States.

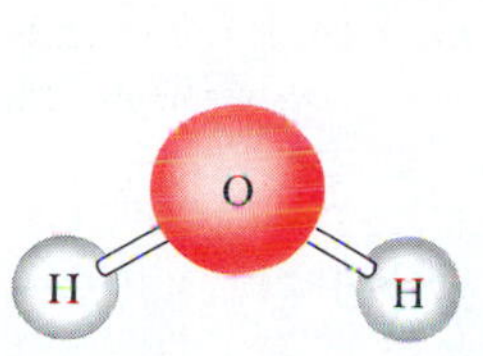

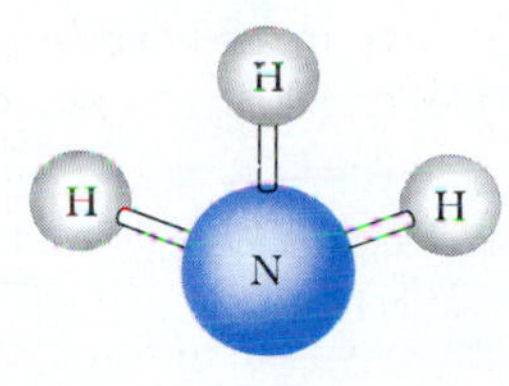

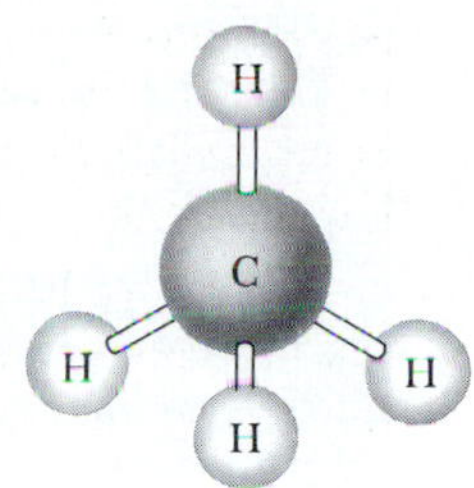

Figure 2.5
Ball-and-stick models of H_2O, NH_3, and CH_4. The "sticks" represent covalent bonds between H atoms and O, N, or C atoms.

Molecules

Two or more atoms may combine with one another to form an uncharged **molecule**. The atoms involved are usually those of nonmetallic elements. Within the molecule, atoms are held to one another by strong forces called *covalent bonds,* which consist of shared pairs of electrons (Chapter 7). Forces between neighboring molecules, in contrast, are quite weak.

"Super glue" is weak compared to covalent bonds

The structures of molecules are sometimes represented by **structural formulas**, which show the bonding pattern within the molecule. The structural formulas of hydrogen chloride, water, ammonia, and methane are

$$\mathrm{H{-}Cl} \qquad \mathrm{H{-}O{-}H} \qquad \begin{matrix} \mathrm{H{-}N{-}H} \\ | \\ \mathrm{H} \end{matrix} \qquad \begin{matrix} \mathrm{H} \\ | \\ \mathrm{H{-}C{-}H} \\ | \\ \mathrm{H} \end{matrix}$$

The dashes represent covalent bonds. The three-dimensional geometries of these molecules are shown in Figure 2.5.

Most commonly, molecular substances are represented by **molecular formulas**, in which the number of atoms of each element is indicated by a subscript written after the symbol of the element. The molecular formulas of the substances just described are

Molecular formula = n(simplest formula), where $n = 1, 2, 3, \ldots$

hydrogen chloride: HCl (1 H atom, 1 Cl atom per molecule)
water: H_2O (2 hydrogen atoms, 1 oxygen atom per molecule)
ammonia: NH_3 (3 hydrogen atoms, 1 nitrogen atom per molecule)
methane: CH_4 (4 hydrogen atoms, 1 carbon atom per molecule)

The **simplest** *(empirical)* **formula** of any compound gives the simplest, whole-number ratio of the atoms present. For a molecular substance, the molecular formula may be identical with the simplest formula or a whole-number multiple of it (Table 2.2, page 34).

Example 2.2 Write the molecular and simplest formula of

(a) glucose, whose molecule contains 6 carbon atoms, 12 hydrogen atoms, and 6 oxygen atoms.
(b) ethanol, with 2 carbon atoms, 6 hydrogen atoms, and 1 oxygen atom per molecule.

Strategy For the molecular formula, show the number of atoms of each type as subscripts. To find the simplest formula, reduce to the simplest whole-number ratio.

Solution

(a) Molecular formula: $C_6H_{12}O_6$ Simplest formula: CH_2O

(b) Molecular and simplest formula: C_2H_6O

Table 2.2 Types of Formulas

Substance	Structural Formula	Molecular Formula	Simplest Formula
Water	H—O—H	H_2O	H_2O
Hydrogen peroxide	H—O—O—H	H_2O_2	HO
Methane	H—C(H)(H)—H	CH_4	CH_4
Ethane	H—C(H)(H)—C(H)(H)—H	C_2H_6	CH_3

GROUP 15	GROUP 16	GROUP 17	GROUP 18
		$H_2(g)$	
$N_2(g)$	$O_2(g)$	$F_2(g)$	
$P_4(s)$	$S_8(s)$	$Cl_2(g)$	
		$Br_2(l)$	
		$I_2(s)$	

Figure 2.6
Molecular elements and their physical states in the periodic table.

Elements as well as compounds can be molecular in nature. In hydrogen gas, the basic building block is a molecule consisting of two hydrogen atoms joined by a covalent bond:

H—H

Other molecular elements are shown in Figure 2.6.

Ions

When an atom loses or gains electrons, charged particles called **ions** are formed. Typically, metal atoms tend to lose electrons to form positively charged ions called **cations**. Examples include the Na^+ and Ca^{2+} ions, formed from atoms of the metals sodium and calcium:

$$\underset{(11\,p^+,\,11\,e^-)}{\text{Na atom}} \longrightarrow \underset{(11\,p^+,\,10\,e^-)}{\text{Na}^+\text{ ion}} + e^-$$

$$\underset{(20\,p^+,\,20\,e^-)}{\text{Ca atom}} \longrightarrow \underset{(20\,p^+,\,18\,e^-)}{\text{Ca}^{2+}\text{ ion}} + 2\,e^-$$

(The arrows separate *reactants,* Na and Ca atoms, from *products,* cations and electrons.)

Nonmetal atoms form negative ions (**anions**) by gaining electrons. Consider, for example, what happens when atoms of the nonmetals chlorine and oxygen acquire electrons:

Ions are formed when metal atoms react with nonmetal atoms: $Na + Cl \rightarrow Na^+ + Cl^-$

$$\underset{(17p^+,\,17e^-)}{\text{Cl atom}} + e^- \longrightarrow \underset{(17p^+,\,18e^-)}{\text{Cl}^-\text{ ion}}$$

$$\underset{(8p^+,\,8e^-)}{\text{O atom}} + 2e^- \longrightarrow \underset{(8p^+,\,10e^-)}{\text{O}^{2-}\text{ ion}}$$

Notice that when an ion is formed, the number of protons in the nucleus is unchanged. It is the number of electrons that increases or decreases.

Example 2.3 Give the number of protons and electrons in Al^{3+}, a cation suspected of playing a role in Alzheimer's disease.

Strategy The atomic number (periodic table) gives the number of protons and electrons in the neutral atom. The positive charge tells you how many electrons have been lost.

Solution The atomic number of aluminum is 13; the charge of the cation is +3.

Hence: no. protons = 13; no. electrons = 13 − 3 = 10.

The ions dealt with to this point (e.g., Na^+, Cl^-) are **monatomic**; that is, they are derived from a single atom by the loss or gain of electrons. Many of the most important ions in chemistry are **polyatomic**, containing more than one atom. Examples include the hydroxide ion (OH^-) and the ammonium ion (NH_4^+). In these and other polyatomic ions, the atoms are held together by covalent bonds, e.g.,

$$(\mathrm{O{-}H})^- \qquad \left(\begin{array}{c} \mathrm{H} \\ | \\ \mathrm{H{-}N{-}H} \\ | \\ \mathrm{H} \end{array}\right)^+$$

In a very real sense, you can think of a polyatomic ion as a "charged molecule."

Since a bulk sample of matter is electrically neutral, ionic compounds always contain both cations (positively charged particles) and anions (negatively charged particles). Ordinary table salt, sodium chloride, is made up of an equal number of Na^+ and Cl^- ions. The structure of sodium chloride is shown in Figure 2.7. Notice that there are no discrete molecules; positive and negative ions are bonded together in a continuous network.

You can't buy a bottle of Na^+ ions

Ionic compounds are held together by strong electrical forces between oppositely charged ions (e.g., Na^+, Cl^-). These forces are referred to as **ionic bonds**. Typically, ionic compounds are solids at room temperature and have

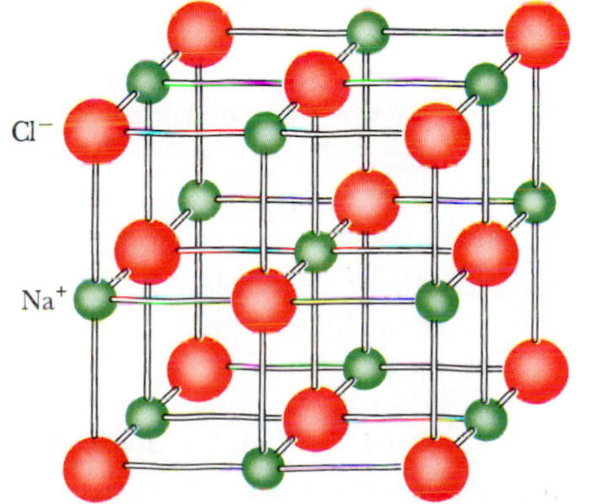

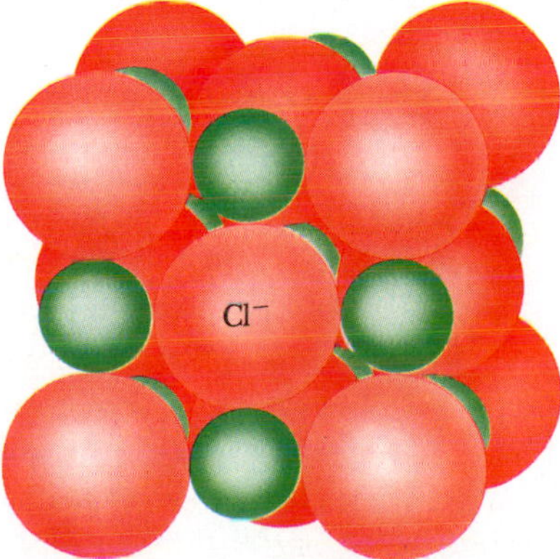

Figure 2.7
Two different ways of showing the structure of NaCl. The small spheres represent Na^+ ions, the large spheres Cl^- ions. Notice that there are equal numbers of Na^+ and Cl^- ions, but no NaCl molecules.

relatively high melting points (mp NaCl = 801°C, $CaCl_2$ = 772°C). To melt an ionic compound requires that oppositely charged ions be separated from one another, thereby breaking ionic bonds.

2.5 FORMULAS OF IONIC COMPOUNDS

The formula of an ionic compound shows the simplest whole-number ratio between cations and anions. To deduce the formula, apply the **principle of electrical neutrality**, which requires that the total positive charge of the cations equal the total negative charge of the anions. Consider, for example, the ionic compound calcium chloride. The ions present are Ca^{2+} and Cl^-. Clearly, for the compound to be electrically neutral, there must be two Cl^- ions for every Ca^{2+} ion. The formula of calcium chloride must be $CaCl_2$, indicating that the simplest ratio of Cl^- to Ca^{2+} ions is 2:1.

The simplest ratio is indicated by the subscripts

To predict the formulas of ionic compounds, you must know the charges of the ions involved. The remainder of this section is devoted to that topic.

Monatomic Ions

Figure 2.8 shows the charges of a large number of monatomic ions superimposed upon the periodic table. The charges of ions formed by main-group atoms, shown in blue in the figure, can be predicted by applying a simple principle:

Atoms which are close to a noble gas (Group 18) *in the periodic table tend to form ions which contain the same number of electrons as the neighboring noble gas atom.*

This is reasonable; noble gas atoms must have an extremely stable electronic structure, since they are so unreactive. Other atoms might be expected to acquire noble gas electronic structures by losing or gaining electrons.

Applying this principle, you can deduce the charges of ions formed by main-group atoms:

Group	No. of Electrons in Atom	Charge of Ion Formed
1	1 more than noble gas atom	+1
2	2 more than noble gas atom	+2
16	2 less than noble gas atom	−2
17	1 less than noble gas atom	−1

Metals form cations, nonmetals form anions; C, P and the metalloids (B, Si, . . .) do not form monatomic ions

Several metals that are farther removed from the noble gases in the periodic table form positive ions. The structures of these ions are not related in any direct way to those of noble gas atoms. Indeed, there is no simple way to predict their charges. Some of the more important of these ions are listed in Figure 2.8. They are derived from the **transition metals** (those in the groups near the center of the periodic table) and the *post-transition* metals in Groups 14 and 15. The most common charge among these ions, you will note, is +2.

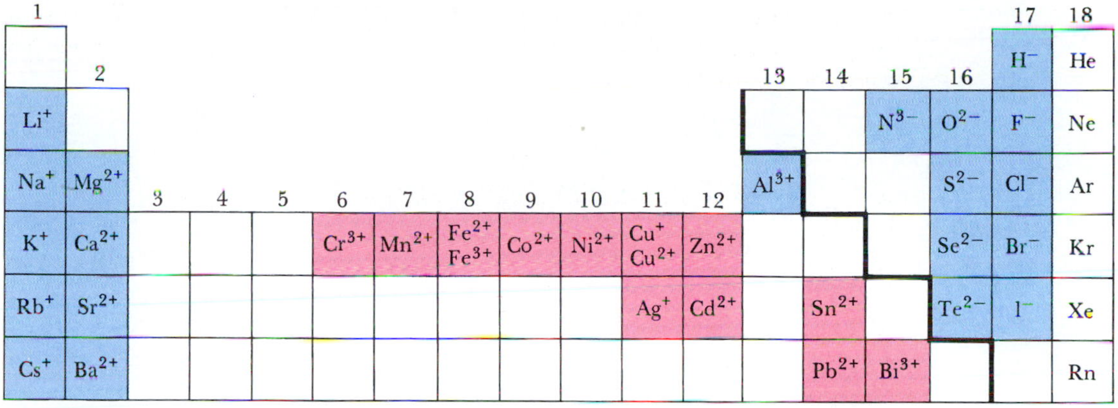

1	2	3	4	5	6	7	8	9	10	11	12	13	14	15	16	17	18
																H^-	He
Li^+														N^{3-}	O^{2-}	F^-	Ne
Na^+	Mg^{2+}											Al^{3+}			S^{2-}	Cl^-	Ar
K^+	Ca^{2+}				Cr^{3+}	Mn^{2+}	Fe^{2+} Fe^{3+}	Co^{2+}	Ni^{2+}	Cu^+ Cu^{2+}	Zn^{2+}				Se^{2-}	Br^-	Kr
Rb^+	Sr^{2+}									Ag^+	Cd^{2+}		Sn^{2+}		Te^{2-}	I^-	Xe
Cs^+	Ba^{2+}												Pb^{2+}	Bi^{3+}			Rn

Figure 2.8
Charges of ions found in solid ionic compounds. The step-like diagonal line separates metals from nonmetals and cations from anions. Ions shown in blue have the same number of electrons as the neighboring noble gas atom. For example, S^{2-}, Cl^-, K^+, and Ca^{2+} all have 18 electrons, as does the Ar atom. The cations shown in red do not have a noble gas structure.

Charges of +1 (Ag^+) and +3(Cr^{3+}, Bi^{3+}) are less common. Several of these metals form more than one cation:

$$Fe^{2+} \text{ and } Fe^{3+} \qquad Cu^+ \text{ and } Cu^{2+}$$

Using Figure 2.8 and the principle of electrical neutrality, it is possible to predict the formulas of a large number of ionic compounds. Note that, *in writing the formula of an ionic compound, the positive ion is always placed first.*

Learn the charges of these cations and anions

Example 2.4 Predict the formulas of the ionic compounds formed by

(a) magnesium and sulfur. (b) cobalt and chlorine.
(c) aluminum and oxygen. (d) bismuth and fluorine.

Strategy First, identify the charges of the cation and anion, using Figure 2.8. Then balance positive with negative charges to arrive at the formula.

Solution

(a) MgS; one Mg^{2+} ion requires one S^{2-} ion.

(b) $CoCl_2$; one Co^{2+} ion requires two Cl^- ions.

(c) Al_2O_3; two Al^{3+} ions (total charge = +6) require three O^{2-} ions (total charge = −6).

(d) BiF_3; one Bi^{3+} ion requires three F^- ions.

Table 2.3 Some Common Polyatomic Ions

+1	−1	−2	−3
NH_4^+ (ammonium)	OH^- (hydroxide)	CO_3^{2-} (carbonate)	PO_4^{3-} (phosphate)
	NO_3^- (nitrate)	SO_4^{2-} (sulfate)	
	ClO_3^- (chlorate)	CrO_4^{2-} (chromate)	
	ClO_4^- (perchlorate)	$Cr_2O_7^{2-}$ (dichromate)	
	CN^- (cyanide)	HPO_4^{2-} (hydrogen phosphate)	
	$C_2H_3O_2^-$ (acetate)		
	MnO_4^- (permanganate)		
	HCO_3^- (hydrogen carbonate)		
	$H_2PO_4^-$ (dihydrogen phosphate)		

Polyatomic Ions

Table 2.3 lists some of the polyatomic ions that you will need to know, along with their names and charges. Notice that

— there is only one polyatomic cation, NH_4^+. *All other cations considered in this text are derived from metal atoms* (e.g., Na^+ from Na, Ca^{2+} from Ca, . . .).

— most of the polyatomic anions contain one or more oxygen atoms; collectively these species are called **oxoanions**.

Example 2.5 Using Figure 2.8 and Table 2.3, predict the formulas of (a) strontium hydroxide. (b) sodium carbonate. (c) ammonium phosphate.

Strategy The reasoning here is entirely similar to that in Example 2.4. The only difference is that you must know the formulas and charges of the polyatomic ions in Table 2.3.

Solution

(a) One Sr^{2+} ion requires two OH^- ions. The formula is $Sr(OH)_2$. Parentheses are used to indicate that there are two polyatomic OH^- ions for every Sr^{2+}.

(b) Two Na^+ ions require one CO_3^{2-} ion. The formula is Na_2CO_3.

(c) Three NH_4^+ ions are required for one PO_4^{3-} ion. The formula is $(NH_4)_3PO_4$.

2.6 NAMES OF COMPOUNDS

To name a compound, you first have to decide whether it's ionic or molecular; the rules are different

A compound can be identified either by its formula (e.g., NaCl) or its name (sodium chloride). In this section, you will learn the rules used to name ionic and simple molecular compounds. To start with, it will be helpful to show how individual ions within ionic compounds are named.

Ions

Monatomic cations take the name of the metal from which they are derived. Examples include:

Na^+ sodium K^+ potassium

There is one complication: Certain metals, notably those in the transition series, form more than one type of cation. An example is iron, which forms both Fe^{2+} and Fe^{3+}. To distinguish between these cations, the charge must be indicated in the name. This is done by putting the charge as a Roman numeral in parentheses after the name of the metal:

Roman numerals are used *only* with cations of the transition and post-transition metals

Fe^{2+} iron(II) Fe^{3+} iron(III)

(An older system used the suffixes *-ic* for the ion of higher charge and *-ous* for the ion of lower charge. These were added to the stem of the Latin name of the metal, so that the Fe^{3+} ion was referred to as ferric, the Fe^{2+} ion as ferrous.)

We seldom use this system nowadays

Monatomic anions are named by adding the suffix *-ide* to the stem of the name of the nonmetal from which they are derived.

				H^-	hydride
N^{3-}	nitride	O^{2-}	oxide	F^-	fluoride
		S^{2-}	sulfide	Cl^-	chloride
		Se^{2-}	selenide	Br^-	bromide
		Te^{2-}	telluride	I^-	iodide

Polyatomic ions, as you have seen (Table 2.3), are given special names. Certain nonmetals in Groups **15–17** of the periodic table form more than one polyatomic ion containing oxygen (oxoanions). The names of several such **oxoanions** are shown in Table 2.4. From the entries in the table, you should be able to deduce the following rules.

1. When a nonmetal forms two oxoanions, the suffix *-ate* is used for the anion with the larger number of oxygen atoms. The suffix *-ite* is used for the anion containing fewer oxygen atoms.
2. When a nonmetal forms more than two oxoanions, the prefixes *per-* (largest number of oxygen atoms) and *hypo-* (fewest oxygen atoms) are used as well.

Only Cl, Br, and I form more than two oxoanions

Table 2.4 Oxoanions of Nitrogen, Sulfur, and Chlorine

Nitrogen	Sulfur	Chlorine
		ClO_4^- *per*chlor*ate*
NO_3^- nitr*ate*	SO_4^{2-} sulf*ate*	ClO_3^- chlor*ate*
NO_2^- nitr*ite*	SO_3^{2-} sulf*ite*	ClO_2^- chlor*ite*
		ClO^- *hypo*chlor*ite*

Potassium dichromate, $K_2Cr_2O_7$, is an ionic compound. (Marna G. Clarke)

Ionic Compounds

The name of an ionic compound consists of two words. The first word names the cation and the second names the anion. This is, of course, the same order in which the ions appear in the formula.

Example 2.6 Name the following ionic compounds:

(a) CaS (b) $Al(NO_3)_3$ (c) $FeCl_2$

Strategy To name an ionic compound, you must know the rules for naming individual ions, as discussed above. Where transition metals are involved, it is customary to show the charge of the cation; main-group metals typically form only one cation.

Solution

(a) calcium sulfide (b) aluminum nitrate (c) iron(II) chloride

Binary Molecular Compounds

At this point, you're not expected to predict the formulas of molecular compounds, but you can match names with formulas

The systematic name of a binary molecular compound, which contains two different nonmetals, consists of two words.

1. The first word gives the name of the element that appears first in the formula; a Greek prefix (Table 2.5) is used to show the number of atoms of that element in the formula.
2. The second word consists of

— the appropriate Greek prefix designating the number of atoms of the second element
— the stem of the name of the second element
— the suffix *-ide*

To illustrate these rules, consider the names of the several oxides of nitrogen:

N_2O_5	*di*nitrogen *penta*ox*ide*	N_2O_3	*di*nitrogen *tri*ox*ide*
N_2O_4	*di*nitrogen *tetra*ox*ide*	NO	nitrogen ox*ide*
NO_2	nitrogen *di*ox*ide*	N_2O	*di*nitrogen ox*ide*

Table 2.5 Greek Prefixes Used in Nomenclature

Number*	Prefix	Number	Prefix	Number	Prefix
2	di	5	penta	8	octa
3	tri	6	hexa	9	nona
4	tetra	7	hepta	10	deca

* The prefix mono (1) is seldom used.

QUESTIONS & PROBLEMS

Symbols, Formulas, and Equations

1. Using information given in this chapter, write the chemical formula for

a. ammonia.
b. water.
c. methane.
d. hydrochloric acid.

2. Using the figures in this chapter, give the color and physical state at room temperature of

a. Cr_2O_3 **b.** CrO_3

3. Using the figures in the chapter, give the color, formula, and physical state of

a. anhydrous copper sulfate.
b. cobalt(II) chloride hexahydrate.

4. Using the figures in the chapter, give the color, formula, and physical state of

a. copper(II) sulfate pentahydrate.
b. cobalt(II) chloride tetrahydrate.

Atomic Theory and Laws

5. Who is the father of atomic theory? State in your own words the law of constant composition.

6. Who first stated the law of conservation of mass? State the law in its modern form.

7. Which of the three laws (if any) listed on p. 27 is illustrated by each of the following statements?

a. Lavoisier found that when mercury(II) oxide, HgO, decomposes, the total mass of mercury and oxygen formed equals the mass of mercury(II) oxide decomposed.

b. Analysis of the calcium carbonate, $CaCO_3$, found in the marble of Carrara, Italy, and in the stalactites of the Carlsbad Caverns of New Mexico gives the same value for the percent calcium.

c. The atom ratio of oxygen to hydrogen is twice as large in one compound as it is in another compound made up of the two elements.

d. Hydrogen occurs as a mixture of two isotopes, one of which is twice as heavy as the other.

8. Which of the three laws (if any) listed on p. 27 is illustrated by each of the following statements?

a. A cold pack has the same mass before and after the seal between the two compartments is broken.

b. It is highly improbable that the formula for carbon monoxide gas found in London, England, is $C_{1.1}O_{2.5}$.

c. The mass of phosphorus, P, combined with one gram of hydrogen, H, in the highly toxic gas phosphine, PH_3, is a little more than twice the mass of nitrogen, N, combined with one gram of hydrogen in ammonia gas, NH_3.

9. When magnesium ribbon, Mg(s), is heated in oxygen gas, magnesium oxide, a white powder, is produced. In one experiment, 3.56 g of magnesium ribbon is completely consumed in reacting with 7.00 g of oxygen to produce 5.93 g of magnesium oxide; some oxygen remains unreacted. In a second experiment, 2.50 g of magnesium ribbon reacts with 1.10 g of oxygen gas. This time, all the oxygen is consumed; some unreacted magnesium remains and 2.75 g of magnesium oxide is produced. Show that these results are consistent with the law of constant composition.

10. Mercury(II) oxide, a red powder, can be decomposed by heating to produce liquid mercury and oxygen gas. When a sample of this compound is decomposed, 3.87 g of oxygen and 48.43 g of mercury are produced. In a second experiment, 15.68 g of mercury is allowed to react with an excess of oxygen; 16.93 g of red mercury(II) oxide is produced. Show that these results are consistent with the law of constant composition.

Nuclear Symbols and Isotopes

11. Who discovered the electron? Describe the experiment that led to the deduction that electrons are negatively charged particles.

12. Who discovered the nucleus? Describe the experiment that led to this discovery.

13. Studies show that there is an inverse relationship between the selenium content of the blood and the incidence of breast cancer in women. $^{80}_{34}Se$ is the most abundant form of naturally occurring selenium. How many protons are there in an Se-80 atom? How many neutrons?

14. The eruption of Mount St. Helens in Washington state produced a considerable amount of a radioactive gas, radon-222. Write the nuclear symbol for this isotope of radon (Rn).

15. **a.** Do the symbols $^{57}_{26}Fe$ and $_{26}Fe$ have the same meaning?

b. Do the symbols $^{57}_{26}Fe$ and ^{57}Fe convey the same information?

16. Explain how the two isotopes of copper, Cu-63 and Cu-65, differ from each other. Write nuclear symbols for each isotope.

17. Lithium is an element that is used by physicians in the treatment of some mental disorders. Lithium-7 is one of its isotopes. How many

a. protons are in its nucleus?
b. neutrons are in its nucleus?
c. electrons are in a lithium atom?
d. neutrons, protons, and electrons are in the Li^+ ion formed from this isotope?

18. An isotope of iodine used in thyroid disorders is $^{131}_{53}I$. How many

a. protons are in its nucleus?
b. neutrons are in its nucleus?
c. electrons are in an iodine atom?
d. neutrons and protons are in the I^- ion formed from this isotope?

19. Complete the table below, using the periodic table when necessary.

Symbol	Charge	Number of Protons	Number of Neutrons	Number of Electrons
_____	0	9	10	_____
P	0	_____	16	_____
_____	+3	26	30	_____
_____	_____	16	16	18

20. Complete the table below, using the periodic table when necessary.

Nuclear Symbol	Charge	Number of Protons	Number of Neutrons	Number of Electrons
$^{79}_{35}Br$	0	_____	_____	_____
_____	−3	7	7	_____
_____	+5	33	42	_____
$^{90}_{40}Zr$	0	_____	_____	_____

21. Give the number of protons and electrons in
a. an S^{2-} ion. **b.** an S_8 molecule.
c. an H_2S molecule. **d.** an H^+ ion.

22. Give the number of protons and electrons in
a. a P_4 molecule. **b.** a PCl_5 molecule.
c. a P^{3-} ion. **d.** a P^{5+} ion.

23. Complete the following table.

Species	Number of Neutrons	Number of Protons	Number of Electrons
$^{55}_{25}Mn^{2+}$	_____	_____	_____
_____	46	34	36
$^{194}_{78}Pt^{3+}$	_____	_____	_____

24. Complete the following table.

Species	Number of Protons	Number of Electrons
As^{5+}	_____	_____
N_2O_4	_____	_____
I^-	_____	_____
CH_4	_____	_____

Elements and the Periodic Table

25. Give the symbols for
a. potassium. **b.** cadmium. **c.** gold.
d. antimony. **e.** rubidium.

26. Name the elements whose symbols are
a. Mn **b.** Na **c.** As **d.** W **e.** P

27. Classify the elements in Question 25 as metals, nonmetals, or metalloids.

28. Classify the elements in Question 26 as metals, nonmetals, or metalloids.

29. How many elements are there in the following groups?
a. Group 12 **b.** Group 13 **c.** Group 18
d. Group 6

30. How many elements are there in the following periods?
a. period 1 **b.** period 2 **c.** period 3
d. period 4 **e.** period 5

Names and Formulas of Ionic and Molecular Compounds

31. Complete the following table of molecular compounds.

Name	Molecular Formula	Simplest Formula
_____	NH_3	_____
diselenium dichloride	_____	_____
_____	XeO_3	_____
_____	$BrCl_3$	_____
_____	P_2O_5	_____
dinitrogen tetraoxide	_____	_____

32. Complete the following table of molecular compounds.

Name	Molecular Formula	Simplest Formula
_____	N_2H_4	_____
hydrogen peroxide	_____	_____
_____	XeF_4	_____
_____	S_4N_4	_____
nitrogen trifluoride	_____	_____
carbon tetrachloride	_____	_____

33. Give the formulas of all the compounds containing no ions other than K^+, Sr^{2+}, Br^-, or O^{2-}.

34. Give the formulas of compounds in which
a. the cation is Li^+; the anion is S^{2-} or N^{3-}.
b. the anion is O^{2-}; the cation is Fe^{2+} or Fe^{3+}.

35. Write the formulas of the following ionic compounds. Use the periodic table and, if necessary, Table 2.3 and Figure 2.8.
a. iron(III) sulfate **b.** potassium acetate
c. lead(II) oxide **d.** barium chlorate
e. calcium sulfate

36. Follow the directions of Question 35 for the following compounds.
- **a.** copper(I) selenide
- **b.** manganese(II) carbonate
- **c.** sulfuric acid
- **d.** sodium hydrogen carbonate
- **e.** nickel(II) perchlorate

37. Follow the directions of Question 35 for the following compounds.
- **a.** sodium dichromate **b.** aluminum phosphate
- **c.** chromium(III) oxide **d.** calcium phosphate
- **e.** sulfurous acid

38. Follow the directions of Question 35 for the following compounds.
- **a.** tin(II) sulfide
- **b.** gold(III) chloride
- **c.** nitric acid
- **d.** calcium perchlorate
- **e.** potassium dihydrogen phosphate

39. Complete the following table of ionic compounds.

Name	Formula
barium nitride	______
copper(II) hydroxide	______
______	Ag_2Te
______	$Fe_2(CO_3)_3$
strontium chromate	______
______	$NaHCO_3$
ammonium perchlorate	______

40. Complete the following table of ionic compounds.

Name	Formula
______	CsOH
sodium permanganate	______
lithium dichromate	______
______	NH_4CN
aluminum sulfate	______
______	$Ba(NO_3)_2$

Oxoanions, Oxoacids

41. Give the names of
a. $HClO_4$ **b.** $HClO_2$ **c.** HIO **d.** HNO_3

42. Give the names of the oxoanions derived from the acids in Question 41.

43. Write the formulas of
- **a.** potassium chlorite **b.** calcium nitrite
- **c.** sodium sulfite **d.** sodium hypochlorite

44. Give the formulas of the oxoacids derived from the anions in Question 43.

Hydrates

45. Give the formula for
- **a.** washing soda **b.** Epsom salts
- **c.** Glauber's salt
- **d.** cobalt(II) chloride hexahydrate

46. Explain what is meant by water of hydration; efflorescence; clathrate.

47. How would you show experimentally that a white solid is $BaCl_2 \cdot H_2O$, not $BaCl_2$?

48. A sample of cobalt(II) chloride hexahydrate contains 2.0×10^{21} Cl^- ions. How many Co^{2+} ions does it contain? How many water molecules?

Unclassified

49. Criticize each of the following statements.
- **a.** In an ionic compound, the number of cations is always twice the number of anions.
- **b.** The molecular formula of calcium chloride is $CaCl_2$.
- **c.** The mass number is always twice the atomic number.
- **d.** For any ion, the number of electrons is always less than the number of protons.

50. Which of the following statements are always true? never true? usually true?
- **a.** A compound with the molecular formula C_6H_6 has the same simplest formula.
- **b.** Since $C_3H_6O_3$ and $C_6H_{12}O_6$ have the same simplest formula, they represent the same compound.
- **c.** A molecule is made up of nonmetallic atoms.
- **d.** An ionic compound always has a met[illegible] atom.

51. Identify each of the following elements.
- **a.** A halogen with 53 protons in the nucleus.
- **b.** A member of the same period as sulfur whose anion has a −1 charge and 18 electrons.
- **c.** A member of Group 9 with 45 protons in the nucleus.
- **d.** A metalloid in Group 16.

52. Write the formulas and names of the following compounds.
- **a.** A molecule made up of one atom of an element with six protons and four atoms of a halogen in period 4.
- **b.** The major inorganic component of bones, made up of calcium and an anion containing four oxygen atoms and one phosphorus atom.
- **c.** The barium salt of the anion derived from the nitrate ion by the loss of one oxygen atom.

Challenge Problems

53. Ethane and ethene are two gases containing only hydrogen and carbon atoms. In a certain sample of ethane, 4.53 g of hydrogen is combined with 18.0 g of carbon. In a sample of ethene, 7.25 g of hydrogen is combined with 43.20 g of carbon.

a. Show how these data illustrate the law of multiple proportions.

b. Suggest reasonable formulas for the two compounds.

54. Calculate the average density of a single Al-27 atom by assuming that it is a sphere with a radius of 0.143 nm. The masses of a proton, electron, and neutron are 1.6727×10^{-24} g, 9.1095×10^{-28} g, and 1.6750×10^{-24} g, respectively. The volume of a sphere is $4\pi r^3/3$, where r is its radius. Express the answer in grams per cubic centimeter. The density of aluminum is experimentally found to be 2.70 g/cm^3. What does that suggest about the packing of aluminum atoms in the metal?

55. The mass of a beryllium atom is 1.4965×10^{-23} g. Using that fact and other information given in this chapter, find the mass of a Be^{2+} ion.

56. Each time you inhale, you take in about 500 mL of air; each milliliter of air contains about 2.5×10^{19} molecules. It has been estimated that Abraham Lincoln, in delivering the Gettysburg Address, inhaled about 200 times.

a. How many molecules did Lincoln take in?

b. In the entire atmosphere, there are about 1.1×10^{44} molecules. What fraction of the molecules in the earth's atmosphere was inhaled by Lincoln at Gettysburg?

c. In the next breath that you take, estimate the number of molecules that were inhaled by Lincoln at Gettysburg.

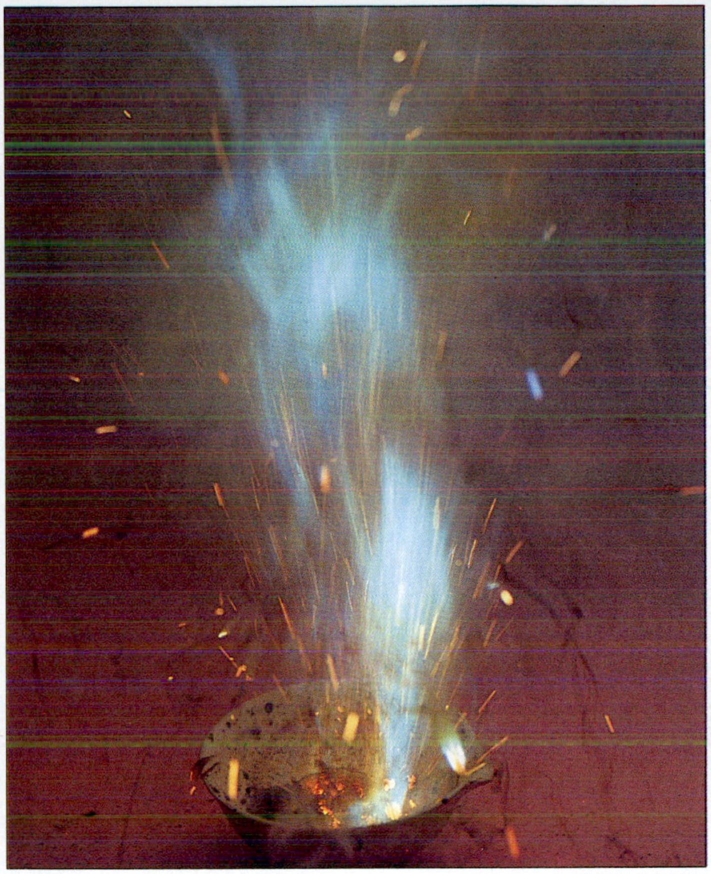

CHARLES D. WINTERS

$$NH_4NO_3(s) \longrightarrow N_2O(g) + 2H_2O(g)$$

(See p. 72)

3
Mass Relations in Chemistry; Stoichiometry

Was not all the knowledge Of the Egyptians writ in mystic symbols?

—BEN JONSON
The Alchemist

CHAPTER OUTLINE

To this point, our study of chemistry has been largely qualitative, involving very few calculations. However, chemistry is a quantitative science. Atoms of elements differ from one another not only in composition (number of protons, electrons, neutrons), but also in mass. Chemical formulas of compounds tell us not only the atom ratios in which elements are present, but also the mass ratios.

The general topic of this chapter is **stoichiometry**, the study of mass relations in chemistry. Whether dealing with atoms and molecules (Section 3.1), molar masses (Section 3.2), chemical formulas (Section 3.3), or chemical reactions (Section 3.4), you will be answering some very practical questions that ask "how much—" or "how many —," e.g.,

How many molecules are there in a glass of water? (Section 3.1)
How much iron can be obtained from a ton of iron ore? (Section 3.3)
How much nitrogen gas is required to form a kilogram of ammonia? (Section 3.4)

3.1 ATOMIC AND FORMULA MASSES

Individual atoms are too small to be seen, let alone weighed. However, as you will soon see, it is possible to determine quite accurately the relative masses of different atoms and molecules. Indeed, it is possible to go a step further and calculate the actual masses of these tiny building blocks of matter.

Atomic Masses; the Carbon-12 Scale

Relative masses of atoms of different elements are expressed in terms of their **atomic masses** (often referred to as atomic weights). The atomic mass of an element indicates how heavy, on the average, one atom of that element is compared to an atom of another element.

In order to set up a scale of atomic masses, it is necessary to establish a standard value for one particular species. For many years, two different standards which differed slightly from one another were in common use. Chemists took the atomic mass of the element oxygen to be exactly 16, while physicists assigned that value to the most common isotope of oxygen, $^{16}_{8}O$. In 1961, this confusing situation was resolved by adopting a single scale based on the most common isotope of carbon, $^{12}_{6}C$. This isotope is assigned a mass of exactly 12 *atomic mass units (amu).*

$$\text{mass of C-12 atom} = 12 \text{ amu (exactly)}$$

It follows that an atom half as heavy as a C-12 atom would weigh 6 amu, an atom twice as heavy as C-12 would have a mass of 24 amu, and so on.

Atomic masses are readily obtained from the periodic table, where they are listed directly below the symbol of the element. In the table on the inside front cover of this text, atomic masses are listed to four significant figures, although more precise values are available (e.g., atomic mass of F = 18.998403 amu). Notice that hydrogen has an atomic mass of 1.008 amu; helium has an atomic mass of 4.003 amu. This means that, on the average, a helium atom has a mass which is about one third that of a C-12 atom:

$$\frac{4.003 \text{ amu}}{12.00 \text{ amu}} = 0.3336$$

or about four times that of a hydrogen atom:

$$\frac{4.003 \text{ amu}}{1.008 \text{ amu}} = 3.971$$

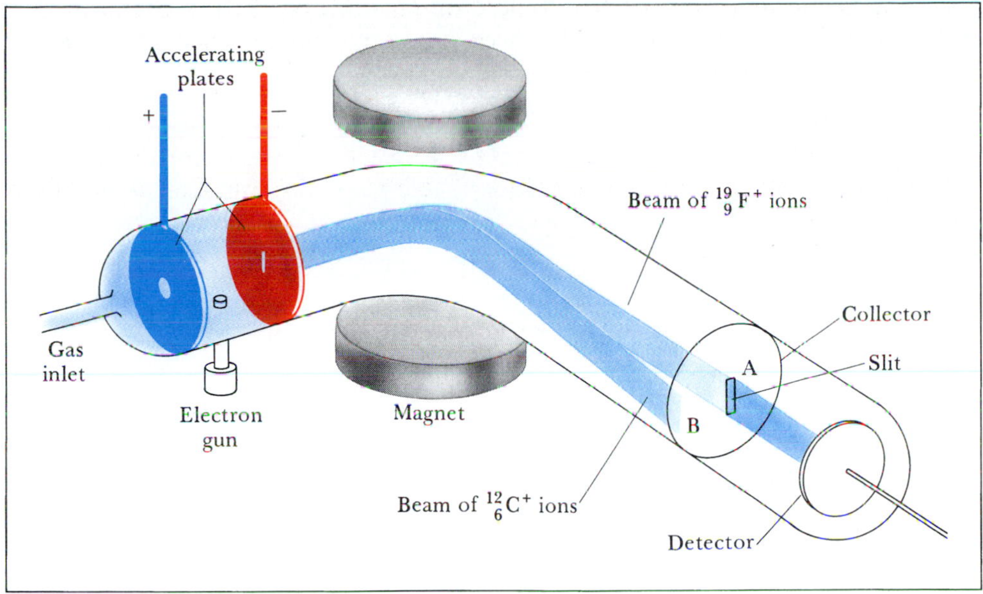

Figure 3.1
The mass spectrometer. A beam of gaseous ions is deflected in the magnetic field toward the collector plate. Light ions are deflected more than heavy ones. By comparing the accelerating voltages required to bring the two ions to the same point, it is possible to determine the relative masses of the ions.

Atomic Masses and Isotopic Abundances

Relative masses of individual atoms can be determined using a mass spectrometer (Fig. 3.1). Here, gaseous atoms or molecules at very low pressures are ionized by removing one or more electrons. The cations formed are accelerated by a potential of 500 to 2000 V toward a magnetic field, which deflects the ions from their straight-line path. The extent of deflection is inversely related to the mass of the ion. By measuring the voltages required to bring two ions of different mass to the same point on the collector, it is possible to determine their relative masses. For example, using a mass spectrometer, it is found that a $^{19}_{9}F$ atom is 1.583 times as heavy as a $^{12}_{6}C$ atom and so has a mass of

Light ions are more readily deflected than heavy ones

$$1.583 \times 12.00 \text{ amu} = 19.00 \text{ amu}$$

As it happens, naturally occurring fluorine consists of a single isotope, $^{19}_{9}F$. It follows that the atomic mass of the element fluorine must be the same as that of F-19, 19.00 amu. The situation with most elements is more complex, since they occur in nature as a mixture of two or more isotopes. In order to determine the atomic mass of such an element, it is necessary to know not only the masses of the individual isotopes but also their percentages *(abundances)* in nature.

Fortunately, isotopic abundances as well as isotopic masses can be determined by mass spectroscopy. The situation with chlorine, where there are two stable isotopes, Cl-35 and Cl-37, is shown in Figure 3.2 on page 52. The atomic masses of the two isotopes are determined in the usual way. The relative abun-

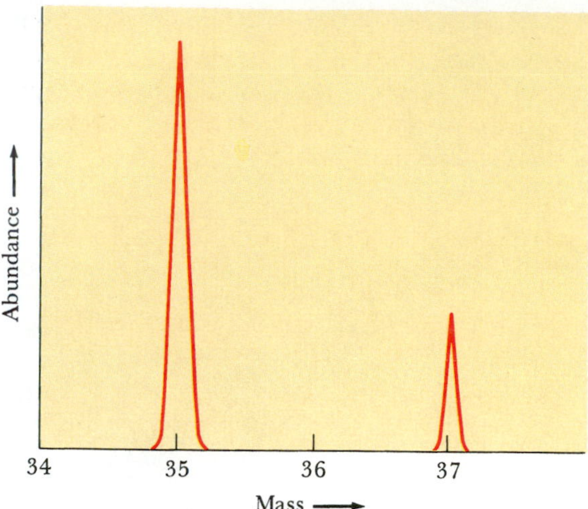

Figure 3.2
Mass spectrum of chlorine, which contains isotopes with atomic masses of 34.97 (75.53%) and 36.97 (24.47%).

dances of these isotopes are proportional to the heights of the recorder peaks, or, more accurately, to the areas under these peaks. For chlorine, the data obtained from the mass spectrometer are

	Atomic Mass	Abundance
Cl-35	34.97 amu	75.53%
Cl-37	36.97 amu	24.47%

Using these data and the general equation

$$\text{atomic mass Y} = (\text{atomic mass Y}_1) \times \frac{\%Y_1}{100\%} + (\text{atomic mass Y}_2) \times \frac{\%Y_2}{100\%} + \cdots$$

where Y is the element in question and $Y_1, Y_2, \ldots$ are its stable isotopes, the atomic mass of the element is readily calculated.

Example 3.1 Using the data cited for chlorine, calculate its atomic mass.

Strategy Substitute the data into the general relation given above and solve for atomic mass.

Solution

$$\text{atomic mass Cl} = 34.97 \text{ amu} \times \frac{75.53\%}{100\%} + 36.97 \text{ amu} \times \frac{24.47\%}{100\%} = 35.46 \text{ amu}$$

The atomic mass of chlorine is a weighted average of those of the two isotopes; it is closer to that of the more abundant isotope, Cl-35.

Calculations of the type shown in Example 3.1, using data obtained with a mass spectrometer, can give answers precise to seven or eight significant figures. The accuracy of tabulated atomic masses is limited mostly by variations in natural abundances. Sulfur is an interesting case in point. It consists largely of two isotopes, $^{32}_{16}S$ and $^{34}_{16}S$. The abundance of sulfur-34 varies from about 4.18% in sulfur deposits in Texas and Louisiana to 4.34% in volcanic sulfur from Italy. This leads to an uncertainty of 0.006 amu in the atomic mass of sulfur.

Ordinarily, atomic masses of elements increase in the same order as atomic number, but there are a few exceptions. The first of these involves the two elements argon and potassium.

	Atomic Number	Atomic Mass
Ar	18	39.95 amu
K	19	39.10 amu

There are three other such reversals; can you find them?

As you might guess, this anomaly is related to the isotopic compositions of the two elements. The major isotope of argon is Ar-40 (99.60%); that of potassium is K-39 (93.22%).

Masses of Individual Atoms; Avogadro's Number

For most purposes in chemistry, it is sufficient to know the relative masses of different atoms. Sometimes, however, it is necessary to go one step further and calculate the mass in grams of individual atoms. Let us consider how this can be done.

To start with, consider the elements helium and hydrogen. A helium atom is about four times as heavy as a hydrogen atom (He = 4.003 amu, H = 1.008 amu). It follows that a sample containing 100 helium atoms weighs four times as much as a sample containing 100 hydrogen atoms. Again, comparing samples of the two elements containing a million atoms each, the masses will be in a 4 (helium) to 1 (hydrogen) ratio. Turning this argument around, it follows that a sample of helium weighing four grams must contain the same number of atoms as a sample of hydrogen weighing one gram. More exactly:

If a nickel weighs twice as much as a dime, there are equal numbers of coins in 1000 g of nickels and 500 g of dimes

$$\text{no. of He atoms in 4.003 g helium} = \text{no. of H atoms in 1.008 g hydrogen}$$

This reasoning is readily extended to other elements. A sample of an element with a mass in grams equal to its atomic mass contains a certain definite number of atoms, N_A, regardless of the identity of the element.

The question now arises as to the numerical value of N_A; that is, how many atoms are there in 4.003 g of helium, 1.008 g of hydrogen, 32.07 g of sulfur, etc.? As it happens, this problem is one that has been studied for at least a century. Several ingenious experiments have been designed to determine this number, known as **Avogadro's number** (see Problem 82, end of chapter). As you can imagine, it is huge. (Remember that atoms are tiny. There must be a lot of them in 4.003 g of He, 1.008 g of H, etc.) To four significant figures,

$$N_A = 6.022 \times 10^{23}$$

To get some idea of how large this number is, suppose the entire population of the world were assigned to counting the atoms in 4.003 g of helium. If each person counted one atom per second and worked a 48-hour week, the task would take more than ten million years.

Most people have better things to do

The importance of Avogadro's number in chemistry should be clear. *It represents the number of atoms in a sample of an element with a mass in grams numerically equal to its atomic mass.* Thus there are

The conversion factor approach described in Chapter 1 comes in handy here

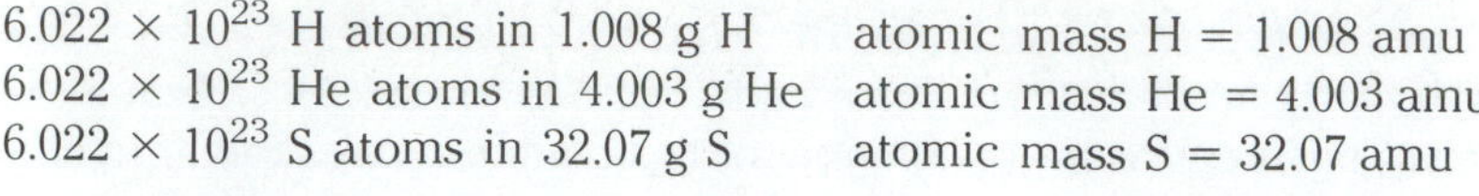

6.022×10^{23} H atoms in 1.008 g H   atomic mass H = 1.008 amu
6.022×10^{23} He atoms in 4.003 g He   atomic mass He = 4.003 amu
6.022×10^{23} S atoms in 32.07 g S   atomic mass S = 32.07 amu

Knowing Avogadro's number and the atomic mass of an element, it is possible to calculate the mass of an individual atom (Example 3.2a). You can also determine the number of atoms in a weighed sample of any element (Example 3.2b).

Glass is made red by adding a compound of selenium to sand. (Charles D. Winters)

Example 3.2 When selenium (Se) is added to glass, it gives the glass a brilliant red color. Taking Avogadro's number to be 6.022×10^{23}, calculate

(a) the mass of a selenium atom.
(b) the number of selenium atoms in a 1.000-g sample of the element.

Strategy The atomic mass of Se, from the periodic table, is 78.96 amu. It follows that

$$6.022 \times 10^{23} \text{ Se atoms} = 78.96 \text{ g Se}$$

This relation yields the required conversion factors.

Solution

$$\text{(a) mass of Se atom} = 1 \text{ Se atom} \times \frac{78.96 \text{ g Se}}{6.022 \times 10^{23} \text{ Se atoms}} = \boxed{1.311 \times 10^{-22} \text{ g}}$$

$$\text{(b) no. of Se atoms} = 1.000 \text{ g} \times \frac{6.022 \times 10^{23} \text{ Se atoms}}{78.96 \text{ g}}$$

$$= \boxed{7.627 \times 10^{21} \text{ Se atoms}}$$

Formula Masses

The development we have gone through to this point has been restricted to atoms and, at least by implication, to elements made up of individual atoms. To extend these ideas to other types of particles and to all types of substances, it is helpful to define a quantity called **formula mass**. Quite simply, the formula mass is the sum of the atomic masses in the formula of a substance. Thus:

Formula	Formula Mass
O	16.00 amu
O_2	2(16.00 amu) = 32.00 amu
H_2O	2(1.008 amu) + 16.00 amu = 18.02 amu
NaCl	22.99 amu + 35.45 amu = 58.44 amu

The "formula unit" of $CaCl_2$ consists of 1 Ca^{2+} and 2 Cl^- ions

The formula mass represents the mass, on the carbon-12 scale, of the unit represented by the formula. The "formula unit" may be an atom (O), a molecule (O_2 or H_2O), or a set of ions (1 Na^+, 1 Cl^- ion). From the data above, you can conclude that an H_2O molecule is about 18/16 as heavy as an O atom; an O_2 molecule is exactly twice as heavy as an O atom.

The statement made earlier relating Avogadro's number to atomic masses can now be generalized:

A sample of a substance which has a mass in grams numerically equal to its formula mass contains Avogadro's number of formula units.

6.022×10^{23} O atoms weigh 16.00 g
6.022×10^{23} O_2 molecules weigh 32.00 g
6.022×10^{23} H_2O molecules weigh 18.02 g
6.022×10^{23} (Na^+ ions + Cl^- ions) weigh 58.44 g

This relationship is particularly useful for elements and molecular compounds (Fig. 3.3).

Example 3.3 How many molecules are there in a drop of water weighing 0.050 g?

Strategy Use the relation 6.022×10^{23} H_2O molecules = 18.02 g H_2O to find the appropriate conversion factor.

Solution

$$0.050 \text{ g } H_2O \times \frac{6.022 \times 10^{23} \text{ } H_2O \text{ molecules}}{18.02 \text{ g } H_2O} = 1.7 \times 10^{21} \text{ } H_2O \text{ molecules}$$

Since molecules are so small, it takes a lot of them to make up a sample large enough to be seen and weighed.

Figure 3.3
One mole of iron, sulfur, and aspirin. The iron nails weigh 55.85 g and contain 6.022×10^{23} Fe atoms. The pile of yellow sulfur weighs 32.07 g and contains 6.022×10^{23} S atoms. The aspirin tablets weigh 180.15 g and contain 6.022×10^{23} $C_9H_8O_4$ molecules. (Marna G. Clarke)

3.2 THE MOLE

The quantity represented by Avogadro's number is so important that it is given a special name, the **mole**. A mole represents 6.022×10^{23} items, whatever they may be.

1 mol H atoms = 6.022×10^{23} H atoms
1 mol O atoms = 6.022×10^{23} O atoms
1 mol H_2 molecules = 6.022×10^{23} H_2 molecules
1 mol H_2O molecules = 6.022×10^{23} H_2O molecules
1 mol electrons = 6.022×10^{23} electrons
1 mol pennies = 6.022×10^{23} pennies

(One mole of pennies is a lot of money. It's enough to pay all the expenses of the United States for the next billion years or so.)

A mole represents not only a specific number of particles but also a definite mass of a substance. In general, *the molar mass, $\mathcal{M}$, in grams per mole, is numerically equal to the formula mass.* Thus:

The size of the mole was chosen to make this relation valid

Formula	Formula Mass	Molar Mass, $\mathcal{M}$
O	16.00 amu	16.00 g/mol
O_2	32.00 amu	32.00 g/mol
H_2O	18.02 amu	18.02 g/mol
NaCl	58.44 amu	58.44 g/mol

Notice that the formula of a substance must be known to find its molar mass. It would be ambiguous, to say the least, to refer to the "molar mass of hydrogen." One mole of hydrogen atoms, represented by the symbol H, weighs

1.008 g; the molar mass of H is 1.008 g/mol. One mole of hydrogen molecules, represented by the formula H_2, weighs 2.016 g; the molar mass of H_2 is 2.016 g/mol.

Mole–Gram Conversions

As you will see later in this chapter, it is often necessary to convert from moles of a substance to mass in grams or vice versa. Such conversions are readily made by using the general relation

$$m = \mathcal{M} \times n$$

where m is the mass in grams, $\mathcal{M}$ is the molar mass (g/mol), and n is the amount in moles.

A bed of calcium carbonate (limestone) along the Verde River in Arizona. (© 1979 James Cowlin)

Example 3.4 Calcium carbonate is the principal ingredient of the chalk used in most classrooms. Determine the number of moles of calcium carbonate in a stick of chalk containing 14.8 g of calcium carbonate.

Strategy Find the molar mass of calcium carbonate and use it to convert 14.8 g to moles. Before calculating the molar mass, you must come up with the formula of calcium carbonate (Chapter 2).

Solution The formula is $CaCO_3$, so the molar mass is

$$\mathcal{M} = [40.08 + 12.01 + 3(16.00)] \text{ g/mol} = 100.09 \text{ g/mol}$$

$$n = 14.8 \text{ g } CaCO_3 \times \frac{1 \text{ mol } CaCO_3}{100.09 \text{ g } CaCO_3} = 0.148 \text{ mol } CaCO_3$$

Example 3.5 Acetylsalicylic acid, $C_9H_8O_4$, is the principal ingredient of aspirin. What is the mass in grams of 0.287 mol of acetylsalicylic acid?

Strategy Find the molar mass of $C_9H_8O_4$ and use it to obtain the conversion factor required to convert 0.287 mol to mass in grams.

Solution The molar mass of $C_9H_8O_4$ is

$$\mathcal{M} = [9(12.01) + 8(1.008) + 4(16.00)] \text{ g/mol} = 180.15 \text{ g/mol}$$

Hence

$$\text{mass } C_9H_8O_4 = 0.287 \text{ mol } C_9H_8O_4 \times \frac{180.15 \text{ g } C_9H_8O_4}{1 \text{ mol } C_9H_8O_4} = 51.7 \text{ g } C_9H_8O_4$$

Almost every calculation you will make in this chapter requires that you know precisely what is meant by a mole

Conversions of the type we have just carried out come up over and over again in chemistry. They will be required in nearly every chapter of this text. Clearly, you must know what is meant by a mole. Remember, a mole always represents a certain number of items, 6.022×10^{23}. Its mass, however, differs with the substance involved: A mole of H_2O, 18.02 g, weighs considerably more than a mole of H_2, 2.016 g, even though they both contain the same number of molecules. In the same way, a dozen bowling balls weigh a lot more than a dozen eggs, even though each involves the same number of items.

3.3 MASS RELATIONS IN CHEMICAL FORMULAS

As you will see shortly, the formula of a compound can be used to determine the mass percents of the elements present. Conversely, if the percentages of the elements are known, the simplest formula can be determined. Knowing the molar mass of the compound, it is possible to go one step further and find the molecular formula. In this section we will consider how these three types of calculations are carried out.

Percent Composition from Formula

The "percent composition" of a compound is specified by citing the mass percents of the elements present. For example, in a one-hundred-gram sample of water there are 11.19 g of hydrogen and 88.81 g of oxygen. Hence, the percentages of the two elements are

$$\frac{11.19 \text{ g H}}{100.00 \text{ g}} \times 100\% = 11.19\% \text{ H} \qquad \frac{88.81 \text{ g O}}{100.00 \text{ g}} \times 100\% = 88.81\% \text{ O}$$

We would say that the percent composition of water is 11.19% H, 88.81% O.

Knowing the formula of a compound, you can readily calculate the mass percents of its constituent elements. It is convenient to start with one mole of the compound (Example 3.6).

Example 3.6 Sodium hydrogen carbonate, commonly called "bicarbonate of soda," is used in many commercial products to relieve an upset stomach. It has the formula $NaHCO_3$. What are the mass percents of Na, H, C, and O in sodium hydrogen carbonate?

Strategy Find the mass in grams of each element in one mole of $NaHCO_3$. Then find

$$\% \text{ element} = \frac{\text{mass element}}{\text{total mass compound}} \times 100\%$$

Solution It is convenient to set up a table to determine the mass of each element in one mole of $NaHCO_3$.

	n	×	$\mathcal{M}$	=	m
Na	1 mol	×	22.99 g/mol	=	22.99 g
H	1 mol	×	1.008 g/mol	=	1.008 g
C	1 mol	×	12.01 g/mol	=	12.01 g
O	3 mol	×	16.00 g/mol	=	48.00 g
					84.01 g $NaHCO_3$

Since 84.01 g of $NaHCO_3$ contains 22.99 g of Na, 1.008 g of H, 12.01 g of C, and 48.00 g of O,

$$\text{mass \% Na} = \frac{22.99 \text{ g}}{84.01 \text{ g}} \times 100\% = 27.36\% \qquad \text{mass \% H} = \frac{1.008 \text{ g}}{84.01 \text{ g}} \times 100\% = 1.200\%$$

$$\text{mass \% C} = \frac{12.01\ \text{g}}{84.01\ \text{g}} \times 100\% = \boxed{14.30\%} \qquad \text{mass \% O} = \frac{48.00\ \text{g}}{84.01\ \text{g}} \times 100\% = \boxed{57.14\%}$$

The percentages add up to 100, as they should:

$$27.36\% + 1.200\% + 14.30\% + 57.14\% = 100.00\%$$

The calculations in Example 3.6 illustrate an important characteristic of formulas. In one mole of $NaHCO_3$, there is 1 mol Na (22.99 g), 1 mol H (1.008 g); 1 mol C (12.01 g), and 3 mol O (48.00 g). In other words, the mole ratio is 1 mol Na : 1 mol H : 1 mol C : 3 mol O. This is the same as the atom ratio in $NaHCO_3$, 1 atom Na : 1 atom H : 1 atom C : 3 atoms O. In general, *the subscripts in a formula represent not only the atom ratio in which the different elements are combined, but also the mole ratio.*

The atom ratio must equal the mole ratio; can you see why?

The formula of a compound can also be used in a straightforward way to find the mass of an element in a weighed sample of the compound (Example 3.7).

Example 3.7 An iron-containing mineral responsible for the red color of soils in many parts of the country is limonite, which has the formula $Fe_2O_3 \cdot \frac{3}{2}H_2O$. What mass of iron can be obtained from a metric ton (10^3 kg = 10^6 g) of limonite?

Mn compounds also produce red soils

Strategy First find the mass percent of iron. Then, using the mass percent, determine how much iron there is in one metric ton of the ore.

Solution In one mole of limonite, there is

$$2\ \text{mol Fe} \times \frac{55.85\ \text{g Fe}}{1\ \text{mol Fe}} = 111.7\ \text{g Fe}$$

The molar mass of limonite is:

$$(111.7 + 3(16.00) + \frac{3}{2}(18.02))\ \text{g/mol} = 186.7\ \text{g/mol}$$

Thus:

$$\%\ \text{Fe} = \frac{111.7\ \text{g}}{186.7\ \text{g}} \times 100\% = 59.83\%$$

In one metric ton of limonite

$$\text{mass Fe} = 1.000 \times 10^6\ \text{g limonite} \times \frac{59.83\ \text{g Fe}}{100.0\ \text{g limonite}} = \boxed{5.983 \times 10^5\ \text{g Fe}}$$

Limonite ore (Paul Silverman/FUNDAMENTAL PHOTOGRAPHS, New York)

Simplest Formula from Chemical Analysis

A major task of chemical analysis is to determine the simplest formulas of compounds. The analytical data obtained may be expressed in various ways. You may know

— the masses of the elements in a weighed sample of the compound
— the mass percents of the elements in the compound
— the masses of products obtained by reaction of a weighed sample of the compound

The strategy used to calculate the simplest formula depends to some extent upon which of these types of information is given. The basic objective in each case is to find the number of moles of each element, then the simplest mole ratio, and finally the simplest formula.

Example 3.8 A 25.0-g sample of an orange compound contains 6.64 g of potassium, 8.84 g of chromium, and 9.52 g of oxygen. Find the simplest formula.

Strategy First (1), convert the masses of the three elements to moles. Knowing the number of moles (n) of K, Cr, and O, you can then (2) calculate the mole ratios. Finally (3), equate the mole ratio to the atom ratio, which gives you the simplest formula.

Mass data → mole ratio = atom ratio → simplest formula

Solution

(1) $n_K = 6.64 \text{ g K} \times \frac{1 \text{ mol K}}{39.10 \text{ g K}} = 0.170 \text{ mol K}$

$n_{Cr} = 8.84 \text{ g Cr} \times \frac{1 \text{ mol Cr}}{52.00 \text{ g Cr}} = 0.170 \text{ mol Cr}$

$n_O = 9.52 \text{ g O} \times \frac{1 \text{ mol O}}{16.00 \text{ g O}} = 0.595 \text{ mol O}$

(2) To find the mole ratios, divide by the smallest number, 0.170 mol K:

$$\frac{0.170 \text{ mol Cr}}{0.170 \text{ mol K}} = \frac{1.00 \text{ mol Cr}}{1 \text{ mol K}} \qquad \frac{0.595 \text{ mol O}}{0.170 \text{ mol K}} = \frac{3.50 \text{ mol O}}{1 \text{ mol K}}$$

The mole ratio is 1 mol of K:1 mol of Cr:3.50 mol of O.

(3) As pointed out earlier, the mole ratio is the same as the atom ratio. To find the simplest whole-number atom ratio, multiply throughout by 2:

2 K:2 Cr:7 O

The simplest formula of the orange compound is $K_2Cr_2O_7$.

Sometimes you will be given the mass percents of the elements in a compound. If that is the case, one extra step is involved. *Assume a 100-g sample and calculate the mass of each element in that sample.*

Suppose, for example, you are told that the percentages of K, Cr, and O in a compound are 26.6%, 35.4%, and 38.0%, respectively. It follows that in a 100-g sample there are

Using 100 g simplifies the math

26.6 g K, 35.4 g Cr, 38.0 g O

Working with these masses, you can go through the same procedure followed in Example 3.8 to arrive at the same answer (try it!).

The most complex problem of this type requires you to determine the simplest formula of a compound, given only the raw data obtained from its analysis. Here, an additional step is involved; you have to determine the masses of the elements present in a fixed mass of the compound (Example 3.9).

Example 3.9 The compound that gives fermented grape juice, malt liquor, and vodka their intoxicating properties is ethyl alcohol, which contains the elements carbon, hydrogen, and oxygen. When a sample of ethyl alcohol is burned in air, it is found that

$$5.00 \text{ g ethyl alcohol} \longrightarrow 9.55 \text{ g } CO_2 + 5.87 \text{ g } H_2O$$

What is the simplest formula of ethyl alcohol?

The "trick" in working this problem is knowing where to start

Strategy The first step here is to calculate the masses of C, H, and O in the 5.00-g sample. To do this, note that all of the carbon has been converted to carbon dioxide and that there is 12.01 g (1 mol) of C in 44.01 g (1 mol) of CO_2. Hence, to find the mass of carbon in 9.55 g of CO_2, you use the conversion factor:

$$12.01 \text{ g C}/44.01 \text{ g } CO_2$$

By the same token, all the hydrogen ends up as water. There is 2.016 g (2 mol) of H in 18.02 g (1 mol) of H_2O. The conversion factor to go from mass of water to mass of hydrogen is

$$2.016 \text{ g H}/18.02 \text{ g } H_2O$$

Why can't you find the mass of oxygen from that of CO_2?

The mass of oxygen is found by difference:

$$\text{mass O} = \text{mass of sample} - (\text{mass of C} + \text{mass of H})$$

Once the masses of the elements are obtained, it's all downhill; follow the same path as in Example 3.8.

Solution

(1) $$\text{mass C} = 9.55 \text{ g } CO_2 \times \frac{12.01 \text{ g C}}{44.01 \text{ g } CO_2} = 2.61 \text{ g C}$$

$$\text{mass H} = 5.87 \text{ g } H_2O \times \frac{2.016 \text{ g H}}{18.02 \text{ g } H_2O} = 0.657 \text{ g H}$$

$$\text{mass O} = 5.00 \text{ g sample} - (2.61 \text{ g C} + 0.657 \text{ g H}) = 1.73 \text{ g O}$$

(2) $$n_C = 2.61 \text{ g C} \times \frac{1 \text{ mol C}}{12.01 \text{ g C}} = 0.217 \text{ mol C}$$

$$n_H = 0.657 \text{ g H} \times \frac{1 \text{ mol H}}{1.008 \text{ g H}} = 0.652 \text{ mol H}$$

$$n_O = 1.73 \text{ g O} \times \frac{1 \text{ mol O}}{16.00 \text{ g O}} = 0.108 \text{ mol O}$$

(3) $$\frac{0.217 \text{ mol C}}{0.108 \text{ mol O}} = \frac{2.01 \text{ mol C}}{1 \text{ mol O}} \qquad \frac{0.652 \text{ mol H}}{0.108 \text{ mol O}} = \frac{6.04 \text{ mol H}}{1 \text{ mol O}}$$

Rounding off to whole numbers, the mole ratios are

$$2 \text{ mol C} : 6 \text{ mol H} : 1 \text{ mol O}$$

(4) The simplest formula of ethyl alcohol is C_2H_6O.

Molecular Formula from Simplest Formula

Chemical analysis always leads to the simplest formula of a compound because it gives only the simplest atom ratio of the elements. As pointed out in Section 2.4, the molecular formula is a whole-number multiple of the simplest formula. That multiple may be 1 as in H_2O, 2 as in H_2O_2, 3 as in C_3H_6, or some other integer. To find the multiple, one more piece of data is needed: the molar mass.

Example 3.10 Vitamin C may (or may not) help prevent the common cold. Its simplest formula is found by analysis to be $C_3H_4O_3$. From another experiment, the molar mass is found to be about 180 g/mol. What is the molecular formula of Vitamin C?

Strategy Calculate the molar mass corresponding to the simplest formula, i.e., $\mathcal{M}_{C_3H_4O_3}$. Then find the multiple by dividing the observed molar mass, 180 g/mol, by $\mathcal{M}_{C_3H_4O_3}$.

Solution

$$\mathcal{M}_{C_3H_4O_3} = 3(12.01\text{ g/mol}) + 4(1.008\text{ g/mol}) + 3(16.00\text{ g/mol}) = 88.06\text{ g/mol}$$

The ratio of the observed molar mass to that of $C_3H_4O_3$ is

You don't need to know the molar mass accurately. Why?

$$\frac{180\text{ g/mol}}{88.06\text{ g/mol}} = 2.04$$

The multiple is 2; the molecular formula is $C_6H_8O_6$.

3.4 MASS RELATIONS IN REACTIONS

A chemist who carries out a reaction in the laboratory needs to know how much **product** can be obtained from a given amount of starting materials (**reactants**). To do this, he or she starts by writing a balanced chemical equation.

Writing and Balancing Equations

Chemical reactions are represented by chemical equations, which identify reactants and products. Formulas of reactants appear on the left side of the equation; those of products are written on the right. In a balanced chemical equation, there are the same number of atoms of a given element on both sides. The same situation holds for a chemical reaction that you carry out in the laboratory; atoms are conserved. For that reason, *any calculation involving a reaction must be based upon the balanced equation for that reaction.*

The total number of moles usually differs on the two sides

A chemical equation is essentially useless unless it's balanced

To illustrate how a relatively simple equation can be written and balanced, consider a reaction used in the Titan rocket motor to launch the Gemini spacecraft (Fig. 3.4). The reactants were two liquids, hydrazine and dinitrogen tetraoxide, whose molecular formulas are N_2H_4 and N_2O_4, respectively. The products of the reaction are gaseous nitrogen, N_2, and water vapor. To write a balanced equation for this reaction, follow the process outlined on page 62.

1. *Write a "skeleton" equation in which the formulas of the reactants appear on the left and those of the products on the right.* In this case,

$$N_2H_4 + N_2O_4 \longrightarrow N_2 + H_2O$$

Using the letters (*s*), (*l*), (*g*), (*aq*) gives physical reality to the equation

2. *Indicate the physical state of each reactant and product, after the formula, by writing*

(*g*) for a gaseous substance
(*l*) for a pure liquid
(*s*) for a solid
(*aq*) for an ion or molecule in water (aqueous) solution,

In this case

$$N_2H_4(l) + N_2O_4(l) \longrightarrow N_2(g) + H_2O(g)$$

3. *Balance the equation.* To accomplish this, start by writing a coefficient of 4 for H_2O, thus obtaining 4 oxygen atoms on both sides:

$$N_2H_4(l) + N_2O_4(l) \longrightarrow N_2(g) + 4H_2O(g)$$

Now consider the hydrogen atoms. There are $4 \times 2 = 8$ H atoms on the right. To obtain 8 H atoms on the left, write a coefficient of 2 for N_2H_4:

$$2N_2H_4(l) + N_2O_4(l) \longrightarrow N_2(g) + 4H_2O(g)$$

Finally, consider nitrogen. There are a total of $(2 \times 2) + 2 = 6$ nitrogen atoms on the left. To balance nitrogen, write a coefficient of 3 for N_2:

$$2N_2H_4(l) + N_2O_4(l) \longrightarrow 3N_2(g) + 4H_2O(g)$$

This is the final balanced equation for the reaction of hydrazine with dinitrogen tetraoxide.

Figure 3.4
Titan rocket motor. (NASA)

Three points concerning the balancing process are worth noting.

6 N is also no good; the formula must be that of the species actually present

1. Equations are balanced by adjusting coefficients in front of formulas, never by changing subscripts within formulas. On paper, the above equation could have been balanced by writing N_6 on the right, but that would have been absurd. Elemental nitrogen exists as diatomic molecules, N_2; there is no such thing as an N_6 molecule.
2. In balancing an equation, it is best to start with an element that appears in only one species on each side of the equation. In this case, either oxygen or hydrogen are good starting points. Nitrogen would have been a poor choice, however, since there are nitrogen atoms in both reactant molecules, N_2H_4 and N_2O_4.
3. In principle, there are an infinite number of balanced equations that can be written for any reaction. The equations

$$4N_2H_4(l) + 2N_2O_4(l) \longrightarrow 6N_2(g) + 8H_2O(g)$$

$$N_2H_4(l) + \tfrac{1}{2}N_2O_4(l) \longrightarrow \tfrac{3}{2}N_2(g) + 2H_2O(g)$$

are balanced in that there are the same number of atoms of each element on both sides. Ordinarily, the equation with the simplest whole number coefficients

$$2N_2H_4(l) + N_2O_4(l) \longrightarrow 3N_2(g) + 4H_2O(g)$$

is preferred. Sometimes, though, we'll use fractional coefficients, as in the next-to-last equation on page 62.

Frequently, when you are asked to balance an equation, the formulas of products and reactants are given. Sometimes, though, you will have to derive the formulas, given only the names (Example 3.11).

Example 3.11 Crystals of sodium hydroxide (lye) react with carbon dioxide of the air to form a white powder, sodium carbonate, and a colorless liquid, water. Write a balanced equation for this chemical reaction.

Strategy To translate names into formulas, recall the discussion in Section 2.6, Chapter 2. The physical states are given or implied. To balance the equation, you could start with either sodium or hydrogen.

Solution The "skeleton" equation is

$$NaOH(s) + CO_2(g) \longrightarrow Na_2CO_3(s) + H_2O(l)$$

Since there are two Na atoms on the right, a coefficient of 2 is written in front of NaOH:

$$2NaOH(s) + CO_2(g) \longrightarrow Na_2CO_3(s) + H_2O(l)$$

Careful inspection shows that all other atoms are now balanced.

Mass Relations from Equations

The principal reason for writing balanced equations is to make it possible to relate the masses of reactants and products. Calculations of this sort are based upon a very important principle:

The coefficients of a balanced equation represent numbers of moles of reactants and products.

To show that this statement is valid, recall the equation

$$2N_2H_4(l) + N_2O_4(l) \longrightarrow 3\ N_2(g) + 4H_2O(g)$$

The coefficients in this equation represent numbers of molecules; that is,

$$2 \text{ molecules } N_2H_4 + 1 \text{ molecule } N_2O_4 \longrightarrow 3 \text{ molecules } N_2 + 4 \text{ molecules } H_2O$$

A balanced equation remains valid if each coefficient is multiplied by the same number, including Avogadro's number, N_A:

$$2N_A \text{ molecules } N_2H_4 + N_A \text{ molecules } N_2O_4 \longrightarrow 3N_A \text{ molecules } N_2 + 4N_A \text{ molecules } H_2O$$

Recall from Section 3.2 that a mole represents Avogadro's number of items, N_A. Thus the equation above can be written

$$2 \text{ mol } N_2H_4 + 1 \text{ mol } N_2O_4 \longrightarrow 3 \text{ mol } N_2 + 4 \text{ mol } H_2O$$

This is the most practical interpretation of the balanced equation

The quantities 2 mol N_2H_4, 1 mol N_2O_4, 3 mol N_2, and 4 mol H_2O are chemically equivalent to each other in this reaction. Hence they can be used in conversion factors such as

$$\frac{2 \text{ mol } N_2H_4}{3 \text{ mol } N_2}, \quad \frac{3 \text{ mol } N_2}{1 \text{ mol } N_2O_4}, \quad \text{or a variety of other combinations}$$

If you wanted to know how many moles of hydrazine were required to form 1.80 mol of elemental nitrogen, the conversion would be

$$n_{N_2H_4} = 1.80 \text{ mol } N_2 \times \frac{2 \text{ mol } N_2H_4}{3 \text{ mol } N_2} = 1.20 \text{ mol } N_2H_4$$

To find the number of moles of nitrogen produced from 2.60 mol of N_2O_4,

$$n_{N_2} = 2.60 \text{ mol } N_2O_4 \times \frac{3 \text{ mol } N_2}{1 \text{ mol } N_2O_4} = 7.80 \text{ mol } N_2$$

Simple mole relationships of the type just discussed are readily extended to relate

— moles of one substance to grams of another (Example 3.12a)
— grams of one substance to grams of another (Examples 3.12b and c)

Example 3.12 Ammonia used to make fertilizers for lawns and gardens is made by reacting nitrogen of the air with hydrogen. The balanced equation for the reaction is

$$3H_2(g) + N_2(g) \longrightarrow 2NH_3(g)$$

Determine

(a) the mass in grams of ammonia, NH_3, formed when 1.34 mol N_2 reacts.
(b) the mass in grams of N_2 required to form 1.00 kg NH_3.
(c) the mass in grams of H_2 required to react with 6.00 g N_2.

To go from moles of A to mass of A, use molar mass

Strategy In each case, you use the mole ratios given by the coefficients of the balanced equation to relate moles of one substance to moles of another. Beyond that, use molar mass to relate moles of one substance to mass in grams of that substance. Before starting, decide upon the path you will follow to go from the quantity given to that required, i.e.,

(a) $n_{N_2} \longrightarrow n_{NH_3} \longrightarrow$ mass of NH_3
(b) mass of $NH_3 \longrightarrow n_{NH_3} \longrightarrow n_{N_2} \longrightarrow$ mass of N_2
(c) mass of $N_2 \longrightarrow n_{N_2} \longrightarrow n_{H_2} \longrightarrow$ mass of H_2

To go from moles of A to moles of B, use the coefficients of the balanced equation

Conversions indicated by colored arrows involve mole ratios given by the coefficients of the balanced equation. The other conversions require only molar masses and are essentially identical to those carried out in Examples 3.4 and 3.5.

Solution

(a) $$\text{mass of } NH_3 = 1.34 \text{ mol } N_2 \times \frac{2 \text{ mol } NH_3}{1 \text{ mol } N_2} \times \frac{17.03 \text{ g } NH_3}{1 \text{ mol } NH_3} = 45.6 \text{ g } NH_3$$

(b) $$\text{mass of } N_2 = 1000 \text{ g } NH_3 \times \frac{1 \text{ mol } NH_3}{17.03 \text{ g } NH_3} \times \frac{1 \text{ mol } N_2}{2 \text{ mol } NH_3} \times \frac{28.02 \text{ g } N_2}{1 \text{ mol } N_2} = 823 \text{ g } N_2$$

(c) $$\text{mass of } H_2 = 6.00 \text{ g } N_2 \times \frac{1 \text{ mol } N_2}{28.02 \text{ g } N_2} \times \frac{3 \text{ mol } H_2}{1 \text{ mol } N_2} \times \frac{2.016 \text{ g } H_2}{1 \text{ mol } H_2} = 1.30 \text{ g } H_2$$

Limiting Reactant and Theoretical Yield

When the two elements antimony and iodine are heated in contact with one another (Fig. 3.5), they react to form antimony(III) iodide.

$$2Sb(s) + 3I_2(s) \longrightarrow 2SbI_3(s)$$

The coefficients in this equation show that two moles of Sb (243.6 g) react with exactly three moles of I_2 (761.4 g) to form two moles of SbI_3 (1005.0 g). Put another way, the maximum quantity of SbI_3 that can be obtained under these conditions, assuming the reaction goes to completion and no product is lost, is 1005.0 g. This quantity is referred to as the **theoretical yield** of SbI_3.

Ordinarily, in the laboratory, reactants are not mixed in exactly the ratio required for reaction. Instead an excess of one reactant, usually the cheaper one, is used. For example, 3.00 mol of Sb could be mixed with 3.00 mol of I_2. In that case, after the reaction is over, 1.00 mol of Sb remains unreacted.

$$\begin{aligned}\text{excess Sb} &= 3.00 \text{ mol Sb originally} - 2.00 \text{ mol Sb consumed}\\ &= 1.00 \text{ mol Sb}\end{aligned}$$

The 3.00 mol of I_2 should be completely consumed in forming the 2.00 mol of SbI_3:

$$n\text{SbI}_3 \text{ formed} = 3.00 \text{ mol I}_2 \times \frac{2 \text{ mol SbI}_3}{3 \text{ mol I}_2} = 2.00 \text{ mol SbI}_3$$

After the reaction is over, the solid obtained would be a mixture of product, 2.00 mol of SbI_3 (1005.0 g), together with 1.00 mol of unreacted Sb (121.8 g).

In situations such as this, a distinction is made between the reactant in excess (Sb) and the other reactant (I_2), called the **limiting reactant**. The amount of product formed is determined (limited) by the amount of limiting reactant. With 3.00 mol of I_2, only 2.00 mol of SbI_3 is obtained, regardless of how large an excess of Sb is used.

Under these conditions, the theoretical yield of product is the amount produced if the limiting reactant is completely consumed. In the case just cited, the theoretical yield of SbI_3 is 2.00 mol, the amount formed from the limiting reactant, I_2.

Often you will be given the amounts of two different reactants and asked to determine which is the limiting reactant and to calculate the theoretical yield of product. To do this, it helps to follow a systematic, three-step procedure:

1. *Calculate the amount of product that would be formed if the first reactant were completely consumed.*
2. *Repeat this calculation for the second reactant: that is, calculate how much product would be formed if all of that reactant were consumed.*
3. *Choose the smaller of the two amounts calculated in* (1) *and* (2). *This is the theoretical yield of product; the reactant that produces the smaller amount is the limiting reactant.* The other reactant is in excess; only part of it is consumed.

Figure 3.5
When antimony (gray powder) comes in contact with iodine (violet vapor), a vigorous reaction takes place (*middle picture*). The product is a red solid, SbI_3. (Charles D. Winters)

Example 3.13 Consider the reaction

$$2Sb(s) + 3I_2(s) \longrightarrow 2SbI_3(s)$$

Determine the limiting reactant and the theoretical yield when

(a) 1.20 mol of Sb and 2.40 mol of I_2 are mixed.
(b) 1.20 g of Sb and 2.40 g of I_2 are mixed.

Strategy Follow the three steps outlined above. In steps (1) and (2), follow the strategy described in Example 3.12 to calculate the amount of product formed. In (a), a simple one-step conversion is required; in (b), the path is longer because you have to go from mass of reactant to mass of product.

Solution

(a) (1) $n\text{SbI}_3 \text{ from Sb} = 1.20 \text{ mol Sb} \times \frac{2 \text{ mol SbI}_3}{2 \text{ mol Sb}} = 1.20 \text{ mol SbI}_3$

(2) $n\text{SbI}_3 \text{ from I}_2 = 2.40 \text{ mol I}_2 \times \frac{2 \text{ mol SbI}_3}{3 \text{ mol I}_2} = 1.60 \text{ mol SbI}_3$

(3) Since 1.20 mol is the smaller amount of product, that is the theoretical yield of SbI_3. This amount of SbI_3 is produced by the antimony, so Sb is the limiting reactant.

You have 1 L of vodka and 1 L of tomato juice. Which is the limiting reactant for "bloody Marys" (1 part vodka, 3 parts tomato juice)?

(b) (1) $\text{mass SbI}_3 \text{ from Sb} = 1.20 \text{ g Sb} \times \frac{1 \text{ mol Sb}}{121.8 \text{ g Sb}} \times \frac{2 \text{ mol SbI}_3}{2 \text{ mol Sb}} \times \frac{502.5 \text{ g SbI}_3}{1 \text{ mol SbI}_3}$

$= 4.95 \text{ g SbI}_3$

(2) $\text{mass SbI}_3 \text{ from I}_2 = 2.40 \text{ g I}_2 \times \frac{1 \text{ mol I}_2}{253.8 \text{ g I}_2} \times \frac{2 \text{ mol SbI}_3}{3 \text{ mol I}_2} \times \frac{502.5 \text{ g SbI}_3}{1 \text{ mol SbI}_3}$

$= 3.17 \text{ g SbI}_3$

(3) The reactant that yields the smaller amount (3.17 g) of SbI_3 is I_2. Hence I_2 is the limiting reactant. The smaller amount, 3.17 g of SbI_3, is the theoretical yield.

The theoretical yield is what you get if everything goes perfectly; it never does

---

Remember that, in deciding upon the theoretical yield of product, you *choose the smaller of the two calculated amounts.* To see why this must be the case, refer back to Example 3.13b. There 1.20 g of Sb was mixed with 2.40 g of I_2. Calculations show that the theoretical yield of SbI_3 is 3.17 g, and 0.43 g of Sb is left over. Thus

$$1.20 \text{ g Sb} + 2.40 \text{ g I}_2 \longrightarrow 3.17 \text{ g SbI}_3 + 0.43 \text{ g Sb}$$

This makes sense: 3.60 g of reactants yields a total of 3.60 g of "products," including the unreacted antimony. Suppose, however, that 4.95 g of SbI_3 was chosen as the theoretical yield. The following nonsensical situation would arise:

$$1.20 \text{ g Sb} + 2.40 \text{ g I}_2 \longrightarrow 4.95 \text{ g SbI}_3$$

This violates the law of conservation of mass; 3.60 g of reactants cannot form 4.95 g of product.

Experimental Yield; Percent Yield

The theoretical yield is the maximum amount of product that can be obtained. In calculating the theoretical yield, it is assumed that the limiting reactant is 100% converted to product. In the real world, that is unlikely to happen. Some of the limiting reactant may be consumed in competing reactions. Some of the product may be lost in separating it from the reaction mixture. For these and other reasons, the experimental yield is ordinarily less than the theoretical yield. Put another way, the *percent yield* is expected to be less than 100%:

$$\text{percent yield} = \frac{\text{experimental yield}}{\text{theoretical yield}} \times 100\%$$

You never do quite as well as you hoped, so you never get 100%

To illustrate how percent yield is calculated, consider again the reaction described in Example 3.13b. Suppose the actual yield in the laboratory is 2.42 g. Since the theoretical yield was 3.17 g,

$$\text{percent yield} = \frac{2.42\text{ g}}{3.17\text{ g}} \times 100\% = 76.3\%$$

PERSPECTIVE • Nonstoichiometric Solids

Throughout this chapter we have assumed, in effect, that all compounds follow the law of constant composition; regardless of their origin or mode of preparation, they contain the same elements in the same atom ratio. For gases, this is always true. However, certain solid compounds contain crystal defects that make them *nonstoichiometric;* the atom ratios of the elements differ slightly from whole-number values.

A classic example of nonstoichiometry involves sodium chloride crystals which have been irradiated with ultraviolet light. Perhaps one chloride ion in a thousand undergoes the reaction:

$$Cl^-(s) \longrightarrow Cl(g) + e^-(s)$$

The Cl atom migrates to the surface of the crystal and escapes; the electron takes the place of the Cl^- ion in the crystal, giving it an intense blue color. The "simplest formula" of the irradiated solid might be written as $Na_{1.000}Cl_{0.999}$.

Many oxides and sulfides of the transition metals can be nonstoichiometric. Stoichiometric nickel(II) oxide, in which the atom ratio is exactly 1:1 (i.e., $Ni_{1.00}O_{1.00}$), is a green, nonconducting solid. If this compound is heated to 1200°C in pure oxygen, it turns black and undergoes a change in composition to something close to $Ni_{0.97}O_{1.00}$. This nonstoichiometric solid was one of the early semiconductors; it has the structure shown in Figure 3.A. A few Ni^{2+} ions are missing from the crystal lattice; to maintain electrical neutrality, two Ni^{2+} ions are converted to Ni^{3+} ions.

Nonstoichiometry is relatively common among "mixed" metal oxides, in which more than one metal is present. In 1986, it was discovered that certain compounds of this type showed the phenomenon of *superconductivity;* upon cooling to about 100 K, their electrical resistance drops to zero. A typical formula here is $YBa_2Cu_3O_x$, where x varies from 6.5 to 7.2, depending upon the method of preparation of the solid.

Potentially, superconductors have a host of practical applications. A power line made of a superconducting material could carry electricity over hundreds of kilometers with virtually no power loss. On a more exotic level, it might be possible to design electric trains traveling at 500 km/h over superconducting tracks. The train would be literally suspended in the air, perhaps

Ni^{2+}	O^{2-}	Ni^{2+}	O^{2-}
O^{2-}	Ni^{2+}	O^{2-}	Ni^{3+}
Ni^{2+}	O^{2-}	□	O^{2-}
O^{2-}	Ni^{3+}	O^{2-}	Ni^{2+}
Ni^{2+}	O^{2-}	Ni^{2+}	O^{2-}

Figure 3.A
In nonstoichiometric nickel(II) oxide, a few Ni^{2+} ions are missing from the crystal lattice. For every Ni^{2+} ion removed, two others are converted to Ni^{3+} ions.

20 cm off the ground (Fig. 3.B). This effect arises because the magnetic field induced when current flows through a superconductor repels that of a magnet above it.

As you might guess, there are some practical problems that have held back the commercial development of superconducting devices. For one thing, mixed metal oxides are brittle, difficult to fabricate into wire or thin film. Perhaps more serious, they are limited as to how much current they can carry. Currents in excess of a critical value produce a tremendous increase in resistance.

Figure 3.B
A pellet of superconducting material, previously cooled to 77 K with liquid nitrogen, floats above a magnet. (Courtesy of Edmund Scientific)

CHAPTER HIGHLIGHTS

KEY CONCEPTS

1. *Relate the atomic mass of an element to isotopic masses and abundances* (Example 3.1; Problems 3–10)
2. *Use Avogadro's number to calculate the mass of an atom or molecule* (Examples 3.2, 3.3; Problems 11–18, 77, 78)
3. *Use molar mass to relate*
   — moles to mass of a substance (Examples 3.4, 3.5; Problems 19–26)
   — molecular formula to simplest formula (Example 3.10; Problems 43, 44)
4. *Use the formula of a compound to find percent composition or mass of an element in a sample* (Examples 3.6, 3.7; Problems 27–34, 79, 81, 83, 84)
5. *Find the simplest formula of a compound from chemical analysis* (Examples 3.8, 3.9; Problems 35–46, 85)
6. *Balance chemical equations by inspection* (Example 3.11; Problems 49–54)

7. *Use a balanced equation to*
   — *relate masses of reactants and products*
   (Example 3.12; Problems 55–64, 80, 86)
   — *find the limiting reactant and theoretical yield*
   (Example 3.13; Problems 65–72)

KEY TERMS

atomic mass unit	molar mass	percent yield
Avogadro's number	mole	simplest formula
isotopic abundance	molecular formula	theoretical yield
limiting reactant	percent composition	

SUMMARY PROBLEM

Consider oxygen gas which is essential to our survival.

a. It consists of three isotopes with atomic masses 16.00, 17.00, and 18.00 amu. Their abundances are 99.76%, 0.04%, and 0.20%, respectively. What is the atomic mass of oxygen?
b. What is the mass in grams of an O_2 molecule? How many molecules are present in 1.000 g of O_2 gas?
c. How many moles are there in 15.00 g of O_2 gas? What is the mass in grams of 0.2500 mol of O_2?
d. When iron reacts with oxygen, iron(III) oxide is formed. Write a balanced equation for this reaction.
e. What is the molar mass of iron(III) oxide?
f. How many grams of oxygen will react with 0.35 mol of iron?
g. If 10.00 g of iron react with an excess of oxygen, how many grams of product are formed?
h. If 10.00 g of iron react with 10.00 g of oxygen, how many grams of product are formed?
i. If 12.61 g of product is produced in the laboratory when 10.00 g of iron react with 10.00 g of oxygen, what is the percent yield?
j. When oxygen is bubbled through a solution of sodium polysulfide, photographic "hypo" is produced. If "hypo" is made up of Na, S, and O, and contains 29.08% Na and 40.57% S, what is its simplest formula?

Answers

a. 16.00 amu **b.** 5.314×10^{-23} g, 1.882×10^{22} molecules
c. 0.4688 mol; 8.000 g **d.** $4Fe(s) + 3O_2(g) \rightarrow 2Fe_2O_3(s)$
e. 159.7 g **f.** 8.4 g **g.** 14.30 g **h.** 14.30 g **i.** 88.18%
j. $Na_2S_2O_3$

QUESTIONS & PROBLEMS

Atomic Masses

1. Arrange the following in order of increasing mass.
a. 2 Na atoms **b.** 3 Mg atoms
c. a K atom **d.** a K^+ ion

2. Calculate the mass ratio of an arsenic (As) atom to an atom of
a. Mg **b.** Sn **c.** Hg

3. Strontium consists of four isotopes with masses of 83.9134 amu (0.5%), 85.9094 amu (9.9%), 86.9089 amu (7.0%), and 87.9056 amu (82.6%). Calculate the atomic mass of strontium.

4. Naturally occurring europium (Eu) consists of two isotopes with masses 150.9199 amu (48.03%) and 152.9212 amu (51.97%). Calculate the atomic mass of Eu.

5. Naturally occurring boron consists of two isotopes with masses 10.013 amu and 11.009 amu. Estimate the abundances of these isotopes. The atomic mass of naturally occurring boron is 10.811 amu.

6. Nitrogen has two isotopes, N-14 and N-15, with masses of 14.0031 amu and 15.001 amu, respectively. What is the abundance of N-15? (atomic mass $N = 14.00674$ amu)

7. Magnesium (atomic mass = 24.305 amu) consists of three isotopes with masses 23.98 amu, 24.98 amu, and 25.98 amu. The abundance of the middle isotope is 10.1%. Estimate the abundances of the other two isotopes.
8. Chromium (atomic mass = 51.9961 amu) has four isotopes. Their masses are 49.9461 amu, 51.9405 amu, 52.9407 amu, and 53.9389 amu. The last two isotopes have abundances of 9.50% and 2.36%, respectively. Estimate the other two abundances.
9. Silicon consists of three isotopes with masses 27.977 amu, 28.977 amu, and 29.974 amu. Their abundances are 92.34%, 4.70%, and 2.96%, respectively. Sketch the mass spectrum for silicon.
10. Chlorine has two isotopes, Cl-35 and Cl-37. Their abundances are 75.53% and 24.47%, respectively. Assuming that the only hydrogen isotope present is H-1
   a. How many different HCl molecules are possible?
   b. What is the sum of the mass numbers of the two atoms in each molecule?
   c. Sketch the mass spectrum for HCl if all the positive ions are obtained by removing a single electron from an HCl molecule.

Avogadro's Number and the Mole

11. The atomic mass of tungsten, W, is 183.9 amu. Calculate the
   a. mass in grams of one tungsten atom.
   b. number of atoms in one milligram of tungsten.
12. Mercury has atomic mass 200.6 amu. Calculate the
   a. mass of 2.0×10^{15} atoms.
   b. number of atoms in one nanogram of mercury.
13. Determine the
   a. mass of one billion gold (Au) atoms.
   b. number of atoms in one ounce of gold.
14. Determine the
   a. mass of a trillion (one million million) lead (Pb) atoms.
   b. number of atoms in one ounce of lead.
15. The national debt in 1990 was 3.2333 trillion dollars. How many moles of pennies would be required to pay off the national debt?
16. Calculate the number of molecules in an eight-ounce glass of water. (1 lb = 16 oz = 453.6 g)
17. How many electrons are there in
   a. an aluminum atom?
   b. a mole of aluminum atoms?
   c. 0.2843 mol of aluminum?
   d. 0.2843 g of aluminum?
18. How many protons are there in
   a. one neon (Ne) atom?
   b. a mole of neon atoms?
   c. 20.02 mol of neon?
   d. 20.02 g of neon?

Molar Masses and Mole–Gram Conversions

19. Calculate the molar masses (g/mol) of
   a. gallium, Ga, a metal that literally melts in your hands.
   b. laughing gas, N_2O, one of the first anesthetics used.
   c. cane sugar, $C_{12}H_{22}O_{11}$.
20. Calculate the molar masses (g/mol) of
   a. ammonia (NH_3).
   b. baking soda ($NaHCO_3$).
   c. osmium (Os) metal.
21. Convert the following to moles.
   a. 0.830 g of strychnine ($C_{21}H_{22}N_2O_2$), present in rat poison
   b. two hundred fifty milligrams of aspirin ($C_9H_8O_4$)
   c. a gram of Vitamin C ($C_6H_8O_6$)
22. Convert the following to moles.
   a. 3.86 g of carbon dioxide, CO_2
   b. 0.485 g of ethyl alcohol (C_2H_5OH)
   c. 6.00×10^3 g of hydrazine, N_2H_4, a rocket propellant
23. Calculate the mass in grams of 5.75 mol of
   a. nitrogen atoms. **b.** nitrogen molecules (N_2).
   c. ammonia molecules.
24. Calculate the mass in grams of 6.83 mol of
   a. vinyl chloride, C_2H_3Cl, the starting material for a plastic.
   b. stannous fluoride, SnF_2, the active ingredient in toothpastes for cavity prevention.
   c. nitrogen dioxide, NO_2, the brown gas responsible for the color of smog.
25. Complete the following table for ethylene glycol, $C_2H_6O_2$, an antifreeze used in cars.

	Number of Grams	Number of Moles	Number of Molecules	Number of C Atoms
a.	0.1245	____	____	____
b.	____	0.0375	____	____
c.	____	____	2.0×10^{25}	____
d.	____	____	____	3.6×10^{12}

26. Complete the following table for acetone, C_3H_6O, the main component of nail polish remover.

	Number of Grams	Number of Moles	Number of Molecules	Number of H Atoms
a.	0.2500	____	____	____
b.	____	0.2000	____	____
c.	____	____	2.5×10^{10}	____
d.	____	____	____	8.0×10^{15}

Percent Composition from Formula

27. Turquoise has the following chemical formula: $CuAl_6(PO_4)_4(OH)_8 \cdot 4H_2O$. Calculate the mass percent of each element in turquoise.
28. The active ingredient of some antiperspirants is alumi-

num chlorohydrate, $Al_2(OH)_5Cl$. Calculate the mass percent of each element in this ingredient.

29. The hormone thyroxine secreted by the thyroid gland has the formula $C_{15}H_{11}NO_4I_4$. How many milligrams of iodine can be extracted from 5.00 g of thyroxine?

30. One of the most common rocks on earth is feldspar. One type of feldspar has the formula $CaAl_2Si_2O_8$. How much aluminum can be obtained from mining and smelting one metric ton (10^3 kg) of feldspar?

31. A 250.0-mg sample of a commercial headache remedy contains 152 mg of aspirin, $C_9H_8O_4$. Assume all the carbon in the sample is in aspirin.

- **a.** What is the mass percent of aspirin in the product?
- **b.** How many grams of carbon are there in the aspirin contained in a tablet of this product weighing 0.611 g?

32. The active ingredient in Pepto-Bismol (an over-the-counter remedy for upset stomach) is bismuth subsalicylate, $C_7H_5BiO_4$. Analysis of a 1.50-g sample of Pepto-Bismol yields 346 mg of bismuth. What percent by mass of the sample is bismuth subsalicylate? (Assume that there is no other bismuth-containing compound in Pepto-Bismol.)

33. Toluene is now widely used instead of benzene, since benzene has been proven to be carcinogenic. Combustion of 1.000 g of toluene, a compound of carbon and hydrogen atoms, gives 3.348 g carbon dioxide. What are the mass percents of carbon and hydrogen in toluene?

34. Hexachlorophene, a compound made up of atoms of carbon, hydrogen, chlorine, and oxygen, is an ingredient in germicidal soaps. Combustion of a 1.000-g sample yields 1.407 g of carbon dioxide, 0.134 g of water, and 0.5228 g of chlorine gas. What are the mass percents of carbon, hydrogen, chlorine, and oxygen in hexachlorophene?

Simplest Formula from Analysis

35. Arsenic reacts with chlorine to form a chloride. If 1.587 g of arsenic reacts with 3.755 g of chlorine, what is the simplest formula of the chloride?

36. If 7.35 g of chromium reacted directly with oxygen to form 10.74 g of a metal oxide, what is the simplest formula of the oxide?

37. Determine the simplest formulas of the following three compounds.

- **a.** citric acid, which has the composition: 37.51% C, 4.20% H, and 58.29% O
- **b.** tetraethyl lead, the gasoline additive, which has the composition: 29.71% C, 6.234% H, and 64.07% Pb
- **c.** saccharin, the artificial sweetener, which has the composition: 45.90% C, 2.75% H, 26.20% O, 17.50% S, and 7.65% N

38. Determine the simplest formulas of the following compounds

- **a.** rubbing alcohol, a compound made up of carbon, hydrogen, and oxygen atoms, which has the composition: 59.96% C and 13.42% H
- **b.** the food enhancer, monosodium glutamate (MSG), which has the composition: 35.51% C, 4.77% H, 37.85% O, 8.29% N, and 13.60% Na
- **c.** the sedative chloral hydrate, which has the composition: 14.52% C, 1.828% H, 64.30% Cl, and 19.35% O

39. Vanillin, a flavoring agent, is made up of carbon, hydrogen, and oxygen atoms. When a sample of vanillin weighing 2.500 g burns in oxygen, 5.79 g of carbon dioxide and 1.18 g water are obtained. What is the simplest formula of vanillin?

40. Methyl salicylate is a common "active ingredient" in liniments such as Ben-Gay. It is also known as oil of wintergreen. It is made up of carbon, hydrogen, and oxygen atoms. When a sample of methyl salicylate weighing 5.287 g is burned in excess oxygen, 12.24 g of carbon dioxide and 2.505 g of water are formed. What is the simplest formula for oil of wintergreen?

41. The hormone norepinephrine is an antidote to the allergic reaction resulting from a bee sting. It is made up of hydrogen, carbon, oxygen, and nitrogen atoms. A 2.587-g sample of norepinephrine burned in oxygen yields 5.387 g of carbon dioxide and 1.526 g of water. Another experiment determines that the hormone contains 8.281% nitrogen by mass. What is the simplest formula of norepinephrine?

42. Dimethyl nitrosamine is a known carcinogen. It can be formed in the intestinal tract when digestive juices react with the nitrite ion in preserved and smoked meats. It is made up of carbon, hydrogen, nitrogen, and oxygen atoms. A 4.319-g sample of dimethyl nitrosamine burned in oxygen yields 5.134 g of carbon dioxide and 3.151 g of water. The compound contains 37.82% by mass of nitrogen. What is the simplest formula of dimethyl nitrosamine?

43. Dimethylhydrazine, the fuel used in the Apollo lunar descent module, has a molar mass of 60.10 g/mol. It is made up of carbon, hydrogen, and nitrogen atoms. The combustion of 2.859 g of the fuel in excess oxygen yields 4.190 g carbon dioxide and 3.428 g of water. What are the simplest and molecular formulas for dimethylhydrazine?

44. Hexamethylenediamine ($\mathcal{M}$ = 116.2 g/mol), a compound made up of carbon, hydrogen, and nitrogen atoms, is used in the production of nylon. When 6.315 g of hexamethylenediamine is burned in oxygen, 14.36 g of carbon dioxide and 7.832 g of water are obtained. What are the simplest and molecular formulas of the compound?

45. A sample of a compound of chlorine and oxygen reacts with excess hydrogen to give 0.3059 g HCl and 0.5287 g of water. What is the simplest formula of the compound?

46. A 2.103-g sample of a copper oxide, when heated in a stream of hydrogen gas, yields 0.476 g of water. What is the formula of the copper oxide?

47. Washing soda is a hydrate of sodium carbonate. Its formula is $Na_2CO_3 \cdot xH_2O$. A 2.714-g sample of washing soda is heated until a constant mass of 1.006 g of Na_2CO_3 is reached. What is x?

48. A certain hydrate of potassium aluminum sulfate (alum) has the formula $KAl(SO_4)_2 \cdot xH_2O$. When a hydrate sample weighing 5.459 g is heated to remove all the water, 2.583 g of $KAl(SO_4)_2$ remains. What is the mass percent of water in the hydrate? What is x?

Balancing Equations

49. Balance the following equations.
   a. $Au_2S_3(s) + H_2(g) \longrightarrow Au(s) + H_2S(g)$
   b. $C_3H_8(g) + O_2(g) \longrightarrow CO_2(g) + H_2O(g)$
   c. $SiO_2(s) + C(s) \longrightarrow Si(s) + CO(g)$

50. Balance the following equations.
   a. $UO_2(s) + HF(l) \longrightarrow UF_4(s) + H_2O(l)$
   b. $PH_3(g) + O_2(g) \longrightarrow P_4O_{10}(s) + H_2O(g)$
   c. $C_2H_3Cl(l) + O_2(g) \longrightarrow CO_2(g) + H_2O(g) + HCl(g)$

51. Write balanced equations for the reaction of sodium, Na(*s*), with the following nonmetals to form ionic solids. (Recall Fig. 2.6, p. 34)
   a. nitrogen b. oxygen, forming O^{2-} ions
   c. sulfur d. bromine e. iodine

52. Write balanced equations for the reaction of chlorine, $Cl_2(g)$, with the following metals to form solids that you can take to be ionic. (Recall Figure 2.8, p. 37).
   a. aluminum b. strontium c. lithium
   d. chromium e. silver

53. Write a balanced equation for
   a. the reaction of boron trifluoride gas with water to give liquid hydrogen fluoride and solid boric acid (H_3BO_3).
   b. the reaction of magnesium oxide with iron to form iron(III) oxide and magnesium.
   c. the decomposition of dinitrogen oxide gas to its elements.
   d. the reaction of calcium carbide (CaC_2) solid with water to form calcium hydroxide and acetylene (C_2H_2) gas.
   e. the reaction of solid calcium cyanamide ($CaCN_2$) with water to form calcium carbonate and ammonia gas.

54. Write a balanced equation for the
   a. decomposition of ammonium nitrate to dinitrogen oxide gas and steam.
   b. reaction of ammonia gas with $O_2(g)$ to form $N_2(g)$ and steam.
   c. decomposition of potassium chlorate to potassium chloride and oxygen.
   d. reaction of solid diborontrioxide with C(*s*) and $Cl_2(g)$ to produce two gases, boron trichloride and carbon monoxide (CO).
   e. combustion of liquid benzene (C_6H_6) to carbon dioxide and water.

Mole–Mass Relations in Reactions

55. One way to remove nitrogen oxide (NO) from smokestack emissions is to react it with ammonia:

$$4NH_3(g) + 6NO(g) \longrightarrow 5N_2(g) + 6H_2O(l)$$

Fill in the blanks below.
   a. 12.3 mol NO reacts with ________ mol ammonia.
   b. 5.87 mol NO yields ________ mol nitrogen.
   c. 0.2384 mol nitrogen requires ________ mol nitrogen oxide.
   d. 13.9 mol NH_3 produces ________ mol water.

56. The reaction of the mineral fluorapatite with sulfuric acid follows the equation

$$Ca_{10}F_2(PO_4)_6(s) + 7H_2SO_4(l) \longrightarrow 2HF(g) + 3Ca(H_2PO_4)_2(s) + 7CaSO_4(s)$$

Fill in the blanks below.
   a. 7.38 mol fluorapatite yields ________ mol calcium sulfate.
   b. 3.98 mol calcium dihydrogen phosphate requires ________ mol H_2SO_4.
   c. 0.379 mol sulfuric acid reacts with ________ mol fluorapatite.
   d. 4.983 mol fluorapatite produces ________ mol $Ca(H_2PO_4)_2$.

57. Using the equation given in Problem 55, calculate
   a. the mass of nitrogen produced from 2.93 mol of nitrogen oxide.
   b. the mass of ammonia required to form 7.65 mol of water.
   c. the mass of nitrogen oxide that yields 0.356 g water.
   d. the mass of ammonia required to react with 20.0 g nitrogen oxide.

58. Using the equation given in Problem 56, calculate
   a. the mass of calcium sulfate formed from 0.349 mol of fluorapatite.
   b. the number of moles of fluorapatite required to form one gram of hydrogen fluoride.
   c. the mass of fluorapatite required to yield 3.79 g of calcium sulfate.
   d. the mass of hydrogen fluoride formed from 13.98 g of sulfuric acid.

59. Iron ore consists mainly of iron(III) oxide. When iron(III) oxide is heated with an excess of coke (carbon), iron metal and carbon monoxide are produced.
   a. Write a balanced equation for the reaction.
   b. How many moles of iron(III) oxide are required to form 12.79 mol of iron?
   c. How many grams of carbon monoxide are formed from 13.68 g of coke?

60. Nitrogen trichloride gas reacts with water to form ammonia and hypochlorous acid, HClO(*aq*), the main component of bleach.

a. Write a balanced equation for this reaction.
b. How many moles of ammonia are produced from 275 mL of water ($d = 1.00\ g/cm^3$)?
c. How many grams of nitrogen trichloride are required to produce 14.8 g of ammonia?

61. A wine cooler contains about 4.5% ethyl alcohol, C_2H_5OH, by mass. Assume that the alcohol in 2.85 kg of wine cooler is produced by the fermentation of the glucose in grapes in the reaction.

$$C_6H_{12}O_6(aq) \longrightarrow 2C_2H_5OH(l) + 2CO_2(g)$$

a. How many grams of glucose are needed to produce the ethyl alcohol in the wine?
b. What volume of carbon dioxide gas ($d = 1.80$ g/L) is produced at the same time?

62. When corn is allowed to ferment, ethyl alcohol is produced from the glucose in corn according to the fermentation reaction given in Problem 61.
a. What volume of ethyl alcohol ($d = 0.789\ g/cm^3$) is produced from one pound of glucose?
b. Gasohol is a mixture of 10 cm^3 of ethyl alcohol per 90 cm^3 of gasoline. How many grams of glucose are required to produce the ethyl alcohol in one gallon of gasohol?

63. Oxygen masks for producing O_2 in emergency situations contain potassium superoxide, KO_2. It reacts with CO_2 and H_2O in exhaled air to produce oxygen:

$$4KO_2(s) + 2H_2O(g) + 4CO_2(g) \longrightarrow 4KHCO_3(s) + 3O_2(g)$$

If a person wearing such a mask exhales 0.702 g CO_2/min, how many grams of KO_2 are consumed in ten minutes?

64. A crude oil burned in electrical generating plants contains about 1.2% sulfur by mass. When the oil burns, the sulfur forms sulfur dioxide gas:

$$S(s) + O_2(g) \longrightarrow SO_2(g)$$

How many liters of SO_2 ($d = 2.60$ g/L) are produced when one metric ton (10^3 kg) of oil burns?

Limiting Reactant; Theoretical Yield

65. Chlorine and fluorine react to form gaseous chlorine trifluoride. You start with 1.75 mol chlorine and 3.68 mol fluorine.
a. Write a balanced equation for the reaction.
b. What is the limiting reactant?
c. What is the theoretical yield of chlorine trifluoride in moles?
d. How many moles of the excess reactant remain unreacted?

66. A gaseous mixture containing 7.50 mol $H_2(g)$ and 9.00 mol $Cl_2(g)$ reacts to form hydrogen chloride gas.
a. Write a balanced equation for the reaction.
b. Which reactant is limiting?
c. If all the limiting reactant is consumed, how many moles of hydrogen chloride are formed?
d. How many moles of the excess reactant remain unreacted?

67. The space shuttle uses aluminum metal and ammonium perchlorate, NH_4ClO_4, in its reusable booster rockets. The products of the reaction are aluminum oxide, aluminum chloride, NO, and steam. The reaction mixture contains 5.75 g of Al and 7.32 g of NH_4ClO_4.
a. What is the theoretical yield of aluminum chloride, $AlCl_3$?
b. If 1.87 g of aluminum chloride is formed, what is the percent yield?

68. Oxyacetylene torches used for welding reach temperatures near 2000°C. The reaction involved is the combustion of acetylene, C_2H_2:

$$2C_2H_2(g) + 5\ O_2(g) \longrightarrow 4CO_2(g) + 2H_2O(g)$$

Starting with 175 g of both acetylene and oxygen, what is the theoretical yield, in grams, of carbon dioxide? If 52.5 L of carbon dioxide ($d = 1.80$ g/L) are produced, what is the percent yield?

69. In the Ostwald process, nitric acid, HNO_3, is produced from ammonia by a three-step process:

$$4NH_3(g) + 5\ O_2(g) \longrightarrow 4NO(g) + 6H_2O(g)$$
$$2NO(g) + O_2(g) \longrightarrow 2NO_2(g)$$
$$3NO_2(g) + H_2O(g) \longrightarrow 2HNO_3(aq) + NO(g)$$

Assuming a 75% yield in each step, how many grams of nitric acid can be made from one hundred liters of ammonia ($d = 0.695$ g/L)?

70. A century ago, $NaHCO_3$ was prepared from Na_2SO_4 by a three-step process:

$$Na_2SO_4(s) + 4C(s) \longrightarrow Na_2S(s) + 4CO(g)$$
$$Na_2S(s) + CaCO_3(s) \longrightarrow CaS(s) + Na_2CO_3(s)$$
$$Na_2CO_3(s) + H_2O(l) + CO_2(g) \longrightarrow 2NaHCO_3(s)$$

How many kilograms of $NaHCO_3$ could be formed from one kilogram of Na_2SO_4, assuming an 82% yield in each step?

71. Aspirin, $C_9H_8O_4$, is prepared by reacting salicylic acid, $C_7H_6O_3$, with acetic anhydride, $C_4H_6O_3$, in the reaction

$$2C_7H_6O_3(s) + C_4H_6O_3(l) \longrightarrow 2C_9H_8O_4(s) + H_2O(l)$$

A student is told to prepare 25.0 g of aspirin. He is also told that he should use a 50.0% excess of acetic anhydride and expect to get a 65.0% yield in the reaction. How many grams of each reactant should he use?

72. A student prepares phosphorous acid, H_3PO_3, by reacting solid phosphorus triiodide with water:

$$PI_3(s) + 3H_2O(l) \longrightarrow H_3PO_3(s) + 3HI(g)$$

The student needs to obtain 0.100 L of H_3PO_3 (d = 1.651 g/cm^3). The procedure calls for a 40.0% excess of water and a yield of 55.0%. How much phosphorus triiodide should she weigh out? What volume of water (d = 1.00 g/cm^3) should she use?

Nonstoichiometric Solids

73. Consider nonstoichiometric nickel oxide, which has the formula $Ni_{0.97}O$. All the anions are O^{2-} ions. What percent of the cations are Ni^{2+} ions? Ni^{3+}?

74. What is the value of x for nonstoichiometric Cu_xS, in which 88% of the copper is in the form of Cu^+, while 12% is present as Cu^{2+}? Assume all the anions are S^{2-} ions.

75. Explain what is meant by nonstoichiometry; superconductivity.

76. List some of the applications for superconductors and state the problems that have held back the commercial development of superconductors.

Unclassified

77. Carbon tetrachloride, CCl_4, was a popular dry-cleaning agent until it was shown to be carcinogenic. It has a density of 1.594 g/cm^3. What volume of carbon tetrachloride will contain a total of 5.00×10^{24} molecules of CCl_4?

78. Some brands of salami contain 0.090% sodium benzoate ($NaC_7H_5O_2$) by mass as a preservative. If you eat 2.52 oz of this salami, how many atoms of sodium will you consume, assuming salami contains no other source of that element?

79. All the fertilizers listed below contribute nitrogen to the soil. If all these fertilizers are sold for the same price per gram of nitrogen, which will cost the least per 50-lb bag?

urea $(NH_2)_2CO$
ammonia NH_3
ammonium nitrate NH_4NO_3
guanidine $HNC(NH_2)_2$

80. Most wine is prepared by the fermentation of the glucose in grape juice by yeast:

$$C_6H_{12}O_6(aq) \longrightarrow 2C_2H_5OH(aq) + 2CO_2(g)$$

How many grams of glucose should there be in grape juice to produce 750.0 mL of wine that is 11.0% ethyl alcohol, C_2H_5OH (d = 0.789 g/cm^3), by volume?

Challenge Problems

81. Chlorophyll, the substance responsible for the green color of leaves, has one magnesium atom per chlorophyll molecule and contains 2.72% magnesium by mass. What is the molar mass of chlorophyll?

82. By x-ray diffraction, it is possible to determine the geometric pattern in which atoms are arranged in a crystal and the distances between atoms. In a crystal of silver, four atoms effectively occupy the volume of a cube 0.409 nm on an edge. Taking the density of silver to be 10.5 g/cm^3, calculate the number of atoms in one mole of Ag.

83. A 5.025-g sample of calcium is burned in air to produce a mixture of two ionic compounds, calcium oxide and calcium nitride. Water is added to this mixture. It reacts with calcium oxide to form 4.832 g of calcium hydroxide. How many grams of calcium oxide are formed? How many grams of calcium nitride?

84. A mixture of potassium chloride and potassium bromide weighing 3.595 g is heated with chlorine, which converts the mixture completely to potassium chloride. The total mass of potassium chloride after the reaction is 3.129 g. What percent of the original mixture is potassium bromide?

85. A sample of an oxide of vanadium weighing 4.589 g was heated with hydrogen gas to form water and another oxide of vanadium weighing 3.782 g. The second oxide was treated further with hydrogen until only 2.573 g of vanadium metal remained.

a. What are the simplest formulas of the two oxides?
b. What is the total mass of water formed in the successive reactions?

86. A sample of cocaine, $C_{17}H_{21}O_4N$, is diluted with sugar, $C_{12}H_{22}O_{11}$. When a 1.00-mg sample of this mixture is burned, 1.00 mL of carbon dioxide (d = 1.80 g/L) is formed. What is the percentage of cocaine in the mixture?

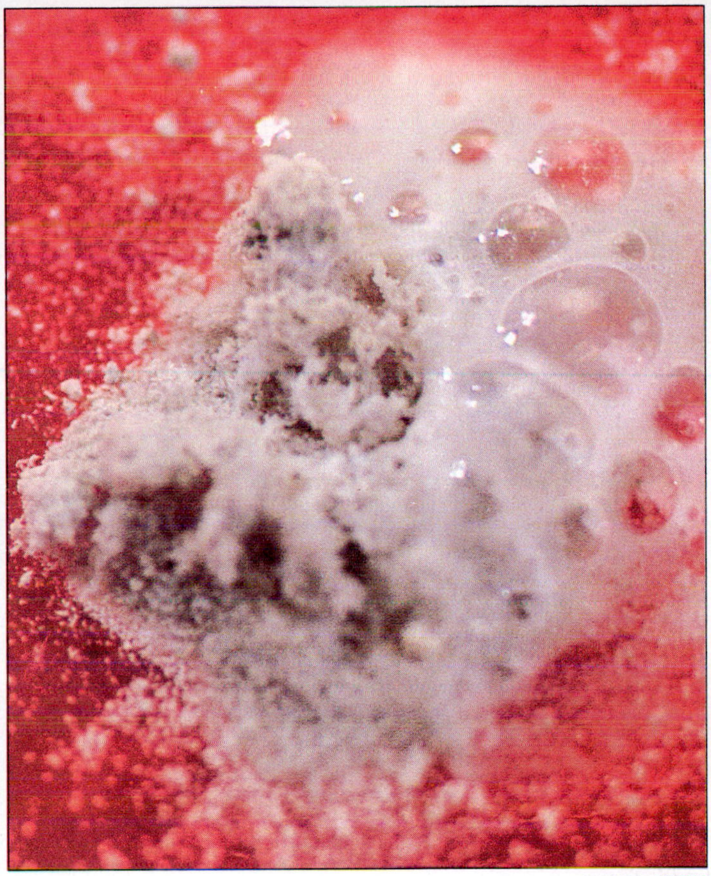

CHARLES D. WINTERS

$CaH_2(s) + 2H_2O \longrightarrow 2H_2(g) + Ca^{2+}(aq) + 2\ OH^-(aq)$

(See p. 528)

4
Reactions in Aqueous Solutions

Surrounded by beakers, by strange coils,
By ovens and flasks with twisted necks,
The chemist, fathoming the whims of attractions,
Artfully imposes on them their precise meetings.

—SULLY-PRUDHOMME
The Naked World
(translated by William Dock)

CHAPTER OUTLINE

All of the reactions considered in Chapter 3 involved pure substances reacting with each other. However, most of the reactions you will carry out in the laboratory or hear about in lecture take place in water solution. Beyond that, most of the reactions that occur in the world around you involve ions or molecules dissolved in water. For these reasons, among others, you need to become familiar with some of the more important types of aqueous reactions. These include

— precipitation reactions (Section 4.1)
— acid-base reactions (Section 4.2)
— oxidation-reduction reactions (Section 4.3)

The emphasis in these sections is upon writing and balancing chemical equations for these reactions.

The last half of this chapter is devoted to mass relations in solution reactions. We consider in turn

— the unit of *molarity*, used to express concentrations of solution species (Section 4.4)
— stoichiometric calculations for solution reactions (Section 4.5)
— the use of solution reactions in quantitative analysis (Section 4.6)

Figure 4.1
When solutions of sodium carbonate, Na_2CO_3, and calcium chloride, $CaCl_2$, are mixed, a white precipitate is obtained. (Charles D. Winters)

4.1 PRECIPITATION REACTIONS

When an ionic solid dissolves in water, the cations and anions separate from one another. This process can be represented by a chemical equation. For the dissolution of sodium carbonate and calcium chloride, the equations are

$$Na_2CO_3(s) \longrightarrow 2Na^+(aq) + CO_3^{2-}(aq)$$

$$CaCl_2(s) \longrightarrow Ca^{2+}(aq) + 2Cl^-(aq)$$

It is important to realize that compounds such as these are completely ionized in water; there is no such thing as a "molecule" of Na_2CO_3 or $CaCl_2$.

Sometimes, when water solutions of two different ionic compounds are mixed, an insoluble solid separates out of solution. This is the case when solutions of Na_2CO_3 and $CaCl_2$ are mixed (Fig. 4.1). The **precipitate** that forms is itself ionic; the cation comes from one solution, the anion from the other. To predict the occurrence of reactions of this type, you must know which ionic substances are insoluble in water.

Ionic solids cover an enormous range of solubilities. At one extreme is lithium chlorate, $LiClO_3$, which dissolves to the extent of 35 mol (3200 g) per liter of water at 25°C. Mercury(II) sulfide, HgS, is at the opposite extreme; the

You must learn these rules to work with precipitation reactions

Table 4.1 Solubility Rules*

NO_3^-	All nitrates are soluble.
Cl^-	All chlorides are soluble except AgCl, Hg_2Cl_2, and $PbCl_2$.
SO_4^{2-}	Most sulfates are soluble; exceptions include $SrSO_4$, $BaSO_4$, and $PbSO_4$.
CO_3^{2-}	All carbonates are insoluble except those of the Group 1 elements and NH_4^+.
OH^-	All hydroxides are insoluble except those of the Group 1 elements, $Sr(OH)_2$, and $Ba(OH)_2$. ($Ca(OH)_2$ is slightly soluble.)
S^{2-}	All sulfides except those of Group 1 and 2 elements and NH_4^+ are insoluble.

* Insoluble compounds are those that precipitate when we mix equal volumes of ~0.1 mol/L solutions of the corresponding ions.

(a)

(b)

$LiClO_3$ (a) is extremely soluble. In contrast, HgS (b) is about as insoluble as it is possible for a compound to be. (Marna G. Clarke)

calculated solubility at 25°C is 10^{-26} mol/L. This means, in principle at least, about 200 L of a solution of HgS would be required to contain a single pair of Hg^{2+} and S^{2-} ions.

They'd be hard to find

Solubility Rules

Using the solubility rules listed in Table 4.1, it is possible to make qualitative predictions concerning solubilities of literally hundreds of different ionic compounds. These rules are quite simple to interpret. For example, the following facts should be evident from the table:

— $Ni(NO_3)_2$ is soluble. (All nitrates are soluble.)
— $BaCl_2$ is soluble. ($BaCl_2$ is not one of the three insoluble chlorides listed.)
— PbS is insoluble. (Pb is not a Group 1 or Group 2 element.)

Figure 4.2 shows how these rules apply to some common ionic compounds.

Using the solubility rules, you can predict whether or not a precipitate will form when two solutions of ionic compounds are mixed. For precipitation to occur, *the cation of one solution must combine with the anion of the other solution to form an insoluble solid.*

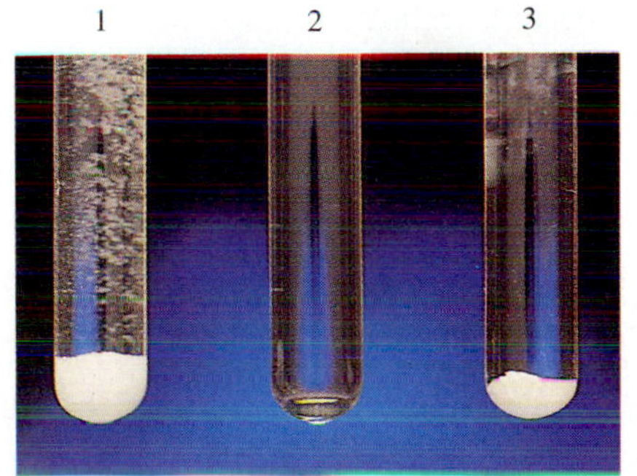

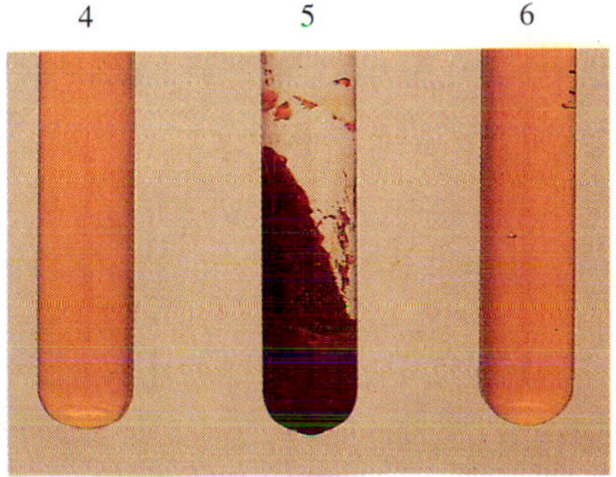

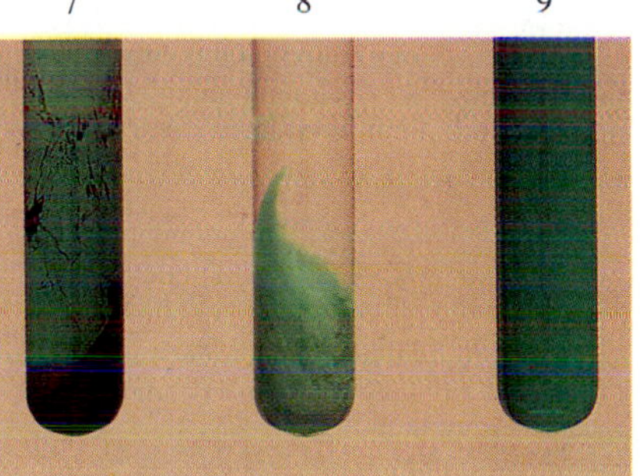

Figure 4.2
Of the compounds of Pb^{2+}, $Pb(NO_3)_2$ (tube 2) is soluble; $PbCl_2$ and $PbSO_4$ (tubes 1, 3) are insoluble white solids. The brownish solids $FeCl_3$ and $Fe(NO_3)_3$ (tubes 4, 6) are water soluble; $Fe(OH)_3$ (tube 5) is a reddish-brown insoluble solid. Two insoluble compounds of Ni^{2+} are black NiS (tube 7) and green $Ni(OH)_2$ (tube 8); $NiSO_4$ (tube 9) dissolves in water to give a deep green solution. All of these observations are consistent with the solubility rules in Table 4.1. (Charles D. Winters)

Example 4.1 Using the solubility rules in Table 4.1, predict what will happen when the following pairs of aqueous solutions are mixed.

(a) $CuSO_4$ and $NaNO_3$ (b) Na_2CO_3 and $CaCl_2$

Strategy First decide what cation and anion are present in each solution. Then write the formulas of the two possible precipitates, combining a cation of one solution with the anion of the other solution. Check the solubility rules to see if one or both of these compounds is insoluble. If so, a precipitation reaction occurs.

Solution

(a) Ions present in first solution: Cu^{2+}, $SO_4{}^{2-}$; second solution: Na^+, $NO_3{}^-$
Possible precipitates: $Cu(NO_3)_2$, Na_2SO_4
From Table 4.1, both of these compounds must be soluble, so

no precipitate forms.

(b) Ions present: Na^+, $CO_3{}^{2-}$; Ca^{2+}, Cl^-
Possible precipitates: NaCl, $CaCO_3$
Sodium chloride is soluble but calcium carbonate is not. When these two solutions are mixed, $CaCO_3$ precipitates (Fig. 4.1).

Net Ionic Equations

The precipitation reaction that occurs when solutions of Na_2CO_3 and $CaCl_2$ are mixed can be represented by a simple equation. The product of the reaction is solid $CaCO_3$, formed by the reaction between Ca^{2+} and $CO_3{}^{2-}$ ions in aqueous solution. The equation for the reaction is

$$Ca^{2+}(aq) + CO_3{}^{2-}(aq) \longrightarrow CaCO_3(s)$$

Notice that this equation includes only those ions that participate in the reaction. To be specific, Na^+ and Cl^- ions do not appear. They are "spectator ions," which are present in solution before and after the precipitation of calcium carbonate.

All kinds of reactions in water solution are described by net ionic equations

Equations such as this, which involve ions and exclude any species that do not take part in the reaction, are referred to as **net ionic equations**. We will use net ionic equations throughout this chapter and indeed the entire text to represent a wide variety of reactions in aqueous solution. Like all equations, net ionic equations must show

— *atom balance.* There must be the same number of atoms of each element on both sides. In the equation above, there is one Ca atom, one C atom, and three oxygen atoms on both sides.
— *charge balance.* There must be the same total charge on both sides. In the equation written above, the total charge is zero on both sides.

Example 4.2 Write a net ionic equation for any precipitation reaction that occurs when solutions of the following ionic compounds are mixed.

(a) NaOH and $Cu(NO_3)_2$ (b) $BaCl_2$ and Ag_2SO_4 (c) $(NH_4)_2S$ and K_2CO_3

Strategy Follow the procedure of Example 4.1 to decide whether or not a precipitate will form. If it does, write its formula, followed by (*s*), on the right side of the equation. On the left (reactant) side, write the formulas of the ions (*aq*) required to produce the precipitate. Finally, balance the equation.

Solution

(a) Ions present: Na^+, OH^-; Cu^{2+}, NO_3^-
Possible precipitates: $NaNO_3$, $Cu(OH)_2$
$NaNO_3$ is soluble, but $Cu(OH)_2$ is not.

Equation: $Cu^{2+}(aq) + 2\,OH^-(aq) \longrightarrow Cu(OH)_2(s)$

(b) Ions present: Ba^{2+}, Cl^-; Ag^+, SO_4^{2-}
Possible precipitates: $BaSO_4$, $AgCl$
Both compounds are insoluble, so two reactions occur.

Equations: $Ba^{2+}(aq) + SO_4^{2-}(aq) \longrightarrow BaSO_4(s)$

$Ag^+(aq) + Cl^-(aq) \longrightarrow AgCl(s)$

(c) Ions present: NH_4^+, S^{2-}; K^+, CO_3^{2-}
Possible precipitates: $(NH_4)_2CO_3$, K_2S
Both compounds are soluble, so there is no precipitation reaction and no equation.

4.2 ACID-BASE REACTIONS

You are probably familiar with a variety of aqueous solutions that are either acidic or basic (Fig. 4.3). Acidic solutions have a sour taste and affect the color of certain organic dyes known as acid-base indicators. For example, litmus turns from blue to red in acidic solution. Basic solutions have a slippery feeling and turn the colors of indicators (e.g., red to blue for litmus).

The species that give these solutions their characteristic properties are called acids and bases. In this chapter, we use the definitions first proposed by Svante Arrhenius over a century ago.

An acid is a species that produces H^+ ions in water solution.

A base is a species that produces OH^- ions in water solution.

Strong and Weak Acids and Bases

There are two types of acids, strong and weak, which differ in the extent of their dissociation in water. **Strong acids** dissociate completely, forming H^+ ions and anions. A typical strong acid is HCl. It undergoes the following reaction upon addition to water:

$$HCl(aq) \longrightarrow H^+(aq) + Cl^-(aq)$$

In a solution prepared by adding 0.1 mol of HCl to water, there is 0.1 mol of H^+ ions, 0.1 mol of Cl^- ions, and no HCl molecules. There are six common strong acids, whose names and formulas are listed in Table 4.2.

All acids other than those listed in Table 4.2 can be taken to be weak. A **weak acid** is only partially dissociated to H^+ ions in water. All of the weak acids considered in this chapter are *molecules containing an ionizable hydrogen*

Figure 4.3
Many common household items, including vinegar, lemon juice, and cola drinks, are acidic. Others, including ammonia and most detergents and cleaning agents, are basic. (Marna G. Clarke)

Table 4.2 Common Strong Acids and Bases

Acid	Name of Acid	Base	Name of Base
HCl	hydrochloric acid	$LiOH$	lithium hydroxide
HBr	hydrobromic acid	$NaOH$	sodium hydroxide
HI	hydriodic acid	KOH	potassium hydroxide
HNO_3	nitric acid	$Ca(OH)_2$	calcium hydroxide
$HClO_4$	perchloric acid	$Sr(OH)_2$	strontium hydroxide
H_2SO_4	sulfuric acid	$Ba(OH)_2$	barium hydroxide

Sulfuric acid dissociation:
$H_2SO_4(aq) \longrightarrow H^+(aq) + HSO_4^-(aq)$

atom. Their general formula can be represented as HB; the general dissociation reaction in water is

$$HB(aq) \rightleftharpoons H^+(aq) + B^-(aq)$$

The double arrow implies that this reaction does not go to completion. Instead, a mixture is formed containing significant amounts of both products and reactants. With the weak acid hydrogen fluoride

$$HF(aq) \rightleftharpoons H^+(aq) + F^-(aq)$$

a solution prepared by adding 0.1 mol of HF to a liter of water contains about 0.01 mol of H^+ ions, 0.01 mol of F^- ions, and 0.09 mol of HF molecules.

Mostly HF molecules

Bases, like acids, are classified as strong or weak. A **strong base** in water solution is completely ionized to OH^- ions and cations. As you can see from Table 4.2, the strong bases are the hydroxides of the Group 1 and Group 2 metals. These are typical ionic solids, completely ionized in both the solid state and in water solution. The equations written to represent the processes by which NaOH and $Ca(OH)_2$ dissolve in water are

With 0.1 mol $Ca(OH)_2$ in water, there is 0.1 mol Ca^{2+} and 0.2 mol OH^-

$$NaOH(s) \longrightarrow Na^+(aq) + OH^-(aq)$$

$$Ca(OH)_2(s) \longrightarrow Ca^{2+}(aq) + 2\,OH^-(aq)$$

In a solution prepared by adding 0.1 mol of NaOH to water, there is 0.1 mol of Na^+ ions, 0.1 mol of OH^- ions, and no NaOH molecules.

Weak bases produce OH^- ions in a quite different manner. They react with H_2O molecules, acquiring H^+ ions and leaving OH^- ions behind. The reaction of ammonia, NH_3, is typical:

$$NH_3(aq) + H_2O \rightleftharpoons NH_4^+(aq) + OH^-(aq)$$

As with all weak bases, this reaction does not go to completion. In a solution prepared by adding 0.1 mol of ammonia to a liter of water, there is about 0.001 mol of NH_4^+, 0.001 mol of OH^-, and nearly 0.099 mol of NH_3.

About 99% NH_3 molecules

A common class of weak bases consists of the organic molecules known as *amines*. An amine can be considered to be a derivative of ammonia in which one or more hydrogen atoms have been replaced by hydrocarbon groups. In the simplest case, methylamine, a hydrogen atom is replaced by a $—CH_3$ group to give the CH_3NH_2 molecule, which reacts with water in a manner very similar to NH_3:

Other amines include $CH_3CH_2NH_2$ and $(CH_3)_2NH$

$$CH_3NH_2(aq) + H_2O \rightleftharpoons CH_3NH_3^+(aq) + OH^-(aq)$$

Equations for Acid-Base Reactions

When an acidic water solution is mixed with a basic solution, an acid-base reaction takes place. The nature of the reaction and hence the equation written for it depend upon whether the acid and base involved are strong or weak.

1. Strong acid–strong base. Consider what happens when a solution of a strong acid such as HNO_3 is added to a solution of a strong base such as NaOH. Since HNO_3 is a strong acid, it is completely dissociated to H^+ and NO_3^- ions in solution. Similarly, with the strong base NaOH, the solution species are the Na^+ and OH^- ions. When the solutions are mixed, the H^+ and OH^- ions react with each other to form H_2O molecules. This reaction, referred to as **neutralization**, is represented by the net ionic equation

$$H^+(aq) + OH^-(aq) \longrightarrow H_2O$$

This same equation applies to the reaction of $HClO_4$ with $Ca(OH)_2$; ClO_4^- and Ca^{2+} ions are spectators.

The Na^+ and NO_3^- ions take no part in the reaction and so do not appear in the equation.

There is considerable evidence to indicate that the above reaction occurs when any strong base reacts with any strong acid in water solution. It follows that the neutralization equation written above applies to any strong acid–strong base reaction.

2. Weak acid–strong base. When a strong base such as NaOH is added to a solution of a weak acid, HB, a two-step reaction occurs. The first step is the dissociation of the HB molecule to H^+ and B^- ions; the second is the neutralization of the H^+ ions produced in the first step by the OH^- ions of the NaOH solution.

$$(1) \quad HB(aq) \rightleftharpoons H^+(aq) + B^-(aq)$$

$$(2) \quad H^+(aq) + OH^-(aq) \longrightarrow H_2O$$

The equation for the overall reaction is obtained by adding the two equations just written:

$$HB(aq) + OH^-(aq) \longrightarrow B^-(aq) + H_2O$$

where HB stands for any weak acid. For the reaction between solutions of sodium hydroxide and hydrogen fluoride, the equation is

$$HF(aq) + OH^-(aq) \longrightarrow H_2O + F^-(aq)$$

HF is in the equation because it is the principal species in aqueous hydrofluoric acid

Here, as always, spectator ions such as Na^+ are not included in the net ionic equation.

3. Strong acid–weak base. As an example of this type of reaction, consider what happens when an aqueous solution of a strong acid like HCl is added to an aqueous solution of ammonia, NH_3. Again, we consider the reaction to take place in two steps. The first step is the reaction of NH_3 with H_2O to form NH_4^+ and OH^- ions. Then, in the second step, the H^+ ions of the strong acid neutralize the OH^- ions formed in the first step.

$$(1) \quad NH_3(aq) + H_2O \rightleftharpoons NH_4^+(aq) + OH^-(aq)$$

$$(2) \quad H^+(aq) + OH^-(aq) \longrightarrow H_2O$$

The overall equation is obtained by summing those for the individual steps. Cancelling species (OH^-, H_2O) which appear on both sides, we obtain the simple equation

When a base is weak, like NH_3, its formula appears directly in the equation

$$H^+(aq) + NH_3(aq) \longrightarrow NH_4^+(aq)$$

In another case, for the reaction of a strong acid such as HNO_3 with methylamine, CH_3NH_2, the equation is

$$H^+(aq) + CH_3NH_2(aq) \longrightarrow CH_3NH_3^+(aq)$$

Table 4.3 summarizes the equations written for the three types of acid-base reactions discussed above. You should find it helpful in writing the equations called for in Example 4.3.

Example 4.3 Write a net ionic equation for each of the following reactions in dilute water solution.

(a) Nitrous acid (HNO_2) with sodium hydroxide (NaOH)
(b) Ethylamine ($C_2H_5NH_2$) with perchloric acid ($HClO_4$)
(c) Hydrobromic acid (HBr) with potassium hydroxide (KOH)

Strategy Decide whether the acid and base are strong or weak. Then decide which of the three types of acid-base reactions is involved. Finally, use Table 4.3 to derive the proper equation.

Solution

(a) HNO_2 is weak; NaOH is strong. This is a weak acid–strong base reaction.

$$HNO_2(aq) + OH^-(aq) \longrightarrow H_2O + NO_2^-(aq)$$

(b) $HClO_4$ is strong (Table 4.2); $C_2H_5NH_2$ is weak. This is a strong acid–weak base reaction.

$$H^+(aq) + C_2H_5NH_2(aq) \longrightarrow C_2H_5NH_3^+(aq)$$

(c) HBr is a strong acid; KOH is a strong base.

$$H^+(aq) + OH^-(aq) \longrightarrow H_2O$$

Table 4.3 Types of Acid-Base Reactions

Reactants	Reacting Species	Net Ionic Equation
Strong acid–Strong base	H^+, OH^-	$H^+(aq) + OH^-(aq) \longrightarrow H_2O$
Weak acid–Strong base	HB, OH^-	$HB(aq) + OH^-(aq) \longrightarrow H_2O + B^-(aq)$
Strong acid–Weak base	H^+, B	$H^+(aq) + B(aq) \longrightarrow BH^+(aq)$

4.3 OXIDATION-REDUCTION REACTIONS

Another common type of reaction in aqueous solution involves an exchange of electrons between two species. Such a reaction is called an **oxidation-reduction** or **redox reaction**. Many familiar reactions fit into this category, including metallic corrosion (Fig. 4.4).

In a redox reaction, one species *loses* electrons and is said to be *oxidized.* The other species, which *gains electrons,* is *reduced.* To illustrate this situation, consider the redox reaction that takes place when zinc pellets are added to hydrochloric acid. The net ionic equation for the reaction is

$$Zn(s) + 2H^+(aq) \longrightarrow Zn^{2+}(aq) + H_2(g)$$

Here, zinc atoms are oxidized to Zn^{2+} ions by *losing* electrons; the *half-reaction* is:

$$\text{oxidation:} \qquad Zn(s) \longrightarrow Zn^{2+}(aq) + 2e^-$$

while H^+ ions are reduced to H_2 molecules by *gaining* electrons; the *half-reaction* is:

$$\text{reduction:} \qquad 2H^+(aq) + 2e^- \longrightarrow H_2(g)$$

From this example, it should be clear that

— *oxidation and reduction occur together,* in the same reaction; you can't have one without the other.

— *there is no net change in the number of electrons in a redox reaction.* Those given off in the oxidation half-reaction are taken on by another species in the reduction half-reaction.

This means that electrons can't appear in the overall redox equation

The two species that exchange electrons in a redox reaction are given special names. The ion or molecule that accepts electrons is called the **oxidiz-**

Figure 4.4
Corrosion is the result of an oxidation-reduction reaction. It is a major problem for an industrialized society and can range in scale from the rusting of screws to the weakening of bridges. (M. D. Ippolito)

ing agent; the species that donates electrons is called the **reducing agent**. In the reaction

$$Zn(s) + 2H^+(aq) \longrightarrow Zn^{2+}(aq) + H_2(g)$$

the H^+ ion is the oxidizing agent; it brings about the oxidation of zinc. By the same token, zinc acts as a reducing agent; it furnishes the electrons required to reduce H^+ ions.

In earlier sections of this chapter, we showed how to write and balance equations for precipitation reactions (Section 4.1) and acid-base reactions (Section 4.2). Here our goal is more modest; we concentrate upon balancing redox equations, given the identity of reactants and products. To do that, it is convenient to introduce a new concept, oxidation number.

Oxidation Number

Oxidation numbers are calculated, not determined experimentally

The concept of **oxidation number** was introduced to simplify the electron bookkeeping in redox reactions. For a monatomic ion (e.g., Na^+, S^{2-}), the oxidation number is, quite simply, the charge of the ion (+1, −2). In a molecule or polyatomic ion, the oxidation number of an element is a "pseudo-charge" obtained in a rather arbitrary way, assigning bonding electrons to the atom with the greater attraction for electrons.

In practice, oxidation numbers in all kinds of species are assigned according to a set of four arbitrary rules.

1. *The oxidation number of an element in an elementary substance is 0.* For example, the oxidation number of chlorine in Cl_2 or of phosphorus in P_4 is 0.

2. *The oxidation number of an element in a monatomic ion is equal to the charge of that ion.* In the ionic compound NaCl, sodium has an oxidation number of +1, chlorine an oxidation number of −1. The oxidation numbers of aluminum and oxygen in Al_2O_3 (Al^{3+}, O^{2-} ions) are +3 and −2, respectively.

Oxidation numbers help us analyze redox reactions

3. *Certain elements have the same oxidation number in all or almost all their compounds.* The Group 1 metals always exist as +1 ions in their compounds and, hence, are assigned an oxidation number of +1. By the same token, Group 2 elements always have oxidation numbers of +2 in their compounds.

Oxygen is ordinarily assigned an oxidation number of −2 in its compounds. (An exception arises in compounds containing the peroxide ion, O_2^{2-}, where the oxidation number of oxygen is −1.)

Hydrogen in its compounds ordinarily has an oxidation number of +1. (The major exception is in metal hydrides such as NaH and CaH_2, where hydrogen is present as the H^- ion and hence is assigned an oxidation number of −1.)

Oxidation number does not have to be integral

4. *The sum of the oxidation numbers in a neutral species is 0; in a polyatomic ion, it is equal to the charge of that ion.* The application of this very useful principle is illustrated in Example 4.4.

Example 4.4 What is the oxidation number of sulfur in Na_2SO_4? Of manganese in MnO_4^-?

Strategy First look for elements whose oxidation number is always or almost always the same (Rule 3). Then solve for the oxidation number of the remaining element by applying Rule 4.

Solution In Na_2SO_4, the oxidation numbers of sodium and oxygen are +1 and −2, respectively. Sodium sulfate, like all compounds, is neutral, so the sum of the oxidation numbers is zero.

$$0 = 2(+1) + \text{oxid. no. S} + 4(-2) \qquad \text{oxid. no. S} = +6$$

In the SO_4^{2-} ion and the SO_3 molecule, the oxidation number of sulfur is also +6

In the MnO_4^- ion, oxygen has an oxidation number of −2. The ion has a charge of −1, so the sum of the oxidation numbers must be −1 (Rule 4).

$$-1 = \text{oxid. no. Mn} + 4(-2) \qquad \text{oxid. no. Mn} = +7$$

The concept of oxidation number leads directly to a working definition of the terms oxidation and reduction. **Oxidation** is defined as *an increase in oxidation number,* and **reduction** as a *decrease in oxidation number.* Examples include

$$2Al(s) + 3Cl_2(g) \longrightarrow 2AlCl_3(s) \qquad \begin{matrix}\text{Al oxidized (oxid. no. } 0 \longrightarrow +3) \\ \text{Cl reduced (oxid. no. } 0 \longrightarrow -1)\end{matrix}$$

$$4As(s) + 5\,O_2(g) \longrightarrow 2As_2O_5(s) \qquad \begin{matrix}\text{As oxidized (oxid. no. } 0 \longrightarrow +5) \\ \text{O reduced (oxid. no. } 0 \longrightarrow -2)\end{matrix}$$

These definitions are compatible with our earlier interpretation of oxidation and reduction in terms of loss and gain of electrons. An element that loses electrons must increase in oxidation number. The gain of electrons always results in a decrease in oxidation number. Sometimes, though, it's not obvious, just looking at a redox equation, which species are gaining or losing electrons, as in the reaction

$$2HCl(g) + 2HNO_3(l) \longrightarrow 2NO_2(g) + Cl_2(g) + 2H_2O(l)$$

Analysis in terms of oxidation number reveals that chlorine is oxidized (oxid. no. = −1 in HCl, 0 in Cl_2). Nitrogen is reduced (oxid. no. = +5 in HNO_3, +4 in NO_2). HCl is the reducing agent, HNO_3 the oxidizing agent.

Balancing Redox Equations

Frequently you will be working with redox reactions where the coefficients in the balanced equation are far from obvious and the "trial and error" method of Chapter 3 doesn't work very well. A general approach for balancing such equations is the **half-equation** (or *ion electron*) **method.** This is perhaps the most straightforward way to balance a variety of redox equations for reactions in aqueous solution. Four steps are involved.

Tough problems are often best handled by breaking them into parts

1. *Split the equation into two half-equations, one for reduction, the other for oxidation.* To decide which species are being reduced and oxidized, it is helpful to assign oxidation numbers.
2. *Balance one of the half-equations with respect to both atoms and charge.* That is, make sure that, on both sides of the equation, there are the same number of atoms of each type and the same total charge.
3. *Balance the other half-equation.*

4. *Combine the two half-equations in such a way as to eliminate electrons.* The final equation should show both atom and charge balance. It should contain no electrons.

Reaction Between Cr^{3+} and Cl^- Ions To illustrate the half-equation method in its simplest form, consider the reaction that occurs when a direct electric current is passed through a water solution of chromium(III) chloride, producing the two elements chromium and chlorine. The unbalanced equation for the reaction is

$$Cr^{3+}(aq) + Cl^-(aq) \longrightarrow Cr(s) + Cl_2(g)$$

To balance this equation, proceed as follows:

1. Chromium is reduced (oxid. no. $+3 \rightarrow 0$); chlorine is oxidized (oxid. no. $-1 \rightarrow 0$).

$$\text{reduction:} \quad Cr^{3+}(aq) \longrightarrow Cr(s) \quad (1a)$$

$$\text{oxidation:} \quad Cl^-(aq) \longrightarrow Cl_2(g) \quad (1b)$$

Since the oxid. no. of Cr goes from +3 to 0, $3e^-$ have to be added to the left

2. Equation 1a is balanced with respect to atoms; there is one atom of Cr on both sides. The charges, though, are unbalanced; the Cr atom on the right has zero charge, while the Cr^{3+} ion on the left has a charge of +3. To correct this, add three electrons to the left of Equation 1a:

$$Cr^{3+}(aq) + 3e^- \longrightarrow Cr(s) \quad (2)$$

3. Equation 1b must first be balanced with respect to atoms by providing two Cl^- ions to give a molecule of Cl_2:

$$2Cl^-(aq) \longrightarrow Cl_2$$

To balance charges, two electrons must be added to the right, giving a charge of -2 on both sides:

$$2Cl^-(aq) \longrightarrow Cl_2(g) + 2e^- \quad (3)$$

4. Having arrived at two balanced half-equations, *combine them so as to make the number of electrons gained in reduction equal to the number lost in oxidation.* In Equation 2, three electrons are gained; in Equation 3, two electrons are given off. To arrive at a final equation in which no electrons appear, multiply Equation 2 by 2, Equation 3 by 3, and add the resulting half-equations:

$$\begin{array}{lll} 2 \times \text{Eqn}(2): & 2Cr^{3+}(aq) + 6e^- & \longrightarrow 2Cr(s) \\ 3 \times \text{Eqn}(3): & 6Cl^-(aq) & \longrightarrow 3Cl_2(g) + 6e^- \\ \hline & 2Cr^{3+}(aq) + 6Cl^-(aq) & \longrightarrow 2Cr(s) + 3Cl_2(g) \end{array}$$

Reaction between MnO_4^- and Fe^{2+} (Acidic Solution) In many redox reactions, elements other than those being reduced or oxidized take part in the reaction. Most commonly, these elements are hydrogen (oxid. no. = +1) and oxygen (oxid. no. = −2). They can be balanced, in acidic solution, using H_2O molecules and/or H^+ ions. To illustrate how this works, consider the reaction shown in Fig. 4.5:

$$MnO_4^-(aq) + Fe^{2+}(aq) \longrightarrow Mn^{2+}(aq) + Fe^{3+}(aq)$$

This equation is balanced by following the four-step path described above.

1. The oxidation number of iron increases from +2 to +3; that of manganese decreases from +7 to +2. Hence

$$\text{oxidation:} \quad Fe^{2+}(aq) \longrightarrow Fe^{3+}(aq) \quad (1a)$$

$$\text{reduction:} \quad MnO_4^-(aq) \longrightarrow Mn^{2+}(aq) \quad (1b)$$

2. Half-equation 1a is balanced by writing

$$Fe^{2+}(aq) \longrightarrow Fe^{3+}(aq) + e^- \quad (2)$$

3. To balance half-equation 1b, first make sure that there are the same number of manganese atoms on both sides: 1. Next, balance oxygen using H_2O molecules. Add four H_2O molecules on the right to account for the four oxygen atoms in the MnO_4^- ion:

$$MnO_4^-(aq) \longrightarrow Mn^{2+}(aq) + 4H_2O$$

To complete the atom balance, add eight H^+ ions on the left to balance the eight hydrogen atoms in the four H_2O molecules:

$$MnO_4^-(aq) + 8H^+(aq) \longrightarrow Mn^{2+}(aq) + 4H_2O$$

In acidic solution, H^+ and H_2O can be reactants or products

This half-equation is now balanced with respect to atoms. However, there is a charge of +7 on the left (8 − 1) as opposed to a charge of +2 on the right. To balance charge, add five electrons to the left:

$$MnO_4^-(aq) + 8H^+(aq) + 5e^- \longrightarrow Mn^{2+}(aq) + 4H_2O \quad (3)$$

4. Finally, half-equations 2 and 3 are combined so as to eliminate electrons from the final equation. To do this, multiply Equation 2 by 5 and add to Equation 3:

$$\begin{array}{lrl} 5 \times \text{Eqn}(2): & 5Fe^{2+}(aq) & \longrightarrow 5Fe^{3+}(aq) + 5e^- \\ \text{Eqn}(3): & MnO_4^-(aq) + 8H^+(aq) + 5e^- & \longrightarrow Mn^{2+}(aq) + 4H_2O \\ \hline & MnO_4^-(aq) + 8H^+(aq) + 5Fe^{2+}(aq) & \longrightarrow Mn^{2+}(aq) + 4H_2O + 5Fe^{3+}(aq) \end{array}$$

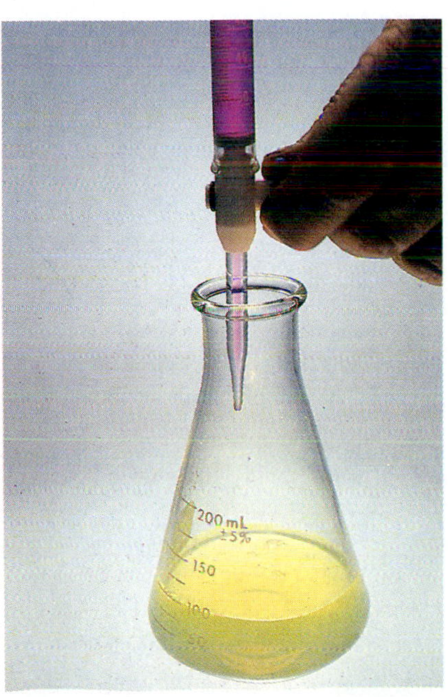

Figure 4.5
When potassium permanganate (buret) is added to an acidic solution containing Fe^{2+} ions (flask), a redox reaction occurs. The equation for the reaction is derived and balanced in the text. As reaction takes place, the purple color characteristic of the MnO_4^- ion fades. (Charles D. Winters)

Turning back to Step 3 for a moment, let us emphasize that to balance a redox half-equation in acidic solution, you balance, in order:

Balance oxygen before hydrogen

— the atoms of the element being oxidized or reduced
— oxygen, using H_2O molecules
— hydrogen, using H^+ ions
— charge, using e^-

Reaction Between MnO_4^- and I^- (Basic Solution) Redox reactions frequently occur in basic solution. In this case, H^+ ions should not be present in the final equation; their concentration in basic solution is very small. Instead, hydrogen should be in the form of OH^- ions or H_2O molecules. A simple way to accomplish this is to eliminate any H^+ ions appearing in the equation, "neutralizing" them by adding an equal number of OH^- ions to both sides. Consider, for example, the oxidation, in basic solution, of iodide by permanganate ions:

$$I^-(aq) + MnO_4^-(aq) \longrightarrow IO_3^-(aq) + MnO_2(s)$$

Proceeding as in the example we have just gone through, you should arrive at the following balanced equation involving H^+ ions and H_2O molecules (try it!).

Start out as though the solution were acidic

$$2MnO_4^-(aq) + 2H^+(aq) + I^-(aq) \longrightarrow 2MnO_2(s) + IO_3^-(aq) + H_2O$$

The H^+ ions must be removed to obtain an equation valid in basic solution. To do this, two OH^- ions are added to both sides; water is formed on the left by "neutralizing" H^+ ions with OH^- ions.

Then add OH^- to get rid of H^+ ions

$$\begin{array}{llll} 2MnO_4^-(aq) + 2H^+(aq) + I^-(aq) & \longrightarrow & 2MnO_2(s) + IO_3^-(aq) + H_2O & \\ \quad + 2\,OH^-(aq) & & & + 2\,OH^-(aq) \\ \hline 2MnO_4^-(aq) + 2H_2O + I^-(aq) & \longrightarrow & 2MnO_2(s) + IO_3^-(aq) + H_2O + 2\,OH^-(aq) & \end{array}$$

Eliminating one H_2O molecule from each side gives the final redox equation in basic solution:

$$2MnO_4^-(aq) + H_2O + I^-(aq) \longrightarrow 2MnO_2(s) + IO_3^-(aq) + 2\,OH^-(aq)$$

Example 4.5 Balance the equation $Cl_2(g) \rightarrow Cl^-(aq) + ClO_3^-(aq)$, first in acidic and then in basic solution.

Strategy To balance in acidic solution, follow the four-step process (1–4) outlined on pp. 85–86.

Don't panic because the same species is reduced and oxidized

(1) Split the overall equation into two half-equations.
(2) Write a balanced reduction half-equation.
(3) Write a balanced oxidation half-equation. (Note that the same species, Cl_2, is both reduced and oxidized).
(4) Combine half-equations so as to cancel electrons.

To find the balanced equation in basic solution, one more step is required. Eliminate H^+ ions by adding OH^- ions to both sides.

Solution

(1) The oxidation numbers of chlorine are 0 in Cl_2, -1 in Cl^-, and $+5$ in ClO_3^-.

$$\text{reduction:} \quad Cl_2(g) \longrightarrow Cl^-(aq)$$

$$\text{oxidation:} \quad Cl_2(g) \longrightarrow ClO_3^-(aq)$$

(2) To balance the reduction half-equation, use two Cl^- ions on the right and add $2e^-$ to the left:

$$Cl_2(g) + 2e^- \longrightarrow 2Cl^-(aq)$$

(3) For the oxidation half-equation

— to balance Cl, use two ClO_3^- ions:

$$Cl_2(g) \longrightarrow 2ClO_3^-(aq)$$

— to balance oxygen, add six H_2O to the left:

$$Cl_2(g) + 6H_2O \longrightarrow 2ClO_3^-(aq)$$

— to balance hydrogen, add $12H^+$ to the right:

$$Cl_2(g) + 6H_2O \longrightarrow 2ClO_3^-(aq) + 12H^+(aq)$$

— to balance charge, add $10e^-$ to the right to give the final half-equation:

$$Cl_2(g) + 6H_2O \longrightarrow 2ClO_3^-(aq) + 12H^+(aq) + 10e^-$$

(4) To cancel electrons, multiply the reduction half-equation by five and add to the oxidation half-equation:

$$\begin{array}{rcl} 5Cl_2(g) + 10e^- & \longrightarrow & 10Cl^-(aq) \\ Cl_2(g) + 6H_2O & \longrightarrow & 2ClO_3^-(aq) + 12H^+(aq) + 10e^- \\ \hline 6Cl_2(g) + 6H_2O & \longrightarrow & 10Cl^-(aq) + 2ClO_3^-(aq) + 12H^+(aq) \end{array}$$

To obtain an equation with the smallest whole-number coefficients, divide through by 2. The final equation in acidic solution is

$$3Cl_2(g) + 3H_2O \longrightarrow 5Cl^-(aq) + ClO_3^-(aq) + 6H^+(aq)$$

(5) To obtain the equation in basic solution, add six OH^- ions to both sides:

$$\begin{array}{l} 3Cl_2(g) + 3H_2O \longrightarrow 5Cl^-(aq) + ClO_3^-(aq) + 6H^+(aq) \\ \quad + 6\,OH^-(aq) \qquad\qquad\qquad\qquad\qquad + 6\,OH^-(aq) \\ \hline 3Cl_2(g) + 3H_2O + 6\,OH^-(aq) \longrightarrow 5Cl^-(aq) + ClO_3^-(aq) + 6H_2O \end{array}$$

Subtracting three H_2O molecules from each side gives the final equation:

$$3Cl_2(g) + 6\,OH^-(aq) \longrightarrow 5Cl^-(aq) + ClO_3^-(aq) + 3H_2O$$

4.4 SOLUTE CONCENTRATIONS; MOLARITY

Before discussing solution stoichiometry, it is necessary to consider how to express the **concentrations** of species in solution. That is, we need to specify how much solute is present in a given volume of solution. It would be meaningless to talk about "250 mL of NaOH solution" without knowing how many grams or moles of NaOH there are per liter of solution.

So far as solution stoichiometry is concerned, it is most convenient to express the concentration of solute in terms of **molarity**:

$$\text{molarity } (M) = \frac{\text{moles of solute}}{\text{liters of solution}}$$

The symbol [ ] is commonly used to represent the molarity of a species in solution. For a solution containing 1.20 mol of substance A in 2.50 L of solution,

$[NH_3]$ means "concentration of ammonia in moles per liter"

Figure 4.6
To prepare one liter of 0.100 M K_2CrO_4, you would start by weighing out 19.4 g of potassium chromate. The solid is transferred to a 1000-mL volumetric flask. Enough water is added to dissolve, by swirling, all of the potassium chromate. More water is then added to bring the level up to the mark on the neck. Finally, the flask is shaken repeatedly until a homogeneous solution is obtained. (Marna G. Clarke)

$$[A] = \frac{1.20 \text{ mol}}{2.50 \text{ L}} = 0.480 \text{ mol/L} = 0.480\ M$$

One liter of such a solution would contain 0.480 mol of A; 100 mL of solution would contain 0.0480 mol of A, and so on.

To prepare a solution to a desired molarity, you first calculate the amount of solute required. This is then dissolved in enough solvent to form the required volume of solution. Suppose, for example, you want to make one liter of 0.100 M K_2CrO_4 solution. First, weigh out 19.4 g (0.100 mol) of K_2CrO_4 ($\mathcal{M}$ = 194.20 g/mol). Then stir with enough water (Fig. 4.6) to form one liter (1000 mL) of solution.

The molarity of a solution can be used to calculate

— the number of moles of solute in a given volume of solution
— the volume of solution containing a given number of moles of solute

Here, as in so many other cases, a conversion factor approach is used (Example 4.6).

Example 4.6 The bottle labeled "concentrated hydrochloric acid" in the lab contains 12.0 mol of HCl per liter of solution. That is, [HCl] = 12.0 M.

(a) How many moles of HCl are there in 25.0 mL of this solution?
(b) What volume of concentrated hydrochloric acid must be taken to contain 1.00 mol of HCl?

Strategy The required conversion factors are

$$\frac{12.0 \text{ mol HCl}}{1 \text{ L}} \quad \text{or} \quad \frac{1 \text{ L}}{12.0 \text{ mol HCl}}$$

Solution

(a) $$n_{HCl} = 25.0 \text{ mL} \times \frac{1 \text{ L}}{1000 \text{ mL}} \times \frac{12.0 \text{ mol HCl}}{1 \text{ L}} = \boxed{0.300 \text{ mol HCl}}$$

(b) $$V = 1.00 \text{ mol HCl} \times \frac{1 \text{ L}}{12.0 \text{ mol HCl}} = \boxed{0.0833 \text{ L (83.3 mL)}}$$

In all the reactions considered in this chapter, at least one of the reactants is an ion in solution. The concentration of that ion frequently differs from that of the corresponding ionic compound. Consider, for example, what happens when magnesium chloride dissolves in water.

$$MgCl_2(s) \longrightarrow Mg^{2+}(aq) + 2Cl^-(aq)$$

Since 1 mol of $MgCl_2$ yields 1 mol of Mg^{2+} and 2 mol of Cl^-, it follows that a 1 *M* solution of $MgCl_2$ is 1 *M* in Mg^{2+} but 2 *M* in Cl^-. More generally, in any solution of magnesium chloride, the molarity of Cl^- is twice that of $MgCl_2$.

In a solution of $Al_2(SO_4)_3$, $[Al^{3+}] = 2x$, $[SO_4^{2-}] = 3x$, where $x = [Al_2(SO_4)_3]$

Example 4.7 Give the concentration, in moles per liter, of each ion in

(a) 0.080 *M* K_2SO_4 (b) 0.40 *M* $LaBr_3$

Strategy To go from concentration of solute to concentration of an individual ion, you must know the conversion factor relating moles of ions to moles of solute. To find this conversion factor, it is helpful to write the equation for the solution process.

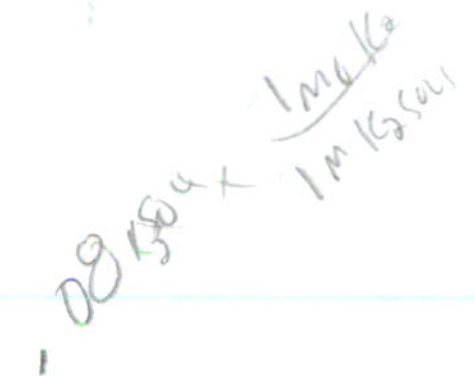

Solution

(a) $K_2SO_4(s) \longrightarrow 2K^+(aq) + SO_4^{2-}(aq)$

The conversion factors are 2 mol K^+/1 mol K_2SO_4 and 1 mol SO_4^{2-}/1 mol K_2SO_4.

$$[K^+] = \frac{0.080 \text{ mol } K_2SO_4}{1 \text{ L}} \times \frac{2 \text{ mol } K^+}{1 \text{ mol } K_2SO_4} = 0.16\ M\ K^+$$

$$[SO_4^{2-}] = \frac{0.080 \text{ mol } K_2SO_4}{1 \text{ L}} \times \frac{1 \text{ mol } SO_4^{2-}}{1 \text{ mol } K_2SO_4} = 0.080\ M\ SO_4^{2-}$$

(b) $LaBr_3(s) \longrightarrow La^{3+}(aq) + 3Br^-(aq)$

$$[La^{3+}] = \frac{0.40 \text{ mol } LaBr_3}{1 \text{ L}} \times \frac{1 \text{ mol } La^{3+}}{1 \text{ mol } LaBr_3} = 0.40\ M\ La^{3+}$$

$$[Br^-] = \frac{0.40 \text{ mol } LaBr_3}{1 \text{ L}} \times \frac{3 \text{ mol } Br^-}{1 \text{ mol } LaBr_3} = 1.2\ M\ Br^-$$

4.5 SOLUTION STOICHIOMETRY

Mass relations in solution reactions are very similar to those considered in Chapter 3. The principal difference is that, for species in solution, the number of moles is calculated by multiplying volume by molarity.

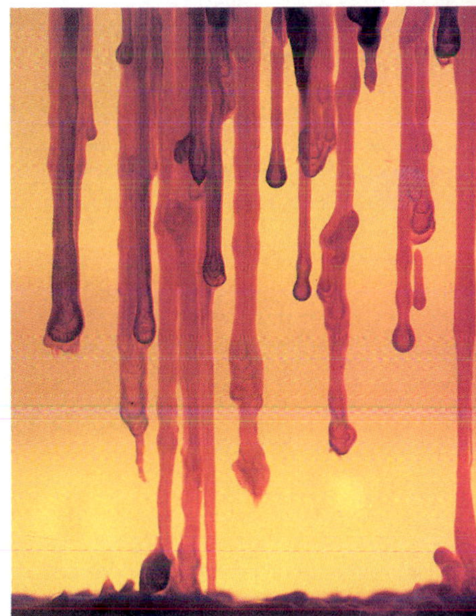

$Fe^{3+}(aq) + 3\,OH^-(aq) \longrightarrow Fe(OH)_3(s)$
When aqueous solutions of sodium hydroxide and iron(III) nitrate are mixed a red gelatinous precipitate forms.

Example 4.8 When aqueous solutions of sodium hydroxide and iron(III) nitrate are mixed, a red gelatinous precipitate forms. Calculate the mass of precipitate formed when 50.00 mL of 0.200 *M* NaOH and 30.00 mL of 0.125 *M* $Fe(NO_3)_3$ are mixed.

Strategy Data are given for two reactants, so this is a limiting reactant problem, similar to those worked in Chapter 3. There are, however, a couple of preliminary steps. First, decide upon the net ionic equation for the precipitation reaction. Second, calculate the number of moles of each reactant. Then apply the three-step procedure described on p. 65.

Always start with the net ionic equation

Solution

(1) Following the procedure outlined in Section 4.1, you should arrive at the net ionic equation $Fe^{3+}(aq) + 3\,OH^-(aq) \rightarrow Fe(OH)_3(s)$.

(2) $$n_{Fe^{3+}} = 0.03000\text{ L Fe(NO}_3)_3 \times \frac{0.125\text{ mol Fe(NO}_3)_3}{1\text{ L Fe(NO}_3)_3} \times \frac{1\text{ mol Fe}^{3+}}{1\text{ mol Fe(NO}_3)_3}$$

$$= 3.75 \times 10^{-3}\text{ mol Fe}^{3+}$$

Note from the formula of iron(III) nitrate that there is one mole of Fe^{3+} per mole of $Fe(NO_3)_3$.

$$n_{OH^-} = 0.05000\text{ L NaOH} \times \frac{0.200\text{ mol NaOH}}{1\text{ L NaOH}} \times \frac{1\text{ mol OH}^-}{1\text{ mol NaOH}} = 1.00 \times 10^{-2}\text{ mol OH}^-$$

(3) If Fe^{3+} is limiting,

$$n_{Fe(OH)_3} = 3.75 \times 10^{-3}\text{ mol Fe}^{3+} \times \frac{1\text{ mol Fe(OH)}_3}{1\text{ mol Fe}^{3+}} = 3.75 \times 10^{-3}\text{ mol Fe(OH)}_3$$

(4) If OH^- is limiting,

$$n_{Fe(OH)_3} = 1.00 \times 10^{-2}\text{ mol OH}^- \times \frac{1\text{ mol Fe(OH)}_3}{3\text{ mol OH}^-} = 3.33 \times 10^{-3}\text{ mol Fe(OH)}_3$$

(5) Since 3.33×10^{-3} is less than 3.75×10^{-3}, OH^- is the limiting reactant. The theoretical yield of $Fe(OH)_3$ is 3.33×10^{-3} mol. The molar mass of $Fe(OH)_3$ is 106.87 g/mol, so

$$\text{mass Fe(OH)}_3 = 3.33 \times 10^{-3}\text{ mol Fe(OH)}_3 \times \frac{106.87\text{ g Fe(OH)}_3}{1\text{ mol Fe(OH)}_3} = 0.356\text{ g Fe(OH)}_3$$

In many solution reactions, you need to know what volume of one solution is required to react with a known amount of the other (Example 4.9).

Example 4.9 As you found in Section 4.3, the balanced equation for the reaction between MnO_4^- and Fe^{2+} in acidic solution is

$$MnO_4^-(aq) + 8H^+(aq) + 5Fe^{2+}(aq) \longrightarrow 5Fe^{3+}(aq) + Mn^{2+}(aq) + 4H_2O$$

What volume of 0.684 *M* $KMnO_4$ solution is required to react completely with 27.50 mL of 0.250 *M* $Fe(NO_3)_2$?

Strategy Start by calculating the number of moles of Fe^{2+}. Then use the coefficients of the balanced equation to find the number of moles of MnO_4^-. Finally, use molarity as a conversion factor to find the volume of $KMnO_4$ solution.

Solution

(1)

$$n_{Fe^{2+}} = 0.02750\text{ L} \times \frac{0.250\text{ mol Fe(NO}_3)_2}{1\text{ L}} \times \frac{1\text{ mol Fe}^{2+}}{1\text{ mol Fe(NO}_3)_2}$$

$$= 6.88 \times 10^{-3}\text{ mol Fe}^{2+}$$

(2) $$n_{MnO_4^-} = 0.00688\text{ mol Fe}^{2+} \times \frac{1\text{ mol MnO}_4^-}{5\text{ mol Fe}^{2+}} = 0.00138\text{ mol MnO}_4^-$$

(3) $$n_{KMnO_4} = n_{MnO_4^-} = 1.38 \times 10^{-3}\text{ mol KMnO}_4$$

$$V = 1.38 \times 10^{-3}\text{ mol KMnO}_4 \times \frac{1\text{ L}}{0.684\text{ mol KMnO}_4} = 2.02 \times 10^{-3}\text{ L} \quad (2.02\text{ mL})$$

4.6 SOLUTION REACTIONS IN QUANTITATIVE ANALYSIS

Reactions taking place in aqueous solution are commonly used in quantitative analysis to determine the concentration of a dissolved species or its percentage in a solid mixture. To do this, you carry out a titration, measuring the volume of a "standard" solution of known concentration required to react with a measured amount of sample. The standard solution is delivered from a buret (Fig. 4.7). The point in the titration at which the reaction is complete is called the **equivalence point**. This point is commonly detected using an indicator that changes color at the *end point*, telling you it's time to quit. If the indicator has been chosen properly, the equivalence point and the end point coincide.

Titrations are typically accurate to ±0.1%

Calculations for titrations are very similar to those described in Section 4.5. You may be asked to calculate the concentration of a species in solution (Example 4.10) or the percentage of a species in a solid mixture (Example 4.11).

Example 4.10 One way to determine blood alcohol level is to titrate with potassium dichromate. The reaction is

$$C_2H_5OH(aq) + Cr_2O_7^{2-}(aq) \longrightarrow CO_2(g) + Cr^{3+}(aq)$$

A "breathalyzer" uses this reaction, measuring the color change from red ($Cr_2O_7^{2-}$) to green (Cr^{3+})

(a) Balance this redox equation.
(b) If 10.0 mL of 0.100 *M* $Cr_2O_7^{2-}$ is required to titrate a 20.0-mL sample of blood plasma, what is the molarity of alcohol, C_2H_5OH?

Strategy The equation is balanced in the usual way, following the four-step procedure described on pp. 85–86. Part (b) is very similar to Example 4.9, except that you are asked to calculate concentration rather than volume. Follow the path

$$n_{Cr_2O_7^{2-}} \longrightarrow n_{C_2H_5OH} \longrightarrow [C_2H_5OH]$$

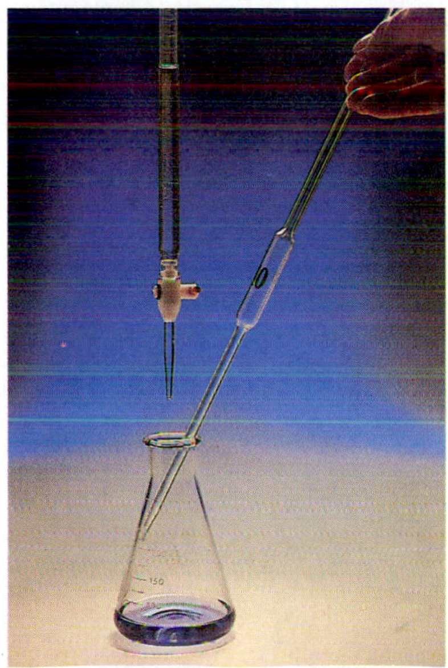

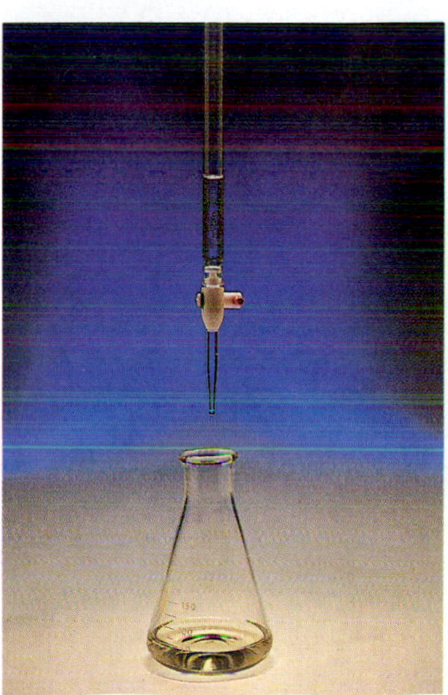

Figure 4.7
Titration of a base (flask) with an acid (buret). Originally, the indicator has the color characteristic of basic solution (blue). At the end point, the color changes sharply to green. With excess acid, we get the acid color, yellow. (Charles D. Winters)

Solution

(a) C_2H_5OH is oxidized (oxid. no. C: $-2 \rightarrow +4$); $Cr_2O_7^{2-}$ is reduced (oxid. no. Cr: $+6 \rightarrow +3$). The balanced half-equations are

$$\text{oxidation:} \quad C_2H_5OH(aq) + 3H_2O \longrightarrow 2CO_2(g) + 12H^+(aq) + 12e^-$$

$$\text{reduction:} \quad Cr_2O_7^{2-}(aq) + 14H^+(aq) + 6e^- \longrightarrow 2Cr^{3+}(aq) + 7H_2O$$

Multiplying the reduction half-equation by 2 and adding it to the oxidation half-equation gives, after simplification,

$$C_2H_5OH(aq) + 2Cr_2O_7^{2-}(aq) + 16H^+(aq) \longrightarrow 2CO_2(g) + 4Cr^{3+}(aq) + 11H_2O$$

(b) $n_{Cr_2O_7^{2-}} = 0.0100\text{ L} \times \dfrac{0.100\text{ mol } Cr_2O_7^{2-}}{1\text{ L}} = 1.00 \times 10^{-3}\text{ mol } Cr_2O_7^{2-}$

$$n_{C_2H_5OH} = 1.00 \times 10^{-3}\text{ mol } Cr_2O_7^{2-} \times \frac{1\text{ mol } C_2H_5OH}{2\text{ mol } Cr_2O_7^{2-}} = 5.00 \times 10^{-4}\text{ mol } C_2H_5OH$$

$$[C_2H_5OH] = \frac{5.00 \times 10^{-4}\text{ mol}}{2.00 \times 10^{-2}\text{ L}} = 2.50 \times 10^{-2}\,M$$

(A concentration of 0.022 M corresponds to 0.10% alcohol, which is the level judged to be evidence of intoxication in many states.)

Other antacids include $Mg(OH)_2$, $Al(OH)_3$, and $NaHCO_3$

Example 4.11 The principal ingredient of certain commercial antacids (Tums, Chooz), is calcium carbonate, $CaCO_3$. A student titrates an antacid tablet weighing 0.542 g with hydrochloric acid; the reaction is

$$CaCO_3(s) + 2H^+(aq) \longrightarrow Ca^{2+}(aq) + CO_2(g) + H_2O$$

If 38.5 mL of 0.200 M HCl is required for complete reaction, what is the percentage of $CaCO_3$ in the antacid tablet?

Strategy From the titration data, the number of moles of H^+ is readily calculated. Then follow the path

$$n_{H^+} \longrightarrow n_{CaCO_3} \longrightarrow \text{mass of } CaCO_3 \longrightarrow \%CaCO_3$$

Solution

(1) $n_{H^+} = 0.0385\text{ L} \times \dfrac{0.200\text{ mol HCl}}{1\text{ L}} \times \dfrac{1\text{ mol } H^+}{1\text{ mol HCl}} = 7.70 \times 10^{-3}\text{ mol } H^+$

(2) $n_{CaCO_3} = 7.70 \times 10^{-3}\text{ mol } H^+ \times \dfrac{1\text{ mol } CaCO_3}{2\text{ mol } H^+} = 3.85 \times 10^{-3}\text{ mol } CaCO_3$

(3) mass of $CaCO_3 = 3.85 \times 10^{-3}\text{ mol } CaCO_3 \times \dfrac{100.09\text{ g } CaCO_3}{1\text{ mol } CaCO_3} = 0.385\text{ g } CaCO_3$

(4) $\%CaCO_3 = \dfrac{0.385\text{ g}}{0.542\text{ g}} \times 100\% = 71.0\%$

The fizz of the antacid is due to CO_2 gas being released. (Charles D. Winters)

PERSPECTIVE • Amines

The organic compounds called amines (p. 80) can be classified according to the number of hydrocarbon groups bonded to nitrogen (Table 4.4). Most amines of low molar mass are volatile compounds with distinctly unpleasant odors. For example, $(CH_3)_3N$ is a gas at room temperature (bp = 3°C) with an odor somewhere between that of ammonia and of spoiled fish.

The reaction of amines with H^+ ions (p. 81) has an interesting practical application. Amines of high molar mass, frequently used as drugs, have very low water solubilities. They can be converted to a water-soluble form by treatment with strong acid, e.g.,

$$\underset{\text{novocaine}}{C_9H_{10}NO_2{-}\underset{\displaystyle C_2H_5}{\underset{|}{N}}{-}C_2H_5(s)} + HCl(aq) \longrightarrow \underset{\text{novocaine hydrochloride}}{[C_9H_{10}NO_2{-}\overset{\displaystyle H}{\overset{|}{\underset{\displaystyle C_2H_5}{\underset{|}{N}}}}{-}C_2H_5]^+(aq)} + Cl^-(aq)$$

Table 4.4 Classification of Amines

Type	General Formula	Example
Primary	RNH_2	$CH_3{-}\underset{\displaystyle H}{\underset{\vert}{N}}{-}H$
Secondary	R_2NH	$CH_3{-}\underset{\displaystyle H}{\underset{\vert}{N}}{-}CH_3$
Tertiary	R_3N	$CH_3{-}\underset{\displaystyle CH_3}{\underset{\vert}{N}}{-}CH_3$

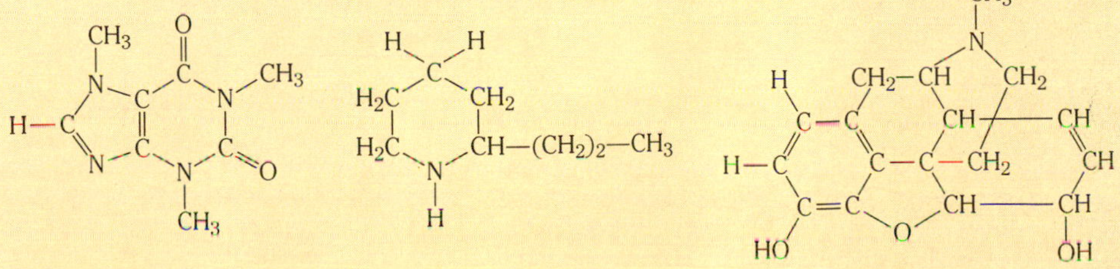

Caffeine Coniine Morphine

Figure 4.A
Structural formulas of three alkaloids. Other amines in this category include nicotine, cocaine, quinine, and strychnine.

Figure 4.B
Kona coffee cherries. Coffee contains the alkaloid caffeine. (Andrew J. Martinez/ Photo Researchers, Inc.)

Novocaine hydrochloride is about 200 times as soluble as novocaine itself. When your dentist injects "Novocaine," the liquid in the syringe is a water solution of novocaine hydrochloride.

Alkaloids such as caffeine, coniine, and morphine (Figs. 4.A–C) form an important class of naturally occurring amines. Caffeine occurs in tea leaves, coffee beans, and cola nuts. Morphine is obtained from unripe opium poppy seed pods. Coniine, extracted from hemlock, is the alkaloid that killed Socrates. He was sentenced to death because of unconventional teaching methods; teacher evaluations had teeth in them in ancient Greece.

During the Civil War (1860–1865) morphine was widely used as a painkiller injected via the hypodermic syringe developed during the 1850s. Soldiers (and desperate physicians at prisoner-of-war camps such as Andersonville, Georgia) discovered that morphine had the side effect of relieving dysentery, a disease as deadly as combat. When the war ended, more than a hundred thousand veterans were addicted to morphine, and the United States faced (and largely ignored) its first drug crisis.

Figure 4.C
Opium poppy flower and unripe seed capsules. The capsule has a milky, gummy substance that contains the alkaloids morphine, heroin, and codeine. (Scott Camazine/ Photo Researchers, Inc.)

CHAPTER HIGHLIGHTS

KEY CONCEPTS

1. *Apply the solubility rules (Table 4.1) to predict precipitation reactions and write net ionic equations for them*
   (Examples 4.1, 4.2; Problems 5–10)
2. *Write net ionic equations for acid-base reactions*
   (Example 4.3; Problems 17–20)
3. *Determine oxidation numbers of elements in compounds*
   (Example 4.4; Problems 21–24)
4. *Balance redox equations by the half-equation method*
   (Example 4.5; Problems 29–36)
5. *Relate molarity of a solute to*
   —number of moles and volume
   (Example 4.6; Problems 37–42)
   —molarities of ions
   (Example 4.7; Problems, 43, 44)
6. *Use balanced net ionic equations to determine*
   —the amount of product formed in a solution reaction
   (Example 4.8; Problems 45, 46, 56, 58)
   —the volume of solution required for reaction
   (Example 4.9; Problems 45, 47, 48, 50–53, 55, 57, 58, 63, 74, 78)
   —the concentration of a reactant
   (Examples 4.10, 4.11; Problems 46, 49, 54, 59–62, 65–68, 75–77, 80)

KEY TERMS

acid
—strong
—weak
base
—strong
—weak
limiting reactant

molarity
mole
net ionic equation
neutralization
oxidation
oxidation number

oxidizing agent
precipitate
reducing agent
reduction
theoretical yield
titration

SUMMARY PROBLEM

Nitric acid, NHO_3, and strontium hydroxide, $Sr(OH)_2$, are a strong acid and a strong base, respectively. They participate in a variety of acid-base reactions with other species. In addition, nitric acid is often involved in redox reactions because the NO_3^- ion is readily reduced. Finally, strontium hydroxide can form precipitates involving either the Sr^{2+} ion or the OH^- ion.

a. Write the net ionic equation for the reaction between solutions of
   (1) strontium hydroxide and aluminum nitrate.
   (2) nitric acid and ammonia.
   (3) strontium hydroxide and nitric acid.
   (4) strontium hydroxide and acetic acid ($HC_2H_3O_2$).
   (5) nitric acid and Cu(*s*) forming nitrogen oxide gas and copper(II) ion.

b. How many grams of strontium hydroxide are required to prepare 350.0 mL of 0.223 *M* solution? What is the concentration of each ion in a solution that is 0.223 *M* $Sr(OH)_2$?

c. A concentrated nitric acid solution is 14.0 *M*. How many milliliters of the solution are needed for a reaction that requires 2.75 mol of HNO_3?

d. When 15.00 mL of 0.2050 *M* strontium hydroxide reacts with 22.00 mL of 0.1380 *M* cobalt(III) sulfate, two precipitates form. Calculate the mass of each precipitate and the concentrations of all ions after reaction, assuming the final volume is the sum of the initial volumes.

e. Nitric acid is used to analyze a sulfide ore; S^{2-} ions are oxidized to sulfur, while NO_3^- ions are reduced to $NO(g)$. It is found that 64.5 mL of a solution 0.125 *M* in NO_3^- is required to react exactly with a 1.00-g sample of ore. Calculate the percentage of sulfur in the ore.

f. To determine the molarity of a solution of hydrobromic acid, it is titrated with 0.2384 *M* strontium hydroxide. It is determined that 20.50 mL of strontium hydroxide is required to react with 33.20 mL of hydrobromic acid. What is the molarity of the hydrobromic acid?

Answers

a. (1) $Al^{3+}(aq) + 3OH^-(aq) \longrightarrow Al(OH)_3(s)$
(2) $H^+(aq) + NH_3(aq) \longrightarrow NH_4^+(aq)$ (3) $H^+(aq) + OH^-(aq) \longrightarrow H_2O$
(4) $HC_2H_3O_2(aq) + OH^-(aq) \longrightarrow H_2O + C_2H_3O_2^-(aq)$
(5) $2NO_3^-(aq) + 8H^+(aq) + 3Cu(s) \longrightarrow 2NO(g) + 4H_2O + 3Cu^{2+}(aq)$

b. 9.49 g; 0.223 *M* Sr^{2+}, 0.446 *M* OH^- **c.** 196 mL

d. 0.5648 g $SrSO_4$, 0.2254 g $Co(OH)_3$; $[Sr^{2+}] = [OH^-] \approx 0$; 0.1631 *M* SO_4^{2-}; 0.1087 *M* Co^{3+} **e.** 38.8% **f.** 0.2944 *M*

QUESTIONS & PROBLEMS

Precipitation Reactions and Solubility

1. Write the formulas of the following compounds. Using Table 4.1, decide which of them are soluble in water.
a. calcium chloride **b.** ammonium sulfide
c. barium sulfate **d.** iron(II) hydroxide

2. Follow the instructions for Question 1 for the following compounds.
a. potassium carbonate **b.** magnesium hydroxide
c. cerium(III) nitrate **d.** nickel(II) sulfate

3. Name the reagent that you would add to a solution of nickel(II) nitrate to precipitate
a. nickel(II) carbonate. **b.** nickel(II) sulfide.
c. nickel(II) hydroxide.

4. Describe how you would prepare
a. copper(II) hydroxide from a solution of copper(II) chloride.
b. strontium sulfate from sulfuric acid.
c. mercury(I) chloride from a solution of mercury(I) nitrate, $Hg_2(NO_3)_2$.

5. Write net ionic equations for the formation of
a. a green precipitate when solutions of nickel(II) nitrate and sodium hydroxide are mixed.
b. a white precipitate (that is soluble in acid) when potassium hydroxide and magnesium chloride are mixed.

6. Write net ionic equations to explain the formation of
a. a red precipitate when solutions of iron(III) chloride and sodium hydroxide are mixed.
b. two different precipitates, one yellow and the other white, when solutions of cadmium(II) sulfate and strontium sulfide are mixed.

7. Using Table 4.1, decide whether a precipitate will form when the following solutions are mixed. If a precipitate forms, write a net ionic equation for the reaction.
a. aluminum sulfate and sodium chloride
b. lead(II) nitrate and ammonium chloride
c. barium nitrate and chromium(III) sulfate
d. potassium nitrate and sodium hydroxide
e. arsenic(III) chloride and sodium sulfide

8. Follow the directions of Question 7 for solutions of
a. copper(II) chloride and sodium hydroxide.
b. copper(II) sulfate and potassium sulfide.
c. mercury(I) nitrate, $Hg_2(NO_3)_2$, and hydrochloric acid.
d. iron(III) sulfate and barium hydroxide.
e. nickel(II) nitrate and ammonium carbonate.

9. Write a net ionic equation for any precipitation reaction that occurs when 0.1 *M* solutions of the following are mixed.
a. sodium carbonate and barium chloride
b. zinc(II) sulfate and ammonium sulfide
c. calcium chloride and silver nitrate
d. strontium sulfide and potassium hydroxide

10. Follow the directions for Question 9 for the following pairs of solutions.
a. scandium(III) chloride and nickel(II) nitrate
b. sodium carbonate and calcium chloride
c. ammonium sulfide and lead(II) nitrate
d. iron(III) nitrate and potassium hydroxide

Acid-Base Reactions

11. For an acid-base reaction, what is the reacting species, i.e., the ion or molecule that appears in the chemical equation, in the following acids?

a. nitric acid b. hydrofluoric acid
c. perchloric acid d. sulfurous acid (H_2SO_3)
e. propionic acid ($HC_3H_5O_2$)

12. For an acid-base reaction, what is the reacting species in the following acids?
a. nitrous acid (HNO_2) b. sulfuric acid
c. acetic acid ($HC_2H_3O_2$) d. hydrochloric acid
e. lactic acid ($HC_3H_5O_3$)

13. For an acid-base reaction, what is the reacting species in the following bases?
a. sodium hydroxide b. ethylamine ($C_2H_5NH_2$)
c. ammonia d. barium hydroxide

14. For an acid-base reaction, what is the reacting species in the following bases?
a. lithium hydroxide
b. dimethylamine, $(CH_3)_2NH$
c. strontium hydroxide
d. pyridine, (C_5H_5N)

15. Classify the following compounds as acids or bases, weak or strong.
a. hydrobromic acid b. cesium hydroxide
c. phosphoric acid d. trimethyl amine, $(CH_3)_3N$

16. Follow the directions for Question 15 for
a. hypochlorous acid
b. barium hydroxide
c. ammonia
d. hydrochloric acid

17. Write a balanced net ionic equation for each of the following acid-base reactions in water.
a. butyric acid ($HC_4H_7O_2$) with lithium hydroxide
b. ammonia with hydriodic acid
c. hydrofluoric acid with strontium hydroxide

18. Write a balanced net ionic equation for each of the following acid-base reactions in water.
a. perchloric acid and potassium hydroxide
b. nitrous acid (HNO_2) and sodium hydroxide
c. aniline ($C_6H_5NH_2$) and sulfuric acid

19. Consider the equation $H^+(aq) + OH^-(aq) \rightarrow H_2O$. For which of the following pairs would this be the correct equation for the acid-base reaction in solution? If it is not correct, write the proper equation for the acid-base reaction between the pair.
a. hydrochloric acid and calcium hydroxide
b. HBr and CH_3NH_2
c. nitric acid and ammonia
d. H_2SO_4 and NaOH
e. perchloric acid and potassium hydroxide
f. HF and barium hydroxide

20. Follow the directions of Question 19 for the following pairs.
a. nitric acid and $C_2H_5NH_2$
b. perchloric acid and cesium hydroxide
c. $HC_2H_3O_2$ and LiOH
d. sulfuric acid and calcium hydroxide
e. barium hydroxide and hydriodic acid

Redox Reactions

21. Give the oxidation number of each atom in
a. hydrogen carbonate ion
b. magnesium sulfate
c. sulfur hexafluoride
d. iodic acid (HIO_3)
e. sodium molybdate (Na_2MoO_4)

22. Give the oxidation number of each atom in
a. nitrite ion (NO_2^-) b. TeF_8^{2-}
c. dinitrogen trioxide d. thiosulfate ion ($S_2O_3^{2-}$)
e. perchlorate ion

23. Give the oxidation number of each atom in
a. Sb_4O_{10} b. HPO_3^{2-} c. RuF_5 d. $C_2H_6O_2$

24. Give the oxidation number of each atom in
a. CaC_2O_4 b. HSO_4^- c. $Na_2Fe_2O_4$ d. NOF
e. N_2H_4

25. Classify each of the following half-equations as oxidation or reduction.
a. $Ca(s) \longrightarrow Ca^{2+}(aq)$
b. $OH^-(aq) \longrightarrow O_2(g)$
c. $NO_3^-(aq) \longrightarrow NO(g)$
d. $AuCl_4^-(aq) \longrightarrow AuCl_2^-(aq)$

26. Classify each of the following half-equations as oxidation or reduction.
a. $Co^{3+}(aq) \longrightarrow Co^{2+}(aq)$
b. $Cl_2(g) \longrightarrow ClO_3^-(aq)$
c. $Fe^{3+}(aq) \longrightarrow Fe^{2+}(aq)$
d. $Hg(l) \longrightarrow Hg_2^{2+}(aq)$

27. For each unbalanced equation given below, identify the species oxidized and the species reduced; identify the oxidizing agent and the reducing agent.
a. $Mg(s) + O_2(g) \longrightarrow MgO(s)$
b. $Cr_2O_7^{2-}(aq) + Sn^{2+}(aq) \longrightarrow Cr^{3+}(aq) + Sn^{4+}(aq)$

28. For each unbalanced equation given below, identify the species oxidized and the species reduced; identify the oxidizing agent and the reducing agent.
a. $FeS(s) + NO_3^-(aq) \longrightarrow NO(g) + SO_4^{2-}(aq) + Fe^{2+}(aq)$
b. $C_2H_4(g) + O_2(g) \longrightarrow CO_2(g) + H_2O(l)$

29. Balance the equations in Question 27.

30. Balance the equations in Question 28.

31. Write balanced equations for the following reactions in acid solution.
a. $Hg^{2+}(aq) + Cu(s) \longrightarrow Hg(l) + Cu^{2+}(aq)$
b. $Zn(s) + VO_3^-(aq) \longrightarrow Zn^{2+}(aq) + V^{2+}(aq)$
c. $H_2O_2(aq) + Cr_2O_7^{2-}(aq) \longrightarrow Cr^{3+}(aq) + O_2(g)$
d. $MnO_2(s) + Cl^-(aq) \longrightarrow Mn^{2+}(aq) + Cl_2(g)$
e. $IO_3^-(aq) + I^-(aq) \longrightarrow I_3^-(aq)$

32. Write balanced equations for the following reactions (a–e) in acid solution.
a. $P_4(s) \longrightarrow PH_3(g) + HPO_3^{2-}(aq)$
b. $H_3AsO_3(aq) + BrO_3^-(aq) \longrightarrow H_3AsO_4(aq) + Br^-(aq)$

c. $MnO_4^-(aq) + HSO_3^-(aq) \longrightarrow Mn^{2+}(aq) + SO_4^{2-}(aq)$
d. $Sn^{2+}(aq) + O_2(g) \longrightarrow Sn^{4+}(aq) + H_2O$
e. $Pt(s) + NO_3^-(aq) + Cl^-(aq) \longrightarrow PtCl_6^{2-}(aq) + NO(g) + H_2O$

33. Write balanced net ionic equations for the following reactions in acid solution.
a. Solid phosphorus (P_4) reacts with hypochlorous acid, HClO, to form phosphoric acid, H_3PO_4, and chloride ion.
b. Tellurium, Te, is oxidized by nitrate ion to form solid tellurium dioxide and NO(*g*).
c. An aqueous solution of bromine is reduced to Br^-; at the same time iodide ions are oxidized to iodate ions, IO_3^-.

34. Write balanced net ionic equations for the following reactions in acid solution.
a. Silver is dissolved in nitric acid, forming aqueous silver nitrate and nitrogen dioxide gas.
b. Solid copper(II) sulfide is dissolved in nitric acid, forming copper(II) nitrate (*aq*), sulfur, and NO(*g*).
c. Tin(II) ion reacts with periodate ion, IO_4^-, yielding iodide ion and tin(IV) ion.

35. Write balanced equations for the following reactions in basic solution.
a. $ClO^-(aq) + CrO_2^-(aq) \longrightarrow Cl^-(aq) + CrO_4^{2-}(aq)$
b. $Al(s) + H_2O \longrightarrow Al(OH)_4^-(aq) + H_2(g)$
c. $Ni^{2+}(aq) + Br_2(l) \longrightarrow NiO(OH)(s) + Br^-(aq)$

36. Write balanced equations for the following reactions in basic solution.
a. $S_2O_3^{2-}(aq) + I_2(s) \longrightarrow SO_4^{2-}(aq) + I^-(aq)$
b. $CN^-(aq) + MnO_4^-(aq) \longrightarrow CNO^-(aq) + MnO_2(s)$
c. $Cr(OH)_3(s) + ClO_3^-(aq) \longrightarrow CrO_4^{2-}(aq) + Cl^-(aq)$

Molarity

37. How would you prepare 425 mL of 0.628 M
a. K_2CrO_4? b. NaI? c. $C_6H_{12}O_6$?

38. Given the pure solid and water, how would you prepare
a. 0.500 L of 1.25 M KOH?
b. 0.750 L of 3.50 M $CuSO_4$?

39. You are given a bottle labeled "3.00 M HNO_3."
a. How many moles of HNO_3 are there in 12.45 mL of this solution?
b. What volume of this solution contains 0.800 mol HNO_3?

40. On a shelf are bottles marked 0.125 M NaCl and 6.00 M acetic acid, $HC_2H_3O_2$.
a. How many grams of solute are there in 35.0 mL of each solution?
b. What volume of each solution must be taken to obtain 0.0854 mol of solute?

41. Complete the table below for aqueous solutions.

Solute	Mass of Solute	Volume	Molarity
a. Na_2CO_3	3.58 g	0.250 L	____
b. CH_3OH	____	0.500 L	6.00
c. $Ba(NO_3)_2$	5.89 g	____	1.21
d. $Al_2(SO_4)_3$	____	0.455 L	0.105

42. Complete the table below for aqueous solutions.

Solute	Mass of Solute	Volume	Molarity
a. $CaCl_2$	45.0 g	0.400 L	____
b. H_2O_2	____	0.300 L	0.155
c. K_2CO_3	5.98 g	____	2.50
d. ZnI_2	____	0.150 L	0.0536

43. How many moles of ions are present in water solutions prepared by dissolving 0.25 mol of
a. calcium bromide? b. magnesium sulfate?
c. iron(III) nitrate? d. nickel(II) sulfate?

44. How many moles of ions are present in water solutions prepared by dissolving 0.33 mol of
a. cobalt(II) nitrate? b. lithium carbonate?
c. cesium sulfate? d. aluminum sulfate?

Solution Stoichiometry

45. A precipitate forms when solutions of silver nitrate and scandium(III) chloride are mixed:

$$Ag^+(aq) + Cl^-(aq) \longrightarrow AgCl(s)$$

a. What volume of 0.0385 *M* scandium(III) chloride is required to completely react with 22.0 mL of 0.130 *M* silver nitrate?
b. What mass in grams of silver chloride is formed?

46. Mixing solutions of sodium oxalate, $Na_2C_2O_4$, and lanthanum(III) chloride precipitates lanthanum(III) oxalate:

$$3C_2O_4^{2-}(aq) + 2La^{3+}(aq) \longrightarrow La_2(C_2O_4)_3(s)$$

a. What is the molarity of a solution of lanthanum(III) chloride if 25.0 mL is required to react with 13.85 mL of 0.0225 *M* sodium oxalate?
b. How many grams of precipitate are formed?

47. What volume of 0.100 *M* lead nitrate is required to precipitate completely
a. 25.0 mL of 0.0832 *M* nickel(II) sulfate?
b. 55.8 mL of 0.222 *M* hydrochloric acid?
c. 18.7 mL of 0.389 *M* potassium chromate?

48. What volume of 0.0750 *M* cobalt(II) sulfate is required to react completely with
a. 28.7 mL of 0.183 *M* sodium hydroxide?
b. 42.5 mL of 0.189 *M* barium nitrate?
c. 68.4 mL of 0.273 *M* ammonium sulfide?

49. A student finds that 38.4 mL of 0.215 *M* hydrochloric acid is required to neutralize a 20.0-mL sample of barium hydroxide. What is the molarity of $Ba(OH)_2$?

50. What volume of 0.317 *M* potassium hydroxide is required to neutralize 32.0 mL of 0.164 *M* nitric acid?

51. Determine the volume of 0.250 *M* hydrochloric acid required to react with

- **a.** 30.0 mL of 0.278 *M* lithium hydroxide.
- **b.** 17.6 mL of 0.0162 *M* strontium hydroxide.
- **c.** 15.0 mL of a solution ($d = 0.958\ g/cm^3$) containing 10.0% by mass of NH_3.

52. Calculate the volume of 0.309 *M* sodium hydroxide required to react with

- **a.** 15.9 mL of 0.190 *M* hydrofluoric acid.
- **b.** 22.9 mL of 0.296 *M* perchloric acid.
- **c.** 10.0 g of concentrated acetic acid ($HC_2H_3O_2$) that is 95.0% pure.

53. Iodine, I_2, reacts with the thiosulfate ion, $S_2O_3^{2-}$, to give the iodide ion and the tetrathionate ion, $S_4O_6^{2-}$.

- **a.** Write a balanced net ionic equation for the reaction.
- **b.** If 10.0 g of iodine is dissolved in enough water to make 3.00×10^{-1} L of solution, what volume of 0.125 *M* sodium thiosulfate will be needed for complete reaction?

54. An acidic solution of potassium permanganate reacts with oxalate ions, $C_2O_4^{2-}$, to form carbon dioxide and manganese(II) ions.

- **a.** Write a balanced equation for the reaction.
- **b.** If 38.4 mL of 0.150 *M* potassium permanganate is required to titrate 25.2 mL of sodium oxalate solution, what was the concentration of oxalate ion?

55. Consider the reaction between copper and nitric acid, for which the unbalanced equation is: $Cu(s) + H^+(aq) + NO_3^-(aq) \rightarrow Cu^{2+}(aq) + NO_2(g) + H_2O$

- **a.** Balance the equation.
- **b.** What volume of 16.0 *M* HNO_3 is needed to furnish enough H^+ ions to react with 5.87 g of copper?

56. Consider the reaction of iron(II) hydroxide with oxygen, for which the unbalanced equation is: $Fe(OH)_2(s) + O_2(g) + H_2O \rightarrow Fe(OH)_3(s)$

- **a.** Balance the equation (basic solution).
- **b.** What mass of $Fe(OH)_3$ is formed from 4.00 g of $Fe(OH)_2$?

57. Consider the acidic reaction between dichromate ions and iron(II) ions. The products are chromium(III) ions and iron(III) ions. It is found that 23.8 mL of a potassium dichromate solution reacts with 29.3 mL of a 0.0325 *M* iron(II) nitrate solution.

- **a.** What is the molarity of the potassium dichromate solution?
- **b.** If a 0.100 *M* solution of an acid furnishing one mole of H^+ per mole of acid is used to acidify the solution, what is the minimum volume of acid required?

58. Consider the acidic reaction between permanganate ions and manganese(II) ions to form manganaese dioxide, $MnO_2(s)$.

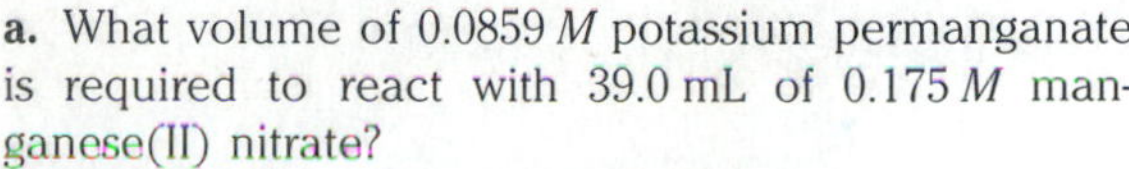

- **a.** What volume of 0.0859 *M* potassium permanganate is required to react with 39.0 mL of 0.175 *M* manganese(II) nitrate?
- **b.** What is the mass of precipitate formed in (a)?

Chemical Analysis

59. 24.41 mL of 0.1645 *M* HCl is required to titrate a 20.00-mL sample of NaOH. What is the molarity of NaOH?

60. A 25.00-mL sample of $Ca(OH)_2$ requires 12.14 mL of 0.1000 *M* HCl in an acid-base titration. What is the molarity of $Ca(OH)_2$?

61. A Vitamin C capsule is analyzed by titrating it with 0.250 *M* sodium hydroxide. It is found that 10.3 mL of base is required to react with a capsule weighing 0.518 g. What is the percentage of Vitamin C, $C_6H_8O_6$, in the capsule? (One mole of Vitamin C reacts with one mole of hydroxide ion.)

62. The percentage of sodium bicarbonate, $NaHCO_3$, in a powder used for stomach upsets is found by titrating with 0.187 *M* hydrochloric acid. If 20.5 mL of hydrochloric acid is required to react with 0.375 g of the powder, what is the percentage of sodium bicarbonate in the sample? The reaction is: $H^+(aq) + HCO_3^-(aq) \rightarrow CO_2(g) + H_2O$.

63. An artificial fruit beverage contains 12.0 g of tartaric acid, $H_2C_4H_4O_6$, to achieve tartness. It is titrated with a basic solution that has a density of 1.045 g/cm^3 and contains 5.00 mass % KOH. What volume of the basic solution is required? (One mole of tartaric acid reacts with two moles of hydroxide ion.)

64. Lactic acid, $C_3H_6O_3$, is the acid present in sour milk. A 0.100-g sample of pure lactic acid requires 12.95 mL of 0.0857 *M* sodium hydroxide for complete reaction. How many moles of hydroxide ion are required to neutralize one mole of lactic acid?

65. A wire weighing 0.100 g and containing 99.78% Fe is dissolved in HCl. The iron is completely oxidized to Fe^{3+} by bromine water. The solution is then treated with tin(II) chloride to bring about the reaction

$$Sn^{2+}(aq) + 2Fe^{3+}(aq) \longrightarrow Sn^{4+}(aq) + 2Fe^{2+}(aq)$$

If 9.47 mL of tin(II) chloride solution is required for complete reaction, what is its molarity?

66. Limonite, an ore of iron, is brought into solution and titrated with potassium permanganate, $KMnO_4$. The unbalanced equation for the redox reaction (acid solution) is

$$Fe^{2+}(aq) + MnO_4^-(aq) \longrightarrow Fe^{3+}(aq) + Mn^{2+}(aq)$$

It is found that a 0.500-g sample of limonite requires 40.0 mL of 0.0187 *M* $KMnO_4$. What is the percentage of iron in the limonite sample? If the iron in limonite is present as Fe_2O_3, what is the percentage of Fe_2O_3 in the ore?

67. Laws passed in some states define a drunk driver as one who drives with a blood alcohol level of 0.1% by mass or higher. The level of alcohol can be determined by titrating blood plasma with potassium dichromate according to the unbalanced equation

$$H^+(aq) + Cr_2O_7^{2-}(aq) + C_2H_5OH(aq) \longrightarrow Cr^{3+}(aq) + CO_2(g) + H_2O$$

Assuming that the only substance that reacts with dichromate in blood plasma is alcohol, is a person legally drunk if 45.02 mL of 0.05000 *M* potassium dichromate is required to titrate a fifty gram sample of blood plasma?

68. The molarity of iodine in solution can be determined by titration with H_3AsO_3; the unbalanced equation for the redox reaction (acid solution) is

$$I_2(aq) + H_3AsO_3(aq) \longrightarrow I^-(aq) + H_3AsO_4(aq)$$

What is the molarity of I_2 if 28.9 mL of the solution reacts exactly with 0.750 g of H_3AsO_3?

Amines

69. Write the general formulas for primary, secondary, and tertiary amines.

70. Write a balanced net ionic equation for the reaction between dimethylethylamine, $(CH_3)_2—N—CH_2CH_3$, and HCl.

71. Give two examples of alkaloids, and name a substance in which they naturally occur.

72. How many mL of 0.0100 *M* HCl are required to completely react with 125 mg of novocaine?

Unclassified

73. Classify each of the following as a precipitation, acid-base, or redox reaction.

a. the reaction between solutions of sulfuric acid and barium nitrate

b. the reaction between solutions of sulfuric acid and calcium hydroxide

c. the reaction of HCl with Al to evolve H_2

d. the reaction of a solution of tin(II) chloride with air to form SnO_2

74. Gold metal will dissolve only in *aqua regia,* a mixture of concentrated hydrochloric acid and concentrated nitric acid. The products of the reaction between gold and aqua regia are $AuCl_4^-(aq)$, $NO(g)$, and H_2O.

a. Write a balanced net ionic equation for the redox reaction, treating HCl and HNO_3 as strong acids.

b. What ratio of hydrochloric acid to nitric acid should be used?

c. What volumes of 12 *M* HCl and 16 *M* HNO_3 are required to furnish the Cl^- and NO_3^- ions to react with 10.0 g of gold?

75. The iron content of hemoglobin is determined by destroying the hemoglobin molecule and producing small water-soluble ions and molecules. The iron in the aqueous solution is reduced to iron(II) ion and then titrated against potassium permanganate. In the titration, iron(II) is oxidized to iron(III) and permanganate is reduced to manganese(II) ion. A 5.00-g sample of hemoglobin requires 32.3 mL of a 0.002100 *M* solution of potassium permanganate. What is the mass percent of iron in hemoglobin?

76. A sample of limestone weighing 0.145 g is dissolved in 50.00 mL of 0.100 *M* hydrochloric acid. The following reaction occurs:

$$CaCO_3(s) + 2H^+(aq) \longrightarrow Ca^{2+}(aq) + CO_2(g) + H_2O$$

It is found that 13.05 mL of 0.175 *M* NaOH is required to titrate the excess HCl left after reaction with the limestone. What is the mass percent of $CaCO_3$ in the limestone?

Challenge Problems

77. Calcium in blood or urine can be determined by precipitation as calcium oxalate, CaC_2O_4. The precipitate is dissolved in strong acid and titrated with potassium permanganate. The products of the reaction are carbon dioxide and manganese(II) ion. A 24-h urine sample is collected from an adult patient, reduced to a small volume, and titrated with 26.2 cm^3 of 0.0946 *M* $KMnO_4$. How many grams of calcium oxalate are in the sample? Normal range for Ca^{2+} output for an adult is 100–300 mg per 24 h. Is the sample within the normal range?

78. Stomach acid is approximately 0.020 *M* HCl. What volume of this acid is neutralized by an antacid tablet that weighs 330 mg and contains 41.0% $Mg(OH)_2$, 36.2% $NaHCO_3$, and 22.8% NaCl? The reactions involved are

$$Mg(OH)_2(s) + 2H^+(aq) \longrightarrow Mg^{2+}(aq) + 2H_2O$$

$$HCO_3^-(aq) + H^+(aq) \longrightarrow CO_2(g) + H_2O$$

79. Several years ago, 20,000 gallons of concentrated nitric acid (72% HNO_3 by mass, $d = 1.42\ g/cm^3$) spilled from a tank car in a Denver railyard. Sodium carbonate was spread on the acid to react with it to form carbon dioxide gas. The reaction is

$$CO_3^{2-}(aq) + 2H^+(aq) \longrightarrow CO_2(g) + H_2O$$

How many grams of sodium carbonate were required?

80. A solution contains both iron(II) and iron(III) ions. A 50.00-mL sample of the solution is titrated with 35.0 mL of 0.0280 *M* MnO_4, which oxidizes Fe^{2+} to Fe^{3+}; the MnO_4^- ion is reduced to Mn^{2+}. Another 50.00-mL sample of the solution is treated with zinc, which reduces all the Fe^{3+} to Fe^{2+}. The resulting solution is again titrated with 0.0280 *M* $KMnO_4$; this time 48.0 mL is required. What are the concentrations of Fe^{2+} and Fe^{3+} in the solution?

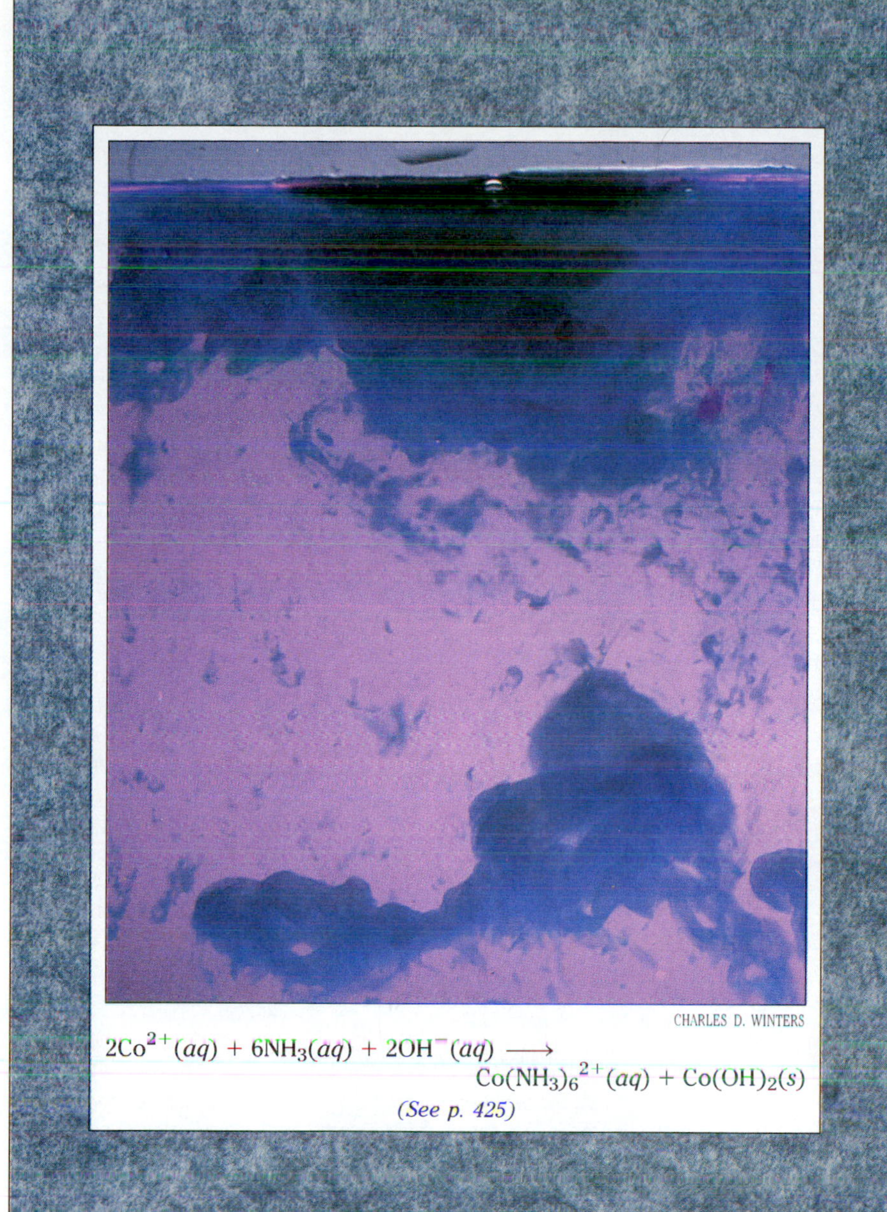

CHARLES D. WINTERS

$$2Co^{2+}(aq) + 6NH_3(aq) + 2OH^-(aq) \longrightarrow Co(NH_3)_6{}^{2+}(aq) + Co(OH)_2(s)$$

(See p. 425)

5
Gases

Religious faith is a most
filling vapor.
It swirls occluded in us
under tight
Compression to uplift us
out of weight—
As in those buoyant bird bones
thin as paper,
To give them still more
buoyancy in flight.
Some gas like helium must
be innate.

ROBERT FROST
Innate Helium

CHAPTER OUTLINE

By far the most familiar gas to all of us is the air we breathe. The Greeks considered air to be one of the four fundamental elements of nature, along with earth, water, and fire. Late in the eighteenth century, Cavendish, Priestley, and Lavoisier studied the composition of air, which is primarily a mixture of nitrogen and oxygen with smaller amounts of argon, carbon dioxide, and water vapor. Today, it appears that the concentrations of some of the minor components of the atmosphere may be changing, with adverse effects on the environ-

Ozone is a vital, but minor, component of the upper atmosphere (Chap. 11)

ment. The depletion of the ozone layer and increases in the amounts of "greenhouse" gases are topics for the evening news.

All gases resemble one another closely in their physical behavior. Their volumes respond in almost exactly the same way to changes in pressure, temperature, or amount of gas. In fact, it is possible to write a simple equation relating these four variables which is valid for all gases. This equation, known as the ideal gas law, is the central theme of this chapter; it is introduced in Section 5.2. The law is applied to

— pure gases in Section 5.3
— gases in chemical reactions in Section 5.4
— gas mixtures in Section 5.5

Section 5.6 considers the kinetic theory of gases, the molecular model upon which the ideal gas law is based. Finally, in Section 5.7, we describe the extent to which real gases deviate from the law.

5.1 MEASUREMENTS ON GASES

To completely describe the state of a gaseous substance, its volume, amount, temperature, and pressure are specified. The first three of these quantities were discussed in earlier chapters and will be reviewed briefly here. Pressure, a somewhat more abstract quantity, will be examined in more detail.

Volume, Amount, and Temperature

A gas expands uniformly to fill any container in which it is placed. This means that the volume of a gas is the volume of its container. Volumes of gases can be expressed in liters, cubic centimeters, or cubic meters:

$$1\ \text{L} = 10^3\ \text{cm}^3 = 10^{-3}\ \text{m}^3$$

Most commonly, the amount of matter in a gaseous sample is expressed in terms of the number of moles (n). In some cases, the mass (m) is given instead. These two quantities are related through the molar mass, $\mathcal{M}$.

$$n = m/\mathcal{M}$$

The temperature of a gas is ordinarily measured using a thermometer marked in degrees Celsius. However, *in any calculation involving the physical behavior of gases, temperatures must be expressed on the Kelvin scale.* To convert between °C and K, use the relation introduced in Chapter 1:

$$T_{\text{K}} = t_{°\text{C}} + 273.15$$

Typically, in gas law calculations, temperatures are expressed only to the nearest degree. In that case, the Kelvin temperature can be found by simply adding 273 to the Celsius temperature.

Pressure

In Canada, atmospheric pressure is commonly expressed in kilopascals

Pressure is defined as force per unit area. The SI unit of pressure (Chap. 1) is the *pascal* (Pa), the pressure exerted by a force of one newton on an area of one square meter. Atmospheric pressure is about 10^5 Pa or 100 *kilopascals* (kPa).

The two pressure units used most frequently in the text are

— the **atmosphere (atm)**, defined by the relation

$$1 \text{ atm} = 1.01325 \times 10^5 \text{ Pa}$$

— the **millimeter of mercury (mm Hg)**, the pressure exerted by a column of mercury one millimeter high. A column of mercury 760 mm high exerts a pressure of one atmosphere*:

$$1 \text{ atm} = 760 \text{ mm Hg}$$

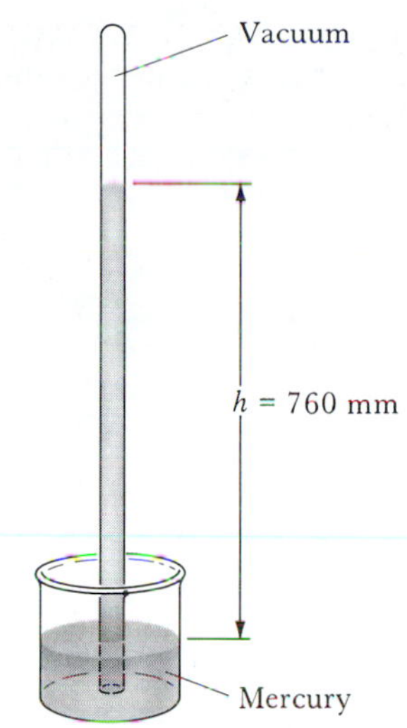

A representation of Torricelli's barometer. The height of the mercury column gives the atmospheric pressure.

Example 5.1 A balloon with a volume of 2.36×10^4 m^3 contains 4.68×10^6 g of helium at 18°C and 120.0 kPa. Express the volume of the balloon in liters, the amount in moles, the temperature in K, and the pressure in both atmospheres and millimeters of mercury.

Strategy Use the following conversion factors:

$$\frac{1 \text{ L}}{10^{-3} \text{ m}^3} \qquad \frac{1 \text{ mol He}}{4.003 \text{ g He}} \qquad \frac{760 \text{ mm Hg}}{101.3 \text{ kPa}} \qquad \frac{1 \text{ atm}}{101.3 \text{ kPa}}$$

For the temperature conversion, use the relation: $T_K = t_C + 273$

Solution

$$V = 2.36 \times 10^4 \text{ m}^3 \times \frac{1 \text{ L}}{10^{-3} \text{ m}^3} = \boxed{2.36 \times 10^7 \text{ L}}$$

$$n_{He} = 4.68 \times 10^6 \text{ g He} \times \frac{1 \text{ mol He}}{4.003 \text{ g He}} = \boxed{1.17 \times 10^6 \text{ mol He}}$$

$$T = 18 + 273 = \boxed{291 \text{ K}}$$

$$P = 120.0 \text{ kPa} \times \frac{1 \text{ atm}}{101.3 \text{ kPa}} = \boxed{1.185 \text{ atm}}$$

$$P = 120.0 \text{ kPa} \times \frac{760 \text{ mm Hg}}{101.3 \text{ kPa}} = \boxed{900.3 \text{ mm Hg}}$$

5.2 THE IDEAL GAS LAW

All gases closely resemble each other in the dependence of volume on amount, temperature, and pressure.

1. *Volume is directly proportional to amount.* Figure 5.1a (p. 106) shows a typical plot of volume (V) versus number of moles (n) for a gas. Notice that the graph is a straight line passing through the origin. The general equation for such a plot is

$$V = k_1 n \qquad \text{(constant } T, P\text{)}$$

* The pressure exerted by a column of mercury depends upon its density, which varies slightly with temperature. To get around this ambiguity, the *torr* was defined to be the pressure exerted by 1 mm of mercury at certain specified conditions, notably 0°C. Over time, the unit torr has become a synonym for millimeter of mercury. Throughout this text, we will use millimeter of mercury rather than torr because the former has a clearer physical meaning.

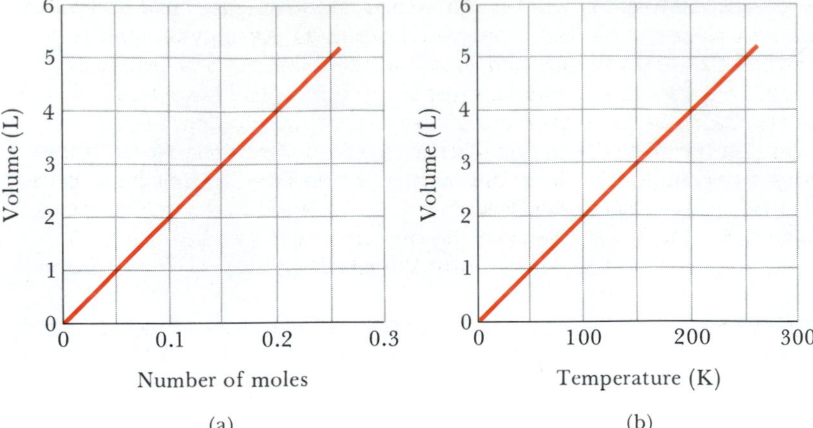

Figure 5.1
At constant pressure, the volume of a gas is directly proportional to the number of moles (a) and to the absolute temperature (b).

where k_1 is a "constant," that is, it is independent of individual values of V and n and of the nature of the gas. This is the equation of a direct proportionality.

2. *Volume is directly proportional to absolute temperature.* The dependence of volume (V) on the Kelvin temperature (T) is shown in Figure 5.1b. Here again, the graph is a straight line through the origin. The equation of the line is

$$V = k_2T \qquad \text{(constant } n, P\text{)}$$

where k_2 is a constant. This relationship was first suggested, in a different form, by two French scientists, Charles and Gay-Lussac. It is often referred to as the *law of Charles and Gay-Lussac.*

3. *Volume is inversely proportional to pressure.* Figure 5.2 shows a typical plot of volume (V) versus pressure (P). Notice that V decreases as P increases. The graph is a hyperbola. The general relation between the two variables is

$$V = k_3/P \qquad \text{(constant } n, T\text{)}$$

The quantity k_3, like k_1 and k_2, is a constant. This is the equation of an inverse proportionality. The fact that volume is inversely proportional to pressure was first established by Robert Boyle; the equation above is one form of *Boyle's law.*

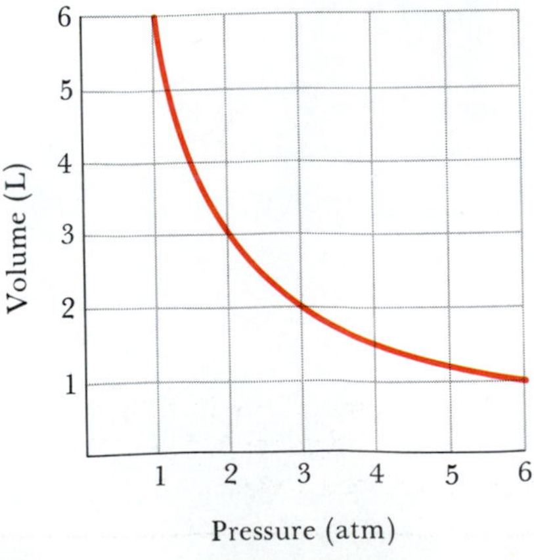

Figure 5.2
At constant temperature, the volume of a gas sample is inversely proportional to pressure. In this case, the volume decreases from 6 L to 1 L when the pressure increases from 1 atm to 6 atm.

The gas laws just stated were the result of the work of several gifted individuals who had a profound impact on the development of chemistry.

Joseph Gay-Lussac
(1778–1850)

Robert Boyle (1627–1691) was born in Ireland, the fourteenth child of the Earl of Cork. He was one of the first experimental scientists. Among other things, Boyle invented a vacuum pump and used it to show that air is necessary for combustion, respiration, and the transmission of sound. His discovery, in 1660, of the law that bears his name grew out of his work on the properties of air.

Jacques Charles (1746–1823) started his career as a minor bureaucrat in the government of France. When dismissed from that post in an economy move, he turned to the study of gases. On December 1, 1783, Charles was on the second balloon ever to lift a human being off the surface of the earth (see Fig. 5.A). This accomplishment so impressed Louis XVI that Charles was given a laboratory at the Sorbonne.

In 1787, Charles discovered that the temperature dependence of gas volume could be expressed by the equation

$$V = V_0(1 + \alpha t)$$

where V_0 is the volume at 0°C, t is the Celsius temperature, and α is a constant. The quantity α is the same for all gases and is approximately equal to 1/273. Notice what happens to the equation when $t = -273$°C; at that temperature, $V = V_0(1 - 273/273) = 0$. In other words, the volume of a gas should go to zero at -273°C = 0 K. This is the basis for our earlier statement that volume is directly proportional to Kelvin temperature: $V = k_2T$. **Joseph Gay-Lussac** (1778–1850) found out about it by accident. He repeated Charles' work and published his results in 1802. Gay-Lussac, like Charles, was fascinated by balloons. In 1804, he ascended to a height of 7 km in a hydrogen-filled balloon. This altitude record remained unbroken for 50 years.

Gay Lussac isolated boron and first prepared HF

The equation

$$V = k_1n \qquad \text{(constant } T, P)$$

can be interpreted to mean that equal volumes of different gases (constant T, P) contain equal numbers of moles. This relationship is a modern form of Avogadro's law, proposed in 1811 by an Italian physicist with the improbable name of Lorenzo Romano Amedeo Carlo Avogadro di Quarequa e di Cerreto. **Avogadro** (1776–1856) suggested that *equal volumes of all gases at the same temperature and pressure contain the same number of molecules.*

This law made possible the first reliable determination of atomic masses

Figure 5.A
Jacques Charles' ascent in a hydrogen balloon at Paris, December 1, 1783. (The Granger Collection, New York)

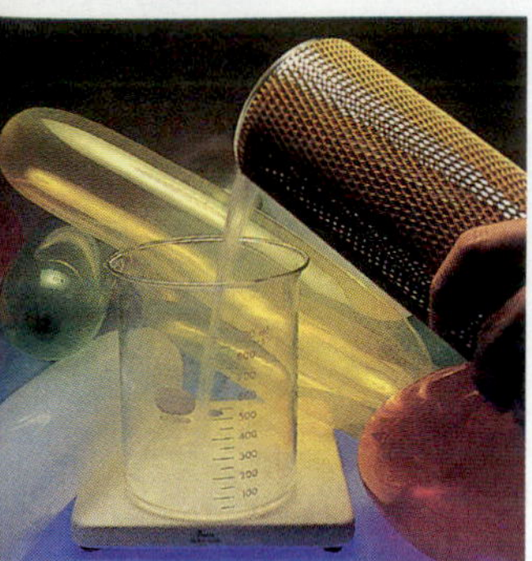
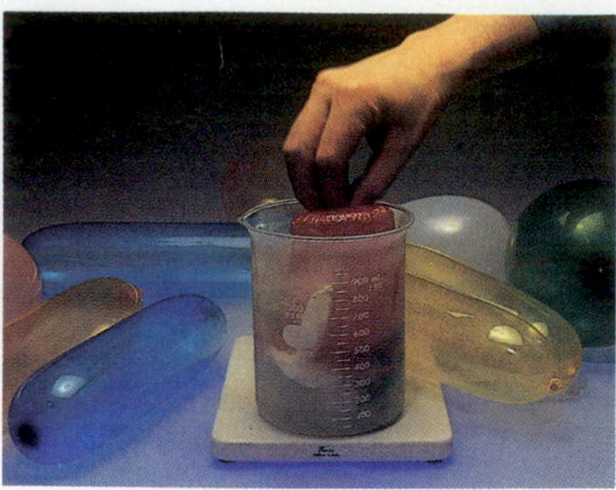
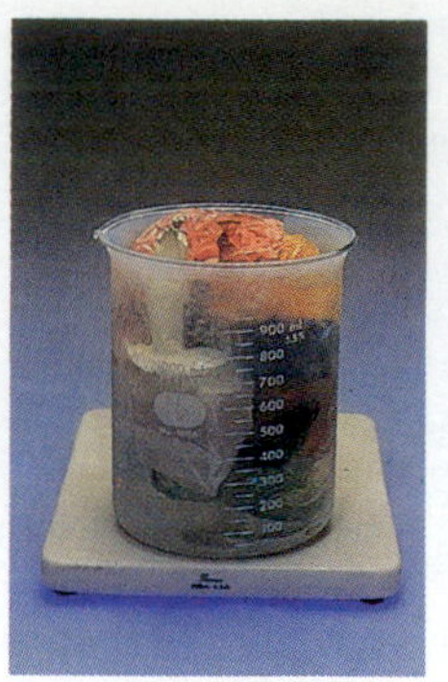

An illustration of Charles' law. When the air-filled balloons are placed in liquid nitrogen ($T = 77$ K), the volume of the gas decreases. When the balloons are removed from the liquid nitrogen, they re-inflate to their original volume as the air in the balloons return to room temperature. (Charles D. Winters)

The three equations relating the volume, pressure, temperature, and amount of a gas can be combined into a single equation. Since V is directly proportional to both n and T

$$V = k_1 n \qquad V = k_2 T$$

and inversely proportional to P

$$V = \frac{k_3}{P}$$

it follows that

$$V = \text{constant} \times \frac{n \times T}{P}$$

R is a constant because it is independent of the values of P, V, n, and T

Ordinarily, the constant is represented by the symbol R. Both sides of the equation are multiplied by P to give the **ideal gas law**

$$PV = nRT$$

where P is the pressure, V the volume, n the number of moles, and T the Kelvin temperature. The quantity R appearing in the ideal gas law is a true constant, independent of P, V, n, T, or the identity of the gas. Experimentally, it is found that the ideal gas law predicts remarkably well the experimental behavior of real gases (e.g., H_2, N_2, O_2, . . .) at ordinary temperatures and pressures.

The value of the gas constant R can be calculated from experimental values of P, V, n, and T. Consider, for example, the situation that applies at 0°C and 1 atm. These conditions are often referred to as *standard temperature and pressure* for a gas (**STP**). At STP, one mole of any gas occupies a volume of 22.4 L. Solving the ideal gas law for R,

$$R = \frac{PV}{nT}$$

Substituting $P = 1.00$ atm, $V = 22.4$ L, $n = 1.00$ mol, and $T = 0 + 273 = 273$ K.

Can you calculate R in L · mm Hg/mol · K?

$$R = \frac{1.00 \text{ atm} \times 22.4 \text{ L}}{1.00 \text{ mol} \times 273 \text{ K}} = 0.0821 \text{ L} \cdot \text{atm/(mol} \cdot \text{K)}$$

The value of R obtained under the most precise conditions at low pressures is 0.082056 L · atm/(mol · K). Note that R has the units of atmospheres, liters, moles, and K. These units must be used for pressure, volume, amount, and temperature in any problem in which this value of R is used.

5.3 GAS LAW CALCULATIONS

The ideal gas law can be used to solve a variety of problems. We will show how you can use it to find

— the final state of a gas, knowing its initial state and the changes in P, V, n, or T that occur
— one of the four variables, P, V, n, or T, given the values of the other three
— the molar mass or density of a gas

Commonly, a gas undergoes a change from an "initial" to a "final" state. Typically, you are asked to determine the effect upon V, P, n, or T of a change in one or more of these variables. For example, starting with a sample of gas at 25°C and 1.00 atm, you might be asked to calculate the pressure developed upon heating to 95°C at constant volume.

The ideal gas law is readily applied to problems of this type. A relationship between the variables involved is derived from this law. In this case, pressure and temperature change, while n and V remain constant.

$$\text{initial state:} \quad P_1V = nRT_1$$
$$\text{final state:} \quad P_2V = nRT_2$$

To obtain a two-point equation, write the gas law twice and divide to eliminate constants

Dividing the second equation by the first cancels V, n, and R, leaving the relation

$$P_2/P_1 = T_2/T_1 \qquad (\text{constant } n, V)$$

Applying this general relation to the problem described above,

$$P_2 = P_1 \times \frac{T_2}{T_1} = 1.00 \text{ atm} \times \frac{368 \text{ K}}{298 \text{ K}} = 1.23 \text{ atm}$$

Similar "two-point" equations can be derived from the ideal gas law to solve any problem of this type.

Example 5.2 On a cold day, a person takes in a breath of 450 mL of air at 756 mm Hg and −10°C. Assuming that amount and pressure remain constant, what is the volume of the air when it warms up to body temperature (37°C) in the lungs?

Strategy Find a "two-point" relation between V and T at constant n and P. Substitute for V_1, T_1, and T_2; solve for V_2.

Solution

$$\text{initial state:} \quad PV_1 = nRT_1$$
$$\text{final state:} \quad PV_2 = nRT_2$$

Dividing the second equation by the first

$$V_2/V_1 = T_2/T_1 \qquad V_2 = V_1 \times \frac{T_2}{T_1} = 450 \text{ mL} \times \frac{310 \text{ K}}{263 \text{ K}} = \boxed{530 \text{ mL}}$$

Calculation of *P*, *V*, *n*, or *T*

Frequently, values are known for three of these quantities (perhaps V, n, and T); the other one (P) must be calculated. This is readily done by direct substitution into the ideal gas law.

The Cl atoms hasten the decomposition of O_3

Example 5.3 Freon, CF_2Cl_2, used as a refrigerant in car air conditioners, is one of the culprits in the depletion of the ozone layer. If 3.00 g of CF_2Cl_2 gas is introduced into an evacuated 500.0-mL container at 10°C, what pressure in atmospheres is developed?

Strategy Substitute directly into the ideal gas law and solve for P. Note that V, n, and T have to be in units consistent with $R = 0.0821 \text{ L} \cdot \text{atm/(mol} \cdot \text{K)}$.

Solution Converting to the appropriate units,

$$V = 500.0 \text{ mL} \times \frac{1 \text{ L}}{1000 \text{ mL}} = 0.5000 \text{ L}$$

$$T = 10 + 273 = 283 \text{ K}$$

The molar mass of CF_2Cl_2 is 120.91 g/mol. Hence

$$n = 3.00 \text{ g } CF_2Cl_2 \times \frac{1 \text{ mol } CF_2Cl_2}{120.91 \text{ g } CF_2Cl_2} = 0.0248 \text{ mol}$$

Substituting into the law,

$$P = \frac{nRT}{V} = \frac{0.0248 \text{ mol} \times 0.0821 \text{ L} \cdot \text{atm/mol} \cdot \text{K} \times 283 \text{ K}}{0.5000 \text{ L}} = \boxed{1.15 \text{ atm}}$$

CFCs used in automotive air conditioners are now being recycled during repairs. Earlier, they were simply released into the atmosphere before the repair work began. (Courtesy of Robinair)

Molar Mass and Density

The ideal gas law offers a simple approach to the experimental determination of the molar mass of a gas. Indeed, this approach can be applied to volatile liquids like acetone (Example 5.4). All you need to know is the mass of a sample confined to a container of fixed volume at a particular temperature and pressure.

Example 5.4 Acetone is widely used as a nail polish remover. A sample of liquid acetone is placed in a 3.00-L flask and vaporized by heating to 95°C at 1.02 atm. The vapor filling the flask at this temperature and pressure weighs 5.87 g. Calculate the molar mass of acetone.

This experiment is commonly done in the general chem lab; make sure you boil off all the liquid

Strategy Perhaps the simplest approach here is to derive from the ideal gas law an expression for the molar mass, $\mathcal{M}$. This is readily done by using the relation $n = m/\mathcal{M}$, where m is the mass in grams.

$$PV = \frac{mRT}{\mathcal{M}}$$

Solution Solving for $\mathcal{M}$,

$$\mathcal{M} = \frac{mRT}{PV}$$

All the quantities required to calculate $\mathcal{M}$ are given in the statement of the problem.

$$\mathcal{M} = \frac{5.87\ \text{g} \times 0.0821\ \text{L} \cdot \text{atm/mol} \cdot \text{K} \times 368\ \text{K}}{3.00\ \text{L} \times 1.02\ \text{atm}} = \boxed{58.0\ \text{g/mol}}$$

The density of a gas is dependent upon

— *pressure.* Compressing a gas increases its density by reducing its volume ($d = m/V$).
— *temperature.* Hot air rises because a gas becomes less dense when its temperature is increased.
— *molar mass.* Hydrogen ($\mathcal{M} = 2.016$ g/mol) has the lowest molar mass and the lowest density (at a given T and P) of all gases.

The general expression for gas density is: $d = \mathcal{M}P/RT$

The ideal gas law can be used to derive a relation for gas density consistent with these observations (Example 5.5).

Example 5.5 Taking the molar mass of dry air to be 29.0 g/mol, calculate the density of air at 27°C and 1 atm.

Strategy Use the ideal gas law to derive a relation between density (d), molar mass ($\mathcal{M}$), pressure (P), and temperature (T). It may be simplest to start with the relation derived in Example 5.4:

Alternatively, you could find the mass of one liter of the gas

$$\mathcal{M} = \frac{mRT}{PV}$$

Substituting d for m/V

$$\mathcal{M} = \frac{dRT}{P}$$

Solving for density

$$d = \frac{\mathcal{M}P}{RT}$$

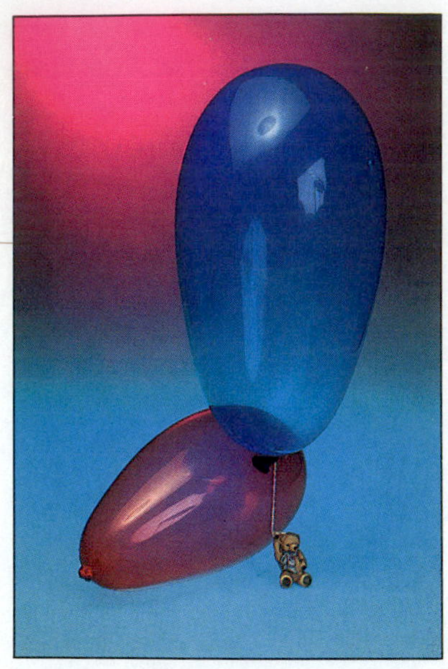

The two balloons are filled with about the same mass of gas at the same temperature and pressure. The blue balloon is filled with helium ($d = 0.16$ g/L) while the red balloon is filled with argon ($d = 1.6$ g/L). (Charles D. Winters)

Solution

$$d = \frac{\mathcal{M}P}{RT} = \frac{29.0\ \text{g/mol} \times 1.00\ \text{atm}}{0.0821\ \text{L} \cdot \text{atm/mol} \cdot \text{K} \times 300\ \text{K}} = \boxed{1.18\ \text{g/L}}$$

As promised, the density equation shows that gas density increases with increasing molar mass and decreases with increasing temperature. Balloons, which must contain a gas less dense than the surrounding air, take advantage of one or the other of these effects. "Hot air" balloons are filled with air at a temperature higher than that of the atmosphere, making the air inside less dense than that outside. First used in France in the eighteenth century, they are now seen in balloon races and other sporting events. Heat is supplied on demand by using a propane burner.

The other type of balloon uses a gas with a molar mass less than that of air. Hydrogen ($\mathcal{M} = 2.016$ g/mol) has the greatest lifting power, because it has the lowest density of all gases. However, it has not been used in manned balloons since 1937, when the Hindenberg, a hydrogen-filled airship, exploded and burned. Helium ($\mathcal{M} = 4.003$ g/mol) is slightly less effective than hydrogen but a lot safer to work with, since it is nonflammable. It is used in a variety of balloons, ranging from the small ones used at parties to meteorological balloons with volumes of a billion liters.

5.4 STOICHIOMETRY OF GASEOUS REACTIONS

As pointed out in Chapter 3, a balanced equation can be used to relate moles or grams of substances taking part in a reaction. Where gases are involved, these relations can be extended to include volumes. To do this, we use the ideal gas law and the conversion factor approach described in Chapter 3.

Example 5.6 Hydrogen peroxide is the active ingredient in commercial preparations for bleaching hair. What mass of hydrogen peroxide must be used to produce 1.00 L of oxygen gas at 25°C and 1.00 atm? The equation for the reaction is

$$2H_2O_2(aq) \longrightarrow O_2(g) + 2H_2O$$

Strategy First convert volume of $O_2(g)$ to moles, using the ideal gas law. Then find the mass of H_2O_2 by the conversion factor approach described in Chapter 3; the path is $n_{O_2} \rightarrow n_{H_2O_2} \rightarrow m_{H_2O_2}$

Solution

$$(1)\ n_{O_2} = \frac{PV}{RT} = \frac{1.00\ \text{atm} \times 1.00\ \text{L}}{0.0821\ \text{L} \cdot \text{atm/mol} \cdot \text{K} \times 298\ \text{K}} = 0.0409\ \text{mol}\ O_2$$

$$(2)\ \text{mass of}\ H_2O_2 = 0.0409\ \text{mol}\ O_2 \times \frac{2\ \text{mol}\ H_2O_2}{1\ \text{mol}\ O_2} \times \frac{34.02\ \text{g}\ H_2O_2}{1\ \text{mol}\ H_2O_2} = \boxed{2.78\ \text{g}\ H_2O_2}$$

Example 5.7 Octane, C_8H_{18}, is one of the hydrocarbons in gasoline. Upon combustion, (burning in oxygen) octane produces carbon dioxide and water. How many liters of oxygen, measured at 0.974 atm and 24°C, are required to burn 1.00 g of octane?

Strategy First write a balanced equation for the reaction. Then calculate the number of moles of oxygen required to burn 1.00 g of C_8H_{18} ($\mathcal{M}$ = 114.22 g/mol). Finally, use the ideal gas law to find the volume of oxygen.

Solution

(1) The balanced equation for the reaction is:

$$2C_8H_{18}(l) + 25\ O_2(g) \longrightarrow 16CO_2(g) + 18H_2O(l)$$

(2) $$n_{O_2} = 1.00\text{ g } C_8H_{18} \times \frac{1\text{ mol } C_8H_{18}}{114.22\text{ g } C_8H_{18}} \times \frac{25\text{ mol } O_2}{2\text{ mol } C_8H_{18}} = 0.109\text{ mol } O_2$$

(3) $$V_{O_2} = \frac{nRT}{P} = \frac{0.109\text{ mol} \times 0.0821\text{ L}\cdot\text{atm/mol}\cdot\text{K} \times 297\text{ K}}{0.974\text{ atm}} = 2.73\text{ L}$$

In going a mile, a typical engine takes in 1000 L of air

Since air is only 21% O_2 by volume, a large volume of air must pass through an automobile engine during combustion.

Perhaps the first stoichiometric relationship to be discovered was the **law of combining volumes**, proposed by Gay-Lussac in 1808: *The volume ratio of any two gases in a reaction at constant temperature and pressure is the same as the reacting mole ratio.*

To illustrate the law, consider the reaction

Similarly, 2 L of H_2 react with 1 L of O_2 to form water

$$2H_2O(l) \longrightarrow 2H_2(g) + O_2(g)$$

As you can see from Figure 5.3, the volume of hydrogen produced is twice that of the other gaseous product, oxygen.

The law of combining volumes, like so many relationships involving gases, is readily explained by the ideal gas law. At constant temperature and pressure, volume is directly proportional to number of moles ($V = k_1n$). It follows that for gaseous species involved in reactions, the volume ratio must be the same as the mole ratio given by the coefficients of the balanced equation.

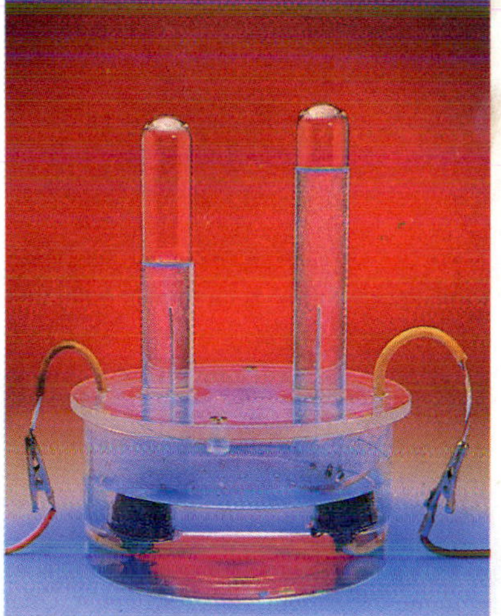

Figure 5.3
When water is electrolyzed, the volume of hydrogen gas formed in the tube at the left is twice that of oxygen (right tube), in accordance with the equation: $2H_2O(l) \longrightarrow 2H_2(g) + O_2(g)$. (Charles D. Winters)

5.5 GAS MIXTURES: PARTIAL PRESSURES AND MOLE FRACTIONS

The ideal gas law applies not only to pure gases but also to gas mixtures. For a mixture of two gases A and B, the total pressure is given by the expression

$$P_{tot} = n_{tot}\frac{RT}{V} = (n_A + n_B)\frac{RT}{V}$$

Separating the two terms on the right,

$$P_{tot} = n_A\frac{RT}{V} + n_B\frac{RT}{V}$$

The terms n_ART/V and n_BRT/V are, according to the ideal gas law, the pressures that gases A and B would exert if they were alone. These quantities are referred to as **partial pressures**, P_A and P_B.

$$P_A = \text{partial pressure A} = n_ART/V$$

$$P_B = \text{partial pressure B} = n_BRT/V$$

Substituting for n_ART/V and n_BRT/V in the equation for P_{tot},

$$P_{tot} = P_A + P_B$$

The relation just derived was first proposed by John Dalton in 1801; it is often referred to as **Dalton's law** of partial pressures:

Dalton didn't believe in the law of combining volumes

The total pressure of a gas mixture is the sum of the partial pressures of the components of the mixture.

To illustrate Dalton's law, consider a gaseous mixture of hydrogen and helium in which

$$P_{H_2} = 2.46 \text{ atm} \qquad P_{He} = 3.69 \text{ atm}$$

It follows from Dalton's law that

$$P_{tot} = 2.46 \text{ atm} + 3.69 \text{ atm} = 6.15 \text{ atm}$$

Wet Gases; Partial Pressure of Water

When a gas such as hydrogen is collected by bubbling through water (Fig. 5.4), it picks up water vapor; molecules of water escape from the liquid and are carried along with the gas. Dalton's law can be applied to the resulting gas mixture:

$$P_{tot} = P_{H_2O} + P_{H_2}$$

In this case, P_{tot} is the measured pressure. The partial pressure of water vapor, P_{H_2O}, is equal to the **vapor pressure** of liquid water. It has a fixed value at a given temperature (see Appendix 1, p. 607). The partial pressure of hydrogen, P_{H_2}, can be calculated by subtraction. The number of moles of hydrogen in the wet gas, n_{H_2}, can then be determined using the ideal gas law.

Example 5.8 A student prepares a sample of hydrogen gas by electrolyzing water at 25°C. She collects 152 mL of H_2 at a total pressure of 758 mm Hg. Using Appendix 1 to find the vapor pressure of water, calculate

This is perhaps the most common application of Dalton's law

(a) the partial pressure of hydrogen.
(b) the number of moles of hydrogen collected.

Figure 5.4
When a gas is collected by displacing water, it becomes saturated with water vapor; the partial pressure of $H_2O(g)$ is equal to the vapor pressure of liquid water at the temperature of the system.
(Marna G. Clarke)

Strategy

(a) Use Dalton's law to find the partial pressure of hydrogen, P_{H_2}.
(b) Use the ideal gas law to calculate n_{H_2}, with P_{H_2} as the pressure.

Solution

(a) From Appendix 1, P_{H_2O} = 23.76 mm Hg at 25°C. The total pressure, P_{tot}, is 758 mm Hg.

$$P_{H_2} = P_{tot} - P_{H_2O} = 758 \text{ mm Hg} - 23.76 \text{ mm Hg} = 734 \text{ mm Hg}$$

$$\text{(b) } n_{H_2} = \frac{(P_{H_2})V}{RT} = \frac{(734/760 \text{ atm})(0.152 \text{ L})}{(0.0821 \text{ L}\cdot\text{atm/mol}\cdot\text{K})(298 \text{ K})} = 0.00600 \text{ mol } H_2$$

Partial Pressure and Mole Fraction

As pointed out earlier, the following relationship applies to a mixture containing gas A (and gas B):

$$P_A = \frac{n_A RT}{V} \qquad P_{tot} = \frac{n_{tot} RT}{V}$$

Dividing P_A by P_{tot} gives

$$\frac{P_A}{P_{tot}} = \frac{n_A}{n_{tot}}$$

The fraction n_A/n_{tot} is referred to as the **mole fraction** of A in the mixture. It is the fraction of the total number of moles that is accounted for by gas A. Using X_A to represent the mole fraction of A (i.e., $X_A = n_A/n_{tot}$),

$$P_A = X_A P_{tot}$$

In other words, *the partial pressure of a gas in a mixture is equal to its mole fraction multiplied by the total pressure.* This relation is commonly used to calculate partial pressures of gases in a mixture when the total pressure and the composition of the mixture are known (Example 5.9).

⅕ of the molecules in air are O_2, so ⅕ of the total pressure is due to O_2

Example 5.9 Chemical analysis of air shows that the mole fractions of nitrogen, oxygen, and argon are 0.781, 0.210, and 0.009, respectively. Calculate the partial pressure of each of these gases on a day when the barometric pressure is 747 mm Hg.

Strategy The barometric pressure is the total pressure; use the equation $P_A = X_A P_{tot}$ to find the partial pressures.

Solution

$$P_{N_2} = 0.781 \times 747 \text{ mm Hg} = 583 \text{ mm Hg}$$

$$P_{O_2} = 0.210 \times 747 \text{ mm Hg} = 157 \text{ mm Hg}$$

$$P_{Ar} = 0.009 \times 747 \text{ mm Hg} = 7 \text{ mm Hg}$$

The partial pressures add to the total pressure, 747 mm Hg, as they should by Dalton's law.

5.6 KINETIC THEORY OF GASES

The fact that the ideal gas law applies to all gases indicates that the gaseous state is a relatively simple one from a molecular standpoint. Gases must have certain common properties that cause them to follow the same natural law. Between about 1850 and 1880, James Maxwell, Rudolf Clausius, Ludwig Boltzmann, and others developed the **kinetic theory** of gases. They based it on the idea that all gases behave similarly so far as particle motion is concerned. Since that time, the kinetic theory has had to be modified only slightly. It is one of the most successful scientific theories, ranking in stature with the atomic theory of matter.

Postulates of the Kinetic Theory

The kinetic theory of gases is based on several assumptions, including the following.

In air, a molecule undergoes about 10 billion collisions per second

1. Gases consist of atoms or molecules in continuous, random motion. These particles undergo frequent collisions with each other and with the walls of their container.
2. Collisions between gas particles are elastic; there is no change in total energy when a collision occurs.
3. The volume occupied by gas particles is negligibly small compared to that of their container.
4. Attractive forces between particles have a negligible effect on their behavior. The atoms or molecules in a gas can be treated as independent particles.

The two most important postulates for our purposes are:

A baseball in motion has translational energy

5. *The average translational kinetic energy, E_t, of a gas particle is directly proportional to the absolute temperature.* That is,

$$E_t = cT$$

where c is a constant.

6. *At a given temperature, all gases have the same average translational kinetic energy.* In other words, the quantity c in the above equation is a universal constant, which has the same value for all gases.

Average Speeds of Gas Particles

A basic law of physics states that the translational kinetic energy (energy of motion in a straight line) of a particle is one half of the product of the particle's mass times the square of its speed. Applied to gas particles, this means that

$$E_t = mu^2/2$$

where E_t is the average translational energy and u is the corresponding speed, which we will call the average speed.* Combining this relation with Postulate 5 of the kinetic theory, $E_t = cT$, it follows that

$$mu^2/2 = cT$$

* More rigorously, u^2 is the average of the squares of the speeds of all molecules.

Solving this equation for u:

$$u = \left(\frac{2cT}{m}\right)^{1/2}$$

E depends only upon T; u depends both on $\mathcal{M}$ and T

where c is a universal constant, with the same value for all gases (Postulate 6).

In case you're curious, $c = 3R/2N_A$

From the equation just written, you can see that the average speed, u, is

— *directly proportional to the square root of the absolute temperature.* For a given gas at two different temperatures, T_2 and T_1, the quantities c and m are constant, and we can write

$$\frac{u_2}{u_1} = \left(\frac{T_2}{T_1}\right)^{1/2}$$

— *inversely proportional to the square root of molecular mass (m) or molar mass ($\mathcal{M}$).* For two different gases A and B at the same temperature (c and T constant):

$$\frac{u_B}{u_A} = \left(\frac{\mathcal{M}_A}{\mathcal{M}_B}\right)^{1/2}$$

Example 5.10 At 25°C, the average speed of an O_2 molecule is 482 m/s.

(a) What is the average speed of an H_2 molecule at 25°C?
(b) What is the average speed of an H_2 molecule at 125°C?

Given u for one gas at one T, you should be able to calculate u for any gas at any T

Strategy Use the two equations cited above to find the ratio of speeds. Remember to use Kelvin temperatures.

Solution

(a) $\frac{u_{H_2}}{u_{O_2}} = \left(\frac{\mathcal{M}_{O_2}}{\mathcal{M}_{H_2}}\right)^{1/2} = \left(\frac{32.00}{2.016}\right)^{1/2} = 3.984$

$u_{H_2} = 3.984 \times 482 \text{ m/s} = 1.92 \times 10^3 \text{ m/s}$

(b) $\frac{u_2}{u_1} = \left(\frac{398}{298}\right)^{1/2} = 1.16$ $\quad u_2 = 1.16 \times 1.92 \times 10^3 \text{ m/s} = 2.23 \times 10^3 \text{ m/s}$

Effusion of Gases; Graham's Law

One way to check the validity of calculations made from kinetic theory is to study the process of **effusion**, the flow of gas particles through tiny pores or pinholes. The relative rates of effusion of different gases depend upon two factors: the pressures of the gases and the relative speeds of their particles. If two different gases A and B are compared at the same pressure, only their speeds are of concern and

$$\frac{\text{rate of effusion A}}{\text{rate of effusion B}} = \frac{u_A}{u_B}$$

where u_A and u_B are average speeds. As pointed out earlier, at a given temperature

$$\frac{u_A}{u_B} = \left(\frac{\mathcal{M}_B}{\mathcal{M}_A}\right)^{1/2}$$

It follows that, at constant pressure and temperature,

$$\frac{\text{rate of effusion A}}{\text{rate of effusion B}} = \left(\frac{\mathcal{M}_B}{\mathcal{M}_A}\right)^{1/2}$$

Among other things, Graham discovered dialysis

This relation in a somewhat different form was discovered experimentally by the Scottish chemist Thomas Graham in 1829. Graham was interested in a wide variety of chemical and physical problems, among them the separation of the components of air. Graham's law can be stated as:

At a given temperature and pressure, the rate of effusion of a gas is inversely proportional to the square root of its molar mass.

Graham's law gives a way of determining molar masses of gases. All that needs to be done is to compare the rate of effusion of the gas in question to that of another gas of known molar mass. Usually, either the distances moved by the two gases in equal times (Fig. 5.5) or the times required to effuse are measured. Since time is inversely related to rate,

$$\text{rate} = \frac{\text{distance}}{\text{time}}$$

it follows that

$$\frac{\text{time A}}{\text{time B}} = \left(\frac{\mathcal{M}_A}{\mathcal{M}_B}\right)^{1/2}$$

In other words, the time required for effusion increases with molar mass; heavy molecules take longer to effuse.

Example 5.11 In an effusion experiment, 45 s was required for a certain number of moles of an unknown gas X to pass through a small opening into a vacuum. Under the same conditions, it took 28 s for the same number of moles of Ar to effuse. Find the molar mass of the unknown gas.

Strategy Use the relation between time of effusion and molar mass.

Solution

$$\frac{\text{time for Ar}}{\text{time for X}} = \left(\frac{\text{molar mass of Ar}}{\text{molar mass of X}}\right)^{1/2}$$

Substituting numbers into this equation and squaring both sides gives

$$\left(\frac{28\text{ s}}{45\text{ s}}\right)^2 = \frac{39.95\text{ g/mol}}{\text{molar mass of X}}$$

$$\text{molar mass of X} = \frac{39.95\text{ g/mol}}{(28\text{ s}/45\text{ s})^2} = 1.0 \times 10^2\text{ g/mol}$$

A practical application of Graham's law arose during World War II, when scientists were studying the fission of uranium atoms as a source of energy. It became necessary to separate $^{235}_{92}U$, which is fissionable, from the more abun-

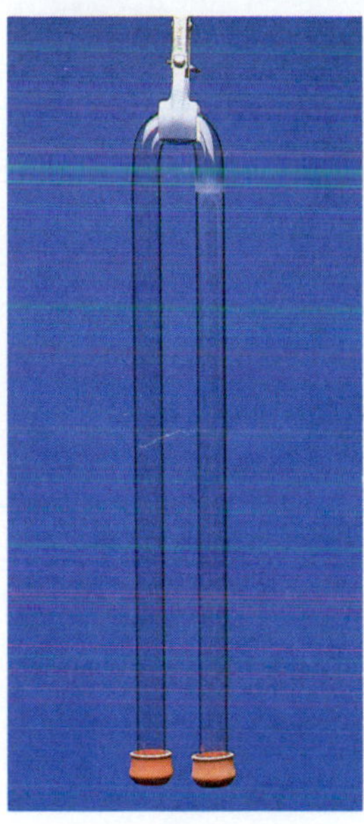

Figure 5.5
When ammonia gas, injected into the left arm of the U-tube, comes in contact with hydrogen chloride (right arm), they react to form a white deposit of ammonium chloride: $NH_3(g) + HCL(g) \longrightarrow NH_4Cl(s)$. Since NH_3 ($\mathcal{M}$ = 17 g/mol) effuses faster than HCl ($\mathcal{M}$ = 36.5 g/mol), the deposit forms closer to the HCl end of the tube. (Marna G. Clarke)

dant isotope of uranium, $^{238}_{92}U$, which is not fissionable. Since the two isotopes have almost identical chemical properties, chemical separation was not feasible. Instead, an effusion process was worked out using uranium hexafluoride, UF_6. This compound is a gas at room temperature and low pressures. Preliminary experiments indicated that $^{235}_{92}UF_6$ could indeed be separated from $^{238}_{92}UF_6$ by effusion. The separation factor is very small, since the rates of effusion of these two species are nearly equal:

$$\frac{\text{rate of effusion of } ^{235}_{92}UF_6}{\text{rate of effusion of } ^{238}_{92}UF_6} = \left(\frac{352}{349}\right)^{1/2} = 1.004$$

so a great many repetitive separations are necessary. An enormous plant was built for this purpose in Oak Ridge, Tennessee. In this process, UF_6 effuses many thousands of times through porous barriers. The lighter fractions move on to the next stage while heavier fractions are recycled through earlier stages. Eventually, a nearly complete separation of the two isotopes is achieved.

Distribution of Molecular Speeds and Energies

As pointed out earlier, the average speed of an O_2 molecule at 25°C is 482 m/s, while that of H_2 is even higher, 1920 m/s. However, not all molecules in these gases have these speeds. The motion of particles in a gas is utterly chaotic. In the course of a second, a particle undergoes millions of collisions with other particles. As a result, the speed and direction of motion of a particle is constantly changing. Over a period of time, the speed will vary from almost zero to some very high value, considerably above the average.

In 1860, James Clerk Maxwell, a Scottish physicist and one of the greatest theoreticians the world has ever known, showed that different possible speeds are distributed among particles in a definite way. Indeed, he developed a mathematical expression for this distribution. His results are shown graphically

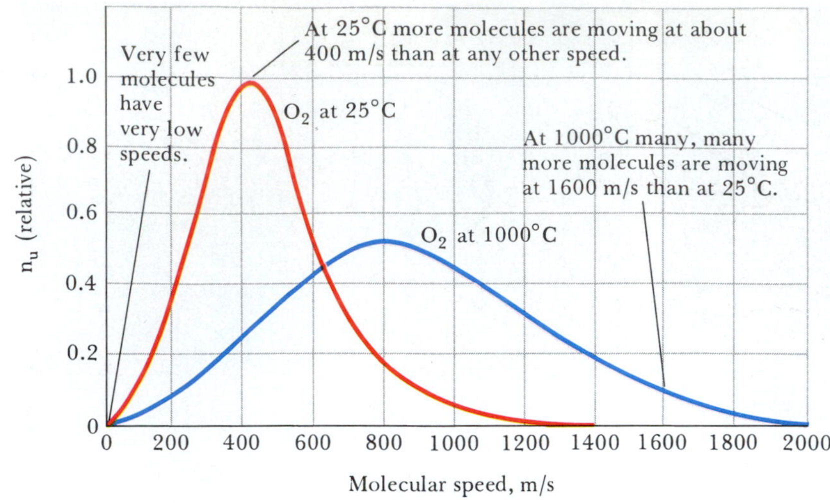

Figure 5.6
The distribution of molecular velocities in oxygen gas at two different temperatures, 25°C and 1000°C. At the higher temperature, the fraction of molecules moving at very high speeds in much greater.

in Figure 5.6 for O_2 at 25 and 1000°C. On the graph, the relative number of molecules having a certain speed is plotted against that speed. At 25°C, this number increases rapidly with the speed, up to a maximum of about 400 m/s. This is the most probable speed of an oxygen molecule at 25°C. Above about 400 m/s, the number of molecules moving at any particular speed decreases. For speeds in excess of about 1200 m/s, the fraction of molecules drops off to nearly zero. In general, most molecules have speeds rather close to the average value.

Sort of like people; most of them go along with the crowd

As temperature increases, the speed of the molecules increases. The distribution curve for molecular speeds (Fig. 5.6) shifts to the right and becomes broader. The chance of a molecule having a very high speed is much greater at 1000°C than it is at 25°C. Note, for example, that a large number of molecules have speeds greater than 1200 m/s at 1000°C.

5.7 REAL GASES

In this chapter, the ideal gas law has been used in all calculations, with the assumption that it applies exactly. Under ordinary conditions, this assumption is a good one; however, all real gases deviate at least slightly from the ideal gas law. Table 5.1 shows the extent to which two gases, O_2 and CO_2, deviate from ideality at different temperatures and pressures. The data compare the experimentally observed molar volume, V_m

The molar volume is the volume when $n = 1$

$$\text{molar volume} = V_m = V/n$$

with the molar volume calculated from the ideal gas law:

$$V_m^\circ = RT/P$$

It should be obvious from Table 5.1 that deviations from ideality become larger at *high pressures* and *low temperatures*. Moreover, the deviations are

Table 5.1 Real vs. Ideal Gases; Percent Deviation* in Molar Volume

	O_2			CO_2		
P(atm)	50°C	0°C	−50°C	50°C	0°C	−50°C
1	−0.0%	−0.1%	− 0.2%	− 0.4%	−0.7%	−1.4%
10	−0.4%	−1.0%	− 2.1%	− 4.0%	−7.1%	Condenses to liquid.
40	−1.4%	−3.7%	− 8.5%	−17.9%		
70	−2.2%	−6.0%	−14.4%	−34.2%		
100	−2.8%	−7.7%	−19.1%	−59.0%		

$$* \text{ \% dev.} = \frac{(V_m - V^\circ_m)}{V^\circ_m} \times 100\%$$

larger for CO_2 than for O_2. All of these effects can be correlated in terms of a simple, common-sense observation:

In general, the closer a gas is to the liquid state, the more it will deviate from the ideal gas law.

A gas is liquefied by going to low temperatures and/or high pressures. Moreover, as you can see from Table 5.1, carbon dioxide is much easier to liquefy than oxygen.

From a molecular standpoint, deviations from the ideal gas law arise because it neglects two factors (recall Postulates 3 and 4 of the kinetic theory):

1. The finite volume of gas particles
2. Attractive forces between gas particles

We will now consider in turn the effect of these two factors on the molar volumes of real gases.

Attractive Forces

Notice that in Table 5.1 all the deviations are negative; the observed molar volume is less than that predicted by the ideal gas law. This effect can be attributed to attractive forces between gas particles. These forces tend to pull the particles toward one another, reducing the space between them. As a result, the particles are crowded into a smaller volume, just as if an additional external pressure were applied. The observed molar volume, V_m, becomes less than V°_m, and the deviation from ideality is *negative:*

Attractive forces make the molar volume smaller than expected

$$\frac{V_m - V^\circ_m}{V^\circ_m} < 0$$

The magnitude of this effect depends upon the strength of the attractive forces and hence upon the nature of the gas. Intermolecular attractive forces are stronger in CO_2 than they are in O_2, which explains why the deviation from ideality of V_m is greater with carbon dioxide and why carbon dioxide is more readily condensed to a liquid than is oxygen.

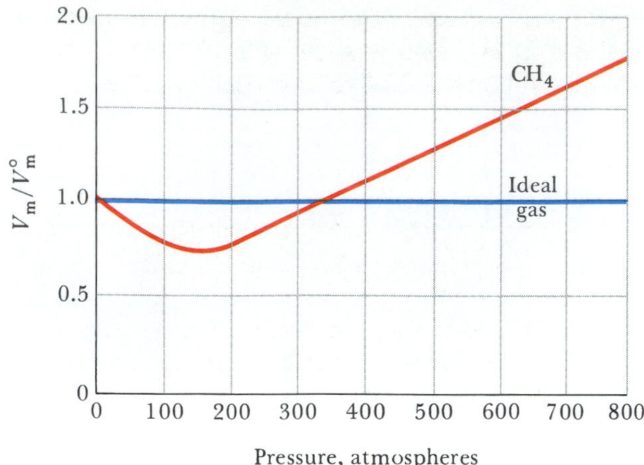

Figure 5.7
Below about 350 atm, attractive forces between CH_4 molecules cause the observed molar volume of methane gas at 25°C to be less than that calculated from the ideal gas law. At 350 atm, the effect of the attractive forces is just balanced by that of the finite volume of CH_4 molecules, and the gas appears to behave ideally. Above 350 atm, the effect of finite molecular volume predominates and $V_m > V_m^\circ$.

Particle Volume

Figure 5.7 shows a plot of V_m/V_m° versus pressure for methane at 25°C. Up to about 150 atm, methane shows a steadily increasing negative deviation from ideality, as might be expected on the basis of attractive forces. At 150 atm, V_m is only about 70% of V_m°.

At very high pressures, methane behaves quite differently. Above 150 atm, the ratio V_m/V_m° *increases*, becoming 1 at about 350 atm. Above that pressure, methane shows a *positive* deviation from the ideal gas law:

Particle volume causes the molar volume to be larger than expected

$$\frac{V_m - V_m^\circ}{V_m^\circ} > 0$$

This effect is by no means unique to methane; it is observed with all gases. If the data in Table 5.1 are extended to very high pressures, oxygen and carbon dioxide behave like methane; V_m becomes larger than V_m°.

An increase in molar volume above that predicted by the ideal gas law is related to the finite volume of gas particles. These particles contribute to the observed volume, making V_m greater than V_m°. Ordinarily, this effect becomes evident only at high pressures, where the particles are quite close to one another.

Table 5.2 van der Waals Constants

Gas	a $\left(\frac{L^2 \cdot atm}{mol^2}\right)$	b (L/mol)
H_2	0.244	0.027
O_2	1.360	0.032
N_2	1.390	0.039
CH_4	2.253	0.043
CO_2	3.592	0.043
SO_2	6.714	0.056
Cl_2	6.493	0.056
H_2O	5.464	0.030

van der Waals Equation

A number of different equations have been derived to relate P, V, and T for gases, taking into account attractions between particles and finite particle volumes. One of these is the van der Waals equation:

$$(P + a/V_m^2)(V_m - b) = RT$$

where V_m is the molar volume of the gas. In this equation, a and b are constants, independent of P, V, and T. They do, however, vary from one gas to another (Table 5.2). The term a/V_m^2 reflects the attractive forces between particles, while the constant b corrects for the effect of particle volume; b is roughly equal to the molar volume of the liquid.

The van der Waals equation is much better than the ideal gas law for predicting the behavior of real gases at moderate to high pressures. Consider, for example, $O_2(g)$ at −50°C and 100 atm. The observed molar volume is 0.148 L. The value calculated from the ideal gas law is 0.183 L. The van der Waals equation predicts a molar volume of 0.140 L, considerably closer to the true value.

PERSPECTIVE • The Greenhouse Effect

This chapter has emphasized the common properties of different gases. Many properties, however, differ tremendously from one gas to another. One of these is the ability to absorb infrared radiation (heat). Among the major components of the atmosphere, only carbon dioxide and water vapor show this behavior. They absorb much of the infrared radiation given off by the warm earth. In this way, CO_2 and H_2O act as an insulating blanket to prevent heat from escaping into outer space; this is often referred to as the greenhouse effect.

Of the two gases, water vapor absorbs more infrared radiation than carbon dioxide because its concentration is higher. This property of water vapor accounts for the fact that the temperature drops less on nights when there is a heavy cloud cover. In desert regions, where there is very little water vapor, large variations between day and night temperatures are common.

Although the concentration of water vapor in the atmosphere varies greatly with location, it remains relatively constant over time. In contrast, the concentration of carbon dioxide has increased by more than 20% over the past century, due to human activities. Increased combustion of fossil fuels is mainly responsible. Every gram of fossil fuel burned releases about three grams of carbon dioxide into the atmosphere. Part of this CO_2 is used by plants in photosynthesis or is absorbed by the oceans, but at least half of it remains. Extensive land clearing, which reduces the amount of carbon dioxide consumed by photosynthesis, is also a factor in raising the CO_2 content of the atmosphere. This is one of the adverse effects of the destruction of tropical rain forests for agricultural purposes.

It has been estimated that, unless preventive action is taken, increasing CO_2 levels could raise the earth's temperature by 3°C over the next century. This could raise sea level by as much as 1 m, flooding many coastal areas, including much of the state of Florida. On a more optimistic note, an increase in CO_2 concentration would promote photosynthesis, perhaps increasing the world's food supply.

Recent studies show that average global temperatures have indeed increased over the past century, by about 0.5°C (1°F). Beyond that, the three years 1989, 1988, and 1987 were, in that order, the warmest on record. Nobody knows whether these data reflect increased concentrations of CO_2 and other greenhouse gases (Fig. 5.B) or statistical fluctuations. The general consensus among atmospheric scientists is that carbon dioxide emissions should be reduced to avoid a worst-case scenario. There are several ways to do this:

— raise fuel efficiency standards for automobiles
— impose a surtax on all carbon-containing fuels
— develop "clean" sources of energy, notably solar

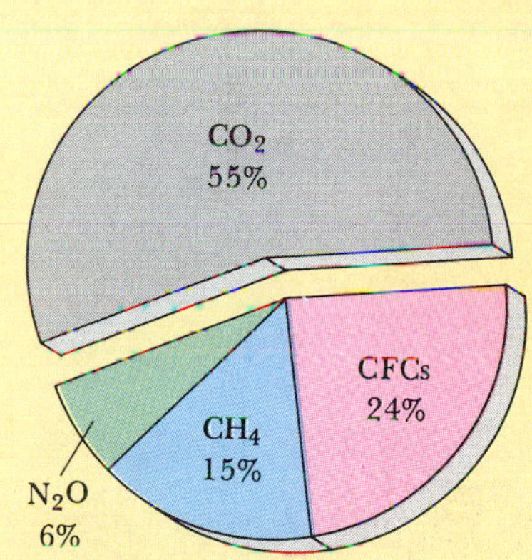

Figure 5.B
Contributions of different gases to global warming (1980–1990). Carbon dioxide is the major factor, but chlorofluorocarbons, such as CF_2Cl_2, make a contribution. These compounds are used in refrigerators, air conditioners, and, until recently, aerosol sprays.

CHAPTER HIGHLIGHTS

KEY CONCEPTS

1. *Convert between different units of pressure, volume, temperature, and amount of gas*
   (Example 5.1; Problems 1–4)
2. *Use the ideal gas law to*
   — *solve initial and final state problems*
   (Example 5.2; Problems 5–16)
   — *calculate P, V, T, or n*
   (Example 5.3, Problems 17–22)
   — *calculate density or molar mass*
   (Examples 5.4, 5.5; Problems 23–32)
   — *relate amounts and volumes of gases in reactions*
   (Examples 5.6, 5.7; Problems 33–38, 66–68, 74, 76)
3. *Use Dalton's law to relate partial pressures to total pressures and mole fractions*
   (Examples 5.8, 5.9; Problems 39–44)
4. *Use the relation $u = (2cT/m)^{1/2}$ to relate average speed to molar mass and temperature*
   (Example 5.10; Problems 49–52)
5. *Use Graham's law to relate rate or time of effusion to molar mass*
   (Example 5.11; Problems 45–48, 70, 72)

KEY EQUATIONS

Ideal gas law	$PV = nRT$	$n = m/\mathcal{M}$
Dalton's law	$P_{tot} = P_A + P_B + \cdots$	$P_A = X_A P_{tot}$
Average speed	$u = (2cT/m)^{1/2}$	
Graham's law	$\frac{\text{rate}_A}{\text{rate}_B} = \frac{\text{time}_B}{\text{time}_A} = \left(\frac{\mathcal{M}_B}{\mathcal{M}_A}\right)^{1/2}$	

KEY TERMS

atmosphere	kilopascal	mole fraction
density	millimeter of mercury	partial pressure
effusion	molar mass	*R* (gas constant)
Kelvin scale	mole	STP

SUMMARY PROBLEM

When aluminum foil is added to a solution of hydrochloric acid, the following reaction occurs:

$$2Al(s) + 6H^+(aq) \longrightarrow 2Al^{3+}(aq) + 3H_2(g)$$

a. If 255 mL of H_2 is produced, measured at 22°C and 740 mm Hg, what is the yield of hydrogen in moles? In grams?

b. What pressure would be required to compress the sample of H_2 referred to in (a) to one half of its volume at 22°C? At what temperature would the volume be doubled, if the pressure is held constant at 740 mm Hg?

c. What is the density of $H_2(g)$ at 22°C and 740 mm Hg?

d. Suppose the hydrogen is collected over water at 25°C and a total pressure of 756 mm Hg. What is the partial pressure of $H_2(g)$? If the volume of the sample is 235 mL, what is the mass of H_2?

e. What is the mole fraction of hydrogen in the wet gas in (d)?

f. What mass of aluminum is required to generate 1.00 L of $H_2(g)$ at 25°C and 1.00 atm?
g. What volume of H_2 at STP could be generated from 250 mL of "dilute" hydrochloric acid, in which the molarity of H^+ is 6.0 mol/L?
h. Compare the rate of effusion of H_2 to that of He; compare the time required for equal numbers of moles of H_2 and CO_2 to effuse.
i. The average speed of a hydrogen molecule at 25°C is 1.92×10^3 m/s. What is the average speed of a helium molecule at the same temperature?

Answers

a. 0.0103 mol; 0.0207 g **b.** 1480 mm Hg; 317°C **c.** 0.0810 g/L
d. 732 mm Hg; 0.0187 g **e.** 0.968 **f.** 0.735 g **g.** 17 L
h. 1.41; 0.214 **i.** 1.36×10^3 m/s

QUESTIONS & PROBLEMS

Measurements on Gases

1. A bedroom 11 ft × 12 ft × 8.0 ft contains 35.41 kg of air at 25°C. Express the volume of the room in liters, the amount of air in moles ($\mathcal{M}$ of air = 29.0 g/mol), and the temperature in K.

2. A three-gallon tank of oxygen contains 0.461 mol O_2 gas at 27°C. Express the volume of the tank in liters, the amount of O_2 in the tank in grams, and the temperature of the tank in K.

3. Complete the following table of pressure conversions.

mm Hg	Atmospheres	Kilopascals
728	____	____
____	1.28	____
____	____	99.8

4. Carry out the indicated conversions between pressure units.

mm Hg	Atmospheres	Kilopascals
158	____	____
____	0.795	____
____	____	128

Ideal Gas Law: Initial and Final States

5. A sample of carbon dioxide gas occupies a volume of 5.75 L at 0.890 atm. If the temperature and the number of moles remain constant, calculate the volume when the pressure is

a. increased to 1.25 atm
b. decreased to 0.350 atm

6. The pressure of a 5.00-L sample of xenon gas is 725 mm Hg. Assuming that the temperature and moles of xenon are unchanged, calculate the new pressure when the volume becomes

a. 8.75 L **b.** 1.35 L

7. A nitrogen sample of 30°C has a volume of 1.75 L. If the pressure and the amount of gas remain unchanged, determine the volume when

a. the Celsius temperature is doubled.
b. the Celsius temperature is halved.

8. A gas is originally at a temperature of 50°C. To what temperature must it be heated to triple its volume (with n and P constant)?

9. A basketball is inflated in a garage at 20°C to a gauge pressure of 8.0 psi. Gauge pressure is the pressure above atmospheric pressure, which is 14.7 psi. The ball is used in the driveway at a temperature of −5°C and feels "flat." What is the actual pressure of the air in the ball? What is the gauge pressure?

10. A tire is inflated to a gauge pressure of 28.0 psi at 67°F. After several hours of driving, the gauge pressure in the tire is 34.0 psi. What is the temperature of the air in the tire in °F? Assume volume changes are negligible.

11. A 2.90-cm^3 air bubble forms in a deep lake at a depth where the temperature is 8°C at a total pressure of 1.98 atm. The bubble rises to a depth where the temperature and pressure are 15°C and 1.50 atm, respectively. Assuming that the amount of air in the bubble has not changed, calculate its new volume.

12. On a cold day, a person takes in a breath of 450 mL of air at 756 mm Hg and −10°C. What is the volume of this air in the lungs at 37°C and 758 mm Hg?

13. An open flask contains 0.200 mol of air. Atmospheric pressure is 745 mm Hg and room temperature is 68°F. How many moles are present in the flask when the pressure is 1.10 atm and the temperature is 33°C?

14. A closed syringe contains 0.01765 mol of the foul-smelling gas hydrogen sulfide, H_2S. The pressure in the container is 725 mm Hg and the temperature is 23°C. An additional 0.00125 mol of H_2S is injected into the syringe. The temperature rises to 35°C. Assuming constant volume, calculate the new pressure in atmospheres.

15. Houses with well-water systems have ballast tanks to hold a supply of water. Typically, water flows into the tank from the well until the gauge pressure (pressure in excess of 15 lb/in^2) of the air in the tank reaches 50 lb/in^2. As water is drawn from the tank, the air above it expands and its pressure drops. When the gauge pressure reaches 20 lb/in^2, the pump delivers more water to the tank. Suppose a 1.50-m^3 tank is 83% full of water when the gauge pressure is 50 lb/in^2. How much water can be withdrawn before the pump turns on at 20 lb/in^2?

16. Frequently, ballast tanks of the type described in Problem 15 lose most of their air and become "water-logged." When this happens, the pump operates more frequently. Suppose the 1.50-m^3 tank referred to in Problem 15 has to be 92% full of water before the gauge pressure reaches 50 lb/in^2. How much water can be withdrawn from the tank before the pump turns on at 20 lb/in^2?

Ideal Gas Law: Calculation of One Variable

17. On a warm day, an amusement park balloon is filled with 47.8 g of helium. The temperature is 33°C and the pressure in the balloon is 2.25 atm. Calculate the volume of the balloon.

18. A ten-liter gas cylinder contains 3.8×10^2 g of nitrogen. What pressure is exerted by the nitrogen at 25°C?

19. A drum used to transport crude oil has a volume of 162 L. How many water molecules, as steam, are required to fill the drum at 1.00 atm and 100°C? What volume of liquid water (d $H_2O(l)$ = 1.00 g/cm^3) is required to produce that amount of steam?

20. How many moles of air are there in a 125-mL Erlenmeyer flask if the pressure is 755 mm Hg and the temperature is 20°C?

21. Use the ideal gas law to complete the following table for ammonia gas.

Pressure	Volume	Temperature	Moles	Grams
2.50 atm	______	0°C	______	32.0
______	75.0 mL	30°C	______	0.385
768 mm Hg	6.0 L	100°C	______	______
195 kPa	58.7 L	______	______	19.8

22. Complete the following table for dinitrogen tetraoxide gas, assuming ideal gas behavior.

Pressure	Volume	Temperature	Moles	Grams
735 mm Hg	______	15°C	______	18.9
______	489 mL	38°C	______	27.8
1.49 atm	0.885 L	45°C	______	______
239 kPa	2.75 L	______	______	45.0

Ideal Gas Law: Density and Molar Mass

23. Calculate the densities of the following gases at 127°C and 763 mm Hg.

a. uranium hexafluoride **b.** carbon monoxide
c. chlorine

24. Calculate the densities of the following gases at 268°F and 115 kPa.

a. nitrogen **b.** argon **c.** iodine chloride

25. Assuming water vapor behaves ideally at 100°C and 1.00 atm, what is the ratio of the density of $H_2O(l)$ to $H_2O(g)$ under these conditions? The density of liquid water at 100°C is 0.958 g/cm^3.

26. Recent measurements show that the atmosphere on Venus consists mostly of carbon dioxide. The temperature at the surface of Venus is 460°C; the pressure is 75 atm. Compare the density of CO_2 on Venus's surface to that on the earth's surface at 25°C and one atmosphere.

27. Cyclopropane mixed in the proper ratio with oxygen can be used as an anesthetic. At 755 mm Hg, and 25°C, it has a density of 1.71 g/L.

a. What is the molar mass of cyclopropane?
b. Cyclopropane is made up of 85.7% C and 14.3% H. What is its molecular formula?

28. Methyl isocyanate is the volatile, highly toxic gas that killed thousands of people in Bhopal, India, in 1984. It is made up of 42.1% C, 5.30% H, 24.6% N, and oxygen. A sample of the volatile liquid is vaporized completely in a 125-mL flask at 99°C and 745 mm Hg. The vapor is condensed and weighs 0.229 g. What is the molecular formula of methyl isocyanate?

29. To prevent a condition called the "bends," deep-sea divers breathe a mixture containing, in mole percent, 10.0% O_2, 10.0% N_2, and 80.0% He.

a. Calculate the molar mass of this mixture.
b. What is the ratio of the density of this gas to that of pure oxygen?

30. Exhaled air contains 74.5% N_2, 15.7% O_2, 3.6% CO_2, and 6.2% H_2O (mole percent).

a. Calculate the molar mass of exhaled air.
b. Calculate the density at 22°C and 750 mm Hg and compare the value obtained to that for ordinary air ($\mathcal{M}$ = 29.0 g/mol).

31. A 2.00-g sample of $SX_6(g)$ has a volume of 329.5 cm^3 at 1.00 atm and 20°C. Identify the element X. Name the compound.

32. A 1.58-g sample of $C_2H_3X_3(g)$ has a volume of 297 mL at 769 mm Hg and 35°C. Identify the element X.

Gases in Reactions

33. When hydrogen sulfide gas, H_2S, reacts with oxygen, sulfur dioxide gas and steam are produced.

a. Write a balanced equation for the reaction.
b. How many liters of sulfur dioxide would be produced from 4.0 L of oxygen? Assume 100% yield and that all gases are measured at the same temperature and pressure.

34. Nitrogen trifluoride gas reacts with steam to form the gases HF, NO, and NO_2.

a. Write a balanced equation for the reaction.
b. What volume of nitrogen oxide, NO, is formed when six liters of nitrogen trifluoride are made to react with

6.00 L of steam? Assume 100% yield and a constant temperature and pressure throughout the reaction.

35. Hydrogen cyanide, HCN, is a poisonous gas that was used in the gas chambers of Hitler's concentration camps. It can be formed by the reaction

$$NaCN(s) + H^+(aq) \longrightarrow HCN(g) + Na^+(aq)$$

What mass of sodium cyanide, NaCN, is required to make 8.53 L of hydrogen cyanide at 22°C and 751 mm Hg?

36. Calcium reacts with water, yielding hydrogen gas and calcium hydroxide. How many mL of water (d = 1.00 g/cm^3) are required to produce 7.00 L of dry hydrogen at 1.05 atm and 30°C?

37. Oxygen can be made by the decomposition of hydrogen peroxide:

$$2H_2O_2(aq) \longrightarrow 2H_2O + O_2(g)$$

a. What volume of oxygen gas at 27°C and 745 mm Hg can be made from 5.50 mL of a solution containing 3.00 mass percent hydrogen peroxide and having a density of 1.01 g/cm^3?

b. The label on the bottle of hydrogen peroxide claims that the solution produces ten times its own volume of oxygen gas. Does it?

38. Diborane, $B_2H_6(s)$, is a highly explosive compound formed by the reaction

$$3NaBH_4(s) + 4BF_3(g) \longrightarrow 2B_2H_6(g) + 3NaBF_4(s)$$

a. What mass of sodium borohydride, $NaBH_4$, is required to form one liter of diborane at STP (0°C and 1.00 atm)?

b. What volume of boron trifluoride at 20°C and 742 mm Hg is required to produce 6.00 g of sodium borofluoride, $NaBF_4$?

Dalton's Law

39. Suppose exhaled air (Problem 30) is at a pressure of 751 mm Hg. Calculate the partial pressures of nitrogen, oxygen, carbon dioxide, and water.

40. A gaseous mixture contains 5.78 g of CH_4, 2.15 g of Ne, and 6.80 g of SO_2. What pressure is exerted by the mixture inside a 75.0-L cylinder at 85°C? Which gas contributes the greatest pressure?

41. A student collects 355 cm^3 of oxygen saturated with water vapor at 27°C. The mixture exerts a total pressure of 775 mm Hg. At 27°C, the vapor pressure of $H_2O(l)$ = 26.7 mm Hg.

a. What is the partial pressure of oxygen in the sample?

b. How many grams of oxygen does the sample contain?

42. To prepare a sample of hydrogen gas, a student reacts zinc with hydrochloric acid. The overall reaction is

$$Zn(s) + 2H^+(aq) \longrightarrow Zn^{2+}(aq) + H_2(g)$$

The hydrogen is collected over water at 24°C and the total pressure is 758 mm Hg (vp $H_2O(l)$ = 22.4 mm Hg).

a. What is the partial pressure of hydrogen?

b. How many grams of hydrogen are there in a 2.00-L sample of wet gas?

43. A sample of oxygen gas is collected over water at 25°C (vp $H_2O(l)$ = 23.8 mm Hg). The wet gas occupies a volume of 12.83 L at a total pressure of 745 mm Hg. If all the water is removed, what volume will the dry oxygen occupy at a pressure of 762 mm Hg and a temperature of 50°C?

44. A sample of gas collected over water at 42°C occupies a volume of one liter. The wet gas has a pressure of 0.986 atm. The gas is dried and the dry gas occupies 1.04 L with a pressure of 1.00 atm at 90°C. What is the vapor pressure of water at 42°C?

Kinetic Theory

45. A gas effuses 1.25 times faster than nitrogen trifluoride at the same temperature and pressure.

a. Is the gas heavier or lighter than nitrogen trifluoride?

b. Calculate the ratio of the molar mass of nitrogen trifluoride to that of the unknown gas.

46. What is the ratio of the rate of effusion of the most dense gas known, uranium hexafluoride, to that of the most abundant gas, nitrogen?

47. There is a tiny leak in a system containing neon gas at 22°C and 760 mm Hg. After one minute, the pressure of Ne drops to 749 mm Hg. The neon is replaced by helium, again at 22°C and 760 mm Hg. What would you expect the pressure of helium to be after one minute?

48. It takes 32.0 s for ammonia to effuse down a capillary tube. How long will it take hydrogen chloride to effuse down an identical capillary tube at the same conditions of temperature and pressure?

49. Consider the sulfur tetrafluoride gas molecule.

a. At what temperature will it have half the average speed it has at 25°C?

b. If the molecule being considered were fluorine, would the answer to part (a) change? How?

50. The average speed of a fluorine molecule at 15°C is 435 m/s. What is the average speed of a xenon tetrafluoride gas molecule at the same temperature?

51. Taking the average speed of an oxygen gas molecule at 25°C to be 482 m/s, calculate the average speed of

a. a Cl_2 molecule at 25°C.

b. an argon atom at −25°C.

52. A professional tennis player can serve a tennis ball traveling at 45 m/s. At 25°C, the average speed of a nitrogen molecule is 515 m/s. At what temperature will a nitrogen molecule have the same average speed as the tennis ball?

Real Gases

53. The normal boiling points of CO and SO_2 are −192°C and −10°C, respectively.

a. At 25°C and 1 atm, which gas would you expect to have a molar volume closest to the ideal value?

b. If you wanted to reduce the deviation from ideal gas behavior, in what direction would you change the temperature? The pressure?

54. A sample of $CH_4(g)$ is at 50°C and 20 atm. Would you expect it to behave more ideally or less ideally if
   a. the pressure were reduced to 1 atm?
   b. the temperature were reduced to −50°C?

55. Using Figure 5.7, estimate the density of methane gas at 200 atm and 25°C and compare to the value calculated from the ideal gas law.

56. Calculate the pressure of one mole of oxygen gas in a 125-mL container at 0°C using the
   a. ideal gas law.
   b. van der Waals equation (see Table 5.2).

Greenhouse Effect

57. Describe the role of CO_2 and H_2O in the greenhouse effect.

58. Assume that an automobile burns octane, C_8H_{18} (d = 0.692 g/cm^3).
   a. Write the equation for the combustion of octane to CO_2 and H_2O.
   b. A car has a fuel efficiency rating of 32 miles/gallon of octane. What volume of CO_2 at 25°C and 1.00 atm is generated when that car goes on a ten-mile trip?

59. Follow the directions of Problem 58 using methanol, CH_3OH (d = 0.791 g/cm^3) as the fuel being burned by a car with the same fuel efficiency rating.

60. What are some of the consequences, both favorable and unfavorable, of an increase in average global temperatures?

Unclassified

61. The pressure exerted by a column of liquid is directly proportional to its density and to its height. If you wanted to fill a barometer with ethyl alcohol (C_2H_5OH, d = 0.789 g/cm^3), how long a tube should you use (d Hg = 13.6 g/cm^3)?

62. For an ideal gas, sketch graphs of
   a. V vs. P at constant T, n.
   b. P vs. n at constant V, T.
   c. u vs. T at constant P, V.
   d. E_{trans} vs. P at constant T, n.

63. For an ideal gas, sketch graphs of
   a. V vs. T at constant P, n.
   b. P vs. T at constant V, n.
   c. n vs. T at constant P, V.
   d. E_{trans} vs. T at constant P, n.

64. A mixture of 3.5 mol neon and 3.9 mol chlorine gas occupies a 5.00-L container at 27°C. Which gas has the larger
   a. average translational energy?
   b. partial pressure?
   c. mole fraction?
   d. effusion rate?

65. Given that 1.00 mol nitrogen and 1.00 mol fluorine gas are in separate containers at the same temperature and pressure, calculate each of the following ratios.
   a. volume F_2/volume N_2
   b. density F_2/density N_2
   c. average translational energy F_2/average translational energy N_2
   d. number of atoms F/number of atoms N

66. A Porsche 928 S4 engine has a cylinder volume of 618 cm^3. The cylinder is full of air at 75°C and 1.00 atm.
   a. How many moles of oxygen are in the cylinder? (Mole percent of oxygen in air = 21.0.)
   b. Assume that the hydrocarbons in gasoline have an average molar mass of 1.0×10^2 g/mol and react with oxygen in a 1:12 mole ratio. How many grams of gasoline should be injected into the cylinder to react with the oxygen?

67. Gasoline is a mixture of hydrocarbons of which octane, C_8H_{18}, is typical. The combustion of octane is represented by the equation

$$2C_8H_{18}(l) + 25\ O_2(g) \longrightarrow 16CO_2(g) + 18H_2O(l)$$

What volume of air (21.0% by volume oxygen) is required at 25°C and 1.00 atm for the combustion of one gallon of octane (d = 0.692 g/cm^3)?

68. An intermediate reaction used in the production of nitrogen-containing fertilizers is that between ammonia and oxygen:

$$4NH_3(g) + 5\ O_2(g) \longrightarrow 4NO(g) + 6H_2O(g)$$

A 150.0-L reaction chamber is charged with reactants to the following partial pressures at 500°C: P_{NH_3} = 1.3 atm; P_{O_2} = 0.80 atm. What is the theoretical yield of NO in grams?

69. Glycine is an amino acid made up of carbon, hydrogen, oxygen, and nitrogen atoms. Combustion of a 0.2036-g sample gives 132.9 mL of CO_2 at 25°C and 1.00 atm and 0.122 g of water. What are the percentages of carbon and hydrogen in glycine? Another sample of glycine weighing 0.2500 g is treated in such a way that all the nitrogen atoms are converted to $N_2(g)$; this gas has a volume of 40.8 mL at 25°C and 1.00 atm. What is the percentage of nitrogen in glycine? The percentage of oxygen? What is the empirical formula of glycine?

70. At 25°C and 380 mm Hg, the density of sulfur dioxide is 1.31 g/L. The rate of effusion of sulfur dioxide through an orifice is 4.48 mL/s. What is the density of a sample of gas that effuses through an identical orifice at the rate of 6.78 mL/s under the same conditions? What is the molar mass of the gas?

71. A research laboratory has two steel cylinders of equal volume; they are at the same temperature. Cylinder I contains carbon dioxide gas, while cylinder II contains nitrogen.
   a. If each cylinder contains one kilogram of gas, which cylinder has the larger pressure?
   b. Suppose cylinder I has a pressure of 7.0 atm at 50°C, while cylinder II has a pressure of 5.5 atm at 10°C.

Which cylinder contains the larger number of molecules?

Challenge Problems

72. A tube 3.0 ft long is originally filled with air. Samples of NH_3 and HCl, at the same temperature and pressure, are introduced simultaneously at opposite ends of the tube. When the two gases meet, a white ring of $NH_4Cl(s)$ forms. How far from the end at which ammonia was introduced will the ring form?

73. The Rankine temperature scale resembles the Kelvin scale in that 0° is taken to be the lowest attainable temperature (0°R = 0 K). However, the Rankine degree is the same size as the Fahrenheit degree, whereas the Kelvin degree is the same size as the Celsius degree. What is the value of the gas law constant in L · atm/(mol · °R)?

74. A 0.2500-g sample of an Al–Zn alloy reacts with HCl to form hydrogen gas:

$$Al(s) + 3H^+(aq) \longrightarrow Al^{3+}(aq) + \tfrac{3}{2}H_2(g)$$

$$Zn(s) + 2H^+(aq) \longrightarrow Zn^{2+}(aq) + H_2(g)$$

The hydrogen produced has a volume of 0.153 L at 25°C and 1.00 atm. What is the mass percent of aluminum in the alloy?

75. The buoyant force on a balloon is equal to the mass of air it displaces. The gravitational force on the balloon is equal to the sum of the masses of the balloon, the gas it contains, and the balloonist. If the balloon and balloonist together weigh 175 kg, what would the diameter of a spherical hydrogen-filled balloon have to be in meters if the rig is to get off the ground at 22°C and 752 mm Hg? (Take $\mathcal{M}_{air}$ = 29.0 g/mol.)

76. A mixture in which the mole ratio of hydrogen to oxygen is 2:1 is used to prepare water by the reaction

$$2H_2(g) + O_2(g) \longrightarrow 2H_2O(g)$$

The total pressure in the container is 0.820 atm at 25°C before the reaction. What is the final pressure in the container at 125°C after reaction, assuming a 75.0% yield and no volume change?

77. The volume fraction of a gas, A, in a mixture is defined by the equation:

$$\text{volume fraction A} = V_A/V$$

where V is the total volume and V_A is the volume that gas A would occupy alone at the same temperature and pressure. Show that, assuming ideal gas behavior, the volume fraction is the same as the mole fraction. Explain why the volume fraction differs from the mass fraction.

6
Electronic Structure and the Periodic Table

Not chaos-like crush'd
and bruis'd,
But, as the world,
harmoniously confus'd,
Where order in variety we see,
And where, though all
things differ,
all agree.

ALEXANDER POPE
Windsor Forest

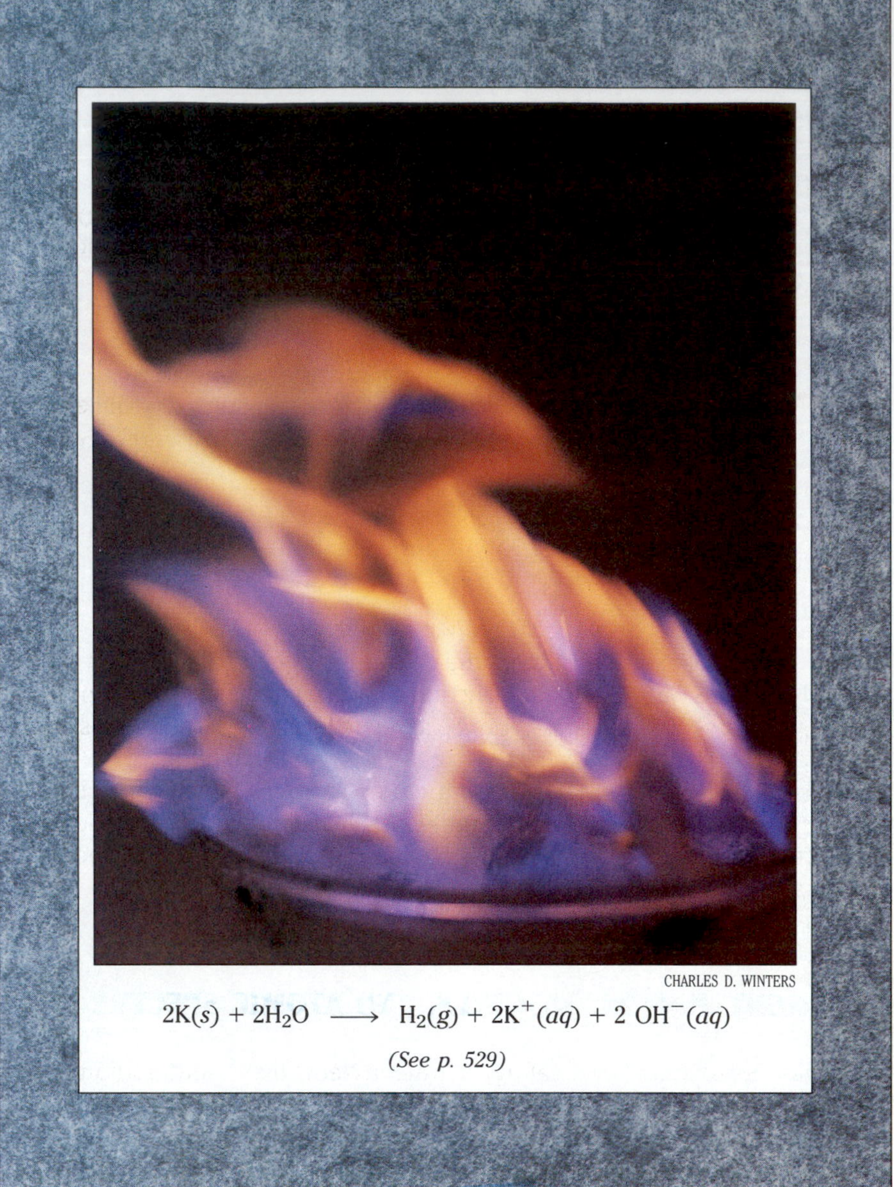

CHARLES D. WINTERS

$$2K(s) + 2H_2O \longrightarrow H_2(g) + 2K^+(aq) + 2\ OH^-(aq)$$

(See p. 529)

CHAPTER OUTLINE

In Chapter 2, we briefly considered the structure of the atom. You will recall that every atom has a tiny, positively charged nucleus, made up of protons and neutrons. The nucleus is surrounded by negatively charged electrons. The number of protons in the nucleus is characteristic of the atoms of a particular element and is referred to as the atomic number. In a neutral atom, the number of electrons is equal to the number of protons and hence to the atomic number.

In this chapter, we focus upon electron arrangements in atoms, paying particular attention to the relative energies of different electrons *(energy levels)* and their spatial locations *(orbitals)*. Specifically, we consider the nature of the energy levels and orbitals available to

Chemical properties of atoms and molecules depend on their electron arrangements

— the single electron in the hydrogen atom (Section 6.2)
— the several electrons in more complex atoms (Section 6.3)

With this background, we show how electron arrangements in multi-electron atoms and the monatomic ions derived from them can be described in terms of

— *electron configurations,* which show the number of electrons in each energy level (Sections 6.4, 6.6)
— *orbital diagrams,* which show the arrangement of electrons within orbitals (Sections 6.5, 6.6)

The electron configuration or orbital diagram of an atom of an element can be deduced from its position in the periodic table. Beyond that, position in the table can be used to predict (Section 6.7) the relative sizes of atoms and ions *(atomic radius, ionic radius),* and the relative tendencies of atoms to give up or acquire electrons *(ionization energy, electronegativity).*

Before dealing with electronic structures as such, it will be helpful, in Section 6.1, to examine briefly the experimental evidence upon which such structures are based. In particular, we need to look at the phenomenon of *atomic spectra.*

6.1 LIGHT, PHOTON ENERGIES, AND ATOMIC SPECTRA

Fireworks displays are fascinating to watch. Neon lights and sodium vapor lamps can transform the skyline of a city with their brilliant colors. The eerie phenomenon of the aurora borealis is an unforgettable experience when you see it for the first time. All of these events relate to the generation of light and its transmission through space.

The Wave Nature of Light; Wavelength and Frequency

Light travels through space as a wave, which is made up of successive crests, which rise above the midline, and troughs, which sink below it. Waves have three primary characteristics (Fig. 6.1), two of which are of particular interest at this point:

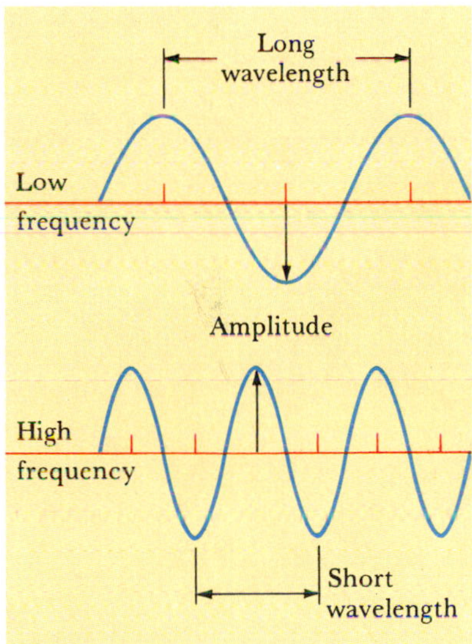

Figure 6.1
Three characteristics of a wave are its amplitude, wavelength, and frequency. The *amplitude* (ψ) is the height of a crest or the depth of a trough. The *wavelength* (λ) is the distance between successive crests or troughs. The *frequency* (ν) is the number of wave cycles (successive crests or troughs) that pass a given point in a given time.

1. Wavelength (λ), the distance between two consecutive crests or troughs, most often measured in meters or nanometers (1 nm = 10^{-9} m).

2. Frequency (ν), the number of wave cycles (successive crests or troughs) that pass a given point in unit time. If 10^8 cycles pass a particular point in one second,

$$\nu = 10^8/\text{s} = 10^8 \text{ Hz}$$

The frequency unit *hertz* (Hz) represents one cycle per second.

The colored lights are generated by Sr, Cu and Na atoms. (Richard Megna/FUNDAMENTAL PHOTOGRAPHS, New York.)

The speed at which a wave moves through space can be found by multiplying the length of a wave cycle (λ) by the number of cycles passing a point in unit time (ν). For light,

$$\lambda\nu = c$$

where c, the speed of light, is a constant, 2.998×10^8 m/s.

Example 6.1 The green light associated with the aurora borealis is emitted by excited (high-energy) oxygen atoms at 557.7 nm. What is the frequency of this light?

Strategy Use the equation: $\lambda\nu = c$, taking $c = 2.998 \times 10^8$ m/s. Note that λ must be expressed in meters.

Solution

$$\nu = \frac{2.998 \times 10^8 \text{ m/s}}{557.7 \times 10^{-9} \text{ m}} = 5.376 \times 10^{14}/\text{s} = 5.376 \times 10^{14} \text{ Hz}$$

Red light (~700 nm) has a longer wavelength than violet (~400 nm)

Light visible to the eye is a tiny portion of the entire electromagnetic spectrum (Fig. 6.2), covering only the narrow wavelength region 400–700 nm. You can see the red glow given off by charcoal on a barbecue grill, but the heat given off is largely in the infrared (IR) region, above 700 nm. Microwave ovens produce radiation at even longer wavelengths. At the other end of the spectrum, below 400 nm, are ultraviolet (UV) radiation, responsible for sunburn, and x-rays, used for diagnostic purposes in medicine.

The Particle Nature of Light; Photon Energies

Einstein developed the quantum theory of radiation, first proposed by Planck

A hundred years ago it was generally supposed that all of the properties of light could be explained in terms of its wave nature. A series of investigations carried out between 1900 and 1910 by Max Planck (blackbody radiation) and Albert Einstein (photoelectric effect) shot down that notion. Today we consider light to be generated as a stream of particles called **photons**, whose energy E is given by the Einstein equation:

$$E = h\nu = hc/\lambda$$

The quantity h appearing in this equation is referred to as Planck's constant.

$$h = 6.626 \times 10^{-34} \text{ J} \cdot \text{s}$$

Notice from this equation that energy is *inversely* related to wavelength. This explains why you put on a sunscreen to protect yourself from UV solar radiation (<400 nm) and a "lead apron" when dental x-rays are being taken. Conversely, IR (>700 nm) and microwave radiation are of relatively low energy (but don't try walking on hot coals).

Example 6.2 Referring back to Example 6.1, calculate

(a) the energy, in joules, of a photon emitted by an excited oxygen atom.
(b) the energy, in kilojoules, of a mole of such photons.

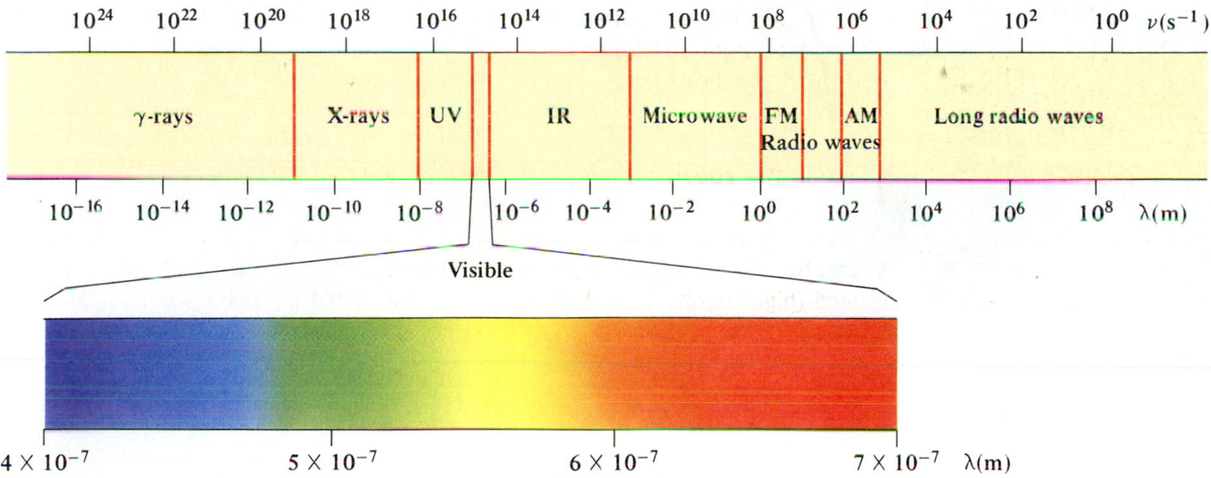

Figure 6.2
The electromagnetic spectrum. Note that only a small fraction is visible to the human eye.

Strategy Use the equation $E = hc/\lambda$. In part (b), remember that 1 mol = 6.022×10^{23} particles.

Solution

$$\text{(a) } E = \frac{hc}{\lambda} = \frac{(6.626 \times 10^{-34}\ \text{J}\cdot\text{s})(2.998 \times 10^{8}\ \text{m/s})}{557.7 \times 10^{-9}\ \text{m}} = 3.562 \times 10^{-19}\ \text{J}$$

This seems like a tiny amount of energy, but remember that it's coming from a single oxygen atom.

$$\text{(b) } E = 3.562 \times 10^{-19}\ \text{J} \times \frac{1\ \text{kJ}}{10^{3}\ \text{J}} \times \frac{6.022 \times 10^{23}}{1\ \text{mol}} = 2.145 \times 10^{2}\ \text{kJ/mol}$$

This is roughly comparable to energy effects in chemical reactions; about 240 kJ of heat is evolved when a mole of hydrogen burns.

Atomic Spectra

In the seventeenth century, Sir Isaac Newton showed that visible (white) light from the sun can be broken down into its various color components by a prism. The *spectrum* obtained is continuous; it contains essentially all wavelengths between 400 and 700 nm. The situation with high-energy atoms of gaseous elements is quite different (Fig. 6.3). Here, the spectrum consists of discrete lines given off at specific wavelengths. Each element has a characteristic spectrum which can be used to identify it. In the case of sodium, there are two strong lines in the yellow region at 589.0 nm and 589.6 nm. These lines account for the yellow color of sodium vapor lamps used to illuminate highways.

One of the simplest of atomic spectra, and the most important from a theoretical standpoint, is that of hydrogen. When energized by a high-voltage discharge, gaseous hydrogen atoms emit radiation at wavelengths which can

Atomic spectroscopy is used to identify metals at concentrations as low as 10^{-7} mol/L

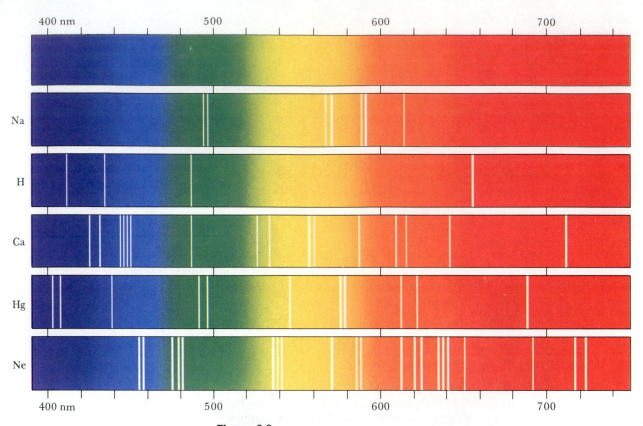

Figure 6.3
Emission spectra. The continuous spectrum at the top contains all visible wavelengths. The atomic spectra of gaseous Na, H, Ca, Hg, and Ne are quite different. They consist of discrete lines at certain definite wavelengths.

be grouped into several different series (Table 6.1). The first of these to be discovered, the Balmer series, lies in the visible region. It consists of a strong line at 656.28 nm followed by successively weaker lines, closer and closer together, at lower wavelengths.

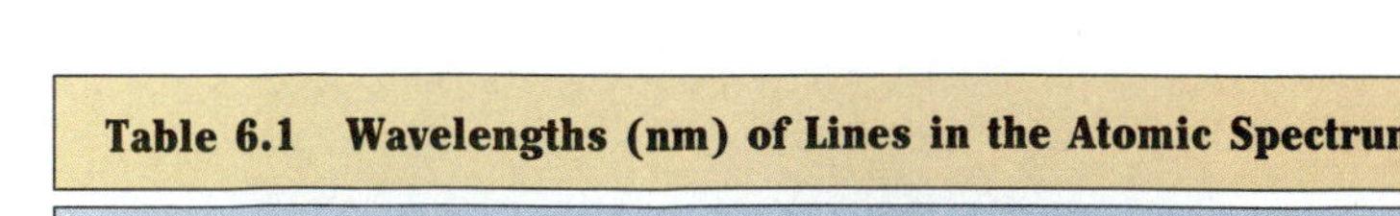
Table 6.1 Wavelengths (nm) of Lines in the Atomic Spectrum of Hydrogen

Ultraviolet (Lyman Series)	Visible (Balmer Series)	Infrared (Paschen Series)
121.53	656.28	1875.09
102.54	486.13	1281.80
97.23	434.05	1093.80
94.95	410.18	1004.93
93.75	397.01	
93.05		

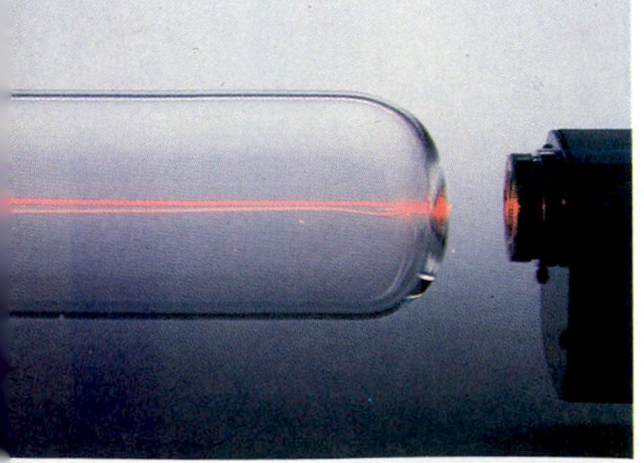
The He–Ne laser gives off orange-red light at 640.2 nm, corresponding to an electronic transition in the neon atom. (Marna G. Clarke)

6.2 THE HYDROGEN ATOM

The hydrogen atom, containing a single electron, has played a major role in the development of models of electronic structure. In 1913, Niels Bohr, a Danish physicist, offered a theoretical explanation of the atomic spectrum of hydrogen; his model was based largely upon classical mechanics. A decade later, a more sophisticated model of the hydrogen atom was developed, using the new discipline of quantum mechanics.

Bohr Model

Bohr assumed that a hydrogen atom consists of a central proton about which an electron moves in a circular orbit. He related the electrostatic force of attraction of the proton for the electron to the centrifugal force due to the circular motion of the electron. In this way, Bohr was able to express the energy of the atom in terms of the radius of the electron's orbit. To this point, his analysis was purely classical, based on Coulomb's law of electrostatic attraction and Newton's laws of motion. To progress beyond this point, Bohr boldly and arbitrarily assumed, in effect, that the electron in the hydrogen atom can have only certain definite energies. Using arguments that we will not go into, Bohr obtained the following equation for the energy of the hydrogen electron:

$$E_{\mathbf{n}} = -R_H/\mathbf{n}^2$$

where $E_{\mathbf{n}}$ is the energy of the electron, R_H is a quantity called the Rydberg constant (modern value = 2.180×10^{-18} J), and **n** is an integer called the principal quantum number. Depending upon the state of the electron, **n** can have any positive, integral value, i.e.,

$$\mathbf{n} = 1, 2, 3, \ldots$$

Before proceeding with the Bohr model, let us make three points:

Bohr, a giant of 20th century physics, was respected by scientists and politicians alike

1. In setting up his model, Bohr designated zero energy as the point at which the proton and electron are completely separated. Energy has to be absorbed to reach that point. This means that the electron, in all its allowed energy states within the atom, must have an energy below zero, i.e., must be negative, hence the minus sign in the equation:

$$E_{\mathbf{n}} = -R_H/\mathbf{n}^2$$

2. Ordinarily the hydrogen electron is in its lowest state, referred to as the **ground state** or ground level, for which $\mathbf{n} = 1$. When an electron absorbs enough energy, it moves to a higher, **excited state.** In a hydrogen atom, the first excited state has $\mathbf{n} = 2$, the second $\mathbf{n} = 3$, and so on.

3. When an excited electron gives off energy as a photon of light, it drops back to a lower energy state. The electron can return to the ground state, (from $\mathbf{n} = 2$ to $\mathbf{n} = 1$, for example) or to a lower excited state (from $\mathbf{n} = 3$ to $\mathbf{n} = 2$). At any rate, the energy of the photon ($h\nu$) evolved is equal to the difference in energy between the two states:

In the Balmer series, $\mathbf{n}_{lo} = 2$; $\mathbf{n}_{hi} = 3, 4, 5, \ldots$

$$\Delta E = h\nu = E_{hi} - E_{lo}$$

where E_{hi} and E_{lo} are the energies of the higher and lower states, respectively.

Using this expression for ΔE and the equation $E_{\mathbf{n}} = -R_H/\mathbf{n}^2$, it is possible to relate the frequency of the light emitted to the quantum numbers, $\mathbf{n}_{hi}$ and $\mathbf{n}_{lo}$, of the two states:

$$h\nu = -R_H\left[\frac{1}{(\mathbf{n}_{hi})^2} - \frac{1}{(\mathbf{n}_{lo})^2}\right]$$

$$\nu = \frac{R_H}{h}\left[\frac{1}{(\mathbf{n}_{lo})^2} - \frac{1}{(\mathbf{n}_{hi})^2}\right]$$

The last equation written is the one Bohr derived in applying his model to the hydrogen atom. Given

$$R_H = 2.180 \times 10^{-18}\ \text{J} \qquad h = 6.626 \times 10^{-34}\ \text{J} \cdot \text{s}$$

you can use the equation to find the frequency or wavelength of any of the lines in the hydrogen spectrum.

Balmer was a Swiss high school teacher

Example 6.3 Calculate the wavelength in nanometers of the line in the Balmer series that results from the transition $\mathbf{n} = 3$ to $\mathbf{n} = 2$.

Strategy Use the above equation to find the frequency; $\mathbf{n}_{lo} = 2$, $\mathbf{n}_{hi} = 3$. Then use the equation $\lambda = c/\nu$ to find the wavelength.

Solution

$$\nu = \frac{2.180 \times 10^{-18}\ \text{J}}{6.626 \times 10^{-34}\ \text{J} \cdot \text{s}}\left[\frac{1}{4} - \frac{1}{9}\right] = 4.570 \times 10^{14}/\text{s}$$

$$\lambda = \frac{c}{\nu} = \frac{2.998 \times 10^{8}\ \text{m/s}}{4.570 \times 10^{14}/\text{s}} \times \frac{10^9\ \text{nm}}{1\ \text{m}} = 656.0\ \text{nm}$$

Compare this value to that listed in Table 6.1 for the first line in the Balmer series.

Quantum Mechanical Model

Bohr's theory for the structure of the hydrogen atom was highly successful. Scientists of the day must have thought they were on the verge of being able to predict the allowed energy levels of all atoms. However, the extension of Bohr's ideas to atoms with two or more electrons gave, at best, only qualitative agreement with experiment. Consider, for example, what happens when Bohr's theory is applied to the helium atom. Here, the errors in calculated energies and wavelengths are of the order of 5% instead of the 0.1% error with hydrogen. There appeared to be no way the theory could be modified to make it work well with helium or other atoms. Indeed, it soon became apparent that there was a fundamental problem with the Bohr model. The idea of an electron moving about the nucleus in a well-defined orbit at a fixed distance from the nucleus had to be abandoned.

Too bad in a way; it was a simple model, readily visualized

Scientists in the 1920s, speculating on this problem, became convinced that an entirely new approach was required to treat electrons in atoms, ions, and molecules. A new discipline, called *quantum mechanics*, was developed to describe the motion of small particles confined to tiny regions of space. In the quantum mechanical atom, no attempt is made to specify the position of an electron at a given instant, nor does quantum mechanics concern itself with the path that an electron takes about the nucleus. (After all, if we can't say where the electron is, we certainly don't know how it got there). Instead, quantum mechanics deals only with the *probability* of finding a particle within a given region of space.

In 1926, Erwin Schrödinger, an Austrian physicist working in Zurich, made a major contribution to quantum mechanics. He wrote down an equation from which one could calculate the amplitude (height) ψ of the electron wave at various points in space. Although we will not use the Schrödinger wave equation in any calculations, we write it down so you will have some idea of its nature. For a single particle of mass m moving in only one dimension (x), the equation has the form

Sort of reassuring

$$\frac{d^2\psi}{dx^2} + \frac{8\pi^2 m}{h^2}(E - V)\psi = 0$$

where E and V are the total and potential energies, respectively, of the particle, and h is Planck's constant. The quantity ψ, commonly called a *wave function*, can be interpreted as the amplitude of the wave associated with the particle. The first term in the equation is the second derivative of ψ with respect to x; for motion in three dimensions, similar terms in y and z must be added.

For the hydrogen electron, the square of the wave function, ψ^2, is directly proportional to the probability of finding the electron at a particular point. If ψ^2 at point A is twice as large as at point B, then we are twice as likely to find the electron at A as at B. Putting it another way, over time the electron will turn up at A twice as often as at B.

Figure 6.4 shows, in two different ways, how ψ^2 for the hydrogen electron in its ground state (**n** = 1) varies moving out from the nucleus. In Figure 6.4a, an *electron cloud* diagram, the depth of the color is directly proportional to ψ^2 and hence to the probability of finding the electron at a point. As you can see, the color fades moving out from the nucleus in any direction; the value of ψ^2 drops accordingly. This feature is emphasized in Figure 6.4b, where ψ_x^2 is plotted against distance along the x-axis. Notice that ψ_x^2 has its greatest value at the nucleus ($x = 0$) and decreases smoothly and rapidly as x increases.

Seems strange, but the electron is more likely to be found at the nucleus than at any other point

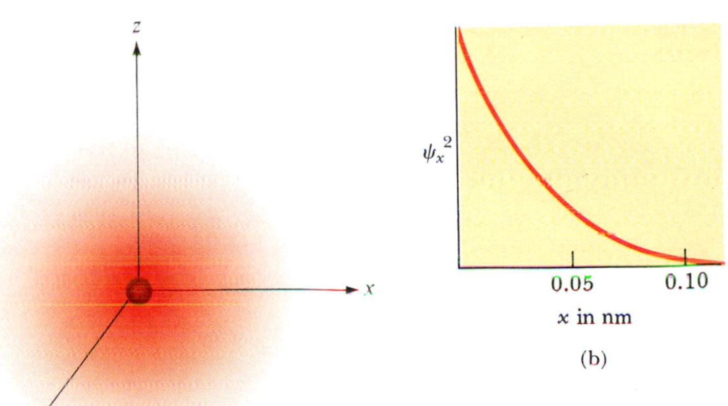

Figure 6.4
Two different ways of showing the relative probability of finding the hydrogen electron in its ground state at various distances from the nucleus. In (a), the depth of color is proportional to ψ^2. In (b), the probability, ψ_x^2, of finding the electron at a given distance along the x-axis is plotted.

6.3 QUANTUM NUMBERS, ENERGY LEVELS, AND ORBITALS

Although the Schrödinger equation is a rather complex one, it can be solved exactly for the hydrogen atom. Indeed, it turns out that there are many solutions for ψ, each associated with a set of numbers called **quantum numbers**. There are three such numbers, given the symbols **n**, ℓ, and $\mathbf{m}_\ell$. A wave function corresponding to a particular set of three quantum numbers (e.g., $\mathbf{n} = 2, \ell = 1, \mathbf{m}_\ell = 0$) is referred to as an **atomic orbital**. Orbitals differ from one another in their energy and in the shape and spatial orientation of their electron cloud.

Quantum numbers serve to classify electrons

For reasons we will discuss later, a fourth quantum number is required to completely describe a specific electron in a multi-electron atom. The fourth quantum number is given the symbol $\mathbf{m}_s$. Each electron in an atom has a set of four quantum numbers, **n**, ℓ, $\mathbf{m}_\ell$, and $\mathbf{m}_s$. We will now discuss the quantum numbers of electrons as they are used in atoms beyond hydrogen.

First Quantum Number, n; Principal Energy Levels

The first quantum number, **n**, comes from the Bohr model of the hydrogen atom, where the energy depends only upon **n** ($E_\mathbf{n} = -R_H/\mathbf{n}^2$). In other atoms, the energy of each electron depends mainly, but not completely, upon the value of **n**. As **n** increases, the energy of the electron increases and, on the average, it is found farther out from the nucleus. The quantum number **n** can take on only integral values, starting with 1:

$$\mathbf{n} = 1, 2, 3, 4, \ldots$$

In an atom, the value of **n** designates what we call a **principal energy level**. Thus, an electron for which $\mathbf{n} = 1$ is said to be in the first principal level. If $\mathbf{n} = 2$, we are dealing with the second principal level, and so on.

Second Quantum Number, ℓ; Sublevels (s, p, d, f)

Each principal level includes one or more **sublevels**. The sublevels are denoted by the second quantum number, ℓ. As we shall see later, the general shape of the electron cloud associated with an orbital is determined by ℓ. Larger values of ℓ produce more complex shapes.

The quantum numbers **n** and ℓ are related; ℓ can take on any integral value starting with 0 and going up to a maximum of $(\mathbf{n} - 1)$. That is,

This relation between ℓ and **n** comes from the Schrödinger equation for the H atom

$$\ell = 0, 1, 2, \ldots, (\mathbf{n} - 1)$$

If $\mathbf{n} = 1$, there is only one possible value of ℓ, namely 0. This means that, in the first principal level, there is only one sublevel, for which $\ell = 0$. If $\mathbf{n} = 2$, two values of ℓ are possible, 0 and 1. In other words, there are two sublevels ($\ell = 0$ and $\ell = 1$) within the second principal energy level. Similarly,

if $\mathbf{n} = 3$: $\ell = 0, 1,$ or 2 (three sublevels)

if $\mathbf{n} = 4$: $\ell = 0, 1, 2,$ or 3 (four sublevels)

The number of sublevels is one greater than the maximum value of ℓ

In general, *in the nth principal level, there are n different sublevels.*

Another method is commonly used to designate sublevels. Instead of giving the quantum number ℓ, the letters s, p, d, or f* indicate the sublevels $\ell = 0$, 1, 2, or 3, respectively. That is,

* These letters come from the adjectives used by spectroscopists to describe spectral lines: *s*harp, *p*rincipal, *d*iffuse and *f*undamental.

quantum number, ℓ	0	1	2	3
type of sublevel	s	p	d	f

Usually, in designating a sublevel, a number is included to indicate the principal level as well. Thus reference is made to a 1s sublevel ($\mathbf{n} = 1, \ell = 0$), a 2s sublevel ($\mathbf{n} = 2, \ell = 0$), a 2p sublevel ($\mathbf{n} = 2, \ell = 1$), and so on.

For the hydrogen atom, the energy of an electron is independent of the value of ℓ. (This is implied by the equation $E = -R_H/\mathbf{n}^2$, which involves only one quantum number). For multi-electron atoms, the situation is quite different. Here, the energy is dependent on ℓ as well as on $\mathbf{n}$. Within a given principal level (same value of $\mathbf{n}$), sublevels increase in energy in the order

$$\mathbf{n}s < \mathbf{n}p < \mathbf{n}d < \mathbf{n}f$$

Thus a 2p sublevel has a slightly higher energy than a 2s sublevel. By the same token, when $\mathbf{n} = 3$, the 3s sublevel has the lowest energy, the 3p is intermediate, and the 3d has the highest energy.

Which would have the higher energy, 4p or 3s?

Third Quantum Number, $\mathbf{m}_\ell$; Orbitals

Each sublevel contains one or more **orbitals**, which differ from one another in the value assigned to the third quantum number, $\mathbf{m}_\ell$. This quantum number determines the direction in space of the electron cloud surrounding the nucleus. The value of $\mathbf{m}_\ell$ is related to that of ℓ. For a given value of ℓ, $\mathbf{m}_\ell$ can have any integral value, including 0, between ℓ and $-\ell$; that is,

Electrons occupy orbitals within a sublevel within a principal level

$$\mathbf{m}_\ell = \ell, \ldots, +1, 0, -1, \ldots, -\ell$$

To illustrate how this rule works, consider an s sublevel ($\ell = 0$). Here $\mathbf{m}_\ell$ can have only one value, 0. This means that an s sublevel contains only one orbital, referred to as an *s orbital*. The electron cloud associated with an s orbital is spherically symmetrical; the density of the cloud varies with distance from the nucleus but is independent of direction. Most commonly, an s orbital is shown as a simple sphere (Fig. 6.5a). The radius of the sphere indicates the region within which there is a specified probability of finding the electron. If we agree that the sphere should contain 90% of the density of the electron cloud, then a 1s orbital in a hydrogen atom has a radius of 0.14 nm. As the value of $\mathbf{n}$ increases, the radius of the orbital becomes larger. This means that an electron in a 2s orbital is more likely to be found far out from the nucleus than one in a 1s orbital.

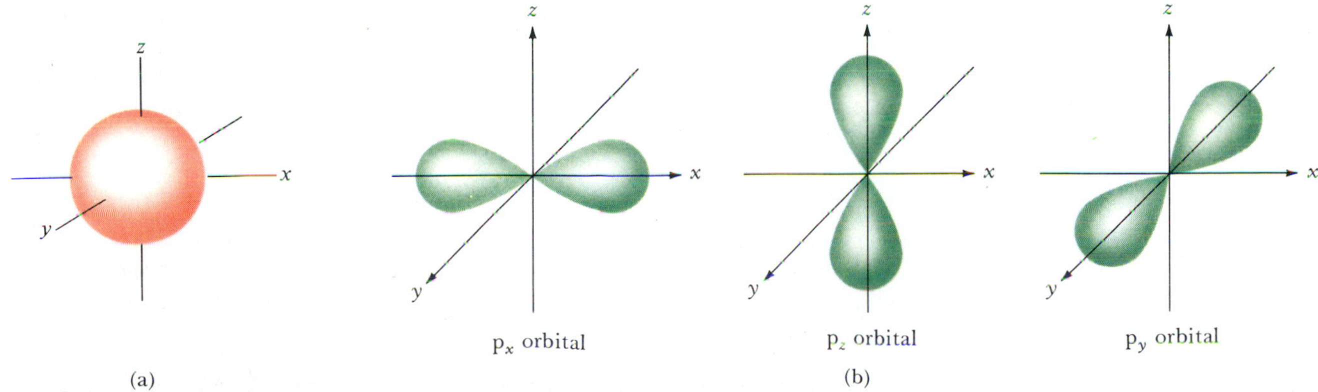

Figure 6.5
All s orbitals (a) are spherically symmetrical; the density of the electron cloud is independent of direction (x, y or z). In contrast, the electron density in p orbitals (b) is concentrated along the x, y or z axes. The three p orbitals are directed at 90° angles to each other.

For a p sublevel ($\ell = 1$), $m_\ell = 1, 0,$ or -1. Within a given p sublevel, there are three different orbitals described by the quantum numbers $m_\ell = 1, 0,$ and -1. Commonly, p orbitals are referred to as p_x, p_y, and p_z orbitals (Fig. 6.5b).

For the d and f sublevels:

d sublevel:	$\ell = 2$	$m_\ell = 2, 1, 0, -1, -2$	5 orbitals
f sublevel:	$\ell = 3$	$m_\ell = 3, 2, 1, 0, -1, -2, -3$	7 orbitals

In general, for a sublevel of quantum number ℓ, there are a total of $2\ell + 1$ orbitals. All of the orbitals in a given sublevel (e.g.. $2p_x$, $2p_y$, $2p_z$) have essentially the same energy.

Fourth Quantum Number, m_s; Electron Spin

The fourth quantum number, m_s, is associated with the spin of the electron. An electron has magnetic properties that correspond to those of a charged particle spinning on its axis. Either of two spins are possible, clockwise or counterclockwise.

The quantum number m_s was introduced to make theory consistent with experiment. In that sense, it differs from the first three quantum numbers, which came from the solution to the Schrödinger wave equation for the hydrogen atom. This quantum number is not related to n, ℓ, or m_ℓ. It can have either of two possible values:

$$m_s = +1/2 \qquad \text{or} \qquad -1/2$$

Electrons that have the same value of m_s (i.e., both $+\frac{1}{2}$ or both $-\frac{1}{2}$) are said to have *parallel* spins. Electrons that have different m_s values (i.e., one $+\frac{1}{2}$ and the other $-\frac{1}{2}$) are said to have *opposed* spins.

Pauli Exclusion Principle

The four quantum numbers that characterize an electron in an atom have now been considered. There is an important rule, called the Pauli exclusion principle, that relates to these numbers. It requires that *no two electrons in an atom can have the same set of four quantum numbers.* This principle was first stated in 1925 by Wolfgang Pauli, a colleague of Bohr, again to make theory consistent with the properties of atoms.

Our model for electronic structure is a pragmatic blend of theory and experiment

The Pauli exclusion principle has an implication that is not obvious at first glance. It requires that no more than two electrons can fit into an orbital. Moreover, if two electrons occupy the same orbital they must have opposed spins. To see that this is the case, consider the 2s orbital. Any electron in this orbital must have

$$n = 2 \qquad \ell = 0 \qquad m_\ell = 0$$

To satisfy the Pauli exclusion principle, the electrons in this orbital must have different m_s values. But there are only two possible values of m_s. Hence only two electrons can enter that orbital. If the orbital is filled, one electron must have $m_s = +\frac{1}{2}$, and the other $m_s = -\frac{1}{2}$. In other words, the two electrons must have opposed spins.

Capacities of Principal Levels, Sublevels, and Orbitals

The Pauli exclusion principle and the rules for quantum numbers fix the capacities of principal levels, sublevels, and orbitals. These rules, applied to the first three principal energy levels, are summarized in Table 6.2. At the bottom of the table, each arrow indicates an electron. In each orbital, there are two electrons of opposed spin.

This table summarizes and illustrates everything we've said in this section

---

Example 6.4

(a) How many electrons can fit into the principal level for which $\mathbf{n} = 4$?
(b) What is the capacity for electrons of the 3d sublevel?

Strategy Recall that

— each principal level of quantum number **n** contains *n* sublevels
— an s sublevel can hold $2e^-$, a p sublevel $6e^-$, a d sublevel $10e^-$, and an f sublevel $14e^-$.

Solution

(a) The sublevels are 4s ($2e^-$), 4p ($6e^-$), 4d ($10e^-$), and 4f ($14e^-$). The total capacity adds to $32e^-$, which is indeed $2\mathbf{n}^2$: $2(4)^2 =$ 32.

(b) Like all d sublevels, the 3d has a capacity of $10e^-$. This is equal to $4\ell + 2$: $4(2) + 2 =$ 10.

Table 6.2 Summary of Principal Level, Sublevel, and Orbital Capacities

Level n Capacity: $2\mathbf{n}^2$	1 $2e^-$	2 $8e^-$				3 $18e^-$								
Sublevel														
No. of sublevels = **n**	1 sublevel	2 sublevels				3 sublevels								
Quantum no. ℓ	0	0		1		0		1				2		
Letter designation	s	s		p		s		p				d		
Capacity $(4\ell + 2)$	$2e^-$	$2e^-$		$6e^-$		$2e^-$		$6e^-$				$10e^-$		
Orbital														
No. of orbitals $(2\ell + 1)$	1 orbital	1 orbital		3 orbitals		1 orbital		3 orbitals				5 orbitals		
Quantum no. $\mathbf{m}_\ell$	0	0	+1	0	−1	0	+1	0	−1	+2	+1	0	−1	−2
Capacity	$2e^-$	$2e^-$	$2e^-$	$2e^-$	$2e^-$	$2e^-$	$2e^-$	$2e^-$	$2e^-$	$2e^-$	$2e^-$	$2e^-$	$2e^-$	$2e^-$
Spin: $\mathbf{m}_s$ $\uparrow\downarrow = +\frac{1}{2}, -\frac{1}{2}$	↑↓	↑↓	↑↓	↑↓	↑↓	↑↓	↑↓	↑↓	↑↓	↑↓	↑↓	↑↓	↑↓	↑↓

6.4 ELECTRON CONFIGURATIONS IN ATOMS

Given the rules referred to in Section 6.3, it is possible to assign quantum numbers to each electron in an atom. Beyond that, electrons can be assigned to specific principal levels, sublevels, and orbitals. There are several ways to do this. Perhaps the simplest way to describe the arrangement of electrons in an atom is to give its **electron configuration**, which shows the number of electrons, indicated by a superscript, in each sublevel. For example, a species with the electron configuration

$$1s^2 2s^2 2p^5$$

has two electrons in the 1s sublevel, two electrons in the 2s sublevel, and five electrons in the 2p sublevel.

In this section, you learn how to predict the electron configurations of atoms of elements. There are a couple of different ways of doing this, which we consider in turn.

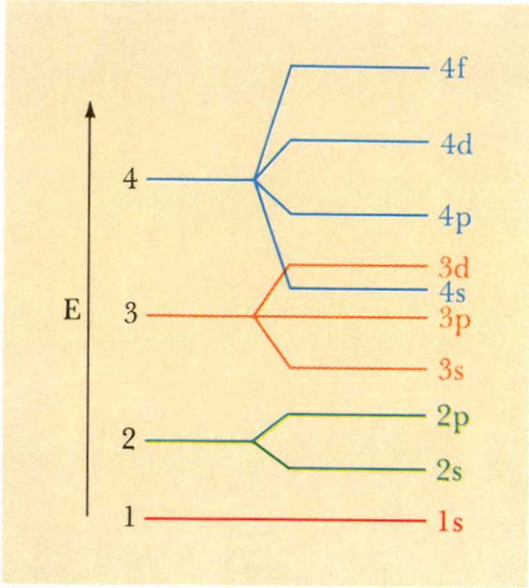

Figure 6.6
Arrangement of sublevels in order of increasing energy. This is also the order of filling, starting at the bottom with 1s, as atomic number increases.

Electron Configuration from Sublevel Energies

Electron configurations are readily obtained if the order of filling sublevels is known. Electrons enter the available sublevels in order of increasing sublevel energy. Ordinarily, a sublevel is filled to capacity before the next one starts to fill. The relative energies of different sublevels can be obtained from experiment. Figure 6.6 is a plot of these energies for atoms through the **n** = 4 principal level.

From Figure 6.6, it is possible to predict the electron configurations of atoms of elements with atomic numbers 1 through 36. Since an s sublevel can hold only two electrons, the 1s is filled at helium ($1s^2$). With lithium ($Z = 3$), the third electron has to enter a new sublevel: this is the 2s, the lowest sublevel of the second principal energy level. Lithium has one electron in this sublevel ($1s^2 2s^1$). With beryllium ($Z = 4$), the 2s sublevel is filled ($1s^2 2s^2$). The next six elements fill the 2p sublevel. Their electron configurations are

These are ground-state configurations; $1s^2 2s^1 2p^2$ would be an excited state for boron

$_5B$	$1s^2 2s^2 2p^1$	$_8O$	$1s^2 2s^2 2p^4$
$_6C$	$1s^2 2s^2 2p^2$	$_9F$	$1s^2 2s^2 2p^5$
$_7N$	$1s^2 2s^2 2p^3$	$_{10}Ne$	$1s^2 2s^2 2p^6$

Beyond neon, electrons enter the third principal level. The 3s sublevel is filled at magnesium:

$$_{12}Mg \quad 1s^2 2s^2 2p^6 3s^2$$

Six more electrons are required to fill the 3p sublevel with argon:

$$_{18}Ar \quad 1s^2 2s^2 2p^6 3s^2 3p^6$$

After argon, an "overlap" of principal energy levels occurs. The next electron enters the *lowest* sublevel of the fourth principal level (4s) instead of the *highest* sublevel of the third principal level (3d). Potassium ($Z = 19$) has one electron in the 4s sublevel; calcium ($Z = 20$) fills it with two electrons:

$$_{20}Ca \quad 1s^2 2s^2 2p^6 3s^2 3p^6 4s^2$$

Now, the 3d sublevel starts to fill with scandium ($Z = 21$). Recall that a d sublevel has a capacity of ten electrons. Hence the 3d sublevel becomes filled at zinc ($Z = 30$):

$$_{30}\text{Zn} \qquad 1s^22s^22p^63s^23p^64s^23d^{10}$$

The next sublevel, 4p, is filled at krypton ($Z = 36$):

$$_{36}\text{Kr} \qquad 1s^22s^22p^63s^23p^64s^23d^{10}4p^6$$

Up to atomic number 36, the order and populations are

$$1s^22s^22p^63s^23p^64s^23d^{10}4p^6$$

Example 6.5 Find the electron configuration of the sulfur atom and the nickel atom.

Strategy Determine the number of electrons in the atom from its atomic number. Fill the appropriate sublevels using the energy diagram in Figure 6.6.

Solution The sulfur atom has atomic number 16, so there are $16e^-$. Two electrons go into the 1s sublevel, two into the 2s sublevel, and six into the 2p sublevel. Two more go into the 3s sublevel, and the remaining four go into the 3p sublevel. The electron configuration of the S atom is

$$1s^22s^22p^63s^23p^4$$

The nickel atom has 28 electrons. Proceeding as above, the electron configuration for nickel is

$$1s^22s^22p^63s^23p^64s^23d^8$$

The 4s sublevel fills before the 3d sublevel.

Often, to save space, electron configurations are shortened; the **abbreviated electron configuration** starts with the preceding noble gas. For the elements sulfur and nickel,

	Complete Electron Configuration	Abbreviated Electron Configuration
$_{16}$S	$1s^22s^22p^63s^23p^4$	[Ne] $3s^23p^4$
$_{28}$Ni	$1s^22s^22p^63s^23p^64s^23d^8$	[Ar] $4s^23d^8$

The symbol [Ne] indicates that the first 10 electrons in the sulfur atom have the neon configuration $1s^22s^22p^6$; similarly, [Ar] represents the first 18 electrons in the nickel atom.

Electron Configuration from the Periodic Table

In principle, a diagram such as Figure 6.6 can be extended to include all sublevels occupied by electrons in the 109 known elements. As a matter of fact, that is a relatively simple thing to do; such a figure is in effect incorporated into the periodic table introduced in Chapter 2. The table can be used to predict the order of filling of all sublevels and hence the electron configurations of all elements from atomic number 1 to 109.

To understand how position in the periodic table relates to electron configuration, consider the metals in the first two groups. Atoms of the Group 1 elements all have one s electron in the outermost principal energy level (Table

Table 6.3 Abbreviated Electron Configurations of Group 1 and 2 Elements

Group 1		Group 2	
$_{3}$Li	[He]**2s^1**	$_{4}$Be	[He]**2s^2**
$_{11}$Na	[Ne]**3s^1**	$_{12}$Mg	[Ne]**3s^2**
$_{19}$K	[Ar]**4s^1**	$_{20}$Ca	[Ar]**4s^2**
$_{37}$Rb	[Kr]**5s^1**	$_{38}$Sr	[Kr]**5s^2**
$_{55}$Cs	[Xe]**6s^1**	$_{56}$Ba	[Xe]**6s^2**

6.3). In each Group 2 atom, there are two s electrons in the outermost level. A similar relationship applies to the elements in any group:

The periodic table "works" because an element's chemical properties depend on the number of outer electrons

The atoms of elements in a group of the periodic table have the same distribution of electrons in the outermost principal energy level.

This means that the order in which electron sublevels are filled is determined by position in the periodic table. Figure 6.7 shows how this works. Notice the following points:

1. *The elements in Groups 1 and 2 are filling an s sublevel.* Thus, Li and Be in the second period fill the 2s sublevel. Na and Mg in the third period fill the 3s sublevel, and so on.

2. *The elements in Groups* 13 *through* 18 *(six elements in each period) fill p sublevels,* which have a capacity of six electrons. In the second period, the 2p sublevel starts to fill with B ($Z = 5$) and is completed with Ne ($Z = 10$). In the third period, the elements Al ($Z = 13$) through Ar ($Z = 18$) fill the 3p sublevel.

3. *The transition metals, in the center of the periodic table, fill d sublevels.* Remember that a d sublevel can hold ten electrons. In the fourth period, the ten elements Sc ($Z = 21$) through Zn ($Z = 30$) fill the 3d sublevel. In the fifth period, the 4d sublevel is filled by the elements Y ($Z = 39$) through Cd ($Z = 48$). The ten transition metals in the sixth period fill the 5d sublevel.

4. *The two sets of 14 elements listed separately at the bottom of the table are filling f sublevels* with a principal quantum number two less than the period number. That is,

These elements are similar chemically because they differ only in the number of inner f electrons

— 14 elements in the sixth period ($Z = 57$ to 70) are filling the 4f sublevel. These elements are sometimes called rare earths or, more commonly nowadays, **lanthanides**, after the name of the first element in the series, lanthanum (La).

— 14 elements in the seventh period ($Z = 89$ to 102) are filling the 5f sublevel. The first element in this series is actinium (Ac); collectively, these elements are referred to as **actinides**.

Figure 6.7 (or any periodic table) can be used to deduce the electron configuration of any element. It is particularly useful for heavier elements such as iodine (Example 6.6).

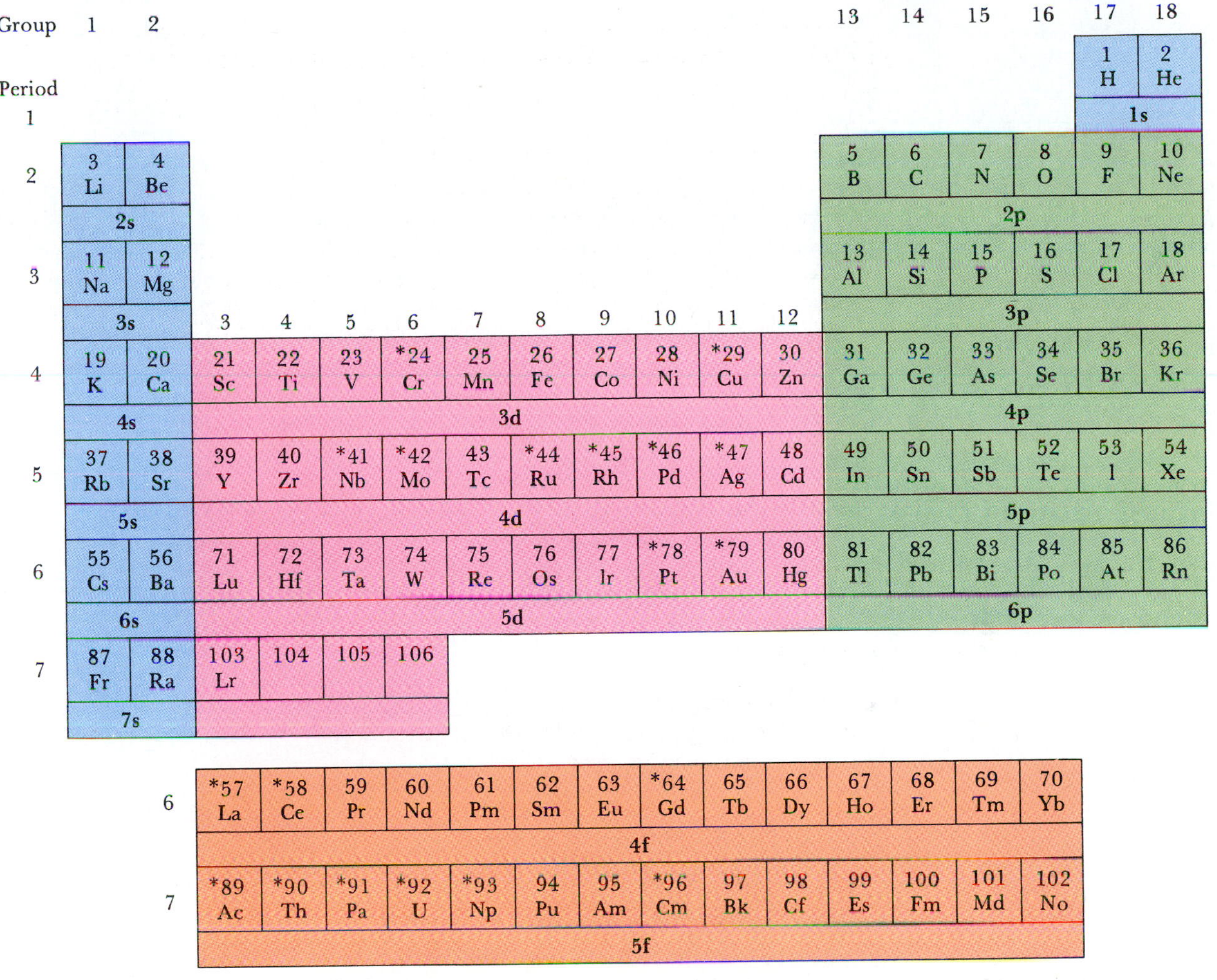

Figure 6.7
The periodic table can be used to deduce the electron configurations of atoms. Elements marked with an asterisk have electron configurations slightly different from that predicted by the periodic table.

As you can see from Figure 6.7, the electron configurations of several elements (marked *) differ slightly from those predicted. In every case, the difference involves a shift of one or, at the most, two electrons from one sublevel to another of very similar energy. For example, in the first transition series, two elements, chromium and copper, have an "extra" electron in the 3d as compared to the 4s orbital.

We don't know why this happens, but then we don't know a lot of things

	Predicted	**Observed**
$_{24}Cr$	[Ar] $4s^2 3d^4$	[Ar] $4s^1 3d^5$
$_{29}Cu$	[Ar] $4s^2 3d^9$	[Ar] $4s^1 3d^{10}$

A similar transposition occurs with six metals (Nb, Mo, Ru, Rh, Pd, Ag) in the second transition series and two more (Pt, Au) in the third.

(LIBRARY OF CONGRESS)

The periodic table, based on chemical similarities between elements, came long before electron configurations. Indeed, the table predated the discovery of the electron itself by 30 years. The person whose name is most closely associated with the periodic table is Dmitri Mendeleev, Professor of Chemistry at the University of St. Petersburg in Tsarist Russia. In writing a textbook of general chemistry, Mendeleev devoted separate chapters to families of elements with similar chemical properties, including the alkali metals (Group 1 in the modern periodic table), the alkaline earth metals (Group 2), and the halogens (Group 17). Reflecting on the behavior of these elements and others, Mendeleev was struck by the fact that their chemical properties vary periodically with atomic mass. Mendeleev went a step further and, in a paper presented to the Russian Chemical Society in March, 1869, proposed a primitive version of today's periodic table (Figure 6.A). The acclaim that greeted this and successive papers lifted Mendeleev from obscurity to fame. His textbook, *Principles of Chemistry,* first appeared in 1870 and was widely adopted through eight editions.

Mendeleev liked to live dangerously

The genius of Mendeleev was that he shrewdly left empty spaces in his periodic table for new elements yet to be discovered. Indeed, he went further than that, predicting detailed properties for three such elements: "ekaboron" (scandium), "ekaaluminum" (gallium), and "ekasilicon" (germanium). By 1886, all of the elements predicted by Mendeleev had been isolated. They were shown to have properties very close to those he predicted (Table 6.A). The spectacular agreement between prediction and experiment removed any doubts about the validity and value of Mendeleev's periodic table.

Mendeleev (as revised, 1871)

I	II	III	IV	V	VI	VII	VIII
R_2O	RO	R_2O_3	RO_2	R_2O_5	RO_3	R_2O_7	RO_4
H							
Li	Be	B	C	N	O	F	
Na	Mg	Al	Si	P	S	Cl	
K	Ca	–	Ti	V	Cr	Mn	Fe, Co, Ni
Cu	Zn	–	–	As	Se	Br	Ru, Rh, Pd
Ag	Cd	In	Sn	Sb	Te	I	
Cs	Ba						

Figure 6.A
An early version of the periodic table. It is in the condensed form used by early chemists, with transition metals placed with main-group elements.

Table 6.A Predicted and Observed Properties of Germanium (Ekasilicon)

Property	Predicted by Mendeleev (1871)	Observed (1886)
Atomic mass	73 amu	72.3 amu
Density	5.5 g/cm^3	5.47 g/cm^3
Specific heat	0.31 J/(g · °C)	0.32 J/(g · °C)
Melting point	very high	960°C
Formula of oxide	RO_2	GeO_2
Formula of chloride	RCl_4	$GeCl_4$
Density of oxide	4.7 g/cm^3	4.70 g/cm^3
Boiling point of chloride	100°C	86°C

Example 6.6 Write the electron configuration of iodine.

Strategy Using Figure 6.7, go across each period in succession, noting the sublevels filled, until you get to iodine.

Solution

Period 1: the 1s sublevel fills ($1s^2$)
Period 2: the 2s and 2p sublevels fill ($2s^22p^6$)
Period 3: the 3s and 3p sublevels fill ($3s^23p^6$)
Period 4: the 4s, 3d, and 4p sublevels fill ($4s^23d^{10}4p^6$)
Period 5: the 5s and 4d sublevels fill; five electrons enter the 5p ($5s^24d^{10}5p^5$)

Putting it all together, the electron configuration of iodine must be

$$1s^22s^22p^63s^23p^64s^23d^{10}4p^65s^24d^{10}5p^5$$

6.5 ORBITAL DIAGRAMS OF ATOMS

For many purposes, electron configurations are sufficient to describe the arrangements of electrons in atoms. Sometimes, however, it is useful to go a step further and show how electrons are distributed among orbitals. In such cases, **orbital diagrams** are used. Each orbital is represented by parentheses ( ), and electrons are shown by arrows written ↑ or ↓, depending upon spin.

There are two possible spins ($m_s = +\frac{1}{2}, -\frac{1}{2}$), so we use up and down arrows

To show how orbital diagrams are obtained from electron configurations, consider the boron atom ($Z = 5$). Its electron configuration is $1s^22s^22p^1$. The pair of electrons in the 1s orbital must have opposed spins ($+\frac{1}{2}, -\frac{1}{2}$ or ↑↓). The same is true of the two electrons in the 2s orbital. There are three orbitals in the 2p sublevel. The single 2p electron in boron could be in any one of these orbitals. Its spin could be either "up" or "down." The orbital diagram is ordinarily written

	1s	2s	2p
$_5B$	(↑↓)	(↑↓)	(↑)( )( )

with the first electron in an orbital arbitrarily designated by an "up" arrow, ↑.

With the next element, carbon, a complication arises. In which orbital should the sixth electron go? It could go in the same orbital as the other 2p electron, in which case it would have to have the opposite spin, ↓. It could go into one of the other two orbitals, either with a parallel spin, ↑, or an opposed spin, ↓. Experiment shows that there is an energy difference between these arrangements. The most stable is the one in which the two electrons are in different orbitals with parallel spins. The orbital diagram of the carbon atom is

	1s	2s	2p
$_6C$	(↑↓)	(↑↓)	(↑)(↑)( )

Similar situations arise frequently. There is a general principle that applies in all such cases. **Hund's rule** predicts that

> *When several orbitals of equal energy are available, as in a given sublevel, electrons enter singly with parallel spins.* Only after all the orbitals are half-filled do electrons "pair up" in orbitals.

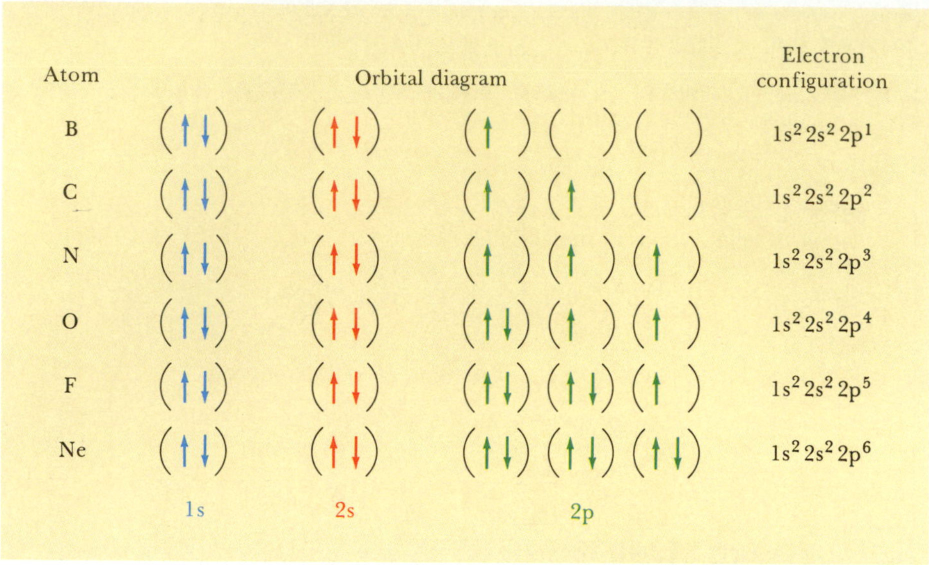

Atom	Orbital diagram 1s	2s	2p	Electron configuration
B	(↑↓)	(↑↓)	(↑) ( ) ( )	$1s^2 2s^2 2p^1$
C	(↑↓)	(↑↓)	(↑) (↑) ( )	$1s^2 2s^2 2p^2$
N	(↑↓)	(↑↓)	(↑) (↑) (↑)	$1s^2 2s^2 2p^3$
O	(↑↓)	(↑↓)	(↑↓) (↑) (↑)	$1s^2 2s^2 2p^4$
F	(↑↓)	(↑↓)	(↑↓) (↑↓) (↑)	$1s^2 2s^2 2p^5$
Ne	(↑↓)	(↑↓)	(↑↓) (↑↓) (↑↓)	$1s^2 2s^2 2p^6$

Figure 6.8
Orbital diagrams showing electron arrangements for atoms with five to ten electrons. Electrons entering orbitals of equal energy remain unpaired as long as possible.

Following this principle, the orbital diagrams for the elements boron through neon are shown in Figure 6.8. Notice that

— in all filled orbitals, the two electrons have opposed spins. Such electrons are often referred to as being *paired.* There are four paired electrons in the B, C, and N atoms, six in the oxygen atom, eight in the fluorine atom, and ten in the neon atom.

Hund's rule keeps electrons as far apart as possible

— in accordance with Hund's rule, within a given sublevel there are as many half-filled orbitals as possible. Electrons in such orbitals are said to be *unpaired.* There is one unpaired electron in atoms of B and F, two unpaired electrons in C and O atoms, and three unpaired electrons in the N atom. When there are two or more unpaired electrons, as in C, N, and O, those electrons have parallel spins.

Hund's rule, like the Pauli exclusion principle, is based upon experiment. It is possible to determine the number of unpaired electrons in an atom. With solids, this is done by studying their behavior in a magnetic field. If there are unpaired electrons present, the solid will be attracted into the field. Such a substance is said to be *paramagnetic.* If the atoms in the solid contain only paired electrons, it is slightly repelled by the field. Substances of this type are called *diamagnetic.* With gaseous atoms, the atomic spectrum can also be used to establish the presence and number of unpaired electrons.

Example 6.7 Construct orbital diagrams for atoms of sulfur and iron.

Strategy Start with the electron configuration, obtained as in Section 6.4. Then write the orbital diagram, recalling the number of orbitals per sublevel, putting two electrons of opposed spin in each orbital within a completed sublevel, and applying Hund's rule where sublevels are partially filled.

Solution Recall from Example 6.5 that the electron configuration of sulfur is $1s^22s^22p^63s^23p^4$. Its orbital diagram is

	1s	2s	2p	3s	3p
$_{16}S$	(↑↓)	(↑↓)	(↑↓)(↑↓)(↑↓)	(↑↓)	(↑↓)(↑)(↑)

The atomic number of iron is 26; its electron configuration is $1s^22s^22p^63s^23p^64s^23d^6$. All the orbitals are filled except those in the 3d sublevel, which is populated according to Hund's rule to give four unpaired electrons.

	1s	2s	2p	3s	3p	4s	3d
$_{26}Fe$	(↑↓)	(↑↓)	(↑↓)(↑↓)(↑↓)	(↑↓)	(↑↓)(↑↓)(↑↓)	(↑↓)	(↑↓)(↑)(↑)(↑)(↑)

6.6 ELECTRON ARRANGEMENTS IN MONATOMIC IONS

The discussion so far in this chapter has focused upon electron configurations and orbital diagrams of neutral atoms. It is also possible to assign electronic structures to monatomic ions, formed from atoms by gaining or losing electrons. In general, when a monatomic ion is formed from an atom, *electrons are added to or removed from sublevels in the highest principal energy level.*

Ions with Noble Gas Structures

As pointed out in Chapter 2, elements close to a noble gas in the periodic table form ions that have the same number of electrons as the noble gas atom. This means that these ions have noble gas electron configurations. Thus the three elements preceding Ne (N, O, and F) and the three elements following neon (Na, Mg, and Al) all form ions with the neon configuration, $1s^22s^22p^6$. The three nonmetal atoms achieve this structure by gaining electrons to form anions:

Important ideas are worth repeating

$$_7N\ (1s^22s^22p^3) + 3e^- \longrightarrow {}_7N^{3-}\ (1s^22s^22p^6)$$
$$_8O\ (1s^22s^22p^4) + 2e^- \longrightarrow {}_8O^{2-}\ (1s^22s^22p^6)$$
$$_9F\ (1s^22s^22p^5) + e^- \longrightarrow {}_9F^-\ (1s^22s^22p^6)$$

The three metal atoms acquire the neon structure by losing electorns to form cations:

$$_{11}Na\ (1s^22s^22p^63s^1) \longrightarrow {}_{11}Na^+\ (1s^22s^22p^6) + e^-$$
$$_{12}Mg\ (1s^22s^22p^63s^2) \longrightarrow {}_{12}Mg^{2+}\ (1s^22s^22p^6) + 2e^-$$
$$_{13}Al\ (1s^22s^22p^63s^23p^1) \longrightarrow {}_{13}Al^{3+}\ (1s^22s^22p^6) + 3e^-$$

The species N^{3-}, O^{2-}, F^-, Ne, Na^+, Mg^{2+} and Al^{3+} are said to be *isoelectronic;* they have the same electron configuration.

"iso" means "the same" in this case

There are a great many monatomic ions that have noble gas configurations; Figure 6.9 shows 24 ions of this type. Note, once again, that ions in a given main group have the same charge (+1 for Group 1, +2 for Group 2, −2 for Group **16**, −1 for Group **17**). This explains, in part, the chemical similarity

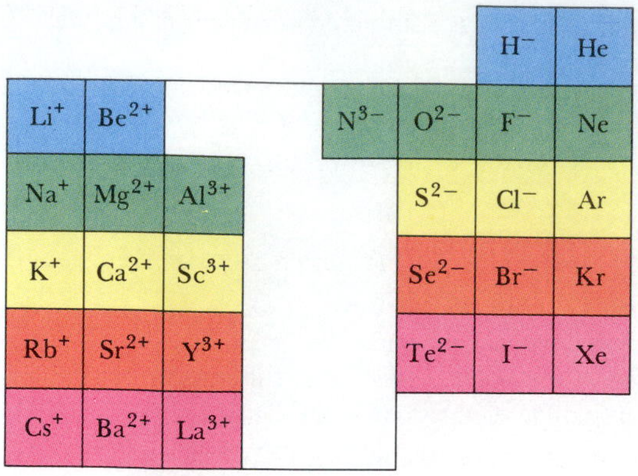

					H^-	He
Li^+	Be^{2+}		N^{3-}	O^{2-}	F^-	Ne
Na^+	Mg^{2+}	Al^{3+}		S^{2-}	Cl^-	Ar
K^+	Ca^{2+}	Sc^{3+}		Se^{2-}	Br^-	Kr
Rb^+	Sr^{2+}	Y^{3+}		Te^{2-}	I^-	Xe
Cs^+	Ba^{2+}	La^{3+}				

Figure 6.9
Species (cations, anions, or atoms) with noble gas structures. The color coding indicates species that are *isoelectronic* (same electronic structure).

between elements in the same main group. In particular, ionic compounds formed by such elements have similar chemical formulas. For example,

— halides of the alkali metals have the general formula MX where M = Li, Na, K, . . . and X = F, Cl, Br . . . .
— halides of the alkaline earth metals have the general formula MX_2 where M = Mg, Ca, Sr, . . . and X = F, Cl, Br, ...
— oxides of the alkaline earth metals have the general formula MO (M = Mg, Ca, Sr, . . .).

Transition Metal Cations

The transition metals to the right of the scandium subgroup do not form ions with noble gas configurations. To do so, they would have to lose four or more electrons. The energy requirement is too high for that to happen. However, as pointed out in Chapter 2, these metals do form cations with charges of +1, +2, or +3. Applying the principle that, in forming cations, electrons are removed from the sublevel of highest **n**, you can predict correctly that *when transition metal atoms form positive ions, the outer s electrons are lost first.* Consider, for example, the formation of the Mn^{2+} ion from the Mn atom:

The *s* electrons go in first with the atoms and come out first with the ions

$$_{25}Mn \quad [Ar]\ 4s^23d^5 \qquad _{25}Mn^{2+} \quad [Ar]\ 3d^5$$

Notice that it is the 4s electrons that are lost rather than the 3d electrons. This is known to be the case because the Mn^{2+} ion has been shown to have five unpaired electrons (the five 3d electrons). If two 3d electrons had been lost, the Mn^{2+} ion would have had only three unpaired electrons.

All the transition metals form cations by a similar process, i.e., loss of outer s electrons. Only after those electrons are lost are electrons removed from the inner d sublevel. Consider, for example, what happens with iron, which, you will recall, forms two different cations. First the 4s electrons are lost to give the Fe^{2+} ion:

$$_{26}Fe\ (Ar\ 4s^23d^6) \longrightarrow {}_{26}Fe^{2+}\ (Ar\ 3d^6) + 2e^-$$

Then an electron is removed from the 3d level to form the Fe^{3+} ion:

$$_{26}Fe^{2+}\ (Ar\ 3d^6) \longrightarrow {}_{26}Fe^{3+}\ (Ar\ 3d^5) + e^-$$

In the Fe^{2+} and Fe^{3+} ions, as in all transition metal ions, there are no outer s electrons.

Recall (Figure 6.6) that for fourth period *atoms*, the 4s level appears to be lower in energy than the 3d. This situation is reversed in the corresponding *ions;* the 4s electrons are higher in energy and hence are removed before the 3d.

Example 6.8 Give the electron configuration of

(a) Zn^{2+} (b) Se^{2-}

Strategy First obtain the electron configuration of the corresponding atom, as in Section 6.4. Then add or remove electrons from sublevels of highest **n**.

Solution

(a) The electron configuration of Zn ($Z = 30$) is $1s^22s^22p^63s^23p^64s^23d^{10}$

Two electrons are removed from the 4s sublevel to form Zn^{2+}:

$$Zn^{2+} \quad 1s^22s^22p^63s^23p^63d^{10}$$

(b) The electron configuration of Se ($Z = 34$) is $1s^22s^22p^63s^23p^64s^23d^{10}4p^4$.

Two electrons are added to the 4p sublevel to form Se^{2-}:

$$Se^{2-} \quad 1s^22s^22p^63s^23p^64s^23d^{10}4p^6$$

This is the electron configuration of the noble gas krypton ($Z = 36$).

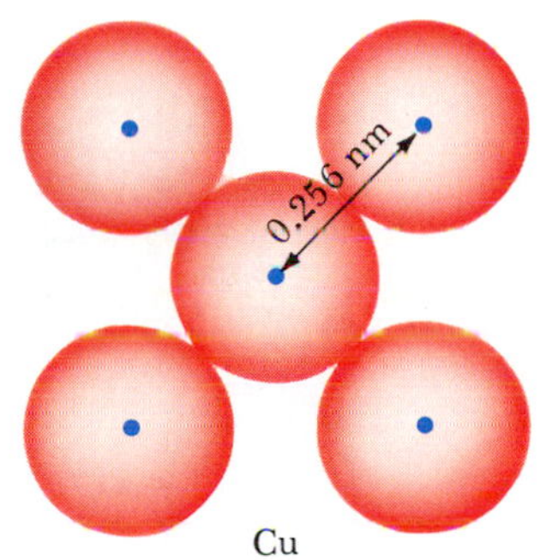

Atomic radius = $\frac{0.256 \text{ nm}}{2} = 0.128$ nm

6.7 PERIODIC TRENDS IN THE PROPERTIES OF ATOMS

In Section 6.4, you saw how the periodic table serves to obtain electron configurations. Beyond that, it can be used for a variety of purposes, including the correlation of properties on an atomic scale.

Atomic Radius

Strictly speaking, the "size" of an atom is a rather nebulous concept. The electron cloud surrounding the nucleus does not have a sharp boundary. However, a quantity called the **atomic radius** can be defined and measured, assuming a spherical atom. Ordinarily, the atomic radius is taken to be one half the distance of closest approach between atoms in an elemental substance (Fig. 6.10).

The atomic radii of the main-group elements are shown at the top of Figure 6.11. Notice that, in general, atomic radii

— decrease across a period from left to right in the periodic table
— increase down a group in the periodic table

These trends can be rationalized in terms of the *effective nuclear charge* experienced by an electron at the outer edge of an atom. For any electron, the effective nuclear charge, Z_{eff}, is approximated by the expression

$$Z_{eff} \approx Z - S$$

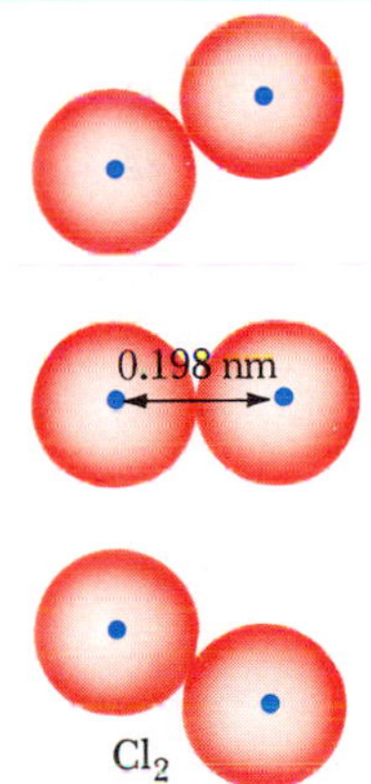

Atomic radius = $\frac{0.198 \text{ nm}}{2} = 0.099$ nm

Figure 6.10
Atomic radii are determined by assuming that atoms in closest contact in an element touch one another. The atomic radius is taken to be one-half of the closest internuclear distance.

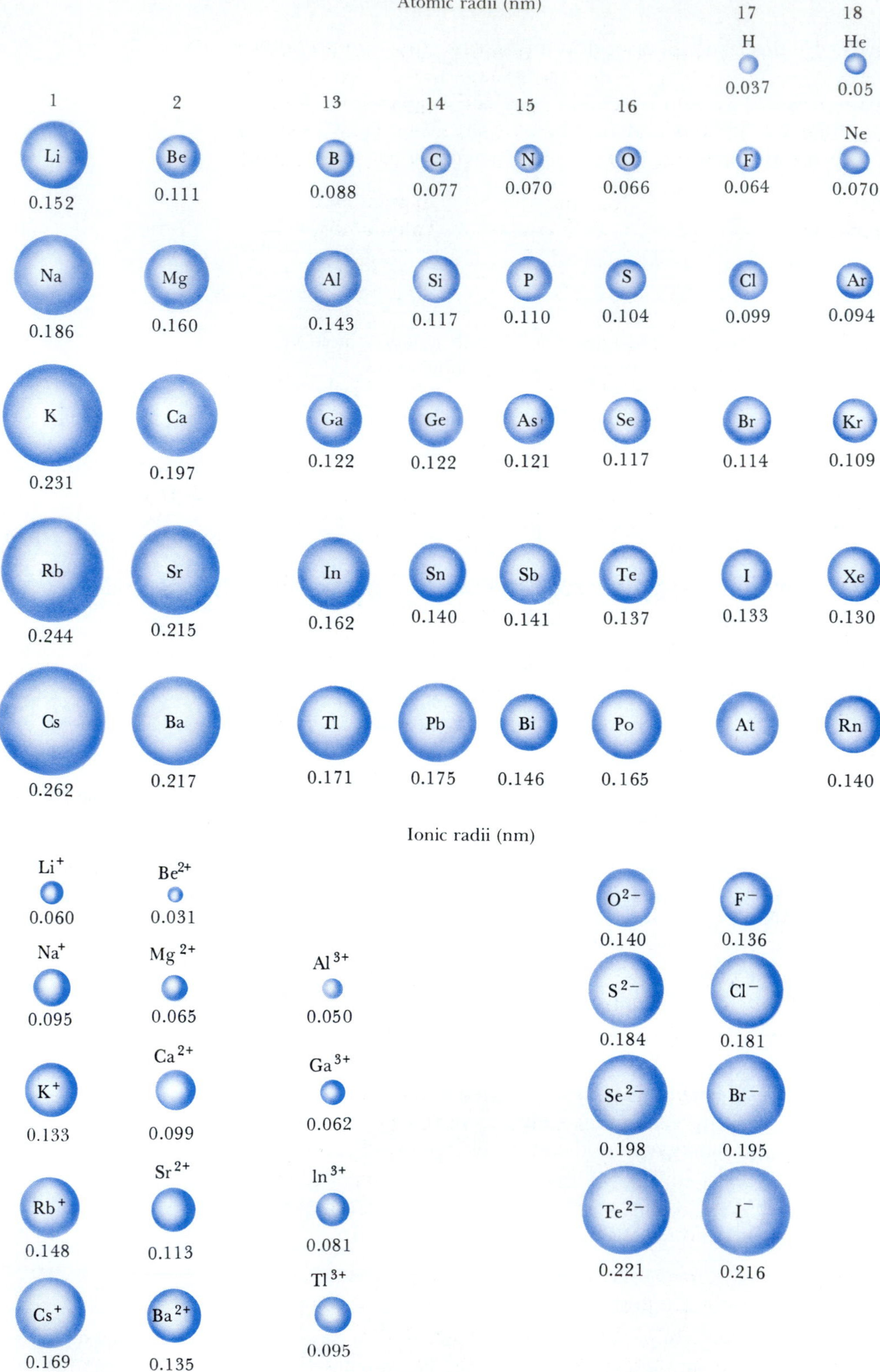

Figure 6.11
Sizes of atoms and ions of the main-group elements. Negative ions are always larger than the atoms from which they derived, whereas positive ions are smaller.

where Z is the actual nuclear charge (the atomic number), and S is the number of core electrons, those in the inner, complete energy levels. Core electrons shield the outer electron from the nucleus, in effect reducing the positive charge attracting that electron toward the nucleus. In contrast, outer electrons do not shield one another effectively. For the three atoms Na, Mg, and Al:

Electron Configuration	Z	Number of Core Electrons	Approximate Value of Z_{eff}
$_{11}$Na $1s^22s^22p^63s^1$	11	2 + 2 + 6 = 10	11 − 10 = 1
$_{12}$Mg $1s^22s^22p^63s^2$	12	2 + 2 + 6 = 10	12 − 10 = 2
$_{13}$Al $1s^22s^22p^63s^23p^1$	13	2 + 2 + 6 = 10	13 − 10 = 3

Thus Z_{eff} is about 1 for sodium, 2 for magnesium, and 3 for aluminum. You can see that effective nuclear charge increases across a period because the number of core electrons, S, remains the same while the atomic number, Z, increases.

The greater the effective nuclear charge, the stronger will be the attractive force pulling an outer electron in toward the nucleus. As this force increases, the electron is drawn in closer to the nucleus and atomic radius decreases. As a result, atoms get smaller from left to right across a period.

Z_{eff} increases →, so atomic radius decreases →

Moving down the periodic table in the same group, the effective nuclear charge experienced by an outer electron should remain essentially constant. For example, for all the metals in Group 1 (Li, Na, K, . . .), Z_{eff} should be 1. However, recall that the distance of the electron from the nucleus increases with the principal quantum number. Hence, atomic radius increases moving from Li (2s electron) to Na (3s electron) to K (4s electron), and so on. A similar argument can be applied to other groups, explaining why atoms get larger moving down a group.

Ionic Radius

The radii of cations and anions derived from atoms of the main-group elements are shown at the bottom of Figure 6.11. The trends referred to above for atomic radii are clearly visible here as well. Notice, for example, that ionic radius increases moving down a group in the periodic table. Moreover, the radii of both cations (left) and anions (right) decrease from left to right across a period.

Comparing the radii of cations and anions to those of the atoms from which they are derived (Fig. 6.11):

— *positive ions are smaller than the metal atoms from which they are formed.* The Na^+ ion has a radius, 0.095 nm, only a little more than half that of the Na atom, 0.186 nm.

— *negative ions are larger than the nonmetal atoms from which they are formed.* The radius of the Cl^- ion, 0.181 nm, is nearly twice that of the Cl atom, 0.099 nm.

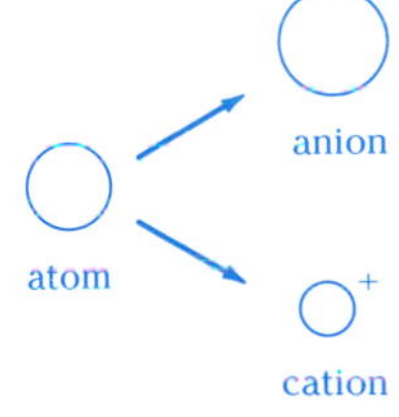

As a result of these effects, anions in general are larger than cations. Compare, for example, the Cl^- ions (radius = 0.181 nm) to the Na^+ ion (radius = 0.095 nm). This means that in sodium chloride, and indeed in the vast majority of all ionic compounds, most of the space in the crystal lattice is taken up by anions.

The differences in radii between atoms and ions can be explained quite simply. A cation is smaller than the corresponding metal atom because the excess of protons in the ion draws the outer electrons in closer to the nucleus. In contrast, an extra electron in an anion adds to the repulsion between outer electrons, making a negative ion larger than the corresponding nonmetal atom.

Example 6.9 Using only the periodic table, arrange each of the following sets of atoms and ions in order of increasing size.

(a) Mg, Al, Ca (b) S, Cl, S^{2-} (c) Fe, Fe^{2+}, Fe^{3+}

Strategy Recall that radius decreases across a period and increases moving down a group. An atom is larger than the corresponding cation but smaller than the corresponding anion.

Solution

(a) Compare the other atoms to Mg. Al, to the right, is smaller than Mg. Ca, below Mg, is larger. The predicted order is $Al < Mg < Ca$.

(b) Compare to the S atom. Cl, to the right, is smaller. The S^{2-} anion is larger than the S atom. The predicted order is $Cl < S < S^{2-}$.

(c) Compare to the Fe^{2+} ion. The Fe atom, with no charge, is larger. The Fe^{3+} ion, with a +3 charge, is smaller. The predicted order is $Fe^{3+} < Fe^{2+} < Fe$.

Ionization Energy

Ionization energy is a measure of how difficult it is to remove an electron from a gaseous atom. Energy must always be *absorbed* to bring about ionization: Ionization energies are always *positive* quantities.

The (first) ionization energy is the energy change for the removal of the outermost electron from a gaseous atom to form a +1 ion:

$$M(g) \longrightarrow M^{+}(g) + e^{-} \qquad \Delta E_1 = \text{first ionization energy}$$

The more difficult it is to remove electrons, the larger the ionization energy.

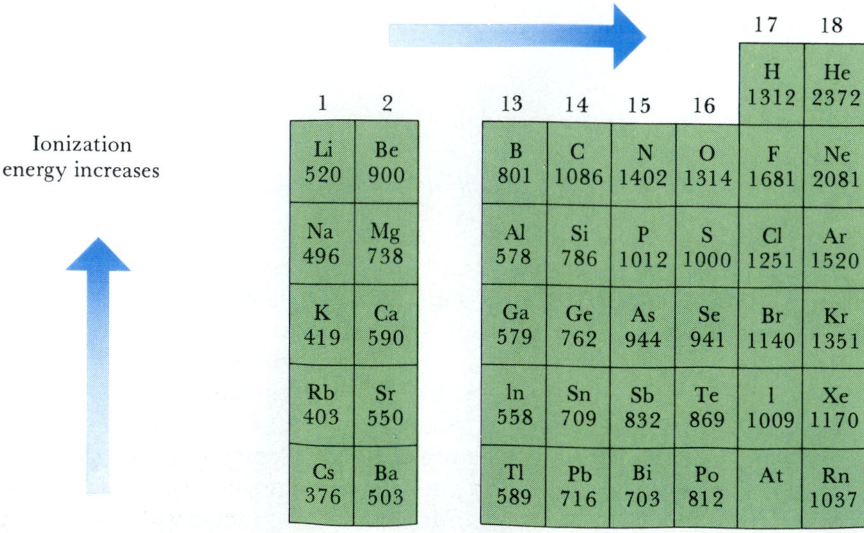

Figure 6.12
First ionization energies of the main-group elements, in kilojoules per mole. In general, ionization energy decreases moving down in the periodic table and increases going across, although there are several exceptions.

Ionization energies of the main-group elements are listed in Figure 6.12. Notice that ionization energy

— increases across the periodic table from left to right
— decreases moving down the periodic table

Comparing Figure 6.11 and 6.12 shows an inverse correlation between ionization energy and atomic radius. The smaller the atom, the more tightly its electrons are held to the nucleus and the more difficult they are to remove.

Example 6.10 Consider the three elements B, C, and Al. Using only the periodic table, predict which of the three elements has

(a) the largest atomic radius; the smallest atomic radius.
(b) the largest ionization energy; the smallest ionization energy.

Strategy Since these three elements form a block,

B C
Al

in the periodic table, it is convenient to compare both carbon and aluminum to boron. Recall the trends for atomic radius and ionization energy.

Solution

(a) B is larger than C but smaller than Al, so Al must be the largest atom and C the smallest.

(b) B has a smaller ionization energy than C, but a larger ionization energy than Al. Hence Al has the smallest ionization energy and C the largest.

If you look carefully at Figure 6.12, you will note a few exceptions to the general trends referred to above and illustrated in Example 6.10. For example, the ionization energy of B (801 kJ/mol) is *less* than that of Be (900 kJ/mol). This happens because the electron removed from the boron atom comes from the 2p as opposed to the 2s sublevel for beryllium. Since 2p is higher in energy than 2s, it is not too surprising that less energy is required to remove an electron from that sublevel.

Electronegativity

The ionization energy of an atom is a measure of its tendency to lose electrons; the larger the ionization energy, the more difficult it is to remove an electron. There are several different ways of comparing the tendencies of different atoms to gain electrons. The most useful of these for our purposes is the **electronegativity**, which measures the ability of an atom to attract to itself the electron pair forming a covalent bond.

The greater the electronegativity of an atom, the greater its affinity for electrons. Table 6.4 shows a scale of electronegativities first proposed by Linus Pauling. Here, each element is assigned a number, ranging from 4.0 for the most electronegative element, fluorine, to 0.8 for cesium, the least electronegative. Among the main-group elements, electronegativity increases moving from

Electronegativity is a positive quantity

left to right in the periodic table. Ordinarily, it decreases moving down a group. You will find Table 6.4 very useful when we discuss covalent bonding in Chapter 7.

Table 6.4 Electronegativity Values

H 2.2						
Li 1.0	Be 1.6	B 2.0	C 2.5	N 3.0	O 3.5	F 4.0
Na 0.9	Mg 1.3	Al 1.6	Si 1.9	P 2.2	S 2.6	Cl 3.2
K 0.8	Ca 1.0	Sc 1.4	Ge 2.0	As 2.2	Se 2.5	Br 3.0
Rb 0.8	Sr 0.9	Y 1.2	Sn 1.9	Sb 2.0	Te 2.1	I 2.7
Cs 0.8	Ba 0.9					

PERSPECTIVE

•

Lanthanides and Actinides

Lanthanides (La–Yb)

The 14 elements of atomic numbers 57–70 are referred to collectively as lanthanides. An earlier term, "rare earths," was something of a misnomer: all except one of these elements are more abundant than iodine. Promethium ($_{61}Pm$) does not occur in nature, although its "discovery" was reported several times in the first half of the twentieth century.

Because atoms of these elements differ only in the number of inner f electrons, they strongly resemble each other chemically. The common oxidation number, both in water and in the solid state, is +3. Salts of the lanthanides are difficult to separate from one another. Until about 40 years ago, the only general technique available was fractional crystallization from water. In 1911, Charles James at the University of New Hampshire carried out 15,000 crystallizations to obtain a small amount of pure $Tm(BrO_3)_3$.

Modern separation techniques, notably chromatography, have greatly increased the availability of compounds of these elements. A brilliant red phosphor used in color TV receivers contains a small amount of europium oxide, Eu_2O_3. This is added to a base of yttrium oxide, Y_2O_3, or gadolinium oxide, Gd_2O_3. Cerium(IV) oxide is used to coat interior surfaces of "self-cleaning" ovens, where it prevents the buildup of tar deposits.

Actinides (Ac–No)

Prior to World War II, the last element listed in the periodic table was uranium, $_{92}U$. Then, in 1940, at the University of California at Berkeley, a group led by Edwin McMillan and Glenn Seaborg discovered two new elements, prepared by bombardment of U-238 with neutrons and deuterons*

$$^{238}_{92}U + ^{1}_{0}n \longrightarrow ^{239}_{93}Np + ^{0}_{-1}e$$

$$^{238}_{92}U + ^{2}_{1}H \longrightarrow ^{238}_{94}Pu + 2^{1}_{0}n + ^{0}_{-1}e$$

These elements were named neptunium ($_{93}Np$) and plutonium ($_{94}Pu$) after the planets Neptune and Pluto, which follow Uranus in the solar system.

In 1944, Seaborg made the revolutionary suggestion that the elements $_{89}Ac$–$_{94}Pu$ were analogous to the lanthanides in that they were filling an inner f level, in this case the 5f. He commented later that he was warned this suggestion might destroy his scientific reputation; as Seaborg put it, "since I didn't have any scientific reputation, I published it anyway."

Between 1944 and 1958, Seaborg and a series of colleagues at Berkeley (notably Albert Ghiorso) prepared the remaining eight actinides—a truly remarkable achievement. All of these elements are radioactive; nuclear stability decreases rapidly with increasing atomic number. The longest-lived isotope of nobelium ($_{102}No$) has a half-life of about 3 min; i.e., in 3 minutes half of the sample decomposes. Nobelium and the preceding element, mendeleevium ($_{101}Md$), were identified in samples containing one to three atoms of No or Md.

* Nuclear equations such as these are discussed in Chapter 18.

CHAPTER HIGHLIGHTS

KEY CONCEPTS

1. *Relate wavelength, frequency, and energy*
   (Examples 6.1, 6.2; Problems 3–10, 69, 70, 83)
2. *Use the Bohr model to identify lines in the hydrogen spectrum*
   (Example 6.3; Problems 11–18)
3. *Identify the quantum numbers and capacities corresponding to energy levels, sublevels, and orbitals*
   (Example 6.4; Problems 21–30, 75, 81, 82)
4. *Write electron configurations of atoms and ions*
   (Examples 6.5, 6.6, 6.8; Problems 33–42, 53–56)
5. *Write orbital diagrams for atoms and ions*
   (Example 6.7; Problems 43–52)
6. *Identify periodic trends in radii, ionization energy, and electronegativity*
   (Examples 6.9, 6.10; Problems 57–64)

KEY EQUATIONS

Frequency–wavelength	$\lambda\nu = c = 2.998 \times 10^{8}$ m/s	
Energy–frequency	$E = h\nu = hc/\lambda$	$h = 6.626 \times 10^{-34}$ J · s
Bohr model	$E_{\mathbf{n}} = -R_H/\mathbf{n}^2$	$R_H = 2.180 \times 10^{-18}$ J

KEY TERMS

atomic orbital (s, p, d, f)
atomic radius
effective nuclear charge
electron configuration
—abbreviated
electron spin
electronegativity
energy level
frequency
ground state
ionization energy
main-group element
orbital diagram
photon
quantum no. ($\mathbf{n}$, ℓ, $\mathbf{m}_\ell$, $\mathbf{m}_s$)
spectrum
sublevel (s, p, d, f)
transition metal
wavelength

SUMMARY PROBLEM

Consider the element cobalt ($Z = 27$).

a. There is a line in the cobalt spectrum at 345.35 nm. In what region of the spectrum (ultraviolet, visible, or infrared) is this line found? What is the frequency of the line? What is the energy difference between the two levels responsible for this line, in kilojoules per mole?
b. The ionization energy of cobalt from the ground state is 758 kJ/mol. Assume that the transition in (a) is from the ground state to an excited state. If that is the case, calculate the ionization energy from the excited state.
c. Give the electron configuration of the Co atom; the Co^{3+} ion.
d. Give the orbital diagram (beyond Ar) for Co and Co^{3+}.
e. How many unpaired electrons are there in the Co atom? In the Co^{3+} ion?
f. How many p electrons are there in the Co atom? In the Co^{3+} ion?
g. Rank Co, Co^{2+}, and Co^{3+} in order of increasing size.

Answers

a. UV; 8.681×10^{14}/s; 346.4 kJ/mol **b.** 412 kJ
c. $1s^22s^22p^63s^23p^64s^23d^7$; $1s^22s^22p^63s^23p^63d^6$
d.

	4s	3d
Co	(↑↓)	(↑↓)(↑↓)(↑)(↑)(↑)
Co^{3+}	( )	(↑↓)(↑)(↑)(↑)(↑)

e. 3; 4 **f.** 12; 12 **g.** $Co^{3+} < Co^{2+} < Co$

QUESTIONS & PROBLEMS

Energy, Wavelength, and Frequency

1. Compare the energies and frequencies of two photons, one with a short wavelength, one with a long wavelength.

2. Compare the energy of an electron in the ground state to that in an excited state. In which state does the electron have the higher energy?

3. A photon of yellow light has a wavelength of 585 nm. Calculate the
a. frequency.
b. energy in joules per particle.
c. energy in kilojoules per mole.

4. An infrared ray has a frequency of 5.71×10^{12}/s. Calculate the
a. wavelength.
b. energy in joules per particle.
c. energy in kilojoules per mole.

5. The ionization energy of potassium is 419 kJ/mol. Do x-rays with a wavelength of 80 nm have sufficient energy to ionize potassium?

6. Carbon dioxide absorbs energy at a frequency of 2.001×10^{13}/s.
a. Calculate the wavelength of this absorption in nanometers.
b. In what spectral range does this absorption occur?
c. What is the energy difference in kilojoules per mole?

7. Two states differ in energy by 5.00×10^2 kJ/mol. When an electron moves from the higher to the lower of these states, calculate
a. λ in nanometers.
b. ν in $(\text{seconds})^{-1}$.

8. Two states differ in energy by 4.08×10^{-18} joules. When an electron moves from the higher to the lower state, calculate
a. λ in meters.
b. λ in nanometers.
c. ν in $(\text{seconds})^{-1}$.

9. When sunlight is scattered by air molecules, the sky appears blue. Blue light has a frequency of 6.6×10^{14} Hz. Calculate
a. the wavelength in nanometers associated with this radiation.
b. the energy in joules associated with this frequency.

10. Compact disc players use lasers with light at a wavelength of 785 nm.

a. What is the frequency of this light?

b. What is the energy associated with one photon of this light?

Bohr Model of the Hydrogen Atom

11. Consider the following transitions

(1) $\mathbf{n} = 1$ to $\mathbf{n} = 3$
(2) $\mathbf{n} = 3$ to $\mathbf{n} = 4$
(3) $\mathbf{n} = 3$ to $\mathbf{n} = 2$
(4) $\mathbf{n} = 5$ to $\mathbf{n} = 3$

a. Which one involves the ground state?

b. Which absorbs the most energy?

c. Which emits the most energy?

12. Consider the transitions in Question 11. For which of the transitions in energy absorbed? For which is energy emitted?

13. According to the Bohr model, the radius of a circular orbit is given by the expression

$$r \text{ (in nm)} = 0.0529\,\mathbf{n}^2$$

Draw successive orbits for the hydrogen electron at $\mathbf{n} = 1$, 2, 3, and 4. Indicate by arrows transitions between orbits that lead to lines in the

a. Lyman series ($\mathbf{n}_{lo} = 1$).

b. Balmer series ($\mathbf{n}_{lo} = 2$).

14. Calculate $E_\mathbf{n}$ for $\mathbf{n} = 1, 2, 3$, and 4. Make a one-dimensional graph showing energy, at different values of **n**, increasing vertically. On this graph, indicate by vertical arrows transitions in the

a. Lyman series ($\mathbf{n}_{lo} = 1$).

b. Balmer series ($\mathbf{n}_{lo} = 2$).

15. For the Brackett series, $\mathbf{n}_{lo} = 4$. Calculate the wavelength in nanometers of a transition from $\mathbf{n} = 6$ to $\mathbf{n} = 4$.

16. The Paschen series lines in the atomic spectrum of hydrogen result from electronic transitions from $\mathbf{n} > 3$ to $\mathbf{n} = 3$. Calculate the wavelength, in nanometers, of a line in this series resulting from the $\mathbf{n} = 7$ to $\mathbf{n} = 3$ transition.

17. A line in the Balmer series ($\mathbf{n}_{lo} = 2$) occurs at 434.05 nm. Calculate the $\mathbf{n}_{hi}$ for the transition associated with this line.

18. In the Pfund series $\mathbf{n}_{lo} = 5$. Calculate the longest wavelength possible for a transition in this series.

Quantum Mechanical Model of the Hydrogen Atom

19. What is the principal difference between the Bohr model and the quantum mechanical model of the hydrogen atom?

20. Explain in your own words the physical meaning of the two graphs in Figure 6.4.

Energy Levels, Sublevels, and Quantum Numbers

21. What are the possible values for $\mathbf{m}_\ell$ for

a. $\ell = 2$. **b.** $\ell = 4$. **c.** $\mathbf{n} = 4$ (all sublevels).

22. What are the possible values of $\mathbf{m}_\ell$ for

a. $\ell = 0$. **b.** $\ell = 3$. **c.** $\mathbf{n} = 3$ (all sublevels).

23. For the following pairs of orbitals, indicate which is higher in energy in a many-electron atom.

a. 3p or 4p **b.** 4s or 4d
c. 2s or 3d **d.** 5s or 4f

24. Which orbital in each of the following pairs is lower in energy in a many-electron atom?

a. 3s or 3p **b.** 4p or 4d
c. 1s or 2s **d.** 4d or 5f

25. What type of electron orbital (i.e., s, p, d, or f) is designated by

a. $\mathbf{n} = 2$, $\ell = 1$, $\mathbf{m}_\ell = 1$?

b. $\mathbf{n} = 4$, $\ell = 3$, $\mathbf{m}_\ell = 0$?

c. $\mathbf{n} = 1$, $\ell = 0$, $\mathbf{m}_\ell = 0$?

26. What type of electron orbital (i.e., s, p, d, or f) is designated by

a. $\mathbf{n} = 4$, $\ell = 1$, $\mathbf{m}_\ell = 0$?

b. $\mathbf{n} = 3$, $\ell = 2$, $\mathbf{m}_\ell = 0$?

c. $\mathbf{n} = 2$, $\ell = 0$, $\mathbf{m}_\ell = 0$?

27. Give the number of orbitals in

a. the principal level $\mathbf{n} = 3$. **b.** a 4d sublevel.

c. an f sublevel.

28. State the total capacity for electrons in

a. the principal level $\mathbf{n} = 4$. **b.** a 4d sublevel.

c. a 4d orbital.

29. State the relationship, if any, between the following pairs of quantum numbers:

a. **n** and ℓ **b.** ℓ and $\mathbf{m}_\ell$ **c.** $\mathbf{m}_\ell$ and $\mathbf{m}_s$

30. What is the

a. minimum **n** value for $\ell = 2$?

b. letter used to designate the sublevel with $\ell = 2$?

c. number of orbitals in a sublevel with $\ell = 3$?

d. number of different sublevels when $\mathbf{n} = 3$?

31. Given the following sets of electron quantum numbers, indicate those that could not occur and explain your answer.

a. 1, 1, 0, $+\frac{1}{2}$

b. 2, 1, 0, $+\frac{1}{2}$

c. 2, 0, 1, $-\frac{1}{2}$

d. 2, 1, 0, 0

e. 3, 2, 0, $-\frac{1}{2}$

32. Arrange the following sets of quantum numbers in order of increasing energy. If they have the same energy, place them together.

a. 4, 2, -1, $+\frac{1}{2}$

b. 1, 0, 0, $-\frac{1}{2}$

c. 3, 1, 1, $-\frac{1}{2}$

d. 2, 0, 0, $+\frac{1}{2}$

e. 2, 1, 0, $+\frac{1}{2}$

f. 3, 1, 1, $+\frac{1}{2}$

Electron Configurations

33. Write the electron configuration (ground state) for
a. Cl **b.** Cr **c.** Se **d.** Si **e.** Te

34. Write the electron configuration (ground state) for
a. Mg **b.** Co **c.** K **d.** In **e.** Pb

35. Write the abbreviated electron configuration for
a. Br **b.** Ti **c.** Cd **d.** F **e.** Hg

36. Write the abbreviated electron configuration for
a. O **b.** Ba **c.** Ni **d.** Te **e.** Bi

37. Give the symbol of the element of lowest atomic number whose ground state has
a. a d electron.
b. three f electrons.
c. a completed s sublevel.
d. six p electrons.

38. Give the symbol of the element of lowest atomic number whose ground state has
a. a completed d sublevel.
b. three 4d electrons.
c. five 3p electrons.
d. one s electron.

39. What fraction of the total number of electrons is in s sublevels in
a. Ar? **b.** Cd? **c.** Ne?

40. What fraction of the total number of electrons is in p sublevels in
a. Ca? **b.** Cs? **c.** Ru?

41. Which of the following electron configurations are for atoms in the ground state? In excited states? Which are impossible?
a. $1s^1 1p^1$
b. $2s^1 2p^1$
c. $1s^2 2s^2 2p^5$
d. $[_{10}Ne]3s^2 3p^3 3d^1$
e. $[_{10}Ne]3s^2 3p^7 4s^2 5d^1$
f. $1s^2 2s^1 2p^3 3s^2$

42. Which of the following electron configurations are for atoms in the ground state? In excited states? Which are impossible?
a. $1s^2 2s^2$ **b.** $1s^2 2s^2 4s^1$ **c.** $[_{10}Ne]3s^2 3p^6 4s^1$
d. $[_{18}Ar]3s^1$ **e.** $[_2He]2s^2 2p^6 2d^3$ **f.** $[_{10}Ne]3p^1$

Orbital Diagrams; Hund's Rule

43. Give the orbital diagram of
a. C **b.** Fe **c.** P **d.** Ar

44. Give the orbital diagram of
a. Si **b.** Nb **c.** Sr **d.** Sb

45. Give the symbol of the atom with the following orbital diagram beyond argon.

	4s	3d	4p
a.	(↓↑)	(↓↑)(↓↑)(↓↑)(↓)(↓)	( )( )( )
b.	(↓↑)	(↓↑)(↓↑)(↓)(↓)(↓)	( )( )( )
c.	(↓↑)	(↓↑)(↓↑)(↓↑)(↓↑)(↓↑)	(↓)(↓)( )

46. Give the symbol of the atom that has the following orbital diagram.

	1s	2s	2p	3s	3p
a.	(↓↑)	(↓↑)	(↓↑)(↓↑)(↓↑)	(↓↑)	( )( )( )
b.	(↓↑)	(↓↑)	(↓↑)(↓↑)(↓↑)	(↓↑)	(↓)(↓)(↓)
c.	(↓↑)	(↓↑)	(↓↑)(↓)(↓)	( )	( )( )( )

47. Give the symbols of
a. all the elements that have filled 3p sublevels.
b. all the metals in the 4th period that have no unpaired electrons.
c. all the nonmetals in the 3rd period that have one unpaired electron.
d. all the elements in the 3rd period in which all sublevels are half-full.

48. Give the symbols of
a. all the elements in which all the 3d orbitals are half-full.
b. all the metals in the 4th period that have one unpaired electron.
c. all the nonmetals in the 4th period that have no unpaired electrons.
d. all the elements in the 5th period in which the 5p sublevel is half-full.

49. Give the number of unpaired electrons in an atom of
a. O **b.** Sr **c.** V

50. How many unpaired electrons are there in an atom of
a. S? **b.** Tc? **c.** B?

51. Give the symbol of the element(s) in the first transition series with the following number of unpaired electrons per atom.
a. 0 **b.** 1 **c.** 2
d. 3 **e.** 4 **f.** 5

52. In what main group(s) of the periodic table do element(s) have the following number of filled p orbitals in the outermost principal level?
a. 0 **b.** 1 **c.** 2 **d.** 3

Electron Arrangement in Ions

53. Write the electron configuration for
a. a calcium atom; a Ca^{2+} ion.
b. a selenium atom; an Se^{2-} ion.
c. a zirconium atom; a Zr^{2+} ion.
d. a Co^{2+} ion; a Co^{3+} ion.

54. Give the electron configuration of
a. a nitrogen atom; an N^{3-} ion.
b. a potassium atom; a K^+ ion.
c. a scandium atom; an Sc^{3+} ion.
d. an Fe^{2+} ion; an Fe^{3+} ion.

55. How many unpaired electrons are there in each of the following ions?
a. Cr^{3+} **b.** Cr^{2+} **c.** Br^- **d.** Cu^+

56. Give the number of unpaired electrons in
a. Sn **b.** Sn^{2+} **c.** Sn^{4+} **d.** Cd^{2+} **e.** Na^+

Trends in the Periodic Table

57. Arrange the elements Rb, Te, and I in order of
a. increasing atomic radius.

b. increasing ionization energy.
c. increasing electronegativity.

58. Arrange the elements Mg, S, and Cl in order of
a. increasing atomic radius.
b. increasing ionization energy.
c. increasing electronegativity.

59. Which of the four atoms Na, Mg, S, or K
a. has the smallest atomic radius?
b. has the lowest ionization energy?
c. is the least electronegative?

60. Which of the following atoms, Ca, As, Br, and Rb,
a. has the largest atomic radius?
b. has the highest ionization energy?
c. is the least electronegative?

61. Select the larger member of each pair.
a. K and K^+ **b.** O and O^{2-}
c. Tl and Tl^{3+} **d.** Ni^{2+} and Ni^{3+}

62. Select the larger member of each pair.
a. N and N^{3-} **b.** Ba and Ba^{2+}
c. Cr and Cr^{3+} **d.** Cu^+ and Cu^{2+}

63. List the following species in order of decreasing radius.
a. K, Ca, Ca^{2+}, Rb **b.** S, Te^{2-}, Se, Te

64. List the following species in order of decreasing radius.
a. Co, Co^{2+}, Co^{3+} **b.** Cl, Cl^-, Br^-

Lanthanides and Actinides

65. Write the abbreviated electron configuration for promethium. What actinide would it most resemble chemically?

66. Predict the formula of
a. lanthanum chloride. **b.** lanthanum oxide.
c. yttrium sulfide.
Explain your reasoning.

67. Write the abbreviated electron configuration for nobelium. What lanthanide would it most resemble chemically?

68. Rank the actinides Pa, Bk, Cf, and Es in order of increasing nuclear stability.

Unclassified

69. A light bulb radiates 7.0% of the energy supplied to it as visible light. If the wavelength of visible light is assumed to be 565 nm, how many photons per second are emitted by a 75-W bulb (1 W = 1 J/s)?

70. A carbon dioxide laser produces radiation of wavelength 10.6 μm (1 μm = 10^{-6} m). If the laser produces about one joule of energy per pulse, how many photons are produced per pulse?

71. Explain the difference between
a. the Bohr model of the atom and the quantum mechanical model.
b. wavelength and frequency.
c. paramagnetism and diamagnetism.
d. the geometries of the three different p orbitals.

72. Explain in your own words what is meant by
a. the Pauli exclusion principle.
b. Hund's rule.
c. a line in an atomic spectrum.
d. the principal quantum number.

73. Indicate whether each of the following statements is true or false. If false, correct the statement.
a. An electron transition from **n** = 3 to **n** = 1 absorbs energy.
b. Light emitted by an **n** = 4 to **n** = 2 transition will have a shorter wavelength than that from an **n** = 5 to **n** = 2 transition.
c. A sublevel of ℓ = 4 has a capacity of 18 electrons.
d. An atom of a Group **16** element has two unpaired electrons.

74. Criticize the following statements.
a. The energy of a photon is inversely proportional to frequency.
b. The energy of the hydrogen electron is inversely proportional to the quantum number **n**.
c. Electrons start to enter the fourth principal level as soon as the third is full.

75. No currently known elements contain electrons in g (ℓ = 4) orbitals in the ground state. If an element is discovered that has electrons in the g orbital, what is the lowest value for **n** in which these g orbitals could exist? What are the possible values of $\mathbf{m}_\ell$? How many electrons could a set of g orbitals hold?

76. Explain why
a. negative ions are larger than their corresponding atoms.
b. scandium, a transition metal, forms an ion with a noble gas structure.
c. atomic radius decreases going across a period in the periodic table.

77. Indicate whether each of the following is true or false.
a. Effective nuclear charge stays about the same going across a period.
b. Group **17** elements have seven outer electrons.
c. Energy is given off when an electron is removed from an atom.

78. Name and give the symbol for the element with the characteristic given below.
a. Electron configuration of $1s^22s^22p^63s^23p^3$.
b. Lowest ionization energy in Group **17**.
c. Its +2 ion has the configuration [$_{18}$Ar] $3d^2$.
d. Alkali metal with the largest atomic radius.
e. Largest ionization energy in the third period.

Challenge Problems

79. The energy of any one-electron species in its **n**th state is given by $E = -BZ^2/\mathbf{n}^2$, where Z is the charge on the nucleus and B is 2.180×10^{-18} J. Find the ionization energy of the Li^{2+} ion in its first excited state in kilojoules per mole.

80. In 1885, Johann Balmer, a numerologist, derived the following relation for the wavelength of lines in the visible spectrum of hydrogen.

$$\lambda = 364.5\ n^2/(n^2 - 4)$$

where λ is in nanometers and **n** is an integer that can be 3, 4, 5, ... Show that this relation follows from the Bohr equation and the equation using the Rydberg constant. Note that in the Balmer series, the electron is returning to the **n** = 2 level.

81. Suppose the rules for assigning quantum numbers were as follows.

$$\mathbf{n} = 1, 2, 3, \ldots$$

$$\ell = 0, 1, 2, \ldots, \mathbf{n}$$

$$\mathbf{m}_\ell = 0, 1, 2, \ldots, \ell + 1$$

$$\mathbf{m}_s = +\tfrac{1}{2} \text{ or } -\tfrac{1}{2}$$

Prepare a table similar to Table 6.2, based on these rules, for **n** = 1 and **n** = 2. Give the electron configuration for an atom with eight electrons.

82. Suppose that the spin quantum number could have the values $\frac{1}{2}$, 0, and $-\frac{1}{2}$. Assuming that the rules governing the values of the other quantum numbers and the order of filling sublevels were unchanged,

a. what would be the electron capacity of an s sublevel? A p sublevel? A d sublevel?

b. how many electrons could fit in the **n** = 3 level?

c. what would be the electron configuration of the element with atomic number 8? 17?

83. In the photoelectric effect, electrons are ejected from a metal surface when light strikes it. A certain minimum energy, E_{min}, is required to eject an electron. Any energy absorbed beyond that minimum gives kinetic energy to the electron. It is found that when light at a wavelength of 540 nm falls on a cesium surface, an electron is ejected with a kinetic energy of 2.60×10^{-20} J. When the wavelength is 400 nm, the kinetic energy is 1.54×10^{-19} J.

a. Calculate E_{min} for cesium, in joules.

b. Calculate the longest wavelength, in nanometers, that will eject electrons from cesium.

7 Covalent Bonding

What immortal hand or eye
Could frame thy fearful
symmetry?

WILLIAM BLAKE
The Tiger

CHARLES D. WINTERS

$$Fe_2O_3(s) + 2Al(s) \longrightarrow 2Fe(s) + Al_2O_3(s)$$

(See p. 211)

CHAPTER OUTLINE

Earlier, we referred to the forces that hold nonmetal atoms to one another, covalent bonds. These bonds consist of an electron pair shared between two atoms. To represent the covalent bond in the H_2 molecule, the structures

$$H:H \qquad \text{or} \qquad H—H$$

are commonly written. These structures can be misleading if they are taken to mean that the two electrons are fixed in position between the two nuclei. A more accurate picture of the electron density in H_2 is shown in Figure 7.1. At a given instant, the two electrons may be located at any of various points about

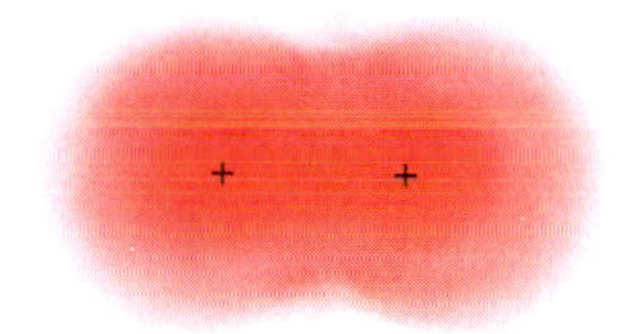

Figure 7.1
Electron density in H_2. The depth of color is proportional to the probability of finding an electron in a particular region.

the two nuclei. However, they are more likely to be found between the nuclei than at the far ends of the molecule.

You may well wonder why the sharing of an electron pair between two nuclei leads to increased stability. Why, for example, is the H_2 molecule more stable than two isolated hydrogen atoms? There are two major factors involved.

1. When two hydrogen atoms approach each other, the electron of one atom is attracted to the nucleus of the other. This lowers the electrostatic energy of the system and so stabilizes it.
2. The two atomic orbitals of the hydrogen atoms overlap one another in the H_2 molecule. This allows for electron exchange between the orbitals and lowers the total energy.

This chapter is devoted to the covalent bond as it exists in molecules and polyatomic ions. We consider

VSEPR stands for Valence Shell Electron Pair Repulsion

— the distribution of outer level *(valence)* electrons in species where atoms are joined by covalent bonds. These distributions are most simply described by *Lewis structures* (Section 7.1)
— molecular geometries. The so-called *VSEPR model* can be used to predict the angles between covalent bonds formed by a central atom (Section 7.2).
— the polarity of covalent bonds and the molecules they form (Section 7.3). Most bonds and many molecules are polar in the sense that they have a positive and a negative pole.
— the distribution of valence electrons among *atomic orbitals*, using the "valence bond" approach (Section 7.4).

7.1 LEWIS STRUCTURES; THE OCTET RULE

Noble gas structures are stable in molecules, as they are in atoms and ions

The idea of the covalent bond was first suggested by the American physical chemist G. N. Lewis in 1916. He pointed out that the electron configuration of the noble gases appears to be a particularly stable one. Noble-gas atoms are themselves extremely unreactive. Lewis suggested that ***atoms, by sharing electrons to form an electron-pair bond, can acquire a stable, noble-gas structure.*** Consider, for example, two hydrogen atoms, each with one electron. The process by which they combine to form an H_2 molecule can be shown as

$$\text{H}\cdot + \text{H}\cdot \longrightarrow \text{H}:\text{H}$$

using dots to represent electrons; the circles emphasize that the pair of electrons in the covalent bond can be considered to occupy the 1s orbital of either hydrogen atom. In that sense, each atom in the H_2 molecule has the electronic structure of the noble gas helium, with the electron configuration $1s^2$.

This idea is readily extended to other simple molecules containing nonmetal atoms. An example is the F_2 molecule. You will recall that a fluorine atom has the electron configuration $1s^22s^22p^5$. It has seven electrons in its outermost principal energy level ($\mathbf{n} = 2$). These are referred to as **valence electrons**. Showing them as dots about the symbol of the element, the fluorine atom can be represented as

$$:\ddot{\underset{..}{F}}\cdot$$

The combination of two fluorine atoms leads to the structure

$$:\ddot{\underset{..}{F}}\cdot + \cdot\ddot{\underset{..}{F}}: \longrightarrow :\ddot{\underset{..}{F}}:\ddot{\underset{..}{F}}:$$

This "electron-dot" structure shows that each fluorine atom "owns" six valence electrons outright and shares two others. Putting it another way, each F atom is surrounded by eight valence electrons. By this model, both F atoms achieve the electron configuration $1s^22s^22p^6$, which is that of the noble gas neon. This, according to Lewis, explains why the F_2 molecule is stable and why F atoms combine to form F_2 rather than F_3, F_4, ...

Shared electrons are counted for both atoms

The structures (without the circles) are referred to as **Lewis structures**. In writing Lewis structures, only the valence electrons written above are shown, since they are the ones that participate in covalent bonding. For the main-group elements, the only ones dealt with here, the number of valence electrons is equal to the last digit of the group number in the periodic table (Table 7.1). Notice that elements in the same main group have the same number of valence electrons. This explains why such elements behave similarly when they react to form covalently bonded species.

Note that with Groups **13–18**, the last digit is boldfaced

In the Lewis structure of a molecule or polyatomic ion, valence electrons ordinarily occur in pairs. There are two kinds of electron pairs.

1. A pair of electrons shared between two atoms is a **covalent bond**, ordinarily shown as a straight line between bonded atoms.
2. An **unshared pair** of electrons, belonging entirely to one atom, is shown as a pair of dots on that atom.

The Lewis structures for the species OH^-, H_2O, NH_3, and NH_4^+ are

OH^-	H_2O	NH_3	NH_4^+
$[:\ddot{\underset{..}{O}}—H]^-$	H—$\ddot{\underset{..}{O}}$—H	H—$\ddot{N}$(—H)—H	$[$H—N(—H)(—H)—H$]^+$

Notice that in each case the oxygen or nitrogen atom is surrounded by eight valence electrons. In each species, a single electron pair is shared between two bonded atoms. These bonds are called **single bonds**. There is one single bond in the OH^- ion, two in the H_2O molecule, three in NH_3, and four in NH_4^+. There are three unshared pairs in the hydroxide ion, two in the water molecule, one in the ammonia molecule, and none in the ammonium ion.

Table 7.1 Lewis Structures of Atoms

Group	1	2	13	14	15	16	17	18
No. of valence e^-	1	2	3	4	5	6	7	8
Lewis structure	H·	·Be·	·$\dot{B}$·	·$\dot{\underset{.}{C}}$·	·$\ddot{\underset{.}{P}}$·	·$\ddot{\underset{..}{O}}$·	:$\ddot{\underset{..}{F}}$·	:$\ddot{\underset{..}{Ne}}$:

Bonded atoms can share more than one electron pair. A **double bond** occurs when bonded atoms share two electron pairs; in a **triple bond**, three pairs of electrons are shared. In ethylene (C_2H_4) and acetylene (C_2H_2), the carbon atoms are linked by a double bond and a triple bond, respectively. Using two parallel lines to represent a double bond and three for a triple bond, we write the structures of these molecules as

```
H       H
 \     /
  C=C
 /     \
H       H
```

ethylene, C_2H_4

H—C≡C—H

acetylene, C_2H_2

Note that each carbon is surrounded by eight valence electrons and each hydrogen by two.

These examples illustrate the principle that atoms in covalently bonded species tend to have noble-gas electronic structures. This generalization is often referred to as the **octet rule**. Nonmetals, except for hydrogen, achieve a noble-gas structure by sharing in an "octet" of electrons (eight). Hydrogen atoms, in molecules or polyatomic ions, are surrounded by two electrons.

Most molecules follow the octet rule

Writing Lewis Structures

For very simple species, Lewis structures can often be written by inspection. Usually, though, you will save time by following the steps listed below.

1. *Count the number of valence electrons.* For a molecule, simply sum up the valence electrons of the atoms present. For a polyatomic anion, one electron is added for each unit of negative charge. For a polyatomic cation, a number of electrons equal to the positive charge must be subtracted.

2. *Draw a skeleton structure for the species, joining atoms by single bonds.* In some cases, only one arrangement of atoms is possible; in others, there are two or more alternative structures. Most of the molecules and polyatomic ions dealt with in this chapter consist of a *central atom* bonded to two or more *terminal atoms,* located at the outer edges of the molecule or ion. For such species (e.g., NH_4^+, SO_2, CCl_4), it is relatively easy to derive the skeleton structure. *The central atom is usually the one written first in the formula* (N in NH_4^+, S in SO_2, C in CCl_4); *put this in the center of the molecule or ion. Terminal atoms are most often hydrogen, oxygen, or a halogen; bond these atoms to the central atom.*

3. *Determine the number of valence electrons still available for distribution.* To do this, deduct two valence electrons for each single bond written in Step 2.

4. *Determine the number of valence electrons required to fill out an octet for each atom* (except H) *in the skeleton.* Remember that shared electrons are counted for both atoms.

a. If the number of electrons available (Step 3) is equal to the number required (Step 4), distribute the available electrons as unshared pairs, satisfying the octet rule for each atom.

b. If the number of electrons available (Step 3) is less than the number required (Step 4), the skeleton structure must be modified by changing single to multiple bonds. If you are two electrons short, convert a single bond to a double bond; if there is a deficiency of four electrons, convert a single bond to a triple bond (or two single bonds to double bonds).

It would be nice if all shortages could be solved so simply

The application of these steps and some further guiding principles are shown in Example 7.1.

Example 7.1 Draw Lewis structures of

(a) the hypochlorite ion, OCl^-. (b) methyl alcohol, CH_4O.

Strategy Follow the steps listed above. For the OCl^- ion, only one skeleton is possible; for methyl alcohol, keep in mind that the carbon atom ordinarily forms four bonds. Hydrogen, since it forms only one bond, must be a terminal atom.

Solution

(a)

(1) The number of valence electrons is 6 (from O) + 7 (from Cl) + 1 (from the −1 charge) = 14.

(2) The skeleton structure is $[O—Cl]^-$.

(3) The number of electrons available for distribution is 14 (originally) − 2 (used in skeleton) = 12.

(4) The number of electrons required to give each atom an octet is 6 (for O) + 6 (for Cl) = 12.

Since the number of available electrons is the same as the number required, the skeleton structure is correct; there are no multiple bonds. The Lewis structure is

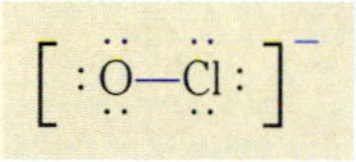

$$\left[:\ddot{\underset{\cdot\cdot}{O}}—\ddot{\underset{\cdot\cdot}{Cl}}:\right]^-$$

(b)

(1) The number of valence electrons is 4 (from C) + 4 (1 from each H) + 6 (from O) = 14.

(2) Since carbon forms four bonds and H must always be a terminal atom, the only reasonable skeleton is

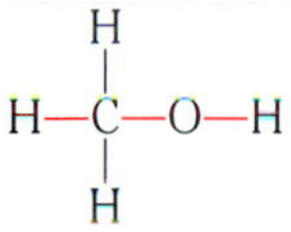

```
     H
     |
  H—C—O—H
     |
     H
```

A molecular model kit helps you to write Lewis structures

(3) The number of electrons available for distribution = 14 − 10 = 4.

(4) The number of electrons needed to satisfy the octet rule is 4 (for O).

Again, the number of electrons available is equal to the number required. The Lewis structure is

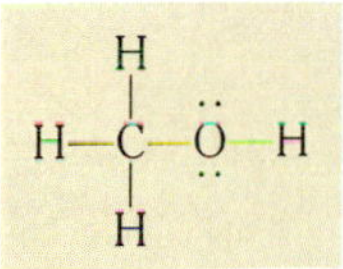

```
     H
     |   ..
  H—C—O—H
     |   ..
     H
```

Example 7.2 Draw Lewis structures for

(a) SO_2 (b) N_2

Strategy Follow the four steps for writing Lewis structures.

Solution

(a)

(1) There are 18 valence electrons. Both sulfur (2 atoms) and oxygen (1 atom) are in Group **16**; $3 \times 6 = 18$

(2) The skeleton structure, with sulfur as the central atom, is

O—S—O

(3) There are $18 - 4 = 14$ electrons available for distribution.

(4) Sixteen electrons are required to give each atom an octet (4 for S and 6 for each O). There is a deficiency of 2 electrons. This means that a single bond in the skeleton must be converted to a double bond. The Lewis structure of SO_2 is

Forming a multiple bond "saves" electrons because bonding pairs are counted for both atoms

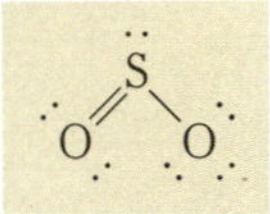

(b)

(1) There are 10 valence electrons; nitrogen is in Group **15**.

(2) The skeleton structure is

N—N

(3) There are $10 - 2 = 8$ electrons available for distribution.

(4) Each N needs 6 electrons for an octet, so 12 electrons are needed. This means that there is a deficiency of $12 - 8 = 4$ electrons. Convert the single bond between the two N atoms to a triple bond. The Lewis structure is

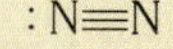

Resonance Forms

In certain cases, the Lewis structure does not adequately describe the properties of the ion or molecule that it represents. Consider, for example, the SO_2 structure derived in Example 7.2. This structure implies that there are two different kinds of sulfur-to-oxygen bonds in SO_2. One of these appears to be a single bond, the other a double bond. Yet experiment shows that there is only one kind of bond in the molecule.

One way to explain this situation is to assume that each of the bonds in SO_2 is intermediate between a single and a double bond. To express this concept, two structures are written:

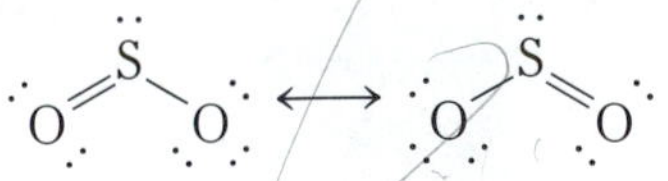

with the understanding that the true structure is intermediate between them. These are referred to as *resonance forms*. The concept of **resonance** is invoked whenever a single Lewis structure does not adequately reflect the properties of a substance.

Another species for which it is necessary to invoke the idea of resonance is the nitrate ion. Here three equivalent structures can be written

$$\left[\begin{array}{c} :\ddot{O}\diagdown\quad\diagup\ddot{O}: \\ N \\ | \\ :\ddot{O}: \end{array}\right]^- \longleftrightarrow \left[\begin{array}{c} :\ddot{O}\diagdown\quad\diagup\ddot{O}: \\ N \\ \| \\ :O: \end{array}\right]^- \longleftrightarrow \left[\begin{array}{c} :O\diagdown\quad\diagup\ddot{O}: \\ N \\ | \\ :\ddot{O}: \end{array}\right]^-$$

to explain the experimental observation that the three nitrogen to oxygen bonds in the NO_3^- ion are identical in all respects.

Gilbert Newton Lewis (1875–1946)

(DR. GLENN SEABORG, UNIVERSITY OF CALIFORNIA, LAWRENCE BERKELEY LABORATORY)

Born in Massachusetts, G. N. Lewis grew up in Nebraska, then came back east to obtain his B.S. (1896) and Ph.D. (1899) at Harvard. Although he stayed on for a few years as an instructor, Lewis seems never to have been happy at Harvard. A precocious student and an intellectual rebel, he was repelled by the highly traditional atmosphere that prevailed in the chemistry department there in his time. Many years later, he refused an honorary degree from his alma mater.

After leaving Harvard, Lewis made his reputation at M.I.T., where he was promoted to full professor in only four years. In 1912, he moved across the country to the University of California at Berkeley as Dean of the College of Chemistry and department chairman. He remained there for the rest of his life. Under his guidance, the chemistry department at Berkeley became perhaps the most prestigious in the country. Among the faculty and graduate students that he attracted were five future Nobel Prize winners: Harold Urey in 1934, William Giauque in 1949, Glenn Seaborg in 1951, Willard Libby in 1960, and Melvin Calvin in 1961.

In administering the chemistry department at Berkeley, Lewis demanded excellence in both research and teaching. Virtually the entire staff was involved in the general chemistry program; at one time eight full professors carried freshman sections.

Lewis' interest in chemical bonding and structure dated from 1902. In attempting to explain "valence" to a class at Harvard, he devised an atomic model to rationalize the octet rule. His model was deficient in many respects; for one thing, Lewis visualized cubic atoms with electrons located at the corners. Perhaps this explains why his ideas of atomic structure were not published until 1916. In that year, Lewis conceived of the electron-pair bond, perhaps his greatest single contribution to chemistry. At that time, it was widely believed that all bonds were ionic; Lewis's ideas were rejected by many well-known organic chemists.

In 1923, Lewis published a classic book (later reprinted by Dover Publications) entitled *Valence and the Structure of Atoms and Molecules.* Here, in Lewis' characteristically lucid style, we find many of the basic principles of covalent bonding discussed in this chapter. Included are electron-dot structures, the octet rule, and the concept of electronegativity. Here too is the Lewis definition of acids and bases (Chap. 15). That same year, Lewis published with Merle Randall a text called *Thermodynamics and the Free Energy of Chemical Substances.* Seventy years later, a revised edition of that text is still widely used in graduate courses in chemistry.

The years from 1923 to 1938 were relatively unproductive for G. N. Lewis insofar as his own research was concerned. The applications of the electron-pair bond came largely in the areas of organic and quantum chemistry; in neither of these fields did Lewis feel at home. In the early 1930s, he published a series of relatively minor papers dealing with the properties of deuterium. Then, in 1939, he began to publish in the field of photochemistry. Of approximately 20 papers in this area, several were of fundamental importance, comparable in quality to the best work of his early years. Retired officially in 1945, Lewis died a year later while carrying out an experiment on fluorescence.

Lewis certainly deserved a Nobel Prize, but he never received one

Example 7.3 Write three resonance forms for SO_3.

Strategy Write a Lewis structure for SO_3 following the four steps for writing Lewis structures. Then write all the resonance forms by changing the position of the multiple bond. Do *not* change the skeleton structure.

Solution The Lewis structure and all the resonance forms for SO_3 are

$$O{=}S(-O)(-O) \longleftrightarrow O{-}S({=}O)(-O) \longleftrightarrow O{-}S(-O)({=}O)$$

$:\ddot{O}-\ddot{S}:$ bonded through O, with O=O below, is *not* a resonance form of sulfur trioxide

Formal Charge

Often, it is possible to write two different Lewis structures for a molecule differing in the arrangement of atoms, i.e.,

A—A—B or A—B—A

Sometimes, both structures represent real compounds, which are *isomers* of one another (Chap. 15, 21). More often, only one structure exists in nature. For example, methanol (CH_4O) has the structure

$$H-\underset{H}{\overset{H}{|}}\!\!C-\ddot{O}-H$$

In contrast, the structure

$$H-\ddot{C}-\ddot{O}-H \quad (\text{H on C and H on O below})$$

does not correspond to any real compound even though it obeys the octet rule.

There are several ways to choose the more plausible of two structures differing in their arrangement of atoms. As pointed out in Example 7.1, the fact that carbon almost always forms four bonds leads to the correct structure for methanol. Another approach involves a concept called **formal charge**, which can be applied to any atom within a Lewis structure. The formal charge, C_f, is defined as

Formal charge is the charge an atom would have if valence electrons were distributed fairly

$$C_f = E_v - (E_u + \tfrac{1}{2}E_b)$$

where E_v = no. valence e^- in the unbonded atom = last digit of group number, E_u = no. of unshared electrons "owned" by the atom, and E_b = no. of bonding electrons shared by the atom.

Empirically, it is found that the more likely Lewis structure is the one in which

— the formal charges are as close to zero as possible
— any negative formal charge is located on a strongly electronegative atom

Sounds reasonable

To show how this works, let us assign formal charges to the carbon and oxygen atoms in the two structures discussed above.

```
    H
    |  ..
H—C—O—H
    |  ..
    H
```
C_f of C = 4 − $\frac{1}{2}$(8) = 0
C_f of O = 6 − [4 + $\frac{1}{2}$(4)] = 0

```
   ..  ..
H—C—O—H
   |   |
   H   H
```
C_f of C = 4 − [2 + $\frac{1}{2}$(6)] = −1
C_f of O = 6 − [2 + $\frac{1}{2}$(6)] = +1

The second structure is frowned upon because the formal charges are not zero; moreover, the negative formal charge is on the *less* electronegative atom (recall from Table 6.4 that EN of C = 2.5, EN of O = 3.5).

The concept of formal charge has a much wider applicability than this short discussion might imply. In particular, it can be used to predict situations where conventional Lewis structures, written in accordance with the octet rule, may be incorrect (Table 7.2). Experimental evidence suggests that the structures for BeF_2 and BF_3 written at the right of the table are valid even though they "violate" the octet rule. (See also Problems 87, 88.)

Exceptions to the Octet Rule

Although most of the molecules and polyatomic ions referred to in general chemistry follow the octet rule, there are some familiar species that do not. Among these are molecules containing an odd number of valence electrons. Nitric oxide, NO, and nitrogen dioxide, NO_2, fall in this category:

NO   no. of valence electrons = 5 + 6 = 11

NO_2 no. of valence electrons = 5 + 6(2) = 17

With an odd number of electrons, there's no way they could all be paired

For such *odd electron* species (sometimes called free radicals) it is impossible to write Lewis structures in which each atom obeys the octet rule. In the NO molecule, the unpaired electron is put on the nitrogen atom, giving both atoms a formal charge of zero:

```
 ..  ..
·N═O:
```

In NO_2, the "best" structure one can write again puts the unpaired electron on the nitrogen atom:

```
      ·
      N
   ..╱ ╲╲ ..
  :O     O:
   ..
```

Table 7.2 Possible Structures for BeF_2 and BF_3

Structure I	C_f	Structure II	C_f
:F̤═Be═F̤:	Be = −2 F = +1	:F̤—Be—F̤:	Be = 0 F = 0
BF_3 with B═F double bond	B = −1 F = +1,0,0	BF_3 with three B—F single bonds	B = 0 F = 0

Figure 7.2
Oxygen is attracted into a magnetic field and can actually be suspended between the poles of an electromagnet. Both the paramagnetism and the blue color are due to the unpaired electrons in the O_2 molecule. (S. Ruven Smith)

We will not discuss molecular orbitals in this text

Elementary oxygen, like NO and NO_2, is paramagnetic (Fig. 7.2). Experimental evidence suggests that the O_2 molecule contains two unpaired electrons *and* a double bond. It is impossible to write a conventional Lewis structure for O_2 that has these two characteristics. A more sophisticated model of bonding, using molecular orbitals (MO theory) is required to explain the properties of oxygen.

There are a few species in which the central atom "violates" the octet rule in the sense that it is surrounded by two or three electron pairs rather than four. Examples include the fluorides of beryllium and boron, BeF_2 and BF_3. Although one could write multiple bonded structures for these molecules in accordance with the octet rule (Table 7.2), experimental evidence suggests the structures

$$:\ddot{\underset{..}{F}}—Be—\ddot{\underset{..}{F}}: \qquad \text{and} \qquad \begin{array}{ccc} :\ddot{\underset{..}{F}} & & \ddot{\underset{..}{F}}: \\ & \diagdown \; \diagup & \\ & B & \\ & | & \\ & :\underset{..}{F}: & \end{array}$$

in which the central atom is surrounded by four and six valence electrons, respectively, rather than eight.

Expanded Octets

An atom has an "expanded octet" if it is surrounded by 5 or 6 electron pairs

The largest class of molecules to "violate" the octet rule consists of species in which the central atom is surrounded by more than four pairs of valence electrons. Typical molecules of this type are phosphorus pentachloride, PCl_5, and sulfur hexafluoride, SF_6. The Lewis structures of these molecules are

$$\begin{array}{c} :\ddot{Cl}: \\ :\ddot{Cl}-P-\ddot{Cl}: \\ :\ddot{Cl}\quad \ddot{Cl}: \end{array} \qquad \begin{array}{c} :\ddot{F}: \\ :\ddot{F}\quad \ddot{F}: \\ S \\ :\ddot{F}\quad \ddot{F}: \\ :\ddot{F}: \end{array}$$

As you can see, the central atoms in these molecules have "expanded octets." In PCl_5, the phosphorus atom is surrounded by 10 valence electrons (5 shared pairs); in SF_6, there are 12 valence electrons (6 shared pairs) around the sulfur atom.

In molecules of this type, the terminal atoms are most often halogens (F, Cl, Br, I); in a few molecules, oxygen is a terminal atom. The central atom is a nonmetal in the third, fourth, or fifth period of the periodic table. Most frequently, it is one of the following elements:

	Group 15	Group 16	Group 17	Group 18
3rd period	P	S	Cl	
4th period	As	Se	Br	Kr
5th period	Sb	Te	I	Xe

All of these atoms have d orbitals available for bonding (3d, 4d, 5d). These are the orbitals in which the "extra" pairs of electrons are located in such species as PCl_5 and SF_6 (Section 7.4).

Since there is no 2d sublevel, C, N, O and F never form expanded octets

Sometimes, as with PCl_5 and SF_6, it is clear from the formula that the central atom has an expanded octet. Often, however, it is by no means obvious that this is the case. At first glance, formulas such as ClF_3 or XeF_4 look completely straightforward. However, when you try to draw the Lewis structure it becomes clear that an expanded octet is involved. The number of electrons available after the skeleton is drawn is *greater* than the number required to give each atom an octet. When that happens, distribute the extra electrons (two or four) around the central atom as unshared pairs.

Example 7.4 Draw the Lewis structure of XeF_4.

Strategy Follow the usual four-step sequence. If there is an electron surplus, add the extra electrons to the central xenon atom as unshared pairs.

Solution

(1) The number of valence electrons is 8 (from Xe) + 28 (from four F atoms) = 36.
(2) The skeleton is

$$\begin{array}{c} F \\ | \\ F-Xe-F \\ | \\ F \end{array}$$

(3) The number of electrons available for distribution is 36 − 8 = 28.
(4) Each fluorine atom needs 6 electrons to complete its octet; the number of electrons required = 4(6) = 24.

There is a surplus of electrons; 28 are available and only 24 are required to fill out octets. The four extra electrons are added to the central Xe atom. The Lewis structure for XeF_4 is

```
      ..
     :F:
      |
 ..   ..   ..
:F — Xe — F:
 ..   ..   ..
      |
     :F:
      ..
```

In this molecule there are six pairs of electrons around the xenon atom.

7.2 MOLECULAR GEOMETRY

The geometry of a diatomic molecule such as Cl_2 or HCl can be described very simply. Since two points define a straight line, the molecule must be linear:

Cl—Cl H—Cl

With molecules containing three or more atoms, the geometry is not so obvious. Here, the angles between bonds, called **bond angles**, must be considered. For example, a molecule of the type YX_2, where Y represents the central atom and X an atom bonded to it, can be

— *linear,* with a bond angle of 180°: X—Y—X

— *bent,* with a bond angle less than 180°: X/Y\X

This is a very useful rule; to use it, you start with the Lewis structure

The major features of molecular geometry can be predicted on the basis of a quite simple principle—electron-pair repulsion. This principle is the essence of the *valence-shell electron-pair repulsion (VSEPR) model,* first suggested by N. V. Sidgwick and H. M. Powell in 1940. It was developed and expanded later by R. J. Gillespie and R. S. Nyholm. According to VSEPR theory, *the valence electron pairs surrounding an atom repel one another, so the orbitals containing those electron pairs are oriented to be as far apart as possible.*

In this section we apply this model to predict the geometry of some rather simple molecules and polyatomic ions. In all of these species, a central atom is surrounded by from two to six pairs of electrons.

Ideal Geometries with Two to Six Electron Pairs on the Central Atom

We begin by considering species in which a central atom, A, is surrounded by from two to six electron pairs, all of which are used to form single bonds with terminal atoms, X. These species have the general formulas AX_2, AX_3, . . . , AX_6. It is understood that there are no unshared pairs around atom A.

Figure 7.3
(Charles D. Winters)

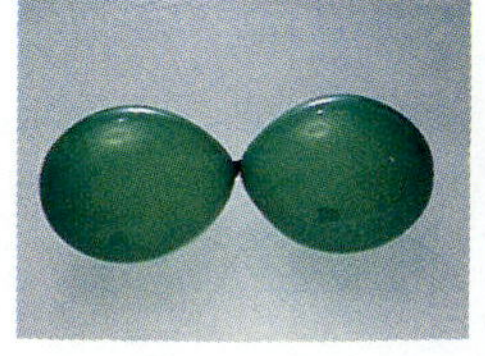

Species type	Orientation of electron pairs	Predicted bond angles	Example	Ball and stick model
AX_2	Linear	180°	BeF_2	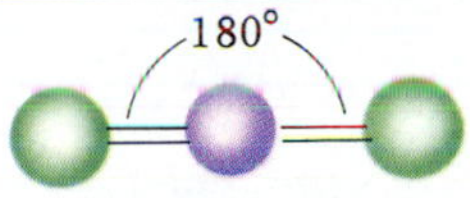 180°
AX_3	Triangular Planar	120°	BF_3	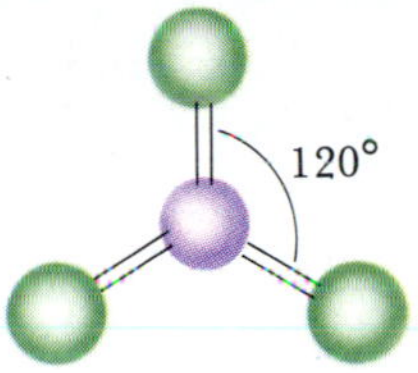 120°
AX_4	Tetrahedron	109.5°	CH_4	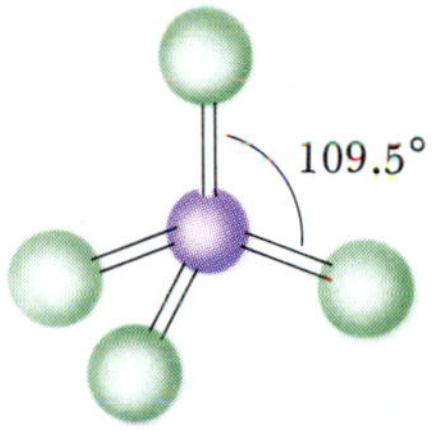 109.5°
AX_5	Triangular Bipyramid	90° 120° 180°	PCl_5	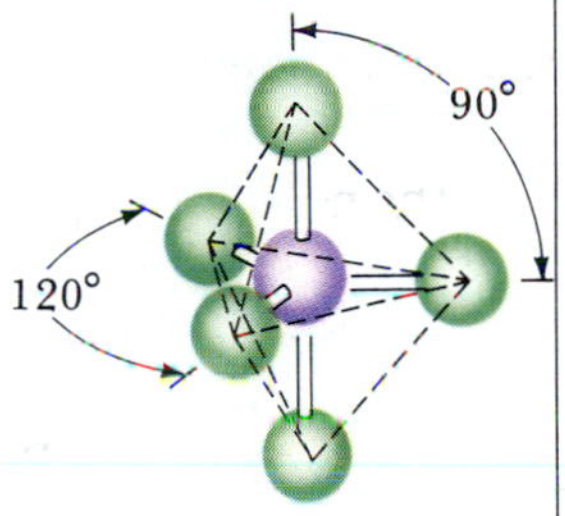 90° 120°
AX_6	Octahedron	90° 180°	SF_6	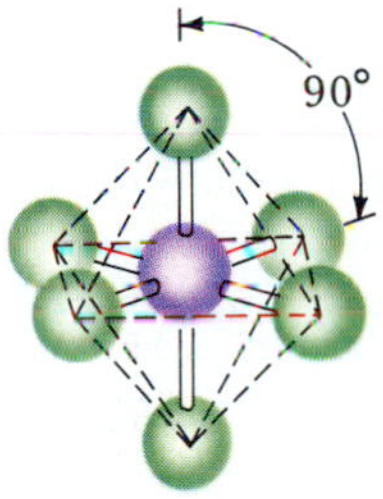 90°

Figure 7.4
Geometries to be expected from the VSEPR model for molecules with 2 to 6 electron pairs around a central atom. (Compare with Fig. 7.3.)

To see how the bonding orbitals surrounding the central atom are oriented with respect to one another, consider Figure 7.3, which shows the positions taken naturally by two to six balloons tied together at the center. The balloons, like the orbitals they represent, arrange themselves to be as far from one another as possible.

Figure 7.4 shows the geometries predicted by the VSEPR model for molecules of the types AX_2 to AX_6. The geometries for two and three electron pairs are those associated with species in which the central atom has less than an octet of electrons. Molecules of this type include BeF_2 and BF_3, which have the Lewis structures

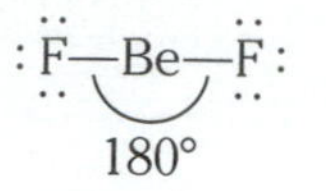

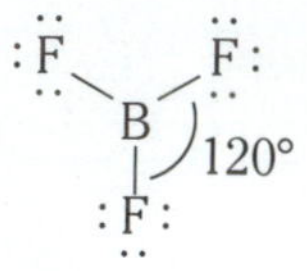

:F—Be—F: 180°

:F—B(—F:)—F: 120°

Two electron pairs are as far apart as possible when they are directed at 180° to one another. This gives BeF_2 a **linear** structure. The three electron pairs around the boron atom in BF_3 are directed toward the corners of an equilateral triangle; the bond angles are 120°. We describe this geometry as **triangular planar**.

In species that follow the octet rule, the central atom is surrounded by four electron pairs. If each of these pairs forms a single bond with a terminal atom, a molecule of the type AX_4 results. The four bonds are directed toward the corners of a regular **tetrahedron**. All the bond angles are 109.5°, the tetrahedral angle. This geometry is found in many polyatomic ions (NH_4^+, SO_4^{2-}, etc.) and in a wide variety of organic molecules, the simplest of which is methane, CH_4.

This puts the four electron pairs as far apart as possible

Molecules of the type AX_5 and AX_6 require that the central atom have an expanded octet. The geometries of these molecules are shown at the bottom of Figure 7.4. In PF_5, the five bonding pairs are directed toward the corners of a **triangular bipyramid**, a figure formed when two triangular pyramids are fused together, base-to-base. Three of the fluorine atoms are located at the corners of an equilateral triangle with the phosphorus atom at the center; the other two fluorine atoms are directly above and below the P atom. In SF_6, the six bonds are directed toward the corners of a regular **octahedron**. An octahedron can be formed by fusing two square pyramids base-to-base. Four of the fluorine atoms in SF_6 are located at the corners of a square with the S atom at the center; one fluorine atom is directly above the S atom, another directly below it.

Effect of Unshared Pairs on Molecular Geometry

In many molecules and polyatomic ions, one or more of the electron pairs around the central atom is unshared. The VSEPR model is readily extended to predict the geometries of these species. In general:

1. The electron-pair geometry is approximately the same as that observed when only single bonds are involved. The bond angles are either equal to the ideal values listed in Figure 7.4 or a little less than these values.

2. The *molecular geometry* is quite different when one or more unshared pairs are present. In describing molecular geometry, we refer only to the positions of the bonded atoms. These positions can be determined experimentally; positions of unshared pairs cannot be established by experiment. Hence the locations of unshared pairs are not specified in describing molecular geometry.

With these principles in mind, consider the NH_3 molecule:

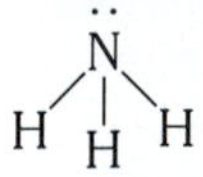

$\ddot{N}$ bonded to H, H, H

Experimentally, we can locate atoms but not unshared pairs

The apparent orientation of the four electron pairs around the N atom in NH_3 is shown at the left of Figure 7.5. Notice that, as in CH_4, the four pairs are directed toward the corners of a regular tetrahedron. The diagram at the right of Figure

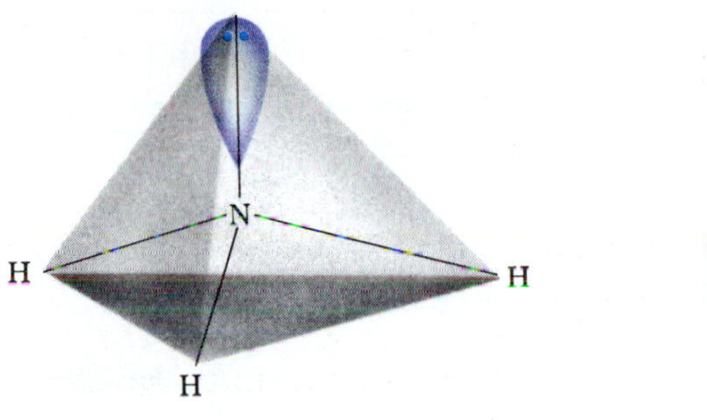

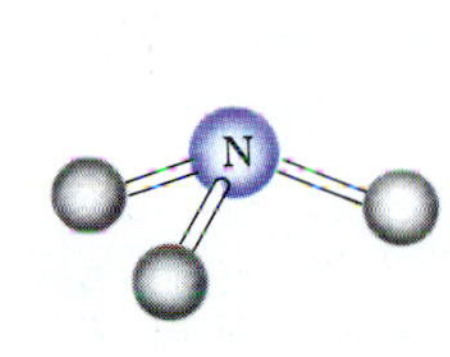

Figure 7.5
Two ways of showing the geometry of the NH_3 molecule. The orientation of the electron pairs, including the unshared pair (blue lobe), is shown at the left. The orientation of the atoms is shown at the right. The nitrogen atom is located directly above the center of the equilateral triangle formed by the three hydrogen atoms. The NH_3 molecule is described as triangular pyramid.

7.5 shows the positions of the atoms in NH_3. The nitrogen atom is located above the center of an equilateral triangle formed by the three hydrogen atoms. The molecular geometry of the NH_3 molecule is described as a **triangular pyramid**. The nitrogen atom is at the apex of the pyramid, while the three hydrogen atoms form its triangular base. The molecule is three-dimensional, as the word "pyramid" implies.

The development we have just gone through for NH_3 is readily extended to the water molecule, H_2O. Here, the Lewis structure shows that the central oxygen atom is surrounded by two single bonds and two unshared pairs:

$$\overset{\cdot\cdot}{\underset{\cdot\cdot}{\text{O}}}\ \text{with H and H}$$

The diagram at the left of Figure 7.6 emphasizes that the four electron pairs are oriented tetrahedrally. At the right, the positions of the atoms are shown. Clearly they are not in a straight line; the H_2O molecule is **bent**.

Unshared pairs reduce bond angles below ideal values

Experiments show that the bond angles in NH_3 and H_2O are slightly less than the ideal value of 109.5°. In NH_3 (three single bonds, one unshared pair around N), the bond angle is 107°. In H_2O (two single bonds, two unshared pairs around O), the bond angle is about 105°.

These effects can be explained in a rather simple way. An unshared pair is attracted by one nucleus, that of the atom to which it belongs. In contrast, a bonding pair is attracted by two nuclei, those of the two atoms it joins. Hence the electron cloud of an unshared pair is expected to spread out over a larger volume than that of a bonding pair. In NH_3, this tends to force the bonding pairs closer to one another, thereby reducing the bond angle. Where there are two unshared pairs, as in H_2O, this effect is more pronounced. In general, the VSEPR model predicts that unshared electron pairs will occupy slightly more space than bonding pairs.

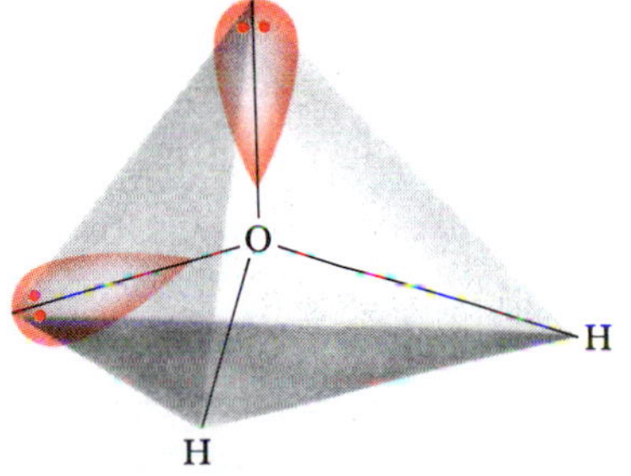

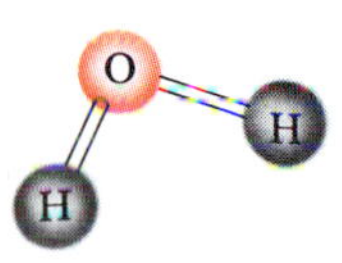

Figure 7.6
Two ways of showing the geometry of the H_2O molecule. At the left, the two unshared pairs are shown. As you can see from the drawing at the right, H_2O is a bent molecule. The bond angle, 105°, is a little smaller than the tetrahedral angle, 109.5°. The unshared pairs spread out over a larger volume than that occupied by the bonding pairs.

Table 7.3 Geometries with 2, 3, or 4 Electron Pairs Around a Central Atom

No. of Terminal Atoms (X) + Unshared Pairs (E)	Species Type	Ideal Bond Angles	Molecular Geometry	Examples
2	AX_2	180°	linear	BeF_2, CO_2
3	AX_3	120°	triangular planar	BF_3, SO_3
	AX_2E	120°*	bent	GeF_2, SO_2
4	AX_4	109.5°	tetrahedron	CH_4
	AX_3E	109.5°*	triangular pyramid	NH_3
	AX_2E_2	109.5°*	bent	H_2O

*In these species, the observed bond angle is ordinarily somewhat less than the ideal value.

AX_3E means that a central atom (A) is bonded to three atoms (X) and has one unshared pair (E)

Table 7.3 summarizes the molecular geometries of species in which a central atom is surrounded by two, three, or four electron pairs. The table is organized in terms of the number of terminal atoms, X, and unshared pairs, E, surrounding the central atom, A.

Example 7.5 Predict the geometries of the following molecules.

(a) BeH_2 (b) OF_2 (c) PF_3

Strategy Draw the Lewis structure as a first step. Then decide what type (AX_2, AX_3, etc.) the molecule is, focusing on the central atom. Remember, X represents a terminal atom, E an unshared pair of electrons.

Solution

(a) The Lewis structure for BeH_2 is

H—Be—H

The central atom has no unshared electron pairs and two bonded atoms. The molecule is of type AX_2. It is linear with a bond angle of 180°.

(b) The Lewis structure of OF_2 is

:Ö: with two :F̤: atoms bonded (bent)

The central oxygen atom has two unshared pairs and two bonds. The molecule is of type AX_2E_2 (like H_2O); it should be bent with a bond angle somewhat less than the ideal value of 109.5°. The observed value is 103°.

(c) The Lewis structure of PF_3 is

P with one unshared pair bonded to three :F̤: atoms

The central phosphorus atom has one unshared pair and three bonded atoms. The molecule is of type AX_3E; it should be a triangular pyramid (like NH_3) with a bond angle somewhat less than 109.5° (actually, 104°).

In many expanded-octet molecules, one or more of the electron pairs around the central atom is unshared. Recall, for example, the Lewis structure of xenon tetrafluoride, XeF_4,

$$\begin{array}{ccc} :\ddot{F} & & \ddot{F}: \\ & \ddot{\underset{..}{Xe}} & \\ :\underset{..}{F} & & \underset{..}{F}: \end{array} \qquad AX_4E_2$$

Here there are six electron pairs around the xenon atom; four of these are covalent bonds to fluorine and the other two pairs are unshared. This molecule is classified as AX_4E_2.

Geometries of molecules such as these can be predicted by the VSEPR model. The results are shown in Figure 7.7. The structures listed include those of all types of molecules having five or six electron pairs around the central atom, one or more of which may be unshared.

Multiple Bonds

The VSEPR model is readily extended to species in which double or triple bonds are present. Here, a simple principle applies: ***Insofar as molecular geometry is concerned, a multiple bond behaves like a single electron pair.*** This makes sense. The four electrons in a double bond, or the six electrons in a triple bond, must be located between the two atoms, as are the two electrons in a single bond. This means that the electron pairs in a multiple bond must occupy the same region of space as those in a single bond. Hence, the "extra" electron pairs in a multiple bond have no effect upon geometry.

To illustrate this principle, consider the CO_2 molecule. Its Lewis structure is

$$:\ddot{O}{=}C{=}\ddot{O}:$$

The central atom, carbon, has two double bonds and no unshared pairs. For purposes of determining molecular geometry, we pretend that the double bonds are single bonds, ignoring the "extra" bonding pairs. The bonds are directed to be as far apart as possible, giving a 180° O—C—O bond angle. The CO_2 molecule, like BeF_2, is linear:

$$\underset{180^\circ}{F{-}Be{-}F} \qquad \underset{180^\circ}{O{=}C{=}O}$$

CO_2, like BeF_2, is an AX_2 molecule; the geometries are the same

This principle can be restated in a somewhat different way for molecules in which there is a single central atom. The geometry of such a molecule depends only upon

— *the number of terminal atoms, X, bonded to the central atom, irrespective of whether the bonds are single, double, or triple*
— *the number of unshared pairs, E, around the central atom*

This means that Table 7.3 can be used in the usual way to predict the geometry of a species containing multiple bonds.

Figure 7.7
Geometries of molecules with expanded octets. The red spheres represent terminal atoms. The open ellipses represent unshared electron pairs.

5 ELECTRON PAIRS

Species type	Structure	Description	Example	Bond angles
AX_5	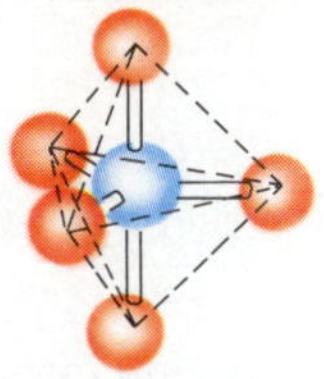	Triangular bipyramidal	PF_5	90°, 120°, 180°
AX_4E	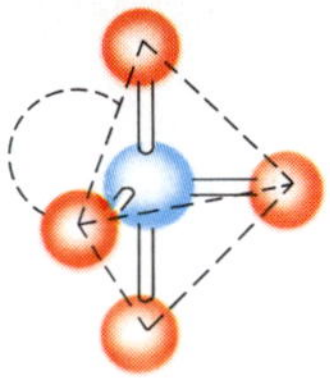	See-saw	SF_4	90°, 120°, 180°
AX_3E_2	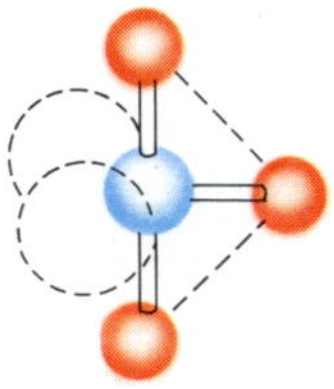	T-shaped	ClF_3	90°, 180°
AX_2E_3	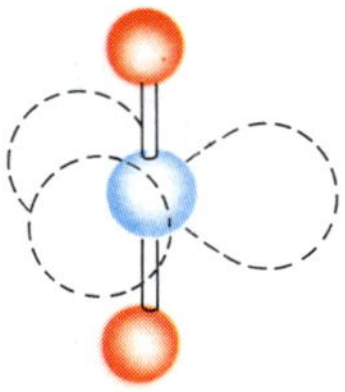	Linear	XeF_2	180°

6 ELECTRON PAIRS

Species type	Structure	Description	Example	Bond angles
AX_6	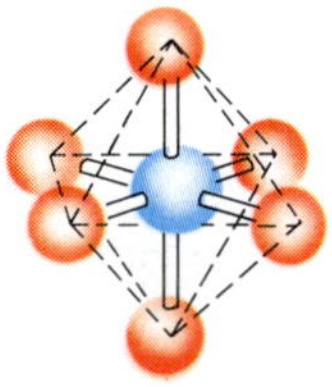	Octahedral	SF_6	90°, 180°
AX_5E	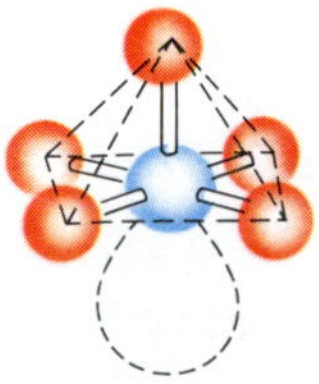	Square pyramidal	ClF_5	90°, 180°
AX_4E_2	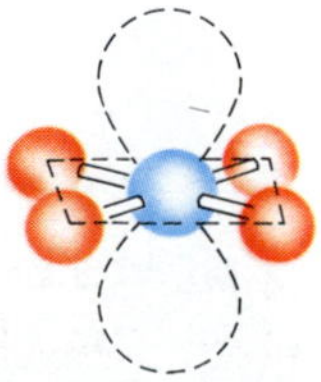	Square planar	XeF_4	90°, 180°

Example 7.6 Predict the geometries of the ClO_3^- ion, the NO_3^- ion, and the N_2O molecule, which have the Lewis structures

(a) $\left[:\ddot{O}-\ddot{Cl}-\ddot{O}:\right]^-$ with $:\ddot{O}:$ bonded below Cl (b) $\left[:\ddot{O}-N-\ddot{O}:\right]^-$ with $:O:$ double-bonded below N (c) $:\ddot{N}=N=\ddot{O}:$

Strategy Classify each species as AX_mE_n and use Table 7.3.

Solution

(a) The central atom, chlorine is bonded to three oxygen atoms; it has one unshared pair. The ClO_3^- ion is of the type AX_3E. It is a triangular pyramid; the ideal bond angle is 109.5°.
(b) The central atom, nitrogen, is bonded to three oxygen atoms; it has no unshared pairs. The NO_3^- ion is of the type AX_3. It has the geometry of an equilateral triangle, with the nitrogen atom at the center; the bond angle is 120°. The ion is triangular planar.
(c) The central nitrogen atom is bonded to two other atoms with no unshared pairs. The molecule, type AX_2, is linear, with a bond angle of 180°.

The VSEPR model applies equally well to molecules in which there is no single central atom. Consider the acetylene molecule, C_2H_2. Recall that here the two carbon atoms are joined by a triple bond:

$$H-C\equiv C-H$$

Each carbon atom behaves as if it were surrounded by two electron pairs. Both of the bond angles (H—C≡C and C≡C—H) are 180°. The molecule is linear; the four atoms are in a straight line. The two "extra" electron pairs in the triple bond do not affect the geometry of the molecule.

In ethylene, C_2H_4, there is a double bond between the two carbon atoms. The molecule has the geometry to be expected if each carbon atom had only three pairs of electrons around it.

$$\begin{matrix} H & & & & H \\ & \diagdown & & \diagup & \\ & & C=C & & \\ & \diagup & & \diagdown & \\ H & & & & H \end{matrix}$$

The six atoms are located in a plane, with bond angles of 120°.

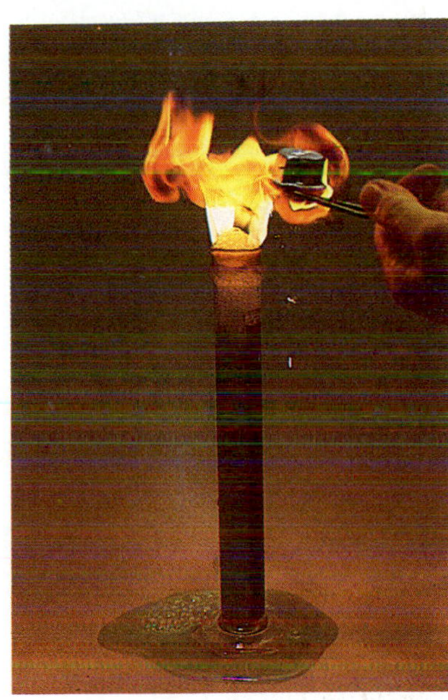

Synthesis and combustion of acetylene, C_2H_2. (Charles D. Winters)

7.3 POLARITY OF MOLECULES

Covalent bonds and molecules held together by such bonds may be

— *polar.* As a result of an unsymmetrical distribution of electrons, the bond or molecule contains a positive and a negative pole and is therefore a **dipole**.
— *nonpolar.* A symmetrical distribution of electrons leads to a bond or molecule with no positive or negative poles.

Polar and Nonpolar Covalent Bonds

The two electrons in the H_2 molecule are shared equally by the two nuclei. Stated another way, a bonding electron is as likely to be found in the vicinity of one nucleus as another. Bonds of this type are described as nonpolar. **Non-**

polar bonds are formed whenever the two atoms joined are identical, as in H_2 and F_2.

All molecules, except those of elements, have polar bonds

In the HF molecule, the distribution of the bonding electrons is somewhat different from that found in H_2 or F_2. Here, the density of the electron cloud is greater about the fluorine atom. The bonding electrons, on the average, are shifted toward fluorine and away from the hydrogen (atom Y in Fig. 7.8). Bonds in which the electron density is unsymmetrical are referred to as **polar bonds**.

Atoms of two different elements always differ at least slightly in their electronegativity (recall Table 6.4, p. 156). Hence covalent bonds between unlike atoms are always polar. Consider, for example, the H—F bond. Since fluorine has a higher electronegativity (4.0) than does hydrogen (2.2), bonding electrons are displaced toward the fluorine atom. The H—F bond is polar, with a partial negative charge at the fluorine atom and a partial positive charge at the hydrogen atom.

The extent of polarity of a covalent bond is related to the difference in electronegativities of the bonded atoms. If this difference is large, as in HF ($\Delta EN = 1.8$), the bond is strongly polar. Where the difference is small, as in H—C ($\Delta EN = 0.3$), the bond is only slightly polar.

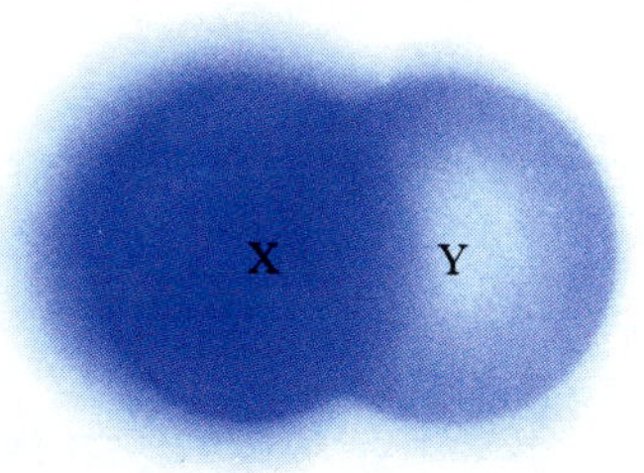

Figure 7.8
If two atoms, X and Y, differ in electronegativity, the bond between them is polar. The electron cloud associated with the bonding electrons is concentrated around the more electronegative atom, in this case, X.

Polarity of Molecules

A polar molecule is one that contains positive and negative poles. There is a partial positive charge (positive pole) at one point in the molecule and a partial negative charge (negative pole) at a different point. As shown in Figure 7.9, polar molecules orient themselves in the presence of an electric field. The positive pole in the molecule tends to align with the external negative charge, and the negative pole with the external positive charge. In contrast, there are no positive and negative poles in a nonpolar molecule. In an electric field, nonpolar molecules, such as H_2, show no preferred orientation.

If a molecule is diatomic, it is easy to decide whether it is polar or nonpolar. A diatomic molecule has only one kind of bond; hence the polarity of the molecule is the same as the polarity of the bond. Hydrogen and fluorine (H_2, F_2) are nonpolar because the bonded atoms are identical and the bond is

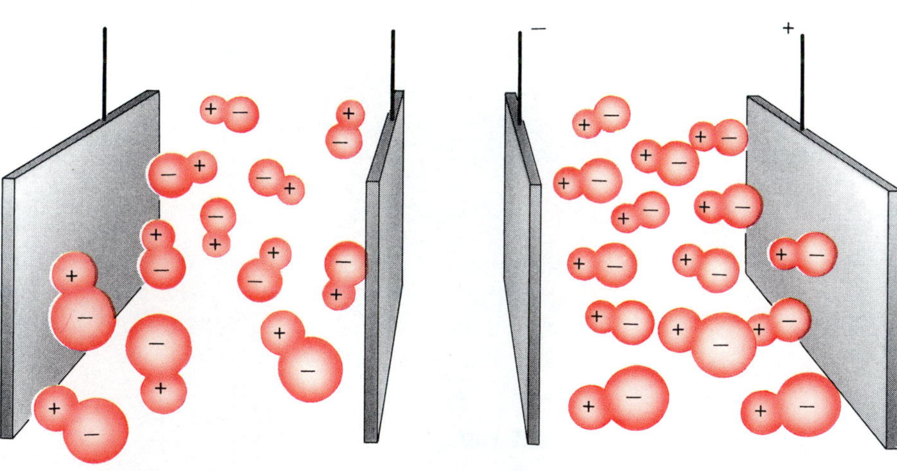

Figure 7.9
Orientation of polar molecules in an electric field. In the absence of the field, polar molecules are randomly oriented. In an electric field, polar molecules such as HF line up as shown ("Field on"). Nonpolar molecules such as H_2 do not line up.

nonpolar. Hydrogen fluoride, HF, on the other hand, has a polar bond, so the molecule is polar. The bonding electrons spend more time near the fluorine atom so that there is a negative pole at that end and a positive pole at the hydrogen end. This is sometimes indicated by writing

$$\text{H} \mapsto \text{F}$$

The arrow points toward the negative end of the polar bond (F atom); the plus sign is at the positive end (H atom). The HF molecule is called a *dipole;* it contains positive and negative poles.

If a molecule contains more than two atoms it is not so easy to decide whether it is polar or nonpolar. In this case, not only bond polarity but also molecular geometry determines the polarity of the molecule. To illustrate what is involved, consider the molecules shown in Figure 7.10.

1. In BeF_2 there are two polar Be—F bonds; in both bonds, the electron density is concentrated around the more electronegative fluorine atom. However, since the BeF_2 molecule is linear, the two Be $\mapsto$ F dipoles are in opposite directions and cancel one another. The molecule has no net dipole and hence is nonpolar. From a slightly different point of view, in BeF_2 the centers of positive and negative charge coincide with each other, at the Be atom. There is no way that a BeF_2 molecule can line up in an electric field.

2. Since oxygen is more electronegative than hydrogen (3.5 vs. 2.2) an O—H bond is polar, with the electron density higher around the oxygen atom. In the bent H_2O molecule, the two H $\mapsto$ O dipoles do not cancel each other. Instead, they add to give the H_2O molecule a net dipole. The center of negative charge is located at the O atom; this is the negative pole of the molecule. The center of positive charge is located midway between the two H atoms; the positive pole of the molecule is at that point. The H_2O molecule is polar. It tends to line up in an electric field with the oxygen atom oriented toward the positive electrode.

Nonpolar molecules: A_2, AX_2 (linear), AX_3 (triangular planar), AX_4 (tetrahedral)

3. Carbon tetrachloride, CCl_4, is another molecule which, like BeF_2, is nonpolar despite the presence of polar bonds. Each of its four bonds is a dipole,

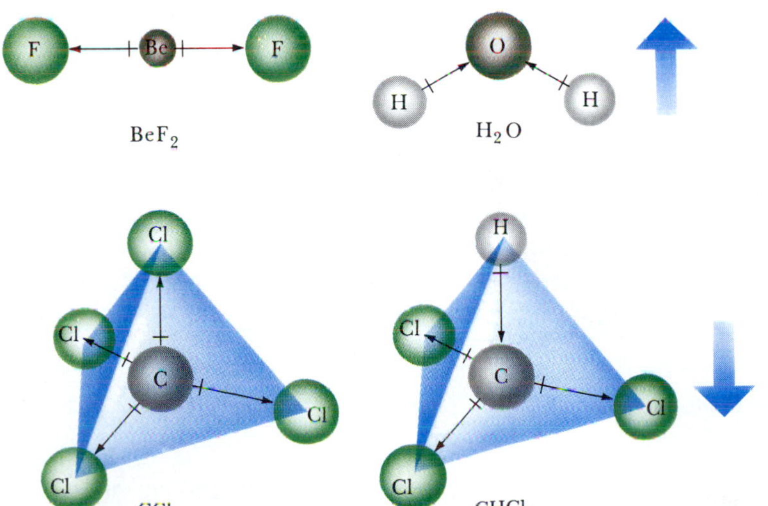

Figure 7.10
Polarity of molecules. In all these molecules, the bonds are polar, as indicated by the $\mapsto$ notation. However, in BeF_2 and CCl_4, the bond dipoles cancel and the molecule is nonpolar. In H_2O and $CHCl_3$, on the other hand, there is a net dipole and the molecule is polar. (The broad arrow beside the molecule points to the negative pole.)

The molecule may be nonpolar if the geometry is symmetrical

C $\leftrightarrow$ Cl. However, because the four bonds are arranged symmetrically around the carbon atom, they cancel. As a result, the molecule has no net dipole; it is nonpolar. If one of the Cl atoms in CCl_4 is replaced by hydrogen, the situation changes. In the $CHCl_3$ molecule, the H $\leftrightarrow$ C dipole does not cancel with the three C $\leftrightarrow$ Cl dipoles. Hence $CHCl_3$ is polar.

There are two criteria for determining the polarity of a molecule; bond polarity and molecular geometry. ***If the polar A—X bonds in a molecule AX_mE_n are arranged symmetrically around the central atom A, the molecule is nonpolar.***

Example 7.7 Determine whether each of the following is polar or nonpolar

(a) SO_2 (b) BF_3 (c) CO_2

Strategy Write the Lewis structure of the molecule and classify it as AX_mE_n. Using Table 7.3 or Figure 7.7, decide upon the geometry of the molecule. Then decide whether the dipoles cancel, in which case it is nonpolar.

Solution

(a) The Lewis structure of SO_2 is shown on p. 168; it is of the type AX_2E. The molecule is bent, so the dipoles do not cancel. The SO_2 molecule is polar.

(b) The Lewis structure of BF_3 is shown on p. 172; it is of the type AX_3. The molecule is an equilateral triangle with the boron atom at the center. The three polar bonds cancel one another; BF_3 is nonpolar.

(c) The Lewis structure of CO_2 is shown on p. 179; it is of the type AX_2. The molecule is linear, so the two C $\leftrightarrow$ O dipoles cancel each other; CO_2 is nonpolar.

7.4 ATOMIC ORBITALS; HYBRIDIZATION

Pauling is a versatile chemist with contributions in many areas

In the 1930s a theoretical treatment of the covalent bond was developed by Linus Pauling and J. C. Slater, among others. It is referred to as the *atomic orbital*, or **valence bond**, model. According to this model, a covalent bond consists of a pair of electrons of opposed spin within an atomic orbital. For example, a hydrogen atom forms a covalent bond by accepting an electron from another atom to complete its 1s orbital. Using orbital diagrams, we could write

	1s
isolated H atom	(↑)
H atom in a stable molecule	(↑↓)

The second electron, shown in color, is contributed by another atom. This could be another H atom in H_2, an F atom in HF, a C atom in CH_4, and so on.

This simple model is readily extended to other atoms. The fluorine atom (electron configuration $1s^22s^22p^5$) has a half-filled p orbital:

	1s	2s	2p		
isolated F atom	(↑↓)	(↑↓)	(↑↓)	(↑↓)	(↑)

By accepting an electron from another atom, F can complete this 2p orbital:

	1s	2s	2p		
F atom in HF, F_2, . . .	(↑↓)	(↑↓)	(↑↓)	(↑↓)	(↑↓)

According to this model, it would seem that in order for an atom to form a covalent bond, it must have an unpaired electron. Indeed, the number of bonds formed by an atom should be determined by its number of unpaired electrons. Since hydrogen has an unpaired electron, an H atom should form one covalent bond, as indeed it does. The same holds for the F atom, which forms only one bond. Noble gas atoms, such as He and Ne, which have no unpaired electrons, should not form bonds at all; they don't.

When this simple idea is extended beyond hydrogen, the halogens, and the noble gases, problems arise. Consider, for example, the three atoms Be ($Z = 4$), B ($Z = 5$), and C ($Z = 6$):

	1s	2s	2p		
Be atom	(↑↓)	(↑↓)	( )	( )	( )
B atom	(↑↓)	(↑↓)	(↑)	( )	( )
C atom	(↑↓)	(↑↓)	(↑)	(↑)	( )

Notice that the beryllium atom has no unpaired electrons, the boron atom has one, and the carbon atom two. Simple valence bond theory would predict that Be, like He, should not form covalent bonds. A boron atom should form one bond, carbon two. Experience tells us that these predictions are wrong. Beryllium forms two bonds in BeF_2; B forms three bonds in BF_3. Carbon ordinarily forms four bonds, not two.

To explain these and other discrepancies, simple valence bond theory must be modified. It is necessary to invoke a new kind of atomic orbital, called a **hybrid orbital**.

Hybrid Orbitals: sp, sp^2, sp^3, sp^3d, sp^3d^2

The formation of the BeF_2 molecule can be explained by assuming that, as two fluorine atoms approach, the atomic orbitals of the beryllium atom undergo a significant change. Specifically, the 2s orbital is mixed or *hybridized* with a 2p orbital to form two new **sp hybrid orbitals**.

one s atomic orbital + one p atomic orbital ⟶ *two* sp hybrid orbitals

Notice (Fig. 7.11) that the number of hybrid orbitals formed is equal to the number of atomic orbitals mixed. This is always true in hybridization of orbitals. Moreover, the energies of the hybrid orbitals are intermediate between those of the atomic orbitals from which they are derived.

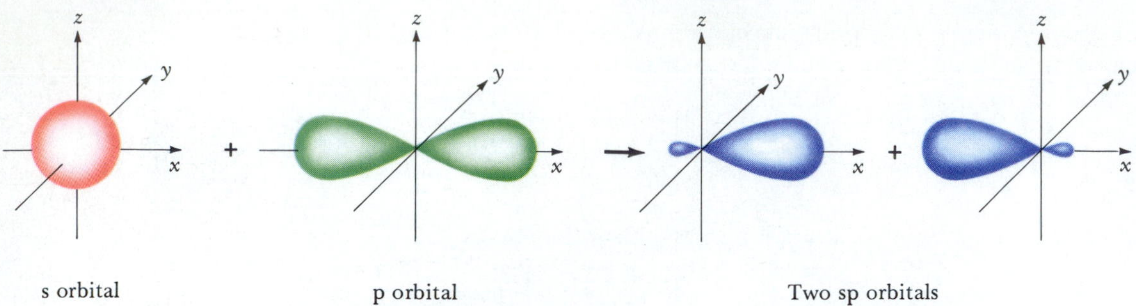

Figure 7.11
Formation of sp hybrid orbitals. The mixing of an s orbital with a p orbital gives two sp hybrid orbitals.

In the BeF_2 molecule, there are two electron-pair bonds. These electron pairs are located in the sp hybrid orbitals just described. In each case, one electron comes from the beryllium atom; the other is the unpaired 2p electron of the fluorine atom. The "orbital diagram" for the beryllium atom in the BeF_2 molecule can be written as:

	1s	2s	2p
Be in BeF_2	(↑↓)	(↑↓)	(↑↓)( )( )

(The colored arrows indicate the two electrons supplied by the fluorine atoms. The horizontal lines enclose the hybrid orbitals involved in bond formation).

A similar argument can be used to explain why boron forms three bonds and carbon four.

In the case of boron:

one s atomic orbital + two p atomic orbitals ⟶ *three* sp^2 hybrid orbitals

With carbon:

The number of orbitals is indicated by the superscript

one s atomic orbital + three p atomic orbitals ⟶ *four* sp^3 hybrid orbitals

Thus we have

	1s	2s	2p
B in BF_3	(↑↓)	(↑↓)	(↑↓) (↑↓) ( )
C in CH_4	(↑↓)	(↑↓)	(↑↓) (↑↓) (↑↓)

Hybridization: sp (AX_2), sp^2(AX_2E, AX_3), sp^3(AX_4, AX_3E, AX_2E_2)

You will recall that the bond angles in NH_3 and H_2O are very close to that in CH_4. This suggests that the four electron pairs surrounding the central atom in NH_3 and H_2O, like those in CH_4, occupy sp^3 hybrid orbitals. In NH_3, three of these orbitals are filled by bonding electrons, the other by the unshared pair on the nitrogen atom. In H_2O, two of the sp^3 orbitals of the oxygen atom contain bonding electron pairs; the other two contain unshared pairs. The situation in NH_3 and H_2O is not unique. In general, we find that *unshared as well as shared electron pairs can be located in hybrid orbitals.*

Table 7.4 Hybrid Orbitals and Their Geometries

Number of Electron Pairs	Atomic Orbitals	Hybrid Orbitals	Orientation	Examples
2	s, p	sp	linear	BeF_2, CO_2
3	s, two p	sp^2	triangular planar	BF_3, SO_3
4	s, three p	sp^3	tetrahedron	CH_4, NH_3, H_2O
5	s, three p, d	sp^3d	triangular bipyramid	PCl_5, SF_4, ClF_3
6	s, three p, two d	sp^3d^2	octahedron	SF_6, ClF_5, XeF_4

The "extra" electron pairs in an expanded octet are accommodated by using d orbitals. The phosphorus atom in PCl_5 and the sulfur atom in SF_6 make use of 3d as well as 3s and 3p orbitals:

		3s	3p	3d
P atom in PCl_5	$[_{10}Ne]$	(↑↓)	(↑↓) (↑↓) (↑↓)	(↑↓) ( ) ( ) ( ) ( )
S atom in SF_6	$[_{10}Ne]$	(↑↓)	(↑↓) (↑↓) (↑↓)	(↑↓) (↑↓) ( ) ( ) ( )

The orbitals used by the five pairs of bonding electrons surrounding the phosphorus atom in PCl_5 are **sp^3d hybrid orbitals**.

one s orbital + three p orbitals + one d orbital ⟶ *five* sp^3d hybrid orbitals

Similarly, in SF_6, the six pairs of bonding electrons are located in **sp^3d^2 hybrid orbitals**:

sp^3d (AX_5, AX_4E, AX_3E_2, AX_2E_3); sp^3d^2 (AX_6, AX_5E, AX_4E_2)

one s orbital + three p orbitals + two d orbitals ⟶ *six* sp^3d^2 hybrid orbital

Table 7.4 gives the orientation in space of hybrid orbitals. The geometries are exactly as predicted by VSEPR theory. In each case, the several hybrid orbitals are directed to be as far apart as possible.

Example 7.8 Give the hybridization of

(a) carbon in CF_4 (b) phosphorus in PF_3 (c) sulfur in SF_4

Strategy Draw the Lewis structure for the molecule and determine the number of electron pairs (single bonds or unshared pairs) around the central atom. The possible hybridizations are sp (two pairs), sp^2 (three pairs), sp^3 (four pairs), sp^3d (five pairs), and sp^3d^2 (six pairs).

Solution

(a) For CF_4, the Lewis structure is

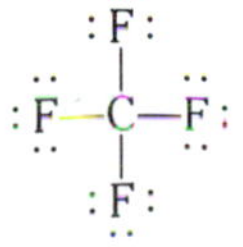

```
       ..
      :F:
 ..    |    ..
:F  — C — F:
 ..    |    ..
      :F:
       ..
```

There are four bonds and no unshared electron pairs around C, the central atom. The hybridization is sp^3.

(b) The Lewis structure of PF_3 is

```
 ..  ..  ..
:F—P—F:
 ..  |  ..
    :F:
     ..
```

There are three bonds and one unshared pair, making a total of four pairs. The hybridization is sp^3.

(c) Sulfur tetrafluoride has the Lewis structure

```
 ..      ..
:F       F:
 .. \ .. / ..
      S
 .. /    \ ..
:F       F:
 ..      ..
```

There are five electron pairs around sulfur: four bonds, and one unshared pair. The hybridization is sp^3d.

Multiple Bonds; Pi and Sigma Bonds

In a multiple bond, only one of the electron pairs is hybridized. This explains why the hybridizations in ethylene, C_2H_4, and acetylene, C_2H_2, are sp^2 and sp, respectively. Using blue lines for hybridized electron pairs, red for unhybridized:

```
H       H
  \
   C=C            H—C≡C—H
  /
H       H
```

ethylene
sp^2; bond angle 120°

acetylene
sp; bond angle 180°

The "extra" electron pairs don't affect the geometry or the hybridization

The "extra" electron pair present in the C_2H_4 molecule is located in a special type of orbital, shown in red at the left of Figure 7.12. This orbital consists of two lobes, one above the bond axis, the other below it. Along the bond axis itself, the electron density is zero. An orbital of this type is called a pi (π) orbital. A bond consisting of two electrons located in such an orbital is referred to as a **pi (π) bond**.

There is one pi bond in the C_2H_4 molecule. In the C_2H_2 molecule, shown at the right of Figure 7.12, there are two pi bonds. The bonding electrons are located in two different pi orbitals at right angles to each other.

The remaining bond orbitals in C_2H_2 and C_2H_4, shown in blue in Figure 7.12, have a quite different shape. They consist of a single lobe in which the electron density is concentrated in the region directly between the two bonded atoms. Bonds of this type are referred to as **sigma (σ) bonds**. There are five sigma bonds in C_2H_4, four C—H bonds and one C—C bond. There are three sigma bonds in C_2H_2, two C—H bonds and one C—C bond. In general,

σ and π orbitals, like s and p orbitals, differ in shape; each orbital can hold $2e^-$

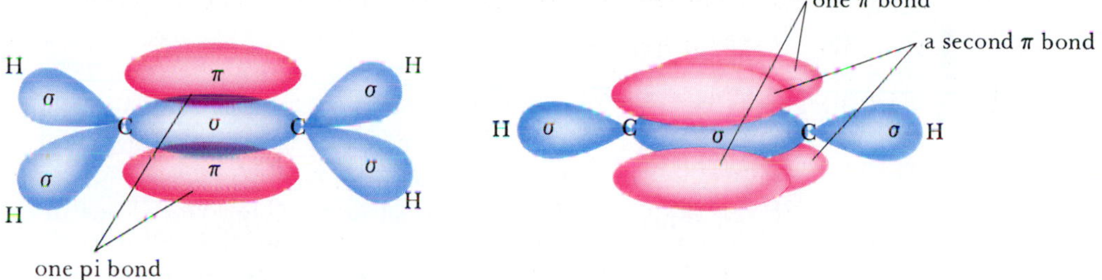

Figure 7.12
Bonding in ethylene, C_2H_4, and acetylene, C_2H_2. The sigma bond backbone is shown in blue; the pi bonds (1 in C_2H_4, 2 in C_2H_2) are shown in red. Note that a pi bonding orbital consists of two lobes.

— all single bonds are sigma bonds
— one of the electron pairs in a multiple bond is a sigma bond; the others are pi bonds.

Example 7.9 Consider the species

(a) N_2 (b) NO_3^-

For each, state the hybridization of nitrogen and give the number of pi and sigma bonds.

Strategy The first step is to draw Lewis structures. To find the hybridization of the nitrogen atom, include

— unshared electron pairs
— electron pairs forming single bonds
— one and only one of the electron pairs in a multiple bond

To find the number of sigma bonds, include

— all single bonds
— one and only one of the electron pairs in a multiple bond

The remaining electron pairs in the multiple bond correspond to pi bonds.

Solution

(a) :N≡N:

For each N atom, hybridize one unshared pair and one of the electron pairs in the triple bond; the hybridization is sp. There are three bonds, one sigma, two pi.

(b) $\left[:\ddot{O}-N-\ddot{O}: \right]^-$ with N double-bonded to $:\ddot{O}:$ below

Three of the four electron pairs around the nitrogen atom are hybridized; the hybridization is sp^2. There are three sigma bonds and one pi bond.

PERSPECTIVE

The Noble Gases

The modern periodic table contains six relatively unreactive gases which were unknown to Mendeleev: the noble gases comprising Group **18** at the far right of the table. The first of these elements to be isolated was argon, which makes up about 0.9% of air. The physicist Lord Rayleigh found that the density of "atmospheric nitrogen," obtained by removing O_2, CO_2, and H_2O from air, was slightly greater than that of chemically pure N_2 ($\mathcal{M}$ = 28.02 g/mol). Following up on that observation, Sir William Ramsey separated argon ($\mathcal{M}$ = 39.95 g/mol) from air. Over a three-year period between 1895 and 1898, this remarkable Scotsman, who never took a formal course in chemistry, isolated three more noble gases: Ne, Kr, and Xe. In effect, Ramsey added a whole new group to the periodic table.

Helium, the first member of the group, was detected in the spectrum of the sun in 1868. Because of its low density (1/7 that of air), helium is used in all kinds of balloons and in synthetic atmospheres to make breathing easier for people suffering from emphysema. Research laboratories use helium as a liquid coolant to achieve very low temperatures (bp He = −269°C). Argon and, more recently, krypton are used to provide an inert atmosphere in light bulbs, thereby extending the life of the tungsten filament. In "neon signs," a high voltage is passed through a glass tube containing neon at very low pressures. The red glow emitted corresponds to an intense line at 640 nm in the neon spectrum.

Until about 30 years ago, these elements were referred to as "inert gases"; they were believed to be entirely unreactive toward other substances. In 1962 Neil Bartlett, a 29-year-old chemist at the University of British Columbia, shook up the world of chemistry by preparing the first noble gas compound. In the course of his research on platinum-fluorine compounds, he isolated a reddish solid that he showed to be $O_2^+(PtF_6^-)$. Bartlett realized that the ionization energy of Xe (1170 kJ/mol) is virtually identical to that of the O_2 molecule (1165 kJ/mol). This encouraged him to attempt to make the analogous compound $XePtF_6$. His success opened up a new era in noble gas chemistry.

Figure 7.A
Crystals of xenon tetrafluoride, XeF_4. (Argonne National Laboratory)

The most stable binary compounds of xenon are the three fluorides, XeF_2, XeF_4, and XeF_6. Xenon difluoride can be prepared quite simply by exposing a 1:1 mol mixture of xenon and fluorine to ultraviolet light; colorless crystals of XeF_2 (mp = 129°C) form slowly.

$$Xe(g) + F_2(g) \longrightarrow XeF_2(s)$$

The higher fluorides are prepared using excess fluorine (Figure 7.A). All of these compounds are stable in dry air at room temperature. However, they react with water to form compounds in which one or more of the fluorine atoms has been replaced by oxygen. Thus xenon hexafluoride reacts rapidly with water to give the trioxide

$$XeF_6(s) + 3H_2O(l) \longrightarrow XeO_3(s) + 6HF(g)$$

Xenon trioxide is highly unstable; it detonates if warmed above room temperature.

The chemistry of xenon is much more extensive than that of any other noble gas. Only one binary compound of krypton, KrF_2, has been prepared.

It is a colorless solid with a positive heat of formation and decomposes at room temperature. The chemistry of radon is difficult to study because all of its isotopes are radioactive. Indeed, the radiation given off is so intense that it decomposes any reagent added to radon in an attempt to bring about a reaction.

Radon infiltrating into home basements can cause a health problem (Chap. 18)

CHAPTER HIGHLIGHTS

KEY CONCEPTS

1. *Draw Lewis structures for molecules and polyatomic ions*
   (Examples 7.1, 7.2, 7.4; Problems 1–20, 80, 87, 88)
2. *Write resonance forms*
   (Example 7.3; Problems 21–26)
3. *Use Table 7.3 and Figure 7.7, applying VSEPR theory, to predict molecular geometry*
   (Examples 7.5, 7.6; Problems 27–48)
4. *Having derived the geometry of a molecule or polyatomic ion, predict whether it will be a dipole*
   (Example 7.7; Problems 49–54)
5. *State the hybridization of an atom in a covalently bonded species and/or the number of pi and sigma bonds*
   (Examples 7.8, 7.9; Problems 55–72)

KEY TERMS

bond
—double
—pi, sigma
—nonpolar
—polar
—triple
dipole

expanded octet
formal charge
hybrid orbital
Lewis structure
octahedron
octet rule
resonance

tetrahedron
triangular bipyramid
triangular planar
triangular pyramid
unshared pair
valence electron

SUMMARY PROBLEM

Consider the molecules SO_3, XeO_3, XeO_4, and XeF_2.

a. Draw the Lewis structures of these molecules.
b. Draw resonance structures for SO_3.
c. Describe the geometries of these molecules, including ideal bond angles.
d. Classify each molecule as being polar or nonpolar.
e. Give the hybridization of the central atom in each molecule.
f. State the number of pi and sigma bonds in each molecule.

Answers

a. 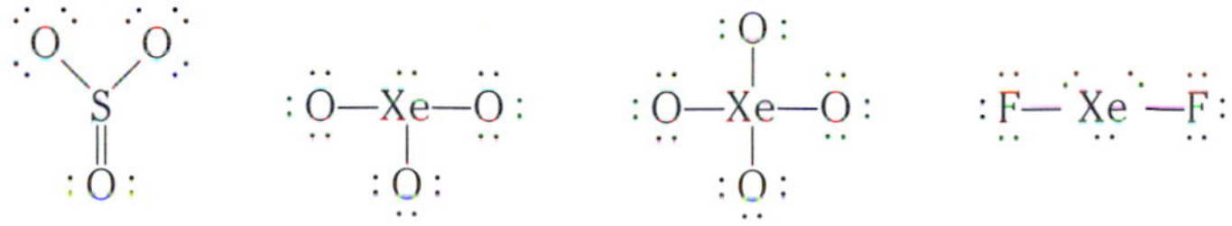 :Ö: single-bonded to S, :Ö: single-bonded to S, S=Ö: ; :Ö—Xe—Ö: with Xe—:Ö: ; :Ö—Xe—Ö: with :Ö: above and :Ö: below Xe ; :F̈—Xe—F̈:

b. 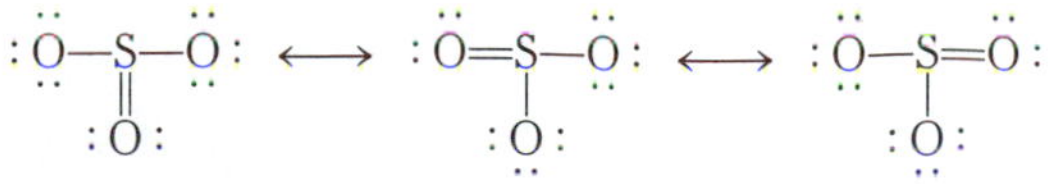 :Ö—S—Ö: (S=O:) ⟷ :O=S—Ö: (S—:Ö:) ⟷ :Ö—S=O: (S—:Ö:)

c. SO_3: triangular planar, 120° XeO_3: triangular pyramid, 109.5°
XeO_4: tetrahedron, 109.5° XeF_2: linear, 180°
d. All nonpolar except XeO_3
e. sp^2 in SO_3; sp^3 in XeO_3 and XeO_4; sp^3d in XeF_2
f. SO_3: 3 σ, 1 π; XeO_3: 3 σ; XeO_4: 4 σ; XeF_2: 2 σ

QUESTIONS & PROBLEMS

Lewis Structures

1. Write the Lewis structure for the following molecules and polyatomic ions. In each case, the first atom listed is the central atom.
a. $POCl_3$ **b.** NF_3 **c.** ClO_4^- **d.** $GeCl_4$

2. Follow the directions of Question 1 for
a. PH_3 **b.** SiF_4 **c.** NHO_2 **d.** NH_4^+

3. Follow the direction of Question 1 for
a. PCl_4^+ **b.** N_3^- **c.** CN^- **d.** ClO_2^-

4. Follow the directions of Question 1 for
a. IO_3^- **b.** ClF_2^+ **c.** SO_3^{2-} **d.** AsO_3^{3-}

5. Follow the directions of Question 1 for
a. SeF_6 **b.** PCl_5 **c.** $TeBr_4$ **d.** KrF_2

6. Follow the directions of Question 1 for
a. BrF_3 **b.** XeO_2F_2 **c.** I_3^- **d.** IF_5

7. Draw Lewis structures for the following species (the skeleton is indicated by the way the molecule is written).
a. $H_2N—OH$ **b.** $H_2C—CO$ **c.** $H_2C—N—N$

8. Follow the directions of Question 7 for the following species.
a. $(HO—CO_2)^-$ **b.** $H_3C—CO_2H$ **c.** $(HO)_2—S—O$

9. Radioastronomers have detected the isoformyl ion, HOC^+, in outer space. Write the Lewis structure for this ion.

10. Formation of dioxirane, H_2CO_2, has been suggested as a factor in smog formation. The molecule contains an oxygen–oxygen bond. Draw its Lewis structure.

11. Formic acid is the irritating substance that gets on your skin when an ant bites. Its formula is HCOOH. Carbon is the central atom and it has no O—O bonds. Write its Lewis structure.

12. Peroxyacetylnitrate (PAN) is the substance in smog that makes your eyes water. Its skeletal structure is

$$H_3C—CO_3—NO_2$$

It has one O—O bond and three O—N bonds. Draw its Lewis structure.

13. Two different molecules have the formula $C_2H_4Br_2$. Draw a Lewis structure for each molecule. (All the H and Br atoms are bonded to carbon.)

14. There are two compounds with the molecular formula C_2H_6O. Draw a Lewis structure for each compound.

15. Give the formula of a polyatomic ion which you would expect to have the same Lewis structure as
a. HCl **b.** F_2 **c.** N_2 **d.** CCl_4

16. Give the formula of a molecule which you would expect to have the same Lewis structure as
a. ClO^- **b.** $H_2PO_4^-$ **c.** PH_4^+ **d.** SiO_4^{4-}

17. Write a Lewis structure for
a. HSO_3^- **b.** SCN^- **c.** $NFCl_2$
d. $P_2O_7^{4-}$ (no O—O bonds)

18. Write Lewis structures for the following species.
a. BCl_4^- **b.** $S_2O_3^{2-}$ **c.** $H_2PO_4^-$ **d.** ClO_3^-

19. Write reasonable Lewis structures for the following species, none of which follow the octet rule.
a. CH_3 **b.** CO^- **c.** NO **d.** BCl_3

20. Write reasonable Lewis structures for the following species, none of which obey the octet rule.
a. NO_2 **b.** BeH_2 **c.** SO_2^- **d.** CO^+

Resonance Forms

21. Draw resonance structures for
a. CO_3^{2-} **b.** SeO_3 **c.** CS_3^{2-}

22. Draw resonance structures for
a. SCN^- **b.** HCO_2^- **c.** $HONO_2$

23. The oxalate ion, $C_2O_4^{2-}$, has the skeleton structure

```
O       O
 \     /
  C — C
 /     \
O       O
```

a. Complete the Lewis structure of this ion.
b. Draw three resonance forms for $C_2O_4^{2-}$, equivalent to the Lewis structure drawn in (a).
c. Is

```
  ..  ..  ..      ..    2-
[ O═C—O—C—O: ]
  ..      ..  ║   ..
             :O:
```

another resonance form of the oxalate ion? Explain.

24. The Lewis structure for hydrazoic acid may be written as

```
H
 \        ..
  N═N═N:
  ..
```

a. Draw two other resonance forms for this molecule.
b. Is

```
      ..
      N
     / \\
 :N ——— N—H
  ..
```

another resonance form of hydrazoic acid? Explain.

25. The skeleton structure for disulfur dinitride, S_2N_2, is

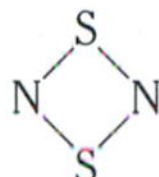

Draw the resonance forms of this molecule.

26. Borazine, $B_3N_3H_6$, has the skeleton

H—B ring: H—N, N—H, H—B, B—H, N—H (six-membered ring of alternating B and N, each bearing one H)

Draw the resonance forms of the molecule.

Molecular Geometry

27. Describe the geometry of each of the following molecules. (The way the Lewis structure is written does not necessarily imply the geometry.)

a. :O—S=O: **b.** :N≡N—O: **c.** :S=C=O: **d.** :Cl—N(H)—H

28. Describe the geometry of each of the following molecules. (The way the Lewis structure is written does not necessarily imply the geometry.)

a. :Cl—Be—Cl: **b.** :Cl—Ge—Cl: **c.** :O=N—Cl: **d.** H—C(H)(Br)—Br:

29. Describe the geometry of the following polyatomic ions.

a. $[:O—N=O]^-$ **b.** $[H—N(H)(H)—H]^+$ **c.** $[:O—N(=O)—O:]^-$ **d.** $[:S=C=N:]^-$

30. Describe the geometry of the following polyatomic ions.

a. $[:O—Mn(O)(O)—O:]^-$ **b.** $[:O—O—O:]^{2-}$ **c.** $[:O—C(=O)—O:]^{2-}$ **d.** $[:O—Cl(O)—O:]^-$

31. Predict the geometry of the following molecules
a. RnF_4 **b.** PCl_5 **c.** SeF_6 **d.** $TeBr_4$

32. Predict the geometry of the following molecules.
a. $SeCl_4$ **b.** SF_5Cl **c.** KrF_2 **d.** IF_3

33. Give all the ideal bond angles (109.5°, 120°, or 180°) in the following molecules and ions.

a. $[:O—N=O]^-$ **b.** H—C(H)(H)—C≡C—H **c.** $[:O—Si(O)(O)—O:]^{4-}$

34. Give all the ideal bond angles (109.5°, 120°, or 180°) in the following molecules and ions.

a. $[:O—C(=O)—O:]^{2-}$ **b.** Cl—C(H)=C(H)—H **c.** H—C(=O)—O—H

35. Draw the Lewis structure and describe the geometry of
a. HOCl **b.** OCN^- **c.** NF_3 **d.** N_3^-

36. Draw the Lewis structure and describe the geometry of
a. C_2HCl **b.** HCO_2^- **c.** CH_3Cl **d.** PCl_4^+

37. Describe the geometry of the species in which there are, around the central atom,
a. three bonds and two unshared pairs.
b. six bonds.
c. four bonds and one unshared pair.

38. Describe the geometry of the species in which there are, around the central atom,
a. four bonds and two unshared pairs.
b. five bonds and one unshared pair.
c. five bonds.

39. Draw the Lewis structure and describe the geometry of
a. SiF_4 **b.** BrO_3^- **c.** HOClO **d.** $AsCl_3$

40. Draw the Lewis structure and describe the geometry of
a. Cl_2CO **b.** NI_3 **c.** PO_3^{3-} **d.** O_3

41. For the structures shown in Question 29, determine
—the number of unshared pairs around the central atom.
—the number of atoms bonded to the central atom.
—the bond angle, using Table 7.3.

42. For the structures shown in Question 30, follow the directions of Question 41.

43. Peroxypropionyl nitrate (PPN) is an eye irritant found in smog. Its skeleton structure is

$H_3C—C(H)(H)—C(O)—O—O—N(O)—O$

Write the Lewis structure for PPN and give all the bond angles.

44. An objectionable component of smoggy air is acetylperoxide, which has the skeleton structure

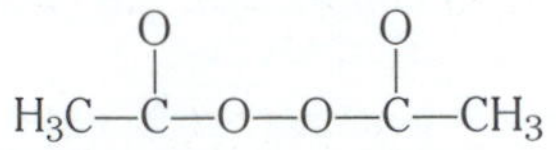

$$H_3C-\overset{\overset{\displaystyle O}{|}}{C}-O-O-\overset{\overset{\displaystyle O}{|}}{C}-CH_3$$

a. Draw the Lewis structure of this compound.
b. Indicate all the bond angles.

45. List all the bond angles in the following hydrocarbons.
a. 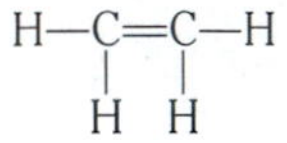 $H-\underset{\underset{\displaystyle H}{|}}{C}=\underset{\underset{\displaystyle H}{|}}{C}-H$ b. C_2H_6
c. $H-C\equiv C-H$

46. List all the bond angles in the following hydrocarbons.
a. CH_4 b. $H_3C-C\equiv C-H$
c. 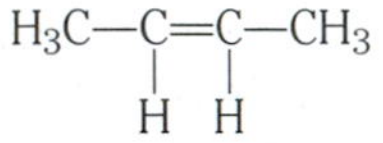 $H_3C-\underset{\underset{\displaystyle H}{|}}{C}=\underset{\underset{\displaystyle H}{|}}{C}-CH_3$

47. In each of the following molecules, a central atom is surrounded by a total of three atoms or unshared pairs: BCl_3, $SnCl_2$, SO_2. In which of these molecules would you expect the bond angle to be less than 120°? Explain your reasoning.

48. Consider the following molecules: SiH_4, PH_3, H_2S. In each case, a central atom is surrounded by four electron pairs. In which of these molecules would you expect the bond angle to be less than 109.5°? Explain your reasoning.

Molecular Polarity

49. Which of the species in Question 27 are dipoles?
50. Which of the species in Question 28 are dipoles?
51. Which of the species in Question 29 are dipoles?
52. Which of the species in Question 30 are dipoles?
53. There are three compounds with the formula $C_2H_2Cl_2$:

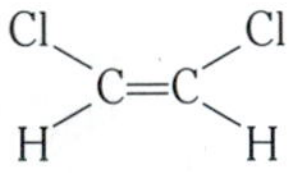

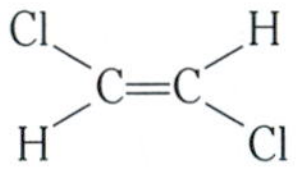

(Cl, H)C=C(Cl, H) (Cl, H)C=C(H, Cl) and (H, H)C=C(Cl, Cl)

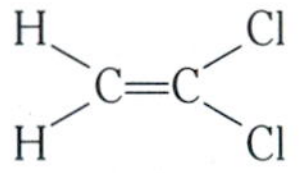

Which of these molecules are polar? Nonpolar?

54. There are two different molecules with the formula N_2F_2:

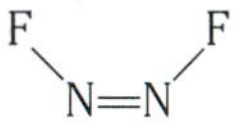

F–N=N–F (both F on the same side) and F–N=N–F (F on opposite sides)

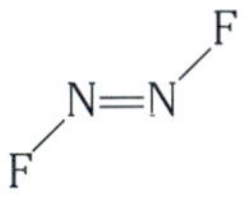

Is either molecule polar? Explain.

Hybridization

55. Give the hybridization of the central atom in each molecule in Question 27.
56. Give the hybridization of the central atom in each molecule in Question 28.
57. Give the hybridization of the central atom in each molecule in Question 29.
58. Give the hybridization of the central atom in each molecule in Question 30.
59. Consider the species in Question 31. How many electron pairs are there around the central atom in each species? What is the hybridization of the central atom?
60. Consider the species in Question 32. Answer the questions in Question 59.
61. In each of the following polyatomic ions, the central atom has an expanded octet. Determine the number of electron pairs around the central atom and the hybridization in
a. ClF_2^- b. $GeCl_4^{2-}$ c. ICl_4^-
62. Follow the directions of Question 61 for the following polyatomic ions.
a. XeF_3^- b. $SiCl_6^{2-}$ c. PCl_4^-
63. Give the hybridization of each C, N, and O atom in nitrobenzene (unshared electron pairs are not shown):

(Benzene ring of six C atoms with alternating double bonds; five C each bonded to H, one C bonded to N; N bonded to O and double-bonded to O.)

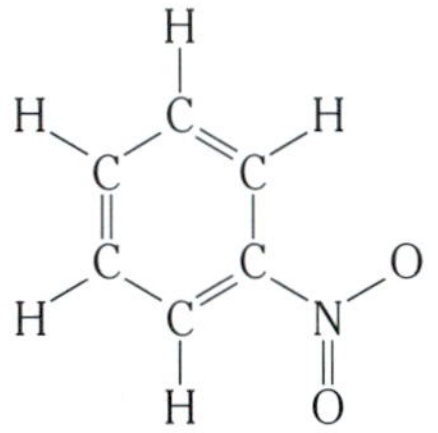

64. Give the hybridization of each C, N, and O atom in 3-pyridine carboxylic acid (unshared electron pairs are not shown):

(Pyridine ring: H–C, H–C, H–C, C–H, N, and a C bearing $-\overset{\overset{\displaystyle O}{\|}}{C}-O-H$.)

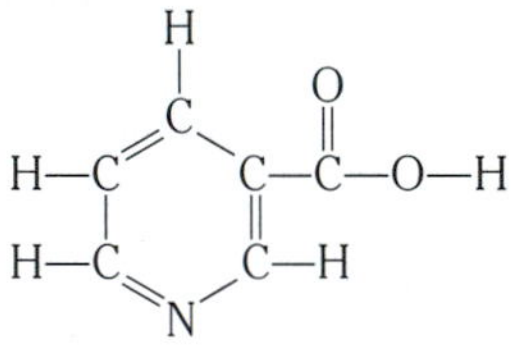

65. What is the hybridization of nitrogen in
a. 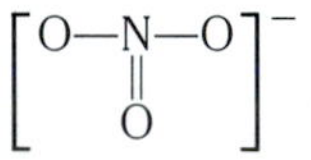 $\left[O-\underset{\underset{\displaystyle O}{\|}}{N}-O\right]^-$ b. 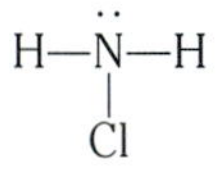 $H-\underset{\underset{\displaystyle Cl}{|}}{\ddot{N}}-H$
c. $:N\equiv N:$ d. $:N\equiv N-O$

66. What is the hybridization of carbon in
a. CH_3Cl b. 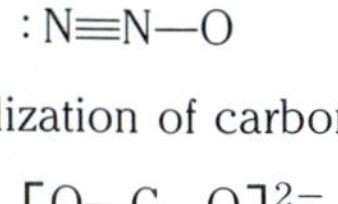 $\left[O-\underset{\underset{\displaystyle O}{\|}}{C}-O\right]^{2-}$
c. $O=C=O$ d. 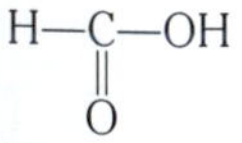 $H-\underset{\underset{\displaystyle O}{\|}}{C}-OH$

67. Give the hybridization of the central atom (underlined) in
a. $F_2\underline{C}O$ b. $\underline{O}F_2$ c. $O\underline{P}Cl_3$
68. Give the hybridization of the central atom (underlined) in
a. $HO\underline{I}O_2$ b. $(H_2N)_2\underline{C}O$ c. $\underline{N}HF_2$

Sigma and Pi Bonds

69. Give the number of sigma and pi bonds in each species listed in Question 65.

70. Give the number of sigma and pi bonds in each species listed in Question 66.

71. Give the formula of an ion or molecule in which an atom of

- **a.** B forms four bonds using sp^3 hybrid orbitals.
- **b.** C forms two pi bonds.
- **c.** N forms a pi and two sigma bonds.
- **d.** S forms a pi bond.

72. Give the formula of an ion or molecule in which an atom of

- **a.** B forms three bonds using sp^2 hybrid orbitals.
- **b.** O forms a pi bond.
- **c.** N forms four bonds using sp^3 hybrid orbitals.
- **d.** C forms four bonds, in three of which it uses sp^2 hybrid orbitals.

Noble Gases

73. Calculate the density at STP of

- **a.** pure nitrogen.
- **b.** the mixture of N_2 and Ar prepared by removing all the other components of air (the mole fractions of N_2 and Ar in air are 0.7808 and 0.0093, respectively).

74. For XeF_2 and XeF_4:

- **a.** Write the Lewis structure of each compound.
- **b.** Discuss the geometry, polarity, hybridization, and number of sigma and pi bonds for each molecule.

75. Give at least one use for helium, neon, argon, and krypton.

76. Write a balanced equation for the reaction of $XeF_6(s)$ with water to form $XeO_3(s)$ and $HF(g)$. If 10.0 g of XeF_6 reacts with excess water in a sealed 2.00-L container, what pressure of HF is developed? Take the final temperature to be 20°C.

Unclassified

77. Two possible structures for the HOCN molecule are

$$\text{H—}\ddot{\underset{\cdot\cdot}{\text{O}}}\text{—C}\equiv\text{N:} \qquad \text{and} \qquad \text{H—}\ddot{\text{O}}\text{=C=}\ddot{\text{N}}\text{:}$$

Assign formal charges to each atom in both structures. Which structure do you think is more likely?

78. A possible structure for $BeCl_2$ which would obey the octet rule is

$$:\ddot{\text{Cl}}\text{=Be=}\ddot{\text{Cl}}:$$

Assign formal charges to each atom in this structure. Explain why this structure seems less plausible than one with only single bonds.

79. Explain the meaning of the following terms.

- **a.** expanded octet
- **b.** resonance
- **c.** unshared electron pair
- **d.** odd-electron species

80. Consider the dichromate ion. It has no metal–metal nor oxygen–oxygen bonds. Write a Lewis structure for the dichromate ion. Consider chromium to have six valence electrons.

81. In which of the following molecules does the sulfur atom have an expanded octet?

a. SO_2 **b.** SF_4 **c.** SO_2Cl_2 **d.** SF_6

82. Consider acetyl salicylic acid, better known as aspirin. Its structure is

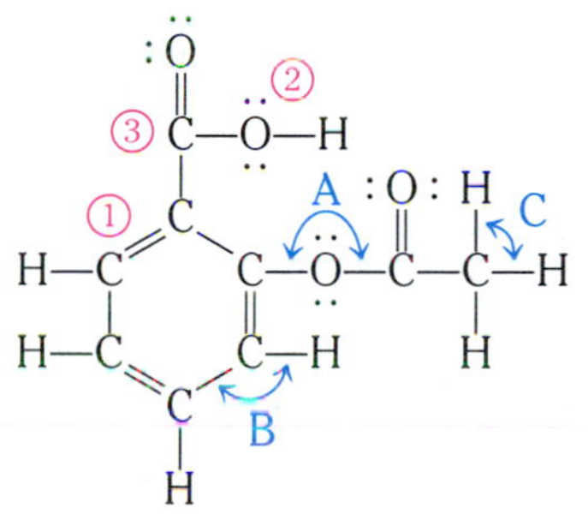

```
                :O:
                 ‖    ②
              ③ C—Ö—H
                 |        A :O: H
              ①  C           ‖   |  C
          H—C      C—Ö—C—C—H
             |      ‖           |
          H—C      C—H          H
                C     B
                |
                H
```

- **a.** How many sigma and pi bonds are there in aspirin?
- **b.** What are the approximate values of the angles marked (in blue) A, B, and C?
- **c.** What is the hybridization of each atom marked (in red) 1, 2, and 3?

83. Complete the following table

Species	Atoms Around Central Atom A	Unshared Pairs Around A	Geometry	Hybridization	Polarity *(Assume all X atoms the same)*
AX_2E_2	____	____	____	____	____
____	3	0	____	____	____
AX_4E_2	____	____	____	____	____
____	____	____	triangular bipyramid	____	____

Challenge Problems

84. A compound of chlorine and fluorine, ClF_x, reacts at about 75°C with uranium to produce uranium hexafluoride and chlorine fluoride, ClF. A certain amount of uranium produced 5.63 g of uranium hexafluoride and 457 mL of chlorine fluoride at 75°C and 3.00 atm. What is x? Describe the geometry, polarity, and bond angles of the compound and the hybridization of chlorine. How many sigma and pi bonds are there?

85. Draw the Lewis structure and describe the geometry of the hydrazine molecule, N_2H_4. Would you expect this molecule to be polar?

86. Consider the polyatomic ion IO_6^{5-}. How many pairs of electrons are there around the central iodine atom? What is its hybridization? Describe the geometry of the ion.

87. It is possible to write a simple Lewis structure for the SO_4^{2-} ion, involving only single bonds, which follows the octet rule. However, Linus Pauling and others have suggested an alternative structure, involving double bonds, in which the sulfur atom is surrounded by six electron pairs.

a. Draw the two Lewis structures.

b. What geometries are predicted for the two structures?

c. What is the hybridization of sulfur in each case?

d. What are the formal charges of the atoms in the two structures?

88. Phosphoryl chloride, $POCl_3$, has the skeleton structure

```
     O
     |
Cl—P—Cl
     |
    Cl
```

Write

a. a Lewis structure for $POCl_3$ following the octet rule. Calculate the formal charges in this structure.

b. a Lewis structure where all the formal charges are zero. (The octet rule need not be followed.)

YOAV LEVY/PHOTOTAKE NYC

$Cu(s) + 2Ag^{+}(aq) \longrightarrow Cu^{2+}(aq) + 2Ag(s)$

(See p. 491)

8
Thermochemistry

Some say the world will end in fire,
Some say in ice.
From what I've tasted of desire
I hold with those who favour fire.

ROBERT FROST
Fire and Ice

CHAPTER OUTLINE

This chapter deals with energy and heat, two terms used widely by both the general public and scientists. Energy, in the vernacular, is equated with pep and vitality. Heat conjures images of blast furnaces and sweltering summer days. Scientifically, these terms have quite different meanings. *Energy* can be defined as the capacity to do work. *Heat* is a particular form of energy that is transferred from a body at a high temperature to one at a lower temperature when they are brought into contact with each other. Two centuries ago, heat was believed to be a material fluid ("caloric"); we still use the phrase "heat flow" to refer to heat transfer or to heat effects in general.

Thermochemistry refers to the study of the heat flow that accompanies chemical reactions. Our discussion of this subject will focus upon

— the basic principles of heat flow (Section 8.1)
— the experimental measurement of the magnitude and direction of heat flow, known as *calorimetry* (Section 8.2)
— the concept of enthalpy (heat content) and *enthalpy change, ΔH* (Section 8.3)
— the calculation of ΔH for reactions, using *thermochemical equations* (Section 8.4) and *heats of formation* (Section 8.5)
— heat effects in the breaking and formation of covalent bonds (Section 8.6)
— the relation between heat and other forms of energy, as expressed by the first law of thermodynamics (Section 8.7).

8.1 PRINCIPLES OF HEAT FLOW

The universe is a big place

In any discussion of heat flow, it is important to distinguish between system and surroundings. The **system** is that part of the universe upon which attention is focused. In a very simple case (Fig. 8.1), it might be a sample of water in contact with a hot plate. The **surroundings**, which exchange energy with the system, comprise in principle the rest of the universe. For all practical purposes, however, they include only those materials in close contact with the system. In Figure 8.1, the surroundings would consist of the hot plate, the beaker holding the water sample, and the air around it.

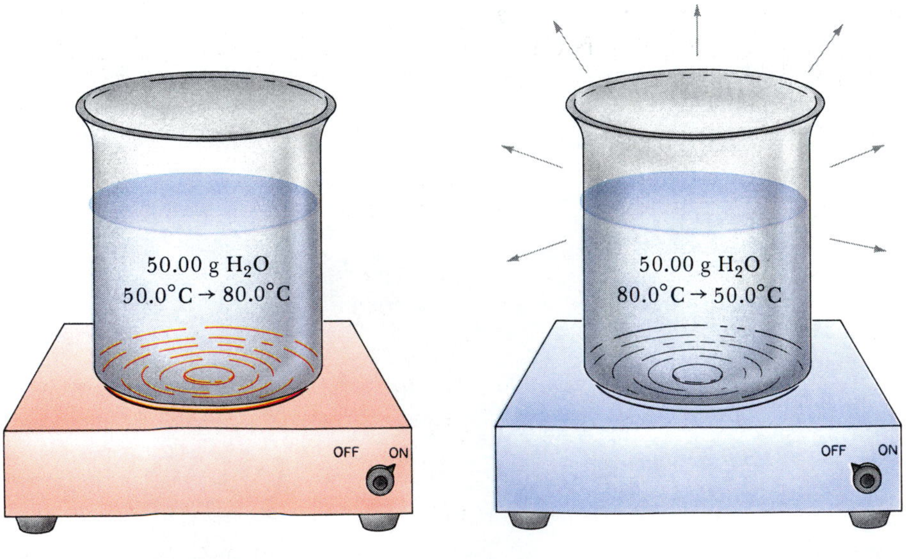

Figure 8.1
When the system (50.00 g H_2O) absorbs heat from the surroundings (hot plate), its temperature increases from 50.0 to 80.0°C. When the hot plate is turned off, the system gives off heat to the surrounding air and its temperature drops.

State Properties

The **state** of a system is described by giving its composition, temperature, and pressure. The system at the left of Figure 8.1 consists of

$$50.0 \text{ g of } H_2O(l) \text{ at } 50.0°C \text{ and } 1 \text{ atm}$$

When this system is heated, its state changes, perhaps to one described as

$$50.0 \text{ g of } H_2O(l) \text{ at } 80.0°C \text{ and } 1 \text{ atm}$$

Certain quantities, called **state properties**, depend only upon the state of the system, not upon the way the system reached that state. Putting it another way, if X is a state property, then

$$\Delta X = X_{\text{final}} - X_{\text{initial}}$$

That is, the change in X is the difference between its values in final and initial states. Most of the quantities that you are familiar with are state properties; volume is a common example. You may be surprised to learn, however, that heat flow is *not* a state property; its magnitude depends upon how a process is carried out (Section 8.7).

The density of water is a state property; its cost isn't

Direction and Sign of Heat Flow

Consider again the setup in Figure 8.1. If the hot plate is turned on, there is a flow of heat from the surroundings into the system, 50.0 g of water. This situation is described by stating that the heat flow, q, for the system is a positive quantity.

q is + when heat flows into the system from the surroundings

Usually, when heat flows into a system, its temperature rises. In this case, the temperature of the 50.0 g water sample might increase from 50.0 to 80.0°C. When the hot plate in Figure 8.1 is shut off, the hot water gives off heat to the surrounding air. In this case, q for the system is a negative quantity.

q is − when heat flows out of the system into the surroundings

Phosphorus reacting with chlorine to give PCl_3 is an exothermic reaction. (Charles D. Winters)

Here, as is usually the case, the temperature of the system drops when heat flows out of it into the surroundings. The 50.0 g water sample might cool from 80.0°C back to 50.0°C.

This same reasoning can be applied to a reaction where the system consists of the reaction mixture (products and reactants). We can distinguish between

— *an **endothermic** process ($q > 0$), in which heat flows from the surroundings into the reaction system.* An example is the melting of ice

endo = into system

$$H_2O(s) \longrightarrow H_2O(l) \qquad q > 0$$

The melting of ice *absorbs* heat from the surroundings, which might be the water in a glass of iced tea. The temperature of the surroundings drops, perhaps from 25 to 0°C, as they give up heat to the system.

— *an **exothermic** process ($q < 0$), in which heat flows from the reaction system into the surroundings.* A familiar example is the combustion of methane gas

exo = out of system

$$CH_4(g) + 2\,O_2(g) \longrightarrow CO_2(g) + 2H_2O(l) \qquad q < 0$$

This reaction *evolves* heat to the surroundings, which might be the air around a Bunsen burner in the laboratory or a potato being baked in a gas oven. In either case, the effect of the heat transfer is to raise the temperature of the surroundings.

Magnitude of Heat Flow

In any process, we are interested not only in the direction of heat flow but also in its magnitude. The magnitude of q is ordinarily cited in joules (J) or kilojoules (kJ).

$$1\ \text{kJ} = 10^3\ \text{J}$$

Most of the remainder of this chapter is devoted to a discussion of the magnitude of the heat flow in chemical reactions or phase changes. Here, however, we will focus upon a simpler process in which the only effect of the heat flow is to change the temperature of a system. In general, the relationship between the magnitude of the heat flow, q, and the temperature change, Δt, is given by the equation

$$q = C \times \Delta t \qquad (\Delta t = t_{\text{final}} - t_{\text{initial}})$$

The quantity C appearing in this equation is known as the **heat capacity** of the system. It represents the amount of heat required to raise the temperature of the system 1°C and has the units J/°C.

For a pure substance of mass m, the expression for q can be written as

C depends upon composition and mass; c depends only upon composition

$$q = m \times c \times \Delta t$$

The quantity c is known as the **specific heat**; it is defined as the amount of heat required to raise the temperature of one gram of a substance 1°C.

Specific heat, like density or melting point, is an intensive property which can be used to identify a substance. Water has an unusually large specific heat, **4.18 J/g · °C**. This explains why swimming is not a popular pastime in northern Minnesota in May. Even if the air temperature rises to 90°F, the water temperature will remain below 60°F. Metals have a relatively low specific heat (Table 8.1). When you warm water in a stainless steel saucepan, nearly all of the heat is absorbed by the water, very little by the steel.

August, maybe

Table 8.1 Specific Heats of Some Elements and Compounds

	c (J/g · °C)*		c (J/g · °C)
$Br_2(l)$	0.474	$CO_2(g)$	0.843
$Cl_2(g)$	0.478	$C_2H_5OH(l)$	2.43
$Cu(s)$	0.382	$C_6H_6(l)$	1.72
$Fe(s)$	0.446	$NaCl(s)$	0.866

* Since a Celsius degree (°C) is equal in magnitude to a kelvin (K), the units of c can equally well be J/g · K.

Example 8.1 How much heat is given off by a 50.0-g sample of copper when it cools from 80.0 to 50.0°C?

Strategy Calculate q by first determining Δt and then substituting into the equation $q = m \times c \times \Delta t$. The specific heat of copper is given in Table 8.1.

Solution The temperature change is

$$\Delta t = t_{\text{final}} - t_{\text{initial}} = 50.0°\text{C} - 80.0°\text{C} = -30.0°\text{C}$$

$$q = 50.0\text{ g} \times 0.382\frac{\text{J}}{\text{g °C}} \times (-30.0°\text{C}) = -573\text{ J}$$

The negative sign indicates that heat flows from copper to the surroundings.

8.2 MEASUREMENT OF HEAT FLOW; CALORIMETRY

To measure the heat flow in a reaction, it is carried out in a device known as a **calorimeter**. The apparatus contains water and/or other materials of known heat capacity. The walls of the calorimeter are insulated so that there is no exchange of heat with the air outside the calorimeter. It follows that the only heat flow is between the reaction system and the calorimeter. The heat flow for the reaction system is equal in magnitude but opposite in sign to that for the calorimeter:

The reaction mixture is the system; the calorimeter is the surroundings

$$q_{\text{reaction}} = -q_{\text{calorimeter}}$$

This is the basic equation of calorimetry. It allows you to calculate the amount of heat absorbed or evolved in a reaction if you know the heat capacity, C_{cal}, and the temperature change, Δt, of the calorimeter.

$$q_{\text{reaction}} = -C_{\text{cal}} \times \Delta t$$

Coffee-Cup Calorimeter

Figure 8.2 shows a simple calorimeter used in the general chemistry laboratory. It consists of a polystyrene foam cup partially filled with water. The cup has a tightly fitting cover through which an accurate thermometer is inserted. Since polystyrene foam is a good insulator, there is very little heat flow through the walls of the cup. Essentially all the heat evolved by a reaction taking place within the calorimeter is absorbed by the water. This means that, to a good degree of approximation, the heat capacity of the coffee-cup calorimeter is that of the water:

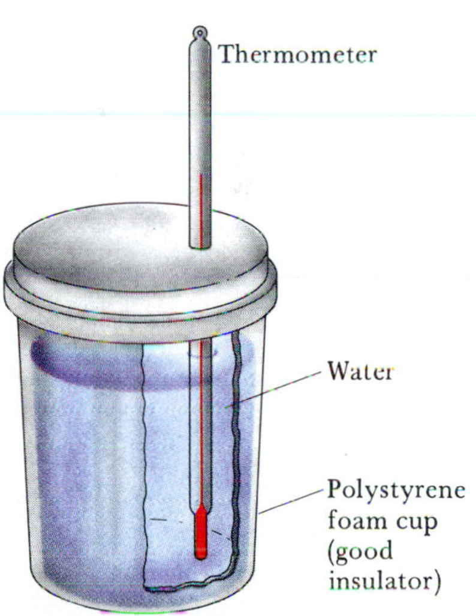

Figure 8.2
Coffee-cup calorimeter. The heat given off by a reaction is absorbed by the water. Knowing the mass of the water, its specific heat (4.18 J/g·°C), and the temperature change as read on the thermometer, you can calculate the heat flow, q, for the reaction.

$$C_{\text{cal}} = m_{\text{water}} \times c_{\text{water}} = m_{\text{water}} \times 4.18\frac{\text{J}}{\text{g} \cdot °\text{C}}$$

and hence

$$q_{\text{reaction}} = -m_{\text{water}} \times 4.18\frac{\text{J}}{\text{g} \cdot °\text{C}} \times \Delta t$$

Example 8.2 When 1.00 g of ammonium nitrate, NH_4NO_3, dissolves in 50.0 g of water in a coffee-cup calorimeter, the following reaction occurs:

$$NH_4NO_3(s) \longrightarrow NH_4^+(aq) + NO_3^-(aq)$$

and the temperature of the water drops from 25.00 to 23.32°C. Assuming that all the heat absorbed by the reaction comes from the water, calculate q for the reaction system.

Strategy Apply the relation $q_{\text{reaction}} = -m_{\text{water}} \times 4.18 \frac{\text{J}}{\text{g} \cdot °\text{C}} \times \Delta t$

Solution

$$q_{\text{reaction}} = -50.0\ \text{g} \times 4.18 \frac{\text{J}}{\text{g} \cdot °\text{C}} \times (23.32°\text{C} - 25.00°\text{C}) = \boxed{+351\ \text{J}}$$

See?

Notice that q is positive; this is an endothermic reaction.

Bomb Calorimeter

A coffee-cup calorimeter is suitable for measuring heat flows for reactions in solution. However, it cannot be used for reactions involving gases, which would escape from the cup. Neither would it be appropriate for reactions in which the products reach high temperatures. The bomb calorimeter, shown in Figure 8.3, is more versatile. To use it, a weighed sample of the reactant(s) is added to the heavy-walled steel container called a "bomb." This is then sealed and lowered into a metal vessel that fits snugly within the insulating walls of the calorimeter. An amount of water sufficient to cover the bomb is added, and the entire apparatus is closed. The initial temperature is measured precisely. The reaction is then started, perhaps by electrical ignition. In an exothermic reaction, the hot products give off heat to the walls of the bomb and to the water. The final temperature is taken to be the highest value read on the thermometer.

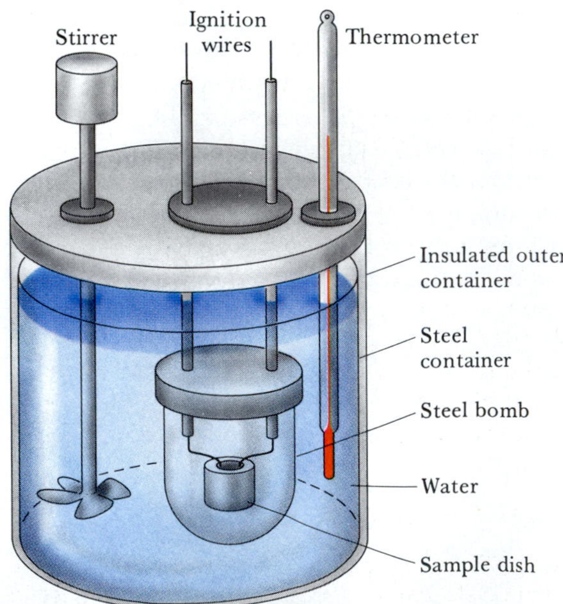

Figure 8.3
Bomb calorimeter. The heat flow, q, for the reaction is calculated from the temperature change multiplied by the heat capacity of the calorimeter, which is determined in a preliminary experiment.

To analyze the heat flow in a bomb calorimeter, the basic equation

$$q_{reaction} = -q_{calorimeter} = -C_{cal} \times \Delta t$$

is used. Since heat is absorbed by both the water and the metal parts of the calorimeter, C_{cal} must be found experimentally. This is ordinarily done by carrying out a reaction with known q in the bomb calorimeter, measuring Δt, and calculating C_{cal}. Suppose, for example, that you carried out a reaction known to evolve 93.3 kJ of heat and found that the temperature of the calorimeter rose from 20.00 to 30.00°C. It follows that

In principle, C_{cal} could be calculated, but it's simpler to use this approach

$$C_{cal} = \frac{93.3 \text{ kJ}}{10.00°\text{C}} = 9.33 \text{ kJ/°C}$$

Knowing the heat capacity of the calorimeter, the heat flow for the reaction can be calculated (Example 8.3).

Example 8.3 The reaction between hydrogen and chlorine, $H_2(g) + Cl_2(g) \rightarrow 2HCl(g)$, can be studied in a bomb calorimeter. It is found that when a 1.00-g sample of H_2 reacts completely, the temperature rises from 20.00 to 29.82°C. Taking the heat capacity of the calorimeter to be 9.33 kJ/°C, calculate the amount of heat evolved in the reaction.

Strategy Use the equation $q_{reaction} = -C_{cal} \times \Delta t$

Solution

$$q_{reaction} = -9.33 \frac{\text{kJ}}{°\text{C}} \times (29.82°\text{C} - 20.00°\text{C}) = -91.6 \text{ kJ}$$

A bomb calorimeter used in the laboratory. (Leon Lewandowski)

8.3 ENTHALPY

We have referred several times to the "heat flow for the reaction system," symbolized as $q_{reaction}$. At this point, you may well find this concept a bit nebulous and wonder if it could be made more concrete by relating $q_{reaction}$ to some property of reactants and products. This can indeed be done; the situation is particularly simple for reactions taking place at constant pressure. Under that condition, the heat flow for the reaction system is equal to the difference in **enthalpy** (H) between products and reactants. That is,

$$q_{reaction} \text{ at constant pressure} = \Delta H = H_{products} - H_{reactants}$$

Enthalpy is a type of chemical energy, sometimes referred to as "heat content." Reactions that occur in the laboratory in an open container or in the world around us take place at a constant pressure, that of the atmosphere. For such reactions, the above equation is valid, making enthalpy a very useful quantity.

Figure 8.4a shows the enthalpy relations for an exothermic reaction such as

$$CH_4(g) + 2\,O_2(g) \longrightarrow CO_2(g) + 2H_2O(l) \qquad \Delta H < 0$$

Here, the products, 1 mol of $CO_2(g)$ and 2 mol of $H_2O(l)$, have a lower enthalpy than the reactants, 1 mol of $CH_4(g)$ and 2 mol of $O_2(g)$. The decrease in

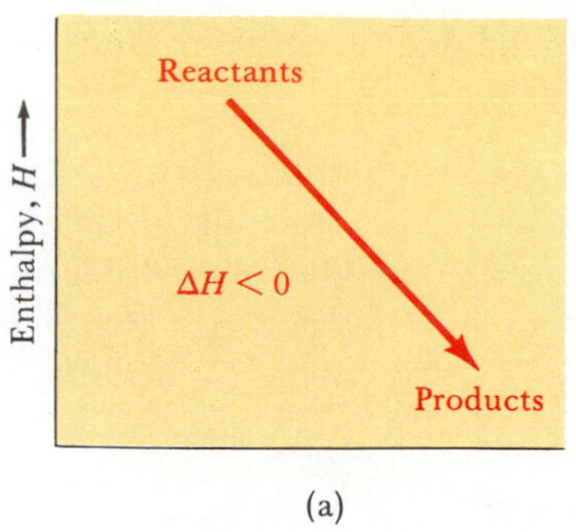

(a)

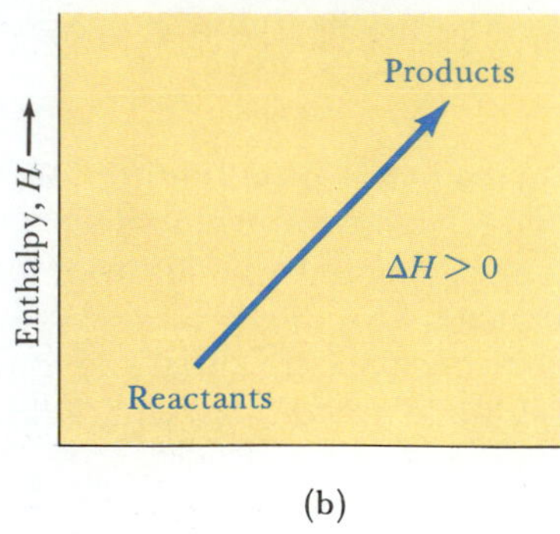

(b)

Figure 8.4
In an exothermic reaction (a), the products have a lower enthalpy than the reactants; thus, ΔH is negative, and heat is given off to the surroundings. In an endothermic reaction (b), the products have a higher enthalpy than the reactants, so ΔH is positive, and heat is absorbed from the surroundings.

enthalpy is the source of the heat evolved to the surroundings. Figure 8.4b shows the situation for an endothermic process such as

$$H_2O(s) \longrightarrow H_2O(l) \qquad \Delta H > 0$$

Since liquid water has a higher enthalpy than ice, heat must be transferred from the surroundings to melt the ice.

In general, the following relations apply for reactions taking place at constant pressure.

When wood burns, ΔH is negative; when ice melts, ΔH is positive

exothermic reaction: $q = \Delta H < 0 \qquad H_{products} < H_{reactants}$

endothermic reaction: $q = \Delta H > 0 \qquad H_{products} > H_{reactants}$

The enthalpy of a substance, like its volume, is a state property. A sample of one gram of liquid water at 25.00°C and 1 atm has a fixed enthalpy, H. In practice, no attempt is made to determine absolute values of enthalpy. Instead, scientists deal with changes in enthalpy, which are readily determined. For the process

$$1.00\text{ g } H_2O(l,\ 25.00°C,\ 1\text{ atm}) \longrightarrow 1.00\text{ g } H_2O(l,\ 26.00°C,\ 1\text{ atm})$$

ΔH is 4.18 J because the specific heat of water is 4.18 J/g · °C.

8.4 THERMOCHEMICAL EQUATIONS

An equation that shows both mass and enthalpy relationships between products and reactants is called a **thermochemical equation**. This type of equation contains, at the right of the balanced chemical equation, the appropriate value and sign for ΔH.

To see where a thermochemical equation comes from, consider again the process by which ammonium nitrate dissolves in water:

$$NH_4NO_3(s) \longrightarrow NH_4^+(aq) + NO_3^-(aq)$$

Recall (Example 8.2) that a simple experiment with a coffee-cup calorimeter shows that when one gram of NH_4NO_3 dissolves, $q_{reaction} = 351$ J. Since the calorimeter is open to the atmosphere, the pressure is constant and

$$\Delta H \text{ for dissolving } 1.00 \text{ g of } NH_4NO_3 = 351 \text{ J} = 0.351 \text{ kJ}$$

When one mole (80.05 g) of NH_4NO_3 dissolves, ΔH should be eighty times as great:

This implies that ΔH is directly proportional to amount, which it is

$$\Delta H \text{ for dissolving 1.00 mol of } NH_4NO_3 = 0.351\frac{\text{kJ}}{\text{g}} \times 80.05 \text{ g} = 28.1 \text{ kJ}$$

The thermochemical equation for this reaction must then be (Fig. 8.5)

$$NH_4NO_3(s) \longrightarrow NH_4^+(aq) + NO_3^-(aq) \qquad \Delta H = 28.1 \text{ kJ}$$

By an entirely analogous procedure, the thermochemical equation for the formation of HCl from the elements (Example 8.3) is found to be

$$H_2(g) + Cl_2(g) \longrightarrow 2HCl(g) \qquad \Delta H = -185 \text{ kJ}$$

In other words, 185 kJ of heat is evolved when two moles of HCl are formed from H_2 and Cl_2.

These thermochemical equations are typical of those used throughout this text. It is important to realize that

— the sign of ΔH indicates whether the reaction, when carried out at constant pressure, is endothermic (positive ΔH) or exothermic (negative ΔH)

— in interpreting a thermochemical equation, the coefficients represent numbers of moles (ΔH is -185 kJ when **1 mol** H_2 + **1 mol** $Cl_2 \rightarrow$ **2 mol** HCl)

— the phases (physical states) of all species must be specified, using the symbols (s), (l), (g), or (aq). The enthalpy of one mole of $H_2O(g)$ at 25°C is 44 kJ larger than that of one mole of $H_2O(l)$; the difference, which represents the heat of vaporization of water, is clearly significant.

This explains why burns from steam are more painful than those from boiling water

— the value quoted for ΔH applies when products and reactants are at the same temperature, ordinarily taken to be 25°C unless specified otherwise

Rules of Thermochemistry

To make effective use of thermochemical equations, three basic rules of thermochemistry are applied.

1. *The magnitude of ΔH is directly proportional to the amount of reactant or product.* This is a common-sense rule, consistent with experience. The amount of heat that must be absorbed to boil a sample of water is directly proportional to its mass. In another case, the more gasoline you burn in your car's engine, the more energy you produce.

Figure 8.5
When the seal separating the compartments in a "cold pack" is broken, the following endothermic reaction occurs: $NH_4NO_3(s) \rightarrow NH_4^+(aq) + NO_3^-(aq)$; $\Delta H = +28.1$ kJ, and the temperature, as read on the thermometer, drops. (Marna G. Clarke)

This rule allows you to find ΔH corresponding to any desired amount of reactant or product. To do this, you follow the conversion factor approach used in Chapter 3 with ordinary chemical equations. Consider, for example,

$$H_2(g) + Cl_2(g) \longrightarrow 2HCl(g) \qquad \Delta H = -185 \text{ kJ}$$

The mass relationships in this equation are such that one mole of hydrogen, one mole of chlorine, and two moles of HCl are chemically equivalent to one another. The thermochemical equation adds the enthalpy relation and hence leads to conversion factors such as

Conversion factors again

$$\frac{-185 \text{ kJ}}{1 \text{ mol } Cl_2} \qquad \frac{1 \text{ mol } H_2}{-185 \text{ kJ}} \qquad \frac{-185 \text{ kJ}}{2 \text{ mol HCl}}$$

Example 8.4 Consider the thermochemical equation for the formation of two moles of HCl from the elements. Calculate ΔH when

(a) 1.00 mol of HCl is formed. (b) 1.00 g of Cl_2 reacts.

Strategy Use the conversion factor approach. In (a), only one conversion is required, from moles of HCl to ΔH in kilojoules. In (b), you first have to convert to moles of Cl_2 and then to ΔH in kilojoules.

Solution

$$\text{(a) } \Delta H = 1.00 \text{ mol HCl} \times \frac{-185 \text{ kJ}}{2 \text{ mol HCl}} = \boxed{-92.5 \text{ kJ}}$$

This means that the thermochemical equation for the formation of one mole of HCl would be

$$\tfrac{1}{2} H_2(g) + \tfrac{1}{2} Cl_2(g) \longrightarrow HCl(g) \qquad \Delta H = -92.5 \text{ kJ}$$

$$\text{(b) } \Delta H = 1.00 \text{ g } Cl_2 \times \frac{1 \text{ mol } Cl_2}{70.90 \text{ g } Cl_2} \times \frac{-185 \text{ kJ}}{1 \text{ mol } Cl_2} = \boxed{-2.61 \text{ kJ}}$$

This relation between ΔH and amounts of substances is equally useful in dealing with chemical reactions or phase changes.

Example 8.5 Using a coffee-cup calorimeter, it is found that when an ice cube weighing 24.6 g melts, it absorbs 8.19 kJ of heat. Calculate ΔH for the phase change represented by the thermochemical equation $H_2O(s) \rightarrow H_2O(l)$ $\Delta H = ?$

Strategy The data given leads directly to the conversion factor

$$\frac{8.19 \text{ kJ}}{24.6 \text{ g ice}}$$

You need to know ΔH when one mole (18.02 g) of ice melts; two conversions are required.

Solution

$$\Delta H = 1 \text{ mol } H_2O \times \frac{18.02 \text{ g } H_2O}{1 \text{ mol } H_2O} \times \frac{8.19 \text{ kJ}}{24.6 \text{ g } H_2O} = \boxed{6.00 \text{ kJ}}$$

This calculation shows that 6.00 kJ of heat must be absorbed to melt one mole of ice:

$$H_2O(s) \longrightarrow H_2O(l) \qquad \Delta H = +6.00 \text{ kJ}$$

What is the heat of fusion of ice per gram? Ans: about 1/3 kJ

The heat absorbed when a solid melts ($s \rightarrow l$) is referred to as the **heat of fusion**; that absorbed when a liquid vaporizes ($l \rightarrow g$) is called the **heat of vaporization**. Heats of fusion (ΔH_{fus}) and vaporization (ΔH_{vap}) are most often expressed in kilojoules per mole (kJ/mol). Values for several different substances are given in Table 8.2.

Example 8.6 Given

$$H_2(g) + \tfrac{1}{2} O_2(g) \longrightarrow H_2O(l) \qquad \Delta H = -285.8 \text{ kJ}$$

calculate ΔH for the equation

$$2H_2O(l) \longrightarrow 2H_2(g) + O_2(g)$$

Strategy Note that for the required equation the coefficients are twice as great as in the given equation; the equation is also reversed. Apply Rules 1 and 2 in succession.

Solution

Applying Rule 1:

$$2H_2(g) + O_2(g) \longrightarrow 2H_2O(l) \qquad \Delta H = 2(-285.8 \text{ kJ}) = -571.6 \text{ kJ}$$

Applying Rule 2:

$$2H_2O(l) \longrightarrow 2H_2(g) + O_2(g) \qquad \Delta H = +571.6 \text{ kJ}$$

Table 8.2 ΔH(kJ/mol) for Phase Changes

Substance		mp(°C)	ΔH_{fus}*	bp(°C)	ΔH_{vap}*
Benzene	C_6H_6	5	9.84	80	30.8
Bromine	Br_2	−7	10.8	59	29.6
Mercury	Hg	−39	2.33	357	59.4
Naphthalene	$C_{10}H_8$	80	19.3	218	43.3
Water	H_2O	0	6.00	100	40.7

* Values of ΔH_{fus} are given at the melting point, values of ΔH_{vap} at the boiling point. The heat of vaporization of water decreases from 44.9 kJ/mol at 0°C to 44.0 kJ/mol at 25°C to 40.7 kJ/mol at 100°C.

2. *ΔH for a reaction is equal in magnitude but opposite in sign to ΔH for the reverse reaction.* Another way to state this rule is to say that the amount of heat evolved in a reaction is exactly equal to the amount of heat absorbed in the reverse reaction. This again is a common-sense rule. If 6.00 kJ of heat is absorbed when a mole of ice melts,

$$H_2O(s) \longrightarrow H_2O(l) \qquad \Delta H = +6.00\text{ kJ}$$

then 6.00 kJ of heat should be evolved when a mole of liquid water freezes.

$$H_2O(l) \longrightarrow H_2O(s) \qquad \Delta H = -6.00\text{ kJ}$$

3. *The value of ΔH for a reaction is the same whether it occurs in one step or in a series of steps.* This rule is also known as **Hess' law**. If a thermochemical equation can be expressed as the sum of two or more equations,

$$\text{equation} = \text{equation (1)} + \text{equation (2)} + \cdots$$

ΔH must be independent of path, since H is a state property

then ΔH for the overall equation is the sum of the ΔH's for the individual equations:

$$\Delta H = \Delta H_1 + \Delta H_2 + \cdots$$

Hess' law is very convenient for obtaining values of ΔH for reactions that are difficult to carry out in a calorimeter. Consider, for example, the formation of the toxic gas carbon monoxide from the elements

$$C(s) + \tfrac{1}{2}O_2(g) \longrightarrow CO(g)$$

It is difficult, essentially impossible, to measure ΔH for this reaction because when carbon burns the major product is always carbon dioxide, CO_2. It is possible, however, to calculate ΔH, using thermochemical data for two other reactions (Example 8.7).

Example 8.7 Given

(1) $C(s) + O_2(g) \longrightarrow CO_2(g) \qquad \Delta H = -393.5\text{ kJ}$
(2) $2CO(g) + O_2(g) \longrightarrow 2CO_2(g) \qquad \Delta H = -566.0\text{ kJ}$

calculate ΔH for the reaction

$$C(s) + \tfrac{1}{2}O_2(g) \longrightarrow CO(g) \qquad \Delta H = ?$$

Strategy The "trick" here is to work with the given information until you arrive at two equations which will add to give the equation you want ($C + \frac{1}{2}O_2 \rightarrow CO$). To do this, it helps to realize that you want *one* mole (not two) of CO on the *right* side (not the left side) of the equation.

Solution To get one mole of CO on the right side, reverse Equation (2) and divide the coefficients by two. Applying Rule 1 and Rule 2 in succession,

$$CO_2(g) \longrightarrow CO(g) + \tfrac{1}{2}O_2(g) \qquad \Delta H = +566.0\text{ kJ}/2 = +283.0\text{ kJ}$$

Now, add Equation (1) and simplify:

$$CO_2(g) \longrightarrow CO(g) + \tfrac{1}{2}O_2(g) \qquad \Delta H = +283.0\text{ kJ}$$

$$(1)\quad C(s) + O_2(g) \longrightarrow CO_2(g) \qquad \Delta H = -393.5\text{ kJ}$$

$$C(s) + \tfrac{1}{2}O_2(g) \longrightarrow CO(g) \qquad \Delta H = -110.5\text{ kJ}$$

Note that thermochemical equations can be added in exactly the same manner as algebraic equations; in this case, 1 CO_2 and $\frac{1}{2}O_2$ "cancelled" when the equations were added.

Summarizing the rules of thermochemistry:

1. ΔH is directly proportional to the amount of reactant or product.
2. ΔH changes sign when a reaction is reversed.
3. ΔH for a reaction has the same value regardless of the number of steps.

8.5 ENTHALPIES OF FORMATION

We have now written several thermochemical equations. In each case, we have cited the corresponding value of ΔH. Literally thousands of such equations would be needed to list the ΔH values for all the reactions that have been studied. Clearly there has to be some more concise way of recording data of this sort. These data should be in a form that can easily be used to calculate ΔH for any reaction. It turns out that there is a simple way to do this, using quantities known as enthalpies of formation.

Meaning of ΔH_f°

The **standard molar enthalpy of formation** of a compound, ΔH_f°, is equal to the enthalpy change when one mole of the compound is formed at a constant pressure of 1 atm and a fixed temperature, ordinarily 25°C, from the elements in their stable states at that pressure and temperature. From the equations

Sometimes called simply "heat of formation"

$$Ag(s, 25°C) + \tfrac{1}{2}Cl_2(g, 25°C, 1 \text{ atm}) \longrightarrow AgCl(s, 25°C) \qquad \Delta H = -127.1 \text{ kJ}$$

$$\tfrac{1}{2}N_2(g, 1 \text{ atm}, 25°C) + O_2(g, 1 \text{ atm}, 25°C) \longrightarrow NO_2(g, 1 \text{ atm}, 25°C) \qquad \Delta H = +33.2 \text{ kJ}$$

it follows that

$$\Delta H_f^\circ\ AgCl(s) = -127.1 \text{ kJ/mol} \qquad \Delta H_f^\circ\ NO_2(g) = +33.2 \text{ kJ/mol}$$

Enthalpies of formation are listed for a variety of compounds in Table 8.3. Notice that, with a few exceptions, enthalpies of formation are negative quantities. This means that the formation of a compound from the elements is ordinarily exothermic. Conversely, when a compound decomposes to the elements, heat usually must be absorbed.

You will note from Table 8.3 that there are no entries for elemental species such as $Br_2(l)$ and $O_2(g)$. This is a consequence of the way in which enthalpies of formation are defined. In effect, the enthalpy of formation of an element in its stable state at 25°C and 1 atm is taken to be zero. That is,

$$\Delta H_f^\circ\ Br_2(l) = \Delta H_f^\circ\ O_2(g) = 0$$

However, $\Delta H_f^\circ\ Br_2(g)$ is +29.6 kJ/mol; explain

Calculation of ΔH°

Enthalpies of formation can be used to calculate ΔH° for a reaction. To do this, apply the general rule:

The standard enthalpy change, ΔH°, for a given thermochemical equation is equal to the sum of the standard enthalpies of formation of the product compounds minus the sum of the standard enthalpies of formation of the reactant compounds.

Enthalpy of formation was defined so as to make this equation valid

Table 8.3 Enthalpies of Formation, ΔH_f° (kJ/mol), of Compounds at 25°C, 1 atm

Compound	ΔH_f°	Compound	ΔH_f°	Compound	ΔH_f°	Compound	ΔH_f°
$AgBr(s)$	−100.4	$CaCl_2(s)$	−795.8	$H_2O(g)$	−241.8	$NH_4NO_3(s)$	−365.6
$AgCl(s)$	−127.1	$CaCO_3(s)$	−1206.9	$H_2O(l)$	−285.8	$NO(g)$	+90.2
$AgI(s)$	−61.8	$CaO(s)$	−635.1	$H_2O_2(l)$	−187.8	$NO_2(g)$	+33.2
$AgNO_3(s)$	−124.4	$Ca(OH)_2(s)$	−986.1	$H_2S(g)$	−20.6	$N_2O_4(g)$	+9.2
$Ag_2O(s)$	−31.0	$CaSO_4(s)$	−1434.1	$H_2SO_4(l)$	−814.0	$NaCl(s)$	−411.2
$Al_2O_3(s)$	−1675.7	$CdCl_2(s)$	−391.5	$HgO(s)$	−90.8	$NaF(s)$	−573.6
$BaCl_2(s)$	−858.6	$CdO(s)$	−258.2	$KBr(s)$	−393.8	$NaOH(s)$	−425.6
$BaCO_3(s)$	−1216.3	$Cr_2O_3(s)$	−1139.7	$KCl(s)$	−436.7	$NiO(s)$	−239.7
$BaO(s)$	−553.5	$CuO(s)$	−157.3	$KClO_3(s)$	−397.7	$PbBr_2(s)$	−278.7
$BaSO_4(s)$	−1473.2	$Cu_2O(s)$	−168.6	$KClO_4(s)$	−432.8	$PbCl_2(s)$	−359.4
$CCl_4(l)$	−135.4	$CuS(s)$	−53.1	$KNO_3(s)$	−369.8	$PbO(s)$	−219.0
$CHCl_3(l)$	−134.5	$Cu_2S(s)$	−79.5	$MgCl_2(s)$	−641.3	$PbO_2(s)$	−277.4
$CH_4(g)$	−74.8	$CuSO_4(s)$	−771.4	$MgCO_3(s)$	−1095.8	$PCl_3(g)$	−287.0
$C_2H_2(g)$	+226.7	$Fe(OH)_3(s)$	−823.0	$MgO(s)$	−601.7	$PCl_5(g)$	−374.9
$C_2H_4(g)$	+52.3	$Fe_2O_3(s)$	−824.2	$Mg(OH)_2(s)$	−924.5	$SiO_2(s)$	−910.9
$C_2H_6(g)$	−84.7	$Fe_3O_4(s)$	−1118.4	$MgSO_4(s)$	−1284.9	$SnO_2(s)$	−580.7
$C_3H_8(g)$	−103.8	$HBr(g)$	−36.4	$MnO(s)$	−385.2	$SO_2(g)$	−296.8
$CH_3OH(l)$	−238.7	$HCl(g)$	−92.3	$MnO_2(s)$	−520.0	$SO_3(g)$	−395.7
$C_2H_5OH(l)$	−277.7	$HF(g)$	−271.1	$NH_3(g)$	−46.1	$ZnI_2(s)$	−208.0
$CO(g)$	−110.5	$HI(g)$	+26.5	$N_2H_4(l)$	+50.6	$ZnO(s)$	−348.3
$CO_2(g)$	−393.5	$HNO_3(l)$	−174.1	$NH_4Cl(s)$	−314.4	$ZnS(s)$	−206.0

Using the symbol Σ to represent "the sum of,"

$$\Delta H^\circ = \sum \Delta H_f^\circ \text{ products} - \sum \Delta H_f^\circ \text{ reactants}$$

In taking these summations, you must take into account the coefficients of products and reactants in the thermochemical equation. For the general reaction

$$a\text{A}(g) + b\text{B}(g) \longrightarrow c\text{C}(g) + d\text{D}(g)$$

where a, b, c, and d are the coefficients of substances A, B, C, and D:

$$\Delta H^\circ = c\Delta H_f^\circ \text{ C}(g) + d\Delta H_f^\circ \text{ D}(g) - [a\Delta H_f^\circ \text{ A}(g) + b\Delta H_f^\circ \text{ B}(g)]$$

Strictly speaking, ΔH° calculated from enthalpies of formation listed in Table 8.3 represents the enthalpy change at 25°C and 1 atm. Actually, ΔH is independent of pressure and varies relatively little with temperature, changing by perhaps 1 to 10 kJ per 100°C.

Example 8.8 Calculate ΔH° for the combustion of one mole of propane, C_3H_8, according to the equation

$$C_3H_8(g) + 5\,O_2(g) \longrightarrow 3CO_2(g) + 4H_2O(l)$$

Strategy Use the general relation $\Delta H° = \Sigma\,\Delta H_f°$ products $- \Sigma\,\Delta H_f°$ reactants. Take enthalpies of formation from Table 8.3. Remember that the values in the table are for one mole of compound; if there are n moles in the equation, multiply the molar enthalpy of formation by n.

Solution Expressing $\Delta H°$ in terms of enthalpies of formation,

$$\Delta H° = 3\Delta H_f°\,CO_2(g) + 4\Delta H_f°\,H_2O(l) - [\Delta H_f°\,C_3H_8(g) + 5\Delta H_f°\,O_2(g)]$$

Taking the enthalpy of formation of $O_2(g)$ to be zero and substituting values for the other substances from Table 8.3,

$$\Delta H° = 3\text{ mol}\left(-393.5\frac{\text{kJ}}{\text{mol}}\right) + 4\text{ mol}\left(-285.8\frac{\text{kJ}}{\text{mol}}\right) - 1\text{ mol}\left(-103.8\frac{\text{kJ}}{\text{mol}}\right)$$

$$= -2219.9\text{ kJ}$$

The relation between $\Delta H°$ and enthalpies of formation is perhaps used more often than any other in thermochemistry. To show its validity, consider the thermite reaction, once used to weld rails (Fig. 8.6):

$$2Al(s) + Fe_2O_3(s) \longrightarrow 2Fe(s) + Al_2O_3(s)$$

This can be considered to be the sum of two different reactions:

(1) $2Al(s) + \frac{3}{2}O_2(g) \longrightarrow Al_2O_3(s)$ $\quad \Delta H_1°$

(2) $Fe_2O_3(s) \longrightarrow 2Fe(s) + \frac{3}{2}O_2(g)$ $\quad \Delta H_2°$

(3) $2Al(s) + Fe_2O_3(s) \longrightarrow 2Fe(s) + Al_2O_3(s)$ $\quad \Delta H_3°$

From the definition of enthalpy of formation,

$$\Delta H_1° = \Delta H_f°\,Al_2O_3(s) \qquad \Delta H_2° = -\Delta H_f°\,Fe_2O_3(s)$$

Applying Hess' law,

Figure 8.6
When a piece of burning magnesium ribbon is used as a fuse, a finely divided mixture of aluminum powder and iron(III) oxide undergoes an exothermic reaction: $2Al(s) + Fe_2O_3(s) \rightarrow 2Fe(s) + Al_2O_3(s)$; $\Delta H° = -851.5$ kJ. Enough heat is generated to produce molten iron. (Charles D. Winters)

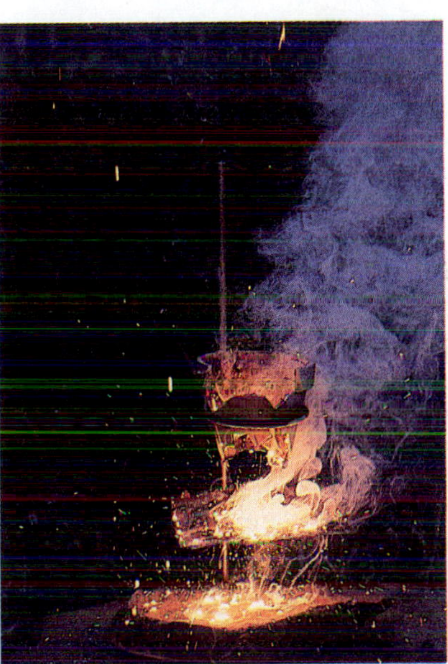

$$\Delta H_3^\circ = \Delta H_1^\circ + \Delta H_2^\circ$$
$$= \Delta H_f^\circ \; Al_2O_3(s) - \Delta H_f^\circ \; Fe_2O_3(s)$$

which is exactly the expression obtained by applying the relation

$$\Delta H^\circ = \sum \Delta H_f^\circ \text{ products} - \sum \Delta H_f^\circ \text{ reactants}$$

directly to the equation

$$2Al(s) + Fe_2O_3(s) \longrightarrow 2Fe(s) + Al_2O_3(s)$$

That's for sure

It's a great deal easier to use the general relation than to go through the tedious, stepwise analysis using Hess' law.

Enthalpies of Formation of Ions in Solution

It is possible to set up a table, very much like Table 8.3, for enthalpies of formation of ions in water solution. There is, however, one problem. The enthalpy of formation of an individual ion cannot be measured. In any reaction involving ions, at least two of them are present, as required by the principle of electrical neutrality. Consider, for example, the reaction that occurs when HCl is added to water:

$$HCl(g) \longrightarrow H^+(aq) + Cl^-(aq)$$

Here, two different ions are formed, H^+ and Cl^-. The same situation applies in all other cases. To get around this dilemma, the enthalpy of formation of the H^+ ion is arbitrarily taken to be zero:

There's nothing unique about H^+, but we have to start somewhere

$$\Delta H_f^\circ \; H^+(aq) = 0$$

Doing this establishes enthalpies of formation for other ions. Take, for instance, the Cl^- ion. For the reaction that occurs when HCl is added to water, ΔH° is found to be -74.9 kJ. That is,

$$HCl(g) \longrightarrow H^+(aq) + Cl^-(aq) \qquad \Delta H^\circ = -74.9 \text{ kJ}$$

Applying the relation between ΔH° and enthalpies of formation,

$$-74.9 \text{ kJ} = \Delta H_f^\circ \; H^+(aq) + \Delta H_f^\circ \; Cl^-(aq) - \Delta H_f^\circ \; HCl(g)$$

Taking the enthalpy of formation of $H^+(aq)$ to be zero and that of $HCl(g)$ to be -92.3 kJ (Table 8.3) gives

$$-74.9 \text{ kJ} = 0 + \Delta H_f^\circ \; Cl^-(aq) + 92.3 \text{ kJ}$$

or

$$\Delta H_f^\circ \; Cl^-(aq) = -74.9 \text{ kJ} - 92.3 \text{ kJ} = -167.2 \text{ kJ}$$

Enthalpies of formation for other ions in water solution can be established in a similar way. Table 8.4 lists values for several common ions. This table can be used, much like Table 8.3, to calculate ΔH for reactions in solution. To do that, use the general relation

$$\Delta H^\circ = \sum \Delta H_f^\circ \text{ products} - \sum \Delta H_f^\circ \text{ reactants}$$

realizing that ΔH_f° for $H^+(aq)$ is zero, as is ΔH_f° for an element in its stable state.

Table 8.4 Enthalpies of Formation, ΔH_f° (kJ/mol), of Aqueous Ions at 25°C, 1 *M*

Cations				Anions			
$Ag^+(aq)$	+105.6	$Hg^{2+}(aq)$	+171.1	$Br^-(aq)$	−121.6	$HPO_4^{2-}(aq)$	−1292.1
$Al^{3+}(aq)$	−531.0	$K^+(aq)$	−252.4	$CO_3^{2-}(aq)$	−677.1	$HSO_4^-(aq)$	−887.3
$Ba^{2+}(aq)$	−537.6	$Mg^{2+}(aq)$	−466.8	$Cl^-(aq)$	−167.2	$I^-(aq)$	−55.2
$Ca^{2+}(aq)$	−542.8	$Mn^{2+}(aq)$	−220.8	$ClO_3^-(aq)$	−104.0	$MnO_4^-(aq)$	−541.4
$Cd^{2+}(aq)$	−75.9	$Na^+(aq)$	−240.1	$ClO_4^-(aq)$	−129.3	$NO_2^-(aq)$	−104.6
$Cu^+(aq)$	+71.7	$NH_4^+(aq)$	−132.5	$CrO_4^{2-}(aq)$	−881.2	$NO_3^-(aq)$	−205.0
$Cu^{2+}(aq)$	+64.8	$Ni^{2+}(aq)$	−54.0	$Cr_2O_7^{2-}(aq)$	−1490.3	$OH^-(aq)$	−230.0
$Fe^{2+}(aq)$	−89.1	$Pb^{2+}(aq)$	−1.7	$F^-(aq)$	−332.6	$PO_4^{3-}(aq)$	−1277.4
$Fe^{3+}(aq)$	−48.5	$Sn^{2+}(aq)$	−8.8	$HCO_3^-(aq)$	−692.0	$S^{2-}(aq)$	+33.1
$H^+(aq)$	0.0	$Zn^{2+}(aq)$	−153.9	$H_2PO_4^-(aq)$	−1296.3	$SO_4^{2-}(aq)$	−909.3

Example 8.9 When hydrochloric acid is added to a solution of sodium carbonate, carbon dioxide gas is formed (Fig. 8.7). The equation for the reaction is

$$2H^+(aq) + CO_3^{2-}(aq) \longrightarrow CO_2(g) + H_2O(l)$$

Calculate ΔH° for this thermochemical equation.

Strategy Apply the general relation between ΔH° and enthalpies of formation. Obtain enthalpies of formation from Tables 8.3 and 8.4. Remember that $\Delta H_f^\circ\ H^+(aq) = 0$.

Solution

$$\Delta H^\circ = \Delta H_f^\circ\ CO_2(g) + \Delta H_f^\circ\ H_2O(l) - [2\Delta H_f^\circ\ H^+(aq) + \Delta H_f^\circ\ CO_3^{2-}(aq)]$$

Taking $\Delta H_f^\circ\ H^+(aq)$ to be zero and obtaining the other enthalpies of formation from Tables 8.3 and 8.4.

$$\Delta H^\circ = 1\ \text{mol}\left(-393.5\frac{\text{kJ}}{\text{mol}}\right) + 1\ \text{mol}\left(-285.8\frac{\text{kJ}}{\text{mol}}\right) - 1\ \text{mol}\left(-677.1\frac{\text{kJ}}{\text{mol}}\right)$$
$$= -2.2\ \text{kJ}$$

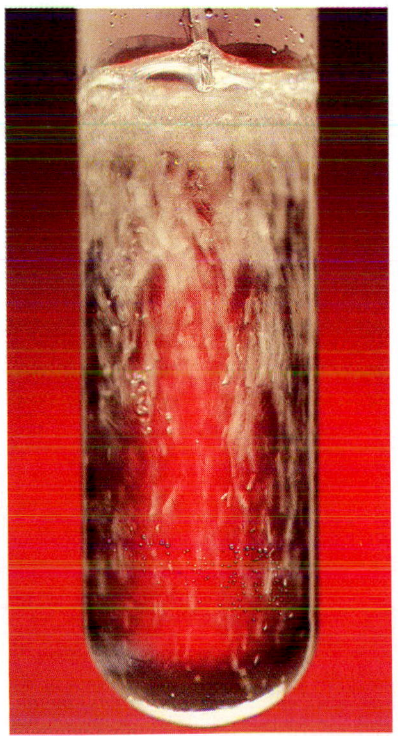

Figure 8.7

8.6 BOND ENERGY

For many reactions, ΔH is a large negative number; the reaction gives off a lot of heat. In other cases, ΔH is positive; heat must be absorbed for the reaction to occur. You may well wonder why the enthalpy change should vary so widely from one reaction to another. Is there some basic property of the molecules involved in the reaction which determines the sign and magnitude of ΔH?

These questions can be answered on a molecular level in terms of a quantity known as bond energy. (More properly, but less commonly, it is called

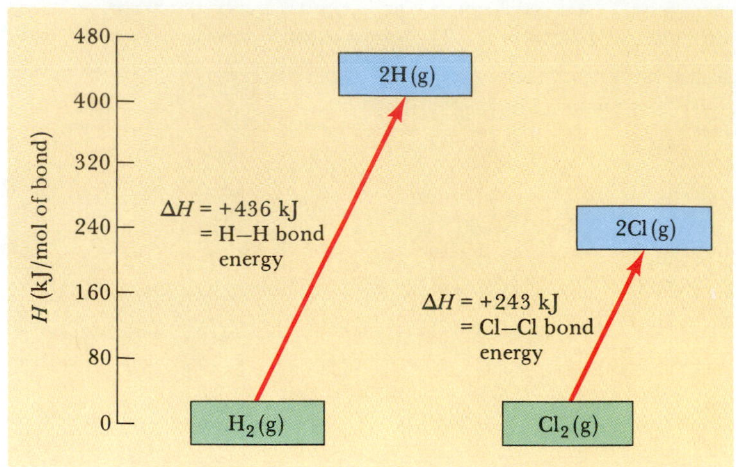

Figure 8.8
Bond energies in H_2 and Cl_2. The H—H bond energy is greater than the Cl—Cl bond energy (+436 kJ/mol vs. +243 kJ/mol). This means that the bond in H_2 is stronger than that in Cl_2.

bond enthalpy). The bond energy is defined as ΔH when one mole of bonds is broken in the gaseous state. From the equations (see also Fig. 8.8)

$$H_2(g) \longrightarrow 2H(g) \qquad \Delta H = +436 \text{ kJ}$$

$$Cl_2(g) \longrightarrow 2Cl(g) \qquad \Delta H = +243 \text{ kJ}$$

it follows that the H—H bond energy is +436 kJ/mol, while that for Cl—Cl is 243 kJ/mol. In both reactions, one mole of bonds (H—H and Cl—Cl) is broken.

The higher the bond energy, the stronger the bond

Table 8.5 Bond Energies

Single Bond Energies (kJ/mol)

	H	C	N	O	S	F	Cl	Br	I
H	436	414	389	464	339	565	431	368	297
C		347	293	351	259	485	331	276	218
N			159	222	—	272	201	243	—
O				138	—	184	205	201	201
S					226	285	255	213	—
F						153	255	255	277
Cl							243	218	209
Br								193	180
I									151

Multiple Bond Energies (kJ/mol)

C=C	612	N=N	418	C≡C	820
C=N	615	N=O	607	C≡N	890
C=O	715	O=O	498	C≡O	1075
C=S	477	S=O	498	N≡N	941

Bond energies for a variety of single and multiple bonds are listed in Table 8.5. Note that bond energy is always a positive quantity; energy is always absorbed when chemical bonds are broken. Conversely, energy is given off when bonds are formed from gaseous atoms. Thus

$$H(g) + Cl(g) \longrightarrow HCl(g) \qquad \Delta H = -431\ kJ$$

$$H(g) + F(g) \longrightarrow HF(g) \qquad \Delta H = -565\ kJ$$

You will note from Table 8.5 that the bond energy is larger for a multiple bond than for a single bond between the same two atoms. Thus

$$\text{bond energy C—C} = 347\ kJ/mol$$

$$\text{bond energy C=C} = 612\ kJ/mol$$

$$\text{bond energy C≡C} = 820\ kJ/mol$$

This effect is a reasonable one; the greater the number of bonding electrons, the more difficult it should be to break the bond between two atoms.

Estimation of ΔH from Bond Energies

A table of bond energies can be used to calculate ΔH for a reaction involving gaseous, molecular species. To see how this is done, consider the reaction

Remember, bond energies apply only in the gas state

$$H_2(g) + Cl_2(g) \longrightarrow 2HCl(g)$$

which can be divided into two steps:

1. The bonds in the reactant molecules are broken

$$H_2(g) + Cl_2(g) \longrightarrow 2H(g) + 2Cl(g)$$

break 1 mol of H—H, 1 mol of Cl—Cl bonds

$$\Delta H_1 = \text{bond energy H—H} + \text{bond energy Cl—Cl}$$
$$= 436\ kJ + 243\ kJ = 679\ kJ$$

This step is endothermic because energy must be absorbed to break bonds.

2. The gaseous atoms formed in Step 1 combine with one another, forming new bonds in HCl:

$$2H(g) + 2Cl(g) \longrightarrow 2HCl(g)$$

form 2 mol of H—Cl bonds

$$\Delta H_2 = -2(\text{bond energy H—Cl})$$
$$= -2(431\ kJ) = -862\ kJ$$

This step is exothermic because energy is given off when bonds are formed.

To determine ΔH for the overall reaction, Hess' law is applied:

$$H_2(g) + Cl_2(g) \longrightarrow 2H(g) + 2Cl(g) \qquad \Delta H_1 = +679\ kJ$$

$$2H(g) + 2Cl(g) \longrightarrow 2HCl(g) \qquad \Delta H_2 = -862\ kJ$$

$$H_2(g) + Cl_2(g) \longrightarrow 2HCl(g) \qquad \Delta H = -183\ kJ$$

This reaction is exothermic ($\Delta H = -183$ kJ) because less energy is absorbed breaking bonds in the reactants ($\Delta H_1 = +679$ kJ) than is evolved when product bonds are formed ($\Delta H_2 = -862$ kJ). In general,

— a reaction in which "weak bonds" are broken and "strong bonds" formed is exothermic ($\Delta H_1 + \Delta H_2 < 0$)
— a reaction in which "strong bonds" are broken and "weak bonds" formed is endothermic ($\Delta H_1 + \Delta H_2 > 0$).

Example 8.10 Using bond energies, estimate ΔH for the reaction of ethylene, C_2H_4, with chlorine:

Suppose only the molecular formulas were given here (C_2H_4, Cl_2, —) Could you still calculate ΔH?

$$\text{H—}\overset{\text{H}}{\overset{|}{\text{C}}}\text{=}\overset{\text{H}}{\overset{|}{\text{C}}}\text{—H}(g) + \text{Cl—Cl}(g) \longrightarrow \text{H—}\underset{\text{Cl}}{\underset{|}{\overset{\text{H}}{\overset{|}{\text{C}}}}}\text{—}\underset{\text{Cl}}{\underset{|}{\overset{\text{H}}{\overset{|}{\text{C}}}}}\text{—H}(g)$$

Strategy First take an inventory of bonds in reactants and products to decide which bonds are broken and which are formed. Then apply Hess' law to find ΔH:

$$\Delta H = \Delta H_1 + \Delta H_2$$

where $\Delta H_1 = \Sigma$ bond energies of reactants and $\Delta H_2 = -\Sigma$ bond energies of products.

Solution

(1) The bonds involved are

reactants: 4 C—H bonds, 1 C=C bond, 1 Cl—Cl bond
products: 4 C—H bonds, 1 C—C bond, 2 C—Cl bonds

Note that the four C—H bonds remain intact and so do not enter into the calculation.

bonds broken: 1 C=C bond, 1 Cl—Cl bond
bonds formed: 1 C—C bond, 2 C—Cl bonds

(2) ΔH_1 = B.E. C=C + B.E. Cl—Cl = 612 kJ + 243 kJ = +855 kJ

B.E. means bond energy

ΔH_2 = −(B.E. C—C + 2 B.E. C—Cl) = −347 kJ − 662 kJ = −1009 kJ

$\Delta H = \Delta H_1 + \Delta H_2$ = +855 kJ − 1009 kJ = **−154 kJ**

We should point out one difficulty with calculating ΔH for reactions from bond energies. In most cases it is not possible to assign an exact value to a bond energy. The bond energy varies to some extent with the species in which the bond is found. Consider, for example, the enthalpy changes for the two reactions

$$\text{H—O—H}(g) \longrightarrow \text{H}(g) + \text{O—H}(g) \qquad \Delta H = +499 \text{ kJ}$$

$$\text{O—H}(g) \longrightarrow \text{H}(g) + \text{O}(g) \qquad \Delta H = +428 \text{ kJ}$$

The ΔH values are quite different, even though both involve breaking an O—H bond. The bond energy quoted in Table 8.5, +464 kJ/mol, is an average, calculated from ΔH values for many different reactions in which O—H bonds are broken.

A similar situation applies with other bond energies; the amount of energy required to break a given type of bond varies from one molecule to another. For that reason, the value of ΔH calculated from tables of bond energies is likely to be in error, often by 10 kJ or more. That is why, in Example 8.10, we asked you to "estimate" the value of ΔH. The accurate value for ΔH of this reaction, calculated from enthalpies of formation as in Section 8.5, is -182 kJ.

8.7 THE FIRST LAW OF THERMODYNAMICS

So far in this chapter, our discussion has focused upon thermochemistry, the study of the heat effects in chemical reactions. Thermochemistry is a branch of *thermodynamics,* which deals with all kinds of energy effects in all kinds of processes. Thermodynamics distinguishes between two types of energy. One of these is heat (q); the other is **work**, represented by the symbol w. The thermodynamic definition of work is quite different from its colloquial meaning. Quite simply, *work includes all forms of energy except heat.*

The law of conservation of energy states that energy (E) can neither be created nor destroyed; it can only be transferred between system and surroundings. That is,

$$\Delta E_{\text{system}} = -\Delta E_{\text{surroundings}}$$

The first law of thermodynamics goes a step further. Taking account of the fact that there are two kinds of energy, heat and work, the first law states:

In any process, the total change in energy of the system, ΔE, is equal to the sum of the heat absorbed, q, and the work, w, done on the system.

$$\Delta E = q + w$$

In applying the first law, note (Fig. 8.9) that q and w are positive when heat or work enters the system from the surroundings. If the transfer is in the opposite direction, from system to surroundings, q and w are negative.

If work is done on the system or heat is added to it, its energy increases

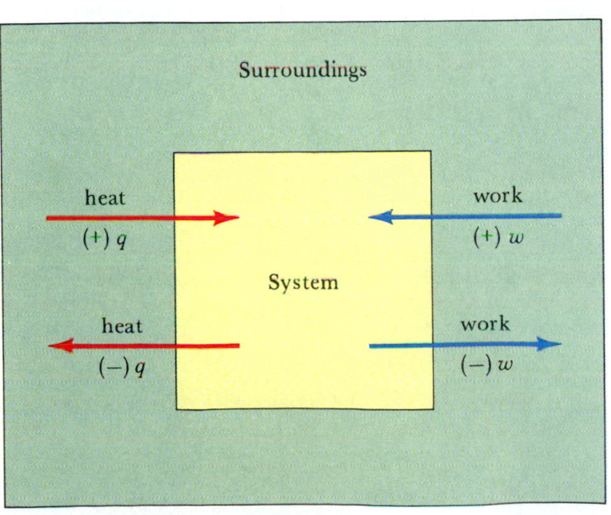

Figure 8.9
Heat and work are positive when they enter a system and negative when they leave.

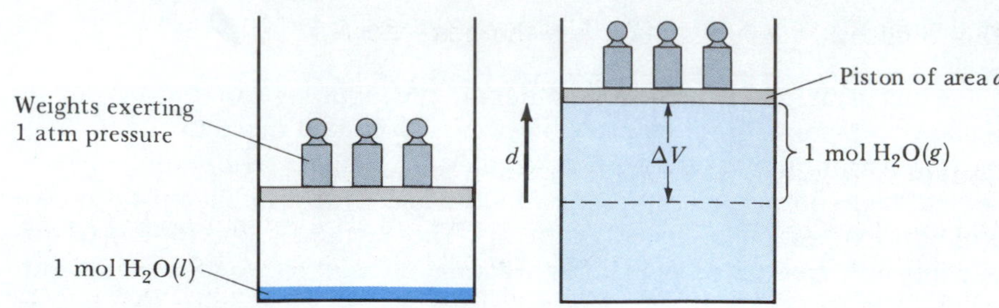

Figure 8.10
In general, the expansion work for a system working against a constant pressure, P, is $-P\Delta V$, where ΔV is the change in volume. In this case, where one mole of liquid water is vaporized at 100°C and 1 atm, w turns out to be -30.6 L · atm or -3.10 kJ.

Expansion Work

When a chemical reaction is carried out directly in the laboratory, only one type of work is ordinarily involved: work of expansion (or contraction) done by a gas against a restraining pressure. Typically, the gases involved are products or reactants in the reaction. The "restraining pressure" is simply the constant pressure of the atmosphere.

To illustrate expansion work, consider what happens when a mole of liquid water is vaporized at 100°C and 1 atm (Fig. 8.10). Work is defined as the product of force (f) times the distance (d) through which the force moves

$$w = -f \times d$$

(The minus sign arises because when expansion occurs, as is the case here, work is done by the system on the surroundings). Since pressure (P) is force per unit area (a), it follows that $f = P \times a$, so

$$w = -P \times a \times d$$

But, as you can see from the figure, the product of area times distance is simply the volume change for the process, ΔV. Hence, the general expression for expansion work occurring against a constant external pressure P is

$$w = -P\Delta V$$

To evaluate the expansion work for the process shown in Figure 8.10,

$$H_2O(l,\ 1\ \text{atm},\ 100°C) \longrightarrow H_2O(g,\ 1\ \text{atm},\ 100°C)$$

note that the volume change is

$$\Delta V = V(H_2O,\ g,\ 1\ \text{atm},\ 100°C) - V(H_2O,\ l,\ 1\ \text{atm},\ 100°C)$$

But the volume of liquid water is much, much smaller than that of the vapor, so

$$\Delta V \approx V(H_2O,\ g) = \frac{nRT}{P}$$

$$-P\Delta V = -nRT = (1\ \text{mol})(0.0821\ \text{L} \cdot \text{atm/mol} \cdot \text{K})(373\ \text{K}) = -30.6\ \text{L} \cdot \text{atm}$$

To convert to more familiar energy units, note that

$$1\ \text{L} \cdot \text{atm} = 0.1013\ \text{kJ}$$

Hence

$$w = -30.6\ \text{L atm} \times \frac{0.1013\ \text{kJ}}{1\ \text{L} \cdot \text{atm}} = -3.10\ \text{kJ}$$

This is relatively small compared to the heat effect:

$$q = \Delta H_{vap}\ H_2O = +40.7\ kJ$$

Constant Volume, Constant Pressure Processes

The quantities q and w, unlike E and H, are *not* state properties. Their value depends upon the path by which a process is carried out. To illustrate this point, let us evaluate the heat and work effects associated with processes carried out at constant volume vs. constant pressure.

1. Constant Volume. In a bomb calorimeter or any other sealed container, there is no change in volume upon reaction (unless, of course, the container explodes!). It follows that there is no expansion work associated with a constant volume process:

$$w_v = 0$$

(The subscript v is used to indicate constant volume). Hence from the first law

$$q_v = \Delta E$$

The heat flow measured in a bomb calorimeter is ΔE for the reaction

2. Constant Pressure. For a process carried out in a coffee-cup calorimeter or any other container open to the atmosphere (beaker, flask, test tube, etc.), the expansion work is given by the expression derived above:

$$w_p = -P\Delta V$$

Ordinarily, there is at least a small change in volume, so w_p has a finite value other than zero and $w_p \neq w_v$.

To find the heat flow for a constant pressure process, we again apply the first law:

$$q_p = \Delta E - w_p$$

$$q_p = \Delta E + P\Delta V$$

Comparing this expression to that for the heat flow at constant volume ($q_v = \Delta E$), you can see that q_p and q_v are *not* the same. Indeed,

$$q_p = q_v + P\Delta V$$

where P and ΔV are the pressure and the volume change, respectively, for the constant pressure process.

Table 8.6 Comparison of Reactions at Constant Volume and Constant Pressure (25°C)

	Constant V		Constant P	
	w_v	q_v	w_p	q_p
$N_2O_4(g) \longrightarrow 2NO_2(g)$	0	+54.7 kJ	−2.5 kJ	+57.2 kJ
$PCl_3(g) + Cl_2(g) \longrightarrow PCl_5(g)$	0	−85.4 kJ	+2.5 kJ	−87.9 kJ
$N_2(g) + 3H_2(g) \longrightarrow 2NH_3(g)$	0	−87.2 kJ	+5.0 kJ	−92.2 kJ
$2HI(g) \longrightarrow H_2(g) + I_2(g)$	0	+7.0 kJ	0	+7.0 kJ

Why is $q_p = q_v$ in this case?

Example 8.11 Calculate the difference in kilojoules between q_p and q_v for the decomposition of one mole of ammonium chloride at 25°C:

$$NH_4Cl(s) \longrightarrow NH_3(g) + HCl(g)$$

Strategy Find ΔV, using the ideal gas law for NH_3 and HCl; the molar volume of the solid is negligibly small by comparison. Then calculate $P\Delta V$, first in liter-atmospheres and then in kilojoules (1 L · atm = 0.1013 kJ).

Solution

$$\Delta V \approx V_m NH_3 + V_m HCl = \frac{RT}{P} + \frac{RT}{P} = \frac{2RT}{P}$$

$$P\Delta V = 2RT$$

$$= 2\text{ mol }(0.0821\text{ L}\cdot\text{atm/mol}\cdot\text{K})(298\text{ K}) \times \frac{0.1013\text{ kJ}}{1\text{ L}\cdot\text{atm}} = 4.96\text{ kJ}$$

ΔH and ΔE

The equation

$$q_p = q_v + P\Delta V$$

can be used to relate the change in enthalpy, ΔH, to the change in energy, ΔE. As you have just seen, $q_v = \Delta E$; earlier, we pointed out that $q_p = \Delta H$. It follows that

$$\Delta H = \Delta E + P\Delta V$$

Since $P\Delta V$ is ordinarily a small quantity, it follows that *the values of ΔH and ΔE for a reaction are close to one another.* For the reaction considered in Example 8.11, recall that $P\Delta V = +4.96$ kJ; ΔH for this reaction, calculated from enthalpies of formation, is +176.0 kJ. Thus

Some people would ignore this small difference, but thermodynamicists are picky

$$\Delta E = \Delta H - P\Delta V = +176.0\text{ kJ} - 4.96\text{ kJ} = +171.0\text{ kJ}$$

The two quantities ΔH and ΔE differ in this case by less than 3%.

PERSPECTIVE

•

Energy Balance in the Human Body

All of us require energy to maintain life processes and do the muscular work associated with such activities as studying, writing, walking, or jogging. The fuel used to produce that energy is the food we eat. In this discussion, we focus upon "energy input" and "energy output" and the balance between them.

Energy Input

Energy values of food can be estimated on the basis of the content of carbohydrate, protein, and fat:

Table 8.A Energy Values of Food Portions*

	kJ	kcal
8 oz glass (250 g) skim milk	330	80
whole milk	710	170
beer	420	100
1 portion vegetable	100	25
1 portion fruit	170	40
1 portion bread or starchy vegetable	300	70
1 oz (30 g) lean meat	230	55
medium-fat meat	320	77
high-fat meat	420	100
1 teaspoon (5 g) fat or oil	190	45
1 teaspoon (5 g) sugar	80	20

* Adapted from E. N. Whitney and E. M. Hamilton, *Understanding Nutrition,* 3rd ed., West Publishing Co., St. Paul, 1984.

carbohydrate: 17 kJ (4.0 kcal) of energy per gram*
protein: 17 kJ (4.0 kcal) of energy per gram
fat: 38 kJ (9.0 kcal) of energy per gram

Alcohol, which doesn't fit into any of these categories, furnishes about 29 kJ (7.0 kcal) per gram.

Using these guidelines, it is possible to come up with the data given in Table 8.A, which lists approximate energy values for standard portions of different types of foods. With a little practice, you can use the table to estimate your energy input within $\pm 10\%$.

Energy Output

Energy output can be divided, more or less arbitrarily, into two categories.

1. Metabolic Energy. This is the largest item in most people's energy budget; it comprises the energy necessary to maintain life. Metabolic rates are ordinarily estimated based on a person's sex, weight, and age. For a young adult female, the metabolic rate is about 100 kJ per day per kilogram of body weight. For a 20-year-old woman weighing 120 lb,

$$\text{metabolic rate} = 120\text{ lb} \times \frac{1\text{ kg}}{2.20\text{ lb}} \times 100\frac{\text{kJ}}{\text{d}\cdot\text{kg}}$$

$$= 5500\text{ kJ/d}\ \ (1300\text{ kcal/day})$$

For a young adult male of the same weight, the figure is about 10% higher.

* In most nutrition texts, energy is expressed in kilocalories rather than kilojoules, although that situation is changing. Remember that 1 kcal = 4.18 kJ; 1 kg = 2.20 lb.

Studying can be tiring, but it doesn't consume much energy

2. Muscular Activity. Energy is consumed whenever your muscles contract, whether in writing, walking, playing tennis, or even sitting in class. Generally speaking, energy expenditure for muscular activity ranges from 50% to 100% of metabolic energy, depending upon life style. If the 20-year-old woman referred to earlier is a student who does nothing more strenuous than lift textbooks (and that not too often), the energy spent on muscular activity might be

$$0.50 \times 5500 \text{ kJ} = 2800 \frac{\text{kJ}}{\text{d}}$$

Her total energy output per day would be 5500 kJ + 2800 kJ = 8300 kJ (2000 kcal).

Energy Balance

Skiing at 10 mph requires approximately 600 kcal/hour or 2500 kJ/hour. (Gordon Wiltsie)

To maintain constant weight, your daily energy input, as calculated from the foods you eat, should be about 700 kJ (170 kcal) greater than output. The difference allows for the fact that about 40 g of protein is required to maintain body tissues and fluids. If the excess of input over output is greater than 700 kJ/day, the unused food, whether it be carbohydrate, protein, or fat, is converted to fatty tissue and stored as such in the body.

Fatty tissue consists of about 85% fat and 15% water; its energy value is

$$0.85 \times 9.0 \frac{\text{kcal}}{\text{g}} \times \frac{454 \text{ g}}{1 \text{ lb}} = 3500 \text{ kcal/lb}$$

This means that to lose one pound of fat, energy input from foods must be decreased by about 3500 kcal. To lose weight at a sensible rate of one pound per week, it is necessary to cut down by 500 kcal (2100 kJ) per day.

It's also possible, of course, to lose weight by increasing muscular activity. Table 8.B shows the amount of energy consumed per hour with various types of exercise. In principle, you can lose a pound a week by climbing mountains for an hour each day, provided you're not already doing that (most people aren't). Alternatively, you could spend $1\frac{1}{2}$ hours a day ice skating or water skiing, depending on the weather.

Table 8.B Energy Consumed by Various Types of Exercise*

kJ/hour	kcal/hour	
1000–1250	240–300	walking (3 mph), bowling, golf (pulling cart)
1250–1500	300–360	volleyball, calisthenics, golf (carrying clubs)
1500–1750	360–420	ice skating, roller skating
1750–2000	420–480	tennis (singles), water skiing
2000–2500	480–600	jogging (5 mph), downhill skiing, mountain climbing
>2500	>600	running (6 mph), handball, squash

* Adapted from Jane Brody: *Jane Brody's Nutrition Book: A Lifetime Guide to Good Eating for Better Health and Weight Control by the Personal Health Columnist for the New York Times.* Norton, New York, 1981.

CHAPTER HIGHLIGHTS

KEY CONCEPTS

1. *Relate heat flow to specific heat, m, and* Δt
   (Example 8.1; Problems 5–8, 70)
2. *Calculate q for a reaction from calorimetric data*
   (Examples 8.2, 8.3; Problems 9–18)
3. *Use a thermochemical equation to relate* ΔH *to mass of reactant or product and/or to* ΔH *of the reverse reaction*
   (Examples 8.4–8.6; Problems 19–28, 68)
4. *Use Hess' law to calculate* ΔH
   (Example 8.7; Problems 29–34, 67)
5. *Calculate* ΔH *from enthalpies of formation*
   (Examples 8.8, 8.9; Problems 35–48)
6. *Estimate* ΔH *from bond energies*
   (Example 8.10; Problems 49–56, 73)
7. *Relate* q_p *to* q_v
   (Example 8.11; Problems 59–62)

KEY EQUATIONS

Heat flow	$q = C \times \Delta t = m \times c \times \Delta t$
Calorimetry	$q_{reaction} = -C_{cal} \times \Delta t$
Enthalpy of formation	$\Delta H° = \sum \Delta H_f° \text{ products} - \sum \Delta H_f° \text{ reactants}$

KEY TERMS

bond energy	exothermic	mole
calorimeter	heat	specific heat
—bomb	—capacity	state property
—coffee cup	—of fusion	surroundings
endothermic	—of vaporization	system
enthalpy	joule	work
enthalpy of formation		

SUMMARY PROBLEM

Consider the thermochemical equation:

$$C_2H_6(g) + \tfrac{7}{2}\,O_2(g) \longrightarrow 2CO_2(g) + 3H_2O(l) \qquad \Delta H = -1559.7 \text{ kJ}$$

a. Calculate ΔH for the thermochemical equations

$$2C_2H_6(g) + 7\,O_2(g) \longrightarrow 4CO_2(g) + 6H_2O(l)$$

$$2CO_2(g) + 3H_2O(l) \longrightarrow C_2H_6(g) + \tfrac{7}{2}\,O_2(g)$$

b. The heat of vaporization of $H_2O(l)$ is +44.0 kJ/mol. Calculate ΔH for the equation

$$C_2H_6(g) + \tfrac{7}{2}\,O_2(g) \longrightarrow 2CO_2(g) + 3H_2O(g)$$

c. The heats of formation of $CO_2(g)$ and $H_2O(l)$ are −393.5 and −285.8 kJ/mol, respectively. Calculate $\Delta H_f°$ of $C_2H_6(g)$.

d. How much heat is evolved when 1.00 g of $C_2H_6(g)$ is burned to give $CO_2(g)$ and $H_2O(l)$ in an open container?

e. When 1.00 g of C_2H_6 is burned in a bomb calorimeter, the temperature rises from 22.00°C to 33.13°C. In a separate experiment, it is found that absorption of 10.0 kJ of heat raises the temperature of the calorimeter by 2.15°C. Calculate $C_{calorimeter}$ and $q_{reaction}$.

f. Calculate $\Delta E = q_v$ at 25°C for the equation

$$C_2H_6(g) + \tfrac{7}{2}O_2(g) \longrightarrow 2CO_2(g) + 3H_2O(l) \qquad \Delta H = q_p = -1559.7 \text{ kJ}$$

g. Using bond energies, estimate ΔH for the reaction

$$2C(g) + 3H_2(g) \longrightarrow C_2H_6(g)$$

Answers

a. −3119.4 kJ; +1559.7 kJ **b.** −1427.7 kJ **c.** −84.7 kJ
d. 51.9 kJ **e.** 4.65 kJ/°C, −51.8 kJ **f.** −1553.5 kJ **g.** −1523 kJ

QUESTIONS & PROBLEMS

Formulas, Symbols, and Equations

1. Using Table 8.2, write thermochemical equations for the following.
  a. benzene boiling **b.** naphthalene melting
  c. bromine freezing

2. Using Table 8.2, write thermochemical equations for the following.
  a. mercury freezing **b.** naphthalene boiling
  c. benzene melting

3. Write chemical equations for the formation of one mole of the following compounds from the elements in their stable state at 25°C and 1 atm.
  a. aluminum oxide (s)
  b. carbon tetrachloride (l)
  c. sodium sulfate (s)
  d. potassium dichromate (s)

4. Write chemical equations for the formation of one mole of the following compounds from the elements in their stable state at 25°C and 1 atm.
  a. potassium oxide (s) **b.** hydrazine (l)
  c. ammonium carbonate (s)
  d. dinitrogen tetraoxide (g)

Specific Heat and Calorimetry

5. The specific heat of solid aluminum is 0.902 J/g · °C. Calculate the amount of heat that must be absorbed to raise the temperature of a 4.75-g aluminum sample from 25.0°C to 26.8°C.

6. The specific heat of solid platinum is 0.133 J/g · °C. Calculate the temperature change in °C if a 37.0-g sample of platinum absorbs 125 J.

7. It is found that 252 J of heat must be absorbed to raise the temperature of 50.0 g of nickel from 20.0 to 31.4°C. What is the specific heat of nickel?

8. The specific heat of iron is 0.449 J/g · °C. A sample of iron absorbs 175 J of energy when the temperature increases from 24.00°C to 29.75°C. What is the mass of the sample?

9. When 2.80 g of calcium chloride dissolves in 100.0 g of water, the temperature of the water rises from 20.50 to 25.40°C. Assume that all the heat is absorbed by the water (specific heat = 4.18 J/g · °C).
  a. Write a balanced equation for the solution process.
  b. What is q for the process described?
  c. Is the solution process exothermic or endothermic?
  d. How much heat is absorbed by the water if one mole of calcium chloride is dissolved?

10. When 1.34 g of potassium bromide dissolves in 74.0 g of water in a coffee-cup calorimeter, the temperature drops from 18.000 to 17.279°C. Assume all the heat absorbed in the solution process comes from the water (specific heat = 4.18 J/g · °C).
  a. Write a balanced equation for the solution process.
  b. Is the process exothermic or endothermic?
  c. What is q when 1.34 g of potassium bromide dissolves?
  d. What is q when one mole of potassium bromide dissolves?

11. A sample of sucrose, $C_{12}H_{22}O_{11}$, weighing 4.50 g is burned in a bomb calorimeter. The heat capacity of the calorimeter is 2.411×10^4 J/°C. The temperature rises from 22.15 to 25.22°C. Calculate q for the combustion of one mole of sucrose.

12. When 5.00 mL of ethyl alcohol, $C_2H_5OH(l)$ (d = 0.789 g/cm^3), is burned in a bomb calorimeter, the temperature in the bomb rises from 19.75 to 28.67°C. The calorimeter heat capacity is 13.24 kJ/°C. Calculate q for the combustion of one mole of ethyl alcohol.

13. When 1.750 g of methane, CH_4, burns in a bomb calorimeter, the temperature in the bomb rises 3.293°C. Under these conditions, 885.3 kJ of heat is evolved per mole of methane burned. Calculate the calorimeter heat capacity in J/°C.

14. A 350.0-mg sample of benzoic acid gave off 9.258 kJ of heat when burned in a bomb calorimeter. The temperature rose 2.152°C. Calculate the calorimeter heat capacity in joules per degree Celsius.

15. A 0.2500-g sample of naphthalene, $C_{10}H_8$, is burned in a bomb calorimeter, heat capacity = 4999 J/°C. If the bomb is initially at 20.00°C, what is the final temperature of the system? (When one mole of naphthalene burns, 5.15×10^3 kJ of heat are evolved.)

16. A 4.00-g sample of salicylic acid, $C_7H_6O_3$, is burned in a bomb calorimeter with a heat capacity of 2.046×10^4 J/°C.

The combustion produces 87.7 kJ of heat, and the temperature after combustion is 26.10°C. Calculate the initial temperature.

17. When 1.00 g of acetylene, $C_2H_2(g)$, burns in a bomb calorimeter, the temperature rises 10.94°C. In a separate experiment, it is found that when 8.16 kJ of heat is added to the calorimeter, the temperature increases by 1.79°C. Calculate

a. the heat capacity of the calorimeter.
b. q for the combustion of one gram of acetylene.
c. q for the combustion of one mole of acetylene.

18. When 1.00 g of ethylene, $C_2H_4(g)$, burns in a bomb calorimeter, the temperature rises 12.92°C. In a separate experiment, it is found that the addition of 6.23 kJ of heat to the calorimeter raises its temperature by 1.61°C. What is

a. the heat capacity of the calorimeter?
b. q for the combustion of one gram of ethylene?
c. q for the combustion of one mole of ethylene?

Rules of Thermochemistry

19. The combustion of one mole of benzene, $C_6H_6(l)$, in oxygen liberates 3.268×10^3 kJ of heat. The products of the reaction are carbon dioxide and water.

a. Write the thermochemical equation for the combustion of benzene.
b. Is the reaction exothermic or endothermic?
c. Draw a diagram, similar to Figure 8.4, for this reaction.
d. Calculate ΔH when 10.00 g of benzene is burned.
e. How many grams of benzene must be burned to evolve one kilojoule of heat?

20. One mole of nickel carbonyl gas, $Ni(CO)_4$, decomposes upon heating to nickel and carbon monoxide. In addition, 160.7 kJ of heat is absorbed.

a. Write a thermochemical equation for this reaction.
b. Is this reaction exothermic or endothermic?
c. Draw a diagram, similar to Figure 8.4, for this reaction.
d. Calculate ΔH when 1.000 g of nickel carbonyl decomposes.
e. How many grams of nickel carbonyl decompose when 10.00 kJ of heat is absorbed?

21. Consider the dissociation of water into ions.

$$H_2O(l) \longrightarrow H^+(aq) + OH^-(aq) \qquad \Delta H = 55.8 \text{ kJ}$$

a. Calculate ΔH when one mole of water is formed from the ions.
b. What is ΔH when 1.00 g of water is formed?

22. Consider the dissolving of silver bromide.

$$AgBr(s) \longrightarrow Ag^+(aq) + Br^-(aq) \qquad \Delta H = 84.4 \text{ kJ}$$

a. Calculate ΔH when one mole of silver bromide solid precipitates from solution.
b. What is ΔH when one gram of silver bromide is precipitated?

23. When ignited, ammonium dichromate decomposes in a fiery display. This is the reaction for the "laboratory volcanoes" often demonstrated in science fair projects. The decomposition of 1.000 g of ammonium dichromate produces 1.19 kJ of energy, nitrogen gas, steam, and chromium(III) oxide.

a. Write a balanced thermochemical equation for the reaction.
b. What is ΔH for the formation of 11.2 L of nitrogen at 0°C and 1.00 atm?

24. When magnesium and carbon dioxide react, 16.7 kJ is evolved per gram of magnesium. The products are solid carbon and magnesium oxide.

a. Write a balanced thermochemical equation for the reaction.
b. How much heat is evolved if one gram of products is formed from this reaction?

25. A typical fat in the body is glyceryl trioleate ($C_{57}H_{104}O_6$). When it is metabolized in the body, it combines with oxygen to produce carbon dioxide, water, and 3.022×10^4 kJ of heat per mole of fat.

a. Write a balanced thermochemical equation for the metabolism of fat.
b. How many grams of fat would have to be burned to heat one liter of water ($d = 1.00$ g/mL) from 25.00°C to 30.00°C? The specific heat of water is 4.18 J/g · °C.

26. Using the metabolism of fat reaction in Problem 25, how many kilojoules of energy must be evolved in the form of heat if you want to get rid of one pound of this fat by combustion? How many calories is this? How many nutritional calories (1 nutritional calorie = 1×10^3 calories)?

27. Which requires the absorption of a greater amount of heat, melting 100.0 g of naphthalene or boiling 100.0 g of water? Use Table 8.2.

28. Which evolves more heat, the freezing of 25.0 g of mercury or the condensation of 100.0 g of benzene vapor? Use Table 8.2.

Hess' Law

29. a. Calculate the amount of heat required to raise the temperature of 25.00 g of water from 25 to 100°C. (c H_2O =4.18 J/g · °C)
b. Calculate the amount of heat involved in the conversion of 25.00 g of water at 100°C to steam at 100°C (Table 8.2).
c. Using Hess' law, calculate the amount of heat involved in the conversion of 25.00 g of water at 25°C to steam at 100°C.

30. Follow the step-wise process outlined in Problem 29 to calculate the amount of heat evolved in freezing 25.00 g of water at 37°C to ice at 0°C.

31. A lead ore, galena, consisting mainly of lead(II) sulfide, is the principal source of lead. To obtain the lead, the ore is first heated in the air to form lead oxide.

$$PbS(s) + \tfrac{3}{2}O_2(g) \longrightarrow PbO(s) + SO_2(g) \qquad \Delta H = -415.4 \text{ kJ}$$

The oxide is then reduced to metal with carbon.

$$PbO(s) + C(s) \longrightarrow Pb(s) + CO(g) \qquad \Delta H = +108.5 \text{ kJ}$$

Calculate ΔH for the reaction of one mole of lead(II) sulfide with oxygen and carbon, forming lead, sulfur dioxide, and carbon monoxide.

32. To produce silicon, used for semiconductors, from sand (SiO_2), a reaction is used which can be broken down into three steps:

$$SiO_2(s) + 2C(s) \longrightarrow Si(s) + 2CO(g) \qquad \Delta H = 689.9 \text{ kJ}$$

$$Si(s) + 2Cl_2(g) \longrightarrow SiCl_4(g) \qquad \Delta H = -657.0 \text{ kJ}$$

$$SiCl_4(g) + 2Mg(s) \longrightarrow 2MgCl_2(s) + Si(s) \qquad \Delta H = -625.6 \text{ kJ}$$

Write the thermochemical equation for the overall reaction for the formation of silicon from silicon dioxide; CO and $MgCl_2$ are byproducts. What is ΔH for the formation of one mole of silicon? Is the reaction endothermic or exothermic?

33. Given the following thermochemical equations

$$2H_2(g) + O_2(g) \longrightarrow 2H_2O(l) \qquad \Delta H = -571.6 \text{ kJ}$$

$$N_2O_5(g) + H_2O(l) \longrightarrow 2HNO_3(l) \qquad \Delta H = -73.7 \text{ kJ}$$

$$\tfrac{1}{2}N_2(g) + \tfrac{3}{2}O_2(g) + \tfrac{1}{2}H_2(g) \longrightarrow HNO_3(l) \qquad \Delta H = -174.1 \text{ kJ}$$

calculate ΔH for the formation of one mole of dinitrogen pentaoxide from its elements in their stable state at 25°C and 1 atm.

34. Given the following thermochemical equations

$$C_2H_2(g) + \tfrac{5}{2}O_2(g) \longrightarrow 2CO_2(g) + H_2O(l) \qquad \Delta H = -1299.5 \text{ kJ}$$

$$C(s) + O_2(g) \longrightarrow CO_2(g) \qquad \Delta H = -393.5 \text{ kJ}$$

$$H_2(g) + \tfrac{1}{2}O_2(g) \longrightarrow H_2O(l) \qquad \Delta H = -285.8 \text{ kJ}$$

calculate ΔH for the decomposition of one mole of acetylene, $C_2H_2(g)$, to its elements in their stable state at 25°C and 1 atm.

$\Delta H°$ and Heats of Formation

35. Given

$$2CuO(s) \longrightarrow 2Cu(s) + O_2(g) \qquad \Delta H° = 314.6 \text{ kJ}$$

a. Determine the heat of formation of CuO.
b. Calculate $\Delta H°$ for the formation of 13.58 g of CuO.

36. Given

$$2Al_2O_3(s) \longrightarrow 4Al(s) + 3O_2(g) \qquad \Delta H° = 3351.4 \text{ kJ}$$

a. What is the heat of formation of aluminum oxide?
b. What is $\Delta H°$ for the formation of 12.50 g of aluminum oxide?

37. Limestone, $CaCO_3$, when subjected to a temperature of 900°C in a kiln, decomposes to calcium oxide and carbon dioxide. How much heat is evolved or absorbed when one gram of limestone decomposes? (Use Table 8.3.)

38. A first-aid hot pack is made up of two compartments: One has solid calcium chloride, the other has water. When heat is required, you are instructed to twist the hot pack. This breaks the diaphragm separating both compartments. Calcium chloride dissolves in water, giving calcium ions, chloride ions, and heat. How much heat is evolved if the hot pack contains five grams of calcium chloride? (Use Table 8.3 and 8.4.)

39. Use Tables 8.3 and 8.4 to obtain $\Delta H°$ for the following thermochemical equations.

a. $2Br^-(aq) + I_2(s) \longrightarrow Br_2(l) + 2I^-(aq)$
b. $Cr(OH)_3(s) + 5\,OH^-(aq) + 3Fe^{3+}(aq) \longrightarrow 3Fe^{2+}(aq) + 4H_2O(l) + CrO_4^{2-}(aq)$
($\Delta H_f°$ for $Cr(OH)_3 = -1064.0$ kJ/mol)
c. $SO_4^{2-}(aq) + 4H^+(aq) + Ni(s) \longrightarrow Ni^{2+}(aq) + SO_2(g) + 2H_2O(l)$

40. Use Tables 8.3 and 8.4 to obtain $\Delta H°$ for the following thermochemical equations.

a. $2Br^-(aq) + 2H_2O(l) \longrightarrow Br_2(l) + H_2(g) + 2\,OH^-(aq)$
b. $Pb^{2+}(aq) + Sn(s) \longrightarrow Pb(s) + Sn^{2+}(aq)$
c. $3Mn^{2+}(aq) + 2NO_3^-(aq) + 2H_2O(l) \longrightarrow 2NO(g) + 3MnO_2(s) + 4H^+(aq)$

41. Use Table 8.3 to calculate $\Delta H°$ for

a. the combustion of one mole of ethyl alcohol, C_2H_5OH, to form carbon dioxide and water (l).
b. the combustion of one mole of ethyl alcohol, C_2H_5OH, to form carbon dioxide and steam.

42. Use Table 8.3 to calculate $\Delta H°$ for

a. the reaction of ammonia and oxygen gas to form one mole of nitrogen and steam.
b. the reaction of ammonia and oxygen gas to form two moles of nitrogen and $H_2O(l)$.

43. When one mole of calcium carbonate reacts with ammonia, solid calcium cyanamide, $CaCN_2$, and liquid water are formed. The reaction absorbs 90.1 kJ of heat.

a. Write a balanced thermochemical equation for the reaction.
b. Using Table 8.3, calculate $\Delta H_f°$ for calcium cyanamide.

44. When one mole of ethylene gas, C_2H_4, is burned in oxygen and hydrogen chloride, the products are liquid ethylene chloride, $C_2H_4Cl_2$, and liquid water; 318.7 kJ of heat is evolved.

a. Write a thermochemical equation for the reaction.
b. Using Table 8.3, calculate the standard heat of formation of ethylene chloride.

45. Use Tables 8.3 and 8.4 to determine the heat of formation of the Co^{2+} ion given $\Delta H° = -1936.2$ kJ for the reaction

$$2MnO_4^-(aq) + 16H^+(aq) + 5Co(s) \longrightarrow 2Mn^{2+}(aq) + 5Co^{2+}(aq) + 8H_2O(l)$$

46. Use Tables 8.3 and 8.4 to determine the heat of formation of lead sulfate from the following equation.

$$PbO_2(s) + SO_4^{2-}(aq) + 4H^+(aq) + 2I^-(aq) \longrightarrow PbSO_4(s) + 2H_2O(l) + I_2(s) \qquad \Delta H° = -194.4 \text{ kJ}$$

47. Glucose, $C_6H_{12}O_6(s)$ ($\Delta H_f° = -1275.2$ kJ/mol) is converted to ethyl alcohol, $C_2H_5OH(l)$, and carbon dioxide in the fermentation of grape juice. What quantity of heat is liberated when a liter of wine containing 12.0% ethyl alcohol by volume ($d = 0.789$ g/cm^3) is produced by the fermentation of grape juice?

48. When ammonia reacts with dinitrogen oxide gas ($\Delta H_f° = 82.05$ kJ/mol), liquid water and nitrogen gas are formed. How much heat is liberated or absorbed by the reaction that produces 275 mL of nitrogen gas at 25°C and 777 mm Hg?

ΔH and Bond Energies

49. Calculate the amount of heat that must be absorbed (kJ/mol) to dissociate into atoms:

a. Cl_2 **b.** C_2H_6 **c.** CCl_4

50. Calculate ΔH per mole for the formation of the following molecules from gaseous atoms.

a. N_2 **b.** CH_4 **c.** NH_3

51. Using bond energies, estimate ΔH for the reaction between hydrogen gas and bromine gas to form one mole of hydrogen bromide gas.

52. Using bond energies, estimate ΔH for the reaction between hydrogen and nitrogen to form one mole of hydrazine gas, N_2H_4.

53. Using bond energies, estimate ΔH for the reaction between carbon monoxide and chlorine gas to form one mole of phosgene, Cl—C(=O)—Cl.

54. Isopropyl alcohol, sold as rubbing alcohol, can be prepared by the reaction between acetone, the primary ingredient in nail polish remover, and hydrogen:

$$H_3C—C(=O)—CH_3(g) + H_2(g) \longrightarrow H_3C—CH(OH)—CH_3(g)$$

Estimate ΔH for this reaction.

55. The first step in the manufacture of acrylonitrile, a chemical widely used in the plastics industry, is

$$H_2C—CH_2(g) \text{ (with O bridging, epoxide)} + HCN(g) \longrightarrow HO—CH_2—CH_2—CN(g)$$

Estimate ΔH for this reaction.

56. The calculation of ΔH using bond energies works only for gaseous molecules. In the reaction for the formation of one mole of hydrogen cyanide, HCN, from its elements, nitrogen and hydrogen gases react with carbon in its solid state. Calculate ΔH for this reaction using bond energies and adding the reaction

$$C(s) \longrightarrow C(g) \qquad \Delta H = 717 \text{ kJ}$$

into your calculations.

First Law of Thermodynamics

57. Calculate

a. q when a system does 72 J of work and its energy decreases by 90 J.

b. ΔE for a gas that releases 35 J of heat and has 128 J of work done on it.

58. Find

a. ΔE when a gas absorbs 45 J of heat and has 32 J of work done on it.

b. q when 62 J of work are done on a system and its internal energy is increased by 84 J.

59. Determine the difference in kilojoules at 25°C between q_p and q_v for

$$2CO(g) + O_2(g) \longrightarrow 2CO_2(g)$$

60. For the vaporization of one mole of water at 100°C, determine

a. $q_p = \Delta H$ **b.** q_v **c.** the expansion work, w_p

61. Consider the combustion of one mole of acetylene, C_2H_2, which yields carbon dioxide and liquid water.

a. Using Table 8.3, calculate ΔH.

b. Calculate ΔE at 25°C.

62. Consider the combustion of propane, C_3H_8, the fuel that is commonly used in portable gas barbecue grills. The products of combustion are carbon dioxide and liquid water.

a. Write a thermochemical equation for the combustion of one mole of propane.

b. How much heat would be evolved if 1.00 g of propane were burned in a bomb calorimeter (25°C)?

Energy Balance

63. A young woman weighing 165 lb is determined to have a metabolic rate of 1.00×10^2 kJ/day/kg of body weight. Her muscular activity is about 75% of her metabolic energy. If the young woman's caloric intake is 2500 kcal/day, will she lose weight? Will she lose a pound a week?

64. A young man weighing 181 lb has a metabolic rate of 115 kJ/day/kg of body weight. If the young man has a strenuous exercise program that results in muscular activity which is 85% of metabolic activity, how many kilocalories can he take in daily without gaining weight?

65. A cup of ice milk contains 6.0 g of protein, 36 g of carbohydrate, and 8.0 g of fat. Approximately how many hours of walking would be required to burn up the energy taken in by eating a cup of ice milk?

66. How is energy balance calculated?

Unclassified

67. Given the following reactions

$$N_2H_4(l) + O_2(g) \longrightarrow N_2(g) + 2H_2O(g) \qquad \Delta H° = -534.2 \text{ kJ}$$

$$H_2(g) + \tfrac{1}{2} O_2(g) \longrightarrow H_2O(g) \qquad \Delta H° = -241.8 \text{ kJ}$$

Calculate the heat of formation of hydrazine.

68. In World War II, the Germans made use of unusable airplane parts by grinding them up into powdered aluminum. This was made to react with ammonium nitrate to produce powerful bombs. The products of this reaction were nitrogen gas, steam, and aluminum oxide. If 10.00 kg of ammonium nitrate is mixed with 10.00 kg of powdered aluminum, how much heat is generated?

69. One way to lose weight is to exercise. Walking briskly at 4.0 mph for an hour consumes 1700 kJ of energy. How many miles would you have to walk to lose one pound of body fat? (One gram of body fat is equivalent to about 32 kJ of energy.)

70. Brass has a density of 8.25 g/cm^3 and a specific heat of 0.362 J/g · °C. A cube of brass 7.50 mm on an edge is heated in a Bunsen burner flame to a temperature of 95.0°C. It is then immersed into 20.0 mL of water (d = 1.00 g/cm^3) at 22.0°C in an insulated container. Assuming no heat loss, what is the final temperature of the water?

71. Some solutes have large heats of solution, and care should be taken in preparing solutions of these substances. The heat evolved when sodium hydroxide dissolves is 44.5 kJ/mol. What is the final temperature of the water, originally at 20.0°C, used to prepare 500.0 cm^3 of 6.00 M NaOH solution? Assume all the heat is absorbed by 500 g of water, specific heat = 4.18 J/g · °C.

72. A 1.00-L sample of a gaseous mixture at 0°C and 1.00 atm (STP) evolves, upon complete combustion at constant pressure, 82.58 kJ of heat. If the gas is a mixture of ethane (C_2H_6) and propane (C_3H_8), what is the mole fraction of ethane in the mixture?

73. Use bond energies to calculate ΔH for the reaction between ammonia and oxygen to form steam and one mole of nitrogen. (Assume a double bond in O_2.) Compare with $\Delta H°$ obtained from heats of formation.

Challenge Problems

74. Biking at 13 mph (moderate pace) uses about 2000 kJ of energy per hour. This comes from the oxidation of foods, which is about 30% efficient. How much energy do you "save" by biking one mile instead of driving a car that gets 20.0 mi/gal of gasoline (d gasoline = 0.68 g/cm^3; heat of combustion = 48 kJ/g)?

75. On a hot day, you take a six-pack of beer on a picnic, cooling it with ice. Each (aluminum) can weighs 38.5 g and contains 12.0 oz of beer. The specific heat of aluminum is 0.902 J/g · °C; take that of beer to be 4.10 J/g · °C.

a. How much heat must be absorbed from the six-pack to lower the temperature from 25.0 to 5.0°C?

b. How much ice must be melted to absorb this amount of heat (ΔH_{fus} of ice is given in Table 8.2)?

76. A cafeteria sets out glasses of tea at room temperature; the customer adds ice. Assuming the customer wants to have some ice left when the tea cools to 0°C, what fraction of the total volume of the glass should be left empty for adding ice? Make any reasonable assumptions needed to work this problem.

77. The thermite reaction was once used to weld rails:

$$2Al(s) + Fe_2O_3(s) \longrightarrow Al_2O_3(s) + 2Fe(s)$$

a. Calculate ΔH for this reaction using heat of formation data.

b. Take the specific heats of Al_2O_3 and Fe to be 0.77 and 0.45 J/g · °C, respectively. Calculate the temperature to which the products of this reaction will be raised, starting at room temperature, by the heat given off in the reaction.

c. Will the reaction produce molten iron (mp Fe = 1535°C, ΔH_{fus} = 270 J/g)?

CHARLES D. WINTERS

$$2Sb(s) + 3Cl_2(g) \longrightarrow 2SbCl_3(s)$$

(See p. 550)

9 Liquids and Solids

When water turns ice does it
remember one time
it was water?
When ice turns back into
water does it remember
it was ice?

CARL SANDBURG
Metamorphoses

CHAPTER OUTLINE

In Chapter 5 we pointed out that at ordinary temperatures and pressures all gases follow the ideal gas law. Unfortunately, there is no simple "equation of state" that can be written to correlate the properties of liquids or solids. There are two reasons for this difference in behavior.

1. Molecules are much closer to one another in liquids and solids. In the gas state, particles are typically separated by 10 molecular diameters or more; in liquids or solids, they touch one another. This explains why liquids and solids have densities so much larger than those of gases. At 100°C and 1 atm, $H_2O(l)$ has a density of 0.95 g/mL; that of $H_2O(g)$ under the same conditions is only 0.00059 g/mL.

2. Intermolecular forces, which are essentially negligible with gases, play a much more important role in liquids and solids. Among the effects of these forces is the phenomenon of *surface tension,* shown by liquids. Molecules at the surface are attracted inward by unbalanced intermolecular forces; as a result, liquids tend to form spherical drops, for which the ratio of surface area to volume is as small as possible (Fig. 9.1).

Figure 9.1
Surface tension causes water to bead on the polished surface of a car. (Richard Megna/FUNDAMENTAL PHOTOGRAPHS, New York)

Our discussion of liquids and solids will focus upon

— the relationships among particle structure, interparticle forces, and physical properties. Section 9.1 explores these relationships for molecular substances; Section 9.2 extends the discussion to nonmolecular solids (*network covalent,* ionic, and metallic). Section 9.3 is devoted to the crystal structure of ionic and metallic solids.

— phase equilibria. Section 9.4 considers several different phenomena related to gas-liquid equilibria, including vapor pressure, boiling point behavior, and *critical properties.* Section 9.5 deals with *phase diagrams,* which describe all three types of phase equilibria (gas-liquid, gas-solid, and liquid-solid).

9.1 MOLECULAR SUBSTANCES; INTERMOLECULAR FORCES

As a class, substances composed of molecules tend to be

1. *Nonconductors of electricity when pure.* Since molecules are uncharged, they cannot carry an electric current. In most cases, (e.g., iodine, I_2, and ethyl alcohol, C_2H_5OH), water solutions of molecular substances are nonconductors. A few polar molecules, including HCl, react with water to form ions:

$$HCl(g) \longrightarrow H^+(aq) + Cl^-(aq)$$

and hence produce a conducting water solution.

Bromine, sulfur, and sugar are substances with covalent bonds. (Richard Megna/FUNDAMENTAL PHOTOGRAPHS, New York)

2. *Insoluble in water but soluble in nonpolar solvents such as CCl_4 or toluene.* Iodine is typical of most molecular substances; it is only slightly soluble in water (0.0013 mol/L at 25°C), much more soluble in toluene (0.5 mol/L). A few molecular substances, including ethyl alcohol, are very soluble in water. As you will see later in this section, such substances have intermolecular forces similar to those in water.

3. *Low melting and boiling.* Many molecular substances are gases at 25°C and 1 atm (for example, N_2, O_2, and CO_2), which means that they have boiling points below 25°C. Others (such as H_2O and CCl_4) are liquids with melting (freezing) points below room temperature. Of the molecular substances that are solids at ordinary temperatures, most are low melting (mp I_2 = 114°C, mp naphthalene = 80°C). The upper limit for melting and boiling points of most molecular substances is about 300°C.

The generally low melting and boiling points of molecular substances reflect the fact that the forces between molecules (**intermolecular forces**) are weak. To melt or boil a molecular substance, the molecules must be set free from one another. This requires only that enough energy be supplied to overcome the weak attractive forces between molecules. The strong covalent bonds within molecules remain intact when a molecular substance melts or boils.

It's a lot easier to separate two molecules than to separate two atoms within a molecule

The boiling points of different molecular substances are directly related to the strength of the intermolecular forces involved. *The stronger the intermolecular forces, the higher the boiling point of the substance.* In the remainder of this section, we examine the nature of the three different types of intermolecular forces: *dispersion forces, dipole forces,* and *hydrogen bonds.*

Dispersion (London) Forces

The most common type of intermolecular force, found in all molecular substances, is referred to as a **dispersion force**. It is basically electrical in nature, involving an attraction between temporary or *induced* dipoles in adjacent molecules. To understand the origin of dispersion forces, consider Figure 9.2.

Once called van der Waals forces

On the average, electrons in a nonpolar molecule, such as H_2, are as close to one nucleus as to the other. However, at a given instant, the electron cloud may be concentrated at one end of the molecule (position 1A in Fig. 9.2). This momentary concentration of the electron cloud on one side of the molecule creates a temporary dipole in H_2. One side of the molecule, shown in deeper color in Figure 9.2, acquires a partial negative charge; the other side has a partial positive charge of equal magnitude.

This temporary dipole induces a similar dipole (an induced dipole) in an adjacent molecule. When the electron cloud in the first molecule is at 1A, the electrons in the second molecule are attracted to 2A. These temporary dipoles, both in the same direction, lead to an attractive force between the molecules. This is the dispersion force.

All molecules have dispersion forces. The strength of these forces depends upon two factors:

In nonpolar molecules, dispersion is the only intermolecular force

— the number of electrons in the atoms that make up the molecule.
— the ease with which electrons are *dispersed* to form temporary dipoles.

Both these factors increase with increasing molecular size. Large molecules are made up of more and/or larger atoms. The outer electrons in larger atoms are

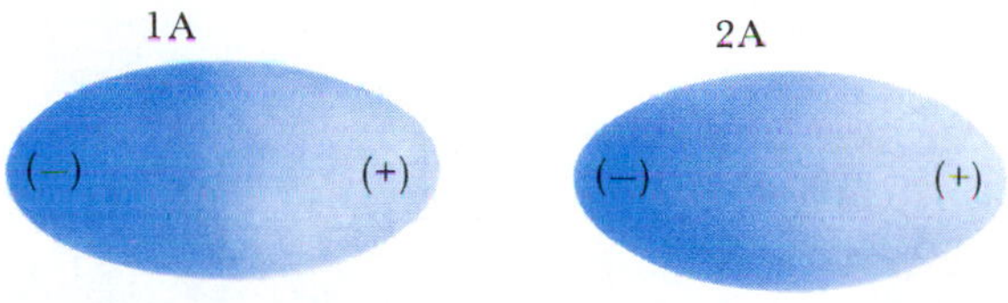

Figure 9.2
Temporary dipoles in adjacent H_2 molecules line up to create an electrical attractive force known as the dispersion force. Deeply shaded areas indicate regions where the electron cloud is momentarily concentrated.

Table 9.1 Effect of Molar Mass on Boiling Points of Molecular Substances

Noble Gases*			Halogens			Hydrocarbons		
	$\mathcal{M}$ (g/mol)	bp (°C)		$\mathcal{M}$ (g/mol)	bp (°C)		$\mathcal{M}$ (g/mol)	bp (°C)
He	4	−269	F_2	38	−188	CH_4	16	−161
Ne	10	−246	Cl_2	71	−34	C_2H_6	30	−88
Ar	40	−186	Br_2	160	59	C_3H_8	44	−42
Kr	84	−152	I_2	254	184	n-C_4H_{10}	58	0

* Strictly speaking, the noble gases are "atomic" rather than molecular. However, like molecules, the noble gas atoms are attracted to one another by dispersion forces.

relatively far from the nucleus and are easier to disperse than the electrons in small atoms. In general, molecular size and molar mass parallel one another. Thus *as molar mass increases, dispersion forces become stronger, and the boiling point of nonpolar molecular substances increases* (Table 9.1).

Dipole Forces

Dipole forces are much weaker than polar covalent bonds

Dipole forces arise as a result of the interaction between polar molecules. The nature of this interaction is illustrated in Figure 9.3, which shows the orientation of polar molecules, such as ICl, in a crystal. Adjacent molecules line up so that the negative pole of one molecule (small Cl atom) is as close as possible to the positive pole (large I atom) of its neighbor. Under these conditions, there is an electrical attractive force, referred to as a dipole force, between adjacent polar molecules.

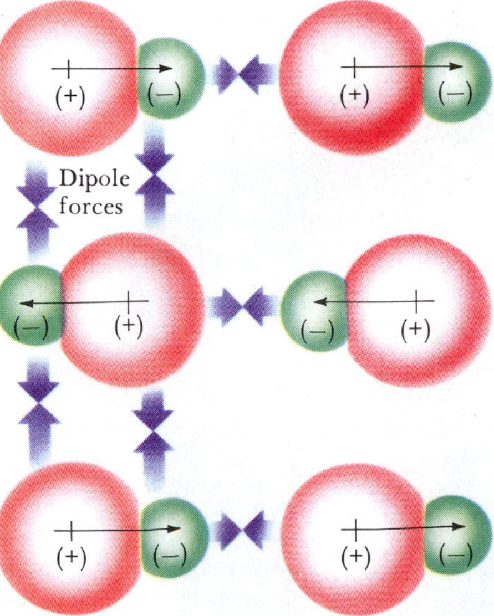

Figure 9.3
Dipole forces in the ICl crystal. The (+) and (−) indicate fractional charges on the I and Cl atoms in the polar molecules. The existence of these partial charges causes the molecules to line up in the pattern shown. Adjacent molecules are attracted to each other by dipole forces between the (+) pole of one molecule and the (−) pole of the other.

Table 9.2 Boiling Points of Nonpolar vs. Polar Substances

Nonpolar			Polar		
Formula	$\mathcal{M}$ (g/mol)	bp (°C)	Formula	$\mathcal{M}$ (g/mol)	bp (°C)
N_2	28	−196	CO	28	−192
SiH_4	32	−112	PH_3	34	−88
GeH_4	77	−90	AsH_3	78	−62
Br_2	160	59	ICl	162	97

When iodine chloride is heated to 27°C, the rather weak dipole forces are unable to keep the molecules rigidly aligned and the solid melts. Dipole forces are still important in the liquid state, since the polar molecules remain close to one another. Only in the gas, where the molecules are far apart, do the effects of dipole forces become negligible. Hence boiling points as well as melting points of polar compounds such as ICl are somewhat higher than those of nonpolar substances of comparable molar mass. This effect is shown in Table 9.2.

Example 9.1 Explain, in terms of intermolecular forces, why

(a) the boiling point of O_2 (−183°C) is higher than that of N_2 (−196°C).
(b) the boiling point of NO (−151°C) is higher than that of either O_2 or N_2.

Strategy Determine whether the molecule is polar or nonpolar and identify the intermolecular forces present. Remember, dispersion forces are always present and increase with molar mass.

Solution

(a) Only dispersion forces are involved with these nonpolar molecules. Since the molar mass of O_2 is greater (32.0 g/mol vs. 28.0 g/mol for N_2), its dispersion forces are stronger, making its boiling point higher.
(b) Dispersion forces in NO (30 g/mol) are comparable in strength to those in O_2 and N_2. The polar NO molecule shows an additional type of intermolecular force not present in N_2 or O_2: the dipole force. As a result, its boiling point is the highest of the three substances.

Hydrogen Bonds

Ordinarily, polarity has a relatively small effect on boiling point. In the series HCl → HBr → HI, boiling point increases steadily with molar mass, even though polarity decreases moving from HCl to HI. However, when hydrogen is bonded to a small, highly electronegative atom (N, O, F), polarity has a much greater effect upon boiling point. Hydrogen fluoride, HF, despite its low molar mass (20 g/mol), has the highest boiling point of all the hydrogen halides. Water ($\mathcal{M}$ = 18 g/mol) and ammonia ($\mathcal{M}$ = 17 g/mol) also have abnormally

Table 9.3 Effect of Hydrogen Bonding on Boiling Point

	bp(°C)		bp(°C)		bp(°C)
NH_3	−33	H_2O	100	HF	19
PH_3	−88	H_2S	−60	HCl	−85
AsH_3	−63	H_2Se	−42	HBr	−67
SbH_3	−18	H_2Te	−2	HI	−35

* Molecules in color show hydrogen bonding.

Water also has an unusually high heat of vaporization

high boiling points (Table 9.3). In these cases, the effect of polarity reverses the normal trend expected from molar mass alone.

The unusually high boiling points of HF, H_2O, and NH_3 result from an unusually strong type of dipole force called a **hydrogen bond**. The hydrogen bond is a force exerted between an H atom bonded to an F, O, or N atom in one molecule and an unshared pair on the F, O, or N atom of a neighboring molecule:

$$\underset{\substack{\uparrow \\ \text{hydrogen bond}}}{\text{X—H - - - - - : X—H}} \qquad X = \text{N, O, or F}$$

There are two reasons that hydrogen bonds are stronger than ordinary dipole forces:

1. The difference in electronegativity between hydrogen (2.2) and fluorine (4.0), oxygen (3.5), or nitrogen (3.0) is quite large. It causes the bonding electrons in molecules such as HF, H_2O, and NH_3 to be primarily associated with the more electronegative atom (F, O, or N). So the hydrogen atom, insofar as its interaction with a neighboring molecule is concerned, behaves almost like a bare proton.

2. The small size of the hydrogen atom allows the unshared pair of an F, O, or N atom of one molecule to approach the H atom in another very closely. It is significant that hydrogen bonding occurs only with these three nonmetals, all of which have small atomic radii.

Hydrogen bonding is important in biochemistry, particularly with proteins

Hydrogen bonds can exist in many molecules other than HF, H_2O, and NH_3. The basic requirement is simply that hydrogen be bonded to a fluorine, oxygen, or nitrogen atom with at least one unshared electron pair.

Example 9.2 Would you expect to find hydrogen bonds in

(a) ethyl alcohol?

```
   H  H
   |  |  ..
H—C—C—O—H
   |  |  ..
   H  H
```

(b) dimethyl ether?

```
   H      H
   |  ..  |
H—C—O—C—H
   |  ..  |
   H      H
```

(c) hydrazine, N_2H_4?

Strategy For hydrogen bonding to occur, hydrogen must be bonded to F, O, or N. In (c), draw the Lewis structure first.

Solution

(a) There should be hydrogen bonds in ethyl alcohol, since an H atom is bonded to oxygen.
(b) Since all the hydrogen atoms are bonded to carbon in dimethyl ether, there should be no hydrogen bonds.
(c) The Lewis structure of hydrazine is

$$\mathrm{H-\overset{\cdot\cdot}{\underset{|}{N}}-\overset{\cdot\cdot}{\underset{|}{N}}-H}$$
$$\mathrm{H\quad H}$$

Since hydrogen is bonded to nitrogen, hydrogen bonding can occur between neighboring N_2H_4 molecules. The boiling point of hydrazine (114°C) is much higher than that of molecular O_2 (−183°C), which has the same molar mass.

Water has many unusual properties in addition to its high boiling point. In particular, water, in contrast to nearly all other substances, expands upon freezing. Ice is one of the very few solids that has a density less than that of the liquid from which it is formed (d ice at 0°C = 0.917 g/cm^3, d water at 0°C = 1.000 g/cm^3). This behavior is an indirect result of hydrogen bonding. When water freezes to ice, an open hexagonal pattern of molecules results (Fig. 9.4). Each oxygen atom in an ice crystal is bonded to four hydrogens. Two of these are attached by ordinary covalent bonds at a distance of 0.099 nm. The other two form hydrogen bonds 0.177 nm in length. The large proportion of "empty space" in the ice structure explains why ice is less dense than water.

If ice were more dense, lakes would freeze from the bottom up; that would tend to discourage skating

---

Example 9.3 What types of intermolecular forces are present in H_2? CCl_4? OCS? NH_3?

Strategy Determine whether the molecules are polar or nonpolar; only polar molecules show dipole forces. Check the Lewis structures for H atoms bonded to F, O, or N. All molecules have dispersion forces.

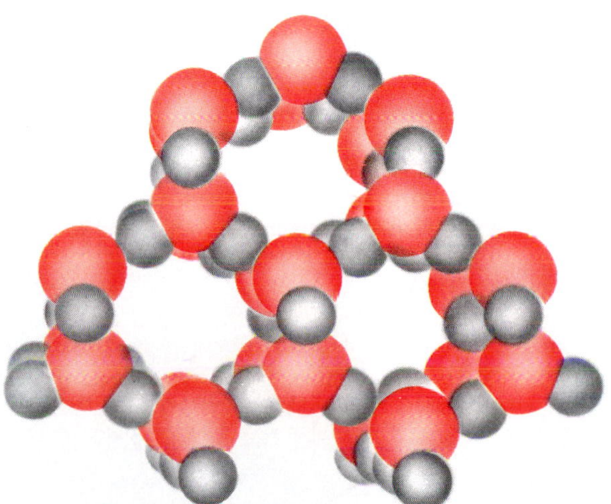

Figure 9.4
In ice, the water molecules are arranged in an open hexagonal pattern that gives ice its low density. Each oxygen atom (red) is bonded covalently to two hydrogen atoms (gray) and forms hydrogen bonds with two other hydrogen atoms.

Solution The H_2 and CCl_4 molecules are nonpolar (Chap. 7). Thus only dispersion forces are present among neighboring molecules. Both OCS and NH_3 are polar molecules:

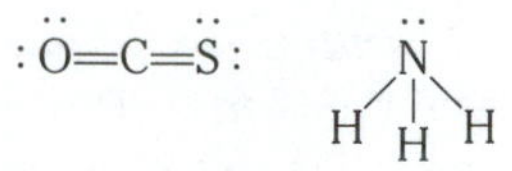

$$:\ddot{O}=C=\ddot{S}: \qquad \ddot{N}H_3$$

There are dipole forces as well as dispersion forces with OCS molecules. In NH_3, there are hydrogen bonds and dispersion forces.

We have now discussed three types of intermolecular forces: dispersion forces, dipole forces, and hydrogen bonds. You should bear in mind that all of these forces are relatively weak compared to ordinary covalent bonds. Consider, for example, the situation in H_2O. The total intermolecular attractive energy in ice is about 50 kJ/mol; the O—H bond energy is an order of magnitude greater, about 464 kJ/mol. This explains why it is a lot easier to melt ice or boil water than it is to decompose water into the elements. Even at a temperature of 1000°C and 1 atm, only about one H_2O molecule in a billion decomposes to hydrogen and oxygen atoms.

9.2 NETWORK COVALENT, IONIC, AND METALLIC SOLIDS

Virtually all substances that are gases or liquids at 25°C and 1 atm are molecular. In contrast, there are at least three types of nonmolecular solids (Fig. 9.5). These are

— **network covalent solids**, in which atoms are joined by a continuous network of covalent bonds. The entire crystal, in effect, consists of one huge molecule.

— **ionic solids**, held together by strong electrical forces (ionic bonds) between oppositely charged ions adjacent to one another

— **metallic solids**, in which the structural units are electrons (e^-) and cations (M^+)

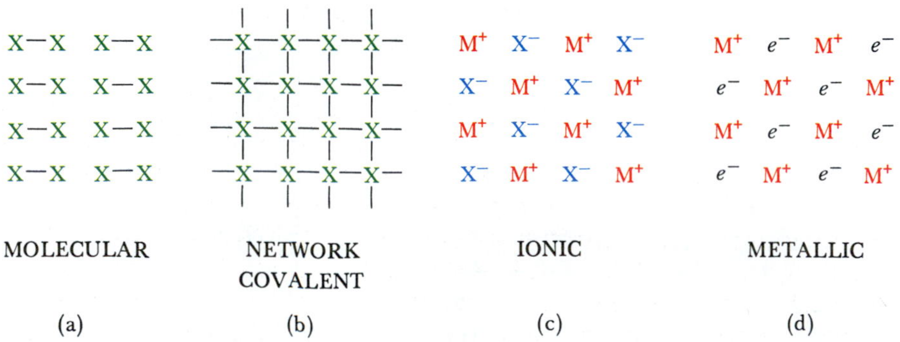

X—X X—X	—X—X—X—X—	M^+ X^- M^+ X^-	M^+ e^- M^+ e^-
X—X X—X	—X—X—X—X—	X^- M^+ X^- M^+	e^- M^+ e^- M^+
X—X X—X	—X—X—X—X—	M^+ X^- M^+ X^-	M^+ e^- M^+ e^-
X—X X—X	—X—X—X—X—	X^- M^+ X^- M^+	e^- M^+ e^- M^+
MOLECULAR	NETWORK COVALENT	IONIC	METALLIC
(a)	(b)	(c)	(d)

Figure 9.5
Schematic diagrams of four types of substances (see text discussion). M^+ represents a cation, X^- an anion, X a nonmetal atom, and e^- an electron. Covalent bonds are shown by straight lines joining atoms.

Network Covalent Solids

As a class, solids of this type are

1. *High melting, often with melting points above 1000°C.* To melt the solid, covalent bonds between atoms must be broken. In this respect, solids of this type differ markedly from molecular solids, which have much lower melting points.

2. *Insoluble in all common solvents.* For solution to occur, covalent bonds throughout the solid would have to be broken.

3. *Poor electrical conductors.* In most network covalent substances (graphite is an exception), there are no mobile electrons to carry a current.

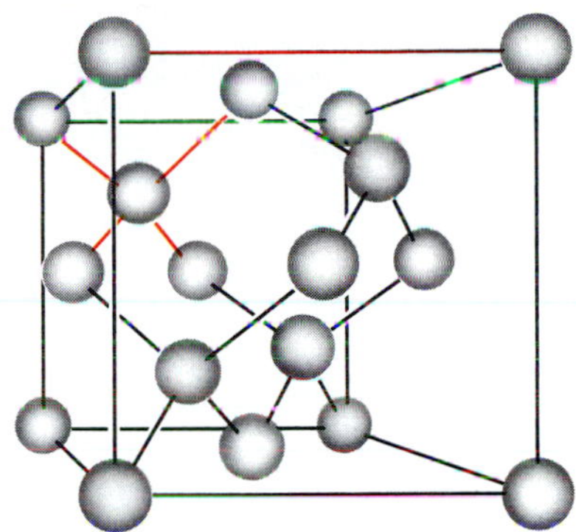

Diamond crystal

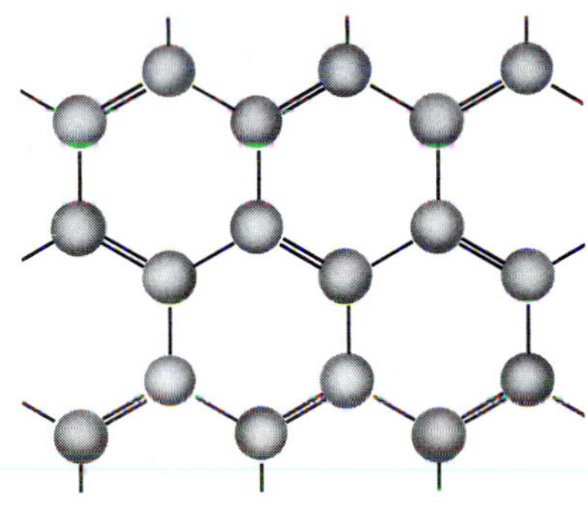

Graphite layer

Figure 9.6
Diamond has a three-dimensional structure in which each carbon atom is surrounded tetrahedrally by four other atoms. Graphite has a two-dimensional layer structure with weak dispersion forces between the layers.

Graphite and Diamond

Several nonmetallic elements have a network covalent structure. The most important of these is carbon, which has two different crystalline forms of the network covalent type. Both graphite and diamond have high melting points, above 3500°C. However, the bonding patterns in the two solids are quite different (Fig. 9.6).

In diamond, each carbon atom forms single bonds with four other carbon atoms arranged tetrahedrally around it. The hybridization in diamond is sp^3. The three-dimensional covalent bonding contributes to diamond's unusual hardness. Diamond is one of the hardest substances known; it is used in cutting tools and quality grindstones.

Graphite is planar, with the carbon atoms arranged in a hexagonal pattern. Each carbon atom is bonded to three others, two by single bonds, one by a double bond. The hybridization is sp^2. The forces between adjacent layers in graphite are of the dispersion type and quite weak. A "lead" pencil really contains a graphite rod, thin layers of which rub off onto the paper as you write.

At room temperature and atmospheric pressure, graphite is the stable form of carbon. Diamond, in principle, should slowly transform to graphite under ordinary conditions. Fortunately for the owners of diamond rings, this transition occurs at zero rate unless the diamond is heated to about 1500°C, at which temperature the conversion occurs rapidly. For understandable reasons, no one has ever become very excited over the commercial possibilities of this process. The more difficult task of converting graphite to diamond has aroused much greater enthusiasm.

Diamond has a higher density than graphite (3.51 vs. 2.26 g/cm^3); as the more dense phase, it should be favored by high pressures. Theoretically, at 25°C and 15,000 atm graphite should turn to diamond. However, under those conditions the reaction has a negligible rate. At higher temperatures it goes faster, but the required pressure goes up too; at 2000°C, a pressure of about 100,000 atm is needed. In 1954, scientists at the General Electric laboratories were able to achieve these high temperatures and pressures and converted graphitic carbon to diamond. Nowadays, all industrial diamonds are synthetic.

Pure research paid off handsomely here

In 1985, a group of chemists at Rice University reported the discovery of a *molecular* form of solid carbon with the formula C_{60}. Five years later, a method was developed to make large quantities of this substance by the laser vaporization of graphite. The cagelike C_{60} molecule, known as buckminsterfullerene or

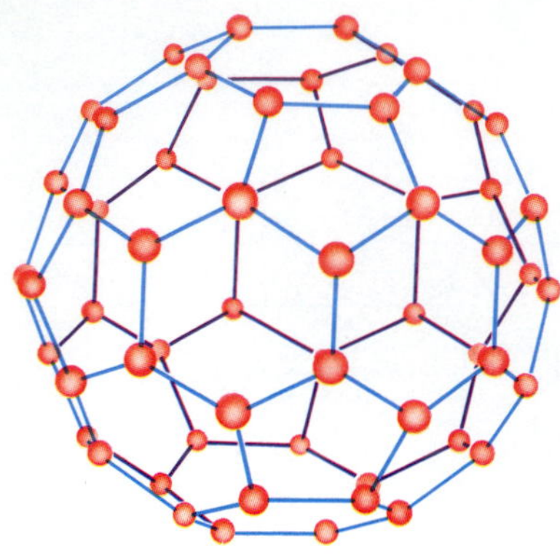

Figure 9.7
A molecular form of solid carbon, molecular formula C_{60}, was discovered in 1985. It has a cagelike structure not unlike that of the molecular hydrates mentioned in Chapter 2 (Fig. 2.C, p. 43).

"buckyball," has the general geometry of a geodesic dome; more to the point, it shows an amazing resemblance to a soccer ball (Fig. 9.7). Studies of the physical and chemical properties of C_{60} and its "cousin" C_{70} comprise one of the hottest research areas of the 1990s.

Quartz

Among the many compounds with a network covalent structure is quartz, the most common form of SiO_2 and the major component of sand. In quartz, each silicon atom bonds tetrahedrally to four oxygen atoms. Each oxygen atom bonds to two silicons, thus linking adjacent tetrahedra to one another (Fig. 9.8). Notice that the network of covalent bonds extends throughout the entire crystal. Unlike most pure solids, quartz does not melt sharply to a liquid. Instead, it turns to a viscous mass over a wide temperature range, first softening at about 1400°C. The viscous fluid probably contains long —Si—O—Si—O— chains, with enough bonds broken to allow flow.

Ionic Solids

An ionic solid consists of cations and anions (e.g., Na^+, Cl^-). No simple, discrete molecules are present in NaCl or other ionic compounds; rather, the ions are held in a regular, repeating arrangement by strong **ionic bonds**, electrostatic interactions between oppositely charged ions. Because of this structure, ionic solids show the following properties:

1. *Ionic solids are nonvolatile and high-melting* (typically at 600°C to 2000°C). Ionic bonds must be broken to melt the solid, separating oppositely charged ions from each other. Only at high temperatures do the ions acquire enough kinetic energy for this to happen.

Charged particles must move to carry a current

2. *Ionic solids do not conduct electricity because the charged ions are fixed in position.* They become good conductors, however, when melted or dissolved in water. In both cases, in the melt or solution, the ions (such as Na^+ and Cl^-) are free to move through the liquid and thus can conduct an electric current.

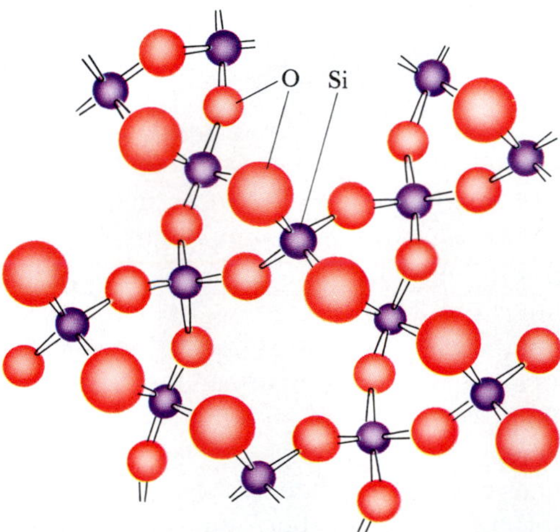

Figure 9.8
Crystal structure of quartz, SiO_2. Every silicon atom (small) is linked tetrahedrally to four oxygen atoms (large).

3. *Many, but not all, ionic compounds* (for example, NaCl but not $CaCO_3$) *are soluble in water, a polar solvent.* In contrast, ionic compounds are insoluble in nonpolar solvents such as benzene (C_6H_6) or carbon tetrachloride (CCl_4).

Metals

Figure 9.5d on p. 236 illustrates a simple model of bonding in metals known as the **electron-sea model**. The metallic crystal is pictured as an array of positive ions, e.g., Na^+, Mg^{2+}. These are anchored in position, like buoys in a mobile "sea" of electrons. These electrons are not attached to any particular positive ion, but rather can wander through the crystal. The electron-sea model explains many of the characteristic properties of metals:

Electrons in metals have no permanent address

1. *High electrical conductivity.* The presence of large numbers of relatively mobile electrons explains why metals have electrical conductivities several hundred times greater than those of typical nonmetals. Silver is the best electrical conductor but is too expensive for general use. Copper, with a conductivity close to that of silver, is the metal most commonly used for electrical wiring.

2. *High thermal conductivity.* Heat is carried through metals by collisions between electrons, which occur frequently. Saucepans used for cooking commonly contain aluminum, copper, or stainless steel; their handles are made of a nonmetallic material that is a good thermal insulator.

A glass containing a metal spoon will not break when boiling hot water is poured into it. Why?

3. *Ductility, malleability.* Most metals are ductile (capable of being drawn out into a wire) and malleable (capable of being hammered into thin sheets; Fig. 9.9). In a metal, the electrons act like a flexible glue holding the atomic nuclei together. As a result, metal crystals can be deformed without shattering.

4. *Luster.* Polished metal surfaces reflect light. Most metals have a silvery white "metallic" color because they reflect light of all wavelengths. Since electrons are not restricted to a particular bond, they can absorb and re-emit light over a wide wavelength range.

5. *Insolubility in water and other common solvents.* No metals dissolve in water; electrons cannot go into solution, and cations cannot dissolve by them-

Substances with ionic bonds. (Richard Megna/FUNDAMENTAL PHOTOGRAPHS, New York)

Copper, gold, and silver are substances with metallic bonds. (Richard Megna/FUNDAMENTAL PHOTOGRAPHS, New York)

Figure 9.9
The iron worker hammers the metal into sheets taking advantage of its malleability. (Mystic Village Photo Archives)

selves. The only liquid metal, mercury, dissolves many metals, forming solutions called *amalgams.* An Ag–Sn–Hg amalgam is used in filling teeth.

In general, the melting points of metals cover a wide range, from −39°C for mercury to 3410°C for tungsten. This variation in melting point corresponds to a similar variation in the strength of the metallic bond. Generally speaking, the lowest-melting metals are those that form +1 cations, like sodium (mp = 98°C) and potassium (mp = 64°C).

Much of what has been said about the four structural types of solids in Sections 9.1 and 9.2 is summarized in Table 9.4 and Example 9.4.

Example 9.4 A certain substance is a liquid at room temperature and is insoluble in water. Suggest which of the four types of basic structural units is present in the substance and list additional experiments that could be carried out to confirm your prediction.

Strategy Use Table 9.4, which shows the general properties of the four different types of substances. Note that a given property (e.g., high electrical conductivity) may be shown by more than one type of substance.

Solution The fact that the substance is a liquid suggests that it is probably molecular. The fact that it is insoluble in water agrees with this classification. To confirm that it is molecular, measure the conductivity, which should be zero. This test eliminates mercury, a liquid metal.

Table 9.4 Structures and Properties of Types of Substances

Type	Structural Particles	Forces Within Particles	Forces Between Particles	Properties	Examples
Molecular	molecules				
	(a) nonpolar	covalent bond	dispersion	low mp, bp; often gas or liquid at 25°C; nonconductors; insoluble in water, soluble in organic solvents	H_2 CCl_4
	(b) polar	covalent bond	dispersion, dipole, H bond	similar to nonpolar but generally higher mp and bp, more likely to be water-soluble	HCl NH_3
Network covalent	atoms	—	covalent bond	hard, very high-melting solids; nonconductors; insoluble in common solvents	C SiO_2
Ionic	ions	—	ionic bond	high mp; conductors in molten state or water solution; often soluble in water, insoluble in organic solvents	NaCl MgO $CaCO_3$
Metallic	cations, mobile electrons	—	metallic bond	variable mp; good conductors in solid; insoluble in common solvents	Na Fe

9.3 CRYSTAL STRUCTURES

Solids tend to crystallize in definite geometric forms that often can be seen by the naked eye. In ordinary table salt, cubic crystals of NaCl are clearly visible. Large, beautifully formed crystals of such minerals as fluorite, CaF_2, are found in nature. It is possible to observe distinct crystal forms of many metals under a microscope.

Crystals have definite geometric forms because the atoms or ions present are arranged in a definite, three-dimensional pattern. The nature of this pattern can be deduced by a technique known as x-ray diffraction. The basic information that comes out of such studies has to do with the dimensions and geometric form of the **unit cell**, the smallest structural unit that, repeated over and over again in three dimensions, generates the crystal.

Metals

Three of the simpler unit cells found in metals, shown in Figure 9.10, are the following:

1. **Simple cubic cell.** This is a cube that consists of eight atoms whose centers are located at the corners of the cell. Atoms at adjacent corners of the cube touch one another.

2. **Face-centered cubic cell (FCC).** Here there is an atom at each corner of the cube and one in the center of each of the six faces of the cube. In this structure, atoms at the corners of the cube do not touch one another; they are forced slightly apart. Instead, contact occurs along a face diagonal. The atom at the center of each face touches atoms at opposite corners of the face.

About 40 metals crystallize in the FCC structure; simple cubic structures are very rare

3. **Body-centered cubic cell (BCC).** This is a cube with atoms at each corner and one in the center of the cube. Here again, corner atoms do not touch each other. Instead, contact occurs along the body diagonal; the atom at the center of the cube touches atoms at opposite corners.

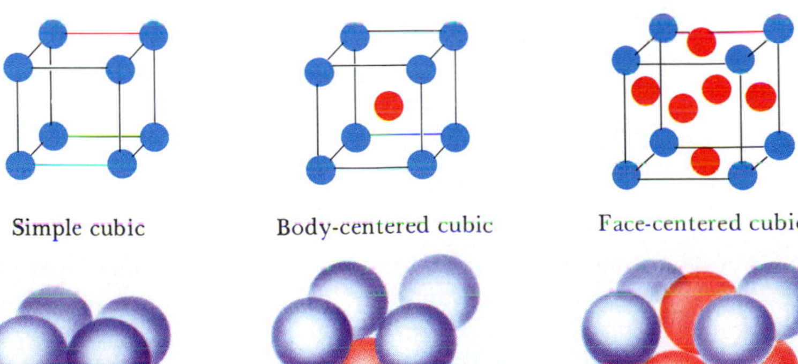

Figure 9.10
Three types of unit cells. In each case, there is an atom at each of the eight corners of the cube. In the body-centered lattice, there is an additional atom in the center of the cube. In the face-centered lattice, there is an atom in the center of each of the six faces.

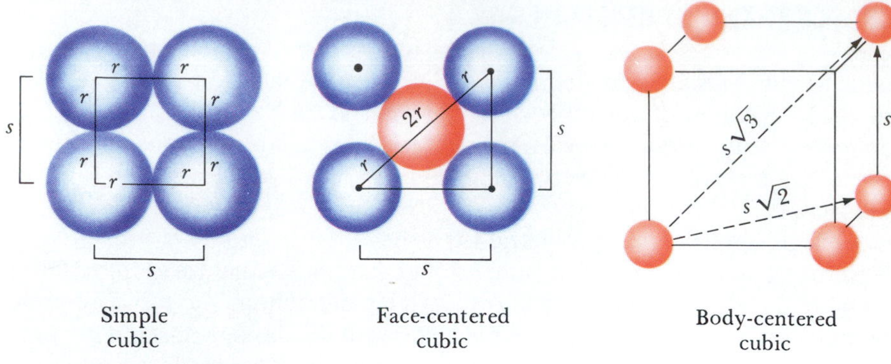

Figure 9.11
Relation between atomic radius, r, and length of edge, s, for cubic cells. Notice that:

- in the FCC cell, the face diagonal has a length of $4r$. Hence: $(4r)^2 = s^2 + s^2 = 2s^2$; $4r = s\sqrt{2}$
- in the BCC cell, the body diagonal has a length of $4r$. It is also the hypotenuse of a right triangle whose other sides are a face diagonal ($s\sqrt{2}$) and an edge (s).

Hence: $(4r)^2 = 2s^2 + s^2 = 3s^2$; $4r = s\sqrt{3}$

If the nature and dimensions of the unit cell of a metallic crystal are known, the atomic radius of the metal can be calculated. It should be evident from Figure 9.11 and the Pythagorean theorem

$$a^2 = b^2 + c^2$$

(a is the hypotenuse of a right triangle; b and c are the other two sides) that the radii (r) and sides (s) of the unit cells are related:

$$\text{simple cubic cell:} \quad 2r = s$$

$$\text{face-centered cubic cell:} \quad 4r = s\sqrt{2}$$

$$\text{body-centered cubic cell:} \quad 4r = s\sqrt{3}$$

Example 9.5 Silver crystallizes with a face-centered cubic (FCC) unit cell 0.407 nm on an edge. Calculate the atomic radius of silver.

Strategy Find the appropriate relation between the radius and the length of a side of an FCC cell. Substitute 0.407 nm for s.

Solution The equation is

$$4r = s\sqrt{2}$$

Substituting $s = 0.407$ nm and solving for r:

$$r = \frac{0.407\text{ nm} \times \sqrt{2}}{4} = \frac{0.407\text{ nm} \times 1.41}{4} = 0.144\text{ nm}$$

Ionic Crystals

The geometry of ionic crystals, in which there are two different kinds of ions, is more difficult to describe than that of metals. However, in many cases the packing can be visualized in terms of the unit cells described above. Lithium chloride, LiCl, is a case in point. Here, the larger Cl^- ions form a face-centered cubic lattice (Fig. 9.12). The smaller Li^+ ions fit into "holes" between the Cl^- ions. This puts an Li^+ ion at the center of each edge of the cube.

In the sodium chloride crystal, the Na^+ ion is slightly too large to fit into holes in a face-centered lattice of Cl^- ions (Fig. 9.12). As a result, the Cl^- ions

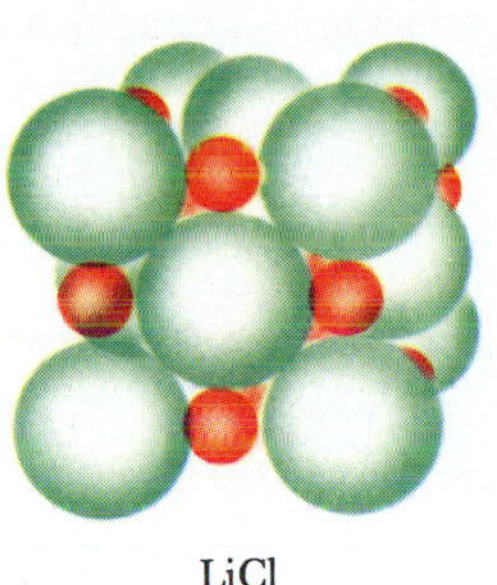

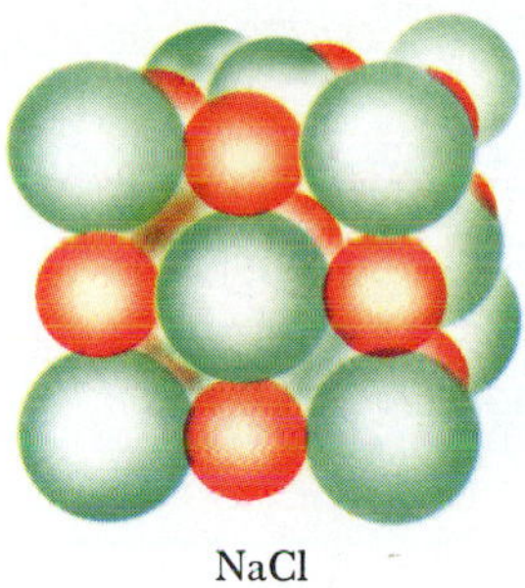

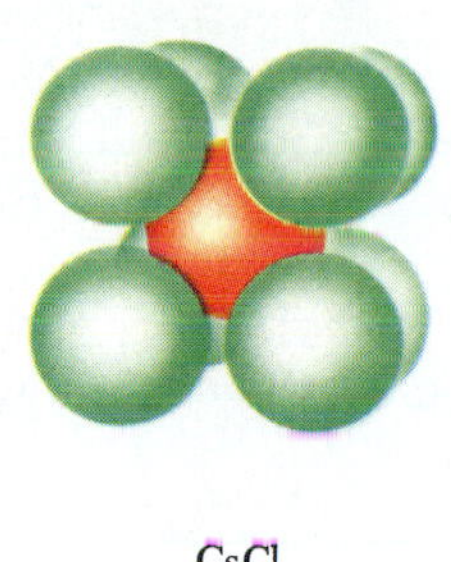

Figure 9.12
Three types of lattices in ionic crystals. In LiCl, the Cl^- ions are in contact with each other, forming a face-centered cubic lattice. In NaCl, the Cl^- ions are forced slightly apart by the larger Na^+ ions. In CsCl, the structure is quite different; the large Cs^+ ion at the center touches Cl^- ions at each corner of the cube.

are pushed slightly apart so that they are no longer touching, and only Na^+ ions are in contact with Cl^- ions. However, the relative positions of positive and negative ions remain the same as in LiCl: Each anion is surrounded by six cations and each cation by six anions.

NaCl is face-centered in both Na^+ and Cl^- ions

The structures of LiCl and NaCl are typical of all the alkali halides (Group 1 cation, Group 17 anion) except those of cesium. Because of the large size of the Cs^+ ion, CsCl crystallizes in a quite different structure. Here, each Cs^+ ion is located at the center of a cube outlined by Cl^- ions. The Cs^+ ion at the center touches all the Cl^- ions at the corners; the Cl^- ions do not touch each other. As you can see, each Cs^+ ion is surrounded by eight Cl^- ions, while each Cl^- ion is surrounded by eight Cs^+ ions.

CsCl is simple cubic in both Cs^+ and Cl^- ions

9.4 LIQUID-VAPOR EQUILIBRIUM

All of us are familiar with the process of **vaporization**, in which a liquid is converted to a vapor. In an open container, *evaporation* continues until all the liquid is gone. If the container is closed, the situation is quite different. At first, the movement of molecules is primarily in one direction, from liquid to vapor. Here, however, the vapor molecules cannot escape from the container. Some of them collide with the surface and re-enter the liquid. As time passes, and the concentration of molecules in the vapor increases, so does the rate of condensation. When the rate of condensation becomes equal to the rate of vaporization, the liquid and vapor are in a state of **dynamic equilibrium**:

$$\text{liquid} \rightleftharpoons \text{vapor}$$

The double arrow implies that the forward and reverse processes are occurring at the same rate, which is characteristic of a dynamic equilibrium.

Vapor Pressure

Once equilibrium between liquid and vapor is reached, the number of molecules per unit volume in the vapor does not change with time. This means that *the pressure exerted by the vapor over the liquid remains constant.* The pressure of vapor in equilibrium with a liquid is called the **vapor pressure**. This quantity is a characteristic property of a given liquid at a particular temperature. It varies from one liquid to another, depending upon the strength of the intermolecular forces. At 25°C, the vapor pressure of water is 24 mm Hg; that of ether, where intermolecular forces are weaker, is 537 mm Hg.

It is important to realize that *so long as both liquid and vapor are present, the pressure exerted by the vapor is independent of the volume of the container.* If a small amount of liquid is introduced into a closed container, some of it will vaporize, establishing its equilibrium vapor pressure. The greater the volume of the container, the greater will be the amount of liquid that vaporizes to establish that pressure. Only if all the liquid vaporizes will the pressure drop below the equilibrium value.

Example 9.6 Consider a sample of 1.00 g of water, which has an equilibrium vapor pressure of 24 mm Hg at 25°C.

(a) If this sample is introduced into a 10.0-L flask, how much liquid water will remain when equilibrium is established?
(b) What would be the minimum volume of a flask in which the 1.00-g sample would be completely vaporized?
(c) Suppose the 1.00-g sample is introduced into a 60.0-L flask. What will be the pressure exerted by the vapor?

Strategy In each case, use the ideal gas law. In (a), calculate the mass of $H_2O(g)$ required to establish a vapor pressure of 24 mm Hg in a 10.0-L flask at 25°C; the mass of $H_2O(l)$ left can then be found by subtraction. In (b), calculate the volume occupied by 1.00 g of $H_2O(g)$ at 25°C and 24 mm Hg. In (c), calculate the pressure exerted by 1.00 g of $H_2O(g)$ at 25°C in a 60.0-L flask.

Solution

$$\text{(a) } n_{H_2O(g)} = \frac{PV}{RT} = \frac{(24/760 \text{ atm})(10.0 \text{ L})}{(0.0821 \text{ L}\cdot\text{atm/mol}\cdot\text{K})(298 \text{ K})} = 0.013 \text{ mol}$$

$$m_{H_2O(g)} = 0.013 \text{ mol} \times \frac{18.02 \text{ g}}{1 \text{ mol}} = 0.23 \text{ g}$$

$$m_{H_2O(l)} = 1.00 \text{ g} - 0.23 \text{ g} = \boxed{0.77 \text{ g}}$$

How would these answers change if 2.00 g of water were used? Ans: (a) 1.77 g (b) 86 L (c) 24 mm Hg

$$\text{(b) } V = \frac{nRT}{P} = \frac{(1.00/18.02 \text{ mol})(0.0821 \text{ L}\cdot\text{atm/mol}\cdot\text{K})(298 \text{ K})}{(24/760 \text{ atm})} = \boxed{43 \text{ L}}$$

$$\text{(c) } P = \frac{nRT}{V} = \frac{(1.00/18.02 \text{ mol})(0.0821 \text{ L}\cdot\text{atm/mol}\cdot\text{K})(298 \text{ K})}{60.0 \text{ L}} = 0.0226 \text{ atm}$$

$$= 0.0226 \text{ atm} \times \frac{760 \text{ mm Hg}}{1 \text{ atm}} = \boxed{17.2 \text{ mm Hg}}$$

This is less than the equilibrium vapor pressure of water, which makes sense. Since 1.00 g of $H_2O(l)$ vaporizes completely in a 43-L flask, there is none left to establish equilibrium in a 60.0-L flask, and the pressure drops below its equilibrium value.

Vapor Pressure vs. Temperature

The vapor pressure of a liquid always increases as temperature rises. Water evaporates more readily on a hot, dry day. Stoppers in bottles of volatile liquids such as ether or gasoline pop out when the temperature rises.

The vapor pressure of water, which is 24 mm Hg at 25°C, becomes 92 mm Hg at 50°C and 1 atm (760 mm Hg) at 100°C. The data for water are plotted at the top of Figure 9.13. As you can see, the graph of vapor pressure vs. temperature is not a straight line, as it would be if pressure were plotted vs. temperature for an ideal gas. Instead, the slope increases steadily as temperature rises, reflecting the fact that more molecules vaporize at higher temperatures. At 100°C, the concentration of H_2O molecules in the vapor in equilibrium with liquid is 25 times as great as at 25°C.

In working with the relationship between two variables, such as vapor pressure and temperature, scientists prefer to deal with linear (straight-line) functions. Straight-line graphs are easier to construct and to interpret. In this case, it is possible to obtain a linear function by making a simple shift in variables. Instead of plotting vapor pressure (P) vs. temperature (T), we plot the *natural logarithm** of the vapor pressure ($\ln P$) vs. the reciprocal of the absolute temperature ($1/T$). Such a plot for water is shown at the bottom of Figure 9.13; as with all other liquids, a plot of $\ln P$ vs. $1/T$ is a straight line.

The general equation of a straight line is

$$y = mx + A \qquad (m = \text{slope}, A = y\text{-intercept})$$

The y-coordinate in Figure 9.13b is $\ln P$, while the x-coordinate is $1/T$. The slope, which is a negative quantity, turns out to be $-\Delta H_{vap}/R$, where ΔH_{vap} is the molar heat of vaporization and R is the gas constant, in the proper units. Hence the equation of the straight line in Figure 9.13b is

$$\ln P = \frac{-\Delta H_{vap}}{RT} + A$$

For many purposes, it is convenient to have a two-point relation between the vapor pressures (P_2, P_1) at two different temperatures (T_2, T_1). Such a relation is obtained by first writing separate equations at each temperature:

$$\text{at } T_2: \qquad \ln P_2 = A - \frac{\Delta H_{vap}}{RT_2}$$

$$\text{at } T_1: \qquad \ln P_1 = A - \frac{\Delta H_{vap}}{RT_1}$$

Upon subtraction, the constant A is eliminated.

$$\ln P_2 - \ln P_1 = -\frac{\Delta H_{vap}}{R}\left[\frac{1}{T_2} - \frac{1}{T_1}\right] = \frac{\Delta H_{vap}}{R}\left[\frac{1}{T_1} - \frac{1}{T_2}\right]$$

This equation is known as the **Clausius-Clapeyron equation**. Rudolph Clausius was a prestigious nineteenth-century German scientist. B. P. E. Clapeyron, a French engineer, first proposed a modified version of the equation in 1834.

In using the Clausius-Clapeyron equation, the units of ΔH_{vap} and R must be consistent. If ΔH is expressed in joules, then R must be expressed in joules per kelvin. To find R in these units, recall (Chapter 8) that

$$1 \text{ L} \cdot \text{atm} = 101.3 \text{ J}$$

Hence

$$R = 0.0821 \frac{\text{L} \cdot \text{atm}}{\text{mol} \cdot \text{K}} \times \frac{101.3 \text{ J}}{1 \text{ L} \cdot \text{atm}} = 8.31 \text{ J/mol} \cdot \text{K}$$

We'll use this value of R frequently in future chapters

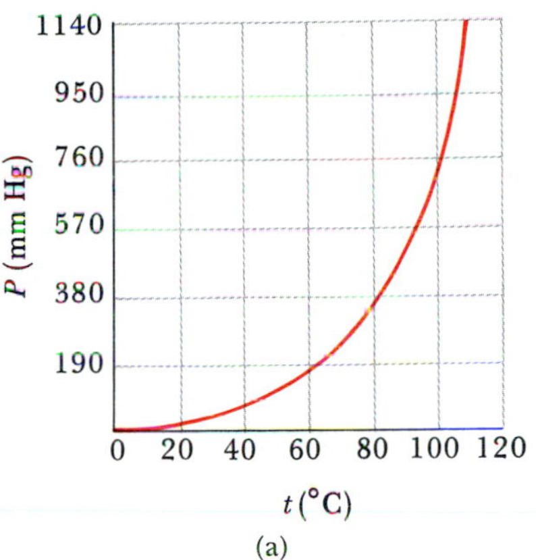

(a)

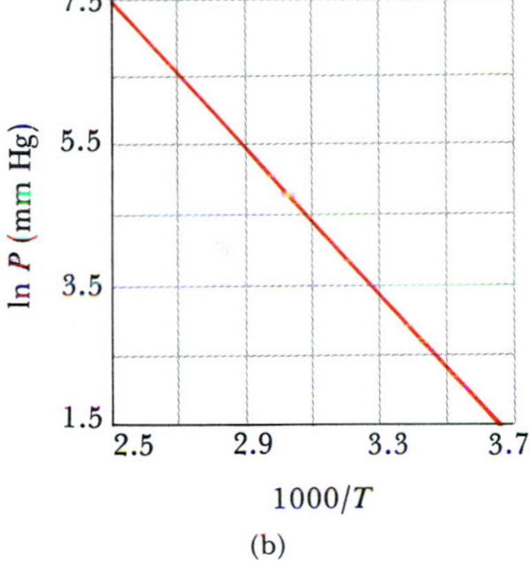

(b)

Figure 9.13
A plot of vapor pressure vs. temperature for water (or any other liquid) is a curve with a steadily increasing slope. On the other hand, a plot of the logarithm of vapor pressure vs. the reciprocal of the absolute temperature is a straight line.

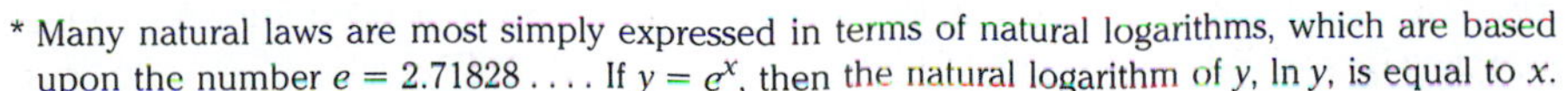

* Many natural laws are most simply expressed in terms of natural logarithms, which are based upon the number $e = 2.71828 \ldots$. If $y = e^x$, then the natural logarithm of y, $\ln y$, is equal to x.

Example 9.7 A certain organic compound has a vapor pressure of 91 mm Hg at 40°C. Calculate its vapor pressure at 25°C, taking the heat of vaporization to be 5.31×10^4 J/mol.

Strategy It is convenient to use the subscript 2 for the higher temperature and pressure. Substitute into the Clausius-Clapeyron equation, solving for P_1. Remember to express temperature in K and take $R = 8.31$ J/mol · K.

Solution

$\ln P_2/P_1 = \ln P_2 - \ln P_1$

$$\ln \frac{P_2}{P_1} = \ln \frac{91 \text{ mm Hg}}{P_1} = \frac{5.31 \times 10^4 \text{ J/mol}}{8.31 \text{ J/mol} \cdot \text{K}}\left[\frac{1}{298} - \frac{1}{313}\right] = 1.03$$

Taking inverse logs,

$$\frac{91 \text{ mm Hg}}{P_1} = e^{1.03} = 2.8$$

(e is the base of natural logarithms; on many calculators, to find the number whose natural logarithm is 1.03, you first enter 1.03, then press in succession the "INV" and "ln x" keys).

Solving for P_1,

$$P_1 = \frac{91 \text{ mm Hg}}{2.8} = 32 \text{ mm Hg}$$

This value is reasonable in the sense that lowering the temperature should reduce the vapor pressure.

Boiling Point

When a liquid is heated in an open container, bubbles form, usually at the bottom, where heat is applied. The first small bubbles are air, driven out of solution by the increase in temperature. Eventually, at a certain temperature, large vapor bubbles form throughout the liquid. These vapor bubbles rise to the surface, where they break. When this happens, the liquid is said to be boiling. For a pure liquid, the temperature remains constant throughout the boiling process.

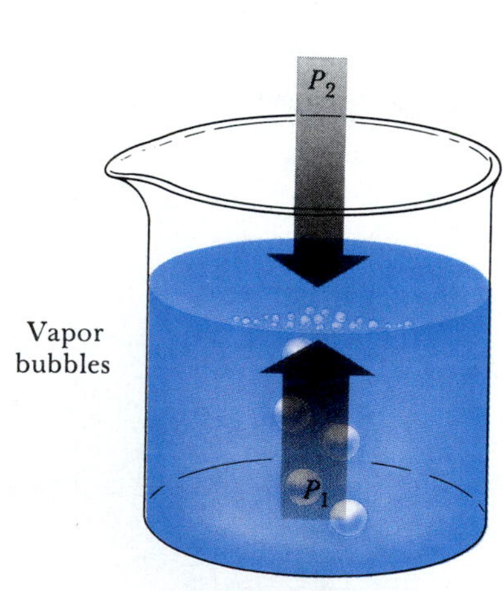

Figure 9.14
A liquid boils when its vapor pressure (P_1) slightly exceeds the pressure above it (P_2).

The temperature at which a liquid boils depends upon the pressure above it. To understand why this is the case, consider Figure 9.14. This shows vapor bubbles rising in a boiling liquid. For a vapor bubble to form, the pressure within it, P_1, must be at least equal to the pressure above it, P_2. Since P_1 is simply the vapor pressure of the liquid, it follows that ***a liquid boils at a temperature at which its vapor pressure is equal to the pressure above its surface.*** If this pressure is 1 atm (760 mm Hg), the temperature is referred to as the normal boiling point. (When the term "boiling point" is used without qualification, normal boiling point is implied.) The normal boiling point of water is 100°C; its vapor pressure is 760 mm Hg at that temperature.

As you might expect, the boiling point of a liquid can be reduced by lowering the pressure above it. Water can be made to boil at 25°C by evacuating the space above it. When a pressure of 24 mm Hg, the equilibrium vapor pressure at 25°C, is reached, the water starts to boil. Chemists often take advantage of this effect in purifying a high-boiling compound that might decompose or oxidize at its normal boiling point. They boil it at a reduced temperature under vacuum and condense the vapor.

If you have been fortunate enough to camp in the high Sierras or the Rockies, you may have noticed that it takes longer at high altitudes to cook foods in boiling water. The reduced pressure lowers the temperature at which water boils in an open container and thus slows down the physical and chemical changes that take place when foods like potatoes or eggs are cooked. In principle, this problem can be solved by using a pressure cooker. In that device, the pressure that develops is high enough to raise the boiling point of water above 100°C. Pressure cookers are indeed used in places like Cheyenne, Wyoming (elevation 1848 m), but not by mountain climbers, who have to carry all their equipment on their backs.

They have enough trouble as it is

Critical Temperature and Pressure

Consider an experiment in which liquid carbon dioxide is introduced into an otherwise evacuated glass tube, which is then sealed (Fig. 9.15). At 0°C, the pressure above the liquid is 34 atm, the equilibrium vapor pressure of $CO_2(l)$ at that temperature. As the tube is heated, some of the liquid is converted to vapor and the pressure rises, to 44 atm at 10°C and 56 atm at 20°C. Nothing spectacular happens (unless there happens to be a weak spot in the tube) until 31°C is reached, where the vapor pressure is 73 atm. Suddenly, as the temperature goes above 31°C, the meniscus between the liquid and vapor disappears! The tube now contains only one phase.

It is impossible to have liquid carbon dioxide at temperatures above 31°C, no matter how much pressure is applied. Even at pressures as high as 1000 atm, carbon dioxide gas does not liquefy at 35 or 40°C. This behavior is typical of all substances. There is a temperature, called the **critical temperature**, above which the liquid phase of a pure substance cannot exist. The pressure that must be applied to cause condensation at that temperature is called the **critical pressure**. Quite simply, the critical pressure is the vapor pressure of the liquid at the critical temperature.

Figure 9.15
When liquid carbon dioxide under pressure is heated, some of it vaporizes (a); vapor bubbles form in the liquid (b). Suddenly, at 31°C, the critical temperature of CO_2, the meniscus disappears (c). Above that temperature, only one phase is present, regardless of the applied pressure. (Marna G. Clarke)

(a)

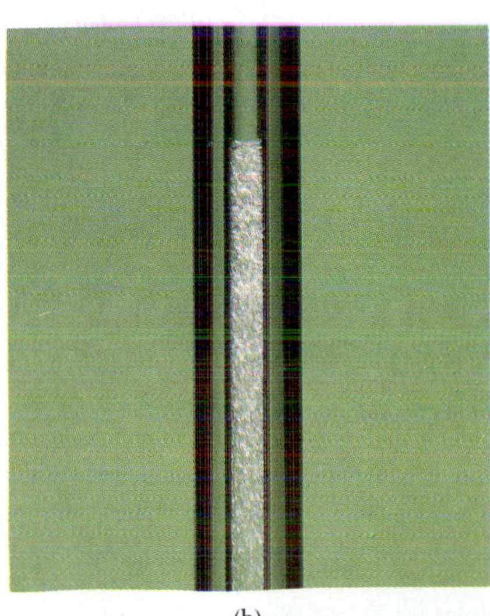
(b)

(c)

Table 9.5 Critical Temperatures (°C)

Permanent Gases		Condensable Gases		Liquids	
Helium	−268	Carbon dioxide	31	Ethyl ether	194
Hydrogen	−240	Ethane	32	Ethyl alcohol	243
Nitrogen	−147	Propane	97	Benzene	289
Argon	−122	Ammonia	132	Bromine	311
Oxygen	−119	Chlorine	144	Water	374
Methane	−82	Sulfur dioxide	158		

Table 9.5 lists the critical temperatures of several common substances. The species in the column at the left all have critical temperatures below 25°C. They are often referred to as "permanent gases." Applying pressure at room temperature will not condense a permanent gas. It must be cooled as well. When you see a truck labeled "liquid nitrogen" on the highway, you can be sure that the cargo trailer is refrigerated to at least −147°C, the critical temperature of N_2.

"Permanent gases" are most often stored and sold in steel cylinders under high pressures, often 150 atm or greater. When the valve on a cylinder of N_2 or O_2 is opened, gas escapes and the pressure drops accordingly. The substances listed in the center column of Table 9.5, all of which have critical temperatures above 25°C, are handled quite differently. They are available commercially as liquids in high-pressure cylinders. When the valve on a cylinder of propane is opened, the gas that escapes is replaced by vaporization of liquid. The pressure quickly returns to its original value. Only when the liquid is completely vaporized does the pressure drop as gas is withdrawn. This indicates that almost all of the propane is gone, and it's time to recharge the tank.

One tank contains methane, another propane. In which tank is the pressure more sensitive to temperature? Explain.

Above the critical temperature and pressure, a substance is referred to as a supercritical fluid. Such fluids have unusual solvent properties that have led to many practical applications. Water at the critical point has a density of about 0.3 g/mL and a heat capacity that approaches infinity. More important, oxygen and most organic molecules are completely soluble in supercritical water. This may offer a safe, inexpensive method of disposing of hazardous organic waste by oxidation. In a quite different area, supercritical CO_2 is now used to extract caffeine from coffee and nicotine from tobacco.

9.5 PHASE DIAGRAMS

In the preceding section, we discussed several features of the equilibrium between a liquid and its vapor. For a pure substance, at least two other types of phase equilibria need to be considered. One is the equilibrium between a solid and its vapor, the other between solid and liquid at the melting (freezing) point. Many of the important relations in all these equilibria can be shown in a **phase diagram**. A phase diagram is a graph that shows the pressures and temperatures at which different phases are in equilibrium with each other. The phase diagram of water is shown in Figure 9.16. This figure, which covers a wide range of temperatures and pressures, is not drawn to scale.

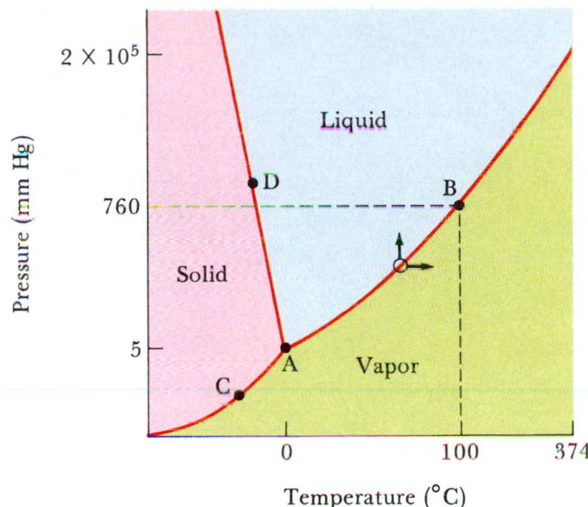

Figure 9.16
Phase diagram for water (not to scale). The triple point is at 0.01°C, 4.56 mm Hg; the critical point is at 374°C, 1.66×10^5 mm Hg (218 atm).

To understand what a phase diagram implies, consider first the three lines AB, AC, and AD in Figure 9.16. Each of these lines shows the pressures and temperatures at which two adjacent phases are at equilibrium.

1. Line AB is a portion of the vapor pressure–temperature curve of liquid water. At any temperature and pressure along this line, liquid water is in equilibrium with water vapor. At point A on the curve, these two phases are in equilibrium at 0°C and about 5 mm Hg (more exactly, 0.01°C and 4.56 mm Hg). At B, corresponding to 100°C, the pressure exerted by the vapor in equilibrium with liquid water is 1 atm. The extension of line AB beyond point B gives the equilibrium vapor pressure of the liquid above the normal boiling point. The line ends at 374°C, the critical temperature of water, where the pressure is 218 atm.

When T and P fall on a line, two phases are in equilibrium

2. Line AC represents the vapor pressure curve of ice. At any point along this line, such as point A (0°C, 5 mm Hg) or point C, which might represent −3°C and 3 mm Hg, ice and vapor are in equilibrium with each other.

3. Line AD gives the temperatures and pressures at which liquid water is in equilibrium with ice.

Point A on the phase diagram is the only one at which all three phases, liquid, solid, and vapor, are in equilibrium with each other. It is called the **triple point**. For water, the triple point temperature is 0.01°C. At this temperature, liquid water and ice have the same vapor pressure, 4.56 mm Hg.

In the three areas of the phase diagram labeled "solid," "liquid," and "vapor," only one phase is present. To understand this, consider what happens to an equilibrium mixture of two phases when the pressure or temperature is changed. Suppose we start at the point on AB indicated by an open circle. Here liquid water and vapor are in equilibrium with each other, let us say at 70°C and 234 mm Hg. If the pressure on this mixture is increased, condensation occurs. The phase diagram confirms this; increasing the pressure at 70°C *(vertical arrow)* puts us in the liquid region. In another experiment, the temperature might be increased at constant pressure. This should cause the liquid to vaporize. The phase diagram shows that this is indeed what happens. An increase in temperature *(horizontal arrow)* shifts us into the vapor region.

Figure 9.17
Solid iodine passes directly to vapor at any temperature below the triple point, 114°C. The vapor condenses back to solid on a cold surface, as in the upper tube, which is filled with dry ice. (Marna G. Clarke)

Example 9.8 Consider a sample of H_2O at point D in Figure 9.16.

(a) What phase(s) is (are) present?
(b) If the temperature of the sample were reduced at constant pressure, what would happen?
(c) How would you convert the sample to vapor without changing the temperature?

Strategy Use the phase diagram in Figure 9.16. Note that P increases moving up vertically; T increases moving to the right.

Solution

(a) Point D is on the solid-liquid equilibrium line. Ice and liquid water are present.
(b) Move to the left to reduce T. This penetrates the solid area, which implies that the solid freezes completely.
(c) Reduce the pressure to below the triple point value, perhaps to 4 mm Hg.

Sublimation

The process by which a solid changes directly to vapor without passing through the liquid phase is called *sublimation.* A solid can sublime only at temperatures below the triple point; above that temperature it will melt to liquid (Fig. 9.16). At temperatures below the triple point, a solid can be made to sublime by reducing the pressure of the vapor above it to less than the equilibrium value. To illustrate what this means, consider the conditions under which ice sublimes. This happens on a cold, dry, winter day when the temperature is below 0°C and the pressure of water vapor in the air is less than the equilibrium value (4.5 mm Hg at 0°C). The rate of sublimation can be increased by evacuating the space above the ice. This is how foods are freeze-dried. The food is frozen, put into a vacuum chamber, and evacuated to a pressure of 1 mm Hg or less. The ice crystals formed upon freezing sublime, which leaves a product whose mass is only a fraction of that of the original food.

By the reverse process, snow is formed from water vapor in the atmosphere

Iodine sublimes more readily than ice because its triple point pressure, 90 mm Hg, is much higher. Sublimation occurs upon heating (Fig. 9.17) below the triple point temperature, 114°C. If the triple point is exceeded, the solid melts. Solid carbon dioxide (dry ice) has a triple point pressure above 1 atm (5.2 atm at −57°C). Liquid carbon dioxide cannot be made by heating dry ice in an open container. Solid CO_2 always passes directly to vapor at 1 atm pressure.

The white "smoke" around a piece of dry ice is $CO_2(g)$, formed by sublimation

Melting Point

For a pure substance, the **melting point** is identical to the **freezing point**. It represents the temperature at which solid and liquid phases are in equilibrium. Melting points are usually measured in an open container, i.e., at atmospheric pressure. For most substances, the melting point at 1 atm (the "normal" melting point) is virtually identical with the triple point temperature. For water, the difference is only 0.01°C.

Although the effect of pressure upon melting point is very small, its direction is still important. To decide whether the melting point will be increased or

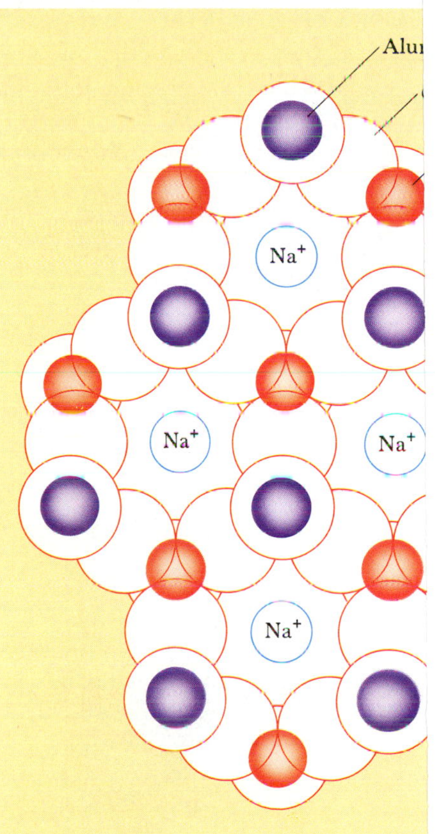

Another class of three-dimensio
zeolites (Fig. 9.C). These contain rel
H_2O molecules or cations such as N
zeolites occur naturally. They can a
case the dimensions of the "holes"

Zeolites are widely used in ho
containing Ca^{2+} (or Mg^{2+}) ions, f
zeolite, NaZ, the following reaction

$$Ca^{2+}(aq) + 2NaZ(s)$$

Sodium ions migrate out of the cavi
replace Ca^{2+} ions in the water by N
properties of hard water (precipitat
etc.) are removed.

Synthetic zeolites are also used
ecules such as H_2O. Anhydrous alc
ment with a zeolite that quantitative
are widely used in the petroleum in
carbon molecules from larger or m
into holes in the structure.

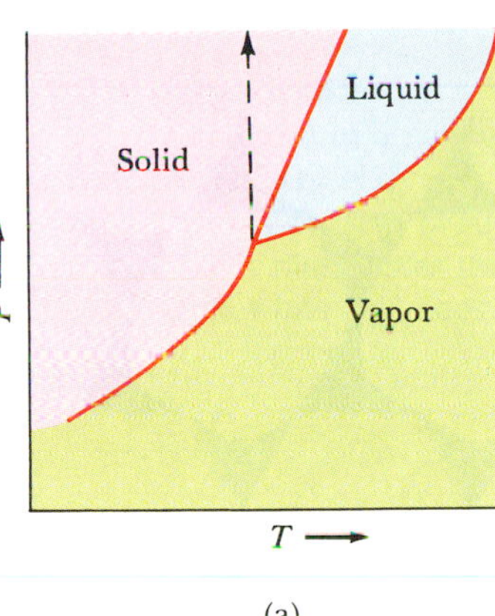

(a)

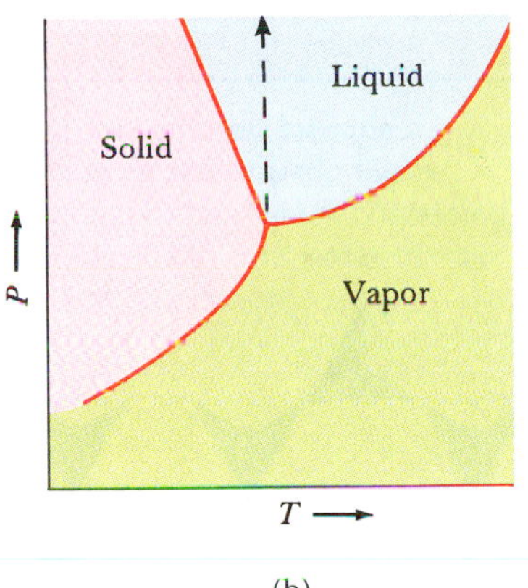

(b)

Figure 9.18
Effect of pressure on the melting point of a solid. (a) When the solid is the more dense phase, an increase in pressure converts liquid to solid; the melting point increases. (b) If the liquid is the more dense phase, an increase in pressure converts solid to liquid and the melting point decreases.

decreased by compression, a simple principle is applied. *An increase in pressure favors the formation of the more dense phase.*

Two types of behavior are shown in Figure 9.18.

1. *The solid is the more dense phase* (Fig. 9.18a). The solid-liquid equilibrium line is inclined to the right, shifting away from the *y*-axis as it rises. At higher pressures, the solid becomes stable at temperatures above the normal melting point. In other words, the melting point is raised by an increase in pressure. This behavior is shown by most substances.

2. *The liquid is the more dense phase* (Fig. 9.18b). The liquid-solid line is inclined to the left, toward the *y*-axis. An increase in pressure favors the formation of liquid; that is, the melting point is decreased by raising the pressure. Water is one of the few substances that behaves this way; ice is less dense than liquid water. The effect is exaggerated for emphasis in Figure 9.18b. Actually, an increase in pressure of 134 atm is required to lower the melting point of ice by 1°C.

PERSPECTIVE • Silicate Minerals

More than 90% of the rocks and minerals found in the earth's crust are classed as silicates; they contain silicon atoms bonded tetrahedrally to four oxygen atoms. Silicates are essentially ionic; the negative ion is an oxoanion in which silicon is the central atom. The anion may be a very simple species (e.g., SiO_4^{4-}) or a more complex species of the network covalent type. The negative charge of the silicon oxoanion is balanced by positive charges on the cations of one or more metals (e.g., Na^+, Mg^{2+}, Al^{3+}) that fit into the crystal lattice.

The simplest silicates are those containing the tetrahedral SiO_4^{4-} anion, found in the semiprecious stone zircon, $ZrSiO_4$, and the garnets, of which $Ca_3Cr_2(SiO_4)_3$ is typical. Most silicate minerals have network covalent structures in which SiO_4^{4-} tetrahedra are linked together in one, two, or three dimensions. In the structure at the left of Figure 9.A, tetrahedra are linked together to form what amounts to a one-dimensional infinite chain. This occurs in fibrous minerals such as diopside, $CaSiO_3 \cdot MgSiO_3$. The best

Figure 9.A Two infinite silicate lattices. Asbestos contains a double anion chain similar to the one shown at the left. The structure of talc is that of the infinite sheet, of which a portion is shown at the right.

Figure 9.B Amethyst crystals. (Charles D. Winters)

CHAPTER HIGHLIGHTS

KEY CONCEPTS

1. *Identify intermolecular forces in different substances and evaluate their effects on boiling point*
(Examples 9.1–9.3; Problems 1–14, 64)
2. *Classify substances as molecular, network covalent, ionic, or metallic on the basis of their properties*
(Example 9.4; Problems 15–24, 65)
3. *Relate atomic radius (r) to side (s) for different unit cells*
(Example 9.5; Problems 25–34)
4. *Use the ideal gas law to determine whether a liquid will completely vaporize*
(Example 9.6; Problems 35–38, 66, 68, 69)
5. *Use the Clausius-Clapeyron equation to relate the vapor pressure of a liquid to temperature*
(Example 9.7; Problems 39–46)
6. *Use a phase diagram to determine the phase(s) present at a given T and P*
(Example 9.8; Problems 49–54)

KEY EQUATIONS

Vapor pressure–temperature $$\ln P_2/P_1 = \frac{\Delta H}{R}\left[\frac{1}{T_1} - \frac{1}{T_2}\right]$$

Unit cell $2r = s$ (SC) $4r = s\sqrt{2}$ (FCC)
$4r = s\sqrt{3}$ (BCC)

KEY TERMS

boiling point	dipole force	phase diagram
critical temperature	dispersion force	sublimation
cubic cell	electron-sea model	triple point
—body-centered (BCC)	hydrogen bond	unit cell
—face-centered (FCC)	melting point	vapor pressure
—simple	network covalent	

SUMMARY PROBLEM

Consider methyl alcohol, an organic compound of structural formula H_3C—OH.

a. What kind of intermolecular forces would you expect to find in methyl alcohol?
b. How would you expect the normal boiling point of CH_3OH to compare to that of O_2? To C_2H_5OH? To NaCl? To SiO_2?
c. Would you expect CH_3OH to conduct electricity? How would its water solubility compare to that of O_2? To SiO_2?
d. Give the symbol of a solid nonmetal that is a better conductor than CH_3OH. Give the formula of a solid compound that, unlike CH_3OH, has a network structure.
e. The normal boiling point of CH_3OH is 65°C, the normal freezing point is −98°C, and the critical temperature is 239°C. Draw a rough sketch of the phase diagram of methyl alcohol, assuming the solid is more dense than the liquid.
f. If you wanted to purify methyl alcohol by sublimation, at what temperature would you operate?
g. A sample of methyl alcohol vapor at 250°C and 1 atm is cooled at constant pressure. At what temperature does liquid first appear?
h. Taking the normal boiling point of CH_3OH to be 65°C and its heat of vaporization to be 38.0 kJ/mol, estimate its normal vapor pressure at 25°C. Estimate the temperature at which its vapor pressure is 0.500 atm.

i. A sample of 1.00 mL of CH_3OH (d = 0.792 g/mL) is placed in a one-liter flask at 35°C, where its vapor pressure is 203 mm Hg. Show by calculation whether there is any liquid left in the flask when equilibrium is established.

Answers

a. dispersion, H bonds **b.** higher than O_2, lower than the others
c. no; higher than O_2 or SiO_2 **d.** C(graphite) or Si; SiO_2
e. See Figure 9.18a **f.** below about −98°C **g.** 65°C
h. 123 mm Hg, 49°C **i.** yes

QUESTIONS & PROBLEMS

Intermolecular Forces

1. Arrange the following in order of increasing boiling point.

a. Ar **b.** He **c.** Ne **d.** Xe

2. Arrange the following in order of decreasing boiling point.

a. I_2 **b.** F_2 **c.** Cl_2 **d.** Br_2

3. Which of the following would you expect to show dispersion forces? Dipole forces?

a. CH_4 **b.** CCl_4 **c.** CO **d.** CO_2

4. Which of the following would you expect to show dispersion forces? Dipole forces?

a. O_2 **b.** H_2O **c.** $CHCl_3$ **d.** CH_3Cl

5. Which of the following would show hydrogen bonding?

a. $CH_3—CH_3$ **b.** $CH_3—OH$
c. CH_3NH_2 **d.** HO—OH

6. Which of the following would show hydrogen bonding?

a. CH_3F **b.** NH_3
c. $H_3C—O—CH_3$ **d.** $[H—F—H]^+$

7. Explain in terms of forces between structural units why

a. ICl has a higher melting point than Br_2.
b. C_2H_6 has a higher boiling point than CH_4.
c. H_2O_2 has a higher boiling point than C_2H_6.
d. C_2H_5OH has a lower boiling point than NaF.

8. Explain in terms of forces between structural units why

a. Br_2 is lower-melting than NaBr.
b. C_2H_5OH is higher-boiling than butane, C_4H_{10}.
c. Water is higher-boiling than hydrogen telluride, H_2Te.
d. Acetic acid, $H_3C—C(=O)—OH$, is lower-boiling than benzoic acid, $C_6H_5C(=O)—OH$.

9. In which of the following processes are covalent bonds broken?

a. melting mothballs made up of napthalene, $C_{10}H_8$
b. boiling water
c. boiling ethyl alcohol
d. dissolving iodine in water

10. In which of the following processes is it necessary to break covalent bonds as opposed to simply overcoming intermolecular forces?

a. electrolysis of water
b. subliming dry ice
c. melting SiO_2
d. decomposing dinitrogen tetraoxide to nitrogen dioxide

11. For each of the following pairs, choose the member with the lower boiling point. Explain your reason in each case.

a. SO_2 or CO_2 **b.** HCl or HBr
c. H_2 or O_2 **d.** F_2 or Cl_2

12. Follow the directions for Question 11 for the following substances.

a. AsH_3 or PH_3 **b.** C_6H_6 or $C_{10}H_8$
c. NH_3 or PH_3 **d.** LiCl or C_3H_8

13. What are the strongest attractive forces that must be overcome to

a. boil silicon hydride (SiH_4)?
b. melt iodine (I_2)?
c. dissolve chlorine in carbon tetrachloride?
d. vaporize calcium chloride?

14. What are the strongest attractive forces that must be overcome to

a. melt ice?
b. boil carbon tetrachloride (CCl_4)?
c. sublime bromine (Br_2)?
d. melt benzene (C_6H_6)?

Types of Substances

15. Classify as metallic, network covalent, ionic, or molecular, each of the following solids.

a. melts below 100°C and is insoluble in water
b. conducts electricity only when melted
c. is insoluble in water and conducts electricity

16. Classify as metallic, network covalent, ionic, or molecular solids a–c.

a. is insoluble in water, melts above 500°C, and does not conduct electricity
b. dissolves in water but is nonconducting as an aqueous solution, as a solid, or when molten
c. dissolves in water, melts above 100°C, and conducts electricity when present in an aqueous solution

17. Of the four general types of solids, which one(s)
 a. are generally low-boiling?
 b. are ductile and malleable?
 c. are generally soluble in nonpolar solvents?

18. Of the four general types of solids, which one(s)
 a. are generally insoluble in water?
 b. are very high-melting?
 c. conduct electricity as solids?

19. Classify each of the following species as molecular, network covalent, ionic, or metallic.
 a. K b. $CaCO_3$ c. C_8H_{18}
 d. C(diamond) e. HCl(*g*)

20. Classify each of the following species as molecular, network covalent, ionic, or metallic.
 a. mercury b. iodine bromide
 c. sand d. chromium(III) sulfate
 e. chlorine gas

21. Give the formula of a solid containing oxygen that is
 a. polar molecular. b. ionic.
 c. network covalent. d. nonpolar molecular.

22. Give the formula of a solid containing carbon that is
 a. molecular. b. ionic.
 c. network covalent. d. metallic.

23. Describe the nature of the structural units in
 a. NaI b. N_2 c. CO_2 d. W

24. Describe the nature of the structural units in
 a. C(graphite) b. SiC c. $FeCl_2$ d. C_2H_2

Crystal Structure

25. Aluminum (at. radius = 0.143 nm) crystallizes with a face-centered cubic unit cell. What is the length of a side of the cell?

26. Xenon crystallizes with a face-centered cubic unit cell. The edge of the unit cell is 0.620 nm. Calculate the atomic radius of xenon.

27. Chromium crystallizes with a body-centered cubic unit cell. The volume of the unit cell is 2.376×10^{-2} nm^3. What is the atomic radius of chromium?

28. Sodium crystallizes with a body-centered cubic unit cell. Its atomic radius is 0.186 nm. What is the volume of the cell?

29. Potassium iodide has a unit cell similar to that of NaCl (Fig. 9.12). The ionic radii of K^+ and I^- are 0.133 nm and 0.216 nm, respectively. How long is
 a. one side of the cube?
 b. the face diagonal of the cube?

30. In the LiCl structure shown in Figure 9.12, the chloride ions form a face-centered cubic unit cell 0.513 nm on an edge. The ionic radius of Cl^- is 0.181 nm.
 a. Along a cell edge, how much space is there between the Cl^- ions?
 b. Would an Na^+ ion (r = 0.095 nm) fit into this space? a K^+ ion (r = 0.133 nm)?

31. Consider the CsCl cell (Fig. 9.12). The ionic radii of Cs^+ and Cl^- are 0.169 and 0.181 nm, respectively. What is the length of
 a. the body diagonal?
 b. the side of the cell?

32. For a cell of the CsCl type (Fig. 9.12), how is the length of one side of the cell, *s*, related to the sum of the radii of the ions, $r_{cation} + r_{anion}$?

33. Consider the CsCl unit cell shown in Figure 9.12. How many Cs^+ ions are there per unit cell? How many Cl^- ions? (Note that each Cl^- ion is shared by 8 cubes).

34. Consider the NaCl unit cell shown in Figure 9.12. Looking only at the front face (5 large Cl^- ions, 4 small Na^+ ions),
 a. how many cubes share each of the Na^+ ions in this face?
 b. how many cubes share each of the Cl^- ions in this face?

Liquid-Vapor Equilibrium

35. A sample of benzene vapor in a flask of constant volume exerts a pressure of 325 mm Hg at 80°C. The flask is slowly cooled.
 a. Assuming no condensation, use the ideal gas law to calculate the pressure of the vapor at 50°C; at 60°C.
 b. Compare your answers in (a) to the equilibrium vapor pressures of benzene: 269 mm Hg at 50°C, 389 mm Hg at 60°C.
 c. On the basis of your answers to (a) and (b), predict the pressure exerted by the benzene at 50°C; at 60°C.

36. The vapor pressure of $I_2(s)$ at 30°C is 0.466 mm Hg.
 a. How many milligrams of iodine will sublime into an evacuated 1.00-L flask?
 b. If 2.0 mg of I_2 is used, what will the final pressure be?
 c. If 10 mg of I_2 is used, what will the final pressure be?

37. Ethyl ether, $C_4H_{10}O$, was a widely used anesthetic in the early days of surgery. It has a vapor pressure of 537 mm Hg at room temperature, 25°C. A ten-mL sample (d = 0.708 g/mL) is placed in a sealed 0.250-L flask.
 a. Will there be any liquid left in the flask at equilibrium at 25°C?
 b. What is the maximum volume the flask can have if equilibrium is to be established between liquid and vapor?
 c. If the flask has a volume of 5.00 L, what is the pressure of ethyl ether in the flask?

38. A humidifier is used in a bedroom kept at 22°C. The bedroom's volume is 4.0×10^4 L. Assume that the air is originally dry and no moisture leaves the room while the humidifier is operating.
 a. If the humidifier has a capacity of 3.0 gallons of water, will there be enough water to saturate the room with water vapor (vp of water at 22°C = 19.83 mm Hg; d water = 1.00 g/cm^3)?
 b. What is the final pressure of water vapor in the room when the humidifier has vaporized two thirds of its water supply?

39. Carbon tetrachloride was once widely used in the dry cleaning industry. Its use was discouraged, however, after it was determined to be carcinogenic. It has a vapor pressure of 9.708 mm Hg at $-20.0°C$ and 90.00 mm Hg at 20.0°C. Use the Clausius-Clapeyron equation to estimate the

a. heat of vaporization.

b. temperature at which the vapor pressure is 0.50 atm.

40. Octane, C_8H_{18}, is a principal component of gasoline. It has a vapor pressure of 145 mm Hg at 75.0°C and 20.0 mm Hg at 32.0°C. Use the Clausius-Clapeyron equation to estimate the

a. heat of vaporization of octane.

b. vapor pressure of octane on a warm day (85°F).

41. Mexico City has an elevation of 7400 ft above sea level. Its normal atmospheric pressure is 589 mm Hg. At what temperature does water boil in Mexico City? (ΔH_{vap} of water = 40.7 kJ/mol.)

42. Mount Kilimanjaro has an altitude of 19,350 ft. Water boils at about 81°C atop Mount Kilimanjaro. What is the normal atmospheric pressure on the mountain?

43. Mercury is used extensively in studying the effect of pressure on the volume of gases. It is fairly inert at the surface, and most gases are insoluble in it. Should we worry about the contribution of the vapor pressure of mercury to the pressure of gases above it? Use the following data to determine the vapor pressure of mercury at 0 and 100°C: ΔH_{vap} of Hg = 59.4 kJ/mol; normal boiling point = 357°C.

44. When water boils inside a pressure cooker, its vapor pressure is 1.500×10^3 mm Hg. Use the Clausius-Clapeyron equation to calculate the temperature of the water. Take the heat of vaporization to be 40.7 kJ/mol.

45. The data below are for the vapor pressure of toluene, a common solvent.

vp (mm Hg)	40	60	80	100
t (°C)	32	40	47	52

Plot ln vp vs. $1/T$ $\left(\ln P = A - \frac{\Delta H_{vap}}{RT}\right)$ and use your graph to estimate the heat of vaporization of toluene.

46. Consider the following data for the vapor pressure of acetone.

vp (mm Hg)	39	116	283	613
t (°C)	-10	10	30	50

Follow the instructions of Problem 45 to estimate the heat of vaporization of acetone.

47. The critical point of CO is $-139°C$, 35 atm. Liquid CO has a vapor pressure of 6 atm at $-171°C$. Which of the following statements must be true?

a. CO is a gas at $-171°C$ and 1 atm.

b. A tank of CO at 20°C can have a pressure of 35 atm.

c. CO gas cooled to $-145°C$ and 40 atm pressure will condense.

d. The normal boiling point of CO lies above $-171°C$.

48. The normal boiling point of SO_2 is $-10°C$. At 32°C its vapor pressure is 5 atm. Which of the following statements concerning sulfur dioxide must be true?

a. A tank of SO_2 at 32°C that has a pressure of 4 atm must contain liquid SO_2.

b. A tank of SO_2 at 32°C that has a pressure of 1 atm cannot contain liquid SO_2.

c. The critical temperature of SO_2 must be greater than 30°C.

Phase Diagrams

49. Referring to Figure 9.16, state what phase(s) is (are) present at

a. $-3°C$, 5 mm Hg. **b.** 25°C, 1 atm.

c. 70°C, 20 mm Hg.

50. Referring to Figure 9.16, state what phase(s) is (are) present at

a. 1 atm, 100°C. **b.** 0.5 atm, 100°C.

c. 0.5 atm, 0°C.

51. Iodine has a triple point at 114°C, 90 mm Hg. Its critical temperature is 535°C. The density of the solid is 4.93 g/cm^3, while that of the liquid is 4.00 g/cm^3. Sketch a phase diagram for iodine, and use it to fill in the blanks below, using either "liquid" or "solid."

a. Iodine vapor at 80 mm Hg condenses to the ______ when cooled sufficiently.

b. Iodine vapor at 125°C condenses to the ________ when enough pressure is applied.

c. Iodine vapor at 700 mm Hg condenses to the ______ when cooled above the triple point temperature.

52. Argon gas has its triple point at $-189.3°C$ and 516 mm Hg. It has a critical point at $-122°C$ and 48 atm. The density of the solid is 1.65 g/cm^3, while that of the liquid is 1.40 g/cm^3. Sketch the phase diagram for argon and use it to fill in the blanks below with the words "boils," "melts," "sublimes," or "condenses."

a. Solid argon at 500 mm Hg ______________ when the temperature is increased.

b. Solid argon at 2 atm ______________ when the temperature is increased.

c. Argon gas at $-150°C$ ______________ when the pressure is increased.

d. Argon gas at $-165°C$ ______________ when the pressure is increased.

53. A pure substance A has a liquid vapor pressure of 320 mm Hg at 125°C, 800 mm Hg at 150°C, and 60 mm Hg at the triple point, 85°C. The melting point of A decreases slightly as pressure increases.

a. Sketch a phase diagram for A.

b. From the phase diagram, estimate the normal boiling point.

c. What changes occur when, at a constant pressure of 320 mm Hg, the temperature drops from 150 to 100°C?

54. A pure substance X has the following properties: mp = 90°C, increasing slightly as pressure increases; normal bp = 120°C; liquid vp = 65 mm Hg at 100°C, 20 mm Hg at the triple point.

a. Draw a phase diagram for X.

b. Label solid, liquid, and vapor regions of the diagram.

c. What changes occur if, at a constant pressure of 100 mm Hg, the temperature is raised from 100°C to 150°C?

Silicates

55. Why is the best-known fibrous silicate being removed from homes and public buildings?

56. Show that in a fibrous silicate (Fig. 9.A, left), the silicon-to-oxygen ratio is 1:3, while in layer minerals (Fig. 9.A, right), the ratio is $1:2\frac{1}{2}$. *Hint:* Consider how many of the oxygen atoms in each tetrahedron are shared with adjacent tetrahedra.

57. Explain the similarity between graphite and talc.

58. Describe how zeolites in home water softeners work. Why are these water softeners potentially hazardous to people suffering from high blood pressure?

Unclassified

59. Criticize each of the following statements.

a. Vapor pressure is inversely proportional to volume.

b. Vapor pressure is directly proportional to temperature.

c. Boiling point increases with molar mass.

d. The melting point of a molecular substance depends upon the strength of the covalent bonds within the molecule.

60. Which of the following statements are true?

a. The critical pressure is the highest vapor pessure that a liquid can have.

b. To sublime a solid, it must be heated above the triple point.

c. NaF melts higher than F_2 because its molar mass is larger.

d. One metal crystallizes in a body-centered cubic cell, another in a face-centered cubic cell of the same size. The two atomic radii must be equal.

61. Explain the difference between

a. A face-centered cubic and a body-centered cubic unit cell.

b. melting and sublimation.

c. a boiling point and a normal boiling point.

d. a phase diagram and a vapor pressure curve.

62. Differentiate between

a. intramolecular forces and intermolecular forces.

b. a molecular solid and a network covalent solid.

c. a dipole force and a hydrogen bond.

d. an ionic solid and a molecular solid.

63. The density of liquid mercury at 20°C is 13.6 g/cm^3; its vapor pressure is 1.2×10^{-3} mm Hg.

a. What volume (cm^3) is occupied by one mole of Hg(l) at 20°C?

b. What volume (cm^3) is occupied by one mole of Hg(g) at 20°C and the equilibrium vapor pressure?

c. The atomic radius of Hg is 0.155 nm. Calculate the volume (cm^3) of one mole Hg atoms ($V = 4\pi r^3/3$).

d. From your answers to (a), (b), and (c), calculate the percentage of the total volume occupied by the atoms in Hg(l) and Hg(g) at 20°C and 1.2×10^{-3} mm Hg.

64. Explain the difference in boiling points for each of the following pairs of substances:

a. ethyl alcohol, C_2H_5OH (79°C), and ethyl ether, $C_2H_5—O—C_2H_5$ (34.6°C)

b. HF (20°C) and HCl (−85°C)

c. LiF (1717°C) and HF (20°C)

65. Elemental boron is almost as hard as diamond. It is insoluble in water, does not conduct electricity at room temperature, and melts at 2300°C. What type of solid is boron?

66. An experiment is performed to determine the vapor pressure of formic acid. A 30.0-L volume of helium gas at 20.0°C is passed through 10.00 g of liquid formic acid (HCOOH) at 20.0°C. After the experiment, 7.50 g of liquid formic acid remains. Assume that the helium gas becomes saturated with formic acid vapor and that the total gas volume and temperature remain constant. What is the vapor pressure of formic acid at 20.0°C?

67. Which of the following substances can be liquefied by applying pressure at 25°C?

Substance	Critical Points
Ethane	32°C, 48 atm
Carbon disulfide	273°C, 76 atm
Krypton	−63°C, 54 atm
Nitrogen oxide	−94°C, 65 atm

Challenge Problems

68. A flask with a volume of 10.0 L contains 0.400 g of hydrogen gas and 3.20 g of oxygen. The mixture is ignited and the reaction

$$2H_2(g) + O_2(g) \longrightarrow 2H_2O$$

goes to completion. The mixture is cooled to 27°C. Assuming 100% yield,

a. what physical state(s) of water are present in the flask?

b. what is the final pressure in the flask?

69. Chloroform ($CHCl_3$) is the liquid that usually saturates the handkerchief that is pressed over the nose of an unsuspecting victim in TV spy dramas. Chloroform has a vapor pressure of 199 mm Hg at 25.0°C. If half a pint of chloroform (d = 1.489 g/mL) is spilled on the floor of a room that is 12 ft × 13 ft × 8.0 ft, will all the chloroform vaporize?

70. It has been suggested that the pressure exerted on a skate blade is sufficient to melt the ice beneath it and form a thin film of water, which makes it easier for the blade to slide over the ice. Assume that a skater weighs 120 lb and the blade has an area of 0.10 in^2. Calculate the pressure exerted on the blade (1 atm = 15 lb/in^2). From information in the text, calculate the decrease in melting point at this pressure. Comment on the plausibility of this explanation, and suggest another mechanism by which the water film might be formed.

71. As shown in Figure 9.12, Li^+ ions fit into a closely packed array of Cl^- ions, but Na^+ ions do not. What is the value of the r_{cation}/r_{anion} ratio at which a cation just fits into a structure of this type?

72. When the temperature drops from 20°C to 10°C, the pressure of a cylinder of compressed N_2 drops by 3.4%. The same temperature change decreases the pressure of a propane (C_3H_8) cylinder by 42%. Explain the difference in behavior.

10
Solutions

See plastic nature working to this end,
The single atoms each to other tend.
Attract, attracted to, the next in place
Form'd and impell'd its neighbor to embrace.

ALEXANDER POPE

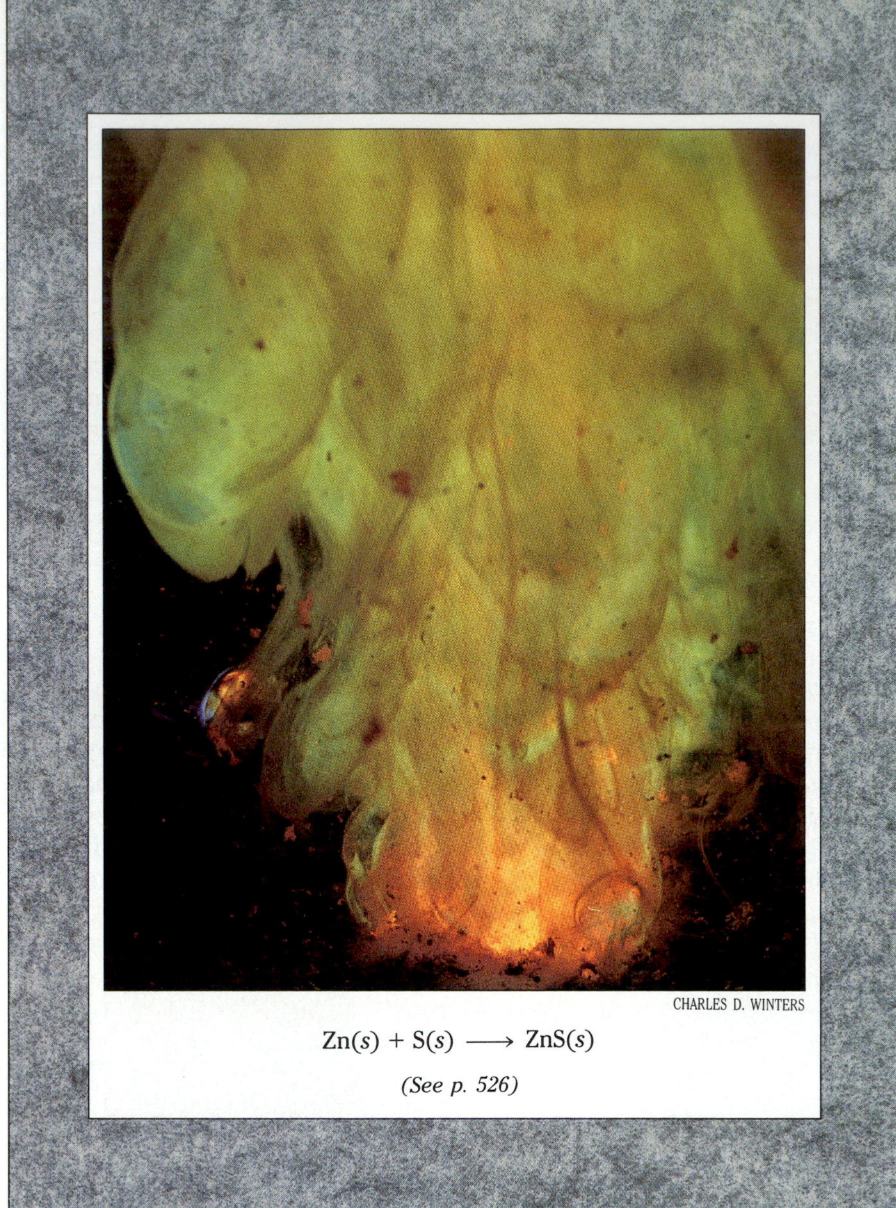

CHARLES D. WINTERS

$Zn(s) + S(s) \longrightarrow ZnS(s)$

(See p. 526)

CHAPTER OUTLINE

In the course of a day, you use or make solutions many times. Your morning cup of coffee is a solution of solids (sugar and coffee) in a liquid (water). The gasoline you fill your gas tank with is a solution of several different liquid hydrocarbons. The soda you drink at a study break is a solution of a gas (carbon dioxide) in a liquid (water).

This chapter covers several of the physical aspects of solutions, including

— types of solutions (Section 10.1)
— methods of expressing solution concentrations by specifying the relative amounts of solute and solvent (Section 10.2)

— factors affecting solubility, including the nature of the solute and the solvent, the temperature, and the pressure (Section 10.3)
— the effect of solutes upon such solvent properties as vapor pressure, freezing point, and boiling point (Section 10.4)

10.1 TYPES OF SOLUTIONS

Solutions can be classified in various ways. For many purposes, it is most convenient to emphasize either the relative amount of solute or the nature of solute particles.

Saturated, Unsaturated, and Supersaturated Solutions

Figure 10.1
This column of solid sodium acetate was formed by slowly pouring a supersaturated solution over a seed crystal. (Charles D. Winters)

A **saturated solution** is one that is, or could be, in equilibrium with undissolved solute. An **unsaturated solution** contains a lower concentration of solute than the saturated solution. The unsaturated solution is not at equilibrium; if solute is added, it dissolves until saturation is reached. A **supersaturated solution** contains more than the equilibrium concentration of solute. It is unstable in the presence of undissolved solute (Fig. 10.1).

To illustrate the meaning of these terms, consider sucrose (table sugar), $C_{12}H_{22}O_{11}$. At 20°C, the solubility of sucrose is 204 g per 100 g of water. A saturated solution of sugar contains precisely this ratio of solute to solvent, i.e., 204 g sucrose/100 g water. Any solution in which the concentration of sucrose at 20°C is less than this value is unsaturated; if more sucrose is added it will dissolve until saturation is reached.

Only a saturated solution is in equilibrium with solute

A supersaturated solution contains more than 204 g of $C_{12}H_{22}O_{11}$ in 100 g of water at 20°C. To prepare such a solution we take advantage of the fact that the solubility of sucrose increases with temperature. At 80°C, 362 g of sucrose dissolves in 100 g of water. If this solution is cooled carefully to 20°C, without shaking or stirring, the excess solute stays in solution, producing a supersaturated solution. If a small seed crystal of sucrose is then added, crystallization quickly takes place. The excess solute (362 g − 204 g = 158 g) comes out of solution, establishing equilibrium between the saturated solution and the sucrose crystals.

The crystallization of excess solute is a common problem in the preparation of candies and in the storage of jam and honey. From these supersaturated solutions, sugar separates either as tiny crystals, causing the "graininess" in fudge, or as large crystals, which often appear in honey kept for a long time (Fig. 10.2).

Figure 10.2
Rock candy is formed by the crystallization of sugar from a saturated solution that is slowly cooled. (Marna G. Clarke)

Nonelectrolytes vs. Electrolytes

Pure water does not conduct an electric current (Fig. 10.3a). Solutes in water can be classified into one of two categories according to the conductivity of the solutions they form.

1. Nonelectrolytes form water solutions that do not conduct an electric current (Fig. 10.3b). Typically, these substances are molecular and dissolve as molecules. Since molecules are neutral, they do not migrate in an electric field. Hence, they do not conduct an electric current. The processes by which

(a)

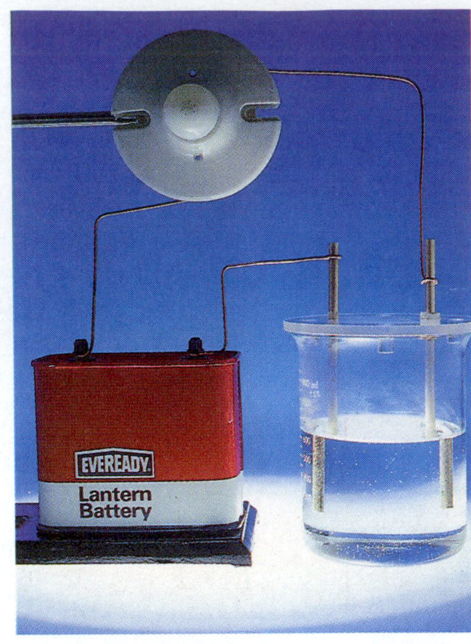

(b)

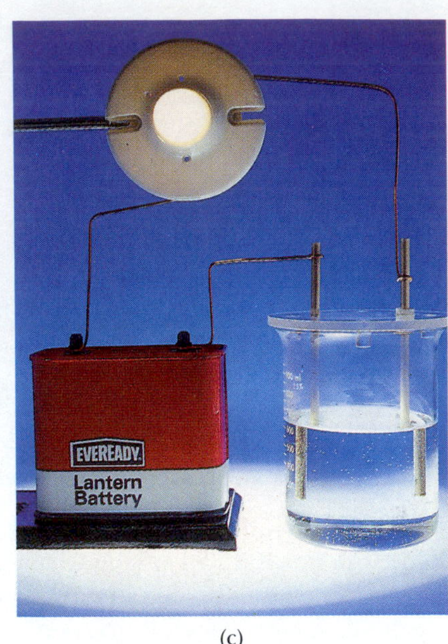

(c)

Figure 10.3
An apparatus for testing electrical conductivity. For an electric current to flow, the solution must contain ions, which carry electric charge. The liquids used are (a) pure water, (b) a solution of pure water and sucrose, and (c) a solution of pure water and NaCl. (Marna G. Clarke)

methyl alcohol, CH_3OH, and sugar, $C_{12}H_{22}O_{11}$, dissolve in water can be represented by the simple equations

$$CH_3OH(l) \longrightarrow CH_3OH(aq)$$

$$C_{12}H_{22}O_{11}(s) \longrightarrow C_{12}H_{22}O_{11}(aq)$$

2. Electrolytes are solutes whose water solutions conduct an electric current (Fig. 10.3c). *These substances produce ions in solution.* The charged ions migrate in an electric field, thereby carrying a current. Sodium chloride (table salt) is an electrolyte; the solid consists of Na^+ and Cl^- ions. When sodium chloride dissolves in water, these ions are set free (Fig. 10.4). The solution process can be represented as

H_2O molecules do not appear in the equation, but they're involved in the process

$$NaCl(s) \longrightarrow Na^+(aq) + Cl^-(aq)$$

Electrolyte systems are important biologically. Blood plasma is an electrolyte solution containing many different ions, among them Na^+, K^+, Mg^{2+}, Ca^{2+}, Cl^-, HCO_3^-, SO_4^{2-}, and HPO_4^{2-}. The concentrations of these ions in blood must stay nearly constant. Slight variations, particularly in K^+, can be critical. Many anorexics and bulimics change the K^+ ion concentration in their blood quite rapidly when they fluctuate between starving and overeating. Fatalities from these eating disorders have been attributed to the rapid change in concentration of K^+.

As pointed out in Chapter 4, *strong acids and bases* are ionized in water:

In dilute hydrochloric acid, there are no HCl molecules

$$HCl(aq) \longrightarrow H^+(aq) + Cl^-(aq)$$

$$NaOH(s) \longrightarrow Na^+(aq) + OH^-(aq)$$

and hence behave as strong electrolytes in water. Like other electrolytes, HCl and NaOH produce more than one mole of particles (ions) per mole of solute (Table 10.1). In contrast, a mole of nonelectrolyte such as I_2 gives one mole of particles (molecules) in solution.

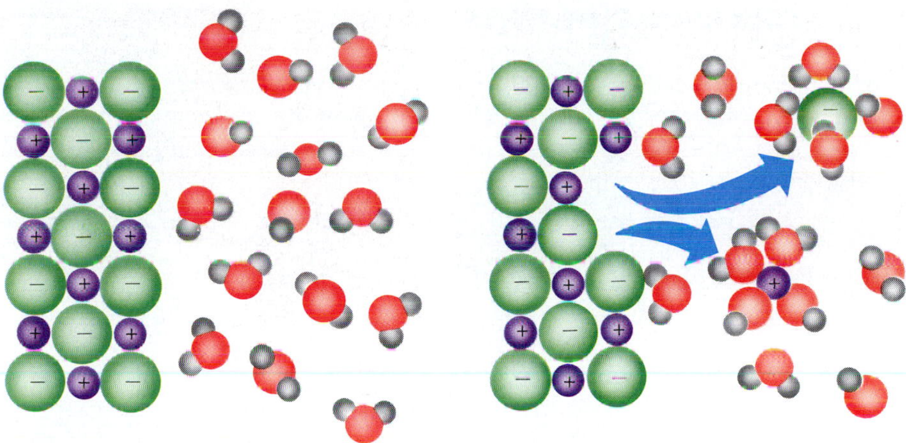

Figure 10.4
Dissolving NaCl in water. The attraction between the oxygen atoms in H_2O molecules and Na^+ ions in NaCl tends to bring the Na^+ ions into solution. At the same time, Cl^- ions, attracted to the hydrogen atoms of H_2O molecules, also enter the solution. The H_2O molecules that surround the ions tend to prevent oppositely charged ions from recombining.

Recall from Chapter 4 that *weak acids and bases* exist in water solution as a mixture of molecules and ions. The behavior of HF is typical of weak acids:

$$HF(aq) \rightleftharpoons H^+(aq) + F^-(aq)$$

while ammonia is typical of weak bases:

$$NH_3(aq) + H_2O \rightleftharpoons NH_4^+(aq) + OH^-(aq)$$

Hydrogen fluoride and ammonia are **weak electrolytes**; the conductivities of their water solutions are slightly greater than those of nonelectrolytes. They produce slightly more than one mole of particles (molecules and ions) per mole of solute.

Table 10.1 Solution Behavior of Some Aqueous Solutes

Solute	Type	Solution Equation	Moles of Solute Particles per Mole Solute
$I_2(s)$	nonelectrolyte	$I_2(s) \longrightarrow I_2(aq)$	1
$C_6H_{12}O_6(s)$ (glucose)	nonelectrolyte	$C_6H_{12}O_6(s) \longrightarrow C_6H_{12}O_6(aq)$	1
$C_{12}H_{22}O_{11}(s)$ (sucrose)	nonelectrolyte	$C_{12}H_{22}O_{11}(s) \longrightarrow C_{12}H_{22}O_{11}(aq)$	1
$HCl(g)$	electrolyte	$HCl(g) \longrightarrow H^+(aq) + Cl^-(aq)$	2 (1 H^+ and 1 Cl^-)
$LiBr(s)$	electrolyte	$LiBr(s) \longrightarrow Li^+(aq) + Br^-(aq)$	2 (1 Li^+ and 1 Br^-)
$K_2SO_4(s)$	electrolyte	$K_2SO_4(s) \longrightarrow 2\ K^+(aq) + SO_4^{2-}(aq)$	3 (2 K^+ and 1 SO_4^{2-})
$LaBr_3(s)$	electrolyte	$LaBr_3(s) \longrightarrow La^{3+}(aq) + 3\ Br^-(aq)$	4 (1 La^{3+} and 3 Br^-)

10.2 CONCENTRATION UNITS

Several different methods are used to express relative amounts of solute and solvent in a solution. Two concentration units, **molarity** and **mole fraction**, were referred to in previous chapters. Two others, **mass percent** and **molality**, are considered here for the first time.

Mass Percent; Parts per Million

The mass percent of solute in solution is expressed quite simply:

$$\text{mass percent of solute} = \frac{\text{mass solute}}{\text{total mass solution}} \times 100\%$$

In a solution prepared by dissolving 24 g of NaCl in 152 g of water,

$$\text{mass percent of NaCl} = \frac{24\text{ g}}{24\text{ g} + 152\text{ g}} \times 100\% = \frac{24}{176} \times 100\% = 14\%$$

When the amount of solute is very small, as with trace impurities in water, concentration is often expressed in **parts per million (ppm)**:

Parts per billion (ppb) is defined similarly

$$\text{ppm solute} = \frac{\text{mass solute}}{\text{total mass solution}} \times 10^6$$

Comparing the defining equations for mass percent and parts per million, it should be clear that

$$\text{ppm} = \text{mass percent} \times 10^4$$

In the United States and Canada, drinking water cannot contain more than 5×10^{-4} mg of mercury per gram of sample. In parts per million that would be

$$\text{ppm Hg} = \frac{5 \times 10^{-4}\text{ mg Hg}}{1 \times 10^3\text{ mg}} \times 10^6 = 0.5$$

Mole fraction (X)

Recall from Chapter 5 the defining equation for mole fraction (X) of a component A:

$$X_A = \frac{\text{moles A}}{\text{total moles}} = \frac{n_A}{n_{tot}}$$

The mole fractions of all components of a solution (A, B, . . .) must add to unity:

$$X_A + X_B + \cdots = 1$$

Example 10.1 What are the mole fractions of CH_3OH and H_2O in a solution prepared by dissolving 1.20 g of methyl alcohol in 16.8 g of water?

Strategy First (1), convert grams to moles for both components. Then (2), using the defining equation, calculate the mole fraction of CH_3OH. Finally (3), obtain the mole fraction of H_2O, most simply by using the relation: $X_{CH_3OH} + X_{H_2O} = 1$.

Solution

(1) $n_{CH_3OH} = 1.20\text{ g } CH_3OH \times \dfrac{1\text{ mol } CH_3OH}{32.04\text{ g } CH_3OH} = 0.0375\text{ mol } CH_3OH$

$n_{H_2O} = 16.8\text{ g } H_2O \times \dfrac{1\text{ mol } H_2O}{18.02\text{ g } H_2O} = 0.932\text{ mol } H_2O$

(2) $X_{CH_3OH} = \dfrac{n_{CH_3OH}}{n_{CH_3OH} + n_{H_2O}} = \dfrac{0.0375}{0.0375 + 0.932} = 0.0387$

(3) $X_{H_2O} = 1 - X_{CH_3OH} = 1 - 0.0387 = 0.9613$

96% of the molecules in this solution are H_2O; about 4% are CH_3OH

Molality (m)

The concentration unit molality, given the symbol m, is defined as the number of moles of solute per kilogram (1000 g) of *solvent.*

$$\text{molality } (m) = \frac{\text{moles solute}}{\text{kilograms solvent}}$$

The molality of a solution is readily calculated if the masses of solute and solvent are known (Example 10.2).

Example 10.2 A solution used for intravenous feeding contains 4.80 g of glucose, $C_6H_{12}O_6$, in 90.0 g of water. What is the molality of glucose?

Strategy First (1), calculate the number of moles of glucose ($\mathcal{M}$ = 180.16 g/mol), then (2), the number of kilograms of water. Finally (3), use the defining equation to calculate molality.

Solution

(1) $n_{glucose} = 4.80\text{ g glucose} \times \dfrac{1\text{ mol glucose}}{180.16\text{ g glucose}} = 0.0266\text{ mol glucose}$

(2) $\text{kg water} = 90.0\text{ g water} \times \dfrac{1\text{ kg water}}{1000\text{ g water}} = 0.0900\text{ kg water}$

(3) $\text{molality} = \dfrac{0.0266\text{ mol glucose}}{0.0900\text{ kg water}} = 0.296\text{ m}$

Molarity (M)

In Chapter 4, molarity was the concentration unit of choice in dealing with solution stoichiometry. You will recall that molarity is defined as

Molality and molarity are concentration units; morality is something else

$$\text{molarity } (M) = \frac{\text{moles solute}}{\text{liters solution}}$$

A solution can be prepared to a specified molarity by weighing out the calculated mass of solute and dissolving in enough solvent to form the desired volume of solution. Alternatively, you can start with a more concentrated solu-

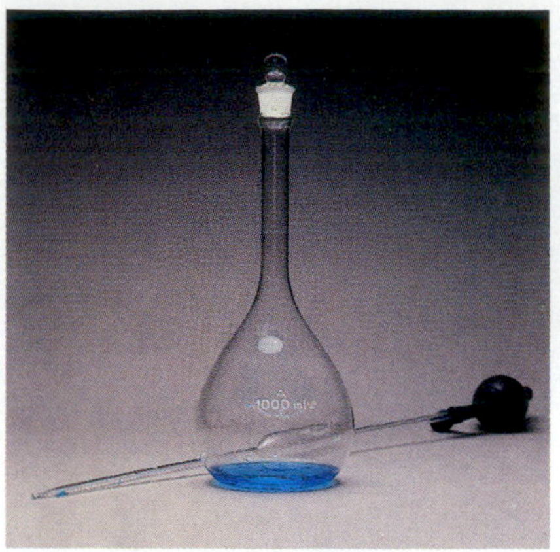

Figure 10.5
To prepare one liter of 0.100 *M* $CuSO_4$ solution, pipet 50.0 mL of 2.00 *M* $CuSO_4$ into a volumetric flask (left). Then add water in stages (center, right), to bring the level to the narrow mark on the neck of the flask, corresponding to 1.00 L. (Marna G. Clarke)

tion and dilute with water to give a solution of the desired molarity (Fig. 10.5). The calculations here are straightforward if you keep a simple point in mind: Adding solvent cannot change the number of moles of solute, B, i.e., n_B.

$$n_B \text{ in concentrated solution} = n_B \text{ in dilute solution}$$

In both solutions, the number of moles of solute can be found by multiplying the molarity, [B], by the volume in liters, *V*. Hence,

$$[B]_c V_c = [B]_d V_d$$

where the subscripts c and d stand for concentrated and dilute solutions, respectively.

Example 10.3 How would you prepare 1.00 L of 0.100 *M* $CuSO_4$ starting with 2.00 *M* $CuSO_4$?

Strategy This question might be restated as: What volume of 2.00 *M* $CuSO_4$ should be diluted with water to give 1.00 L of 0.100 *M* $CuSO_4$? To do this, you use the equation $[CuSO_4]_c V_c = [CuSO_4]_d V_d$ to calculate V_c, the volume of the more concentrated solution.

Solution Solving for V_c:

$$V_c = \frac{[CuSO_4]_d \times V_d}{[CuSO_4]_c} = \frac{0.100\,M \times 1.00\text{ L}}{2.00\,M} = 0.0500\text{ L} = 50.0\text{ mL}$$

Measure out 50.0 mL of 1.00 *M* $CuSO_4$ and dilute with enough water (about 950 mL) to form 1.00 L of 0.100 *M* solution.

Molarity, unlike mass %, mole fraction, or molality, requires volume measurements

The advantage of preparing solutions by the method illustrated in Example 10.3 and Figure 10.5 is that only volume measurements are involved. If you wander into the general chemistry storeroom, you're likely to find concentrated "stock" solutions of various chemicals. Storeroom personnel prepare the more dilute solutions that you use in the laboratory on the basis of calculations like those shown in the example.

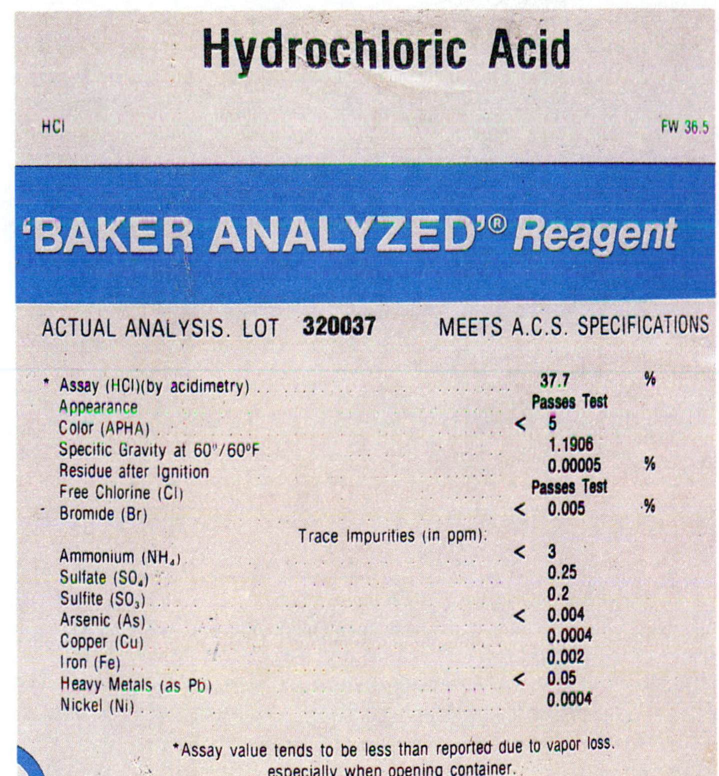

Figure 10.6
The label on a bottle of concentrated hydrochloric acid typically gives the mass percent of HCl ("assay") and the density (or "specific gravity") of the solution. Given that information, it is possible to calculate the molality, molarity, and mole fraction of HCl. (Marna G. Clarke)

Conversions Between Concentration Units

Frequently it is necessary to convert from one concentration unit to another. This problem arises, for example, in making up solutions of hydrochloric acid. Typically, the analysis or assay that appears on the label (Fig. 10.6) does not give the molarity or molality of the acid. Instead, it lists the mass percent of solute and the density of the solution.

Conversions between concentration units are relatively straightforward provided you *first decide upon a fixed amount of solution*. The amount chosen depends upon the unit in which concentration is originally expressed.

Most problems can be solved if you get off to a good start

When the Original Concentration Is	Start with
Mass percent	100 g solution
Molarity (M)	1.00 L solution
Molality (m)	1000 g solvent
Mole fraction (X)	1 mol (solute + solvent)

To illustrate the approach used, let us consider how the molarity of HCl can be calculated from the data shown on the label (Fig. 10.6).

Example 10.4 Calculate the molarity of a concentrated solution of hydrochloric acid that is 37.7% by mass HCl; the solution has a density of 1.19 g/mL.

Strategy Start with 100.0 g of solution. First (1), knowing the percent by mass of HCl (37.7%) and its molar mass (36.46 g/mol), you can readily calculate the number of moles of HCl in 100.0 g of solution. Then (2), knowing the density of the solution, you should be able to calculate the volume occupied by 100.0 g of solution. Finally (3), knowing both the number of moles and the volume in liters, use the defining equation to calculate the molarity of HCl.

Solution

(1) $n_{HCl} = 100.0 \text{ g soln.} \times \frac{37.7 \text{ g HCl}}{100.0 \text{ g soln.}} \times \frac{1 \text{ mol HCl}}{36.46 \text{ g HCl}} = 1.03 \text{ mol HCl}$

(2) $V = 100.0 \text{ g} \times \frac{1 \text{ mL}}{1.19 \text{ g}} \times \frac{1 \text{ L}}{1000 \text{ mL}} = 0.0840 \text{ L}$

(3) $[HCl] = 1.03 \text{ mol}/0.0840 \text{ L} = 12.3\ M$

This same approach can be used to convert between molarity and molality (Example 10.5).

Example 10.5 A 1.13 *M* solution of KOH has a density of 1.05 g/mL. Calculate its molality.

Strategy Start with one liter of solution, which contains 1.13 mol of KOH. Then find, in order, (1) the total mass of solution, knowing the density, (2) the mass of KOH, knowing the molarity, and (3) the mass of water. Finally (4), use the defining equation to calculate molality.

Solution

(1) total mass of 1 L of soln. $= 1.00 \text{ L} \times \frac{1000 \text{ mL}}{1 \text{ L}} \times \frac{1.05 \text{ g}}{1 \text{ mL}} = 1050 \text{ g}$

(2) mass of KOH in 1 L of soln. $= 1.13 \text{ mol KOH} \times \frac{56.11 \text{ g KOH}}{1 \text{ mol KOH}} = 63.4 \text{ g KOH}$

(3) mass of water in 1 L of soln. $= 1050 \text{ g} - 63.4 \text{ g} = 987 \text{ g}$

(4) molality $= \frac{1.13 \text{ mol KOH}}{0.987 \text{ kg water}} = 1.14\ m$

You will notice from Example 10.5 that the molarity and molality of the KOH solutions are very close to one another (1.13 *M*, 1.14 *m*). This is generally true of *dilute water solutions;* one liter of a dilute aqueous solution contains approximately one kilogram of water. For concentrated or nonaqueous solutions, molarity and molality ordinarily differ considerably from each other.

10.3 PRINCIPLES OF SOLUBILITY

The extent to which a solute dissolves in a particular solvent depends upon several factors. The most important of these are

— the nature of solvent and solute particles and the interactions between them

— the temperature at which the solution is formed
— the pressure of a gaseous solute

In this section we consider in turn the effect of each of these factors upon solubility.

Solute-Solvent Interactions

In discussing solubility, it is sometimes stated that "like dissolves like." A more meaningful way to express this idea is to say that two substances with intermolecular forces of about the same type and magnitude are likely to be very soluble in one another. To illustrate, consider the hydrocarbons pentane, C_5H_{12}, and hexane, C_6H_{14}, which are completely miscible with each other. Molecules of these nonpolar substances are held together by dispersion forces of about the same magnitude. A pentane molecule experiences little or no change in intermolecular forces when it goes into solution in hexane.

Most nonpolar substances have very small water solubilities. Petroleum, a mixture of hydrocarbons, spreads out in a thin film on the surface of a body of water rather than dissolving. The mole fraction of pentane, C_5H_{12}, in a saturated water solution is only 0.0001. These low solubilities are readily understood in terms of the structure of liquid water. To dissolve appreciable amounts of pentane in water, it would be necessary to break the hydrogen bonds holding H_2O molecules together. There is no comparable attractive force between C_5H_{12} and H_2O to supply the energy required to break into the water structure.

Of the relatively few organic compounds that dissolve readily in water, most contain —OH groups. Three familiar examples are methyl alcohol (methanol), ethyl alcohol (ethanol), and ethylene glycol, all of which are infinitely soluble in water.

```
   H              H  H              H  H
   |              |  |              |  |
H—C—OH        H—C—C—OH        H—C—C—H
   |              |  |              |  |
   H              H  H             HO  OH
methyl alcohol   ethyl alcohol    ethylene glycol
```

In these compounds, as in water, the principal intermolecular forces are hydrogen bonds. When a substance like methyl alcohol dissolves in water, it forms hydrogen bonds with H_2O molecules. These hydrogen bonds, joining a CH_3OH molecule to an H_2O molecule, are about as strong as those in the pure substances.

The presence or absence of H bonds has a pronounced effect on solubility

As we noted in Chapter 4, the solubility of ionic compounds in water varies tremendously from one solid to another. The extent to which solution occurs depends upon a balance between two forces, both electrical in nature:

1. The force of attraction between H_2O molecules and the ions, which tends to bring the solid into solution. If this factor predominates, the compound is very soluble in water, as is the case with NaCl, NaOH, and many other ionic solids.

We can't predict which factor will dominate, so we can't predict solubility

2. The force of attraction between oppositely charged ions, which tends to keep them in the solid state. If this is the major factor, the water solubility is very low. The fact that $CaCO_3$ and $BaSO_4$ are almost insoluble in water implies that interionic attractive forces predominate with these ionic solids.

A warm glass rod placed in a glass of ginger ale increases the temperature of the carbonated solution. The increase in temperature decreases the solubility of CO_2 in water, causing the CO_2 to bubble out of solution. (Charles D. Winters)

Effect of Temperature upon Solubility

When an excess of a solid such as sodium chloride is shaken with water, it forms a saturated solution. An equilibrium is established between the solid and its ions in solution.

$$NaCl(s) \rightleftharpoons Na^+(aq) + Cl^-(aq)$$

A similar type of equilibrium is established when a gas such as carbon dioxide is bubbled through water:

$$CO_2(g) \rightleftharpoons CO_2(aq)$$

The effect of a temperature change upon solubility equilibria such as these can be predicted by applying a simple principle. *An increase in temperature always shifts the position of an equilibrium so as to favor an endothermic process.* This means that if the solution process absorbs heat, ($\Delta H_{soln.} > 0$), an increase in temperature increases the solubility. Conversely, if the solution process is exothermic ($\Delta H < 0$), an increase in temperature decreases the solubility.

Dissolving a solid in a liquid is usually an endothermic process; heat must be absorbed to break down the crystal lattice.

$$\text{solid} + \text{liquid} \rightleftharpoons \text{solution} \qquad \Delta H_{soln.} > 0$$

Consistent with this effect, the solubilities of solids usually increase as the temperature rises. More sugar dissolves in hot coffee than in cold coffee. Cooling a saturated solution usually causes a solid to crystallize out of solution, indicating that it is less soluble at the lower temperature.

When a gas condenses to a liquid, heat is always evolved. By the same token, heat is usually evolved when a gas dissolves in a liquid:

$$\text{gas} + \text{liquid} \rightleftharpoons \text{solution} \qquad \Delta H_{soln.} < 0$$

An open can of soda loses its carbonation faster at room temperature

This means that the reverse process (gas coming out of solution) is endothermic. Hence it is favored by an increase in temperature; typically, gases become less soluble as the temperature rises. This rule is followed by all gases in water. You have probably noticed this effect when heating water in an open pan or beaker. Bubbles of air are driven out of the water by an increase in temperature. The reduced solubility of oxygen in water at high temperatures (Fig. 10.7a) may explain why trout congregate at the bottom of deep pools on hot summer days when the surface water is depleted in dissolved oxygen.

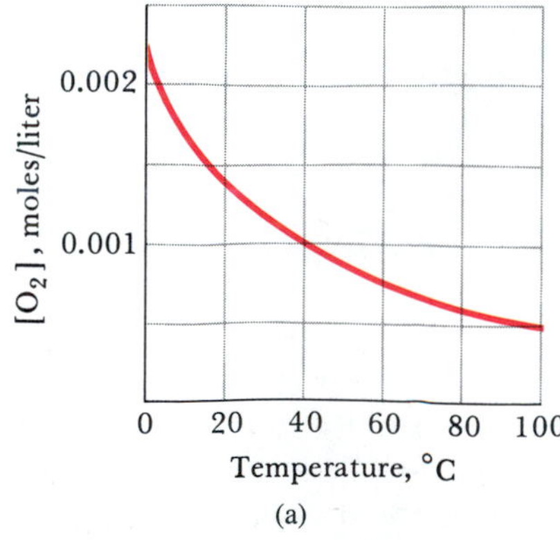

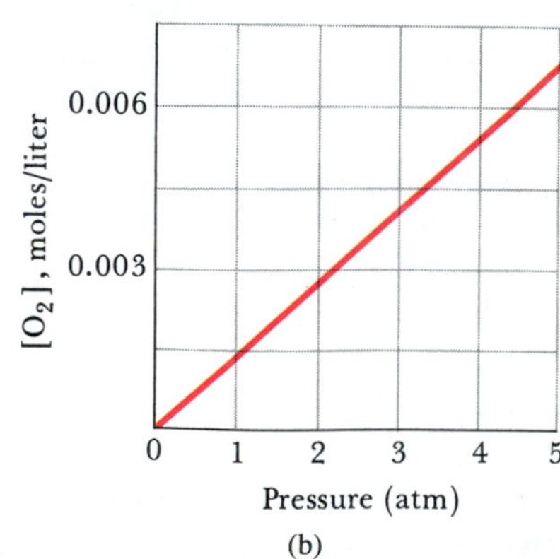

Figure 10.7
The solubility of $O_2(g)$ in water decreases as temperature rises (a) and increases as pressure increases (b). In (a), the pressure is held constant at 1 atm; in (b), the temperature is held constant at 25°C.

Effect of Pressure upon Solubility

Pressure has a major effect on solubility only for gas-liquid systems. At a given temperature, a rise in pressure increases the solubility of a gas. Indeed, at low to moderate pressures, gas solubility is directly proportional to pressure (**Henry's law**; Figure 10.7b).

$$C_g = kP_g$$

where P_g is the partial pressure of the gas over the solution, C_g is its concentration in the solution, and k is a constant characteristic of the particular gas-liquid system. This effect arises because increasing the pressure raises the concentration of molecules in the gas phase. To balance this change and maintain equilibrium, more gas molecules enter the solution, increasing their concentration in the liquid phase.

Example 10.6 The solubility of pure nitrogen in blood at body temperature, 37°C, and one atmosphere, is 6.2×10^{-4} *M*. If a diver breathes air ($X_{N_2} = 0.78$) at a depth where the total pressure is 2.5 atm, calculate the concentration of nitrogen in his blood.

Strategy Use Henry's law in any problem involving gas solubility and pressure. Perhaps the simplest approach here is to use the data at one atmosphere to calculate k. Then apply Henry's law to calculate C_g at the higher pressure. Note that it is the *partial* pressure of N_2 that is required; use the relation $P_{N_2} = X_{N_2} \times P_{tot}$ to find it.

Solution At 1 atm:

$$k = \frac{\text{concentration of } N_2}{\text{pressure of } N_2} = \frac{6.2 \times 10^{-4}\,M}{1.00 \text{ atm}} = 6.2 \times 10^{-4}\,M/\text{atm}$$

At the higher pressure:

$$P_{N_2} = X_{N_2} \times P_{tot} = 0.78 \times 2.5 \text{ atm} = 2.0 \text{ atm}$$

$$[N_2] = kP_{N_2} = 6.2 \times 10^{-4} \frac{M}{\text{atm}} \times 2.0 \text{ atm} = 1.2 \times 10^{-3}\,M$$

The influence of partial pressure on gas solubility is used in making carbonated beverages such as beer, sparkling wines, and many soft drinks. These beverages are bottled under pressures of CO_2 as high as 4 atm. When the bottle or can is opened, the pressure above the liquid drops to 1 atm and the carbon dioxide bubbles rapidly out of solution. Pressurized containers for shaving cream, whipped cream, and cheese spreads work on a similar principle. Pressing a valve reduces the pressure on the dissolved gas, causing it to rush from solution, carrying liquid with it as a foam.

Sometimes, as with beer, too rapidly

Another consequence of the effect of pressure on gas solubility is the painful, sometimes fatal, affliction known as the "bends." This occurs when a person goes rapidly from deep water (high pressure) to the surface (lower pressure). The rapid decompression causes air, dissolved in blood and other body fluids, to bubble out of solution. These bubbles impair blood circulation and affect nerve impulses. To minimize these effects, deep-sea divers and aquanauts breathe a helium–oxygen mixture rather than compressed air (nitrogen–

Scuba divers have to worry about this

oxygen). Helium is only about one third as soluble as nitrogen, and hence much less gas comes out of solution upon decompression.

10.4 COLLIGATIVE PROPERTIES

The properties of a solution differ considerably from those of the pure solvent. Those solution properties which depend primarily upon the *concentration of solute particles* rather than their nature are called **colligative properties**. Such properties include vapor pressure lowering, osmotic pressure, boiling point elevation, and freezing point depression. This section considers the relations between colligative properties and solute concentration, starting with nonelectrolytes and then going on (briefly) to electrolytes.

The relationships among colligative properties and solute concentration are best regarded as limiting laws. They are approached more closely as the solution becomes more dilute. In practice, the relationships discussed here are valid, for nonelectrolytes, to within a few percent at concentration as high as 1 *M*. At higher concentrations, solute-solute interactions lead to larger deviations.

Vapor Pressure Lowering (Nonelectrolytes)

You may have noticed that concentrated aqueous solutions evaporate more slowly than does pure water. This reflects the fact that the vapor pressure of water over the solution is less than that of pure water (Fig. 10.8).

Vapor pressure lowering is a true colligative property; that is, it is independent of the nature of the solute but directly proportional to its concentration. For example, the vapor pressure of water above a 0.10 *M* solution of either glucose or sucrose at 0°C is the same, about 0.008 mm Hg less than that of pure water. In 0.30 *M* solution, the vapor pressure lowering is almost exactly three times as great, 0.025 mm Hg.

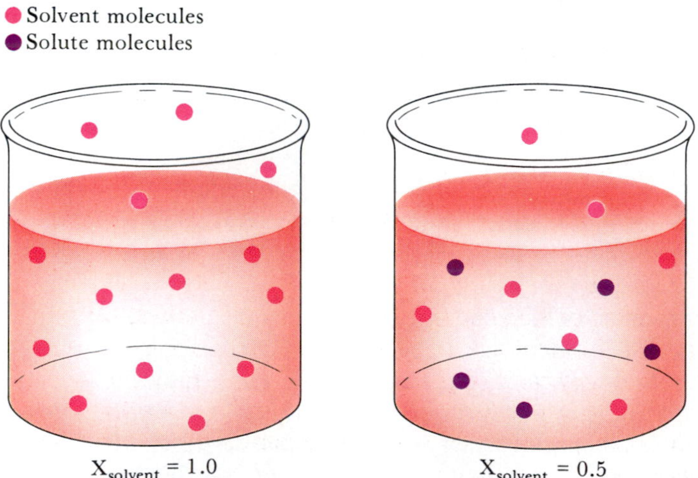

Figure 10.8
Raoult's law. Adding a solute lowers the concentration of solvent molecules in the liquid phase. To maintain equilibrium, the concentration of solvent molecules in the gas phase must decrease, thereby lowering the solvent vapor pressure.

The relationship between solvent vapor pressure and concentration is ordinarily expressed in the form of **Raoult's law**:

$$P_1 = X_1 P_1^\circ$$

In this equation, P_1 is the vapor pressure of solvent over the solution, P_1° is the vapor pressure of the pure solvent at the same temperature, and X_1 is the mole fraction of solvent. Note that since X_1 in a solution must be less than 1, P_1 must be less than P_1°.

To obtain a direct expression for vapor pressure lowering, note that $X_1 = 1 - X_2$, where X_2 is the mole fraction of solute. Substituting $1 - X_2$ for X_1 in Raoult's law,

$$P_1 = (1 - X_2)P_1^\circ$$

Rearranging,

$$P_1^\circ - P_1 = X_2 P_1^\circ$$

The quantity $(P_1^\circ - P_1)$ is the vapor pressure lowering (ΔP). It is the difference between the solvent vapor pressure in the pure solvent and in solution.

$$\Delta P = X_2 P_1^\circ$$

Example 10.7 A solution contains 102 g of sugar, $C_{12}H_{22}O_{11}$, in 375 g of water. Calculate the vapor pressure lowering at 25°C (vp of pure water = 23.76 mm Hg).

Strategy First (1), calculate the number of moles of sugar ($\mathcal{M}$ = 342.30 g/mol) and water ($\mathcal{M}$ = 18.02 g/mol). That information allows you to calculate (2) the mole fraction of sugar. Finally (3), use Raoult's law to find the vapor pressure lowering.

Solution

(1) $n_{\text{sugar}} = 102 \text{ g sugar} \times \dfrac{1 \text{ mol sugar}}{342.30 \text{ g sugar}} = 0.298 \text{ mol sugar}$

$n_{\text{water}} = 375 \text{ g water} \times \dfrac{1 \text{ mol water}}{18.02 \text{ g water}} = 20.8 \text{ mol water}$

(2) $X_{\text{sugar}} = \dfrac{0.298}{0.298 + 20.8} = \dfrac{0.298}{21.1} = 0.0141$

(3) $\Delta P = X_{\text{sugar}} \times P^\circ_{H_2O} = 0.0141 \times 23.76 \text{ mm Hg} = 0.335 \text{ mm Hg}$

Osmotic Pressure (Nonelectrolytes)

Consider the apparatus shown in Figure 10.9, where a sugar solution in the tall tube is separated from water in the beaker by a "semipermeable" membrane. This may be an animal bladder, a slice of vegetable tissue, or a synthetic material. The membrane, by a mechanism that is not well understood, allows water molecules to pass through it, but not sugar molecules. Experimentally, it is found that water moves from the beaker, through the membrane, and into the tube. The liquid level in the tube rises; the sugar solution becomes more dilute. This process, in which a solvent moves through a semipermeable membrane into a solution, is called **osmosis**.

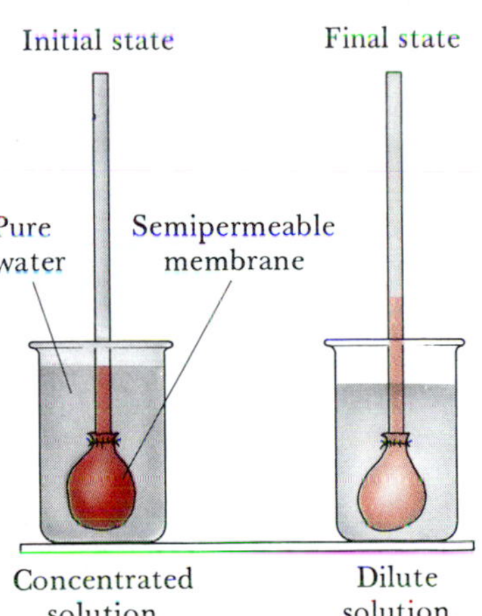

Figure 10.9
In osmosis, water molecules move through a semipermeable membrane from a region of high vapor pressure (pure water) to a region of lower vapor pressure (solution).

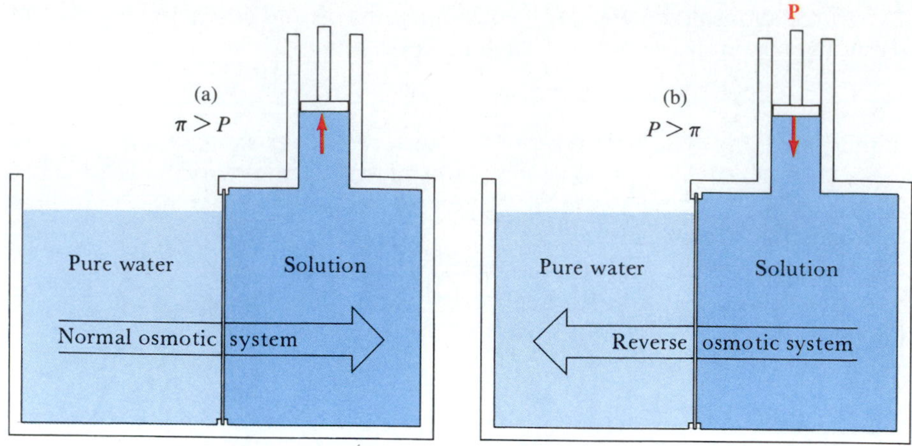

Figure 10.10
Osmosis can be prevented by applying to the solution a pressure P that just balances the osmotic pressure, π. If $P < \pi$, normal osmosis occurs. If $P > \pi$, water flows in the opposite direction, producing reverse osmosis, a process that can be used to obtain fresh water from seawater.

The driving force behind osmosis is the vapor pressure difference between solvent and solution. The vapor pressure of pure water in the beaker is greater than that in the solution inside the tube. Water and other liquids tend to move from a region of high vapor pressure to one of low vapor pressure, just as a compressed gas flows spontaneously from a high-pressure cylinder when the valve is opened.

The osmotic pressure, π, is equal to the external pressure, P, just sufficient to prevent osmosis (Fig. 10.10). If P is less than π, osmosis takes place in the normal way and water moves through the membrane into the solution (Fig. 10.10a). By making the external pressure large enough, it is possible to reverse this process (Fig. 10.10b). When $P > \pi$, water molecules move through the membrane from the solution to pure water. This process, called **reverse osmosis**, is used to obtain fresh water from seawater in arid regions of the world, including Saudi Arabia.

Osmotic pressure, like vapor pressure lowering, is a colligative property. For any nonelectrolyte B, π is directly proportional to molarity, [B]. The equation relating these two quantities is very similar to the ideal gas law:

$$\pi = \frac{n_B RT}{V} = [B]RT$$

where R is the gas law constant, 0.0821 L · atm/mol · K, and T is the Kelvin temperature. Even in dilute solution, the osmotic pressure is quite large. Consider, for example, a 0.10 M solution at 25°C:

$$\pi = (0.10 \text{ mol/L})\left(0.0821 \frac{\text{L} \cdot \text{atm}}{\text{mol} \cdot \text{K}}\right)(298 \text{ K}) = 2.4 \text{ atm}$$

A pressure of 2.4 atm is equivalent to a column of water 25 m (more than 80 ft) high.

If you soak in a hot tub long enough, your skin will look "pickled" too

If a cucumber is placed in a concentrated brine solution, it shrinks and assumes the wrinkled skin of a pickle. The skin of the cucumber acts as a semipermeable membrane. The water solution inside the cucumber is more dilute than the solution surrounding it. As a result, water flows out of the cucumber into the brine (Fig. 10.11).

Figure 10.11
When a cucumber is pickled, water moves out of the cucumber by osmosis into the concentrated brine solution. A prune placed in pure water swells as water moves into the prune, again by osmosis. (Marna G. Clarke)

When a dried prune is placed in water, the skin also acts as a semipermeable membrane. This time the solution inside the prune is more concentrated than the water, so that water flows into the prune, making the prune less wrinkled.

Nutrient solutions used in intravenous feeding must be *isotonic* with blood; that is, they must have the same osmotic pressure as blood. If the solution is too dilute, its osmotic pressure will be less than that of the fluids inside blood cells; in that case, water will flow into the cell until it bursts. Conversely, if the nutrient solution has too high a concentration of solutes, water will flow out of the cell until it shrivels and dies.

Boiling Point Elevation and Freezing Point Lowering (Nonelectrolytes)

When a solution of a nonvolatile solute is heated, it does not begin to boil until the temperature exceeds the boiling point of the solvent. The difference in temperature is called the **boiling point elevation**, ΔT_b.

$$\Delta T_b = T_b - T_b^\circ$$

where T_b and T_b° are the boiling points of the solution and the pure solvent, respectively. As boiling continues, pure solvent distils off, the concentration of solute increases, and the boiling point continues to rise (Fig. 10.12).

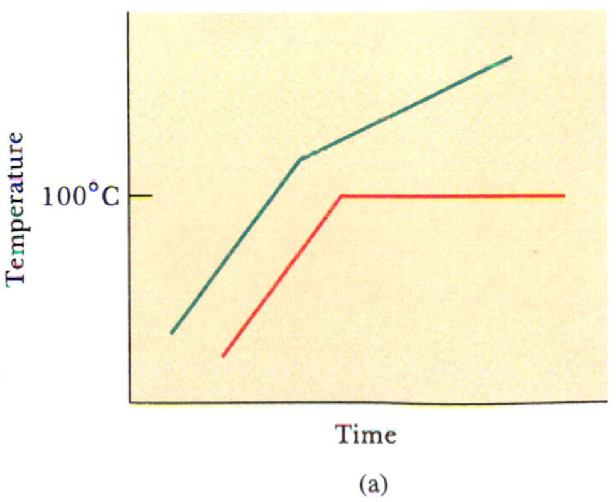

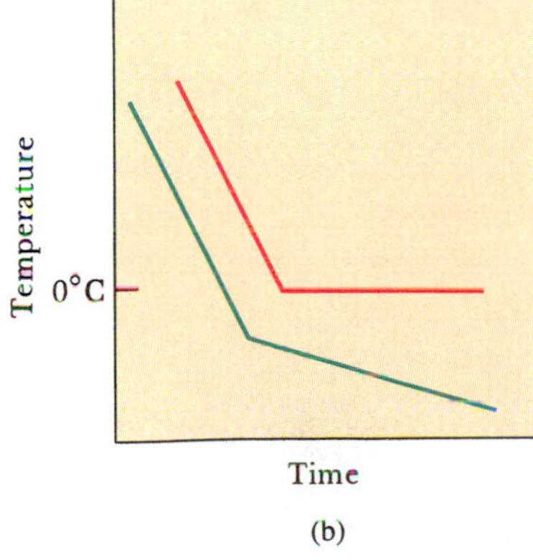

Figure 10.12
Heating (a) and cooling (b) curves for pure water (red) and an aqueous solution (green). For pure water, the temperature remains constant during boiling or freezing. For the solution, the temperature changes constantly during the phase change because water is being removed, increasing the concentration of solute.

Table 10.2 Molal Freezing Point and Boiling Point Constants

Solvent	fp (°C)	k_f (°C/m)	bp (°C)	k_b (°C/m)
Water	0.00	1.86	100.00	0.52
Acetic acid	16.66	3.90	117.90	2.53
Benzene	5.50	5.10	80.10	2.53
Cyclohexane	6.50	20.2	80.72	2.75
Camphor	178.40	40.0	207.42	5.61
p-Dichlorobenzene	53.1	7.1	174.1	6.2
Naphthalene	80.29	6.94	217.96	5.80

Solutes raise the boiling point and lower the freezing point

When a solution is cooled, it does not begin to freeze until a temperature below the freezing point of the pure solvent is reached. The **freezing point lowering**, ΔT_f, is defined to be a positive quantity:

$$\Delta T_f = T_f^\circ - T_f$$

where T_f°, the freezing point of the solvent, lies above T_f, the freezing point of the solution. As freezing takes place, pure solvent freezes out, the concentration of solute increases, and the freezing point continues to drop.

Boiling point elevation is a direct result of vapor pressure lowering. At any given temperature, a solution of a nonvolatile solute* has a vapor pressure *lower* than that of the pure solvent. Hence, a *higher* temperature must be reached before the solution boils—that is, before its vapor pressure becomes equal to the external pressure. Figure 10.13 illustrates this reasoning graphically.

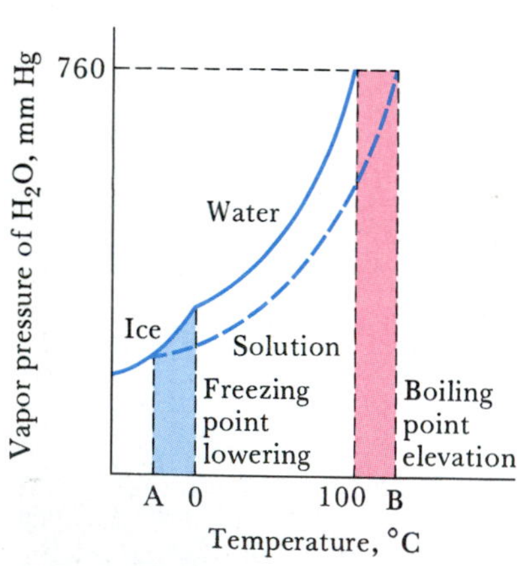

Figure 10.13
Since a nonvolatile solute lowers the vapor pressure of a solvent, the boiling point of a solution will be higher and the freezing point lower than the corresponding values for the pure solvent. Water solutions freeze *below* 0°C at point A and boil *above* 100°C at point B.

The freezing point lowering, like the boiling point elevation, is a direct result of the lowering of the solvent vapor pressure by the solute. Notice from Figure 10.13 that the freezing point of the solution is the temperature at which the solvent in solution has the same vapor pressure as the pure solid solvent. This implies that it is pure solvent (e.g., ice) that separates when the solution freezes.

Boiling point elevation and freezing point lowering, like vapor pressure lowering, are colligative properties. They are directly proportional to solute concentration, generally expressed as molality, m. The relevant equations are

$$\Delta T_b = k_b \text{ (molality)}$$

$$\Delta T_f = k_f \text{ (molality)}$$

The proportionality constants in these equations, k_b and k_f, are called the *molal boiling point constant* and the *molal freezing point constant*, respectively. Their magnitudes depend upon the nature of the solvent (Table 10.2). Note that when the solvent is water

$$k_b = 0.52°\text{C}/m \qquad k_f = 1.86°\text{C}/m$$

* Volatile solutes ordinarily lower the boiling point because they contribute to the total vapor pressure of the solution.

Example 10.8 An antifreeze solution is prepared containing 50.0 cm^3 of ethylene glycol, $C_2H_6O_2$ (d = 1.12 g/cm^3), in 50.0 g of water. Calculate the freezing point of this "50-50" mixture.

Strategy First (1), calculate the number of moles of $C_2H_6O_2$ ($\mathcal{M}$ = 62.07 g/mol). Then (2), apply the defining equation to calculate the molality. Finally (3), use the equation $\Delta T_f = (1.86°C/m) \times$ molality to find the freezing point lowering.

Solution

$$(1)\ n_{C_2H_6O_2} = 50.0\ cm^3 \times \frac{1.12\ g}{1\ cm^3} \times \frac{1\ mol}{62.04\ g} = 0.903\ mol$$

$$(2)\ \text{molality} = \frac{\text{moles of } C_2H_6O_2}{\text{kg of water}} = \frac{0.903\ mol}{0.0500\ kg} = 18.1\ m$$

$$(3)\ \Delta T_f = k_f\ (\text{molality}) = 1.86°C/m \times 18.1\ m = 33.7°C$$

The freezing point of the solution is 33.7°C below that of pure water (0°C). Hence the solution should freeze at −33.7°C. Actually, the freezing point is somewhat lower, about −37°C (−35°F). The deviation reminds us that the equation used here, $\Delta T_f = k_f$(molality) is a limiting law.

You take advantage of freezing point lowering when you add antifreeze to your automobile radiator in winter. Ethylene glycol is the solute commonly used in so-called "permanent" antifreeze. It has a high boiling point (197°C), is virtually nonvolatile at 100°C, and raises the boiling point of water. Hence antifreeze that contains ethylene glycol does not boil away in summer driving.

It is, however, toxic

Commercial grade ethylene glycol is added to a car's radiator to prevent the water from freezing in the winter and boiling over in the summer. (Union Carbide)

Determination of Molar Masses of Nonelectrolytes from Colligative Properties

Colligative properties, particularly freezing point depression, can be used to determine molar masses of a wide variety of nonelectrolytes. The approach used is illustrated in Example 10.9.

Example 10.9 A student dissolves 1.50 g of a newly prepared compound in 75.0 g of cyclohexane. She measures the freezing point of the solution to be 2.70°C; that of pure cyclohexane is 6.50°C. Cyclohexane has a k_f of 20.2°C/m. Using these data, calculate the molar mass of the compound.

Strategy First (1), knowing the freezing point of the solution and that of the pure solvent, calculate ΔT_f. Then (2), use the equation $\Delta T_f = k_f$ (molality) to find the molality. Finally (3), obtain the molar mass using the defining equation for molality.

Solution

$$(1)\ \Delta T_f = T_f^\circ - T_f = 6.50°C - 2.70°C = 3.80°C$$

$$(2)\ \Delta T_f = k_f\ (\text{molality})$$

$$\text{molality} = \frac{\Delta T_f}{k_f} = \frac{3.80°C}{20.2°C/m} = 0.188\ m$$

(3) $\text{molality} = \dfrac{\text{moles solute}}{\text{kilograms solvent}} = \dfrac{\text{grams solute}/\mathcal{M}}{\text{kilograms solvent}}$

Solving for the molar mass:

$$\mathcal{M} = \frac{\text{grams solute}}{(\text{molality})(\text{kilograms solvent})}$$

All the quantities on the right side of this equation are known; 1.50 g of solute is dissolved in 75.0 g (0.0750 kg) of solvent. The molality was calculated to be 0.188 m. So

$$\mathcal{M} = \frac{1.50 \text{ g}}{\left(\dfrac{0.188 \text{ mol}}{\text{kg solvent}}\right) \times 0.0750 \text{ kg solvent}} = \boxed{106 \text{ g/mol}}$$

Molar masses of polymers like polyethylene can be measured this way

Molar masses can also be determined using other colligative properties. Osmotic pressure measurements are often used, particularly for solutes of high molar mass, where the concentration is likely to be quite low. The advantage of using osmotic pressure is that the effect is relatively large. Consider, for example, a 0.0010 M aqueous solution, for which

$$\pi \text{ at } 25°\text{C} = 0.024 \text{ atm} = 18 \text{ mm Hg}$$

$$\Delta T_f \approx 1.86 \times 10^{-3}°\text{C}$$

$$\Delta T_b \approx 5.2 \times 10^{-4}°\text{C}$$

A pressure of 18 mm Hg can be measured relatively accurately; temperature differences of the order of 0.001°C are essentially impossible to measure accurately.

Example 10.10 A solution contains 1.0 g of hemoglobin dissolved in enough water to form 100 cm^3 of solution. The osmotic pressure at 20°C is found to be 2.75 mm Hg. Calculate

(a) the molarity of hemoglobin.
(b) the molar mass of hemoglobin.

Strategy The molarity of hemoglobin, [Hem], is readily calculated from the relation $\pi = [\text{Hem}]RT$. Then the molar mass can be obtained from the defining equation for molarity.

Solution

(a) $[\text{Hem}] = \pi/RT$

$\pi = (2.75/760) \text{ atm} \qquad R = 0.0821 \text{ L} \cdot \text{atm/mol} \cdot \text{K} \qquad T = 293 \text{ K}$

$$[\text{Hem}] = \frac{(2.75/760)\text{atm}}{(0.0821 \text{ L} \cdot \text{atm/mol} \cdot \text{K})(293 \text{ K})} = \boxed{1.50 \times 10^{-4} \text{ mol/L}}$$

(b) From the defining equation for molarity,

$$\text{molarity} = \frac{\text{moles solute}}{\text{liters solution}} = \frac{\text{grams solute}/\mathcal{M}}{\text{liters solution}}$$

Solving for the molar mass,

$$\mathcal{M} = \frac{\text{grams solute}}{(\text{molarity})(\text{liters solution})}$$

Recall that there is 1.0 g of hemoglobin in 100 cm^3 of solution. Hence

$$\mathcal{M} = \frac{1.0 \text{ g}}{(1.50 \times 10^{-4} \text{ mol/L})(0.100 \text{ L})} = 6.7 \times 10^4 \text{ g/mol}$$

Colligative Properties of Electrolytes

As noted earlier, colligative properties of solutions are directly proportional to the concentration of solute *particles.* On this basis, it is reasonable to suppose that, at a given concentration, an electrolyte should have a greater effect upon these properties than does a nonelectrolyte. When one mole of a nonelectrolyte such as glucose dissolves in water, one mole of solute molecules is obtained. On the other hand, one mole of the electrolyte NaCl yields two moles of ions. With calcium chloride, $CaCl_2$, three moles of ions are produced per mole of solute.

This reasoning is confirmed experimentally. Compare, for example, the vapor pressure lowerings for 1 *M* solutions of glucose, sodium chloride, and calcium chloride at 25°C.

	Glucose	NaCl	$CaCl_2$
ΔP	0.42 mm Hg	0.77 mm Hg	1.3 mm Hg

With many electrolytes, ΔP is so large that the solid, when exposed to moist air, picks up water *(deliquesces).* This occurs with calcium chloride, whose saturated solution has a vapor pressure only 30% that of pure water. If dry $CaCl_2$ is exposed to air in which the relative humidity is greater than 30%, it absorbs water and forms a saturated solution. Deliquescence continues until the vapor pressure of the solution becomes equal to that of the water in the air.

The freezing points of electrolyte solutions, like their vapor pressures, are lower than those of nonelectrolytes at the same concentration. Sodium chloride and calcium chloride are used to lower the melting point of ice on highways; their aqueous solutions can have freezing points as low as −21 and −55°C, respectively. The equation for the freezing point lowering of an electrolyte is similar to that for nonelectrolytes, except for the introduction of a multiplier, *i*, called the *Van't Hoff factor.* For aqueous solutions of electrolytes,

$CaCl_2$ is better for the environment than NaCl; urea, $CO(NH_2)_2$ is still better

$$\Delta T_f = 1.86°\text{C}/m \times \text{molality} \times i$$

Similar equations apply for other colligative properties.

$$\Delta T_b = 0.52°\text{C}/m \times \text{molality} \times i$$

$$\pi = \text{molarity} \times R \times T \times i$$

If we assume that the ions of an electrolyte behave independently, *i* should be equal to *the number of moles of ions per mole of electrolyte.* Thus we would predict for the freezing points of 0.50 *m* solutions:

For a nonelectrolyte, $i = 1$

Electrolyte	Moles Ions	Predicted i	Predicted ΔT_f
NaCl	$1 + 1 = 2$	2	−1.86°C
$MgSO_4$	$1 + 1 = 2$	2	−1.86°C
Na_2SO_4	$2 + 1 = 3$	3	−2.79°C
$Al(NO_3)_3$	$1 + 3 = 4$	4	−3.72°C

The data in Table 10.3 suggest that the situation is not as simple as this discussion implies. The observed freezing point lowerings of NaCl and $MgSO_4$ are smaller than would be predicted with $i = 2$. For example, 0.50 *m* solutions of NaCl and $MgSO_4$ freeze at −1.68 and −0.995°C, respectively; the predicted freezing point is −1.86°C. Only in very dilute solution does the multiplier i approach the predicted value of 2.

This behavior is generally typical of electrolytes. Their colligative properties deviate considerably from "ideal" values, even at concentrations below 1 *m*. There are at least a couple of reasons for this effect.

1. Because of electrostatic attraction, an ion in solution tends to surround itself with more ions of opposite than of like charge (Fig. 10.14). The existence of this *ionic atmosphere*, first proposed by Peter Debye in 1923, prevents ions from acting as completely independent solute particles. The result is to make an ion somewhat less effective than a nonelectrolyte molecule in its influence on colligative properties.

2. Oppositely charged ions may interact strongly enough to form a discrete species called an *ion pair*. This effect is essentially nonexistent with electrolytes such as NaCl, in which the ions have low charges (+1, −1). However, with $MgSO_4$ (+2, −2 ions), ion pairing plays a major role. Even at concentrations as low as 0.1 *m*, there are more $MgSO_4$ ion pairs than "free" Mg^{2+} and SO_4^{2-} ions.

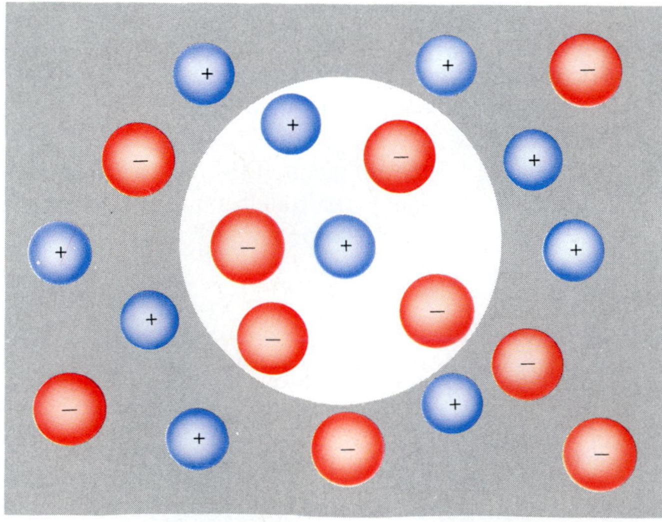

Figure 10.14
Ion atmosphere. An ion on the average is surrounded by more ions of opposite charge than of like charge.

Table 10.3 Freezing Point Lowerings of Solutions

	ΔT_f Observed (°C)		i (Calc from ΔT_f)	
Molality	NaCl	$MgSO_4$	NaCl	$MgSO_4$
0.005	0.0182	0.0160	1.96	1.72
0.01	0.0360	0.0285	1.94	1.53
0.02	0.0714	0.0534	1.92	1.44
0.05	0.176	0.121	1.89	1.30
0.10	0.348	0.225	1.87	1.21
0.20	0.685	0.418	1.84	1.12
0.50	1.68	0.995	1.81	1.07

PERSPECTIVE • Maple Syrup

The collection of maple sap and its conversion to syrup or sugar illustrate many of the principles covered in this chapter. Moreover, in northern New England, making maple syrup is an interesting way to spend the month of March (Fig. 10.A), which separates midwinter from "mud season."

The driving force behind the flow of maple sap is by no means obvious. The calculated osmotic pressure of sap, a 2% solution of sucrose ($\mathcal{M}$ = 342 g/mol) is

$$\pi = \frac{20/342 \text{ mol}}{1 \text{ L}} \times 0.0821 \frac{\text{L} \cdot \text{atm}}{\text{mol} \cdot \text{K}} \times 280 \text{ K} = 1.3 \text{ atm}$$

This is sufficient to push water to a height of about 45 ft; many maple trees are taller than that. Besides, a maple tree continues to bleed sap for several

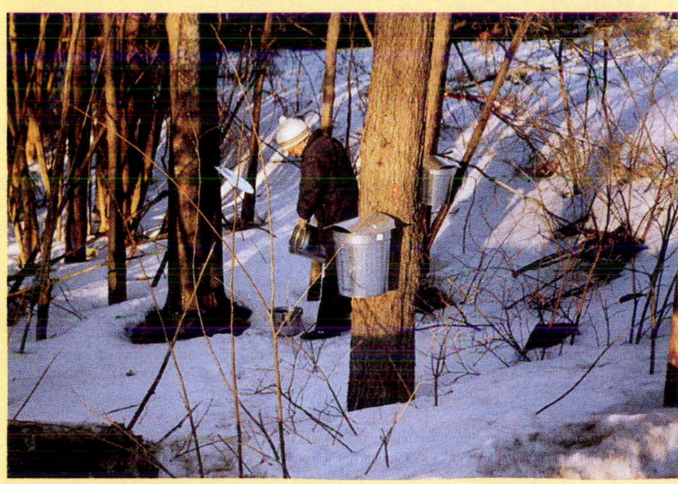

Figure 10.A
William Masterton collecting maple syrup.

days after it has been cut down. An alternative theory suggests that sap is forced out of the tree by bubbles of $CO_2(g)$, produced by respiration. When the temperature drops at night, the carbon dioxide goes into solution and the flow of sap ceases. This would explain the high sensitivity of sap flow to temperature; the aqueous solubility of carbon dioxide doubles when the temperature falls by 15°C.

If you want to make your own maple syrup on a small scale, perhaps 10–20 L per season, there are a few principles to keep in mind.

1. Make sure the trees you tap are *maples;* hemlocks would be a particularly poor choice. Identify the trees to be tapped in the fall, before the leaves fall. If possible, select sugar maples, which produce about one liter of maple syrup per tree per season. Other types of maples are less productive.

2. It takes 20–40 L of sap to yield one liter of maple syrup. To remove the water, you could freeze the sap, as the colonists did 200 years ago. The ice that forms is pure water (p. 276); by discarding it, you increase the concentration of sugar in the remaining solution. Large-scale operators today use reverse osmosis (p. 274) to remove about half of the water. The remainder must be boiled off. The characteristic flavor of maple syrup is caused by compounds formed upon heating, such as

acetol　　cyclotene

It's best to boil off the water outdoors. If you do this in the kitchen, you may not be able to open the doors and windows for a couple of weeks. Wood tends to swell when it absorbs a few hundred liters of water.

3. When the concentration of sugar reaches 66%, you are at the maple syrup stage. The calculated boiling point elevation (660 g of sugar, 340 g of water),

$$\Delta T_b = 0.52°C \times \frac{660/342}{0.340} = 3.0°C$$

is somewhat less than the observed value, about 4°C. The temperature rises very rapidly around 104°C (Fig. 10.B), which makes thermometry the method of choice for detecting the "end point." Shortly before that point, add a drop or two of vegetable oil to prevent foaming; perhaps this is the source of the phrase "spreading oil on troubled waters."

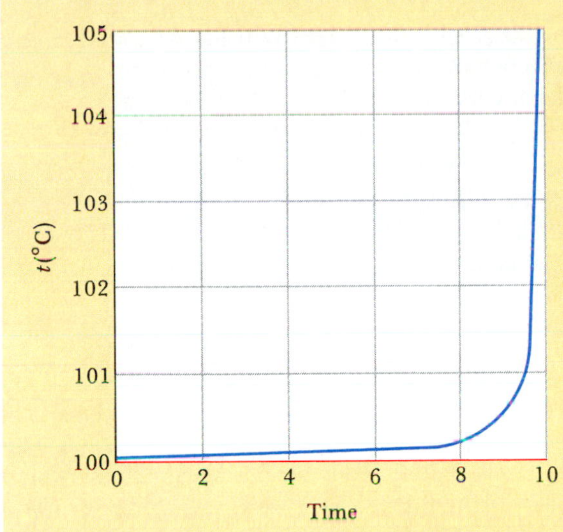

Figure 10.B
It seems to take forever to boil down maple sap until you reach 101°C. Then the temperature rises rapidly to the "endpoint," 104°C. Further *careful* heating produces maple sugar.

CHAPTER HIGHLIGHTS

KEY CONCEPTS

1. *Calculate the concentration of a solute (mass percent, mole fraction, molality, molarity) given mass and/or density data*
   (Examples 10.1, 10.2, 10.4, 10.5; Problems 7–16, 21–24, 73)
2. *Make dilution calculations*
   (Example 10.3; Problems 17–20, 70)
3. *Apply Henry's law to relate gas solubility to partial pressure*
   (Example 10.6; Problems 29, 30)
4. *Apply Raoult's law to calculate the vapor pressure of a solution*
   (Example 10.7; Problems 31–34)
5. *Relate the freezing point, boiling point, or osmotic pressure of a solution to solute concentration*
   (Examples 10.8–10.10; Problems 35–54, 72)

KEY EQUATIONS

Henry's law	$C_g = kP_g$
Raoult's law	$P_1 = X_1P_1^\circ$
Osmotic pressure	$\pi = MRT$
Boiling point, freezing point	$\Delta T_b = 0.52°C \times m \times i$ $\quad \Delta T_f = 1.86°C \times m \times i$ (i = no. of moles of particles per mole of solute)

KEY TERMS

boiling point
colligative property
electrolyte
mass percent
melting point
molality
molar mass
molarity
mole fraction
nonelectrolyte
osmosis
osmotic pressure
partial pressure
saturated solution
solvent
supersaturated solution
vapor pressure

SUMMARY PROBLEM

Consider camphor, $C_{10}H_{16}O$ ($\mathcal{M}$ = 152 g/mol), a substance obtained from the Formosa camphor tree. It has considerable use in the polymer and drug industries.

A solution of camphor is prepared by mixing 30.0 g of camphor with 1.25 L of ethanol, C_2H_5OH (d = 0.789 g/mL). Assume no change in volume when the solution is prepared.

1. What is the mass percent of camphor in the solution?
2. What is the concentration of camphor in parts per million?
3. What is the molarity of the solution?
4. What is the molality of the solution?
5. The vapor pressure of pure ethanol at 25°C is 59.0 mm Hg. What is the vapor pressure of ethanol in the solution at this temperature?
6. What is the osmotic pressure of the solution at 25°C?
7. What is the boiling point of the solution? The normal boiling point of ethanol is 78.26°C (k_b = 1.22°C/m).
8. Camphor can also be used as a solvent. It is used in the Rast method for determining molar mass. The molar mass of cortisone acetate is determined by dissolving 2.50 g in 50.0 g camphor (k_f = 40.0°C/m). The freezing point of the mixture is determined to be 173.44°C; that of pure camphor is 178.40°C. What is the molar mass of cortisone acetate?

Answers

1. 2.95% **2.** 2.95×10^4 ppm **3.** 0.158 M **4.** 0.200 m
5. 58.5 mm Hg **6.** 3.87 atm **7.** 78.50°C **8.** 403 g/mol

QUESTIONS & PROBLEMS

Symbols, Formulas, and Equations

1. Write the formulas of the following compounds.
a. table sugar **b.** nitric acid
c. calcium chloride

2. Write the formulas of the following compounds.
a. glucose **b.** sulfuric acid
c. potassium chromate

3. Write an equation for dissolving each of the following electrolytes in water.
a. chromium(III) nitrate **b.** calcium phosphate
c. magnesium iodide
d. rubidium hydrogen carbonate

4. Write an equation for dissolving each of the following electrolytes in water.
a. potassium perchlorate **b.** scandium(III) sulfate
c. calcium bromide **d.** nickel(II) chlorate

5. Write an equation to represent the dissolving of the following species in water.
a. table sugar **b.** hydrogen chloride gas
c. aluminum nitrate **d.** iodine

6. Write a reaction to represent the dissolving of the following species in water.
a. bromine **b.** potassium permanganate
c. ammonium phosphate **d.** glucose

Concentrations of Solutions

7. A solution is prepared by dissolving 4.87 g of sodium dichromate in 25.0 mL of water (d = 1.00 g/mL). Calculate
a. the mass percent of sodium dichromate in the solution.
b. the mass percent of water in the solution.
c. the mole fraction of sodium dichromate in the solution.

8. A solution is made by adding 3.50 mL of ethyl acetate, $C_4H_8O_2$ (d = 0.901 g/mL), to 25.00 mL of water (d = 1.00 g/mL). Assuming volumes are additive, calculate
a. the mass percent of ethyl acetate in solution.
b. the volume percent of water in solution.
c. the mole fraction of ethyl acetate in solution.

9. The "proof" of an alcoholic beverage is twice the volume percent of ethyl alcohol, C_2H_5OH, in solution. For a 90-proof vodka, what is the molality of the ethyl alcohol? Take the densities of ethyl alcohol and water to be 0.789 g/mL and 1.00 g/mL, respectively.

10. Household chlorine bleach contains 5.00% by mass of sodium hypochlorite, NaClO. If the only other component were water, what would be the molality of ClO^-?

11. The Dead Sea contains 58 moles of bromide ion in 1.0×10^3 kg of water. Calculate the concentration, in ppm, of bromide ion in the Dead Sea.

12. The Salton Sea in California contains a relatively large amount of lithium ions. Its concentration of Li^+ is 1.9 ppm. How many moles of lithium ions are present in ten kilograms of water from the Salton Sea?

13. Complete the following table for water solutions of sodium permanganate, $NaMnO_4$.

	Mass Solute	Moles Solute	Volume Solution	Molarity
a.	10.3 g	______	315 mL	______
b.	______	2.65	______	0.832
c.	______	______	3.85 L	1.58

14. Complete the following table for water solutions of oxalic acid, $H_2C_2O_4$.

	Mass Solute	Moles Solute	Volume Solution	Molarity
a.	12.5 g	______	456 mL	______
b.	______	0.0375	______	0.138
c.	______	______	1.75 L	0.496

15. Complete the following table for water solutions of acetic acid, $HC_2H_3O_2$.

	Molality	Mass Percent Solute	ppm of Solute	Mole Fraction Solute
a.	0.257	______	______	______
b.	______	5.00	______	______
c.	______	______	1542	______
d.	______	______	______	0.387

16. Complete the following table for water solutions of copper(II) sulfate.

	Molality	Mass Percent Solute	ppm of Solute	Mole Fraction Solute
a.	______	12.00	______	______
b.	______	______	586	______
c.	0.389	______	______	______
d.	______	______	______	0.0534

17. Describe in detail how you would prepare 350.0 mL of 0.250 *M* potassium chromate solution starting with

a. solid potassium chromate.

b. 1.50 *M* potassium chromate solution.

18. Describe how you would prepare 5.00 L of 0.250 *M* formic acid, HCOOH, starting with

a. 0.750 *M* formic acid. **b.** 0.350 *M* formic acid.

19. A solution is made by diluting 175 mL of 0.238 *M* aluminum nitrate, $Al(NO_3)_3$, solution with water to a final volume of 5.00×10^2 mL. Calculate

a. the molarity of aluminum nitrate, aluminum ion, and nitrate ion in the diluted solution.

b. the number of moles of nitrate ion in the original solution.

20. A solution is made by diluting 125 mL of 0.230 *M* potassium phosphate solution with water to a final volume of 7.50×10^2 mL. Calculate

a. The molarities of potassium phosphate, potassium ion, and phosphate ion in the diluted solution.

b. the molarities of potassium ion and phosphate ion in the original solution.

21. The concentrated sulfuric acid available in the laboratory is 98.0% sulfuric acid by mass. Its density is 1.83 g/mL.

a. Calculate the molarity of concentrated sulfuric acid.

b. How would you prepare 1.50 L of 3.00 *M* sulfuric acid solution from concentrated sulfuric acid?

22. A bottle of phosphoric acid is labeled "85.0% H_3PO_4 by mass; density = 1.689 g/cm^3." Calculate the molarity, molality, and mole fraction of the phosphoric acid.

23. Complete the following table for potassium hydroxide solutions.

	Density (g/cm^3)	Molarity	Molality	Mass Percent of Solute
a.	1.05	1.13	______	______
b.	1.29	______	______	30.0
c.	1.43	______	14.2	______

24. Complete the following table for ammonium sulfate solutions.

	Density (g/cm^3)	Molarity	Molality	Mass Percent of Solute
a.	1.06	0.886	______	______
b.	1.15	______	______	26.0
c.	1.23	______	3.11	______

Solubilities

25. Choose the member of each set that you would expect to be more soluble in water. Explain your answer.

a. ethane, CH_3CH_3, or methyl alcohol, CH_3OH

b. potassium chloride or carbon tetrachloride

c. methyl fluoride, CH_3F, or hydrogen fluoride

d. benzene, C_6H_6, or hydrogen peroxide, H—O—O—H

26. Choose the member of each set (a–d) that you would expect to be more soluble in water. Explain your answer.

a. nitrogen trifluoride or sodium fluoride

b. ammonia or methane

c. carbon dioxide or silicon dioxide

d. methyl alcohol, CH_3OH, or methyl ether, $H_3C—O—CH_3$

27. Consider the process by which ammonium chloride dissolves in water:

$$NH_4Cl(s) \longrightarrow NH_4^+(aq) + Cl^-(aq)$$

a. Using data from tables in Chapter 8, calculate ΔH for this reaction.
b. Would you expect the solubility of NH_4Cl to increase or decrease if the temperature is lowered?

28. A certain gaseous solute dissolves in water, evolving 4.8 kJ/mol of heat. Its solubility at 25°C and 2.00 atm is 0.010 *M*. Would you expect the solubility to be greater or less than 0.010 *M* at

a. 0°C and 5 atm? **b.** 50°C and 1 atm?
c. 15°C and 2 atm? **d.** 25°C and 1 atm?

29. The concentration of hydrogen sulfide, H_2S, in hot springs is relatively high. This accounts for the rotten-egg smell around the "mud pots" in Yellowstone National Park. The solubility of hydrogen sulfide in water at 25°C is 0.0932 *M* at 1.00 atm. If the partial pressure of H_2S at Yellowstone is 0.12 atm, calculate the molarity of H_2S in the mud pots.

30. A carbonated beverage is made by saturating water with carbon dioxide at 0°C and a pressure of 3.0 atm. The bottle is then opened at room temperature (25°C), and comes to equilibrium with air in the room containing CO_2 ($P_{CO_2} = 3.4 \times 10^{-4}$ atm).

a. What is the concentration of carbon dioxide in the bottle before it is opened?
b. What is the concentration of carbon dioxide in the bottle after it has been opened and come to equilibrium with the air?

The Henry's law constant for the solubility of CO_2 in water is 0.0769 *M*/atm at 0°C and 0.0313 *M*/atm at 25°C.

Colligative Properties

31. Calculate the vapor pressure lowering in an aqueous sucrose solution at 22°C (vp pure water = 19.83 mm Hg) if the mole fraction of sucrose is

a. 0.0100 **b.** 0.100 **c.** 0.120

What is the vapor pressure of water over each of these solutions?

32. Repeat the calculations called for in Problem 31 at 90°C (vp = 525.8 mm Hg).

33. What is the vapor pressure at 20°C of a solution of 3.50 g naphthalene, $C_{10}H_8$, in 28.75 g of benzene, C_6H_6 (vp pure C_6H_6 = 74.7 mm Hg)? Assume the vapor pressure of napththalene is negligible.

34. The vapor pressure of pure chloroform at 70.0°C is 1.34 atm. How much iodine should be dissolved in one liter of chloroform, $CHCl_3$ ($d = 1.49$ g/cm^3), to lower the vapor pressure by 1.00×10^2 mm Hg?

35. Calculate the osmotic pressure at 25°C in solutions of urea, $CO(NH_2)_2$, containing the following masses of solute per liter of solution.

a. 10.0 g **b.** 50.0 g **c.** 100.0 g

36. Lysozyme is an enzyme that cleaves bacterial cell walls. A sample of lysozyme extracted from egg white has a molar mass of 13,930 g/mol. If 15.0 mg of this enzyme is dissolved in 175 mL of water at 25°C, what is the osmotic pressure of this solution?

37. Calculate the freezing point and normal boiling point of each of the following solutions.

a. 25.0 g of propylene glycol, $C_3H_8O_2$, in 250.0 mL of water ($d = 1.00$ g/cm^3)
b. 25.0 mL of methanol, CH_3OH ($d = 0.792$ g/cm^3), in 325 mL of water ($d = 1.00$ g/cm^3)

38. How many grams of the following nonelectrolyte solutes would have to be dissolved in 100.0 mL of water ($d = 1.00$ g/cm^3) to give a solution freezing at −1.50°C? What would be the normal boiling point of the solution?

a. glucose, $C_6H_{12}O_6$ **b.** citric acid, $C_6H_8O_7$

39. What is the freezing point and normal boiling point of a solution made by adding 20.0 mL of isopropyl alcohol, C_3H_7OH, to 80.0 mL of water? The density of isopropyl alcohol is 0.785 g/cm^3, while the density of water is 1.00 g/cm^3.

40. An automobile radiator is filled with an antifreeze solution prepared by mixing equal volumes of ethylene glycol, $C_2H_6O_2$ ($d = 1.12$ g/cm^3) and water ($d = 1.00$ g/cm^3). Estimate the freezing point of the mixture. Will this mixture protect automobile engines in Connecticut if the lowest temperature expected is −20°F?

41. Using Table 10.2, calculate the freezing point and the normal boiling point of solutions of 12.50 g of naphthalene, $C_{10}H_8$, in 100.0 g of

a. *p*-dichlorobenzene. **b.** benzene.
c. cyclohexane.

42. When 20.25 g of lactic acid, $C_3H_6O_3$, is dissolved in 250.0 mL of acetone ($d = 0.791$ g/mL), the resulting solution boils at 57.89°C. If the boiling point of pure acetone is 55.95°C, what is the boiling point constant for acetone?

43. A compound contains 42.9% C, 2.4% H, 16.6% N, and 38.1% O. The addition of 3.16 g of this compound to 75.0 mL of cyclohexane ($d = 0.779$ g/cm^3) gives a solution with a freezing point at 0.0°C. Using Table 10.2, determine the molecular formula of the compound.

44. Lauryl alcohol is obtained from the coconut and is an ingredient in many hair shampoos. Its empirical formula is $C_{12}H_{26}O$. A solution of 5.00 g of lauryl alcohol in 100.0 g of benzene boils at 80.78°C. Using Table 10.2, find the molecular formula of lauryl alcohol.

45. The freezing point of *p*-dichlorobenzene is 53.1°C; its k_f value is 7.10°C/*m*. A solution of 1.52 g of sulfanilamide (a sulfa drug) in 10.0 g of *p*-dichlorobenzene freezes at 46.7°C. What is the molar mass of sulfanilamide?

46. A student dissolved menthol ($C_{10}H_{19}OH$), a crystalline nonelectrolyte with the taste of peppermint, in 100.0 g of cyclohexane. The solution froze at −1.95°C. Using Table 10.2, calculate the percent by mass of menthol in the solution.

47. Aqueous solutions introduced into the blood stream by injection must have the same osmotic pressure as blood, i.e., they must be "isotonic" with blood. At 25°C, the average osmotic pressure of blood is 7.7 atm. What is the molarity of a glucose solution isotonic with blood?

48. Refer to Problem 47 and determine the concentration of an isotonic saline solution (NaCl in H_2O). Recall that NaCl is an electrolyte; assume complete conversion to Na^+ and Cl^- ions.

49. A biochemist isolated a new protein and determined its molar mass by osmotic pressure measurements. She used 0.270 g of the protein in 50.0 mL of solution and observed an osmotic pressure of 3.86 mm Hg for this solution at 25°C. What should she report as the molar mass of the new protein?

50. The molar mass of a type of hemoglobin was determined by osmotic pressure measurement. A student measured an osmotic pressure of 4.60 mm Hg for a solution at 20°C containing 3.27 g of hemoglobin in 0.200 L of solution. What is the molar mass of the hemoglobin?

51. Arrange 0.10 *m* solutions of the following solutes in order of decreasing freezing point.

a. $C_2H_6O_2$ **b.** $CrCl_3$
c. $Al_2(SO_4)_3$ **d.** Na_2CO_3

52. Estimate the freezing points of 0.10 *m* solutions of

a. K_2SO_4 **b.** $CsNO_3$ **c.** $Al(NO_3)_3$

53. The freezing point of 0.10 *m* $KHSO_3$ is −0.38°C. Which of the following equations best represents what happens when $KHSO_3$ dissolves in water?

a. $KHSO_3(s) \longrightarrow KHSO_3(aq)$
b. $KHSO_3(s) \longrightarrow K^+(aq) + HSO_3^-(aq)$
c. $KHSO_3(aq) \longrightarrow K^+(aq) + H^+(aq) + SO_3^{2-}(aq)$

54. The freezing point of 0.20 *m* HF is −0.38°C. Is HF primarily nonionized in this solution (HF molecules), or is it dissociated to H^+ and F^- ions?

Maple Syrup

55. What is the osmotic pressure of a 5.0% sucrose solution at 10°C? To what height could water be pushed up by that pressure? (*Hint:* Mercury is 13.6 times heavier than water.)

56. Give an explanation for the high sensitivity of sap flow to temperature.

57. Assume that 30 L of sap yields one kilogram of maple syrup (66% sucrose). What is the molality of the sucrose solution after half of the water content of the sap has been removed?

58. What is the freezing point of maple syrup (66% sucrose)?

Unclassified

59. How would you prepare a saturated solution of $CO_2(g)$ in water? A supersaturated solution?

60. You are given a clear water solution containing KNO_3. How would you determine experimentally whether the solution is unsaturated, saturated, or supersaturated?

61. Explain why

a. the freezing point of 0.10 *m* $CaCl_2$ is lower than that of 0.10 *m* $MgSO_4$.
b. a water solution of NaCl conducts an electric current but NaCl(*s*) does not.
c. the solubility of solids in water usually increases as temperature increases.
d. pressure must be applied to cause reverse osmosis to occur.

62. Explain why

a. 0.10 *M* $CaCl_2$ has a higher electrical conductivity than 0.10 *M* NaCl.
b. the solubility of gases in water decreases as temperature increases.
c. a water solution of HCl conducts an electric current but HCl(*l*) does not.
d. molality and molarity are nearly the same in dilute water solution.

63. In your own words, explain

a. why the concentrations of solutions used for intravenous feeding must be controlled carefully.
b. why fish in a lake (and fishermen) seek deep, shaded places during summer afternoons.
c. why champagne "fizzes" in a glass.
d. the difference between molality and molarity.

64. In your own words, explain

a. why seawater has a lower freezing point than fresh water.
b. why we believe that vapor pressure lowering is a colligative property.
c. why one often obtains a "grainy" product when making fudge (a supersaturated sugar solution).
d. what causes the "bends" in divers.

65. Criticize the following statements.

a. A saturated solution is always a concentrated solution.
b. The water solubility of a solid always decreases with a drop in temperature.
c. For aqueous solutions, molarity and molality are equal.
d. The freezing point depression of a 0.10 *m* $CaCl_2$ solution is twice that of a 0.10 *m* KCl solution.
e. A 0.10 *m* sucrose solution and a 0.10 *m* NaCl solution have the same osmotic pressure.

66. Explain, in your own words,

a. how to determine experimentally whether a pure substance is an electrolyte or nonelectrolyte.
b. why a cold glass of beer goes "flat" upon warming.
c. why the molality of a solute is ordinarily larger than its mole fraction.
d. why the boiling point is raised by the presence of a solute.

67. A water solution containing 275.9 g of sugar, $C_{12}H_{22}O_{11}$, per liter has a density of 1.104 g/cm^3 at 20°C. Calculate the

a. molarity **b.** molality

c. vapor pressure (vp H_2O at 20°C = 17.54 mm Hg)

d. freezing point

68. A water solution containing 28% by mass of iron(III) chloride has a density of 1.271 g/cm^3. Assuming ideal behavior, calculate the

a. molarity. **b.** molality.

c. osmotic pressure at 25°C. **d.** normal boiling pt.

69. So far as its colligative properties are concerned, seawater behaves like a 0.60 M solution of NaCl, with i = 1.9. Calculate the minimum pressure that must be applied to obtain pure water from seawater by reverse osmosis at 25°C.

Challenge Problems

70. A solution contains 158.2 g KOH per liter; its density is 1.13 g/cm^3. A lab technician wants to prepare 0.250-molal KOH, starting with 100.0 mL of this solution. How much water or solid KOH should be added to the 100.0-mL portion?

71. Show that the following relation is generally valid for any solution:

$$\text{molality} = \frac{\text{molarity}}{d - \dfrac{\mathcal{M}(\text{molarity})}{1000}}$$

where d is solution density (g/mL), and $\mathcal{M}$ is the molar mass of the solute. Using this equation, explain why molality approaches molarity in dilute solution when water is the solvent, but not with other solvents.

72. The water-soluble nonelectrolyte X has a molar mass of 410 g/mol. A 0.100-g mixture containing this substance and sugar ($\mathcal{M}$ = 342 g/mol) is added to 1.00 g of water to give a solution freezing at −0.500°C. Estimate the mass percent of X in the mixture.

73. A martini, weighing about 5.0 oz (142 g), contains 30% by mass of alcohol. About 15% of the alcohol in the martini passes directly into the bloodstream (7.0 L for an adult). Estimate the concentration of alcohol in the blood (g/cm^3) of a person who drinks two martinis before dinner. (A concentration of 0.0010 g/cm^3 or more is frequently considered indicative of intoxication in a "normal" adult.)

74. When water is added to a mixture of aluminum metal and sodium hydroxide, hydrogen gas is produced; this reaction is used in commercial drain cleaners:

$$2Al(s) + 6H_2O(l) + 2\,OH^-(aq) \longrightarrow 2Al(OH)_4{}^-(aq) + 3H_2(g)$$

A sufficient amount of water is added to 49.92 g of NaOH to make 0.600 L of solution; 41.28 g Al is added to this solution and hydrogen gas is formed.

a. Calculate the molarity of the initial NaOH solution.

b. How many moles of hydrogen were formed?

c. The hydrogen was collected over water at 25°C and 758.6 mm Hg. The vapor pressure of water at this temperature is 23.8 mm Hg. What volume of hydrogen was generated?

75. It is found experimentally that the volume of a gas that dissolves in a given amount of water is independent of the pressure of the gas; that is, if 5 cm^3 of a gas dissolves in 100 g of water at 1 atm pressure, 5 cm^3 will dissolve at a pressure of 2 atm, 5 atm, 10 atm, ... Show that this relationship follows logically from Henry's law and the ideal gas law.

11
Rate of Reaction

Not every collision,
not every punctilious trajectory
by which billiard-ball complexes
arrive at their calculable
meeting places leads
to reaction

....

Men (and women) are not
as different from molecules
as they think.

ROALD HOFFMANN
Men and Molecules

CHARLES D. WINTERS

$NH_4Cl(s) \longrightarrow NH_3(g) + HCl(g) \longrightarrow NH_4Cl(g)$

(See p. 119)

CHAPTER OUTLINE

For a chemical reaction to be feasible, it must occur at a reasonable rate. Consequently, it is important to be able to control the rate of reaction. Most often, this means making it occur more rapidly. When you carry out a reaction in the general chemistry laboratory, you want it to take place quickly. A research chemist trying to synthesize a new drug has the same objective. Sometimes, though, it is desirable to reduce the rate of reaction. The aging process, a complex series of biological oxidations, believed to involve "free radicals" such as

$$\cdot\ddot{\underset{\cdot\cdot}{O}}\text{—H} \qquad \text{and} \qquad \cdot\ddot{\underset{\cdot\cdot}{O}}\text{—}\ddot{\underset{\cdot\cdot}{O}}:$$

Amen! (WLM)

is one we would all like to slow down.

This chapter sets forth the principles of **chemical kinetics**, the study of reaction rates. The main emphasis is upon those factors that influence rate. These include

— the concentrations of reactants (Sections 11.2, 11.3)
— the nature of the reaction (Section 11.4)
— the temperature (Section 11.5)
— the reaction mechanism (Section 11.6)
— the presence of a catalyst (Section 11.7)

11.1 MEANING OF REACTION RATE

To discuss reaction rate meaningfully, it must be defined precisely. *The rate of reaction is a positive quantity that expresses how the concentration of a reactant or product changes with time.* To illustrate what this means, consider the reaction

$$N_2O_5(g) \longrightarrow 2NO_2(g) + \tfrac{1}{2}O_2(g)$$

As you can see from Figure 11.1, the concentration of N_2O_5 decreases with time; the concentrations of NO_2 and O_2 increase. Because these species have different coefficients in the balanced equation, their concentrations do not change at the same rate. When *one* mole of N_2O_5 decomposes, *two* moles of NO_2 and *one-half* mole of O_2 are formed. This means that

$$-\Delta[N_2O_5] = \frac{\Delta[NO_2]}{2} = \frac{\Delta[O_2]}{\frac{1}{2}}$$

where $\Delta[\ \ ]$ refers to the change in concentration in moles per liter. The minus sign in front of the N_2O_5 term takes account of the fact that $[N_2O_5]$ decreases as the reaction takes place; the numbers in the denominator of the terms on the right ($2, \frac{1}{2}$) are the coefficients of these species in the balanced equation. The rate of reaction can now be defined by dividing by the change in time, Δt:

$$\text{rate} = \frac{-\Delta[N_2O_5]}{\Delta t} = \frac{\Delta[NO_2]}{2\Delta t} = \frac{\Delta[O_2]}{\frac{1}{2}\Delta t}$$

More generally, for the reaction

$$aA + bB \longrightarrow cC + dD$$

where A, B, C, and D represent substances in the gas phase (g) or in aqueous solution (aq) and a, b, c, d are their coefficients in the balanced equation:

This is the defining equation for reaction rate

$$\text{rate} = \frac{-\Delta[A]}{a\Delta t} = \frac{-\Delta[B]}{b\Delta t} = \frac{\Delta[C]}{c\Delta t} = \frac{\Delta[D]}{d\Delta t}$$

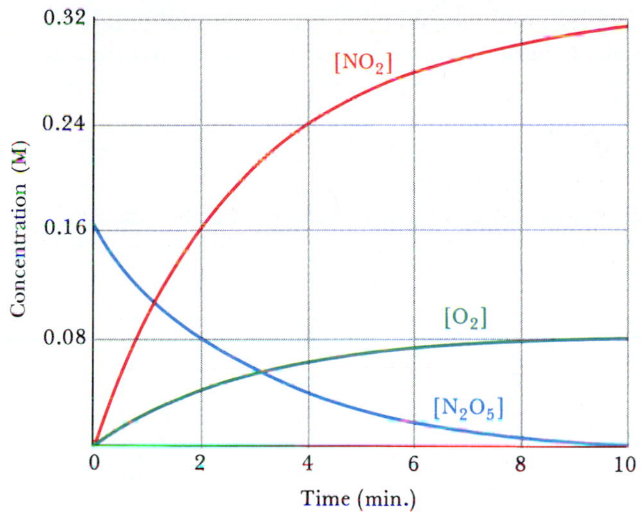

Figure 11.1
For the reaction: $N_2O_5(g) \longrightarrow 2NO_2(g) + \frac{1}{2}O_2(g)$, the concentrations of NO_2 and O_2 increase with time, while that of N_2O_5 decreases. The reaction rate is defined as:

$$\frac{-\Delta[N_2O_5]}{\Delta t} = \frac{\Delta[NO_2]}{2\Delta t} = \frac{2\Delta[O_2]}{\Delta t}$$

Measurement of Rate

For the reaction

$$N_2O_5(g) \longrightarrow 2NO_2(g) + \tfrac{1}{2}O_2(g)$$

the rate could be determined by measuring

— the absorption of visible light by the NO_2 formed; this species has a reddish-brown color, while N_2O_5 and O_2 are colorless

— the change in pressure that results from the increase in the number of moles of gas (1 mol reactant → $2\frac{1}{2}$ mol product)

The graphs shown in Figure 11.1 were plotted from data obtained by measurements of this type.

To find the rate of decomposition of N_2O_5, it is convenient to use Figure 11.2, which is a magnified version of a portion of Figure 11.1. If a tangent is

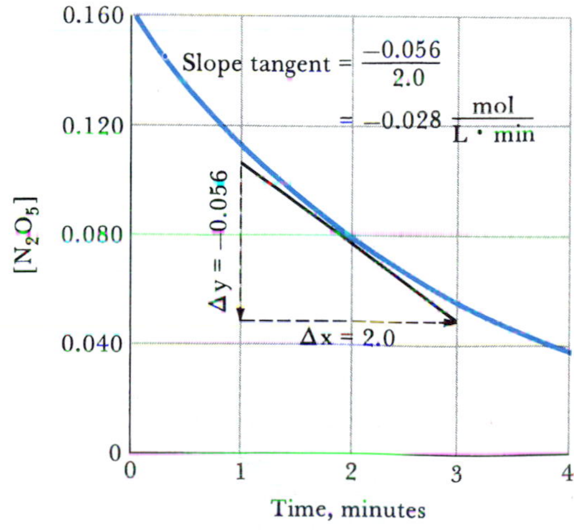

Figure 11.2
To determine the rate of a reaction, plot concentration vs. time and take the tangent to the curve at the desired point. For the reaction: $N_2O_5(g)$ $2NO_2(g) + \frac{1}{2}O_2(g)$, it appears that the reaction rate at $[N_2O_5] = 0.080\ M$ is 0.028 mol/L · min.

drawn to the curve of concentration vs. time, its slope at that point must equal $\Delta[N_2O_5]/\Delta t$. But since the reaction rate is $-\Delta[N_2O_5]/\Delta t$, it follows that

$$\text{instantaneous rate} = -\text{slope of tangent}$$

From Figure 11.2 it appears that the slope of the tangent at $t = 2$ min is -0.028 mol/L · min. Hence

$$\text{rate at 2 min} = -(-0.028 \text{ mol/L} \cdot \text{min}) = 0.028 \text{ mol/L} \cdot \text{min}$$

11.2 REACTION RATE AND CONCENTRATION

Ordinarily, reaction rate is directly related to reactant concentration. The higher the concentration of starting materials, the more rapidly a reaction takes place. Pure hydrogen peroxide, in which the concentration of H_2O_2 molecules is about 40 mol/L, is an extremely dangerous substance to work with. In the presence of trace impurities, it decomposes explosively

$$H_2O_2(l) \longrightarrow H_2O(g) + \tfrac{1}{2}\,O_2(g)$$

at a rate too rapid to measure. The hydrogen peroxide you buy in the drugstore is a dilute aqueous solution in which $[H_2O_2] \approx 1\ M$. At this relatively low concentration, decomposition is so slow that the solution is stable for several months.

The dependence of reaction rate upon concentration is readily explained. Ordinarily, *reactions occur as the result of collisions between reactant molecules.* The higher the concentration of molecules, the greater the number of collisions in unit time and hence the faster the reaction. As reactants are consumed, their concentrations drop, collisions occur less frequently, and reaction rate decreases. This explains the common observation that reaction rate drops off with time, eventually going to zero when all the reactants are consumed.

Rate Expression and Rate Constant

The dependence of reaction rate upon concentration is readily determined for the decomposition of N_2O_5. Figure 11.3 shows what happens when reaction rate is plotted versus $[N_2O_5]$. As you would expect, rate increases as concentration increases, going from zero when $[N_2O_5] = 0$ to about 0.06 mol/L · min when $[N_2O_5] = k0.16\ M$. Moreover, as you can see from the figure, the plot of

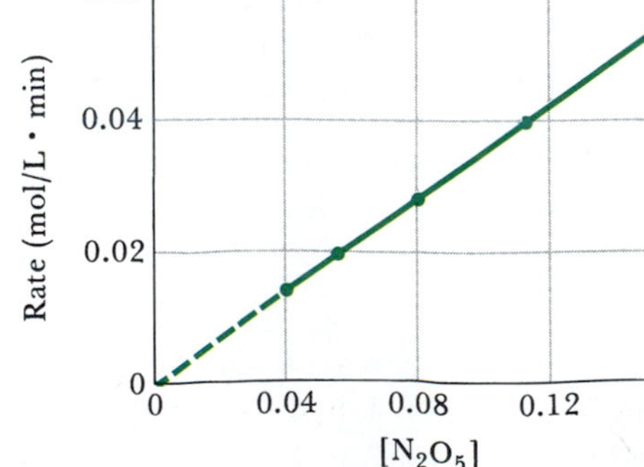

Figure 11.3
For the decomposition of N_2O_5, a plot of rate vs. conc. N_2O_5 is a straight line. The line, if extrapolated, passes through the origin. This means that rate is directly proportional to concentration; that is, rate $= k[N_2O_5]$.

rate vs. concentration is a straight line through the origin, which means that rate must be directly proportional to concentration.

$$\text{rate} = k[N_2O_5]$$

This equation is referred to as the **rate expression** for the decomposition of N_2O_5. It tells how the rate of the reaction

Rate depends upon concentration, but rate constant does not

$$N_2O_5(g) \longrightarrow 2NO_2(g) + \tfrac{1}{2}O_2(g)$$

depends upon the concentration of reactant. The proportionality constant k is called a **rate constant**. It is independent of the other quantities in the equation.

As we will see shortly, the rate expression can take various forms, depending upon the nature of the reaction. It can be quite simple, as in the N_2O_5 decomposition, or exceedingly complex. The value of the rate constant, k, varies with the nature of the reaction and with the temperature (Section 11.5).

Order of Reaction Involving a Single Reactant

Rate expressions have been determined by experiment for a large number of reactions. For the process

$$A \longrightarrow \text{products}$$

the rate expression has the general form

$$\text{rate} = k[A]^m$$

The power to which the concentration of reactant A is raised in the rate expression is called the **order of the reaction**, m. If m is 0, the reaction is said to be "zero-order." If $m = 1$, the reaction is "first-order"; if $m = 2$, it is "second-order," and so on.

The order of a reaction must be determined experimentally; *it cannot be deduced from the coefficients of the balanced equation.* This must be true because there is only one reaction order, but there are many different ways in which the equation for the reaction can be balanced. For example, although we wrote

$$N_2O_5(g) \longrightarrow 2NO_2(g) + \tfrac{1}{2}O_2(g)$$

to describe the decomposition of N_2O_5, it could have been written

$$2N_2O_5(g) \longrightarrow 4NO_2(g) + O_2(g)$$

The reaction is still first-order no matter how the equation is written.

One way to find the order of a reaction is to measure the **initial rate** (i.e., the rate at $t = 0$) as a function of the concentration of reactant. Suppose, for example, that the rate of decomposition of a species A is measured at two different concentrations, 1 and 2. From the rate expression,

$$\text{rate}_2 = k[A]_2{}^m \qquad \text{rate}_1 = k[A]_1{}^m$$

Dividing the second rate by the first,

$$\frac{\text{rate}_2}{\text{rate}_1} = \frac{[A]_2{}^m}{[A]_1{}^m} = \left(\frac{[A]_2}{[A]_1}\right)^m$$

Since all the quantities in this equation are known except m, the reaction order can be calculated (Example 11.1).

Example 11.1 The initial rate of decomposition of acetaldehyde, CH_3CHO, at 600°C

$$CH_3CHO(g) \longrightarrow CH_4(g) + CO(g)$$

was measured at a series of concentrations with the following results:

$[CH_3CHO]$	0.10 *M*	0.20 *M*	0.30 *M*	0.40 *M*
rate (mol/L · s)	0.085	0.34	0.76	1.4

Using these data, determine the reaction order; that is, determine the value of *m* in the equation: $\text{rate} = k[CH_3CHO]^m$.

Strategy Choose the first two concentrations, 0.10 *M* and 0.20 *M*. Calculate the ratio of the rates, the ratio of the concentrations, and finally the order of reaction, using the general relation derived above.

Solution

$$\frac{\text{rate}_2}{\text{rate}_1} = \frac{0.34 \text{ mol/L} \cdot \text{s}}{0.085 \text{ mol/L} \cdot \text{s}} = 4.0$$

$$\frac{[CH_3CHO]_2}{[CH_3CHO]_1} = \frac{0.20\ M}{0.10\ M} = 2.0$$

Hence, the general relation becomes $4.0 = (2.0)^m$. Clearly, $m = 2$; the reaction is second-order.

Once the order of the reaction is known, the rate constant is readily calculated. Consider, for example, the decomposition of acetaldehyde, where we have shown that the rate expression is

There are two variables in this equation, rate and concentration, and two constants, *k* and reaction order

$$\text{rate} = k[CH_3CHO]^2$$

The data in Example 11.1 show that the rate at 600°C is 0.085 mol/L · s when the concentration is 0.10 mol/L. It follows that

$$k = \frac{\text{rate}}{[CH_3CHO]^2} = \frac{0.085 \text{ mol/L} \cdot \text{s}}{(0.10 \text{ mol/L})^2} = 8.5 \text{ L/mol} \cdot \text{s}$$

The same value of *k* would be obtained, within experimental error, using any other data pair.

Having established the value of *k* and the reaction order, the rate is readily calculated at any concentration. Again, using the decomposition of acetaldehyde as an example, we have established that

$$\text{rate} = 8.5 \frac{\text{L}}{\text{mol} \cdot \text{s}} [CH_3CHO]^2$$

If the concentration of acetaldehyde were 0.50 *M*,

$$\text{rate} = 8.5 \frac{\text{L}}{\text{mol} \cdot \text{s}} (0.50 \text{ mol/L})^2 = 2.1 \text{ mol/L} \cdot \text{s}$$

The rate of reaction of zinc with aqueous sulfuric acid depends on the concentration of H_2SO_4. The beaker with the dilute solution (left) reacts slowly while the beaker with the concentrated solution (right) reacts rapidly. (Charles Steele)

Order of Reaction with More Than One Reactant

Many reactions (indeed, most reactions) involve more than one reactant. For a reaction between two species A and B,

$$aA + bB \longrightarrow \text{products}$$

the general form of the rate expression is

$$\text{rate} = k[A]^m \times [B]^n$$

Here m is referred to as "the order of the reaction with respect to A." Similarly, n is "the order of the reaction with respect to B." The **overall order** of the reaction is the sum of the exponents, $m + n$. If $m = 1$, $n = 2$, then the reaction is first-order in A, second-order in B, and third-order overall.

When more than one reactant is involved, the order can be determined by holding the initial concentration of one reactant constant while varying that of the other reactant. From rates measured under these conditions, it is possible to deduce the order of the reaction with respect to the reactant whose initial concentration is varied.

This way the concentration terms cancel for that reactant

Example 11.2 Consider the reaction at 55°C

$$(CH_3)_3CBr(aq) + OH^-(aq) \longrightarrow (CH_3)_3COH(aq) + Br^-(aq)$$

A series of experiments is carried out with the following results:

	Expt. 1	Expt. 2	Expt. 3	Expt. 4	Expt. 5
$[(CH_3)_3CBr]$	0.50	1.0	1.5	1.0	1.0
$[OH^-]$	0.050	0.050	0.050	0.10	0.20
rate (mol/L · s)	0.0050	0.010	0.015	0.010	0.010

Find the order of the reaction with respect to both $(CH_3)_3CBr$ and OH^-.

Strategy To find the order of the reaction with respect to $(CH_3)_3CBr$, choose two experiments, perhaps 1 and 3, where $[OH^-]$ is constant. Under these conditions, the relation $\text{rate}_3/\text{rate}_1 = ([(CH_3)_3CBr]_3/[(CH_3)_3CBr]_1)^m$ applies because $[OH^-]^n$ has the same value in the two experiments regardless of the value of n. To find m, proceed as in Example 11.1. A similar approach can be used to find n; compare experiments 2 and 5, where $[(CH_3)_3CBr]$ is constant.

Solution

$$(1)\ \frac{\text{rate}_3}{\text{rate}_1} = \left(\frac{[(CH_3)_3CBr]_3}{[(CH_3)_3CBr]_1}\right)^m \qquad \frac{0.015}{0.0050} = \left(\frac{1.5}{0.50}\right)^m \qquad 3 = 3^m$$

Clearly, $m = 1$.

$$(2)\ \frac{\text{rate}_5}{\text{rate}_2} = \left(\frac{[OH^-]_5}{[OH^-]_2}\right)^n \qquad \frac{0.010}{0.010} = \left(\frac{0.20}{0.050}\right)^n \qquad 1 = 4^n$$

Anything raised to the power zero equals one

In this case, $n = 0$; the rate is independent of the concentration of OH^-.

11.3 REACTANT CONCENTRATION AND TIME

The rate expression

$$\text{rate} = k[N_2O_5]$$

shows how the rate of decomposition of N_2O_5 changes with concentration. From a practical standpoint, however, it is more important to know the relation between concentration and *time* rather than between concentration and rate. Suppose, for example, you are studying the decomposition of dinitrogen pentaoxide. Most likely, you would want to know how much N_2O_5 is left after 5 min, 1 h, or several days. An equation relating rate to concentration does not answer that purpose.

Using calculus, it is possible to develop **integrated rate equations** relating reactant concentration to time. We now examine several such equations, starting with first-order reactions.

First-Order Reactions

Radioactivity (Chap. 18) is a first-order reaction

For the decomposition of N_2O_5 and other first-order reactions of the type

$$A \longrightarrow \text{products} \qquad \text{rate} = k[A]$$

it can be shown by using calculus that the relationship between concentration and time is*

$$\ln \frac{[A]_o}{[A]} = kt$$

* Throughout Section 11.3, balanced chemical equations are written in such a way that the coefficient of the reactant is 1. In general, if the coefficient of the reactant is a, where a may be 2, or 3, or . . . , then k in each integrated rate equation must be replaced by the product ak. (See Problem 76.)

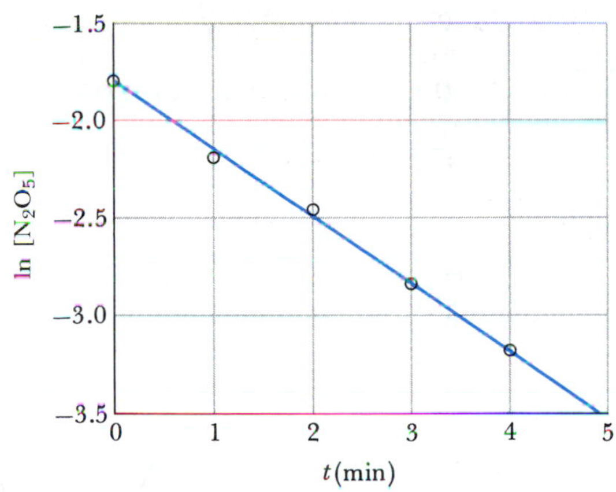

Figure 11.4
The rate constant for a first-order reaction can be determined from the slope of a plot of ln[A] vs. time. From the graph of data for the reaction: $N_2O_5(g) \longrightarrow 2NO_2(g) + \frac{1}{2}O_2(g)$ at 67°C, it appears that the first-order rate constant is about 0.35/min.

where $[A]_o$ is the original concentration of reactant, [A] is its concentration at time t, k is the first-order rate constant, and the abbreviation "ln" refers to the natural logarithm.

Since $\ln a/b = \ln a - \ln b$, the first-order equation can be written in the form

$$\ln\,[A]_o - \ln\,[A] = kt$$

Solving for ln [A],

$$\ln\,[A] = \ln\,[A]_o - kt$$

Comparing this equation to the general equation of a straight line,

$$y = b + mx \qquad (b = y\text{-intercept},\ m = \text{slope})$$

it is clear that a plot of ln [A] vs. t should be a straight line with a y-intercept of ln $[A]_o$ and a slope of $-k$. This is indeed the case, as you can see from Figure 11.4 where we have plotted ln $[N_2O_5]$ vs. t for the decomposition of N_2O_5. Drawing the best straight line through the points and taking the slope based upon the two points on the y- and x-axes,

$$\text{slope} = \frac{-3.5 - (-1.8)}{4.9 - 0} = -0.35$$

It follows that the rate constant is 0.35/min; the integrated first-order equation for the decomposition of N_2O_5 is

$$\ln \frac{[N_2O_5]_o}{[N_2O_5]} = \frac{0.35}{\text{min}}t \qquad (\text{at } 67°\text{C})$$

Example 11.3 For the decomposition of N_2O_5 at 67°C, calculate

(a) the concentration after 4.0 min, starting at 0.160 M.
(b) the time required for the concentration to drop from 0.160 to 0.100 M.
(c) the time required for half a sample of N_2O_5 to decompose.

Strategy In each case, the equation ln $[N_2O_5]_o/[N_2O_5]$ = (0.35/min)t is used. In (a) and (b), two of the three variables, $[N_2O_5]_o$, $[N_2O_5]$, t, are known; the other is readily calculated. In (c), you should be able to find the ratio $[N_2O_5]_o/[N_2O_5]$; knowing that ratio, the time can be calculated.

Solution

(a) Substituting in the integrated first-order equation,

$$\ln \frac{0.160\ M}{[N_2O_5]} = \frac{0.35}{\text{min}}(4.0\ \text{min}) = 1.4$$

Taking inverse logarithms,

If $\ln x = y$, then $x = e^y$

$$\frac{0.160\ M}{[N_2O_5]} = e^{1.4} = 4.0 \qquad [N_2O_5] = \frac{0.160\ M}{4.0} = \boxed{0.040\ M}$$

(b) Solving the concentration-time relation for t,

$$t = \frac{1}{k}\ln\frac{[A]_o}{[A]} = \frac{1}{0.35/\text{min}}\ln\frac{0.160\ M}{0.100\ M} = \frac{\ln 1.60}{0.35/\text{min}}$$

$$= 0.47\ \text{min}/0.35 = \boxed{1.3\ \text{min}}$$

(c) When half of the sample has decomposed,

$$[N_2O_5] = [N_2O_5]_o/2 \qquad [N_2O_5]_o = 2[N_2O_5] \qquad [N_2O_5]_o/[N_2O_5] = 2$$

Using the equation in (b),

$$t = \frac{1}{k}\ln 2 = \frac{0.693}{k}$$

With $k = 0.35$/min, we have

$$t = \frac{0.693}{0.35} = \boxed{2.0\ \text{min}}$$

The analysis of Example 11.3c reveals an important feature of a first-order reaction: *The time required for one half of a reactant to decompose via a first-order reaction has a fixed value, independent of concentration.* This quantity, called the **half-life**, is given by the expression

This expression holds only for a first-order reaction

$$t_{1/2} = \frac{\ln 2}{k} = \frac{0.693}{k} \qquad \text{first-order reaction}$$

where k is the rate constant. For the decomposition of N_2O_5, where $k =$ 0.35/min, $t_{1/2} = 2.0$ min. Thus, every 2 minutes, one half of a sample of N_2O_5 decomposes (Table 11.1).

Table 11.1 Decomposition of N_2O_5 at 67°C ($t_{1/2}$ = 2.0 min)

t (min)	0.0	2.0	4.0	6.0	8.0
$[N_2O_5]$	0.160	0.080	0.040	0.020	0.010
fraction of N_2O_5 decomposed	0	$\frac{1}{2}$	$\frac{3}{4}$	$\frac{7}{8}$	$\frac{15}{16}$
fraction of N_2O_5 left	1	$\frac{1}{2}$	$\frac{1}{4}$	$\frac{1}{8}$	$\frac{1}{16}$
number of half-lives	0	1	2	3	4

Zero- and Second-Order Reactions

For a **zero-order reaction**,

$$A \longrightarrow \text{products} \qquad \text{rate} = k[A]^0 = k$$

since any quantity raised to the zero power, including [A], is equal to one. In other words, the rate of a zero-order reaction is constant, independent of concentration. As you might expect from our earlier discussion, zero-order reactions are relatively rare. Most of them take place at solid surfaces, where the rate is independent of concentration in the gas phase. A typical example is the thermal decomposition of hydrogen iodide on gold:

They are also the simplest to treat mathematically

$$HI(g) \xrightarrow{Au} \tfrac{1}{2}H_2(g) + \tfrac{1}{2}I_2(g)$$

When the gold surface is completely covered with HI molecules, increasing the concentration of HI(g) has no effect on reaction rate.

It can readily be shown (Problem 75) that the concentration–time relation for a zero-order reaction is

$$[A] = [A]_o - kt$$

Comparing this equation to that for a straight line

$$y = b + mx \qquad (b = y\text{-intercept}, m = \text{slope})$$

it should be clear that a plot of [A] vs. t should be a straight line with a slope of $-k$. Putting it another way, if a plot of concentration vs. time is linear, the reaction must be zero-order; the rate constant k is numerically equal to the slope of that line but has the opposite sign.

For a **second-order reaction** involving a single reactant, such as acetaldehyde (recall Example 11.1):

$$A \longrightarrow \text{products} \qquad \text{rate} = k[A]^2$$

it is again necessary to resort to calculus* to obtain the concentration–time relationship

$$\frac{1}{[A]} - \frac{1}{[A]_o} = kt$$

where the symbols [A], $[A]_o$, t, and k have their usual meanings. For a second-order reaction, a plot of $1/[A]$ vs. t should be linear.

The characteristics of zero-, first-, and second-order reactions are summarized in Table 11.2. To determine reaction order, the properties in either of the last two columns at the right of the table can be used (Example 11.4).

* If you are taking a course in calculus, you may be surprised to learn how useful it can be in the real world (e.g., chemistry). The general rate expressions for zero-, first-, and second-order reactions are

$$-d[A]/dt = k \qquad -d[A]/dt = k[A] \qquad -d[A]/dt = k[A]^2$$

Integrating these equations from 0 to t and from $[A]_o$ to [A], you should be able to derive the equations for zero-, first-, and second-order reactions.

Notice that, for zero- and second-order reactions, $t_{1/2}$ depends on initial concentration

Table 11.2 Characteristics of Zero-, First-, and Second-Order Reactions of the Form $A(g) \rightarrow$ products; [A], $[A]_o$ = conc. A at t and $t = 0$, respectively

Order	Rate Expression	Conc.–Time Relation	Half-Life	Linear Plot
0	rate = k	$[A]_o - [A] = kt$	$[A]_o/2k$	[A] vs. t
1	rate = $k[A]$	$\ln \frac{[A]_o}{[A]} = kt$	$0.693/k$	ln [A] vs. t
2	rate = $k[A]^2$	$\frac{1}{[A]} - \frac{1}{[A]_o} = kt$	$1/k[A]_o$	$\frac{1}{[A]}$ vs. t

Example 11.4 The following data were obtained for the gas-phase decomposition of hydrogen iodide:

time (h)	0	2	4	6
[HI]	1.00	0.50	0.33	0.25

Is this reaction zero-, first-, or second-order in HI?

Strategy Note from the last column of Table 11.2 that

— if the reaction is zero-order, a plot of [A] vs. t should be linear
— if the reaction is first-order, a plot of ln [A] vs. t should be linear
— if the reaction is second-order, a plot of 1/[A] vs. t should be linear

Using these data, prepare these plots and determine which one is a straight line.

Solution It is useful to prepare a table listing [HI], ln [HI], and 1/[HI] as a function of time.

t(h)	[HI]	ln [HI]	1/[HI]
0	1.00	0	1.0
2	0.50	−0.69	2.0
4	0.33	−1.10	3.0
6	0.25	−1.39	4.0

If it is not obvious from this table that the only linear plot is that of 1/[HI] vs. time, that point should be clear from Figure 11.5. This is a second-order reaction.

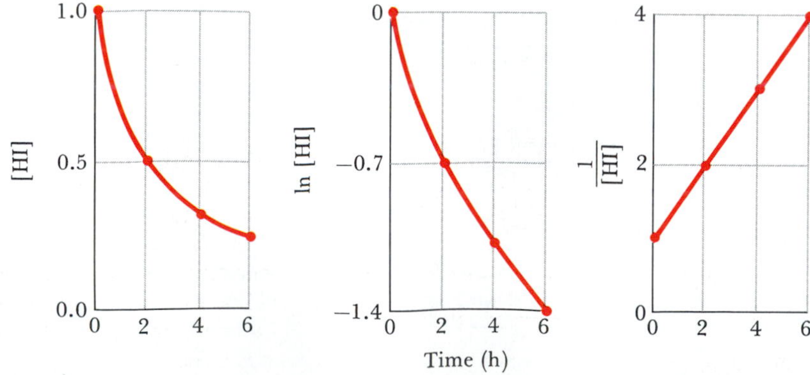

Figure 11.5
One way to determine reaction order is to search for a linear relation between some function of concentration and time (Table 11.2). Since a plot of 1/[HI] vs. t is linear, the decomposition of HI must be second-order.

11.4 ACTIVATION ENERGY

Consider the reaction

$$CO(g) + NO_2(g) \longrightarrow CO_2(g) + NO(g)$$

Above 600 K, this reaction takes place as a direct result of collisions between CO and NO_2 molecules. When the concentration of CO doubles (Figure 11.6) the number of these collisions in a given time increases by a factor of two; doubling the concentration of NO_2 has the same effect. Assuming that reaction rate is directly proportional to the collision rate, the following relation should hold:

$$\text{reaction rate} = k[CO] \times [NO_2]$$

Experimentally, this prediction is confirmed; the reaction is first-order in both carbon monoxide and nitrogen dioxide.

There is a restriction on this simple model for the CO–NO_2 reaction. According to the kinetic theory of gases, for a reaction mixture at 700 K and concentrations of 0.10 *M*, every CO molecule should collide with about 10^9 NO_2 molecules in one second. If every collision were effective, the reaction should be over in a fraction of a second. In reality, this does not happen; under these conditions, the half-life is about 10 s. This implies that not every CO–NO_2 collision leads to reaction.

There would be an explosion

There are a couple of reasons why collision between reactant molecules does not always lead to reaction. For one thing, the molecules have to be properly oriented with respect to one another when they collide. Suppose, for example, that the carbon atom of a CO molecule strikes the nitrogen atom of an NO_2 molecule (Fig. 11.7). This is very unlikely to result in the transfer of an oxygen atom from NO_2 to CO, which is required for reaction to occur.

A more important factor that reduces the number of effective collisions has to do with the kinetic energy of reactant molecules. These molecules are held together by strong chemical bonds. Only if the colliding molecules are moving very rapidly will the kinetic energy be large enough to supply the energy required to break these bonds. Molecules with small kinetic energies bounce off one another without reacting. As a result, only a small fraction of collisions are effective.

The high-energy molecules get the job done

For every reaction, there is a certain minimum energy that molecules must possess for collision to be effective. This is referred to as the **activation energy**.

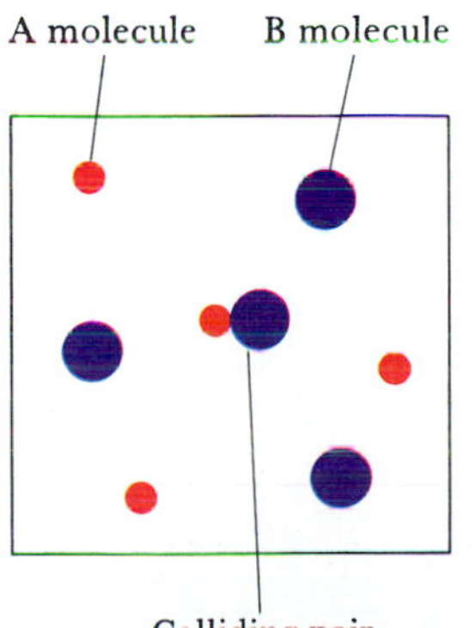

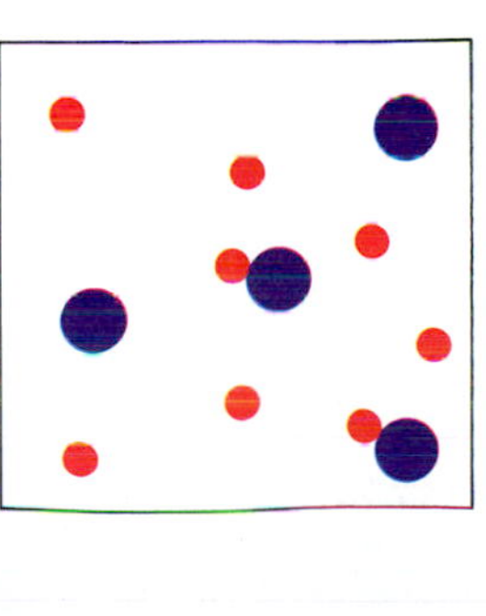

Figure 11.6
Consider the reaction: $A + B \longrightarrow$ products, where reaction occurs as the result of collision. If we double the number of A molecules (square at right), there will be twice as many collisions per unit time, so the rate will be doubled. In general, the rate will be directly proportional to the concentration of A. The same reasoning applies to B, so the rate law is rate = k(conc. A) × (conc. B).

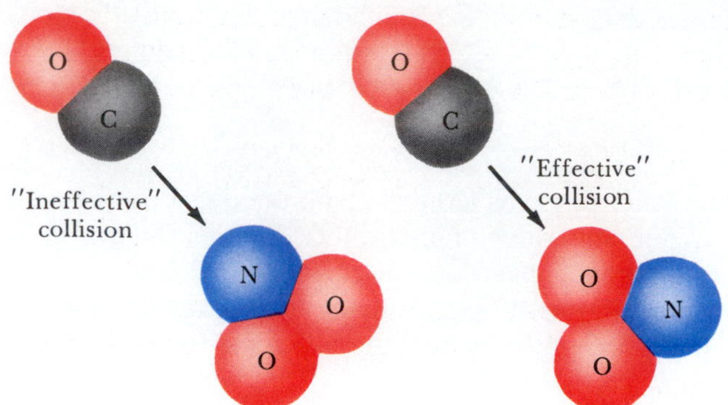

Figure 11.7
For a collision to result in reaction, the molecules must be properly oriented. For the reaction: $CO(g) + NO_2(g) \longrightarrow CO_2(g) + NO(g)$, the carbon atom of the CO molecule must strike an oxygen atom of the NO_2 molecule, forming CO_2 as one product, NO as the other.

It has the symbol E_a and is expressed in kilojoules. For the reaction between one mole of CO and one mole of NO_2, E_a is 134 kJ. The colliding molecules (CO and NO_2) must have a total kinetic energy of at least 134 kJ/mol if they are to react.

The activation energy for a reaction

— *is a positive quantity* ($E_a > 0$)
— *depends upon the nature of the reaction.* Other factors being equal, we expect "fast" reactions to have a small activation energy. A reaction with a large activation energy takes place slowly under ordinary conditions. The larger the value of E_a, the smaller will be the fraction of molecules having enough kinetic energy to react when they collide.
— *is independent of temperature or concentration*

Activation Energy Diagrams

Figure 11.8 is an energy diagram for the CO–NO_2 reaction. Reactants, CO and NO_2, are shown at the left. Products, CO_2 and NO, are at the right. They have an energy 226 kJ less than that of the reactants; ΔH for the reaction is -226 kJ. In the center of the figure is an intermediate called an *activated complex.* This is an unstable, high-energy species that must be formed before the reaction can occur. It has an energy 134 kJ greater than that of the reactants and 360 kJ greater than that of the products.

In the activated complex, electrons are often in excited states

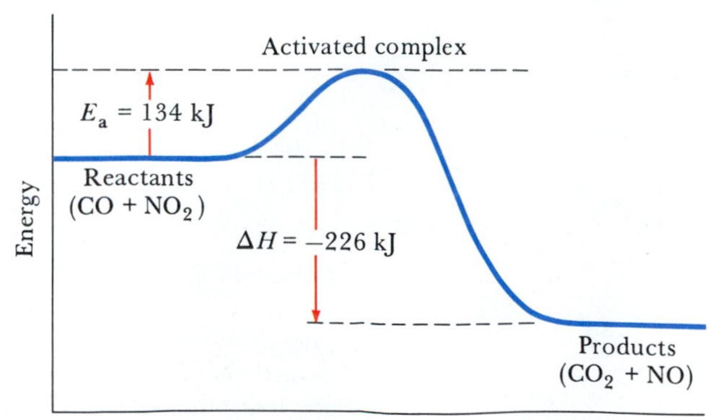

Figure 11.8
During the reaction initiation, 134 kJ—the activation energy E_a—must be furnished to the reactants for every mole of CO that reacts. This energy activates each CO–NO_2 complex to the point where reaction can proceed.

The exact nature of the activated complex is difficult to determine. For this reaction, the activated complex might be a "pseudomolecule" made up of CO and NO_2 molecules in close contact. The path of the reaction might be more or less as follows:

$$\underset{\text{reactants}}{O{\equiv}C + O{-}N{=}O} \longrightarrow \underset{\text{activated complex}}{O{\equiv}C\cdots O\cdots N{=}O} \longrightarrow \underset{\text{products}}{O{=}C{=}O + N{=}O}$$

The dotted lines stand for "partial bonds" in the activated complex. The N—O bond in the NO_2 molecule has been partially broken. A new bond between carbon and oxygen has started to form.

11.5 REACTION RATE AND TEMPERATURE

The rates of most reactions increase as the temperature rises. A person in a hurry to prepare dinner applies this principle in using a pressure cooker to raise the temperature for cooking potatoes, apples, or a pot roast (not all at the same time, we trust). By storing the leftovers in a refrigerator, the chemical reactions responsible for food spoilage are slowed down. As a general and very approximate rule, it is often stated that an increase in temperature of 10°C doubles the reaction rate. If this rule holds, foods should cook twice as fast in a pressure cooker at 110°C as in an open saucepan and deteriorate four times as rapidly at room temperature (25°C) as they do in a refrigerator at 5°C.

Hibernating animals lower their body temperatures, slowing down life processes

The effect of temperature on reaction rate can be explained in terms of the kinetic theory of gases. Recall from Chapter 5 that raising the temperature greatly increases the fraction of molecules having very high kinetic energies. These are the molecules that are most likely to react when they collide. The higher the temperature, the larger the fraction of molecules that can provide the activation energy required for reaction. This effect is apparent from Figure 11.9, where the distribution of kinetic energies is shown at two different temperatures. Notice that the fraction of molecules having a kinetic energy equal to or greater than the activation energy E_a (shaded area) is considerably larger at the higher temperature. Hence the fraction of effective collisions increases; this is the major factor causing reaction rate to increase with temperature.

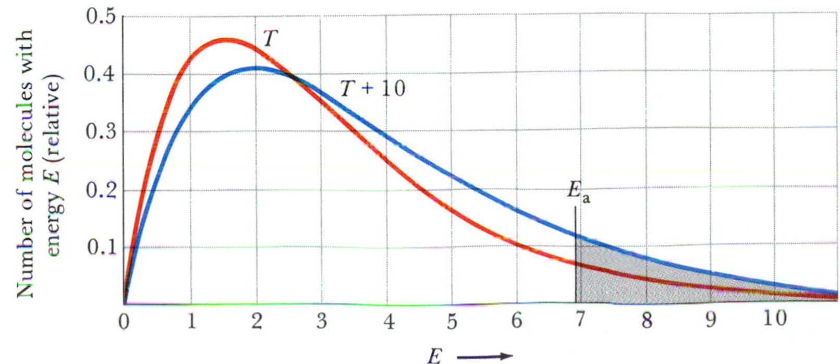

Figure 11.9
When T is increased to $T + 10$, the fraction of molecules with very high energies increases sharply. Hence, many more molecules possess the activation energy, E_a, and reaction occurs more rapidly. If E_a is of the order of 50 kJ/mol, an increase in temperature of 10°C approximately doubles the number of molecules having energy E_a or greater, and thus doubles the reaction rate.

The Arrhenius Equation

The argument just developed can be made quantitative. According to the kinetic theory of gases, the fraction, f, of molecules having an energy equal to or greater than E_a is

$$f = e^{-E_a/RT}$$

where e is the base of natural logarithms, R is the gas constant, E_a is the activation energy, and T is the absolute temperature in K. Assuming that the rate constant, k, is directly proportional to f (which is approximately true),

$$k = Af = Ae^{-E_a/RT}$$

where A is a constant. Taking the natural logarithm of both sides of this equation,

$$\ln k = \ln A - E_a/RT$$

This equation was first shown to be valid by the Swedish physical chemist Svante Arrhenius in 1889 and is referred to as the **Arrhenius equation**.

Comparing the Arrhenius equation to the general equation for a straight line,

$$y = b + mx$$

it should be clear that a plot of $\ln k$ ("y") vs. $1/T$ ("x") should be linear. The slope of the straight line, m, is equal to $-E_a/R$. Figure 11.10 shows such a plot for the CO–NO_2 reaction. The slope appears to be about -1.61×10^4 K. Hence

$$\frac{-E_a}{R} = -1.61 \times 10^4 \text{ K}$$

$$E_a = 8.31 \frac{\text{J}}{\text{mol} \cdot \text{K}}(1.61 \times 10^4 \text{ K}) = 1.34 \times 10^5 \text{ J/mol} = 134 \text{ kJ/mol}$$

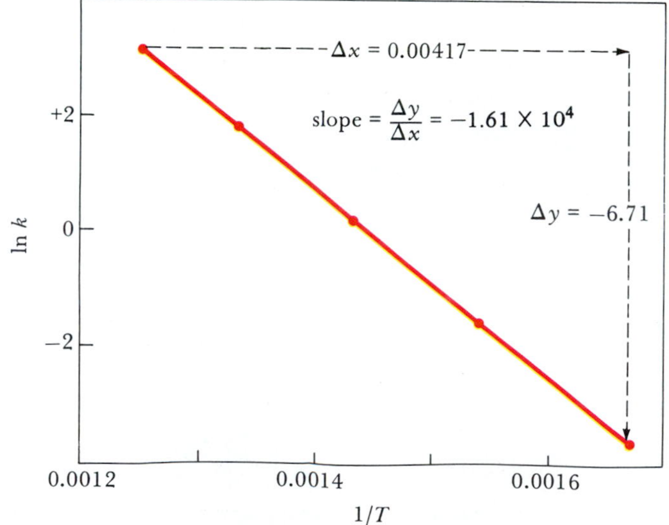

Figure 11.10
A plot of $\ln k$ vs. $1/T$ (k = rate constant, T = Kelvin temperature) is a straight line. From the slope of this line, the activation energy can be determined: $E_a = -R \times$ slope. For the CO–NO_2 reaction shown here, $E_a = -(8.31 \text{ J/mol} \cdot \text{K})(-1.61 \times 10^4 \text{ K}) = 1.34 \times 10^5 \text{ J/mol} = 134$ kJ/mol.

"Two-Point" Equation Relating *k* and *T*

The Arrhenius equation can be expressed in a different form by following the procedure used with the Clausius-Clapeyron equation in Chapter 9. At two different temperatures, T_2 and T_1,

$$\ln k_2 = \ln A - \frac{E_a}{RT_2}$$

$$\ln k_1 = \ln A - \frac{E_a}{RT_1}$$

Subtracting the second equation from the first,

$$\ln k_2 - \ln k_1 = \frac{-E_a}{R}\left[\frac{1}{T_2} - \frac{1}{T_1}\right]$$

or

$$\ln \frac{k_2}{k_1} = \frac{E_a}{R}\left[\frac{1}{T_1} - \frac{1}{T_2}\right]$$

Taking $R = 8.31$ J/mol · K, the activation energy is expressed in joules per mole.

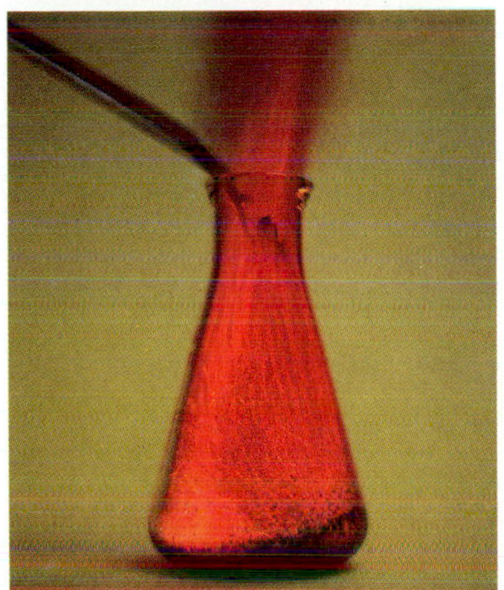

Antimony powder reacts more rapidly with bromine at high temperatures. The reaction is: $2Sb(s) + 3Br_2(l) \longrightarrow 2SbBr_3(s)$. (James Morgenthaler)

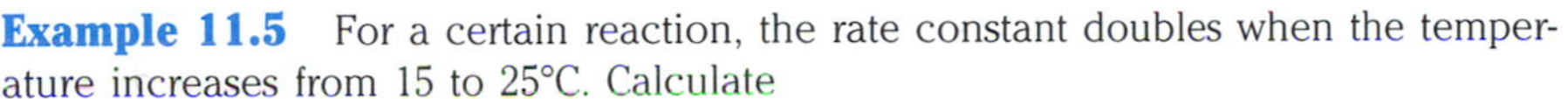

Example 11.5 For a certain reaction, the rate constant doubles when the temperature increases from 15 to 25°C. Calculate

(a) the activation energy, E_a.
(b) the rate constant at 100°C, taking k at 25°C to be 1.2×10^{-2} L/mol · s.

Strategy Use the two-point form of the Arrhenius equation. In (a), the two temperatures and the ratio k_2/k_1 are known, so E_a is readily calculated. That value of E_a is used in (b) to calculate k_2, knowing k_1, T_2, and T_1. Remember to express temperatures in K.

Solution

(a) $k_2/k_1 = 2.00$ $\quad \ln k_2/k_1 = \ln 2.00 = 0.693$ $\quad T_2 = 298$ K $\quad T_1 = 288$ K
Substituting into the two-point form of the Arrhenius equation,

$$0.693 = \frac{E_a}{8.31 \text{ J/mol} \cdot \text{K}}\left[\frac{1}{288 \text{ K}} - \frac{1}{298 \text{ K}}\right]$$

Solving,

$$E_a = \boxed{49 \text{ kJ/mol}}$$

(b) $\ln \frac{k_2}{k_1} = \frac{49{,}000 \text{ J/mol}}{8.31 \text{ J/mol} \cdot \text{K}}\left[\frac{1}{298 \text{ K}} - \frac{1}{373 \text{ K}}\right] = 3.98 \qquad k_2/k_1 = e^{3.98} = 53$

Remember, E_a must be expressed in joules

Taking $k_1 = 1.2 \times 10^{-2}$ L/mol · s

$$k_2 = 53(1.2 \times 10^{-2} \text{ L/mol} \cdot \text{s}) = \boxed{0.64 \text{ L/mol} \cdot \text{s}}$$

The E.F. Smith Memorial Collection in the History of Chemistry, Dept. of Special Collections, Van Pelt-Dietrich Library, University of Pennsylvania

Benjamin Franklin was wrong

Today, Arrhenius is probably best known for his contributions to chemical kinetics. At the age of 30, he derived the relation for the temperature dependence of the rate constant that is now referred to as the *Arrhenius equation.* Beyond that, he introduced the concepts of activation energy and the activated complex.

Arrhenius had an extremely productive scientific career, but it was not an easy one. In his Ph.D. thesis, written at the University of Uppsala in Sweden in 1884, he proposed that salts, strong acids, and strong bases are completely dissociated in dilute water solution. Today, it seems quite reasonable that solutions of NaCl, HCl, and NaOH contain, respectively, Na^+ and Cl^- ions, H^+ and Cl^- ions, and Na^+ and OH^- ions. It did not seem nearly so obvious to the chemistry faculty at Uppsala in 1884. Arrhenius's dissertation received the lowest passing grade "approved not without praise."

A few years later, when the colligative properties of electrolyte solutions were shown to be in agreement with the Arrhenius model, the situation changed. But the old ideas died hard in Uppsala. In 1901, Arrhenius was narrowly elected to the Swedish Academy of Science, over strong opposition. Two years later, he received the Nobel Prize in chemistry.

In his later years, Arrhenius turned his attention to popularizing chemistry. He wrote several different textbooks that were well received. In 1925, under pressure from his publisher to submit a manuscript, Arrhenius started getting up at 4 A.M. to write. As might be expected, rising at such an early hour had an adverse effect on his health. Arrhenius suffered a physical breakdown in 1925, from which he never really recovered, dying two years later.

11.6 REACTION MECHANISMS

A **reaction mechanism** is a description of a path, or a sequence of steps, by which a reaction occurs at the molecular level. In the simplest case, only a single step is involved. This is a collision between two reactant molecules. This is the "mechanism" for the reaction of CO with NO_2 at high temperatures, above about 600 K:

$$CO(g) + NO_2(g) \longrightarrow NO(g) + CO_2(g)$$

At low temperatures, the reaction between carbon monoxide and nitrogen dioxide takes place by a quite different mechanism. Two steps are involved:

Most reactions occur in a series of steps

$$\begin{array}{l} NO_2(g) + NO_2(g) \longrightarrow NO_3(g) + NO(g) \\ \underline{CO(g) + NO_3(g) \longrightarrow CO_2(g) + NO_2(g)} \\ CO(g) + NO_2(g) \longrightarrow NO(g) + CO_2(g) \end{array}$$

Notice that the overall reaction, obtained by summing the individual steps, is identical with that for the one-step process. The rate expressions are quite different, however:

high temperatures: $\text{rate} = k[CO] \times [NO_2]$

low temperatures: $\text{rate} = k[NO_2]^2$

In general, the nature of the rate expression and hence *the reaction order depend upon the mechanism by which the reaction takes place.*

Elementary Steps

The individual steps that constitute a reaction mechanism are referred to as **elementary steps**. These steps may be *unimolecular*

$$A \longrightarrow B + C \qquad \text{rate} = k[A]$$

bimolecular

$$A + A \longrightarrow B + C \qquad \text{rate} = k[A] \times [A] = k[A]^2$$

or, in rare cases, *termolecular*

$$A + B + C \longrightarrow D + E \qquad \text{rate} = k[A] \times [B] \times [C]$$

(Color is used to distinguish elementary steps from overall reactions.)

Notice from the rate expressions just written that *the rate of an elementary step is equal to a rate constant k multiplied by the concentration of each reactant molecule*. This rule is readily explained. Consider, for example, a step in which two molecules, A and B, collide effectively with each other to form C and D. As pointed out earlier, the rate of collision and hence the rate of reaction will be directly proportional to the concentration of each reactant.

Slow Steps

Often, one step in a mechanism is much slower than any other. If this is the case, *the slow step is rate-determining*. That is, the rate of the overall reaction can be taken to be that of the slow step. Consider, for example, a three-step reaction:

Step 1:	$A \longrightarrow B$	fast
Step 2:	$B \longrightarrow C$	slow
Step 3:	$C \longrightarrow D$	fast
	$A \longrightarrow D$	

The rate at which A is converted to D (the overall reaction) is approximately equal to the rate of conversion of B to C (the slow step).

To understand the rationale behind this rule, and its limitations, consider an analogous situation. Suppose three people (A, B, and C) are assigned to grade general chemistry examinations that contain three questions. On the average, A spends 10 s grading Question 1 and B spends 15 s grading Question 2. In contrast, C, the ultimate procrastinator, takes 5 min to grade Question 3. The rate at which exams are graded is

Maybe Questions 1 and 2 are multiple choice

$$\frac{1 \text{ exam}}{10 \text{ s} + 15 \text{ s} + 300 \text{ s}} = \frac{1 \text{ exam}}{325 \text{ s}} = 0.00308 \text{ exam/s}$$

This is approximately equal to the rate of the slow grader:

$$\frac{1 \text{ exam}}{300 \text{ s}} = 0.00333 \text{ exam/s}$$

Extrapolating from general chemistry exams to chemical reactions, we can say that

— the overall rate of the reaction cannot exceed that of the slowest step
— if that step is by far the slowest, its rate will be approximately equal to that of the overall reaction

Deducing a Rate Expression from a Proposed Mechanism

The question to be considered here is: *Given a mechanism for a several-step reaction, how can you deduce the rate expression corresponding to that mechanism?* In principle, at least, all you have to do is to apply the rules just cited.

1. *Find the slowest step and equate the rate of the overall reaction to the rate of that step.*
2. *Find the rate expression for the slowest step.*

To illustrate this process, consider the two-step mechanism for the low-temperature reaction between CO and NO_2:

$$\begin{array}{llr} NO_2(g) + NO_2(g) \longrightarrow NO_3(g) + NO(g) & & \text{slow} \\ CO(g) + NO_2(g) \longrightarrow CO_2(g) + NO_2(g) & & \text{fast} \\ \hline CO(g) + NO_2(g) \longrightarrow CO_2(g) + NO(g) & & \end{array}$$

Applying the above rules in order,

$$\text{rate of overall reaction} = \text{rate of 1st step} = k[NO_2]^2$$

This analysis explains why the rate expression for the two-step mechanism is different from that for the direct, one-step reaction.

Elimination of Intermediates

If [intermediate] can't be measured, it better not be in the rate expression

Sometimes the rate expression obtained by the process just described involves a reactive intermediate, i.e., a species produced in one step of the mechanism and consumed in a later step. Ordinarily, concentrations of such species are too small to be determined experimentally. Hence they must be eliminated from the rate expression if it is to be compared with experiment. *The final rate expression must include only those species that appear in the balanced equation for the overall reaction.*

To illustrate this situation, consider the reaction between nitric oxide and chlorine, which is believed to proceed by a two-step mechanism:

$$\text{Step 1:} \quad NO(g) + Cl_2(g) \underset{k_{-1}}{\overset{k_1}{\rightleftharpoons}} NOCl_2(g) \qquad \text{fast}$$

$$\begin{array}{ll} \text{Step 2:} & NOCl_2(g) + NO(g) \xrightarrow{k_2} 2NOCl(g) \qquad \text{slow} \\ \hline & 2NO(g) + Cl_2(g) \longrightarrow 2NOCl(g) \end{array}$$

The first step occurs rapidly and reversibly; a dynamic equilibrium is set up in which the rates of forward and reverse reactions are equal. Since the second step is slow and rate-determining, it follows that

$$\text{rate of overall reaction} = \text{rate of Step 2} = k_2[NOCl_2] \times [NO]$$

The rate expression just written is unsatisfactory in that it cannot be checked against experiment. The species $NOCl_2$ is a reactive intermediate whose concentration is too small to be measured accurately, if at all. To eliminate the $[NOCl_2]$ term from the rate expression, recall that the rates of forward and reverse reactions in Step 1 are equal, which means that

$$k_1[NO] \times [Cl_2] = k_{-1}[NOCl_2]$$

Solving for [$NOCl_2$] and substituting in this rate expression,

$$[NOCl_2] = \frac{k_1[NO] \times [Cl_2]}{k_{-1}}$$

$$\text{rate of overall reaction} = k_2[NOCl_2] \times [NO] = \frac{k_2k_1[NO]^2 \times [Cl_2]}{k_{-1}}$$

The quotient k_2k_1/k_{-1} is the experimentally observed rate constant for the reaction, which is found to be second-order in NO and first-order in Cl_2, as predicted by this mechanism.

Steel wool glows when heated in air. If it is then immersed in a bottle of pure oxygen, where [O_2] is five times greater than in air, it bursts into flame.

Example 11.6 The decomposition of ozone, O_3, to diatomic oxygen, O_2, is believed to occur by a two-step mechanism:

$$\text{Step 1:} \quad O_3(g) \underset{k_{-1}}{\overset{k_1}{\rightleftharpoons}} O_2(g) + O(g) \quad \text{fast}$$

$$\text{Step 2:} \quad O_3(g) + O(g) \xrightarrow{k_2} 2\,O_2(g) \quad \text{slow}$$

$$2\,O_3(g) \longrightarrow 3\,O_2(g)$$

Obtain the rate expression corresponding to this mechanism.

Strategy Write the rate expression for the rate-determining second step. This will involve the unstable intermediate, O. To get rid of [O], use the fact that reactants and products are in equilibrium in Step 1, so forward and reverse reactions occur at the same rate.

Solution

(1) Overall rate = rate Step 2 = $k_2[O_3] \times [O]$

(2) $k_1[O_3] = k_{-1}[O_2] \times [O]$

$$[O] = \frac{k_1[O_3]}{k_{-1}[O_2]}$$

(3) Overall rate $= \dfrac{k_1k_2[O_3]^2}{k_{-1}[O_2]} = \dfrac{k[O_3]^2}{[O_2]}$

Notice that the concentration of O_2, a product in the reaction, appears in the denominator. The rate is inversely proportional to the concentration of molecular oxygen, a feature that you would never have predicted from the balanced equation for the reaction.

It is important to point out one of the limitations of mechanism studies. Usually more than one mechanism is compatible with the same experimentally obtained rate expression. To make a choice between alternative mechanisms, other evidence must be considered. A classic example of this situation is the reaction between hydrogen and iodine

$$H_2(g) + I_2(g) \longrightarrow 2HI(g)$$

for which the observed rate expression is

$$\text{rate} = k[H_2] \times [I_2]$$

For many years, it was assumed that the H_2–I_2 reaction occurs in a single step, a collision between an H_2 molecule and an I_2 molecule. That would, of course, be compatible with the rate expression on p. 309. However, there is now evidence to indicate that a quite different and more complex mechanism is involved (see Problem 61).

11.7 CATALYSIS

The rate of a reaction can be increased by raising the temperature. However, this is not always feasible or practical. Some reactants and products decompose at high temperatures. From an economic standpoint, raising the temperature means increased energy costs. Fortunately, there are certain substances called *catalysts* that offer an alternative approach to speeding up a reaction.

A catalyst increases the rate of a reaction without being consumed by it. A catalyst operates by changing the reaction mechanism to one with a lower activation energy. Consider, for example, the decomposition of N_2O:

$$N_2O(g) \longrightarrow N_2(g) + \tfrac{1}{2}\,O_2(g)$$

For the direct reaction, taking place in the gas phase, the activation energy is 250 kJ/mol. On a gold surface, which acts as a catalyst for this reaction, the activation energy is only 120 kJ/mol (Fig. 11.11). Hence, the reaction occurs more rapidly on a gold surface.

Heterogeneous Catalysis

A *heterogeneous catalyst* is one that is in a different phase from the reaction mixture. Most commonly, the catalyst is a solid that increases the rate of a gas phase or liquid phase reaction. An example is the reaction previously cited:

$$N_2O(g) \xrightarrow{\text{Au}} N_2(g) + \tfrac{1}{2}\,O_2(g)$$

In the catalyzed decomposition, N_2O is chemically adsorbed on the surface of the solid. A chemical bond is formed between the oxygen atom of an N_2O molecule and a gold atom at the surface. This weakens the bond joining nitro-

Many industrial processes use heterogeneous catalysis

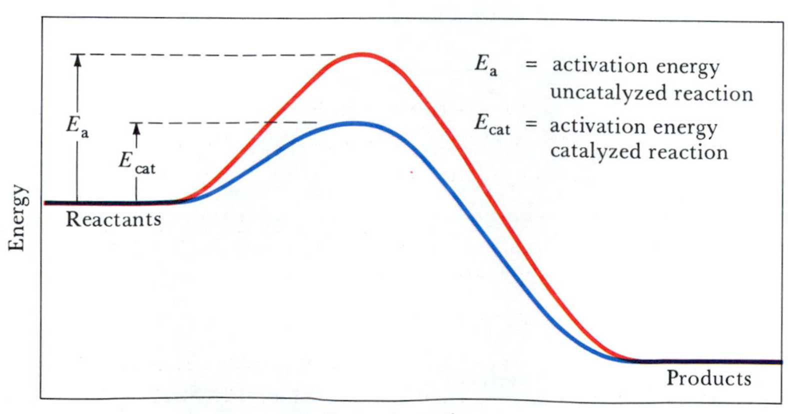

Figure 11.11
By changing the path by which a reaction occurs, a catalyst can lower the activation energy that is required and so speed up the reaction.

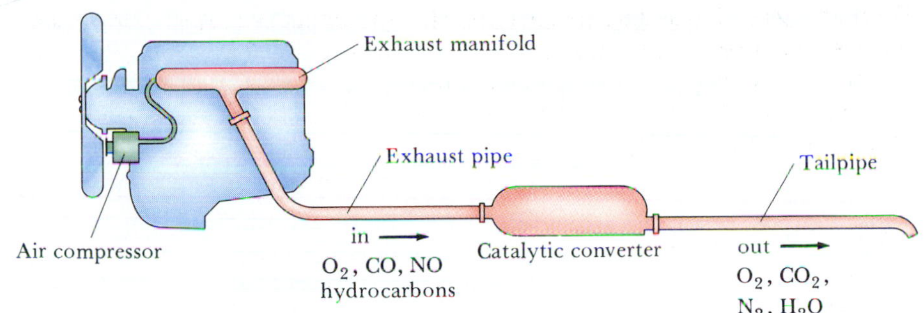

Figure 11.12
Catalytic converters contain a "three-way" catalyst designed to convert CO to CO_2, unburned hydrocarbons to CO_2 and H_2O, and NO to N_2. The active components of the catalysts are the precious metals platinum and rhodium; palladium is sometimes used as well.

gen to oxygen, making it easier for the N_2O molecule to break apart. Symbolically, this process can be shown as

$$\mathrm{N{\equiv}N{-}O}(g) + \mathrm{Au}(s) \longrightarrow \mathrm{N{\equiv}N{-}{-}{-}O{-}{-}{-}Au}(s) \longrightarrow \mathrm{N{\equiv}N}(g) + \mathrm{O}(g) + \mathrm{Au}(s)$$

where the broken lines represent weak covalent bonds.

Perhaps the most familiar example of heterogeneous catalysis is the series of reactions that occur in the catalytic converter of an automobile (Fig. 11.12). Typically this device contains 1–3 g of platinum metal mixed with rhodium. The platinum catalyzes the oxidation of carbon monoxide and unburned hydrocarbons such as benzene, C_6H_6:

$$2\mathrm{CO}(g) + \mathrm{O_2}(g) \xrightarrow{\mathrm{Pt}} 2\mathrm{CO_2}(g)$$

$$\mathrm{C_6H_6}(g) + \tfrac{15}{2}\,\mathrm{O_2}(g) \xrightarrow{\mathrm{Pt}} 6\mathrm{CO_2}(g) + 3\mathrm{H_2O}(l)$$

The rhodium acts as a catalyst to convert oxides of nitrogen to the free element:

$$\mathrm{CO}(g) + \mathrm{NO}(g) \xrightarrow{\mathrm{Rh}} \mathrm{CO_2}(g) + \tfrac{1}{2}\mathrm{N_2}(g)$$

One problem with heterogeneous catalysis is that the solid catalyst is easily "poisoned." Foreign materials deposited on the catalytic surface during the reaction reduce or even destroy its effectiveness. A major reason for using unleaded gasoline is that lead metal poisons the Pt–Rh mixture in the catalytic converter.

Homogeneous Catalysis

A *homogeneous catalyst* is one that is present in the same phase as the reactants. It speeds up the reaction by forming a reactive intermediate that decomposes to give products. In this way, the catalyst provides an alternative mechanism of lower activation energy.

An example of a reaction that is subject to homogeneous catalysis is the decomposition of hydrogen peroxide in aqueous solution:

$$2\mathrm{H_2O_2}(aq) \longrightarrow 2\mathrm{H_2O} + \mathrm{O_2}(g)$$

Under ordinary conditions, this reaction occurs very slowly. However, if a solution of sodium iodide, NaI, is added, reaction occurs almost immediately; you can see the bubbles of oxygen forming.

The catalyzed decomposition of hydrogen peroxide is believed to take place by a two-step mechanism:

$$
\begin{array}{lll}
\text{Step 1:} & H_2O_2(aq) + I^-(aq) & \longrightarrow H_2O + IO^-(aq) \\
\text{Step 2:} & H_2O_2(aq) + IO^-(aq) & \longrightarrow H_2O + O_2(g) + I^-(aq) \\
\hline
 & 2H_2O_2(aq) & \longrightarrow 2H_2O + O_2(g)
\end{array}
$$

Notice that the end result is the same as in the direct reaction.

The I^- ions are not consumed in the reaction. For every I^- ion used up in the first step, one is produced in the second step. The activation energy for this two-step process is much smaller than for the uncatalyzed reaction.

Enzymes

Enzymes often contain more than a thousand atoms

Many reactions that take place slowly under ordinary conditions occur readily in living organisms in the presence of catalysts called **enzymes**. Enzymes are protein molecules of high molar mass. An example of an enzyme-catalyzed reaction is the decomposition of hydrogen peroxide:

$$2H_2O_2(aq) \longrightarrow 2H_2O + O_2(g)$$

In blood or tissues, this reaction is catalyzed by an enzyme called catalase. When 3% hydrogen peroxide is used to treat a fresh cut or wound, oxygen gas is given off rapidly (Fig. 11.13). The function of catalase in the body is to prevent the buildup of hydrogen peroxide, a powerful oxidizing agent.

Many enzymes are extremely specific. For example, the enzyme maltase catalyzes the hydrolysis of maltose:

$$\underset{\text{maltose}}{C_{12}H_{22}O_{11}(aq)} + H_2O \xrightarrow{\text{maltase}} \underset{\text{glucose}}{2C_6H_{12}O_6(aq)}$$

It's nice to be needed

This is the only function of maltase, but it is one that no other enzyme can perform. There are many such digestive enzymes required for the metabolism of carbohydrates, proteins, and fats. It has been estimated that without enzymes, it would take upwards of 50 years to digest a meal.

Figure 11.13
An enzyme present in liver is a catalyst in the rapid decomposition of H_2O_2 into water and oxygen gas. Hydrogen peroxide without the catalyst still decomposes, but at a much slower rate. (Richard Megna/ FUNDAMENTAL PHOTOGRAPHS, New York)

Enzymes, like all other catalysts, lower the activation energy for reaction. They can be enormously effective; it is not uncommon for the rate constant to increase by a factor of 10^{12} or more. However, from a commercial standpoint, enzymes have some drawbacks. A particular enzyme operates best over a narrow range of temperature. An increase in temperature frequently deactivates an enzyme by causing the molecule to "unfold," changing its characteristic shape. Recently, chemists have discovered that this effect can be prevented if the enzyme is immobilized by bonding to a solid support. Among the solids that have been used are synthetic polymers, porous glass, and even stainless steel. The development of immobilized enzymes has led to a host of new products. The sweetener aspartame, used in many diet soft drinks, is made using the enzyme aspartase in immobilized form.

PERSPECTIVE

•

The Ozone Story

In recent years, a minor component of the atmosphere, ozone, has received a great deal of attention. Ozone, molecular formula O_3, is a pale blue gas with a characteristic odor that can be detected after lightning activity, in the vicinity of electric motors, or near a subway train.

Depending upon its location in the atmosphere, ozone can be a villain or a beleaguered hero. In the lower atmosphere (the *troposphere*), ozone is bad news; it is a major component of photochemical smog, formed by the reaction sequence

$$\begin{array}{rcl} NO_2(g) & \longrightarrow & NO(g) + O(g) \\ O_2(g) + O(g) & \longrightarrow & O_3(g) \\ \hline NO_2(g) + O_2(g) & \longrightarrow & NO(g) + O_3(g) \end{array}$$

Ozone is an extremely powerful oxidizing agent, which explains its toxicity to animals and humans. At partial pressures as low as 10^{-7} atm, it can cut in half the rate of photosynthesis by plants.

In the upper atmosphere (the *stratosphere*), the situation is quite different. There the partial pressure of ozone goes through a maximum of about 10^{-5} atm at an altitude of 30 km. From 95 to 99% of sunlight in the wavelength range 200–300 nm is absorbed by ozone in this region, commonly referred to as the "ozone layer." If this ultraviolet radiation were to reach the surface of the earth, it could have several adverse effects. A decrease in ozone concentration of 5% could increase the incidence of skin cancer by 20%. Ultraviolet radiation is also a factor in diseases of the eye, including cataract formation.

Ozone molecules in the stratosphere can decompose by the reaction

$$O_3(g) + O(g) \longrightarrow 2\,O_2(g)$$

This reaction takes place rather slowly by direct collision between an O_3 molecule and an O atom. It can occur more rapidly by a two-step process in which a chlorine atom acts as a catalyst:

$$\begin{array}{lll} O_3(g) + Cl(g) \longrightarrow O_2(g) + ClO(g) & & k = 5.2 \times 10^{9}\ \text{L/mol} \cdot \text{s at 220 K} \\ ClO(g) + O(g) \longrightarrow Cl(g) + O_2(g) & & k = 2.6 \times 10^{10}\ \text{L/mol} \cdot \text{s at 220 K} \\ \hline O_3(g) + O(g) \longrightarrow 2\,O_2(g) & & \end{array}$$

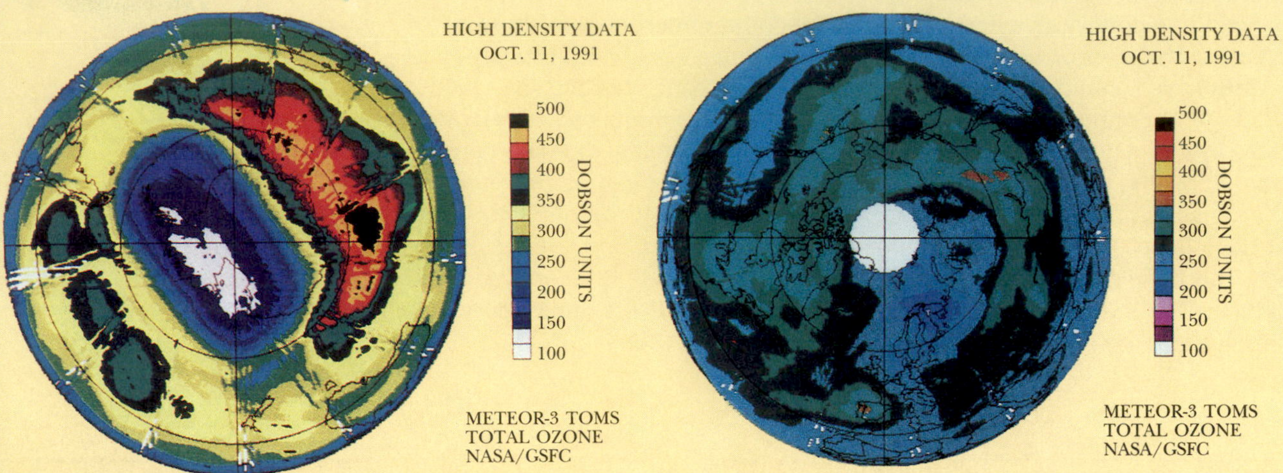

Figure 11.A
Maps of the Southern Hemisphere (left) and Northern Hemisphere (right) show ozone concentrations measured October 11, 1991. The Antarctic ozone hole is located in the lavender and dark purple area at the center of the map on the left. Successively higher ozone concentrations are shown in yellow, green, blue, red, and black (highest). (NASA)

The activation energy for the catalyzed reaction is only about 2 kJ, as compared to nearly 14 kJ for the direct reaction between O_3 and O.

The chlorine atoms that catalyze the decomposition of ozone come from "chlorofluorocarbons" (CFCs), which are widely used in refrigerators and air conditioners. A major culprit is CF_2Cl_2, which forms Cl atoms when exposed to ultraviolet radiation at 200 nm*:

$$CF_2Cl_2(g) \longrightarrow CF_2Cl(g) + Cl(g)$$

The Cl-atom-catalyzed decomposition of ozone is known to be responsible for the ozone hole (Fig. 11.A) that develops in Antarctica each year in September and October, at the end of winter in the Southern Hemisphere. No ozone is generated during the long, dark Antarctic winter. Meanwhile, a heterogeneous reaction occurring on clouds of ice crystals at −85°C produces species such as Cl_2 and HClO. When the sun reappears in September, these molecules decompose photochemically to form Cl atoms, which catalyze the decomposition of ozone, decreasing its partial pressure to less than half of its normal value.

In January 1992, record-high levels of ClO were measured over Canada and northern New England

In 1987, an international treaty was signed in Montreal to cut back on CFC production. A year later, Du Pont, the major producer of CF_2Cl_2, called for a complete phase-out of this compound. As you can imagine, the development of alternative refrigerants is an active area of research. The best bet at this point seems to be

```
     F   H
     |   |
  F—C—C—F
     |   |
     F   H
```

which contains no chlorine atoms.

* It is ironic that CF_2Cl_2 (CFC-12) was developed 50 years ago to replace SO_2 and NH_3 as refrigerants. The use of this nontoxic gas made domestic refrigeration possible; prior to World War II, your grandparents used "ice chests" to keep foods and beverages cold.

CHAPTER HIGHLIGHTS

KEY CONCEPTS

1. *Determine the rate expression (reaction order) from*
   —initial rate data
   (Examples 11.1, 11.2; Problems 21–28, 78)
   —concentration-time data, using Table 11.2
   (Example 11.4; Problems 29–32)
   —reaction mechanism
   (Example 11.6; Problems 59–64)
2. *Relate concentration to time for a first-order reaction*
   (Example 11.3; Problems 33–38, 73)
3. *Use the Arrhenius equation to relate rate constant to temperature*
   (Example 11.5; Problems 49–58, 74)

KEY EQUATIONS

Zero-order reaction	rate $= k$	$[A] = [A]_o - kt$	
First-order reaction	rate $= k[A]$	$\ln [A]_o/[A] = kt$	$t_{1/2} = 0.693/k$
Second-order reaction	rate $= k[A]^2$	$1/[A] - 1/[A]_o = kt$	
Rate constant vs. temperature	$\ln k_2/k_1 = \frac{E_a}{R}\left[\frac{1}{T_1} - \frac{1}{T_2}\right]$		

KEY TERMS

activation energy	mechanism	rate expression
catalyst	order	reaction rate
half-life	rate constant	

SUMMARY PROBLEM

Hydrogen peroxide decomposes to water and oxygen according to the following reaction

$$H_2O_2(aq) \longrightarrow H_2O + \tfrac{1}{2}\,O_2(g)$$

Its rate of decomposition is measured by titrating samples of the solution with potassium permanganate ($KMnO_4$) at certain intervals.

a. Initial rate determinations at 40°C for the decomposition give the following data:

$[H_2O_2]$	Initial Rate (mol/L · min)
0.1000	1.93×10^{-4}
0.2000	3.86×10^{-4}
0.3000	5.79×10^{-4}

What is the order of the reaction? Write the rate equation for the decomposition. Calculate the rate constant and the half-life for the reaction at 40°C.

b. Hydrogen peroxide is sold commercially as a 30.0% solution. If the solution is kept at 40°C, how long will it take for the solution to become 10.0% H_2O_2?

c. It has been determined that at 50°C, the rate constant for the reaction is 4.32×10^{-3}/min. Calculate the activation energy for the decomposition of H_2O_2.

d. Manufacturers recommend that solutions of hydrogen peroxide be kept in a refrigerator at 4°C. How long will it take for a 30.0% solution to decompose to 10.0% if the solution is kept in a refrigerator at 4°C?

e. The rate constant for the uncatalyzed reaction at 25°C is 5.21×10^{-4}/min. The rate constant for the catalyzed reaction at 25°C is 2.95×10^{8}/min. What is the half-life of the uncatalyzed reaction at 25°C? What is the half-life of the catalyzed reaction?

f. Hydrogen peroxide in basic solution oxidizes iodide ions to iodine. The proposed mechanism for this reaction is

$$H_2O_2(aq) + I^-(aq) \longrightarrow HOI(aq) + OH^-(aq) \quad \text{slow}$$

$$HOI(aq) + I^-(aq) \rightleftharpoons I_2(aq) + OH^-(aq) \quad \text{fast}$$

Write the overall redox reaction. Write a rate law consistent with this proposed mechanism.

Answers

a. first-order; rate $= k[H_2O_2]$; $k = 1.93 \times 10^{-3}$/min; $t_{1/2} = 5.98$ h
b. 9.49 h c. 68 kJ/mol d. 2.8×10^2 h
e. $t_{1/2}$ uncatalyzed = 22.2 h; $t_{1/2}$ catalyzed = 2.35×10^{-9} min
f. $H_2O_2(aq) + 2I^-(aq) \rightarrow I_2(aq) + 2\,OH^-(aq)$; rate $= k[H_2O_2] \times [I^-]$

QUESTIONS & PROBLEMS

Meaning of Reaction Rates

1. Express the rate of the reaction

$$C_2H_6(g) \longrightarrow C_2H_4(g) + H_2(g)$$

a. in terms of $\Delta[C_2H_6]$.
b. in terms of $\Delta[C_2H_4]$.

2. Express the rate of the reaction

$$2HI(g) \longrightarrow H_2(g) + I_2(g)$$

a. in terms of $\Delta[H_2]$.
b. in terms of $\Delta[HI]$, if you want the same rate as in (a).
(a).

3. Consider the combustion of ethane:

$$2C_2H_6(g) + 7\,O_2(g) \longrightarrow 4CO_2(g) + 6H_2O(g)$$

If the ethane is burning at a rate of 0.20 mol/L · s, at what rates are CO_2 and H_2O being produced?

4. For the reaction

$$5Br^-(aq) + BrO_3^-(aq) + 6H^+(aq) \longrightarrow 3Br_2(aq) + 3H_2O$$

it was found that at a particular instant bromine was being formed at the rate of 0.039 mol/L per second. At that instant,

a. at what rate was water being formed?
b. at what rate was bromide ion being oxidized?
c. at what rate was H^+ being consumed?

5. Nitrosyl chloride (NOCl) decomposes according to the equation

$$2NOCl(g) \longrightarrow 2NO(g) + Cl_2(g)$$

a. Write an expression for the rate of reaction in terms of $\Delta[NOCl]$.
b. The concentration of NOCl drops from 0.580 *M* to 0.238 *M* in 8.00 min. Calculate the rate of reaction over this time interval.

6. Dinitrogen pentaoxide decomposes according to the following equation:

$$2N_2O_5(g) \longrightarrow 4NO_2(g) + O_2(g)$$

a. Write an expression for reaction rate in terms of $\Delta[N_2O_5]$.
b. The concentration of N_2O_5 decreases by 20.0% in one minute, starting at 0.384 *M*. Calculate the rate of reaction over this one-minute interval.

7. Experimental data are listed for the hypothetical reaction

$$A + B \longrightarrow C + D$$

Time (min)	0	1	2	3	4	5
[A]	0.200	0.166	0.139	0.117	0.100	0.089

Plot these data as in Figure 11.2. Draw a tangent to the curve to find the rate at 3 minutes.

8. Using the data in Problem 7, find the rate at 4 minutes using the plot referred to in Problem 7.

Rate Expressions

9. What is the order with respect to each reactant and the overall order of the reactions described by the following rate expressions?

a. rate $= k_1[A]^2 \times [B]$
b. rate $= k_2[A]$
c. rate $= k_3[A]^2 \times [B]^2$
d. rate $= k_4$

10. What is the order with respect to each reactant and the overall order of the reactions described by the following rate expressions?

a. rate = $k_1[A]^3$
b. rate = $k_2[A] \times [B]$
c. rate = $k_3[A] \times [B]^2$
d. rate = $k_4[B]$

11. What will the units of the rate constants in Question 9 be if rate is expressed in mol/L · s?

12. What will the units of each of the rate constants in Question 10 be if the rate is expressed in mol/L · s?

13. Complete the following table for the first-order reaction

$$A(g) \longrightarrow \text{products}$$

[A]	k (min^{-1})	Rate (mol/L · min)
0.750	2.0×10^{-3}	______
0.035	______	6.82
______	0.0372	7.98×10^{-3}

14. Complete the following table for the reaction

$$A(g) + B(g) \longrightarrow \text{products}$$

which is first-order in both reactants.

[A]	[B]	k (L/mol · s)	Rate (mol/L · s)
0.200	0.300	1.5	______
______	0.029	0.78	0.025
0.450	0.520	______	0.033

15. At 550 K, the decomposition of nitrogen dioxide is second-order, with a rate constant of 0.31 L/mol · s.

a. Write the rate expression.
b. Calculate the rate when the concentration of nitrogen dioxide is 0.050 *M*.
c. At what concentration of nitrogen dioxide will the rate be 1.5×10^{-3} mol/L · s?

16. The decomposition of ammonia on tungsten at 1100°C is zero-order, with a rate constant of 2.5×10^{-4} mol/L · min.

a. Write the rate expression.
b. Calculate the rate when the concentration of ammonia is 0.080 *M*.
c. At what concentration of ammonia is the rate equal to the rate constant?

17. The reaction

$$NO(g) + \tfrac{1}{2}Br_2(g) \longrightarrow NOBr(g)$$

is second-order in nitrogen oxide and first-order in bromine. The rate of the reaction is 1.6×10^{-8} mol/L · min when the nitrogen oxide concentration is 0.020 *M* and the bromine concentration is 0.030 *M*.

a. What is the value of k?
b. At what concentration of bromine is the rate 2.0×10^{-6} mol/L · min when the nitrogen oxide concentration is 0.064 *M*?
c. At what concentration of nitrogen oxide is the rate 4.5×10^{-7} mol/L · min if the bromine concentration is one third of the nitrogen oxide concentration?

18. The reaction

$$ICl(g) + \tfrac{1}{2}H_2(g) \longrightarrow \tfrac{1}{2}I_2(g) + HCl(g)$$

is first-order in both reactants. The rate of the reaction is 4.89×10^{-5} mol/L · s when the ICl concentration is 0.100 *M* and that of hydrogen gas is 0.0300 *M*.

a. What is the value of k?
b. At what concentration of hydrogen gas is the rate 3.00×10^{-3} mol/L · s when the ICl concentration is 0.150 *M*?
c. At what concentration of iodine chloride is the rate 7.85 mol/L · s if the hydrogen gas concentration is twice that of ICl?

19. The greatest decrease in reaction rate for the reaction between A and D where rate = $k \times [A]^2 \times [D]$ is caused by

a. halving conc. D. **b.** halving conc. A
c. doubling conc. D.
d. halving conc. A and conc. D.

20. The greatest increase in rate for the reaction between X and Z where rate = $k[X] \times [Z]^2$ will be caused by

a. doubling conc. Z. **b.** doubling conc. X.
c. tripling conc. X. **d.** lowering the temperature.

Determination of Reaction Order

21. For a reaction involving a single reactant A, the following rate data are obtained.

Rate (mol/L · min)	0.020	0.020	0.019	0.021
[A]	0.100	0.090	0.080	0.070

Determine the order of the reaction.

22. For a reaction involving a single reactant A, the following data are obtained.

Rate (mol/L · min)	0.020	0.016	0.013	0.010
[A]	0.100	0.090	0.080	0.070

Determine the order of the reaction.

23. The rate of the reaction

$$HgCl_2(aq) + \tfrac{1}{2}C_2O_4^{2-}(aq) \longrightarrow Cl^-(aq) + CO_2(g) + \tfrac{1}{2}Hg_2Cl_2(s)$$

is followed by measuring the number of moles of Hg_2Cl_2 that precipitate per liter per second. The following data are obtained:

$[HgCl_2]$	$[C_2O_4^{2-}]$	Initial Rate (mol/L · s)
0.10	0.10	1.3×10^{-7}
0.10	0.20	5.2×10^{-7}
0.20	0.20	1.0×10^{-6}
0.20	0.10	2.6×10^{-7}

a. What is the order of the reaction with respect to $HgCl_2$, with respect to $C_2O_4^{2-}$, and overall?
b. Write the rate expression for the reaction.
c. Calculate k for the reaction.
d. When the concentrations of both mercury(II) chloride and oxalate ion are 0.30 *M*, what is the rate of the reaction?

24. In solution at constant H^+ concentration, I^- reacts with H_2O_2 to produce I_2:

$$H^+(aq) + I^-(aq) + \tfrac{1}{2}H_2O_2(aq) \longrightarrow \tfrac{1}{2}I_2(aq) + H_2O$$

The reaction rate can be followed by monitoring iodine production. The following data apply:

$[I^-]$	$[H_2O_2]$	Initial Rate (mol/L · s)
0.020	0.020	3.3×10^{-5}
0.040	0.020	6.6×10^{-5}
0.060	0.020	9.0×10^{-5}
0.040	0.040	1.3×10^{-4}

a. What is the order with respect to I^-?
b. What is the order with respect to H_2O_2?
c. What is the rate when $[I^-] = 0.010$ *M* and $[H_2O_2] = 0.030$ *M*?

25. The following data refer to the reaction between A and B.

[A]	[B]	Initial Rate (mol/L · s)
0.10	0.10	3.0×10^{-3}
0.10	0.30	3.0×10^{-3}
0.30	0.30	2.7×10^{-2}
0.30	0.60	2.7×10^{-2}

a. What is the order of the reaction with respect to A, with respect to B, and overall?
b. Calculate k for the reaction.
c. At what concentration of A will the rate be 5.0×10^{-3} mol/L · s?

26. The following data refer to the equation

$$A + B + C \longrightarrow \text{products}$$

[A]	[B]	[C]	Initial Rate (mol/L · s)
0.100	0.100	0.100	2.50×10^{-4}
0.300	0.100	0.100	2.25×10^{-3}
0.300	0.200	0.100	2.25×10^{-3}
0.300	0.200	0.400	9.00×10^{-3}

a. What is the order of the reaction with respect to each of the three reactants, and overall?
b. Write the rate expression for the reaction.
c. Calculate k.
d. Calculate the rate of the reaction when $[A] = 2[B] = \tfrac{1}{3}[C] = 0.600$ *M*.

27. The following initial rate data were obtained for the reaction

$$ClO_2(aq) + OH^-(aq) \longrightarrow \tfrac{1}{2}ClO_3^-(aq) + \tfrac{1}{2}ClO_2^-(aq) + \tfrac{1}{2}H_2O$$

$[ClO_2]$	$[OH^-]$	Initial Rate (mol/L · s)
0.0575	0.0216	8.21×10^{-3}
0.0713	0.0216	1.26×10^{-2}
0.0575	0.0333	1.26×10^{-2}
0.0713	0.0333	1.95×10^{-2}

a. Write the rate expression.
b. Calculate k.
c. What is the pH when the concentration of ClO_2 is 0.100 *M* and the rate is 3.56×10^{-2} mol/L · s?

28. Consider the reaction

$$Cr(H_2O)_6^{3+}(aq) + SCN^-(aq) \longrightarrow Cr(H_2O)_5SCN^{2+}(aq) + H_2O$$

The following data were obtained:

$[Cr(H_2O)_6^{3+}]$	$[SCN^-]$	Rate (mol/L · s)
0.028	0.040	8.1×10^{-6}
0.028	0.055	1.1×10^{-5}
0.037	0.055	1.5×10^{-5}
0.037	0.040	1.1×10^{-5}

a. Write the rate expression for the reaction.
b. Calculate k.
c. What is the initial rate if 12 mg of KSCN is added to two liters of a solution 0.087 *M* in $Cr(H_2O)_6^{3+}$?

29. The following data are obtained for the decomposition of CH_3NO_2 at 500 K. Following the procedure used in Example 11.4 find the reaction order.

t (s)	0	300	600	900	1200
$[CH_3NO_2]$	0.200	0.145	0.105	0.076	0.055

30. The following data are obtained for the decomposition of acetaldehyde, CH_3CHO, at 700 K. Following the procedure used in Example 11.4, find the reaction order.

t (s)	0	50	100	150	200
$[CH_3CHO]$	0.400	0.333	0.286	0.250	0.222

31. For the reaction: A → products, it is found that the half-life is 27 s when the initial concentration is 0.50 M and 50 s when the initial concentration is 0.27 M. Is the reaction zero-order, first-order, or second-order?

32. Consider the following data for the decomposition of a species B:

[B]	0.160	0.080	0.040	0.020
t (s)	0	10.0	15.0	17.5

What is the half-life, starting at 0.160 M? 0.080 M? 0.040 M? What is the order of the reaction?

First-Order Reaction

33. The following data apply to the first-order decomposition of ethane, $C_2H_6(g) \rightarrow$ products:

t (s)	$[C_2H_6]$
0	0.01000
200	0.00916
400	0.00839
600	0.00768
800	0.00703

a. From a plot of $[C_2H_6]$ vs. time, estimate the rate at $t = 800$ s.

b. From a plot of ln $[C_2H_6]$ vs. time, determine k, the rate constant. Use k to obtain the rate at $t = 800$ s.

c. Which of the rates just calculated in (a) or (b) do you think is more accurate? Explain.

34. The following data are for the gas-phase decomposition of ethyl chloride, $C_2H_5Cl(g) \rightarrow$ products at 740 K:

t (min)	$[C_2H_5Cl]$	t (min)	$[C_2H_5Cl]$
0	0.200	4	0.187
1	0.197	8	0.175
2	0.193	16	0.153
3	0.190		

a. By plotting the data, show that the reaction is first-order.

b. From the graph, determine k.

c. Using k, find the time for the concentration to drop to one fourth of the original concentration.

35. In the first-order decomposition of cyclobutane at 750 K,

$$C_4H_8(g) \longrightarrow \text{products}$$

it is found that 25% of a sample has decomposed in eighty seconds. Calculate

a. the rate constant. **b.** the half-life.

36. In the first-order decomposition of acetone at 500°C,

$$H_3C{-}CO{-}CH_3(g) \longrightarrow \text{products}$$

it is found that the concentration is 0.0300 M after 200 min and 0.0200 M after 400 min. Calculate

a. the rate constant. **b.** the half-life.

c. the initial concentration.

37. The isomerization reaction $A(g) \rightarrow A^*(g)$ is first-order with a half-life of 64.2 min.

a. What is the rate constant for the isomerization?

b. How many hours are required for [A] to drop from 0.500 M to 0.125 M?

c. How long will it take to reduce [A] to 5.00% of $[A]_o$?

38. The decomposition of dinitrogen pentaoxide

$$N_2O_5(g) \longrightarrow 2NO_2(g) + \tfrac{1}{2}O_2(g)$$

is first-order. Its half-life is 0.363 h.

a. What is the rate constant for the decomposition?

b. What fraction of N_2O_5 will have decomposed after one hour?

c. How long will it take to decompose 15.0% of the dinitrogen pentaoxide?

Zero- and Second-Order Reactions

39. The decomposition of nitrogen dioxide

$$NO_2(g) \longrightarrow NO(g) + \tfrac{1}{2}O_2(g)$$

is a second-order reaction. It takes 125 s for the concentration of NO_2 to go from 0.800 M to 0.0104 M.

a. What is k for the reaction?

b. What is the half-life of the reaction when its initial concentration is 0.500 M?

40. The decomposition of hydrogen iodide

$$HI(g) \longrightarrow \tfrac{1}{2}H_2(g) + \tfrac{1}{2}I_2(g)$$

is second-order. Its half-life is 85 s when the initial concentration is 0.15 M.

a. What is k for the reaction?

b. How long will it take to go from 0.300 M to 0.100 M?

41. The rate constant for the second-order reaction

$$NOBr(g) \longrightarrow NO(g) + \tfrac{1}{2}Br_2(g)$$

is 48 L/mol · min at a certain temperature. How long will it take to decompose 90.0% of a 0.0200 M solution of nitrosyl bromide?

42. The decomposition of nitrosyl chloride

$$NOCl(g) \longrightarrow NO(g) + \tfrac{1}{2}Cl_2(g)$$

is a second-order reaction. If it takes 0.20 min to decompose 15% of a 0.300 *M* solution of nitrosyl chloride, what is *k* for the reaction?

43. The rate constant for the zero-order decomposition of HI on a gold surface

$$HI(g) \xrightarrow{Au} \tfrac{1}{2}H_2(g) + \tfrac{1}{2}I_2(g)$$

is 0.050 mol/L · s. How long will it take for the concentration of HI to drop from 1.00 *M* to 0.20 *M*?

44. For the zero-order decomposition of ammonia on tungsten,

$$NH_3(g) \xrightarrow{W} \tfrac{1}{2}N_2(g) + \tfrac{3}{2}H_2(g)$$

it takes 32 min for the concentration of NH_3 to drop from 0.500 *M* to 0.100 *M*. How long will it take for the rest of the ammonia to decompose? If the original concentration were 0.800 *M*, how long would it take for half of the NH_3 to decompose?

Activation Energy, Catalysis and ΔH

45. Three reactions have the following activation energies:

Reaction	A	B	C
E_a (kJ)	145	210	48

Which reaction would you expect to be the fastest? The slowest? Explain.

46. Consider the data for several systems for the conversion of reactants to products at the same temperature.

System	E_a (kJ)	ΔH (kJ)
1	50	+20
2	85	−20
3	12	−30

a. Which system has the highest forward rate?

b. In each case, determine the difference between the energy of the activated complex and that of the products.

47. For a certain reaction, E_a is 120 kJ and ΔH is 15 kJ. In the presence of a catalyst, the activation energy is lowered to 75 kJ. Draw a diagram similar to those in Figures 11.8 and 11.11 for this reaction.

48. The activation energy of a certain reaction is 85 kJ. In the presence of a catalyst, it is reduced to 40 kJ. For the reaction, ΔH is −50 kJ. Draw a diagram similar to those in Figures 11.8 and 11.11 for this reaction.

Reaction Rate and Temperature

49. The following data were obtained for the reaction

$$SiH_4(g) \longrightarrow Si(s) + 2H_2(g)$$

k (s^{-1})	0.048	2.3	49	590
t (°C)	500	600	700	800

Plot these data (ln k vs. $1/T$) and find the activation energy for the reaction.

50. The following data are for the gas-phase decomposition of acetaldehyde:

$k\left(\frac{L}{mol \cdot s}\right)$	0.0105	0.101	0.60	2.92
T (K)	700	750	800	850

Plot these data and determine the activation energy for the reaction.

51. The activation energy of a certain enzyme-catalyzed reaction is 50.2 kJ/mol. By what percent is the rate of this reaction increased if you have a fever of 104.0°F, assuming that the enzyme works at a normal temperature of 98.6°F?

52. Raw milk will sour in about four hours at 28°C (82°F) but will last up to 48 hours in a refrigerator at 5°C. Assuming the rate constant to be inversely related to souring time, what is the activation energy for the reaction involving the souring of milk?

53. The chirping rate of a cricket, *X*, in chirps per minute, near room temperature is given by

$$X = 7.2t - 32$$

where t is the temperature in °C. Calculate the chirping rates at 25 and 35°C, and use them to estimate the activation energy for this reaction.

54. Cold-blooded animals decrease their body temperature in cold weather to match that of their environment. The activation energy of a certain enzyme-catalyzed reaction in a cold-blooded animal is 65 kJ/mol. By what percentage is the rate of this reaction decreased if the body temperature of the animal drops from 35 to 25°C?

55. For the reaction

$$HI(g) + CH_3I(g) \longrightarrow CH_4(g) + I_2(g)$$

the rate constant is 0.28 L/mol · s at 300°C and 3.9×10^{-3} L/mol · s at 227°C.

a. What is the activation energy of the reaction?

b. What is k at 400°C?

56. For the reaction

$$C_2H_4(g) + H_2(g) \longrightarrow C_2H_6(g)$$

the activation energy is 181 kJ/mol. The rate constant at 500°C is 2.5×10^{-2} L/mol · s.

a. At what temperature is the rate constant twice its value at 500°C?
b. What is the rate constant at 1000°C?

57. The decomposition of ethyl chloride is a first-order reaction. Its activation energy is 248 kJ/mol. At 470°C, it has a half-life of 40.8 min. What is its half-life at 400°C?

58. The decomposition of ethyl bromide is a first-order reaction. Its activation energy is 226 kJ/mol. At 720 K, its half-life is 78 s. What is its half-life at 800 K?

Reaction Mechanisms

59. Write the rate expression for each of the following elementary steps:
a. $N_2O_2 + H_2 \longrightarrow N_2O + H_2O$
b. $ClCO + Cl_2 \longrightarrow Cl_2CO + Cl$
c. $2NO_2 \longrightarrow NO_3 + NO$

60. Write the rate expression for each of the following elementary steps:
a. $K + HCl \longrightarrow KCl + H$
b. $NO_3 + CO \longrightarrow NO_2 + CO_2$
c. $2NO_2 \longrightarrow 2NO + O_2$

61. For the reaction between hydrogen and iodine,

$$H_2(g) + I_2(g) \longrightarrow 2HI(g)$$

The rate expression is rate = $k[H_2] \times [I_2]$. Show that this expression is consistent with the mechanism

$I_2(g) \rightleftharpoons 2I(g)$ (fast)
$H_2(g) + I(g) + I(g) \longrightarrow 2HI(g)$ (slow)

62. For the reaction

$$2H_2(g) + 2NO(g) \longrightarrow N_2(g) + 2H_2O(g)$$

the experimental rate expression is rate = $k[NO]^2 \times [H_2]$. The following mechanism is proposed:

$2NO \rightleftharpoons N_2O_2$ (fast)
$N_2O_2 + H_2 \longrightarrow H_2O + N_2O$ (slow)
$N_2O + H_2 \longrightarrow N_2 + H_2O$ (fast)

Show that the mechanism is consistent with the rate expression.

63. Two mechanisms are proposed for the reaction

$$2NO(g) + O_2(g) \longrightarrow 2NO_2(g)$$

Mechanism 1: $NO + O_2 \rightleftharpoons NO_3$ (fast)
$NO_3 + NO \longrightarrow 2NO_2$ (slow)
Mechanism 2: $NO + NO \rightleftharpoons N_2O_2$ (fast)
$N_2O_2 + O_2 \longrightarrow 2NO_2$ (slow)

Show that each of these mechanisms is consistent with the observed rate law, rate = $k[NO]^2 \times [O_2]$.

64. At low temperatures, the rate law for the reaction $CO(g) + NO_2(g) \longrightarrow CO_2(g) + NO(g)$ is as follows: rate = constant × $[NO_2]^2$. Which of the following mechanisms is consistent with this rate law?

a. $CO + NO_2 \longrightarrow CO_2 + NO$
b. $2NO_2 \rightleftharpoons N_2O_4$ (fast)
$N_2O_4 + 2CO \longrightarrow 2CO_2 + 2NO$ (slow)
c. $2NO_2 \longrightarrow NO_3 + NO$ (slow)
$NO_3 + CO \longrightarrow NO_2 + CO_2$ (fast)
d. $2NO_2 \longrightarrow 2NO + O_2$ (slow)
$2CO + O_2 \longrightarrow 2CO_2$ (fast)

Ozone

65. Write equations for the two-step Cl-catalyzed decomposition of ozone. In what part of the atmosphere does this reaction take place?

66. Using information given in the text, calculate the rates of the two steps in the Cl-catalyzed decomposition of ozone. Take $[O_3] = 5.3 \times 10^{-9}\,M$, $[Cl] = 1.7 \times 10^{-16}\,M$, $[ClO] = 1.1 \times 10^{-13}\,M$, $[O] = 8.2 \times 10^{-17}\,M$.

67. For the reaction

$$O_3(g) + O(g) \longrightarrow 2\,O_2(g)$$

$\Delta H° = -391.2$ kJ. The activation energies for the direct and Cl-catalyzed reactions are given in the text. Draw activation energy diagrams for both direct and Cl-catalyzed reactions.

68. The following reaction is involved in smog formation:

$$O_3(g) + NO(g) \longrightarrow O_2(g) + NO_2(g)$$

This reaction has been shown to be first-order in both ozone and NO with a rate constant of 1.2×10^7 L/mol · s. Calculate the concentration of NO_2 formed per second in polluted air, where O_3 and NO concentrations are both 2×10^{-8} mol/L. From the magnitude of your answer, would you expect the conversion of NO to be rapid or slow?

69. For the first-order thermal decomposition of ozone

$$O_3(g) \longrightarrow O_2(g) + O(g)$$

$k = 3 \times 10^{-26}\,s^{-1}$ at 25°C. What is the half-life for this reaction in years? Comment on the likelihood that this reaction contributes to the depletion of the ozone layer.

Unclassified

70. How does each of the following affect reaction rate?
a. the passage of time
b. decreasing container size for a gas phase reaction
c. adding a catalyst

71. In your own words explain why
a. a decrease in temperature slows the rate of a reaction.
b. doubling the concentration of a reactant does not always double the rate of the reaction.
c. a flame lights a cigarette but the cigarette continues to burn after the flame has been removed.

72. The decomposition of nitrosyl bromide

$$NOBr(g) \longrightarrow NO(g) + \tfrac{1}{2} Br_2(g)$$

is a second-order reaction. Its rate constant is 0.80 L/mol · s; ΔH_f° for NOBr is 82.2 kJ/mol; ΔH_f° for NO is 90.2 kJ/mol and for $Br_2(g)$, 30.9 kJ/mol. If the initial concentration of NOBr is 0.100 *M*, how much heat is evolved or absorbed per liter of solution in one minute?

73. Consider the decomposition of dinitrogen pentaoxide in CCl_4:

$$N_2O_5(g) \longrightarrow N_2O_4(g) + \tfrac{1}{2} O_2(g)$$

The rate constant for this first-order reaction is 0.037/min. A sample weighing 68.0 g is dissolved in CCl_4 and allowed to decompose.

a. How long will it take to reduce the amount of N_2O_5 to 10.0 g?

b. What total volume of O_2 at 45°C and 1.00 atm is produced if the reaction is stopped after 25 minutes?

74. How much faster would a reaction proceed at 25°C than at 10°C if the activation energy of the reaction is 1.00×10^2 kJ/mol?

75. Derive the integrated rate law, $[A] = [A]_o - kt$ for a zero-order reaction. (*Hint:* Start with the relation $-\Delta[A] = k\Delta t$.)

Challenge Problems

76. For a first-order reaction $aA \rightarrow$ products, where *a* is a number other than 1, the rate is equal to $-\Delta[A]/a\Delta t$, or, in derivative notation, $-d[A]/adt$. Derive the integrated rate law for the first-order decomposition of *a* moles of reactant.

77. Using calculus, derive the equation for

a. the concentration–time relation for a second-order reaction (see Table 11.2).

b. the concentration–time relation for a third-order reaction, $A \rightarrow$ products.

78. The following data apply to the reaction

$$A(g) + 3B(g) + 2C(g) \longrightarrow \text{products}$$

[A]	[B]	[C]	Rate
0.20	0.40	0.10	X
0.40	0.40	0.20	8X
0.20	0.20	0.20	X
0.40	0.40	0.10	4X

Determine the rate law for the reaction.

79. In a first-order reaction, let us suppose that a quantity, X, of reactant is added at regular intervals of time, Δt. At first the amount of reactant in the system builds up; eventually, however, it levels off at a "saturation value" given by the expression

$$\text{saturation value} = \frac{X}{1 - 10^{-a}}, \text{ where } a = 0.30\frac{\Delta t}{t_{1/2}}$$

This analysis applies to prescription drugs, where you take a certain amount each day. Suppose you take 0.100 g of a drug three times a day and that the half-life for elimination is 2.0 days. Using this equation, calculate the mass of the drug in the body at saturation. Suppose further that side effects show up when 0.500 g of the drug accumulates in the body. As a pharmacist, what is the maximum dosage you could assign to a patient for an 8-hour period without his suffering side effects?

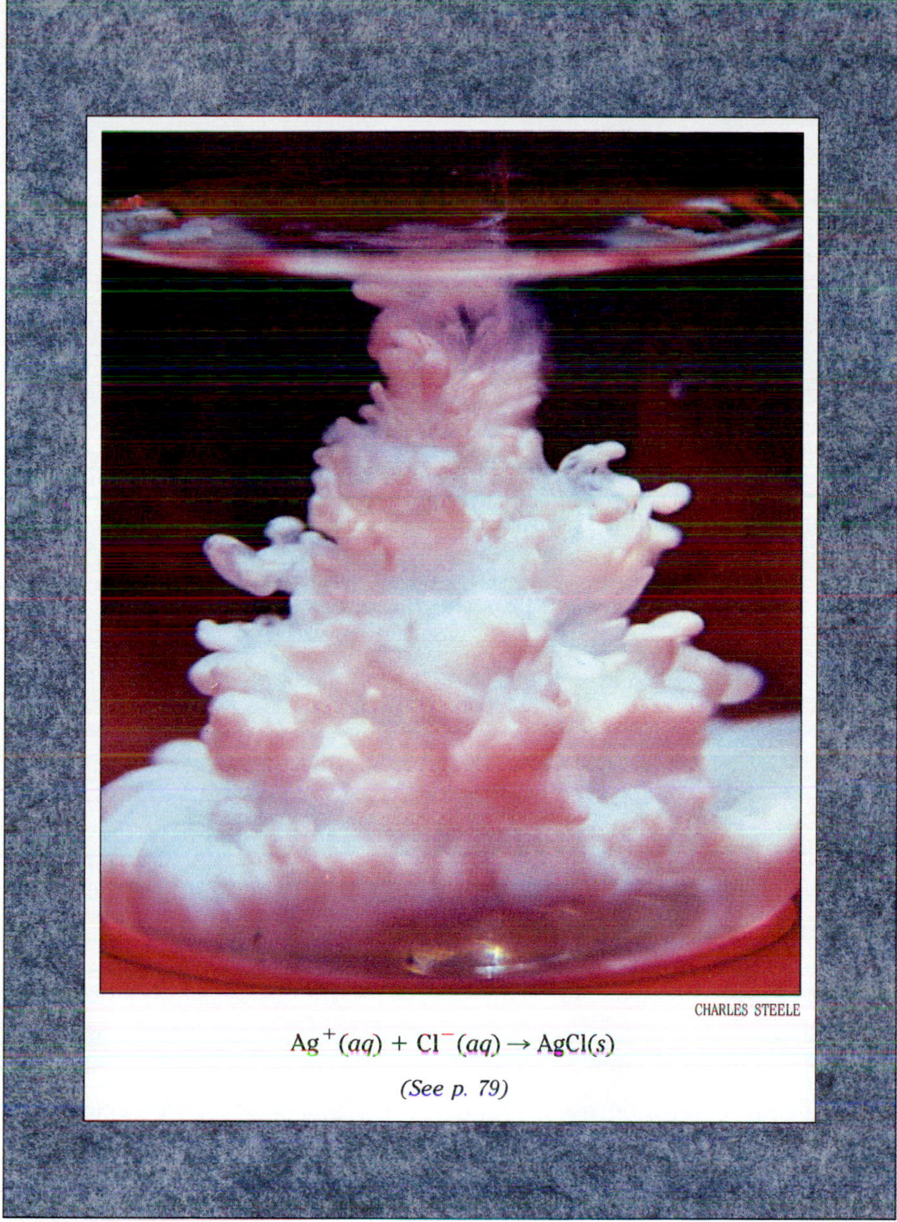

CHARLES STEELE

$Ag^+(aq) + Cl^-(aq) \rightarrow AgCl(s)$

(See p. 79)

12 Gaseous Chemical Equilibrium

The cumbrous elements, Earth,
Flood, Aire, Fire
Flew upward, spirited
various forms,
Each had his place
appointed, each
his course
The rest in circuit walles
this Universe.

JOHN MILTON
Paradise Lost

CHAPTER OUTLINE

Among the topics covered in Chapter 9 was the equilibrium between liquid and gaseous water:

$$H_2O(l) \rightleftharpoons H_2O(g)$$

The state of this equilibrium system at a given temperature can be described in a simple way by citing the equilibrium pressure of water vapor: 0.034 atm at 25°C, 1.00 atm at 100°C, and so on.

They can be reversed by changing the external conditions

Chemical reactions involving gases carried out in closed containers resemble in many ways the $H_2O(l)$–$H_2O(g)$ system. The reactions are reversible; reactants are not completely consumed. Instead, an equilibrium mixture containing both products and reactants is obtained. At equilibrium, forward and reverse reactions take place at the same rate. As a result, the amounts of all species at equilibrium remain constant with time.

There is, however, one complicating factor where gaseous chemical equilibria are involved. Ordinarily, more than one gaseous substance is present. We might, for example, have a system

$$aA(g) + bB(g) \rightleftharpoons cC(g) + dD(g)$$

in which the equilibrium mixture contains two gaseous products (C and D) and two reactants (A and B). This means that

— the **partial pressures** of the gases in the mixture (P_C, P_D, P_A, P_B) must be specified. Recall from Chapter 5 that the partial pressure of a gas, P_i, can be calculated from the ideal gas law:

$$P_i = n_iRT/V$$

— the equilibrium condition cannot be expressed in terms of a single partial pressure, e.g., P_D or P_A

Fortunately, it turns out that there is a relatively simple relationship between the partial pressures of different gases present at equilibrium in a reaction system. This relationship is expressed in terms of a quantity called the **equilibrium constant**, symbol *K*. In this chapter, you will learn how to

— write the expression for *K* corresponding to any chemical equilibrium (Section 12.2)
— calculate the value of *K* from experimental data for the equilibrium system (Section 12.3)
— use the value of *K* to predict the extent to which a reaction will take place (Section 12.4)
— use *K*, along with Le Chatelier's principle, to predict the result of disturbing an equilibrium system (Section 12.5)

The last section of this chapter (12.6) covers some important industrial syntheses (NH_3, HNO_3, H_2SO_4), which apply not only the principles of equilibrium but also those of chemical kinetics (Chapter 11).

12.1 THE N_2O_4–NO_2 EQUILIBRIUM SYSTEM

Consider what happens when a sample of N_2O_4, a colorless gas, is placed in a closed, evacuated container at 100°C. Instantly, a reddish-brown color develops. This color is due to nitrogen dioxide, NO_2, formed by decomposition of part of the N_2O_4 (Fig. 12.1):

$$N_2O_4(g) \longrightarrow 2NO_2(g)$$

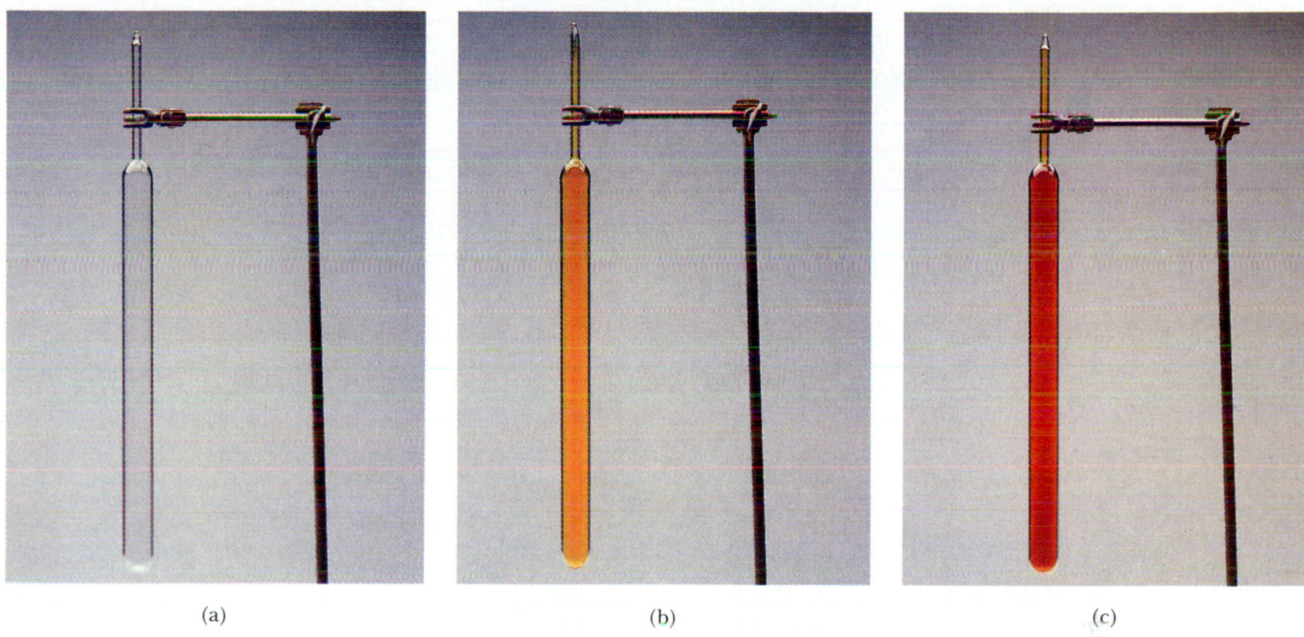

(a) (b) (c)

Figure 12.1
Dinitrogen tetraoxide, N_2O_4 (a) is colorless; nitrogen dioxide, NO_2 (c) is a brownish-red gas. The partial pressure of NO_2 in an equilibrium mixture of the two gases (b) is directly related to the depth of the brown color. (Marna G. Clarke)

At first, this is the only reaction taking place. As soon as some NO_2 is formed, however, the reverse reaction can occur:

$$2NO_2(g) \longrightarrow N_2O_4(g)$$

As time passes, the partial pressure of N_2O_4 drops, and the forward reaction slows down. Conversely, the partial pressure of NO_2 increases, so the rate of the reverse reaction increases. Soon, these rates become equal. A dynamic equilibrium has been established:

$$N_2O_4(g) \rightleftharpoons 2NO_2(g)$$

At equilibrium, appreciable amounts of both gases are present. As time passes, their amounts or partial pressures remain constant, as long as the volume of the container and the temperature remain unchanged.

The approach to equilibrium in this system is illustrated by the data in Table 12.1 and Figure 12.2. Originally, only N_2O_4 is present; its pressure is 1.00 atm. Since there is no NO_2 around, its original pressure is zero. As equilibrium is approached, the *net reaction* is

$$N_2O_4(g) \longrightarrow 2NO_2(g)$$

The amount and hence the partial pressure of N_2O_4 drop, rapidly at first, then more slowly. The partial pressure of NO_2 increases. Finally, both partial pressures level off and become constant. At equilibrium, at 100°C:

Their pressures don't change because forward and reverse reactions occur at the same rate

$$P_{N_2O_4} = 0.22 \text{ atm} \qquad P_{NO_2} = 1.56 \text{ atm}$$

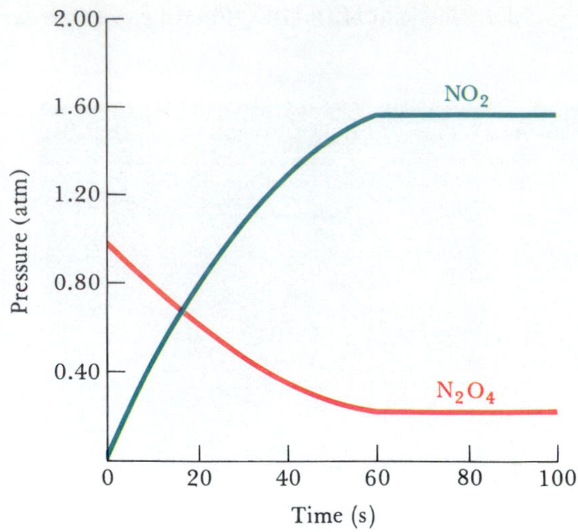

Figure 12.2
Approach to equilibrium in the N_2O_4–NO_2 system. The partial pressure of N_2O_4 starts off at 1.00 atm, drops sharply at first, and finally levels off at the equilibrium value of 0.22 atm. Meanwhile, the partial pressure of NO_2 rises from zero to its equilibrium value, 1.56 atm.

There are many ways to approach equilibrium in the N_2O_4–NO_2 system. Table 12.2 gives data for three experiments in which the original conditions are quite different. Experiment 1 is that just described, starting with pure N_2O_4. Experiment 2 starts instead with pure NO_2 at a partial pressure of 1.00 atm. As equilibrium is approached, some of the NO_2 reacts to form N_2O_4:

$$2NO_2(g) \longrightarrow N_2O_4(g)$$

Finally, in Experiment 3, both N_2O_4 and NO_2 are present originally, each at a partial pressure of 1.00 atm.

Looking at the data in Table 12.2, you might wonder whether these three experiments have anything in common. Specifically, is there any relationship between the equilibrium partial pressures of NO_2 and N_2O_4 that is valid for all the experiments? It turns out that there is, although it is not an obvious one. The value of the quotient $(P_{NO_2})^2/P_{N_2O_4}$ is the same, about 11, in each case:

This relation holds only at equilibrium; initial pressures can have any values

$$\text{Expt. 1} \qquad \frac{(P_{NO_2})^2}{P_{N_2O_4}} = \frac{(1.56)^2}{0.22} = 11$$

$$\text{Expt. 2} \qquad \frac{(P_{NO_2})^2}{P_{N_2O_4}} = \frac{(0.86)^2}{0.07} = 11$$

$$\text{Expt. 3} \qquad \frac{(P_{NO_2})^2}{P_{N_2O_4}} = \frac{(2.16)^2}{0.42} = 11$$

This relationship holds for any equilibrium mixture containing N_2O_4 and NO_2 at 100°C. More generally, it is found that, at any temperature, the quantity

$$\frac{(P_{NO_2})^2}{P_{N_2O_4}}$$

Table 12.1 Establishment of Equilibrium in the System $N_2O_4(g) \rightleftharpoons 2NO_2(g)$ (100°C)

Time (s)	0	20	40	60	80	100
$P_{N_2O_4}$ (atm)	1.00	0.60	0.35	**0.22**	**0.22**	**0.22**
P_{NO_2} (atm)	0.00	0.80	1.30	**1.56**	**1.56**	**1.56**

Table 12.2 Equilibrium Measurements in the N_2O_4–NO_2 System at 100°C

		Original Pressure (atm)	Equilibrium Pressure (atm)
Expt. 1	N_2O_4	1.00	0.22
	NO_2	0.00	1.56
Expt. 2	N_2O_4	0.00	0.07
	NO_2	1.00	0.86
Expt. 3	N_2O_4	1.00	0.42
	NO_2	1.00	2.16

where P_{NO_2} and $P_{N_2O_4}$ are equilibrium pressures in atmospheres, is a constant, independent of the original composition, the volume of the container, or the total pressure. This constant is referred to as the **equilibrium constant *K*** for the system

$$N_2O_4(g) \rightleftharpoons 2NO_2(g)$$

At 100°C, *K* for this system is 11; at 150°C, it has a different value, about 110. Any mixture of NO_2 and N_2O_4 at 100°C will react in such a way that the ratio $(P_{NO_2})^2/P_{N_2O_4}$ becomes equal to 11. At 150°C, reaction occurs until this ratio becomes 110.

12.2 THE EQUILIBRIUM CONSTANT EXPRESSION

For every gaseous chemical system, an equilibrium constant expression can be written stating the condition that must be attained at equilibrium. For the general system involving only gases,

$$a\text{A}(g) + b\text{B}(g) \rightleftharpoons c\text{C}(g) + d\text{D}(g)$$

where A, B, C, D represent different substances and *a, b, c, d* are their coefficients in the balanced equation

$$K = \frac{(P_C)^c \times (P_D)^d}{(P_A)^a \times (P_B)^b}$$

Remember, these are equilibrium pressures in atmospheres

where P_C, P_D, P_A, and P_B are the partial pressures in **atmospheres** of the four gases at equilibrium. Notice that in the expression for *K**

* This equilibrium constant is often given the symbol K_p to emphasize that it involves partial pressures. Other equilibrium expressions for gases are sometimes used, including K_c:

$$K_c = \frac{[C]^c \times [D]^d}{[A]^a \times [B]^b}$$

where the square brackets represent equilibrium concentrations in moles per liter. The two constants, K_c and K_p, are simply related (see Problem 66):

$$K_p = K_c(RT)^{\Delta n_g}$$

where Δn_g is the change in the number of moles of gas in the equation, *R* is the gas law constant, 0.0821 L · atm/mol · K, and *T* is the Kelvin temperature. Throughout this chapter, we deal exclusively with K_p, referring to it simply as *K*.

— the equilibrium partial pressures of **products** (right side of equation) appear in the **numerator**
— the equilibrium partial pressures of **reactants** (left side of equation) appear in the **denominator**
— each partial pressure is raised to a **power** equal to its **coefficient** in the balanced equation

---

Example 12.1 Ammonia can be made from hydrogen and nitrogen (Section 12.6):

$$N_2(g) + 3H_2(g) \rightleftharpoons 2NH_3(g)$$

Write the equilibrium expression for this system.

Strategy Since NH_3 is a product, P_{NH_3} appears in the numerator; the partial pressures of N_2 and H_2 are in the denominator. The exponents are the coefficients in the balanced equation: 2 for NH_3, 1 for N_2, 3 for H_2.

Solution NH_3 is a product; H_2 and N_2 are reactants. The expression is

$$K = \frac{(P_{NH_3})^2}{P_{N_2} \times (P_{H_2})^3}$$

Changing the Chemical Equation

It is important to realize that *the expression for K depends upon the form of the chemical equation written to describe the equilibrium system.* To illustrate what this statement means, consider the N_2O_4–NO_2 system:

K is meaningless unless accompanied by a chemical equation

$$N_2O_4(g) \rightleftharpoons 2NO_2(g) \qquad K = \frac{(P_{NO_2})^2}{P_{N_2O_4}}$$

Many other equations could be written for this system, for example,

$$\tfrac{1}{2}N_2O_4(g) \rightleftharpoons NO_2(g)$$

In this case, the expression for the equilibrium constant would be

$$K' = \frac{P_{NO_2}}{(P_{N_2O_4})^{1/2}}$$

Comparing the expressions for K and K', it is clear that K' is the square root of K. This illustrates a general rule, sometimes referred to as the **coefficient rule**, which states:

If the coefficients in a balanced equation are multiplied by a factor n, the equilibrium constant is raised to the nth power:

$$K' = K^n$$

In this particular case, $n = \frac{1}{2}$ because the coefficients were divided by 2. Hence K' is the square root of K. If $K = 11$, then $K' = (11)^{1/2} = 3.3$.

Another equation that might be written to describe the N_2O_4–NO_2 system is

$$2NO_2(g) \rightleftharpoons N_2O_4(g) \qquad K'' = \frac{P_{N_2O_4}}{(P_{NO_2})^2}$$

For the system: $2A \rightleftharpoons 3B$, $K = 4.0$. What is K for: $3/2B \rightleftharpoons A$? Ans: 0.50

This chemical equation is simply the reverse of that written originally; N_2O_4 and NO_2 switch sides of the equation. Notice that K'' is the reciprocal of K; the numerator and denominator have been inverted. This illustrates the **reciprocal rule**:

The equilibrium constants for forward and reverse reactions are the reciprocals of each other.

$$K'' = 1/K$$

If, for example, $K = 11$, then $K'' = 1/11 = 0.091$.

Adding Chemical Equations

A property of K that you will find very useful in this and succeeding chapters is expressed by the **rule of multiple equilibria**, which states:

If a reaction can be expressed as the sum of two or more reactions, K for the overall reaction is the product of the equilibrium constants of the individual reactions.

That is, if

Reaction 3 = Reaction 1 + Reaction 2

Then

K (Reaction 3) = K (Reaction 1) × K (Reaction 2)

To illustrate the application of this rule, consider the following reactions at 700°C:

$$SO_2(g) + \tfrac{1}{2}O_2(g) \rightleftharpoons SO_3(g) \qquad K = 2.2$$

$$NO_2(g) \rightleftharpoons NO(g) + \tfrac{1}{2}O_2(g) \qquad K = 4.0$$

For the system: $A + B \rightleftharpoons C$, $K = 3$; for $A + B \rightleftharpoons D$, $K = 6$. What is K for: $C \rightleftharpoons D$? Ans: 2

Adding these equations eliminates $\tfrac{1}{2}O_2$; the result is

$$SO_2(g) + NO_2(g) \rightleftharpoons SO_3(g) + NO(g)$$

For this reaction, $K = 2.2 \times 4.0 = 8.8$.

The validity of this rule can be demonstrated by writing the expression for K for the individual reactions and multiplying:

$$\frac{P_{SO_3}}{P_{SO_2} \times (P_{O_2})^{1/2}} \times \frac{P_{NO} \times (P_{O_2})^{1/2}}{P_{NO_2}} = \frac{P_{SO_3} \times P_{NO}}{P_{SO_2} \times P_{NO_2}}$$

The product expression, at the right, is clearly the equilibrium constant for the overall reaction, as it should be according to the rule of multiple equilibria.

Heterogeneous Equilibria

In the reactions described so far, all the reactants and products have been gaseous; the equilibrium systems are *homogeneous*. In certain reactions, one or more of the substances involved is a pure liquid or solid; the others are gases.

Table 12.3 Equilibrium Constant Expressions for $CO_2(g) + H_2(g) \rightleftharpoons CO(g) + H_2O(l)$

	Expt. 1	Expt. 2	Expt. 3	Expt. 4	Expt. 5
Mass $H_2O(l)$	8 g	6 g	4 g	2 g	0
P_{H_2O} (atm)	3×10^{-2}	3×10^{-2}	3×10^{-2}	3×10^{-2}	1×10^{-2}
K_I	9×10^{-6}	9×10^{-6}	9×10^{-6}	9×10^{-6}	9×10^{-6}
K_{II}	3×10^{-4}	3×10^{-4}	3×10^{-4}	3×10^{-4}	9×10^{-4}

$$K_I = \frac{P_{CO} \times P_{H_2O}}{P_{CO_2} \times P_{H_2}} \qquad K_{II} = \frac{P_{CO}}{P_{CO_2} \times P_{H_2}}$$

Such a system is *heterogeneous,* since more than one phase is present. Examples include

$$CO_2(g) + H_2(g) \rightleftharpoons CO(g) + H_2O(l)$$

$$I_2(s) \rightleftharpoons I_2(g)$$

Experimentally, it is found that for such systems

— *the position of the equilibrium is independent of the amount of solid or liquid, as long as some is present*

— *terms for pure liquids or solids need not appear in the expression for K*

To see why this is the case, consider the data shown in Table 12.3 for the system

$$CO_2(g) + H_2(g) \rightleftharpoons CO(g) + H_2O(l)$$

at 25°C. In Experiments 1–4, there is liquid water present, so $P_{H_2O} = 0.03$ atm, the equilibrium vapor pressure of water at 25°C. In Experiment 5, there is no liquid water, so the partial pressure of water vapor drops below 0.03 atm. Notice that

1. In all cases, the quantity

$$K_I = \frac{P_{CO} \times P_{H_2O}}{P_{CO_2} \times P_{H_2}}$$

is constant at 9×10^{-6}.

2. *So long as liquid water is present* (Expt. 1–4), the quantity

P_{H_2O} is constant, so it can be incorporated into the equilibrium constant

$$K_{II} = \frac{P_{CO}}{P_{CO_2} \times P_{H_2}}$$

is constant at 3×10^{-4}. This must be true because

$$K_{II} = K_I/P_{H_2O}$$

and P_{H_2O} is constant at 3×10^{-2} atm. It follows that K_{II} is a valid equilibrium constant for the system

$$CO_2(g) + H_2(g) \rightleftharpoons CO(g) + H_2O(l)$$

Since the expression for K_{II} is simpler than that for K_I, it is the equilibrium constant of choice for the heterogeneous system.

A similar argument can be applied to the equilibrium involved in the sublimation of iodine, $I_2(s) \rightleftharpoons I_2(g)$, where $K = P_{I_2}$. At a given temperature, the pressure of iodine vapor is constant, independent of the amount of solid iodine or any other factor.

Example 12.2 Write the expression for K for:

(a) the reduction of black solid copper(II) oxide (1 mol) with hydrogen to form copper metal and steam.
(b) the reaction of 1 mol of steam with red hot coke (carbon) to form a mixture of hydrogen and carbon monoxide, called water gas.

Strategy First, write the equation for the equilibrium system. Then write the expression for K, deleting pure liquids and solids, and following the usual rules for gases.

Solution

(a) The reaction is $CuO(s) + H_2(g) \rightleftharpoons Cu(s) + H_2O(g)$.

The equilibrium expression is $K = P_{H_2O}/P_{H_2}$

If the product were $H_2O(l)$, $K = 1/P_{H_2}$

(b) The reaction is $C(s) + H_2O(g) \rightleftharpoons H_2(g) + CO(g)$

The equilibrium expression is $K = \dfrac{P_{H_2} \times P_{CO}}{P_{H_2O}}$

In this and succeeding chapters, a wide variety of different types of equilibria will be covered. They may involve gases, pure liquids or solids, and species in aqueous solution. It will always be true that in the expression for the equilibrium constant

To express K for any system, follow these rules

— *gases enter as their partial pressures in atmospheres*
— *pure liquids or solids do not appear; neither does the solvent for a reaction in dilute solution*
— *species (ions or molecules) in water solution enter as their molar concentrations*

Thus for the equilibrium attained when zinc metal reacts with acid:

$$Zn(s) + 2H^+(aq) \rightleftharpoons Zn^{2+}(aq) + H_2(g)$$

$$K = \frac{P_{H_2} \times [Zn^{2+}]}{[H^+]^2}$$

We will have more to say in later chapters about equilibria involving species in aqueous solution.

12.3 DETERMINATION OF K

Numerical values of equilibrium constants can be calculated if the partial pressures of products and reactants at equilibria are known. Sometimes you will be given equilibrium partial pressures directly (Example 12.3). At other times, you will be given the original partial pressures and the equilibrium partial pressure of one species (Example 12.4). In that case, the calculation of K is a bit more difficult.

Example 12.3 Ammonium chloride is sometimes used as a flux in soldering because it decomposes upon heating:

$$NH_4Cl(s) \rightleftharpoons NH_3(g) + HCl(g)$$

The HCl formed removes oxide films from metals to be soldered. In a certain equilibrium system at 400°C, 12.0 g of NH_4Cl is present; the partial pressures of NH_3 and HCl are 3.0 atm and 5.0 atm respectively. Calculate K at 400°C.

Strategy First write the expression for K. Then substitute equilibrium partial pressures into that expression and solve.

There's no term for $NH_4Cl(s)$

Solution $K = (P_{NH_3}) \times (P_{HCl}) = (3.0) \times (5.0) =$ 15

Example 12.4 Consider the equilibrium system

$$2HI(g) \rightleftharpoons H_2(g) + I_2(g)$$

Originally, a system contains only HI at a pressure of 1.00 atm at 520°C. The equilibrium partial pressure of H_2 is found to be 0.10 atm. Calculate

(a) P_{I_2} at equilibrium. (b) P_{HI} at equilibrium. (c) K

Strategy In order to solve this problem, you must

— carefully distinguish between *original* and *equilibrium* partial pressures
— learn how to relate the equilibrium partial pressures of H_2, I_2, and HI, using the coefficients of the balanced equation

Solution

(a) Note from the balanced equation that one mole of I_2 is formed for every mole of H_2. Putting it another way, as reaction proceeds from left to right, the increase in the number of moles of I_2 is the same as that for H_2:

H_2 and I_2 come from the decomposition of HI

$$\Delta n_{I_2} = \Delta n_{H_2}$$

But, since $P = nRT/V$, it follows that for an equilibrium system at constant T and V, pressure is directly proportional to amount in moles, and the above relationship must hold for partial pressures as well:

$$\Delta P_{I_2} = \Delta P_{H_2}$$

Since both partial pressures start at zero, the equilibrium partial pressures must be equal. The equilibrium partial pressure of I_2, like that of H_2, is 0.10 atm.

(b) Two moles of HI are required to form one mole of H_2. More generally,

$$\Delta n_{HI} = -2\Delta n_{H_2}$$

where the minus sign takes account of the fact that when the amount of hydrogen increases, the amount of HI decreases. But, since P is directly proportional to n,

$$\Delta P_{HI} = -2\Delta P_{H_2} = -2(0.10 \text{ atm}) = -0.20 \text{ atm}$$

In other words, the partial pressure of HI decreases by 0.20 atm. Since it starts at 1.00 atm, its equilibrium value must be 1.00 atm − 0.20 atm = 0.80 atm.

Summarizing this reasoning in the form of a table:

	$2HI(g)$ $\rightleftharpoons$	$H_2(g)$ +	$I_2(g)$
P_o (atm)	1.00	0.00	0.00
ΔP (atm)	−0.20	+0.10	+0.10
P_{eq} (atm)	0.80	0.10	0.10

(P_o = initial pressure; P_{eq} = equilibrium pressure)

Note that the numbers in the center row (ΔP) are related through the coefficients of the balanced equation ($2HI$, $1H_2$, $1I_2$).

(c) $K = \dfrac{P_{H_2} \times P_{I_2}}{(P_{HI})^2} = \dfrac{0.10 \times 0.10}{(0.80)^2} = 0.016$

Example 12.4 illustrates a principle that you will find very useful in solving equilibrium problems throughout this (and later) chapters. *Changes in partial pressures of reactants and products as a system approaches equilibrium, like changes in molar amounts, are related to one another through the coefficients of the balanced equation.*

For: $A + 2B \rightleftharpoons 3C$, $\Delta P_C = -3\,\Delta P_A = -3/2\,\Delta P_B$

12.4 APPLICATIONS OF THE EQUILIBRIUM CONSTANT

Sometimes, knowing only the magnitude of the equilibrium constant, it is possible to decide upon the feasibility of a reaction. Consider, for example, a possible method for "fixing" atmospheric nitrogen—converting it to a compound—by reaction with oxygen:

$$N_2(g) + O_2(g) \rightleftharpoons 2NO(g)$$

$$K = \frac{(P_{NO})^2}{P_{N_2} \times P_{O_2}} = 1 \times 10^{-30} \text{ at } 25°C$$

Since K is so small, the partial pressure of NO in equilibrium with N_2 and O_2, and hence the amount of NO, must be extremely small, approaching zero. Clearly, this would not be a suitable way to fix nitrogen, at least at 25°C.

If K is very small, the reaction won't go very far to the right

An alternative approach to nitrogen fixation involves reacting it with hydrogen:

$$N_2(g) + 3H_2(g) \rightleftharpoons 2NH_3(g)$$

$$K = \frac{(P_{NH_3})^2}{P_{N_2} \times (P_{H_2})^3} = 6 \times 10^5 \text{ at } 25°C$$

Here, the equilibrium system must contain mostly ammonia. A mixture of N_2 and H_2 should be almost completely converted to NH_3 at equilibrium (see, however, the discussion of the Haber process in Section 12.6).

If K is very large, the reaction goes virtually to completion

In general, if K is a very small number, the equilibrium mixture will contain mostly unreacted starting materials; for all practical purposes, the forward reaction does not "go." Conversely, a large K implies a reaction which, at least in principle, is feasible; products should be formed in high yield. Frequently, K has an intermediate value, in which case you must make quantitative calculations concerning the *direction* or *extent* of reaction.

Direction of Reaction; the Reaction Quotient (Q)

Consider the general gas phase reaction

$$aA(g) + bB(g) \rightleftharpoons cC(g) + dD(g)$$

for which

$$K = \frac{(P_C)^c \times (P_D)^d}{(P_A)^a \times (P_B)^b}$$

As we have seen, for a given system at a particular temperature, the value of K is fixed.

In contrast, the *actual* pressure ratio, Q

$$Q = \frac{(P_C)^c \times (P_D)^d}{(P_A)^a \times (P_B)^b}$$ (color is used to distinguish actual from equilibrium pressures)

can have any value at all. If you start with pure reactants (A and B), P_C and P_D are zero and the value of Q is 0:

$$Q = \frac{0 \times 0}{(P_A)^a \times (P_B)^b} = 0$$

Conversely, starting with only C and D:

$$Q = \frac{(P_C)^c \times (P_D)^d}{0 \times 0} \longrightarrow \infty$$

If all four substances are present, Q can have any value between zero and infinity.

The value of K is fixed; Q can have any value

The expression for Q, known as the **reaction quotient**, is the same as that for the equilibrium constant, K. The difference is that the partial pressures that appear in Q are those that apply at a particular moment, not necessarily when the system is at equilibrium. By comparing the numerical value of Q to that of K, it is possible to decide in which direction the system will move to achieve equilibrium.

1. If $Q < K$

the reaction proceeds from left to right:

$$aA(g) + bB(g) \longrightarrow cC(g) + dD(g)$$

In this way, the partial pressures of products increase, while those of reactants decrease. As this happens, the reaction quotient Q increases and eventually at equilibrium becomes equal to K.

2. If $Q > K$

the partial pressures of products are "too high" and those of the reactants "too low" to meet the equilibrium condition. Reaction proceeds in the reverse direction:

$$cC(g) + dD(g) \longrightarrow aA(g) + bB(g)$$

increasing the partial pressures of A and B while reducing those of C and D. This lowers Q to its equilibrium value, K.

3. If, perchance, $Q = K$, the system is already at equilibrium, and nothing happens.

Example 12.5 Consider the following system at 100°C:

$$N_2O_4(g) \rightleftharpoons 2NO_2(g) \qquad K = 11$$

Predict the direction in which reaction will occur to reach equilibrium starting with 0.20 mol of N_2O_4 and 0.20 mol of NO_2 in a 4.0-L container.

$Q \rightarrow K$ as the system goes to equilibrium

Strategy First, calculate the partial pressures of N_2O_4 and NO_2, using the ideal gas law, as applied to mixtures: $P_i = n_iRT/V$. Then, calculate Q. Finally, compare Q to K to predict the direction of reaction.

Solution

1. $P_{N_2O_4} = P_{NO_2} = \dfrac{nRT}{V} = \dfrac{(0.20 \text{ mol})(0.0821 \text{ L}\cdot\text{atm/mol}\cdot\text{K})(373 \text{ K})}{4.0 \text{ L}} = 1.5 \text{ atm}$

2. $Q = \dfrac{(P_{NO_2})^2}{P_{N_2O_4}} = \dfrac{(1.5)^2}{1.5} = 1.5$

3. Since $Q < 11$, reaction must proceed in the forward direction. This way, the partial pressure of NO_2 will increase, that of N_2O_4 will decrease, and Q will become equal to K.

Equilibrium Partial Pressures

The equilibrium constant for a chemical system can be used to calculate the partial pressures of the species present at equilibrium. In the simplest case, one equilibrium pressure can be calculated, knowing all the others. Consider, for example, the system

$$N_2(g) + O_2(g) \rightleftharpoons 2NO(g)$$

$$K = \frac{(P_{NO})^2}{P_{N_2} \times P_{O_2}} = 1 \times 10^{-30} \text{ at } 25°C$$

In air, the partial pressures of N_2 and O_2 are about 0.78 and 0.21 atm, respectively. The expression for K can be used to calculate the equilibrium partial pressure of NO under these conditions:

$$(P_{NO})^2 = P_{N_2} \times P_{O_2} \times K$$

$$= (0.78)(0.21)(1 \times 10^{-30}) = 1.6 \times 10^{-31}$$

$$P_{NO} = (1.6 \times 10^{-31})^{1/2} = 4 \times 10^{-16} \text{ atm}$$

This is an extremely small pressure, as you might have guessed from the small magnitude of the equilibrium constant.

That's just as well; if atmospheric N_2 and O_2 were converted to NO, we'd be in big trouble

More commonly, K is used to determine the equilibrium partial pressures of all species, reactants and products, knowing their original pressures. To do this, you follow what amounts to a four-step path.

1. *Using the balanced equation for the reaction, write the expression for K.*
2. *Express the equilibrium partial pressures of all species in terms of a single unknown, x.* To do this, apply the principle mentioned earlier: *The changes in partial pressures of reactants and products are related through the coefficients of the balanced equation.* To keep track of these values, make an equilibrium table, like the one illustrated in Example 12.6.
3. *Substitute the equilibrium terms into the expression for K. This gives an algebraic equation which must be solved for x.*
4. *Having found x, refer back to the table and calculate the equilibrium partial pressures of all species.*

Example 12.6 For the system

$$CO_2(g) + H_2(g) \rightleftharpoons CO(g) + H_2O(g)$$

K is 0.64 at 900 K. Originally, only CO_2 and H_2 are present, each at a partial pressure of 1.00 atm. What are the equilibrium partial pressures of all species?

Strategy Follow the four-step procedure described above. Be particularly careful about step (2), the most critical one in the process.

Solution

(1) $K = \dfrac{P_{CO} \times P_{H_2O}}{P_{CO_2} \times P_{H_2}}$

(2) You must express all equilibrium partial pressures in terms of a single unknown, x. A reasonable choice for x is

$$x = \text{change in partial pressure of CO as the system goes to equilibrium}$$
$$= \Delta P_{CO}$$

All the coefficients in the balanced equation are the same, 1. It follows that all of the partial pressures change by the same amount, x.

$$\Delta P_{H_2O} = \Delta P_{CO} = x$$
$$\Delta P_{H_2} = \Delta P_{CO_2} = -x$$

(The minus sign arises because H_2 and CO_2 are consumed, while CO and H_2O are formed.) The equilibrium table becomes

	$CO_2(g)$ +	$H_2(g)$ $\rightleftharpoons$	$CO(g)$ +	$H_2O(g)$
P_o (atm)	1.00	1.00	0	0
ΔP (atm)	$-x$	$-x$	$+x$	$+x$
P_{eq} (atm)	$1.00 - x$	$1.00 - x$	x	x

(3) Substituting into the expression for K,

$$0.64 = \frac{(x)(x)}{(1.00 - x)(1.00 - x)}$$

Taking the square root of both sides,

Keep the math simple when you can

$$(0.64)^{1/2} = 0.80 = \frac{x}{1.00 - x}$$

Solving for x:

$$x = 0.80 - 0.80x \qquad 1.80x = 0.80 \qquad x = 0.44$$

(4) Referring back to the equilibrium table,

$$P_{CO} = P_{H_2O} = x = \boxed{0.44 \text{ atm}}$$

$$P_{CO_2} = P_{H_2} = 1.00 - x = \boxed{0.56 \text{ atm}}$$

To check the validity of the calculation, note that

$$\frac{P_{CO} \times P_{H_2O}}{P_{CO_2} \times P_{H_2}} = \frac{(0.44)^2}{(0.56)^2} = 0.64 = K$$

The arithmetic involved in equilibrium calculations can be relatively simple, as it was in Example 12.6. Sometimes it is more complex (Example 12.7). The reasoning involved, however, is the same. It is always helpful to set up an equilibrium table to summarize the analysis of the problem.

Example 12.7 Consider the system

$$N_2O_4(g) \rightleftharpoons 2NO_2(g) \qquad K = 11 \text{ at } 100°C$$

Starting with pure N_2O_4 at a pressure of 1.00 atm, what will be the equilibrium partial pressures of NO_2 and N_2O_4?

Strategy Again, follow the four-step path. The second step is a bit trickier here because the coefficients in the balanced equation are not equal to one another. The third step is more difficult algebraically, as you will find.

Solution

(1) $K = \dfrac{(P_{NO_2})^2}{P_{N_2O_4}}$

(2) Some of the N_2O_4 will decompose; let x be the *decrease* in the partial pressure of N_2O_4, i.e., $\Delta P_{N_2O_4} = -x$. The balanced equation shows that 1 N_2O_4 yields 2 NO_2; it follows that the *increase* in partial pressure of NO_2 must be $2x$. That is: $\Delta P_{NO_2} = 2x$. The equilibrium table becomes

	$N_2O_4(g)$ $\rightleftharpoons$	$2NO_2(g)$
P_o (atm)	1.00	0.00
ΔP (atm)	$-x$	$+2x$
P_{eq} (atm)	$1.00 - x$	$2x$

(3) $K = \dfrac{(2x)^2}{(1.00 - x)} = \dfrac{4x^2}{(1.00 - x)} = 11$

This time, you cannot solve for x as simply as in Example 12.6; the denominator on the left side is not a perfect square. Use the general method of solving a quadratic equation, discussed in Appendix 3. This involves rearranging to the form Damn!

$$ax^2 + bx + c = 0$$

and applying the quadratic formula

$$x = \frac{-b \pm \sqrt{b^2 - 4ac}}{2a}$$

Putting the equilibrium constant expression in the desired form,

$$4x^2 + 11x - 11 = 0$$

So

$$a = 4,\ b = 11,\ c = -11$$

$$x = \frac{-11 \pm \sqrt{121 + 16(11)}}{8} = \frac{-11 \pm \sqrt{297}}{8}$$

Noting that the square root of 297 is 17.2,

$$x = \frac{-11 \pm 17.2}{8} = \frac{6.2}{8} \text{ or } \frac{-28.2}{8} = 0.78 \text{ or } -3.52$$

One root will always be physically impossible

Of the two answers, only 0.78 is plausible; a value of −3.52 would imply a negative partial pressure of NO_2, which is ridiculous.

(4) $P_{NO_2} = 2(0.78 \text{ atm}) = $ **1.56 atm**

$P_{N_2O_4} = 1.00 \text{ atm} - 0.78 \text{ atm} = $ **0.22 atm**

One way to check the validity of this answer is to refer back to Table 12.1, where you find that, sure enough, starting with an N_2O_4 pressure of 1.00 atm, the values of the equilibrium partial pressures are those just calculated.

You may be curious as to what would have happened in Example 12.7 with a different choice of variable. Suppose, for example, x had been chosen to be the *increase* in the partial pressure of NO_2. In that case, the partial pressure of N_2O_4 at equilibrium would have decreased by $\frac{1}{2}x$, so

$$P_{NO_2} = x \qquad P_{N_2O_4} = 1.00 - \tfrac{1}{2}x$$

This leads to the algebraic equation

$$\frac{x^2}{(1.00 - \frac{1}{2}x)} = 11$$

Solving this equation, you should find that $x = 1.56$, so $P_{NO_2} = 1.56$ atm, $P_{N_2O_4} = 1.00 \text{ atm} - \frac{1}{2}(1.56 \text{ atm}) = 0.22$ atm. These are, of course, the same values obtained with a different choice of unknown in Example 12.7. In general, it doesn't matter what choice of unknown you make, provided you are consistent in relating equilibrium pressures.

12.5 EFFECT OF CHANGES IN CONDITIONS UPON AN EQUILIBRIUM SYSTEM

Once a system has attained equilibrium, it is possible to change the ratio of products to reactants by changing the external conditions. We will consider three ways in which a chemical equilibrium can be disturbed:

1. Adding or removing a gaseous reactant or product
2. Changing the volume of the system
3. Changing the temperature

You can deduce the direction in which an equilibrium will shift when one of these changes is made by applying **Le Chatelier's principle**.

If a system at equilibrium is disturbed by some change, the system will, if possible, shift so as to partially counteract the effect of the change.

Equilibrium systems, like people, resist change

Adding or Removing a Gaseous Species

According to Le Chatelier's principle, *if a chemical system at equilibrium is disturbed by adding a gaseous* species (reactant or product), the reaction will proceed in such a direction as to consume part of the added species. Conversely, if a gaseous species is removed, the system shifts to restore part of that species.*

To apply this general rule, consider the reaction

$$N_2O_4(g) \rightleftharpoons 2NO_2(g)$$

Suppose this system has reached equilibrium at a certain temperature. This equilibrium could be disturbed by

— *adding N_2O_4*. Here, reaction will occur in the forward direction (left to right). In this way, part of the N_2O_4 will be consumed.
— *adding NO_2*, which causes the reverse reaction (right to left) to occur, using up part of the NO_2 added.
— *removing N_2O_4*. Here, reaction occurs in the reverse direction to restore part of the N_2O_4.
— *removing NO_2*, which causes the forward reaction to occur, restoring part of the NO_2 removed.

It is possible to use K to calculate the extent to which reaction occurs when an equilibrium is disturbed by adding or removing a product or reactant. To show how this is done, consider the effect of adding hydrogen iodide to the $HI–H_2–I_2$ system (Example 12.8).

$$2HI(g) \rightleftharpoons H_2(g) + I_2(g)$$

Example 12.8 In Example 12.4 you found that this system is in equilibrium at 520°C when $P_{HI} = 0.80$ atm and $P_{H_2} = P_{I_2} = 0.10$ atm. Suppose enough HI is added to raise its pressure temporarily to 1.00 atm. When equilibrium is restored, what are P_{HI}, P_{H_2}, and P_{I_2}?

Strategy This problem is entirely analogous to Examples 12.6 and 12.7; the only real difference is that in this case the "original pressures" are those that prevail immediately after equilibrium is disturbed. Note that

$$Q = \frac{(0.10) \times (0.10)}{(1.00)^2} = 0.010 < K = 0.016$$

so some HI must dissociate to establish equilibrium.

Solution

(1) Recall from Example 12.4 that

$$K = \frac{P_{H_2} \times P_{I_2}}{(P_{HI})^2} = 0.016 \text{ at } 520°C$$

* Adding a pure liquid or solid has no effect on the system, as is implied by the discussion of heterogeneous equilibrium. p. 330.

(2) The partial pressure of HI must decrease; those of H_2 and I_2 must increase. Let x be the increase in partial pressure of H_2. Then

$$\Delta P_{I_2} = \Delta P_{H_2} = x \qquad \Delta P_{HI} = -2x$$

The equilibrium table becomes

	$2HI(g) \rightleftharpoons$	$H_2(g)$ +	$I_2(g)$
P_o (atm)	1.00	0.10	0.10
ΔP (atm)	$-2x$	$+x$	$+x$
P_{eq} (atm)	$1.00 - 2x$	$0.10 + x$	$0.10 + x$

3. Substituting into the expression for K,

$$0.016 = \frac{(0.10 + x)^2}{(1.00 - 2x)^2}$$

To solve this equation, take the square root of both sides:

$$0.13 = \frac{0.10 + x}{1.00 - 2x}$$

from which you should find that $x = 0.024$

4. $P_{H_2} = P_{I_2} = 0.10 + x =$ **0.12 atm**

$P_{HI} = 1.00 - 2x =$ **0.95 atm**

Note that the equilibrium partial pressure of HI is intermediate between its value before equilibrium was disturbed (0.80 atm) and that immediately afterward (1.00 atm). This is exactly what Le Chatelier's principle predicts; *part* of the added HI is consumed to re-establish equilibrium.

Changes in Volume

To understand how a change in container volume can change the position of an equilibrium, consider again the N_2O_4–NO_2 system

$$N_2O_4(g) \rightleftharpoons 2NO_2(g)$$

Le Chatelier was a French chemist (1850–1936)

Suppose the volume of the system is decreased (Fig. 12.3). The immediate effect is to increase the number of molecules per unit volume. According to Le Chatelier's principle, the system will shift so as to partially counteract this change. There is a simple way that this can happen. Some of the NO_2 molecules combine with each other to form N_2O_4 (diagram at right of Fig. 12.3). That is, reaction occurs in the reverse direction. Since two molecules of NO_2 form only one molecule of N_2O_4, this reduces the number of molecules.

It is possible, using the value of K, to calculate the extent to which NO_2 is converted to N_2O_4 when the volume is decreased. The results of such calculations are given in Table 12.4. As the volume is decreased from 10.0 to 1.0 L, more and more of the NO_2 is converted to N_2O_4. Notice that the total number of moles of gas decreases steadily as a result of this conversion.

The analysis we have gone through for the N_2O_4–NO_2 system can be applied to any chemical equilibrium involving gases. ***When the volume of an equilibrium system is decreased, reaction takes place in the direction that***

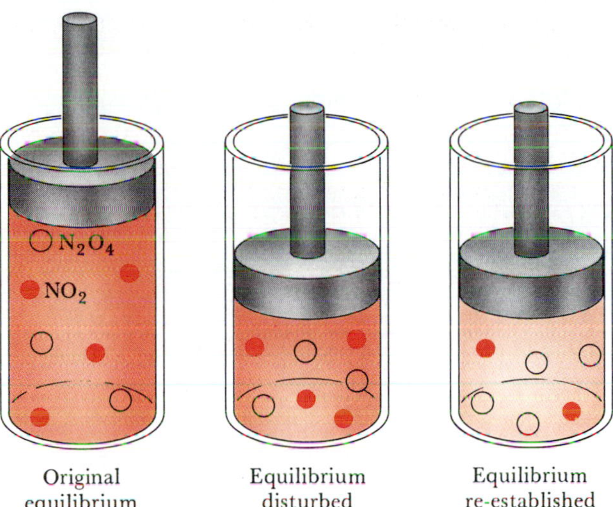

Figure 12.3
Effect of a decrease in volume upon the $N_2O_4(g) \rightleftharpoons 2NO_2(g)$ system at equilibrium. The immediate effect (*middle cylinder*) is to crowd the same number of moles of gas into a smaller volume and so to increase the total pressure. This is partially compensated for by the conversion of some of the NO_2 to N_2O_4, thereby reducing the total number of moles of gas.

decreases the total number of moles of gas. When the volume is increased, the reaction that increases the total number of moles of gas takes place.

The increased volume is filled by an increase in the number of molecules

The application of this principle to several different systems is shown in Table 12.5. In System 2, the number of moles of gas decreases, from $\frac{3}{2}$ to 1, as the reaction goes to the right. Hence, decreasing the volume causes the forward reaction to occur; an increase in volume has the reverse effect. Notice that it is the change in the number of moles of *gas* that determines which way the equilibrium shifts (System 4). When there is no change in the number of moles of gas (System 5), a change in volume has no effect on the position of the equilibrium.

Changes in volume of gaseous systems at equilibrium ordinarily result from changes in applied pressure. The volume of the N_2O_4–NO_2 system could be cut in half by increasing the pressure on the piston in Figure 12.3. Instead of saying that the shift in the position of the equilibrium comes about because of a change in volume, the shift is often attributed to the pressure change that accompanies the volume decrease. The statement is made that an *increase in pressure shifts the position of the equilibrium in such a way as to decrease the number of moles of gas.*

Table 12.4 Effect of Change in Volume on the Equilibrium System $N_2O_4(g) \rightleftharpoons 2NO_2(g)$; $K = 11$ at 100°C

V(L)	Moles NO_2	Moles N_2O_4	Moles Total
10.0	1.20	0.40	1.60
5.0	0.96	0.52	1.48
2.0	0.68	0.66	1.34
1.0	0.52	0.74	1.26

When V decreases, NO_2 is converted to N_2O_4

Table 12.5 Effect of a Change in Volume upon the Position of Gaseous Equilibria

System	*V* Increases	*V* Decreases
1. $N_2O_4(g) \rightleftharpoons 2NO_2(g)$	→	←
2. $SO_2(g) + \frac{1}{2}O_2(g) \rightleftharpoons SO_3(g)$	←	→
3. $N_2(g) + 3H_2(g) \rightleftharpoons 2NH_3(g)$	←	→
4. $C(s) + H_2O(g) \rightleftharpoons CO(g) + H_2(g)$	→	←
5. $N_2(g) + O_2(g) \rightleftharpoons 2NO(g)$	0	0

This statement is valid provided it is understood that the effect of the pressure increase is to compress the system, i.e., reduce the volume. There are, however, many other ways in which pressure can be changed without changing the volume. One way is to add an unreactive gas like helium at constant volume. This increases the total number of moles and hence the total pressure. It has no effect, however, upon the position of the equilibrium. In general, the position of an equilibrium is not shifted by adding an unreactive gas, since that does not change the concentrations or partial pressures of reactants or products.

Example 12.9 The pressure on each of the following systems is reduced so that the volume increases from 2 L to 10 L. Which way does the equilibrium shift?

(a) $2CO_2(g) \rightleftharpoons 2CO(g) + O_2(g)$
(b) $H_2(g) + I_2(g) \rightleftharpoons 2HI(g)$
(c) $H_2(g) + I_2(s) \rightleftharpoons 2HI(g)$

Strategy Using the coefficients of the balanced equation, determine what happens to the number of moles of gas. Then apply Le Chatelier's principle,

Solution

(a) → (2 mol gas → 3 mol gas). The immediate effect of the volume increase is to decrease the number of molecules per unit volume. This is partially compensated for by increasing the number of molecules.

(b) No effect. Same number of moles of gas on both sides.

(c) → (1 mol gas → 2 mol gas). Note that it is the number of moles of gas that is important; solids or liquids don't count.

Changes in Temperature

Le Chatelier's principle can be used to predict the effect of a change in temperature upon the position of an equilibrium. In general, *an increase in temperature causes the endothermic reaction to occur.* Applying this principle to the N_2O_4–NO_2 system,

$$N_2O_4(g) \rightleftharpoons 2NO_2(g) \qquad \Delta H° = +57.2 \text{ kJ}$$

it follows that raising the temperature causes the forward reaction to occur, since that reaction absorbs heat. This is confirmed by experiment (Fig. 12.4). At higher temperatures, more NO_2 is produced, and the reddish-brown color of that gas becomes more intense.

For the synthesis of ammonia,

$$N_2(g) + 3H_2(g) \rightleftharpoons 2NH_3(g) \qquad \Delta H° = -92.2 \text{ kJ}$$

an increase in temperature shifts the equilibrium to the left; some ammonia decomposes to the elements. This reflects the fact that the reverse reaction is endothermic:

$$2NH_3(g) \longrightarrow N_2(g) + 3H_2(g) \qquad \Delta H° = +92.2 \text{ kJ}$$

As pointed out earlier, the equilibrium constant of a system changes with temperature. The form of the equation relating K to T is a familiar one,

$$\ln\frac{K_2}{K_1} = \frac{\Delta H°}{R}\left[\frac{1}{T_1} - \frac{1}{T_2}\right]$$

where K_2 and K_1 are the equilibrium constants at T_2 and T_1, respectively, $\Delta H°$ is the standard enthalpy change for the forward reaction, and R is the gas constant, 8.31 J/mol · K.

To illustrate how this equation is used, let us apply it to calculate the equilibrium constant at 100°C for the system

$$N_2(g) + 3H_2(g) \rightleftharpoons 2NH_3(g) \qquad \Delta H° = -92.2 \text{ kJ}$$

given that $K = 6 \times 10^5$ at 25°C. The relation is

$$\ln\frac{K \text{ at } 100°\text{C}}{6 \times 10^5} = \frac{-92{,}200 \text{ J/mol}}{8.31 \text{ J/mol} \cdot \text{K}}\left[\frac{1}{298 \text{ K}} - \frac{1}{373 \text{ K}}\right] = -7.5$$

Taking inverse logarithms:

$$\frac{K \text{ at } 100°\text{C}}{6 \times 10^5} = 6 \times 10^{-4}$$

$$K \text{ at } 100°\text{C} = 4 \times 10^2$$

Notice that the equilibrium constant becomes smaller as the temperature increases. In general,

— if the forward reaction is exothermic, as is the case here ($\Delta H° = -92.2$ kJ), K decreases as T increases

— if the forward reaction is endothermic, as in the decomposition of N_2O_4:

$$N_2O_4(g) \rightleftharpoons 2NO_2(g) \qquad \Delta H° = +57.2 \text{ kJ}$$

K increases as T increases. For this reaction, K is 0.11 at 25°C and 11 at 100°C.

Figure 12.4
Effect of temperature on the N_2O_4–NO_2 system. At 0°C (*tube at right*), N_2O_4, which is colorless, predominates. At 50°C (*tube at left*), some of the N_2O_4 has dissociated to give the deep brown color of NO_2.
(Marna G. Clarke)

Example 12.10 Consider the system $I_2(g) \rightleftharpoons 2I(g)$; $\Delta H° = +151$ kJ. Suppose the system is at equilibrium at 1000°C. In which direction will reaction occur if

(a) I atoms are added? (b) the system is compressed?
(c) the temperature is increased?

Strategy Apply Le Chatelier's principle to predict the effect of adding a product (a), reducing the volume (b), or increasing the temperature (c).

Equilibrium between liquid bromine and its gas.
$Br_2(l) \rightleftharpoons Br_2(g)$

Solution

(a) $2I(g) \rightarrow I_2(g)$

(b) $2I(g) \rightarrow I_2(g)$ This decreases the number of particles per unit volume.

(c) $I_2(g) \rightarrow 2I(g)$ This reaction absorbs heat.

We should emphasize that of the three changes in conditions described in this section

— adding or removing a gaseous species
— changing the volume
— changing the temperature

the only one that changes the value of the equilibrium constant is a change in temperature. In the other two cases, K remains constant.

12.6 INDUSTRIAL APPLICATIONS OF GASEOUS EQUILIBRIA

A wide variety of materials, both pure substances and mixtures, are made by processes which involve one or more gas-phase reactions. Among these are three of the most important industrial chemicals: ammonia, nitric acid, and sulfuric acid.

Industrial Chemical	Annual U.S. Production, 1990
NH_3	1.5×10^{10} kg
HNO_3	0.70×10^{10} kg
H_2SO_4	4.0×10^{10} kg

The processes used to make these chemicals apply many of the principles of chemical equilibrium, discussed in this chapter, and chemical kinetics (Chap. 11).

Haber Process for Ammonia

Before Haber, all fertilizers were "organic"

Combined ("fixed") nitrogen in the form of protein is essential to all forms of life. There is more than enough elementary nitrogen in the air, about 4×10^{18} kg, to meet all our needs. The problem is to convert the element to compounds that can be used by plants to make proteins. At room temperature and atmospheric pressure, N_2 does not react with any nonmetal. However, in 1908 in Germany, Fritz Haber showed that nitrogen does react with hydrogen at high temperatures and pressures to form ammonia in good yield.

$$N_2(g) + 3H_2(g) \rightleftharpoons 2NH_3(g) \qquad \Delta H° = -92.2 \text{ kJ}$$

The Haber process, represented by this equation, is now the main source of fixed nitrogen. Its feasibility depends upon choosing conditions under which nitrogen and hydrogen react rapidly to give a high yield of ammonia. At 25°C and atmospheric pressure, the position of the equilibrium favors the formation of $NH_3 (K = 6 \times 10^5)$. Unfortunately, however, the rate of reaction is

The scientific career of Fritz Haber was characterized by theoretical studies in areas of emerging practical importance. Training as an organic chemist and self-development as a physical chemist permitted Haber to relate chemistry to engineering. His most important achievement was the synthesis of ammonia from nitrogen and hydrogen.

Haber's approach to the synthesis of ammonia was not original. The French chemist Le Chatelier first studied this reaction, but an explosion led him to abandon the project. The German chemist Walter H. Nernst, a contemporary of Haber, was the first to accomplish the high-pressure synthesis. Disagreement with Nernst's data led Haber to seek the optimum temperature, pressure, and catalyst. After obtaining about 8% ammonia at 600°C and 200 atm, Haber turned the process over to engineers. In 1918 he received the Nobel Prize for his work.

During World War I, Haber directed chemical warfare in Germany, introducing the use of chlorine and mustard gas into warfare. After the war, he attempted unsuccessfully to recover gold from the oceans. In this way, he hoped to repay Germany's war debts (which were never paid).

In 1912, Haber became director of the new Kaiser Wilhelm Institute for Chemistry. Under his influence, the Institute became world famous, attracting outstanding students and professors. In 1933 Haber was ordered by the Nazis to dismiss all Jewish workers at the Institute. He resigned instead. In his resignation letter he wrote, "For more than 40 years I have selected my collaborators on the basis of their intelligence and their character, not on the basis of their grandmothers. I am unwilling for the rest of my life to change this method which I have found so good."

Fritz Haber
(1868–1934)

(Supplied by Professor John Stock, University of Connecticut, Storrs, Connecticut)

virtually zero. Equilibrium is reached more rapidly by raising the temperature. However, since the synthesis of ammonia is exothermic, high temperatures reduce K and hence the yield of ammonia. High pressures, on the other hand, have a favorable effect on both the rate of the reaction and the position of the equilibrium (Table 12.6).

The goal of Haber's research was to find a catalyst to synthesize ammonia at a reasonable rate without going to very high temperatures. Nowadays, the catalyst used is a special mixture of iron, potassium oxide (K_2O), and aluminum oxide (Al_2O_3). The hydrogen and nitrogen gases must be carefully puri-

Table 12.6 Effect of Temperature and Pressure Upon the Yield of Ammonia in the Haber Process ($P_{H_2} = 3P_{N_2}$)

		Mole Percent NH_3 in Equilibrium Mixture				
°C	K	10 atm	50 atm	100 atm	300 atm	1000 atm
200	0.4	51	74	82	90	98
300	4×10^{-3}	15	39	52	71	93
400	2×10^{-4}	4	15	25	47	80
500	2×10^{-5}	1	6	11	26	57
600	3×10^{-6}	0.5	2	5	14	31

fied to remove traces of sulfur compounds, which "poison" the catalyst. Reaction takes place at 450°C and at a pressure of 200 to 600 atm. The ammonia formed is passed through a cooling chamber. Since ammonia boils at −33°C, it is condensed out as a liquid, separating it from unreacted nitrogen and hydrogen. The yield is typically less than 50%, so the reactants are recycled to produce more ammonia.

Ostwald Process for Nitric Acid

Ostwald was a pacifist and visionary who developed the science of color

About 15% of the ammonia made by the Haber process is later converted to nitric acid. The process by which this is done was developed by a German chemist, Wilhelm Ostwald. The reaction takes place in three steps. All of these are carried out at a relatively low pressure, 1 to 10 atm.

1. Ammonia is burned with ten times its volume of air. The desired reaction is

$$4NH_3(g) + 5\,O_2(g) \longrightarrow 4NO(g) + 6H_2O(g)$$

This reaction is carried out at 900°C; a gauze made of a platinum–rhodium alloy (90% Pt, 10% Rh) is used as a catalyst. Under these conditions, more than 90% of the ammonia is converted to nitrogen oxide, NO.

2. The gaseous mixture produced is mixed with more air. This lowers the temperature and brings about the reaction

$$2NO(g) + O_2(g) \rightleftharpoons 2NO_2(g)$$

3. The nitrogen dioxide produced in the second step is passed through water. In this way a solution of nitric acid is formed:

$$3NO_2(g) + H_2O(l) \longrightarrow NO(g) + 2HNO_3(aq)$$

The aqueous solution formed by this reaction contains about 60 mass percent of HNO_3. To obtain the anhydrous acid, H_2SO_4 is added, and the mixture is distilled. Nearly pure nitric acid (bp = 83°C) boils off and is condensed.

Example 12.11 Consider the second step in the Ostwald process:

$$2NO(g) + O_2(g) \rightleftharpoons 2NO_2(g) \qquad \Delta H° = -114.0 \text{ kJ}$$

What would you expect to happen to the rate of NO_2 formation and the yield of NO_2 at equilibrium if

(a) the pressure were increased by compressing the system?
(b) the temperature were increased?
(c) a catalyst were used?

Strategy Recall (Chap. 11) that rate increases with increases in concentration (pressure) and temperature. To deduce the effect on the equilibrium yield, apply Le Chatelier's principle.

Solution

(a) An increase in pressure should increase the rate (higher concentration of reactants). It should also increase the yield of NO_2 (3 mol gas → 2 mol gas).
(b) One would expect an increase in temperature to increase the rate. An in-

crease in temperature would decrease the yield of NO_2 since the forward reaction is exothermic.
(c) A suitable catalyst would speed up the reaction but would have no effect upon the yield of NO_2.

Contact Process for Sulfuric Acid

Sulfuric acid is made commercially by the three-step **"contact" process**:

1. Elemental sulfur is burned in air to form sulfur dioxide:

$$S(s) + O_2(g) \longrightarrow SO_2(g)$$

2. Sulfur dioxide is converted to sulfur trioxide by bringing it into "contact" with oxygen on the surface of a solid catalyst:

$$SO_2(g) + \tfrac{1}{2}O_2(g) \rightleftharpoons SO_3(g) \qquad \Delta H = -98.9 \text{ kJ}$$

This is the key step in the process and the most difficult to carry out. The catalyst used today is an oxide of vanadium, V_2O_5. Platinum is equally effective but is more expensive and more easily "poisoned" (made ineffective) by impurities in the gases.

Since the conversion of SO_2 to SO_3 is exothermic, the equilibrium constant decreases as the temperature rises (Table 12.7). Hence, a low temperature is preferred on equilibrium principles. Too low a temperature, however, reduces the rate to the point at which the reaction becomes impractical. The temperature used in the contact process represents a compromise between equilibrium and rate considerations. A mixture of SO_2 and O_2 is first passed over the catalyst at 600°C; about 80% of the SO_2 is quickly converted to SO_3. The sulfur trioxide is removed from the gas mixture and the remaining SO_2 and O_2 are recycled over a second catalyst bed. This time the temperature is lower, about 450°C. It takes a while under these conditions, but eventually the yield of SO_3 is raised to above 99%.

3. The sulfur trioxide formed in Step 2 is converted to sulfuric acid by reaction with water:

$$SO_3(g) + H_2O \longrightarrow H_2SO_4(aq)$$

This reaction cannot be carried out directly. If SO_3 is bubbled through water, H_2SO_4 is formed as a fog of tiny particles that are difficult to condense. Instead, SO_3 is absorbed in concentrated sulfuric acid to form an intermediate product, $H_2S_2O_7$, called pyrosulfuric acid. Subsequent dilution with water forms H_2SO_4:

$$\begin{array}{rcl} SO_3(g) + H_2SO_4(l) & \longrightarrow & H_2S_2O_7(l) \\ H_2S_2O_7(l) + H_2O & \longrightarrow & 2H_2SO_4(aq) \\ \hline SO_3(g) + H_2O & \longrightarrow & H_2SO_4(aq) \end{array}$$

Table 12.7 Equilibrium Constants for the Reaction $SO_2(g) + \tfrac{1}{2}O_2(g) \rightleftharpoons SO_3(g)$

t(°C)	200	300	400	500	600
K	1.0×10^6	1.3×10^4	5.9×10^2	60	10

CHAPTER HIGHLIGHTS

KEY CONCEPTS

1. *Write the expression for K corresponding to a chemical equation*
   (Examples 12.1, 12.2; Problems 5–12)
2. *Calculate K, knowing*
   —*appropriate K's for other reactions*
   (Problems 13–18)
   —*all the equilibrium partial pressures*
   (Example 12.3; Problems 19–22, 61)
   —*all the original and one equilibrium partial pressure*
   (Example 12.4; Problems 23, 24, 53, 54)
3. *Use the value of K to determine*
   —*the direction of reaction*
   (Example 12.5; Problems 25–28)
   —*equilibrium concentrations of all species*
   (Examples 12.6–12.8; Problems 31–44, 68, 71)
4. *Apply Le Chatelier's principle to predict the direction in which an equilibrium shifts when conditions are changed*
   (Examples 12.9–12.11; Problems 45–52, 55)

KEY EQUATIONS

Expression for K	$K = \dfrac{(P_C)^c \times (P_D)^d}{(P_A)^a \times (P_B)^b}$
Coefficient rule	$K' = K^n$
Reciprocal rule	$K'' = 1/K$
Multiple equilibrium	$K_3 = K_1 \times K_2$

KEY TERMS

endothermic	exothermic	partial pressure
equilibrium constant	mole	reaction quotient

SUMMARY PROBLEM

Consider bromine chloride, an interhalogen compound. It is formed by the reaction between red-orange bromine vapor and yellow chlorine gas; BrCl is itself a gas. The reaction is endothermic.

a. Write the chemical equation for the formation of BrCl, using simplest whole number coefficients.

b. Write the equilibrium expression for K.

c. At 400°C, after the reaction has reached equilibrium, the partial pressures of BrCl, Br_2, and Cl_2 are 1.87 atm, 1.00 atm, and 0.50 atm, respectively. Calculate K.

d. Initially, a 2.00-L flask contains Cl_2 at partial pressure 0.51 atm and Br_2 at partial pressure 0.34 atm. After equilibrium is established at 400°C, the partial pressure of BrCl is 0.46 atm. Calculate the equilibrium partial pressures of Cl_2 and Br_2.

e. Initially, a reaction vessel contains Br_2, Cl_2, and BrCl, all with partial pressures of 0.41 atm. In what direction will the reaction proceed at 400°C? What are the partial pressures of Br_2, Cl_2, and BrCl when equilibrium is established? Suppose enough BrCl is added to raise its pressure temporarily to 0.80 atm. In what direction will equilibrium shift? When equilibrium is restored, what are the equilibrium partial pressures of Br_2, Cl_2, and BrCl?

f. In what direction will this system shift if at equilibrium
  1. the volume is increased?
  2. helium gas is added?
  3. the temperature is increased?

Answers

a. $Br_2(g) + Cl_2(g) \rightleftharpoons 2BrCl(g)$ **b.** $K = \frac{(P_{BrCl})^2}{(P_{Br_2})(P_{Cl_2})}$

c. 7.0 **d.** 0.28 atm; 0.11 atm

e. →; $P_{Cl_2} = P_{Br_2} = 0.26$ atm; $P_{BrCl} = 0.70$ atm; ←; $P_{Cl_2} = P_{Br_2} = 0.28$ atm, $P_{BrCl} = 0.75$ atm

f. 1. no change; **2.** no change; **3.** →.

QUESTIONS & PROBLEMS

Establishment of Equilibrium

1. The following data are for the system

$$A(g) \rightleftharpoons 2B(g)$$

Time(s)	0	30	45	60	75	90
P_A (atm)	0.500	0.390	0.360	0.340	0.325	0.325
P_B (atm)	0.000	0.220	0.280	0.320	0.350	0.350

At what time(s) is the system in equilibrium?

2. For the system in Question 1,
  a. What is the partial pressure of B after 2 minutes?
  b. How does the rate of the forward reaction compare with the rate of the reverse reaction at 45 s? at 90 s?

3. Complete the table below for the reaction

$$A(g) + 2B(g) \rightleftharpoons C(g)$$

Time(s)	0	20	40	60	80	100
P_A (atm)	0.200	0.150	____	____	0.100	____
P_B (atm)	0.250	____	____	0.070	____	____
P_C (atm)	0.000	____	0.075	____	____	0.100

4. The following data apply to the *unbalanced* equation

$$A(g) \rightleftharpoons B(g)$$

Time (min)	0	2	4	6	8
P_A (atm)	0.600	0.480	0.390	0.345	0.315
P_B (atm)	0.000	0.040	0.070	0.085	0.095

  a. Based on the data, balance the equation (simplest whole-number coefficients).
  b. Has the system reached equilibrium? Explain.

Expressions for *K*

5. Write equilibrium constant (K) expressions for the following reactions.
  a. $CS_2(g) + 4H_2(g) \rightleftharpoons CH_4(g) + 2H_2S(g)$
  b. $2H_2O_2(g) \rightleftharpoons 2H_2O(g) + O_2(g)$
  c. $C_3H_8(g) + 5\,O_2(g) \rightleftharpoons 3CO_2(g) + 4H_2O(g)$

6. Write equilibrium constant expressions (K) for the following reactions.
  a. $IF(g) \rightleftharpoons \frac{1}{2}I_2(g) + \frac{1}{2}F_2(g)$
  b. $2C_5H_6(g) \rightleftharpoons C_{10}H_{12}(g)$
  c. $P_4O_{10}(g) + 6PCl_5(g) \rightleftharpoons 10POCl_3(g)$

7. Write equilibrium constant expressions (K) for the following reactions
  a. $ZnO(s) + CO(g) \rightleftharpoons Zn(l) + CO_2(g)$
  b. $2HgO(s) \rightleftharpoons 2Hg(l) + O_2(g)$
  c. $2PbS(s) + 3\,O_2(g) \rightleftharpoons 2PbO(s) + 2SO_2(g)$

8. Write equilibrium constant expressions (K) for the following reactions.
  a. $Ni(s) + 4CO(g) \rightleftharpoons Ni(CO)_4(g)$
  b. $2Ag_2O(s) \rightleftharpoons 4Ag(s) + O_2(g)$
  c. $4KO_2(s) + 2H_2O(g) \rightleftharpoons 4KOH(s) + 3O_2(g)$

9. Given the following descriptions of reversible reactions, write a balanced equation (simplest whole-number coefficients) and the equilibrium constant expression (K) for each.
  a. Liquid acetone, C_3H_6O, is in equilibrium with its vapor.
  b. Hydrogen gas reduces nitrogen dioxide to form ammonia and steam.
  c. Chlorine gas reacts with liquid carbon disulfide to produce the liquids carbon tetrachloride and disulfur dichloride.

10. Given the following descriptions of reversible reactions, write a balanced equation (simplest whole-number coefficients) and the equilibrium expression (K) for each.
  a. Nitrogen gas reacts with solid sodium carbonate and solid carbon to produce carbon monoxide gas and solid sodium cyanide.
  b. Solid magnesium nitride reacts with water vapor to form magnesium hydroxide solid and ammonia gas.
  c. Solid barium carbonate decomposes to solid barium oxide and carbon dioxide gas.

11. Write a chemical equation for an equilibrium system that would lead to the following expressions (a–e) for K:
  a. $\frac{(P_{O_2})^3}{(P_{O_3})^2}$ **b.** $\frac{(P_{SO_3})^2}{(P_{O_2}) \times (P_{SO_2})^2}$

c. $\frac{(P_{NH_3})^2}{(P_{N_2}) \times (P_{H_2})^3}$ d. $\frac{(P_{H_2O})^2 \times (P_{Cl_2})^2}{(P_{O_2}) \times (P_{HCl})^4}$ e. $\frac{P_{H_2O}}{P_{H_2}}$

12. Write a chemical equation for an equilibrium system that would lead to the following expressions for K:

a. $\frac{(P_{H_2}) \times (P_{Br_2})}{(P_{HBr})^2}$ b. $\frac{(P_{CH_3OH})}{(P_{CO}) \times (P_{H_2})^2}$

c. $\frac{(P_{C_2H_4}) \times (P_{H_2})}{(P_{C_2H_6})}$ d. $\frac{(P_{SO_3})}{(P_{SO_2}) \times (P_{O_2})^{1/2}}$

e. $\frac{(P_{CO_2}) \times (P_{H_2})}{(P_{CO})}$

Calculation of K

13. At 25°C, $K = 2.2 \times 10^{-3}$ for the reaction

$$ICl(g) \rightleftharpoons \tfrac{1}{2}I_2(g) + \tfrac{1}{2}Cl_2(g)$$

Calculate K at 25°C for
a. $2ICl(g) \rightleftharpoons I_2(g) + Cl_2(g)$
b. $2I_2(g) + 2Cl_2(g) \rightleftharpoons 4ICl(g)$

14. At 627°C, $K = 0.76$ for the reaction

$$2SO_2(g) + O_2(g) \rightleftharpoons 2SO_3(g)$$

Calculate K at 627°C for
a. $2SO_3(g) \rightleftharpoons 2SO_2(g) + O_2(g)$
b. $4SO_2(g) + 2\,O_2(g) \rightleftharpoons 4SO_3(g)$

15. Given that K for the reactions

$$SnO_2(s) + 2H_2(g) \rightleftharpoons Sn(s) + 2H_2O(g)$$

$$CO(g) + H_2O(g) \rightleftharpoons CO_2(g) + H_2(g)$$

is 21 and 0.034, respectively, calculate K for the reaction $SnO_2(s) + 2CO(g) \rightleftharpoons Sn(s) + 2CO_2(g)$

16. Given that K for the reactions

$$H_2(g) + S(s) \rightleftharpoons H_2S(g)$$

$$S(s) + O_2(g) \rightleftharpoons SO_2(g)$$

is 1.0×10^{-3} and 5.0×10^6, respectively, calculate K for the reaction $H_2(g) + SO_2(g) \rightleftharpoons H_2S(g) + O_2(g)$

17. Given the following data at 25°C

$$2NO(g) \rightleftharpoons N_2(g) + O_2(g) \qquad K = 1 \times 10^{30}$$

$$2NO(g) + Br_2(g) \rightleftharpoons 2NOBr(g) \qquad K = 8 \times 10^1$$

Calculate K for the formation of one mole of NOBr from its elements in the gaseous state at 25°C.

18. Given the following data at a certain temperature

$$2N_2(g) + O_2(g) \rightleftharpoons 2N_2O(g) \qquad K = 1.2 \times 10^{-35}$$

$$N_2O_4(g) \rightleftharpoons 2NO_2(g) \qquad K = 4.6 \times 10^{-3}$$

$$\tfrac{1}{2}N_2(g) + O_2(g) \rightleftharpoons NO_2(g) \qquad K = 4.1 \times 10^{-9}$$

Calculate K for the reaction between one mole of dinitrogen oxide gas and oxygen gas to give dinitrogen tetraoxide gas.

19. Carbon dioxide reacts with solid carbon to give carbon monoxide. At 700°C, the equilibrium partial pressures are $P_{CO} = 0.43$ atm, $P_{CO_2} = 0.092$ atm. At equilibrium, the flask also contains 0.44 mol of carbon.
a. Write a balanced equation for the reaction of one mole of carbon dioxide with carbon.
b. Calculate K for the reaction at 700°C.

20. At 227°C, carbon monoxide gas reacts with hydrogen gas to form one mole of methanol, $CH_3OH(g)$. At equilibrium and 227°C, the partial pressures of the gases in the flask are $P_{CO} = 0.295$ atm, $P_{H_2} = 1.25$ atm, $P_{CH_3OH} = 2.18$ mm Hg.
a. Write a balanced equation for the reaction.
b. Calculate K for the reaction at 227°C.

21. Calculate K at 1000°C for

$$2CO_2(g) \rightleftharpoons 2CO(g) + O_2(g)$$

given that the equilibrium concentrations of carbon monoxide, oxygen, and carbon dioxide at that temperature are $2.0 \times 10^{-6}\,M$, $1.0 \times 10^{-6}\,M$, and $0.25\,M$ respectively.

22. For the system

$$CO(g) + 3H_2(g) \rightleftharpoons CH_4(g) + H_2O(g)$$

analysis shows that at 1400 K, the concentrations of methane, steam, hydrogen gas, and carbon monoxide gas are 0.150 M, 0.233 M, 0.259 M, and 0.513 M, respectively. Calculate K at 1400 K.

23. When nitrogen oxide gas and bromine gas with partial pressures of 1.577 atm and 0.427 atm, respectively, are sealed in a one-liter container at 77°C, the following equilibrium is established.

$$2NO(g) + Br_2(g) \rightleftharpoons 2NOBr(g)$$

Given that the equilibrium partial pressure of NOBr at 77°C is 0.624 atm, calculate K at 77°C.

24. Consider the decomposition of nitrosyl chloride, NOCl.

$$2NOCl(g) \rightleftharpoons 2NO(g) + Cl_2(g)$$

At 220°C in a sealed flask, originally, $P_{NOCl} = 1.000$ atm, $P_{NO} = 0.009$ atm, $P_{Cl_2} = 0.467$ atm. When equilibrium is established at 220°C it is found that the nitrosyl chloride partial pressure has dropped by 11.6%. Calculate K for the decomposition at 220°C.

K; Direction and Extent of the Reaction

25. A gaseous reaction mixture contains 0.30 atm SO_2, 0.16 atm Cl_2, and 0.50 atm SO_2Cl_2 in a 2.0-L container. $K = 0.011$ for the equilibrium system

$$SO_2Cl_2(g) \rightleftharpoons SO_2(g) + Cl_2(g)$$

a. Is the system at equilibrium? Explain.
b. If it is not at equilibrium, in which direction will the system move to reach equilibrium?

26. K is 26 at 300°C for the system

$$PCl_5(g) \rightleftharpoons PCl_3(g) + Cl_2(g)$$

In a 5.0-L flask, a gaseous mixture consists of all three gases with partial pressures as follows: $P_{PCl_5} = 0.012$ atm, $P_{Cl_2} = 0.45$ atm, $P_{PCl_3} = 0.90$ atm.

a. Is the mixture at equilibrium? Explain.
b. If it is not at equilibrium, which way will the system shift to establish equilibrium?

27. For the reaction

$$2NO_2(g) \rightleftharpoons 2NO(g) + O_2(g)$$

K at a certain temperature is 0.50. Predict the direction in which the reaction will occur to establish equilibrium if one starts with

a. $P_{NO_2} = P_{NO} = P_{O_2} = 0.10$ atm.
b. $P_{NO_2} = 0.0848$ atm, $P_{O_2} = 0.0116$ atm.
c. $P_{NO_2} = 0.20$ atm, $P_{NO} = 0.040$ atm, $P_{O_2} = 0.010$ atm.

28. The reversible reaction between hydrogen chloride gas and one mole of oxygen produces steam and chlorine gas: $4HCl(g) + O_2(g) \rightleftharpoons 2H_2O(g) + 2Cl_2(g)$; $K = 1.6$. Predict the direction in which the system will move to reach equilibrium if one starts with

a. $P_{HCl} = P_{O_2} = P_{H_2O} = 0.20$ atm.
b. $P_{HCl} = 0.30$ atm, $P_{O_2} = 0.15$ atm, $P_{H_2O} = 0.35$ atm, $P_{Cl_2} = 0.2$ atm.

29. For the reaction

$$Cl_2(g) \rightleftharpoons 2Cl(g)$$

$K = 1.4 \times 10^{-4}$ at 1100°C. If one starts with 0.5 atm Cl_2 at 1100°C, will the equilibrium system contain

a. mostly Cl_2? b. mostly Cl atoms?
c. about equal amounts of Cl_2 and Cl?

30. At another temperature, K for the reaction in Problem 29 is 3.6×10^{-1}. At this temperature, will the equilibrium system contain mostly Cl_2, mostly Cl, or appreciable amounts of both species?

K; Equilibrium Pressures

31. For the reaction

$$2NO_2(g) \rightleftharpoons 2NO(g) + O_2(g)$$

K is 1.74×10^{-4} at 500 K. Calculate P_{NO} if $P_{O_2} = 0.127$ atm and $P_{NO_2} = 0.384$ atm at equilibrium.

32. Consider the reaction

$$2NO(g) + Cl_2(g) \rightleftharpoons 2NOCl(g)$$

At a certain temperature, $K = 0.30$. Calculate P_{Cl_2} if $P_{NO} = 0.65$ atm and $P_{NOCl} = 0.15$ atm at equilibrium.

33. At a certain temperature, K is 1.3×10^5 for the reaction

$$2H_2(g) + S_2(g) \rightleftharpoons 2H_2S(g)$$

What is the equilibrium pressure of hydrogen sulfide if those of hydrogen and sulfur gases are 0.0200 atm and 0.0400 atm, respectively?

34. At a certain temperature, K is 0.040 for the decomposition of two moles of bromine chloride to its elements. An equilibrium mixture at this temperature contains bromine and chlorine gases at equal partial pressures of 0.0325 atm. What is the equilibrium partial pressure of bromine chloride?

35. For the system

$$PCl_5(g) \rightleftharpoons PCl_3(g) + Cl_2(g)$$

$K = 3.33$ at a certain temperature. In an equilibrium mixture at that temperature, $P_{PCl_3} = 4.0(P_{PCl_5})$. What is the equilibrium partial pressure of Cl_2?

36. For the system

$$2HI(g) \rightleftharpoons H_2(g) + I_2(g)$$

$K = 0.016$ at 800 K. If, at 800 K, $P_{HI} = 0.50$ atm, and $P_{H_2} = P_{I_2}$, calculate the equilibrium partial pressure of H_2.

37. At 460°C, the reaction

$$SO_2(g) + NO_2(g) \rightleftharpoons NO(g) + SO_3(g)$$

has $K = 85.0$. What will be the equilibrium partial pressures of the four gases if a mixture of SO_2 and NO_2 is prepared in which they both have initial partial pressures of 0.075 atm?

38. At 650°C, the reaction

$$H_2(g) + CO_2(g) \rightleftharpoons CO(g) + H_2O(g)$$

has a K value of 0.771. If hydrogen and carbon dioxide, each at a partial pressure of 0.500 atm, are placed in a 4.00-L container and allowed to react, what will be the equilibrium partial pressures of all four gases?

39. Hydrogen gas and cyanogen gas, C_2N_2, react at a certain temperature to form hydrogen cyanide. The equation for the reaction is

$$H_2(g) + C_2N_2(g) \rightleftharpoons 2HCN(g)$$

K for this system is 1.50. A mixture of hydrogen at a partial pressure of 0.200 atm, cyanogen with a partial pressure of 0.150 atm, and hydrogen cyanide with a partial pressure of 0.0100 atm is sealed in a 1.00-L flask. Calculate equilibrium pressures for all gases.

40. Carbonyl bromide, $COBr_2$, can be formed by reacting carbon monoxide with bromine gas. The equation for the reaction is

$$CO(g) + Br_2(g) \rightleftharpoons COBr_2(g)$$

Take K for this system to be 5.26. A mixture of carbon monoxide, bromine, and carbonyl bromide with partial pressures 0.0800 atm, 0.0600 atm, and 0.0040 atm, respectively, is sealed in a flask. Calculate equilibrium pressures for all gases.

41. For the reaction

$$CO(g) + H_2O(g) \rightleftharpoons CO_2(g) + H_2(g)$$

equilibrium is established at a certain temperature when the partial pressures of CO, H_2O, CO_2, and H_2 are 0.010, 0.020, 0.012, and 0.012 atm, respectively.

a. Calculate K.
b. If enough CO is added to raise its partial pressure temporarily to 0.020 atm, what will be the partial pressures of all species after equilibrium has been re-established?

42. Water gas, a commercial fuel, is made by reacting hot coke with steam:

$$C(s) + H_2O(g) \rightleftharpoons CO(g) + H_2(g)$$

When equilibrium is established at 1000 K, the partial pressures of CO, H_2, and H_2O are 0.40 atm, 0.40 atm, and 0.062 atm, respectively.

a. Calculate K.

b. If enough steam is added to raise its partial pressure temporarily to 0.60 atm, what will the partial pressures of CO and H_2 be when equilibrium is re-established?

43. For the formation of nitrogen oxide,

$$N_2(g) + O_2(g) \rightleftharpoons 2NO(g)$$

K is 2.5×10^{-3} at 2400°C. In a certain experiment, the original partial pressures of N_2 and O_2 are 0.35 atm. What is the partial pressure of NO when equilibrium is established?

44. Sodium bicarbonate is often used in household fire extinguishers. It decomposes to sodium carbonate, carbon dioxide, and steam according to the equation

$$2NaHCO_3(s) \rightleftharpoons Na_2CO_3(s) + CO_2(g) + H_2O(g)$$

At 125°C, K for the reaction is 0.25. What are the partial pressures of CO_2 and H_2O when equilibrium is established after decomposition?

Le Chatelier's Principle

45. Consider the system

$$SO_3(g) \rightleftharpoons SO_2(g) + \tfrac{1}{2}O_2(g)$$

ΔH for the forward reaction is 98.9 kJ. Predict whether the forward or reverse reaction will occur when the equilibrium is disturbed by

a. adding oxygen gas.

b. compressing the system at constant temperature.

c. adding argon gas.

d. decreasing the temperature.

e. removing sulfur dioxide gas.

46. Consider the system

$$4NH_3(g) + 3\,O_2(g) \rightleftharpoons 2N_2(g) + 6H_2O(l) \qquad \Delta H = -1530.4\text{ kJ}$$

How will the amount of ammonia at equilibrium be affected by

a. removing oxygen gas?

b. adding nitrogen gas?

c. adding water?

d. expanding the container at constant temperature?

e. increasing the temperature?

47. Predict the direction in which each of the following equilibria will shift if the pressure on the system is decreased by expansion.

a. $Ni(s) + 4CO(g) \rightleftharpoons Ni(CO)_4(g)$

b. $ClF_5(g) \rightleftharpoons ClF_3(g) + F_2(g)$

c. $HBr(g) \rightleftharpoons \tfrac{1}{2}H_2(g) + \tfrac{1}{2}Br_2(g)$

48. Predict the direction in which each of the following equilibria will shift if the pressure on the system is increased by compression.

a. $H_2O(g) + C(s) \rightleftharpoons CO(g) + H_2(g)$

b. $SbCl_5(g) \rightleftharpoons SbCl_3(g) + Cl_2(g)$

c. $CO(g) + H_2O(g) \rightleftharpoons CO_2(g) + H_2(g)$

49. The system

$$2X(g) + Z(g) \rightleftharpoons Q(g)$$

is at equilibrium when the partial pressure of X is 0.60 atm. Sufficient X is added to raise its partial pressure temporarily to 1.0 atm. When equilibrium is re-established, the partial pressure of X could be which of the following?

a. 0.80 atm **b.** 0.60 atm **c.** 0.90 atm
d. 1.0 atm **e.** 1.2 atm

50. For the system in Problem 49, the initial partial pressure of Z was 0.40 atm. When equilibrium is re-established, the partial pressure of Z could be

a. 0.40 atm **b.** 0.50 atm **c.** 0.35 atm
d. 0.30 atm **e.** 0.00 atm

51. What is the value of $\Delta H°$ for a reaction if K doubles when the temperature increases from 25 to 35°C?

52. For the system $CaCO_3(s) \rightarrow CaO(s) + CO_2(g)$, $K = 1.0$ at 1100 K. Using information in Chapter 8, estimate K at 25°C.

Industrial Applications

53. One of the steps in the manufacture of sulfuric acid is

$$SO_2(g) + \tfrac{1}{2}O_2(g) \rightleftharpoons SO_3(g)$$

At 627°C, O_2 and SO_2 with partial pressures of 0.61 atm and 0.65 atm, respectively, are placed in a sealed container. When equilibrium is established, the equilibrium pressure of SO_3 is found to be 0.49 atm. Calculate K for the reaction at 627°C.

54. A mixture of N_2 and H_2 with original partial pressures of 0.206 atm and 0.318 atm, respectively, is allowed to come to equilibrium; at equilibrium, the partial pressure of NH_3 is 0.0122 atm.

a. Calculate K for the system $N_2(g) + 3H_2(g) \rightleftharpoons 2NH_3(g)$.

b. Using Table 12.6, estimate the temperature at which this equilibrium is established.

55. For the Haber process,

$$N_2(g) + 3H_2(g) \rightleftharpoons 2NH_3(g)$$

K is 6×10^5 at 25°C and 4×10^{-4} at 367°C. Is the forward reaction exothermic or endothermic? Based only on equilibrium considerations, what are the optimal conditions (T and P) for the Haber process?

56. The second step of the Ostwald process for making HNO_3 is

$$2NO(g) + O_2(g) \rightleftharpoons 2NO_2(g)$$

An equilibrium mixture consists of 1.40 mol O_2, 0.40 mol NO, and 0.019 mol NO_2. Calculate K for the reaction, given that the total pressure is 0.50 atm.

Acid Rain

57. Almost everywhere in North America and Europe, except in the western United States, sulfuric acid is the major component of acid rain.

a. Write an equation for the reaction that changes SO_3 to H_2SO_4.
b. Most of the SO_3 present in the atmosphere comes from SO_2. What are the two major sources of SO_2?
c. Write an equation for the conversion of SO_2 to SO_3.

58. Explain
a. the adverse effect that acid rain has on trees.
b. the effect of acid rain on marble statuary.

59. How can sulfur dioxide emissions from power plants and smelters be brought under control?

60. The reaction for the "scrubber" operation of a smokestack is

$$Ca^{2+}(aq) + 2\,OH^-(aq) + SO_2(g) + \tfrac{1}{2}\,O_2(g) \longrightarrow CaSO_4(s) + H_2O$$

Smokestack gas contains 0.85% SO_2 by volume at 75°C and a total pressure of 1.00 atm. How many grams of $CaSO_4$ are produced when 200.0 L of the stack gas are "scrubbed" with a solution of $Ca(OH)_2$?

Unclassified

61. Consider the equilibrium

$$C(s) + CO_2(g) \rightleftharpoons 2CO(g)$$

When this system is at equilibrium at 700°C in a 2.0-L container, there are 0.10 mol CO, 0.20 mol CO_2, and 0.40 mol C present. When the system is cooled to 600°C, an additional 0.040 mol C(*s*) forms. Calculate *K* at 700°C and again at 600°C.

62. An examination question asked students to calculate *K* for the reaction

$$CS_2(g) + 4H_2(g) \rightleftharpoons CH_4(g) + 2H_2S(g)$$

given that, at equilibrium in a 5.0-L vessel, the partial pressures are $P_{CH_4} = 0.55$ atm, $P_{H_2S} = 0.125$ atm, $P_{CS_2} = 0.15$ atm, $P_{H_2} = 0.15$ atm. Four students gave the following answers, all wrong. Explain the error(s) in their reasoning.

a. $K = \dfrac{(0.15)(0.15)^4}{(0.55)(0.125)^2}$ **b.** $K = \dfrac{(0.55)(0.25)^2}{(0.15)(0.60)^4}$

c. $K = \dfrac{(0.55) + (0.125)^2}{(0.15) + (0.15)^4}$ **d.** $K = \dfrac{(0.11)(0.025)^2}{(0.03)(0.03)^4}$

63. Phosphorus pentachloride dissociates to phosphorus trichloride and chlorine:

$$PCl_5(g) \rightleftharpoons PCl_3(g) + Cl_2(g)$$

A sealed vessel at 250°C originally contains PCl_5 at 0.27 atm. What is the total pressure at 250°C when equilibrium is reached? *K* at 250°C is 2.1.

64. Consider the statement, "The equilibrium constant of a mixture of graphite, carbon dioxide, and carbon monoxide is 5.28." What information is missing from this statement?

65. At 1800 K, oxygen dissociates into gaseous atoms:

$$O_2(g) \rightleftharpoons 2\,O(g)$$

K for the system is 1.7×10^{-8}. If one mole of O_2 is placed in a 5.0-L flask and heated to 1800 K, what percent by mass of the oxygen dissociates? How many O atoms are there in the flask?

Challenge Problems

66. Derive the relationship

$$K = K_c \times (RT)^{\Delta n_g}$$

where K_c is the equilibrium constant using molarities and Δn_g is the change in the number of moles of gas in the reaction (see p. 327). (*Hint:* Recall that $P_A = n_ART/V$ and $n_A/V = [A]$.)

67. Hydrogen iodide gas decomposes to hydrogen gas and iodine gas:

$$2HI(g) \rightleftharpoons H_2(g) + I_2(g)$$

To determine the equilibrium constant of the system, identical one-liter glass bulbs are filled with 3.20 g of HI and maintained at a certain temperature. Each bulb is periodically opened and analyzed for iodine formation by titration with sodium thiosulfate, $Na_2S_2O_3$.

$$I_2(aq) + 2S_2O_3^{2-}(aq) \longrightarrow S_4O_6^{2-}(aq) + 2I^-(aq)$$

It is determined that when equilibrium is reached, 37.0 mL of 0.200 *M* $Na_2S_2O_3$ is required to titrate the iodine. What is *K* at the temperature of the experiment?

68. For the system

$$SO_3(g) \rightleftharpoons SO_2(g) + \tfrac{1}{2}\,O_2(g)$$

at 1000 K, $K = 0.45$. Sulfur trioxide, originally at 1.00 atm pressure, partially dissociates to SO_2 and O_2 at 1000 K. What is its partial pressure at equilibrium?

69. At a certain temperature, the reaction

$$Xe(g) + 2F_2(g) \rightleftharpoons XeF_4(g)$$

gives a 50.0% yield of XeF_4, starting with Xe ($P_{Xe} = 0.20$ atm) and F_2 ($P_{F_2} = 0.40$ atm). Calculate *K* at this temperature. What must the initial pressure of F_2 be to convert 75.0% of the xenon to XeF_4?

70. A student studies the equilibrium

$$I_2(g) \rightleftharpoons 2I(g)$$

at a high temperature. He finds that the total pressure at equilibrium is 40% greater than it was originally, when only I_2 was present. What is *K* for this reaction at that temperature?

71. Benzaldehyde, a flavoring agent, is obtained by the dehydrogenation of benzyl alcohol

$$C_6H_5CH_2OH(g) \rightleftharpoons C_6H_5CHO(g) + H_2(g)$$

K for the reaction at 250°C is 0.56. If 1.50 g of benzyl alcohol is placed in a 2.0-L flask and heated to 250°C,
a. what is the partial pressure of the benzaldehyde when equilibrium is established?
b. How many grams of benzyl alcohol remain at equilibrium?

13
Acids and Bases

There is nothing in the Universe but alkali and acid, From which Nature composes all things.

OTTO TACHENIUS (1671)

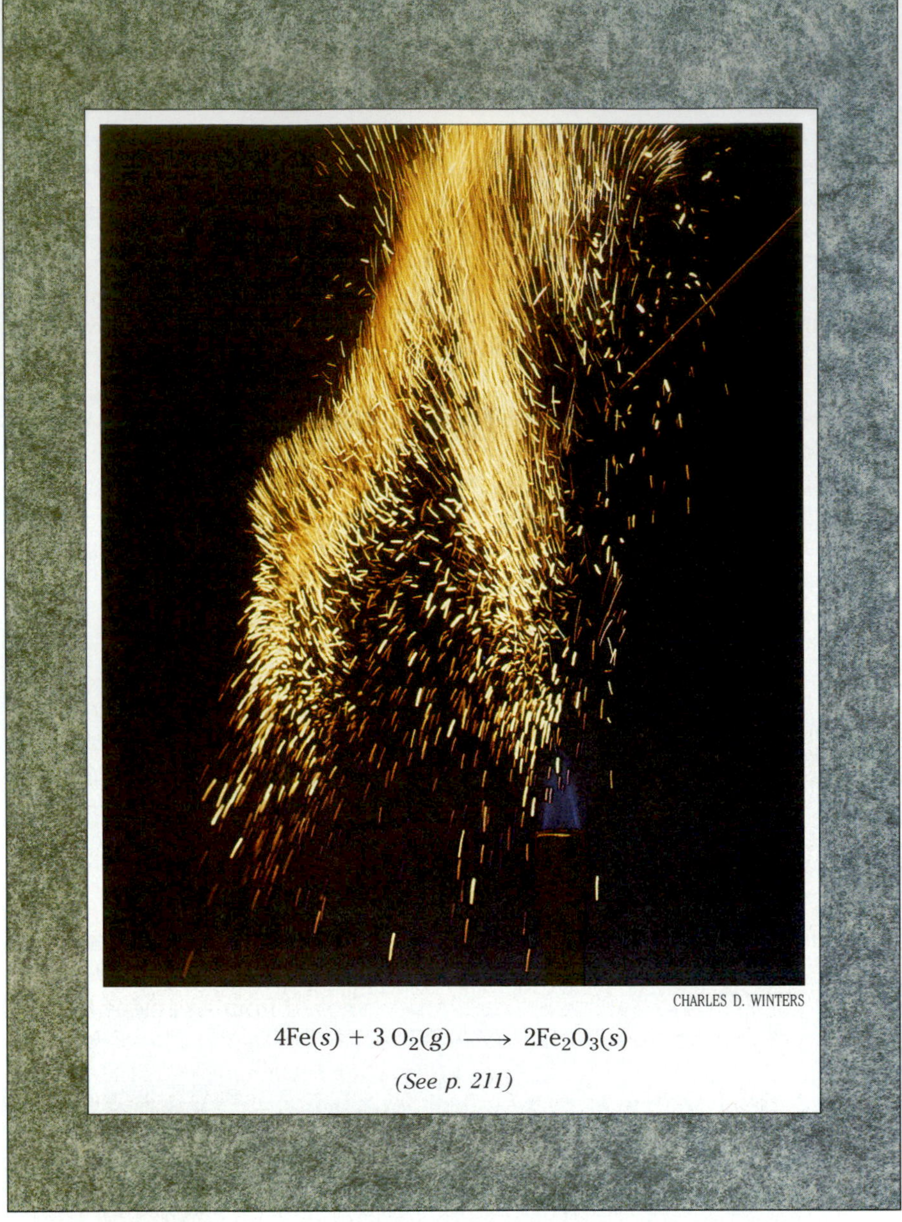

CHARLES D. WINTERS

$$4Fe(s) + 3\,O_2(g) \longrightarrow 2Fe_2O_3(s)$$

(See p. 211)

CHAPTER OUTLINE

Among the solution reactions considered in Chapter 4 were those between acids and bases. In this chapter, we take a closer look at the properties of acidic and basic water solutions. In particular, we examine

— the dissociation of water and the equilibrium between hydrated H^+ ions (H_3O^+ ions) and OH^- ions in water solution (Section 13.2)

— the quantities pH and pOH, used to describe the acidity and/or basicity of water solutions (Section 13.3)
— the types of species that act as weak acids and weak bases and the equilibria that apply in their water solutions (Sections 13.4, 13.5)
— the acid-base properties of salt solutions (Section 13.6)

13.1 BRØNSTED-LOWRY ACID-BASE MODEL

You will recall that in Chapter 4 we considered acids and bases to be species that supply H^+ ions and OH^- ions, respectively, to water. The model of acids and bases used in this chapter is a somewhat more general one developed independently by Johannes Brønsted in Denmark and Thomas Lowry in England in 1923. The **Brønsted-Lowry** model considers that

Sometimes called the Lowry-Brønsted model, at least in England

— an acid is a proton donor
— a base is a proton acceptor
— in an acid-base reaction, a proton is transferred from an acid to a base.

A proton is an H^+ ion

A Brønsted-Lowry acid-base reaction might be represented as

$$HB(aq) + A^-(aq) \rightleftharpoons HA(aq) + B^-(aq)$$

The species shown in red, HB and HA, act as Brønsted-Lowry acids in the forward and reverse reactions, respectively; A^- and B^- (blue) act as Brønsted-Lowry bases. (We will use this color coding consistently throughout the chapter in writing Brønsted-Lowry acid-base equations.)

In connection with the Brønsted-Lowry model, there are some terms which are used frequently.

1. The species formed when a proton is removed from an acid is referred to as the **conjugate base** of that acid; B^- is the conjugate base of HB. The species formed when a proton is added to a base is called the **conjugate acid** of that base; HA is the conjugate acid of A^-.

F^- is the conjugate base of HF, which is the conjugate acid of F^-

2. A species which can either accept or donate a proton is referred to as **amphiprotic**. An example is the H_2O molecule, which can gain a proton to form the hydronium ion, H_3O^+, or lose a proton, leaving the hydroxide ion, OH^-.

HCO_3^- is also amphiprotic; it can form CO_3^{2-} or H_2CO_3

$$[:\ddot{\underset{..}{O}}—H]^- \xleftarrow{-H^+} H—\ddot{\underset{..}{O}}—H \xrightarrow{+H^+} [H—\underset{..}{\overset{H}{\overset{|}{O}}}—H]^+$$

hydroxide ion — water molecule — hydronium ion

13.2 WATER DISSOCIATION CONSTANT

The acidic and basic properties of aqueous solutions are dependent upon an equilibrium that involves the solvent, water. The reaction involved can be regarded as a Brønsted-Lowry acid-base reaction in which the H_2O molecule shows its amphiprotic nature:

$$H_2O + H_2O \rightleftharpoons H_3O^+(aq) + OH^-(aq)$$

Alternatively, and somewhat more simply, the reaction can be regarded as the dissociation of a single H_2O molecule:

$$H_2O \rightleftharpoons H^+(aq) + OH^-(aq)$$

Recall (p. 331, Ch. 12) that in the equilibrium constant expression for reactions in solution

— solutes enter as their molarity, [ ]
— the solvent, H_2O in this case, does not appear

Hence, for the dissociation reaction, the equilibrium constant expression is

$$K_w = [H^+] \times [OH^-]$$

where K_w, referred to as the **dissociation constant of water**, has a very small value. At 25°C°,

$$K_w = 1.0 \times 10^{-14}$$

The concentrations of H^+ ($[H^+] = [H_3O^+]$) ions and OH^- ions in *pure water* at 25°C are readily calculated from K_w. Notice from the equation for dissociation that these two ions are formed in equal numbers. Hence in pure H_2O

There are very few H^+ and OH^- ions in pure water

$$[H^+] = [OH^-]$$
$$K_w = [H^+] \times [OH^-] = [H^+]^2 = 1.0 \times 10^{-14}$$
$$[H^+] = 1.0 \times 10^{-7}\,M = [OH^-]$$

An aqueous solution in which $[H^+]$ *equals* $[OH^-]$ is called a **neutral** solution. It has a $[H^+]$ of $1.0 \times 10^{-7}\,M$ at 25°C.

In most water solutions, the concentrations of H^+ and OH^- are not equal. The equation $[H^+] \times [OH^-] = 1.0 \times 10^{-14}$ indicates that these two quantities are inversely proportional to one another (Fig. 13.1). When $[H^+]$ is very high, $[OH^-]$ is very low, and vice versa.

Example 13.1 In a certain tap-water sample, $[H^+] = 3.0 \times 10^{-7}\,M$. What is the concentration of OH^-?

Strategy Simply substitute into the expression for K_w and solve for $[OH^-]$.

Solution

$$[OH^-] = \frac{K_w}{[H^+]} = \frac{1.0 \times 10^{-14}}{3.0 \times 10^{-7}} = 3.3 \times 10^{-8}\,M$$

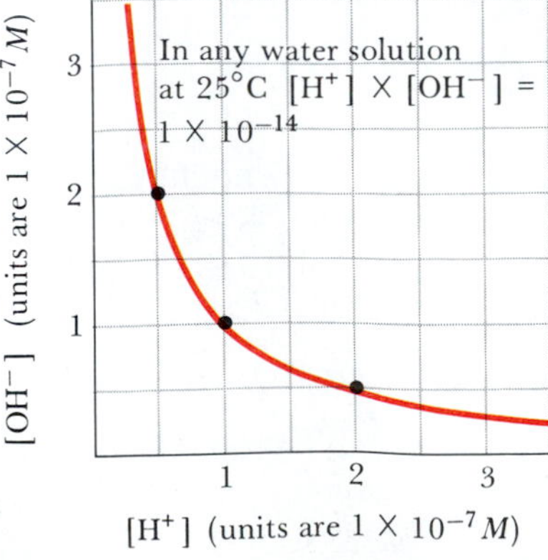

Figure 13.1
A graph of $[OH^-]$ vs. $[H^+]$ looks very much like a graph of gas volume vs. pressure. In both cases, the two variables are inversely proportional to one another. When $[H^+]$ gets larger, $[OH^-]$ gets smaller.

An aqueous solution in which $[H^+]$ is greater than $[OH^-]$ is termed acidic. An aqueous solution in which $[OH^-]$ is greater than $[H^+]$ is basic **(alkaline)**. Therefore,

if $[H^+] > 1.0 \times 10^{-7}\,M$, $[OH^-] < 1.0 \times 10^{-7}\,M$, solution is acidic

if $[OH^-] > 1.0 \times 10^{-7}\,M$, $[H^+] < 1.0 \times 10^{-7}\,M$, solution is basic

The solution in Example 13.1 is acidic

13.3 pH AND pOH

As just pointed out, the acidity or basicity of a solution can be described in terms of its H^+ concentration. In 1909, Søren Sørensen, a biochemist working at the Carlsberg Brewery in Copenhagen, proposed an alternative method of specifying the acidity of a solution. He defined a term called **pH** (for "power of the hydrogen ion"):

It is simpler to use numbers (pH = 4) than exponents ($[H^+] = 10^{-4}\,M$) to describe acidity

$$pH = -\log_{10}[H^+] = -\log_{10}[H_3O^+]$$

or

$$[H^+] = [H_3O^+] = 10^{-pH}$$

Figure 13.2 shows the relationship between pH and $[H^+]$. Notice that, as the defining equation implies, pH increases by one unit when the concentration of H^+ decreases by a power of 10. In general, ***the higher the pH, the less acidic the solution.*** Most aqueous solutions have hydrogen ion concentrations between 1 and $10^{-14}\,M$ and hence have pH's between 0 and 14.

A 10 *M* HCl solution has a pH of −1

The pH, like $[H^+]$ or $[OH^-]$, can be used to differentiate acidic, neutral, and basic solutions. At 25°C:

if $pH < 7.0$, solution is acidic

if $pH = 7.0$, solution is neutral

if $pH > 7.0$, solution is basic

Table 13.1 shows the pH of some common solutions.

A similar approach is used for the hydroxide ion concentration. The **pOH** of a solution is defined as

$$pOH = -\log_{10}[OH^-]$$

Since $[H^+] \times [OH^-] = 1.0 \times 10^{-14}$ at 25°C, it follows that at this temperature

$$pH + pOH = 14.00$$

Thus a solution that has a pH of 6.20 must have a pOH of 7.80, and vice versa.

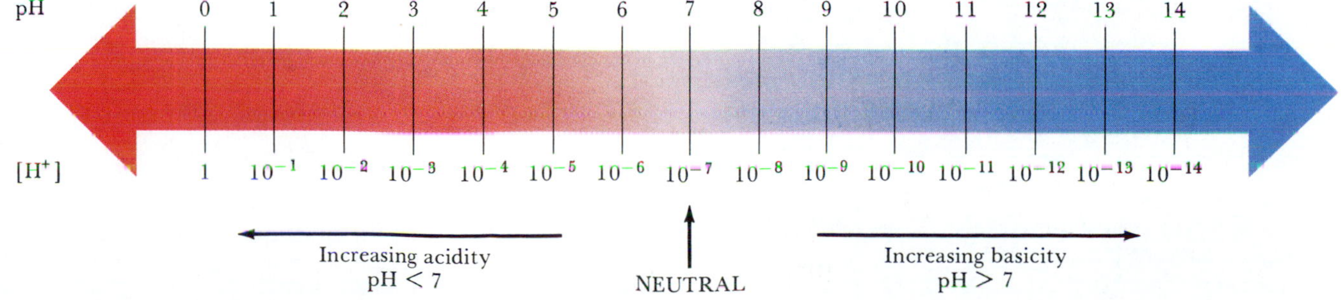

Figure 13.2
Acidity is inversely related to pH; the higher the H^+ ion concentration, the lower the pH. In neutral solution, $[H^+] = [OH^-] = 1.0 \times 10^{-7}\,M$; pH = 7.00 at 25°C.

Table 13.1 pH of Some Common Materials

Lemon juice	2.2–2.4	Urine, human	4.8–8.4
Wine	2.8–3.8	Cow's milk	6.3–6.6
Vinegar	3.0	Saliva, human	6.5–7.5
Tomato juice	4.0	Drinking water	5.5–8.0
Beer	4–5	Blood, human	7.3–7.5
Cheese	4.8–6.4	Seawater	8.3

Example 13.2 Calculate, at 25°C,

(a) the pH and pOH of a lemon juice solution in which $[H^+]$ is $5.0 \times 10^{-3}\,M$.
(b) the $[H^+]$ and $[OH^-]$ of human blood at pH 7.40.

Knowing $[H^+]$, $[OH^-]$, pH or pOH, you can calculate the other 3 quantities

Strategy These calculations can be carried out by substitution into the basic relations

$$\text{pH} = -\log_{10}[H^+] \qquad [H^+] \times [OH^-] = 1.0 \times 10^{-14} \qquad \text{pH} + \text{pOH} = 14.00$$

Use a calculator to find logarithms and inverse logarithms, following the rules for significant figures discussed on p. 619.

Solution

(a) You should find on your calculator that

$$\log_{10}(5.0 \times 10^{-3}) = -2.30$$

Hence $\quad \text{pH} = -(-2.30) = \boxed{2.30} \qquad \text{pOH} = 14.00 - \text{pH} = \boxed{11.70}$

(b) Since the pH is 7.40, $[H^+] = 10^{-7.40}$. To find $[H^+]$, enter -7.40 on your calculator. Then either

— punch the 10^x key, if you have one, or
— punch the INV and then the LOG key

Either way, you should find that $[H^+] = \boxed{4.0 \times 10^{-8}\,M}$.

Knowing $[H^+]$, the concentration of OH^- is calculated as in Example 13.1:

$$[OH^-] = \frac{1.0 \times 10^{-14}}{4.0 \times 10^{-8}} = \boxed{2.5 \times 10^{-7}\,M}$$

Since pH > 7 and $[OH^-] > [H^+]$, blood must be (slightly) basic.

pH of Strong Acids and Strong Bases

As pointed out in Chapter 4, the following acids are strong

HCl	**HBr**	**HI**
$HClO_4$	**HNO_3**	**H_2SO_4**

in the sense that they are completely ionized in water. The reaction of the strong acid HCl with water can be represented by the equation (Fig. 13.3)

$$HCl(aq) + H_2O \longrightarrow H_3O^+(aq) + Cl^-(aq)$$

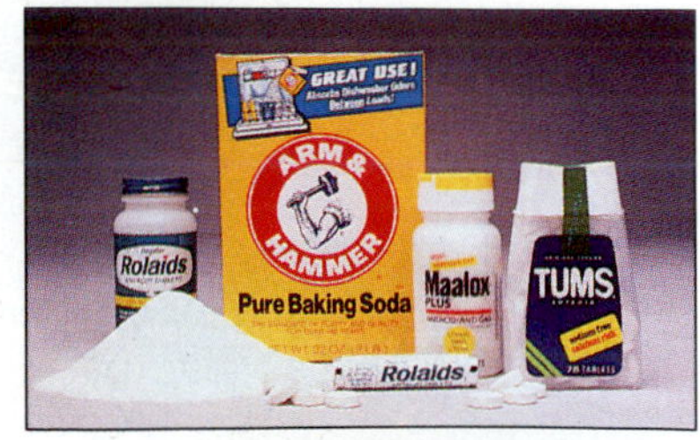

Antacids react with H^+ ions in the stomach to relieve heartburn. (Robert Mathena/FUNDAMENTAL PHOTOGRAPHS, New York)

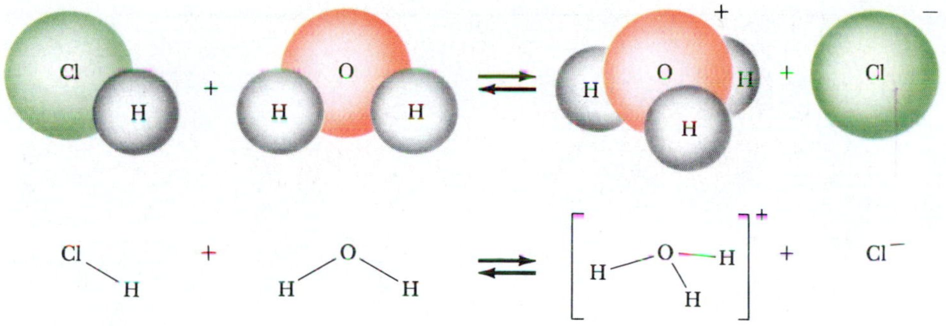

Figure 13.3
When HCl is added to water, there is a proton transfer from an HCl to an H_2O molecule, forming a Cl^- ion and an H_3O^+ ion. In this reaction, HCl acts as a Brønsted-Lowry acid, H_2O as a Brønsted-Lowry base.

Since this reaction goes to completion, it follows that a 0.10 *M* solution of HCl is 0.10 *M* in both H_3O^+ (i.e., H^+) and Cl^- ions.

The strong bases

LiOH	**NaOH**	**KOH**
$\mathbf{Ca(OH)_2}$	$\mathbf{Sr(OH)_2}$	$\mathbf{Ba(OH)_2}$

are completely ionized in dilute water solution. A 0.10 *M* solution of NaOH is 0.10 *M* in both Na^+ and OH^- ions.

The fact that strong acids and bases are completely ionized in water makes it relatively easy to calculate the pH or pOH of their solutions (Example 13.3).

1 *M* HCl has a pH of 0, 1 *M* NaOH a pH of 14

Example 13.3 What is the pH of

(a) 0.025 *M* HNO_3?
(b) a solution prepared by dissolving 10.0 g of $Ba(OH)_2$ per liter?

Strategy To calculate pH, you can either use the defining equation (a), or the relation between pH and pOH (b). Note that both HNO_3 and $Ba(OH)_2$ are completely dissociated in water.

Solution

(a) Since HNO_3 is a strong acid, $[H^+] = 0.025\,M$

$$\text{pH} = -\log_{10}(0.025) = \boxed{1.60}$$

The pH of a 10^{-10} *M* solution of HNO_3 is not 10. Can you see why?

(b) $$[Ba(OH)_2] = \frac{10.0\text{ g}}{1\text{ L}} \times \frac{1\text{ mol}}{171.3\text{ g}} = 0.0584\text{ mol/L}$$

Since 1 mol of $Ba(OH)_2$ yields 2 mol of OH^- ions,

$$[OH^-] = 2 \times 0.0584\text{ mol/L} = 0.117\text{ mol/L}$$

$$\text{pOH} = -\log_{10}(0.117) = 0.932 \qquad \text{pH} = 14.00 - 0.93 = \boxed{13.07}$$

Measuring pH

The pH of a solution can be measured accurately by an instrument called a pH-meter. A pH-meter translates the H^+ ion concentration of a solution into an electrical signal that is converted into either a digital display or a deflection on a meter that reads pH directly (Fig. 13.4).

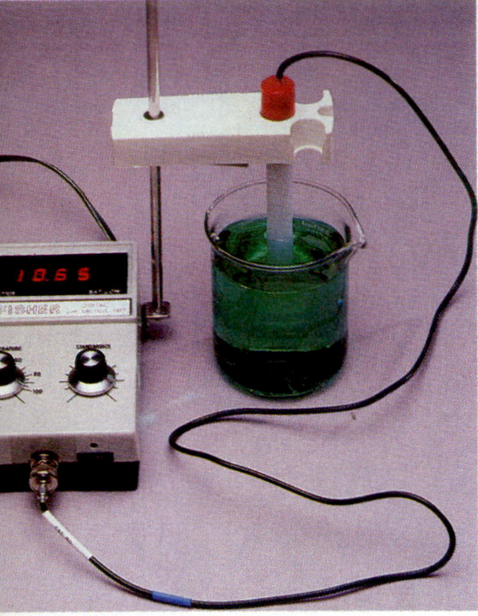

Figure 13.4
A pH meter with digital read-out. The pH of the solution in the beaker appears to be about 10.65. What do you suppose is in the solution? (Marna G. Clarke)

Figure 13.5
Universal indicator is deep red in strongly acidic solution (upper left). It changes to yellow and green at pH 6–8 and then to deep violet in strongly basic solution (lower right). (Charles D. Winters)

A less accurate but more colorful way to measure pH uses a "universal indicator," which is a mixture of *acid-base indicators* that (Fig. 13.5) shows changes in color at different pH values. A similar principle is used with "pH paper." Strips of this paper are coated with a mixture of pH-sensitive dyes; these strips are widely used to test the pH of biological fluids, ground water, and foods. Depending upon the indicators used, a test strip can measure pH over a wide or narrow range.

To get red hydrangeas, add lime

Home gardeners frequently measure and adjust soil pH in an effort to improve the yield and quality of grass, vegetables, and flowers. Interestingly enough, the colors of many flowers depend upon pH; among these are dahlias, delphiniums, and, in particular, hydrangeas (Fig. 13.6). The first research in this area was carried out by Robert Boyle of gas law fame, who published a paper on the relation between flower color and acidity in 1664.

Figure 13.6
Hydrangeas grown in strongly acidic soil (below pH 5) are blue. In weakly acidic soil ($5 < pH < 7$), they are mauve. If the soil is neutral or basic, the flowers are rose-pink. (Michael Dalton/FUNDAMENTAL PHOTOGRAPHS, New York)

13.4 WEAK ACIDS AND THEIR DISSOCIATION CONSTANTS

A wide variety of solutes behave as weak acids; that is, they react reversibly with water to form H_3O^+ ions. Using HB to represent a weak acid, its Brønsted-Lowry reaction with water is

$$HB(aq) + H_2O \rightleftharpoons H_3O^+(aq) + B^-(aq)$$

Typically, this reaction occurs to a very small extent; usually, fewer than 1% of the HB molecules are converted to ions.

Most weak acids fall into one of two categories:

1. Molecules containing an ionizable hydrogen atom. This type of weak acid was discussed in Chapter 4. There are literally thousands of molecular weak acids, most of them organic in nature (see p. 377). Among the molecular inorganic acids is nitrous acid:

$$HNO_2(aq) + H_2O \rightleftharpoons H_3O^+(aq) + NO_2^-(aq)$$

Any acid you encounter except HCl, HBr, HI, HNO_3, $HClO_4$, and H_2SO_4 will be weak

2. Cations. The ammonium ion, NH_4^+, behaves as a weak acid in water; a 0.10 *M* solution of NH_4Cl has a pH of about 5. The process by which the NH_4^+ ion lowers the pH of water can be represented by the (Brønsted-Lowry) equation:

$$NH_4^+(aq) + H_2O \rightleftharpoons H_3O^+(aq) + NH_3(aq)$$

Comparing the equation just written to that cited above for HNO_2, you can see that they are very similar. In both cases, a weak acid (HNO_2, NH_4^+) is converted to its conjugate base (NO_2^-, NH_3). The fact that the NH_4^+ ion has a +1 charge while the HNO_2 molecule is neutral is really irrelevant here.

You may be surprised to learn that most metal cations, except those of Periodic Groups 1 and 2, are weak acids. A 0.10 *M* solution of $Al_2(SO_4)_3$ has a pH close to 3; you can change the color of hydrangeas from red to blue by adding aluminum salts to soil. At first glance it is not at all obvious how a cation such as Al^{3+} can make a water solution acidic. However, the aluminum cation in water solution is really a hydrated species, $Al(H_2O)_6^{3+}$, in which six water molecules are bonded to the central Al^{3+} ion. This species can transfer a proton to a solvent water molecule to form an H_3O^+ ion:

$$Al(H_2O)_6^{3+}(aq) + H_2O \rightleftharpoons H_3O^+(aq) + Al(H_2O)_5(OH)^{2+}(aq)$$

More simply: $Al(H_2O)_6^{3+}(aq) \rightarrow H^+(aq) + Al(H_2O)_5(OH)^{2+}(aq)$

Figure 13.7 shows the structures of the $Al(H_2O)_6^{3+}$ cation and its conjugate base, $Al(H_2O)_5(OH)^{2+}$.

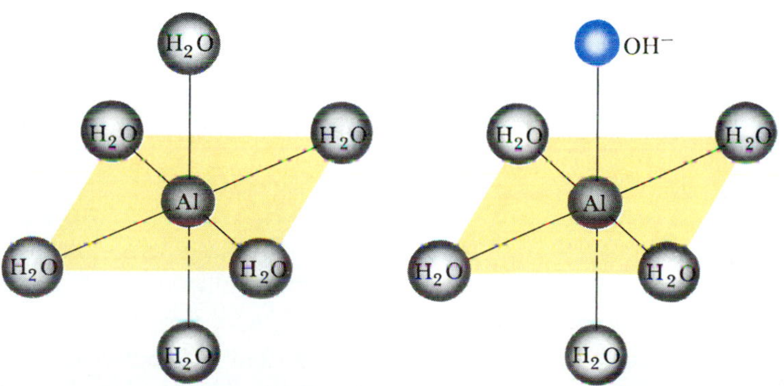

Figure 13.7
The Al^{3+} ion is bonded to six H_2O molecules in the $Al(H_2O)_6^{3+}$ ion (left); in the $Al(H_2O)_5(OH)^{2+}$ ion (right), one of the H_2O molecules has been replaced by an OH^- ion.

A very similar equation can be written to explain why solutions of zinc salts are acidic. Here it appears that the hydrated cation in water solution contains four H_2O molecules bonded to a central Zn^{2+} ion. The Brønsted-Lowry equation is

$$Zn(H_2O)_4{}^{2+}(aq) + H_2O \rightleftharpoons H_3O^+(aq) + Zn(H_2O)_3(OH)^+(aq)$$

Example 13.4 Using the Brønsted-Lowry model, write equations to show why the following species behave as weak acids in water.

(a) acetic acid, $HC_2H_3O_2$, a weak acid found in vinegar
(b) the hydrated iron(II) cation, $Fe(H_2O)_6{}^{2+}$

Strategy In each case, a proton is transferred to a solvent water molecule, which acts as a Brønsted base. The products are an H_3O^+ ion and the conjugate base of the weak acid.

Solution

The formula of acetic acid is often written as CH_3COOH

(a) $HC_2H_3O_2(aq) + H_2O \rightleftharpoons H_3O^+(aq) + C_2H_3O_2{}^-(aq)$

(b) $Fe(H_2O)_6{}^{2+}(aq) + H_2O \rightleftharpoons H_3O^+(aq) + Fe(H_2O)_5(OH)^+(aq)$

The Equilibrium Constant for a Weak Acid

In discussing the equilibria involved when a weak acid is added to water, it is convenient to represent the proton transfer

$$HB(aq) + H_2O \rightleftharpoons H_3O^+(aq) + B^-(aq)$$

as a simple dissociation

$$HB(aq) \rightleftharpoons H^+(aq) + B^-(aq)$$

in which case the expression for the equilibrium constant becomes

$$K_a = \frac{[H^+] \times [B^-]}{[HB]}$$

The equilibrium constant K_a is called the *ionization constant* or **acid dissociation constant** of the weak acid HB.

The K_a values of some weak acids (in order of decreasing strength) are given in Table 13.2, along with values of **pK_a**:

$$pK_a = -\log_{10}K_a$$

In general, the *weaker* the acid, the *smaller* the value of K_a and the *larger* the value of pK_a.

There are several ways to determine the dissociation constant, K_a, of a weak acid. A simple approach involves measuring $[H^+]$ or pH in a solution prepared by dissolving a known amount of the weak acid to form a given volume of solution.

Example 13.5 Aspirin is a weak organic acid whose molecular formula may be written as $HC_9H_7O_4$. A water solution of aspirin is prepared by dissolving 3.60 g per liter. The pH of this solution is found to be 2.60. Calculate K_a for aspirin.

Strategy The approach used here is very similar to that used in Chapter 12, except that partial pressures are replaced by concentrations in moles per liter. Note that you can readily calculate

— the original concentration of $HC_9H_7O_4$ ($\mathcal{M}$ = 180.15 g/mol)

$$[HC_9H_7O_4]_o = \frac{3.60\text{ g}}{1\text{ L}} \times \frac{1\text{ mol}}{180.15\text{ g}} = 2.00 \times 10^{-2}\, M$$

— the equilibrium concentration of H^+

$$[H^+]_{eq} = 10^{-2.60} = 2.5 \times 10^{-3}\, M$$

Your task is to calculate the acid dissociation constant:

$$HC_9H_7O_4(aq) \rightleftharpoons H^+(aq) + C_9H_7O_4^-(aq) \qquad K_a = \frac{[H^+] \times [C_9H_7O_4^-]}{[HC_9H_7O_4]}$$

Solution From the dissociation equation it should be clear that *1 mol* of $C_9H_7O_4^-$ is *produced,* and *1 mol* of $HC_9H_7O_4$ is *consumed* for every mole of H^+ produced. It follows that

$$\Delta[C_9H_7O_4^-] = \Delta[H^+] \qquad \Delta[HC_9H_7O_4] = -\Delta[H^+]$$

Originally, there is essentially no H^+ (ignoring the slight dissociation of water). The same holds for the anion $C_9H_7O_4^-$; the only species present originally is the weak acid, $HC_9H_7O_4$, at a concentration of 0.0200 *M*.

Putting this information together in the form of a table:

	$HC_9H_7O_4(aq)$ $\rightleftharpoons$	$H^+(aq)$ +	$C_7H_9O_4^-(aq)$
$[\ ]_o$	0.0200	0.0000	0.0000
$\Delta[\ ]$	−0.0025	+0.0025	+0.0025
$[\ ]_{eq}$	0.0175	0.0025	0.0025

(Numbers in color are those given or implied in the statement of the problem; the other numbers are deduced using the dissociation equation printed above the table. The symbols $[\ ]_o$ and $[\ ]_{eq}$ refer to original and equilibrium concentrations, respectively.)

All the information needed to calculate K_a is now available.

$$K_a = \frac{(2.5 \times 10^{-3})^2}{0.0175} = 3.6 \times 10^{-4}$$

Table 13.2 Dissociation Constants of Weak Acids at 25°C

Name	Formula	K_a	pK_a
Sulfurous acid	H_2SO_3	1.7×10^{-2}	1.77
Hexaaquairon(III) ion	$Fe(H_2O)_6^{3+}$	6.7×10^{-3}	2.17
Hydrofluoric acid	HF	6.9×10^{-4}	3.16
Nitrous acid	HNO_2	6.0×10^{-4}	3.22
Formic acid	$HCHO_2$	1.9×10^{-4}	3.72
Benzoic acid	$HC_7H_5O_2$	6.6×10^{-5}	4.18
Acetic acid	$HC_2H_3O_2$	1.8×10^{-5}	4.74
Hexaaquaaluminum(III) ion	$Al(H_2O)_6^{3+}$	1.2×10^{-5}	4.92
Hypochlorous acid	HClO	2.8×10^{-8}	7.55
Hydrocyanic acid	HCN	5.8×10^{-10}	9.24
Ammonium ion	NH_4^+	5.6×10^{-10}	9.25
Tetraaquazinc(II) ion	$Zn(H_2O)_4^{2+}$	3.3×10^{-10}	9.48
Diaquasilver(I) ion	$Ag(H_2O)_2^+$	1.2×10^{-12}	11.92

Table 13.3 Percent Dissociation of Weak Acids

Weak Acid	K_a	Percent Dissociation: Orig. Conc. 1.0 *M*	Percent Dissociation: Orig. Conc. 0.10 *M*
Lactic acid	1.4×10^{-4}	1.2	3.7
Acetic acid	1.8×10^{-5}	0.42	1.3
Carbonic acid	4.4×10^{-7}	0.066	0.21
Hypochlorous acid	2.8×10^{-8}	0.017	0.053
Ammonium ion	5.6×10^{-10}	0.0024	0.0075
Hydrocyanic acid	5.8×10^{-10}	0.0024	0.0076

decreasing acid strength

What happens to K_a when the concentration drops? Ans: nothing

In discussing the dissociation of a weak acid,

$$HB(aq) \rightleftharpoons H^+(aq) + B^-(aq)$$

we often refer to the **percent dissociation**:

$$\% \text{ dissociation} = \frac{[H^+]_{eq}}{[HB]_o} \times 100\%$$

For the aspirin solution referred to in Example 13.5,

Aspirin is more acidic than vinegar

$$\% \text{ dissociation} = \frac{2.5 \times 10^{-3}}{2.0 \times 10^{-2}} \times 100\% = 12\%$$

This is a relatively high percent of dissociation; typically values are of the order of 1% or less (Table 13.3).

Calculation of $[H^+]$ in a Water Solution of a Monoprotic Weak Acid

Given the dissociation constant of a weak acid and its original concentration, the H^+ concentration in solution is readily calculated. The reasoning is simplest when the acid is **monoprotic**, containing one ionizable hydrogen atom. The approach used is the inverse of that followed in Example 13.5, where K_a was calculated knowing $[H^+]$; here K_a is known and $[H^+]$ must be calculated.

Example 13.6 Nicotinic acid, $HC_6H_4O_2N$ ($K_a = 1.4 \times 10^{-5}$), is another name for niacin, an important member of the Vitamin B group. Determine $[H^+]$ in a solution prepared by dissolving 0.10 mol of nicotinic acid, HNic, in water to form one liter of solution.

Strategy To start with, set up an equilibrium table similar to the one in the previous example. To do that, note that

— the original concentrations of HNic, H^+ and Nic^- are 0.10 *M*, 0.00 *M*, and 0.00 *M*, ignoring, tentatively at least, the H^+ ions from the dissociation of water.

— the changes in concentration are related by the coefficients of the balanced equation, all of which are 1:

$$\Delta[\text{Nic}^-] = \Delta[\text{H}^+] \qquad \Delta[\text{HNic}] = -\Delta[\text{H}^+]$$

Letting $\Delta[H^+] = x$, it follows that $\Delta[\text{Nic}^-] = x$; $\Delta[\text{HNic}] = -x$. This information should enable you to express the equilibrium concentrations of all species in terms of x. The rest is algebra; substitute into the expression for K_a and solve for $x = [H^+]$.

Solution Setting up the table:

	$\text{HNic}(aq)$ $\rightleftharpoons$	$H^+(aq)$ +	$\text{Nic}^-(aq)$
$[\ ]_o$	0.10	0.00	0.00
$\Delta[\ ]$	$-x$	$+x$	$+x$
$[\ ]_{eq}$	$0.10 - x$	x	x

Equilibrium tables always help; be sure to include the equation

Substituting into the expression for the dissociation constant,

$$K_a = \frac{(x)(x)}{0.10 - x} = 1.4 \times 10^{-5}$$

This is a quadratic equation. It could be rearranged to the form $ax^2 + bx + c = 0$ and solved for x, using the quadratic formula. Such a procedure is time-consuming, and, in this case, unnecessary. Nicotinic acid is a weak acid, only slightly dissociated in water. The equilibrium concentration of HNic, $0.10 - x$, is probably only very slightly less than its original concentration, 0.10 M. So let's make the approximation $0.10 - x \approx 0.10$. This simplifies the equation

$$\frac{x^2}{0.10} = 1.4 \times 10^{-5}$$

You can't discard x in the numerator, since that's what you want to find

$$x^2 = 1.4 \times 10^{-6}$$

Taking square roots:

$$x = 1.2 \times 10^{-3}\,M = [H^+]$$

Note that the concentration of H^+, 0.0012 M, is

— much smaller than the original concentration of weak acid, 0.10 M. In this case, then, the approximation $0.10 - x \approx 0.10$ is justified. This will usually, but not always, be the case (see Example 13.7).

— much larger than $[H^+]$ in pure water, $1 \times 10^{-7}\,M$, justifying the assumption that the dissociation of water can be neglected. *This will always be the case, provided $[H^+]$ from the weak acid is $\geq 10^{-6}\,M$.*

In general, the value of K_a is seldom known to better than ±5%. Hence, in the expression

$$K_a = \frac{x^2}{a - x}$$

where $x = [H^+]$ and a = *original* concentration of weak acid, you can neglect the x in the denominator if doing so does not introduce an error of more than 5%. In other words,

$$\text{if } \frac{x}{a} \leq 0.05$$

This is known as the "5% rule"

then

$$a - x = a.$$

In most of the problems you will work, the approximation $a - x \approx a$ is valid and you can solve for $[H^+]$ quite simply, as in Example 13.6, where $x = 0.012a$. Sometimes, though, you will find that the calculated $[H^+]$ is greater than 5% of the original concentration of weak acid. In that case, you can solve for x either by using the *quadratic formula* or *the method of successive approximations*.

Example 13.7 Calculate $[H^+]$ in a 0.100 *M* solution of nitrous acid, HNO_2, for which $K_a = 6.0 \times 10^{-4}$.

Strategy The setup here is identical with that in Example 13.6. However, you will find, upon solving for x, that $x > 0.050a$, so the approximation $a - x \approx a$ fails. The simplest way to proceed here is to use the calculated value of x to obtain a better estimate of $[HNO_2]$, then solve again for $[H^+]$. An alternative is to use the quadratic formula. (This is a particularly shrewd choice if you have a calculator that can be programmed to solve quadratic equations.)

Solution Proceeding as in Example 13.6, you arrive at the equation

$$K_a = \frac{x^2}{0.100 - x} = 6.0 \times 10^{-4}$$

Making the same approximation as before, $0.100 - x \approx 0.100$,

$$x^2 = 0.100 \times 6.0 \times 10^{-4} = 6.0 \times 10^{-5}$$

$$x = 7.7 \times 10^{-3} \approx [H^+]$$

To check the validity of the approximation, note that

$$\frac{x}{a} = \frac{7.7 \times 10^{-3}}{1.0 \times 10^{-1}} = 0.077 > 0.050$$

So the approximation is not valid.

To obtain a better value for x, you have a choice of two approaches.

(1) *The method of successive approximations.* In the first approximation, you ignored the x in the denominator of the equation. This time, you use the approximate value of x just calculated, 0.0077, to find a more exact value for the concentration of HNO_2

WLM likes this method a lot

$$[HNO_2] = 0.100 - 0.0077 = 0.092\ M$$

Substituting in the expression for K_a,

$$K_a = \frac{x^2}{0.092} = 6.0 \times 10^{-4} \qquad x^2 = 5.5 \times 10^{-5}$$

$$x = 7.4 \times 10^{-3}\ M \approx [H^+]$$

This value is closer to the true $[H^+]$, since 0.092 *M* is a better approximation for $[HNO_2]$ than was 0.100 *M*. If you're still not satisfied, you can go one step further. Using 7.4×10^{-3} for x instead of 7.7×10^{-3}, you can recalculate $[HNO_2]$ and solve again for x. If you do, you will find that your answer does not change. In other words, "you have gone about as far as you can go."

(2) *The quadratic formula.* This gives an exact solution for x but is more time-consuming. Here, rewrite the equation

CNH thinks this method is just as fast (but she's wrong)

$$\frac{x^2}{0.100 - x} = 6.0 \times 10^{-4}$$

in the form $ax^2 + bx + c = 0$. Doing this:

$$x^2 + (6.0 \times 10^{-4}\,x) - (6.0 \times 10^{-5}) = 0$$

Thus, $a = 1$; $b = 6.0 \times 10^{-4}$; $c = -6.0 \times 10^{-5}$. Applying the quadratic formula,

$$x = \frac{-b \pm \sqrt{b^2 - 4ac}}{2a}$$

$$= \frac{-6.0 \times 10^{-4} \pm \sqrt{(6.0 \times 10^{-4})^2 + (24.0 \times 10^{-5})}}{2}$$

If you carry out the arithmetic properly, you should get two answers for x:

$$x = 7.4 \times 10^{-3}\,M \qquad \text{and} \qquad -8.0 \times 10^{-3}\,M$$

The second answer is physically ridiculous; the concentration of H^+ cannot be a negative quantity. The first answer is the same one obtained by the method of successive approximations.

Dissociation of Polyprotic Acids

Certain weak acids are *polyprotic;* they contain more than one ionizable hydrogen atom. Such acids dissociate in steps, with a separate dissociation constant for each step. Oxalic acid, a weak organic acid sometimes used to remove bloodstains, is *diprotic:*

$$H_2C_2O_4(aq) \rightleftharpoons H^+(aq) + HC_2O_4^-(aq) \qquad K_{a1} = 5.9 \times 10^{-2}$$

$$HC_2O_4^-(aq) \rightleftharpoons H^+(aq) + C_2O_4^{2-}(aq) \qquad K_{a2} = 5.2 \times 10^{-5}$$

Phosphoric acid, a common ingredient of cola drinks, is *triprotic:*

$$H_3PO_4(aq) \rightleftharpoons H^+(aq) + H_2PO_4^-(aq) \qquad K_{a1} = 7.1 \times 10^{-3}$$

$$H_2PO_4^-(aq) \rightleftharpoons H^+(aq) + HPO_4^{2-}(aq) \qquad K_{a2} = 6.2 \times 10^{-8}$$

$$HPO_4^{2-}(aq) \rightleftharpoons H^+(aq) + PO_4^{3-}(aq) \qquad K_{a3} = 4.5 \times 10^{-13}$$

The behavior of these acids is typical of that of all polyprotic acids in that

- *the anion formed in one step* (e.g., $HC_2O_4^-$, $H_2PO_4^-$) *dissociates in the next step*
- *the dissociation constant becomes smaller with each successive step.*

$$K_{a1} > K_{a2} > K_{a3}$$

Putting it another way, *the acids formed in successive steps become progressively weaker.* This is reasonable; it should be more difficult to remove a positively charged proton, H^+, from a negatively charged species like $H_2PO_4^-$ than from a neutral molecule like H_3PO_4.

Paper made many years ago used acids in the manufacturing process. (Paul Silverman/FUNDAMENTAL PHOTOGRAPHS, New York)

Table 13.4 Dissociation Constants for Some Polyprotic Acids at 25°C

Acid	Formula	K_{a1}	K_{a2}	K_{a3}
Sulfuric acid	H_2SO_4	∞	1.0×10^{-2}	
Carbonic acid*	H_2CO_3	4.4×10^{-7}	4.7×10^{-11}	
Citric acid	$H_3C_5H_5O_7$	8.7×10^{-4}	1.8×10^{-5}	4.0×10^{-7}
Oxalic acid	$H_2C_2O_4$	5.9×10^{-2}	5.2×10^{-5}	
Phosphoric acid	H_3PO_4	7.1×10^{-3}	6.2×10^{-8}	4.5×10^{-13}
Sulfurous acid	H_2SO_3	1.7×10^{-2}	6.0×10^{-8}	

* Carbonic acid is a water solution of carbon dioxide:

$$CO_2(g) + H_2O \rightleftharpoons H_2CO_3(aq)$$

The dissociation constants listed here are calculated assuming that all the carbon dioxide that dissolves is in the form of H_2CO_3.

Ordinarily, successive dissociation constants of polyprotic acids decrease by a factor of at least 100 (Table 13.4). In that case, *essentially all the hydrogen ions in the solution come from the first dissociation of the acid.* This makes it relatively easy to calculate the pH of a solution of a polyprotic acid.

Example 13.8 The distilled water you use in the laboratory is slightly acidic because of dissolved CO_2, which reacts to form carbonic acid, H_2CO_3. Calculate the pH of a 0.0010 *M* solution of H_2CO_3.

Strategy In principle, there are two different sources of H^+ ions from H_2CO_3:

$$H_2CO_3(aq) \rightleftharpoons H^+(aq) + HCO_3^-(aq) \qquad K_{a1} = 4.4 \times 10^{-7}$$

$$HCO_3^-(aq) \rightleftharpoons H^+(aq) + CO_3^{2-}(aq) \qquad K_{a2} = 4.7 \times 10^{-11}$$

In practice, essentially all the H^+ ions come from the first reaction, since K_{a1} is so much larger than K_{a2}. In other words, H_2CO_3 can be treated as if it were a weak monoprotic acid, and this example is similar to Example 13.6.

Solution For the first dissociation: $H_2CO_3(aq) \rightleftharpoons H^+(aq) + HCO_3^-(aq)$

$$K_{a1} = \frac{x^2}{0.0010 - x} = 4.4 \times 10^{-7} \qquad (x = [H^+] = [HCO_3^-])$$

Making the approximation $0.0010 - x \approx 0.0010$ and solving gives

$$x = (4.4 \times 10^{-10})^{1/2} = 2.1 \times 10^{-5}$$

$$pH = -\log_{10}(2.1 \times 10^{-5}) = 4.68$$

So far as pH is concerned, you need only consider K_{a1}

Note that the solution is indeed acidic, with a pH considerably less than 7.

The approximation $0.0010 - x \approx 0.0010$ is justified; $[H^+]$ is only about 2% of the original concentration of H_2CO_3. To check the assumption that virtually all the H^+ ions come from the first dissociation, take y to be the concentration of H^+ ions produced by the second dissociation:

$$HCO_3^-(aq) \rightleftharpoons H^+(aq) + CO_3^{2-}(aq)$$

Then $[H^+] = [H^+]$ 1st dissociation + $[H^+]$ 2nd dissociation $= x + y$
$[CO_3^{2-}] = [H^+]$ 2nd dissociation $= y$
$[HCO_3^-] = [H^+]$ 1st dissociation $-$ $[H^+]$ 2nd dissociation $= x - y$

Taking $x = 2.1 \times 10^{-5}$:

$$\frac{(2.1 \times 10^{-5} + y)(y)}{(2.1 \times 10^{-5} - y)} = 4.7 \times 10^{-11}$$

Assuming $y << 2.1 \times 10^{-5}$, $2.1 \times 10^{-5} + y \approx 2.1 \times 10^{-5} - y$. Hence,

$$y = 4.7 \times 10^{-11}$$

Notice that y (4.7×10^{-11}) is indeed much less than x (2.1×10^{-5}), so the H^+ ions from the second dissociation can safely be ignored.

13.5 WEAK BASES AND THEIR DISSOCIATION CONSTANTS

Like the weak acids, there are a large number of solutes that act as weak bases. It is convenient to classify weak bases into two groups, molecules and anions.

1. Molecules. As pointed out in Chapter 4, there are many molecular weak bases, including the organic compounds known as amines. The simplest weak base is ammonia, whose reversible, Brønsted-Lowry reaction with water is represented by the equation

$$NH_3(aq) + H_2O \rightleftharpoons NH_4^+(aq) + OH^-(aq)$$

This reaction proceeds to only a very small extent. In 0.10 M NH_3, the concentrations of NH_4^+ and OH^- are only about 0.0013 M.

2. Anions. *Any anion derived from a weak acid is itself a weak base.* A typical example is the fluoride ion, F^-, which is the conjugate base of the weak acid HF. The reaction of the F^- ion with water is

$$F^-(aq) + H_2O \rightleftharpoons HF(aq) + OH^-(aq)$$

Only a few HF molecules dissociate, so $[OH^-] > [H^+]$

The OH^- ions formed make the solution basic. A 0.10 M solution of sodium fluoride, NaF, has a pH of about 8.1.

Notice the similarity between the equation just written and that for the NH_3 molecule cited above:

— both weak bases (NH_3, F^-) accept protons to form the conjugate acid (NH_4^+, HF)
— water acts as a Brønsted-Lowry acid in each case, donating a proton to form the OH^- ion

Example 13.9 Write an equation to explain why each of the following species produces a basic water solution.

(a) NO_2^- (b) CO_3^{2-} (c) HCO_3^-

Strategy In each case, the weak base reacts reversibly with a water molecule, picking up a proton from it. Two species are formed. One is an OH^- ion, which makes the solution basic. The other product is the conjugate acid of the weak base.

Solution

(a) $NO_2^-(aq) + H_2O \rightleftharpoons HNO_2(aq) + OH^-(aq)$

(b) $CO_3^{2-}(aq) + H_2O \rightleftharpoons HCO_3^-(aq) + OH^-(aq)$

(c) $HCO_3^-(aq) + H_2O \rightleftharpoons H_2CO_3(aq) + OH^-(aq)$

The Equilibrium Constant for a Weak Base

For an anion B^- which acts as a weak base

$$B^-(aq) + H_2O \rightleftharpoons HB(aq) + OH^-(aq)$$

the equilibrium constant is written in the usual way:

$$K_b = \frac{[HB] \times [OH^-]}{[B^-]}$$

The constant K_b is often called the **base dissociation constant**. For a molecular weak base such as ammonia,

$$NH_3(aq) + H_2O \rightleftharpoons NH_4^+(aq) + OH^-(aq)$$

a very similar expression can be written:

$$K_b = \frac{[NH_4^+] \times [OH^-]}{[NH_3]}$$

Table 13.5 Dissociation Constants for Some Weak Bases at 25°C

Base	Formula	K_b	pK_b*
Phosphate ion	PO_4^{3-}	2.2×10^{-2}	1.66
Methyl amine	CH_3NH_2	4.2×10^{-4}	3.38
Carbonate ion	CO_3^{2-}	2.1×10^{-4}	3.68
Ammonia	NH_3	1.8×10^{-5}	4.74
Cyanide ion	CN^-	1.7×10^{-5}	4.77
Hypochlorite ion	ClO^-	3.6×10^{-7}	6.44
Sulfite ion	SO_3^{2-}	1.7×10^{-7}	6.77
Hydrogen phosphate ion	HPO_4^{2-}	1.6×10^{-7}	6.80
Hydrogen carbonate ion	HCO_3^-	2.3×10^{-8}	7.64
Acetate ion	$C_2H_3O_2^-$	5.6×10^{-10}	9.25
Aniline	$C_6H_5NH_2$	3.8×10^{-10}	9.42
Formate ion	CHO_2^-	5.3×10^{-11}	10.28
Nitrite ion	NO_2^-	1.7×10^{-11}	10.77
Fluoride ion	F^-	1.4×10^{-11}	10.85
Dihydrogen phosphate ion	$H_2PO_4^-$	1.4×10^{-12}	11.85
Hydrogen sulfite ion	HSO_3^-	5.9×10^{-13}	12.23

* $pK_b = -\log_{10} K_b$.

Values of K_b are listed in Table 13.5. In general, the larger the value of K_b, the stronger the base; base strength decreases moving down the table. Since K_b for NH_3 (1.8×10^{-5}) is larger than K_b for the acetate ion, $C_2H_3O_2^-$ (5.6×10^{-10}), ammonia is a stronger base than the acetate ion.

Calculation of $[OH^-]$ in a Water Solution of a Weak Base

We showed in Section 13.3 how K_a of a weak acid can be used to calculate $[H^+]$ in a solution of that acid. In a very similar way, K_b can be used to find $[OH^-]$ in a solution of a weak base.

Example 13.10 Calculate the pH of 0.10 *M* NaF ($K_b\,F^- = 1.4 \times 10^{-11}$).

Strategy The procedure here is entirely analogous to that followed in Example 13.6. Take $[OH^-] = x$ and set up an equilibrium table, ignoring the OH^- ions present in pure water. Solve for x, making the usual approximation. Then calculate pOH and finally pH. Hopefully, you will obtain a pH greater than 7; a solution of sodium fluoride had better be basic!

Solution The equilibrium table is

	$F^-(aq) + H_2O$ $\rightleftharpoons$	$HF(aq)$ +	$OH^-(aq)$
$[\ ]_o$	0.10	0.00	0.00
$\Delta[\ ]$	$-x$	$+x$	$+x$
$[\ ]_{eq}$	$0.10 - x$	x	x

Substituting into the equilibrium constant expression:

$$K_b = \frac{x^2}{0.10 - x} = 1.4 \times 10^{-11}$$

Note the resemblance to Example 13.6; here you calculate $[OH^-]$

Assuming $0.10 - x \approx x$ and solving for x,

$$x^2 = 1.4 \times 10^{-12} \qquad x = 1.2 \times 10^{-6}$$

Clearly $x \ll 0.10$, so the approximation is justified.

$$[OH^-] = 1.2 \times 10^{-6}\,M$$

$$pOH = -\log_{10}(1.2 \times 10^{-6}) = 5.92$$

$$pH = 14.00 - 5.92 = 8.08$$

Relation Between K_a and K_b

Dissociation constants of weak bases can be measured in the laboratory by procedures very much like those used for weak acids. In practice, though, it is simpler to take advantage of a simple mathematical relationship between K_b for a weak base and K_a for its conjugate acid. This relationship can be derived by adding together the equations for the dissociation of the weak acid HB and the weak base B^-.

$$
\begin{array}{llr}
(1) & HB(aq) \rightleftharpoons H^+(aq) + B^-(aq) & K_I = K_a \text{ of HB} \\
(2) & B^-(aq) + H_2O(aq) \rightleftharpoons HB(aq) + OH^-(aq) & K_{II} = K_b \text{ of } B^- \\
\hline
(3) & H_2O \rightleftharpoons H^+(aq) + OH^-(aq) & K_{III} = K_w
\end{array}
$$

Since Equation (1) + Equation (2) = Equation (3), we have, according to the rule of multiple equilibria (Chap. 12),

$$K_I \times K_{II} = K_{III}$$

or

$$(K_a \text{ of HB}) \times (K_b \text{ of } B^-) = K_w = 1.0 \times 10^{-14}$$

To show how this relation is used, suppose you want to calculate K_b of the F^- ion, knowing that K_a of HF is 6.9×10^{-4}:

$$K_b \text{ of } F^- = \frac{1.0 \times 10^{-14}}{K_a \text{ of HF}} = \frac{1.0 \times 10^{-14}}{6.9 \times 10^{-4}} = 1.4 \times 10^{-11}$$

Since HCN is a very weak acid, the CN^- ion is a relatively strong base

From the general relation between K_a and K_b, it should be clear that these two quantities are inversely related to one another. This makes sense; the stronger the acid, the weaker the conjugate base. Table 13.6 shows this effect and some other interesting features. In particular:

1. Brønsted-Lowry acids (left column) can be divided into three categories:
   (a) strong acids ($HClO_4$, . . .), which are stronger proton donors than the H_3O^+ ion
   (b) weak acids (HF, . . .), which are weaker proton donors than the H_3O^+ ion, but stronger than the H_2O molecule
   (c) species such as C_2H_5OH, which are weaker proton donors than the H_2O molecule and hence do not form acidic water solutions

Table 13.6 Relative Strengths of Brønsted-Lowry Acids and Bases

Species shown in deep color are unstable; they react with H_2O molecules

K_a	Conjugate Acid	Conjugate Base	K_b
very large	$HClO_4$	ClO_4^-	very small
very large	HCl	Cl^-	very small
very large	HNO_3	NO_3^-	very small
	H_3O^+	**H_2O**	
6.9×10^{-4}	HF	F^-	1.4×10^{-11}
1.8×10^{-5}	$HC_2H_3O_2$	$C_2H_3O_2^-$	5.6×10^{-10}
1.4×10^{-5}	$Al(H_2O)_6^{3+}$	$Al(H_2O)_5(OH)^{2+}$	7.1×10^{-10}
4.4×10^{-7}	H_2CO_3	HCO_3^-	2.3×10^{-8}
2.8×10^{-8}	HClO	ClO^-	3.6×10^{-7}
5.6×10^{-10}	NH_4^+	NH_3	1.8×10^{-5}
4.7×10^{-11}	HCO_3^-	CO_3^{2-}	2.1×10^{-4}
	H_2O	**OH^-**	
very small	C_2H_5OH	$C_2H_5O^-$	very large
very small	OH^-	O^{2-}	very large
very small	H_2	H^-	very large

2. Brønsted-Lowry bases (right column) can be divided similarly:
   (a) strong bases (H^-, . . .), which are stronger proton acceptors than the OH^- ion
   (b) weak bases (F^-, . . .), which are weaker proton acceptors than the OH^- ion, but stronger than the H_2O molecule
   (c) the anions of strong acids (ClO_4^-, . . .), which are weaker proton acceptors than the H_2O molecule and hence do not form basic water solutions

13.6 ACID-BASE PROPERTIES OF SALT SOLUTIONS

A salt is an ionic solid containing a cation other than H^+ and an anion other than OH^- or O^{2-}. When a salt such as NaCl, K_2CO_3, or $Al(NO_3)_3$ dissolves in water, the cation and anion separate from one another.

$$NaCl(s) \longrightarrow Na^+(aq) + Cl^-(aq)$$

$$K_2CO_3(s) \longrightarrow 2K^+(aq) + CO_3^{2-}(aq)$$

$$Al(NO_3)_3(s) \longrightarrow Al^{3+}(aq) + 3NO_3^-(aq)$$

To predict whether a given salt solution will be acidic, basic, or neutral, you consider three factors in turn.

1. *Decide what effect, if any, the cation has on the pH of water.*
2. *Decide what effect, if any, the anion has on the pH of water.*
3. *Combine the two effects to decide upon the behavior of the salt.*

Cations: Weak Acids or Spectator Ions?

As we pointed out in Section 13.4, certain cations act as weak acids in water solution because of reactions such as

$$NH_4^+(aq) + H_2O \rightleftharpoons H_3O^+(aq) + NH_3(aq)$$

$$Zn(H_2O)_4^{2+}(aq) + H_2O \rightleftharpoons H_3O^+(aq) + Zn(H_2O)_3(OH)^+(aq)$$

Essentially all transition metal ions behave like Zn^{2+}, forming a weakly acidic solution. Among the main-group cations, Al^{3+} and, to a lesser extent, Mg^{2+}, act as weak acids.

The cations derived from strong bases, e.g.,

— the alkali metal cations (Li^+, Na^+, K^+, . . .)
— the heavier alkaline earth cations (Ca^{2+}, Ba^{2+})

do not react with water. In that sense, these ions are "spectator" ions and do not affect pH (Table 13.7).

Anions: Weak Bases or Spectator Ions?

As pointed out in Section 13.5, anions which are the conjugate bases of weak acids act themselves as weak bases in water. They accept a proton from a water molecule, leaving an OH^- ion which makes the solution basic. The reactions of the fluoride and carbonate ions are typical:

$$F^-(aq) + H_2O \rightleftharpoons HF(aq) + OH^-(aq)$$

$$CO_3^{2-}(aq) + H_2O \rightleftharpoons HCO_3^-(aq) + OH^-(aq)$$

Table 13.7 Acid-Base Properties of Some Common Ions in Water Solution

	Spectator		Basic		Acidic
Anion	Cl^- Br^- I^-	NO_3^- ClO_4^- SO_4^{2-}	$C_2H_3O_2^-$ F^- CO_3^{2-} S^{2-} PO_4^{3-}	CN^- NO_2^-	
Cation	Li^+ Na^+ K^+	Ca^{2+} Ba^{2+}			Mg^{2+} Al^{3+} NH_4^+ transition metal ions

Learn the spectator ions, not the players

Actually, the SO_4^{2-} ion is very, very slightly basic. Can you see why?

Most anions behave this way, except those derived from the six strong acids (Cl^-, Br^-, I^-, NO_3^-, ClO_4^-, SO_4^{2-}). These anions do not react with water; in that sense they behave as "spectator ions" so far as their effect on pH is concerned (Table 13.7).

Salts: Acidic, Basic, or Neutral?

If you know how the cation and anion of a salt affect the pH of water, it is a relatively simple matter to decide what the net effect will be (Example 13.11).

Example 13.11 Consider water solutions of the four salts:

$$NH_4I,\ Zn(NO_3)_3,\ KClO_4,\ Na_3PO_4$$

Which of these solutions are acidic? basic? neutral?

Strategy First, decide what ions are present in the solution. Then classify each cation and anion as acidic, basic, or neutral, using Table 13.7. Finally, consider the combined effects of the two ions in each salt.

Solution To implement this strategy, it is convenient to prepare a table.

Salt	Cation	Anion	Solution of Salt
NH_4I	NH_4^+ (acidic)	I^- (spectator)	acidic
$Zn(NO_3)_2$	Zn^{2+} (acidic)	NO_3^- (spectator)	acidic
$KClO_4$	K^+ (spectator)	ClO_4^- (spectator)	neutral
Na_3PO_4	Na^+ (spectator)	PO_4^{3-} (basic)	basic

These predictions are confirmed by experiment (Figure 13.8).

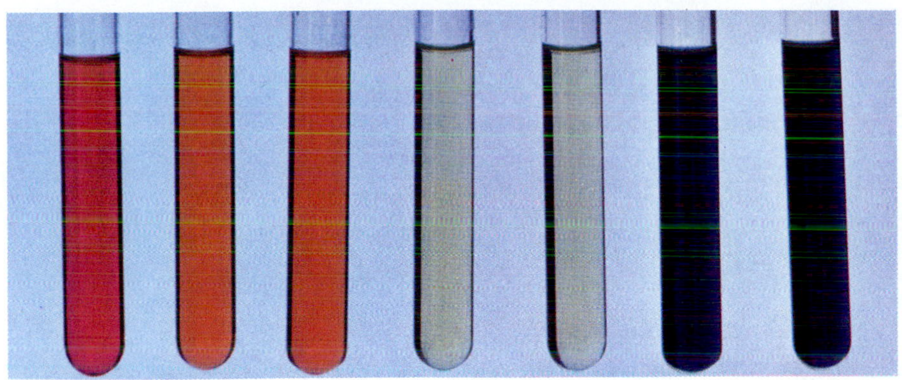

Figure 13.8
Colors of salt solutions with universal indicator, which is red in strong acid *(far left)* and purple in strong base *(far right)*. NH_4I and $Zn(NO_3)_2$ are weakly acidic at about pH 5 (orange); $KClO_4$, like pure water, has a pH of 7 (yellow-green); Na_3PO_4 is basic at about pH 12 (purple).

The procedure described in Example 13.11 does not cover one combination—an acidic cation (e.g., NH_4^+) and a basic anion (e.g., F^-). Here, K_a of the cation must be compared to K_b of the anion. In general:

1. *If $K_a > K_b$, the salt is acidic.* An example is NH_4F (K_a NH_4^+ = 5.6×10^{-10}; K_b F^- = 1.4×10^{-11}), where the pH is about 6.2.
2. *If $K_b > K_a$, the salt is basic.* An example is NH_4ClO (K_a NH_4^+ = 5.6×10^{-10}; $K_b ClO^-$ = 3.6×10^{-7}) where the pH is 8.4.

A similar argument can be used to decide whether an amphiprotic species such as HCO_3^- or $H_2PO_4^-$ is acidic or basic. In principle, these ions can act as either weak acids or weak bases:

$$HCO_3^-(aq) + H_2O \rightleftharpoons H_3O^+(aq) + CO_3^{2-}(aq) \qquad K_a = 4.7 \times 10^{-11}$$

$$HCO_3^-(aq) + H_2O \rightleftharpoons H_2CO_3(aq) + OH^-(aq) \qquad K_b = 2.3 \times 10^{-8}$$

Since $K_b > K_a$, you would predict, correctly, that the HCO_3^- ion produces a basic water solution. Sodium hydrogen carbonate, $NaHCO_3$ ("bicarbonate of soda") is used as an antacid because its water solution is basic with a pH of about 8.3. The reverse situation applies with the $H_2PO_4^-$ ion (K_a = 6.2×10^{-8}; K_b = 1.4×10^{-12}). Since $K_a > K_b$, the $H_2PO_4^-$ ion gives an acidic solution. The compound NaH_2PO_4 is weakly acidic with a solution pH of about 4.7.

Most anions of this type, like HCO_3^-, are basic ($K_b > K_a$)

PERSPECTIVE
•
Organic Acids

As pointed out earlier in this chapter, most weak acids are organic in nature, i.e., they contain carbon and hydrogen atoms. Table 13.8 lists some of the organic acids found in foods. All of these compounds contain the *carboxyl* group:

$$-\underset{\underset{O}{\|}}{C}-O-H$$

Certain drugs, both prescription and over-the-counter, contain organic acids. Two of the most popular products of this type are the analgesics aspirin and ibuprofen (Advil, Nuprin, etc.).

H
H—C—C—COOH
H—C—C—$OCOCH_3$
H

aspirin

H H
C=C CH_3
$(CH_3)_2$—CH—CH_2—C C—C—COOH
C—C H
H H

ibuprofen

Because these compounds are acidic, they can cause stomach irritation unless taken with food or water.

To the general public, probably the best-known organic acid is ascorbic acid (Vitamin C):

O
HO—C—C
O
HO—C—CH
HO—C—H
CH_2OH

ascorbic acid (Vitamin C)

Table 13.8 Some Naturally Occurring Organic Acids

Name		Source
Acetic acid	$CH_3—COOH$	vinegar
Citric acid	OH $HOOC—CH_2—C—CH_2—COOH$ COOH	citrus fruits
Lactic acid	$CH_3—CH—COOH$ OH	sour milk
Malic acid	$HOOC—CH_2—CH—COOH$ OH	apples, watermelons, grape juice, wine
Oxalic acid	HOOC—COOH	rhubarb, spinach, tomatoes
Quinic acid	H OH H C—CH_2 COOH C C HO C—CH_2 OH H OH	cranberries
Tartaric acid	HOOC—CH—CH—COOH OH OH	grape juice, wine

Vitamin C plays a variety of roles in human nutrition, some of which have been discovered only recently. For one thing, it is involved in the formation of collagen, the principal protein in connective tissue. Beyond that, Vitamin C is required for the metabolism of several amino acids and for the absorption of iron. Moreover, since Vitamin C is easily oxidized, it serves to protect other vitamins (A and E) from oxidation. The recommended minimum daily allowance of Vitamin C is 60 mg/day.

Many people, including Nobel laureate Linus Pauling, advocate taking megadoses of Vitamin C—two or more grams per day. In such large doses Vitamin C supposedly helps prevent colds and acts as an anticarcinogen. However, large doses of Vitamin C can be toxic to people with an enzyme deficiency that makes them susceptible to strong reducing agents. Perhaps 2 to 5% of the population falls into this category; for them megadoses of Vitamin C pose a very real health hazard.

CHAPTER HIGHLIGHTS

KEY CONCEPTS

1. *Given $[H^+]$, $[OH^-]$ or pH, calculate the other two quantities*
   (Examples 13.1–13.3; Problems 7–24)
2. *Using the Brønsted-Lowry model, write an equation to explain why a species acts as a*
   —weak acid
   (Example 13.4; Problems 25–28)
   —weak base
   (Example 13.9; Problems 47, 48)
3. *Given pH and $[HB]_o$, calculate K_a*
   (Example 13.5; Problems 33–36)
4. *Given K_a and $[HB]_o$, calculate $[H^+]$*
   (Examples 13.6–13.8; Problems 37–46, 68, 70, 75)
5. *Given K_b and $[B^-]_o$, calculate $[OH^-]$*
   (Example 13.10; Problems 55–58)
6. *Predict whether a given salt solution is acidic, basic, or neutral*
   (Example 13.11; Problems 59–66, 72)

KEY EQUATIONS

Dissociation of water $\quad K_w = [H^+] \times [OH^-] = 1.0 \times 10^{-14}$

Expressions for K_a, K_b

$$K_a = \frac{[H^+] \times [B^-]}{[HB]}$$

$$K_b = \frac{[OH^-] \times [HB]}{[B^-]}$$

$$K_a \times K_b = K_w$$

KEY TERMS

acid	conjugate acid	pK_a
—strong	conjugate base	pOH
—weak	dissociation constant (K_w, K_a, K_b)	percent dissociation
amphiprotic	indicator	polyprotic acid
base	molarity	quadratic formula
—strong	neutral solution	salt
—weak	pH	successive approximations

SUMMARY PROBLEM

Consider the salts ammonium chloride, NH_4Cl, and potassium cyanide, KCN.

1. Hydrochloric acid and ammonia can be used to prepare ammonium chloride. Potassium hydroxide and hydrogen cyanide can be used to prepare potassium cyanide. Classify each reactant as strong or weak, acid or base.
2. Using the Brønsted-Lowry model, write equations to explain the acidity or basicity of HCl, NH_3, and HCN.
3. What is the conjugate base of NH_4^+? the conjugate acid of CN^-?
4. Calculate the pH of 0.10 *M* solutions of HCl, KOH, NH_4Cl, and KCN (Tables 13.2, 13.5).
5. Classify the salts NH_4Cl and KCN as acidic or basic. Write net ionic equations to explain your answers.
6. Would ammonium cyanide, NH_4CN, be acidic or basic?

Answers

1. HCl, strong acid; KOH, strong base; ammonia, weak base; HCN, weak acid
2. $HCl(aq) + H_2O \longrightarrow H_3O^+(aq) + Cl^-(aq)$
   $NH_3(aq) + H_2O \rightleftharpoons NH_4^+(aq) + OH^-(aq)$
   $HCN(aq) + H_2O \rightleftharpoons H_3O^+(aq) + CN^-(aq)$
3. NH_3; HCN
4. HCl, pH = 1.00; KOH, pH = 13.00; NH_4Cl, pH = 5.13; KCN, pH = 11.11
5. NH_4Cl is acidic; $NH_4^+(aq) + H_2O \rightleftharpoons NH_3(aq) + H_3O^+(aq)$
   KCN is basic; $CN^-(aq) + H_2O \rightleftharpoons HCN(aq) + OH^-(aq)$
6. NH_4CN is basic.

QUESTIONS & PROBLEMS

Brønsted-Lowry Acid-Base Model

1. For each of the following reactions, indicate the Brønsted-Lowry acids and bases. What are the conjugate acid-base pairs?

a. $H_3O^+ + HSO_3^-(aq) \rightleftharpoons H_2SO_3(aq) + H_2O$
b. $HF(aq) + OH^-(aq) \rightleftharpoons F^-(aq) + H_2O$
c. $NH_4^+(aq) + H_2O \rightleftharpoons NH_3(aq) + H_3O^+$

2. Follow the directions for Question 1 for the following reactions.

a. $CN^-(aq) + H_2O \rightleftharpoons HCN(aq) + OH^-(aq)$
b. $HCO_3^-(aq) + H_3O^+(aq) \rightleftharpoons H_2CO_3(aq) + H_2O$
c. $HC_2H_3O_2(aq) + HS^-(aq) \rightleftharpoons C_2H_3O_2^-(aq) + H_2S(aq)$

3. According to the Brønsted-Lowry theory, which of the following would you expect to act as an acid? Which as a base?

a. HNO_2 **b.** OCl^- **c.** NH_2^-

4. According to the Brønsted-Lowry theory, which of the following would you expect to be an acid? Which a base?

a. NH_4^+ **b.** $CH_3NH_3^+$ **c.** $C_2H_3O_2^-$

5. Give the formula of the conjugate acid of

a. H_2O **b.** HCO_3^- **c.** CH_3NH_2
d. S^{2-} **e.** $Fe(H_2O)_5(OH)^+$

6. Give the formula of the conjugate base of

a. $HC_2H_3O_2$ **b.** $Fe(H_2O)_5(OH)^{2+}$ **c.** HSO_3^-
d. $CH_3NH_3^+$ **e.** H_2S

[H⁺], [OH⁻], pH, and pOH

7. Find the pH of solutions with the following $[H^+]$. Classify each as acidic or basic.

a. 1×10^{-2} *M* **b.** 0.00010 *M*
c. 4.0×10^{-5} *M* **d.** 6.2×10^{-10} *M*

8. Find the pH of the solutions with the following $[H^+]$. Classify each as acidic or basic.

a. 0.1 *M* **b.** 10 *M*
c. 7.0×10^{-3} *M* **d.** 8.2×10^{-9} *M*

9. Calculate $[H^+]$ and $[OH^-]$ in solutions with the following pH.

a. 4.0 **b.** 8.52 **c.** 0.00 **d.** 12.60

10. Find $[H^+]$ and $[OH^-]$ in solutions with the following pOH.

a. 9.0 **b.** 3.20 **c.** −1.05 **d.** 7.46

11. Solution A has $[OH^-] = 3.2 \times 10^{-4}$ *M*. Solution B has $[H^+] = 6.9 \times 10^{-9}$ *M*. Which solution is more basic? Which has the lower pH?

12. Solution 1 has pH 4.3. Solution 2 has $[OH^-] = 3.4 \times 10^{-7}$ *M*. Which solution is more acidic? Which has the higher pOH?

13. One solution has a pH of 3.2. What must be the pH of another solution so that $[H^+]$ is four times as large? One fourth as large?

14. One solution has a pH of 3.5. Another has a pH of 5.5.

What is the ratio of the H^+ ion concentrations in the two solutions? The ratio of OH^- ion concentrations?

15. Milk of magnesia has a pH of 10.5.
 a. Calculate $[H^+]$.
 b. Calculate the ratio of the $[H^+]$ concentration of gastric juice, pH 1.5, to that of milk of magnesia.

16. Unpolluted rain water has a pH of about 5.5. Acid rain has been shown to have a pH as low as 3.0. Calculate the $[H^+]$ ratio of acid rain to unpolluted rain.

17. Find $[H^+]$ and the pH of the following solutions.
 a. 0.30 *M* HBr
 b. a solution made by diluting 10.0 mL of 6.00 *M* HCl to 0.300 L with water

18. Find $[H^+]$ and the pH of the following solutions.
 a. 0.60 *M* HCl
 b. a solution made by dissolving 75 g HNO_3 in water to make 2.0 L of solution

19. What is the pH of a solution obtained by mixing 245 mL of 0.0235 *M* HI and 438 mL of 0.554 *M* HBr? Assume that volumes are additive.

20. What is the pH of a solution obtained by mixing 38.2 g $HClO_4$ and 28.9 g HCl in enough water to make 0.750 L of solution?

21. Find $[OH^-]$, $[H^+]$, and the pOH of the following solutions.
 a. 0.50 *M* KOH
 b. a solution made by dissolving 100.0 g NaOH in enough water to make 500.0 mL of solution

22. Find $[OH^-]$, $[H^+]$, and the pOH of the following solutions.
 a. 0.80 *M* NaOH
 b. a solution made by diluting 8.0 mL of 6.0 *M* CsOH to a volume of 4.80×10^2 mL with water

23. What is the pH of a solution obtained by adding 32.1 g of NaOH and 56.3 g of KOH to enough water to make 1.75 L of solution?

24. What is the pH of a solution obtained by adding 25.0 mL of 0.125 *M* rubidium hydroxide, RbOH, to 35.0 mL of 0.027 *M* barium hydroxide? Assume that volumes are additive.

Dissociation Expressions, Weak Acids

25. Using the Brønsted-Lowry model, write equations to show why the following species behave as weak acids in water.
 a. $Ni(H_2O)_5(OH)^+$ **b.** $Al(H_2O)_6{}^{3+}$ **c.** H_2S
 d. $H_2PO_4{}^-$ **e.** $Cr(H_2O)_5(OH)^{2+}$ **f.** $HClO_2$

26. Follow the directions of Question 25 for the following species.
 a. $Zn(H_2O)_3(OH)^+$ **b.** $HSO_4{}^-$
 c. HNO_2 **d.** $Fe(H_2O)_6{}^{2+}$
 e. $Mn(H_2O)_6{}^{2+}$ **f.** $HC_2H_3O_2$

27. Write the dissociation equation and the K_a expression for each of the following acids.
 a. $HSO_3{}^-$ **b.** $HPO_4{}^{2-}$ **c.** HNO_2

28. Write the dissociation equation and the K_a expression for each of the following acids.
 a. $PH_4{}^+$ **b.** HS^- **c.** $HC_2O_4{}^-$

29. Referring to Table 13.4, give the two pK_a values for H_2CO_3.

30. Calculate the K_a for the weak acids that have the following pK_a values.
 a. 3.0 **b.** 5.8 **c.** 9.67

31. Consider these acids:

Acid	A	B	C	D
K_a	3×10^{-4}	5×10^{-6}	2×10^{-3}	8×10^{-2}

 a. Arrange the acids in order of increasing acid strength.
 b. Which acid has the smallest pK_a value?

32. Consider these acids:

Acid	A	B	C	D
pK_a	8.3	2.7	12.9	5.6

 a. List the acids in order of decreasing acid strength.
 b. Which acid has the smallest K_a value?

Equilibrium Calculations, Weak Acids

33. Para-aminobenzoic acid (PABA), $HC_7H_6NO_2$, is used in some sunscreen agents. A solution is made by dissolving 0.030 mol PABA to form a liter of solution. The solution has $[H^+] = 8.1 \times 10^{-4}$ *M*. Calculate K_a for para-aminobenzoic acid.

34. Caproic acid, $HC_6H_{11}O_2$, is found in coconut oil and is used in the making of artificial flavors. A solution prepared by dissolving 0.14 mol $HC_6H_{11}O_2$ to form 1.5 L of solution has $[H^+] = 1.1 \times 10^{-3}$ *M*. Calculate K_a for caproic acid.

35. Benzoic acid, $HC_7H_5O_2$, is present in many berries. A benzoic acid solution prepared by dissolving 1.00 g of benzoic acid in 350.0 mL of solution has a pH of 2.91. What is the K_a for benzoic acid?

36. Phenol, HC_6H_5O, is a weak organic acid used in the manufacture of plastics. A solution prepared by dissolving 0.385 g of phenol in water to form a volume of 2.00 L has a pH of 6.29. What is the K_a of phenol?

37. Formic acid, $HCHO_2$, is the irritant in nettles and ants. Its K_a is 1.9×10^{-4}. Calculate the $[H^+]$ in solutions prepared by adding the following number of moles of formic acid to form one liter of solution.
 a. 2.0 **b.** 0.33

38. Follow the directions for Problem 37 for hypobromous acid, HBrO ($K_a = 2.6 \times 10^{-9}$).

39. Lactic acid ($K_a = 1.4 \times 10^{-4}$) is present in sore muscles after vigorous exercise. For a 1.3 *M* solution of lactic acid, calculate
 a. $[H^+]$. **b.** $[OH^-]$. **c.** pH.
 d. % dissociation.

40. Follow the directions of Problem 39 for a 1.25 *M* solution of ammonium chloride ($K_aNH_4^+ = 5.6 \times 10^{-10}$).

41. Chloroacetic acid, $ClCH_2COOH$, has a K_a of 1.4×10^{-3}. Calculate the pH of a 0.20 *M* solution of chloroacetic acid.

42. Chlorous acid, $HClO_2$, has a K_a of 1.0×10^{-2}. Calculate the pH of a 0.100 *M* solution of chlorous acid.

43. Using the K_a values listed in Table 13.4, calculate the pH of 0.10 *M* H_2CO_3.

44. Using the K_a values listed in Table 13.4, calculate the pH of a solution 0.10 *M* in citric acid.

45. For the solution referred to in Problem 43, estimate $[HCO_3^-]$ and $[CO_3^{2-}]$.

46. For the solution referred to in Problem 44, estimate $[H_2C_5H_5O_7^-]$ and $[HC_5H_5O_7^{2-}]$.

Dissociation Expressions, Weak Bases

47. Using the Brønsted-Lowry model, write an equation to show why each of the following species produces a basic water solution.

a. NH_3 **b.** NO_2^- **c.** $C_6H_5NH_2$
d. CO_3^{2-} **e.** F^- **f.** HCO_3^-

48. Follow the directions of Question 47 for the following species.

a. $(CH_3)_3N$ **b.** PO_4^{3-} **c.** HPO_4^{2-}
d. $H_2PO_4^-$ **e.** HS^- **f.** $C_2H_5NH_2$

49. Using the dissociation constants listed in Table 13.5, arrange the following solutions in order of decreasing pH.

a. 0.10 *M* NH_3 **b.** 0.10 *M* KOH
c. 0.10 *M* NaF **d.** 0.10 *M* $C_6H_5NH_2$

50. Follow the directions of Question 49 for 0.10 *M* solutions of

a. HCl **b.** NaOH **c.** $NaNO_2$ **d.** NaClO

51. Which of the following are true regarding a 1 *M* solution of a strong base MOH?

a. The M^+ concentration is 1 *M*.
b. The concentration of MOH molecules is 1 *M*.
c. The sum of $[M^+]$ and $[OH^-]$ is 2 *M*.
d. The H^+ concentration is 1 *M*.
e. The pH is 14.0.

52. Which of the following are true regarding a 0.10 *M* solution of a weak base B^-?

a. The HB concentration is 0.10 *M*.
b. The $[OH^-] \approx [HB]$.
c. $[B^-] \gg [HB]$.
d. The pH is 13.
e. The H^+ concentration is 0.10 *M*.

Equilibrium Calculations, Weak Bases

53. Find the value of K_a for the conjugate acid of the following organic bases.

a. dimethylamine, used to remove hair from hides; $K_b = 5.2 \times 10^{-4}$
b. aniline, an important dye intermediate; $K_b = 3.8 \times 10^{-10}$

54. Find the value of K_b for the conjugate base of the following acids.

a. gallic acid, present in tea; $K_a = 3.9 \times 10^{-5}$
b. mandelic acid, obtained from bitter almonds; $K_a = 4.4 \times 10^{-4}$

55. Write the net ionic equation for the reaction that makes solutions of sodium benzoate, $NaC_7H_5O_2$, basic. K_a for benzoic acid is 6.6×10^{-5}. Find

a. K_b for the reaction.
b. the pH of a 0.23 *M* solution of sodium benzoate.

56. Use Table 13.5 to determine the pH of a 0.50 *M* solution of sodium cyanide.

57. Using Table 13.5, calculate the pH of a household cleaning ammonia solution prepared by dissolving enough ammonia gas in water to make a 0.30 *M* aqueous solution.

58. Cocaine is a weak base. Its ionization in water can be represented (Coc) by the equation

$$Coc(aq) + H_2O \rightleftharpoons CocH^+(aq) + OH^-(aq)$$

A 0.0010 *M* solution of cocaine has a pH of 9.70. Calculate K_b for cocaine.

Salt Solutions

59. State whether 1 *M* solutions of the following salts in water would be acidic, basic, or neutral.

a. NH_4Cl **b.** NH_4CN
c. Na_3PO_4 **d.** KNO_3
e. $KHCO_3$ **f.** NaF

60. State whether 1 *M* solutions of the following salts in water would be acidic, basic, or neutral.

a. $Al(NO_3)_3$ **b.** NH_4NO_2
c. NaClO **d.** NH_4NO_3
e. Na_2CO_3 **f.** $NaHSO_4$

61. Write net ionic equations to explain the acidity or basicity of the various salts listed in Question 59.

62. Write net ionic equations to explain the acidity or basicity of the various salts listed in Question 60.

63. Arrange the following 0.1 *M* aqueous solutions in order of increasing pH: KOH, $ZnCl_2$, NH_3, HBr, BaI_2.

64. Arrange the following 0.1 *M* aqueous solutions in order of decreasing pH: NH_4Br, $Ba(OH)_2$, $HClO_4$, K_2SO_4, LiCN.

65. Write formulas for four salts that

a. contain Fe^{3+} and are acidic.
b. contain NO_3^- and are neutral.
c. contain Ba^{2+} and are basic.
d. contain Cs^+ and are neutral.

66. Write formulas for four salts that

a. contain Li^+ and are basic.
b. contain Li^+ and are neutral.
c. contain SO_4^{2-} and are neutral.
d. contain SO_4^{2-} and are acidic.

Organic Acids

67. Consider Vitamin C.

a. State its other name.
b. What is its role in human nutrition?
c. What is the recommended daily allowance of Vitamin C?

68. The successive dissociation constants of oxalic acid are 5.9×10^{-2} and 5.2×10^{-5}. What is the pH of a 1.00 *M* solution of oxalic acid?

69. Consider the following naturally occurring organic acids: acetic acid, lactic acid, citric acid, and oxalic acid.

a. List sources for these acids.

b. Using the various tables in the chapter, rank 0.1 *M* aqueous solutions of these acids in terms of increasing acidity.

70. There are 324 mg of acetylsalicylic acid ($\mathcal{M}$ = 180.15 g/mol) per aspirin tablet. If two tablets are dissolved to give two ounces ($\frac{1}{16}$ quart) of solution, estimate the pH. K_a of acetylsalicylic acid is 3.6×10^{-4}.

Unclassified

71. Give two examples of

a. an acid stronger than HF.

b. a base weaker than NaOH.

c. a salt that dissolves in water to yield an acidic solution.

d. a salt that dissolves in water to yield a basic solution.

e. a salt that dissolves in water to yield a solution with a pH of 7.

72. Using data in Tables 13.2 and 13.5, classify solutions of the following salts as acidic, basic, or neutral.

a. $NH_4C_2H_3O_2$ **b.** $NH_4H_2PO_4$

c. $AgNO_2$ **d.** CH_3NH_3F

73. Consider the process

$$H_2O \rightleftharpoons H^-(aq) + OH^-(aq) \qquad \Delta H° = 55.8 \text{ kJ}$$

Will $[H^+]$ in pure water be greater or smaller than 1.0×10^{-7} at 40°C? Explain.

74. The K_w for water at 37°C, human body temperature, is 2.40×10^{-14}. Is a saline solution (0.1 *M* NaCl) acidic, basic, or neutral at 37°C?

75. Calculate the pH of a 0.025 *M* solution of aluminum nitrate. K_a for $Al(H_2O)_6{}^{3+}$ is 1.2×10^{-5}.

Challenge Problems

76. Using Table 13.4 and the law of multiple equilibria, determine the equilibrium constant for the reaction

$$H_2CO_3(aq) \rightleftharpoons 2H^+(aq) + CO_3{}^{2-}(aq)$$

77. Silver hydroxide, AgOH, is insoluble in water. Describe a simple qualitative experiment that would enable you to determine whether AgOH is a strong or weak base.

78. Using the tables in Appendix 1, calculate ΔH for the reaction of

a. 1.00 L of 0.100 *M* NaOH with 1.00 L of 0.100 *M* HCl.

b. 1.00 L of 0.100 *M* NaOH with 1.00 L of 0.100 *M* HF, taking the heat of formation of HF(*aq*) to be −320.1 kJ/mol.

79. Show by calculation that when the concentration of a weak acid decreases by a factor of 10, its percent dissociation increases by a factor of $10^{1/2}$.

80. What is the freezing point of vinegar, which is an aqueous solution of 5.00% acetic acid, $HC_2H_3O_2$, by mass (d = 1.006 g/cm^3)?

14
Acid-Base and Precipitation Equilibria

No single thing abides;
but all things flow
Fragment to fragment clings—
the things thus grow
Until we know and name them.
By degrees
They melt, and are no more the
things we know.

TITUS LUCRETIUS CARUS
(92–52 B.C.)
No Single Thing Abides
(Translated by W. H. Mallock)

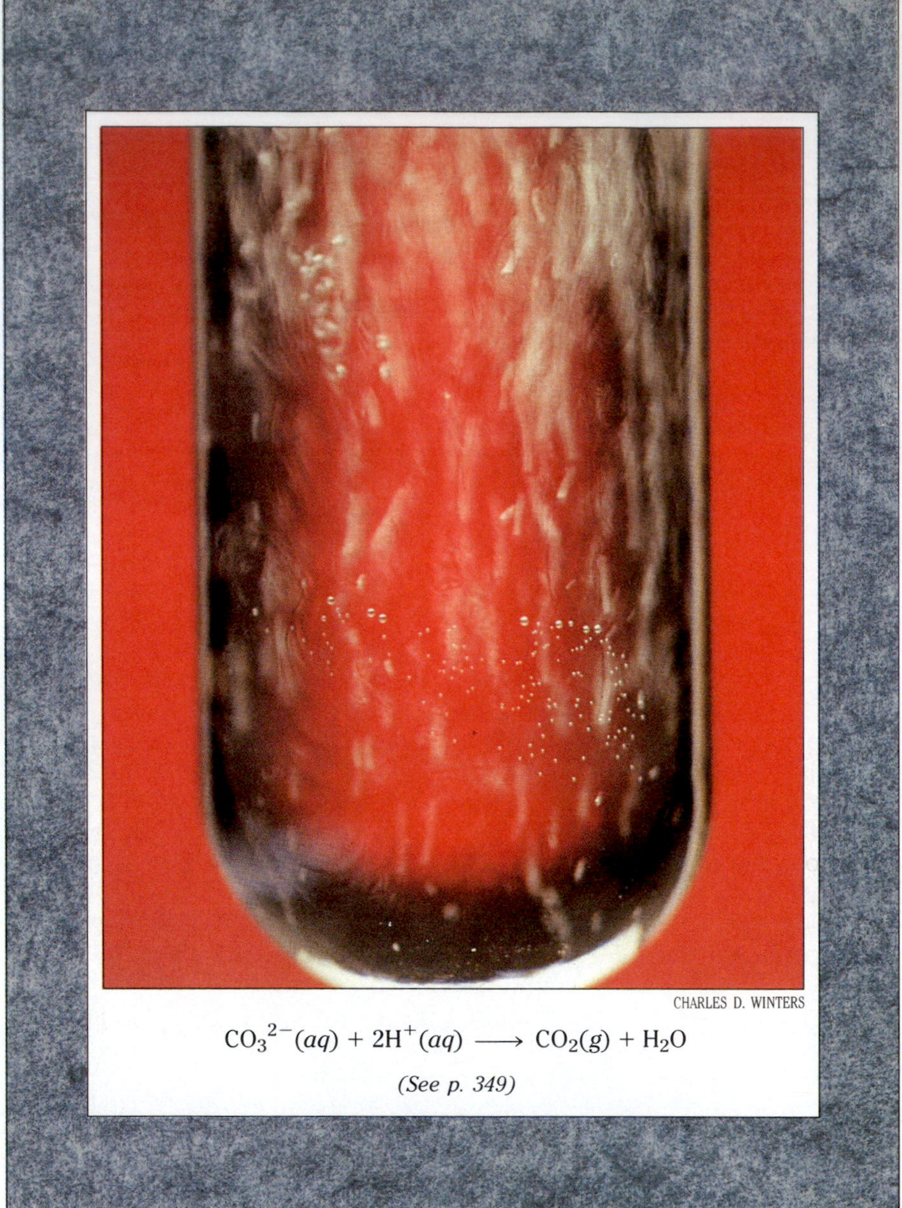

CHARLES D. WINTERS

$$CO_3^{2-}(aq) + 2H^+(aq) \longrightarrow CO_2(g) + H_2O$$

(See p. 349)

CHAPTER OUTLINE

In Chapter 13, we dealt with the equilibrium established when a single solute, either a weak acid or a weak base, is added to water. This chapter focuses upon the equilibrium established when two different solutes are mixed in water solution. These solutes may be

1. *An acid and a base used in an acid-base titration.* This type of reaction was discussed in Chapter 4. Section 14.1 examines the equilibria involved, the way pH changes during the titration, and the choice of indicator for the titration.

2. *A weak acid, HB, and its conjugate base, B^-.* Solutions containing roughly equal amounts of these two species, called *buffers,* are used to establish constant pH. The equilibria involved in buffer solutions are considered in Section 14.2.
3. *Two ionic solutes reacting to form a precipitate.* Again, this type of reaction was discussed qualitatively in Chapter 4. Section 14.3 considers the equilibrium established between a precipitate and its ions in solution.

The equilibrium principles developed in this chapter are of considerable practical importance. Geologists use them to reconstruct the processes by which minerals are deposited in the earth's crust or the bottom of the ocean. Nutritionists use them to recommend diets to patients suffering from arthritis or osteoporosis. Biochemists use the principles of buffer action to control pH and hence reaction rate in living systems.

14.1 ACID-BASE TITRATIONS

In Chapter 4, we considered the general topic of acid-base titrations. In carrying out an acid-base titration, you measure the volume of a standard acidic or basic solution required to just react with a known volume of a base or acid of unknown concentration. The point at which the reaction is complete is referred to as the **equivalence point**; equivalent quantities of acid and base have reacted. This point is detected using an *acid-base indicator.* It is important that the point at which the indicator changes color (the **end point**) coincide with the equivalence point.

The discussion in Chapter 4 focused upon the stoichiometry of acid-base titrations. Here, the emphasis is on the equilibrium principles that apply to the acid-base reactions involved. It is convenient to distinguish between titrations involving

— a strong acid (e.g., HCl) and a strong base (e.g., NaOH, $Ba(OH)_2$)
— a weak acid (e.g., $HC_2H_3O_2$) and a strong base (e.g., NaOH)
— a strong acid (e.g., HCl) and a weak base (e.g., NH_3)

Strong Acid–Strong Base

As pointed out in Chapter 13, strong acids dissociate completely in water to form H_3O^+ ions; strong bases dissolve in water to form OH^- ions. The neutralization reaction that takes place when *any* strong acid reacts with *any* strong base can be represented by a net ionic equation of the Brønsted-Lowry type:

$$H_3O^+(aq) + OH^-(aq) \longrightarrow 2H_2O$$

or, more simply, substituting H^+ ions for H_3O^+ ions, as:

$$H^+(aq) + OH^-(aq) \longrightarrow H_2O$$

Since the equation just written is the reverse of that for the dissociation of water, the equilibrium constant can be calculated by using the reciprocal rule (Chap. 12).

$$K = 1/K_w = 1/(1.0 \times 10^{-14}) = 1.0 \times 10^{14}$$

K must be large if a titration is to work

The enormous value of K means that for all practical purposes this reaction goes to completion, consuming the limiting reactant, H^+ or OH^-.

Consider now what happens when HCl, a typical strong acid, is titrated with NaOH. Figure 14.1 shows how the pH changes during the titration. Two features of this curve are of particular importance:

1. At the equivalence point, when all the HCl has been neutralized by NaOH, a solution of NaCl, a neutral salt, is present. The pH at the equivalence point is 7.

2. Near the equivalence point, the pH rises very rapidly. Indeed, the pH may increase by as much as 6 units (from 4 to 10) when a single drop, ≈0.02 mL, of NaOH is added (Example 14.1).

Example 14.1 When 50.00 mL of 1.000 *M* HCl is titrated with 1.000 *M* NaOH, the pH increases. Find the pH of the solution after the following volumes of 1.000 *M* NaOH have been added.

(a) 49.99 mL (b) 50.01 mL

Strategy This is essentially a stoichiometry problem of the type discussed in Chapter 4. The reaction is $H^+(aq) + OH^-(aq) \longrightarrow H_2O$. The number of moles of H^+ in 50.00 mL of 1.000 *M* HCl is

$$n_{H^+} = 50.00 \times 10^{-3}\,\text{L} \times \frac{1.000\ \text{mol H}^+}{1\ \text{L}} = 50.00 \times 10^{-3}\ \text{mol H}^+$$

In each part of the problem, you start (1) by calculating the number of moles of OH^- added. Then (2), find the number of moles of H^+ or OH^- in excess. Finally (3), calculate $[H^+]$ and the pH.

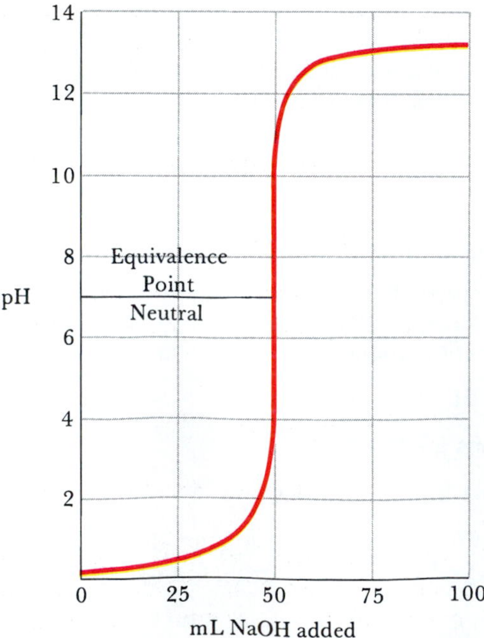

Figure 14.1
Titration of 50.00 mL of 1.000 *M* HCl with 1.000 *M* NaOH. The solution at the equivalence point is neutral (pH = 7). The pH rises very rapidly near the equivalence point.

Solution

(a) (1) $n_{OH^-} = 49.99 \times 10^{-3}\ L \times \dfrac{1.000\ mol\ OH^-}{1\ L} = 49.99 \times 10^{-3}\ mol\ OH^-$

(2) Since H^+ and OH^- react in a 1:1 mole ratio,

$$n_{H^+\,reacted} = 49.99 \times 10^{-3}\ mol$$

$$n_{H^+\,excess} = 50.00 \times 10^{-3}\ mol - 49.99 \times 10^{-3}\ mol$$

$$= 0.01 \times 10^{-3}\ mol = 1 \times 10^{-5}\ mol$$

(3) The total volume is almost exactly 100 mL

$$[H^+] = \frac{1 \times 10^{-5}\ mol\ H^+}{1 \times 10^{-1}\ L} = 1 \times 10^{-4}\ M$$

$$pH = -\log_{10}[H^+] = 4.0$$

(b) (1) $n_{OH^-} = 50.01 \times 10^{-3}\ L \times \dfrac{1.000\ mol\ OH^-}{1\ L} = 50.01 \times 10^{-3}\ mol\ OH^-$

(2) $n_{H^+\,reacted} = 50.00 \times 10^{-3}\ mol$

$$n_{OH^-\,excess} = 50.01 \times 10^{-3}\ mol - 50.00 \times 10^{-3}\ mol$$

$$= 0.01 \times 10^{-3}\ mol = 1 \times 10^{-5}\ mol$$

(3) $[OH^-] = \dfrac{1 \times 10^{-5}\ mol\ OH^-}{1 \times 10^{-1}\ L} = 1 \times 10^{-4}\ M$

$$pOH = -\log_{10}[OH^-] = 4.0$$

$$pH = 14.00 - pOH = 10.0$$

Note that the pH changes from 4 to 10 upon adding (50.01 − 49.99)mL = 0.02 mL of titrant.

Weak Acid–Strong Base

A typical weak acid–strong base titration is that of acetic acid with sodium hydroxide. The net ionic equation for the reaction is

$$HC_2H_3O_2(aq) + OH^-(aq) \longrightarrow C_2H_3O_2^-(aq) + H_2O$$

Notice that this reaction is the reverse of the reaction of the weak base $C_2H_3O_2^-$ (the acetate ion) with water (Chap. 13). It follows from the reciprocal rule that for this reaction

$$K = \frac{1}{(K_b\ C_2H_3O_2^-)} = \frac{1}{(5.6 \times 10^{-10})} = 1.8 \times 10^9$$

Here again, K is a very large number; the reaction of acetic acid with a strong base goes essentially to completion.

Figure 14.2 shows how the pH changes as 50 mL of 1 M $HC_2H_3O_2$ is titrated with 1 M NaOH. Notice that the pH starts at a somewhat higher value than with a strong acid (2.4 vs. 0.0) because acetic acid is only slightly dissociated in

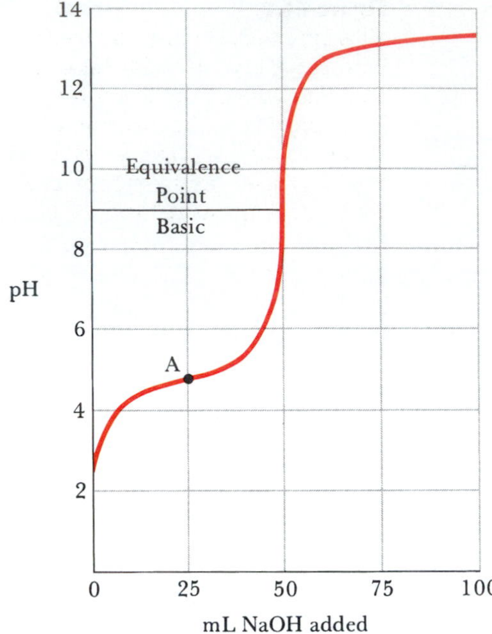

Figure 14.2
Titration of 50.00 mL of the weak acid $HC_2H_3O_2$ (1.000 *M*) with 1.000 *M* NaOH. The solution at the equivalence point is basic (pH = 9.22). Notice that when the titration is half over (point A), the pH is rising very slowly; at this point $[HC_2H_3O_2] = [C_2H_3O_2^-]$.

water. Beyond that, there are a couple of important differences between Figures 14.1 and 14.2.

1. The pH at the equivalence point is greater than 7, actually about 9.2. This is reasonable; at this point the reaction between $HC_2H_3O_2$ and NaOH is complete, forming a solution of sodium acetate, $NaC_2H_3O_2$. The $C_2H_3O_2^-$ ion reacts with water to make the solution weakly basic.

Note also that pH levels off at A, halfway through the titration

2. The pH changes comparatively slowly near the equivalence point. It takes about 3–4 mL of 1 *M* NaOH to change the pH from 6 to 12; a similar shift of 6 pH units with a strong acid requires only 0.02 mL of 1 *M* NaOH (recall Example 14.1).

Strong Acid–Weak Base

The reaction between solutions of hydrochloric acid and ammonia can be represented by the Brønsted-Lowry equation:

$$H_3O^+(aq) + NH_3(aq) \longrightarrow NH_4^+(aq) + H_2O$$

Again, we simplify the equation by replacing H_3O^+ with H^+:

$$H^+(aq) + NH_3(aq) \longrightarrow NH_4^+(aq)$$

To find the equilibrium constant, note that this equation is the reverse of that for the dissociation of the NH_4^+ ion. Hence

$$K = \frac{1}{K_a\ NH_4^+} = \frac{1}{5.6 \times 10^{-10}} = 1.8 \times 10^9$$

Since *K* is so large, the reaction goes virtually to completion.

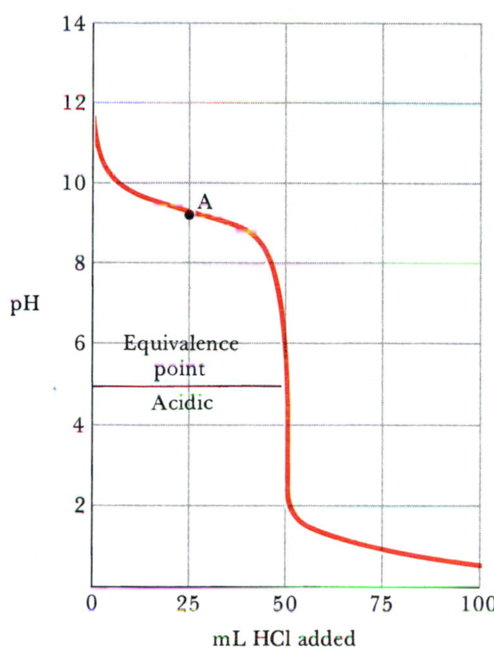

Figure 14.3
Titration of 50.00 mL of the weak base NH_3 (1.000 *M*) with 1.000 *M* HCl. The solution at the equivalence point is acidic, because of the NH_4^+ ion. At point A, half of the NH_3 molecules have been converted to NH_4^+ ions: $[NH_3] = [NH_4^+]$.

Figure 14.3 shows how pH changes when 50 mL of 1 *M* NH_3 is titrated with 1 *M* HCl. Originally, the solution is basic with a pH of about 11.6, that of 1 *M* NH_3. As strong acid is added, moving to the right in Figure 14.3, the pH drops steadily. Notice that

— the pH at the equivalence point is <7, actually about 4.8, due to the presence of the NH_4^+ ion, which is a weak acid

— the pH changes relatively slowly near the equivalence point (compare with Figs. 14.2 and 14.1)

Choice of Indicator

An acid-base indicator is typically a weak acid, HIn,

$$HIn(aq) \rightleftharpoons H^+(aq) + In^-(aq) \qquad K_a = \frac{[H^+] \times [In^-]}{[HIn]}$$

where the two species, HIn and In^-, have different colors (Fig. 14.4). The color that you see when a drop of indicator solution is added in an acid-base titration depends upon the ratio

$$\frac{[HIn]}{[In^-]}$$

For litmus, HIn is red, In^- is blue

Three cases can be distinguished:

1. If $\dfrac{[HIn]}{[In^-]} \geqslant 10$

the principal species is HIn; you see the "acid" color, that of the HIn molecule.

2. If $\dfrac{[HIn]}{[In^-]} \leqslant 0.1$

the principal species is In^-; you see the "base" color, that of the In^- ion.

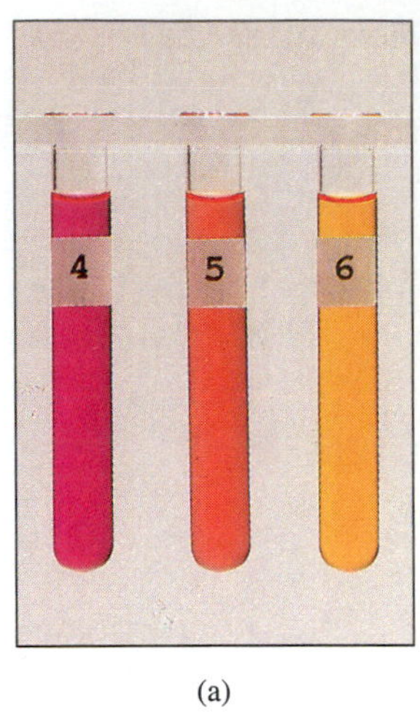

(a)

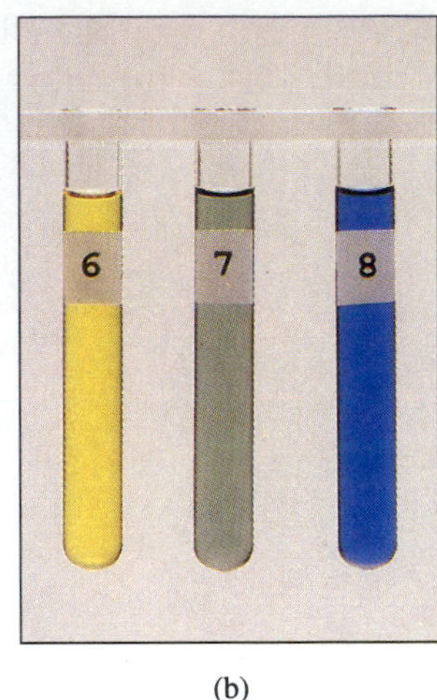

(b)

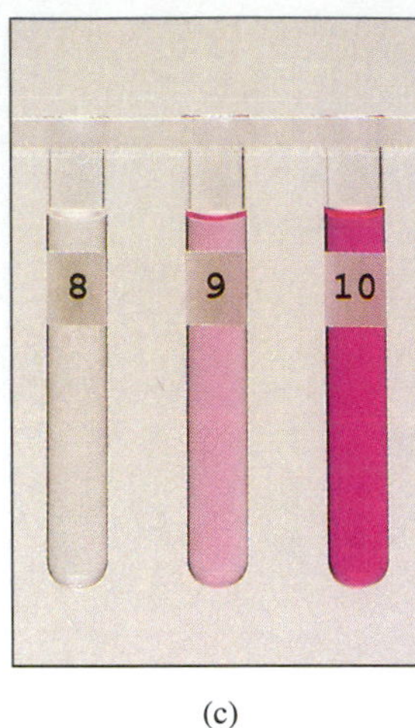

(c)

Figure 14.4
Methyl red (a) goes from red at low pH to orange at about pH 5 to yellow at high pH. Bromthymol blue (b) is yellow at low pH, blue at high pH, and green at about pH 7. Phenolphthalein (c) goes from colorless to pink at about pH 9. (Marna G. Clarke)

3. If $\dfrac{[\text{HIn}]}{[\text{In}^-]} \approx 1$
the color observed is intermediate between those of the two species, HIn and In^-.

The expression for the dissociation of the indicator molecule, HIn,

$$K_a = \frac{[\text{H}^+] \times [\text{In}^-]}{[\text{HIn}]}$$

The indicator measures $[\text{H}^+]$ without affecting it

can be rearranged to give

$$\frac{[\text{HIn}]}{[\text{In}^-]} = \frac{[\text{H}^+]}{K_a}$$

From this expression, it follows that the ratio $[\text{HIn}]/[\text{In}^-]$ and hence the color of an indicator depends upon two factors.

1. $[\text{H}^+]$ or the pH of the solution. At high $[\text{H}^+]$ (low pH), you see the color of the HIn molecule; at low $[\text{H}^+]$ (high pH), the colo: of the In^- ion dominates.

2. K_a of the indicator. Since the dissociation constant K_a varies from one indicator to another, different indicators change colors at different pH's. A color change occurs when

$$[\text{H}^+] \approx K_a \qquad \text{pH} \approx \text{p}K_a$$

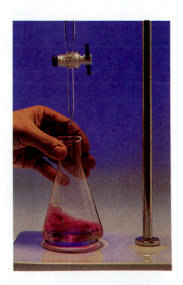

Titrating a solution of HCl with NaOH using phenolphthalein as the indicator. (Charles D. Winters)

Table 14.1 Colors and End Points of Indicators

	Color [HIn]	Color $[\text{In}^-]$	K_a	pH at End Point
Methyl red	red	yellow	1×10^{-5}	5
Bromthymol blue	yellow	blue	1×10^{-7}	7
Phenolphthalein	colorless	pink	1×10^{-9}	9

Table 14.1 shows the characteristics of three indicators: methyl red, bromthymol blue, and phenolphthalein. These indicators change colors at pH 5, 7, and 9, respectively. Keeping this property in mind and glancing back at Figures 14.1–14.3 to refresh your memory, the following points should be clear:

1. In a strong acid–strong base titration, the pH changes so rapidly near the equivalence point (4 to 10 with one drop of base) that any of these indicators works. Phenolphthalein is most often used because the change is visible even if you are colorblind.

2. In a weak acid–strong base titration, only phenolphthalein (color change at pH 9) works. Methyl red and bromthymol blue would change color before the equivalence point is reached.

The reaction would be incomplete; some of the acid would remain

3. In a strong acid–weak base titration, only methyl red (color change at pH 5) works. The end points with bromthymol blue and phenolphthalein would come long before reaching the equivalence point.

Table 14.2 summarizes our discussion of acid-base titrations. Notice that for these three types of titrations

— *the equations (second column) written to describe the reactions are quite different.* Strong acids and bases are represented by H^+ and OH^- ions,

Table 14.2 Characteristics of Acid-Base Titrations

Example	Equation	K	Species Equiv. Pt.	pH Equiv. Pt.	Indicator*
Strong Acid–Strong Base					
		$K = 1/K_w$			
NaOH–HCl	$H^+(aq) + OH^-(aq) \longrightarrow H_2O$	1.0×10^{14}	Na^+, Cl^-	7.00	MR, BB, PP
$Ba(OH)_2$–HNO_3	$H^+(aq) + OH^-(aq) \longrightarrow H_2O$	1.0×10^{14}	Ba^{2+}, NO_3^-	7.00	MR, BB, PP
Weak Acid–Strong Base					
		$K = 1/K_b$			
$HC_2H_3O_2$–NaOH	$HC_2H_3O_2(aq) + OH^-(aq) \longrightarrow C_2H_3O_2^-(aq) + H_2O$	1.8×10^9	Na^+, $C_2H_3O_2^-$	9.22†	PP
HF–KOH	$HF(aq) + OH^-(aq) \longrightarrow F^-(aq) + H_2O$	6.9×10^{10}	K^+, F^-	9.42†	PP
Strong Acid–Weak Base					
		$K = 1/K_a$			
NH_3–HCl	$NH_3(aq) + H^+(aq) \longrightarrow NH_4^+(aq)$	1.8×10^9	NH_4^+, Cl^-	4.78†	MR
ClO^-–HCl	$ClO^-(aq) + H^+(aq) \longrightarrow HClO(aq)$	3.6×10^7	HClO, Cl^-	3.93†	MR

* MR = methyl red (end point pH = 5); BB = bromthymol blue (end point pH = 7); PP = phenolphthalein (end point pH = 9).

† When 1 M acid is titrated with 1 M base.

respectively; weak acids and weak bases are represented by their chemical formulas.

These equations were discussed in Chapter 4; better review them

— *the equilibrium constants (K) for all of these reactions are very large, indicating that the reactions go essentially to completion.*
— *the pH at the equivalence point is determined by what species are present.* When only "spectator ions" are present, as in a strong acid–strong base titration, the pH at the equivalence point is 7. Basic anions (F^-, $C_2H_3O_2^-$), formed during a weak acid–strong base titration, make the pH at the equivalence point greater than 7. Conversely, acidic species (NH_4^+, HClO) formed during a strong acid–weak base titration make the pH at the equivalence point less than 7.

14.2 BUFFERS

Let us turn back to Figures 14.2 and 14.3, focusing now on point A on the titration curves. Notice that

— the pH is changing only slowly at A; the slope of the curve at that point is close to zero.
— at A, the titration is half over, which means that the concentrations of weak acid and its conjugate base are equal:

$$[HC_2H_3O_2] = [C_2H_3O_2^-] \qquad [NH_4^+] = [NH_3]$$

and

$$[H^+] = K_a \qquad pH = pK_a$$

These observations have a significance that goes well beyond acid-base titrations. In general, any solution containing approximately equal amounts of a weak acid and its conjugate base

— *is highly resistant to changes in pH brought about by addition of strong acid or strong base*
— *has a pH approximately equal to pK_a of the weak acid*

A solution showing these properties is given a special name; it is called a **buffer solution**.

Buffers are widely used to maintain nearly constant pH in a variety of commercial and laboratory products (Fig. 14.5). For these applications and many others, it is essential to be able to determine

— the pH of a buffer system made by mixing a weak acid with its conjugate base
— the (small) change in pH that occurs when a strong acid or base is added to a buffer
— the appropriate buffer system to maintain a desired pH

Determination of $[H^+]$ in a Buffer System

The concentration of H^+ ion or pH of a buffer can be calculated using the dissociation constant of the weak acid HB:

$$HB(aq) \rightleftharpoons H^+(aq) + B^-(aq) \qquad K_a = \frac{[H^+] \times [B^-]}{[HB]}$$

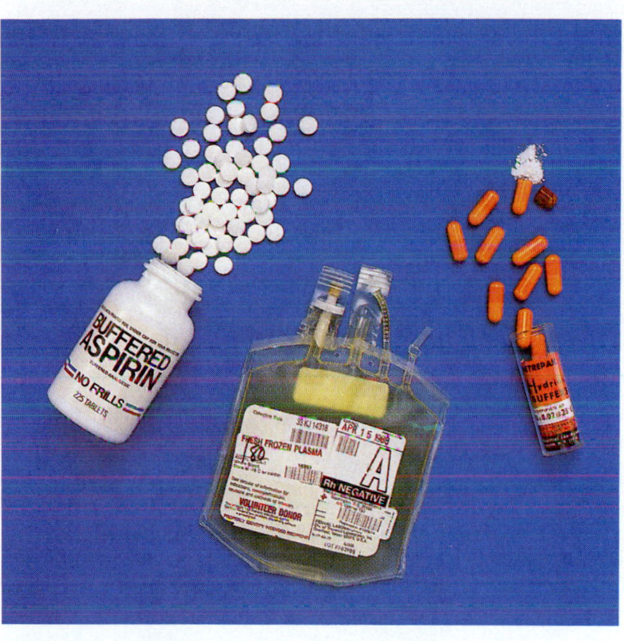

Figure 14.5
Many products, including aspirin and blood plasma, are buffered. Buffer tablets are also available in the laboratory *(far right)* to make up a solution to a specified pH. (Marna G. Clarke)

Solving for $[H^+]$,

$$[H^+] = K_a \times \frac{[HB]}{[B^-]} \qquad (1)^*$$

This is a completely general equation, applicable to all buffer systems. The calculation of $[H^+]$, and hence pH, can be simplified if you keep two points in mind.

1. *You can always assume that equilibrium is established without appreciably changing the original concentrations of either HB or* B^-. Letting $x =$ $[H^+]$ produced by dissociation of HB,

$$[HB]_{eq} = [HB]_o - x \approx [HB]_o$$

$$[B^-]_{eq} = [B^-]_o + x \approx [B^-]_o$$

B^- comes from adding a salt such as NaB to the solution

The justification here is quite simple. Recall from Chapter 13 that when a weak acid is added to water, very little of it dissociates. That is,

$$[HB]_o \gg x \qquad \text{so} \qquad [HB]_{eq} \approx [HB]_o$$

Here, there are a lot of B^- ions to begin with. This shifts the equilibrium

$$HB(aq) \rightleftharpoons H^+(aq) + B^-(aq)$$

to the left, making x very small indeed in comparison to [HB] or $[B^-]$.

* By taking the logarithms of both sides of Equation 1 and multiplying through by −1, you can show (Problem 73) that, for any buffer system,

$$pH = pK_a + \log_{10} [B^-]/[HB]$$

This relation, known as the Henderson-Hasselbach equation, is often used in biology and biochemistry to calculate the pH of buffers.

2. *Since the two species HB and B^- are present in the same solution, the ratio of their concentrations is also their mole ratio.* That is,

$$\frac{[HB]}{[B^-]} = \frac{\text{no. moles HB}/V}{\text{no. moles B}^-/V} = \frac{n_{HB}}{n_{B^-}} \qquad (n = \text{no. moles})$$

Hence Equation 1 can be rewritten as

If you start with 0.20 mol HB and 0.10 mol NaB, $[H^+] = 2K_a$

$$[H^+] = K_a \times \frac{n_{HB}}{n_{B^-}} \tag{2}$$

Frequently, Equation 2 is easier to work with than Equation 1.

Example 14.2 Lactic acid, $C_3H_6O_3$, is a weak organic acid present in both sour milk and buttermilk. It is also a product of carbohydrate metabolism and is found in the blood after vigorous muscular activity. A buffer is prepared by dissolving 1.00 mol of lactic acid, HLac ($K_a = 1.4 \times 10^{-4}$) and 1.00 mol of sodium lactate, NaLac, in enough water to form 550 mL of solution. Calculate $[H^+]$ and the pH of the buffer.

Strategy You are given K_a and n_{HLac} in the statement of the problem. Since sodium lactate, like all salts, is completely dissociated in water, $n_{Lac^-} = n_{NaLac} = 1.00$. Thus you have all the quantities required to calculate $[H^+]$ by Equation 2.

Solution

$$[H^+] = K_a \times \frac{n_{HLac}}{n_{Lac^-}}$$

$$= 1.4 \times 10^{-4} \times \frac{1.00}{1.00} = \boxed{1.4 \times 10^{-4}}$$

$$pH = -\log_{10}(1.4 \times 10^{-4}) = \boxed{3.85}$$

Notice in Example 14.2 that the total volume of solution is irrelevant. All that is required to solve for $[H^+]$ is the number of moles of lactic acid and sodium lactate. The pH of the buffer is 3.85 whether the volume of solution is 550 mL, 1 L, or even 10 L. From a slightly different point of view, *diluting a buffer system with water does not change its pH.* The mole ratio of HB to B^- stays the same on dilution, as does the concentration ratio.

Effect of Added H^+ or OH^- on Buffer Systems

A buffer consisting of a weak acid HB and its conjugate base B^- can react with either strong base (OH^-) or strong acid (H^+).

$$HB(aq) + OH^-(aq) \longrightarrow B^-(aq) + H_2O$$

$$B^-(aq) + H^+(aq) \longrightarrow HB(aq)$$

A buffer "works" because it contains species which react with both H^+ and OH^-

As pointed out in Section 14.1, these reactions go to completion. As a result, the added H^+ or OH^- ions are consumed and do not directly affect the pH. This is the principle of buffer action, explaining why a buffered solution is much more resistant to a change in pH than one which is unbuffered (Fig. 14.6).

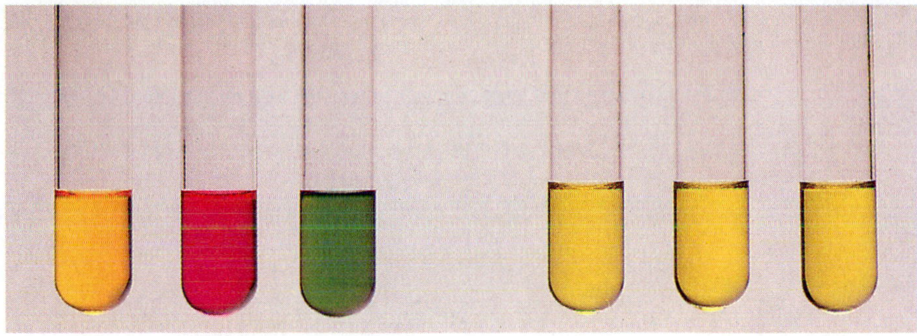

Figure 14.6
The three tubes at the left show the effect of adding a few drops of strong acid or strong base to water. The pH changes drastically, giving a pronounced color change with universal indicator. This experiment is repeated in the three tubes at the right, using a buffer of pH 7 instead of water. This time the pH changes only very slightly, and there is no change in the color of the indicator. (Marna G. Clarke)

The pH of a buffer does change slightly upon addition of a strong acid or strong base. When OH^- ions are added, a small amount of weak acid, HB, is converted to B^- ions. Addition of a strong acid converts a small amount of B^- to HB. In either case, the ratio $[HB]/[B^-]$ changes; this in turn changes the H^+ ion concentration and pH of the buffer. The effect is ordinarily small, as illustrated in Example 14.3.

Example 14.3 Consider the buffer described in Example 14.2, where no. moles HLac = no. moles Lac^- = 1.00 (K_a HLac = 1.4×10^{-4}). You will recall that in this buffer the pH is 3.85. Calculate the pH after addition of

(a) 0.10 mol HCl. (b) 0.10 mol NaOH.

Strategy First (1), write the chemical equation for the reaction that occurs when strong acid or base is added. Then (2), apply the principles of stoichiometry to find the numbers of moles of weak acid and weak base remaining after reaction. Finally (3), apply the relation $[H^+] = K_a \times n_{HB}/n_{B^-}$ to find $[H^+]$ and then the pH.

Solution

(a) (1) $H^+(aq) + Lac^-(aq) \longrightarrow HLac(aq)$

(2) The addition of 0.10 mol of H^+ produces 0.10 mol of HLac and consumes 0.10 mol of Lac^-. Originally, there was 1.00 mol of both HLac and Lac^-. Hence

$$n_{HLac} = (1.00 + 0.10)\text{ mol} = 1.10\text{ mol}$$

$$n_{Lac^-} = (1.00 - 0.10)\text{ mol} = 0.90\text{ mol}$$

To work this problem, you have to combine principles of stoichiometry and equilibrium; that's not easy

(3) $$[H^+] = 1.4 \times 10^{-4} \times \frac{n_{HLac}}{n_{Lac^-}} = 1.4 \times 10^{-4} \times \frac{1.10}{0.90} = 1.7 \times 10^{-4}\,M$$

$\text{pH} = -\log_{10}[H^+] =$ **3.77** (about 0.1 unit less than the original pH)

(b) (1) $HLac(aq) + OH^-(aq) \longrightarrow Lac^-(aq) + H_2O$

(2) The reaction produces 0.10 mol of Lac^- and consumes 0.10 mol of HLac

$$n_{HLac} = (1.00 - 0.10)\text{ mol} = 0.90\text{ mol}$$

$$n_{Lac^-} = (1.00 + 0.10)\text{ mol} = 1.10\text{ mol}$$

$$(3)\ [H^+] = 1.4 \times 10^{-4} \times \frac{n_{HLac}}{n_{Lac^-}} = 1.4 \times 10^{-4} \times \frac{0.90}{1.10} = 1.1 \times 10^{-4}\ M$$

$$pH = -\log_{10} [H^+] = \boxed{3.94}$$

Again, the pH changes by less than 0.1 unit. In contrast, addition of 0.10 mol of HCl or NaOH to a liter of pure water would change the pH by 6 units.

When H^+ is added, weak acid forms; addition of OH^- forms weak base

The general rule, illustrated by Example 14.3, is:

— if x moles of H^+ are added to a buffer, $\Delta n_{HB} = x$; $\Delta n_{B^-} = -x$
— if y moles of OH^- are added to a buffer, $\Delta n_{B^-} = y$; $\Delta n_{HB} = -y$

Choosing a Buffer System

Suppose you want to make up a buffer in the laboratory with a specified pH (e.g., 4.0, 7.0, 10.0, . . .). Looking at the equation

$$[H^+] = K_a \times \frac{[HB]}{[B^-]} = K_a \times \frac{n_{HB}}{n_{B^-}}$$

it is evident that the pH of the buffer depends upon two factors:

1. *The dissociation constant of the weak acid, K_a.* This has the greatest influence on buffer pH. Since HB and B^- are likely to be present in nearly equal amounts,

$$[H^+] \approx K_a \qquad pH \approx pK_a$$

Thus, to make up a buffer of pH close to 7, you should start with a weak acid–weak base pair in which the dissociation constant of the weak acid is about 10^{-7}.

2. *The ratio of the concentrations or amounts of HB and B^-.* Small variations in pH can be achieved by adjusting this ratio. To obtain a slightly more acidic buffer, add more weak acid, HB; addition of more weak base, B^-, will make the buffer a bit more basic.

Example 14.4 To prepare a buffer with a pH of 9.00, which of the systems in Table 14.3 would you use? What should be the ratio of concentrations, $[HB]/[B^-]$?

Strategy Choose the buffer whose pK_a is closest to 9. Clearly, this is the NH_4^+–NH_3 buffer. Then use the equation $[H^+] = K_a \times [NH_4^+]/[NH_3]$ to calculate the necessary ratio. Note that since the pH is 9.00, $[H^+] = 1.0 \times 10^{-9}\ M$.

This is how you prepare a buffer of pH 9

Solution For the NH_4^+–NH_3 buffer,

$$[H^+] = K_a \times \frac{[NH_4^+]}{[NH_3]}$$

Solving for the ratio:

$$\frac{[NH_4^+]}{[NH_3]} = \frac{[H^+]}{K_a} = \frac{1.0 \times 10^{-9}}{5.6 \times 10^{-10}} = \boxed{1.8}$$

You might add 1.8 mol of NH_4Cl to 1.0 mol of NH_3, or 9 mol of NH_4Cl to 5 mol of NH_3, or . . . .

Table 14.3 Buffer Systems at Different pH Values

Desired pH	Buffer System: Weak Acid	Weak Base	K_a(Weak Acid)	pK_a
4	lactic acid (HLac)	lactate ion (Lac^-)	1.4×10^{-4}	3.85
5	acetic acid ($HC_2H_3O_2$)	acetate ion ($C_2H_3O_2^-$)	1.8×10^{-5}	4.74
6	carbonic acid (H_2CO_3)	hydrogen carbonate ion (HCO_3^-)	4.4×10^{-7}	6.36
7	dihydrogen phosphate ion ($H_2PO_4^-$)	hydrogen phosphate ion (HPO_4^{2-})	6.2×10^{-8}	7.21
8	hypochlorous acid (HClO)	hypochlorite ion (ClO^-)	2.8×10^{-8}	7.55
9	ammonium ion (NH_4^+)	ammonia (NH_3)	5.6×10^{-10}	9.25
10	hydrogen carbonate ion (HCO_3^-)	carbonate ion (CO_3^{2-})	4.7×10^{-11}	10.32

Buffer Capacity

A buffer has a limited capacity to react with H^+ or OH^- ions without undergoing a drastic change in pH. To see what happens when a large amount of strong acid or base is added to a buffer, refer again to Figure 14.2. Notice what happens as NaOH is added to the buffer, moving to the right of point A. The pH rises very slowly at first, then more rapidly, and finally "takes off" vertically when all the acetic acid molecules are consumed, destroying the buffer. A similar effect is observed in Figure 14.3, except that here it is H^+ ions that are added from the strong acid, eventually reacting with all the NH_3 molecules and destroying the buffer.

Diluting a buffer doesn't change its pH but does affect its capacity.

The capacity of a buffer solution is directly related to the amounts of weak acid (n_{HB}) and weak base (n_{B^-}) present. In general,

— the larger the number of moles of HB, the greater the capacity for absorbing OH^- ions by the reaction:

$$HB(aq) + OH^-(aq) \longrightarrow B^-(aq) + H_2O$$

— the larger the number of moles of B^-, the greater the capacity for absorbing H^+ ions by the reaction:

$$H^+(aq) + B^-(aq) \longrightarrow HB(aq)$$

One liter of a buffer 1.0 *M* in HB and B^- has a capacity ten times that of an equal volume of a buffer 0.10 *M* in HB and B^-.

14.3 PRECIPITATION EQUILIBRIA

Precipitation reactions, like all reactions, reach a position of equilibrium. Suppose, for example, solutions of $Sr(NO_3)_2$ and K_2CrO_4 are mixed. In this case, Sr^{2+} ions combine with CrO_4^{2-} ions to form a yellow precipitate of strontium chromate, $SrCrO_4$ (Fig. 14.7). Very quickly, an equilibrium is established between the solid and the corresponding ions in solution:

$$SrCrO_4(s) \rightleftharpoons Sr^{2+}(aq) + CrO_4^{2-}(aq)$$

Figure 14.7
A solution prepared by mixing solutions of $Sr(NO_3)_2$ and K_2CrO_4 is in equilibrium with yellow $SrCrO_4(s)$. In such a solution: $[Sr^{2+}] \times [CrO_4^{2-}] = 3.6 \times 10^{-5}$ (Charles D. Winters)

K_{sp} Expression

The equilibrium constant expression for the dissolving of $SrCrO_4$ can be written following the rules cited in Chapters 12 and 13. In particular, the solid does not appear in the expression; the concentration of each ion is raised to a power equal to its coefficient in the chemical equation.

$$K_{sp} = [Sr^{2+}] \times [CrO_4^{2-}]$$

Although the solid doesn't appear in the expression for K_{sp}, it must be present for equilibrium

The symbol K_{sp} represents a particular type of equilibrium constant known as the **solubility product constant**. Like all equilibrium constants, K_{sp} has a fixed value for a given system at a particular temperature. At 25°C, K_{sp} for $SrCrO_4$ is about 3.6×10^{-5}; that is,

$$[Sr^{2+}] \times [CrO_4^{2-}] = 3.6 \times 10^{-5}$$

This relation says that the product of the two ion concentrations at equilibrium must be 3.6×10^{-5}, regardless of how equilibrium is established.

For other slightly soluble ionic solids, K_{sp} expressions analogous to that for strontium chromate can be written. For $PbCl_2$ and Ag_2CrO_4,

$$PbCl_2(s) \rightleftharpoons Pb^{2+}(aq) + 2Cl^-(aq)$$
$$K_{sp}\ PbCl_2 = [Pb^{2+}] \times [Cl^-]^2 = 1.7 \times 10^{-5}$$

$$Ag_2CrO_4(s) \rightleftharpoons 2Ag^+(aq) + CrO_4^{2-}(aq)$$
$$K_{sp}\ Ag_2CrO_4 = [Ag^+]^2 \times [CrO_4^{2-}] = 1 \times 10^{-12}$$

K_{sp} and the Equilibrium Concentrations of Ions

The relation

$$K_{sp}\ SrCrO_4 = [Sr^{2+}] \times [CrO_4^{2-}] = 3.6 \times 10^{-5}$$

can be used to calculate the equilibrium concentration of one ion, knowing that of the other. Suppose, for example, the concentration of CrO_4^{2-} in a certain solution in equilibrium with $SrCrO_4$ is known to be $2.0 \times 10^{-3}\ M$. It follows that

$$[Sr^{2+}] = \frac{K_{sp}\ SrCrO_4}{[CrO_4^{2-}]} = \frac{3.6 \times 10^{-5}}{2.0 \times 10^{-3}} = 1.8 \times 10^{-2}\ M$$

If in another case, $[Sr^{2+}] = 1.0 \times 10^{-4}\ M$,

$$[CrO_4^{2-}] = \frac{K_{sp}\ SrCrO_4}{[Sr^{2+}]} = \frac{3.6 \times 10^{-5}}{1.0 \times 10^{-4}} = 3.6 \times 10^{-1}\ M$$

Example 14.5 illustrates the same kind of calculation for a different electrolyte; the math is a bit more difficult, but the principle is the same.

Example 14.5 Calcium phosphate, $Ca_3(PO_4)_2$, is a water-insoluble mineral, large quantities of which are used to make commercial fertilizers. Taking its K_{sp} value from Table 14.4, calculate

(a) the concentration of PO_4^{3-} in equilibrium with the solid if $[Ca^{2+}] = 1 \times 10^{-9}\ M$.
(b) the concentration of Ca^{2+} in equilibrium with the solid if $[PO_4^{3-}] = 1 \times 10^{-5}\ M$.

Table 14.4 Solubility Product Constants at 25°C

		K_{sp}			K_{sp}
Acetates	$AgC_2H_3O_2$	1.9×10^{-3}	Hydroxides	$Al(OH)_3$	2×10^{-31}
				$Fe(OH)_2$	5×10^{-17}
Bromides	$AgBr$	5×10^{-13}		$Fe(OH)_3$	3×10^{-39}
	Hg_2Br_2	6×10^{-23}		$Mg(OH)_2$	6×10^{-12}
	$PbBr_2$	6.6×10^{-6}		$Tl(OH)_3$	2×10^{-44}
				$Zn(OH)_2$	4×10^{-17}
Carbonates	Ag_2CO_3	8×10^{-12}			
	$BaCO_3$	2.6×10^{-9}	Iodides	AgI	1×10^{-16}
	$CaCO_3$	4.9×10^{-9}		Hg_2I_2	5×10^{-29}
	$MgCO_3$	6.8×10^{-6}		PbI_2	8.4×10^{-9}
	$SrCO_3$	5.6×10^{-10}			
	$PbCO_3$	1×10^{-13}			
			Phosphates	Ag_3PO_4	1×10^{-16}
Chlorides	$AgCl$	1.8×10^{-10}		$AlPO_4$	1×10^{-20}
	Hg_2Cl_2	1×10^{-18}		$Ca_3(PO_4)_2$	1×10^{-33}
	$PbCl_2$	1.7×10^{-5}		$Mg_3(PO_4)_2$	1×10^{-24}
Chromates	Ag_2CrO_4	1×10^{-12}	Sulfates	$BaSO_4$	1.1×10^{-10}
	$BaCrO_4$	1.2×10^{-10}		$CaSO_4$	7.1×10^{-5}
	$PbCrO_4$	2×10^{-14}		$PbSO_4$	1.8×10^{-8}
	$SrCrO_4$	3.6×10^{-5}		$SrSO_4$	3.4×10^{-7}
Fluorides	BaF_2	1.8×10^{-7}	Sulfides	Ag_2S	1×10^{-49}
	CaF_2	1.5×10^{-10}		Bi_2S_3	1×10^{-99}
	MgF_2	7×10^{-11}		CdS	1×10^{-29}
	PbF_2	7.1×10^{-7}		CuS	1×10^{-36}
				Cu_2S	1×10^{-48}
				FeS	2×10^{-19}
				HgS	1×10^{-52}
				MnS	5×10^{-14}
				NiS	1×10^{-21}
				PbS	1×10^{-28}
				SnS	3×10^{-28}

Strategy The first step is to write down the K_{sp} expression:

$$Ca_3(PO_4)_2(s) \rightleftharpoons 3Ca^{2+}(aq) + 2PO_4^{3-}(aq)$$

$$K_{sp} = [Ca^{2+}]^3 \times [PO_4^{3-}]^2 = 1 \times 10^{-33}$$

Now substitute the concentration of one ion and solve for that of the other.

Solution

(a) $[PO_4^{3-}]^2 = \dfrac{1 \times 10^{-33}}{[Ca^{2+}]^3} = \dfrac{1 \times 10^{-33}}{(1 \times 10^{-9})^3} = 1 \times 10^{-6}$ $\quad [PO_4^{3-}] = 1 \times 10^{-3}\,M$

(b) $[Ca^{2+}]^3 = \dfrac{1 \times 10^{-33}}{[PO_4^{3-}]^2} = \dfrac{1 \times 10^{-33}}{(1 \times 10^{-5})^2} = 1 \times 10^{-23}$ $\quad [Ca^{2+}] = 2 \times 10^{-8}\,M$

(To find a cube root on your calculator, use the y^x key, where $x = 1/3 = 0.333333\ldots$)

K_{sp} and Water Solubility

The water solubility of an ionic compound, *s*, in moles per liter, is quite different from its solubility product constant, K_{sp}. For example:

Compound	Water Solubility (mol/L)	Solubility Product Constant, K_{sp}
$SrCrO_4$	6.0×10^{-3}	3.6×10^{-5}
$CaCO_3$	7.0×10^{-5}	4.9×10^{-9}
BaF_2	3.6×10^{-3}	1.8×10^{-7}

In principle, at least, K_{sp} can be calculated from solubility

It is possible, however, to estimate the water solubility, knowing K_{sp}.*

Example 14.8 Calculate the water solubility of

(a) $CaCO_3$ ($K_{sp} = 4.9 \times 10^{-9}$) in mol/L. (b) BaF_2 ($K_{sp} = 1.8 \times 10^{-7}$) in g/L.

Strategy To relate solubility, *s*, to K_{sp}, start by writing the equation for the dissolving of the solid in water:

$$CaCO_3(s) \longrightarrow Ca^{2+}(aq) + CO_3^{2-}(aq)$$

$$BaF_2(s) \longrightarrow Ba^{2+}(aq) + 2F^-(aq)$$

K_{sp} is a constant; *P* can

Now relate the equilibrium concentration of each ion to the solubility of the compound (mol/L). Finally, relate K_{sp} to solubility and solve for *s*.

Solution

(a) For every mole of $CaCO_3$ that dissolves, 1 mol of Ca^{2+} and 1 mol of CO_3^{2-} form.

$[Ca^{2+}] = s \qquad [CO_3^{2-}] = s$

$K_{sp} = [Ca^{2+}] \times [CO_3^{2-}] = s \times s = s^2$

$s = (K_{sp})^{1/2} = (4.9 \times 10^{-9})^{1/2} = 7.0 \times 10^{-5}$ mol/L

(b) For every mole of BaF_2 that dissolves, 1 mol of Ba^{2+} and 2 mol of F^- form.

$[Ba^{2+}] = s \qquad [F^-] = 2s$

$K_{sp} = [Ba^{2+}] \times [F^-]^2 = s \times (2s)^2 = 4s^3$

$s = (K_{sp}/4)^{1/3} = (1.8 \times 10^{-7}/4)^{1/3} = 3.6 \times 10^{-3}$ mol/L

To find the solubility in grams per liter, note that the molar mass of BaF_2 is 175.3 g/mol. So

$$\text{solubility} = \frac{3.6 \times 10^{-3} \text{ mol } BaF_2}{1 \text{ L}} \times \frac{175.3 \text{ g } BaF_2}{1 \text{ mol } BaF_2} = 0.63 \text{ g } BaF_2/\text{L}$$

Stalactites *(upper)* and stalagmites *(lower)* consist of calcium carbonate. They are formed when a water solution containing Ca^{2+} and HCO_3^- ions enters a cave. Carbon dioxide is released and calcium carbonate precipitates: $Ca^{2+}(aq) + 2HCO_3^-(aq) \rightarrow CaCO_3(s) + H_2O + CO_2(g)$. (Dick George, Tom Stack & Associates)

* Experimentally, we usually find that the solubility is slightly greater than that predicted from K_{sp}. For example, the measured solubility of PbI_2 in water at 25°C is 1.7×10^{-3} *M*. This compares to a value of 1.4×10^{-3} *M* calculated from the K_{sp} of PbI_2. The reason for this is that some of the lead in PbI_2 goes into solution in the form of species other than Pb^{2+}. For example, we can detect ions such as $Pb(OH)^+$ and PbI^+ in a water solution of lead iodide.

K_{sp} and the Common Ion Effect

How would you expect the solubility of $CaCO_3$ in water,

$$CaCO_3(s) \rightleftharpoons Ca^{2+}(aq) + CO_3^{2-}(aq)$$

to compare to that in a 0.10 *M* solution of Na_2CO_3, which contains the same (**common**) anion, CO_3^{2-}? A moment's reflection should convince you that the solubility in 0.10 *M* Na_2CO_3 must be *less* than that in pure water. Recall (Example 14.8) that in pure water $[CO_3^{2-}]$ is only 7×10^{-5} *M*. Increasing the concentration of CO_3^{2-} to 0.10 *M* should, by Le Chatelier's principle, drive the above equilibrium to the *left*, repressing the solubility of calcium carbonate. This is, indeed, the case (Example 14.9).

A "common" ion comes from two sources (e.g., $CaCO_3$, Na_2CO_3)

Example 14.9 Taking K_{sp} of $CaCO_3$ to be 4.9×10^{-9}, estimate its solubility (moles per liter) in 0.10 *M* Na_2CO_3 solution.

Strategy Let s = solubility = no. of moles per liter of $CaCO_3$ that dissolves. Relate $[Ca^{2+}]$ and $[CO_3^{2-}]$ to s; notice that in this case *there are two different sources of* CO_3^{2-}. Now relate K_{sp} to s and, finally, solve for s.

Solution Every mole of $CaCO_3$ that dissolves forms 1 mol of Ca^{2+} and 1 mol of CO_3^{2-}. Since there was no Ca^{2+} present originally, its concentration becomes s:

$$[Ca^{2+}] = s$$

Since the concentration of CO_3^{2-} was originally 0.10 *M*, and it increases by s mol/L,

$$[CO_3^{2-}] = 0.10 + s$$

Hence $K_{sp} = 4.9 \times 10^{-9} = [Ca^{2+}] \times [CO_3^{2-}] = s(0.10 + s)$. This is a quadratic equation; as usual, we look for ways to avoid solving it by brute force. A suitable approximation here would appear to be

$$s + 0.10 \approx 0.10$$

The solubility should be much less than 0.10 mol/L, since $CaCO_3$ is quite insoluble. With that assumption,

$$s(0.10) = 4.9 \times 10^{-9}$$

$$s = \frac{4.9 \times 10^{-9}}{1.0 \times 10^{-1}} = 4.9 \times 10^{-8}\ \text{mol/L}$$

The solubility is indeed much less than 0.10 *M*, so the approximation is justified. More important, the solubility in 0.10 *M* Na_2CO_3 is much less than in pure water (Example 14.8):

$$4.9 \times 10^{-8} \ll 7.0 \times 10^{-5}$$

which is the point we set out to prove.

(a)

(b)

Figure 14.9
Sodium chloride can be precipitated from its saturated solution (6 *M*) by adding a solution containing either Na^+ or Cl^- ions at a concentration greater than 6 *M*. Both 12 *M* NaOH (a) and 12 *M* HCl (b) will bring about this reaction, illustrating the common ion effect. (Marna G. Clarke)

The effect illustrated in Example 14.9 is a general one. *An ionic solid is always less soluble in a solution containing a common ion than it is in water* (Fig. 14.9).

PERSPECTIVE

•

pH Control in the Body

Blood, like many natural fluids, is buffered. Indeed, there are three different buffer systems in blood plasma that hold the pH at about 7.40. By far the most important of these involves H_2CO_3 (an aqueous solution of CO_2), and its conjugate base, the HCO_3^- ion. This system consumes H^+ or OH^- ions by the reactions

$$HCO_3^-(aq) + H^+(aq) \longrightarrow H_2CO_3(aq)$$

$$H_2CO_3(aq) + OH^-(aq) \longrightarrow HCO_3^-(aq) + H_2O$$

Since the pH of blood is 7.40, we have

$$[H^+] = 10^{-7.40} = 4.0 \times 10^{-8}\,M$$

The dissociation constant of carbonic acid is 4.4×10^{-7}; it follows that the concentration of HCO_3^- in the blood is more than ten times that of H_2CO_3:

$$\frac{[HCO_3^-]}{[H_2CO_3]} = \frac{K_a\,H_2CO_3}{[H^+]} = \frac{4.4 \times 10^{-7}}{4.0 \times 10^{-8}} = 11$$

Consequently, blood has a much greater capacity for absorbing H^+ ions than for OH^- ions. It needs this capacity; life processes tend to produce H^+ ions rather than OH^- ions. For example, heavy exercise produces lactic acid, a weak organic acid.

If the pH of the blood drops significantly below 7.4, a condition called *acidosis* is created. The nervous system is depressed; fainting and even coma can result. Most often, acidosis is caused by a buildup of carbon dioxide concentration in the blood. You can produce a very mild case of acidosis by holding your breath. A more serious case can be the result of lung diseases such as asthma or emphysema. Diabetics are also susceptible to acidosis because their metabolic processes form organic acids such as

$$CH_3{-}\underset{\substack{\|\\ O}}{C}{-}CH_2{-}COOH \qquad \text{acetoacetic acid}$$

If acetoacetic acid is produced in amounts greater than the HCO_3^- ions in the blood can react with, a diabetic coma can result.

Alkalosis, which results when the pH of the blood rises significantly above its normal value of 7.4, is much less common than acidosis (fortunately, since blood has a relatively small capacity to absorb OH^- ions). The symptoms of alkalosis are the inverse of those for acidosis; the nervous system is overstimulated, leading to muscle cramps and ultimately convulsions. Most often, alkalosis is caused by rapid or heavy breathing (hyperventilation). This can result from fever, infection, or the action of certain drugs. When a person breathes too deeply, carbon dioxide is expelled in large quantities from the bloodstream and the pH rises.

CHAPTER HIGHLIGHTS

KEY CONCEPTS

1. *Calculate the pH during an acid-base titration*
   (Example 14.1; Problems 11–16, 66, 67)
2. *Choose the proper indicator for an acid-base titration*
   (Problems 9, 10, 69)
3. *Calculate the pH of a buffer, before or after adding H^+ or OH^-*
   (Examples 14.2, 14.3; Problems 17–20, 25–34, 37, 38, 59, 60)
4. *Choose a buffer system to get a specified pH*
   (Example 14.4; Problems 21–24)
5. *Use the value of K_{sp} to*
   —calculate the concentration of one ion, knowing that of the other
   (Example 14.5; Problems 43–46)
   —determine whether a precipitate will form
   (Examples 14.6, 14.7; Problems 49–52, 64)
   —calculate the solubility
   (Examples 14.8, 14.9; Problems 55, 56, 68, 70)

KEY EQUATIONS

$[H^+]$ in buffer $\quad [H^+] = K_a \times \frac{n_{HB}}{n_{B^-}}$

KEY TERMS

acid	common ion	ion product
—strong	conjugate acid	pH
—weak	conjugate base	precipitate
base	dissociation constant	solubility
—strong	equivalence point	solubility product constant
—weak	indicator	titration
buffer		

SUMMARY PROBLEM

Consider the weak acid hydrogen fluoride, $K_a = 6.9 \times 10^{-4}$, and its conjugate base, F^- ($K_b = 1.4 \times 10^{-11}$).

1. Write net ionic equations and calculate K for the reactions between
   a. HF and KOH.
   b. F^- and HNO_3.
   c. HF and NH_3 (K_a $NH_4^+ = 5.6 \times 10^{-10}$).
2. Fifty mL of 1.000 *M* HF is titrated with 0.500 *M* KOH.
   a. What is the pH of the HF solution before titration?
   b. At the equivalence point, is the solution acidic, basic, or neutral?
   c. What indicator could be used for this titration?
3. A buffer is made by dissolving 25.0 g of sodium fluoride, NaF, in one liter of a solution of 0.500 *M* HF.
   a. Calculate the pH of the buffer.
   b. Calculate the pH of the buffer after 0.100 mol HCl is added.
   c. Calculate the pH of the buffer after 0.100 mol NaOH is added.
   d. Calculate the pH of the buffer after 0.500 mol NaOH is added.

4. Calcium fluoride ($K_{sp} = 1.5 \times 10^{-10}$) is formed when solutions of calcium nitrate and sodium fluoride are mixed.
   a. Will a precipitate form if 10.00 mL of 0.200 *M* calcium nitrate is added to 25.00 mL of 0.100 *M* sodium fluoride?
   b. Calculate the solubility of calcium fluoride (grams per liter) in water and in 0.100 *M* sodium fluoride.

Answers

1. a. $HF(aq) + OH^-(aq) \rightleftharpoons F^-(aq) + H_2O$ $K = 7.1 \times 10^{10}$
   b. $F^-(aq) + H^+(aq) \rightleftharpoons HF(aq)$ $K = 1.4 \times 10^3$
   c. $HF(aq) + NH_3(aq) \rightleftharpoons NH_4^+(aq) + F^-(aq)$ $K = 1.2 \times 10^6$
2. a. 1.58 b. basic c. phenolphthalein
3. a. 3.24 b. 3.08 c. 3.40 d. 8.59 (pH of 1.095 *M* NaF)
4. a. yes b. 0.026 g/L; 1.2×10^{-6} g/L

QUESTIONS & PROBLEMS

Equilibrium constants required to solve these problems are listed in Table 13.2 (K_a of weak acids), 13.4 (polyprotic acids), 13.5 (K_b of weak bases), or 14.4 (solubility product constants). They are also found in Appendix 1.

Symbols, Formulas, and Equations

1. Write a net ionic equation for the reaction between solutions of
   a. nitric acid and lithium hydroxide.
   b. ammonia and hydrogen iodide.
   c. hydrogen fluoride and potassium cyanide.
   d. calcium hydroxide and nitrous acid, HNO_2.

2. Follow the directions of Question 1 for
   a. HCN and sodium hydroxide.
   b. NaClO and hydrochloric acid.
   c. ammonium chloride and potassium hydroxide.
   d. methylamine, CH_3NH_2, and hypochlorous acid, HClO.

3. Write a balanced net ionic equation for the reaction of each of the following aqueous solutions with H^+ ions.
   a. sodium formate ($NaCHO_2$)
   b. calcium hydroxide c. ammonia

4. Write a balanced net ionic equation for the reaction of each of the following aqueous solutions with H^+ ions.
   a. cesium hydroxide b. potassium cyanide
   c. aniline, $C_6H_5NH_2$

Acid-Base Reaction and Titrations

5. Calculate *K* for the reactions given in Question 1.

6. Calculate *K* for the reactions given in Question 2.

7. Calculate *K* for the reactions given in Question 3.

8. Calculate *K* for the reactions in Question 4.

9. Given three acid-base indicators: methyl orange (end point at pH 4), bromthymol blue (end point at pH 7), and phenolphthalein (end point at pH 9), which would you select for the following acid-base titrations?
   a. formic acid with sodium hydroxide
   b. ethylamine, $C_2H_5NH_2$, with hydrochloric acid
   c. sodium acetate with hydrochloric acid
   d. perchloric acid with lithium hydroxide

10. Given the acid-base indicators in Question 9, select a suitable indicator for the following titrations.
   a. sulfuric acid with potassium hydroxide
   b. ammonia with hydrobromic acid
   c. hydrogen cyanide with barium hydroxide
   d. sodium nitrite with hydriodic acid

11. 35.00 mL of 0.2500 *M* sodium hydroxide is titrated with 0.4375 *M* HCl. Calculate the pH of the resulting solution (assuming volumes are additive) when the following amounts of acid are added.
   a. 10.00 mL b. 20.00 mL c. 30.00 mL

12. 75.00 mL of 0.1350 *M* perchloric acid is titrated with 0.3375 *M* KOH. Calculate the pH of the resulting solution (assuming volumes are additive) when the following amounts of base are added.
   a. 10.00 mL b. 20.00 mL c. 30.00 mL

13. Consider the titration of HNO_2 with KOH.
   a. What species are present at the equivalence point?
   b. Is the solution acidic, basic, or neutral at the equivalence point?

14. Consider the titration of ammonia with HCl.
   a. What species are present at the equivalence point?
   b. Is the solution acidic, basic, or neutral at the equivalence point?

15. Twenty-five mL of 0.100 *M* formic acid, $HCHO_2$, is titrated with KOH.
   a. What is the pH of the formic acid solution before titration starts?
   b. How many moles of KOH are required to react with the formic acid?

c. Write the net ionic equation for the reaction.

d. If the total volume at the equivalence point is 41.7 mL, what is the pH of the solution at this point?

16. Thirty mL of 0.100 *M* acetic acid, $HC_2H_3O_2$, is titrated with NaOH.

a. What is the pH of the acetic acid solution before titration?

b. How many moles of NaOH are required to react with the acetic acid?

c. Write a net ionic equation for the reaction.

d. If the total volume of the solution at the equivalence point is 45.0 mL, what is the pH of the solution?

Buffers

17. Calculate $[H^+]$ and pH in a solution in which $[NH_4^+]$ is 0.20 *M* and $[NH_3]$ is

a. 0.50 *M* **b.** 0.20 *M* **c.** 0.10 *M* **d.** 0.010 *M*

18. Calculate $[OH^-]$ and pH in a solution in which [HClO] is 0.25 *M* and $[ClO^-]$ is

a. 0.50 *M* **b.** 0.25 *M* **c.** 0.050 *M* **d.** 0.010 *M*

19. A solution is prepared by dissolving 0.020 mol of sodium nitrite, $NaNO_2$, in 250.0 mL of 0.040 *M* nitrous acid. Calculate the pH of this buffer.

20. A solution is prepared by dissolving 0.050 mol of potassium fluoride in 150.0 mL of 0.0275 *M* hydrogen fluoride. Calculate the pH of this buffer.

21. Consider the weak acids and their conjugate bases listed in Table 14.3. Which acid-base pair would be best for a buffer at a pH of

a. 3.5 **b.** 7.3 **c.** 9.3

22. Follow the instructions for Problem 21 for a pH of

a. 2.0 **b.** 4.0 **c.** 8.5

23. To make a buffer using $HCHO_2$ and CHO_2^- in which the desired pH is 3.00:

a. What must be the ratio $[HCHO_2]/[CHO_2^-]$?

b. How many moles of $HCHO_2$ must be added to a liter of 0.200 *M* $NaCHO_2$ to give this pH?

c. How many grams of $NaCHO_2$ must be added to a liter of 0.100 *M* $HCHO_2$ to give this pH?

24. A $NaHCO_3$–Na_2CO_3 buffer is to be prepared with a pH of 10.00.

a. What must be the ratio $[HCO_3^-]/[CO_3^{2-}]$?

b. What volume of 1.00 *M* $NaHCO_3$ should be added to a liter of 1.00 *M* Na_2CO_3 to form this buffer?

25. If a buffer solution is made of 15.00 g sodium acetate ($NaC_2H_3O_2$) and 12.50 g acetic acid ($HC_2H_3O_2$) in 5.00×10^2 mL, what is the pH of the buffer? If the buffer is diluted to 1.50 L, what is the pH of the diluted buffer?

26. A buffer solution is prepared by adding 5.50 g of ammonium chloride to 150.0 mL of 0.125 *M* ammonia. What is the pH of the solution? If the buffer is diluted with water to a final volume of 2.00 L, what is the pH of the diluted buffer?

27. How many grams of ammonium nitrate should be added to 250.0 mL of 0.150 *M* ammonia to produce a buffer with pH 9.00?

28. It is desired to convert 250.0 mL of white vinegar (d = 1.006 g/mL), which is 5.00% acetic acid, $HC_2H_3O_2$, by mass, to a buffer. How many grams of sodium acetate, $NaC_2H_3O_2$, are required to produce a buffer of pH 4.00?

29. A solution prepared from 0.050 mol/L of a weak acid, HX, has a pH of 2.75. What is the pH of the solution after 0.035 mol of solid LiX has been dissolved in it?

30. A 0.045 *M* solution of weak acid HX has a pH of 3.65. What is the pH of the solution after 0.015 mol/L of solid NaX has been dissolved in it?

31. A buffer is made up of one liter each of 0.20 *M* NH_3 and 0.20 *M* NH_4Cl. Calculate

a. the pH of the buffer.

b. the pH of the buffer after the addition of 0.0100 mol HCl.

c. the pH of the buffer after the addition of 0.0200 mol KOH.

32. A buffer is made up of one liter each of 0.100 *M* $HC_2H_3O_2$ and 0.150 *M* $NaC_2H_3O_2$. Calculate

a. the pH of the buffer.

b. the pH of the buffer after the addition of 0.0100 mol HCl.

c. the pH of the buffer after the addition of 0.0200 mol NaOH.

33. A buffer is prepared in which the ratio $[HCO_3^-]/[CO_3^{2-}]$ is 4.0.

a. What is the pH of this buffer (K_a $HCO_3^- = 4.7 \times 10^{-11}$)?

b. Enough strong acid is added to make the pH of the buffer 9.40. What is the ratio $[HCO_3^-]/[CO_3^{2-}]$ at this point?

34. A buffer is prepared using the conjugate acid-base pair, NH_4^+–NH_3. The pH of the prepared buffer is 9.00.

a. What is the ratio $[NH_4^+]/[NH_3]$?

b. What does the pH become if 20% of the NH_4^+ ions are converted to ammonia?

c. What does the pH become if 25% of the NH_3 molecules are converted to NH_4^+?

35. Which of the following would form a buffer if added to one liter of 0.20 *M* NaOH?

a. 0.10 mol $HC_2H_3O_2$ **b.** 0.30 mol $HC_2H_3O_2$

c. 0.10 mol $NaC_2H_3O_2$ **d.** 0.20 mol HCl

Explain your reasoning in each case.

36. Which of the following would form a buffer if added to one liter of 0.20 *M* $HC_2H_3O_2$?

a. 0.10 mol $NaC_2H_3O_2$ **b.** 0.10 mol NaOH

c. 0.30 mol NaOH **d.** 0.10 mol HCl

Explain your answers.

37. Calculate the pH of a solution that is prepared by mixing 2.00 g propionic acid, $HC_3H_5O_2$, and 0.45 g NaOH in water (K_a propionic acid $= 1.4 \times 10^{-5}$).

38. Calculate the pH of a solution that is prepared by mixing 1.35 g sodium nitrite, $NaNO_2$, with 100.0 mL of 0.095 *M* HCl (K_a nitrous acid = 6.0×10^{-4}).

Expression for K_{sp}

39. Write the equilibrium equation and the K_{sp} expression for each of the following electrolytes.
 a. $Zr(OH)_4$ b. $PbBr_2$
 c. K_2SiF_6 d. Bi_2S_3

40. Write the equilibrium equation and the K_{sp} expression for each of the following electrolytes.
 a. $Cu_2P_2O_7$ b. $Ni_3(AsO_4)_2$
 c. $Fe(OH)_3$ d. $Mg(NbO_3)_2$

41. Write the equilibrium equations on which the following K_{sp} expressions are based.
 a. $[Tl^+] \times [I^-]$ b. $[Eu^{3+}] \times [OH^-]^3$
 c. $[Pb^{2+}]^3 \times [PO_4^{3-}]^2$ d. $[Zn^{2+}] \times [OH^-]^2$

42. Write the equilibrium equations on which the following K_{sp} expressions are based.
 a. $[Ca^{2+}] \times [IO_3^-]^2$ b. $[Ag^+]^2 \times [SO_3^{2-}]$
 c. $[Mn^{2+}]^3 \times [AsO_4^{3-}]^2$ d. $[Pb^{2+}] \times [C_2O_4^{2-}]$

K_{sp} and Precipitation

43. Complete the following table (K_{sp} $Mg(OH)_2 = 6 \times 10^{-12}$):

	$[Mg^{2+}]$	$[OH^-]$
a.	______	1×10^{-4} *M*
b.	6×10^{-3} *M*	______
c.	______	8×10^{-5} *M*
d.	2×10^{-5} *M*	______

44. Complete the following table (K_{sp} $Ag_2CO_3 = 8 \times 10^{-12}$):

	$[Ag^+]$	$[CO_3^{2-}]$
a.	1×10^{-3} *M*	______
b.	______	5×10^{-3} *M*
c.	5×10^{-4} *M*	______
d.	______	1×10^{-5} *M*

45. Using Table 14.4, calculate the concentration of each of the following ions in equilibrium with 1.0×10^{-4} *M* PO_4^{3-}.
 a. Ag^+ b. Ca^{2+} c. Al^{3+}

46. Using Table 14.4, calculate the concentration of each of the following ions in equilibrium with 1.0×10^{-5} *M* S^{2-}.
 a. Bi^{3+} b. Cu^{2+} c. Cu^+

47. Silver nitrate is added to a solution of 1.00×10^{-2} *M* sodium acetate (K_{sp} $AgC_2H_3O_2 = 1.9 \times 10^{-3}$).
 a. At what concentration of Ag^+ does a precipitate start to form?
 b. Enough silver nitrate is added to make $[Ag^+]$ = 0.500 *M*. What is $[C_2H_3O_2^-]$? What percentage of the acetate ion originally present remains in solution?

48. A solution contains 0.0100 *M* Pb^{2+}. Potassium iodide is added to precipitate lead iodide ($K_{sp} = 8.4 \times 10^{-9}$).
 a. At what concentration of I^- does a precipitate start to form?
 b. When $[I^-] = 1.0 \times 10^{-3}$ *M*, what is $[Pb^{2+}]$? What percentage of the Pb^{2+} originally present remains in solution?

49. Lead chromate, $PbCrO_4$, is the yellow paint pigment known as "chrome yellow." A solution is prepared by mixing a solution that is 1×10^{-4} *M* in lead ion with a solution that is 5×10^{-3} *M* in chromate ion. Ignoring dilution effects, would you expect a precipitate to form?

50. A ground-water source contains 2.5 mg iodide ion per liter. Will lead iodide ($K_{sp} = 8.4 \times 10^{-9}$) precipitate when enough Pb^{2+} is added to make its concentration 4.0×10^{-3} *M*?

51. A solution is prepared by mixing 25.0 mL of 0.100 *M* iron(II) nitrate with 50.0 mL of an NaOH solution of pH 9.00. Will a precipitate form?

52. 15.0 mg magnesium nitrate is added to 50.0 mL 0.025 *M* sodium phosphate. Will a precipitate form?

Solubility

53. Calculate K_{sp} for $MnCO_3$ if 1.07 mg/L are required to make a saturated solution.

54. Calculate K_{sp} for strontium fluoride if 0.073 g/L are required to make a saturated solution.

55. Calculate the solubility (grams per liter) of magnesium phosphate ($K_{sp} = 1 \times 10^{-24}$) in
 a. pure water
 b. 0.010 *M* $Mg(NO_3)_2$
 c. 0.020 *M* Na_3PO_4

56. Calculate the solubility (grams per liter) of copper(II) sulfide in
 a. pure water
 b. 0.010 *M* Na_2S
 c. 0.020 *M* $CuSO_4$

pH of Blood

57. Carbonic acid, H_2CO_3, is a solution of CO_2 in H_2O. Explain, using Le Chatelier's principle, the buildup of CO_2 in the blood when acidosis occurs.

58. What are the symptoms and possible causes of alkalosis?

59. Blood is buffered mainly by the HCO_3^-–H_2CO_3 system (K_a $H_2CO_3 = 4.4 \times 10^{-7}$). The normal pH of blood is 7.40.
 a. What is the ratio $[H_2CO_3]/[HCO_3^-]$?
 b. What does the pH become if 10% of the HCO_3^- ions are converted to H_2CO_3?
 c. What does the pH become if 10% of the H_2CO_3 molecules are converted to HCO_3^-?

60. There is another buffer system ($H_2PO_4^-$–HPO_4^{2-}) in blood that helps keep the blood pH at about 7.40. (K_a $H_2PO_4^- = 6.2 \times 10^{-8}$)
 a. Calculate the ratio of $H_2PO_4^-/HPO_4^{2-}$ at the normal pH of blood.
 b. What percentage of the HPO_4^{2-} ions are converted to $H_2PO_4^-$ when the pH goes down to 7.00?

c. What percentage of the $H_2PO_4^-$ ions are converted to HPO_4^- when the pH goes up to 7.50?

Unclassified

61. Explain why

a. the pH increases when sodium benzoate is added to a benzoic acid solution.

b. the pH of 0.10 M HNO_2 is greater than 1.0.

c. a buffer resists changes in pH caused by addition of H^+ or OH^-.

62. Indicate whether each of the following statements is true or false; if it is false, correct it.

a. The acetate concentration in 0.10 M $HC_2H_3O_2$ is the same as in 0.10 M $NaC_2H_3O_2$.

b. A buffer can be destroyed by adding too much strong acid.

c. If $K_a = 3.1 \times 10^{-5}$, $K_b = 3.2 \times 10^{-9}$.

63. Predict what effect each of the following has on the position of the equilibrium

$$PbCl_2(s) \rightleftharpoons Pb^{2+}(aq) + 2Cl^-(aq); \qquad \Delta H = +23.4 \text{ kJ}$$

a. addition of $Pb(NO_3)_2$ solution

b. increase in temperature

c. addition of Ag^+, forming AgCl

d. addition of hydrochloric acid

64. A town adds 1.0 ppm of F^- ion to fluoridate its water supply (fluoridation of water reduces the incidence of dental caries). If the concentration of Ca^{2+} in the water is 2.0×10^{-4} M, will a precipitate of CaF_2 form when the water is fluoridated?

65. Using K_{sp} data in Table 14.4 to calculate K for the reaction, decide whether this reaction is likely to go to completion:

$$Ag_2CrO_4(s) + 2Cl^-(aq) \longrightarrow 2AgCl(s) + CrO_4^{2-}(aq)$$

66. A solution contains 30.0 mL of 0.100 M HCl. Calculate the pH after 50.0 mL of 0.100 M NaOH is added.

Challenge Problems

67. If 50.00 cm^3 of 1.000 M $HC_2H_3O_2$ ($K_a = 1.8 \times 10^{-5}$) is titrated with 1.000 M NaOH, what is the pH of the solution after the following volumes of NaOH have been added.

a. 0.00 cm^3 **b.** 25.00 cm^3 **c.** 49.90 cm^3

d. 50.00 cm^3 **e.** 50.10 cm^3 **f.** 100.00 cm^3

Use your data to construct a plot similar to that shown in Figure 14.2 (pH vs. volume NaOH added).

68. Ammonium chloride solutions are slightly acidic, so they are better solvents than water for insoluble hydroxides such as $Mg(OH)_2$. Find the solubility of $Mg(OH)_2$ in moles/liter in 0.2 M NH_4Cl and compare with the solubility in water. *Hint:* Find K for the reaction

$$Mg(OH)_2(s) + 2NH_4^+(aq) \longrightarrow Mg^{2+}(aq) + 2NH_3(aq) + 2H_2O$$

69. In a titration of 50.0 mL of 1.00 M $HC_2H_3O_2$ with 1.00 M NaOH, a student used bromcresol green as an indicator ($K_a = 1.0 \times 10^{-5}$). About how many mL of NaOH would it take to reach the end point with this indicator? What would be a better indicator for this titration?

70. What is the solubility of CaF_2 in a buffer solution containing 0.30 M $HCHO_2$ and 0.20 M $NaCHO_2$? (*Hint:* Consider the equation

$$CaF_2(s) + 2H^+(aq) \longrightarrow Ca^{2+}(aq) + 2HF(aq)$$

and solve the equilibrium problem.)

71. What is the I^- concentration just as AgCl begins to precipitate when 1.0 M $AgNO_3$ is slowly added to a solution containing 0.020 M Cl^- and 0.020 M I^-?

72. The concentrations of various cations in seawater, in moles per liter, are

Ion	Na^+	Mg^{2+}	Ca^{2+}	Al^{3+}	Fe^{3+}
Molarity (M)	0.46	0.056	0.01	4×10^{-7}	2×10^{-7}

a. At what $[OH^-]$ does $Mg(OH)_2$ start to precipitate?

b. At this concentration, will any of the other ions precipitate?

c. If enough OH^- is added to precipitate 50% of the Mg^{2+}, what percentage of each of the other ions will be precipitated?

d. Under the conditions in (c), what mass of precipitate will be obtained from one liter of seawater?

73. Starting with the relation

$$[H^+] = K_a \frac{[HB]}{[B^-]}$$

derive the Henderson-Hasselbach equation:

$$pH = pK_a + \log_{10} \frac{[B^-]}{[HB]}$$

15
Complex Ions; Coordination Compounds

Chromium ions are
sensitive to their
Chemical environment.
They make the ruby red.
The emerald green.

The ruby and emerald
are similar.
You say red, I say green.

ANN RAE JONAS
The Causes of Color

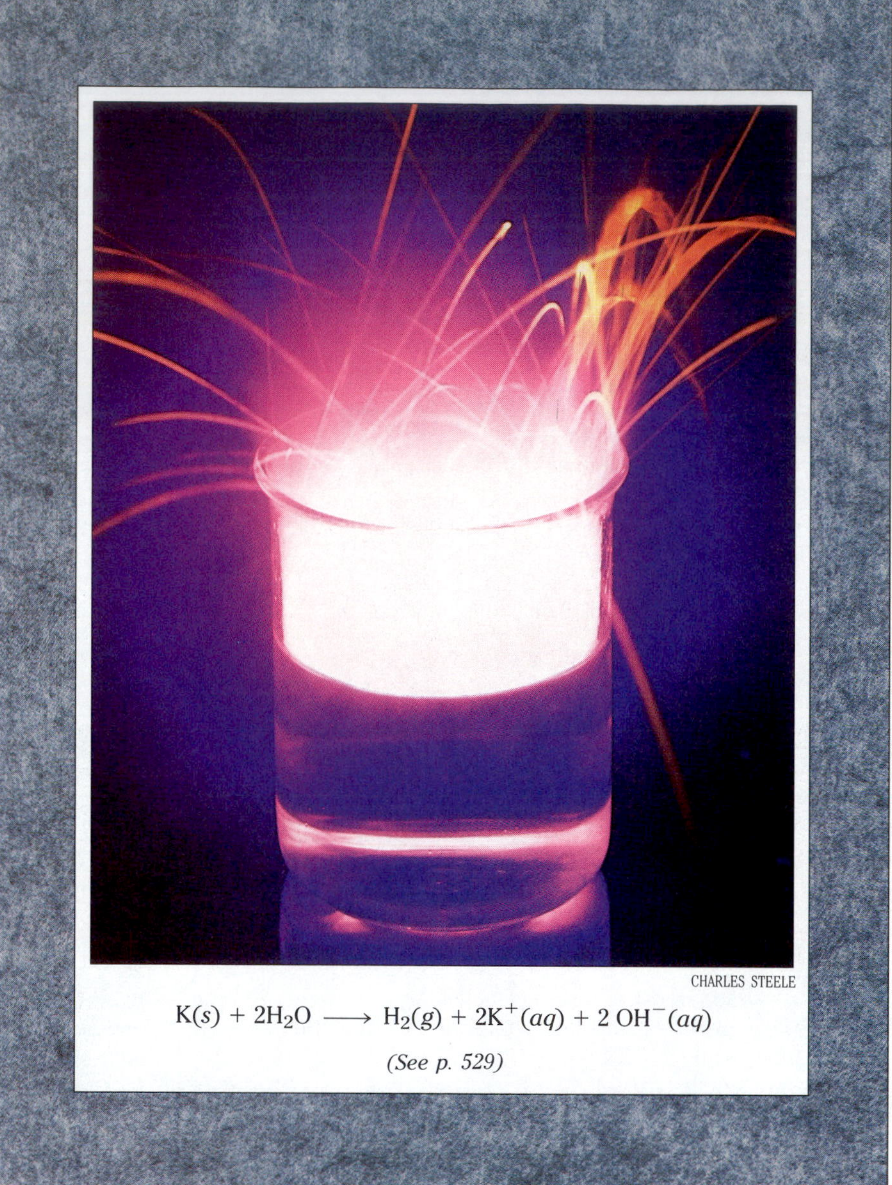

CHARLES STEELE

$$K(s) + 2H_2O \longrightarrow H_2(g) + 2K^+(aq) + 2\,OH^-(aq)$$

(See p. 529)

CHAPTER OUTLINE

In previous chapters we have referred from time to time to compounds of the transition metals. Many of these have relatively simple formulas such as $CuSO_4$, $CrCl_3$, and $Fe(NO_3)_3$. These compounds are ionic: The transition metal is present as a simple cation (Cu^{2+}, Cr^{3+}, Fe^{3+}). In that sense, they resemble the ionic compounds formed by the main-group metals, such as $CaSO_4$ and $Al(NO_3)_3$.

It has been known for more than a century, however, that transition metals also form a variety of ionic compounds with more complex formulas such as

$$[Cu(NH_3)_4]SO_4 \qquad [Cr(NH_3)_6]Cl_3 \qquad K_3[Fe(CN)_6]$$

In these so-called *coordination compounds*, the transition metal is present as a *complex ion*, enclosed within the square brackets.

Cu^{2+} and Fe^{3+} are simple cations; $Cu(NH_3)_4{}^{2+}$ and $Fe(CN)_6{}^{3-}$ are complex ions

This chapter is devoted to complex ions and the important role they play in inorganic chemistry. We consider in turn

— the compositions and names of complex ions and the coordination compounds they form (Sections 15.1 and 15.2)
— the geometry of complex ions (Section 15.3)
— the electronic structure of the central metal atom in a complex (Section 15.4)

15.1 COMPOSITION OF COMPLEX IONS

Figure 15.1
The $Cu(NH_3)_4{}^{2+}$ has an intense blue, almost violet, color, in contrast to the light blue $Cu(H_2O)_4{}^{2+}$ ion. (Charles D. Winters)

When ammonia is added to an aqueous solution of a copper(II) salt, a deep, almost opaque, blue color develops (Fig. 15.1). This color is due to the formation of the $Cu(NH_3)_4{}^{2+}$ ion, in which four NH_3 molecules are bonded to a central Cu^{2+} ion. The formation of this species can be represented in electron-dot notation as

$$Cu^{2+} + 4\,:NH_3 \longrightarrow \left[\begin{array}{c} H_3N \\ | \\ H_3N-Cu-NH_3 \\ | \\ NH_3 \end{array} \right]^{2+}$$

The nitrogen atom of each NH_3 molecule contributes a pair of unshared electrons to form a covalent bond with the Cu^{2+} ion. This bond and others like it, where both electrons are contributed by the same atom, is referred to as a **coordinate** covalent bond.

The $Cu(NH_3)_4{}^{2+}$ ion is commonly referred to as a **complex ion**, a charged species in which a **central metal atom** is bonded to molecules and/or anions referred to collectively as **ligands**. The number of bonds formed by the central metal atom is referred to as its **coordination number**. In the $Cu(NH_3)_4{}^{2+}$ complex ion,

— the central metal atom is copper(II)
— the ligands are NH_3 molecules
— the coordination number is four

Complex ions are commonly formed by transition metals, particularly those toward the right of a transition series ($_{24}Cr \rightarrow {}_{30}Zn$ in the first transition series). Nontransition metals, including Al, Sn, and Pb, form a more limited number of stable complex ions.

Cations of these metals invariably exist in water solution as complex ions, formed by reactions such as

$$Cu^{2+}(aq) + 4NH_3(aq) \longrightarrow Cu(NH_3)_4{}^{2+}(aq)$$

$$Zn^{2+}(aq) + 4\,OH^-(aq) \longrightarrow Zn(OH)_4{}^{2-}(aq)$$

$$Cr^{3+}(aq) + 4NH_3(aq) + 2Cl^-(aq) \longrightarrow Cr(NH_3)_4Cl_2{}^+(aq)$$

Often, hydrated cations such as $Cr(H_2O)_6{}^{3+}$ are present

Reactions such as these, in which one species supplies the electron pair required for bond formation, are referred to as **Lewis acid-base reactions**. In particular,

— a **Lewis base** is a species that donates a pair of electrons. Ligands such as the NH_3 molecule, OH^- ion, and Cl^- ion in the reactions on p. 411 are acting as Lewis bases.

— a **Lewis acid** is a species that accepts a pair of electrons. Cations such as Cu^{2+}, Zn^{2+}, and Cr^{3+} act as Lewis acids when they form complex ions.

Charges of Complexes

The charge of a complex is readily determined by applying a simple principle:

charge of complex = oxid. no. central atom + charge of ligands

The application of this principle is shown in Table 15.1. Notice that one of the complexes listed, $Pt(NH_3)_2Cl_2$, is neutral; the charges of the two Cl^- ions just cancel that of the Pt^{2+} central atom.

Example 15.1 Consider the $Co(H_2O)_4Cl_2^+$ ion.

(a) What is the oxidation number of cobalt?
(b) What is the formula of the coordination compound containing this cation and the Br^- anion? the SO_4^{2-} anion?

Strategy To find the oxidation number, apply the relation

charge complex ion = oxid. no. central atom + charge of ligands

You know the charge of the complex and those of the ligands. To find the formulas of the coordination compounds, note that this ion behaves like any other +1 ion; compare NaCl and Na_2SO_4.

Solution

(a) +1 = oxid. no. Co − 2 oxid. no. Co = +3

(b) $[Co(H_2O)_4Cl_2]Br$ $[Co(H_2O)_4Cl_2]_2SO_4$

The square brackets enclose the formula of the complex ion.

$Pt(NH_3)_2Cl_2$ is a complex but not an ion

Table 15.1 Complexes of Pt^{2+} with NH_3 and Cl^-

Complex	Oxid. No. of Pt	Ligands	Charge of Ligands	Charge Complex
$Pt(NH_3)_4^{2+}$	+2	$4NH_3$	0	+2
$Pt(NH_3)_3Cl^+$	+2	$3NH_3$, $1Cl^-$	−1	+1
$Pt(NH_3)_2Cl_2$	+2	$2NH_3$, $2Cl^-$	−2	0
$Pt(NH_3)Cl_3^-$	+2	$1NH_3$, $3Cl^-$	−3	−1
$PtCl_4^{2-}$	+2	$4Cl^-$	−4	−2

Ligands; Chelating Agents

In principle, any molecule or anion with an unshared pair of electrons can donate them to a metal ion to form a coordinate covalent bond. In practice, a ligand usually contains an atom of one of the more electronegative elements (C, N, O, S, F, Cl, Br, I). Several hundred different ligands are known. Those most commonly encountered in general chemistry are NH_3 and H_2O molecules and Cl^- and OH^- ions.

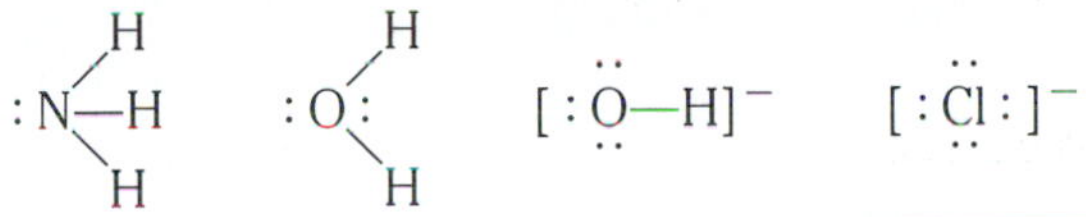

$$:NH_3 \qquad :\ddot{O}H_2 \qquad [:\ddot{O}\text{—}H]^- \qquad [:\ddot{Cl}:]^-$$

Some ligands have more than one atom with an unshared pair of electrons and hence can form more than one bond with a central metal atom. Ligands of this type are referred to as **chelating agents**; the complexes formed are referred to as **chelates** (from the Greek *chela*, crab's claw). Two of the most common chelating agents are the oxalate anion (abbr. *ox*) and the ethylenediamine molecule (abbr. *en*), whose Lewis structures are

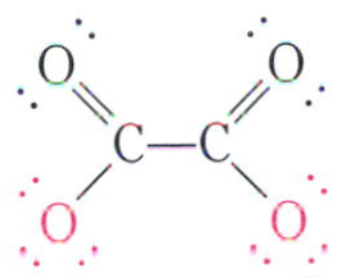

oxalate ion, $C_2O_4^{2-}$

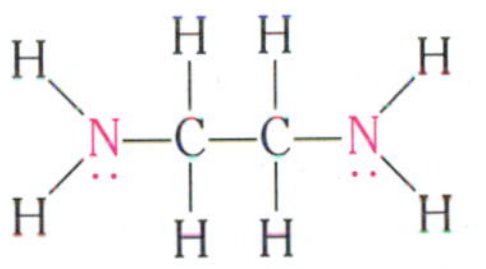

ethylenediamine molecule $H_2N\text{—}(CH_2)_2\text{—}NH_2$

(The atoms which form bonds with the central metal atom are shown in color.)

Figure 15.2 shows the structure of the chelates formed by copper(II) with these ligands. Notice that both the oxalate ion and the ethylenediamine molecule form five-membered rings with Cu^{2+}. Most chelates contain five- or six-membered rings. Four-membered rings are less stable; three-membered rings do not exist in coordination chemistry.

Coordination Number

As shown in Table 15.2, the most common coordination number is 6. A coordination number of 4 is less common. A value of 2 is restricted largely to Cu^+, Ag^+, and Au^+.

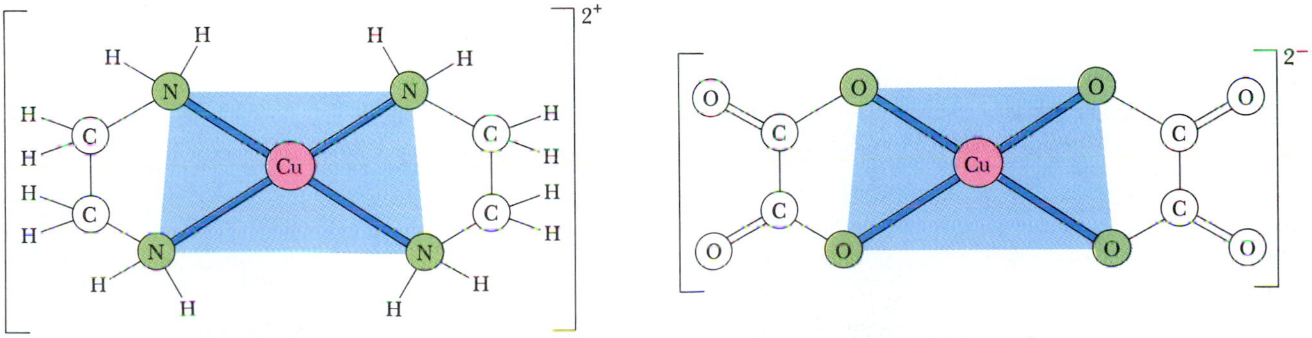

Figure 15.2
Structures of the chelates formed by Cu^{2+} with the ethylenediamine molecule (en) and the oxalate ion ($C_2O_4^{2-}$).

$Cu(en)_2^{2+}$

$Cu(C_2O_4)_2^{2-}$

Table 15.2 Coordination Number and Geometry of Complex Ions*

Metal Ion	Coordination Number	Geometry	Example
Ag^+, Au^+, Cu^+	2	linear	$Ag(NH_3)_2^+$
Cu^{2+}, Ni^{2+}, **Pd^{2+}**, **Pt^{2+}**	4	square planar	$Pt(NH_3)_4^{2+}$
Al^{3+}, Au^+, **Cd^{2+}**, Co^{2+}, Cu^+, Ni^{2+}, **Zn^{2+}**	4	tetrahedral	$Zn(NH_3)_4^{2+}$
Al^{3+}, Co^{2+}, **Co^{3+}**, **Cr^{3+}**, Cu^{2+}, **Fe^{2+}**, **Fe^{3+}**, Ni^{2+}, **Pt^{4+}**	6	octahedral	$Co(NH_3)_6^{3+}$

* The most common coordination numbers are indicated by bold type.

Odd coordination numbers are quite rare

A few cations show only one coordination number in their complexes. Thus, Co^{3+} always shows a coordination number of 6, as in

$$Co(NH_3)_6^{3+} \qquad Co(NH_3)_4Cl_2^{+} \qquad Co(en)_3^{3+}$$

Other cations, such as Al^{3+} and Ni^{2+}, have variable coordination numbers, depending upon the nature of the ligand.

Example 15.2 Write formulas (including charges) of all the complex ions formed by cobalt(III) with NH_3 and/or *en* molecules as ligands.

Strategy As indicated in Table 15.2, the coordination number of Co^{3+} is 6. This means that Co^{3+} forms six bonds with ligands, *including two with each en molecule.* The complex ions can contain 0 to 3 en molecules. Since both ligands are neutral, each complex ion has a charge of +3.

Solution There are four in all:

$$Co(NH_3)_6^{3+} \qquad Co(NH_3)_4(en)^{3+} \qquad Co(NH_3)_2(en)_2^{3+} \qquad Co(en)_3^{3+}$$

15.2 NAMING COMPLEX IONS

To name a complex ion, it is necessary to show

— the number and identity of each ligand attached to the central metal atom
— the identity and oxidation number of the central metal atom
— whether the complex is a cation or an anion

To accomplish this, a simple set of rules is followed.

1. The names of anionic ligands are obtained by substituting the suffix *o* for the normal ending. Examples include

Cl^-	chlor*o*	SO_4^{2-}	sulfat*o*
OH^-	hydrox*o*	CO_3^{2-}	carbonat*o*

Ordinarily, the names of molecular ligands are unchanged. Two important exceptions are

H_2O aqua NH_3 ammine

2. The number of ligands of a particular type is ordinarily indicated by the Greek prefixes *di, tri, tetra, penta, hexa:*

$Cu(H_2O)_4{}^{2+}$ *tetra*aquacopper(II)

$Cr(NH_3)_6{}^{3+}$ *hexa*amminechromium(III)

If the name of the ligand is itself complex (e.g., ethylenediamine), the number of such ligands is indicated by the prefixes *bis, tris,* . . . . The name of the ligand is enclosed in parentheses:

$Cr(en)_3^{3+}$ *tris*(ethylenediamine)chromium(III)

3. If more than one type of ligand is present, they are named in alphabetical order (without regard for prefixes):

$Cu(NH_3)_2(H_2O)_2{}^{2+}$ di*ammine*di*aqua*copper(II)

$Cr(NH_3)_5Cl^{2+}$ penta*amminechloro*chromium(III)

4. As you can deduce from the preceding examples, the oxidation number of the central metal atom is indicated by a Roman numeral written at the end of the name.

5. If the complex is an anion, the suffix *ate* is inserted between the name of the metal and the oxidation number:

Sometimes the Latin name is used for the metal, e.g., $Fe(CN)_6{}^{4-}$ is called hexacyanoferrate(II)

$Zn(OH)_4{}^{2-}$ tetrahydroxozinc*ate*(II)

Example 15.3 Name the following complex ions.

(a) $Cr(NH_3)_4Cl_2{}^+$ (b) $PtCl_6{}^{2-}$ (c) $Co(NH_3)_2(en)_2{}^{3+}$

Strategy Follow the rules just cited. Be careful about the oxidation number of the central metal atom!

Solution

(a) The ligands are four NH_3 molecules and two Cl^- ions. Since the charge of the ion is +1, the oxidation number of chromium must be +3:

$$\text{oxid. no. Cr} + 2(-1) = +1 \qquad \text{oxid. no. Cr} = +3$$

Since *ammine* comes in the alphabet before *chloro,* the NH_3 ligands are named first. The name is tetraamminedichlorochromium(III).

(b) There are six Cl^- ion ligands; the oxidation number of platinum must be +4. Since the complex is an anion, the suffix *ate* is used. The name is

hexachloroplatinate(IV).

(c) diamminebis(ethylenediamine)cobalt(III).

Coordination compounds, like simple ionic compounds, are named in a straightforward way. The cation is named first, followed by the anion. Compounds derived from the complex ions dealt with in Example 15.3 are named as follows:

The entire complex ion is written as one long word

$[Cr(NH_3)_4Cl_2]NO_3$	tetraamminedichlorochromium(III) nitrate
$K_2[PtCl_6]$	potassium hexachloroplatinate(IV)
$[Co(NH_3)_2(en)_2]Br_3$	diamminebis(ethylenediamine)cobalt(III) bromide

15.3 GEOMETRY OF COMPLEX IONS

The physical and chemical properties of complex ions and of the coordination compounds they form depend upon the spatial orientation of ligands around the central metal atom. Here we consider the geometries associated with the common coordination numbers: 2, 4, and 6. With that background, we then examine the phenomenon of **geometric isomerism**, where two or more complex ions have the same chemical formula but different properties because of their different geometries.

Coordination Number = 2

Complex ions in which the central metal atom forms only two bonds to ligands are *linear;* that is, the two bonds are directed at 180° angles. The structures of $CuCl_2^-$, $Ag(NH_3)_2^+$, and $Au(CN)_2^-$ may be represented as

$$(Cl{-}Cu{-}Cl)^- \qquad \left[\begin{matrix} & H & & & & H & \\ H{-} & N & {-}Ag{-} & & & N & {-}H \\ & H & & & & H & \end{matrix}\right]^+ \qquad (N{\equiv}C{-}Au{-}C{\equiv}N)^-$$

Coordination Number = 4

Four-coordinate metal complexes may have either of two different geometries (Fig. 15.3). The four bonds from the central metal atom may be directed toward the corners of a regular tetrahedron. Two common *tetrahedral* complexes

Figure 15.3
Complexes with a coordination number of four can be tetrahedral, as is $Zn(NH_3)_4^{2+}$, or square planar, like $Pt(NH_3)_4^{2+}$.

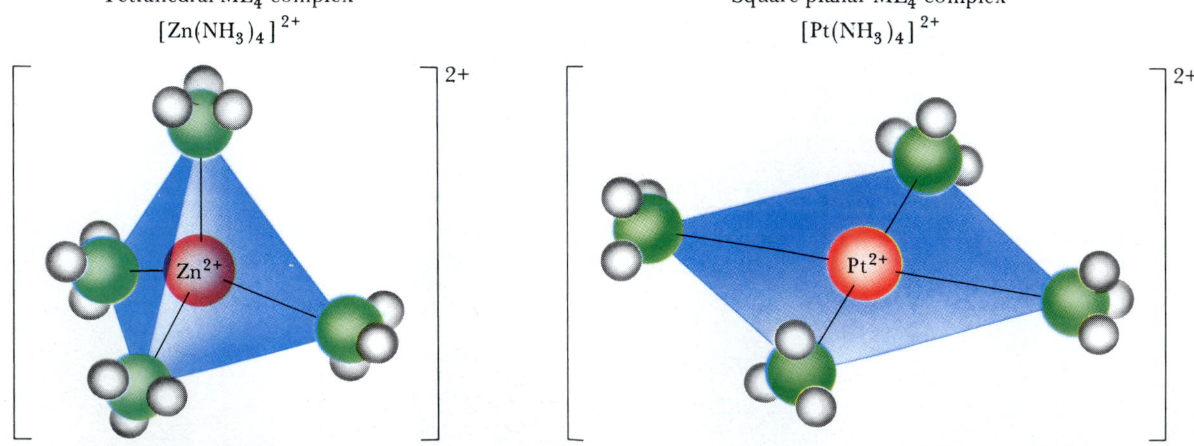

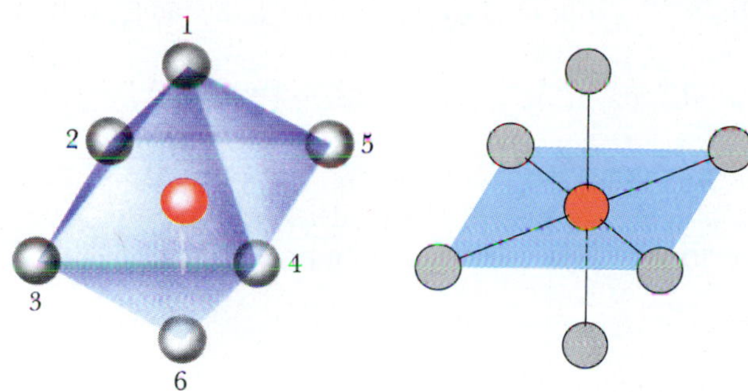

Figure 15.4
The drawing at the left shows six ligands at the corners of an octahedron with a metal atom at the center. A simpler way to represent an octahedral complex is shown at the right.

are $Zn(NH_3)_4^{2+}$ and $CoCl_4^{2-}$. In the other geometry, known as *square planar,* the four bonds are directed toward the corners of a square. The $Pt(NH_3)_4^{2+}$ and $Ni(CN)_4^{2-}$ ions are of this type.

Coordination Number = 6

Octahedral geometry is characteristic of this coordination number. The six ligands surrounding the central metal atom in complexes such as $Fe(CN)_6^{3-}$ and $Co(NH_3)_6^{3+}$ are located at the corners of a regular octahedron. An octahedron is a geometric figure with six corners and eight faces, each of which is an equilateral triangle. The metal atom is located at the center of the octahedron (Fig. 15.4). The six ligands are *equidistant from the metal atom at the center.* The skeleton structure at the right of Figure 15.4 is easier to draw and more commonly used. From this skeleton, we see that an octahedral complex can be regarded as a derivative of a square planar complex. The two extra ligands are located above and below the square, on a line perpendicular to the square at its center.

The structure is often represented simply as

Geometric Isomerism

Two or more species with different physical and chemical properties but the same formula are said to be **isomers** of one another. Complex ions can show many different kinds of isomerism, only one of which is considered here. *Geometric isomers* are ones that differ only in the spatial orientation of ligands around the central metal atom.

Geometric isomerism can exist in two different kinds of complexes.

1. *Square planar.* There are two compounds with the formula $Pt(NH_3)_2Cl_2$, differing in water solubility, melting point, chemical behavior, and biological activity.* Their structures are

```
Cl      NH3        NH3      Cl
  \    /              \    /
   Pt                  Pt
  /    \              /    \
Cl      NH3        Cl       NH3
    cis                trans
```

* The *cis* isomer ("cisplatin") is an effective anticancer drug. This reflects the ability of the two Cl atoms to interact with nitrogen atoms of DNA, a molecule responsible for cell reproduction. The *trans* isomer is ineffective in chemotherapy, presumably because the Cl atoms are too far apart to react with a DNA molecule.

cis = close together; *trans* = far apart

The complex in which the two chlorine atoms are at adjacent corners of the square, as close to one another as possible, is called the **cis** isomer. In the **trans** isomer, the two chlorine atoms are at opposite corners of the square, as far apart as possible.

Geometric isomerism can occur with any square complex of the type Ma_2b_2 or Ma_2bc, where M refers to the central metal atom and a, b, c are different ligands. Thus there are two different square planar complexes with the formula $Pt(NH_3)_2ClBr$ but only one with the formula $Pt(NH_3)_3Cl^+$.

$Br\ \ NH_3$ / Pt / $Cl\ \ NH_3$
cis-diamminebromo chloroplatinum(II)

$NH_3\ \ Br$ / Pt / $Cl\ \ NH_3$
trans-diamminebromo chloroplatinum(II)

$[NH_3\ \ Cl$ / Pt / $NH_3\ \ NH_3]^+$
triamminechloro platinum(II)

2. *Octahedral.* To understand how geometric isomerism can arise in octahedral complexes, refer back to Figure 15.4. Notice that for any given position of a ligand, four positions are equivalent, while a fifth is at a greater distance. Taking position 1 as a point of reference, you can see that groups at 2, 3, 4, and 5 are equidistant from 1; 6 is farther away. In other words, positions 1 and 2, 1 and 3, 1 and 4, 1 and 5 are *cis* to one another; positions 1 and 6 are *trans*. Hence, a complex ion like $Co(NH_3)_4Cl_2^+$ (Fig. 15.5) can exist in two different isomeric forms. In the *cis* isomer, the two Cl^- ions are at adjacent corners of the octahedron, as close together as possible. In the *trans* isomer they are at opposite corners, as far away from one another as possible.

It doesn't matter where you start; the octahedron is symmetrical

---

Example 15.4 How many geometric isomers are possible for the neutral complex $[Co(NH_3)_3Cl_3]$?

Unless you're careful, you'll come up with too many isomers

Strategy It's best to approach a problem of this type systematically. You might start by putting two NH_3 molecules *trans* to one another and see how many different structures can be obtained by placing the third NH_3 molecule at different positions. Then start over with two NH_3 molecules *cis* to one another, and follow the same procedure.

Figure 15.5
Cis and *trans* isomers of the $Co(NH_3)_4Cl_2^+$ ion. Note that the two Cl^- ions are closer to one another in the *cis* isomer (*left*) than in the *trans* isomer (*right*).

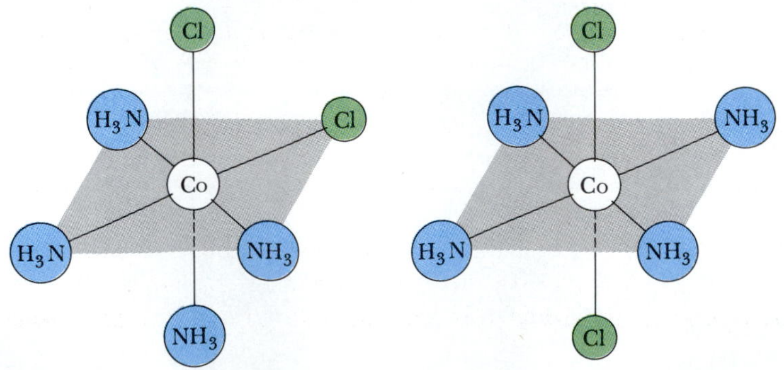

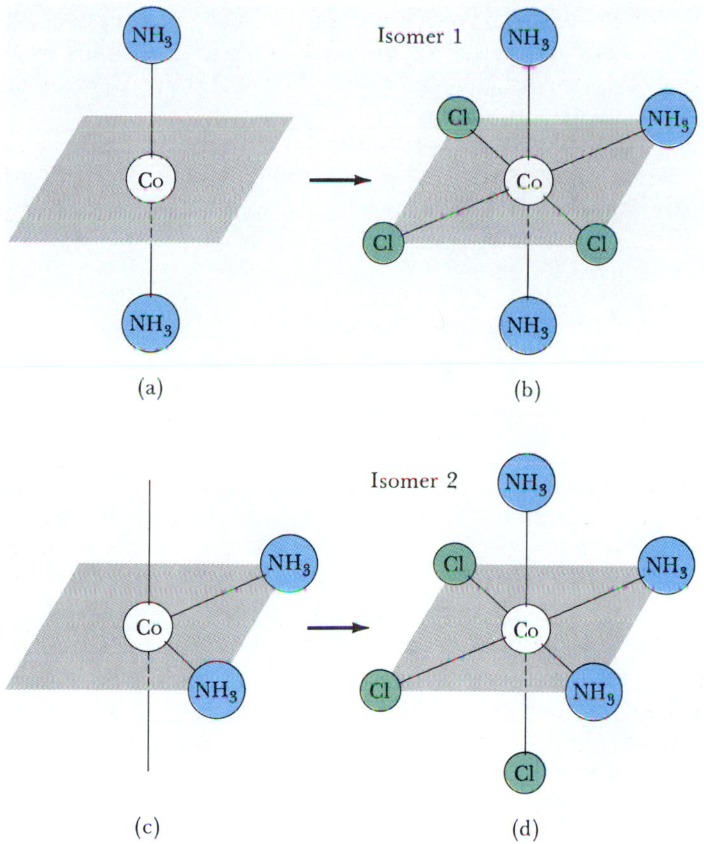

Figure 15.6
Isomers of $Co(NH_3)_3Cl_3$ (Example 15.4). You may be tempted to write down additional structures, but you will find that they are equivalent to one of the two isomers shown here.

Solution

(1) Starting with two NH_3 molecules *trans* to one another (Fig. 15.6a), it clearly doesn't matter where you put the third NH_3 molecule. All the available vacancies are equivalent in the sense that they are *cis* to the two NH_3 molecules already in place. Choosing one of these positions arbitrarily for NH_3 gives the first isomer (Fig. 15.6b); the Cl^- ions occupy the three remaining positions.

(2) Start again with the two NH_3 molecules *cis* to one another (Fig. 15.6c). If you place the third NH_3 molecule at either of the other corners of the square, you will simply reproduce the first isomer. (Remember that the symmetry of a regular octahedron requires that the distance across the diagonal of the square be the same as that from "top" to "bottom".) The only way you can get a "new" isomer is to place the third NH_3 molecule above or below the square so that all three NH_3 molecules are *cis* to one another. Putting that molecule arbitrarily at the top gives the second isomer (Fig. 15.6d). That's all there are.

Geometric isomerism can occur in chelated octahedral complexes (Fig. 15.7). Notice that an *ethylenediamine molecule,* here and indeed in all complexes, *can only bridge cis positions.* It is not long enough to connect to *trans* positions.

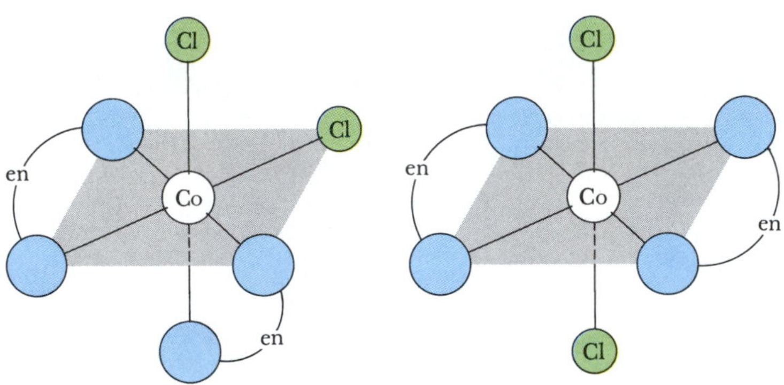

Figure 15.7
There are two complex ions with the formula $Co(en)_2Cl_2^+$. Compounds containing the *cis* cation are purple; those derived from the *trans* cation are green. (Photo, Charles D. Winters)

15.4 ELECTRONIC STRUCTURE AND BONDING IN COMPLEX IONS

"Crystal field" isn't a very descriptive term, but it's part of the jargon

Several different models have been developed to describe the bonding in complex ions. We discuss one of the simpler of these, the crystal-field model, which emphasizes the ionic component of the bonds between the central atom and the ligands. An advantage of this model is that it offers a rational explanation of the brilliant colors that are characteristic of coordination compounds (Fig. 15.8). Before discussing the crystal-field model, it may be helpful to review the electronic structure of monatomic transition metal cations, covered originally in Chapter 6.

Transition Metal Cations

Recall (pp. 150–151) that in transition metal cations:

In transition metal ions, 3d is lower in energy than 4s

1. There are no outer s electrons. Electrons beyond the preceding noble gas are located in an inner d sublevel (3d for the first transition series).
2. Electrons are distributed among the five d orbitals in accordance with Hund's rule (p. 147), giving the maximum number of unpaired electrons.

To illustrate these rules, consider the Fe^{2+} ion. Since the atomic number of iron is 26, this +2 ion must contain $26 - 2 = 24e^-$. Of these electrons, the first

Figure 15.8
Most coordination compounds are brilliantly colored, a property that can be explained readily by the crystal-field model. (Marna G. Clarke)

18 have the argon structure; the remaining 6 are located in the 3d sublevel. The electron configuration is

$$Fe^{2+} \qquad [Ar]3d^6$$

These six electrons are spread over all five orbitals; the orbital diagram is

$$Fe^{2+} \qquad [Ar] \quad \overset{3d}{(\uparrow\downarrow)(\uparrow)(\uparrow)(\uparrow)(\uparrow)}$$

The Fe^{2+} ion is *paramagnetic,* with four unpaired electrons.

Example 15.5 For the Cr^{3+} ion, derive the electron configuration, orbital diagram, and number of unpaired electrons.

Strategy First (1), find the total number of electrons (Z Cr = 24). Then (2), find the electron configuration; the first 18 electrons form the argon core, and the remaining electrons enter the 3d sublevel. Finally (3), apply Hund's rule to obtain the orbital diagram.

Solution

(1) no. of electrons = 24 − 3 = 21

(2) electron configuration $[Ar]3d^3$

(3) orbital diagram (beyond Ar) $\overset{3d}{(\uparrow)(\uparrow)(\uparrow)(\ )(\ )}$ 3 unpaired electrons

---

Crystal-Field Model (Octahedral Complexes)

The **crystal-field** model assumes that the bonding in complexes is primarily ionic rather than covalent. It further assumes that the only effect of the ligands is to create an electrostatic field around the central metal ion, which changes the relative energies of different d orbitals. In the isolated metal ion, before the ligands approach, all the d orbitals in a sublevel such as 3d have the same

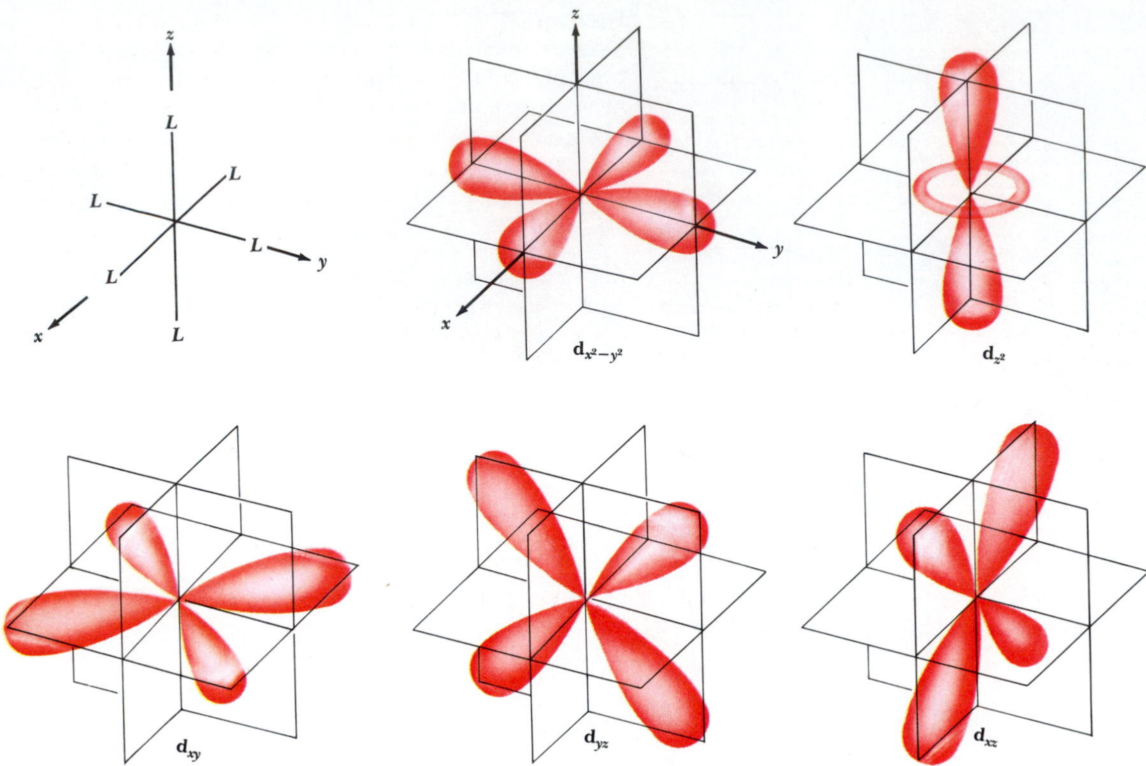

Figure 15.9
The five d orbitals and their relation to ligands on the *x, y,* and *z* axes.

energy. After interaction, the d orbitals are split into two groups with different energies.

To see why this splitting occurs, consider what happens when six ligands (e.g., H_2O, CN^-, NH_3) approach a central metal cation along the *x*-, *y*-, and *z*-axes (Fig. 15.9). The unshared electron pairs on these ligands repel the electrons in the d orbitals of the cation. The repulsion is greatest for the $d_{x^2-y^2}$ and d_{z^2} orbitals, which have their maximum electron density directly along the *x*-, *y*-, and *z*-axes, respectively. Electrons in the other three orbitals (d_{xy}, d_{yz}, d_{xz}) are less affected because their densities are concentrated between the axes rather than along them. The overall effect is to split the d orbitals into two sets with different energies:

The five d orbitals are referred to as d_{xy}, d_{yz}, d_{xz}, d_{z^2}, and $d_{x^2-y^2}$

1. A higher-energy pair, the $d_{x^2-y^2}$ and d_{z^2} orbitals
2. A lower-energy trio, comprising the d_{xy}, d_{yz}, and d_{xz} orbitals

This splitting is shown schematically at the top of Figure 15.10. The energy difference between the two sets of orbitals is known as the **crystal-field splitting energy** and given the symbol Δ_0. Depending upon the magnitude of Δ_0, a cation may form either of two different types of complexes. This is illustrated for the Fe^{2+} ion in the two diagrams at the bottom of Figure 15.10.

When Δ_0 is large, electrons can't jump the gap, so they crowd into the lower levels

1. In the $Fe(CN)_6{}^{4-}$ ion shown at the left, Δ_0 is so large that the six 3d electrons of the Fe^{2+} ion pair up in the lower-energy orbitals. The resulting complex is diamagnetic with no unpaired electrons.

2. In the $Fe(H_2O)_6{}^{2+}$ ion shown at the right, Δ_0 is relatively small, and the six 3d electrons of the Fe^{2+} ion spread out over all the orbitals in accordance

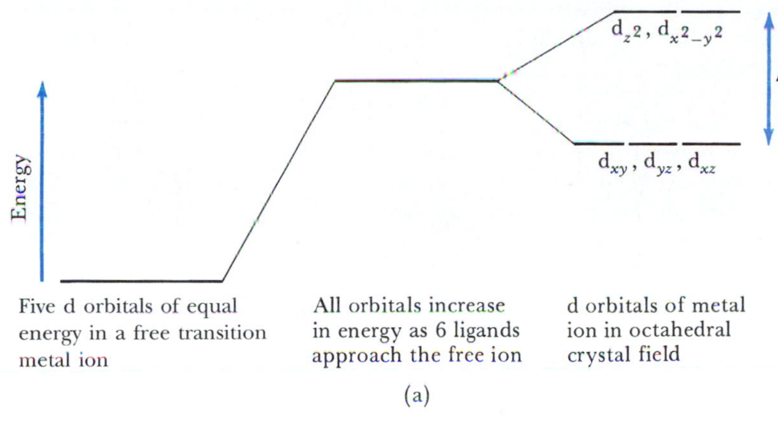

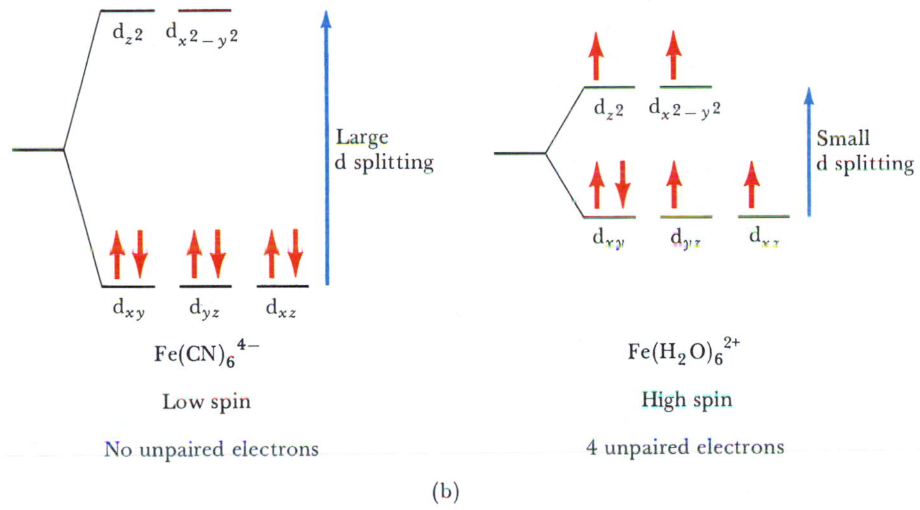

Figure 15.10
In an octahedral field, the five d orbitals are split into a high-energy pair and a low-energy trio (a). Depending upon the magnitude of Δ_0, either of two different complexes may be formed (b).

with Hund's rule. In other words, the electron distribution in the complex is the same as in the Fe^{2+} ion itself. The $Fe(H_2O)_6^{2+}$ complex is paramagnetic with four unpaired electrons.

When Δ_0 is small, the electrons spread out over as many orbitals as possible

More generally, it is possible to distinguish between

— **low spin complexes**, containing the minimum number of unpaired electrons, formed when Δ_0 is large; electrons are concentrated in the lower-energy orbitals.
— **high spin complexes**, containing the maximum number of unpaired electrons, formed when Δ_0 is small; electrons are spread out among the d orbitals in accordance with Hund's rule. In a high spin complex, the electron distribution is the same as in the simple cation.

For a given central metal ion, the value of Δ_0 is determined by the nature of the ligand. So-called "strong-field" ligands (e.g., CN^-), which interact strongly with d-orbital electrons, produce a large Δ_0. Conversely, "weak-field" ligands (e.g., H_2O) interact weakly with d-orbital electrons, producing a small Δ_0. Unfortunately, there is no simple way to predict how a particular ligand will behave with a given cation; that has to be deduced from experiment.

Example 15.6 Derive the structure of the Co^{2+} ion in low spin and high spin octahedral complexes.

Strategy First, find out how many 3d electrons there are in Co^{2+}. Now, consider how those electrons will be distributed if Hund's rule is followed; that corresponds to the high spin complex. Finally, sprinkle as many electrons as possible into the lower three orbitals; this is the low spin complex.

Solution

(1) no. of electrons = 27 − 2 = 25; the electron configuration is $[Ar]3d^7$.

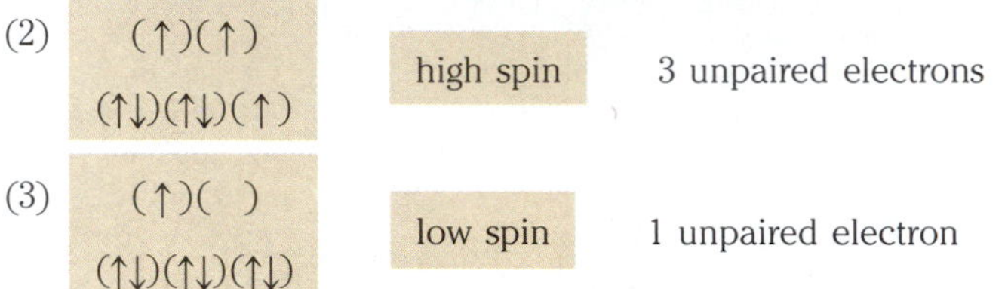

Both types of complexes are known. The $Co(CN)_6{}^{4-}$ ion is of the low spin type; $Co(H_2O)_6{}^{2+}$ is a high spin complex.

The crystal-field model offers a simple explanation of the fact that high spin and low spin complexes occur only with ions that have 4 to 7 d electrons (d^4, d^5, d^6, d^7). With three or fewer electrons, only one distribution is possible; the same is true with eight or more electrons.

( )( )	(↑)(↑)
(↑)(↑)(↑)	(↑↓)(↑↓)(↑↓)
3 electrons	8 electrons

Color

Most coordination compounds are brilliantly colored, with the exception of those formed by ions like Sc^{3+}, which has no d electrons, or Zn^{2+}, where the d orbitals are completely filled. Crystal-field theory explains colors of complex ions in terms of electron transitions between energy levels. The energy difference between two sets of d orbitals in a complex is, in many cases, equivalent to a wavelength in the visible region. Hence, by absorbing visible light, an electron may be able to move from the lower energy set of d orbitals to the higher one. This removes some of the component wavelengths of white light, so that the light reflected or transmitted by the complex is colored. The relation between absorbed and transmitted light can be deduced from the "color wheel" shown in Figure 15.11. For example, a complex that absorbs blue-green light appears orange, and vice versa.

The color of an object is that of the light transmitted or reflected

From the color of a complex ion, it is sometimes possible to deduce the value of Δ_0, the crystal-field splitting energy. The situation is particularly simple in ${}_{22}Ti^{3+}$, where there is only one 3d electron. Consider, for example, the $Ti(H_2O)_6{}^{3+}$ ion, which has an intense purple color. This ion absorbs at 510 nm, in the green region. The purple (red-violet) color of a solution of $Ti(H_2O)_6{}^{3+}$ is what is left over when the green component is subtracted from the visible

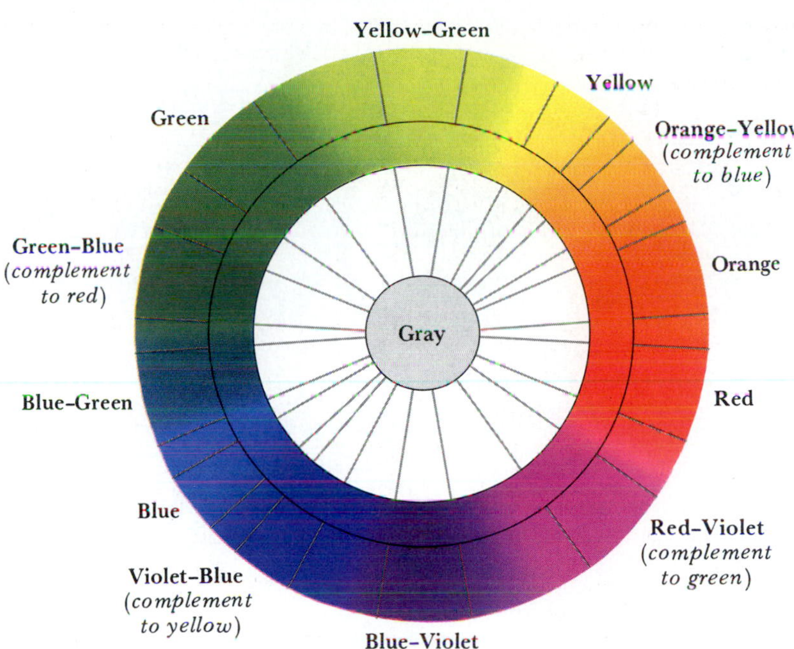

Figure 15.11
The color of a complex is complementary to (180° across from) the color of the light absorbed.

spectrum. Using the Einstein equation (Chap. 6), it is possible to calculate the energy of a photon with a wavelength of 510 nm:

$$E = \frac{hc}{\lambda} = \frac{(6.626 \times 10^{-34}\ \text{J} \cdot \text{s})(2.998 \times 10^{8}\ \text{m/s})}{510 \times 10^{-9}\ \text{m}} = 3.90 \times 10^{-19}\ \text{J}$$

The energy in kilojoules per mole is

$$E = 3.90 \times 10^{-19}\ \text{J} \times \frac{1\ \text{kJ}}{10^{3}\ \text{J}} \times \frac{6.022 \times 10^{23}}{1\ \text{mol}} = 2.35 \times 10^{2}\ \text{kJ/mol}$$

This is the energy absorbed in raising the 3d electron from a lower to a higher orbital. In other words, in the $Ti(H_2O)_6{}^{3+}$ ion, the two sets of d orbitals are separated by this amount of energy. The splitting energy, Δ_0, is 235 kJ/mol.

The smaller the value of Δ_0, the longer the wavelength of the light absorbed. This effect is shown in Table 15.3 and Figure 15.12. Notice that when

Table 15.3 Colors of Complex Ions of Co^{3+}

Complex	Color Observed	Color Absorbed	Approximate Wavelength (nm) Absorbed
$Co(NH_3)_6{}^{3+}$	yellow	violet	430
$Co(NH_3)_5NCS^{2+}$	orange	blue-green	470
$Co(NH_3)_5H_2O^{3+}$	red	green-blue	500
$Co(NH_3)_5Cl^{2+}$	purple	yellow-green	530
trans-$Co(NH_3)_4Cl_2{}^{+}$	green	red	680

Figure 15.12
Compounds containing the five cations listed in Table 15.3 form a spectrochemical series. They vary in color from yellow $[Co(NH_3)_6]Cl_3$ (left) to green $[t\text{-}Co(NH_3)_4Cl_2]Cl$ (right). (Charles D. Winters)

This series is based on experiment, not theory

weak-field ligands like NCS^-, H_2O, or Cl^- are substituted for NH_3, the light absorbed shifts to longer wavelengths (lower energies). On the basis of observations like this, ligands can be arranged in order of decreasing tendency to split the d orbitals. A short version of such a *spectrochemical series* is

$$CN^- > NO_2^- > en > NH_3 > NCS^- > H_2O > F^- > Cl^-$$

$$\xrightarrow{\hspace{2em}}$$

strong field — decreasing Δ_0 — weak field

PERSPECTIVE

•

Chelates: Natural and Synthetic

Chelating agents, capable of bonding to metal atoms at more than one position, are abundant in nature. Certain species of soybeans synthesize and secrete organic chelating agents that extract iron from insoluble compounds in the soil, thereby making it available to the plant. Mosses and lichens growing on boulders use a similar process to obtain the metal ions they need for growth.

Many important natural products are chelates in which a central metal atom is bonded into a large organic molecule. In chlorophyll, the green coloring matter of plants, the central atom is magnesium; in the essential Vitamin B_{12}, it is cobalt. The structure of heme, the pigment responsible for the red color of blood, is shown in Figure 15.A. There is an Fe^{2+} ion at the center of the complex surrounded by four nitrogen atoms at the corners of a square. A fifth coordination position around the iron is occupied by an organic molecule, which, in combination with heme, gives the protein referred to as hemoglobin.

The sixth coordination position of Fe^{2+} in heme is occupied by a water molecule which can be replaced reversibly by oxygen to give a derivative known as oxyhemoglobin, which has the bright red color characteristic of arterial blood.

$$\text{hemoglobin} + O_2 \rightleftharpoons \text{oxyhemoglobin} + H_2O$$

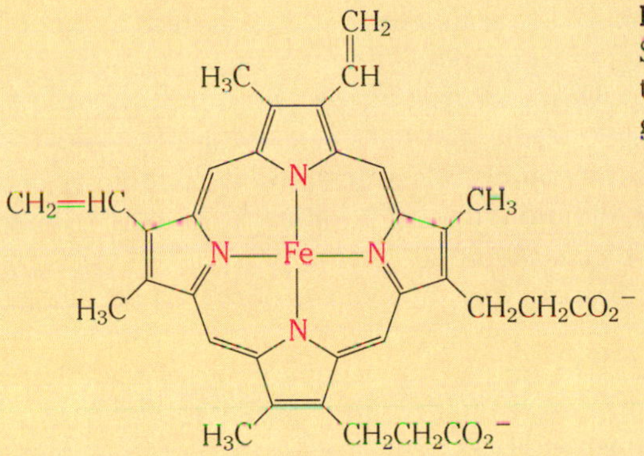
CH2
H3C
CH
CH2=HC
CH3
N
Fe
H3C
CH2CH2CO2−
H3C
CH2CH2CO2−

Figure 15.A
Structure of heme. In the hemoglobin molecule, the Fe^{2+} ion is at the center of an octahedron, surrounded by four nitrogen atoms, a globin molecule, and a water molecule.

The position of this equilibrium is sensitive to the pressure of oxygen. In the lungs, where the blood is saturated with air (P_{O_2} = 0.2 atm), the hemoglobin is almost completely converted to oxyhemoglobin. In the tissues, the partial pressure of oxygen drops, and the oxyhemoglobin breaks down to release O_2 essential for metabolism. In this way, hemoglobin acts as an oxygen carrier, absorbing oxygen in the lungs and liberating it to the tissues.

Unfortunately, hemoglobin forms a complex with carbon monoxide that is considerably more stable than oxyhemoglobin. The equilibrium constant for the reaction

$$\text{hemoglobin} \cdot O_2(aq) + CO(g) \rightleftharpoons \text{hemoglobin} \cdot CO(aq) + O_2(g)$$

is about 200 at body temperature. Consequently, the carbon monoxide complex is formed preferentially in the lungs even at CO concentrations as low as one part per thousand. When this happens, the flow of oxygen to the tissues is cut off, resulting eventually in muscular paralysis and death.

Many chelating agents have been synthesized in the laboratory to take advantage of their ability to tie up *(sequester)* metal cations. Perhaps the best-known synthetic chelating agent is ethylenediaminetetraacetic acid (EDTA), shown in Figure 15.B. An EDTA molecule can bond at as many as

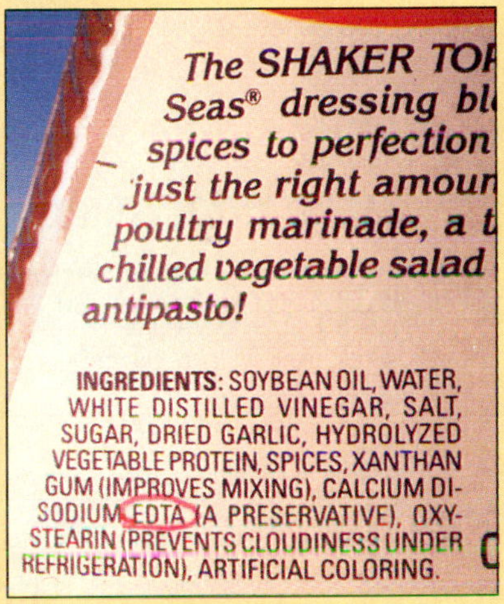
The SHAKER TO
Seas® dressing bl
spices to perfection
just the right amour
poultry marinade, a t
chilled vegetable salad
antipasto!
INGREDIENTS: SOYBEAN OIL, WATER,
WHITE DISTILLED VINEGAR, SALT,
SUGAR, DRIED GARLIC, HYDROLYZED
VEGETABLE PROTEIN, SPICES, XANTHAN
GUM (IMPROVES MIXING), CALCIUM DI-
SODIUM EDTA (A PRESERVATIVE), OXY-
STEARIN (PREVENTS CLOUDINESS UNDER
REFRIGERATION), ARTIFICIAL COLORING.

One of the uses of the chelate EDTA is as a preservative.
(Paul Silverman/FUNDAMENTAL PHOTOGRAPHS, New York)

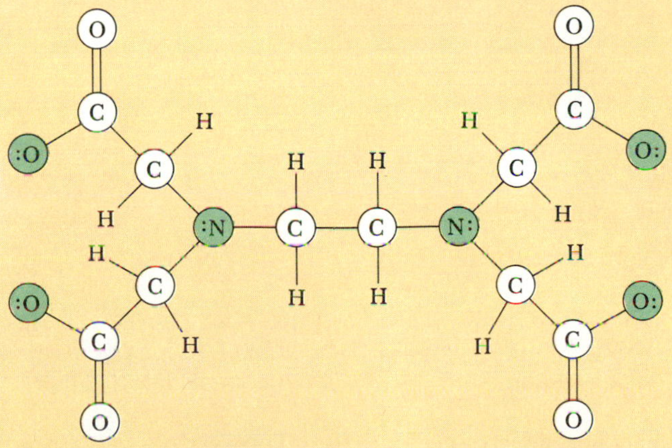
ethylenediaminetetraacetate ion (hexadentate)

Figure 15.B
Structure of the EDTA ion; the chelating atoms, six in all, are shown in color.

six positions, thereby occupying all the octahedral sites surrounding a central metal atom.

EDTA forms 1:1 complexes with a large number of cations, including those of some of the main-group metals. The complex formed by calcium with EDTA is used to treat lead poisoning. When a solution containing the Ca–EDTA complex is given by injection, the calcium is displaced by lead. The more stable Pb–EDTA complex is eliminated in the urine. EDTA has also been used to remove radioactive isotopes of metals, notably plutonium, from body tissues.

You may have noticed Ca–EDTA on the list of ingredients of many prepared foods, ranging from beer to mayonnaise. Here, EDTA acts as a scavenger to pick up traces of metal ions that catalyze the chemical reactions responsible for flavor deterioration, loss of color, or rancidity. Typically, Ca–EDTA is added at a level of 30 to 800 ppm.

CHAPTER HIGHLIGHTS

KEY CONCEPTS

1. *Relate the composition of a complex ion to its charge, the coordination number, and the oxidation number of the central atom*
   (Examples 15.1, 15.2; Problems 1–8, 47–50)
2. *Name complex ions*
   (Example 15.3; Problems 11–16)
3. *Sketch the geometry of complex ions and identify geometric isomers*
   (Example 15.4; Problems 17–24)
4. *Give the electron configuration and/or orbital diagram of a transition metal ion*
   (Example 15.5; Problems 25–28)
5. *Distinguish between high spin and low spin complexes*
   (Example 15.6; Problems 29–36)

KEY TERMS

central atom
chelate
cis-
complex ion
coordination number
geometric isomerism
ligand
octahedral
spin
—high
—low
square planar
splitting energy
tetrahedral
trans-
transition metal

SUMMARY PROBLEM

Consider the complex ion $Co(en)_2NH_3Cl^{2+}$.

a. Identify the ligands and their charges.
b. What is the oxidation number of cobalt?
c. What is the name of this ion? The formula of its sulfate salt?
d. Suppose the ethylenediamine molecules are replaced by OH^- ions. Write the formula of the resulting complex and the formula of the potassium salt of that complex.
e. What is the coordination number of cobalt in the complex?
f. Describe the geometry of the complex.

g. How many possible geometric isomers are there?
h. Give the electron configuration of cobalt(III).
i. Using the crystal-field model, show the electron distribution in the high spin and low spin complexes of cobalt(III).

Answers

a. en = 0; NH_3 = 0; Cl = −1 **b.** +3
c. amminechlorobis(ethylenediamine)cobalt(III); $[Co(en)_2NH_3Cl]SO_4$
d. $Co(NH_3)Cl(OH)_4^{2-}$; $K_2[Co(NH_3)Cl(OH)_4]$
e. 6 **f.** octahedral **g.** two **h.** $[Ar]3d^6$
i. low spin: ( )( ) high spin: (↓)(↓)

(↑↓)(↑↓)(↑↓) (↑↓)(↓)(↓)

QUESTIONS & PROBLEMS

Composition of Complex Ions and Coordination Compounds

1. Consider the complex ion $Co(en)_2(SCN)Cl^+$.
a. Identify the ligands and their charges.
b. What is the oxidation number of cobalt?
c. What is the formula of the sulfide salt of this ion?

2. Consider the complex ion $Co(NH_3)_4Cl_2^+$.
a. Identify the ligands and their charges.
b. What is the oxidation number of cobalt?
c. What would be the formula and charge of the complex ion if the ammonia molecules were replaced by $C_2O_4^{2-}$ ions?

3. Pt^{2+} forms many complexes, among them those with the following ligands. Give the formula and charge of each complex.
a. two ammonia molecules and one oxalate ion, $C_2O_4^{2-}$
b. two ammonia molecules, one thiocyanate ion, SCN^-, and one bromide ion
c. one ethylenediamine molecule and two nitrite ions, NO_2^-

4. Chromium(III) forms many complexes, among them those with the following ligands. Give the formula and charge of each chromium complex ion described below.
a. two oxalate ions ($C_2O_4^{2-}$), and two water molecules
b. five ammonia molecules and one sulfate ion
c. one ethylenediamine molecule (en), two ammonia molecules, and two iodide ions

5. What is the coordination number of the central metal atom in the following complexes?
a. $Cu(en)_2(NH_3)_2^{2+}$ **b.** $Ni(CN)_4^{2-}$
c. $Zn(en)_2^{2+}$ **d.** VCl_6^{4-}

6. What is the coordination number of the central metal atom in the following complexes?
a. $Fe(H_2O)_6^{3+}$ **b.** $Pt(NH_3)Br_3^-$
c. $Ag(NH_3)_2^+$ **d.** $Co(en)_2(SCN)Cl^+$

7. Refer to Table 15.2 to predict the formula of the complex formed by
a. Co^{3+} with NH_3 **b.** Zn^{2+} with OH^-
c. Ag^+ with CN^- **d.** Fe^{3+} with $C_2O_4^{2-}$

8. Refer to Table 15.2 to predict the formula of the complex formed by
a. Ag^+ with en **b.** Fe^{2+} with H_2O
c. Zn^{2+} with CN^- **d.** Pt^{4+} with en

9. What is the mass percent of nitrogen in the $Co(en)_3^{3+}$ complex ion?

10. What is the mass percent of chromium in the chloride salt of $Cr(H_2O)_5(OH)^{2+}$?

Nomenclature

11. Write formulas for the following species.
a. hexachloroplatinate(IV)
b. dichlorobis(ethylenediamine)nickel(II)
c. diamminetriaquahydroxochromium(III)
d. pentachlorohydroxoferrate(III)

12. Write formulas for the following species.
a. tetraammineaquachlorocobalt(III)
b. pentaamminesulfatocobalt(III)
c. tetracyanonickelate(II)
d. pentaamminenitratocobalt(III)

13. Name the following species.
a. $Ru(NH_3)_5Cl^{2+}$ **b.** $Mn(NH_2CH_2CH_2NH_2)_3^{2+}$
c. $PtCl_4^{2-}$ **d.** $Cr(NH_3)_5I^{2+}$

14. Name the following species.
a. $Al(OH)_4^-$ **b.** $Co(C_2O_4)_2(H_2O)_2^-$
c. $Ir(NH_3)_3Cl_3$ **d.** $Cr(en)(NH_3)_2Br_2^+$

15. Name the ions in Question 5.

16. Name the ions in Question 6 (SCN^- = thiocyanate).

Geometry of Complex Ions

17. Sketch the geometry of
a. *cis*-$Cu(H_2O)_2Br_4^{2-}$
b. $Zn(en)Cl_2$(tetrahedral)
c. *trans*-$Ni(NH_3)_2(en)_2^{2+}$
d. $Cu(C_2O_4)_3^{4-}$
e. *trans*-$Ni(H_2O)_2Cl_2$

18. Sketch the geometry of

a. $Ag(CN)_2^-$

b. $Co(H_2O)_2Cl_2$ (tetrahedral)

c. *cis*-$Ni(H_2O)_2Cl_2$

d. *cis*-$Pt(en)_2Br_2^{2+}$

e. *trans*-$Cr(H_2O)_4Cl_2^+$

19. The acetylacetonate ion ($acac^-$)

$$\left(CH_3{-}\overset{\overset{:\ddot{O}:}{\|}}{C}{-}CH{=}\overset{\overset{:\ddot{O}:}{|}}{C}{-}CH_3 \right)^-$$

forms complexes with many metal ions. Sketch the geometry of $Fe(acac)_3$.

20. The compound 1,2-diaminocyclohexane

[cyclohexane ring with two adjacent $\ddot{N}H_2$ groups]

(abbreviated "dech") is a ligand in the promising anticancer complex *cis*-$Pd(H_2O)_2(dech)^{2+}$. Sketch the geometry of this complex.

21. Which of the following octahedral complexes show geometric isomerism? If geometric isomers are possible, draw their structures.

a. $Cr(NH_3)_2(SCN)_4^-$ **b.** $Co(NH_3)_3(NO_2)_3$

c. $Co(en)(NH_3)_2Cl_2^+$

22. Follow the directions of Question 21 for

a. $Co(en)Cl_4^-$ **b.** $Co(en)_2ClBr^+$

c. $Co(NH_3)_2(C_2O_4)_2^-$

23. How many different octahedral complexes of Fe^{3+} can you write using only $C_2O_4^{2-}$ and/or Cl^- as ligands?

24. Draw all the structural formulas for octahedral complexes in compounds of formula $Fe(NH_3)_4Cl_2Br$.

Electronic Structure of Metal Ions

25. Give the electron configuration for

a. Ti^{3+} **b.** Cr^{2+} **c.** Ni^{3+}

d. Co^{2+} **e.** Ru^{4+}

26. Give the electron configuration for

a. Fe^{2+} **b.** V^{2+} **c.** Zn^{2+}

d. Cu^+ **e.** Nb^{2+}

27. Write an orbital diagram and determine the number of unpaired electrons in each species in Question 25.

28. Write an orbital diagram and determine the number of unpaired electrons in each species in Question 26.

Crystal-Field Model

29. Using the crystal-field model, give the electron distribution in low spin and high spin complexes of

a. Co^{3+} **b.** Mn^{2+}

30. Follow the direction of Question 29 for

a. Fe^{2+} **b.** Mn^{3+}

31. Using the crystal-field model, explain why Mn^{3+} forms high spin and low spin octahedral complexes but Mn^{4+} does not.

32. V^{3+} does not form high and low spin octahedral complexes. Use crystal-field theory to explain why this is so.

33. $Cr(CN)_6^{4-}$ is less paramagnetic than $Cr(H_2O)_6^{2+}$. Account for this using electron distribution and crystal-field theory.

34. Using the crystal-field model, account for the fact that $Co(NH_3)_6^{3+}$ is diamagnetic, while CoF_6^{3-} is paramagnetic.

35. Give the number of unpaired electrons in octahedral complexes with strong-field ligands for

a. Mn^{3+} **b.** Co^{3+} **c.** Rh^{3+}

d. Ti^{2+} **e.** Mo^{2+}

36. For the species in Question 35, indicate the number of unpaired electrons with weak-field ligands.

37. The 460-nm absorption in MnF_6^{2-} corresponds to its crystal-field splitting energy, Δ_o. Find Δ_o in kJ/mol.

38. $Ti(NH_3)_6^{3+}$ has a d orbital electron transition where ΔE is 300 kJ/mol. Calculate the wavelength (nm) for this transition.

Natural and Synthetic Chelates

39. What is the central atom in the following chelates?

a. hemoglobin **b.** chlorophyll **c.** Vitamin B_{12}

40. Oxyhemoglobin is red, while hemoglobin is blue. Explain, using Le Chatelier's principle, why venal blood is blue while arterial blood is bright red.

41. For the system

$$\text{hemoglobin} \cdot O_2(aq) + CO(g) \rightleftharpoons \text{hemoglobin} \cdot CO(aq) + O_2(g)$$

$K = 2.0 \times 10^2$. What must be the ratio of P_{CO}/P_{O_2} if 5.0% of the hemoglobin in the bloodstream is converted to the CO complex?

42. A child eats 10.0 g of paint containing 5.0% Pb. How many grams of the sodium salt of EDTA, $Na_4(EDTA)$ should he receive to bring the lead into solution as Pb · EDTA?

Unclassified

43. In your own words, explain why

a. $H_2N(CH_2)_3NH_2$ is a chelating agent.

b. AgCl dissolves in NH_3.

c. there are no geometric isomers of tetrahedral complexes.

44. Determine whether each of the following is true or false. If the statement is false, correct it.

a. The coordination number of iron(III) in $Fe(H_2O)_4(C_2O_4)^+$ is 5.

b. Cu^+ has two unpaired electrons.

c. $Co(CN)_6^{3-}$ is expected to absorb at a longer wavelength than $Co(NH_3)_6^{3+}$.

45. Indicate each of the following statements as true or false. If false, correct the statement to make it true.

a. In $Pt(NH_3)_4Cl_4$, platinum has a +4 charge and a coordination number of 6.

b. Complexes of Cu^{2+} are brightly colored, while those of Zn^{2+} are colorless.

c. Ions with seven or more d electrons cannot form high spin and low spin complexes.

46. Explain why

a. oxalic acid solution removes rust stains.

b. cations such as Ni^{2+} act as Lewis acids.

c. a pale green solution of nickel turns blue when NH_3 is added.

47. Analysis of a coordination compound gives the following results: 22.0% Co, 31.4% N, 6.78% H, and 39.8% Cl. One mole of the compound dissociates in water to form four moles of ions.

a. What is the formula of the compound?

b. Write an equation for its dissociation in water.

48. A chemist synthesizes two coordination compounds. One compound decomposes at 280°C, the other at 240°C. Analysis of the compounds gives the same mass percent data: 52.6% Pt, 7.6% N 1.63% H, and 38.2% Cl. Both compounds contain a +4 central ion.

a. What is the simplest formula of the compounds?

b. Draw structural formulas for the complexes present.

Challenge Problems

49. A certain coordination compound has the simplest formula $PtN_2H_6Cl_2$. It has a molar mass of about 600 g/mol and contains both a complex cation and a complex anion. What is its structure?

50. Two coordination compounds decompose at different temperatures but have the same mass percent analysis data: 20.25% Cu, 15.29% C, 7.07% H, 26.86% N, 10.23% S, and 20.39% O. Each contains Cu^{2+}.

a. Determine the simplest formula of the compounds.

b. Draw the structural formulas of the complex ion in each case.

51. When solid aluminum hydroxide is stirred with a concentrated solution of NaOH, it dissolves, forming the complex ion tetrahydroxoaluminate(III). The equilibrium constant for the formation of the complex ion from Al^{3+} and OH^- ions is 1×10^{33}.

a. Write an equation for the reaction in which solid aluminum hydroxide dissolves in NaOH(*aq*).

b. Write an equation for the formation of the complex ion from Al^{3+} and OH^- ions.

c. Calculate the equilibrium constant for the reaction in (a).

52. When a small amount of aqueous ammonia is added to a solution of $CuSO_4$, a blue precipitate of $Cu(OH)_2$ forms. Addition of more ammonia dissolves the precipitate to form a deep blue, almost violet solution. Treatment of this solution with hydrochloric acid gives a green solution containing the $CuCl_4{}^{2-}$ complex ion. Write balanced net ionic equations for each reaction in this sequence.

16
Spontaneity of Reaction

I am a sleepless
Slowfaring eater,
Maker of rust and rot
In your bastioned fastenings,

I am the crumbler: tomorrow.

CARL SANDBURG
Under

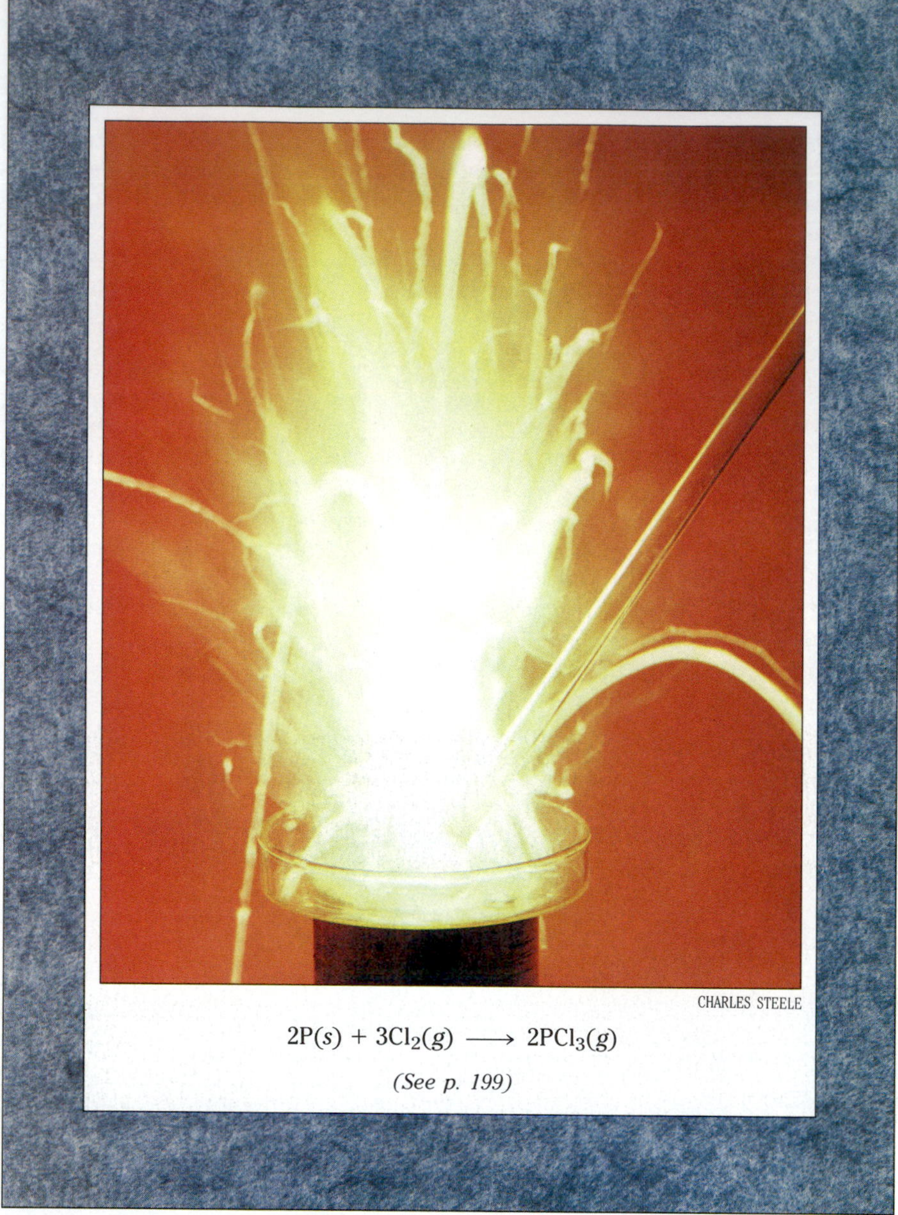

CHARLES STEELE

$$2P(s) + 3Cl_2(g) \longrightarrow 2PCl_3(g)$$

(See p. 199)

CHAPTER OUTLINE

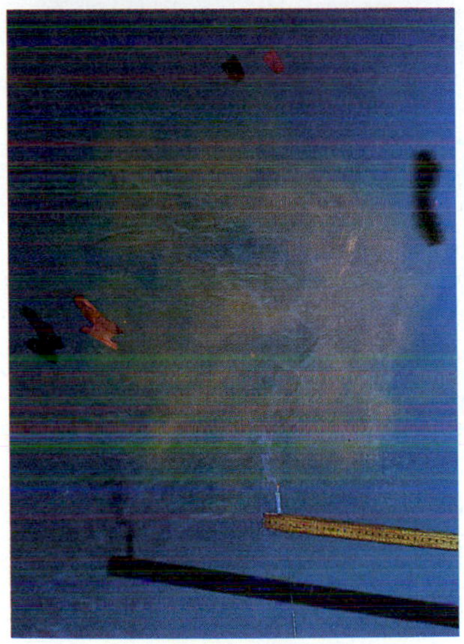

When a hydrogen-filled balloon is ignited from a distance (caution!), it explodes. The reaction $2H_2(g) + O_2(g) \rightarrow 2H_2O(l)$ is spontaneous. (Charles D. Winters)

The goal of this chapter is to answer a basic question: Will a given reaction occur "by itself" at a particular temperature and pressure, without the exertion of any outside force? In that sense, is the reaction *spontaneous?* This is a critical question in just about every area of science and technology. A synthetic organic chemist looking for a source of acetylene, C_2H_2, would like to know whether this compound can be made by heating the elements together. A metallurgist trying to produce titanium metal from TiO_2 would like to know what reaction to use; would hydrogen, carbon, or aluminum be feasible reducing agents?

To develop a general criterion for spontaneity, we will apply the principles of *thermodynamics,* the science that deals with heat and energy effects. Three different thermodynamic functions are of value here.

1. ΔH, the change in enthalpy (Chap. 8); a *negative* value of ΔH tends to make a reaction spontaneous.
2. ΔS, the change in entropy (Section 16.2); a *positive* value of ΔS tends to make a reaction spontaneous.
3. ΔG, the change in free energy (Sections 16.3, 16.4); a reaction at constant temperature and pressure will be *spontaneous* if ΔG is *negative,* no ifs, ands, or buts.

Besides serving as a general criterion for spontaneity, the free energy change can be used to

— determine the effect of temperature, pressure, and concentration on reaction spontaneity (Section 16.5)
— calculate the equilibrium constant for a reaction (Section 16.6)
— determine whether coupled reactions will be spontaneous (Section 16.7)

16.1 SPONTANEOUS PROCESSES

All of us are familiar with certain spontaneous processes. For example,

— an ice cube melts when added to a glass of water at room temperature

$$H_2O(s) \longrightarrow H_2O(l)$$

— a mixture of hydrogen and oxygen burns if ignited by a spark

$$2H_2(g) + O_2(g) \longrightarrow 2H_2O(l)$$

— an iron (steel) tool exposed to moist air rusts

$$2Fe(s) + \tfrac{3}{2}O_2(g) + 3H_2O(l) \longrightarrow 2Fe(OH)_3(s)$$

In other words, these three reactions are spontaneous at 25°C and 1 atm.

The word "spontaneous" does not imply anything about how rapidly a reaction occurs. Some spontaneous reactions, notably the rusting of iron, are quite slow. Often a reaction that is spontaneous does not occur without some sort of stimulus to provide the required activation energy. A mixture of hydrogen and oxygen shows no sign of reaction in the absence of a spark or match. Once started, though, a spontaneous reaction continues by itself without further input of energy from the outside.

A spark is OK, but a continuous input of energy isn't

The Energy Factor

Sounds reasonable, but it isn't necessarily so

Many spontaneous processes proceed with a decrease of energy. Boulders roll downhill, not uphill. A storage battery discharges when you leave your car's headlights on. Extrapolating to chemical reactions, one might guess that spontaneous reactions would be exothermic ($\Delta H < 0$). A century ago, P. M. Berthelot in Paris and Julius Thomsen in Copenhagen proposed this as a general principle, applicable to all reactions.

It turns out that almost all exothermic chemical reactions are indeed spontaneous at 25°C and 1 atm. Consider, for example, the formation of water from the elements and the rusting of iron:

$$2H_2(g) + O_2(g) \longrightarrow 2H_2O(l) \qquad \Delta H = -571.6 \text{ kJ}$$

$$2Fe(s) + \tfrac{3}{2}O_2(g) + 3H_2O(l) \longrightarrow 2Fe(OH)_3(s) \qquad \Delta H = -780.6 \text{ kJ}$$

For both of these spontaneous reactions, ΔH is a negative quantity.

On the other hand, this simple rule fails for many familiar phase changes. An example is the melting of ice. This takes place spontaneously at 1 atm above 0°C, even though it is endothermic:

$$H_2O(s) \longrightarrow H_2O(l) \qquad \Delta H = +6.0 \text{ kJ}$$

There is still another basic objection to using the sign of ΔH as a general criterion for spontaneity. Endothermic reactions that are nonspontaneous at room temperature often become spontaneous when the temperature is raised. Consider, for example, the decomposition of limestone:

$$CaCO_3(s) \longrightarrow CaO(s) + CO_2(g) \qquad \Delta H = +178.3 \text{ kJ}$$

At 25°C and 1 atm, this reaction is nonspontaneous. Witness the existence of the white cliffs of Dover and other limestone deposits over eons of time. However, if the temperature is raised to about 1100 K, the limestone decomposes to give off carbon dioxide gas at 1 atm. In other words, this endothermic reaction becomes spontaneous at high temperatures. This is true despite the fact that ΔH remains about 178 kJ, nearly independent of temperature.

The Randomness Factor

Clearly, the direction of a spontaneous change is not always determined by the tendency for a system to go to a state of lower energy. There is another natural tendency that must be taken into account to predict the direction of spontaneity. *Nature tends to move spontaneously from a state of lower probability to one of higher probability.* Or, as G. N. Lewis put it,

> Each system which is left to its own will, over time, change toward a condition of maximum probability.

To illustrate what these statements mean, consider a pastime far removed from chemistry: tossing dice. If you've ever shot craps (and maybe even if you haven't), you know that when a pair of dice is thrown, a 7 is much more likely to come up than a 12. Figure 16.1 shows why this is the case. There are six different ways to throw a 7 and only one way to throw a 12. Over time, dice will come up 7 six times as often as 12. A 7 is a state of "high" probability; a 12 is a state of "low" probability.

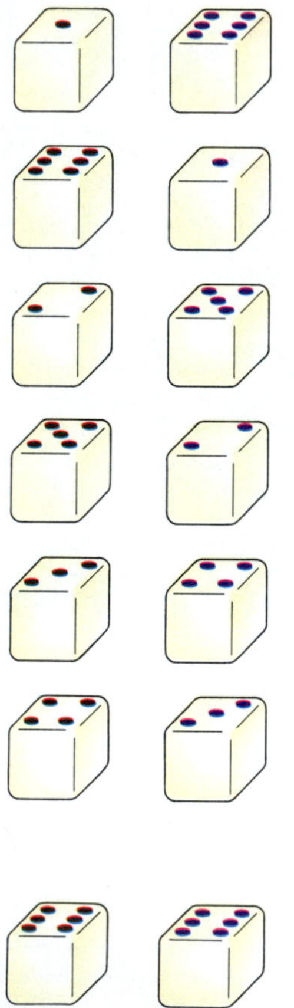

Figure 16.1
Certain "states" are more probable than others. For example, when you toss a pair of dice, a 7 is much more likely to come up than a 12.

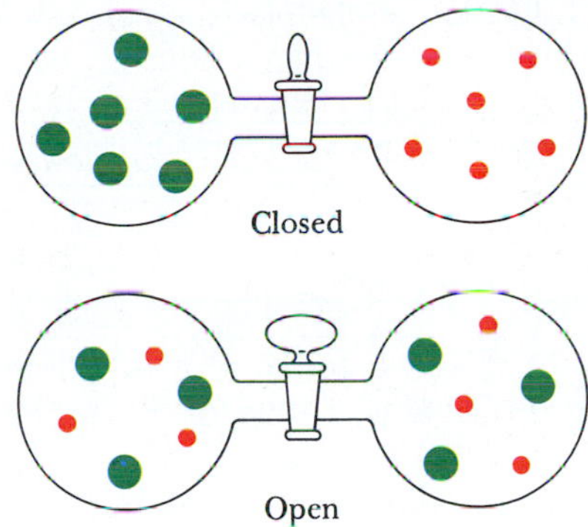

Figure 16.2
Different kinds of gas molecules mix spontaneously, going from a more ordered to a more random state.

Figure 16.3
Some states are much more probable than others. If you shake red and black marbles with each other, the random distribution at the left is much more probable than the highly ordered distribution at the right. (Charles Steele)

Now let's consider a process a bit closer to chemistry (Fig. 16.2). Two different gases, let us say H_2 and N_2, are originally contained in different glass bulbs, separated by a stopcock. When the stopcock is opened, the two different kinds of molecules distribute themselves evenly between the two bulbs. Eventually, half of the H_2 molecules will end up in the left bulb and half in the right; the same holds for the N_2 molecules. Each gas achieves its own most probable distribution, independent of the presence of the other gas.

We could explain the results of this experiment the way we did before; the final distribution is clearly much more probable than the initial distribution. There is, however, another useful way of looking at this process. The system has gone from a highly ordered state (all the H_2 molecules on the left, all the N_2 molecules on the right) to a more disordered, or random, state where the molecules are distributed evenly between the two bulbs. The same situation holds when marbles rather than molecules are mixed (Fig. 16.3). ***In general, nature tends to move spontaneously from more ordered to more random states.***

This statement is quite easy for parents or students to understand. Your room tends to get messy because an ordered room has few options for objects to be moved around (socks on the floor is not an option for an orderly room). The comedian Bill Cosby insists that with an army of 80 two-year-olds he could take over any country in the world, because they have a remarkable ability for disorganization.

Example 16.1 Choose the state that is more random and thus more probable.

(a) A chessboard before the first move or while a game is in progress.
(b) A solved jigsaw puzzle or the unassembled pieces in a box.
(c) Humpty Dumpty after the fall, or Humpty Dumpty together again.

Solution

(a) The chess game in progress is more random. There are many more ways the pieces on the board could be arranged when the game is going on.

(b) A solved jigsaw puzzle is quite orderly. A box has all the pieces at random.
(c) Humpty Dumpty after the fall is more probable. Putting him together again is nonspontaneous; all the king's horses and all the king's men couldn't do it.

16.2 ENTROPY, *S*

The randomness factor can be treated quantitatively in terms of a function called **entropy**, symbol S. In general, the more probable a state or the more random the distribution of molecules, the greater the entropy.

Entropy, like enthalpy (Chap. 8) is a state property. That is, the entropy depends only upon the state of a system, not upon its history. The entropy change is determined by the entropies of the final and initial states, not upon the path followed from one state to another.

$$\Delta S = S_{\text{final}} - S_{\text{initial}}$$

Several factors influence the amount of entropy that a system has in a particular state. In general,

One way or another, heating a substance increases its entropy

— *a liquid has a higher entropy than the solid from which it is formed.* In a solid, the atoms, molecules, or ions are fixed in position; in the liquid, these particles are free to move past one another. In that sense, the liquid structure is more random, the solid more ordered.

— *a gas has a higher entropy than the liquid from which it is formed.* Upon vaporization, the particles acquire greater freedom to move about. They are distributed throughout the entire container instead of being restricted to a small volume.

— *increasing the temperature of a substance increases its entropy.* Raising the temperature increases the kinetic energy of the molecules (or atoms or ions) and hence their freedom of motion. In the solid, the molecules vibrate with a greater amplitude at higher temperatures. In a liquid or a gas, they move about more rapidly.

These effects are shown in Figure 16.4, where the entropy of ammonia, NH_3, is plotted vs. temperature. Note that the entropy of solid ammonia at 0 K is

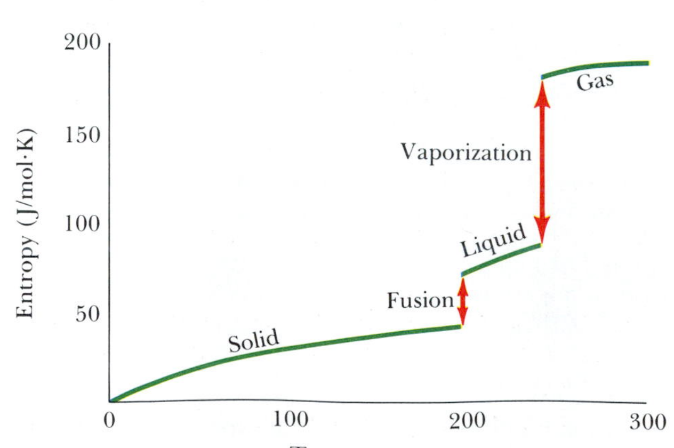

Figure 16.4
Molar entropy of NH_3 as a function of temperature. Note the large increases upon fusion and vaporization.

zero. This reflects the fact that molecules are completely ordered in the solid state at this temperature; there is only one way in which this system can be attained. More generally, the *third law of thermodynamics* tells us that *a completely ordered pure crystalline solid has an entropy of zero at 0 K.*

No disorder $\longrightarrow$ 0 entropy

Notice from the figure that an increase in temperature alone has relatively little effect on entropy. The slope of the curve is small in regions where only one phase is present. In contrast, there is a large jump in entropy when the solid melts and an even larger one when the liquid vaporizes. This behavior is typical of all substances; fusion and vaporization are accompanied by relatively large increases in entropy.

Example 16.2 Predict whether ΔS is positive or negative for each of the following processes:

(a) Taking dry ice from a freezer where its temperature is $-80°C$ and allowing it to warm to room temperature.
(b) Dissolving bromine in hexane.
(c) Condensing gaseous bromine to liquid bromine.

What is the sign of ΔS for: egg $\longrightarrow$ omelet?

Strategy Consider the relative disorder of final and initial states; remember that entropy increases in the order solid < liquid < gas.

Solution

(a) This process involves an increase in temperature and a phase change from solid to gas: $\Delta S > 0$.

(b) The solution is more random than the original liquids: $\Delta S > 0$.

(c) This is a phase change from gas to liquid, from disorder to order; $\Delta S < 0$.

Standard Molar Entropies

The entropy of a substance, unlike its energy or enthalpy, can be evaluated exactly. The details of how this is done are beyond the level of this text, but Figure 16.4, referred to earlier, shows the results for one substance, ammonia. From such a plot you can read off the **standard molar entropy** at 1 atm pressure and any given temperature, most often 25°C. This quantity is given the symbol $S°$ and has the units of joules per mole per kelvin (J/mol · K). From Figure 16.4, it appears that

$$S° \ NH_3(g) \text{ at } 25°C = 192 \text{ J/mol} \cdot \text{K}$$

Standard molar entropies of elements, compounds, and aqueous ions are listed in Table 16.1. Notice that

— *elements have nonzero standard entropies.* This means that in calculating the standard entropy change for a reaction, $\Delta S°$, elements as well as compounds must be taken into account.
— *standard molar entropies of pure substances* (elements and compounds) *are always positive quantities* ($S° > 0$).

Table 16.1 Standard Entropies at 25°C (J/mol · K) of Elements and Compounds at 1 atm, Aqueous Ions at 1 *M*.

Elements

$Ag(s)$	42.6	$Cl_2(g)$	223.0	$I_2(s)$	116.1	$O_2(g)$	205.0
$Al(s)$	28.3	$Cr(s)$	23.8	$K(s)$	64.2	$Pb(s)$	64.8
$Ba(s)$	62.8	$Cu(s)$	33.2	$Mg(s)$	32.7	$P_4(s)$	164.4
$Br_2(l)$	152.2	$F_2(g)$	202.7	$Mn(s)$	32.0	$S(s)$	31.8
$C(s)$	5.7	$Fe(s)$	27.3	$N_2(g)$	191.5	$Si(s)$	18.8
$Ca(s)$	41.4	$H_2(g)$	130.6	$Na(s)$	51.2	$Sn(s)$	51.6
$Cd(s)$	51.8	$Hg(l)$	76.0	$Ni(s)$	29.9	$Zn(s)$	41.6

Compounds

$AgBr(s)$	107.1	$CaCl_2(s)$	104.6	$H_2O(g)$	188.7	$NH_4NO_3(s)$	151.1
$AgCl(s)$	96.2	$CaCO_3(s)$	92.9	$H_2O(l)$	69.9	$NO(g)$	210.7
$AgI(s)$	115.5	$CaO(s)$	39.8	$H_2O_2(l)$	109.6	$NO_2(g)$	240.0
$AgNO_3(s)$	140.9	$Ca(OH)_2(s)$	83.4	$H_2S(g)$	205.7	$N_2O_4(g)$	304.2
$Ag_2O(s)$	121.3	$CaSO_4(s)$	106.7	$H_2SO_4(l)$	156.9	$NaCl(s)$	72.1
$Al_2O_3(s)$	50.9	$CdCl_2(s)$	115.3	$HgO(s)$	70.3	$NaF(s)$	51.5
$BaCl_2(s)$	123.7	$CdO(s)$	54.8	$KBr(s)$	95.9	$NaOH(s)$	64.5
$BaCO_3(s)$	112.1	$Cr_2O_3(s)$	81.2	$KCl(s)$	82.6	$NiO(s)$	38.0
$BaO(s)$	70.4	$CuO(s)$	42.6	$KClO_3(s)$	143.1	$PbBr_2(s)$	161.5
$BaSO_4(s)$	132.2	$Cu_2O(s)$	93.1	$KClO_4(s)$	151.0	$PbCl_2(s)$	136.0
$CCl_4(l)$	216.4	$CuS(s)$	66.5	$KNO_3(s)$	133.0	$PbO(s)$	66.5
$CHCl_3(l)$	201.7	$Cu_2S(s)$	120.9	$MgCl_2(s)$	89.6	$PbO_2(s)$	68.6
$CH_4(g)$	186.2	$CuSO_4(s)$	107.6	$MgCO_3(s)$	65.7	$PCl_3(g)$	311.7
$C_2H_2(g)$	200.8	$Fe(OH)_3(s)$	106.7	$MgO(s)$	26.9	$PCl_5(g)$	364.5
$C_2H_4(g)$	219.5	$Fe_2O_3(s)$	87.4	$Mg(OH)_2(s)$	63.2	$SiO_2(s)$	41.8
$C_2H_6(g)$	229.5	$Fe_3O_4(s)$	146.4	$MgSO_4(s)$	91.6	$SnO_2(s)$	52.3
$C_3H_8(g)$	269.9	$HBr(g)$	198.6	$MnO(s)$	59.7	$SO_2(g)$	248.1
$CH_3OH(l)$	126.8	$HCl(g)$	186.8	$MnO_2(s)$	53.0	$SO_3(g)$	256.7
$C_2H_5OH(l)$	160.7	$HF(g)$	173.7	$NH_3(g)$	192.3	$ZnI_2(s)$	161.1
$CO(g)$	197.6	$HI(g)$	206.5	$N_2H_4(l)$	121.2	$ZnO(s)$	43.6
$CO_2(g)$	213.6	$HNO_3(l)$	155.6	$NH_4Cl(s)$	94.6	$ZnS(s)$	57.7

Cations				Anions			
$Ag^+(aq)$	72.7	$Hg^{2+}(aq)$	−32.2	$Br^-(aq)$	82.4	$HPO_4^{2-}(aq)$	−33.5
$Al^{3+}(aq)$	−321.7	$K^+(aq)$	102.5	$CO_3^{2-}(aq)$	−56.9	$HSO_4^-(aq)$	131.8
$Ba^{2+}(aq)$	9.6	$Mg^{2+}(aq)$	−138.1	$Cl^-(aq)$	56.5	$I^-(aq)$	111.3
$Ca^{2+}(aq)$	−53.1	$Mn^{2+}(aq)$	−73.6	$ClO_3^-(aq)$	162.3	$MnO_4^-(aq)$	191.2
$Cd^{2+}(aq)$	−73.2	$Na^+(aq)$	59.0	$ClO_4^-(aq)$	182.0	$NO_2^-(aq)$	123.0
$Cu^+(aq)$	40.6	$NH_4^+(aq)$	113.4	$CrO_4^{2-}(aq)$	50.2	$NO_3^-(aq)$	146.4
$Cu^{2+}(aq)$	−99.6	$Ni^{2+}(aq)$	−128.9	$Cr_2O_7^{2-}(aq)$	261.9	$OH^-(aq)$	−10.8
$Fe^{2+}(aq)$	−137.7	$Pb^{2+}(aq)$	10.5	$F^-(aq)$	−13.8	$PO_4^{3-}(aq)$	−222
$Fe^{3+}(aq)$	−315.9	$Sn^{2+}(aq)$	−17.4	$HCO_3^-(aq)$	91.2	$S^{2-}(aq)$	−14.6
$H^+(aq)$	0.0	$Zn^{2+}(aq)$	−112.1	$H_2PO_4^-(aq)$	90.4	$SO_4^{2-}(aq)$	20.1

— *aqueous ions may have negative $S°$ values.* This is a consequence of the arbitrary way in which ionic entropies are defined, taking

$$S°\ H^+(aq) = 0$$

The fluoride ion has a standard entropy 13.8 units less than that of H^+; hence $S°\ F^-(aq) = -13.8$ J/mol K.

You will also notice that gases, as a group, have higher entropies than liquids or solids. Moreover, among substances of similar structure and physical state, entropy usually increases with molar mass. Compare, for example, the hydrocarbons

$$CH_4(g) \qquad S° = 186.2\ \text{J/mol}\cdot\text{K}$$
$$C_2H_6(g) \qquad S° = 229.5\ \text{J/mol}\cdot\text{K}$$
$$C_3H_8(g) \qquad S° = 269.9\ \text{J/mol}\cdot\text{K}$$

$\Delta S°$ for Reactions

Table 16.1 can be used to calculate the standard entropy change, $\Delta S°$, for reactions, using the relation

$$\Delta S° = \sum S°_{\text{products}} - \sum S°_{\text{reactants}}$$

In taking these sums, the standard molar entropies are multiplied by the number of moles specified in the balanced chemical equation.

To show how this relation is used, consider the reaction

$$CaCO_3(s) \longrightarrow CaO(s) + CO_2(g)$$

$$\Delta S° = 1\ \text{mol}\left(39.8\frac{\text{J}}{\text{mol}\cdot\text{K}}\right) + 1\ \text{mol}\left(213.6\frac{\text{J}}{\text{mol}\cdot\text{K}}\right) - 1\ \text{mol}\left(92.9\frac{\text{J}}{\text{mol}\cdot\text{K}}\right)$$
$$= 39.8\ \text{J/K} + 213.6\ \text{J/K} - 92.9\ \text{J/K} = +160.5\ \text{J/K}$$

Values of $\Delta S°$ calculated this way, using Table 16.1, have the units joules per kelvin (J/K). To convert to kilojoules per kelvin (kJ/K), all you need do is divide by 1000. For the decomposition of calcium carbonate,

$$\Delta S° = +160.5\frac{\text{J}}{\text{K}} \times \frac{1\ \text{kJ}}{1000\ \text{J}} = +0.1605\ \text{kJ/K}$$

You will notice that $\Delta S°$ for the decomposition of calcium carbonate is a positive quantity. This is reasonable since the gas formed, CO_2, has a much higher molar entropy than either of the solids, CaO or $CaCO_3$. As a matter of fact, it is almost always found that *a reaction that results in an increase in the number of moles of gas is accompanied by an increase in entropy. Conversely, if the number of moles of gas decreases, $\Delta S°$ is a negative quantity.* Consider, for example, the reaction

$$2H_2(g) + O_2(g) \longrightarrow 2H_2O(l)$$
$$\Delta S° = 2S°\ H_2O(l) - 2S°\ H_2(g) - S°\ O_2(g)$$
$$= 139.8\ \text{J/K} - 261.2\ \text{J/K} - 205.0\ \text{J/K} = -326.4\ \text{J/K}$$

Example 16.3 Calculate $\Delta S°$ for the process by which calcium hydroxide dissolves in water: $Ca(OH)_2(s) \rightarrow Ca^{2+}(aq) + 2\ OH^-(aq)$.

Strategy Apply the relation

You can't predict the sign of ΔS for processes in water solution

$$\Delta S° = \sum S°_{products} - \sum S°_{reactants}$$

Use data in Tables 16.1; take account of the coefficients in the balanced equation.

Solution

$$\Delta S° = S°\ Ca^{2+}(aq) + 2S°\ OH^-(aq) - S°\ Ca(OH)_2(s)$$

$$= -53.1\ J/K + 2(-10.8\ J/K) - 83.4\ J/K = -158.1\ J/K$$

The Second Law of Thermodynamics

A philosopher would say that the universe is running down

The relationship between entropy change and spontaneity can be expressed through a basic principle of nature known as the second law of thermodynamics. One way to state this law is to say that, *in a spontaneous process, there is a net increase in entropy, taking into account both system and surroundings.* That is,

$$\Delta S_{universe} = (\Delta S_{system} + \Delta S_{surroundings}) > 0 \qquad \text{spontaneous process}$$

(Recall from Chapter 8 that the system is that portion of the universe upon which attention is focused; the surroundings include everything else.)

Notice that the second law refers to the total entropy change, involving both system and surroundings. For many spontaneous processes, the entropy change for the system is a *negative* quantity. Consider, for example, the rusting of iron, a spontaneous process:

$$2Fe(s) + \tfrac{3}{2}O_2(g) + 3H_2O(l) \longrightarrow 2Fe(OH)_3(s)$$

The reaction mixture is taken to be the system

$\Delta S°$ for this system at 25°C and 1 atm can be calculated from a table of standard entropies; it is found to be −358.4 J/K. The negative sign of $\Delta S°$ is entirely consistent with the second law. All the law requires is that the entropy change of the surroundings be greater than 358.4 J/K, so that $\Delta S_{universe} > 0$.

In principle, the second law can be used to determine whether or not a reaction is spontaneous. To do that, however, requires calculating the entropy change for the surroundings, which is not easy. We follow a conceptually simpler approach (Section 16.3), which deals only with the thermodynamic properties of chemical *systems*.

16.3 FREE ENERGY, *G*

As pointed out earlier, there are two thermodynamic quantities that affect reaction spontaneity. One of these is the enthalpy, *H*; the other is the entropy, *S*. The problem is to put these two quantities together in such a way as to arrive at a single function whose sign will determine whether a reaction is spontaneous. This problem was first solved a century ago by J. Willard Gibbs, (p. 442), who introduced a new quantity, now called the **Gibbs free energy** and given the symbol *G*. Gibbs showed that for a reaction taking place at constant pressure

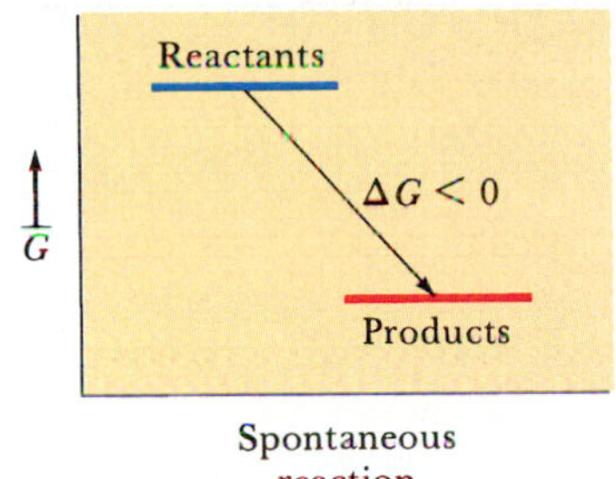

Spontaneous reaction

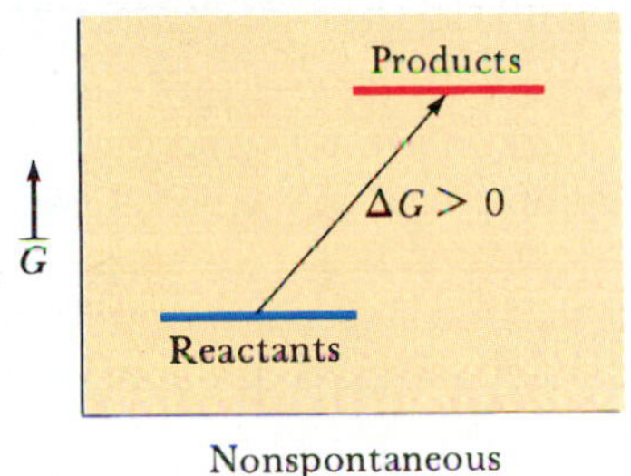

Nonspontaneous reaction

Figure 16.5
For a spontaneous reaction, the free energy of the products is less than that of the reactants: $\Delta G < 0$. For a nonspontaneous reaction, the reverse is true: $\Delta G > 0$.

and temperature, ΔG represents that fraction of the total energy change that is "free" to do useful work.

A spontaneous process can produce useful work

The basic definition of the Gibbs free energy is

$$G = H - TS$$

where T is the absolute (Kelvin) temperature. The free energy of a substance, like its enthalpy and entropy, is a state property; its value is determined only by the state of a system, not by how it got there. Putting it another way, ΔG for a reaction depends only upon the nature of products and reactants, their concentrations or pressures, and the temperature. It does *not* depend upon the path by which the reaction is carried out.

The sign of the free energy change can be used to determine the spontaneity of a reaction carried out at constant temperature and pressure.

1. *If ΔG is negative, the reaction is spontaneous.*
2. *If ΔG is positive, the reaction will not occur spontaneously.* Instead, the reverse reaction will occur.
3. *If ΔG is 0, the system is at equilibrium;* there is no tendency for reaction to occur in either direction.

Putting it another way, ΔG is a measure of the driving force of a reaction. *Reactions, at constant pressure and temperature, go in such a direction as to decrease the free energy of the system.* This means that the direction in which a reaction takes place is determined by the relative free energies of products and reactants. If the products have a lower free energy than the reactants ($G_{products} < G_{reactants}$), the forward reaction will occur (Fig. 16.5). If the reverse is true ($G_{reactants} < G_{products}$), the reverse reaction is spontaneous. Finally, if $G_{products} = G_{reactants}$, there is no driving force to make the reaction go in either direction.

A system has its minimum free energy at equilibrium

Relation Among ΔG, ΔH, and ΔS

From the defining equation for free energy, it follows that at constant temperature

$$\Delta G = \Delta H - T\Delta S$$

where ΔG, ΔH, and ΔS are the changes in free energy, enthalpy, and entropy for a reaction. This relation, known as the **Gibbs-Helmholtz equation**, is perhaps the most important equation in chemical thermodynamics. As you can see, there are two factors that tend to make ΔG negative and hence lead to a spontaneous reaction:

1. *A negative value of* ΔH. Exothermic reactions ($\Delta H < 0$) tend to be spontaneous inasmuch as they contribute to a negative value of ΔG. On the molecular level, this means that there will be a tendency to form "strong" bonds at the expense of "weak" ones.
2. *A positive value of* ΔS. If the entropy change is positive ($\Delta S > 0$), the term $-T\Delta S$ will make a negative contribution to ΔG. Hence, there will be a tendency for a reaction to be spontaneous if the products are less ordered than the reactants.

J. Willard Gibbs (yearbook portrait as a graduating senior at Yale, Class of 1858, Burndy Library.)

Two theoreticians working in the latter half of the nineteenth century changed the very nature of chemistry by deriving the mathematical laws that govern the behavior of matter undergoing physical or chemical change. One of these was James Clerk Maxwell, whose contributions to kinetic theory were discussed in Chapter 5. The other was J. Willard Gibbs, Professor of Mathematical Physics at Yale from 1871 until his death in 1903.

In 1876 Gibbs published the first portion of a remarkable paper in the *Transactions of the Connecticut Academy of Sciences* entitled "On the Equilibrium of Heterogeneous Substances." When the paper was completed in 1878 (it was 323 pages long), the foundation was laid for the science of chemical thermodynamics. Here, for the first time, the concept of free energy appeared. Included as well were the basic principles of chemical equilibrium (Chap. 12), phase equilibrium (Chap. 9), and the relations governing energy changes in electrical cells (Chap 16).

If Gibbs had never published another paper, this single contribution would have placed him among the greatest theoreticians in the history of science. Generations of experimental scientists have established their reputations by demonstrating in the laboratory the validity of the relationships that Gibbs derived at his desk. Many of these relationships were rediscovered by others; an example is the Gibbs-Helmholtz equation developed in 1882 by Helmholtz, a prestigious German physiologist and physicist who was completely unaware of Gibbs' work.

J. Willard Gibbs is often cited as an example of the "prophet without honor in his own country." His colleagues in New Haven and elsewhere in the United States seem not to have realized the significance of his work until late in his life. During his first 10 years as a professor at Yale he received no salary. In 1920, when he was first proposed for the Hall of Fame of Distinguished Americans at New York University, he received 9 votes out of a possible 100. Not until 1950 was he elected to that body. Even today the name of J. Willard Gibbs is generally unknown among educated Americans outside of those interested in the natural sciences.

Maxwell was perhaps the first to recognize Gibbs' genius

Admittedly, Gibbs himself was largely responsible for the fact that for many years his work did not attract the attention it deserved. He made little effort to publicize it; the *Transactions of the Connecticut Academy of Sciences* was hardly the leading scientific journal of its day. Gibbs was one of those rare individuals who seem to have no inner need for recognition by contemporaries. His satisfaction came from solving a problem in his mind; having done so, he was ready to proceed to other problems. His papers are not easy to read; he seldom cites examples to illustrate his abstract reasoning. Frequently, the implications of the laws that he derives are left for the readers to grasp on their own. One of his colleagues at Yale confessed many years later that none of the members of the Connecticut Academy of Sciences understood his paper on thermodynamics; as he put it, "We knew Gibbs and took his contributions on faith."

In many physical changes, the entropy increase is the major driving force. This situation applies when two liquids with similar intermolecular forces, such as benzene (C_6H_6) and toluene (C_7H_9), are mixed. There is no change in enthalpy, but the entropy increases because the pure substances become diluted when they form a solution.*

What is the driving force behind a falling brick? a balloon deflating?

In certain reactions, ΔS is nearly zero and ΔH is the only important component of the driving force for spontaneity. An example is the synthesis of hydrogen fluoride from the elements:

$$\tfrac{1}{2}H_2(g) + \tfrac{1}{2}F_2(g) \longrightarrow HF(g)$$

For this reaction, ΔH is a large negative number, −271.1 kJ, showing that the bonds in HF are stronger than those in the H_2 and F_2 molecules. As you might expect for a gaseous reaction in which there is no change in the number of moles, ΔS is very small, about 0.0070 kJ/K. The free energy change, ΔG, at 1 atm is −273.2 kJ at 25°C, almost identical to ΔH. Even at very high temperatures, the difference between ΔG and ΔH is small, amounting to only about 14 kJ at 2000 K.

16.4 STANDARD FREE ENERGY CHANGE, $\Delta G°$

Although the Gibbs-Helmholtz equation is valid under all conditions, we will apply it only under standard conditions, where any gas involved in the reaction is at 1 atm pressure and all species in aqueous solution are at a concentration of 1 *M*. In other words, we will use the equation in the form

$$\Delta G° = \Delta H° - T\Delta S°$$

where $\Delta G°$ is the standard free energy change (1 atm, 1 *M*); $\Delta H°$ is the standard enthalpy change, which can be calculated from heats of formation, $\Delta H_f°$ (listed in Tables 8.3 and 8.4, Chapter 8); and $\Delta S°$ is the standard entropy change (Table 16.1).

Calculation of $\Delta G°$ at 25°C; Free Energies of Formation

To illustrate the use of the Gibbs-Helmholtz equation, let us apply it first to find $\Delta G°$ at 25°C (Example 16.4).

Example 16.4 For the reaction $CaSO_4(s) \longrightarrow Ca^{2+}(aq) + SO_4^{2-}(aq)$, calculate

(a) $\Delta H°$. (b) $\Delta S°$. (c) $\Delta G°$ at 25°C.

Strategy $\Delta H°$ is calculated from Tables 8.3 and 8.4, Chapter 8, using the relation

$$\Delta H° = \sum \Delta H_{f\,\text{products}}° - \sum \Delta H_{f\,\text{reactants}}°$$

$\Delta S°$ is obtained from Table 16.1 using the relation

$$\Delta S° = \sum S°_{\text{products}} - \sum S°_{\text{reactants}}$$

* Formation of a *water* solution is often accompanied by a *decrease* in entropy because of hydrogen bonding or hydration effects that lead to a highly ordered solution structure. Examples include

$$C_2H_5OH(l) \longrightarrow C_2H_5OH(aq) \qquad \Delta S° = -12.2 \text{ J/K}$$
$$CaCl_2(s) \longrightarrow Ca^{2+}(aq) + 2Cl^-(aq) \qquad \Delta S° = -44.7 \text{ J/K}$$

For consistency, $\Delta S°$ should be converted to kilojoules per kelvin; i.e., 1 J/K = 10^{-3} kJ/K. Finally, $\Delta G°$ is calculated from the Gibbs-Helmholtz equation with T = 298 K.

Solution

(a) $\Delta H° = \Delta H_f° \, Ca^{2+}(aq) + \Delta H_f° \, SO_4^{2-}(aq) - \Delta H_f° \, CaSO_4(s)$

$$= -542.8 \text{ kJ} - 909.3 \text{ kJ} - (-1434.1 \text{ kJ}) = -18.0 \text{ kJ}$$

(b) $\Delta S° = S° \, Ca^{2+}(aq) + S° \, SO_4^{2-}(aq) - S° \, CaSO_4(s)$

$$= -53.1 \text{ J/K} + 20.1 \text{ J/K} - 106.7 \text{ J/K} = -139.7 \text{ J/K}$$

$$= -0.1397 \text{ kJ/K}$$

(c) $\Delta G° = \Delta H° - T\Delta S°$

$$= -18.0 \text{ kJ} - 298 \text{ K} \times (-0.1397 \text{ kJ/K}) = +23.6 \text{ kJ}$$

Since $\Delta G°$ is positive, this reaction at standard conditions

$$CaSO_4(s) \longrightarrow Ca^{2+}(aq, 1\,M) + SO_4^{2-}(aq, 1\,M)$$

should not be spontaneous. In other words, calcium sulfate should not dissolve in water to give a 1 M solution. This is indeed the case. The solubility of $CaSO_4$ at 25°C is considerably less than 1 mol/L, only about 0.02 mol/L.

The Gibbs-Helmholtz equation can be used to calculate the *standard free energy of formation* of a compound. This quantity, $\Delta G_f°$, is analogous to the heat of formation, $\Delta H_f°$. It is defined as the free energy change per mole when a compound is formed from the elements in their stable states at 1 atm.

Example 16.5 Calculate the standard free energy of formation, $\Delta G_f°$, at 25°C, for $CH_4(g)$, using data in Tables 8.3 and 16.1.

Strategy By definition, $\Delta G_f°$ for CH_4 is $\Delta G°$ for the reaction $C(s) + 2H_2(g) \rightarrow CH_4(g)$. The calculations are entirely analogous to those in Example 16.4.

Solution

$$\Delta H° = \Delta H_f° \, CH_4(g) = -74.8 \text{ kJ}$$

$$\Delta S° = S° \, CH_4(g) - S° \, C(s) - 2S° \, H_2(g)$$

$$= 186.2 \text{ J/K} - 5.7 \text{ J/K} - 2(130.6 \text{ J/K}) = -80.7 \text{ J/K} = -0.0807 \text{ kJ/K}$$

At 25°C, that is, 298 K,

$$\Delta G° = \Delta H° - 298\,\Delta S° = -74.8 \text{ kJ} - 298 \text{ K}(-0.0807 \text{ kJ/K}) = -50.7 \text{ kJ}$$

Hence, the free energy of formation of $CH_4(g)$ at 25°C is −50.7 kJ/mol.

$\Delta G_f°$ for an element, like $\Delta H_f°$, is zero

Tables of standard free energies of formation at 25°C of compounds and ions in solution are given in Appendix 1 (along with standard heats of formation and standard entropies). Notice that, for most compounds, $\Delta G_f°$ is a negative quantity, which means that the compound can be formed spontaneously from the elements. This is true for water:

$$H_2(g) + \tfrac{1}{2}O_2(g) \longrightarrow H_2O(l) \qquad \Delta G_f^\circ\ H_2O(l) = -237.2\text{ kJ}$$

and, at least in principle, for methane:

$$C(s) + 2H_2(g) \longrightarrow CH_4(g) \qquad \Delta G_f^\circ\ CH_4(g) = -50.7\text{ kJ}$$

A few compounds, including acetylene, have positive free energies of formation ($\Delta G_f^\circ\ C_2H_2 = +217.1$ kJ). These compounds cannot be made from the elements; indeed, they are potentially unstable with respect to the elements. In the case of acetylene, the reaction

$$C_2H_2(g) \longrightarrow 2C(s) + H_2(g) \qquad \Delta G^\circ \text{ at } 25°\text{C} = -217.1\text{ kJ}$$

occurs with explosive violence unless special precautions are taken.

Values of ΔG_f° can be used to calculate free energy changes for reactions. The relationship here is entirely analogous to that given for enthalpies in Chapter 8:

$$\Delta G^\circ_{\text{reaction}} = \sum \Delta G^\circ_{f\,\text{products}} - \sum \Delta G^\circ_{f\,\text{reactants}}$$

ΔG_f° tables have very limited uses

If you calculate ΔG° in this way, you should keep in mind an important limitation of this equation. *It is valid only at the temperature at which ΔG_f° data are tabulated, in this case 25°C.* Since ΔG° varies considerably with temperature, this approach is not even approximately valid at other temperatures.

Calculation of ΔG° at Other Temperatures

To a good degree of approximation, the temperature variation of ΔH° and ΔS° can be neglected.* This means that, to apply the Gibbs-Helmholtz equation at temperatures other than 25°C, you need only change the value of T. The quantities ΔH° and ΔS° can be calculated in the usual way from tables of standard enthalpies and entropies.

Example 16.6 Calculate ΔG° for the reaction $Cu(s) + H_2O(g) \rightarrow CuO(s) + H_2(g)$ at 500 K.

Strategy Calculate ΔH° using Table 8.3 and ΔS° using Table 16.1. Then apply the Gibbs-Helmholtz equation with $T = 500$ K.

Solution From Table 8.3

$$\Delta H^\circ = \Delta H_f^\circ\ CuO(s) - \Delta H_f^\circ\ H_2O(g) = -157.3\text{ kJ} + 241.8\text{ kJ} = +84.5\text{ kJ}$$

From Table 16.1

$$\Delta S^\circ = S^\circ\ CuO(s) + S^\circ\ H_2(g) - S^\circ\ Cu(s) - S^\circ\ H_2O(g)$$
$$= 42.6\text{ J/K} + 130.6\text{ J/K} - 33.2\text{ J/K} - 188.7\text{ J/K}$$
$$= -48.7\text{ J/K} = -0.0487\text{ kJ/K}$$
$$\Delta G^\circ = \Delta H^\circ - T\Delta S^\circ$$

Hence, $\Delta G^\circ = +84.5\text{ kJ} - 500\text{ K} \times (-0.0487\text{ kJ/K}) = +108.9\text{ kJ}$

Since ΔG° is positive, the reaction is not spontaneous. Instead, the reverse reaction occurs at 1 atm and 500 K (227°C). Under these conditions, CuO is reduced to copper in a stream of H_2 gas.

* As far as the Gibbs-Helmholtz equation is concerned, there is another reason for ignoring the temperature dependence of ΔH and ΔS. These two quantities always change in the same direction as the temperature changes (that is, if ΔH becomes more positive, so does ΔS). Hence, the two effects tend to cancel each other.

From Example 16.6 and the preceding discussion, it should be clear that $\Delta G°$, unlike $\Delta H°$ and $\Delta S°$, is strongly dependent upon temperature. This comes about, of course, because of the T in the Gibbs-Helmholtz equation:

$$\Delta G° = \Delta H° - T\Delta S°$$

Comparing this equation to that of a straight line,

$$y = b + mx$$

it is clear that a plot of $\Delta G°$ vs. T should be linear, with a slope of $-\Delta S°$ and a y-intercept (at 0 K) of $\Delta H°$.

16.5 EFFECT OF TEMPERATURE, PRESSURE, AND CONCENTRATION ON REACTION SPONTANEITY

A change in reaction conditions can, and often does, change the direction in which a reaction occurs spontaneously. The Gibbs-Helmholtz equation in the form

$$\Delta G° = \Delta H° - T\Delta S°$$

is readily applied to deduce the effect of temperature upon reaction spontaneity. To consider the effect of pressure or concentration, we need an analogous expression for the dependence of ΔG upon these factors.

Temperature

When the temperature of a reaction system is increased, the sign of $\Delta G°$, and hence the direction in which the reaction proceeds spontaneously, may or may not change. Whether it does or not depends upon the signs of $\Delta H°$ and $\Delta S°$. The four possible situations, deduced from the Gibbs-Helmholtz equation, are summarized in Figure 16.6.

If $\Delta H°$ and $\Delta S°$ have opposite signs (Fig. 16.6, I and III), it is impossible to reverse the direction of spontaneity by a change in temperature alone. The two terms $\Delta H°$ and $-T\Delta S°$ reinforce one another. Hence, $\Delta G°$ has the same sign at all temperatures. Reactions of this type are rather uncommon. One such reaction is that discussed in Example 16.6:

$$Cu(s) + H_2O(g) \longrightarrow CuO(s) + H_2(g)$$

Here, $\Delta H°$ is +84.5 kJ and $\Delta S°$ is −0.0487 kJ/K. Hence,

$$\Delta G° = \Delta H° - T\Delta S°$$
$$= +84.5 \text{ kJ} + T(0.0487 \text{ kJ/K})$$

Copper metal doesn't react with water vapor, period

Clearly, $\Delta G°$ is positive at all temperatures. The reaction cannot take place spontaneously at 1 atm regardless of temperature.

It is more common to find that $\Delta H°$ and $\Delta S°$ have the same sign (Fig. 16.6, II and IV). When this happens, the enthalpy and entropy factors oppose each other. $\Delta G°$ changes sign as temperature increases, and the direction of spontaneity reverses. At low temperatures, $\Delta H°$ predominates and the exothermic reaction, which may be either the forward or the reverse reaction, occurs. As

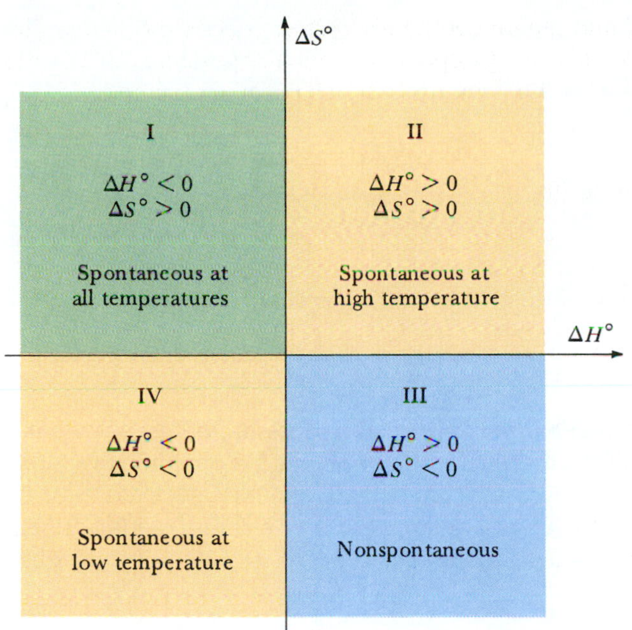

Figure 16.6
Spontaneity of reaction depends upon the sign and magnitude of $\Delta H°$ and $T\Delta S°$. Most commonly, $\Delta H°$ and $\Delta S°$ have the same sign and the direction of spontaneity depends upon the temperature.

the temperature rises, the quantity $T\Delta S°$ increases in magnitude and eventually exceeds $\Delta H°$. At high temperatures, the reaction that leads to an increase in entropy occurs. In most cases, 25°C is a "low" temperature, at least at a pressure of 1 atm. This explains why exothermic reactions are usually spontaneous at room temperature and atmospheric pressure.

An example of a reaction for which $\Delta H°$ and $\Delta S°$ have the same sign is the decomposition of calcium carbonate:

$$CaCO_3(s) \longrightarrow CaO(s) + CO_2(g)$$

Here, $\Delta H° = +178.3$ kJ and $\Delta S° = +160.5$ J/K. Hence,

$$\Delta G° = +178.3 \text{ kJ} - T(0.1605 \text{ kJ/K})$$

For this reaction at "low" temperatures, e.g., 298 K,

$$\Delta G° = +178.3 \text{ kJ} - 47.8 \text{ kJ} = +130.5 \text{ kJ}$$

the $\Delta H°$ term predominates, $\Delta G°$ is positive, and the reaction is nonspontaneous at 1 atm. Conversely, at "high" temperatures, e.g., 2000 K,

$$\Delta G° = +178.3 \text{ kJ} - 321.0 \text{ kJ} = -142.7 \text{ kJ}$$

the $T\Delta S°$ term predominates, $\Delta G°$ is negative, and the reaction is spontaneous at 1 atm.

CaO is made commercially by heating $CaCO_3$

Example 16.7 At what temperature does $\Delta G°$ become zero for the reaction

$$CaCO_3(s) \longrightarrow CaO(s) + CO_2(g)?$$

Strategy This is another application of the Gibbs-Helmholtz equation. Setting $\Delta G° = 0$ gives $\Delta H° = T\Delta S°$. Solving for T gives $T = \Delta H°/\Delta S°$. Knowing $\Delta H°$ and $\Delta S°$, T is readily calculated.

Solution From the preceding discussion,

$$\Delta H° = +178.3 \text{ kJ} \qquad \Delta S° = 0.1605 \text{ kJ/K}$$

Hence

$$T = \frac{\Delta H°}{\Delta S°} = \frac{178.3 \text{ kJ}}{0.1605 \text{ kJ/K}} = 1110 \text{ K}$$

At this temperature, the reaction is at equilibrium at 1 atm pressure. If some $CaCO_3$ is placed in a container and heated to 1110 K, the pressure of CO_2 developed is 1 atm.

The development presented in Example 16.7 is an important one from a practical standpoint. It tells us the temperature at which the direction of spontaneity changes. In this case, one can deduce that calcium carbonate must be heated to about 1110 K ($\approx$840°C) to decompose it to calcium oxide and carbon dioxide. In another case, for the reaction

$$Fe_2O_3(s) + 3H_2(g) \longrightarrow 2Fe(s) + 3H_2O(g)$$

it turns out that $\Delta G°$, which is positive at low temperatures, becomes zero at about 400°C. This means that to reduce hematite ore, Fe_2O_3, to iron in a stream of hydrogen, the system must be heated to 400°C.

Pressure, Concentration

All of the free energy calculations to this point have involved the standard free energy change, $\Delta G°$. It is possible, however, to write a general relation for the free energy change, ΔG, valid under any conditions:

We'll use this equation but we're not in a position to derive it

$$\Delta G = \Delta G° + RT \ln Q$$

The quantity Q which appears in this equation is the "reaction quotient" referred to in Chapter 12. It has the same mathematical form as the equilibrium constant, K; the difference is that the terms that appear in Q are arbitrary, instantaneous pressures or concentrations rather than equilibrium values.

To use this general expression for ΔG, you need to express T in kelvins and R in kilojoules per kelvin (8.31×10^{-3} kJ/K). So far as Q is concerned, the general rules for expressing K are followed:

— gases enter as their partial pressures in atmospheres
— species in aqueous solution enter as their molar concentrations
— pure liquids or solids do not appear; neither does the solvent in a dilute solution

---

Example 16.8 Consider the reaction of zinc with strong acid:

$$Zn(s) + 2H^+(aq) \longrightarrow Zn^{2+}(aq) + H_2(g)$$

(a) Write the expression for Q.
(b) Calculate $\Delta G°$ at 25°C.
(c) Find ΔG at 25°C when $P_{H_2} = 750$ mm Hg, $[Zn^{2+}] = 0.10\ M$, $[H^+] = 1.0 \times 10^{-4}\ M$.

Strategy In (a), follow the rules repeated above for Q. In (b), calculate $\Delta G°$ using the Gibbs-Helmholtz equation, as in Examples 16.4–16.7. Finally (c), determine ΔG using the relation $\Delta G = \Delta G° + RT \ln Q$. Note that the pressure of H_2 must be expressed in atmospheres.

Solution

(a) $Q = \dfrac{[Zn^{2+}] \times P_{H_2}}{[H^+]^2}$

(b) $\Delta H° = \Delta H_f° \, Zn^{2+} = -153.9 \text{ kJ}$

$\Delta S° = S° \, Zn^{2+} + S° \, H_2 - S° \, Zn = -23.1 \text{ J/K} = -0.0231 \text{ kJ/K}$

$\Delta G° = \Delta H° - T\Delta S° = -153.9 \text{ kJ} + 298 \text{ K} \times (0.0231 \text{ kJ/K}) = -147.0 \text{ kJ}$

(c) $\Delta G = \Delta G° + RT \ln Q$

$$= -147.0 \text{ kJ} + (8.31 \times 10^{-3} \text{ kJ/K})\left[\ln \frac{(0.10)(750/760)}{(1.0 \times 10^{-4})^2}\right]$$

$$= -147.0 \text{ kJ} + 39.9 \text{ kJ} = -107.1 \text{ kJ}$$

Figure 16.7
A saturated solution of $SrCrO_4$ contains 0.006 mol/L of Sr^{2+} and CrO_4^{2-}. For the reaction at 25°C: $SrCrO_4(s) \rightleftharpoons Sr^{2+}(aq, 0.006\ m) + CrO_4^{2-}(aq, 0.006\ m)$, $\Delta G = 0$. (Charles D. Winters)

As Example 16.8 implies, changes in pressure and/or concentration can have a considerable effect upon ΔG. Sometimes, when $\Delta G°$ is close to zero, changing the pressure from 1 atm to some other value can change the direction of spontaneity. Consider, for example, the reaction at 300°C:

$$NH_4Cl(s) \longrightarrow NH_3(g) + HCl(g) \qquad \Delta G° = +13.0 \text{ kJ}$$

The positive sign of $\Delta G°$ implies that ammonium chloride will not decompose at 300°C to give NH_3 and HCl, both at 1 atm pressure. However, notice what happens if $P_{NH_3} = P_{HCl} = 0.10$ atm, in which case $Q = (0.10)^2$

If $Q < 1$, $\Delta G < \Delta G°$. Explain why

$$\Delta G = \Delta G° + RT \ln(0.10)^2$$

$$= +13.0 \text{ kJ} + (8.31 \times 10^{-3})(573) \text{ kJ} \times \ln 0.010$$

$$= +13.0 \text{ kJ} - 21.9 \text{ kJ} = -8.9 \text{ kJ}$$

This means that NH_3 and HCl *can* be formed, each at 0.10 atm pressure, by heating ammonium chloride to 300°C.

Another example of how a change in concentration can change the direction of spontaneity involves the reaction

$$SrCrO_4(s) \longrightarrow Sr^{2+}(aq) + CrO_4^{2-}(aq) \qquad \Delta G° = +25.3 \text{ kJ at } 25°C$$

Since $\Delta G°$ is a positive quantity, $SrCrO_4$ does not dissolve spontaneously to form a 1 M solution at 25°C (Fig. 16.7). Suppose, however, the concentrations of Sr^{2+} and CrO_4^{2-} are reduced to 0.0010 M:

$$\Delta G = \Delta G° + RT \ln [Sr^{2+}] \times [CrO_4^{2-}]$$

$$= +23.6 \text{ kJ} + (8.31 \times 10^{-3})(298) \text{ kJ} \times [\ln (1.0 \times 10^{-3})^2]$$

$$= +25.3 \text{ kJ} - 34.2 \text{ kJ} = -8.9 \text{ kJ}$$

Strontium chromate should, and does, dissolve spontaneously to form a 0.0010 M water solution. As you might expect, its solubility at 25°C lies between 1 M and 0.0010 M, at about 0.006 M.

$$\Delta G \text{ at } 0.006\ M = +25.3 \text{ kJ} + RT \ln (0.006)^2 \approx 0$$

16.6 THE FREE ENERGY CHANGE AND THE EQUILIBRIUM CONSTANT

Table 16.2 Values of K Corresponding to $\Delta G°$ Values at 25°C

$\Delta G°$ (kJ)	K
−200	1×10^{35}
−100	3×10^{17}
−50	6×10^{8}
−25	2×10^{4}
0	1
+25	4×10^{-5}
+50	2×10^{-9}
+100	3×10^{-18}
+200	1×10^{-35}

Throughout this chapter, we have stressed the relation between the free energy change and reaction spontaneity. For a reaction to be spontaneous at standard conditions (1 atm, 1 M), $\Delta G°$ must be negative. Another indicator of reaction spontaneity is the equilibrium constant, K; if K is greater than 1, the reaction is spontaneous at standard conditions. As you might suppose, the two quantities $\Delta G°$ and K are related. The nature of that relationship is readily deduced by starting with the general equation

$$\Delta G = \Delta G° + RT \ln Q$$

Suppose now that reaction takes place until equilibrium is established, at which point $Q = K$ and $\Delta G = 0$.

$$0 = \Delta G° + RT \ln K$$

or

$$\Delta G° = -RT \ln K$$

The equation just written is generally applicable to any system. The equilibrium constant may be the K referred to in our discussion of gaseous equilibrium (Chap. 12), or any of the solution equilibrium constants (K_w, K_a, K_b, K_{sp}, . . .) discussed in subsequent chapters. Notice that $\Delta G°$ is the *standard* free energy change (gases at *1 atm*, species in solution at *1 M*). That is why, in the expression for K, gases enter as their partial pressures in *atmospheres* and ions or molecules in solution as their *molarities*.

Example 16.9 Using $\Delta G_f°$ tables in Appendix 1, calculate the solubility product constant, K_{sp}, for AgBr at 25°C.

Strategy Write out the chemical equation for the solution reaction $AgBr(s) \rightleftharpoons Ag^+(aq) + Br^-(aq)$. Now apply the equation

$$\Delta G° = \sum \Delta G°_{f\,products} - \sum \Delta G°_{f\,reactants}$$

to calculate $\Delta G°$. Then, use the equation $\Delta G° = -RT \ln K_{sp}$ to find K_{sp}.

To find K_{sp} at 100°C, you'd have to find $\Delta H°$ and $\Delta S°$

Solution

(1) $\Delta G° = \Delta G_f°\, Ag^+ + \Delta G_f°\, Br^- - \Delta G_f°\, AgBr$

$= 77.1\text{ kJ} - 104.0\text{ kJ} + 96.9\text{ kJ} = 70.0\text{ kJ}$

(2) $\ln K_{sp} = \dfrac{-\Delta G°}{RT} = \dfrac{-70.0\text{ kJ}}{(8.31 \times 10^{-3}\text{ kJ/K})(298\text{ K})} = -28.3$

$K_{sp} = 5 \times 10^{-13}$

This is the value listed in Chapter 14 for K_{sp} of AgBr.

The relation $\Delta G° = -RT \ln K$ (and Table 16.2) allow us to give meaning to the magnitude of the free energy change. If $\Delta G°$ is a large negative number, K will be much greater than 1. This means that the forward reaction will go virtually to completion. Conversely, if $\Delta G°$ is a large positive number, K will be much less than 1, and the reverse reaction will go nearly to completion. Only if $\Delta G°$ is relatively small, perhaps between +50 and −50 kJ, will the reaction yield an equilibrium mixture containing significant amounts of both products and reactants.

16.7 ADDITIVITY OF FREE ENERGY CHANGES; COUPLED REACTIONS

Free energy changes for reactions, like enthalpy or entropy changes, are additive. That is,

if Reaction 3 = Reaction 1 + Reaction 2

then $\Delta G_3 = \Delta G_1 + \Delta G_2$

This relation can be regarded as the free energy equivalent of Hess' law (Chap. 8). To illustrate its application, consider the synthesis of $CuCl_2$ from the elements:

$$\begin{array}{rcll} Cu(s) + \frac{1}{2}Cl_2(g) & \longrightarrow & CuCl(s) & \Delta G° \text{ at } 25°C = -119.9 \text{ kJ} \\ CuCl(s) + \frac{1}{2}Cl_2(g) & \longrightarrow & CuCl_2(s) & \Delta G° \text{ at } 25°C = -55.8 \text{ kJ} \\ \hline Cu(s) + Cl_2(g) & \longrightarrow & CuCl_2(s) & \end{array}$$

For the overall reaction, $\Delta G°$ at 25°C = −119.9 kJ − 55.8 kJ = −175.7 kJ.

Since free energy changes are additive, it is often possible to bring about a nonspontaneous reaction by coupling it with a reaction for which $\Delta G°$ is a large negative number. As an example, consider the preparation of iron metal from hematite ore. The reaction

$$Fe_2O_3(s) \longrightarrow 2Fe(s) + \tfrac{3}{2}O_2(g) \qquad \Delta G° \text{ at } 25°C = +742.2 \text{ kJ}$$

is clearly nonspontaneous; even at temperatures as high as 2000°C, $\Delta G°$ is a positive quantity. Suppose, though, that this reaction is "coupled" with the spontaneous oxidation of carbon monoxide:

$$CO(g) + \tfrac{1}{2}O_2(g) \longrightarrow CO_2(g) \qquad \Delta G° \text{ at } 25°C = -257.1 \text{ kJ}$$

The overall reaction is spontaneous:

$$\begin{array}{rcl} Fe_2O_3(s) & \longrightarrow & 2Fe(s) + \frac{3}{2}O_2(g) \\ 3CO(g) + \frac{3}{2}O_2(g) & \longrightarrow & 3CO_2(g) \\ \hline Fe_2O_3(s) + 3CO(g) & \longrightarrow & 2Fe(s) + 3CO_2(g) \end{array}$$

For the coupled reaction at 25°C:

$$\Delta G° = +742.2 \text{ kJ} + 3(-257.1 \text{ kJ}) = -29.1 \text{ kJ}$$

The negative sign of $\Delta G°$ implies that although Fe_2O_3 does not spontaneously decompose, it can be converted to iron by reaction with carbon monoxide. This is in fact the reaction used in the blast furnace when iron ore consisting mainly of Fe_2O_3 is reduced to iron (Chap. 19).

Many industrial processes are based on coupled reactions

Coupled reactions are common in human metabolism. Spontaneous processes, such as the oxidation of glucose,

$$C_6H_{12}O_6(aq) + 6\,O_2(g) \longrightarrow 6CO_2(g) + 6H_2O \qquad \Delta G° = -2870 \text{ kJ at } 25°C$$

ordinarily do not serve directly as a source of energy. Instead these reactions are used to bring about a nonspontaneous reaction:

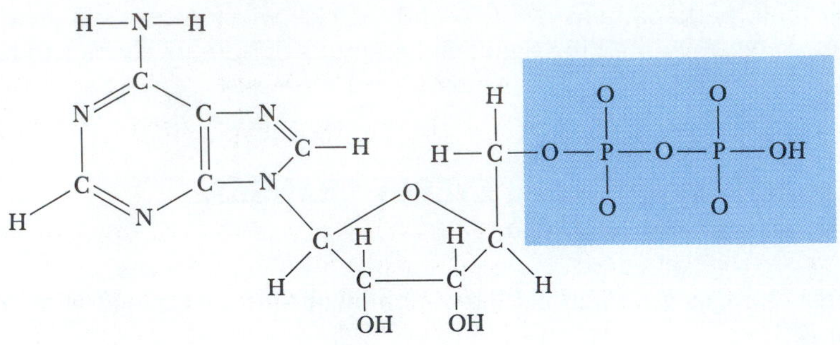

Adenosine diphosphate (ADP)

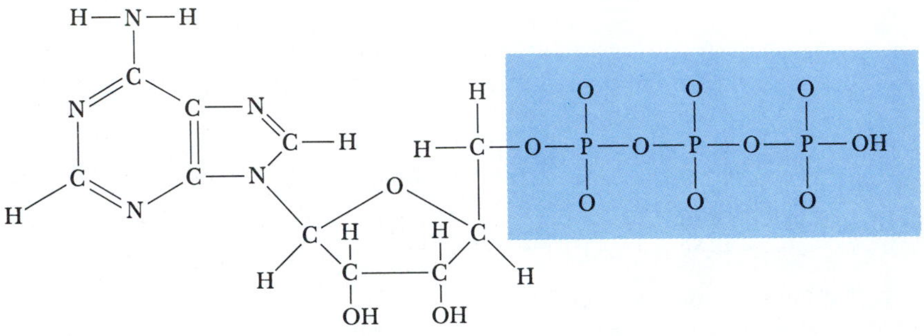

Adenosine triphosphate (ATP)

Figure 16.8
Structures of ADP and ATP.

$$ADP(aq) + HPO_4^{2-}(aq) + 2H^+(aq) \longrightarrow ATP(aq) + H_2O$$

$$\Delta G° = +31 \text{ kJ at } 25°C$$

ADP (adenosine diphosphate) and ATP (adenosine triphosphate) are complex organic molecules (Fig. 16.8) that, in essence, differ only by the presence of an extra phosphate group in ATP. In the coupled reaction with glucose, about 38 mol of ATP are synthesized for every mole of glucose consumed. This gives an overall free energy change for the coupled reaction of

$$-2870 \text{ kJ} + 38(+31 \text{ kJ}) \approx -1700 \text{ kJ}$$

This is the way your body gets energy in a hurry

In a very real sense, your body "stores" energy available from the metabolism of foods in the form of ATP. This molecule in turn supplies the energy required for all sorts of biochemical reactions taking place in the body. It does this by reverting to ADP, i.e., by reversing the above reaction. The amount of ATP consumed is amazingly large; a competitive sprinter may hydrolyze as much as 500 g (about 1 lb) of ATP per minute.

Example 16.10 The lactic acid ($C_3H_6O_3(aq)$, $\Delta G_f° = -559$ kJ) produced in muscle cells after vigorous exercise eventually is absorbed into the bloodstream, where it is metabolized back to glucose ($\Delta G_f° = -919$ kJ) in the liver. The reaction is

$$2C_3H_6O_3(aq) \longrightarrow C_6H_{12}O_6(aq)$$

(a) Calculate $\Delta G°$ for this reaction, using free energies of formation.
(b) If the hydrolysis of ATP to ADP is coupled with this reaction, how many moles of ATP must react to make the process spontaneous?

Strategy Use the relation

$$\Delta G° = \sum \Delta G^{\circ}_{f\,products} - \sum \Delta G^{\circ}_{f\,reactants}$$

to find $\Delta G°$. To solve part (b), note that 31 kJ of energy is produced per mole of ATP consumed.

Solution

(a) $\Delta G° = \Delta G^{\circ}_{f}\ C_6H_{12}O_6(aq) - 2\Delta G^{\circ}_{f}\ C_3H_6O_3(aq)$

$= -919\text{ kJ} + 2(559\text{ kJ}) = \boxed{+199\text{ kJ}}$

(b) $199\text{ kJ} \times \dfrac{1\text{ mol ATP}}{31\text{ kJ}} \approx \boxed{7\text{ mol ATP}}$

PERSPECTIVE • Solar Energy

All of our energy, "free" or otherwise, comes ultimately from the sun. The solar energy we use has been stored by natural processes—for millions of years in the case of petroleum, a century or less for wood, a few days for hydroelectric power. The direct, immediate utilization of the sun's energy comprises less than 0.01% of our energy budget.

The simplest direct application of solar energy is in heating water. Solar water heaters are common in Israel, Japan, and Australia. In the United States, a lot of swimming pools and a few houses are heated by solar energy. For home heating, solar collectors are mounted on a roof with a southern exposure. The base of the collector is painted black to absorb as much sunlight as possible. Water, circulating through the collector, is heated to about 65°C. The hot water enters a storage tank, from which heat is transferred throughout the house.

A major problem with solar heating of homes is that the supply of heat is cut off in periods of cold, cloudy weather and during long winter nights. Hence, an auxiliary heating system of the ordinary type (oil, gas, electric) must be available. This makes the economics of solar heating systems unattractive, at least so far as initial cost is concerned.

If solar energy is to make a major contribution to our energy needs, it will almost certainly be via conversion to electrical energy. Solar cells designed to accomplish this have been available for at least 25 years. Typically, they are sold as discs, perhaps 2 cm in diameter and 0.5 mm thick, cut from a single crystal of ultra-pure silicon. The crystal is impregnated with a trace of an electron-rich impurity such as As or P (Fig. 16.A). At the surface of the crystal, there is a thin, transparent layer of an electron-poor element, B or Al. This creates an "n-p junction," which generates a flow of electrons upon exposure to sunlight.

The major drawback to this type of solar cell is economic; electrical energy produced this way costs five to ten times as much as that from conventional sources. Two approaches are available to solve this problem:

1. *Reduce the cost of the cell,* perhaps by using amorphous silicon, in which Si atoms are not arranged in a fixed geometric pattern. If you have a solar-powered calculator, the light-absorbing material is probably a very thin

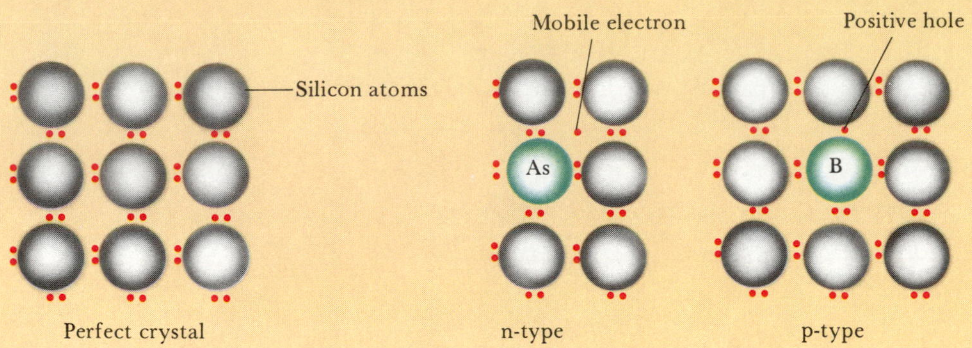

Figure 16.A
Semiconductors derived from silicon. In n-type semiconductors, the impurity atoms furnish mobile electrons to the crystal. In p-type semiconductors there is a deficiency of electrons, since the impurity atoms have three rather than four valence electrons.

film of amorphous silicon. Unfortunately, the efficiency of conversion of solar to electrical energy by amorphous silicon is only 5–10%, considerably less than that for crystalline silicon.

2. *Use a more efficient semiconductor.* Gallium arsenide, is a leading possibility here. Cells made from GaAs have shown efficiencies as high as 25%. However, (you guessed it) GaAs semiconductors are considerably more expensive to make than those based upon silicon.

CHAPTER HIGHLIGHTS

KEY CONCEPTS

1. *Deduce the sign of ΔS based on randomness*
   (Examples 16.1, 16.2; Problems 5–10)
2. *Calculate $\Delta S°$ from tables of standard entropies*
   (Example 16.3; Problems 11–16)
3. *Calculate $\Delta G°$ from*
   —the Gibbs-Helmholtz equation
   (Examples 16.4, 16.6; Problems 17–20, 27–30)
   —free energies of formation
   (Example 16.5; Problems 21–26)
4. *Calculate the temperature at which $\Delta G° = 0$*
   (Example 16.7; Problems 33–38, 41–44, 77, 78)
5. *Calculate ΔG from $\Delta G°$ and Q*
   (Example 16.8; Problems 45–50)
6. *Relate $\Delta G°$ to K*
   (Example 16.9; Problems 55–60, 69, 71, 72, 74)
7. *Calculate $\Delta G°$ for coupled reactions*
   (Example 16.10; Problems 51–54, 70)

KEY EQUATIONS

Entropy change $\Delta S^\circ = \sum S^\circ_{products} - \sum S^\circ_{reactants}$

Free energy change $\Delta G = \Delta H - T\Delta S$

$\Delta G^\circ = \sum \Delta G^\circ_{f\,products} - \sum \Delta G^\circ_{f\,reactants}$

$\Delta G^\circ = -RT \ln K$

$\Delta G = \Delta G^\circ + RT \ln Q$

KEY TERMS

atmosphere	exothermic	molarity
endothermic	free energy	Q
enthalpy	—of formation	ΔS°
—of formation	ΔG°	spontaneous
entropy	ΔH°	system

SUMMARY PROBLEM

Consider acetic acid, CH_3COOH, the main component of vinegar. Its name comes from the Latin name for vinegar, *acetum*. When wine is exposed to air, bacterial oxidation converts alcohol to acetic acid:

$$C_2H_5OH(aq) + O_2(g) \longrightarrow CH_3COOH(aq) + H_2O(l)$$

a. Calculate ΔH° and ΔS° for this process. Use the following data in combination with tables of entropies and heats of formation: ΔH°_f $CH_3COOH(aq) = -485.8$ kJ/mol, ΔH°_f $C_2H_5OH(aq) = -288.3$ kJ/mol, S° $CH_3COOH(aq) = 178.7$ J/mol · K, S° $C_2H_5OH(aq) = +148.5$ J/mol · K.
b. Is the reaction spontaneous at 25°C? At 100°C?
c. The heat of fusion of acetic acid is 11.7 kJ/mol; its freezing point at 1 atm, where $\Delta G^\circ = 0$, is 16.6°C. Calculate ΔS° for the reaction $CH_3COOH(l) \rightarrow CH_3COOH(s)$.
d. What is the standard molar entropy of $CH_3COOH(s)$, taking S° $CH_3COOH(l) = 159.8$ J/mol · K?
e. Calculate ΔG at 25°C for the reaction in (a) when $[CH_3COOH] = 0.100\,M$, $P_{O_2} = 0.50$ atm, and $[C_2H_5OH] = 0.0250\,M$.
f. Calculate ΔG° for the ionization of acetic acid at 25°C ($K_a = 1.8 \times 10^{-5}$).

Answers

a. −483.3 kJ, −104.9 J/K **b.** yes, yes **c.** −40.4 J/K
d. 119.4 J/mol · K **e.** −446.9 kJ **f.** 27 kJ

QUESTIONS & PROBLEMS

Spontaneity

1. Which of the following processes are spontaneous?
  a. glass shattering when it is dropped
  b. outlining your chemistry notes
  c. perfume aroma from an open bottle filling the air

2. Which of the following processes are spontaneous?
  a. building a house of cards
  b. sugar dissolving in water
  c. raking leaves into a pile

3. Based on your experience, predict whether the following reactions are spontaneous.
  a. $C_2H_2(g) + \frac{5}{2}O_2(g) \longrightarrow 2CO_2(g) + H_2O(l)$
  b. $NaCl(s) + H_2O(l) \longrightarrow NaOH(s) + HCl(g)$
  c. $C(s) + O_2(g) \longrightarrow CO_2(g)$
  d. $C_2H_5OH(l) \longrightarrow C_2H_5OH(s)$ at 25°C

4. Follow the directions for Question 3 for
  a. $Na^+(aq) + OH^-(aq) + \frac{1}{2}H_2(g) \longrightarrow Na(s) + H_2O$.
  b. $H_2O(s) \longrightarrow H_2O(l)$ at −10°C.
  c. $Mg(s) + \frac{1}{2}O_2(g) \longrightarrow MgO(s)$.
  d. $Zn(s) + 2H^+(aq) \longrightarrow Zn^{2+}(aq) + H_2(g)$.

Entropy; $\Delta S°$

5. Predict the sign of ΔS for
  a. the freezing of water.
  b. evaporation of a seawater sample to dryness.
  c. ammonia vapor condensing.
  d. weeding a garden.

6. Predict the sign of ΔS for
  a. a candle burning.
  b. butter melting.
  c. separating air into its components.
  d. tea dissolving in water.

7. Predict the sign of $\Delta S°$ for each of the following reactions.
  a. $N_2(g) + 3H_2(g) \longrightarrow 2NH_3(g)$
  b. $C(s) + H_2O(g) \longrightarrow CO(g) + H_2(g)$
  c. $2H_2(g) + O_2(g) \longrightarrow 2H_2O(l)$
  d. $S(s) + O_2(g) \longrightarrow SO_2(l)$

8. Predict the sign of $\Delta S°$ for each of the following four reactions.
  a. $H_2(g) + O_2(g) \longrightarrow H_2O_2(l)$
  b. $CO(g) + 3H_2(g) \longrightarrow CH_4(g) + H_2O(g)$
  c. $NH_3(g) + HCl(g) \longrightarrow NH_4Cl(s)$
  d. $K(s) + O_2(g) \longrightarrow KO_2(s)$

9. Predict the sign of $\Delta S°$ for each of the following reactions.
  a. $H_2(g) + Cu^{2+}(aq) \longrightarrow 2H^+(aq) + Cu(s)$
  b. $2Cl(g) \longrightarrow Cl_2(g)$
  c. $CaCl_2(s) + 6H_2O(g) \longrightarrow CaCl_2 \cdot 6H_2O(s)$

10. Predict the sign of $\Delta S°$ for each of the following reactions.
  a. $CuSO_4 \cdot 5H_2O(s) \longrightarrow CuSO_4(s) + 5H_2O(g)$
  b. $6H^+(aq) + 2Al(s) \longrightarrow 2Al^{3+}(aq) + 3H_2(g)$
  c. $2SO_3(g) \longrightarrow 2SO_2(g) + O_2(g)$

11. Use Table 16.1 to calculate $\Delta S°$ for each of the following reactions.
  a. $4NH_3(g) + 7\,O_2(g) \longrightarrow 4NO_2(g) + 6H_2O(g)$
  b. $2H_2O_2(l) + N_2H_4(l) \longrightarrow N_2(g) + 4H_2O(g)$
  c. $C(s) + O_2(g) \longrightarrow CO_2(g)$
  d. $CH_4(g) + 3Cl_2(g) \longrightarrow CHCl_3(l) + 3HCl(g)$

12. Use Table 16.1 to calculate $\Delta S°$ for each of the following reactions.
  a. $CO(g) + 2H_2(g) \longrightarrow CH_3OH(l)$
  b. $N_2(g) + O_2(g) \longrightarrow 2NO(g)$
  c. $BaCO_3(s) \longrightarrow BaO(s) + CO_2(g)$
  d. $2NaCl(s) + F_2(g) \longrightarrow 2NaF(s) + Cl_2(g)$

13. Use Table 16.1 to calculate $\Delta S°$ for each of the following reactions.
  a. $Zn(s) + 2H^+(aq) \longrightarrow Zn^{2+}(aq) + H_2(g)$
  b. $H^+(aq) + OH^-(aq) \longrightarrow H_2O(l)$
  c. $NH_3(g) + H_2O(l) \longrightarrow NH_4^+(aq) + OH^-(aq)$

14. Use Table 16.1 to calculate $\Delta S°$ for each of the following reactions.
  a. $2Na(s) + 2H_2O(l) \longrightarrow 2Na^+(aq) + 2\,OH^-(aq) + H_2(g)$
  b. $Zn(s) + 2Ag^+(aq) \longrightarrow 2Ag(s) + Zn^{2+}(aq)$
  c. $2NO_3^-(aq) + 8H^+(aq) + 3Cu(s) \longrightarrow 3Cu^{2+}(aq) + 2NO(g) + 4H_2O(l)$

15. Use Table 16.1 to calculate $\Delta S°$ for each of the following reactions.
  a. $2H_2S(g) + 3\,O_2(g) \longrightarrow 2H_2O(g) + 2SO_2(g)$
  b. $Ag(s) + 2H^+(g) + NO_3^-(aq) \longrightarrow Ag^+(aq) + H_2O(l) + NO_2(g)$

16. Use Table 16.1 to calculate $\Delta S°$ for each of the following reactions.
  a. $2HNO_3(l) + 3H_2S(g) \longrightarrow 4H_2O(l) + 2NO(g) + 3S(s)$
  b. $PCl_5(g) + 4H_2O(l) \longrightarrow 6H^+(aq) + 5Cl^-(aq) + H_2PO_4^-(aq)$
  c. $MnO_4^-(aq) + 3Fe^{2+}(aq) + 4H^+(aq) \longrightarrow 3Fe^{3+}(aq) + MnO_2(s) + 2H_2O(l)$

$\Delta G°$ and the Gibbs-Helmholtz Equation

17. Calculate $\Delta G°$ at 25°C for reactions for which
  a. $\Delta H° = +210$ kJ; $\Delta S° = +32.5$ J/K.
  b. $\Delta H° = +638$ kJ; $\Delta S° = -215.2$ J/K.
  c. $\Delta H° = +7.34$ kJ; $\Delta S° = +0.337$ kJ/K.

18. Calculate $\Delta G°$ at 25°C for reactions for which
  a. $\Delta H° = +79.6$ kJ; $\Delta S° = +433.1$ J/K.
  b. $\Delta H° = -837.4$ kJ; $\Delta S° = +173.8$ J/K.
  c. $\Delta H° = -34.9$ kJ; $\Delta S° = +0.039$ kJ/K.

19. Calculate $\Delta G°$ at 400°C for each of the reactions in Problem 11. State whether the reactions are spontaneous or not.

20. Calculate $\Delta G°$ at 400 K for each of the reactions in Problem 12. State whether the reactions are spontaneous or not.

21. Using values of $\Delta G_f°$ given in Appendix 1, calculate $\Delta G°$ at 25°C for each of the reactions in Problem 11.

22. Follow the directions of Problem 21 for each of the reactions in Problem 12.

23. Calculate $\Delta G_f°$ at 25°C, using standard entropies and heats of formation, for
  a. calcium carbonate
  b. manganese(II) oxide
  c. phosphorus pentachloride

Which of these compounds are stable with respect to the elements?

24. Follow the directions of Problem 23 for the following compounds.
  a. tin(IV) oxide (s)
  b. silver nitrate (s)
  c. hydrogen peroxide (l)

25. A student warned his friends not to swim in a river close to an electric plant. He claimed that the ozone produced by the plant turned the river water to hydrogen peroxide, which would bleach hair. The reaction is

$$O_3(g) + H_2O(l) \longrightarrow H_2O_2(aq) + O_2(g)$$

Show by calculation whether his claim is plausible, assuming the river water is at 25°C, and all species are at standard concentrations. Take ΔG_f° $O_3(g)$ at 25°C to be +163.2 kJ/mol and ΔG_f° $H_2O_2(aq)$ = −134.0 kJ/mol.

26. It has been proposed that wood alcohol, CH_3OH, a relatively inexpensive fuel to produce, be decomposed to produce methane. Methane is a natural gas commonly used for heating homes. Is the decomposition of wood alcohol to methane and oxygen thermodynamically feasible at 25°C and 1 atm?

27. Sodium carbonate, also called "washing soda," can be made by heating sodium hydrogen carbonate:

$$2NaHCO_3(s) \longrightarrow Na_2CO_3(s) + CO_2(g) + H_2O(g)$$

$$\Delta H^\circ = +135.6 \text{ kJ}; \ \Delta G^\circ = +34.6 \text{ kJ at } 25°C$$

a. Calculate ΔS° for this reaction. Is the sign reasonable?
b. Calculate ΔG° at 0 K; at 1000 K.

28. Oxygen can be made in the laboratory by reacting sodium peroxide and water:

$$2Na_2O_2(s) + 2H_2O(l) \longrightarrow 4NaOH(s) + O_2(g)$$

$$\Delta H^\circ = -109.0 \text{ kJ}; \ \Delta G^\circ = -148.4 \text{ kJ at } 25°C$$

a. Calculate ΔS° for this reaction. Is this sign reasonable?
b. Calculate S° for Na_2O_2.
c. Calculate ΔH_f° for Na_2O_2.

29. Consider the reaction

$$2CuCl(s) + 2\,OH^-(aq) \longrightarrow Cu_2O(s) + 2Cl^-(aq) + H_2O(l)$$

$$\Delta H^\circ = -54.3 \text{ kJ}; \ \Delta S^\circ = +125.1 \text{ J/K}$$

a. Calculate ΔG° for this reaction at 25°C.
b. Determine ΔH_f° for $CuCl(s)$.
c. Calculate S° for $CuCl(s)$.

30. Phosgene, $COCl_2$, can be formed by the reaction of chloroform, $CHCl_3(l)$, with oxygen:

$$2CHCl_3(l) + O_2(g) \longrightarrow 2COCl_2(g) + 2HCl(g)$$

$$\Delta H^\circ = -353.2 \text{ kJ}; \ \Delta G^\circ = -452.4 \text{ kJ at } 25°C$$

a. Calculate ΔS° for the reaction. Is the sign reasonable?
b. Calculate S° for phosgene.
c. Calculate ΔH_f° for phosgene.

Temperature Dependence of Spontaneity

31. Discuss the effect of temperature change upon the spontaneity of the following reactions at 1 atm.
a. $Al_2O_3(s) + 2Fe(s) \longrightarrow 2Al(s) + Fe_2O_3(s)$
$\Delta H^\circ = 851.5$ kJ; $\Delta S^\circ = 38.5$ J/K
b. $N_2H_4(l) \longrightarrow N_2(g) + 2H_2(g)$
$\Delta H^\circ = -50.6$ kJ; $\Delta S^\circ = 0.3315$ kJ/K
c. $SO_3(g) \longrightarrow SO_2(g) + \frac{1}{2}O_2(g)$
$\Delta H^\circ = 98.9$ kJ; $\Delta S^\circ = 0.0939$ kJ/K

32. Discuss the effect of temperature change upon the spontaneity of the following reactions at 1 atm.
a. $2PbO(s) + 2SO_2(g) \longrightarrow 2PbS(s) + 3\,O_2(g)$
$\Delta H^\circ = +830.8$ kJ; $\Delta S^\circ = 168$ J/K
b. $2As(s) + 3F_2(g) \longrightarrow 2AsF_3(l)$
$\Delta H^\circ = -1643$ kJ; $\Delta S^\circ = -0.316$ kJ/K
c. $CO(g) \longrightarrow C(s) + \frac{1}{2}O_2(g)$
$\Delta H^\circ = 110.5$ kJ; $\Delta S^\circ = -89.4$ J/K

33. At what temperature does ΔG° become zero for each of the reactions in Problem 31? Explain the significance of your answers.

34. At what temperature does ΔG° become zero for each of the reactions in Problem 32? Explain the significance of your answers.

35. For the reaction

$$NH_4Cl(s) \longrightarrow NH_3(g) + HCl(g)$$

Calculate the temperature at which $\Delta G^\circ = 0$.

36. For the reaction

$$SO_2(g) + 2H_2S(g) \longrightarrow 3S(s) + 2H_2O(g)$$

Calculate the temperature at which $\Delta G^\circ = 0$.

37. For the decomposition of Ag_2O,

$$2Ag_2O(s) \longrightarrow 4Ag(s) + O_2(g)$$

a. Using Tables 8.3 and 16.1, obtain an expression for ΔG° as a function of temperature. Use it to prepare a table of ΔG° values at 100 K intervals between 100 and 500 K.
b. Calculate the temperature at which ΔG° becomes zero.

38. Earlier civilizations smelted iron from ore by heating it with charcoal from a wood fire:

$$2Fe_2O_3(s) + 3C(s) \longrightarrow 4Fe(s) + 3CO_2(g)$$

a. Obtain an expression for ΔG° as a function of temperature. Prepare a table of ΔG° values at 200 K intervals between 200 and 1000 K.
b. Calculate the lowest temperature at which the smelting could be carried out.

39. Two possible ways of producing iron from iron ore are
a. $Fe_2O_3(s) + \frac{3}{2}C(s) \longrightarrow 2Fe(s) + \frac{3}{2}CO_2(g)$
b. $Fe_2O_3(s) + 3H_2(g) \longrightarrow 2Fe(s) + 3H_2O(g)$
Which of these reactions would proceed spontaneously at the lower temperature?

40. It is desired to produce tin from its ore, cassiterite, SnO_2, at as low a temperature as possible. The ore could be
a. decomposed by heating, producing tin and oxygen.
b. heated with hydrogen gas, producing tin and water vapor.
c. heated with carbon, producing tin and carbon dioxide.

Based solely on thermodynamic principles, which method would you recommend? Show calculations.

41. Red phosphorus is formed by heating white phosphorus. Calculate the temperature at which the two forms are at equilibrium, given

white P: $\Delta H_f^\circ = 0.00$ kJ/mol; $S^\circ = 41.09$ J/mol · K

red P: $\Delta H_f^\circ = -17.6$ kJ/mol; $S^\circ = 22.80$ J/mol · K

42. Tin organ pipes in unheated churches develop tin "disease," in which white tin is converted to gray tin. Given

white Sn: $\Delta H_f^\circ = 0.00$ kJ/mol; $S^\circ = 51.55$ J/mol · K

gray Sn: $\Delta H_f^\circ = -2.09$ kJ/mol; $S^\circ = 44.14$ J/mol · K

calculate the equilibrium temperature for the transition

$$Sn_{white}(s) \longrightarrow Sn_{gray}(s)$$

43. Given the following data for mercury

Hg(*l*): $S^\circ = 76.0$ J/mol · K

Hg(*g*): $S^\circ = 175.0$ J/mol · K; $\Delta H_f^\circ = 61.32$ kJ/mol

estimate the normal boiling point ($\Delta G^\circ = 0$) of Hg.

44. Diethyl ether, $(C_2H_5)_2O$, boils at 35.0°C and 1 atm. Its heat of vaporization is 26.0 kJ/mol. Calculate ΔS° for the reaction

$$(C_2H_5)_2O(l) \longrightarrow (C_2H_5)_2O(g)$$

Effect of Concentration/Pressure on Spontaneity

45. Is the reaction

$$HF(aq) \longrightarrow H^+(aq) + F^-(aq); \qquad \Delta G^\circ = +18.0 \text{ kJ}$$

spontaneous at 25°C when

a. $[H^+] = [F^-] = [HF] = 1.0\ M$?
b. $[H^+] = [F^-] = 1.0 \times 10^{-3}\ M$; $[HF] = 1.0\ M$?

46. Show by calculation, using ΔG_f° values, whether the reaction

$$CaCO_3(s) \longrightarrow Ca^{2+}(aq) + CO_3^{2-}(aq)$$

is spontaneous at 25°C when

a. $[Ca^{2+}] = [CO_3^{2-}] = 1.0\ M$.
b. $[Ca^{2+}] = [CO_3^{2-}] = 1.0 \times 10^{-10}\ M$.

47. For the reaction

$$2H_2O + 2Cl^-(aq) \longrightarrow H_2(g) + Cl_2(g) + 2\,OH^-(aq)$$

a. calculate ΔG° at 25°C.
b. calculate ΔG at 25°C when $P_{H_2} = P_{Cl_2} = 0.200$ atm, $[Cl^-] = 0.500\ M$, and the pH of the solution is 12.25.

48. For the reaction

$$O_2(g) + 4H^+(aq) + 4Fe^{2+}(aq) \longrightarrow 2H_2O(l) + 4Fe^{3+}(aq)$$

a. calculate ΔG° at 25°C.
b. calculate ΔG at 25°C when $[Fe^{2+}] = [Fe^{3+}] = 0.200\ M$, $P_{O_2} = 0.800$ atm, and the pH of the solution is 2.56.

49. Consider the reaction

$$AgCl(s) \longrightarrow Ag^+(aq) + Cl^-(aq)$$

a. Calculate ΔG° at 25°C.
b. What should the concentrations of Ag^+ and Cl^- be so that $\Delta G = -1.0$ kJ (i.e., just spontaneous)? Take $[Ag^+] = [Cl^-]$
c. The K_{sp} for AgCl is 1.8×10^{-10}. Is the answer to (b) reasonable? Explain.

50. Consider the reaction

$$2SO_2(g) + O_2(g) \longrightarrow 2SO_3(g)$$

a. Calculate ΔG° for the reaction at 25°C.
b. If the partial pressures of SO_2 and SO_3 are kept at 0.500 atm, what partial pressure should O_2 have so that the reaction just becomes nonspontaneous (i.e., $\Delta G = +1.0$ kJ)?

Additivity of ΔG; Coupled Reactions

51. Given that, at 25°C,

$$Fe(s) + Cl_2(g) \longrightarrow FeCl_2(s); \qquad \Delta G^\circ = -302.3 \text{ kJ}$$

$$Fe(s) + \tfrac{3}{2}Cl_2(g) \longrightarrow FeCl_3(s); \qquad \Delta G^\circ = -334.0 \text{ kJ}$$

Calculate ΔG° at 25°C for the reaction

$$2FeCl_2(s) + Cl_2(g) \longrightarrow 2FeCl_3(s)$$

52. Given that, at 25°C,

$$2Cu(s) + \tfrac{1}{2}O_2(g) \longrightarrow Cu_2O(s); \qquad \Delta G^\circ = -146.0 \text{ kJ}$$

$$Cu_2O(s) + \tfrac{1}{2}O_2(g) \longrightarrow 2CuO(s); \qquad \Delta G^\circ = -113.4 \text{ kJ}$$

Calculate ΔG_f° CuO(*s*) at 25°C.

53. How many moles of ATP must be converted to ADP by the reaction

$$ATP(aq) + H_2O \longrightarrow ADP(aq) + HPO_4^{2-}(aq) + 2H^+(aq) \qquad \Delta G^\circ = -31 \text{ kJ}$$

to bring about a nonspontaneous biochemical reaction in which $\Delta G^\circ = +372$ kJ?

54. Consider the following reactions at 25°C:

$$C_6H_{12}O_6(aq) + 6\,O_2(g) \longrightarrow 6CO_2(g) + 6H_2O \qquad \Delta G^\circ = -2870 \text{ kJ}$$

$$ADP(aq) + HPO_4^{2-}(aq) + 2H^+(aq) \longrightarrow ATP(aq) + H_2O \qquad \Delta G^\circ = 31 \text{ kJ}$$

Write a balanced equation for a coupled reaction between glucose, $C_6H_{12}O_6$, and ADP in which $\Delta G^\circ = -390$ kJ.

Free Energy and Equilibrium

55. Consider the reaction

$$H_2O(l) \rightleftharpoons H^+(aq) + OH^-(aq)$$

Using the appropriate tables, calculate

a. ΔG° at 25°C. **b.** K_w at 25°C.

56. Consider the reaction

$$CaCO_3(s) \rightleftharpoons CaO(s) + CO_2(g)$$

Using the appropriate tables, calculate

a. ΔG° at 500°C. **b.** K at 500°C.

57. Calculate $\Delta G°$ at 25°C for the following reactions.

a. $NH_4Cl(s) \rightleftharpoons NH_3(g) + HCl(g)$; $K = 1.1 \times 10^{-16}$
b. $2NO_2(g) \rightleftharpoons 2NO(g) + O_2(g)$; $K = 4.5 \times 10^{-13}$
c. $H_2(g) + I_2(g) \rightleftharpoons 2HI(g)$; $K = 6.2 \times 10^{2}$

58. For the reaction

$$Cl_2(g) \rightleftharpoons 2Cl(g)$$

K is 1.0×10^{-37} at 25°C and 4.2×10^{-5} at 1000°C. Calculate $\Delta G°$ at each of these temperatures.

59. Given that $\Delta H_f°$ for HF(aq) is −320.1 kJ/mol and $S°$ for HF(aq) is 88.7 J/mol · K, find K_a for HF at 25°C.

60. Using values of $\Delta G_f°$ in Appendix 1, calculate K_{sp} for barium sulfate at 25°C. Compare with the value given in Chapter 14, Table 14.4.

Solar Energy

61. Describe the principle upon which a solar cell works.

62. Referring to Figure 16.A, explain why arsenic (Group 15) gives an n-type semiconductor, while boron (Group 13) gives a p-type.

63. Amorphous silicon is cheaper than crystalline silicon. Why hasn't it completely replaced the crystalline form for semiconductor applications?

64. Why are solar units for heating water more efficient in the Mojave Desert than in Juneau, Alaska?

Unclassified

65. Which of the following quantities can be taken to be independent of temperature? independent of pressure?

a. ΔH for a reaction **b.** ΔS for a reaction
c. ΔG for a reaction **d.** S for a substance

66. Criticize each of the following statements.

a. An exothermic reaction is spontaneous.
b. When $\Delta G°$ is positive, the reaction cannot occur.
c. $\Delta S°$ is positive for a reaction in which there is an increase in the number of moles.
d. If $\Delta H°$ and $\Delta S°$ are both positive, $\Delta G°$ will be positive.

67. In your own words, explain why

a. $\Delta S°$ is negative for a reaction in which the number of moles of gas decreases.
b. we take $\Delta S°$ to be independent of T, even though entropy increases with T.
c. a solid has lower entropy than its corresponding liquid.

68. Fill in the blanks.

a. $\Delta H°$ and $\Delta G°$ become equal at ____________ K.
b. At equilibrium, ΔG is ____________.
c. In a spontaneous reaction, ΔG is ____________.
d. $S°$ for ice is ____________ than $S°$ for liquid water.

69. For the equilibrium system at 1000 K:

$$CO_2(g) + H_2(g) \rightleftharpoons CO(g) + H_2O(g)$$

the partial pressures of CO, H_2O, CO_2, and H_2 are 2.65 atm, 1.28 atm, 1.84 atm, and 2.81 atm, respectively. Calculate $\Delta G°$ at 1000 K.

70. Some bacteria use light energy to convert carbon dioxide and water to glucose and oxygen:

$$6CO_2(g) + 6H_2O(l) \longrightarrow C_6H_{12}O_6(aq) + 6\,O_2(g)$$
$$\Delta G° = +2870 \text{ kJ at } 25°C$$

Other bacteria, those that do not have light available to them, couple the reaction

$$H_2S(g) + \tfrac{1}{2}O_2(g) \longrightarrow H_2O(l) + S(s)$$

to the glucose synthesis above. Coupling the two reactions, the overall reaction is

$$24H_2S(g) + 6CO_2(g) + 6\,O_2(g) \longrightarrow C_6H_{12}O_6(s) + 18H_2O(l) + 24S(s)$$

Show that the overall reaction is spontaneous at 25°C.

71. How many grams of $PbCl_2$ will dissolve in 200.0 mL of water at 100°C? (*Hint:* Use the appropriate tables to calculate the K_{sp} of $PbCl_2$ at 100°C.)

72. For acetic acid(aq), $\Delta H_f° = -485.8$ kJ/mol and $S° = 178.7$ J/mol · K. For the acetate ion, $\Delta H_f° = -486.0$ kJ/mol and $S° = 86.6$ J/mol · K. Calculate K_a for acetic acid at 100°C. Compare with the K_a at 25°C.

Challenge Problems

73. The normal boiling point of benzene is 80.1°C. Thermodynamic data for benzene are

$$S°\ C_6H_6(l) = 172.80 \text{ J/mol} \cdot \text{K}$$
$$S°\ C_6H_6(g) = 268.2 \text{ J/mol} \cdot \text{K}$$

At what temperature is the vapor pressure of benzene 1.50×10^2 mm Hg?

74. $\Delta H_f°$ for iodine gas is 62.4 kJ/mol, while $S°$ is 260.7 J/mol · K. Calculate the equilibrium partial pressures of $I_2(g)$, $H_2(g)$, and HI(g) for the system

$$2HI(g) \rightleftharpoons H_2(g) + I_2(g)$$

at 520°C if the initial partial pressures are all 0.100 atm.

75. The heat of fusion of ice is 333 J/g. For the process $H_2O(s) \rightarrow H_2O(l)$, determine

a. $\Delta H°$ **b.** $\Delta G°$ at 0°C
c. $\Delta S°$ **d.** $\Delta G°$ at −10°C
e. $\Delta G°$ at 10°C

76. The overall reaction that occurs when sugar is metabolized is

$$C_{12}H_{22}O_{11}(s) + 12\,O_2(g) \longrightarrow 12CO_2(g) + 11H_2O(l)$$

For this reaction, $\Delta H°$ is −5650 kJ and $\Delta G°$ is −5790 kJ at 25°C.

a. If 30% of the free energy change is actually converted to useful work, how many kilojoules of work could be obtained when one gram of sugar is metabolized at body temperature, 37°C?
b. How many grams of sugar would you have to eat to get the energy to climb a mountain 1610 meters high? ($w = 9.79 \times 10^{-3}\,mh$, where w = work in kilojoules, m is body mass in kilograms, and h is height in meters.)

77. Hydrogen has been suggested as the fuel of the future. One way to store it is to convert it to a compound that can then be heated to release the hydrogen. One such compound is calcium hydride, CaH_2. This compound has a heat of formation of -186.2 kJ/mol and a standard entropy of 42.0 J/mol · K. What is the minimum temperature to which calcium hydride would have to be heated to produce hydrogen at one atmosphere pressure?

78. When a copper wire is exposed to air at room temperature, it becomes coated with a black oxide, CuO. If the wire is heated above a certain temperature, the black oxide is converted to a red oxide, Cu_2O. At a still higher temperature, the oxide coating disappears. Explain these observations in terms of the thermodynamics of the reactions

$$2CuO(s) \longrightarrow Cu_2O(s) + \tfrac{1}{2}\,O_2(g)$$

$$Cu_2O(s) \longrightarrow 2Cu(s) + \tfrac{1}{2}\,O_2(g)$$

and estimate the temperatures at which the changes occur.

17
Electrochemistry

If by fire
Of sooty coal th' empiric
Alchymist
Can turn, or holds it possible
to turn
Metals of drossiest ore to
perfect gold.

JOHN MILTON

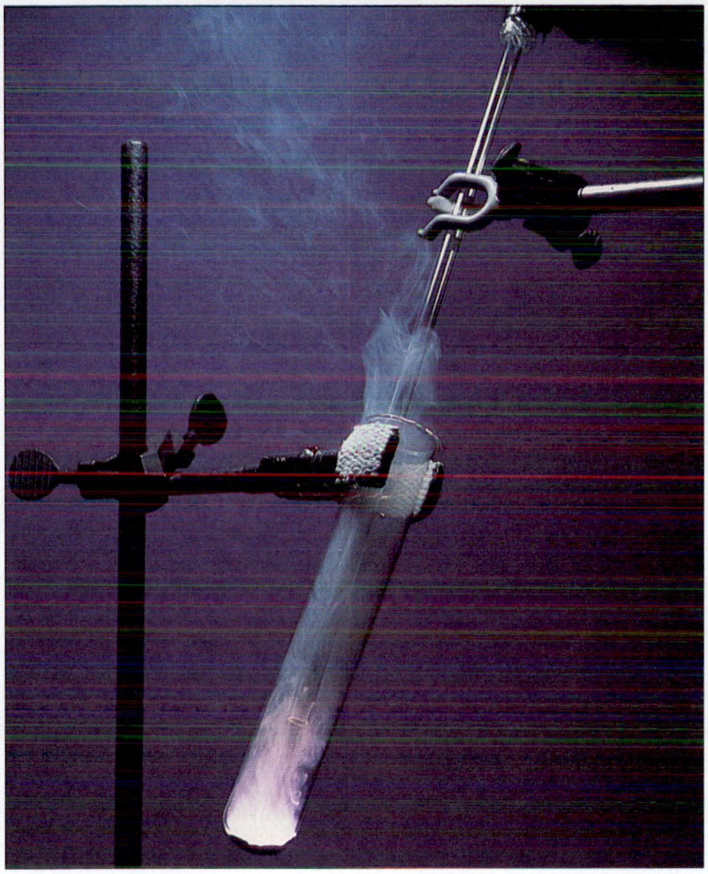

CHARLES STEELE

$$2K(s) + Cl_2(g) \longrightarrow 2KCl(s)$$

(See p. 529)

CHAPTER OUTLINE

Electrochemistry is the study of the interconversion of electrical and chemical energy. This conversion takes place in an electrochemical cell which may be a(n)

— **voltaic cell** (Section 17.1), where a spontaneous reaction generates electrical energy
— **electrolytic cell** (Section 17.5), where electrical energy is used to bring about a nonspontaneous reaction

Cathode = reduction; anode = oxidation

The reaction taking place in an electrochemical cell is of the oxidation-reduction type (Chap. 4). At one electrode, called the **cathode**, a **reduction** half-reaction occurs; electrons are consumed. Typical examples include

$$Cu^{2+}(aq) + 2e^- \longrightarrow Cu(s)$$

$$Cl_2(g) + 2e^- \longrightarrow 2Cl^-(aq)$$

At the other electrode, called the **anode**, an **oxidation** half-reaction occurs. Electrons are produced, as in the half-reactions

$$Zn(s) \longrightarrow Zn^{2+}(aq) + 2e^-$$

$$2I^-(aq) \longrightarrow I_2(s) + 2e^-$$

The number of electrons produced at the anode is exactly equal to the number of electrons consumed at the cathode. In any cell, *anions* move to the *anode; cations* move to the *cathode.*

One of the most important characteristics of a cell is its **voltage**, which is a measure of reaction spontaneity. Cell voltages depend upon the nature of the half-reactions occurring at the electrodes (Section 17.2) and upon the concentrations of species involved (Section 17.4). From the voltage measured at standard concentrations, it is possible to calculate the standard free energy change and the equilibrium constant (Section 17.3) of the reaction involved.

The principles discussed in this chapter have a host of practical applications. Whenever you start your car, turn on a flashlight, or take a logarithm on a calculator, you are making use of a voltaic cell. Many of our most important elements, including hydrogen and chlorine, are made in electrolytic cells. These applications, among others, are discussed in Section 17.6.

17.1 VOLTAIC CELLS

Figure 17.1
When a strip of zinc is placed in a solution containing Cu^{2+} ions (*left*), a spontaneous redox reaction occurs. The final result is shown at the right. Copper metal plates out and the blue color due to Cu^{2+} fades. (Marna G. Clarke)

In principle at least, any spontaneous redox reaction can serve as a source of energy in a voltaic cell. The cell must be designed in such a way that oxidation occurs at one electrode (anode) with reduction at the other electrode (cathode). The electrons produced at the anode must be transferred to the cathode, where they are consumed. To do this, the electrons move through an external circuit, where they do electrical work.

To understand how a voltaic cell operates, let us start with some simple cells that are readily made in the general chemistry laboratory.

The Zn–Cu^{2+} Cell ($Zn \mid Zn^{2+} \parallel Cu^{2+} \mid Cu$)

When a piece of zinc is added to a water solution containing Cu^{2+} ions, the following redox reaction takes place:

$$Zn(s) + Cu^{2+}(aq) \longrightarrow Zn^{2+}(aq) + Cu(s)$$

In this reaction, copper metal plates out on the surface of the zinc. The blue color of the aqueous Cu^{2+} ion fades as it is replaced by the colorless aqueous Zn^{2+} ion (Fig. 17.1). Clearly this redox reaction is spontaneous; it involves electron transfer from a Zn atom to a Cu^{2+} ion.

Otherwise, the cell is "shorted" out

To design a voltaic cell using the Zn–Cu^{2+} reaction as a source of electrical energy, the electron transfer must occur indirectly; that is, the electrons

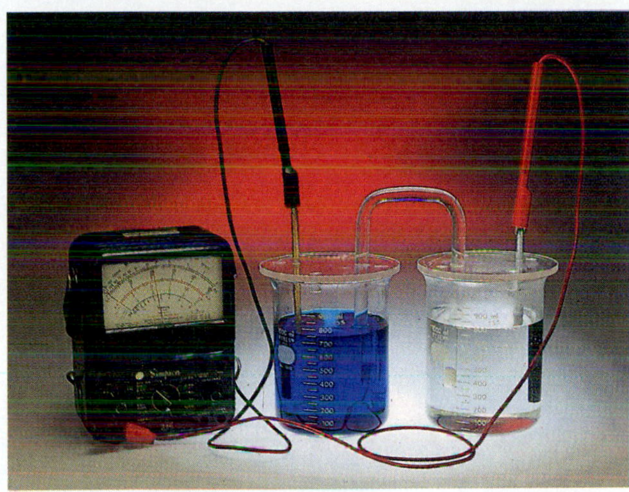

Figure 17.2
In this voltaic cell, the following spontaneous redox reaction takes place: $Zn(s) + Cu^{2+}(aq) \longrightarrow Zn^{2+}(aq) + Cu(s)$. The salt bridge allows ions to pass from one solution to the other to complete the circuit. At the same time, it prevents direct contact between Zn atoms and Cu^{2+} ions. (Charles D. Winters)

given off by zinc atoms must be made to pass through an external electric circuit before they reduce Cu^{2+} ions to copper atoms. One way to do this is shown in Figure 17.2. The voltaic cell consists of two half-cells:

— a Zn anode dipping into a solution containing Zn^{2+} ions, shown in the beaker at the far right
— a Cu cathode dipping into a solution containing Cu^{2+} ions (blue), shown in the beaker at the center of Figure 17.2

The "external circuit" consists of a voltmeter with leads (red and black) to the anode and cathode.

Let us trace the flow of electric current through this cell.

1. At the zinc *anode*, electrons are produced by the *oxidation* half-reaction

$$Zn(s) \longrightarrow Zn^{2+}(aq) + 2e^-$$

This electrode, which "pumps" electrons into the external circuit, is ordinarily marked as the negative pole of the cell.

2. Electrons generated at the anode move through the external circuit (right to left in Fig. 17.2) to the copper *cathode*. There they are consumed, reducing Cu^{2+} ions present in the solution around the electrode:

$$Cu^{2+}(aq) + 2e^- \longrightarrow Cu(s)$$

The copper electrode, which "pulls" electrons from the external circuit, is considered to be the positive pole of the cell.

3. As the above half-reactions proceed, a surplus of positive ions (Zn^{2+}) tends to build up around the zinc electrode. The region around the copper electrode tends to become deficient in positive ions as Cu^{2+} ions are consumed. To maintain electrical neutrality, cations must move toward the copper cathode or, alternatively, anions must move toward the zinc anode. In practice, both migrations occur.

In the cell shown in Figure 17.2, movement of ions occurs through a *salt bridge*. In its simplest form, a salt bridge may consist of an inverted U-tube, plugged with glass wool at each end. The tube is filled with a solution of a salt that takes no part in the electrode reactions; potassium nitrate, KNO_3, is frequently used. As current is drawn from the cell, K^+ ions move from the salt bridge into the cathode half-cell. At the same time, NO_3^- ions move into the anode half-cell. In this way, electrical neutrality is maintained without Cu^{2+} ions coming in contact with the Zn electrode, which would short-circuit the cell.

The cell shown in Figure 17.2 is often abbreviated as

Anode || Cathode

$$Zn \mid Zn^{2+} \parallel Cu^{2+} \mid Cu$$

In this notation,

— the **anode** reaction (**oxidation**) is shown at the left. Zn atoms are oxidized to Zn^{2+} ions.
— the salt bridge (or other means of separating the half-cells) is indicated by the symbol $\parallel$.
— the **cathode** reaction (**reduction**) is shown at the right. Cu^{2+} ions are reduced to Cu atoms.

Oxidation and anode both start with vowels; reduction and cathode begin with consonants

Notice that the anode half-reaction comes first in the cell notation, just as the letter *a* comes before *c*.

Other Salt Bridge Cells

Cells similar to that shown in Figure 17.2 can be set up for many different spontaneous redox reactions. Consider, for example, the reaction

$$Ni(s) + Cu^{2+}(aq) \longrightarrow Ni^{2+}(aq) + Cu(s)$$

This reaction, like that between Zn and Cu^{2+}, can serve as a source of electrical energy in a voltaic cell. The cell is similar to that shown in Figure 17.2 except that, in the anode compartment, a nickel electrode is surrounded by a

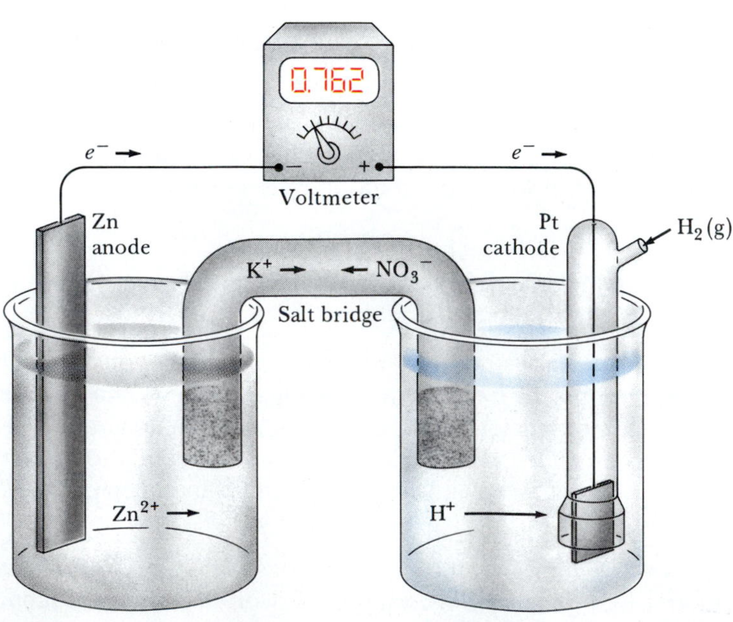

Figure 17.3
A voltaic cell in which the reaction: $Zn(s) + 2H^+(aq) \longrightarrow Zn^{2+}(aq) + H_2(g)$ occurs. Hydrogen gas is bubbled over a specially prepared platinum electrode, which is surrounded by a solution containing H^+ ions.

solution of a nickel(II) salt, such as $NiCl_2$ or $NiSO_4$. The cell notation is $Ni \mid Ni^{2+} \parallel Cu^{2+} \mid Cu$.

Another spontaneous redox reaction that can serve as a source of electrical energy is that between zinc metal and H^+ ions:

$$Zn(s) + 2H^+(aq) \longrightarrow Zn^{2+}(aq) + H_2(g)$$

A voltaic cell using this reaction is similar to the $Zn–Cu^{2+}$ cell; the $Zn \mid Zn^{2+}$ half-cell and the salt bridge are the same. Since no metal is involved in the cathode half-reaction, an *inert* electrode that conducts an electric current is used. Frequently, the cathode is made out of platinum (Fig. 17.3). Hydrogen gas is bubbled over the cathode, which is surrounded by H^+ ions from a solution of HCl.

Nichrome or graphite can also be used

The half-reactions occurring in the cell are

anode: $Zn(s) \longrightarrow Zn^{2+}(aq) + 2e^-$ (oxidation)

cathode: $2H^+(aq) + 2e^- \longrightarrow H_2(g)$ (reduction)

The cell notation is $Zn \mid Zn^{2+} \parallel H^+ \mid H_2 \mid Pt$. The symbol (Pt) is used to indicate the presence of an inert platinum electrode.

Example 17.1 When chlorine gas is bubbled through a water solution of NaBr, a spontaneous redox reaction occurs:

$$Cl_2(g) + 2Br^-(aq) \longrightarrow 2Cl^-(aq) + Br_2(l)$$

This reaction can serve as a source of electrical energy in the voltaic cell shown in Figure 17.4. In this cell,

(a) what is the cathode reaction? the anode reaction?
(b) which way do electrons move in the external circuit?
(c) which way do anions move within the cell? cations?

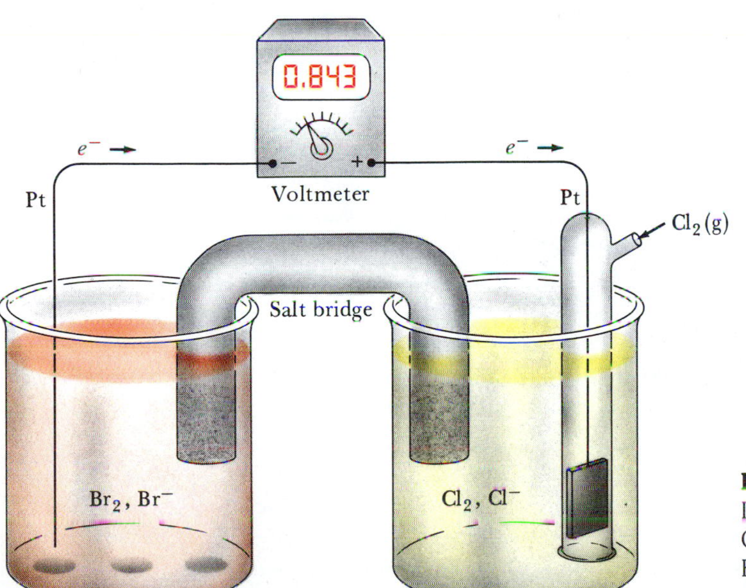

Figure 17.4
In this voltaic cell, the spontaneous redox reaction is $Cl_2(g) + 2Br^-(aq) \longrightarrow 2Cl^-(aq) + Br_2(l)$. Both electrodes are made of platinum; the one at the left is the anode.

Strategy Split the reaction into two half-reactions. Remember that oxidation occurs at the anode, reduction at the cathode. Anions move to the anode, cations to the cathode. Electrons are produced at the anode and transferred through the external circuit to the cathode, where they are consumed.

Solution

(a) cathode: $Cl_2(g) + 2e^- \longrightarrow 2Cl^-(aq)$ (reduction)

anode: $2Br^-(aq) \longrightarrow Br_2(l) + 2e^-$ (oxidation)

(b) From anode to cathode (left to right in Fig. 17.4).

(c) Anions move to the anode (right to left); cations move to the cathode (left to right).

17.2 STANDARD VOLTAGES

The driving force behind the spontaneous reaction in a voltaic cell is measured by the cell voltage, which is an *intensive* property, independent of the number of electrons passing through the cell. Cell voltage depends upon the nature of the redox reaction and the concentrations of the species involved; for the moment, we'll concentrate upon the first of these factors.

The standard voltage for a given cell is that measured when the current flow is essentially zero, ***all ions and molecules in solution are at a concentration of 1 M, and all gases are at a pressure of 1 atm.*** To illustrate, consider the $Zn{-}H^+$ cell referred to earlier. Let us suppose that the half-cells are set up in such a way that the concentrations of Zn^{2+} and H^+ are both 1 M and the pressure of $H_2(g)$ is 1 atm. Under these conditions, the cell voltage at very low current flow is +0.762 V. This quantity is referred to as the **standard voltage** and is given the symbol $\boldsymbol{E°}$.

$$Zn(s) + 2H^+(aq,\ 1\ M) \longrightarrow Zn^{2+}(aq,\ 1\ M) + H_2(g,\ 1\ \text{atm}) \quad E° = +0.762\ \text{V}$$

$E°_{ox}$ and $E°_{red}$

Any redox reaction can be split into two half-reactions, an oxidation and a reduction. It is possible to associate standard voltages, shown as $\boldsymbol{E°_{ox}}$ and $\boldsymbol{E°_{red}}$, with the oxidation and reduction half-reactions. The standard voltage for the overall reaction, $E°$, is the sum of these two quantities

$$E° = E°_{ox} + E°_{red}$$

To illustrate, consider the reaction between Zn and H^+ ions, where the standard voltage is +0.762 V.

$$+0.762\ \text{V} = E°_{ox}(Zn \rightarrow Zn^{2+}) + E°_{red}(H^+ \rightarrow H_2)$$

We need two half-cells to measure a voltage

There is no way to measure the standard voltage for a half-reaction; only $E°$ can be measured directly. To obtain values for $E°_{ox}$ and $E°_{red}$, the value zero is arbitrarily assigned to the standard voltage for reduction of H^+ ions to H_2 gas:

$$2H^+(aq, 1\ M) + 2e^- \longrightarrow H_2(g, 1\ \text{atm}) \qquad E^\circ_{red}(H^+ \rightarrow H_2) = 0.000\ V$$

Using this convention, it follows that the standard voltage for the oxidation of zinc must be +0.762 V, that is,

$$Zn(s) \longrightarrow Zn^{2+}(aq, 1\ M) + 2e^- \qquad E^\circ_{ox}(Zn \rightarrow Zn^{2+}) = +0.762\ V$$

As soon as one half-reaction voltage is established, others can be determined. For example, the standard voltage for the Zn–Cu^{2+} cell shown in Figure 17.2 is found to be +1.101 V. Knowing that E°_{ox} for zinc is +0.762 V, it follows that

$$E^\circ_{red}(Cu^{2+} \rightarrow Cu) = E^\circ - E^\circ_{ox}(Zn \rightarrow Zn^{2+})$$
$$= 1.101\ V - 0.762\ V = +0.339\ V$$

Standard potential = E°_{red}

Standard half-cell voltages are ordinarily obtained from a list of **standard potentials** such as those in Table 17.1, pp. 468–469. The potentials listed are the standard voltages for reduction half-reactions. For example, since the standard potentials listed in the table for $Zn^{2+} \longrightarrow Zn$ and $Cu^{2+} \longrightarrow Cu$ are −0.762 V and +0.339 V, respectively,

$$Zn^{2+}(aq) + 2e^- \longrightarrow Zn(s) \qquad E^\circ_{red} = -0.762\ V$$
$$Cu^{2+}(aq) + 2e^- \longrightarrow Cu(s) \qquad E^\circ_{red} = +0.339\ V$$

Standard voltages for oxidation half-reactions are obtained by changing the sign of the standard potential listed in Table 17.1. Thus,

$E^\circ_{ox} = -E^\circ_{red}$ for a half-reaction

$$Zn(s) \longrightarrow Zn^{2+}(aq) + 2e^- \qquad E^\circ_{ox} = +0.762\ V$$
$$Cu(s) \longrightarrow Cu^{2+}(aq) + 2e^- \qquad E^\circ_{ox} = -0.339\ V$$

In general, standard voltages for forward and reverse half-reactions are equal in magnitude but opposite in sign.

Strength of Oxidizing and Reducing Agents

As pointed out in Chapter 4, an oxidizing agent is a species that can gain electrons; it is a reactant in a reduction half-reaction. Since Table 17.1 lists reduction half-reactions from left to right, it follows that oxidizing agents are located in the left column of the table. All of the species listed in that column (Li^+, . . . , F_2) are, in principle, oxidizing agents. A "strong" oxidizing agent is one that has a strong attraction for electrons and hence *can readily oxidize other species*. In contrast, a "weak" oxidizing agent does not gain electrons readily. It is capable of reacting only with those species that are very easily oxidized.

Strong oxidizing agents (large E°_{red}) at bottom of left column

The strength of an oxidizing agent is directly related to the standard voltage for its reduction, E°_{red}. *The more positive E°_{red} is, the stronger the oxidizing agent.* Looking at Table 17.1, you can see that *oxidizing strength increases moving down the left column.* The Li^+ ion, at the top of that column, is a very weak oxidizing agent with a large negative reduction voltage (E°_{red} = −3.040 V). In practice, cations of the Group 1 metals (Li^+, Na^+, K^+, . . .) and the Group 2 metals (Mg^{2+}, Ca^{2+}, . . .) never act as oxidizing agents in water solution.

Table 17.1 Standard Potentials in Water Solution at 25°C

Oxidizing Agent	Reducing Agent	E°_{red}(V)
	Acidic Solution	
$Li^+(aq) + e^-$	$\rightarrow Li(s)$	−3.040
$K^+(aq) + e^-$	$\rightarrow K(s)$	−2.936
$Ba^{2+}(aq) + 2e^-$	$\rightarrow Ba(s)$	−2.906
$Ca^{2+}(aq) + 2e^-$	$\rightarrow Ca(s)$	−2.869
$Na^+(aq) + e^-$	$\rightarrow Na(s)$	−2.714
$Mg^{2+}(aq) + 2e^-$	$\rightarrow Mg(s)$	−2.357
$Al^{3+}(aq) + 3e^-$	$\rightarrow Al(s)$	−1.68
$Mn^{2+}(aq) + 2e^-$	$\rightarrow Mn(s)$	−1.182
$Zn^{2+}(aq) + 2e^-$	$\rightarrow Zn(s)$	−0.762
$Cr^{3+}(aq) + 3e^-$	$\rightarrow Cr(s)$	−0.744
$Fe^{2+}(aq) + 2e^-$	$\rightarrow Fe(s)$	−0.409
$Cr^{3+}(aq) + e^-$	$\rightarrow Cr^{2+}(aq)$	−0.408
$Cd^{2+}(aq) + 2e^-$	$\rightarrow Cd(s)$	−0.402
$PbSO_4(s) + 2e^-$	$\rightarrow Pb(s) + SO_4^{2-}(aq)$	−0.356
$Tl^+(aq) + e^-$	$\rightarrow Tl(s)$	−0.336
$Co^{2+}(aq) + 2e^-$	$\rightarrow Co(s)$	−0.282
$Ni^{2+}(aq) + 2e^-$	$\rightarrow Ni(s)$	−0.236
$AgI(s) + e^-$	$\rightarrow Ag(s) + I^-(aq)$	−0.152
$Sn^{2+}(aq) + 2e^-$	$\rightarrow Sn(s)$	−0.141
$Pb^{2+}(aq) + 2e^-$	$\rightarrow Pb(s)$	−0.127
$2H^+(aq) + 2e^-$	$\rightarrow H_2(g)$	0.000
$AgBr(s) + e^-$	$\rightarrow Ag(s) + Br^-(aq)$	0.073
$S(s) + 2H^+(aq) + 2e^-$	$\rightarrow H_2S(aq)$	0.144
$Sn^{4+}(aq) + 2e^-$	$\rightarrow Sn^{2+}(aq)$	0.154
$SO_4^{2-}(aq) + 4H^+(aq) + 2e^-$	$\rightarrow SO_2(g) + 2H_2O$	0.155
$Cu^{2+}(aq) + e^-$	$\rightarrow Cu^+(aq)$	0.161
$Cu^{2+}(aq) + 2e^-$	$\rightarrow Cu(s)$	0.339
$Cu^+(aq) + e^-$	$\rightarrow Cu(s)$	0.518
$I_2(s) + 2e^-$	$\rightarrow 2I^-(aq)$	0.534
$Fe^{3+}(aq) + e^-$	$\rightarrow Fe^{2+}(aq)$	0.769
$Hg_2^{2+}(aq) + 2e^-$	$\rightarrow 2Hg(l)$	0.796
$Ag^+(aq) + e^-$	$\rightarrow Ag(s)$	0.799
$2Hg^{2+}(aq) + 2e^-$	$\rightarrow Hg_2^{2+}(aq)$	0.908
$NO_3^-(aq) + 4H^+(aq) + 3e^-$	$\rightarrow NO(g) + 2H_2O$	0.964
$AuCl_4^-(aq) + 3e^-$	$\rightarrow Au(s) + 4Cl^-(aq)$	1.001
$Br_2(l) + 2e^-$	$\rightarrow 2Br^-(aq)$	1.077
$O_2(g) + 4H^+(aq) + 4e^-$	$\rightarrow 2H_2O$	1.229
$MnO_2(s) + 4H^+(aq) + 2e^-$	$\rightarrow Mn^{2+}(aq) + 2H_2O$	1.229
$Cr_2O_7^{2-}(aq) + 14H^+(aq) + 6e^-$	$\rightarrow 2Cr^{3+}(aq) + 7H_2O$	1.33
$Cl_2(g) + 2e^-$	$\rightarrow 2Cl^-(aq)$	1.360
$ClO_3^-(aq) + 6H^+(aq) + 5e^-$	$\rightarrow \frac{1}{2}Cl_2(g) + 3H_2O$	1.458

Table 17.1 (*Continued*)

Oxidizing Agent	Reducing Agent	E°_{red}(V)
	Acidic Solution	
$Au^{3+}(aq) + 3e^-$	$\rightarrow Au(s)$	1.498
$MnO_4^-(aq) + 8H^+(aq) + 5e^-$	$\rightarrow Mn^{2+}(aq) + 4H_2O$	1.512
$PbO_2(s) + SO_4^{2-}(aq) + 4H^+(aq) + 2e^-$	$\rightarrow PbSO_4(s) + 2H_2O$	1.687
$H_2O_2(aq) + 2H^+(aq) + 2e^-$	$\rightarrow 2H_2O$	1.763
$Co^{3+}(aq) + e^-$	$\rightarrow Co^{2+}(aq)$	1.953
$F_2(g) + 2e^-$	$\rightarrow 2F^-(aq)$	2.889
	Basic Solution	
$Fe(OH)_2(s) + 2e^-$	$\rightarrow Fe(s) + 2\,OH^-(aq)$	−0.891
$2H_2O + 2e^-$	$\rightarrow H_2(g) + 2\,OH^-(aq)$	−0.828
$Fe(OH)_3(s) + e^-$	$\rightarrow Fe(OH)_2(s) + OH^-(aq)$	−0.547
$S(s) + 2e^-$	$\rightarrow S^{2-}(aq)$	−0.445
$NO_3^-(aq) + 2H_2O + 3e^-$	$\rightarrow NO(g) + 4\,OH^-(aq)$	−0.140
$NO_3^-(aq) + H_2O + 2e^-$	$\rightarrow NO_2^-(aq) + 2\,OH^-(aq)$	0.004
$ClO_4^-(aq) + H_2O + 2e^-$	$\rightarrow ClO_3^-(aq) + 2\,OH^-(aq)$	0.398
$O_2(g) + 2H_2O + 4e^-$	$\rightarrow 4\,OH^-(aq)$	0.401
$ClO_3^-(aq) + 3H_2O + 6e^-$	$\rightarrow Cl^-(aq) + 6\,OH^-(aq)$	0.614
$ClO^-(aq) + H_2O + 2e^-$	$\rightarrow Cl^-(aq) + 2\,OH^-(aq)$	0.890

Further down the left column of Table 17.1, the H^+ ion ($E^\circ_{red} = 0.000$ V) is a reasonably strong oxidizing agent. It is capable of oxidizing many metals, including magnesium and zinc:

$$Mg(s) + 2H^+(aq) \longrightarrow Mg^{2+}(aq) + H_2(g) \qquad E^\circ = +2.357\text{ V}$$

$$Zn(s) + 2H^+(aq) \longrightarrow Zn^{2+}(aq) + H_2(g) \qquad E^\circ = +0.762\text{ V}$$

The strongest oxidizing agents are those at the bottom of the left column. Species such as $Cr_2O_7^{2-}$ ($E^\circ_{red} = +1.33$ V) and Cl_2 ($E^\circ_{red} = +1.360$ V) are commonly used as oxidizing agents in the laboratory. The fluorine molecule, at the bottom of the left column, should be the strongest of all oxidizing agents. In practice, fluorine is never used as an oxidizing agent in water solution. The F_2 molecule takes electrons away from just about anything, including water, often with explosive violence.

The argument we have just gone through can be applied, in reverse, to reducing agents. These species are listed in the column to the right of Table 17.1 (Li, . . . , F^-). In principle, at least, all of them can supply electrons to another species in a redox reaction. Their strength as reducing agents is directly related to their E°_{ox} values. *The more positive E°_{ox} is, the stronger the reducing agent.* Looking at the values, remembering that $E^\circ_{ox} = -E^\circ_{red}$,

$$Li(s) \longrightarrow Li^+(aq) + e^- \qquad E^\circ_{ox} = +3.040\text{ V}$$

$$H_2(g) \longrightarrow 2H^+(aq) + 2e^- \qquad E^\circ_{ox} = 0.000\text{ V}$$

$$2F^-(aq) \longrightarrow F_2(g) + 2e^- \qquad E^\circ_{ox} = -2.889\text{ V}$$

Strong reducing agents (large E°_{ox}) at top of right column

it should be clear that *reducing strength decreases moving down the table.* Actually, the five metals at the top of the right column (Li, K, Ba, Ca, Na) cannot be used as reducing agents in water solution because they react directly with water, reducing it to hydrogen gas:

$$Li(s) + H_2O \longrightarrow Li^+(aq) + OH^-(aq) + \tfrac{1}{2}H_2(g)$$

$$Ba(s) + 2H_2O \longrightarrow Ba^{2+}(aq) + 2\,OH^-(aq) + H_2(g)$$

Example 17.2 Consider the following species in acidic solution: $Cr_2O_7^{2-}$, NO_3^-, Br^-, Mg, and Sn^{2+}. Using Table 17.1,

(a) classify each of these as an oxidizing and/or reducing agent.
(b) arrange the oxidizing agents in order of increasing strength.
(c) do the same with the reducing agents.

Strategy Remember that oxidizing agents are located in the left column of Table 17.1, reducing agents in the right column. Large positive values of E°_{red} and E°_{ox} are associated with strong oxidizing and reducing agents, respectively.

Solution

(a) Oxidizing agents:

Sn^{2+} ($E^\circ_{red} = -0.141$ V), NO_3^- ($E^\circ_{red} = +0.964$ V), $Cr_2O_7^{2-}$ ($E^\circ_{red} = +1.33$ V),

Reducing agents:

Mg ($E^\circ_{ox} = +2.357$ V), Sn^{2+} ($E^\circ_{ox} = -0.154$ V), Br^- ($E^\circ_{ox} = -1.077$ V)

Note that the Sn^{2+} ion can act as either an oxidizing agent, when it is reduced to Sn, or a reducing agent, when it is oxidized to Sn^{4+}.
(b) Comparing values of E°_{red}:

$$Sn^{2+} < NO_3^- < Cr_2O_7^{2-}$$

This ranking correlates with the positions of these species in the left column of Table 17.1; $Cr_2O_7^{2-}$ is near the bottom, Sn^{2+} closest to the top.

(c) Comparing values of E°_{ox}:

$$Br^- < Sn^{2+} < Mg$$

Calculation of E° from E°_{red} and E°_{ox}

As pointed out earlier, the standard voltage for a redox reaction is the sum of the standard voltages of the two half-reactions, reduction and oxidation; that is,

$$E^\circ = E^\circ_{red} + E^\circ_{ox}$$

This simple relation makes it possible, using Table 17.1, to calculate standard voltages for more than 3000 different redox reactions.

Example 17.3 Using Table 17.1, calculate E° for a voltaic cell in which the reaction is

$$2Fe^{3+}(aq) + 2I^-(aq) \longrightarrow 2Fe^{2+}(aq) + I_2(s)$$

Strategy Split the reaction into two half-reactions, find the appropriate values of E°_{red} and E°_{ox} from Table 17.1, and add to obtain E°.

Solution

reduction:	$2Fe^{3+}(aq) + 2e^- \longrightarrow 2Fe^{2+}(aq)$	$E^\circ_{red} = +0.769$ V
oxidation:	$2I^-(aq) \longrightarrow I_2(s) + 2e^-$	$E^\circ_{ox} = -0.534$ V
		$E^\circ = +0.235$ V

Two general points concerning cell voltages are illustrated by Example 17.3.

1. The calculated voltage, E°, is always a positive quantity for a reaction taking place in a voltaic cell.
2. The quantities E°, E°_{ox}, and E°_{red} are independent of how the equation for the cell reaction is written. You *never* multiply the voltage by the coefficients of the balanced equation.

Spontaneity of Redox Reactions

To determine whether a given redox reaction is spontaneous, apply a simple principle:

If the calculated voltage for a redox reaction is a positive quantity, the reaction will be spontaneous. If the calculated voltage is negative, the reaction will not occur.

Ordinarily, this principle is applied at standard concentrations (1 atm for gases, 1 *M* for species in aqueous solution). Here, it is the sign of E° that serves as the criterion for spontaneity. To show how this works, consider the problem of oxidizing nickel metal to Ni^{2+} ions. This *cannot* be accomplished using Zn^{2+} ions:

$$Ni(s) + Zn^{2+}(aq, 1\,M) \longrightarrow Ni^{2+}(aq, 1\,M) + Zn(s)$$

$$E^\circ = E^\circ_{ox}\,Ni + E^\circ_{red}\,Zn^{2+} = +0.236\text{ V} - 0.762\text{ V} = -0.526\text{ V}$$

Sure enough, if you immerse a piece of nickel in a solution of $ZnSO_4$, nothing happens. Suppose, however, you add a bar of nickel to a solution of $CuSO_4$ (Fig. 17.5). Now the nickel is oxidized through the spontaneous redox reaction:

$$Ni(s) + Cu^{2+}(aq, 1\,M) \longrightarrow Ni^{2+}(aq, 1\,M) + Cu(s)$$

$$E^\circ = E^\circ_{ox}\,Ni + E^\circ_{red}\,Cu^{2+} = +0.236\text{ V} + 0.339\text{ V} = +0.575\text{ V}$$

Figure 17.5
Nickel metal reacts spontaneously with Cu^{2+} ions, producing Cu metal and Ni^{2+} ions. Copper plates out on the surface of the nickel, and the blue color of Cu^{2+} is replaced by the green color of Ni^{2+}. (Marna G. Clarke)

Example 17.4 Using standard potentials listed in Table 17.1, decide whether, at standard concentrations,

(a) Fe(*s*) will be oxidized to Fe^{2+} by treatment with hydrochloric acid (HCl).
(b) Cu(*s*) will be oxidized to Cu^{2+} by treatment with hydrochloric acid.
(c) Cu(*s*) will be oxidized to Cu^{2+} by treatment with nitric acid (HNO_3).

Strategy The oxidation half-reaction is given in each case; you must decide upon the nature of the reduction half-reaction. Once that is done, look up the appropriate standard potentials and combine them to find out whether E° is positive or negative.

Figure 17.6
Finely divided iron in the form of steel wool reacts with hydrochloric acid to evolve hydrogen: $Fe(s) + 2H^+(aq) \longrightarrow Fe^{2+}(aq) + H_2(g)$.
(Charles D. Winters)

Solution

(a) The oxidation half-reaction is

$$Fe(s) \longrightarrow Fe^{2+}(aq) + 2e^- \qquad E^\circ_{ox} = +0.409 \text{ V}$$

Hydrochloric acid consists of H^+ and Cl^- ions. Of these two ions, only H^+ is listed in the left column of Table 17.1; the Cl^- ion cannot be reduced. The reduction half-reaction must be

$$2H^+(aq) + 2e^- \longrightarrow H_2(g) \qquad E^\circ_{red} = 0.000 \text{ V}$$

Since the calculated voltage is positive,

$$E^\circ = +0.409 \text{ V} + 0.000 \text{ V} = +0.409 \text{ V}$$

The following redox reaction

$$Fe(s) + 2H^+(aq) \longrightarrow Fe^{2+}(aq) + H_2(g)$$

should and does occur (Fig. 17.6).

(b) Proceeding in the same way,

$$\begin{array}{rll} Cu(s) & \longrightarrow Cu^{2+}(aq) + 2e^- & E^\circ_{ox} = -0.339 \text{ V} \\ 2H^+(aq) + 2e^- & \longrightarrow H_2(g) & E^\circ_{red} = 0.000 \text{ V} \\ \hline Cu(s) + 2H^+(aq) & \longrightarrow Cu^{2+}(aq) + H_2(g) & E^\circ = -0.339 \text{ V} \end{array}$$

As predicted, no reaction occurs when copper is added to 1 *M* hydrochloric acid.

(c) Here, there is another possible oxidizing agent, the NO_3^- ion. Looking at Table 17.1, you should find that E°_{red} for the NO_3^- ion is +0.964 V. It follows that E° is positive, so the following reaction occurs (Fig. 17.7):

$$\begin{array}{rl} 3[Cu(s) & \longrightarrow Cu^{2+}(aq) + 2e^-] \\ 2[NO_3^-(aq) + 4H^+(aq) + 3e^- & \longrightarrow NO(g) + 2H_2O] \\ \hline 3Cu(s) + 2NO_3^- + 8H^+(aq) & \longrightarrow 3Cu^{2+}(aq) + 2NO(g) + 4H_2O \end{array}$$

$$E^\circ = E^\circ_{ox} + E^\circ_{red} = -0.339 \text{ V} + 0.964 \text{ V} = +0.625 \text{ V}$$

(a)

(b)

The oxidation-reduction reaction between zinc and hydrochloric acid. Zinc metal (a) loses electrons and is oxidized to $Zn^{2+}(aq)$. Hydrogen gas is evolved (b) when H^+ (from HCl) is reduced to H_2.
(Richard Megna/FUNDAMENTAL PHOTOGRAPHS, New York)

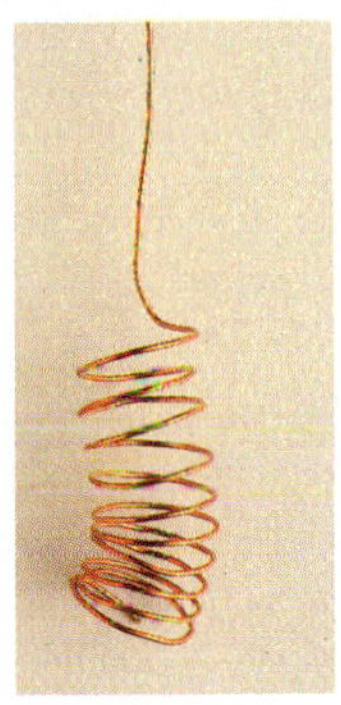

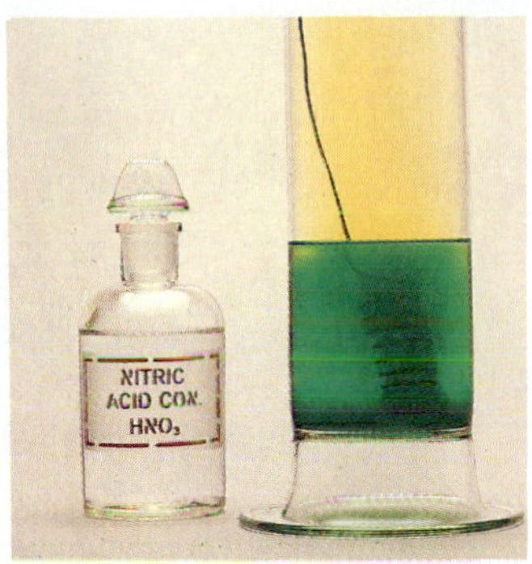

Figure 17.7
Copper metal is comparatively inactive, but it reacts with concentrated nitric acid. The brown fumes are $NO_2(g)$, a reduction product of HNO_3. The copper is oxidized to Cu^{2+} ions, which impart their color to the solution. (Marna G. Clarke)

17.3 RELATIONS BETWEEN $E°$, $\Delta G°$, AND K

As pointed out previously, the value of the standard cell voltage, $E°$, is a measure of the spontaneity of a cell reaction. In Chapter 16, we showed that the standard free energy change, $\Delta G°$, is a general criterion for reaction spontaneity. As you might suppose, these two quantities are simply related to one another and to the equilibrium constant, K, for the cell reaction.

$E°$ and $\Delta G°$

It was pointed out in Chapter 16 (p. 441) that the free energy change is a measure of the amount of useful work that can be obtained from a reaction carried out at constant temperature and pressure. The relation between these two quantities is

$$\Delta G = w_{max}$$

where w_{max} is the maximum amount of useful work done *on* the reaction system. For a voltaic cell, the "useful work" produced *by* the cell reaction, $-w_{max}$, is the electrical energy generated. This in turn is the product of the charge, Q, in coulombs, times the voltage, E:

$$-w_{max} = QE$$

Hence

$$\Delta G = -QE$$

We would like to relate ΔG to the number of moles of electrons taking part in the reaction, n. To do this, note that*

$$1 \text{ mol } e^- = 9.648 \times 10^4 \text{ C}$$

$$Q = 9.648 \times 10^4 \frac{\text{C}}{\text{mol}} \times n \qquad (n = \text{no. moles } e^- \text{ exchanged in reaction})$$

Substituting for Q in the expression for ΔG:

$$\Delta G = -9.648 \times 10^4 \frac{\text{C}}{\text{mol}} \times n \times E$$

* The charge of an electron is 1.602×10^{-19} C. The charge on one mole of electrons must then be $(6.022 \times 10^{23})(1.602 \times 10^{-19} \text{ C}) = 9.648 \times 10^4$ C.

This equation is usually written in the form

$$\Delta G = -nFE$$

$F = 9.648 \times 10^4$ C/mol $= 9.648 \times 10^4$ J/mol · V

where F is the **Faraday constant**, 9.648×10^4 C/mol. Noting that 1 J = 1 V × 1 C, it follows that an alternative expression for the Faraday constant is

$$F = 9.648 \times 10^4 \frac{\text{C}}{\text{mol}} \times \frac{1\text{ J}}{1\text{V} \cdot 1\text{ C}} = 9.648 \times 10^4 \frac{\text{J}}{\text{mol} \cdot \text{V}}$$

At standard concentrations

$$\Delta G° = -nFE°$$

Notice from this equation that $\Delta G°$ and $E°$ have opposite signs. This is reasonable; a spontaneous reaction is one for which $\Delta G°$ is *negative* but $E°$ is *positive.*

Example 17.5 Calculate $\Delta G°$ for the reaction

$$Cl_2(g) + 2Br^-(aq) \longrightarrow 2Cl^-(aq) + Br_2(l)$$

using data in Table 17.1.

Strategy Split the reaction into two half-reactions; find $E°_{ox}$ and $E°_{red}$ from Table 17.1, and add to obtain $E°$. Then use the relation $\Delta G° = -nFE°$.

Solution

(1)

$$\begin{array}{rll} Cl_2(g) + 2e^- & \longrightarrow 2Cl^-(aq) & E°_{red} = +1.360\text{ V} \\ 2Br^-(aq) & \longrightarrow Br_2(l) + 2e^- & E°_{ox} = -1.077\text{ V} \\ \hline Cl_2(g) + 2Br^-(aq) & \longrightarrow 2Cl^-(aq) + Br_2(l) & E° = +0.283\text{ V} \end{array}$$

(2) Note from the half-equations that $n = 2$:

$$\Delta G° = -2\text{ mol }(9.648 \times 10^4\text{ J/mol} \cdot \text{V})(0.283\text{ V}) = -5.46 \times 10^4\text{ J} = -54.6\text{ kJ}$$

Notice that $E°_{tot}$ is positive, while $\Delta G°$ is negative, indicating a spontaneous reaction. This reaction can serve as a basis for a voltaic cell (Fig. 17.4) or as a way of testing for Br^- ions in solution (Fig. 17.8).

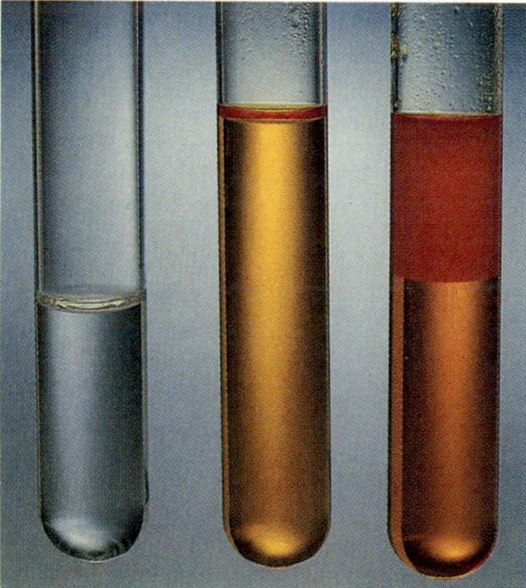

Figure 17.8
When a water solution saturated with $Cl_2(g)$ is added to a solution containing Br^- ions (left), a redox reaction occurs. The Br_2 formed gives the water solution a light orange color (center). Extraction with a small amount of an organic solvent gives the characteristic reddish-orange color of Br_2 (right). (Charles D. Winters)

$E°$ and K

Recall from Chapter 16 that

$$\Delta G° = -RT \ln K$$

We have just shown that $\Delta G° = -nFE°$. It follows that

$$RT \ln K = nFE°$$

or

$$E° = \frac{RT}{nF} \ln K$$

The quantity RT/F is readily evaluated at 25°C, the temperature at which standard potentials are tabulated.

$$\frac{RT}{F} = \frac{8.31 \text{ J/mol} \cdot \text{K} \times 298 \text{ K}}{9.648 \times 10^4 \text{ J/mol} \cdot \text{V}} = 0.0257 \text{ V}$$

Hence

$$E° = \frac{(0.0257 \text{ V})}{n} \ln K \qquad \text{(at 25°C)}$$

$$E° = \frac{(0.0591 \text{ V})}{n} \log_{10} K$$

Notice that if the standard voltage is positive, $\ln K$ is also positive and K is greater than one. Conversely, if the standard voltage is negative, $\ln K$ is also negative and K is less than one.

Example 17.6 For the reaction

$$3Ag(s) + NO_3^-(aq) + 4H^+(aq) \longrightarrow 3Ag^+(aq) + NO(g) + 2H_2O$$

calculate the equilibrium constant K, using data in Table 17.1.

Strategy Use Table 17.1 to obtain the standard voltage. Then use the equation

$E° = \frac{(0.0257 \text{ V})}{n} \ln K$ to find K.

Solution

(1) $E° = E°_{red}\ NO_3^- + E°_{ox}\ Ag = +0.964 \text{ V} - 0.799 \text{ V} = +0.165 \text{ V}$
(2) To find n, it is helpful to break the equation into two half-equations:

$$3Ag(s) \longrightarrow 3Ag^+(aq) + 3e^-$$

$$NO_3^-(aq) + 4H^+(aq) + 3e^- \longrightarrow NO(g) + 2H_2O$$

Clearly, $n = 3$.

$$\ln K = \frac{nE°}{(0.0257 \text{ V})} = \frac{3(+0.165 \text{ V})}{(0.0257 \text{ V})} = 19.3;\ K = 2 \times 10^8$$

Table 17.2 lists values of K corresponding to various values of $E°$ with $n = 2$. Notice that if the standard voltage is greater than about 0.2 V, K is very large; if $E°$ is smaller than -0.2 V, K is very small. Only if the standard voltage falls in a rather narrow range, say 0.2 to -0.2 V, will the value of the equilibrium constant be such that the redox reaction will produce an equilibrium mixture containing appreciable amounts of both reactants and products.

Most redox reactions go either to completion or not at all

Table 17.2 Relation between $E°$ and K ($n = 2$)

$E°$	ln K	K	$E°$	ln K	K
1.00	77.8	6×10^{33}	−0.05	−3.9	0.02
0.80	62	1×10^{27}	−0.10	−7.8	0.0004
0.60	47	2×10^{20}	−0.20	−16	2×10^{-7}
0.40	31	3×10^{13}	−0.40	−31	3×10^{-14}
0.20	16	6×10^{6}	−0.60	−47	5×10^{-21}
0.10	7.8	2×10^{3}	−0.80	−62	9×10^{-28}
0.05	3.9	50	−1.00	−77.8	2×10^{-34}
0	0	1			

17.4 EFFECT OF CONCENTRATION UPON VOLTAGE

To this point, we have dealt only with "standard" voltages, i.e., voltages when all gases are at 1 atm pressure and all species in aqueous solution are at a concentration of 1 *M*. When the concentration of a reactant or product changes, the voltage changes as well. Qualitatively, the direction in which the voltage shifts is readily predicted:

1. Voltage will *increase* if the concentration of a reactant is increased or that of a product is decreased. Either of these changes increases the driving force behind the redox reaction, making it more spontaneous.
2. Voltage will *decrease* if the concentration of a reactant is decreased or that of a product is increased. Either of these changes makes the redox reaction less spontaneous.

When a voltaic cell operates, supplying electrical energy, the concentration of reactants decreases and that of the products increases. As time passes, the voltage drops steadily. Eventually it becomes zero, and we say that the cell is "dead." At that point, the redox reaction taking place within the cell is at equilibrium and there is no driving force to produce a voltage.

That's what happens when you leave your car lights on

Nernst Equation

To obtain a quantitative relation between cell voltage and concentration, it is convenient to start with the general expression for the free energy change discussed in Chapter 16:

$$\Delta G = \Delta G° + RT \ln Q$$

Substituting for ΔG and $\Delta G°$ from the relations obtained in Section 17.3,

$$\Delta G = -nFE \qquad \Delta G° = -nFE°$$

yields $-nFE = -nFE° + RT \ln Q$.

Solving for the cell voltage E,

$$E = E° - \frac{RT}{nF} \ln Q$$

This relationship is generally known as the **Nernst equation**. Recalling that, at 25°C, the quantity RT/F is 0.0257 V,

Nernst was a German physical chemist (1864–1941)

$$E = E° - \frac{(0.0257 \text{ V})}{n} \ln Q$$

In terms of base 10 logarithms ($\ln x = 2.303 \log_{10} x$),

$$E = E° - \frac{(0.0591 \text{ V})}{n} \log_{10} Q$$

In these equations, E is the cell voltage, $E°$ the standard voltage, n is the number of moles of electrons exchanged in the reaction, and Q is the reaction quotient.

If $Q > 1$, the reaction is less spontaneous and $E < E°$

Remember that gases enter Q as their partial pressures in atmospheres. Species in water solution enter as their molar concentrations. Pure liquids and solids do not appear in the expression for Q. For example,

$$a\text{A}(s) + b\text{B}(aq) \longrightarrow c\text{C}(aq) + d\text{D}(g) \qquad Q = \frac{[\text{C}]^c \times (P_\text{D})^d}{[\text{B}]^b}$$

Example 17.7 Consider a voltaic cell in which the following reaction occurs:

$$O_2(g) + 4H^+(aq) + 4Br^-(aq) \longrightarrow 2H_2O + 2Br_2(l)$$

Calculate the cell voltage, E, when O_2 is at 1.0 atm pressure, $[H^+] = [Br^-] = 0.10\ M$.

Strategy First, set up the Nernst equation, following the rules for Q listed above. Then calculate $E°$, using standard potentials in Table 17.1. Finally, using the Nernst equation, calculate E.

Solution

(1) $Q = \dfrac{1}{(P_{O_2}) \times [H^+]^4 \times [Br^-]^4}$

To find n, break the reaction down into two half-reactions:

$$O_2(g) + 4H^+(aq) + 4e^- \longrightarrow 2H_2O$$

$$4Br^-(aq) \longrightarrow 2Br_2(l) + 4e^-$$

Clearly, $n = 4$. The Nernst equation must then be

$$E = E° - \frac{(0.0257 \text{ V})}{4} \ln \frac{1}{(P_{O_2}) \times [H^+]^4 \times [Br^-]^4}$$

(2) $E° = E°_{red}\, O_2 + E°_{ox}\, Br^- = +1.229 \text{ V} - 1.077 \text{ V} = +0.152 \text{ V}$

(3) $E = +0.152 \text{ V} - \dfrac{(0.0257 \text{ V})}{4} \ln \dfrac{1}{1.0(0.10)^4(0.10)^4}$

$= +0.152 \text{ V} - \dfrac{(0.0257 \text{ V})}{4} \ln (1.0 \times 10^8)$

$= +0.152 \text{ V} - \dfrac{(0.0257 \text{ V})(18.4)}{4} = +0.034 \text{ V}$

The Nernst equation can also be used to determine the effect of changes in concentration upon the voltage of an individual half-cell, E°_{red} or E°_{ox}. Consider, for example, the half-reaction

$$MnO_4^-(aq) + 8H^+(aq) + 5e^- \longrightarrow Mn^{2+}(aq) + 4H_2O \qquad E^\circ_{red} = +1.512\ V$$

Here the Nernst equation takes the form

$$E_{red} = +1.512\ V - \frac{(0.0257\ V)}{5}\ln\frac{[Mn^{2+}]}{[MnO_4^-] \times [H^+]^8}$$

where E_{red} is the observed reduction voltage corresponding to any given concentrations of Mn^{2+}, MnO_4^-, and H^+.

Use of the Nernst Equation to Determine Ion Concentrations

This approach is particularly useful for ions at low concentrations

In chemistry, the most important use of the Nernst equation lies in the experimental determination of the concentration of ions in solution. Suppose you measure the cell voltage E and know the concentrations of all but one species in the two half-cells. It should then be possible to calculate the concentration of that species by using the Nernst equation (Example 17.8).

Example 17.8 Consider a voltaic cell in which the reaction is

$$Zn(s) + 2H^+(aq) \longrightarrow Zn^{2+}(aq) + H_2(g)$$

It is found that the cell voltage is +0.460 V when $[Zn^{2+}] = 1.0\ M$, $P_{H_2} = 1.0$ atm. What must be the concentration of H^+ in the H_2–H^+ half-cell?

Strategy This example is handled exactly like Example 17.7, except that in the last step you solve for $[H^+]$ instead of E.

Solution

(1) Setting up the Nernst equation with $n = 2$,

$$E = E^\circ - \frac{(0.0257\ V)}{2}\ln\frac{[Zn^{2+}] \times (P_{H_2})}{[H^+]^2}$$

(2) $E^\circ = E^\circ_{ox}\ Zn + E^\circ_{red}\ H^+ = +0.762\ V$

(3) All that remains is to substitute for E, E°_{tot}, $[Zn^{2+}]$, P_{H_2}, and solve for $[H^+]$:

$$+0.460\ V = +0.762\ V - \frac{0.0257\ V}{2}\ln\frac{1 \times 1}{[H^+]^2}$$

Solving:

$$\ln\frac{1}{[H^+]^2} = \frac{2(+0.460\ V - 0.762\ V)}{-0.0257\ V} = 23.5$$

$$\frac{1}{[H^+]^2} = 1.6 \times 10^{10} \qquad [H^+] = 8 \times 10^{-6}\ M$$

As Example 17.8 implies, the $Zn\,|\,Zn^{2+}\,\|\,H^+\,|\,H_2\,|\,Pt$ cell can be used to measure the concentration of H^+ or pH of a solution. Indeed, cells of this type are commonly used to measure pH; Figure 17.9 shows a schematic diagram of a cell used with a pH-meter. The pH-meter, referred to in Chapter 13, is actually a high-resistance voltmeter calibrated to read pH rather than voltage. The cell connected to the pH-meter consists of two half-cells. One of these is a reference half-cell of known voltage. The other half-cell contains a solution of known pH separated by a thin fragile *glass electrode* from a solution whose pH is to be determined. The voltage of this cell is a linear function of the pH of the solution in the beaker.

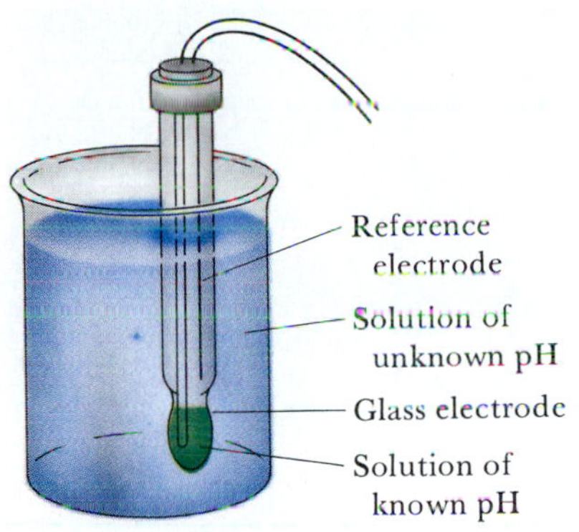

Figure 17.9
The pH of a solution can be determined with the aid of a "glass electrode." The voltage between the glass electrode and the reference electrode is directly related to pH. The leads from the electrode are connected to a pH meter of the type discussed in Chapter 13.

Specific ion electrodes, similar in design to the glass electrode, have been developed to analyze for a variety of cations and anions. One of the first to be used extensively was a fluoride ion electrode which is sensitive to F^- at concentrations as low as 0.1 part per million and hence is ideal for monitoring fluoridated water supplies. An electrode which is specific for Cl^- ions is used to diagnose for cystic fibrosis. Attached directly to the skin, it detects the abnormally high concentrations of sodium chloride in sweat that are a characteristic symptom of this disorder. Diagnoses that used to require an hour or more can now be carried out in a few minutes; as a result, large numbers of children can be rapidly and routinely screened.

The general approach illustrated by Example 17.8 is widely used to determine equilibrium constants for solution reactions. The pH-meter in particular can be used to determine acid or base dissociation constants by measuring the pH of solutions containing known concentrations of weak acids or bases. Specific ion electrodes are readily adapted to the determination of solubility product constants. For example, a chloride ion electrode can be used to find $[Cl^-]$ in equilibrium with AgCl(*s*) and a known $[Ag^+]$. From that information, K_{sp} of AgCl can be calculated.

Most K_{sp} values are determined this way

17.5 ELECTROLYTIC CELLS

In an electrolytic cell, a nonspontaneous redox reaction is made to occur by pumping electrical energy into the system. A generalized diagram for such a cell is shown in Figure 17.10. The storage battery at the left provides a source of direct electric current. From the terminals of the battery, two wires lead to the electrolytic cell. This consists of two electrodes, A and C, dipping into a solution containing ions M^+ and X.

The battery acts as an electron pump, pushing electrons into the *cathode,* C, and removing them from the *anode,* A. To maintain electrical neutrality, some process within the cell must consume electrons at C and liberate them at A. This process is an oxidation-reduction reaction; when carried out in an electrolytic cell, it is called **electrolysis**.

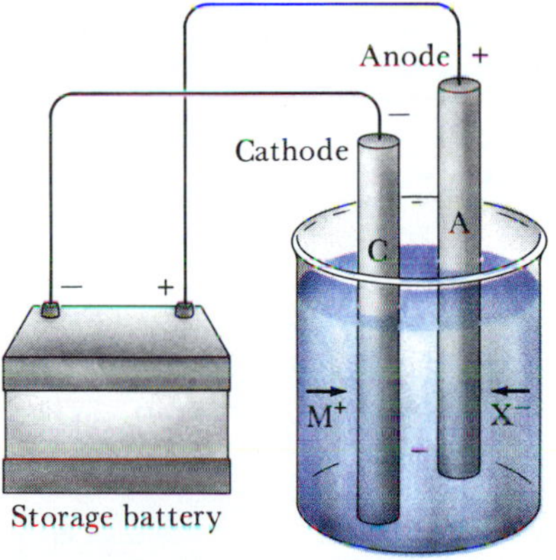

Figure 17.10
Schematic diagram of an electrolytic cell.

There is a simple relationship between the amount of electricity passed through an electrolytic cell and the amounts of substances produced by oxidation or reduction at the electrodes. From the balanced half-equations

$$Ag^+(aq) + e^- \longrightarrow Ag(s)$$

$$Cu^{2+}(aq) + 2e^- \longrightarrow Cu(s)$$

$$Au^{3+}(aq) + 3e^- \longrightarrow Au(s)$$

you can deduce that

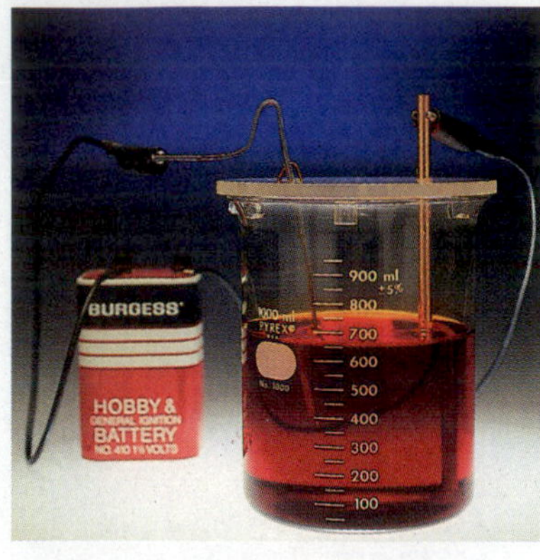

$$1 \text{ mol } e^- \longrightarrow 1 \text{ mol Ag } (107.9 \text{ g Ag})$$
$$2 \text{ mol } e^- \longrightarrow 1 \text{ mol Cu } (63.55 \text{ g Cu})$$
$$3 \text{ mol } e^- \longrightarrow 1 \text{ mol Au } (197.0 \text{ g Au})$$

Relations of this type, obtained from balanced half-equations, can be used in many practical calculations involving electrolytic cells. Frequently, the relations between electrical units provided in Table 17.3 will be required as well.

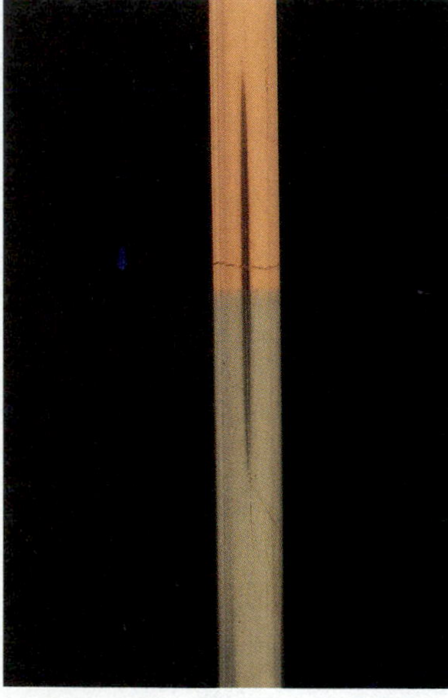

Figure 17.11
Chromium metal can be plated on copper by electrolysis of a deep red water solution of CrO_3. (Charles D. Winters)

Example 17.9 Chromium metal can be electroplated from an acidic solution of CrO_3 (Fig. 17.11).

(a) How many grams of chromium will be plated by 1.00×10^4 C?
(b) How long will it take to plate one gram of chromium using a current of 6.00 A?

Strategy The first thing to do is to write the half-equation for the reduction. Following the rules cited in Chapter 4, you should arrive at

$$CrO_3(s) + 6H^+(aq) + 6e^- \longrightarrow Cr(s) + 3H_2O$$

From this equation, it should be clear that 6 mol e^- = 1 mol Cr (52.0 g Cr). That relationship, along with the relations 1 mol e^- = 9.648×10^4 C and 1 A = 1 C/s can be used to calculate the required quantities. Follow a conversion factor approach.

Solution

(a) $1.00 \times 10^4 \text{ C} \times \dfrac{1 \text{ mol } e^-}{9.648 \times 10^4 \text{ C}} \times \dfrac{52.0 \text{ g Cr}}{6 \text{ mol } e^-} = \boxed{0.898 \text{ g Cr}}$

(b) The most straightforward approach here is to calculate the number of coulombs first and then the time.

(1) $1.00 \text{ g Cr} \times \dfrac{6 \text{ mol } e^-}{52.0 \text{ g Cr}} \times \dfrac{9.648 \times 10^4 \text{ C}}{1 \text{ mol } e^-} = 1.11 \times 10^4 \text{ C}$

(2) Since 1 A = 1 C/s,

$$\text{time (s)} = \frac{\text{charge (C)}}{\text{current (A)}} = \frac{1.11 \times 10^4 \text{ C}}{6.00 \text{ C/s}} = \boxed{1.85 \times 10^3 \text{ s}} \quad \text{(about a half-hour)}$$

Table 17.3 Electrical Units

Quantity	Unit	Defining Relation	Conversion Factors
Charge	coulomb (C)	1 C = 1 A · s = 1 J/V	1 mol e^- = 9.648×10^4 C
Current	ampere (A)	1 A = 1 C/s	
Potential	volt (V)	1 V = 1 J/C	
Power	watt (W)	1 W = 1 J/s	
Energy	joule (J)	1 J = 1 V · C	1 kWh = 3.600×10^6 J

1 kilowatt hour = 1 kWh

Example 17.10 Consider the electroplating of chromium, referred to in Example 17.9. If the applied voltage is 4.5 V, calculate the amount of electrical energy absorbed in plating 1.00 g Cr, first in joules and then in kilowatt-hours.

Strategy Recall that, in Example 17.9b, you calculated the number of coulombs required to plate one gram of Cr. Multiplying that by the number of volts will give the energy in joules, which can then be converted to kilowatt-hours.

Solution

$$\text{energy (J)} = \text{voltage (V)} \times \text{charge (C)} = 4.5\text{ V} \times 1.1 \times 10^4\text{ C} = \boxed{5.0 \times 10^4\text{ J}}$$

$$\text{energy in kWh} = 5.0 \times 10^4\text{ J} \times \frac{1\text{ kWh}}{3.6 \times 10^6\text{ J}} = \boxed{1.4 \times 10^{-2}\text{ kWh}}$$

The laws of electrolysis were discovered by Michael Faraday, perhaps the most talented experimental scientist of the nineteenth century. Faraday lived his entire life in what is now greater London. The son of a blacksmith, he had no formal education beyond the rudiments of reading, writing, and arithmetic. Apprenticed to a bookbinder at the age of 13, Faraday educated himself by reading virtually every book that came into the shop. Anxious to escape a life of drudgery as a tradesman, Faraday wrote to Sir Humphrey Davy at the Royal Institution, requesting employment. Shortly afterwards, a vacancy arose and Faraday was hired as a laboratory assistant.

Faraday at about 30 years old. (Oesper Collection in the History of Chemistry, University of Cincinnati)

Davy quickly recognized Faraday's talents and as time passed allowed him to work more and more independently. In his years with Davy, Faraday published papers covering almost every field of chemistry. They included studies on the condensation of gases (he was the first to liquefy ammonia), the reaction of silver compounds with ammonia, and the isolation of several organic compounds, the most important of which was benzene. In 1816, Faraday began a series of lectures at the Royal Institution that were brilliantly successful. In 1825 he succeeded Davy as director of the laboratory. As Faraday's reputation grew, it was said that "Humphry Davy's greatest discovery was Michael Faraday." Perhaps it was witticisms of this sort that led to an estrangement between master and protégé. Late in his life, Davy opposed Faraday's nomination as a Fellow of the Royal Society and is reputed to have cast the only vote against him.

To Michael Faraday, science was an obsession; one of his biographers describes him as a "work maniac." An observer (Faraday had no students) said of him " . . . if he had to cross the laboratory for anything, he did not walk, he ran; the quickness of his perception was equalled by the calm rapidity of his movements." In 1839, he suffered a nervous breakdown, the result of overwork. For much of the rest of his life, Faraday was in poor health. He gradually gave up more and more of his social engagements but continued to do research at the same pace as before.

Faraday developed the laws of electrolysis between 1831 and 1834. In mid-December of 1833, he began a quantitative study of the electrolysis of several metal cations, including Sn^{2+}, Pb^{2+}, and Zn^{2+}. Despite taking a whole day off for Christmas, he managed to complete these experiments, write up the results of 3 years' work, and get his paper published in the *Philosophic Transactions of the Royal Society* on January 9, 1834. In this paper, Faraday introduced the basic vocabulary of electrochemistry, using for the first time the terms "anode," "cathode," "ion," "electrolyte," and "electrolysis."

In his spare time, Faraday invented the electric motor and generator

Nobody is 100% efficient

In working Examples 17.9 and 17.10, we have in effect assumed that the electrolyses were 100% efficient in converting electrical energy into chemical energy. In practice, this is almost never the case. Some electrical energy is wasted in side reactions at the electrodes and in the form of heat. This means that the actual yield of products is less than the "theoretical yield."

17.6 COMMERCIAL CELLS

To a chemist, electrochemical cells are of interest primarily for the information they yield concerning the spontaneity of redox reactions, the strengths of oxidizing and reducing agents, and the concentrations of trace species in solution. The viewpoint of an engineer is somewhat different; here, applications of electrolytic cells in electroplating and electrosynthesis are of particular importance. To the layman, electrochemistry is important primarily because of commercial voltaic cells, which supply the electrical energy for instruments ranging in size from pacemakers to automobiles.

Electrolysis of Aqueous NaCl

From a commercial standpoint, the most important electrolysis carried out in water solution is that of sodium chloride (Fig. 17.12). At the anode, Cl^- ions are oxidized to chlorine gas:

$$\text{anode:} \quad 2Cl^-(aq) \longrightarrow Cl_2(g) + 2e^-$$

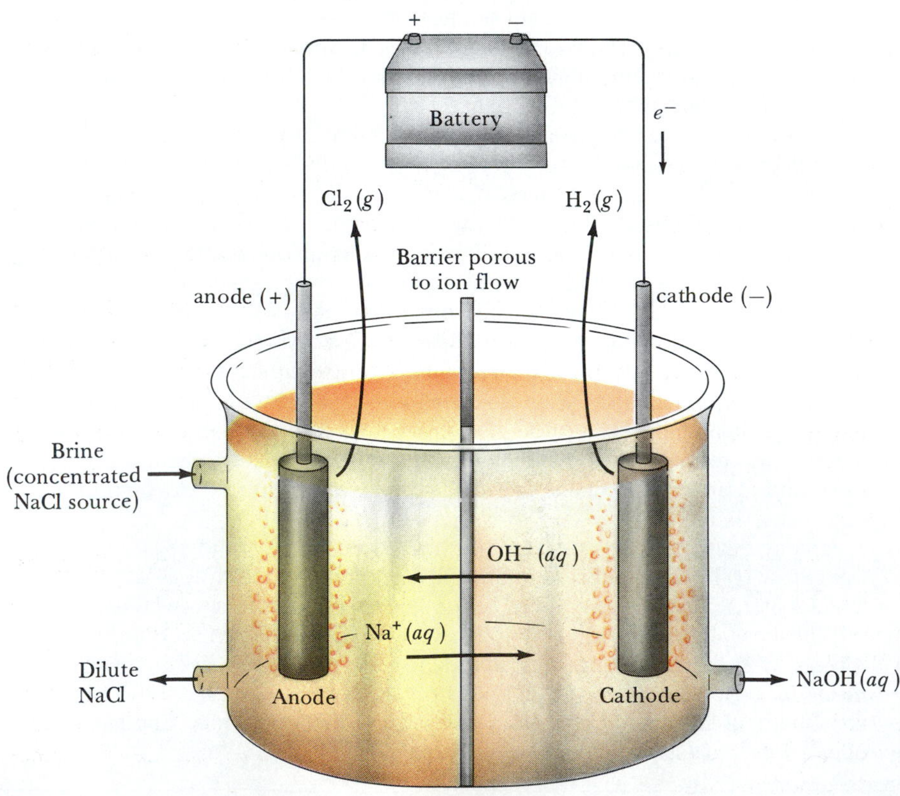

Figure 17.12
Schematic diagram for the electrolysis of aqueous NaCl (brine). Migration of ions through the membrane maintains charge balance.

At the cathode, the half-reaction involves H_2O molecules, which are easier to reduce ($E^\circ_{red} = -0.828$ V) than Na^+ ions ($E^\circ_{red} = -2.714$ V).

$$\text{cathode:} \qquad 2H_2O + 2e^- \longrightarrow H_2(g) + 2\,OH^-(aq)$$

The overall cell reaction is obtained by summing the half-reactions:

$$2Cl^-(aq) + 2H_2O \longrightarrow Cl_2(g) + H_2(g) + 2\,OH^-(aq)$$

For the reaction, $E < 0$ and $\Delta G > 0$

Chlorine gas bubbles out of solution at the anode. At the cathode, hydrogen gas is formed, and the solution around the electrode becomes strongly basic.

The products of electrolysis have a variety of uses. Chlorine is used to purify drinking water; large quantities of it are consumed in making plastics such as polyvinyl chloride (PVC). Hydrogen, prepared in this and many other industrial processes, is used chiefly in the synthesis of ammonia (Chap. 12). Sodium hydroxide (lye), obtained on evaporation of the electrolyte, is used in processing pulp and paper, in the purification of aluminum ore, in the manufacture of glass and textiles, and for many other purposes.

Primary (Nonrechargeable) Voltaic Cells

The construction of the ordinary dry cell (Leclanché cell) used in flashlights is shown in Figure 17.13. The zinc wall of the cell is the anode. The graphite rod through the center of the cell is the cathode. The space between the electrodes is filled with a moist paste. This contains MnO_2, $ZnCl_2$, and NH_4Cl. When the cell operates, the half-reaction at the anode is

A flashlight draws about 1 A and runs for about an hour before "dying"

$$Zn(s) \longrightarrow Zn^{2+}(aq) + 2e^-$$

At the cathode, manganese dioxide is reduced to species in which Mn is in the +3 oxidation state, such as Mn_2O_3:

$$2MnO_2(s) + 2NH_4^+(aq) + 2e^- \longrightarrow Mn_2O_3(s) + 2NH_3(aq) + H_2O$$

The overall reaction occurring in this voltaic cell is

$$Zn(s) + 2MnO_2(s) + 2NH_4^+(aq) \longrightarrow Zn^{2+}(aq) + Mn_2O_3(s) + 2NH_3(aq) + H_2O$$

If too large a current is drawn from a Leclanché cell, the ammonia forms a gaseous insulating layer around the carbon cathode. When this happens, the voltage drops sharply and then returns slowly to its normal value of 1.5 V. This problem can be avoided by using an "alkaline" dry cell, in which the paste between the electrodes contains KOH rather than NH_4Cl. In this case the overall cell reaction is simply

$$Zn(s) + 2MnO_2(s) \longrightarrow ZnO(s) + Mn_2O_3(s)$$

No gas is produced. The alkaline dry cell, although more expensive than the Leclanché cell, has a longer shelf-life and provides more current.

Another important primary battery is the mercury cell. It usually comes in very small sizes and is used in hearing aids, watches, cameras, and some calculators. The anode of this cell is a zinc–mercury amalgam; the reacting species is zinc. The cathode is a plate made up of mercury(II) oxide, HgO. The electrolyte is a paste containing HgO and sodium or potassium hydroxide. The electrode reactions are

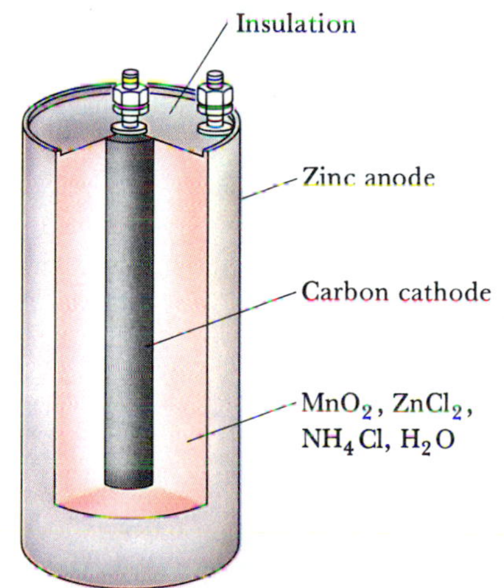

Figure 17.13
Section of an ordinary Zn–MnO_2 dry cell. This cell produces 1.5 V and will deliver a current of about half an ampere for 6 hours.

$$
\begin{array}{rlcl}
\text{anode:} & Zn(s) + 2\,OH^-(aq) & \longrightarrow & Zn(OH)_2(s) + 2e^- \\
\text{cathode:} & HgO(s) + H_2O + 2e^- & \longrightarrow & Hg(l) + 2\,OH^-(aq) \\
\hline
 & Zn(s) + HgO(s) + H_2O & \longrightarrow & Zn(OH)_2(s) + Hg(l)
\end{array}
$$

Notice that the overall reaction does not involve any ions in solution, so there are no concentration changes when current is drawn. As a result, the battery maintains a constant voltage of about 1.3 V throughout its life.

Storage (Rechargeable) Voltaic Cells

A storage cell, unlike an ordinary dry cell, can be recharged repeatedly. This can be accomplished because the products of the reaction are deposited directly on the electrodes. By passing a current through a storage cell, it is possible to reverse the electrode reactions and restore the cell to its original condition.

A 12-V storage battery can deliver 300 A for a minute or so

The best-known voltaic cell of this type is the lead storage battery. The 12-V battery used in automobiles consists of six voltaic cells of the type shown in Figure 17.14. A group of lead plates, the grills of which are filled with spongy gray lead, forms the anode of the cell. The multiple cathode consists of another group of plates of similar design filled with lead(IV) oxide, PbO_2. These two sets of plates alternate through the cell. They are immersed in a water solution of sulfuric acid, H_2SO_4, which acts as the electrolyte.

When a lead storage battery is supplying current, the lead in the anode grids is oxidized to Pb^{2+} ions. These immediately react with SO_4^{2-} ions in the electrolyte, precipitating $PbSO_4$ (lead sulfate) on the plates. At the cathode, lead dioxide is reduced to Pb^{2+} ions, which also precipitate as $PbSO_4$:

$$
\begin{array}{rcl}
Pb(s) + SO_4^{2-}(aq) & \longrightarrow & PbSO_4(s) + 2e^- \\
PbO_2(s) + 4H^+(aq) + SO_4^{2-}(aq) + 2e^- & \longrightarrow & PbSO_4(s) + 2H_2O \\
\hline
Pb(s) + PbO_2(s) + 4H^+(aq) + 2SO_4^{2-}(aq) & \longrightarrow & 2PbSO_4(s) + 2H_2O
\end{array}
$$

Deposits of lead sulfate slowly build up on the plates, partially covering and replacing the lead and lead dioxide. As the cell discharges, the concentration of sulfuric acid decreases. For every mole of lead reacting, two moles of H_2SO_4

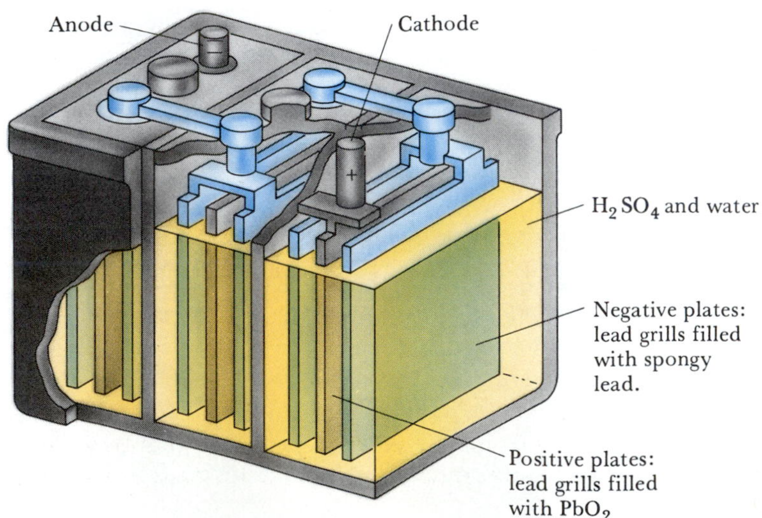

Figure 17.14
One cell of a lead storage battery. Three advantages of the lead storage battery are its ability to deliver large amounts of energy for a short time, the ease of recharging, and a nearly constant voltage from full charge to discharge. A disadvantage is its high mass/energy ratio.

$(4H^+, 2SO_4{}^{2-})$ are replaced by two moles of water. The state of charge of a storage battery can be checked by measuring the density of the electrolyte. When the battery is fully charged, the density is in the range of 1.25 to 1.30 g/cm^3. A density below 1.20 g/cm^3 indicates a low sulfuric acid concentration and hence a partially discharged cell.

A lead storage battery can be recharged and thus restored to its original condition. To do this, a direct current is passed through the cell in the reverse direction. While a storage battery is being recharged, it acts as an electrolytic cell. The overall cell reaction is the reverse of that occurring when the battery discharges:

$$2PbSO_4(s) + 2H_2O \longrightarrow Pb(s) + PbO_2(s) + 4H^+(aq) + 2SO_4{}^{2-}(aq)$$

The electrical energy required to bring about this nonspontaneous reaction in an automobile is furnished by an alternator equipped with a rectifier to convert alternating to direct current.

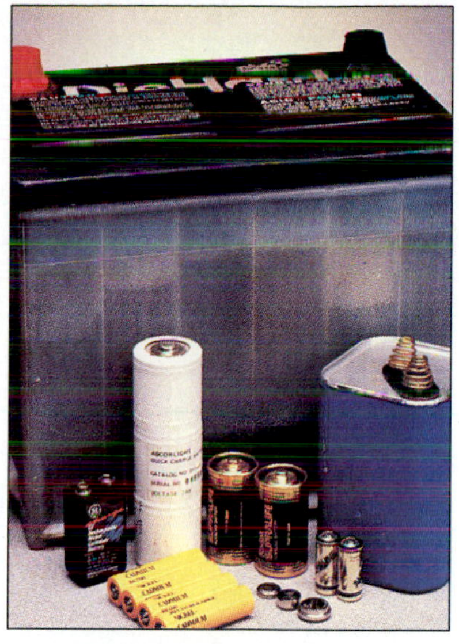

Four types of batteries. (Richard Megna/FUNDAMENTAL PHOTOGRAPHS, New York)

As you may have found from experience, lead storage batteries do not endure forever, particularly if they are allowed to stand for some time when discharged. Repeated quick-charging can cause Pb, PbO_2, and $PbSO_4$ to flake off the electrodes. This collects as a sludge at the bottom of the battery, often short-circuiting one or more cells. Discharged batteries are also susceptible to freezing, since the sulfuric acid concentration is low. If freezing occurs, the electrodes may warp and come in contact with one another.

Another type of rechargeable voltaic cell is the "Nicad" storage battery, used for small appliances, tools, and calculators. The anode in this cell is made of cadmium metal, and the cathode contains nickel(IV) oxide, NiO_2. The electrolyte is a concentrated solution of potassium hydroxide. The discharge reactions are

$$\begin{array}{lrcl} \text{anode:} & Cd(s) + 2OH^-(aq) & \longrightarrow & Cd(OH)_2(s) + 2e^- \\ \text{cathode:} & NiO_2(s) + 2H_2O + 2e^- & \longrightarrow & Ni(OH)_2(s) + 2\,OH^-(aq) \\ \hline & Cd(s) + NiO_2(s) + 2H_2O & \longrightarrow & Cd(OH)_2(s) + Ni(OH)_2(s) \end{array}$$

The insoluble hydroxides of cadmium and nickel deposit on the electrodes. Hence, the half-reactions are readily reversed during recharging. Nicad batteries are more expensive than lead storage batteries for a given amount of electrical energy delivered but also have a longer life.

PERSPECTIVE • Corrosion of Metals

Most metals corrode when exposed to the atmosphere, reacting with oxygen, water vapor, or carbon dioxide. Gold and platinum are among the few metals that retain their shiny appearance indefinitely when exposed to air; these metals are very difficult to oxidize (E°_{ox} Pt = −1.320 V, Au = −1.498 V).

Aluminum (E°_{ox} = +1.68 V) reacts readily with oxygen of the air:

$$4Al(s) + 3\,O_2(g) \longrightarrow 2Al_2O_3(s)$$

However, the Al_2O_3 coating, which is only about 10^{-8} m thick, adheres tightly to the surface of the metal. This prevents further corrosion and explains why aluminum cookware does not disintegrate upon exposure to air.

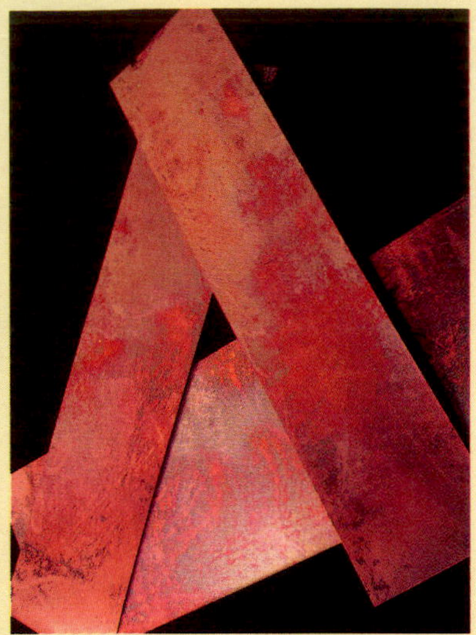

Figure 17.A
Copper strips exposed to moist air. (Paul Silverman/FUNDAMENTAL PHOTOGRAPHS, New York)

Copper in moist air (Fig. 17.A) slowly acquires a dull green coating. The green material is a 1:1 mole mixture of $Cu(OH)_2$ and $CuCO_3$:

$$2Cu(s) + H_2O(g) + CO_2(g) + O_2(g) \longrightarrow Cu(OH)_2 \cdot CuCO_3(s)$$

Several other elements, including zinc and lead, react similarly. The products, $Zn(OH)_2 \cdot ZnCO_3$ and $Pb(OH)_2 \cdot PbCO_3$, are white and adhere tightly to the metal, preventing further corrosion. In the case of lead, the protective coating dissolves in acetic acid, primarily because Pb^{2+} forms a very stable complex with the acetate ion. It has been suggested that the ancient Romans suffered from lead poisoning because they stored wine (containing some acetic acid) in pottery vessels glazed with lead compounds.

From an economic standpoint, the most important corrosion reaction is that involving iron and steel. About 20% of all the iron produced each year goes to replace products whose usefulness has been destroyed by rust. When a piece of iron is exposed to water containing dissolved oxygen, the half-reactions of oxidation

$$Fe(s) \longrightarrow Fe^{2+}(aq) + 2e^-$$

and reduction

$$\tfrac{1}{2}O_2(g) + H_2O + 2e^- \longrightarrow 2\,OH^-(aq)$$

occur at different locations. The surface of a piece of corroding iron consists of a series of tiny voltaic cells. At *anodic areas,* iron is oxidized to Fe^{2+} ions; at *cathodic areas,* elementary oxygen is reduced to OH^- ions. Electrons are transferred through the iron, which acts like the external conductor of an ordinary voltaic cell. The electrical circuit is completed by the flow of ions through the water solution or film covering the iron.

Many characteristics of corrosion are most readily explained in terms of an electrochemical mechanism. A perfectly dry metal surface is not attacked by oxygen; iron exposed to dry air does not corrode. This seems plausible if corrosion occurs through a voltaic cell, which requires a water solution through which ions can move to complete the circuit. The fact that corrosion occurs more readily in seawater than in fresh water has a similar explanation. The dissolved salts in seawater supply the ions necessary for the conduction of current.

The existence of discrete cathodic and anodic areas on a piece of corroding iron requires that adjacent surface areas differ from each other chem-

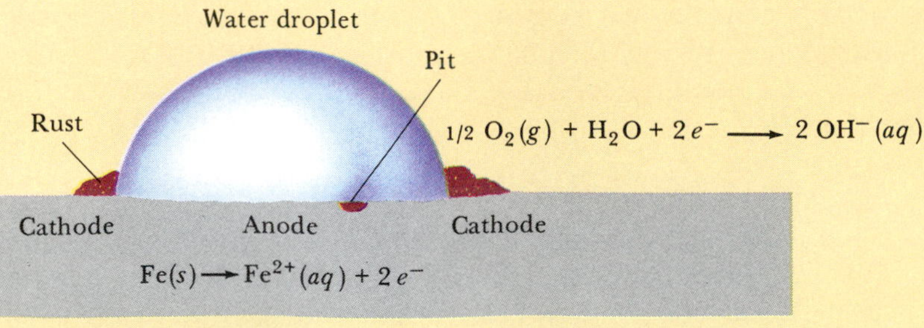

Figure 17.B
Corrosion of iron under a drop of water. The Fe^{2+} ions migrate toward the edge of the drop, where they precipitate as $Fe(OH)_2$, which later forms $Fe(OH)_3$.

ically. This can happen if there are differences in oxygen concentration along the metal surface, as when a drop of water adheres to the surface of a piece of iron exposed to the air (Fig. 17.B). The metal around the edges of the drop is in contact with water containing a high concentration of dissolved oxygen. The water touching the metal beneath the center of the drop is depleted in oxygen, since it is cut off from contact with air. As a result, a small oxygen concentration cell is set up. The area around the edge of the drop, where the oxygen concentration is high, becomes cathodic; oxygen molecules are reduced there. Directly beneath the drop is an anodic area where the iron is oxidized. A particle of dirt on the surface of an iron object can act in much the same way as a drop of water to cut off the supply of oxygen to the area beneath it and thereby establish anodic and cathodic areas. This explains why garden tools left covered with soil are particularly susceptible to rusting.

CHAPTER HIGHLIGHTS

KEY CONCEPTS

1. *Draw a diagram for a voltaic cell, labeling electrodes and direction of current flow*
   (Example 17.1; Problems 3–6)
2. *Use standard potentials (Table 17.1) to*
   —rank oxidizing and reducing agents
   (Example 17.2; Problems 7–14)
   —calculate $E°$ and/or reaction spontaneity
   (Examples 17.3, 17.4; Problems 15–32)
3. *Relate $E°$ to $\Delta G°$ and K*
   (Examples 17.5, 17.6; Problems 33–42)
4. *Use the Nernst equation to relate voltage to concentration*
   (Examples 17.7, 17.8; Problems 43–54, 71, 75–78)
5. *Relate mass to coulombs or joules in electrolysis reactions*
   (Examples 17.9, 17.10; Problems 55–60, 65, 66, 69, 70)

KEY EQUATIONS

Standard voltage $$E° = E°_{ox} + E°_{red}$$

$E°$, $\Delta G°$, K $$E° = \frac{\Delta G°}{-nF} = \frac{RT \ln K}{nF}$$

Nernst equation $$E = E° - \frac{RT}{nF} \ln Q$$

KEY TERMS

ampere
anode
cathode
coulomb
$E°$, $E°_{ox}$, $E°_{red}$
electrolytic cell
$\Delta G°$
oxidation
oxidizing agent
reducing agent
reduction
salt bridge
standard potential
volt
voltaic cell

SUMMARY PROBLEM

A voltaic cell consists of two half-cells, one of which contains a Pt electrode surrounded by Cr^{3+} and $Cr_2O_7^{2-}$ ions. The other half-cell contains a Pt electrode surrounded by Mn^{2+} ions and $MnO_2(s)$. Assume the cell reaction, which produces a positive voltage, involves both Cr^{3+} and Mn^{2+} ions. Take $T = 25°C$.

a. Write the anode half-reaction, the cathode half-reaction, and the overall equation for the cell.
b. Write the cell description in abbreviated notation.
c. Calculate $E°$ for the cell.
d. For the redox reaction in (a), calculate K and $\Delta G°$.
e. Calculate the voltage of the cell when all ionic species except H^+ have a concentration of 0.300 M and the solution has a pH of 3.5.

An electrolytic cell contains a solution of $Cr(NO_3)_3$. Assume that chromium metal plates out at one electrode and oxygen gas is evolved at the other electrode.

f. Write the anode half-reaction, the cathode half-reaction, and the overall reaction for the cell.
g. How long will it take to deposit 15.0 g of chromium metal, using a current of 4.50 A?
h. A current of 4.50 A is passed through the cell for 30.0 min. Starting out with 250 mL of 1.00 M $Cr(NO_3)_3$, what is the concentration of Cr^{3+} after electrolysis? The pH of the solution? Assume 100% current efficiency and no change in volume during electrolysis.

Answers

a. anode: $Mn^{2+}(aq) + 2H_2O \longrightarrow MnO_2(s) + 4H^+(aq) + 2e^-$
cathode: $Cr_2O_7^{2-}(aq) + 14H^+(aq) + 6e^- \longrightarrow 2Cr^{3+}(aq) + 7H_2O$
overall: $3Mn^{2+}(aq) + Cr_2O_7^{2-}(aq) + 2H^+(aq) \longrightarrow 2Cr^{3+}(aq) + 3MnO_2(s) + H_2O$
b. (Pt) $Mn^{2+} | MnO_2 \| Cr_2O_7^{2-} | Cr^{3+} |$ Pt
c. 0.10 V **d.** $K = 1.4 \times 10^{10}$; $\Delta G° = -58$ kJ **e.** 0.02 V
f. anode: $Cr^{3+}(aq) + 3e^- \longrightarrow Cr(s)$
cathode: $2H_2O \longrightarrow 4H^+(aq) + O_2(g) + 4e^-$
overall: $4Cr^{3+}(aq) + 6H_2O \longrightarrow 4Cr(s) + 12H^+(aq) + 3\,O_2(g)$
g. 5.15 h **h.** 0.888 M $Cr(NO_3)_3$; pH = 0.47

QUESTIONS & PROBLEMS

Voltaic Cells

1. Write a balanced chemical equation for the overall cell reaction represented as
a. (Pt) $H_2 | H^+ \| Fe^{3+} | Fe^{2+} |$ Pt
b. $Cd | Cd^{2+} \| Ni^{2+} | Ni$
c. (Pt) $Cl^- | Cl_2 \| MnO_4^- | Mn^{2+} |$ Pt

2. Write a balanced chemical equation for the overall cell reaction represented as
a. $Zn | Zn^{2+} \| Cr^{3+} | Cr$
b. $Sn | Sn^{2+} \| O_2 | H_2O |$ Pt
c. $Al | Al^{3+} \| I_2 | I^- |$ Pt

3. Draw a diagram for a salt bridge cell for each of the following reactions. Label the anode and cathode, and indicate the direction of current flow throughout the circuit.
a. $Sn(s) + 2Ag^+(aq) \longrightarrow Sn^{2+}(aq) + 2Ag(s)$
b. $H_2(g) + Hg_2Cl_2(s) \longrightarrow 2H^+(aq) + 2Cl^-(aq) + 2Hg(l)$
c. $Pb(s) + PbO_2(s) + 4H^+(aq) + 2SO_4^{2-}(aq) \longrightarrow 2PbSO_4(s) + 2H_2O$

4. Follow the directions for Question 3 for the following.
a. $Zn(s) + Cd^{2+}(aq) \longrightarrow Zn^{2+}(aq) + Cd(s)$
b. $2AuCl_4^-(aq) + 3Cu(s) \longrightarrow 2Au(s) + 8Cl^-(aq) + 3Cu^{2+}(aq)$
c. $Fe(s) + Cu(OH)_2(s) \longrightarrow Cu(s) + Fe(OH)_2(s)$

5. Consider a salt bridge cell in which the anode is a manganese rod immersed in an aqueous manganese(II) sulfate solution. The cathode is a chromium strip immersed in an aqueous chromium(III) sulfate solution. Sketch a diagram of the cell, indicating the flow of current throughout. Write the half-equations for the electrode reactions, the overall equation, and the abbreviated notation for the cell.

6. Follow the directions for Question 5 for a salt bridge cell in which the anode is a platinum rod immersed in an aqueous solution of sodium iodide containing solid iodine crystals. The cathode is another platinum rod immersed in a solution of sodium bromide with bromine liquid.

Strength of Oxidizing and Reducing Species

7. Which species in each pair is the better oxidizing agent?
a. Br_2 or H_2O_2 **b.** SO_4^{2-} or MnO_4^-
c. $AgBr(s)$ or Ag^+
d. O_2 in acidic solution or O_2 in basic solution

8. Which species in each pair is the better reducing agent?
a. Au or Ag **b.** Cl^- or I^-
c. Mn^{2+} or Fe^{2+}
d. H_2 in acidic solution or H_2 in basic solution

9. Using Table 17.1, arrange the following reducing agents in order of increasing strength.

Br^-, Zn, Co, $PbSO_4$, H_2S (acidic solution)

10. Using Table 17.1, arrange the following oxidizing agents in order of increasing strength.

Al^{3+}, AgBr, F_2, ClO_3^- (acidic), Ni^{2+}

11. Consider the following species.

Cu^+, Zn, Ni^{2+}, Fe^{2+}, H^+ (acidic solution)

Classify each species as oxidizing agent, reducing agent, or both. Arrange the oxidizing agents in order of increasing strength. Do the same for the reducing agents.

12. Follow the directions for Question 11 for the following species.

Cr^{3+}, $Hg(l)$, H_2, Sn^{2+}, Br_2 (acidic solution)

13. Use Table 17.1 to select
a. a reducing agent that will convert Pb^{2+} to Pb but not Tl^+ to Tl.
b. an oxidizing agent to convert Fe to Fe^{2+} but not Co to Co^{2+}.
c. a reducing agent that converts Au^{3+} to Au but not $AuCl_4^-$ to Au.

14. Use Table 17.1 to select
a. an oxidizing agent that converts I^- to I_2 but not Cl^- to Cl_2.
b. a reducing agent capable of converting Co^{2+} to Co but not Zn^{2+} to Zn.
c. an oxidizing agent capable of converting Cl^- to Cl_2 but not F^- to F_2.

Calculation of $E°$

15. Calculate $E°$ for the following voltaic cells.
a. $Pb(s) + 2Ag^+(aq) \longrightarrow Pb^{2+}(aq) + 2Ag(s)$
b. $O_2(g) + 4Fe^{2+}(aq) + 4H^+(aq) \longrightarrow 2H_2O + 4Fe^{3+}(aq)$
c. a Cd–Cd^{2+} half-cell and a Zn–Zn^{2+} half-cell

16. Calculate $E°$ for the following voltaic cells.
a. $MnO_2(s) + 4H^+(aq) + 2I^-(aq) \longrightarrow Mn^{2+}(aq) + 2H_2O + I_2(s)$
b. $H_2(g) + 2\,OH^-(aq) + S(s) \longrightarrow 2H_2O + S^{2-}(aq)$
c. an Ag–Ag^+ half-cell and an Au–$AuCl_4^-$ half-cell.

17. Using Table 17.1, calculate $E°$ for
a. the reaction of chromium(II) ions with tin(IV) ions to produce chromium(III) ions and tin(II) ions.
b. the reaction between manganese(II) ions and hydrogen peroxide to produce solid manganese dioxide (MnO_2).
c. the reaction between iron and oxygen in base.

18. Using Table 17.1, calculate $E°$ for the reaction between
a. iron and water to produce iron(II) hydroxide and hydrogen gas.
b. iron and iron(III) ions to give iron(II) ions.
c. iron(II) hydroxide with oxygen in basic solution.

19. Calculate $E°$ for the cells
a. $Al \mid Al^{3+} \parallel NO_3^- \mid NO \mid Pt$
b. $(Pt)\ Cr^{2+} \mid Cr^{3+} \parallel O_2(\text{acidic}) \mid H_2O \mid Pt$
c. $Cu \mid Cu^{2+} \parallel I_2 \mid I^- \mid Pt$

20. Calculate $E°$ for the cells
a. $Pb \mid Pb^{2+} \parallel Sn^{4+} \mid Sn^{2+} \mid Pt$
b. $Cu \mid Cu^{2+} \parallel NO_3^- \mid NO \mid Pt$
c. $Mn \mid Mn^{2+} \parallel Ni^{2+} \mid Ni$

21. Suppose $E°_{red}$ for $H^+ \longrightarrow H_2$ were taken to be 0.500 V instead of 0.000 V. What would be
a. $E°_{ox}$ for H_2?
b. $E°_{red}$ for $Cu^{2+} \longrightarrow Cu$?
c. $E°$ for the cell $Zn \mid Zn^{2+} \parallel Cu^{2+} \mid Cu$?

22. Suppose $E°_{red}$ of $Ag^+ \longrightarrow Ag$, instead of that of $H^+ \longrightarrow H_2$, were set equal to zero. What would be
a. $E°_{red}$ for $H^+ \longrightarrow H_2$?
b. $E°_{ox}$ of $Ca \longrightarrow Ca^{2+}$?
c. $E°$ for the cell $Zn \mid Zn^{2+} \parallel Cu^{2+} \mid Cu$?

Spontaneity and $E°$

23. Which of the following reactions are spontaneous at standard conditions?
a. $AuCl_4^-(aq) + 3Fe^{2+}(aq) \longrightarrow Au(s) + 4Cl^-(aq) + 3Fe^{3+}(aq)$
b. $Zn(s) + 2Fe^{3+}(aq) \longrightarrow Zn^{2+}(aq) + 2Fe^{2+}(aq)$
c. $Cu(s) + 2H^+(aq) \longrightarrow Cu^{2+}(aq) + H_2(g)$

24. Which of the following reactions are spontaneous at standard conditions?
a. $O_2(g) + 4H^+(aq) + 4Cl^-(aq) \longrightarrow 2H_2O + 2Cl_2(g)$
b. $2NO_3^-(aq) + 8H^+(aq) + 6Cl^- \longrightarrow 2NO(g) + 4H_2O + 3Cl_2(g)$
c. $I_2(s) + 2Br^-(aq) \longrightarrow Br_2(l) + 2I^-(aq)$

25. Using Table 17.1, calculate $E°$ and decide whether the following ions will oxidize chloride ions to chlorine gas in acidic solution at standard concentrations.
a. permanganate ion **b.** dichromate ion
c. nitrate ion

26. Using Table 17.1, calculate $E°$ and decide whether the following ions will reduce chlorate ion (ClO_3^-) to chlorine gas in acidic solution at standard concentrations.
a. iodide ion **b.** fluoride ion **c.** copper(I) ion

27. Under standard conditions, write the equation for the reaction that occurs, if any, when each of the following experiments is performed.

a. Crystals of iodine are added to a solution of sodium bromide.

b. Bromine is added to a solution of sodium chloride.

c. A chromium wire is placed into a solution of nickel(II) chloride.

28. Under standard conditions, write the equation for the reaction that occurs, if any, when each of the following experiments is performed.

a. Sulfur is added to a solution of iron(II) nitrate.

b. Manganese dioxide in acidic solution is added to liquid mercury.

c. Aluminum metal is added to a solution of potassium ions.

29. Which of the following metals will react with 1 *M* HCl?

a. Au **b.** Fe **c.** Cu **d.** Pb

30. Which of the following will be oxidized by 1 *M* HNO_3?

a. I^- **b.** Mg **c.** Ag **d.** F^-

31. Using Table 17.1, decide what reaction, if any, will occur when the following are mixed (standard concentrations).

a. Co^{2+}, Co, Fe^{2+}

b. Fe^{2+}, Fe^{3+}, Ag^+

c. ClO_3^-, H^+, $Hg(l)$

32. Predict what reaction, if any, will occur when liquid bromine is added to an acidic aqueous solution of each of the following (standard concentrations).

a. $Ca(NO_3)_2$ **b.** FeI_2 **c.** AgF

$E°$, $\Delta G°$, and K

33. Calculate $\Delta G°$ and K at 25°C for cell reactions in which $n = 3$ and $E°$ is

a. 0.00 V **b.** +0.500 V **c.** −0.500 V

34. Calculate $\Delta G°$ and K at 25°C for cell reactions in which $n = 6$ and $E°$ is

a. −1.00 V **b.** 1.00 V **c.** −2.00 V

35. For a certain cell, $\Delta G°$ is −38.7 kJ. Calculate $E°$ if n is

a. 1 **b.** 2 **c.** 4

36. For a certain cell, $n = 3$. Calculate $E°$ if $\Delta G°$ is

a. −42.3 kJ **b.** +42.3 kJ

37. Calculate $E°$, $\Delta G°$, and K at 25°C for the reaction

$$NO_3^-(aq) + H_2O + 2Fe(OH)_2(s) \longrightarrow 2Fe(OH)_3(s) + NO_2^-(aq)$$

38. Calculate $E°$, $\Delta G°$, and K at 25°C for the reaction

$$4ClO_3^-(aq) \longrightarrow Cl^-(aq) + 3ClO_4^-(aq)$$

in basic solution.

39. Calculate $\Delta G°$ for each of the reactions referred to in Question 15 (assume smallest whole-number coefficients in part c).

40. Calculate $\Delta G°$ for each of the reactions referred to in Question 16 (assume smallest whole-number coefficients in part c).

41. Calculate K at 25°C for each of the reactions referred to in Question 17 (assume smallest whole-number coefficients).

42. Calculate K at 25°C for each of the reactions referred to in Question 18 (assume smallest whole-number coefficients).

Nernst Equation

43. Consider a voltaic cell in which the following reaction takes place.

$$2Cr(s) + 6H^+(aq) \longrightarrow 2Cr^{3+}(aq) + 3H_2(g)$$

a. Write the Nernst equation for this cell.

b. Calculate $E°$.

c. Calculate E under the following conditions: $[H^+] = 1.00 \times 10^{-3}$ *M*, $[Cr^{3+}] = 1.00$ *M*, $P_{H_2} = 1.00$ atm.

44. Consider a voltaic cell in which the following reaction occurs.

$$Br_2(l) + 2I^-(aq) \longrightarrow 2Br^-(aq) + I_2(s)$$

a. Write the Nernst equation for this cell.

b. Calculate $E°$.

c. Calculate E when the concentration of iodide ion is twice the concentration of bromide ion.

45. Consider the half-reaction for the reduction of dichromate ion to chromium(III) ion. Using the Nernst equation, calculate E_{red} when all the ionic species, except H^+, are at 0.100 *M* and $[H^+] = 1.00 \times 10^{-4}$ *M*.

46. Consider the half-reaction for the oxidation of lead sulfate to lead dioxide, sulfate ion, and hydrogen ion. Using the Nernst equation, calculate E_{ox} when the sulfate ion concentration is 0.0100 *M* and the hydrogen ion concentration is 2.00×10^{-3} *M*.

47. Calculate the voltages of cells under the following conditions.

a. $Zn \mid Zn^{2+}$ (0.50 *M*) $\|$ Cd^{2+} (0.020 *M*) $\mid$ Cd

b. $Cu \mid Cu^{2+}$ (0.0010 *M*) $\|$ H^+ (0.010 *M*) $\mid$ H_2 (1 atm)

Are the cell reactions spontaneous?

48. Calculate the voltages of cells under the following conditions.

a. $Fe \mid Fe^{2+}$ (0.010 *M*) $\|$ Cu^{2+} (0.10 *M*) $\mid$ Cu

b. $Pt \mid Sn^{2+}$ (0.10 *M*) $\mid$ Sn^{4+} (0.010 *M*) $\|$ Co^{2+} (0.10 *M*) $\mid$ Co

Are the cell reactions spontaneous?

49. Consider the reaction

$$O_2(g) + 4H^+(aq) + 4Cl^-(aq) \longrightarrow 2Cl_2(g) + 2H_2O$$

a. Calculate $E°$.

b. At what pressure of Cl_2 is the voltage zero, if all other species are at standard concentrations?

50. Consider the reaction

$$2Fe^{3+}(aq) + 2I^-(aq) \longrightarrow 2Fe^{2+}(aq) + I_2(s)$$

a. Calculate $E°$.

b. At what concentration of I^- is the voltage zero, if all other species are at standard concentrations?

51. Consider the cell

$$Zn(s) + 2H^+(aq) \longrightarrow Zn^{2+}(aq) + H_2(g)$$

At what pH is the voltage 0.500 V, taking $[Zn^{2+}] = 1.0\,M$, $P_{H_2} = 1.0$ atm?

52. Consider the cell

$$Sn^{2+}(aq) + \tfrac{1}{2}O_2(g) + 2H^+(aq) \longrightarrow Sn^{4+}(aq) + H_2O$$

At what pH is the voltage 0.820 V, assuming all species other than H^+ are at standard concentrations?

53. Consider a cell in which the reaction is

$$Pb(s) + 2H^+(aq) \longrightarrow Pb^{2+}(aq) + H_2(g)$$

a. Calculate $E°$.

b. Chloride ions are added to the Pb | Pb^{2+} half-cell to precipitate $PbCl_2$. The voltage is measured to be +0.210 V. Taking $[H^+] = 1.0\,M$ and $P_{H_2} = 1.0$ atm, calculate $[Pb^{2+}]$

c. Taking $[Cl^-]$ in (b) to be 0.10 M, calculate K_{sp} for $PbCl_2$.

54. Consider a cell in which the reaction is

$$2Ag(s) + Cu^{2+}(aq) \longrightarrow 2Ag^+(aq) + Cu(s)$$

a. Calculate $E°$ for this cell.

b. Chloride ions are added to the Ag | Ag^+ half-cell to precipitate AgCl. The measured voltage is +0.060 V. Taking $[Cu^{2+}] = 1.0\,M$, calculate $[Ag^+]$.

c. Taking $[Cl^-]$ in (b) to be 0.10 M, calculate K_{sp} of AgCl.

Electrolytic Cells

55. The electrolysis of an aqueous solution of NaCl has the overall reaction

$$2H_2O + 2Cl^-(aq) \longrightarrow H_2(g) + Cl_2(g) + 2\,OH^-(aq)$$

During electrolysis, 2.6×10^{22} electrons pass through the cell.

a. How many coulombs does this represent?

b. What masses of $H_2(g)$, $Cl_2(g)$, and OH^- are produced, assuming 100% yield?

56. An electrolytic cell is producing aluminum from Al_2O_3 at the rate of one kilogram per day. Assuming a yield of 100%,

a. how many electrons must pass through the cell in one day?

b. what is the current passing through the cell?

c. how much oxygen is being produced simultaneously?

57. A spoon with an area of 4.00 cm^2 is plated with silver from a $Ag(CN)_2^-$ solution, using a current of 0.500 A for two hours.

a. If the current efficiency is 80.0%, how many grams of silver are plated?

b. What is the thickness of the silver plate formed (d Ag = 10.5 g/cm^3)?

58. It is desired to plate a coin, which has a diameter of 1.05 in. and a thickness of 0.0600 in., with a layer of gold 0.0010 in. thick.

a. How many grams of gold (d = 19.3 g/cm^3) are required?

b. How long will it take to plate the coin from AuCN, using a current of 0.200 A and assuming 100% yield?

59. A lead storage battery delivers a current of 2.00 A for one hour at a voltage of 12.0 V.

a. How many grams of Pb are converted to $PbSO_4$?

b. How much electrical energy is produced (kilowatt-hours)?

60. Aluminum is produced by the electrolysis of Al_2O_3, using a voltage of 6.0 V.

a. How many joules of electrical energy are required to form 1.00 kg Al?

b. What is the cost of the electrical energy in (a) at the rate of 6.0¢ per kilowatt-hour?

Corrosion

61. Explain why aluminum cookware does not disintegrate upon exposure to air.

62. Write chemical equations to represent the half-reactions taking place in the anodic and cathodic areas of corroding iron.

63. Explain

a. why iron exposed to dry air does not rust.

b. why garden tools left covered with soil are susceptible to rusting.

64. When exposed to an atmosphere containing SO_2, copper metal is oxidized to $Cu(OH)_2 \cdot CuSO_4$. Write a balanced equation for the reaction involved.

Unclassified

65. Consider the electrolysis of a solution of $CuCl_2$ to form $Cu(s)$ and $Cl_2(g)$. Calculate the minimum voltage required to carry out this reaction. If a voltage of 1.50 V is actually used, how many kilojoules of electrical energy is consumed in producing 1.00 g of Cu?

66. When a 1.50 V dry cell is used to "pump" $2.01 \times 10^{23}\,e^-$ through an electrolytic cell, determine

a. the number of coulombs passed through the cell.

b. the current in amperes if the time required is 10.0 min.

c. the energy in kilojoules and kilowatt-hours.

67. Explain why

a. sulfuric acid is used in a lead storage battery.

b. a salt bridge is used in a voltaic cell.

c. Cu reacts with HNO_3 but not with HCl.

68. Identify each of the following statements as true or false. If false, correct it to make it true.

a. Sn^{2+} disproportionates in acid solution.

b. K^+ ions from a salt bridge move to the anode during operation of a voltaic cell.

c. When a lead storage battery is recharged, lead sulfate is deposited.

69. A current of 2.00 A is drawn from a dry cell for a period of ten minutes. Assuming that the only reaction taking place is

$$Zn(s) + 2MnO_2(s) + 2NH_4^+(aq) \longrightarrow Zn^{2+}(aq) + Mn_2O_3(s) + 2NH_3(aq) + H_2O$$

a. How many grams of zinc are consumed?
b. What volume of ammonia gas at 25°C and 1.00 atm would be produced if the solution were heated to drive off all the ammonia?

70. A cell used to electrolyze sodium chloride solution is filled with 1.00 L of 2.00 *M* NaCl. A current of two amperes is passed through this solution for 8.00 h. Assuming that the only reaction that occurs is

$$2H_2O + 2Cl^-(aq) \longrightarrow H_2(g) + Cl_2(g) + 2\,OH^-(aq)$$

a. What volume of chlorine gas is produced at 25°C and 1.00 atm?
b. What is the pH of the solution after eight hours?

71. A voltaic cell consists of a copper electrode immersed in 1.0 *M* $CuSO_4$ connected via a salt bridge to a solution of 1.0 *M* silver nitrate, in which is immersed a silver electrode. By what amount will the voltage change if

a. the concentration of Cu^{2+} is decreased to 0.00100 *M*?
b. enough Cl^- is added to precipitate AgCl ($K_{sp} = 1.8 \times 10^{-10}$) and leave $[Cl^-] = 1.0$ *M*?
c. the area of the copper electrode is doubled?

72. Atomic masses can be determined by electrolysis. In one hour, a current of 0.600 A deposits 2.42 g of a certain metal, *M*, which is present in solution as M^+ ions. What is the atomic mass of the metal?

73. The standard potential for the reduction of AgSCN is 0.0895 V.

$$AgSCN(s) + e^- \longrightarrow Ag(s) + SCN^-(aq)$$

Find another electrode potential to use together with the above value and calculate K_{sp} for AgSCN.

74. Consider the following reaction at 25°C.

$$O_2(g) + 4H^+(aq) + 4Br^-(aq) \longrightarrow 2H_2O + 2Br_2(l)$$

If $[H^+]$ is adjusted by adding a buffer that is 0.100 *M* in sodium acetate and 0.100 *M* in acetic acid, the pressure of O_2 is 1.00 atm, and the bromide concentration is 0.100 *M*, what is the calculated cell voltage? (K_a acetic acid $= 1.8 \times 10^{-5}$.)

Challenge Problems

75. In a fully charged lead storage battery, the electrolyte consists of 38% sulfuric acid by mass. The solution has a density of 1.286 g/cm^3. Calculate *E* for the cell. (Assume all the H^+ ions come from the first dissociation of H_2SO_4, which is complete; K_a $HSO_4^- = 1.0 \times 10^{-2}$.)

76. Consider a voltaic cell in which the following reaction occurs.

$$Zn(s) + Sn^{2+}(aq) \longrightarrow Zn^{2+}(aq) + Sn(s)$$

a. Calculate $E°$ for the cell.
b. When the cell operates, what happens to the concentration of Zn^{2+}? the concentration of Sn^{2+}?
c. When the cell voltage drops to zero, what is the ratio of the concentration of Zn^{2+} to that of Sn^{2+}?
d. If the concentration of both cations is 1.0 *M* originally, what are their concentrations when the voltage drops to zero?

77. In biological systems, acetate ion is converted to ethyl alcohol in a two-step process:

$$CH_3COO^-(aq) + 3H^+(aq) + 2e^- \longrightarrow CH_3CHO(aq) + H_2O; \quad E°' = -0.581\ V$$

$$CH_3CHO(aq) + 2H^+(aq) + 2e^- \longrightarrow C_2H_5OH(aq); \quad E°' = -0.197\ V$$

($E°'$ is the standard reduction voltage at 25°C and a pH of 7.00.)

a. Calculate $\Delta G°'$ for each step and for the overall conversion.
b. Calculate $E°'$ for the overall conversion.

78. Consider the cell $(Pt)H_2 \mid H^+ \parallel H^+ \mid H_2 \mid Pt$. In the anode half-cell, hydrogen gas at 1.0 atm is bubbled over a platinum electrode dipping into a solution that has a pH of 7.0. The other half-cell is identical to the first except that the solution around the platinum electrode has a pH of 0.0. What is the cell voltage?

18
Nuclear Reactions

The soul, perhaps, is a
gust of gas
And wrong is a form of right—
But we know that Energy
equals Mass
By the Square of the
Speed of Light.

MORRIS BISHOP
E = MC²

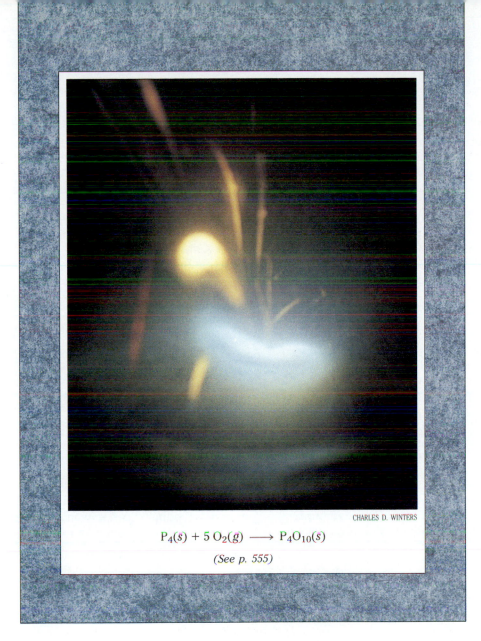

CHARLES D. WINTERS

$P_4(s) + 5\ O_2(g) \longrightarrow P_4O_{10}(s)$

(See p. 555)

CHAPTER OUTLINE

The "ordinary chemical reactions" discussed to this point involve changes in the outer electronic structure of atoms or molecules. In contrast, nuclear reactions result from changes taking place within atomic nuclei. You will recall (Chap. 2) that atomic nuclei are represented by symbols such as

$$^{12}_{6}C \qquad ^{14}_{6}C$$

Here, the atomic number *Z* (number of protons in the nucleus), is shown as a left subscript. The mass number, *A* (number of protons + number of neutrons in the nucleus), appears as a left superscript. Nuclei with the same number of protons but different numbers of neutrons are called *isotopes*. The symbols written above represent two isotopes of the element carbon ($Z = 6$). One isotope has 6 neutrons in the nucleus and hence has a mass number of $6 + 6 = 12$. The heavier isotope has 8 neutrons and hence a mass number of 14.

Nuclear reactions usually involve transmutation of elements

We start this chapter by discussing briefly the factors that contribute to nuclear stability (Section 18.1). Unstable nuclei, whether formed in nature or in the laboratory, decompose by a type of nuclear reaction called radioactivity (Section 18.2). One of the most important features of this process is the rate at which radioactive nuclei decompose or "decay" (Section 18.3). Another feature is the amount of energy evolved, which is directly related to the change in mass (Section 18.4). This factor is particularly important in the nuclear reactions known as fission (Section 18.5) and fusion (Section 18.6).

18.1 NUCLEAR STABILITY

According to classical electrostatics, we would expect the protons in a nucleus to repel one another, causing the nucleus to fly apart. It turns out, however, that at the very short distances of separation characteristic of atomic nuclei, about 10^{-15} m, there are strong attractive forces between nuclear particles. The stability of a nucleus depends upon the balance between these forces and those of electrostatic repulsion.

Our knowledge of nuclear stability today leaves a great deal to be desired; all we really have is a series of empirical rules for predicting which nuclei will be most stable. The more important of these rules are listed below.

1. The neutron-to-proton ratio required for stability varies with atomic number (Fig. 18.1). For light elements, this ratio is close to one. For example,

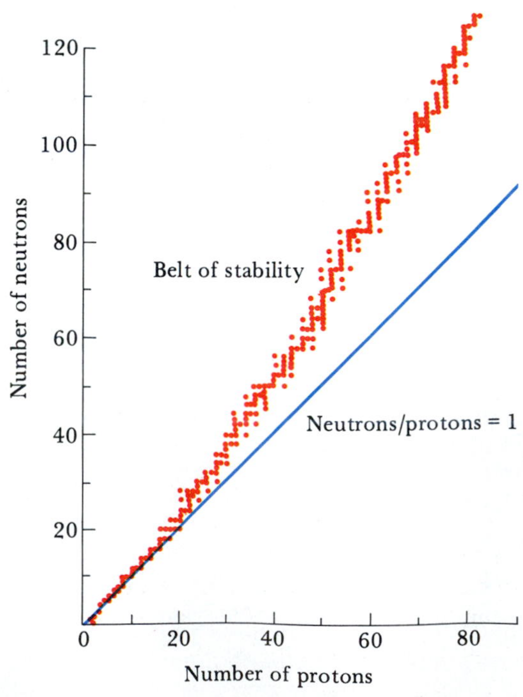

Figure 18.1
Stable isotopes (red dots) have neutron-to-proton ratios that fall within a narrow range, referred to as a "belt of stability." For light isotopes of small atomic number, the stable ratio is 1.0; with heavier isotopes it increases to about 1.5. There are no stable isotopes for elements of atomic number greater than 83 (Bi).

the isotopes $^{12}_{6}C$, $^{14}_{7}N$, and $^{16}_{8}O$ are stable. As atomic number increases, the ratio increases; the "belt of stability" shifts to higher numbers of neutrons. With very heavy isotopes such as $^{206}_{82}Pb$, the stable neutron-to-proton ratio is about 1.5:

$$(206 - 82)/82 = 124/82 = 1.51$$

2. Nuclei containing more than 83 protons are unstable. Putting it another way, no element beyond bismuth ($Z = 83$) has a stable isotope; all the isotopes of such elements are radioactive.

3. Nuclei with an even number of *nucleons,* either protons or neutrons, tend to be more stable than those with an odd number of nuclear particles. Consider, for example, the stable isotopes of the elements of atomic numbers 50–53:

Nature prefers evens to odds, here and elsewhere

	At. No.	Mass No. of Stable Isotope
Sn	50	112, 114, 115, 116, 117, 118, 119, 120, 122, 124
Sb	51	121, 123
Te	52	120, 122, 123, 124, 125, 126, 128, 130
I	53	127

Of the 21 stable isotopes of these elements, 18 have an even number of protons, 16 an even number of neutrons. Thirteen of 21 have an even number of both protons and neutrons; statistically, we would expect only 5 to fall in this category.

4. Certain numbers of protons and neutrons appear to be particularly stable: 2, 8, 20, 28, 50, 82, and 126. These *magic numbers* were first suggested by Maria Mayer, who developed a nuclear shell model similar to, but much less regular than, the electron shell model discussed in Chapter 6. Mayer won a Nobel Prize for her work in physics in 1963.

18.2 RADIOACTIVITY

An unstable nucleus undergoes a reaction called radioactive *decomposition* or *decay.* A few such nuclei occur in nature; their decomposition is referred to as *natural radioactivity.* Many more unstable nuclei have been made in the laboratory; the process by which such nuclei decompose is called *induced radioactivity.*

Natural Radioactivity

Natural radioactivity was discovered, almost accidentally, by Henri Becquerel and his colleagues Marie and Pierre Curie at the Sorbonne in Paris in 1896. The radiation given off in natural radioactivity can be separated by an electric or magnetic field into three distinct parts (Fig. 18.2):

1. Alpha radiation consists of a stream of positively charged particles (alpha particles) with a charge of +2 and a mass of 4 on the atomic mass scale. These particles are identical with the nuclei of ordinary helium atoms, $^{4}_{2}He$.

When an alpha particle is ejected from the nucleus, the atomic number decreases by two units; the mass number decreases by four units. Consider, for example, the loss of an alpha particle by a uranium-238 atom:

$$^{238}_{92}U \longrightarrow {}^{4}_{2}He + {}^{234}_{90}Th$$

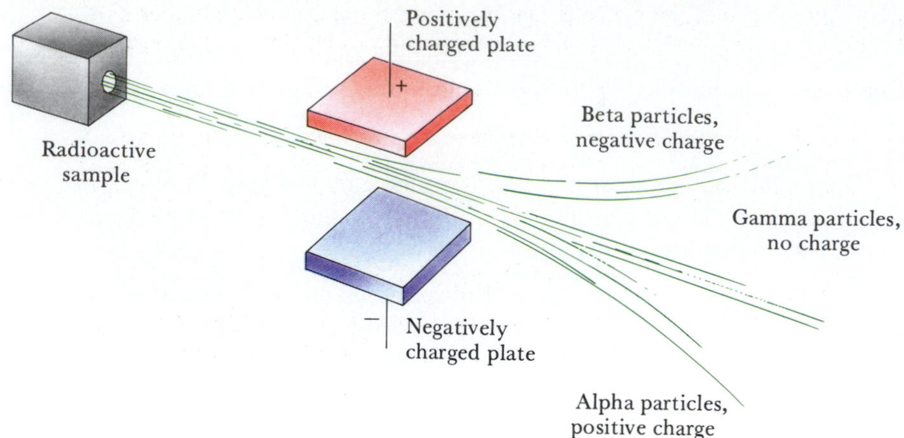

Figure 18.2
The direction in which beta particles are deflected shows that they are negatively charged. Alpha particles move so as to indicate that they carry a positive charge. Gamma rays are undeflected and so must be uncharged.

Here, as in all nuclear equations, there is a balance of both atomic number (90 + 2 = 92) and mass number (4 + 234 = 238) on the two sides.

2. Beta radiation is made up of a stream of negatively charged particles (beta particles) identical in their properties to electrons. The ejection of a beta particle (mass ≈ 0, charge = −1) leaves the mass number unchanged but increases the atomic number by one unit. An example of beta emission is the radioactive decay of thorium-234:

$$^{234}_{90}\text{Th} \longrightarrow {}^{0}_{-1}e + {}^{234}_{91}\text{Pa}$$

Effectively, an electron has an atomic number of −1

The symbol ${}^{0}_{-1}e$ is written to stand for a beta particle (electron). Again, there is a balance of mass number (234) and atomic number (90) on both sides of the equation.

Beta emission comes about when a neutron in the nucleus is converted to a proton:

$$^{1}_{0}n \longrightarrow {}^{1}_{1}\text{H} + {}^{0}_{-1}e$$

This type of reaction occurs with unstable nuclei containing "too many" neutrons, that is, nuclei lying above the belt of stability shown in Figure 18.1.

3. Gamma radiation consists of high-energy photons of very short wavelength (λ = 0.0005 to 0.1 nm). The emission of gamma radiation accompanies most nuclear reactions. Since gamma radiation changes neither the atomic number nor the mass number, it is ordinarily omitted in writing nuclear equations.

Frequently, several steps are required to reach a stable isotope

Example 18.1 Thorium-232 undergoes radioactive decay in a three-step process:

(a) $^{232}_{90}\text{Th} \longrightarrow \text{Q} + {}^{4}_{2}\text{He}$ (b) $\text{Q} \longrightarrow \text{R} + {}^{0}_{-1}e$ (c) $\text{R} \longrightarrow \text{T} + {}^{228}_{90}\text{Th}$

Write a balanced nuclear equation for each step, identifying Q, R, and T by their nuclear symbols.

Strategy The basic principle is that mass number and atomic number must be "conserved," i.e., they must have the same sum on both sides of the equation.

Solution

(a) Alpha emission by $^{232}_{90}Th$ decreases the mass number by four and the atomic number by two. The mass number of Q is 232 − 4 = 228; its atomic number is 90 − 2 = 88. Referring to the periodic table, element 88 is radium, symbol Ra.

$$^{232}_{90}Th \longrightarrow ^{4}_{2}He + ^{228}_{88}Ra$$

(b) When $^{228}_{88}Ra$ emits an electron, the mass number is unchanged, but the atomic number increases by 1 to 89 (element 89 = actinium, Ac).

$$^{228}_{88}Ra \longrightarrow ^{0}_{-1}e + ^{228}_{89}Ac$$

(c) The unbalanced equation in this case is

$$^{228}_{89}Ac \longrightarrow T + ^{228}_{90}Th$$

For the equation to balance, T must have a mass number of 0 and an atomic number of −1; T must be an electron. The balanced equation is

$$^{228}_{89}Ac \longrightarrow ^{0}_{-1}e + ^{228}_{90}Th$$

Induced Radioactivity; Bombardment Reactions

During the past 60 years, more than 1500 radioactive isotopes have been prepared in the laboratory. The number of such isotopes per element ranges from one (hydrogen and boron) to 34 (indium). They are all prepared by bombardment reactions in which a stable nucleus is converted to one that is radioactive. A typical reaction is the one that occurs when the stable isotope of aluminum, $^{27}_{13}Al$, absorbs a neutron to form $^{28}_{13}Al$. The latter is unstable, decaying by electron emission to give a stable isotope of silicon, $^{28}_{14}Si$. The two steps involved in the process are

neutron bombardment: $^{27}_{13}Al + ^{1}_{0}n \longrightarrow ^{28}_{13}Al$

radioactive decay: $^{28}_{13}Al \longrightarrow ^{28}_{14}Si + ^{0}_{-1}e$

The first radioactive isotopes to be made in the laboratory were prepared in 1934 by Irene Curie and her husband, Frederic Joliot. They achieved this by bombarding certain stable isotopes with high-energy alpha particles. One reaction was

$$^{27}_{13}Al + ^{4}_{2}He \longrightarrow ^{30}_{15}P + ^{1}_{0}n$$

The product, phosphorus-30, is radioactive. It decays by emitting a **positron**:

$$^{30}_{15}P \longrightarrow ^{30}_{14}Si + ^{0}_{1}e$$

The positron is a particle that has the same mass as the electron, but a charge of +1 rather than −1. Positron emission converts a proton in the nucleus to a neutron (15 p, 15 n in P-30; 14 p, 16 n in Si-30). This same result can be achieved by **K-electron capture**. Here an electron in the innermost energy level (***n*** = 1) "falls" into the nucleus:

$$^{82}_{37}Rb + ^{0}_{-1}e \longrightarrow ^{82}_{36}Kr$$

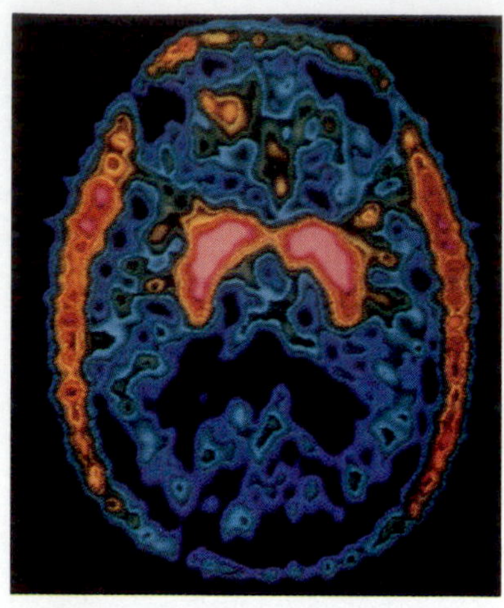

PET scan (positron emission tomography) of normal human brain. (Cheodoke-McMaster Hospitals)

Positron emission or K-electron capture occurs with unstable nuclei which have "too many" protons, i.e., nuclei which fall below the belt of stability shown in Figure 18.1. Electron capture is more common with heavy isotopes, presumably because the $n = 1$ shell is closer to the nucleus.

Uses of Radioactive Isotopes

A large number of radioactive isotopes have been used both in industry and in many areas of basic and applied research. A few of these are discussed below.

Medicine

Radioactive isotopes are commonly used in cancer therapy, usually to eliminate any malignant cells left after surgery. Cobalt-60 is most often used; γ-rays from this source are focused at small areas where cancer is suspected. Certain types of cancer can be treated internally with radioactive isotopes. If a patient suffering from cancer of the thyroid drinks a solution of NaI containing radioactive iodide ions (^{131}I or ^{123}I), the iodine moves preferentially to the thyroid gland. There, the radiation destroys malignant cells without affecting the rest of the body.

More commonly, radioactive isotopes are used for diagnosis (Table 18.1). Positron emission tomography (PET) is a technique used to study brain disorders. The patient is given a dose of glucose ($C_6H_{12}O_6$) containing a small amount of carbon-11, a positron emitter. The brain is then scanned to detect positron emission from the radioactive, "labeled" glucose. In this way, differences in glucose uptake and metabolism in the brains of normal and abnormal patients are established. For example, PET scans have determined that the brain of a schizophrenic metabolizes only about 20% as much glucose as that of a normal individual.

Table 18.1 Diagnostic Uses of Radioactive Isotopes

Isotope	Use
$^{11}_{6}C$	brain scan (PET); see text
$^{24}_{11}Na$	circulatory disorders
$^{32}_{15}P$	detection of eye tumors
$^{59}_{26}Fe$	anemia
$^{67}_{31}Ga$	scan for lung tumors, abscesses
$^{75}_{34}Se$	pancreas scan
$^{99}_{43}Tc$	imaging of brain, liver, kidneys, bone marrow
$^{133}_{54}Xe$	lung imaging
$^{201}_{81}Tl$	heart disorders

Chemistry

Radioactive isotopes are used extensively in chemical analysis. One technique of particular importance is *neutron activation analysis.* This procedure depends upon the phenomenon of induced radioactivity. A sample is bombarded by neutrons, bringing about such reactions as

$$^{84}_{38}Sr + ^{1}_{0}n \longrightarrow ^{85}_{38}Sr$$

Ordinarily the element retains its chemical identity, but the isotope formed is radioactive, decaying by gamma emission. The magnitude of the energy change and hence the wavelength of the gamma ray varies from one element to another and so can serve for the qualitative analysis of the sample. The intensity of the radiation depends upon the amount of the element present in the sample; this permits quantitative analysis of the sample. Neutron activation analysis can be used to analyze for 50 different elements in amounts as small as one picogram (10^{-12} g).

Radioactive isotopes are also used to trace the path of photosynthesis

One application of neutron activation analysis is in the field of archaeology. By measuring the amount of strontium in the bones of prehistoric humans, it is possible to get some idea of their diet. Plants contain considerably more strontium than animals do, so a high strontium content suggests a largely vegetarian diet. Strontium analyses of bones taken from ancient farming communities consistently show a difference by sex; women have higher strontium levels than men. Apparently, in those days, women did most of the farming; men spent a lot of time away from home hunting and eating their kill.

The history of radiochemistry is in no small measure the story of two remarkable women, Marie and Irene Curie, and their husbands, Pierre Curie and Frederic Joliot. Marie Curie was born Maria Sklodowska in 1867 in Warsaw, Poland, then a part of the Russian empire. In 1891 she emigrated to Paris to study at the Sorbonne, where she met and married a French physicist, Pierre Curie. The Curies were associates of Henri Becquerel, the man who discovered that uranium salts are radioactive. They showed that thorium, like uranium, is radioactive and that the amount of radiation emitted is directly proportional to the amount of uranium or thorium in the sample.

In 1898, Marie and Pierre Curie isolated two new radioactive elements, which they named radium and polonium. To obtain a few milligrams of these elements, they started with several tons of pitchblende ore and carried out a long series of tedious separations. Their work was done in a poorly equipped, unheated shed where the temperature reached 6°C (43°F) in winter. Four years later, in 1902, Marie determined the atomic mass of radium to within 0.5%, working with a tiny sample.

In 1903, the Curies received the Nobel Prize in physics (with Becquerel) for the discovery of radioactivity. Three years later, Pierre Curie died at the age of 46, the victim of a tragic accident. He stepped from behind a carriage in a busy Paris street and was run down by a truck. That same year, Marie became the first woman instructor at the Sorbonne. In 1911, she narrowly missed election to the French Academy of Science for quite irrelevant reasons. She did, however, receive the Nobel Prize in chemistry for the discovery of radium and polonium, thereby becoming the first person to win two Nobel Prizes.

When Europe exploded into war in 1914, scientists largely abandoned their studies to go to the front. Marie Curie, with her daughter Irene, then 17 years old, organized medical units equipped with x-ray machinery. These were used to locate foreign metallic objects in wounded soldiers. Many of the wounds were to the head; French soldiers came out of the trenches without head protection because their government had decided that helmets looked too German. In November of 1918, the Curies celebrated the end of World War I; France was victorious and Marie's beloved Poland was free again.

In 1921, Irene Curie began research at the Radium Institute. Five years later she married Frederic Joliot, a brilliant young physicist who was also an assistant at the Institute. In 1931, they began a research program in nuclear chemistry that led to several important discoveries and at least one near miss. The Joliot-Curies were the first to demonstrate induced radioactivity. They also discovered the positron, a particle that scientists had been seeking for many years. They narrowly missed finding another, more fundamental particle, the neutron. That honor went to Chadwick in England. In 1935, Irene and Frederic Joliot-Curie received the Nobel Prize in physics. The award came too late for Irene's mother, who had died of leukemia a year earlier.

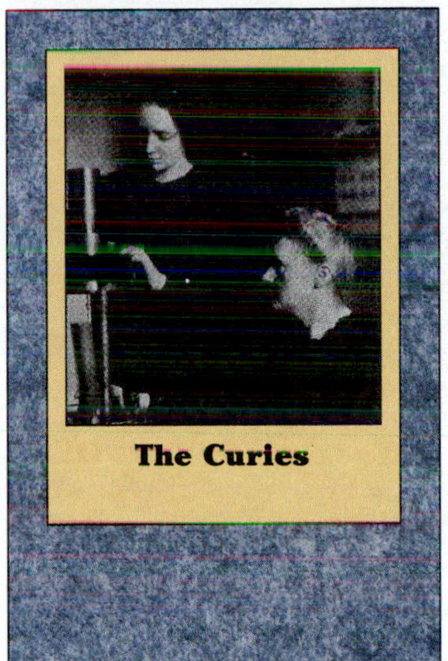

MARIE AND IRENE CURIE. (E. F. SMITH MEMORIAL COLLECTION IN THE HISTORY OF CHEMISTRY, DEPT. OF SPECIAL COLLECTIONS, VAN PELT-DIETRICH LIBRARY, UNIVERSITY OF PENNSYLVANIA)

Commercial Applications

Most smoke alarms (Fig. 18.3) use a radioactive isotope, typically americium-241. A tiny amount of this isotope is placed in a small ionization chamber; decay of Am-241 ionizes air molecules within the chamber. Under the influence of a potential applied by a battery, these ions move across the chamber, producing an electric current. If smoke particles get into the chamber, the flow of ions is impeded and the current drops. This is detected by electronic circuitry, and an alarm sounds. The alarm also goes off if the battery voltage drops, indicating that it needs to be replaced.

Figure 18.3
Most smoke detectors use a tiny amount of a radioactive isotope to produce a current flow that drops off sharply in the presence of smoke particles, emitting an alarm in the process. (Marna G. Clarke)

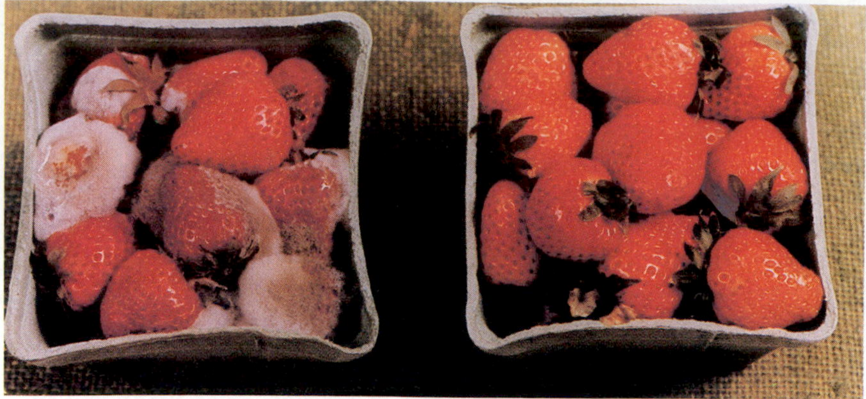

Figure 18.4
Strawberries irradiated with gamma rays from radioactive isotopes are still fresh after 15 days storage at 4°C (*right*), while those not irradiated are moldy (*left*). (International Atomic Energy Agency)

Another potential application of radioactive isotopes is in food preservation (Fig. 18.4). It is well known that gamma rays can kill insects, larvae, and parasites like trichina that cause trichinosis in pork. Radiation can also inhibit sprouting of onions and potatoes. Perhaps most important from a commercial standpoint, it can extend the shelf lives of many foods for weeks or even months. Since many chemicals used to preserve foods have later been shown to have adverse health effects, irradiation would seem to be an attractive alternative.

There are, however, some problems with food irradiation. For one thing, the taste of some fruits and all milk products is affected adversely. More generally, there is concern that irradiation produces a variety of reaction products, at least a few of which could be harmful.

18.3 KINETICS OF RADIOACTIVE DECAY

Nuclear decay is a first-order reaction. That is, the rate of decay is directly proportional to the amount of radioactive isotope present. Mathematically,

$$\text{rate} = kN_t$$

In this equation:

> $\boldsymbol{N_t}$ represents the amount of radioactive isotope present at time t. Most often, N_t is taken to be the number of atoms in the sample.
>
> $\boldsymbol{k}$ is the first-order rate constant and has the units of reciprocal time (e.g., s^{-1}, d^{-1}, yr^{-1}).

The **rate of decay**, often referred to as the **activity** of the sample, is most often expressed in terms of the number of atoms decaying in unit time. This may be stated in becquerels (Bq):

$$1 \text{ Bq} = 1 \text{ atom/s}$$

or, more commonly, in curies (Ci)

$1 \text{ Ci} = 3.700 \times 10^{10} \text{ atoms/s}$

$$1 \text{ Ci} = 3.700 \times 10^{10} \text{ Bq}$$

If a counter indicates that 750 atoms disintegrate over a ten-second interval, the rate of decay in curies is

$$\frac{7.50 \times 10^2 \text{ atoms}}{10.0 \text{ s}} \times \frac{1 \text{ Ci}}{3.700 \times 10^{10} \text{ atoms/s}} = 2.03 \times 10^{-9} \text{ Ci}$$

By measuring the decay rate of a known mass of a radioactive isotope, it is possible to calculate the decay constant, k.

Example 18.2 A 1.00-g sample of cobalt-60 (molar mass = 59.92 g/mol) has an activity of 1.1×10^3 Ci. Find the rate constant k for the decay of cobalt-60.

Strategy Use the relation: rate = kN_t. First, calculate the number of atoms, N_t, in the one-gram sample, using Avogadro's number (Chap. 3). Then find the decay rate in atoms per second (1 Ci = 3.700×10^{10} atoms/s), and finally the rate constant, k.

k = (no. atoms decaying per second)/(total no. atoms)

Solution

(1) $N_t = 1.00 \text{ g Co-60} \times \frac{1 \text{ mol}}{59.92 \text{ g}} \times \frac{6.022 \times 10^{23} \text{ atoms}}{1 \text{ mol}}$

$= 1.01 \times 10^{22}$ atoms

(2) rate $= 1.1 \times 10^3 \text{ Ci} \times \frac{3.700 \times 10^{10} \text{ atoms/s}}{1 \text{ Ci}} = 4.1 \times 10^{13}$ atoms/s

(3) $k = \frac{\text{rate}}{N_t} = \frac{4.1 \times 10^{13} \text{ atoms/s}}{1.01 \times 10^{22} \text{ atoms}} = 4.1 \times 10^{-9}\text{/s}$

Notice from Example 18.2 that the expression for the first-order rate constant, k, is

$$k = \frac{\text{rate}}{N_t} = \frac{\text{no. of atoms decomposing per second}}{\text{total no. of atoms}}$$

From a slightly different point of view, the rate constant represents *the fraction of the sample that decomposes in unit time*. When the rate constant is 0.010/s, 1% of the atoms in the sample are disintegrating per second.

For radioactive decay, the first-order rate law has the form

$$\ln \frac{X_o}{X} = kt$$

where X_o is the original amount of radioactive material, X is the amount left after time t, and k is the first-order rate constant referred to above. The amount of a radioactive isotope can be expressed in terms of moles, grams, or number of atoms. Indeed, X_o and X can even refer to counting rates (counts per minute, counts per second, etc.), since the rate of decay is directly proportional to amount.

Recall the discussion of first-order reactions (Chap. 11)

Decay rates of radioactive isotopes are most often expressed in terms of their *half-lives*, $t_{1/2}$, rather than the first-order rate constant, k. As noted in Chapter 11, these two quantities are related by the equation

$$k = \frac{0.693}{t_{1/2}}$$

Example 18.3 Plutonium-240, produced in nuclear reactors, has a half-life of 6.58×10^3 years. Calculate

(a) the first-order rate constant for the decay of plutonium-240.
(b) the fraction of a sample that will remain after 100 (1.00×10^2) years.

Strategy This is a standard type of first-order rate calculation, similar to those done in Chapter 11. Use the equation $k = 0.693/t_{1/2}$ to calculate k and then the equation $\ln X_o/X = kt$ to find the fraction left, X/X_o.

Solution

(a) $k = 0.693/(6.58 \times 10^3\ \text{yr}) =$ $1.05 \times 10^{-4}/\text{yr}$

(b) $\ln \dfrac{X_o}{X} = 1.05 \times 10^{-4}\ \text{yr}^{-1} \times 100\ \text{yr} = 1.05 \times 10^{-2}$

Taking the antilog, $X_o/X = 1.01$.

Hence, $X/X_o = 1/1.01 =$ 0.99. That is, 99% remains.

This problem illustrates the difficulty inherent in storing or disposing of a long-lived radioactive species. Such isotopes cannot be released into the environment in the hope that they will decompose rapidly. In the case of plutonium-240, its radiation level would be virtually unchanged a century from now.

Half-lives can be interpreted in terms of the level of radiation of the corresponding isotopes. Since uranium has a very long half-life (4.5×10^9 yr), it gives off radiation very slowly. At the opposite extreme is fermium-258, which decays with a half-life of 3.8×10^{-4} s. You would expect the rate of decay to be quite high. Within a second all the radiation from fermium-258 is gone. Species such as this produce a very high level of radiation during their brief existence.

Age of Rocks

As pointed out earlier, uranium-238 is radioactive:

$$^{238}_{92}\text{U} \longrightarrow {}^{234}_{90}\text{Th} + {}^{4}_{2}\text{He} \qquad t_{1/2} = 4.5 \times 10^9\ \text{yr}$$

The product, thorium-234, is itself radioactive; decay continues until a stable product, lead-206, is formed. The overall, 14-step process can be represented by the nuclear equation

$$^{238}_{92}\text{U} \longrightarrow {}^{206}_{82}\text{Pb} + 8\,{}^{4}_{2}\text{He} + 6\,{}^{0}_{-1}e \qquad t_{1/2} = 4.5 \times 10^9\ \text{yr}$$

Since all the other steps have short half-lives, the slow first step is rate-determining.

By analyzing a uranium-containing rock to determine the relative amounts of U-238 and Pb-206, it is possible to determine the "age" of the rock, i.e., the amount of time elapsed since it solidified. Suppose, for example, that equal numbers of atoms of U-238 and Pb-206 are present in the rock. Assuming there was no Pb-206 present originally, this means that one half-life, or 4.5×10^9 years, has passed; the rock is 4.5 billion years old.

Ages of rocks determined by this method range from 3 to 4.5×10^9 years. The larger number is often taken as an approximate value for the age of the

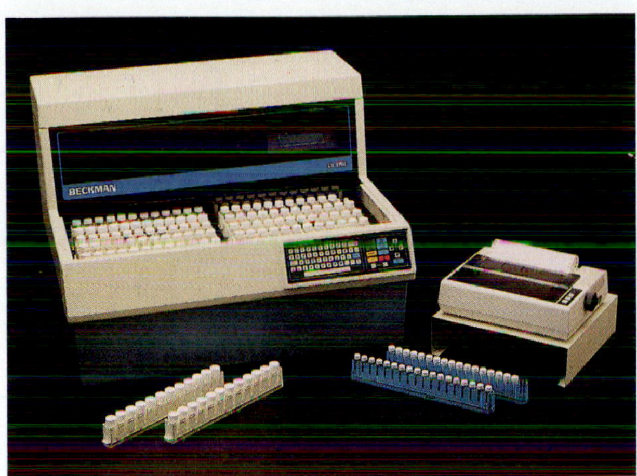

Liquid scintillation counter used to detect radiation and measure disintegrations per minute quickly and accurately. (Beckman)

earth. Analyses of rock samples from the moon indicate ages in the same range. This argues against the once prevalent idea that the moon was torn from the earth's surface by a violent event a long time after the earth solidified.

Several other radioactive "clocks" can be used to find the age of the earth

Age of Organic Material

During the 1950s, Professor W. F. Libby of the University of Chicago and others worked out a method for determining the age of organic material. It is based upon the decay rate of carbon-14. The method can be applied to objects from a few hundred up to 50,000 years old. It has been used to determine the authenticity of canvases of Renaissance painters and to check the ages of relics left by prehistoric cavemen.

Carbon-14 is produced in the atmosphere by the interaction of neutrons from cosmic radiation with ordinary nitrogen atoms:

$$^{14}_{7}N + ^{1}_{0}n \longrightarrow ^{14}_{6}C + ^{1}_{1}H$$

The carbon-14 formed by this nuclear reaction is eventually incorporated into the carbon dioxide of the air. A steady-state concentration, amounting to about one atom of carbon-14 for every 10^{12} atoms of carbon-12, is established in atmospheric CO_2. A living plant, taking in carbon dioxide, has this same $^{14}C/^{12}C$ ratio, as do plant-eating animals or human beings.

When a plant or animal dies, the intake of radioactive carbon stops. Consequently, the radioactive decay of carbon-14

$$^{14}_{6}C \longrightarrow ^{14}_{7}N + ^{0}_{-1}e \qquad t_{1/2} = 5720 \text{ yr}$$

takes over, and the ratio $^{14}C/^{12}C$ drops (Example 18.4).

Example 18.4 A tiny piece of paper taken from the Dead Sea Scrolls, believed to date back to the first century A.D., was found to have a $^{14}C/^{12}C$ ratio of 0.795 times that in a living plant. Estimate the age of the scrolls.

Strategy First calculate the rate constant, knowing that the half-life is 5720 yr. Then find t, using the equation $\ln X_o/X = kt$. Note that the amount of C-14 left, X, is 0.795 times that originally present, X_o.

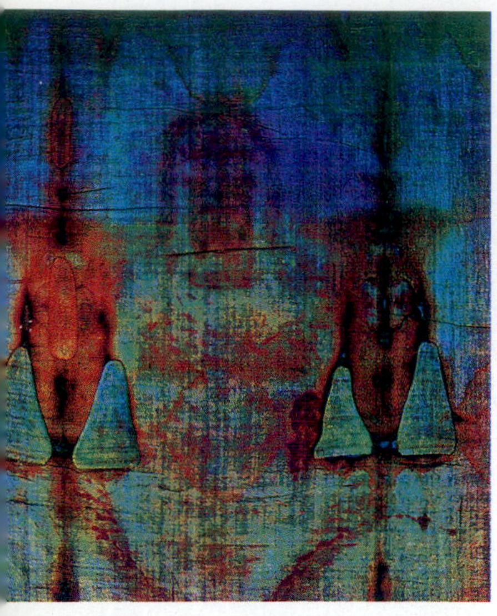

The Shroud of Turin has been represented to be the burial cloth of Jesus Christ. (Jean Lorre, Donald Lynn, and the Shroud of Turin Project)

Solution

(1) $k = 0.693/5720 \text{ yr} = 1.21 \times 10^{-4}/\text{yr}$

(2) $\ln \frac{X_o}{X} = kt;\ t = \frac{1}{k} \ln \frac{X_o}{X}$

$$t = \frac{1}{1.21 \times 10^{-4}/\text{yr}} \ln \frac{1.000}{0.795} = \frac{0.229}{1.21 \times 10^{-4}/\text{yr}} = 1.89 \times 10^3 \text{ yr}$$

This calculation assumes that the $^{14}C/^{12}C$ ratio has remained constant over the years.

In practice, the $^{14}C/^{12}C$ ratio is determined by measuring the activity of the sample and calculating the number of atoms decaying per minute per gram of carbon. In a living plant, the $^{14}C/^{12}C$ ratio is such that, in a one-gram sample of carbon, about 15.3 atoms of C-14 decay per minute. If, in a particular sample, the activity is 13.0 atoms/min · g C, then the $^{14}C/^{12}C$ ratio must be

$$13.0/15.3 = 0.850$$

or about 85% of that in a living plant.

As you can imagine, it is not easy to determine accurately activities of the order of 15 atoms decaying per minute, about one "event" every four seconds. Elaborate precautions have to be taken to exclude background radiation. Moreover, relatively large samples must be used to increase the counting rate. Recently, a technique has been developed whereby C-14 atoms can be counted very accurately in a specially designed mass spectrometer. This method was used to date the Shroud of Turin, using six samples with a total mass of about 0.1 g. Analysis by an international team of scientists showed that the linen of which the Shroud is composed grew in the fourteenth century A.D. Clearly, this remarkable burial garment could not have been used for the body of Christ.

The results were announced in 1988

18.4 MASS–ENERGY RELATIONS

The energy change accompanying a nuclear reaction can be calculated from the relation

$$\Delta E = c^2 \Delta m$$

where Δm is the change in mass*, ΔE is the change in energy, and c is the speed of light. In an "ordinary" chemical reaction, Δm is immeasurably small. In a nuclear reaction, on the other hand, Δm is appreciable, amounting to 0.002% or more of the mass of reactants. The change in mass can readily be calculated from a table of nuclear masses (Table 18.2).

To obtain a more useful form of the equation $\Delta E = c^2 \Delta m$, we substitute for c its value in meters per second:

$$c = 3.00 \times 10^8 \text{ m/s}$$

* Specifically, Δm = mass of products − mass of reactants; ΔE = energy of products − energy of reactants. In spontaneous nuclear reactions, the products weigh less than the reactants (Δm negative). In this case, the energy of the products is less than that of the reactants (ΔE negative), and energy is evolved to the surroundings.

Table 18.2 Nuclear Masses on the ^{12}C Scale*

	At. No.	Mass No.	Mass (amu)		At. No.	Mass No.	Mass (amu)
e	0	0	0.00055				
n	0	1	1.00867	Br	35	79	78.8992
H	1	1	1.00728		35	81	80.8971
	1	2	2.01355		35	87	86.9028
	1	3	3.01550	Rb	37	89	88.8913
He	2	3	3.01493	Sr	38	90	89.8869
	2	4	4.00150	Mo	42	99	98.8846
Li	3	6	6.01348	Ru	44	106	105.8832
	3	7	7.01436	Ag	47	109	108.8790
Be	4	9	9.00999	Cd	48	109	108.8786
	4	10	10.01134		48	115	114.8791
B	5	10	10.01019	Sn	50	120	119.8748
	5	11	11.00656	Ce	58	144	143.8817
C	6	11	11.00814		58	146	145.8868
	6	12	11.99671	Pr	59	144	143.8809
	6	13	13.00006	Sm	62	152	151.8857
	6	14	13.99995	Eu	63	157	156.8908
O	8	16	15.99052	Er	68	168	167.8951
	8	17	16.99474	Hf	72	179	178.9065
	8	18	17.99477	W	74	186	185.9138
F	9	18	17.99601	Os	76	192	191.9197
	9	19	18.99346	Au	79	196	195.9231
Na	11	23	22.98373	Hg	80	196	195.9219
Mg	12	24	23.97845	Pb	82	206	205.9295
	12	25	24.97925		82	207	206.9309
	12	26	25.97600		82	208	207.9316
Al	13	26	25.97977	Po	84	210	209.9368
	13	27	26.97439		84	218	217.9628
	13	28	27.97477	Rn	86	222	221.9703
Si	14	28	27.96924	Ra	88	226	225.9771
S	16	32	31.96329	Th	90	230	229.9837
Cl	17	35	34.95952	Pa	91	234	233.9934
	17	37	36.95657	U	92	233	232.9890
Ar	18	40	39.95250		92	235	234.9934
K	19	39	38.95328		92	238	238.0003
	19	40	39.95358		92	239	239.0038
Ca	20	40	39.95162	Np	93	239	239.0019
Ti	22	48	47.93588	Pu	94	239	239.0006
Cr	24	52	51.92734		94	241	241.0051
Fe	26	56	55.92066	Am	95	241	241.0045
Co	27	59	58.91837	Cm	96	242	242.0061
Ni	28	59	58.91897	Bk	97	245	245.0129
Zn	30	64	63.91268	Cf	98	248	248.0186
	30	72	71.91128	Es	99	251	251.0255
Ge	32	76	75.90380	Fm	100	252	252.0278
As	33	79	78.90288		100	254	254.0331

* Note that these are *nuclear masses*. The masses of the corresponding atoms can be calculated by adding the masses of each extranuclear electron (0.000549). For example, for an *atom* of $^{4}_{2}He$ we have 4.00150 + 2(0.000549) = 4.00260. Similarly, for an atom of $^{12}_{6}C$, 11.99671 + 6(0.000549) = 12.00000

Thus $$\Delta E = 9.00 \times 10^{16} \frac{\text{m}^2}{\text{s}^2} \times \Delta m.$$

But $$1\ \text{J} = 1\ \text{kg} \cdot \frac{\text{m}^2}{\text{s}^2} \qquad 1\frac{\text{m}^2}{\text{s}^2} = 1\frac{\text{J}}{\text{kg}}$$

So $$\Delta E = 9.00 \times 10^{16} \frac{\text{J}}{\text{kg}} \times \Delta m.$$

This equation can be used to calculate the energy change in *joules,* knowing Δm in *kilograms.* Ordinarily, Δm is expressed in *grams,* while ΔE is calculated in *kilojoules.* The relationship between ΔE and Δm in these units can be found by using conversion factors:

$$\Delta E = 9.00 \times 10^{16} \frac{\text{J}}{\text{kg}} \times \frac{1\ \text{kg}}{10^3\ \text{g}} \times \frac{1\ \text{kJ}}{10^3\ \text{J}} \times \Delta m$$

This is the most useful form of the mass-energy relation

$$\Delta E = 9.00 \times 10^{10} \frac{\text{kJ}}{\text{g}} \times \Delta m$$

Example 18.5 For the radioactive decay of radium, $^{226}_{88}\text{Ra} \rightarrow {}^{222}_{86}\text{Rn} + {}^{4}_{2}\text{He}$, calculate ΔE in kilojoules when 1.02 g of radium decays.

Strategy Using Table 18.2, calculate Δm when one mole of radium decays. Then find Δm when 1.02 g of radium decays (1 mol Ra = 226.0 g Ra). Finally, find ΔE using the relation $\Delta E = 9.00 \times 10^{10}\ \Delta m$ kJ/g.

Solution

(1) Using Table 18.2, for one mole of Ra-226,

$$\Delta m = \text{mass 1 mol He-4} + \text{mass 1 mol Rn-222} - \text{mass 1 mol Ra-226}$$
$$= 4.0015\ \text{g} + 221.9703\ \text{g} - 225.9771\ \text{g}$$
$$= -0.0053\ \text{g/mol Ra}$$

(2) When 1.02 g of radium decays,

$$\Delta m = \frac{-0.0053\ \text{g}}{\text{mol Ra}} \times \frac{1\ \text{mol Ra}}{226.0\ \text{g Ra}} \times 1.02\ \text{g Ra} = -2.4 \times 10^{-5}\ \text{g}$$

(3) $\Delta E = 9.00 \times 10^{10} \dfrac{\text{kJ}}{\text{g}} \times (-2.4 \times 10^{-5}\ \text{g}) = -2.2 \times 10^{6}\ \text{kJ}$

In ordinary chemical reactions, the energy change is of the order of 50 kJ/g or less. The energy change in this nuclear reaction is greater by a factor of

$$2.2 \times 10^6/50 = 4 \times 10^4$$

In other words, about 40,000 times as much energy is evolved.

Nuclear Binding Energy

It is always true that *a nucleus weighs less than the individual protons and neutrons of which it is composed.* Consider, for example, the $^{6}_{3}\text{Li}$ nucleus, which contains three protons and three neutrons. According to Table 18.2, one mole of Li-6 nuclei weighs 6.01348 g. In contrast, the total mass of three moles of neutrons and three moles of protons is

$$3(1.00867\text{ g}) + 3(1.00728\text{ g}) = 6.04785\text{ g}$$

Clearly, one mole of Li-6 weighs *less* than the corresponding protons and neutrons. For the process

$$^{6}_{3}\text{Li} \longrightarrow 3\,^{1}_{1}\text{H} + 3\,^{1}_{0}n$$

$\Delta m = 6.04785\text{ g} - 6.01348\text{ g} = 0.03437\text{ g/mol Li}$.

The quantity just calculated is referred to as the *mass defect*. The corresponding energy difference is

$$\Delta E = 9.00 \times 10^{10}\text{ kJ} \times 0.03437\frac{\text{g}}{\text{mol Li}} = 3.09 \times 10^{9}\frac{\text{kJ}}{\text{mol Li}}$$

This energy is referred to as the **binding energy**. It follows that 3.09×10^9 kJ of energy would have to be absorbed to decompose one mole of Li-6 nuclei into protons and neutrons:

It takes a lot of energy to blow a nucleus apart

$$^{6}_{3}\text{Li} \longrightarrow 3\,^{1}_{0}n + 3\,^{1}_{1}\text{H} \qquad \Delta E = 3.09 \times 10^{9}\frac{\text{kJ}}{\text{mol Li}}$$

By the same token, 3.09×10^9 kJ of energy would be evolved when one mole of Li-6 is formed from protons and neutrons.

Example 18.6 Calculate the binding energy of Be-9 in kilojoules per mole.

Strategy Use Table 18.2 to find Δm for the decomposition of Be-9 into neutrons and protons. Then calculate $\Delta E = 9.00 \times 10^{10}\text{ kJ/g} \times \Delta m$.

Solution

(1) $^{9}_{4}\text{Be} \longrightarrow 5\,^{1}_{0}n + 4\,^{1}_{1}\text{H}$

(2) $\Delta m = 5(1.00867\text{ g}) + 4(1.00728\text{ g}) - 9.00999\text{ g} = 0.06248\text{ g/mol Be}$

$$\Delta E = 9.00 \times 10^{10}\frac{\text{kJ}}{\text{g}} \times 0.06248\frac{\text{g}}{\text{mol Be}} = 5.62 \times 10^{9}\text{ kJ/mol Be}$$

The binding energy of a nucleus is, in a sense, a measure of its stability. The greater the binding energy, the more difficult it would be to decompose the nucleus into protons and neutrons. As you might expect, the binding energy as calculated above increases steadily as the nucleus gets heavier, containing more protons and neutrons. A better measure of the relative stabilities of different nuclei is the *binding energy per mole of nuclear particles (nucleons)*. This quantity is calculated by dividing the binding energy per mole of nuclei by the number of particles per nucleus. Thus

$$\text{Li-6:} \qquad 3.09 \times 10^{9}\frac{\text{kJ}}{\text{mol Li-6}} \times \frac{1\text{ mol Li-6}}{6\text{ mol nucleons}} = 5.15 \times 10^{8}\frac{\text{kJ}}{\text{mol}}$$

$$\text{Be-9:} \qquad 5.62 \times 10^{9}\frac{\text{kJ}}{\text{mol Be-9}} \times \frac{1\text{ mol Be-9}}{9\text{ mol nucleons}} = 6.24 \times 10^{8}\frac{\text{kJ}}{\text{mol}}$$

Figure 18.5 shows a plot of this quantity, binding energy per mole of nucleons, vs. mass number. Notice that the curve has a broad maximum in the vicinity of

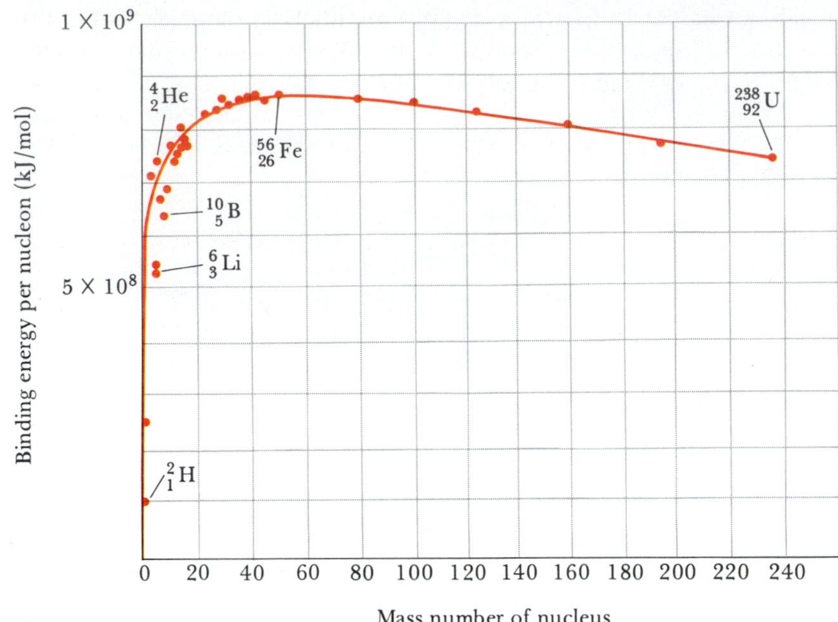

Figure 18.5
The binding energy per nucleon is a measure of nuclear stability. It has its maximum value for nuclei of intermediate mass, falling off for very heavy or very light nuclei. The form of this curve accounts for the fact that both fission and fusion give off large amounts of energy.

The higher the binding energy per nucleon, the more stable the nucleus

mass numbers 50 to 80. Consider what would happen if a heavy nucleus such as $^{235}_{92}U$ were to split into smaller nuclei with mass numbers near the maximum. This process, referred to as *nuclear fission,* should result in an evolution of energy. The same effect would be obtained if very light nuclei such as $^{2}_{1}H$ were to combine with one another. Indeed, this process, called *nuclear fusion,* should evolve even more energy, since the binding energy per nucleon increases very sharply at the beginning of the curve.

18.5 NUCLEAR FISSION

By Otto Hahn and Lise Meitner

The process of nuclear fission was discovered more than half a century ago in 1938 in Germany. With the outbreak of World War II a year later, interest focused on the enormous amount of energy released in the process. At Los Alamos, in the mountains of New Mexico, a group of scientists led by J. Robert Oppenheimer worked feverishly to produce the fission, or "atomic," bomb. Many of the members of this group were exiles from Nazi Germany. They were spurred on by the fear that Hitler would obtain the bomb first. Their work led to the explosion of the first atomic bomb in the New Mexico desert at 5:30 A.M. on July 16, 1945. Less than a month later (August 6, 1945), the world learned of this new weapon when another bomb was exploded over Hiroshima. This bomb killed 70,000 people and completely devastated an area of 10 square kilometers. Three days later Nagasaki and its inhabitants met a similar fate. On August 14, Japan surrendered and World War II was over.

The Fission Process ($^{235}_{92}U$)

Several isotopes of the heavy elements undergo fission if bombarded by neutrons of high enough energy. In practice, attention has centered upon two particular isotopes, $^{235}_{92}U$ and $^{239}_{94}Pu$. Both of these can be split into fragments by low-energy neutrons.

Our discussion concentrates upon the uranium-235 isotope. It makes up only about 0.7% of naturally occurring uranium. The more abundant isotope, uranium-238, does not undergo fission. The first process used to separate these isotopes, and until recently the only one available, was that of gaseous effusion (Chap. 5). The volatile compound uranium hexafluoride, UF_6, which sublimes at 56°C, is used for this purpose.

Fission Products When a uranium-235 atom undergoes fission, it splits into two unequal fragments and a number of neutrons and beta particles. The fission process is complicated by the fact that different uranium-235 atoms split up in many different ways. For example, while one atom of $^{235}_{92}U$ is splitting to give isotopes of rubidium ($Z = 37$) and cesium ($Z = 55$), another may break up to give isotopes of bromine ($Z = 35$) and lanthanum ($Z = 57$), while still another atom yields isotopes of zinc ($Z = 30$) and samarium ($Z = 62$):

$$^{1}_{0}n + ^{235}_{92}U \longrightarrow \begin{cases} ^{90}_{37}Rb + ^{144}_{55}Cs + 2\,^{1}_{0}n \\ ^{87}_{35}Br + ^{146}_{57}La + 3\,^{1}_{0}n \\ ^{72}_{30}Zn + ^{160}_{62}Sm + 4\,^{1}_{0}n \end{cases}$$

Fission occurs when a U-235 nucleus absorbs a neutron

More than 200 isotopes of 35 different elements have been identified among the fission products of uranium-235.

The stable neutron-to-proton ratio near the middle of the periodic table, where the fission products are located, is considerably smaller (~1.2) than that of uranium-235 (1.5). Hence, the immediate products of the fission process contain too many neutrons for stability; they decompose by beta emission. In the case of rubidium-90, three steps are required to reach a stable nucleus:

$$^{90}_{37}Rb \longrightarrow ^{90}_{38}Sr + ^{0}_{-1}e \qquad t_{1/2} = 2.8 \text{ min}$$

$$^{90}_{38}Sr \longrightarrow ^{90}_{39}Y + ^{0}_{-1}e \qquad t_{1/2} = 29 \text{ yr}$$

$$^{90}_{39}Y \longrightarrow ^{90}_{40}Zr + ^{0}_{-1}e \qquad t_{1/2} = 64 \text{ h}$$

The radiation hazard associated with nuclear fallout arises from the formation of radioactive isotopes such as these. One of the most dangerous is strontium-90. In the form of strontium carbonate, $SrCO_3$, it is incorporated into the bones of animals and human beings.

Notice from the fission equations written above that two to four neutrons are produced by fission for every one consumed. Once a few atoms of uranium-235 split, the neutrons produced can bring about the fission of many more uranium-235 atoms. This creates the possibility of a *chain reaction*, whose rate increases exponentially with time. This is precisely what happens in the atomic bomb. The energy evolved in successive fissions escalates to give a tremendous explosion within a few seconds.

For nuclear fission to result in a chain reaction, the sample must be large enough so that most of the neutrons are captured internally. If the sample is too small, most of the neutrons escape, breaking the chain. The *critical mass* of uranium-235 required to maintain a chain reaction in a bomb appears to be about 1 to 10 kg. In the bomb dropped on Hiroshima, the critical mass was achieved by using a conventional explosive to fire one piece of uranium-235 into another.

Fission Energy The evolution of energy in nuclear fission is directly related to the decrease in mass that takes place. About 80,000,000 kJ of energy is given off

The Hiroshima bomb was equivalent to 20,000 tons of TNT

for every gram of $^{235}_{92}U$ that reacts. This is about 40 times as great as the energy change for simple nuclear reactions such as radioactive decay. The heat of combustion of coal is only about 30 kJ/g; the energy given off when TNT explodes is still smaller, about 2.8 kJ/g. Putting it another way, the fission of one gram of $^{235}_{92}U$ produces as much energy as the combustion of 2700 kg of coal or the explosion of 30 metric tons (3×10^4 kg) of TNT.

Nuclear Reactors

Even before the first atomic bomb exploded, scientists and political leaders began to speculate on the use of fission as a peacetime energy source. Twenty-five years ago, it was widely believed that nuclear fission would replace fossil fuels (oil, natural gas, coal), at least for the generation of electrical energy. During the 1970s, disillusionment set in with nuclear power in the United States. Economic factors, along with the problem of disposing of nuclear waste (p. 511), were largely responsible. The newest of the 110 nuclear reactors now operating in the United States was started when Richard Nixon was president. The incident at Three Mile Island (p. 511) in 1979 seemed to be the last straw, stirring fears of nuclear disaster.

Within the past few years, there has been a perceptible shift of public opinion in favor of nuclear power. Mostly, this reflects deficiencies and uncertainties associated with conventional power plants. Coal-fired plants are the major source of acid rain (p. 348). The 1991 war in the Persian Gulf emphasized once again the danger of depending upon oil imported from the Middle East. Finally, the combustion of all fossil fuels is principally responsible for the greenhouse effect. Nuclear energy has none of these shortcomings.

Choose your poison

The so-called "light water reactor" (LWR) used in the United States today is shown in Figure 18.6. Fuel rods alternate with control rods in a containment

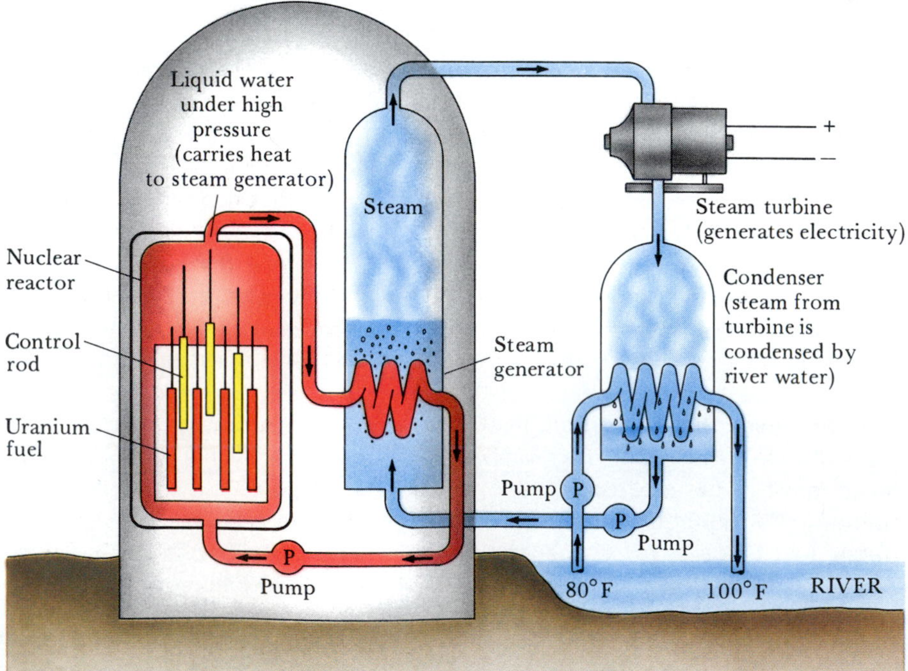

Figure 18.6
Nuclear reactor of the pressurized water type. The control rods are made of a material such as cadmium or boron, which absorb neutrons effectively. The fuel rods contain uranium oxide, usually enriched in U-235.

chamber. The *fuel rods* are cylinders that contain fissionable material, uranium dioxide (UO_2) pellets, in a zirconium alloy tube. The uranium in these reactors is "enriched" so that it contains about 3% U-235, the fissionable isotope. The *control rods* are cylinders composed of substances, such as boron and cadmium, that absorb neutrons. Increased absorption of neutrons slows down the chain reaction. By varying the depth of the control rods within the fuel-rod assembly, the speed of the chain reaction can be controlled. Water at a pressure of 140 atm is passed through the reactor to absorb the heat given off by fission. The water, coming out of the reactor core at 320°C, circulates through a closed loop containing a heat exchanger which produces steam at 270°C. This steam is used to drive a turbogenerator that produces electrical energy.

The nuclear power plant at Chernobyl shortly after the fire and explosion in 1986. (© Peter A. Simon/ Phototake)

In the light water reactor, the circulating water serves another purpose in addition to heat transfer. It acts to slow down or *moderate* the neutrons given off by fission. This is necessary if the chain reaction is to continue; fast neutrons are not readily absorbed by U-235. Reactors in Canada use "heavy water," D_2O, which has an important advantage over H_2O. Its moderating properties are such that naturally occurring uranium can be used as a fuel; enrichment in U-235 is not necessary.

Nuclear Accidents: Three Mile Island and Chernobyl To operate a nuclear reactor safely, it is necessary to prevent significant amounts of radioactive fission products from escaping into the environment. This can happen if the temperature of the nuclear fuel rises to the melting point ("meltdown"), leading to an explosion.

The most likely cause of meltdown is a loss of cooling water. If this happens in a light water reactor, fission stops because the unmoderated neutrons move too fast to continue the chain reaction. However, large amounts of heat are given off by the radioactive decay of fission products (recall Example 18.5). This happened at Three Mile Island in Pennsylvania in March of 1979. Through operator error, water was lost from the cooling system. For some time, the fuel rods were uncovered and there was a real danger of meltdown. Fortunately, this did not happen. Virtually all of the radioactive isotopes released were confined within a complete containment building surrounding the reactor.

A much more severe nuclear accident took place at Chernobyl, near Kiev in the Ukraine in April of 1986. In the absence of a containment building, large amounts of radiation, estimated at 100 million curies, were released to the atmosphere. Much of this was carried westward by prevailing winds to other European countries. There were 31 fatalities from acute radiation exposure at the reactor; 116,000 people had to be permanently evacuated from the Chernobyl area. It is estimated that the number of delayed cancer fatalities from lower-level radiation will be in the vicinity of 25,000.

The worst nuclear accident occurred 40 years ago at Chelyabinsk, when radioactive waste from Pu production was dumped into a lake

Nuclear Waste When a nuclear reactor operates, the fuel rods undergo physical and chemical changes due to the enormous amount of radiation to which they are exposed. Each year, on the average, one fourth of these intensely radioactive rods must be replaced. A typical nuclear reactor produces about 20 metric tons (2×10^4 kg) of spent fuel rods per year. Ideally, these rods should be reprocessed to recover uranium and plutonium, producing a relatively small amount of high-level radioactive waste. In 1982, a federal program was established for the disposal of such waste, which is supposed to be buried in an underground site.

Calculations indicate that waste from reprocessed fuel requires 20,000 years to decay to a "safe" level. Unprocessed waste takes 100 times longer. Data like these contribute to the "nimby" syndrome (*n*ot *i*n *m*y *b*ack *y*ard) where nuclear wastes are concerned. The latest estimate is that a repository for high-activity waste may be available by the year 2010, perhaps within a remote mountain in Nevada.

Maybe

18.6 NUCLEAR FUSION

Recall (Fig. 18.5) that very light isotopes, such as those of hydrogen, are unstable with respect to fusion into heavier isotopes. Indeed, the energy available from nuclear fusion is considerably greater than that given off in the fission of an equal mass of a heavy element (Example 18.7).

Example 18.7 Calculate the amount of energy evolved, in kilojoules per gram of reactants, in

(a) a fusion reaction, $^{2}_{1}H + ^{2}_{1}H \longrightarrow ^{4}_{2}He$.
(b) a fission reaction, $^{235}_{92}U \longrightarrow ^{90}_{38}Sr + ^{144}_{58}Ce + ^{1}_{0}n + 4\ ^{0}_{-1}e$.

Strategy This problem is entirely analogous to Example 18.5. First, find Δm for the equation as written, and then find Δm for one gram of reactant. Finally, calculate $\Delta E = 9.00 \times 10^{10}$ kJ/g $\times \Delta m$.

Solution

(a) (1) using Table 18.2,

$$\Delta m = 4.00150\text{ g} - 2(2.01355\text{ g}) = -0.02560\text{ g}$$

(2) The total mass of reactant, $^{2}_{1}H$, is 4.027 g:

$$\Delta m = \frac{-0.02560\text{ g}}{4.027\text{ g reactant}} = -6.357 \times 10^{-3}\text{ g/g reactant}$$

(3) $$\Delta E = 9.00 \times 10^{10}\frac{\text{kJ}}{\text{g}} \times \frac{(-6.357 \times 10^{-3}\text{ g})}{\text{g reactant}} = \frac{-5.72 \times 10^{8}\text{ kJ}}{\text{g reactant}}$$

(b) (1) $$\Delta m = 89.8869\text{ g} + 143.8817\text{ g} + 1.0087\text{ g} + 4(0.00055\text{ g}) - 234.9934\text{ g}$$
$$= -0.2139\text{ g}$$

(2) The total mass of reactant, U-235, is 235.0 g:

$$\Delta m = \frac{-0.2139\text{ g}}{235.0\text{ g reactant}} = 9.102 \times 10^{-4}\text{ g/g reactant}$$

(3) $$\Delta E = 9.00 \times 10^{10}\frac{\text{kJ}}{\text{g}} \times \left(\frac{-9.102 \times 10^{-4}\text{ g}}{\text{g reactant}}\right) = \frac{-8.19 \times 10^{7}\text{ kJ}}{\text{g reactant}}$$

Comparing the answers to (a) and (b), it appears that the fusion reaction produces about seven times as much energy per gram of reactant (57.2×10^{7} vs. 8.19×10^{7} kJ) as does the fission reaction. This factor varies from about three to ten, depending upon the particular reactions chosen to represent the fusion and fission processes.

As an energy source, nuclear fusion possesses several additional advantages over nuclear fission. In particular, light isotopes suitable for fusion are far more abundant than the heavy isotopes required for fission. You can calculate, for example (Problem 67), that the fusion of only $2 \times 10^{-9}\%$ of the deuterium ($^{2}_{1}H$) in seawater would meet the total annual energy requirements of the world.

Unfortunately, fusion processes, unlike neutron-induced fission, have very high activation energies. In order to overcome the electrostatic repulsion between two deuterium nuclei and cause them to react, they have to be accelerated to velocities of about 10^6 m/s, about 10,000 times greater than ordinary molecular velocities at room temperature. The corresponding temperature for fusion, as calculated from kinetic theory, is of the order of 10^9 °C. In the hydrogen bomb, temperatures of this magnitude were achieved by using a fission reaction to trigger nuclear fusion. If fusion reactions are to be used to generate electricity, it will be necessary to develop equipment in which very high temperatures can be maintained long enough to allow fusion to occur and give off energy. In any conventional container, the reactant nuclei would quickly lose their high kinetic energies by collisions with the walls.

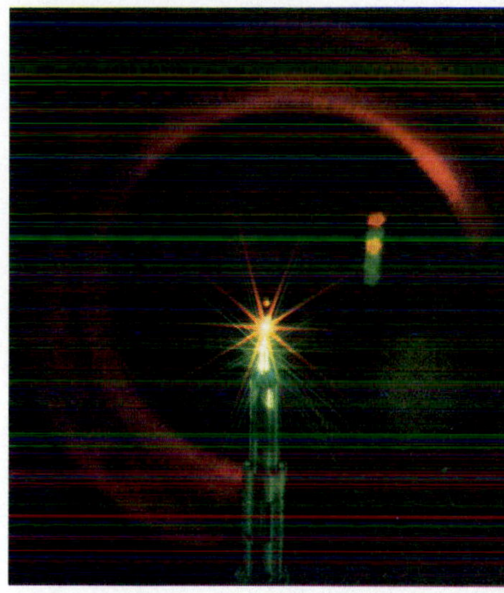

Figure 18.7
Laser fusion. (University of California, Lawrence Livermore Laboratory)

One fusion reaction currently under study is a two-step process involving deuterium and lithium as the basic starting materials:

$$\begin{array}{l} ^{2}_{1}H + ^{3}_{1}H \longrightarrow ^{4}_{2}He + ^{1}_{0}n \\ ^{6}_{3}Li + ^{1}_{0}n \longrightarrow ^{4}_{2}He + ^{3}_{1}H \\ \hline ^{2}_{1}H + ^{6}_{3}Li \longrightarrow 2\,^{4}_{2}He \end{array}$$

This process is attractive because it has a lower activation energy than other fusion reactions.

Promising results have been obtained with nuclear fusion using magnetic fields to confine the reactant nuclei and prevent them from touching the walls of the container, where they would quickly slow down below the velocity required for fusion. Using 400-ton magnets, it is possible to sustain the reaction for a fraction of a second. To achieve a net evolution of energy, this time must be extended to about one second. A practical fusion reactor would have to produce 20 times as much energy as it consumes. It is likely to take at least 20 years to achieve that goal with magnetic confinement.

Nuclear fusion has been the "fuel of the future" for at least 40 years

Another approach to nuclear fusion is shown in Figure 18.7. Tiny glass pellets (about 0.1 mm in diameter) filled with frozen deuterium and tritium serve as a target. The pellets are illuminated by a powerful laser beam, which delivers 10^{12} kilowatts of power in one nanosecond (10^{-9} s). The reaction is the same as with magnetic confinement; unfortunately, at this point energy breakeven seems several years away.

PERSPECTIVE
•
Biological Effects of Radiation

The harmful effects of radiation result from its high energy, sufficient to break the chemical bonds holding organic molecules together in body cells. The extent of damage depends mainly upon two factors. These are the amount of radiation absorbed and the type of radiation. The former is commonly expressed in *rads* (*r*adiation *a*bsorbed *d*ose). A rad corresponds to the absorption of 10^{-2} J of energy per kilogram of tissue:

$$1 \text{ rad} = 10^{-2} \text{ J/kg}$$

It would be better to say reps (*r*adiation *e*quivalent for *p*eople)

The total biological effect of radiation is expressed in *rems* (*r*adiation *e*quivalent for *m*an). The number of rems is found by multiplying the number of rads by an appropriate factor, n, for the particular type of radiation:

$$\text{no. rems} = n \text{ (no. rads)}$$

$$n = 1 \text{ for } \beta, \gamma, \text{x-rays}$$

$$n = 10 \text{ for } \alpha \text{ rays, high-energy neutrons}$$

Table 18.A lists some of the effects to be expected when a person is exposed to a single dose of radiation at various levels.

Small doses of radiation repeated over long periods of time can have very serious consequences. Many of the early workers in the field of radioactivity developed cancer in this way. Cases are known in which cancers developed as long as 40 years after initial exposure. Studies have shown an abnormally large number of cases of leukemia among the survivors of the atomic bombs dropped on Hiroshima and Nagasaki.

Table 18.B lists average exposures to radiation of people living in the United States. Notice that about two thirds of the radiation exposure comes from natural sources. The level depends upon location. Cosmic radiation is much more intense at high elevations. A person living in Denver is exposed to about 100 millirems/year from this source. This is twice the national average.

Radiation listed in Table 18.B as coming from the earth is mostly in the form of $^{222}_{86}Rn$, a radioactive isotope of radon, which is a decay product of uranium-238. Because it is gaseous and chemically inert, radon seeps through cracks in concrete and masonry from the ground into houses. There its concentration builds up, particularly if the house is tightly insulated.

Inhalation of radon-222 can cause health problems because its decay products, including Po-218 and Po-214, are intensely radioactive and readily adsorbed in lung tissue. The Environmental Protection Agency estimates that radon inhalation causes between 5000 and 20,000 deaths from lung cancer annually in the United States. To reduce these numbers, EPA recom-

Table 18.A Effect of Exposure to a Single Dose of Radiation

Dose (rems)	Probable Effect
0 to 25	no observable effect
25 to 50	small decrease in white blood cell count
50 to 100	lesions, marked decrease in white blood cells
100 to 200	nausea, vomiting, loss of hair
200 to 500	hemorrhaging, ulcers, possible death
500+	fatal

Table 18.B Typical Radiation Exposures in the United States (1 millirem = 10^{-3} rem)

Sources	Millirems/yr
I. Natural	
A. External to the body	
1. From cosmic radiation	50
2. From the earth	47
3. From building materials	3
B. Inside the body	
1. Inhalation of air	5
2. In human tissues (mostly $^{40}_{19}K$)	21
Total from natural sources	**126**
II. Man-made	
A. Medical procedures	
1. Diagnostic x-rays	50
2. Radiotherapy x-rays, radioisotopes	10
3. Internal diagnosis, therapy	1
B. Nuclear power industry	0.2
C. Luminous watch dials, TV tubes, industrial wastes	2
D. Radioactive fallout (nuclear tests)	4
Total from man-made sources	**67**
Total	**193**

mends that special ventilation devices be used to remove radon from basements if the radiation level exceeds 4×10^{-12} Ci/L. This corresponds to a radon concentration of only about 10^{-19} mol/L.

CHAPTER HIGHLIGHTS

KEY CONCEPTS

1. *Write balanced nuclear equations*
   (Example 18.1; Problems 1–10)
2. *Relate activity to mass and rate constant*
   (Example 18.2; Problems 21–24, 54–58, 61, 64)
3. *Relate time to mass and rate constant*
   (Examples 18.3, 18.4; Problems 15–20, 25–32, 62)
4. *Relate Δm to ΔE*
   (Examples 18.5, 18.7; Problems 33, 34, 37–46)
5. *Calculate binding energies*
   (Example 18.6; Problems 35, 36)

KEY EQUATIONS

Rate of decay $\ln X_o/X = kt$ $k = \text{rate}/N_t$

Mass–energy $\Delta E = 9.00 \times 10^{10} \frac{\text{kJ}}{\text{g}} \times \Delta m$

KEY TERMS

activity
alpha particle
atomic number
beta particle
binding energy
curie (Ci)
fission
fusion
half-life
isotope
K-electron capture
mass number
positron

SUMMARY PROBLEM

Consider the isotopes of copper.

a. Write the nuclear symbol for Cu-64, which is used medically to scan for brain tumors. How many protons are there in the nucleus? How many neutrons?
b. Write the reaction for the β-decay of Cu-64.
c. When Cu-65 is bombarded with carbon-12, three neutrons and another particle are produced. Write the equation for this reaction.
d. Calculate ΔE in kilojoules when 1.00 g of Cu-63 (molar mass = 62.91367 g/mol) is formed, along with a proton, by neutron bombardment of Zn-63 (molar mass = 62.91674 g/mol).
e. What is the mass defect and binding energy of Cu-63?
f. A one-milligram sample of Cu-64 has an activity of 3.82×10^3 Ci. How many atoms are there in the sample (molar mass Cu-64 = 63.91 g/mol)? What is the decay rate in atoms per second? What is the rate constant (s^{-1})? the half-life in seconds?
g. How long will it take for 20.0% of a sample of Cu-64 to decay?

Solution

a. $^{64}_{29}Cu$, 29 protons, 35 neutrons **b.** $^{64}_{29}Cu \rightarrow {}^{0}_{-1}e + {}^{64}_{30}Zn$
c. $^{65}_{29}Cu + {}^{12}_{6}C \rightarrow 3\,{}^{1}_{0}n + {}^{74}_{35}Br$ **d.** -6.38×10^6 kJ
e. mass defect = 0.5922 g; binding energy = 5.33×10^{10} kJ
f. 9.42×10^{18}; 1.41×10^{14}; 1.50×10^{-5}/s; 4.62×10^4 s
g. 1.49×10^4 s

QUESTIONS & PROBLEMS

Nuclear Equations

1. Lead-210 is used to prepare eyes for corneal transplants. Its decay product is bismuth-210. Identify the emission from lead-210.

2. The blood volume of a patient can be measured by using chromium-51, a positron emitter, administered as a solution of sodium chromate. Write the nuclear equation for the decay of chromium-51.

3. Write balanced nuclear equations for
a. the loss of an alpha particle by Th-230.
b. the loss of an electron by lead-210.
c. the fusion of two C-12 nuclei to give another nucleus and a neutron.
d. the fission of U-235 to give Ba-140, another nucleus, and an excess of two neutrons.

4. Write balanced nuclear equations for
a. the alpha emission resulting in the formation of Pa-233.
b. the loss of a positron by Y-85.
c. the fusion of two C-12 nuclei to give sodium-23 and another particle.
d. the fission of Pu-239 to give tin-130, another nucleus, and an excess of three neutrons.

5. Rubidium-87, a beta emitter, is the product of positron emission. Identify
a. the product of ^{87}Rb decay.
b. the parent nucleus from which ^{87}Rb is formed.

6. Thorium-231 is the product of alpha emission and is radioactive, emitting beta radiation. Determine

a. the parent nucleus of ^{231}Th.
b. the product of ^{231}Th decay.

7. Write balanced nuclear equations for the bombardment of

a. Fe-54 with an alpha particle to produce another nucleus and two protons.
b. Mo-96 with deuterium ($^2_1$H) to produce a neutron and another nucleus.
c. Ar-40 with an unknown particle to produce potassium-43 and a proton.
d. a nucleus with a neutron to produce a proton and P-31.

8. Write balanced nuclear equations for the bombardment of

a. U-238 to produce Fm-249 and five neutrons.
b. Al-26 with an alpha particle to produce P-30.
c. Cu-63 to produce Zn-63 and a neutron.
d. Al-27 with deuterium ($^2_1$H) to produce an alpha particle and another nucleus.

9. Balance the following nuclear equations by filling in the blanks.

a. $^{121}_{51}\text{Sb} + ^4_2\text{He} \longrightarrow$ ________ $+ ^1_1\text{H}$
b. $^{238}_{92}\text{U} + ^1_0n \longrightarrow ^{\ 0}_{-1}e +$ ________
c. $^{14}_{7}\text{N} +$ ________ $\longrightarrow ^{17}_{8}\text{O} + ^1_1\text{H}$
d. ________ $+ ^4_2\text{He} \longrightarrow ^{27}_{14}\text{Si} + ^1_0n$

10. Balance the following nuclear equations by filling in the blanks.

a. $^{238}_{92}\text{U} + ^1_1\text{H} \longrightarrow ^{238}_{93}\text{Np} +$ ________
b. $^{241}_{95}\text{Am} + ^4_2\text{He} \longrightarrow$ ________ $+ 2\,^1_0n$
c. ________ $+ ^4_2\text{He} \longrightarrow ^1_0n + ^{12}_{6}\text{C}$
d. $^{27}_{13}\text{Al} +$ ________ $\longrightarrow ^{24}_{11}\text{Na} + ^4_2\text{He}$

Nuclear Stability

11. Which isotope in each of the following pairs should be more stable?

a. $^{12}_{6}$C or $^{13}_{6}$C **b.** $^{19}_{9}$F or $^{20}_{9}$F **c.** $^{16}_{7}$N or $^{14}_{7}$N

12. Which isotope in each of the following pairs is more stable?

a. $^{28}_{14}$Si or $^{29}_{14}$Si **b.** $^{6}_{3}$Li or $^{8}_{3}$Li **c.** $^{23}_{11}$Na or $^{20}_{11}$Na

13. For each pair of elements listed, predict which one has more stable isotopes.

a. Co or Fe **b.** F or Ge **c.** Ag or Pd

14. For each pair of elements listed, predict which one has more stable isotopes.

a. Ni or Cu **b.** Se or Sb **c.** Cd or Au

Rate of Nuclear Decay

15. Thallium-206 decays to lead-206 and has a half-life of 4.20 minutes. How many milligrams of thallium-206 remain after four half-lives, starting with 10.0 mg?

16. Strontium-90 has a half-life of 28.8 years. How much strontium-90 was present initially, if after 144 years 10.0 g remain?

17. Argon-41 is used to measure the rate of gas flow. It has a half-life of 110 minutes.

a. Calculate k.
b. How long will it take before only 15.0% of the original amount of Ar-41 remains?

18. Strontium-90 was one of the isotopes present after the Chernobyl nuclear disaster. Cow's milk in Amsterdam was found to have high levels of Sr-90 after the explosion. Suppose that a year-old toddler had drunk contaminated milk. Calculate the fraction of Sr-90 left in his body when he reaches 80 years of age, assuming no loss of Sr-90 except by radioactive decay. (k for Sr-90 = 0.024 yr^{-1}.)

19. A sample of sodium-24 chloride containing 0.075 mg of Na-24 is used to study sodium balance in an animal. After 8.0 h, it is determined that there is 0.052 mg of Na-24 left. What is the half-life of Na-24?

20. A sample of Br-82 was found to have an activity of 8.7 × 10^4 disintegrations per minute at 1:00 P.M., November 20, 1991. At 10:15 A.M. November 22, 1991, its activity was redetermined and found to be 3.6 × 10^4 disintegrations/minute. Calculate the half-life of Br-82.

21. I-131 is used to locate tumors in the thyroid glands. A 1.00-g sample of I-131 (at. mass = 130.9 amu) has an activity of 1.3 × 10^5 Ci. Calculate its decay constant and half-life in days.

22. Technetium-99 (at. mass = 98.9 amu) is used for bone scans. It has a half-life of 6.0 h. How many disintegrations/s can you expect from 1.00 mg technetium-99?

23. Np-237 (at. mass = 237.0 amu) has a half-life of 2.20 × 10^6 years. This isotope decays by alpha particle emission. Write a balanced nuclear equation for the decay of Np-237 and calculate the activity in millicuries of a 0.500-g sample.

24. Lead-210 has a half-life of 20.4 years. This isotope decays by beta particle emission. Write a balanced nuclear equation for the decay of Pb-210 and calculate the activity in millicuries of a 0.500-g sample (at. mass = 210.0 amu).

25. A sample of a wooden artifact gives 5.0 disintegrations/min/g of carbon. The half-life of C-14 is 5720 years and the activity of wood just cut down from a tree is 15.3 disintegrations/min/g of carbon. How old is the wooden artifact?

26. A sample of a beam from the tomb of an ancient Egyptian king was analyzed in 1992 and gave 7.0 counts per minute (cpm) in a scintillation counter. A sample of freshly cut wood containing the same amount of carbon gave 15.3 cpm. In what year, approximately, did the king die? ($t_{1/2}$ C-14 = 5720 yr)

27. The radioactive isotope tritium, $^3_1$H, is produced in nature in much the same way as $^{14}_{6}$C. Its half-life is 12.3 yr. Estimate the age of a sample of Scotch whiskey that has a tritium content 0.59 times that of the water in the area where the whiskey was produced.

28. An oil painting supposed to be by Rembrandt (1606–1669 A.D.) is checked by ^{14}C dating. The ^{14}C content ($t_{1/2}$ = 5720 yr) of the canvas is 0.961 times that in a living plant. Could the painting have been by Rembrandt?

29. What is the approximate age of a rock in which the mole ratio of U-238 to Pb-206 is 1.10? Take the half-life of U-238 to be 4.5×10^9 yr.

30. What is the approximate age of a rock in which the mass ratio (grams) of U-238 to Pb-206 is 1.10?

31. One way of dating rocks is to determine the relative amounts of ^{40}K and ^{40}Ar; the decay of ^{40}K has a half-life of 1.26×10^9 yr. Analysis of a certain lunar sample gives the following results in mole ratios.

$^{40}Ar/^{40}K = 4.13$ $\quad t_{1/2}\ ^{40}K = 1.26 \times 10^9$ yr

$^{206}Pb/^{238}U = 0.66$ $\quad t_{1/2}\ ^{238}U = 4.5 \times 10^9$ yr

$^{87}Sr/^{87}Rb = 0.049$ $\quad t_{1/2}\ ^{87}Rb = 4.8 \times 10^{10}$ yr

Using these data, obtain the best possible value for the age of the sample. Can you suggest why the K–Ar method gives a low result?

32. The ^{87}Rb to ^{87}Sr method of dating rocks was used to analyze lunar samples from the Apollo-15 mission. Estimate the age of the lunar sample in which

a. the mole ratio of ^{87}Rb to ^{87}Sr is 25.0 (see Problem 31).

b. the mole ratio of ^{87}Rb to ^{87}Sr is 20.0.

Mass–Energy Changes

33. For the reaction $^{230}_{90}Th \rightarrow ^{226}_{88}Ra + ^{4}_{2}He$.

a. calculate Δm in grams when one mole of $^{230}_{90}Th$ decays.

b. calculate ΔE in joules when one mole of $^{230}_{90}Th$ decays; one gram of $^{230}_{90}Th$ decays.

34. Bk-245 decays by alpha emission. For one gram of ^{245}Bk, calculate Δm (grams) and ΔE (kilojoules).

35. For carbon-14, calculate

a. the mass defect.

b. the binding energy.

36. Which has the larger binding energy, fluorine-19 or oxygen-17?

37. Some of the sun's energy comes from the reaction

$$4\ ^{1}_{1}H \longrightarrow ^{4}_{2}He + 2\ ^{0}_{1}e$$

Calculate the energy change in this reaction per gram of hydrogen.

38. Compare the energies given off per gram of reactant in the two fusion processes considered to occur during the formation of a star:

a. $^{2}_{1}H + ^{1}_{1}H \longrightarrow ^{3}_{2}He$

b. $2\ ^{3}_{2}He \longrightarrow ^{4}_{2}He + 2\ ^{1}_{1}H$

39. The sun radiates energy into space at the rate of 3.9×10^{26} J/s. Calculate the rate of mass loss by the sun.

40. Calculate the change in mass when 1.00 kg of carbon burns to form carbon dioxide, taking the heat of formation of $CO_2(g)$ to be -393.5 kJ/mol.

41. For the fission reaction

$$^{1}_{0}n + ^{235}_{92}U \longrightarrow ^{89}_{37}Rb + ^{144}_{58}Ce + 3\ _{-1}^{0}e + 3\ ^{1}_{0}n$$

a. how much energy (in kJ) is given off per gram of $^{235}_{92}U$?

b. how many kilograms of TNT must be detonated to produce the same amount of energy? ($\Delta E = -2.76$ kJ/g.)

42. Consider the fission reaction

$$^{239}_{94}Pu + ^{1}_{0}n \longrightarrow ^{146}_{58}Ce + ^{90}_{38}Sr + 2\ _{-1}^{0}e + 4\ ^{1}_{0}n$$

a. How many grams of $^{239}_{94}Pu$ would have to react to produce 1.00 kJ of energy?

b. How many atoms of $^{239}_{94}Pu$ would have to react to produce 1.00 kJ of energy?

43. Show by calculation whether the following reaction is spontaneous.

$$^{1}_{1}H + ^{18}_{8}O \longrightarrow ^{19}_{9}F$$

44. Will Al-26 decay spontaneously by positron emission? Show by calculation.

45. Consider the fission reaction where U-235 is bombarded with a neutron. Cerium-144, bromine-87, electrons, and neutrons are produced.

a. Write a balanced nuclear equation for the reaction.

b. Calculate ΔE when one gram of U-235 undergoes fission.

c. The decomposition of ammonium nitrate, an explosive, evolves 37.0 kJ/mol. How many kilograms of NH_4NO_3 are required to produce the same amount of energy as one milligram of U-235?

46. Using the equation in Problem 45, compare the energy produced by the fission of one gram of U-235 to the energy produced by one gram of He-3 in the fusion reaction

$$2\ ^{3}_{2}He \longrightarrow ^{4}_{2}He + 2\ ^{1}_{1}H$$

Biological Effects of Radiation

47. Consider the interaction of beta and slow neutron radiation with human cells.

a. Calculate the number of rems absorbed by a human being exposed to 15 rads of slow neutrons ($n = 3$) and 10 rads of beta radiation ($n = 1$). Assume the dosage is additive.

b. Using Table 18.A, describe the effect of the absorption of the radiation in part (a).

48. Why does radon-222 cause health problems? How does it get into houses?

49. What two factors determine the extent of radiation damage to an individual?

50. How many disintegrations per second occur in a basement that is $40 \times 20 \times 10$ feet if the radiation level from radon is the allowed 4×10^{-12} Ci/L?

Unclassified

51. Classify each of the following statements as true or false. If false, correct the statement to make it true.

a. The mass number increases in beta emission.

b. A radioactive species with a large rate constant, k, decays very slowly.

c. Fusion gives off less energy per gram of fuel than fission.

52. Explain how

a. alpha and beta radiation are separated by an electric field.

b. radioactive ^{11}C can be used as a tracer to study brain disorders.

c. a self-sustaining chain reaction occurs in nuclear fission.

53. Suppose the $^{14}C/^{12}C$ ratio in plants a thousand years ago was 10% higher than it is today. What effect, if any, would this have on the calculated age of an artifact found by the C-14 method to be a thousand years old?

54. The amount of oxygen dissolved in a sample of water can be determined by using thallium metal containing a small amount of the isotope Tl-204. When excess thallium is added to oxygen-containing water, the following reaction occurs.

$$2Tl(s) + \tfrac{1}{2} O_2(g) + H_2O \longrightarrow 2Tl^+(aq) + 2\,OH^-(aq)$$

It is found that, after reaction, the activity of a 25.0 mL water sample is 745 counts per minute (cpm), caused by the presence of Tl^+-204 ions. The activity of Tl-204 is 5.53×10^5 cpm per gram of thallium metal. Assuming O_2 is the limiting reactant in the above equation, calculate its concentration in moles per liter.

55. A 35-mL sample of 0.050 M $AgNO_3$ is mixed with 35 mL of 0.050 M NaI labeled with I-131. The following reaction occurs.

$$Ag^+(aq) + I^-(aq) \longrightarrow AgI(s)$$

The filtrate is found to have an activity of 2.50×10^3 counts per minute per milliliter. The 0.050 M NaI solution had an activity of 1.25×10^{10} counts per minute per milliliter. Calculate K_{sp} for AgI.

56. A 50.0-g sample of water containing tritium, ^3_1H, emits 2.89×10^3 beta particles per second. Tritium has a half-life of 12.3 years. What percentage of all the hydrogen atoms in the water sample is tritium?

57. Using the half-life of tritium in Problem 56, calculate the activity in curies of 1.00 mL of $^3_1H_2$ at STP.

58. In order to measure the volume of the blood in an animal's circulatory system, the following experiment was performed. A 5.0-mL sample of an aqueous solution containing 1.7×10^5 counts per minute (cpm) of tritium was injected into the bloodstream. After an adequate period of time to allow for the complete circulation of the tritium, a 5.0-mL sample of blood was withdrawn and found to have 1.3×10^3 cpm on the scintillation counter. What is the volume of the animal's circulatory system, assuming that only a negligible amount of tritium has decayed during the experiment?

59. One of the causes of the explosion at Chernobyl may have been the reaction between zirconium, which coated the fuel rods, and steam.

$$Zr(s) + 2H_2O(g) \longrightarrow ZrO_2(s) + 2H_2(g)$$

If half a metric ton of zirconium reacted, what pressure was exerted by the hydrogen gas produced at 55°C in the containment chamber of volume 2.0×10^4 L?

60. Consider the fission reaction

$$^1_0n + ^{235}_{92}U \longrightarrow ^{89}_{37}Rb + ^{144}_{58}Ce + ^{0}_{-1}e + 3\,^1_0n$$

How many liters of methane at 25°C and 1.00 atm pressure must be burned to $CO_2(g)$ and $H_2O(l)$ to produce as much energy as the fission of one gram of U-235 fuel?

61. A chelate of Cr^{3+} and $C_2O_4{}^{2-}$ is made by a reaction that involves Na_2CrO_4, a reducing agent, and oxalic acid, $H_2C_2O_4$. The sodium chromate has an activity of 765 counts per minute per gram, from Cr-51. The oxalic acid has an activity of 512 counts per minute per gram; it is labeled with C-14. Since Cr-51 and C-14 emit different particles during decay, their activities can be counted independently. A sample of the chelate was found to have a Cr-51 count of 314 cpm and a C-14 count of 235 cpm. How many oxalate ions are bound to each Cr^{3+} ion?

62. Polonium-210 decays to Pb-206 by alpha emission. Its half-life is 138 days. What volume of helium at 25°C and 755 mm Hg would be obtained form a 10.00-g sample of Po-210 left to decay for 62 hours?

63. Radium-226 decays by alpha emission to radon-222. Suppose that 15.0% of the energy given off by one gram of radium is converted to electrical energy. What is the minimum mass of chromium that would be needed for the voltaic cell $Cr|Cr^{3+} \| Cu^{2+}|Cu$, at standard concentrations, to produce the same amount of electrical work ($\Delta G°$)?

Challenge Problems

64. An activity of 20 picocuries (20×10^{-12} Ci) of radon-222 per liter of air in a house constitutes a health hazard to anyone living in the house. Taking the half-life of radon-222 to be 3.82 d, calculate the concentration of radon in air (moles per liter) corresponding to this activity.

65. Plutonium-239 decays by the reaction $^{239}_{94}Pu \rightarrow ^{235}_{92}U + ^4_2He$, with a rate constant of 5.5×10^{-11}/min. In a one-gram sample of ^{239}Pu.

a. how many grams decompose in 10 min?

b. how much energy in kilojoules is given off in 10 min?

c. what radiation dosage in rems (p. 514) is received by a 70-kg man exposed to a gram of ^{239}Pu for 10 min?

66. It is possible to estimate the activation energy for fusion by calculating the energy required to bring two deuterons close enough to one another to form an alpha particle. This energy can be obtained by using Coulomb's law in the form $E = 8.99 \times 10^9\, q_1q_2/r$, where q_1 and q_2 are the charges of the deuterons (1.60×10^{-19} C), r is the radius of the He nucleus, about 2×10^{-15} m, and E is the energy in joules.

a. Estimate E in joules per alpha particle.

b. Using the equation $E = mv^2/2$, estimate the velocity (meters per second) each deuteron must have if a collision between two of them is to supply the activation energy for fusion (m is the mass of the deuteron in kilograms).

67. Consider the reaction $2\ {}^{2}_{1}H \rightarrow {}^{4}_{2}He$.

a. Calculate ΔE in kilojoules per gram of deuterium fused.

b. How much energy is potentially available from the fusion of all the deuterium atoms in seawater? The percentage of deuterium in water is about 0.0017%. The total mass of water in the oceans is 1.3×10^{24} g.

c. What fraction of the deuterium in the oceans would have to be consumed to supply the annual energy requirements of the world (2.3×10^{17} kJ)?

CHARLES D. WINTERS

$$4Fe^{3+}(aq) + 3Fe(CN)_6{}^{4-}(aq) \longrightarrow Fe_4[Fe(CN)_6]_3(s)$$

(See p. 426)

19 Chemistry of the Metals

The fields of Nature long
prepared and fallow
the silent, cyclic chemistry
The slow and steady
ages plodding,
the unoccupied surface ripening,
the rich ores forming beneath.

WALT WHITMAN
Song of the Redwood Tree

CHAPTER OUTLINE

The diagonal line or "stairway" that runs from the left to the lower right of the periodic table separates metals from nonmetals; the 80 or more elements below and to the left of the line are metals. In discussing the descriptive chemistry of the metals, we concentrate upon

— the **main-group metals in Groups 1 and 2** at the far left of the periodic table. These are commonly referred to as the *alkali* metals (Group 1) and *alkaline earth* metals (Group 2). The group names reflect the strongly basic

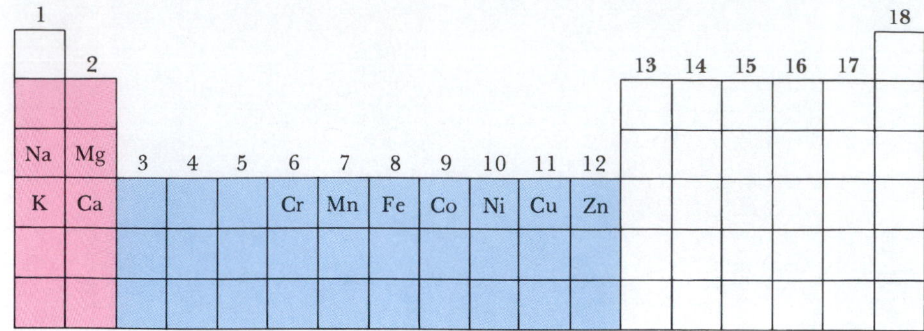

Figure 19.1
The metals considered in this chapter are those in Groups 1 and 2 (red) and the transition metals (blue). Symbols are shown for the more common metals.

nature of the oxides (K_2O, CaO, . . .) and hydroxides (KOH, $Ca(OH)_2$, . . .) of these elements.

— **the transition metals**, located in the center of the periodic table. There are three series of transition metals, each consisting of ten elements, located in the fourth, fifth, and sixth periods. We focus on a few of the more important transition metals (Fig. 19.1), particularly those toward the right of the first series.

Section 19.1 deals with the processes by which these metals are obtained from their principal ores. Section 19.2 describes the reactions of the alkali and alkaline earth metals, particularly those with hydrogen, oxygen, and water. Section 19.3 considers the redox chemistry of the transition metals, their cations (e.g., Fe^{2+}, Fe^{3+}) and their oxoanions (e.g., CrO_4^{2-}). The last two sections of the chapter deal with two somewhat more general topics:

— qualitative analysis of metal cations (Section 19.4)
— alloys of the metals (Section 19.5)

19.1 METALLURGY

An ore is a deposit from which a metal can be extracted profitably

The processes by which metals are extracted from their ores fall within the science of metallurgy. As you might expect, the chemical reactions involved depend upon the type of ore (Fig. 19.2). We consider some typical processes used to obtain metals from chloride, oxide, sulfide, or "native" ores.

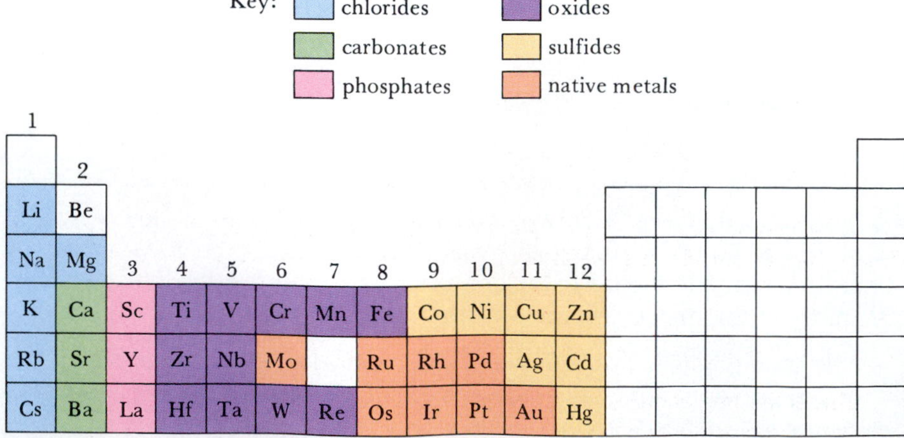

Figure 19.2
Principal ores of the Group 1, Group 2, and transition metals.

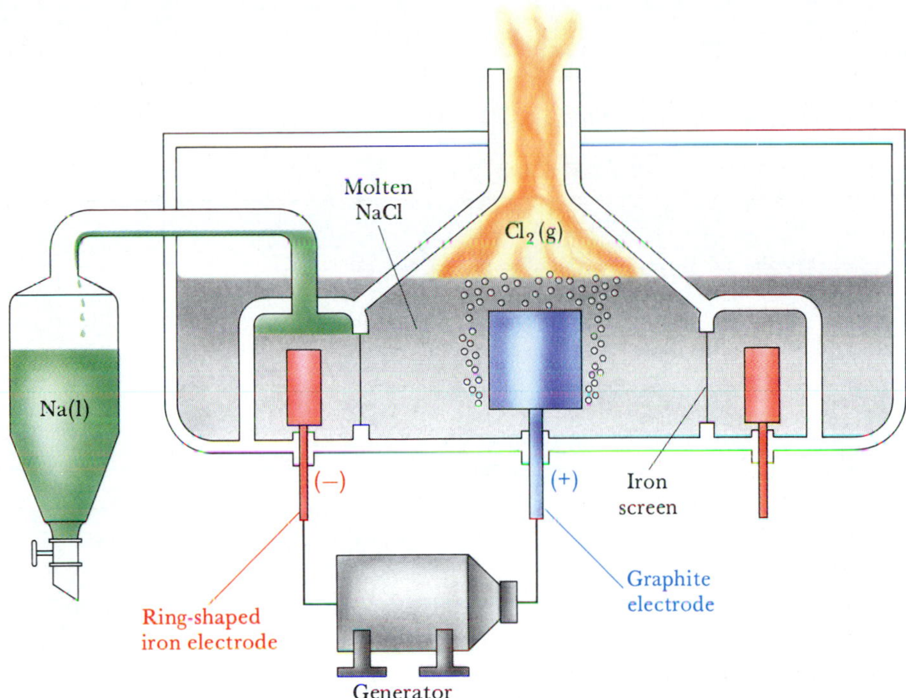

Figure 19.3
Electrolysis of molten sodium chloride, containing some $CaCl_2$ to lower the melting point. The iron screen is used to prevent sodium and chlorine from coming in contact with each other.

Chloride Ores: Na from NaCl

Sodium metal is obtained by the electrolysis of molten sodium chloride (Fig. 19.3). The electrode reactions are quite simple:

$$\begin{array}{rl} \text{cathode:} & 2Na^+(l) + 2e^- \longrightarrow 2Na(l) \\ \text{anode:} & 2Cl^-(l) \longrightarrow Cl_2(g) + 2e^- \\ \hline & 2NaCl(l) \longrightarrow 2Na(l) + Cl_2(g) \end{array}$$

The cell is operated at about 600°C to keep the electrolyte molten; calcium chloride is added to lower the melting point. About 14 kJ of electrical energy is required to produce one gram of sodium (Example 19.1), which is drawn off as a liquid (mp of Na = 98°C). The chlorine gas produced at the anode is a valuable by-product.

It's cheaper to electrolyze $NaCl(aq)$, but you don't get Na(s) that way

Example 19.1 Taking $\Delta H°$ and $\Delta S°$ for the reaction

$$2NaCl(l) \longrightarrow 2Na(l) + Cl_2(g)$$

to be +820 kJ and +0.180 kJ/K respectively, calculate

(a) $\Delta G°$ at the electrolysis temperature, 600°C.
(b) the voltage required to carry out the electrolysis.

Strategy To find $\Delta G°$, apply the Gibbs-Helmholtz equation, $\Delta G° = \Delta H° - T\Delta S°$ (Chap. 16). To find $E°$, use the relation $\Delta G° = -nFE°$ (Chap. 17).

Solution

(a) $\Delta G° = +820\text{ kJ} - 873\text{ K}(0.180\text{ kJ/K}) = +663\text{ kJ}$

This is the free energy change for the formation of two moles of sodium (45.98 g); the electrical energy required per gram is 663 kJ/45.98 g = 14.4 kJ/g.

$$\text{(b) } E° = \frac{-\Delta G°}{nF} = \frac{-6.63 \times 10^5 \text{ J}}{(2 \text{ mol})(9.648 \times 10^4 \text{ J/mol} \cdot \text{V})} = -3.44 \text{ V}$$

At least 3.44 V must be applied to carry out the electrolysis. (In water solution, the calculated $E°$ for the electrolysis of NaCl is −2.19 V.)

Oxide Ores: Al from Al_2O_3, Fe from Fe_2O_3

Oxides of very reactive metals such as calcium or aluminum are reduced by electrolysis. In the case of aluminum, the reaction is

$$2Al_2O_3(l) \longrightarrow 4Al(l) + 3\,O_2(g)$$

Chemistry majors can be very productive

This process was worked out in 1886 by Charles Hall, a chemistry student at Oberlin College. He added cryolite, Na_3AlF_6, to lower the melting point of the electrolyte from 2000 to 1000°C. Curiously enough, the same process was developed by Heroult in France, also in 1886; each man was unaware of the other's work.

With less active metals, a chemical reducing agent can be used to reduce a metal cation to the element. The most common reducing agent in metallurgical processes is carbon, in the form of coke, or, more exactly, carbon monoxide formed from the coke.

The most important metallurgical process involving carbon is the reduction of hematite ore, which consists largely of iron(III) oxide, Fe_2O_3, mixed with silicon dioxide, SiO_2 (sand). Reduction occurs in a blast furnace (Fig. 19.4a) typically 30 m high and 10 m in diameter. The solid charge admitted at

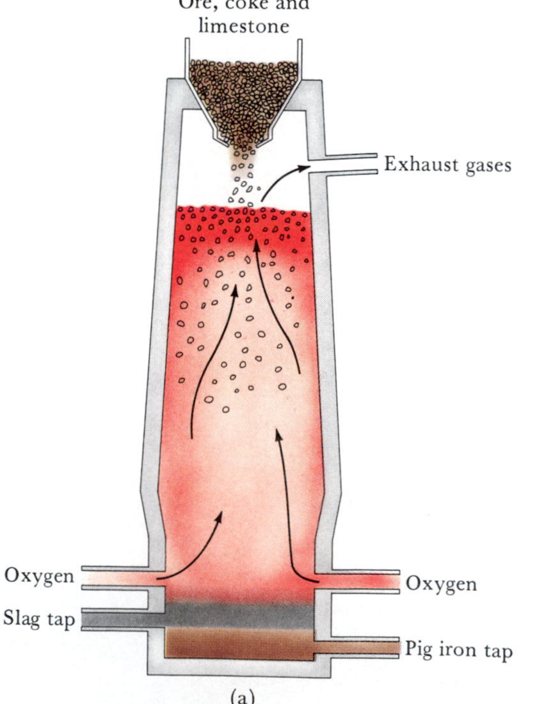

(a)

(b)

Figure 19.4
Blast furnace for the production of pig iron (a). The carbon content of the crude iron is lowered by heating with oxygen in a basic oxygen furnace (b), forming steel.

the top of the furnace consists of iron ore, coke, and limestone ($CaCO_3$). To get the process started, a blast of compressed air or pure O_2 at 500°C is blown into the furnace through nozzles near the bottom. Several different reactions occur, of which three are most important:

1. *Conversion of carbon to carbon monoxide.* In the lower part of the furnace coke burns to form carbon dioxide, CO_2. As the CO_2 rises through the solid mixture, it reacts further with the coke to form carbon monoxide, CO. The overall reaction is

$$2C(s) + O_2(g) \longrightarrow 2CO(g) \qquad \Delta H = -221 \text{ kJ}$$

The heat given off by this reaction maintains a high temperature within the furnace.

2. *Reduction of Fe^{3+} ions to Fe.* The carbon monoxide reacts with the iron(III) oxide in the ore:

$$Fe_2O_3(s) + 3CO(g) \longrightarrow 2Fe(l) + 3CO_2(g)$$

This is the key reaction; it produces molten iron

Molten iron, formed at a temperature of 1600°C, collects at the bottom of the furnace. Four or five times a day, it is drawn off. The daily production of iron from a single blast furnace is about 1000 metric tons. This is enough to make 500 Cadillacs or 1000 Hyundais.

3. *Formation of slag.* The limestone added to the furnace decomposes at about 800°C:

$$CaCO_3(s) \longrightarrow CaO(s) + CO_2(g)$$

The calcium oxide formed reacts with impurities in the iron ore to form a glassy material called **slag**. The main reaction is with SiO_2 to form calcium silicate, $CaSiO_3$:

$$CaO(s) + SiO_2(s) \longrightarrow CaSiO_3(l)$$

The slag, which is less dense than molten iron, forms a layer on the surface of the metal. This makes it possible to draw off the slag through an opening in the furnace above that used to remove the iron. The slag is used to make cement and as a base in road construction.

The product that comes out of the blast furnace, called "pig iron," is highly impure. On the average, it contains about 4% carbon along with lesser amounts of silicon, manganese, and phosphorus. To make steel from pig iron, the carbon content must be lowered below 2%. Most of the steel produced in the world today is made by the basic oxygen process (Fig. 19.4b). The "converter" is filled with a mixture of about 70% molten iron from the blast furnace, 25% scrap iron or steel, and 5% limestone. Pure oxygen under a pressure of about 10 atm is blown through the molten metal. The reaction

Where do you suppose the phrase "pig iron" came from?

$$C(s) + O_2(g) \longrightarrow CO_2(g)$$

occurs rapidly; at the same time, impurities such as silicon are converted to oxides, which react with limestone to form a slag. When the carbon content drops to the desired level, the supply of oxygen is cut off. At this stage, the steel is ready to be poured. The whole process takes from 30 min to 1 h and yields about 200 metric tons of steel in a single "blow."

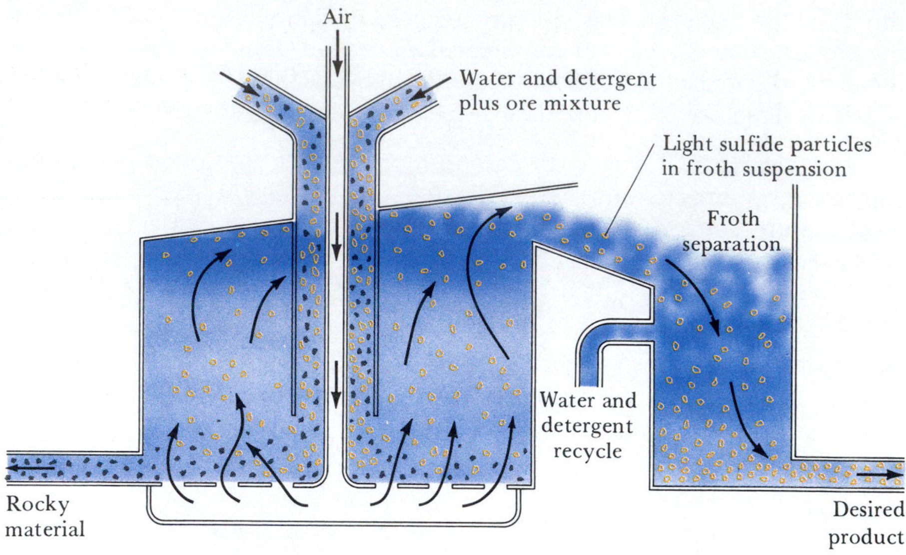

Figure 19.5
Low-grade sulfide ores, including Cu_2S, are often concentrated by flotation. The finely divided sulfide particles are trapped in soap bubbles, while the rocky waste sinks to the bottom and is discarded.

Sulfide Ores: Cu from Cu_2S

Sulfide ores, after preliminary treatment, most often undergo roasting, i.e., heating with air or pure oxygen. With a relatively reactive transition metal such as zinc, the product is the oxide

$$2ZnS(s) + 3\,O_2(g) \longrightarrow 2ZnO(s) + 2SO_2(g)$$

which can then be reduced to the metal with carbon. With sulfides of less reactive metals such as copper or mercury, the free metal is formed directly upon roasting. The reaction with cinnabar, the sulfide ore of mercury, is

$$HgS(s) + O_2(g) \longrightarrow Hg(g) + SO_2(g)$$

Among the several ores of copper, one of the most important is chalcocite, which contains copper(I) sulfide, Cu_2S, in highly impure form. Rocky material typically lowers the fraction of copper in the ore to 1% or less. The Cu_2S is concentrated by a process called *flotation* (Fig. 19.5), which raises the fraction of copper to 20–40%. The concentrated ore is then converted to the metal by blowing air through it at a high temperature, typically above 1000°C. (Pure O_2 is often used instead of air.) The overall reaction that occurs is a simple one:

Because roasting produces Cu, it was one of the first metals known

$$Cu_2S(s) + O_2(g) \longrightarrow 2Cu(s) + SO_2(g)$$

The solid produced is called "blister copper." It has an irregular appearance due to air bubbles that enter the copper while it is still molten. Blister copper is impure, containing small amounts of several other metals.

Copper is purified by electrolysis. The anode, which may weigh as much as 300 kg, is made of blister copper. The electrolyte is 0.5 to 1.0 M $CuSO_4$, adjusted to a pH of about 0 with sulfuric acid. The cathode is a piece of pure copper, weighing perhaps 150 kg. The half-reactions are

$$\text{oxidation:} \quad Cu(s,\ \text{impure}) \longrightarrow Cu^{2+}(aq) + 2e^-$$

$$\text{reduction:} \quad Cu^{2+}(aq) + 2e^- \longrightarrow Cu(s,\ \text{pure})$$

The overall reaction, obtained by adding these two half-reactions, is

$$\text{Cu}(s,\ \text{impure}) \longrightarrow \text{Cu}(s,\ \text{pure})$$

Thus, the net effect of electrolysis is to transfer copper metal from the impure blister copper used as one electrode to the pure copper sheet used as the other electrode. Electrolytic copper is 99.95% pure.

Example 19.2 A major ore of bismuth is bismuth(III) sulfide, Bi_2S_3. Upon roasting in air, it is converted to the corresponding oxide; sulfur dioxide is a by-product. What volume of SO_2, at 25°C and 1.00 atm, is formed from one metric ton (10^6 g) of ore containing 1.25% Bi_2S_3?

Strategy First, write a balanced equation for the reaction, which is very similar to that for ZnS, except that Zn^{2+} is replaced by Bi^{3+}. Using the balanced equation, calculate the number of moles of SO_2. Finally, use the ideal gas law to calculate the volume of SO_2.

Solution

(1) The product is $Bi_2O_3(s)$; the equation is

$$2\text{Bi}_2\text{S}_3(s) + 9\,\text{O}_2(g) \longrightarrow 2\text{Bi}_2\text{O}_3(s) + 6\text{SO}_2(g)$$

(2) $n_{SO_2} = 0.0125 \times 10^6 \text{ g Bi}_2\text{S}_3 \times \dfrac{1 \text{ mol Bi}_2\text{S}_3}{514.2 \text{ g Bi}_2\text{S}_3} \times \dfrac{6 \text{ mol SO}_2}{2 \text{ mol Bi}_2\text{S}_3} = 72.9 \text{ mol SO}_2$

(3) $V_{SO_2}(g) = \dfrac{nRT}{P} = \dfrac{(72.9 \text{ mol})(0.0821 \text{ L} \cdot \text{atm/mol} \cdot \text{K})(298 \text{ K})}{1.00 \text{ atm}} = 1.78 \times 10^3 \text{ L}$

The roasting of sulfide ores is a major source of the gaseous pollutant SO_2, which contributes to acid rain (Chap. 12).

Native Metals: Au

A few very unreactive metals, notably silver and gold, are found in nature in elemental form, mixed with large amounts of rocky material. For countless centuries, people have extracted gold by taking advantage of its high density (19.3 g/mL). In ancient times, gold-bearing sands were washed over sheepskins which retained the gold; this is believed to be the source of the "Golden Fleece" of Greek mythology. The forty-niners in California obtained gold by swirling gold-bearing sands with water in a pan. Less dense impurities were washed away, leaving gold nuggets or flakes at the bottom of the pan.

Nowadays, the gold content of ores is much too low for these simple mechanical separation methods to be effective. Instead, the ore is treated with very dilute (0.01 *M*) sodium cyanide solution, through which air is blown. The following redox reaction takes place:

$$4\text{Au}(s) + 8\text{CN}^-(aq) + \text{O}_2(g) + 2\text{H}_2\text{O} \longrightarrow 4\text{Au(CN)}_2^-(aq) + 4\,\text{OH}^-(aq)$$

Gold at 10 g to the ton can be extracted profitably

The oxidizing agent is O_2, which takes gold to the +1 state. The cyanide ion acts as a complexing ligand, forming the stable $Au(CN)_2^-$ ion. Metallic gold is recovered from solution by adding zinc; the gold in the complex ion is reduced to the metal.

$$\text{Zn}(s) + 2\text{Au(CN)}_2^-(aq) \longrightarrow \text{Zn(CN)}_4^{2-}(aq) + 2\text{Au}(s)$$

19.2 REACTIONS OF THE ALKALI AND ALKALINE EARTH METALS

Not under water!

The metals in Groups 1 and 2 are among the most reactive of all elements (Table 19.1). Their low ionization energies and high E°_{ox} values explain why they are so readily oxidized to cations, +1 cations for Group 1, +2 for Group 2. The alkali metals and the heavier alkaline earth metals (Ca, Sr, Ba) are commonly stored under dry mineral oil or kerosene to prevent them from reacting with oxygen or water vapor in the air. Magnesium is less reactive; it is commonly available in the form of ribbon or powder. Beryllium, as one would expect from its position in the periodic table, is the least metallic element in these two groups. It is also the least reactive toward water, oxygen, or other nonmetals.

Example 19.3 Write balanced equations for the reaction of

(a) sodium with hydrogen. (b) barium with oxygen.

Strategy The formulas of the products can be deduced from Table 19.1. Note that barium forms two products with oxygen, BaO and BaO_2.

Solution

(a) $2Na(s) + H_2(g) \longrightarrow 2NaH(s)$

(b) $2Ba(s) + O_2(g) \longrightarrow 2BaO(s)$
$Ba(s) + O_2(g) \longrightarrow BaO_2(s)$

Hydrogen

The compounds formed by the reaction of hydrogen with the alkali and alkaline earth metals contain H^- ions; for example, sodium hydride consists of Na^+ and H^- ions. These white crystalline solids are often referred to as saline hydrides because of their physical resemblance to NaCl. Chemically, they behave quite differently from sodium chloride; for example, they react with water to produce hydrogen gas. Typical reactions are

$$NaH(s) + H_2O \longrightarrow H_2(g) + Na^+(aq) + OH^-(aq)$$

$$CaH_2(s) + 2H_2O \longrightarrow 2H_2(g) + Ca^{2+}(aq) + 2\,OH^-(aq)$$

A tank of compressed $H_2(g)$ would not be practical

In this way, saline hydrides can serve as compact, portable sources of hydrogen gas for inflating life rafts and balloons.

Water

The alkali metals react vigorously with water (Fig. 19.6) to evolve hydrogen and form a water solution of the alkali hydroxide. The reaction of sodium is typical:

$$2Na(s) + 2H_2O(l) \longrightarrow 2Na^+(aq) + 2\,OH^-(aq) + H_2(g) \quad \Delta H^\circ = -368.6 \text{ kJ}$$

The heat evolved in the reaction frequently causes the hydrogen to ignite.

Table 19.1 Reactions of the Alkali and Alkaline Earth Metals

Reactant	Product	Comments
	Alkali Metals (M)	
$H_2(g)$	$MH(s)$	Upon heating in hydrogen gas
$X_2(g)$	$MX(s)$	X = F, Cl, Br, I
$N_2(g)$	$M_3N(s)$	Only Li reacts; product contains N^{3-} ion
$S(s)$	$M_2S(s)$	Upon heating
$O_2(g)$	$M_2O(s)$	Li; product contains O^{2-} ion
	$M_2O_2(s)$	Na; product contains O_2^{2-} ion
	$MO_2(s)$	K, Rb, Cs; product contains O_2^- ion
$H_2O(l)$	$H_2(g)$, M^+, OH^-	Violent reaction with Na, K
	Alkaline Earth Metals (M)	
$H_2(g)$	$MH_2(s)$	All except Be; heating required
$X_2(g)$	$MX_2(s)$	Any halogen
$N_2(g)$	$M_3N_2(s)$	All except Be; heating required
$S(s)$	$MS(s)$	Upon heating
$O_2(g)$	$MO(s)$	All; product contains O^{2-} ion
	$MO_2(s)$	Ba; product contains O_2^{2-} ion
$H_2O(l)$	$H_2(g)$, M^{2+}, OH^-	Ca, Sr, Ba

Figure 19.6
When a *very small* piece of sodium is added to water, it reacts violently: $Na(s) + H_2O \longrightarrow Na^+(aq) + OH^-(aq) + \frac{1}{2}H_2(g)$. The OH^- ions formed turn the color of the organic dye phenolphthalein from colorless to pink. (Marna G. Clarke)

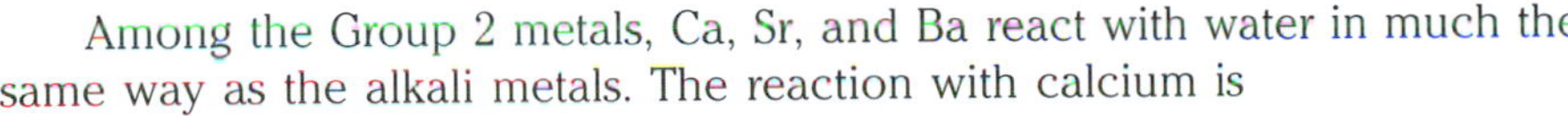

Among the Group 2 metals, Ca, Sr, and Ba react with water in much the same way as the alkali metals. The reaction with calcium is

$$Ca(s) + 2H_2O \longrightarrow Ca^{2+}(aq) + 2\,OH^-(aq) + H_2(g)$$

Beryllium does not react with water at all. Magnesium reacts very slowly with boiling water but reacts more readily with steam at high temperatures:

$$Mg(s) + H_2O(g) \longrightarrow MgO(s) + H_2(g)$$

This reaction, like that of sodium with water, produces enough heat to ignite the hydrogen. Firefighters who try to put out a magnesium fire by spraying water on it have discovered this reaction, often with tragic results. The best way to extinguish burning magnesium is to dump dry sand on it.

Oxygen

Note from Table 19.1 that several different products are possible when an alkali or alkaline earth metal reacts with oxygen. The product may be a normal oxide (O^{2-} ion), a peroxide (O_2^{2-} ion), or superoxide (O_2^- ion).

$$[:\ddot{\underset{..}{O}}:]^{2-} \qquad [:\ddot{\underset{..}{O}}-\ddot{\underset{..}{O}}:]^{2-} \qquad [:\ddot{\underset{..}{O}}-\ddot{\underset{..}{O}}\cdot]^{-}$$

oxide ion — peroxide ion — superoxide ion

The superoxide ion has an unpaired electron; KO_2 is paramagnetic

Lithium is the only Group 1 metal that forms the normal oxide in good yield by direct reaction with oxygen. The other Group 1 oxides (Na_2O, K_2O,

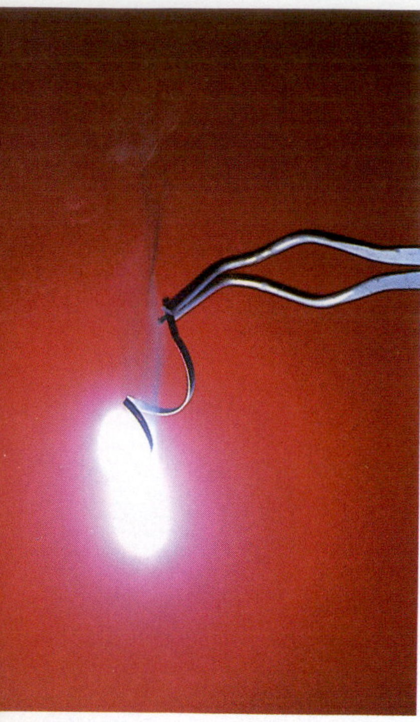

Figure 19.7
A piece of magnesium ribbon bursts into flame when heated in air; the product of the reaction is $MgO(s)$. (Charles D. Winters)

Rb_2O, Cs_2O) must be prepared by other means. In contrast, the Group 2 metals usually react with oxygen to give the normal oxide. Beryllium and magnesium must be heated strongly to give BeO and MgO (Fig. 19.7). Calcium and strontium react more readily to give CaO and SrO. Barium, the most reactive of the Group 2 metals, catches fire when exposed to moist air. The product is a mixture of the normal oxide, BaO, and the peroxide, BaO_2.

The oxides of these metals react with water to form hydroxides:

$$Li_2O(s) + H_2O(l) \longrightarrow 2LiOH(s)$$

$$CaO(s) + H_2O(l) \longrightarrow Ca(OH)_2(s)$$

The reaction with CaO is referred to as the "slaking" of lime; it gives off 65 kJ of heat per mole of $Ca(OH)_2$ formed. A similar reaction with MgO takes place slowly to form $Mg(OH)_2$, the antacid commonly referred to as "milk of magnesia."

When sodium burns in air, the principal product is yellowish sodium peroxide, Na_2O_2:

$$2Na(s) + O_2(g) \longrightarrow Na_2O_2(s)$$

Addition of sodium peroxide to water gives hydrogen peroxide, H_2O_2:

$$Na_2O_2(s) + 2H_2O \longrightarrow 2Na^+(aq) + 2\,OH^-(aq) + H_2O_2(aq)$$

Through this reaction, sodium peroxide finds use as a bleaching agent in the pulp and paper industry.

The heavier alkali metals (K, Rb, Cs) form the superoxide when they burn in air. For example,

$$K(s) + O_2(g) \longrightarrow KO_2(s)$$

Potassium superoxide is used in self-contained breathing apparatus for firefighters and miners. It reacts with the moisture in exhaled air to generate oxygen:

$$4KO_2(s) + 2H_2O(g) \longrightarrow 3\,O_2(g) + 4KOH(s)$$

The carbon dioxide in the exhaled air is removed by reaction with the KOH formed:

$$KOH(s) + CO_2(g) \longrightarrow KHCO_3(s)$$

A person using a mask charged with KO_2 can rebreathe the same air for an extended period of time. This allows that person to enter an area where there are poisonous gases or oxygen-deficient air.

Example 19.4 Consider the compounds strontium hydride, radium peroxide, and cesium superoxide.

(a) Give the formulas of these compounds.
(b) Write equations for the formation of these compounds from the elements.
(c) Write equations for the reactions of strontium hydride and radium peroxide with water.

Strategy The formulas can be deduced from the charges of the ions (Sr^{2+}, Ra^{2+}, Cs^+; H^-, O_2^{2-}, O_2^-). If you know the formulas, the equations are readily written. In part (c), note that

— hydrides on reaction with water give $H_2(g)$ and a solution of the metal hydroxide
— peroxides on reaction with water give $H_2O_2(aq)$ and a solution of the metal hydroxide

Solution

(a) SrH_2, RaO_2, CsO_2

(b) $Sr(s) + H_2(g) \longrightarrow SrH_2(s)$
$Ra(s) + O_2(g) \longrightarrow RaO_2(s)$
$Cs(s) + O_2(g) \longrightarrow CsO_2(s)$

(c) $SrH_2(s) + 2H_2O \longrightarrow 2H_2(g) + Sr^{2+}(aq) + 2\,OH^-(aq)$
$RaO_2(s) + 2H_2O \longrightarrow H_2O_2(aq) + Ra^{2+}(aq) + 2\,OH^-(aq)$

19.3 REDOX CHEMISTRY OF THE TRANSITION METALS

The transition metals, unlike those in Groups 1 and 2, typically show several different oxidation numbers in their compounds. This tends to make their redox chemistry more complex (and more colorful). Only in the lower oxidation states (+1, +2, +3) are the transition metals present as cations (e.g., Ag^+, Zn^{2+}, Fe^{3+}). In higher oxidation states (+4 to +7) a transition metal is covalently bonded to a nonmetal atom, most often oxygen.

In CrO_4^{2-} and MnO_4^-, the metal atom is covalently bonded to oxygen

Reaction of Transition Metals With Acids

Any metal with a positive standard oxidation voltage, E°_{ox}, can be oxidized by the H^+ ions present in a 1 *M* solution of a strong acid. All of the transition metals in the left column of Table 19.2 react spontaneously with dilute solutions of such strong acids as HCl, HBr, and H_2SO_4. The products are hydrogen gas and a cation of the transition metal. A typical reaction is that of nickel:

Most of these reactions are very slow

$$Ni(s) + 2H^+(aq) \longrightarrow Ni^{2+}(aq) + H_2(g)$$
$$E^\circ = E^\circ_{ox}\,Ni = +0.236\text{ V} \qquad \Delta G^\circ = -45.5\text{ kJ}$$

Table 19.2 Ease of Oxidation of Transition Metals

Metal		Cation	E°_{ox} (V)	Metal		Cation	E°_{ox} (V)
Mn	$\longrightarrow$	Mn^{2+}	+1.182	Cu	$\longrightarrow$	Cu^{2+}	−0.339
Cr	$\longrightarrow$	Cr^{2+}	+0.912	Ag	$\longrightarrow$	Ag^+	−0.799
Zn	$\longrightarrow$	Zn^{2+}	+0.762	Hg	$\longrightarrow$	Hg^{2+}	−0.852
Fe	$\longrightarrow$	Fe^{2+}	+0.409	Au	$\longrightarrow$	Au^{3+}	−1.498
Cd	$\longrightarrow$	Cd^{2+}	+0.402				
Co	$\longrightarrow$	Co^{2+}	+0.282				
Ni	$\longrightarrow$	Ni^{2+}	+0.236				

Figure 19.8
Nickel reacts slowly with hydrochloric acid to form $H_2(g)$ and Ni^{2+} ions in solution. Evaporation of the solution formed gives green crystals of $NiCl_2 \cdot 6H_2O$. (Charles D. Winters)

With metals that can form more than one cation, such as iron, the product upon reaction with H^+ in the absence of air is ordinarily the cation of lower charge, e.g., Fe^{2+}:

$$Fe(s) + 2H^+(aq) \longrightarrow Fe^{2+}(aq) + H_2(g)$$

Metals with negative values of E°_{ox}, listed at the right of Table 19.2, are too inactive to react with hydrochloric acid. The H^+ ion is not a strong enough oxidizing agent to convert a metal such as copper ($E^\circ_{ox} = -0.339$ V) to a cation. However, copper can be oxidized by nitric acid. Here, the oxidizing agent is the nitrate ion, NO_3^-, which may be reduced to NO_2 or NO:

$$3Cu(s) + 8H^+(aq) + 2NO_3^-(aq) \longrightarrow 3Cu^{2+}(aq) + 2NO(g) + 4H_2O$$

$$E^\circ = E^\circ_{ox}\,Cu + E^\circ_{red}\,NO_3^- = -0.339\text{ V} + 0.964\text{ V} = +0.625\text{ V}$$

Example 19.5 Write balanced equations for the reactions, if any, at standard concentrations, of

(a) chromium with hydrochloric acid. (b) silver with nitric acid.

Strategy Use standard potentials to decide whether or not reaction will occur. Note that with HCl only the H^+ ion can be reduced (to H_2); in HNO_3, the NO_3^- ion can be reduced to NO ($E^\circ_{red} = +0.964$ V).

Solution

(a) $Cr(s) \longrightarrow Cr^{2+}(aq) + 2e^-$ $\qquad E^\circ_{ox} = +0.912$ V

$2H^+(aq) + 2e^- \longrightarrow H_2(g)$ $\qquad E^\circ_{red} = 0.000$ V

The reaction is

$$Cr(s) + 2H^+(aq) \longrightarrow Cr^{2+}(aq) + H_2(g) \qquad E^\circ = +0.912\text{ V}$$

In the presence of O_2, Cr^{2+} is oxidized to Cr^{3+}

If chromium metal is added to hydrochloric acid in the absence of air, it slowly reacts, forming blue Cr^{2+} and bubbles of hydrogen gas.

(b) $Ag(s) \longrightarrow Ag^+(aq)$ $\qquad E^\circ_{ox} = -0.799$ V

$NO_3^-(aq) \longrightarrow NO(g)$ $\qquad E^\circ_{red} = +0.964$ V

Since E° is a positive quantity, +0.165 V, a redox reaction should occur. The balanced half-equations are oxidation: $Ag(s) \longrightarrow Ag^+(aq) + e^-$

reduction: $NO_3^-(aq) + 4H^+(aq) + 3e^- \longrightarrow NO(g) + 2H_2O$

To obtain the final balanced equation, multiply the first half-equation by 3 and add to the second. The result is

$$3Ag(s) + NO_3^-(aq) + 4H^+(aq) \longrightarrow 3Ag^+(aq) + NO(g) + 2H_2O$$

Although gold is not oxidized by nitric acid, it can be brought into solution in *aqua regia,* a 3:1 mixture by volume of 12 *M* HCl and 16 *M* HNO_3:

$$Au(s) + 4H^+(aq) + 4Cl^-(aq) + NO_3^-(aq) \longrightarrow AuCl_4^-(aq) + NO(g) + 2H_2O$$

The nitrate ion of the nitric acid acts as the oxidizing agent. The function of the hydrochloric acid is to furnish Cl^- ions to form the very stable complex ion $AuCl_4^-$.

Equilibria Between Different Cations of a Transition Metal

Several transition metals form more than one cation. For example, chromium forms Cr^{2+} and Cr^{3+}; copper forms Cu^+ and Cu^{2+}. Table 19.3 lists standard reduction potentials for several such systems. Using the data in this table and in Table 17.1 (p. 468), it is possible to decide upon the relative stabilities of different transition metal cations in water solution.

Cations for which E°_{red} is a large, positive number are readily reduced and hence tend to be unstable in water solution. A case in point is the Mn^{3+} ion, which reacts spontaneously with water:

$$2Mn^{3+}(aq) + H_2O \longrightarrow 2Mn^{2+}(aq) + \tfrac{1}{2}O_2(g) + 2H^+(aq)$$

$$E^\circ = E^\circ_{red}\,Mn^{3+} + E^\circ_{ox}\,H_2O = +1.559\text{ V} - 1.229\text{ V} = +0.330\text{ V}$$

As a result of this reaction, Mn^{3+} cations are never found in water solution. Manganese(III) occurs only in insoluble oxides and hydroxides such as Mn_2O_3 and MnO(OH).

Co^{3+} is also rare, except in complexes

Table 19.3 Ease of Reduction of Transition Metal Cations

Chromium	Cr^{3+}	$\xrightarrow{-0.408\text{ V}}$	Cr^{2+}	$\xrightarrow{-0.912\text{ V}}$	Cr
Manganese	Mn^{3+}	$\xrightarrow{+1.559\text{ V}}$	Mn^{2+}	$\xrightarrow{-1.182\text{ V}}$	Mn
Iron	Fe^{3+}	$\xrightarrow{+0.769\text{ V}}$	Fe^{2+}	$\xrightarrow{-0.409\text{ V}}$	Fe
Cobalt	Co^{3+}	$\xrightarrow{+1.953\text{ V}}$	Co^{2+}	$\xrightarrow{-0.282\text{ V}}$	Co
Copper	Cu^{2+}	$\xrightarrow{+0.161\text{ V}}$	Cu^+	$\xrightarrow{+0.518\text{ V}}$	Cu
Gold	Au^{3+}	$\xrightarrow{+1.400\text{ V}}$	Au^+	$\xrightarrow{+1.695\text{ V}}$	Au
Mercury	Hg^{2+}	$\xrightarrow{+0.908\text{ V}}$	Hg_2^{2+}	$\xrightarrow{+0.796\text{ V}}$	Hg

The Hg_2^{2+} ion has the structure $(Hg—Hg)^{2+}$

The cations in the center column of Table 19.3 (Cr^{2+}, Mn^{2+}, . . .) are in an intermediate oxidation state. They can either be oxidized to a cation of higher charge ($Cr^{2+} \rightarrow Cr^{3+}$) or reduced to the metal ($Cr^{2+} \rightarrow Cr$). With certain cations of this type, these two half-reactions occur simultaneously. Consider, for example, the Cu^+ ion. In water solution, copper(I) salts **disproportionate**, undergoing reduction (to copper metal) and oxidation (to Cu^{2+}) at the same time:

$$2Cu^+(aq) \longrightarrow Cu(s) + Cu^{2+}(aq);\ E° = +0.518\ V - 0.161\ V = +0.357\ V$$

As a result, the only stable copper(I) species are insoluble compounds such as CuCN or complex ions such as $Cu(CN)_2^-$.

The Cu^+ ion is one of the few species in the center column of Table 19.3 that disproportionates in water. However, cations of this type may be unstable for a quite different reason. Water ordinarily contains dissolved air; the O_2 in air may oxidize the cation. When a blue solution of a chromium(II) salt is exposed to air, the color quickly changes to violet or green as the Cr^{3+} ion is formed by the reaction

$$\begin{array}{rll} 2Cr^{2+}(aq) & \longrightarrow 2Cr^{3+}(aq) + 2e^- & E°_{ox} = +0.408\ V \\ \frac{1}{2}O_2(g) + 2H^+(aq) + 2e^- & \longrightarrow H_2O & E°_{red} = +1.229\ V \\ \hline 2Cr^{2+}(aq) + \frac{1}{2}O_2(g) + 2H^+(aq) & \longrightarrow 2Cr^{3+}(aq) + H_2O & E° = +1.637\ V \end{array}$$

As a result of this reaction, chromium(II) salts are difficult to prepare and even more difficult to store.

The Fe^{2+} ion ($E°_{ox} = -0.769$ V) is much more stable toward oxidation than Cr^{2+}. However, iron(II) salts in water solution are slowly converted to iron(III) by dissolved oxygen. In acidic solution, the reaction is

Solid Fe^{2+} salts are often contaminated with Fe^{3+}

$$2Fe^{2+}(aq) + \frac{1}{2}O_2(g) + 2H^+(aq) \longrightarrow 2Fe^{3+}(aq) + H_2O$$

$$E° = E°_{ox}\,Fe^{2+} + E°_{red}\,O_2 = -0.769\ V + 1.229\ V = +0.460\ V$$

A similar reaction takes place in basic solution. Iron(II) hydroxide is pure white when first precipitated but, in the presence of air, it turns first green and then brown as it is oxidized by O_2:

$$2Fe(OH)_2(s) + \frac{1}{2}O_2(g) + H_2O \longrightarrow 2Fe(OH)_3(s)$$

Example 19.6 Using Table 19.3, find

(a) three different cations, in addition to Mn^{3+}, which react with H_2O to form $O_2(g)$ ($E°_{ox}\,H_2O = -1.229$ V).
(b) another cation, in addition to Cu^+, that disproportionates in water.
(c) two other cations, in addition to Cr^{2+} and Fe^{2+}, that are oxidized by $O_2(g)$ dissolved in water ($E°_{red}\,O_2 = +1.229$ V).

Strategy In each case, look for a reaction in which $E°$ is a positive quantity. In (a), $E°_{red}$ for the cation must exceed +1.229 V. In (b), $E°_{red} + E°_{ox}$ for the cation must be a positive quantity. In (c), the cation must have an $E°_{ox}$ value no smaller than −1.229 V.

Solution

(a) Co^{3+} ($E°_{red} = +1.953$ V), Au^{3+} ($E°_{red} = +1.400$ V) and Au^+ ($E°_{red} = +1.695$ V)

(b) Au^+ ($E^\circ_{red} + E^\circ_{ox} = +1.695\ V - 1.400\ V = 0.295\ V$)

(c) Cu^+ ($E^\circ_{ox} = -0.161\ V$) and Hg_2^{2+} ($E^\circ_{ox} = -0.908\ V$)

Figure 19.9
Oxidation States of Chromium. The first two containers (green and violet) contain +3 chromium. The $Cr(H_2O)_4(H_2O)_2^+$ ion is green; $Cr(H_2O)_6^{3+}$ is violet. The two bottles at the right contain +6 chromium; the CrO_4^{2-} ion is yellow while the $Cr_2O_7^{2-}$ ion is red. (Charles D. Winters)

Oxoanions of the Transition Metals (CrO_4^{2-}, $Cr_2O_7^{2-}$, MnO_4^-)

Chromium in the +6 state forms two different oxoanions, the yellow chromate ion, CrO_4^{2-}, and the red dichromate ion, $Cr_2O_7^{2-}$ (Fig. 19.9). The chromate ion is stable in basic or neutral solution; in acid, it is converted to the dichromate ion:

$$\underset{\text{yellow}}{2CrO_4^{2-}(aq)} + 2H^+(aq) \rightleftharpoons \underset{\text{red}}{Cr_2O_7^{2-}(aq)} + H_2O \qquad K = 3 \times 10^{14}$$

The dichromate ion in acidic solution is a powerful oxidizing agent,

$$Cr_2O_7^{2-}(aq) + 14H^+(aq) + 6e^- \longrightarrow 2Cr^{3+}(aq) + 7H_2O \qquad E^\circ_{red} = +1.33\ V$$

As you might expect from the half-equation for its reduction, the oxidizing strength of the dichromate ion decreases as the concentration of H^+ decreases (increasing pH).

The $Cr_2O_7^{2-}$ ion can act as an oxidizing agent in the solid state as well as in water solution. In particular, it can oxidize the NH_4^+ ion to molecular nitrogen. When a pile of ammonium dichromate is ignited, a spectacular reaction occurs (Fig. 19.10).

$$\underset{\text{red}}{(NH_4)_2Cr_2O_7(s)} \longrightarrow N_2(g) + 4H_2O(g) + \underset{\text{green}}{Cr_2O_3(s)}$$

This makes a great lecture demonstration

The ammonium dichromate resembles a tiny volcano as it burns, emitting hot gases, sparks, and a voluminous green dust of chromium(III) oxide.

Chromates and dichromates are gradually disappearing from chemistry teaching laboratories because of concern about their toxicity. Long-term exposure of industrial workers to dust containing chromates has, in a few cases, been implicated in lung cancer. More commonly, repeated contact with chromate salts leads to skin disorders; a few people are extremely allergic to CrO_4^{2-} and $Cr_2O_7^{2-}$ ions, breaking into a rash upon first exposure.

Figure 19.10
Ammonium dichromate, $(NH_4)_2Cr_2O_7$, has a reddish color due to the presence of the $Cr_2O_7^{2-}$ ion. When ignited, it decomposes to give finely divided Cr_2O_3, which is green, nitrogen gas, and water vapor. (Charles D. Winters)

The permanganate ion, MnO_4^-, has an intense purple color (Fig. 19.11), easily visible even in very dilute solution.* Crystals of solid potassium permanganate, $KMnO_4$, have a deep purple, almost black color. This compound is used to treat such diverse ailments as "athlete's foot" and rattlesnake bites. These applications depend upon the fact that the MnO_4^- ion is a very powerful oxidizing agent. This is especially true in acidic solution, where MnO_4^- is reduced to Mn^{2+}:

Also poison ivy and ringworm

$$MnO_4^-(aq) + 8H^+(aq) + 5e^- \longrightarrow Mn^{2+}(aq) + 4H_2O \qquad E^\circ_{red} = +1.512 \text{ V}$$

In basic solution, MnO_4^- is reduced to MnO_2, with a considerably smaller value of E°_{red}:

$$MnO_4^-(aq) + 2H_2O + 3e^- \longrightarrow MnO_2(s) + 4\,OH^-(aq) \qquad E^\circ_{red} = +0.596 \text{ V}$$

However, even in basic solution, MnO_4^- can oxidize water:

$$4MnO_4^-(aq) + 2H_2O \longrightarrow 4MnO_2(s) + 3\,O_2(g) + 4\,OH^-(aq)$$

$$E^\circ = E^\circ_{red}\,MnO_4^- + E^\circ_{ox}\,H_2O = +0.596 \text{ V} - 0.401 \text{ V} = +0.195 \text{ V}$$

This reaction accounts for the fact that laboratory solutions of $KMnO_4$ slowly decompose, producing a brownish solid (MnO_2) and gas bubbles (O_2).

Figure 19.11
The MnO_4^- ion has an intense purple color. (Charles D. Winters)

19.4 QUALITATIVE ANALYSIS OF METAL CATIONS

In the general chemistry laboratory, you will most likely carry out at least one experiment dealing with the qualitative analysis of metal cations. The objective here is to separate and identify the cations present in an "unknown" solution. A scheme of analysis for 22 different cations is shown in Figure 19.12. As you can see, the general approach is to use precipitation reactions to divide the ions into four different groups. The ions within a group are then brought into solution, separated from one another, and identified.

Group I contains the only three common cations that form insoluble chlorides: Ag^+, Pb^{2+}, and Hg_2^{2+}. Addition of HCl precipitates AgCl, $PbCl_2$, and Hg_2Cl_2, all of which are white solids. To separate the cations, lead chloride is brought into solution in hot water

$$PbCl_2(s) \longrightarrow Pb^{2+}(aq) + 2Cl^-(aq)$$

while AgCl is dissolved in ammonia through complex ion formation

Hg_2Cl_2 reacts with NH_3 to form a gray precipitate

$$AgCl(s) + 2NH_3(aq) \longrightarrow Ag(NH_3)_2^+(aq) + Cl^-(aq)$$

Group II consists of six different cations, all of which form very insoluble sulfides with characteristic colors (Fig. 19.13 on p. 538). These compounds are precipitated by adding hydrogen sulfide, a toxic, foul-smelling gas, at a pH of 0.5. At this rather high H^+ ion concentration, the equilibrium

$$H_2S(aq) \rightleftharpoons 2H^+(aq) + S^{2-}(aq)$$

lies far to the left. The concentration of S^{2-} is extremely low but sufficient to precipitate Group II sulfides such as CuS ($K_{sp} = 1 \times 10^{-36}$) and Bi_2S_3 ($K_{sp} = 1 \times 10^{-99}$).

* The purple color of old bottles exposed to the sun for a long time is due to MnO_4^- ions. These are formed when ultraviolet light oxidizes manganese compounds in the glass.

$$Cu^{2+}(aq) + H_2S(aq) \longrightarrow CuS(s) + 2H^+(aq)$$

$$2Bi^{3+}(aq) + 3H_2S(aq) \longrightarrow Bi_2S_3(s) + 6H^+(aq)$$

Group III cations form sulfides which are considerably more soluble than those of Group II (e.g., K_{sp} NiS = 1×10^{-21}). Consequently, they do not precipitate at pH 0.5, allowing for their separation from Group II. However, at pH 9, where the concentration of S^{2-} is considerably higher, five Group III cations

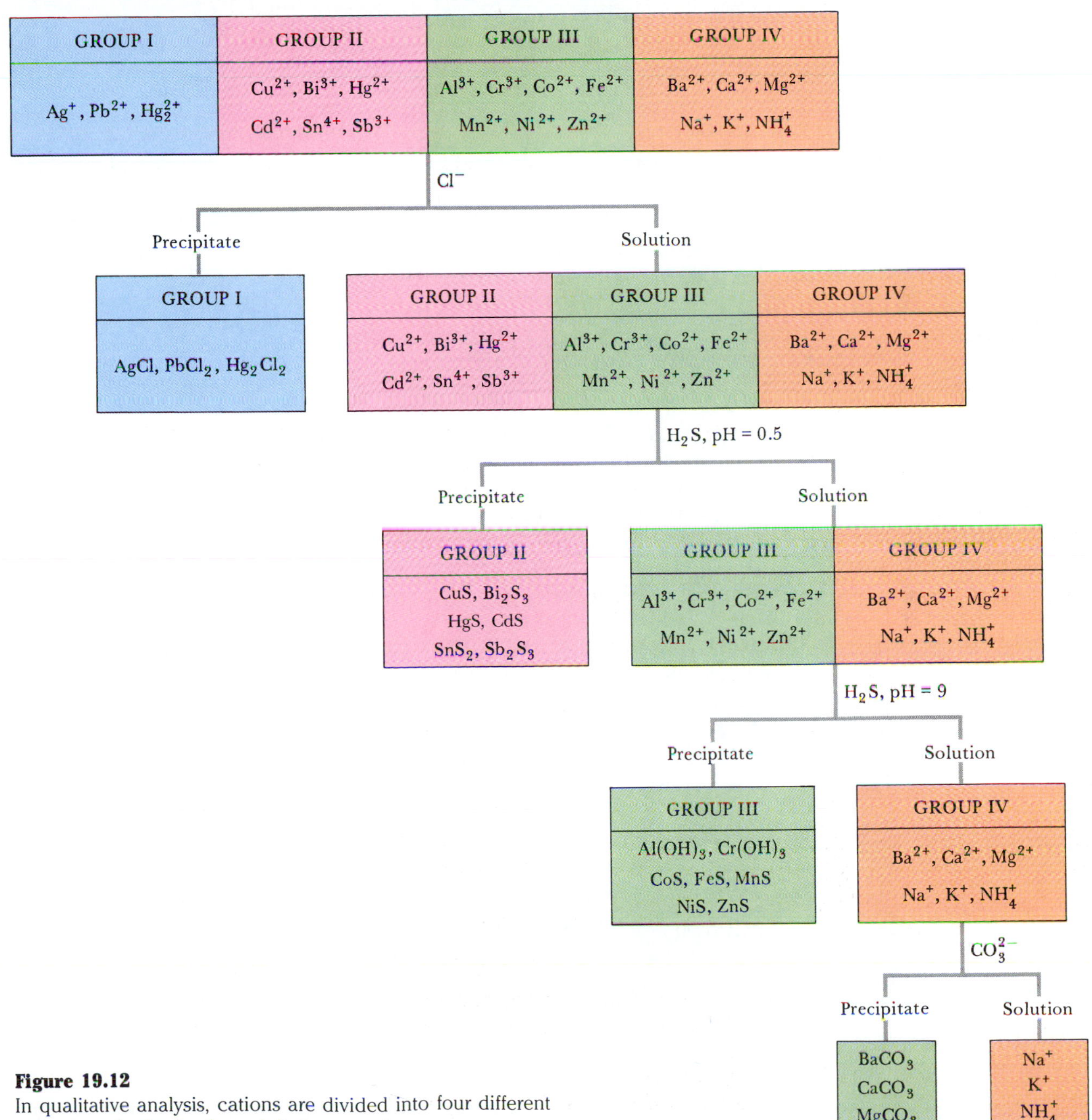

Figure 19.12
In qualitative analysis, cations are divided into four different groups on the basis of precipitation reactions.

Figure 19.13
The Group II sulfides show a variety of colors. CuS, Bi_2S_3, and HgS are black; CdS is orange-yellow, SnS_2 is pale yellow, and Sb_2S_3 is a brilliant red-orange.

precipitate as sulfides; Al^{3+} and Cr^{3+} come down as hydroxides in this basic solution (Fig. 19.14).

The sulfides of Al^{3+} and Cr^{3+} are unstable

Group IV cations have soluble chlorides and sulfides. However, the alkaline earth cations in this group (Mg^{2+}, Ca^{2+}, Ba^{2+}) can be precipitated as carbonates, thereby separating them from the other three cations in this group.

The alkali metal cations (Na^+, K^+) are identified by flame tests (Fig. 19.15). The $NH_4{}^+$ ion is heated with strong base to form NH_3, the only gas that turns red litmus blue:

$$NH_4{}^+(aq) + OH^-(aq) \longrightarrow NH_3(g) + H_2O$$

Example 19.7 A cation "unknown" contains the ions $Hg_2{}^{2+}$, Cd^{2+}, and Ca^{2+}. Write balanced equations for the precipitation reactions by which these ions are brought out of solution in qualitative analysis.

Strategy First, decide upon the formulas of the precipitates; Figure 19.12 should be helpful here. The precipitating agents are Cl^- for Group I, H_2S for Group II, and $CO_3{}^{2-}$ for Group IV. With this information, the net ionic equations follow logically.

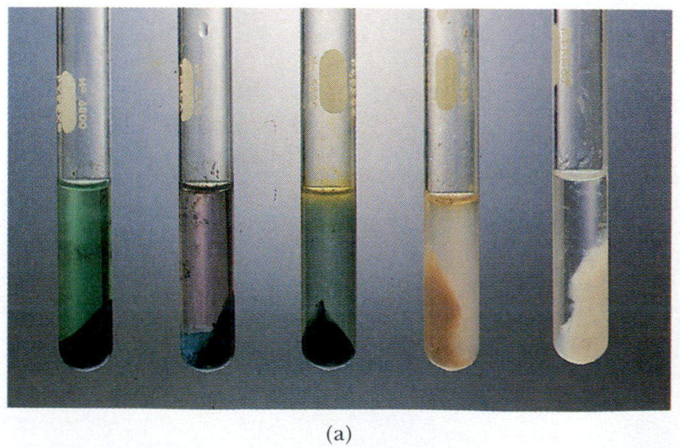

(a)

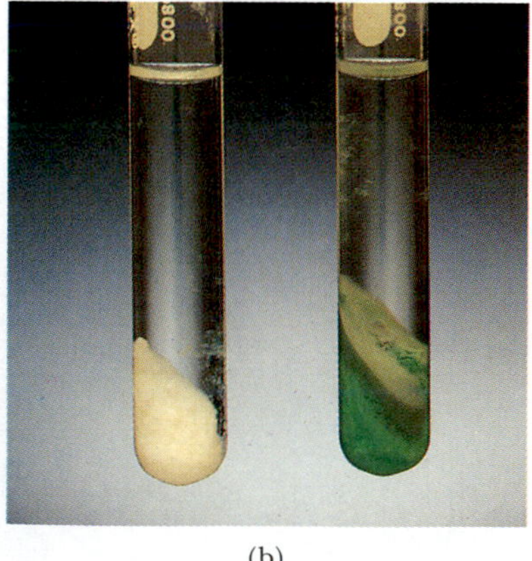

(b)

Figure 19.14
Five Group III cations precipitate as sulfides. These are NiS, CoS and FeS, all of which are black, MnS (light pink), and ZnS (white). Two cations precipitate as hydroxides, $Al(OH)_3$ (white) and $Cr(OH)_3$ (gray-green). (Charles D. Winters)

Solution

$$Hg_2^{2+}(aq) + 2Cl^-(aq) \longrightarrow Hg_2Cl_2(s)$$
$$Cd^{2+}(aq) + H_2S(aq) \longrightarrow CdS(s) + 2H^+(aq)$$
$$Mg^{2+}(aq) + CO_3^{2-}(aq) \longrightarrow MgCO_3(s)$$

19.5 ALLOYS

An important characteristic of metallic elements is their ability to form alloys. An **alloy** is a material with metallic properties that contains two or more elements, at least one of which is a metal. Solid alloys are ordinarily prepared by melting the elements together, stirring the molten mixture until it is homogeneous, and allowing it to cool. Many alloys, notably bronze, brass, and pewter, have been made for centuries by this method.

The type of alloy formed by two metals A and B can be deduced from the phase diagram for the system (Fig. 19.16). For a two-component phase diagram, temperature is plotted along the vertical axis, composition along the horizontal axis. Pure A is at the far left of the diagram, pure B at the far right. The dotted vertical lines show what happens when a molten solution of A and B is cooled; the upper circle shows the temperature at which freezing first occurs, the lower circle the temperature at which solidification is complete. Regions in which two phases are present are shown in color in the phase diagrams.

The idealized phase diagrams shown in Figure 19.16 correspond to two different structural types of alloys:

1. Solid solutions (Fig. 19.16a). Here, A and B are completely soluble in one another in the solid as well as the liquid phase. The solid alloy formed

Figure 19.15
Flame tests are used for Na^+ (yellow) and K^+ (violet). A drop of solution is picked up on a platinum loop and immersed in the flame. The test for K^+ is best done with a filter that hides the strong Na^+ color.

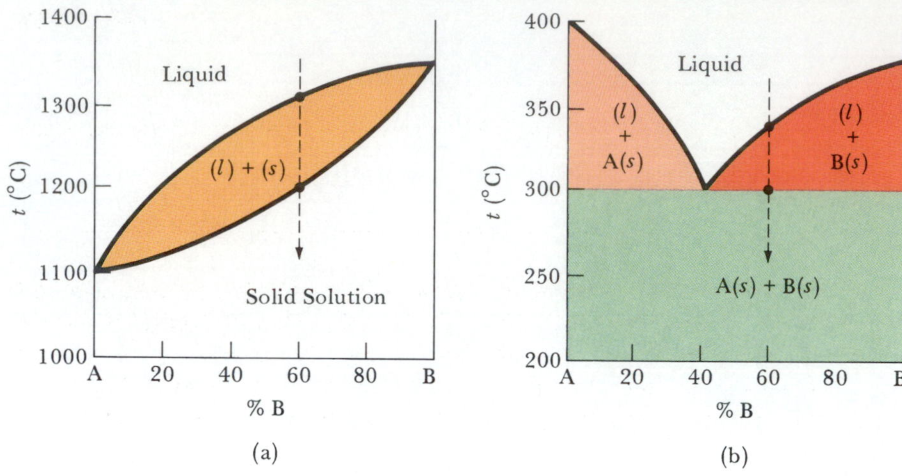

Figure 19.16
Phase diagrams for two-component alloys that are completely soluble in the solid state (a) or completely insoluble (b).

Alloys can be coarse mixtures, solid solutions, or compounds

upon cooling is homogeneous. It is referred to as a *substitutional* solid solution; atoms of A can substitute for B in any proportions.

Systems of this type are relatively rare; complete solubility in the solid requires that the two metals have the same type of unit cell (Chap. 9) and very similar atomic radii. Monel, a Cu–Ni alloy, has such a structure. Both copper and nickel crystallize in the face-centered cubic system; their atomic radii are 0.128 nm and 0.124 nm, respectively.

More commonly, two elements show limited solubility in the solid state. Iron can dissolve a maximum of 2% carbon. The carbon atoms, which are much smaller than iron (at. rad. C = 0.077 nm; at. rad. Fe = 0.126 nm) form an *interstitial* solid solution. They fit into holes in the face-centered cubic structure of iron, stable between 900 and 1400°C. The phase formed, called martensite, makes high-carbon steel hard and brittle.

2. Heterogeneous mixtures (Fig. 19.16b). Here, the two metals A and B are completely insoluble in each other in the solid phase. When the melt is cooled, the first solid that separates may be either A or B, depending upon the composition of the melt. At a lower temperature, an intimate but heterogeneous mixture of tiny crystals of A and B forms. Solder, an Sn–Pb alloy, approximates this behavior. If you look at solder under the microscope, you will see a mixture of nearly pure crystals of the two metals.

Example 19.8 Referring to the phase diagram shown in Figure 19.16b, for a mixture containing 60% B, state

(a) the temperature at which crystals first form on cooling; the temperature at which solidification is complete.
(b) the phase(s) present at 400°C; 310°C; 250°C.

Strategy A composition of 60% B corresponds to the dotted line shown in the figure. Freezing starts at the upper circle, where the temperature appears to be about 330°C; solid B precipitates out of the melt at that point. The liquid is all gone at the lower circle, about 300°C. At that temperature, solid A separates for the first time.

Solution

(a) 330°C; 300°C

(b) At any temperature above 330°C, including 400°C, the system is completely liquid. Between 330°C and 300°C (e.g., at 310°C), liquid is in equilibrium with solid B. Below 300°C (at 250°C, for example), there is a heterogeneous mixture of the two solids A and B.

Many alloys contain **intermetallic compounds** in which two elements A and B are present in a fixed atom ratio, e.g., AB, AB_2. Iron and carbon form a compound called cementite, which has the composition Fe_3C. Copper and zinc form a series of different compounds of which the simplest is CuZn. This 1 : 1 compound has a body-centered cubic structure very similar to that of CsCl (Chap. 9).

The properties of alloys are quite different from those of the component metals. The effect of alloying is ordinarily to

— *lower the melting point.* As you can see from Figure 19.16, an alloy usually (but not always!) melts at a lower temperature than the pure metals. An extreme example is Wood's metal (Table 19.4), which melts at 70°C.

Table 19.4 Some Commercially Important Alloys

Common Name	Composition, Element (Mass Percent)	Uses
Alnico	Fe(50), Al(20), Ni(20), Co(10)	magnets
Aluminum bronze	Cu(90), Al(10)	crankcases, hinges
Brass	Cu(67–90), Zn(10–33)	plumbing, hardware
Bronze	Cu(70–95), Zn(1–25), Sn(1–18)	bearings, bells, medals
Cast iron	Fe(96–97), C(3–4)	castings
Coinage, U.S.	Cu(75), Ni(25)	5¢, 10¢, 25¢, 50¢ coins
Dental amalgam	Hg(50), Ag(35), Sn(15)	dental fillings
Duriron	Fe(84), Si(14), C(1), Mn(1)	pipes, kettles, condensers
German silver	Cu(60), Zn(25), Ni(15)	tea pots, jugs, faucets
Gold, 18 carat	Au(75), Ag(10–20), Cu(5–15)	jewelry
Gold, 10 carat	Au(42), Ag(12–20), Cu(38–46)	jewelry
Gunmetal	Cu(88), Sn(10), Zn(2)	gun barrels, machine parts
Monel	Ni(60–70), Cu(25–35), Fe, Mn	instruments, machine parts
Nichrome	Ni(60), Fe(25), Cr(15)	electrical resistance wire
Pewter	Sn(70–95), Sb(5–15), Pb(0–15)	tableware
Silver solder	Ag(63), Cu(30), Zn(7)	high-melting solder
Solder	Pb(67), Sn(33)	joining metals
Spiegeleisen	Fe(80–95), Mn(5–20), C(0–1)	safes, armor plate, rails
Stainless steel	Fe(73–79), Cr(14–18), Ni(7–9)	instruments, sinks
Steel	Fe(98–99.5), C(0.5–2)	structural metal
Sterling silver	Ag(92.5), Cu(7.5)	tableware, jewelry
Vitallium	Co(62), Cr(26), Mo, Ni	artificial hip, knee joints
White gold	Au(90), Pd(10)	jewelry
Wood's metal	Bi(50), Pb(25), Sn(13), Cd(12)	automatic sprinkler systems

Wood's metal is used in fusible plugs that melt to set off automatic sprinkler systems.

— *increase the hardness.* A small amount of copper is present in sterling silver, which is much harder than pure silver. Gold is alloyed with silver and copper to form a metal hard enough to be used in jewelry. The lead plates used in storage batteries contain small amounts of antimony to prevent them from bending under stress.

Many alloys, notably stainless steel, resist corrosion

— *lower the electrical and thermal conductivity.* Copper used in electrical wiring must be extremely pure; as little as 0.03% of arsenic can lower its conductivity by 15%. Sometimes we take advantage of this effect. High-resistance nichrome wire (Ni, Cr) is used in the heating elements of hair dryers and electric toasters.

PERSPECTIVE

•

Essential Metals in Nutrition

Four of the main-group cations are essential in human nutrition (Table 19.A). Of these, the most important is Ca^{2+}. About 90% of the calcium in the body is found in bones and teeth, largely in the form of hydroxyapatite, $Ca(OH)_2 \cdot 3Ca_3(PO_4)_2$. Calcium ions in bones and teeth exchange readily with those in the blood; about 0.6 g of Ca^{2+} enters and leaves your bones every day. In a normal adult this exchange is in balance, but in elderly people, particularly women, there is sometimes a net loss of bone calcium, leading to the disease known as osteoporosis. It is generally supposed that osteoporosis is related to Ca^{2+} deficiency in early adult years, although the evidence is far from conclusive.

Of the ten trace elements known to be essential to human nutrition, seven are transition metals. For the most part, transition metals in biochemical compounds are present as complex ions, chelated by organic ligands. You will recall (Chap. 15) that hemoglobin has such a structure with Fe^{2+} as the central ion of the complex. The Co^{3+} ion occupies a similar position in Vitamin B_{12}, octahedrally coordinated to organic molecules somewhat similar to those in hemoglobin.

As you can judge from Table 19.A, transition metal cations are frequently found in enzymes. The Zn^{2+} ion alone is known to be a component of at least 70 different enzymes. One of these, referred to as "alcohol dehydrogenase" is concentrated in the liver, where it acts to break down alcohols. Another zinc-containing enzyme is involved in the normal functioning of oil glands in the skin, which accounts for the use of Zn^{2+} compounds in the treatment of acne.

Although Zn^{2+} is essential to human nutrition, compounds of the two elements below zinc in the periodic table, Cd and Hg, are extremely toxic. This reflects the fact that Cd^{2+} and Hg^{2+}, in contrast to Zn^{2+}, form very stable complexes with ligands containing sulfur atoms. As a result, these two cations react with and thereby deactivate enzymes containing —SH groups.

Over the years, compounds of the transition metals have been used for a variety of medicinal purposes. Calomel, Hg_2Cl_2, was the wonder drug of the early nineteenth century, prescribed for everything from constipation to pneumonia. It probably killed more patients than it cured. Quite recently, gold salts have been found effective in the treatment of rheumatoid arthritis. On a more prosaic level, iron(II) compounds such as $FeSO_4$ are used in treating anemia.

Table 19.A Essential Metal Ions

Ion	Amount in Body	Daily Requirement	Function in Body	Rich Sources*
Major Species (Main-Group Cations)				
Na^+	63 g	~2 g	principal cation *outside* cell fluid	table salt, many processed and preserved foods
K^+	150 g	~2 g	principal cation *within* cell fluid	fruits, nuts, fish, instant coffee, wheat bran
Mg^{2+}	21 g	0.35 g	activates enzymes for body processes	chocolate, nuts, instant coffee, wheat bran
Ca^{2+}	1160 g	0.8 g	bone and tooth formation	milk, cheeses, broccoli, canned salmon (with bones)
Trace Species (Transition Metal Cations)				
Ion	Amount in Body	Daily Requirement	Function in Body	Rich Sources*
Fe^{2+}, Fe^{3+}	5000 mg	15 mg	component of hemoglobin, myoglobin	liver, meat, clams, spinach
Zn^{2+}	3000 mg	15 mg	component of many enzymes, hormones	oysters, crab, meat, nuts
Cu^{2+}, Cu^+	100 mg	3 mg	iron metabolism, component of enzymes	liver, lobster, cherries
Mn^{2+}	15 mg	3 mg	metabolism of carbohydrates, lipids	beet greens, nuts, blueberries
Mo(IV,V,VI)	10 mg	0.3 mg	Fe, N metabolism; component of enzymes	legumes, green vegetables
Cr^{3+}	5 mg	0.1 mg	glucose metabolism; affects action of insulin	corn, clams, whole grains
Co^{2+}, Co^{3+}	1 mg	0.005 mg	component of Vitamin B_{12}	liver, shellfish

*Other, more mundane, sources can be found in any nutrition textbook.

CHAPTER HIGHLIGHTS

KEY CONCEPTS

1. *Calculate $\Delta G°$ from thermodynamic data*
   (Example 19.1; Problems 5, 6, 51, 60)
2. *Write balanced equations to represent*
   —*metallurgical processes*
   (Example 19.2; Problems 1–4, 7, 8, 54)
   —*reactions of Group 1, Group 2 metals*
   (Examples 19.3, 19.4; Problems 15, 16, 48)
   —*redox reactions of transition metals*
   (Example 19.5; Problems 19–24, 55)
   —*reactions taking place in qualitative analysis*
   (Example 19.7; Problems 37, 38)
3. *Determine $E°$ and reaction spontaneity from standard potentials*
   (Example 19.6; Problems 25–30)
4. *Interpret phase diagrams for alloy systems*
   (Example 19.8; Problems 39, 40)

KEY TERMS

alkali metal
alkaline earth metal
alloy
disproportionation
E°_{ox}, E°_{red}
electrolysis
main-group metal
metallurgy
oxidation
oxoanion
peroxide
qualitative analysis
reduction
superoxide
transition metal

SUMMARY PROBLEM

Consider the alkaline earth strontium and the transition metal manganese.

a. Write a balanced equation for the reaction of strontium with oxygen; with water.
b. Give the formulas of strontium hydride, strontium nitride, and strontium sulfide.
c. Electrolysis of strontium chloride gives strontium metal and chlorine gas. What mass of the chloride must be electrolyzed to form one kilogram of strontium? One liter of $Cl_2(g)$ at STP?
d. Give the formulas of three manganese compounds containing manganese in different oxidation states.
e. Write a balanced equation for the reaction of manganese metal with hydrochloric acid; Mn^{3+} with water; Mn^{2+} with H_2S.
f. Write a balanced equation for the reduction of the principal ore of manganese, pyrolusite, MnO_2, with CO.
g. For the reaction $MnO(s) + H_2(g) \rightarrow Mn(s) + H_2O(g)$, $\Delta H^\circ = +143$ kJ, $\Delta S^\circ = +0.0297$ kJ/K, would it be feasible to reduce MnO to the metal by heating with hydrogen?
h. Given that E°_{red} $Mn^{3+} \rightarrow Mn^{2+} = +1.559$ V, E°_{red} $Mn^{2+} \rightarrow Mn = -1.182$ V, and E°_{red} $O_2(g) \rightarrow H_2O = +1.229$ V, show by calculation whether Mn^{2+} in water solution will disproportionate; whether it will be oxidized to Mn^{3+} by dissolved oxygen; whether it will be reduced by hydrogen gas to the metal.
i. Given $MnO_4^-(aq) + 2H_2O + 3e^- \rightarrow MnO_2(s) + 4\,OH^-(aq)$, $E^\circ_{red} = +0.59$ V, write a balanced equation for the reaction of MnO_4^- in basic solution with Fe^{2+} to form $Fe(OH)_3$. Calculate E_{red} for the MnO_4^- ion at pH 9.0, taking conc. $MnO_4^- = 1\,M$.
j. At what $[S^{2-}]$ does MnS precipitate in qualitative analysis, using an unknown in which $[Mn^{2+}] = 0.02\,M$? (K_{sp} MnS $= 5 \times 10^{-14}$)

Answers

a. $2Sr(s) + O_2(g) \longrightarrow 2SrO(s)$
$Sr(s) + 2H_2O \longrightarrow Sr^{2+}(aq) + 2\,OH^-(aq) + H_2(g)$
b. SrH_2; Sr_3N_2; SrS
c. 1.809 kg; 7.07 g
d. $MnCl_2$, Mn_2O_3, $KMnO_4$
e. $Mn(s) + 2H^+(aq) \longrightarrow Mn^{2+}(aq) + H_2(g)$
$2Mn^{3+}(aq) + H_2O \longrightarrow 2Mn^{2+}(aq) + \frac{1}{2}O_2(g) + 2H^+(aq)$
$Mn^{2+}(aq) + H_2S(aq) \longrightarrow MnS(s) + 2H^+(aq)$
f. $MnO_2(s) + 2CO(g) \longrightarrow Mn(s) + 2CO_2(g)$
g. no; $T_{calc} \approx 4800$ K **h.** no; in all cases, E° is negative
i. $MnO_4^-(aq) + 3Fe^{2+}(aq) + 5\,OH^-(aq) + 2H_2O \longrightarrow MnO_2(s) + 3Fe(OH)_3(s)$ $E_{red} = +0.98$ V
j. $2 \times 10^{-12}\,M$

QUESTIONS & PROBLEMS

Metallurgy

1. Write a balanced equation to represent the electrolysis of molten sodium chloride. What volume of Cl_2 at STP is formed at the anode when 1.00 g of sodium is formed at the cathode?

2. Write a balanced equation to represent the electrolysis of aluminum oxide. If 2.00 L of O_2 at 25°C and 751 mm Hg is formed at the anode, what mass of Al is formed at the cathode?

3. Write a balanced equation to represent
- **a.** the roasting of nickel(II) sulfide to form nickel(II) oxide.
- **b.** the reduction of nickel(II) oxide by carbon monoxide.

4. Write a balanced equation to represent the roasting of copper(I) sulfide to form "blister copper."

5. Show by calculation whether the reaction in Problem 3(b) is spontaneous at 25°C and 1 atm.

6. Calculate $\Delta G°$ at 200°C for the reaction in Problem 4.

7. Write a balanced equation for the reaction that occurs when
- **a.** finely divided gold is treated with $CN^-(aq)$ in the presence of $O_2(g)$.
- **b.** finely divided zinc metal is added to the solution formed in (a).

8. Write a balanced equation for the reaction that occurs when
- **a.** iron(III) oxide is reduced with carbon monoxide.
- **b.** the excess carbon in pig iron is removed by the basic oxygen process.

9. How many cubic feet of air (assume 21% by volume of oxygen in air) at 25°C and 1.00 atm are required to react with coke to form the CO needed to convert one metric ton of hematite ore (92% Fe_2O_3) to iron?

10. Zinc is produced by electrolytic refining. The electrolytic process, which is similar to that for copper, can be represented by the two half-reactions

$$Zn(\text{impure}, s) \longrightarrow Zn^{2+} + 2e^-$$

$$Zn^{2+} + 2e^- \longrightarrow Zn(\text{pure}, s)$$

For this process, a voltage of 3.0 V is used. How many kilowatt hours are needed to produce one metric ton of pure zinc?

11. When 2.876 g of a certain metal sulfide is roasted in air, 2.368 g of the metal oxide are formed. If the metal has an oxidation number of +2, what is its molar mass?

12. Chalcopyrite, $CuFeS_2$, is an important source of copper. A typical chalcopyrite ore contains about 0.75% Cu. What volume of sulfur dioxide at 25°C and 1.00 atm pressure is produced when one boxcar load (4.00×10^3 ft^3) of chalcopyrite ore ($d = 2.6$ g/cm^3) is roasted? Assume all the sulfur in the ore is converted to SO_2 and no other source of sulfur is present.

Reactions of Alkali Metals and Alkaline Earth Metals

13. Give the formula and name of the compound formed when strontium reacts with

a. nitrogen **b.** bromine
c. water **d.** oxygen

14. Give the formula and name of the compound formed when potassium reacts with

a. nitrogen **b.** iodine **c.** water
d. hydrogen **e.** sulfur

15. Write a balanced equation and give the names of the products for the reaction of
- **a.** magnesium with chlorine.
- **b.** barium peroxide with water.
- **c.** lithium with sulfur.
- **d.** sodium with water.

16. Write a balanced equation and give the names of the products for the reaction of
- **a.** sodium peroxide and water.
- **b.** calcium and oxygen.
- **c.** rubidium and oxygen.
- **d.** strontium hydride and water.

17. To inflate a life raft with hydrogen to a volume of 25.0 L at 25°C and 1.10 atm, what mass of calcium hydride must react with water?

18. What mass of KO_2 is required to remove 90.0% of the CO_2 from a sample of 1.00 L of exhaled air (37°C, 1.00 atm) containing 5.00 mole percent CO_2?

Redox Chemistry of Transition Metals

19. Write a balanced equation to show
- **a.** the reaction of chromate ion with strong acid.
- **b.** the oxidation of water to oxygen gas by permanganate ion in basic solution.
- **c.** the reduction half-reaction of chromate ion to chromium(III) hydroxide in basic solution.

20. Write a balanced equation to show
- **a.** the formation of gas bubbles when cobalt reacts with hydrochloric acid.
- **b.** the reaction of copper with nitric acid.
- **c.** the reduction half-reaction of dichromate ion to Cr^{3+} in acidic solution.

21. Write a balanced redox equation for the reaction of mercury with aqua regia, assuming the products include $HgCl_4^{2-}$ and $NO_2(g)$.

22. Write a balanced redox equation for the reaction of cadmium with aqua regia, assuming the products include $CdCl_4^{2-}$ and $NO(g)$.

23. Balance the following redox equations.
- **a.** $Cu(s) + NO_3^-(aq) \longrightarrow Cu^{2+}(aq) + NO_2(g)$ (acidic)
- **b.** $Cr(OH)_3(s) + ClO^-(aq) \longrightarrow CrO_4^{2-}(aq) + Cl^-(aq)$ (basic)

24. Balance the following redox equations.
- **a.** $Fe(s) + NO_3^-(aq) \longrightarrow Fe^{3+}(aq) + NO_2(g)$ (acidic)
- **b.** $Cr(OH)_3(s) + O_2(g) \longrightarrow CrO_4^{2-}(aq)$ (basic)

25. Show by calculation which of the following metals will react with hydrochloric acid (standard concentrations).

a. Cd **b.** Cr **c.** Co
d. Ag **e.** Au

26. Show by calculation which of the metals in Problem 25 will react with nitric acid to form NO (standard concentrations).

27. Of the cations listed in Table 19.3, show by calculation which one (besides Cu^+) will disproportionate at standard conditions.

28. Using Table 17.1 (Chap. 17) calculate $E°$ for

a. $2Co^{3+}(aq) + H_2O \longrightarrow 2Co^{2+}(aq) + \frac{1}{2}O_2(g) + 2H^+(aq)$

b. $2Cr^{2+}(s) + I_2(s) \longrightarrow 2Cr^{3+}(aq) + 2I^-(aq)$

29. Using Table 19.3, calculate, for the disproportionation of Fe^{2+},

a. the equilibrium constant, K.

b. the concentration of Fe^{3+} in equilibrium with 0.10 M Fe^{2+}.

30. Using Table 19.3, calculate, for the disproportionation of Au^+,

a. K.

b. the concentration of Au^+ in equilibrium with 0.10 M Au^{3+}.

Qualitative Analysis

31. Complete the following table for cations.

Cation	Analytical Group	Precipitating Agent	Precipitate Formed
Ag^+	______	______	______
Bi^{3+}	______	______	______
Co^{2+}	______	______	______
Mg^{2+}	______	______	______

32. Complete the following table for cations.

Species	Test/Reagent	Response to Test/Reagent
______	flame test	yellow
Ag^+	______	precipitate
______	______	dissolves
K^+	flame test	______

33. Give the symbol for the cation that

a. is in Group IV and gives a violet flame.

b. is in Group II and forms a bright orange sulfide precipitate.

c. forms the only white sulfide in Groups II and III.

34. Identify the general characteristic of

a. Group I cations that allows for their separation from cations of Groups II through IV.

b. Group IV cations that separates them from those of Groups I through III.

35. What would happen if a solution containing Ni^{2+} and Cu^{2+}, both at 0.01 M, were treated with S^{2-} at 1×10^{-20} M (Group II)? (K_{sp} NiS = 1×10^{-21}, CuS = 1×10^{-36}).

36. What would happen if a solution containing Cu^{2+} and Ni^{2+}, both at 0.01 M, were treated with S^{2-} at 1×10^{-3} M (Group III)?

37. Write balanced net ionic equations for the precipitation of

a. Silver chloride in Group I.

b. Cadmium sulfide in Group II.

c. Calcium carbonate in Group IV.

38. Write a balanced net ionic equation to show how

a. Pb^{2+} is detected in Group I.

b. Al^{3+} is detected in Group III.

c. NH_4^+ is detected in Group IV.

Alloys

39. Referring to the phase diagram shown in Figure 19.16a, state what phase(s) is (are) present at 1200°C for a sample containing 0% B? 50% B? 100% B? At what temperature does solid first form when a liquid mixture containing 40% B is cooled?

40. Referring to Figure 19.16b, describe what happens when

a. a liquid containing 20% B is cooled from 400 to 200°C.

b. a liquid containing 80% B is cooled from 400 to 200°C.

41. Referring to Table 19.4, calculate

a. the mole percent of copper in the U.S. dime.

b. the number of atoms of silver in 1.00 g of sterling silver.

42. Give the name(s) of

a. a low-melting alloy containing bismuth.

b. an alloy containing two metals in Group **14**.

c. two different alloys of manganese.

Essential Metals

43. List four main-group cations essential to nutrition, and give their functions in the body.

44. Where is most of the calcium in the human body found? In what form?

45. Why are zinc compounds used to treat acne?

46. Why are Cd^{2+} and Hg^{2+} toxic? With which ligands do they complex?

Unclassified

47. A sample of sodium liberates 2.73 L hydrogen at 752 mm Hg and 22°C when it is added to a large amount of water. How much sodium is used?

48. A self-contained breathing apparatus contains 248 g of potassium superoxide. A firefighter exhales 116 L of air at 37°C and 748 mm Hg. The volume percent of water is exhaled air is 6.2. What mass of potassium superoxide is left after the water in the exhaled air reacts with it?

49. Taking K_{sp} $PbCl_2 = 1.7 \times 10^{-5}$ and assuming $[Cl^-] = 0.20$ M in Group I precipitation, calculate the concentration of Pb^{2+} carried over to Group II.

50. The equilibrium constant for the reaction

$$2CrO_4^{2-}(aq) + 2H^+(aq) \rightleftharpoons Cr_2O_7^{2-}(aq) + H_2O$$

is 3×10^{14}. What must the pH be so that the concentrations of chromate and dichromate ion are both 0.10 M?

51. Using data in Appendix 1, estimate the temperature at which Fe_2O_3 can be reduced to iron, using hydrogen gas as a reducing agent (assume $H_2O(g)$ is the other product).

52. A 0.500-g sample of zinc–copper alloy was treated with dilute hydrochloric acid. The hydrogen gas evolved was collected by water displacement at 27°C and a total pressure of 755 mm Hg. The volume of the water displaced by the gas is 105.7 mL. What is the percent composition, by mass, of the alloy? (Vapor pressure of H_2O at 27°C is 26.74 mm Hg.) Assume only the zinc reacts.

53. One type of stainless steel contains 22% by mass nickel. How much nickel sulfide ore, NiS, is required to produce one metric ion of stainless steel?

54. Silver is obtained in much the same manner as gold, using NaCN solution and O_2. Describe with appropriate equations the extraction of silver from argentite ore, Ag_2S. (The products are SO_2 and $Ag(CN)_2^-$, which is reduced with zinc.)

55. Iron(II) can be oxidized to iron(III) by permanganate ion in acidic solution. The permanganate ion is reduced to manganese(II) ion.

- **a.** Write the oxidation half-reaction, the reduction half-reaction, and the overall redox equation.
- **b.** Calculate $E°$ for the reaction.
- **c.** Calculate the percent Fe in an ore if a 0.3500-g sample is dissolved and the Fe^{2+} formed requires for titration 55.63 mL of a 0.0200 *M* solution of $KMnO_4$.

56. Of the cations listed on the center column of Table 19.3, which one is the

- **a.** strongest reducing agent?
- **b.** strongest oxidizing agent?
- **c.** weakest reducing agent?
- **d.** weakest oxidizing agent?

Challenge Problems

57. A sample of 20.00 g barium reacts with oxygen to form 22.38 g of a mixture of barium oxide and barium peroxide. Determine the composition of the mixture.

58. Rust, which you can take to be $Fe(OH)_3$, can be dissolved by treating it with oxalic acid. An acid-base reaction occurs, and a complex ion is formed.

- **a.** Write a balanced equation for the reaction.
- **b.** What volume of 0.10 *M* $H_2C_2O_4$ would be required to remove a rust stain weighing 1.0 g?

59. A 0.500-g sample of steel is analyzed for manganese. The sample is dissolved in acid and the manganese is oxidized to permanganate ion. A measured excess of Fe^{2+} is added to reduce MnO_4^- to Mn^{2+}. The excess Fe^{2+} is determined by titration with $K_2Cr_2O_7$. If 75.00 mL of 0.125 *M* $FeSO_4$ is added and the excess requires 13.50 mL of 0.100 *M* $K_2Cr_2O_7$ to oxidize Fe^{2+}, calculate the percent of Mn in the sample.

60. Calculate the temperature in °C at which the equilibrium constant (K) for the following reaction is 1.00.

$$MnO_2(s) \longrightarrow Mn(s) + O_2(g)$$

61. A solution of potassium dichromate is made basic with sodium hydroxide; the color changes from red to yellow. Addition of silver nitrate to the yellow solution gives a precipitate. This precipitate dissolves in concentrated ammonia but re-forms when nitric acid is added. Write balanced net ionic equations for all the reactions in this sequence.

20 Chemistry of the Nonmetals

For what can so fire us,
Enrapture, inspire us,
As Oxygen? What so delicious
to quaff?
It is so animating,
And so titillating,
E'en grey-beards turn
freshy, dance,
caper, and laugh.

JOHN SHIELD
Oxygen Gas

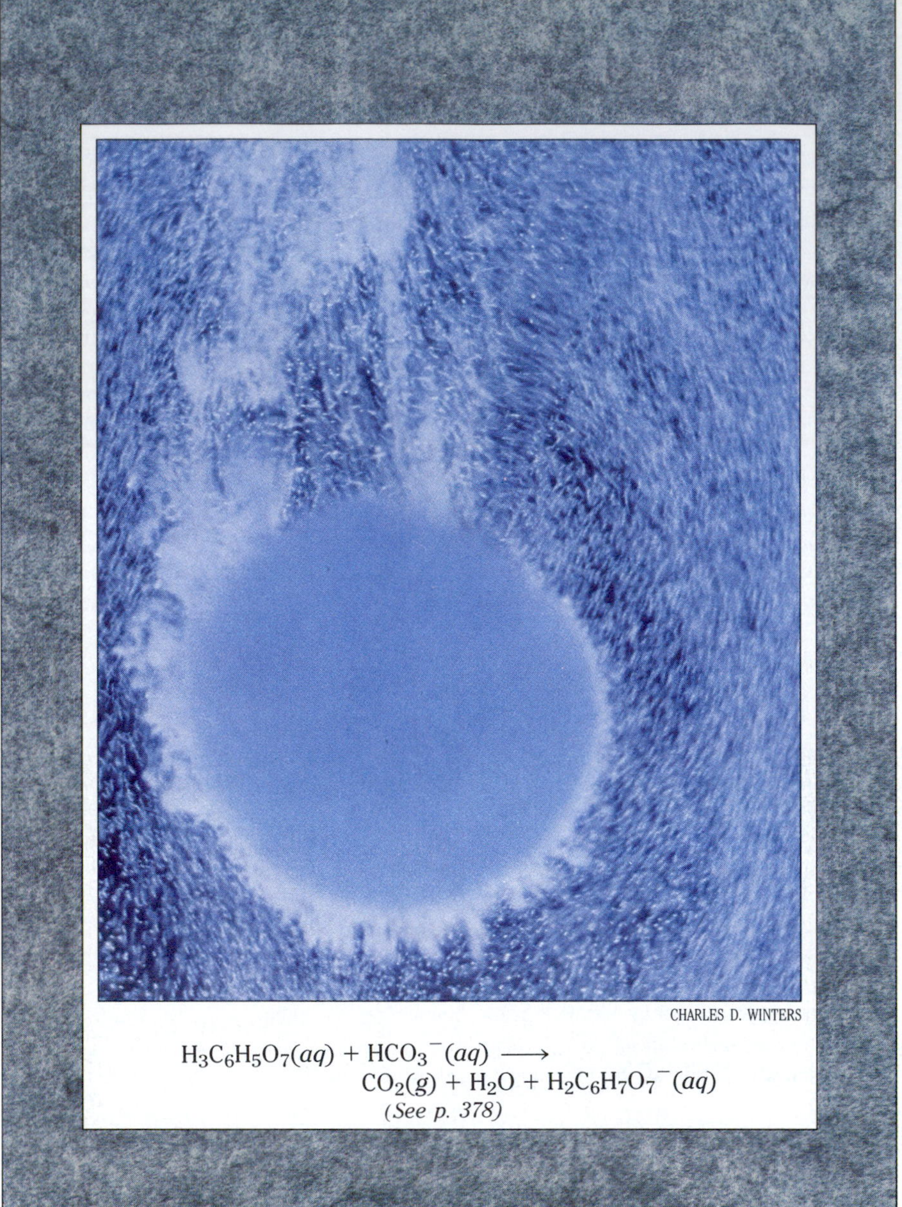

CHARLES D. WINTERS

$$H_3C_6H_5O_7(aq) + HCO_3^-(aq) \longrightarrow CO_2(g) + H_2O + H_2C_6H_7O_7^-(aq)$$

(See p. 378)

CHAPTER OUTLINE

Approximately 18 elements are classified as nonmetals; they lie above and to the right of the "stairway" that runs diagonally across the periodic table (Fig. 20.1). As the word "nonmetal" implies, these elements do not show metallic properties; in the solid state they are brittle as opposed to ductile, insulators rather than conductors. Most of the nonmetals, particularly those in Groups **15–17** of the periodic table, are molecular in nature (e.g., N_2, O_2, F_2). The noble gases (Group **18**) consist of individual atoms attracted to each other by

																17	18
													14	15	16		
														N	O	F	
														P	S	Cl	
																Br	
																I	

Figure 20.1
The 18 nonmetals are shown in color; the symbols indicate those elements whose chemistry is discussed in this chapter.

weak dispersion forces. Carbon and silicon in Group **14** have a network covalent structure.

As indicated in Figure 20.1, this chapter concentrates upon the more common and/or more reactive nonmetals, namely

— nitrogen and phosphorus in Group **15**
— oxygen and sulfur in Group **16**
— the halogens (F, Cl, Br, I) in Group **17**

We will consider

— the properties of these elements and methods of preparing them (Section 20.1)
— their hydrogen compounds (Section 20.2)
— their oxides (Section 20.3)
— their oxoacids and oxoanions (Section 20.4)

20.1 THE ELEMENTS AND THEIR PREPARATION

Table 20.1 lists some of the properties of the eight nonmetals considered in this chapter. Notice that all of these elements are molecular; those of low molar mass (N_2, O_2, F_2, Cl_2) are gases at room temperature and atmospheric pressure

Table 20.1 Properties of Nonmetallic Elements

	Nitrogen	Phosphorus	Oxygen	Sulfur	Fluorine	Chlorine	Bromine	Iodine
Outer electron configuration	$2s^2 2p^3$	$3s^2 3p^3$	$2s^2 2p^4$	$3s^2 3p^4$	$2s^2 2p^5$	$3s^2 3p^5$	$4s^2 4p^5$	$5s^2 5p^5$
Molecular formula	N_2	P_4	O_2	S_8	F_2	Cl_2	Br_2	I_2
Molar mass (g/mol)	28	124	32	257	38	71	160	254
State (25°C, 1 atm)	gas	solid	gas	solid	gas	gas	liquid	solid
Melting point (°C)	−210	44	−218	119	−220	−101	−7	114
Boiling point (°C)	−196	280	−183	444	−188	−34	59	184
Bond energy* (kJ/mol)	941	200	498	266	153	243	193	151
E°_{red}					+2.889 V	+1.360 V	+1.077 V	+0.534 V

* In the element (triple bond in N_2, double bond in O_2).

Figure 20.2
Flasks containing Cl_2, Br_2, and I_2 show a gradation in color from greenish-yellow through deep red to violet. The colors shown for bromine and iodine are those of the vapors in equilibrium with $Br_2(l)$ and $I_2(s)$. (Marna G. Clarke)

(Fig. 20.2). Stronger dispersion forces cause the nonmetals of higher molar mass to be either liquids (Br_2) or solids (I_2, P_4, S_8).

Chemical Reactivity

B.E. = bond energy

Of the eight nonmetals listed in Table 20.1, *nitrogen* is by far the least reactive. Its inertness is due to the strength of the triple bond holding the N_2 molecule together (B.E. N≡N = 941 kJ/mol). This same factor explains why virtually all chemical explosives are compounds of nitrogen (e.g., nitroglycerine, trinitrotoluene, ammonium nitrate, lead azide). These compounds detonate exothermically to form molecular nitrogen. The reaction with lead azide is

$$Pb(N_3)_2(s) \longrightarrow Pb(s) + 3N_2(g) \qquad \Delta H° = -476 \text{ kJ}$$

Sodium azide, NaN_3, decomposes more smoothly; it is used in automobile airbags which inflate upon impact with another car.

F_2 reacts with almost anything

Fluorine is the most reactive of all elements, in part because of the weakness of the F—F bond (B.E. F—F = 153 kJ/mol), but mostly because it is such a powerful oxidizing agent ($E°_{red} = +2.889$ V). Fluorine combines with every element in the periodic table except He, Ne, and Ar. With a few metals, it forms a surface film of metal fluoride, which adheres tightly enough to prevent further reaction. This is the case with nickel, where the product is NiF_2. Fluorine gas is ordinarily stored in containers made of a nickel alloy, such as stainless steel (Fe, Cr, Ni) or Monel (Ni, Cu).

Chlorine is somewhat less reactive than fluorine. Although it reacts with nearly all metals (Fig. 20.3), heating is often required. This reflects the relatively strong bond in the Cl_2 molecule (B.E. Cl—Cl = 243 kJ/mol). Chlorine disproportionates in water, forming Cl^- ions (oxid. no. Cl = −1) and HClO molecules (oxid. no. Cl = +1).

Cl_2 is very soluble in NaOH solution; why?

$$Cl_2(g) + H_2O \rightleftharpoons Cl^-(aq) + H^+(aq) + HClO(aq)$$

The hypochlorous acid, HClO, formed by this reaction is a powerful oxidizing agent ($E^\circ_{red} = +1.630$ V); it kills bacteria, apparently by destroying certain enzymes essential to their metabolism. The taste and odor that we associate with "chlorinated water" is actually due to compounds such as CH_3NHCl, produced by the action of hypochlorous acid on bacteria.

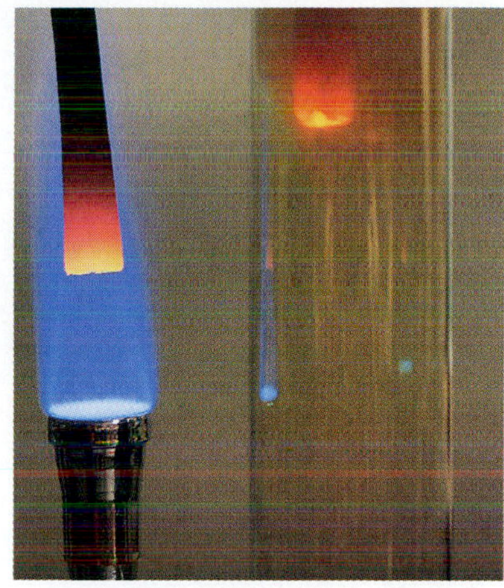

Figure 20.3
When a heated piece of copper foil is plunged into a cylinder containing chlorine gas, it reacts vigorously, giving off sparks. The equation for the reaction is: $Cu(s) + Cl_2(g) \rightarrow CuCl_2(s)$. (Charles D. Winters)

Example 20.1 For the reaction $Cl_2(g) + H_2O \rightleftharpoons Cl^-(aq) + H^+(aq) + HClO(aq)$

(a) Write the expression for the equilibrium constant, *K*.
(b) Given that $K = 2.7 \times 10^{-5}$, calculate the concentration of HClO in equilibrium with $Cl_2(g)$ at 1.0 atm.

Strategy To answer (a), note that, in the expression for *K*, gases enter as their partial pressures in atmospheres, species in aqueous solution as their molarities; water does not appear, since it is the solvent. To answer (b), note that H^+ ions, Cl^- ions, and HClO molecules are formed in equimolar amounts.

Solution

(a) $$K = \frac{[HClO] \times [H^+] \times [Cl^-]}{P_{Cl_2}}$$

(b) Let $x = [HClO]$. Since H^+, Cl^-, and HClO all have coefficients of 1 in the balanced equation, $[H^+] = [Cl^-] = x$. Substituting in the expression for *K*:

$$K = \frac{[HClO] \times [H^+] \times [Cl^-]}{P_{Cl_2}} \qquad 2.7 \times 10^{-5} = \frac{x^3}{1.0}$$

Solving,

$$x = (2.7 \times 10^{-5})^{1/3} = 0.030\,M$$

In other words, the concentration of hypochlorous acid in a solution formed by bubbling chlorine gas through water should be about 0.03 mol/L.

The oxidizing power of the halogens makes them hazardous to work with. Fluorine is the most dangerous, but it is very unlikely that you will ever come across it in a teaching laboratory. You are most likely to encounter chlorine as its saturated water solution, called "chlorine water." Remember that the pressure of chlorine gas over this solution (if it is freshly prepared) is 1 atm and that chlorine was used as a poison gas in World War I. Use small quantities of chlorine water and don't breathe the vapors. Bromine, although not as strong an oxidizing agent as chlorine, can cause severe burns if it comes in contact with your skin, particularly if it gets under your fingernails.

Of the four halogens, iodine is the weakest oxidizing agent. When the authors of this textbook were young, "tincture of iodine," a 10% solution of I_2 in alcohol, was widely used as an antiseptic. Today, hospitals use a product called "povidone-iodine," a quite powerful iodine-containing antiseptic and disinfectant, which can be diluted with water to the desired strength. These applications of molecular iodine should not delude you into thinking that the solid is harmless. On the contrary, if $I_2(s)$ is allowed to remain in contact with your skin, it can cause painful burns that are slow to heal.

Correction: when WLM was young, before CNH was born

Occurrences and Preparation

All the noble gases except He come from liquid air

Of the eight nonmetals considered here, three (nitrogen, oxygen, and sulfur) occur in nature in elemental form. *Nitrogen* and *oxygen* are obtained from air, where their mole fractions are 0.7808 and 0.2095, respectively. When liquid air at −200°C is allowed to warm, the first substance that boils off is nitrogen (bp N_2 = 77 K). After most of the nitrogen has been removed, further warming gives oxygen (bp O_2 = 90 K). About 2×10^{10} kg of O_2 and lesser amounts of N_2 are produced annually in the United States from liquid air.

At the close of the Civil War in 1865, oil prospectors in Louisiana discovered (to their disgust) elemental *sulfur* in the caprock of vast salt domes up to 20 km^2 in area. The sulfur lies 60 to 600 m below the surface of the earth. The process used to mine sulfur is named after its inventor, Herman Frasch, an American chemical engineer. A diagram of the Frasch process is shown in Figure 20.4. The sulfur is heated to its melting point (119°C) by pumping superheated water at 165°C down one of three concentric pipes. Compressed air is used to bring the sulfur to the surface. The air and sulfur form a frothy mixture that rises through the middle pipe. Upon cooling, the sulfur solidifies, filling huge vats that may be 0.5 km long. The sulfur obtained in this way has a purity approaching 99.9%.

The *halogens* are far too reactive to occur in nature as the free elements. Instead, they are found as anions:

— F^- in the mineral calcium fluoride, CaF_2 (fluorite)
— Cl^- in huge underground deposits of sodium chloride, NaCl (rock salt), underlying parts of Oklahoma, Texas, and Kansas
— $Br^-(aq)$ and $I^-(aq)$ in brine wells in Arkansas (conc. Br^- = 0.05 *M*) and Michigan (conc. I^- = 0.001 *M*), respectively

Br_2 and I_2 can also be obtained from seawater

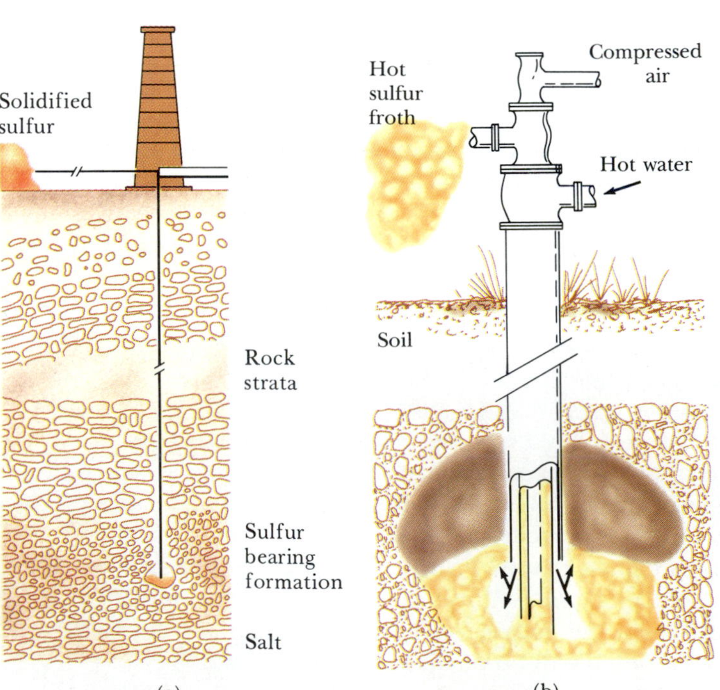

Figure 20.4
Frasch process for mining sulfur. Superheated water at 165°C is sent down through the outer pipe to form a pool of molten sulfur (mp = 119°C) at the base. Compressed air, pumped down the inner pipe, brings the sulfur to the surface. Sulfur deposits are often 100 m or more beneath the earth's surface, covered with quicksand and rock.

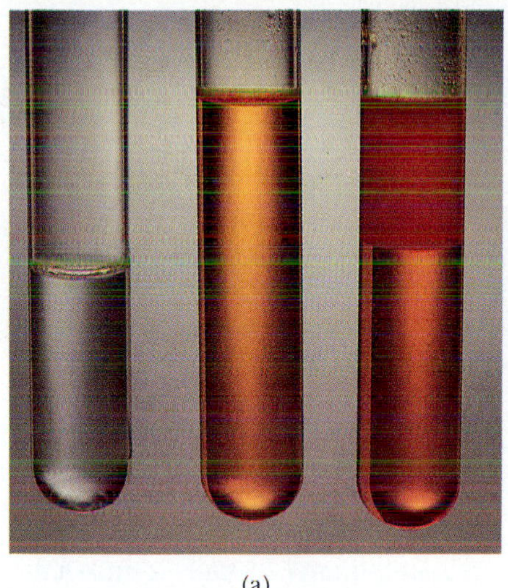

(a)

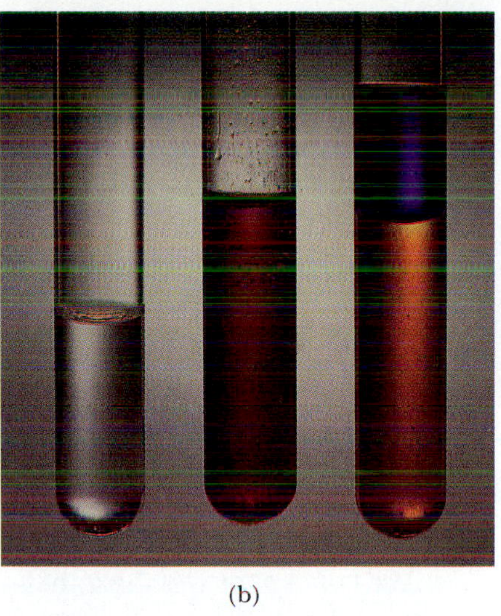

(b)

Figure 20.5
The reaction of chlorine with Br^- and I^- ions is shown in (a) and (b), respectively. The anions themselves (test tubes at *left*) are colorless. Oxidation by chlorine yields the free halogens, Br_2 and I_2, which are colored in water solution (*center* tubes) and more intensely colored in an organic solvent (upper layer in the tubes at the *right*). (Charles D. Winters)

The fluoride and chloride ions are very difficult to oxidize (E°_{ox} F^- = −2.889 V; E°_{ox} Cl^- = −1.360 V). Hence the elements fluorine and chlorine are ordinarily prepared by electrolytic oxidation, using a high voltage. As pointed out in Chapter 17, chlorine is prepared by the electrolysis of aqueous sodium chloride:

$$2Cl^-(aq) + 2H_2O \longrightarrow Cl_2(g) + H_2(g) + 2\,OH^-(aq)$$

The process used to prepare fluorine was developed by Henri Moissan in Paris more than a century ago; it won him one of the early (1906) Nobel Prizes in chemistry. The electrolyte is a mixture of HF and KF in a 2:1 mole ratio. At 100°C, fluorine is generated by the decomposition of hydrogen fluoride:

$$2HF(l) \longrightarrow H_2(g) + F_2(g)$$

Potassium fluoride furnishes the ions required to carry the electric current.

Since bromide and iodide ions are easier to oxidize (E°_{ox} Br^- = −1.077 V; E°_{ox} I^- = −0.534 V), bromine and iodine can be prepared by chemical oxidation. Commonly, the oxidizing agent is chlorine gas (E°_{red} = +1.360 V):

$$Cl_2(g) + 2Br^-(aq) \longrightarrow 2Cl^-(aq) + Br_2(l)$$

$$Cl_2(g) + 2I^-(aq) \longrightarrow 2Cl^-(aq) + I_2(s)$$

In the general chemistry laboratory, these same reactions are often used to test for Br^- and I^- ions (Fig. 20.5).

The first three elements in Group 15: nitrogen gas, phosphorus, and antimony. (Richard Megna/FUNDAMENTAL PHOTOGRAPHS, New York)

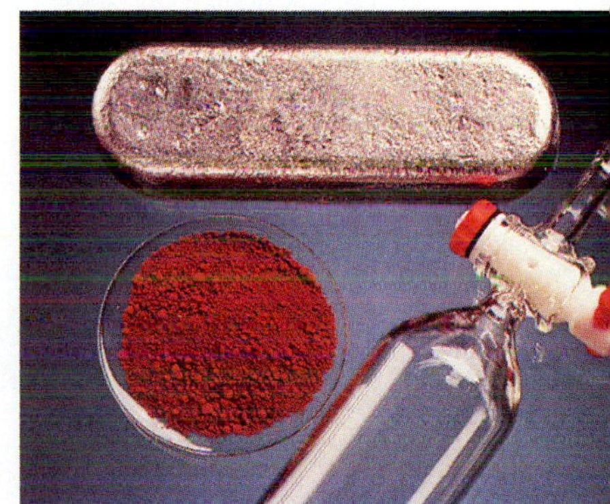

Allotropy

Two or more different structural forms of an element in the same phase (gas, liquid, or solid) are referred to as *allotropes*. One of the simplest forms of allotropy is shown by *oxygen*, which can exist in the gas phase as either O_2 or O_3 (ozone). Commercially, ozone is prepared by passing O_2 gas through a high-voltage (10^4 V) electric discharge. At atmospheric pressure, the reaction

$$2\,O_3(g) \longrightarrow 3\,O_2(g) \qquad \Delta H^\circ = -285.4\text{ kJ} \quad \Delta S^\circ = +137.5\text{ J/K}$$

is thermodynamically spontaneous at all temperatures. Kinetically, however, ozone stays around long enough to find some application as a substitute for chlorine in disinfecting municipal water supplies.

Phosphorus forms several allotropes in the solid state, of which white and red phosphorus are the most common.

1. *White phosphorus* consists of P_4 molecules with the structure shown in Figure 20.6. It is a soft waxy substance with a low melting point (44°C) and boiling point (280°C). Like most molecular substances, white phosphorus is readily soluble in such nonpolar solvents as CCl_4. The chemical reactivity of white phosphorus is so great that it is stored under water to protect it from O_2. A piece of P_4 exposed to air in a dark room glows because of the light given off upon oxidation (Fig. 20.7). White phosphorus is extremely toxic. As little as 0.1 g taken internally can be fatal. Direct contact with the skin produces painful burns.

White phosphorus is used in napalm, a really nasty chemical weapon

2. *Red phosphorus* is the form in which the element is usually found in the laboratory. This allotrope has properties quite different from those of white phosphorus. It is much higher-melting (mp = 590°C at 43 atm) and insoluble in common solvents. The low volatility of red phosphorus makes it much less toxic than the white form. It is also less reactive and must be heated to 250°C to burn in air. These properties are consistent with the structure of red phosphorus, which is known to be network covalent (Fig. 20.6). This allotrope can be made by heating white phosphorus in the absence of air to about 300°C.

Before the undesirable properties of white phosphorus were known, it was used in matches. Today, two different kinds of matches are available, neither of which contains white phosphorus. The heads of "strike-anywhere" matches contain a mixture of a sulfide of phosphorus, P_4S_3, potassium chlorate, $KClO_3$, and powdered glass. When struck against a rough surface, the mixture ignites. Safety matches contain sulfur and potassium chlorate; the special surface against which they are struck contains red phosphorus and powdered glass. Friction sets off a reaction between red phosphorus and $KClO_3$.

Figure 20.6
At the top of the figure are shown the two most common allotropes of phosphorus, the white and red forms. Below are their structures. White phosphorus is molecular, formula P_4. Red phosphorus has a network structure, shown here in simplified form. (photo, Charles D. Winters)

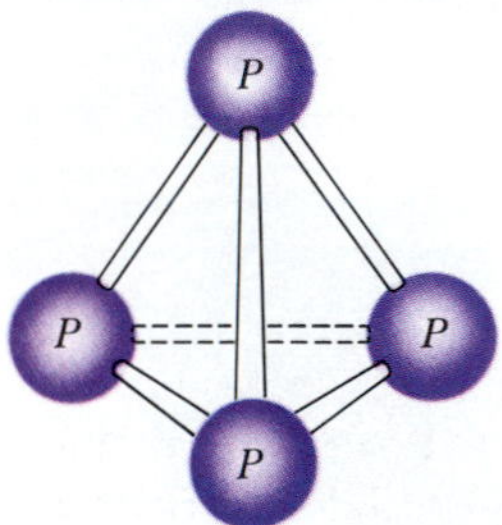

White phosphorus

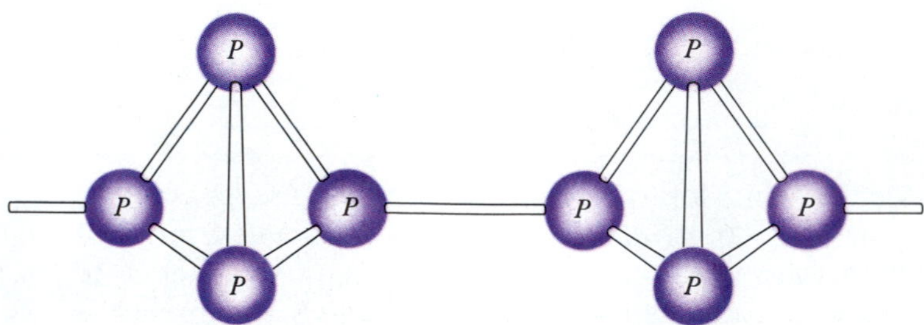

Red phosphorus

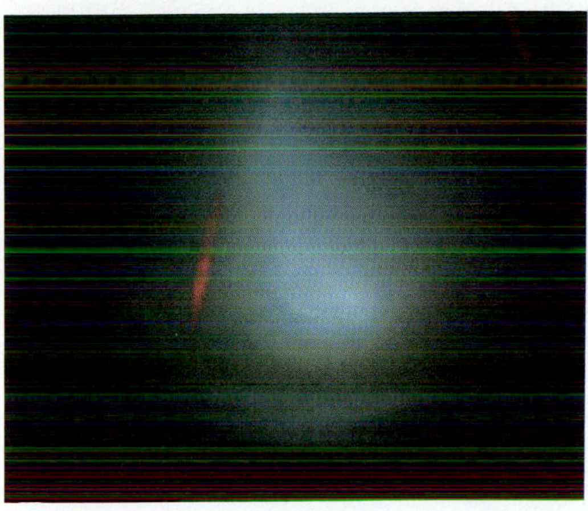

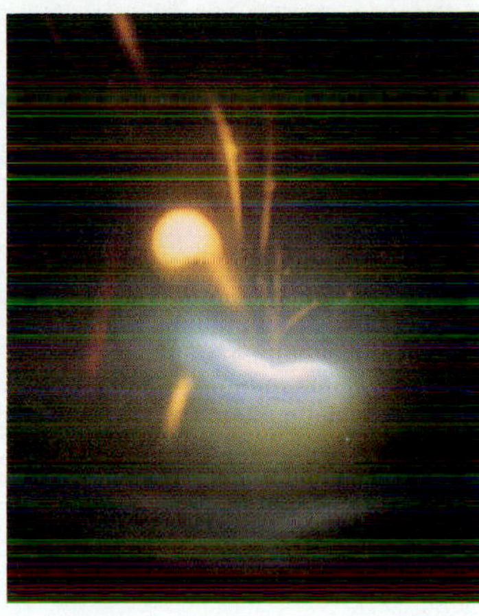

Figure 20.7
When white phosphorus is exposed to oxygen gas, it first glows (phosphorescence) and then bursts into flame. The reaction is: $P_4(s) + 5O_2(g) \rightarrow P_4O_{10}(s)$. (Charles D. Winters)

Example 20.2 For the allotropic conversion P(white) → P(red), $\Delta H°$ is −17.6 kJ and $\Delta S°$ is −18.3 J/K. At what temperature are the two allotropes in equilibrium at 1 atm?

Strategy A system is at equilibrium at 1 atm when $\Delta G° = 0$. To find the temperature at which that happens, use the Gibbs-Helmholtz equation: $\Delta G° = \Delta H° - T\Delta S°$.

Solution Setting $\Delta G° = 0$, it follows that $\Delta H° = T\Delta S°$
Hence

$$T = \frac{\Delta H°}{\Delta S°} = \frac{-17.6 \text{ kJ}}{-0.0183 \text{ kJ/K}} = 962 \text{ K} \approx 690°\text{C}$$

At any temperature below about 700°C, red phosphorus is the stable allotrope.

In the solid state, *sulfur* can have more than 20 different allotropic forms. You will no doubt be relieved to learn that we will only talk about two of these, rhombic and monoclinic sulfur (Fig. 20.8). Both consist of ring molecules of formula S_8. They differ only in the way that molecules are packed in the solid, reflected in different crystal structures.

At room temperature, rhombic sulfur is the stable allotrope. However, since the process

$$S_8(\text{rhombic}) \longrightarrow S_8(\text{monoclinic}) \qquad \Delta H° = +2.6 \text{ kJ}$$

Figure 20.8
Naturally occurring crystals of monoclinic sulfur *(left)* and rhombic sulfur *(right)*. (left, W.K. Fletcher, Photo Researchers, Inc.; right, Allen B. Smith, Tom Stack and Assoc.)

is endothermic, the monoclinic form is stable at high temperatures. The two forms are in equilibrium at 96°C; if rhombic sulfur is heated to that temperature, it slowly converts to monoclinic sulfur. More commonly, the monoclinic allotrope is prepared by freezing liquid sulfur at the melting point (119°C) and then cooling quickly to room temperature. Typically, at 25°C, monoclinic crystals stay around for a day or more before converting to the rhombic form.

The free-flowing, pale yellow liquid formed when sulfur melts contains S_8 molecules. However, upon heating to 160°C, a striking change occurs. The liquid becomes so viscous that it cannot be poured readily. At the same time its color changes to a deep reddish-brown. These effects reflect a change in molecular structure. The S_8 rings break apart and then link to one another to form long chains such as

Under these conditions, sulfur is a polymer

$$\cdot S - S - S - S - S - S - S - S - S - S - S - S - S - S - S - S \cdot$$

Liquid sulfur between 160 and 250°C contains a high proportion of such chains. They vary in length from eight to perhaps a million atoms. The chains become tangled, producing a highly viscous liquid. The deep color is due to the absorption of light by the unpaired electrons at the ends of the chains.

If liquid sulfur at 200°C is quickly poured into water, a rubbery mass results. This is referred to as "plastic sulfur." It consists of long-chain molecules that did not have time to rearrange to the S_8 molecules stable at room temperature. Within a few hours, the plastic sulfur loses its elasticity as it converts to rhombic crystals.

20.2 HYDROGEN COMPOUNDS OF NONMETALS

Table 20.2 lists some of the more important nonmetal hydrides (the hydrogen compounds of carbon are discussed in Chap. 21). The physical states listed are those observed at 25°C and 1 atm. The remainder of this section is devoted to a discussion of the chemical properties of the compounds shown in boldface in the table.

Table 20.2 Hydrogen Compounds of the Nonmetals

Group 15	Group 16	Group 17
Ammonia $NH_3(g)$	Water $H_2O(l)$	**Hydrogen fluoride $HF(g)$**
Hydrazine $N_2H_4(l)$	**Hydrogen peroxide $H_2O_2(l)$**	
Hydrazoic acid $HN_3(l)$		
Phosphine $PH_3(g)$	**Hydrogen sulfide $H_2S(g)$**	**Hydrogen chloride $HCl(g)$**
Diphosphine $P_2H_4(l)$		
		Hydrogen bromide $HBr(g)$
		Hydrogen iodide $HI(g)$

Ammonia, NH_3

Ammonia is one of the most important industrial chemicals; more than ten million tons of NH_3 are produced annually in the United States. You will recall (Chap. 12) that it is made by the Haber process

$$N_2(g) + 3H_2(g) \rightleftharpoons 2NH_3(g) \qquad 450°C,\ 200–600\ atm,\ solid\ catalyst$$

Ammonia is used to make fertilizers and a host of different nitrogen compounds, notably nitric acid, HNO_3.

The NH_3 molecule acts as a *Brønsted-Lowry base* in water, accepting a proton from a water molecule:

$$NH_3(aq) + H_2O \rightleftharpoons NH_4^+(aq) + OH^-(aq)$$

Ammonia can also act as a *Lewis base* when it reacts with a metal cation to form a complex ion:

$$2NH_3(aq) + Ag^+(aq) \longrightarrow Ag(NH_3)_2^+(aq)$$

The NH_3 molecule donates an electron pair to Ag^+

Ammonia is often used to precipitate insoluble metal hydroxides such as $Al(OH)_3$. The OH^- ions formed when ammonia reacts with water precipitate the cation from solution as the hydroxide. The overall equation for the reaction is

$$Al^{3+}(aq) + 3NH_3(aq) + 3H_2O \longrightarrow Al(OH)_3(s) + 3NH_4^+(aq)$$

Nitrogen cannot have an oxidation number lower than −3, which means that when NH_3 takes part in a redox reaction, it always acts as a *reducing agent.* Ammonia may be oxidized to elementary nitrogen or to a compound of nitrogen. An important redox reaction of ammonia is that with hypochlorite ion:

Ammonia has a sharp, irritating odor and is toxic at high concentrations

$$2NH_3(aq) + ClO^-(aq) \longrightarrow N_2H_4(aq) + Cl^-(aq) + H_2O$$

Hydrazine, N_2H_4, is made commercially by this process. Certain by-products of this reaction, notably NH_2Cl and $NHCl_2$, are both toxic and explosive, so solutions of household bleach and ammonia should never be mixed with one another.

Hydrogen Sulfide, H_2S

In water solution, hydrogen sulfide acts as a *Brønsted-Lowry acid*; it can donate a proton to a water molecule:

$$H_2S(aq) + H_2O \rightleftharpoons HS^-(aq) + H_3O^+(aq)$$

Like ammonia, hydrogen sulfide can act as a precipitating agent toward metal cations (recall the discussion of qualitative analysis in Chap. 19). The reaction with Cd^{2+} is typical:

$$Cd^{2+}(aq) + H_2S(aq) \longrightarrow CdS(s) + 2H^+(aq)$$

Like ammonia, hydrogen sulfide (oxid. no. S = −2) can act only as a reducing agent when it takes part in redox reactions. Most often the H_2S is oxidized to elementary sulfur, as in the reaction

$$2H_2S(aq) + O_2(g) \longrightarrow 2S(s) + 2H_2O$$

The sulfur formed is often very finely dispersed, which explains why aqueous solutions of hydrogen sulfide in contact with air have a milky appearance.

If you've worked with H_2S in the laboratory, you won't soon forget its rotten-egg odor. In a sense, it's fortunate that hydrogen sulfide has such a distinctive odor. The gas is highly toxic, as poisonous as HCN. At a concentration of 10 parts per million, H_2S can cause headaches and nausea; at 100 ppm it can be fatal.

Example 20.3 When a solution containing Cu^{2+} is treated with hydrogen sulfide, a black precipitate forms. When another portion of the solution is treated with ammonia, a blue precipitate forms. This precipitate dissolves in excess ammonia to form a deep blue solution containing the $Cu(NH_3)_4{}^{2+}$ ion. Write balanced net ionic equations to explain these observations.

Strategy Before you can write the equations, you must identify the products. The black precipitate is CuS. The blue precipitate must be copper(II) hydroxide, $Cu(OH)_2$. The identity of the final product is given: $Cu(NH_3)_4{}^{2+}$. The equations are readily written, knowing the formulas of the products.

Solution

$$Cu^{2+}(aq) + H_2S(aq) \longrightarrow CuS(s) + 2H^+(aq)$$

$$Cu^{2+}(aq) + 2NH_3(aq) + 2H_2O \longrightarrow Cu(OH)_2(s) + 2NH_4{}^+(aq)$$

$$Cu(OH)_2(s) + 4NH_3(aq) \longrightarrow Cu(NH_3)_4{}^{2+}(aq) + 2\,OH^-(aq)$$

Hydrogen Peroxide

In hydrogen peroxide, oxygen has an oxidation number of -1, intermediate between the extremes for the element, 0 and -2. This means that H_2O_2 can act as either an oxidizing agent, in which case it is reduced to H_2O, or as a reducing agent, where it is oxidized to O_2. In practice, hydrogen peroxide is an extremely strong oxidizing agent:

$$H_2O_2(aq) + 2H^+(aq) + 2e^- \longrightarrow 2H_2O \quad E^\circ_{red} = +1.763\text{ V}$$

but a very weak reducing agent:

$$H_2O_2(aq) \longrightarrow O_2(g) + 2H^+(aq) + 2e^- \quad E^\circ_{ox} = -0.695\text{ V}$$

Hydrogen peroxide tends to decompose in water, which explains why its solutions soon lose their oxidizing power. The reaction involved is *disproportionation,* combining the two half-reactions referred to above:

$$\begin{array}{ll} H_2O_2(aq) + 2H^+(aq) + 2e^- \longrightarrow 2H_2O & E^\circ_{red} = +1.763\text{ V} \\ H_2O_2(aq) \longrightarrow O_2(g) + 2H^+(aq) + 2e^- & E^\circ_{ox} = -0.695\text{ V} \\ \hline 2H_2O_2(aq) \longrightarrow O_2(g) + 2H_2O & E^\circ = +1.068\text{ V} \end{array}$$

H_2O_2 is stored in brown bottles to prevent this reaction

This reaction is catalyzed by a wide variety of materials, including I^- ions, MnO_2, and metal surfaces (Pt, Ag), and even by traces of OH^- ions dissolved from glass.

You are most likely to come across hydrogen peroxide as its water solution. Two concentrations are available; one of these, containing 3 mass percent

H_2O_2, is sold in drugstores. The other solution contains 30 mass percent H_2O_2. Both solutions contain stabilizers to prevent disproportionation from taking place during storage. Hydrogen peroxide is used as a disinfectant (cuts, sore throats) or as a bleach (cloth, paper, hair, etc.).

Hydrogen Fluoride and Hydrogen Chloride

The most common hydrogen halides are HF (annual U.S. production = 3×10^8 kg) and HCl (3×10^9 kg/yr). They are most familiar as water solutions, referred to as hydrofluoric acid and hydrochloric acid, respectively. Recall (Chap. 13) that hydrofluoric acid is weak, incompletely dissociated in water, while HCl is a strong acid.

$$HF(aq) \rightleftharpoons H^+(aq) + F^-(aq) \qquad K_a = 6.9 \times 10^{-4}$$

$$HCl(aq) \longrightarrow H^+(aq) + Cl^-(aq) \qquad K_a \longrightarrow \infty$$

Hydrofluoric and hydrochloric acids undergo very similar reactions with bases such as OH^- or CO_3^{2-} ions. The equations for these reactions look somewhat different because of the difference in acid strength. Thus, for the reaction of hydrochloric acid with a solution of sodium hydroxide, the equation is simply

$$H^+(aq) + OH^-(aq) \longrightarrow H_2O$$

With hydrofluoric acid, the equation is

$$HF(aq) + OH^-(aq) \longrightarrow H_2O + F^-(aq)$$

HF appears in this equation because hydrofluoric acid is weak, containing many more HF molecules than H^+ ions. Similarly, for the reaction of these two acids with a solution of sodium carbonate:

$$\text{hydrochloric acid: } 2H^+(aq) + CO_3^{2-}(aq) \longrightarrow CO_2(g) + H_2O$$

$$\text{hydrofluoric acid: } 2HF(aq) + CO_3^{2-}(aq) \longrightarrow CO_2(g) + H_2O + 2F^-(aq)$$

Hydrochloric acid contains H^+ and Cl^- ions; hydrofluoric acid, mostly HF molecules

Concentrated hydrofluoric acid reacts with glass, which can be considered to be a mixture of SiO_2 and ionic silicates such as calcium silicate, $CaSiO_3$:

$$SiO_2(s) + 4HF(aq) \longrightarrow SiF_4(g) + 2H_2O$$

$$CaSiO_3(s) + 6HF(aq) \longrightarrow SiF_4(g) + CaF_2(s) + 3H_2O$$

As you might guess, HF solutions are never stored in glass bottles; plastic is used instead. Hydrofluoric acid is sometimes used to etch glass. The glass object is first covered with a thin protective coating of wax or plastic. Then the coating is removed from the area to be etched and the glass is exposed to the HF solution. Thermometer stems and burets can be etched or light bulbs frosted in this way.

Hydrogen fluoride is a very unpleasant chemical to work with. If spilled on the skin, it removes Ca^{2+} ions from the tissues, forming insoluble CaF_2. A white patch forms which is agonizingly painful to the touch. To make matters worse, HF is a local anaesthetic, so a person may be unaware of what's happening until it's too late.

You won't encounter HF, HBr, or HI in the general chem lab, but you use HCl frequently

Example 20.4 Write balanced equations to explain why

(a) aluminum hydroxide dissolves in hydrochloric acid.
(b) carbon dioxide gas is evolved when calcium carbonate is treated with hydrochloric acid.
(c) carbon dioxide gas is evolved when calcium carbonate is treated with hydrofluoric acid.

Strategy In each case, decide upon the nature of the products. You can reason by analogy with the equations cited in the text for the reactions with OH^- and CO_3^{2-} ions in solution.

Solution

(a) $Al(OH)_3(s) + 3H^+(aq) \longrightarrow Al^{3+}(aq) + 3H_2O$

(b) $CaCO_3(s) + 2H^+(aq) \longrightarrow Ca^{2+}(aq) + CO_2(g) + H_2O$

(c) $CaCO_3(s) + 2HF(aq) \longrightarrow Ca^{2+}(aq) + CO_2(g) + H_2O + 2F^-(aq)$

20.3 OXYGEN COMPOUNDS OF NONMETALS

Table 20.3 lists some of the more familiar nonmetal oxides. Curiously enough, only 5 of the 21 compounds shown are thermodynamically stable at 25°C and 1 atm (P_4O_{10}, P_4O_6, SO_3, SO_2, I_2O_5). The others, including all of the oxides of nitrogen and chlorine, have positive free energies of formation at 25°C and 1 atm. For example, $\Delta G_f^\circ NO_2(g) = +51.3$ kJ; $\Delta G_f^\circ ClO_2(g) = +120.6$ kJ. Kinetically, these compounds stay around long enough to have an extensive chemistry. Nitrogen dioxide, a reddish-brown gas, is a major factor in the formation of photochemical smog (Fig. 20.9). Chlorine dioxide, a yellow gas, is widely used as an industrial bleach and water purifier, even though it tends to explode at partial pressures higher than 50 mm Hg.

All such compounds are potentially unstable

Table 20.3 Nonmetal Oxides*

Group 15	Group 16	Group 17
$N_2O_5(s)$, $N_2O_4(g)$, $N_2O_3(d)$ **$N_2O(g)$, $NO_2(g)$, $NO(g)$**		$OF_2(g)$, $O_2F_2(g)$
$P_4O_{10}(s)$, $P_4O_6(s)$	**$SO_3(l)$, $SO_2(g)$**	$Cl_2O_7(l)$, $Cl_2O_6(l)$ $Cl_2O(g)$, $ClO_2(g)$ $BrO_2(d)$, $Br_2O(d)$ $I_4O_9(s)$, $I_2O_5(s)$, $I_2O_4(s)$

* The states listed are those observed at 25°C and 1 atm. Compounds that decompose below 25°C are listed as (*d*). Oxides shown in boldface are discussed in the text.

Molecular Structures of Nonmetal Oxides

The Lewis structures of the oxides of nitrogen are shown in Figure 20.10. Two of these species, NO and NO_2, are paramagnetic, with one unpaired electron. When nitrogen dioxide is cooled, it dimerizes; the unpaired electrons combine to form a single bond between the two nitrogen atoms:

$$2NO_2(g) \rightleftharpoons N_2O_4(g)$$

A similar reaction occurs when an equimolar mixture of NO and NO_2 is cooled. Two odd electrons, one from each molecule, pair off to form an N—N bond:

$$NO_2(g) + NO(g) \rightleftharpoons N_2O_3(g)$$

At −20°C, dinitrogen trioxide separates from the mixture as a blue liquid.

Perhaps the best-known oxide of nitrogen is N_2O, commonly called nitrous oxide or "laughing gas." Nitrous oxide is frequently used as an anaesthetic, particularly in dentistry. It is also the propellant gas used in whipped cream containers; N_2O is nontoxic, virtually tasteless, and quite soluble in vegetable oils. The N_2O molecule, like all those in Figure 20.10, can be represented as a resonance hybrid.

Figure 20.9
The brown haze covering this city is pollution caused by NO_2. (National Center for Atmospheric Research)

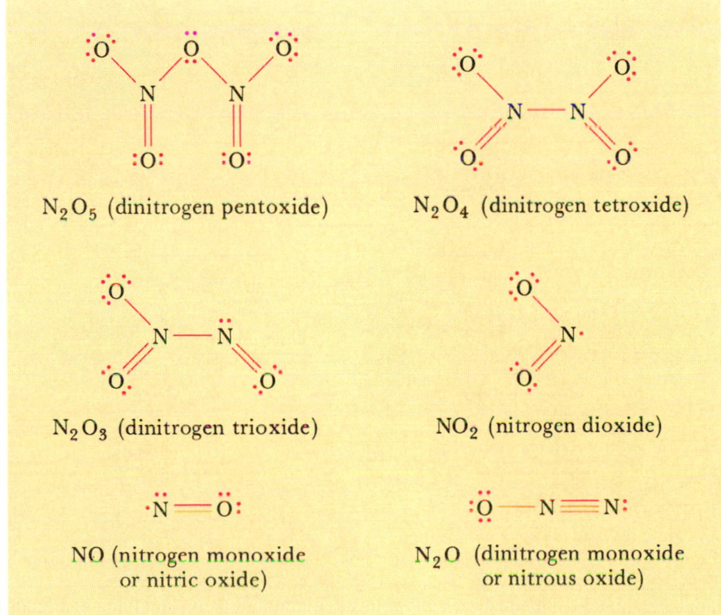

Figure 20.10
Lewis structures of the oxides of nitrogen. Many other resonance forms are possible.

Example 20.5 Consider the N_2O molecule shown in Figure 20.10.

(a) Draw another resonance form of N_2O.
(b) What is the bond angle in N_2O?
(c) Is the N_2O molecule polar or nonpolar?

Strategy Recall the discussion in Chapter 7, Sections 7.1–7.3, where the principles of resonance, molecular geometry, and polarity were considered.

Solution

(a) $:\ddot{O}=N=\ddot{N}:$ or $:O\equiv N-\underset{\cdot\cdot}{\ddot{N}}:$

(b) In any of the resonance forms, the central nitrogen atom, insofar as geometry is concerned, would behave as if it were surrounded by two electron pairs.

The bond angle is 180°; the molecule is linear, like BeF_2.

(c) Polar (unsymmetrical).

The structures of SO_2 and SO_3 were referred to in Chapter 7. These molecules are often cited as examples of resonance; sulfur trioxide, for example, has three equivalent resonance structures:

$$O=S(-O)(-O) \longleftrightarrow O-S(=O)(-O) \longleftrightarrow O-S(-O)(=O)$$

As you might expect, the SO_3 molecule is nonpolar with 120° bond angles.

Of the two oxides of phosphorus, P_4O_6 and P_4O_{10}, the latter is the more stable; it is formed when white phosphorus burns in air:

In a limited supply of air, some P_4O_6 is formed

$$P_4(s) + 5\,O_2(g) \longrightarrow P_4O_{10}(s)$$

Note from Figure 20.11 that in both oxides, as in P_4 itself, the four phosphorus atoms are at the corners of a tetrahedron. The P_4O_6 molecule can be visualized as derived from P_4 by inserting an oxygen atom between each pair of phosphorus atoms. In P_4O_{10}, an extra oxygen atom is bonded to each phosphorus.

Reactions of Nonmetal Oxides with Water

Many nonmetal oxides react with water to form acids. Compounds which behave in this way are referred to as **acid anhydrides**. Looking at the reaction

$$SO_3(g) + H_2O(l) \longrightarrow H_2SO_4(l)$$

you can see that sulfur trioxide is the acid anhydride of sulfuric acid. Notice that, in this reaction, the nonmetal does not change oxidation number; sulfur is in the +6 state in both SO_3 and H_2SO_4. Other acid anhydrides include N_2O_5 and N_2O_3:

What is the acid derived from CO_2? Ans: H_2CO_3

$$+5 \text{ nitrogen:} \quad N_2O_5(s) + H_2O(l) \longrightarrow 2HNO_3(l)$$

$$+3 \text{ nitrogen:} \quad N_2O_3(g) + H_2O(l) \longrightarrow 2HNO_2(aq)$$

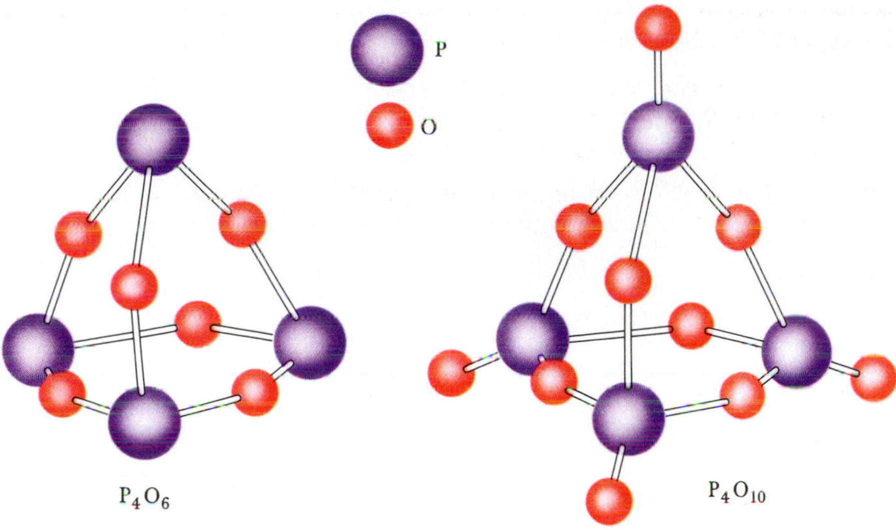

Figure 20.11
Structures of the oxides of phosphorus, simplest formulas P_2O_3 and P_2O_5.

The products here are nitric acid, HNO_3, and an aqueous solution of nitrous acid, HNO_2.

One of the most important reactions of this type involves the +5 oxide of phosphorus, P_4O_{10}. Here the product is phosphoric acid, H_3PO_4:

$$P_4O_{10}(s) + 6H_2O(l) \longrightarrow 4H_3PO_4(s)$$

This reaction is used to prepare high-purity phosphoric acid and salts of that acid for use in food products. Phosphoric acid, H_3PO_4, is added to cola and root beer to give them a tart taste. Much larger amounts of phosphoric acid are prepared by the reaction of calcium phosphate, $Ca_3(PO_4)_2$, with sulfuric acid. The phosphoric acid made this way is highly impure but costs only half as much as that from P_4O_{10}; its major use is in the manufacture of fertilizers.

Example 20.6 Give the formula of the acid anhydride of

(a) H_3PO_3 (b) HClO (c) H_2SO_3

Strategy Referring to Table 20.3, find an oxide in which the central nonmetal atom has the same oxidation number as in the acid.

Solution

(a) +3 P: H_3PO_3 and P_4O_6

(b) +1 Cl: HClO and Cl_2O

(c) +4 S: H_2SO_3 and SO_2

Table 20.4 Oxoacids of the Nonmetals

Group 15	Group 16	Group 17
HNO_3, HNO_2*		
H_3PO_4, H_3PO_3	H_2SO_4, H_2SO_3*	$HClO_4$, $HClO_3$*, $HClO_2$*, $HClO$*
		$HBrO_4$*, $HBrO_3$*, $HBrO$*
		HIO_4, H_5IO_6, HIO_3, HIO*

* These compounds cannot be isolated from water solution.

20.4 OXOACIDS AND OXOANIONS

Table 20.4 lists some of the more important oxoacids of the nonmetals. In all of these compounds, the ionizable hydrogen atoms are bonded to oxygen, not to the central nonmetal atom. Dissociation of one or more protons from the oxoacid gives the corresponding oxoanion (Fig. 20.12).

In this section, we discuss the principles that allow you to predict the relative acid strengths of oxoacids such as those listed in Table 20.4. Then we consider their strengths as oxidizing and/or reducing agents. Finally, we take a closer look at the chemistry of the two most familiar oxoacids, HNO_3 and H_2SO_4.

Acid Strength

The dissociation constants of the oxoacids of the halogens are listed in Table 20.5. Notice that the value of the dissociation constant, K_a, increases with

— *increasing oxidation number of the central atom* ($HClO < HClO_2 < HClO_3$)
— *increasing electronegativity of the central atom* ($HIO < HBrO < HClO$)

These trends are general ones, observed with other oxoacids of the nonmetals. Recall, for example, that nitric acid, HNO_3 (oxid. no. N = +5), is a

Table 20.5 Dissociation Constants of Oxoacids of the Halogens

Oxid. State		K_a		K_a		K_a
+7	$HClO_4$	$\sim 10^7$	$HBrO_4$	$\sim 10^6$	HIO_4*	1.4×10^1
+5	$HClO_3$	$\sim 10^3$	$HBrO_3$	3.0	HIO_3	1.6×10^{-1}
+3	$HClO_2$	1.0×10^{-2}				
+1	$HClO$	2.8×10^{-8}	$HBrO$	2.6×10^{-9}	HIO	2.4×10^{-11}

* Estimated; in water solution the stable species is H_5IO_6, whose first dissociation constant is 5×10^{-4}.

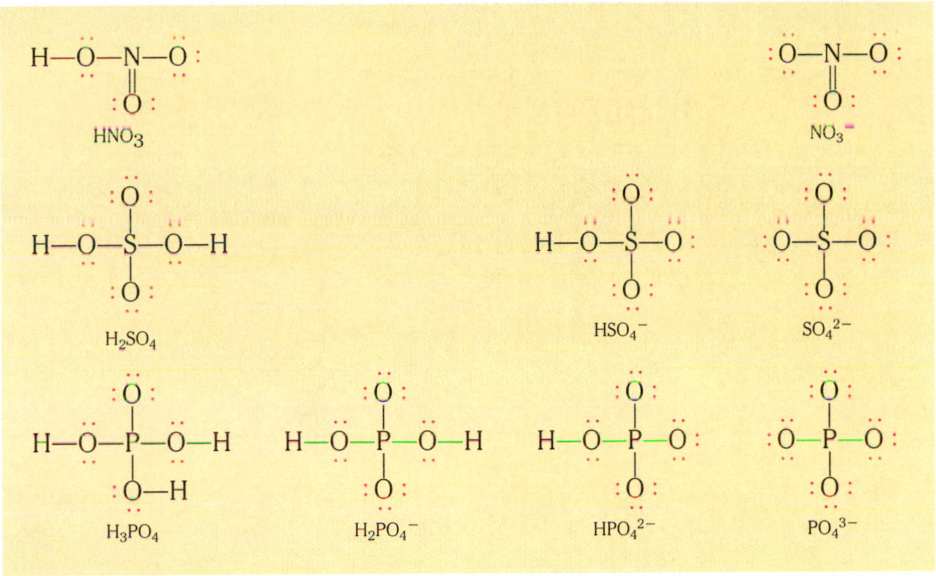

HNO_3 NO_3^-

H_2SO_4 HSO_4^- SO_4^{2-}

H_3PO_4 $H_2PO_4^-$ HPO_4^{2-} PO_4^{3-}

Figure 20.12
Lewis structures of the oxoacids HNO_3, H_2SO_4, H_3PO_4 and the oxoanions derived from them.

strong acid, completely dissociated in water. In contrast, nitrous acid, HNO_2 (oxid. no. N = +3) is a weak acid ($K_a = 6.0 \times 10^{-4}$). The electronegativity effect shows up with the strengths of the oxoacids of sulfur and selenium:

K_{a1} $H_2SO_3 = 1.7 \times 10^{-2}$ K_{a1} $H_2SeO_3 = 2.7 \times 10^{-3}$ (E.N. S = 2.6, Se = 2.5)

E.N. = electronegativity

Trends in acid strength can be explained in terms of molecular structure. In an oxoacid molecule, the hydrogen atom that dissociates is bonded to oxygen, which in turn is bonded to a nonmetal atom, X. The dissociation in water of an oxoacid H—O—X can be represented as

$$H{-}O{-}X(aq) \rightleftharpoons H^+(aq) + XO^-(aq)$$

For a proton, with its +1 charge, to separate from the molecule, the electron density around the oxygen should be as low as possible. This will weaken the O—H bond and favor dissociation. The electron density around the oxygen atom is decreased when

— *X is a highly electronegative atom such as Cl.* This draws electrons away from the oxygen atom and makes hypochlorous acid stronger than hypoiodous acid.

— *Additional, strongly electronegative oxygen atoms are bonded to X.* These tend to draw electrons away from the oxygen atom bonded to H. Thus, we would predict that the ease of dissociation of a proton, and hence K_a, should increase in the following order, from left to right:

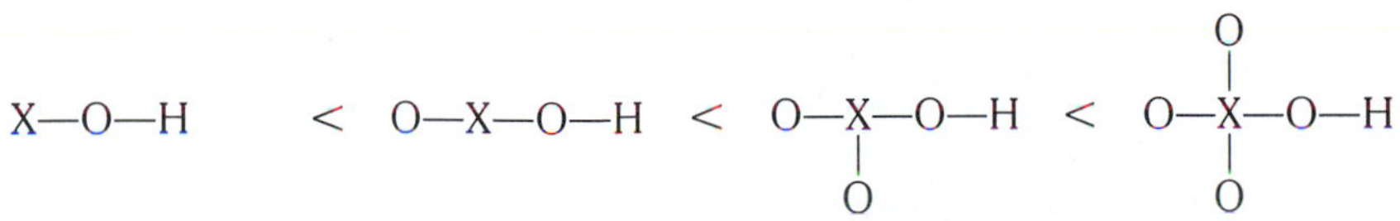

X—O—H < O—X—O—H < O—X(—O)—O—H < O—X(—O)(—O)—O—H

oxid. no. X = +1   +3   +5   +7

Example 20.7 Consider sulfurous acid, H_2SO_3.

(a) Show its Lewis structure and that of the HSO_3^- and SO_3^{2-} ions.
(b) How would its acid strength compare to that of H_2SO_4? H_2TeO_3?

Strategy The structure can be obtained by removing an oxygen atom from H_2SO_4 (Fig. 20.12). Relative acid strengths can be predicted on the basis of the electronegativity and oxidation number of the central nonmetal atom, following the rules cited above.

Solution

(a) H—Ö—S̈—Ö—H with :Ö: below S (sulfurous acid); (H—Ö—S̈—Ö:)$^-$ with :Ö: below S (hydrogen sulfite ion); (:Ö—S̈—Ö:)$^{2-}$ with :Ö: below S (sulfite ion)

(b) $H_2SO_3 < H_2SO_4$ (oxid. no. S = +4, +6)

$H_2SO_3 > H_2TeO_3$ (S more electronegative than Te)

Oxidizing and Reducing Strength

Many of the reactions of oxoacids and oxoanions involve oxidation and reduction. There are certain general principles that apply here, regardless of the particular species involved.

Many oxoanions are very powerful oxidizing agents

1. *A species in which a nonmetal is in its highest oxidation state can act only as an oxidizing agent, never as a reducing agent.* Consider, for example, the ClO_4^- ion, in which chlorine is in its highest oxidation state, +7. In any redox reaction in which this ion takes part, chlorine must be reduced to a lower oxidation state. When that happens, the ClO_4^- ions act as an oxidizing agent, taking electrons away from something else. The same argument applies to

— the NO_3^- ion (highest oxid. no. N = +5)
— the SO_4^{2-} ion (highest oxid. no. S = +6)

2. *A species in which a nonmetal is in an intermediate oxidation state can act as either an oxidizing agent or a reducing agent.* Consider, for example, the ClO_3^- ion (oxid. no. Cl = +5). It can be oxidized to the perchlorate ion, in which case ClO_3^- acts as a reducing agent:

$$ClO_3^-(aq) + H_2O \longrightarrow ClO_4^-(aq) + 2H^+(aq) + 2e^- \qquad E^\circ_{ox} = -1.226\ V$$

Alternatively, the ClO_3^- ion can be reduced, perhaps to a Cl^- ion. When that occurs, ClO_3^- acts as an oxidizing agent:

$$ClO_3^-(aq) + 6H^+(aq) + 6e^- \longrightarrow Cl^-(aq) + 3H_2O \qquad E^\circ_{red} = +1.442\ V$$

3. *Sometimes, with a species such as ClO_3^-, oxidation and reduction occur together, resulting in disproportionation:*

$$4ClO_3^-(aq) \longrightarrow 3ClO_4^-(aq) + Cl^-(aq) \qquad E^\circ = +0.216\ V$$

In general, a *species in an intermediate oxidation state is expected to disproportionate if the sum $E^\circ_{ox} + E^\circ_{red}$ is a positive number.*

4. *The oxidizing strength of an oxoacid or oxoanion is greatest at high $[H^+]$ (low pH). Conversely, its reducing strength is greatest at low $[H^+]$ (high pH).*

Any oxoanion is a stronger oxidizing agent in acidic solution

This principle has a simple explanation. Looking back at the half-equations written on p. 566, you can see that

— when ClO_3^- acts as an oxidizing agent, the H^+ ion is a reactant, so increasing its concentration makes the process more spontaneous.
— when ClO_3^- acts as a reducing agent, the H^+ ion is a product; to make the process more spontaneous, $[H^+]$ should be lowered.

Example 20.8 Calculate E_{red} and E_{ox} for the ClO_3^- ion in neutral solution, at pH 7.00, assuming all other species are at standard concentrations (E°_{red} = +1.442 V; E°_{ox} = −1.226 V). Will the ClO_3^- ion disproportionate at pH 7.00?

Strategy First (1), set up the Nernst equation for the reduction half-reaction and calculate E_{red}. (It's convenient here to use the base 10 form of the Nernst equation, since pH is involved.) Then (2), repeat the calculation for the oxidation half-reaction, finding E_{ox}. Finally (3), add $E_{red} + E_{ox}$; if the sum is positive, disproportionation should occur.

Solution

(1) $ClO_3^-(aq) + 6H^+(aq) + 6e^- \longrightarrow Cl^-(aq) + 3H_2O$

$$E_{red} = +1.442\text{ V} - \frac{0.0592}{6}\log\frac{1}{[H^+]^6}$$

$$= +1.442\text{ V} + 0.0592\log[H^+]$$

$$= +1.442\text{ V} - 0.0592(\text{pH}) = +1.028\text{ V}$$

$$\log\frac{1}{[H^+]^6} = -6\log[H^+]$$

As expected, decreasing the concentration of H^+ makes the half-reaction less spontaneous and hence makes E_{red} less positive.

(2) $ClO_3^-(aq) + H_2O \longrightarrow ClO_4^-(aq) + 2H^+(aq) + 2e^-$

$$E_{ox} = -1.226\text{ V} - \frac{0.0592}{2}\log[H^+]^2$$

$$= -1.226\text{ V} - 0.0592\log[H^+]$$

$$= -1.226\text{ V} + 0.0592(\text{pH}) = -0.822\text{ V}$$

(3) $E = +1.028\text{ V} - 0.822\text{ V} = +0.216\text{ V}$; disproportionation should occur.

Notice that E is the same as E°, +0.216 V. You could have predicted that (and saved a lot of work!), since the overall equation for disproportionation

$$4ClO_3^-(aq) \longrightarrow 3ClO_4^-(aq) + Cl^-(aq)$$

does not involve H^+ or OH^- ions.

Nitric Acid, HNO_3

Nitric acid is a strong acid, completely dissociated to H^+ and NO_3^- ions in dilute water solution:

$$HNO_3(aq) \longrightarrow H^+(aq) + NO_3^-(aq)$$

Many of the reactions of nitric acid are those associated with all strong acids. For example, dilute (6 *M*) nitric acid can be used to dissolve aluminum hydroxide:

$$Al(OH)_3(s) + 3H^+(aq) \longrightarrow Al^{3+}(aq) + 3H_2O$$

or to generate carbon dioxide gas from calcium carbonate:

$$CaCO_3(s) + 2H^+(aq) \longrightarrow Ca^{2+}(aq) + CO_2(g) + H_2O$$

Referring back to Example 20.4, you will find that these equations are identical with those written for the reactions of hydrochloric acid with $Al(OH)_3$ and $CaCO_3$. It is the H^+ ion that reacts in either case: Cl^- and NO_3^- ions take no part in the reactions and hence do not appear in the equation.

Concentrated (16 *M*) nitric acid is a strong oxidizing agent; here the nitrate ion is reduced to nitrogen dioxide. This happens when 16 *M* HNO_3 reacts with copper metal (Fig. 20.13):

$$Cu(s) + 4H^+(aq) + 2NO_3^-(aq) \longrightarrow Cu^{2+}(aq) + 2NO_2(g) + 2H_2O$$

Dilute nitric acid (6 *M*) is a weaker oxidizing agent than 16 *M* HNO_3. It also gives a wider variety of reduction products, depending upon the nature of the reducing agent. With inactive metals such as copper ($E^\circ_{ox} = -0.339$ V), the major product is usually NO (oxid. no. N = +2):

$$3Cu(s) + 2NO_3^-(aq) + 8H^+(aq) \longrightarrow 3Cu^{2+}(aq) + 2NO(g) + 4H_2O$$

With very dilute acid and a strong reducing agent such as zinc ($E^\circ_{ox} = +0.762$ V) reduction may go all the way to the NH_4^+ ion (oxid. no. N = −3):

$$4Zn(s) + NO_3^-(aq) + 10H^+(aq) \longrightarrow 4Zn^{2+}(aq) + NH_4^+(aq) + 3H_2O$$

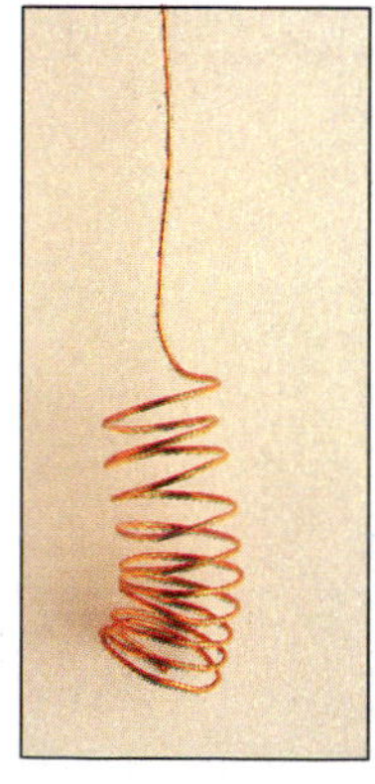

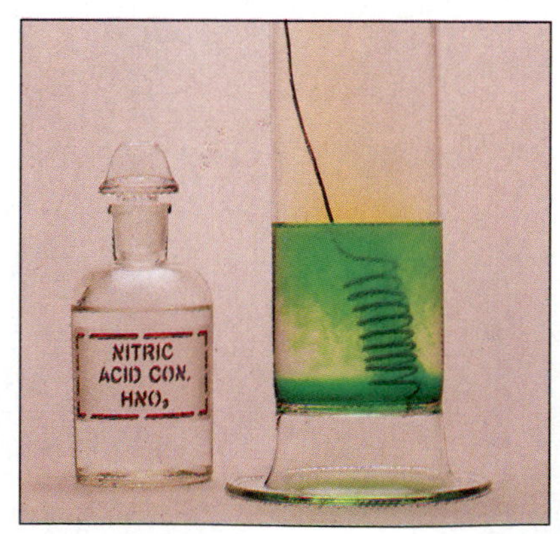

Figure 20.13
Copper metal is comparatively inactive, but it reacts with concentrated nitric acid. The brown fumes are $NO_2(g)$, a reduction product of HNO_3. The copper is oxidized to Cu^{2+} ions, which impart their color to the solution. (Marna G. Clarke)

Example 20.9 Write a balanced net ionic equation for the reaction of nitric acid with insoluble copper(II) sulfide; the products include Cu^{2+}, S(*s*), and $NO_2(g)$.

Strategy Follow the general procedure for writing and balancing redox equations in Chapter 4 and reviewed in Chapter 17. Note that since nitric acid is strong, it should be represented as H^+ and NO_3^- ions.

Solution

(1) The "skeleton" half-equations are

$$\text{oxidation:} \quad CuS(s) \longrightarrow S(s)$$

$$\text{reduction:} \quad NO_3^-(aq) \longrightarrow NO_2(g)$$

(2) The balanced half-equations are

$$\text{oxidation:} \quad CuS(s) \longrightarrow Cu^{2+}(aq) + S(s) + 2e^-$$

$$\text{reduction:} \quad NO_3^-(aq) + 2H^+(aq) + e^- \longrightarrow NO_2(g) + H_2O$$

(3) Multiplying the reduction half-equation by two and adding to the oxidation half-equation, you should obtain, after simplification,

$$CuS(s) + 2NO_3^-(aq) + 4H^+(aq) \longrightarrow Cu^{2+}(aq) + S(s) + 2NO_2(g) + 2H_2O$$

Concentrated nitric acid (16 *M*) is colorless when pure. In sunlight, it turns yellow (Fig. 20.14) because it decomposes to $NO_2(g)$:

$$4HNO_3(aq) \longrightarrow 4NO_2(g) + 2H_2O + O_2(g)$$

You can always smell NO_2 over 16 *M* HNO_3

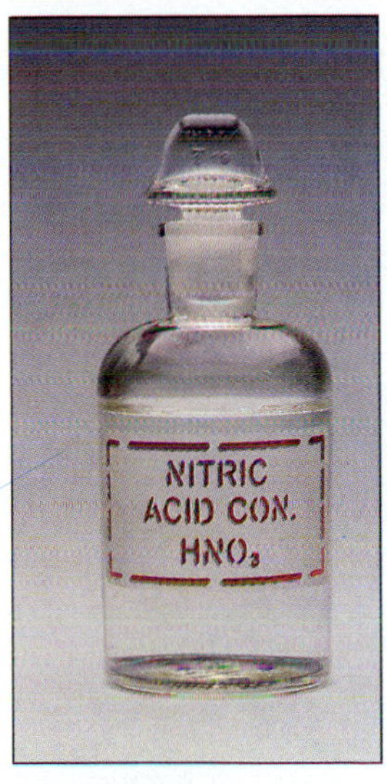

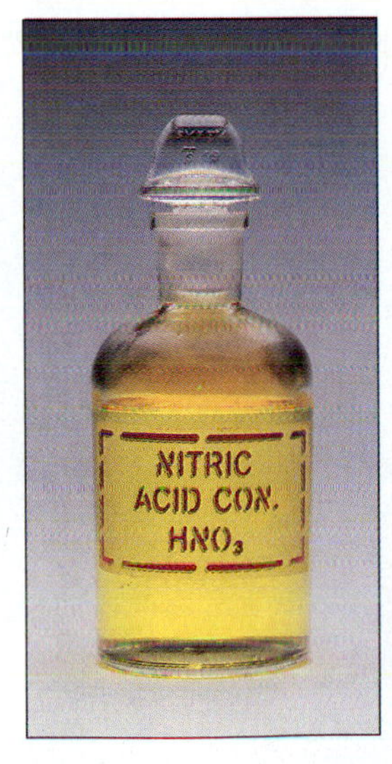

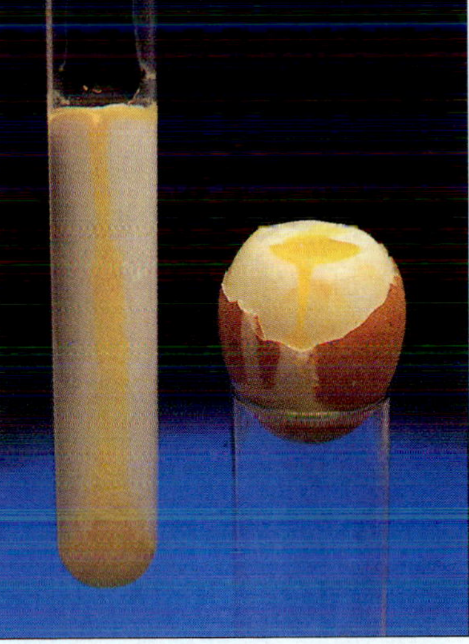

Figure 20.14
A water solution of nitric acid slowly turns yellow because of the $NO_2(g)$ formed by decomposition. Nitric acid also reacts with proteins (casein in milk (test tube) and albumen in eggs) to give a characteristic yellow color. (a,b, Marna G. Clarke; c, Charles D. Winters)

The yellow color that appears on your skin if it comes in contact with nitric acid has quite a different explanation. Nitric acid reacts with proteins to give a yellow material called xanthoprotein.

Sulfuric Acid, H_2SO_4

You will recall from Chapter 4 that H_2SO_4 is a strong acid, completely dissociated to H^+ and HSO_4^- ions in dilute water solution. The HSO_4^- ion dissociates further to give H^+ and SO_4^{2-} ions:

$$H_2SO_4(aq) \longrightarrow H^+(aq) + HSO_4^-(aq) \qquad K_a \longrightarrow \infty$$

$$HSO_4^-(aq) \rightleftharpoons H^+(aq) + SO_4^{2-}(aq) \qquad K_a = 1.0 \times 10^{-2}$$

The dissociation constant of the HSO_4^- ion is relatively large. This explains why in writing equations for the reactions of sulfuric acid, we often consider it to consist of $2H^+(aq) + SO_4^{2-}(aq)$.

Sulfuric acid is a relatively weak oxidizing agent; most often it is reduced to sulfur dioxide:

$$SO_4^{2-}(aq) + 4H^+(aq) + 2e^- \longrightarrow SO_2(g) + 2H_2O \qquad E^\circ_{red} = +0.155\text{ V}$$

This is the case when copper metal is oxidized by hot concentrated sulfuric acid:

$$Cu(s) + 4H^+(aq) + SO_4^{2-}(aq) \longrightarrow Cu^{2+}(aq) + 2H_2O + SO_2(g)$$

When dilute sulfuric acid (3 M) reacts with metals, it is ordinarily the H^+ ion rather than the SO_4^{2-} ion which is reduced. For example, zinc reacts with dilute sulfuric acid to form hydrogen gas:

$$Zn(s) + 2H^+(aq) \longrightarrow Zn^{2+}(aq) + H_2(g)$$

Concentrated sulfuric acid, in addition to being an acid and an oxidizing agent, is also a dehydrating agent. Small amounts of water can be removed

Figure 20.15
When sulfuric acid is added to sugar, sucrose, an exothermic reaction occurs. The elements of water are removed from the sugar, $C_{12}H_{22}O_{11}$, leaving a mass of black carbon. (Charles D. Winters)

from organic liquids such as gasoline by extraction with sulfuric acid. Sometimes it is even possible to remove the elements of water from a compound by treating it with 18 *M* H_2SO_4. This happens with table sugar, $C_{12}H_{22}O_{11}$; the product is a black char that is mostly carbon (Fig. 20.15).

$$C_{12}H_{22}O_{11}(s) \longrightarrow 12C(s) + 11H_2O(l)$$

When concentrated sulfuric acid dissolves in water, a great deal of heat is given off, nearly 100 kJ per mole of H_2SO_4. Sometimes enough heat is evolved to bring the solution to the boiling point. To prevent this and to avoid splattering, the acid should be added slowly to water, with constant stirring. If it comes in contact with the skin, concentrated sulfuric acid can cause painful chemical burns.

Never add water to conc. H_2SO_4

PERSPECTIVE • Essential Nonmetals

The ten nonmetals which are essential in human nutrition can be divided into three categories:

- the four elements oxygen, carbon, hydrogen, and nitrogen together make up about 96% of body weight (65% for O, 18% for C, 10% for H, 3% for N). The first three of these elements are present in all foods; nitrogen occurs in all proteins. No one worries about deficiencies of these elements.
- the three elements *phosphorus, sulfur,* and *chlorine* are classified as "major elements"; they make up 1.0%, 0.25%, and 0.15% of body weight, respectively. Most of the phosphorus is found in bones and teeth; smaller amounts are present in DNA, the genetic code material found in every body cell. Sulfur is contained in all proteins, especially those making up skin, hair, and nails. Chlorine is present as Cl^- ions in body fluids; hydrochloric acid is secreted in the stomach during digestion. Deficiencies of these three elements are virtually unknown.
- the three elements **fluorine**, **iodine**, and **selenium** (0.0043%, 0.00004%, and 0.00001% of body weight, respectively) are classified as trace elements.

One enzyme in the body that destroys harmful peroxides contains selenium. Moreover, there is considerable evidence that selenium compounds are anticarcinogens. For one thing, tests with laboratory animals show that the incidence and size of malignant tumors is reduced when a solution containing Na_2SeO_3 is injected at the part per million level. Beyond that evidence, statistical studies show an inverse correlation between selenium levels in the soil and the incidence of certain types of cancer.

Iodide ions concentrate in the thyroid gland, where they are essential for the synthesis of iodine-containing growth hormones such as thyroxine. Iodine deficiency leads to abnormal enlargement of the thyroid, a condition known as goiter. This disease has been virtually wiped out in the United States and Canada by the use of iodized salt (0.01% KI), but it is estimated that goiter still affects 200 million people worldwide.

The principal mineral component of bones and teeth is hydroxyapatite, whose formula may be written as $Ca(OH)_2 \cdot 3Ca_3(PO_4)_2$. An OH^- ion in this compound is readily replaced by a F^- ion, which has the same charge as OH^- and nearly the same size. The product is fluorapatite, $CaF_2 \cdot 3Ca_3(PO_4)_2$. On the average, about 2% of the OH^- ions in hydroxyapatite in bone and teeth are replaced by F^- ions. The fluoride ions come mostly from drinking water, where they occur naturally or are added at the 1 ppm level in the form of sodium fluoride. About half of the people in the United States and Canada drink fluoridated water. This has been shown to reduce the incidence of tooth decay in children by 50–70%. In areas where fluoridated water is not available, other approaches are possible. Most toothpastes contain a small amount of a metal fluoride like tin(II) fluoride, SnF_2, or sodium fluoride, NaF.

As you might expect, fluorapatite is much more resistant to attack by acid than is hydroxyapatite. The F^- ion is a much weaker base than the OH^- ion:

$$F^-(aq) + H^+(aq) \longrightarrow HF(aq) \qquad K = 1.4 \times 10^3$$

$$OH^-(aq) + H^+(aq) \longrightarrow H_2O \qquad K = 1.0 \times 10^{14}$$

This may explain why F^- ions are effective in preventing tooth decay. Acids produced by plaque bacteria are a major culprit in destroying tooth enamel. Interestingly enough, drinking fluoridated water has also been found to reduce the incidence of osteoporosis, a degenerative bone disease that affects many older people.

If the concentration of F^- ions in drinking water is too high, perhaps 5–10 ppm, children's permanent teeth develop with mottled enamel. Chalky white patches form along with yellowish stains. There is no structural damage but the stains are unsightly and difficult to remove. At concentrations above 50 ppm, the F^- ion is toxic.

CHAPTER HIGHLIGHTS

KEY CONCEPTS

1. *Carry out equilibrium calculations for solution reactions*
   (Example 20.1; Problems 47–52)
2. *Apply the Gibbs-Helmholtz equation*
   (Example 20.2; Problems 53–58)
3. *Write balanced equations for solution reactions*
   (Examples 20.3, 20.4, 20.9; Problems 15–26, 73)
4. *Draw Lewis structures for compounds of the nonmetals*
   (Examples 20.5, 20.7; Problems 31–38)
5. *Relate oxoacids to acid anhydrides and compare their acid strengths*
   (Examples 20.6, 20.7; Problems 7, 8, 71)
6. *Carry out electrochemical calculations involving $E°$, the Nernst equation, and/or electrolysis*
   (Example 20.8; Problems 59–66)

KEY TERMS

acid	E°_{ox}, E°_{red}	oxidizing agent
acid anhydride	electronegativity	oxoacid
allotrope	Lewis structure	oxoanion
base	nonmetal	reducing agent
disproportionation	oxidation number	resonance

SUMMARY PROBLEM

Consider bromine, whose chemistry is quite similar to that of chlorine.

a. Write equations for the reaction of bromine with I^- ions; for the reaction of Br_2 with water (disproportionation).
b. Write equations for the preparation of bromine by the electrolysis of aqueous NaBr; by the reaction of chlorine with bromide ions in aqueous solution.
c. Write equations for the reaction of hydrobromic acid with OH^- ions; with CO_3^{2-} ions; with ammonia.
d. Consider the species Br^-, Br_2, BrO^-, BrO_4^-. In a redox reaction, which of these species can act only as an oxidizing agent? only as a reducing agent? Which can act as either oxidizing or reducing agents?
e. When aqueous sodium bromide is heated with a concentrated solution of sulfuric acid, the products include liquid bromine and sulfur dioxide. Write a balanced equation for the redox reaction involved.
f. Write Lewis structures for the following species: Br^-, BrO^-, BrO_4^-. What is the bond angle in the BrO_4^- ion? For which of these ions is the conjugate acid the weakest?
g. Give the formula of the acid anhydride of HBrO. Knowing that K_a of HBrO is 2.6×10^{-9}, calculate the ratio [HBrO]/[BrO^-] at pH 10.00.
h. For the reduction of HBrO to Br^-, what is the change in voltage when the pH increases by one unit?

Answers

a. $Br_2(l) + 2I^-(aq) \longrightarrow 2Br^-(aq) + I_2(s)$
$Br_2(l) + H_2O \longrightarrow HBrO(aq) + H^+(aq) + Br^-(aq)$
b. $2Br^-(aq) + 2H_2O \longrightarrow Br_2(l) + H_2(g) + 2\,OH^-(aq)$
$2Br^-(aq) + Cl_2(aq) \longrightarrow Br_2(l) + 2Cl^-(aq)$
c. $H^+(aq) + OH^-(aq) \longrightarrow H_2O$; $2H^+(aq) + CO_3^{2-}(aq) \longrightarrow CO_2(g) + H_2O$
$H^+(aq) + NH_3(aq) \longrightarrow NH_4^+(aq)$
d. BrO_4^-; Br^-; Br_2, BrO^-
e. $2Br^-(aq) + SO_4^{2-}(aq) + 4H^+(aq) \longrightarrow Br_2(l) + SO_2(g) + 2H_2O$

f. [:B̈r:]⁻   [:B̈r—Ö:]⁻   [:Ö—Br(—Ö:)(—Ö:)—Ö:]⁻   109.5°; BrO^-

g. Br_2O; 0.038   **h.** −0.0296 V

QUESTIONS & PROBLEMS

Formulas, Equations, and Reactions

1. Name the following species.
a. HIO_4 **b.** BrO_2^- **c.** HIO **d.** $NaClO_3$

2. Name the following compounds.
a. $HBrO_3$ **b.** KIO **c.** $NaClO_2$ **d.** $NaBrO_4$

3. Write the formula for each of the following compounds.
a. chloric acid
b. periodic acid
c. hypobromous acid
d. hydriodic acid

4. Write the formula for each of the following compounds.
a. potassium bromite **b.** calcium bromide
c. sodium periodate **d.** magnesium hypochlorite

5. Write the formula of a compound of each of the following elements which *cannot* act as an oxidizing agent.
a. N **b.** S **c.** Cl

6. Write the formula of an oxoanion of each of the following elements which *cannot* act as a reducing agent.
a. N **b.** S **c.** Cl

7. Give the formula of the acid anhydride of
a. HNO_3 **b.** HNO_2 **c.** H_2SO_4

8. Write the formula of the acid formed when each of the following anhydrides reacts with water.
a. SO_2 **b.** Cl_2O **c.** P_4O_6

9. Write the formulas of the following compounds.
a. ammonia **b.** laughing gas
c. hydrogen perioxide **d.** sulfur trioxide

10. Write the formula for the following compounds.
a. sodium azide **b.** sulfurous acid
c. hydrazine **d.** sodium dihydrogen phosphate

11. Write the formula of a compound of hydrogen with
a. nitrogen which is a gas at 25°C and 1 atm.
b. phosphorus which is a liquid at 25°C and 1 atm.
c. oxygen which contains an O—O bond.

12. Write the formula of a compound of hydrogen with
a. sulfur.
b. nitrogen which is a liquid at 25°C and 1 atm.
c. phosphorus which is a poisonous gas at 25°C and 1 atm.

13. Give the formula of
a. an anion in which S has an oxidation number of −2.
b. two anions in which S has an oxidation number of +4.
c. two different acids of sulfur.

14. Give the formula of a compound of nitrogen that is
a. a weak base. **b.** a strong acid.
c. a weak acid. **d.** capable of oxidizing copper.

15. Write a balanced net ionic equation for
a. the electrolytic decomposition of hydrogen fluoride.
b. the oxidation of iodide ion to iodine by hydrogen peroxide in acidic solution. Hydrogen peroxide is reduced to water.

16. Write a balanced net ionic equation for
a. the oxidation of iodide to iodine by sulfate ion in acidic solution. Sulfur dioxide gas is also produced.
b. The preparation of iodine from an iodide salt and chlorine gas.

17. Write a balanced net ionic equation for the disproportionation reaction
a. of iodine to give iodate and iodide ions in basic solution.
b. of chlorine gas to chloride and perchlorate ions in basic solution.

18. Write a balanced net ionic equation for the disproportionation reaction of
a. hypochlorous acid to chlorine gas and chlorous acid in acidic solution.
b. chlorate ion to perchlorate and chlorite ions.

19. Complete and balance the following equation. If no reaction occurs, write NR.
a. $Cl_2(g) + I^-(aq) \longrightarrow$ **b.** $F_2(g) + Br^-(aq) \longrightarrow$
c. $I_2(s) + Cl^-(aq) \longrightarrow$ **d.** $Br_2(l) + I^-(aq) \longrightarrow$

20. Complete and balance the following equations. If no reaction occurs, write NR.
a. $Cl_2(g) + Br^-(aq) \longrightarrow$ **b.** $I_2(s) + Cl^-(aq) \longrightarrow$
c. $I_2(s) + Br^-(aq) \longrightarrow$ **d.** $Br_2(l) + Cl^-(aq) \longrightarrow$

21. Write a balanced equation for the preparation of
a. F_2 from HF. **b.** Br_2 from NaBr.
c. NH_4^+ from NH_3.

22. Write a balanced equation for the preparation of
a. N_2 from $Pb(N_3)_2$. **b.** O_2 from O_3.
c. S from H_2S.

23. Write a balanced equation for the reaction of ammonia with
a. Cu^{2+} **b.** H^+ **c.** Al^{3+}

24. Write a balanced equation for the reaction of hydrogen sulfide with
a. Cd^{2+} **b.** OH^- **c.** $O_2(g)$

25. Write a balanced net ionic equation for the reaction of nitric acid with
a. a solution of $Ca(OH)_2$
b. Ag(*s*); assume the nitrate ion is reduced to $NO_2(g)$.
c. Cd(*s*); assume the nitrate ion is reduced to $N_2(g)$.

26. Write a balanced net ionic equation for the reaction of sulfuric acid with
a. $CaCO_3(s)$.
b. a solution of NaOH.
c. Cu; assume the SO_4^{2-} ion is reduced to SO_2.

Allotropy

27. Which of the following elements show allotropy?
a. F **b.** O **c.** S **d.** Cl

28. Which of the following elements show allotropy?
a. Br **b.** N **c.** P **d.** I

29. Describe the structural difference between
a. ozone and ordinary oxygen.
b. white phosphorus and red phosphorus.
c. liquid sulfur at 120° and 180°.

30. Describe how
a. ozone can be prepared from O_2.
b. monoclinic sulfur can be prepared from rhombic sulfur.
c. red phosphorus can be made from white phosphorus.

Molecular Structure

31. Give the Lewis structure of
a. NO_2 **b.** NO **c.** SO_2 **d.** SO_3

32. Give the Lewis structure of
a. Cl_2O **b.** N_2O **c.** P_4 **d.** N_2

33. Which of the molecules in Question 31 are polar?
34. Which of the molecules in Question 32 are polar?
35. Give the Lewis structure of

a. HNO_3 **b.** H_2SO_4 **c.** H_3PO_4

36. Give the Lewis structures of the conjugate bases of the species in Question 35.
37. Give the Lewis structure of

a. the strongest oxoacid of bromine.
b. a hydride of nitrogen in which there is an —N—N— bond.
c. an acid added to cola drinks.

38. Give the Lewis structure of

a. an oxide of nitrogen in the +5 state.
b. the strongest oxoacid of nitrogen.
c. a tetrahedral oxoanion of sulfur.

Stoichiometry

39. The average concentration of bromine (as bromide) in seawater is 65 ppm. Calculate

a. the volume of seawater ($d = 64.0\ \text{lb/ft}^3$) in cubic feet required to produce one kilogram of liquid bromine.
b. the volume of chlorine gas in liters, measured at 20°C and 762 mm Hg, required to react with this volume of seawater.

40. A 425-gallon tank is filled with water containing 175 g sodium iodide. How many liters of chlorine gas at 758 mm Hg and 25°C will be required to oxidize all the iodide to iodine?
41. Iodine can be prepared by allowing an aqueous solution of hydrogen iodide to react with manganese dioxide, MnO_2. The reaction is

$$2I^-(aq) + 4H^+(aq) + MnO_2(s) \longrightarrow Mn^{2+}(aq) + 2H_2O + I_2(s)$$

If an excess of hydrogen iodide is added to 0.200 g MnO_2, how many grams of iodine are obtained, assuming 100% yield?
42. When a solution of hydrogen bromide is prepared, 1.283 L of HBr gas at 25°C and 0.974 atm is bubbled into 250.0 mL of water. Assuming all the HBr dissolves with no volume change, what is the molarity of the hydrobromic acid solution produced?
43. When ammonium nitrate explodes, nitrogen, steam, and oxygen gas are produced. If the explosion is carried out by heating one kilogram of ammonium nitrate sealed in a rigid bomb with a volume of one liter, what is the total pressure produced by the gases before the bomb ruptures? Assume the reaction goes to completion and the final temperature is 500°C.
44. Sulfur dioxide can be removed from the smokestack emissions of power plants by reacting it with hydrogen sulfide, producing sulfur and water. What volume of hydrogen sulfide at 27°C and 755 mm Hg is required to remove the sulfur dioxide produced by a power plant that burns one metric ton of coal containing 5.0% sulfur by mass? How many grams of sulfur are produced by the reaction of H_2S with SO_2?
45. A 1.500-g sample containing sodium nitrate was heated to form $NaNO_2$ and O_2. The oxygen evolved was collected over water at 23°C and 752 mm Hg; its volume was 125.0 mL. Calculate the percentage of $NaNO_3$ in the sample. The vapor pressure of water at 23°C is 21.07 mm Hg.
46. Chlorine can remove the foul smell of H_2S in water. The reaction is

$$H_2S(aq) + Cl_2(aq) \longrightarrow 2H^+(aq) + 2Cl^-(aq) + S(s)$$

If the contaminated water has 5.0 ppm hydrogen sulfide by mass, what volume of chlorine gas at STP is required to remove all the H_2S from 1.00×10^3 gallons of water ($d = 1.00$ g/mL)? What is the pH of the solution after treatment with chlorine?

Equilibria

47. The equilibrium constant at 25°C for the reaction

$$Br_2(l) + H_2O \rightleftharpoons H^+(aq) + Br^-(aq) + HBrO(aq)$$

is 1.2×10^{-9}. This is the system present in a bottle of "bromine water." Assuming that HBrO does not ionize appreciably, what is the pH of the bromine water?
48. Calculate the pH and the equilibrium concentration of HClO in a 0.10 M solution of hypochlorous acid. K_a HClO $= 2.8 \times 10^{-8}$.
49. At equilibrium, a gas mixture has a partial pressure of 0.7324 atm for HBr and 2.80×10^{-3} atm for both hydrogen and bromine gases. What is K for the formation of two moles of HBr from H_2 and Br_2?
50. Given

$$HF(aq) \rightleftharpoons H^+(aq) + F^-(aq) \qquad K_a = 6.9 \times 10^{-4}$$
$$HF(aq) + F^-(aq) \rightleftharpoons HF_2^-(aq) \qquad K = 2.7$$

Calculate K for the reaction

$$2HF(aq) \rightleftharpoons H^+(aq) + HF_2^-(aq)$$

51. What is the concentration of fluoride ion in a water solution saturated with BaF_2, $K_{sp} = 1.8 \times 10^{-7}$?
52. Calculate the solubility in grams per 100 mL of BaF_2 in 0.10 M $BaCl_2$ solution.

Thermodynamics

53. Determine whether the following redox reaction is spontaneous at 25°C:

$$2KIO_3(s) + Cl_2(g) \longrightarrow 2KClO_3(s) + I_2(s)$$

Use data in Appendix 1 and the following information: ΔH_f° $KIO_3(s) = -501.4$ kJ/mol, S° $KIO_3(s) = 151.5$ J/mol · K. What is the lowest temperature at which the reaction is spontaneous?
54. Follow the directions for Problem 53 for the reaction

$$2KBrO_3(s) + Cl_2(g) \longrightarrow 2KClO_3(s) + Br_2(l)$$

The following thermodynamic data may be useful:

ΔH_f° $KBrO_3$ = −360.2 kJ/mol; $S^\circ KBrO_3$ = 149.2 J/mol · K

55. Consider the equilibrium system

$$HF(aq) \rightleftharpoons H^+(aq) + F^-(aq)$$

Given $H_f^\circ HF$ = −320.1 kJ/mol;
$\Delta H_f^\circ F^-(aq)$ = −332.6 kJ/mol; $S^\circ F^-(aq)$ = −13.8 J/mol · K;
$K_a HF = 6.9 \times 10^{-4}$ at 25°C

calculate S° for HF(*aq*).

56. Applying the Tables in Appendix 1 to

$$4HCl(g) + O_2(g) \longrightarrow 2Cl_2(g) + 2H_2O(l)$$

determine

a. whether the reaction is spontaneous at 25°C.
b. *K* for the reaction at 25°C.

57. Consider the reaction

$$4NH_3(g) + 5\,O_2(g) \longrightarrow 4NO(g) + 6H_2O(g)$$

a. Calculate ΔH° for this reaction. Is it exothermic or endothermic?
b. Would you expect ΔS° to be positive or negative? Calculate ΔS°.
c. Is the reaction spontaneous at 25°C and 1 atm?
d. At what temperature, if any, is the reaction at equilibrium at 1 atm pressure?

58. Data is given in Appendix 1 for white phosphorus, $P_4(s)$. $P_4(g)$ has the following thermodynamic values: ΔH_f° = 58.9 kJ/mol, S° = 280.0 J/K · mol. What is the temperature at which white phosphorus sublimes at 1 atm pressure?

Electrochemistry

59. In the electrolysis of a KI solution, using 5.00 V, how much electrical energy in kilojoules is consumed when one mole of I_2 is formed?

60. If an electrolytic cell producing fluorine uses a current of 7.00×10^3 A (at 10.0 V), how many grams of fluorine gas can be produced in two days (assuming that the cell operates continuously at 95% efficiency)?

61. Sodium hypochlorite is produced by the electrolysis of cold sodium chloride solution. How long must a cell operate to produce 1.500×10^3 L of 5.00% NaClO by mass if the cell current is 2.00×10^3 A? Assume that the density of the solution is 1.00 g/cm^3.

62. Sodium perchlorate is produced by the electrolysis of sodium chlorate. If a current of 1.50×10^3 A passes through an electrolytic cell, how many kilograms of sodium perchlorate are produced in an eight-hour run?

63. Taking E°_{ox} H_2O_2 = −0.695 V, determine which of the following species will be reduced by hydrogen peroxide (use Table 17.1 to find E°_{red} values).

a. $Cr_2O_7^{2-}$ **b.** Fe^{2+} **c.** I_2 **d.** Br_2

64. Taking E°_{red} H_2O_2 = +1.763 V, determine which of the following species will be oxidized by hydrogen peroxide (use Table 17.1 to find E°_{ox} values).

a. Co^{2+} **b.** Cl^- **c.** Fe^{2+} **d.** Sn^{2+}

65. Consider the reduction of nitrate ion in acidic solution to nitrogen oxide (E°_{red} = 0.964 V) by sulfur dioxide which is oxidized to sulfate ion (E°_{red} = 0.155 V). Calculate the voltage of a cell involving this reaction in which all the gases have pressures of 1.00 atm, all the ionic species (except H^+) are at 0.100 *M*, and the pH is 4.30.

66. For the reaction in Problem 65 if gas pressures are at 1.00 atm and ionic species are at 0.100 *M* (except H^+), at what pH will the voltage be 1.000 V?

Essential Nonmetals

67. Why is selenium an essential element for human metabolism? Discuss the role of selenium compounds an anticarcinogens.

68. Where are iodide ions concentrated in the body? What is their role? What condition results from an iodine deficiency?

69. Explain how fluoride ions are effective in preventing tooth decay.

70. What role do the "major elements," phosphorus, sulfur, and chlorine, play in human metabolism?

Unclassified

71. Choose the strongest acid from each group.

a. HClO, HBrO, HIO **b.** HIO, HIO_3, HIO_4
c. HIO, $HBrO_2$, $HBrO_4$

72. What intermolecular forces are present in the following?

a. Cl_2 **b.** HBr **c.** HF **d.** $HClO_4$ **e.** MgI_2

73. Write a balanced equation for the reaction of hydrofluoric acid with SiO_2. What volume of 2.0 *M* HF is required to react with one gram of silicon dioxide?

74. State the oxidation number of N in

a. NO_2^- **b.** NO_2 **c.** HNO_3 **d.** NH_4^+

75. The density of sulfur vapor at one atmosphere pressure and 973 K is 0.8012 g/L. What is the molecular formula of the vapor?

76. Give the formula of a substance discussed in this chapter that is used

a. to disinfect water. **b.** in safety matches.
c. to prepare hydrazine. **d.** to etch glass.

77. Why does concentrated nitric acid often have a yellow color even though pure HNO_3 is colorless?

78. Explain why

a. acid strength increases as the oxidation number of the central nonmetal atom increases.
b. nitrogen dioxide is paramagnetic.
c. the oxidizing strength of an oxoanion is inversely related to pH.
d. sugar turns black when treated with concentrated sulfuric acid.

Challenge Problems

79. Suppose you wish to calculate the mass of sulfuric acid that can be obtained from an underground deposit of sulfur 1.00 km^2 in area. What additional information do you need to make this calculation?

80. The reaction

$$4HF(aq) + SiO_2(aq) \longrightarrow SiF_4(aq) + 2H_2O$$

can be used to release gold that is distributed in certain quartz (SiO_2) veins of hydrothermal origin. If the quartz contains $1.0 \times 10^{-3}\%$ Au by weight and the gold has a market value of $425 per troy ounce, would the process be economically feasible if commercial HF (50% by weight, $d = 1.17\ g/cm^3$) costs 75¢ a liter? (1 troy ounce = 31.1 g.)

81. The amount of sodium hypochlorite in a bleach solution can be determined by using a given volume of bleach to oxidize excess iodide ion to iodine; ClO^- is reduced to Cl^-. The amount of iodine produced by the redox reaction is determined by titration with sodium thiosulfate, $Na_2S_2O_3$; I_2 is reduced to I^-. The sodium thiosulfate is oxidized to sodium tetrathionate, $Na_2S_4O_6$. In this analysis, potassium iodide was added in excess to 5.00 mL of bleach ($d = 1.00\ g/cm^3$). If 25.00 mL of 0.0700 M $Na_2S_2O_3$ was required to reduce all the iodine produced by the bleach back to iodide, what is the mass percent of NaClO in the bleach?

82. What is the minimum amount of sodium azide that can be added to an automobile airbag to give a volume of 20.0 L upon inflation? Make any reasonable assumptions required to obtain an answer, but state what these assumptions are.

21 Organic Chemistry

He takes up the waters
of the sea in
his hand, leaving the salt;
He disperses it in mist
through the skies;
He recollects and
sprinkles it like
grain in six-rayed snowy stars
over the earth,
There to lie till he dissolves
its bonds again.

HENRY DAVID THOREAU
Journal January 5, 1856

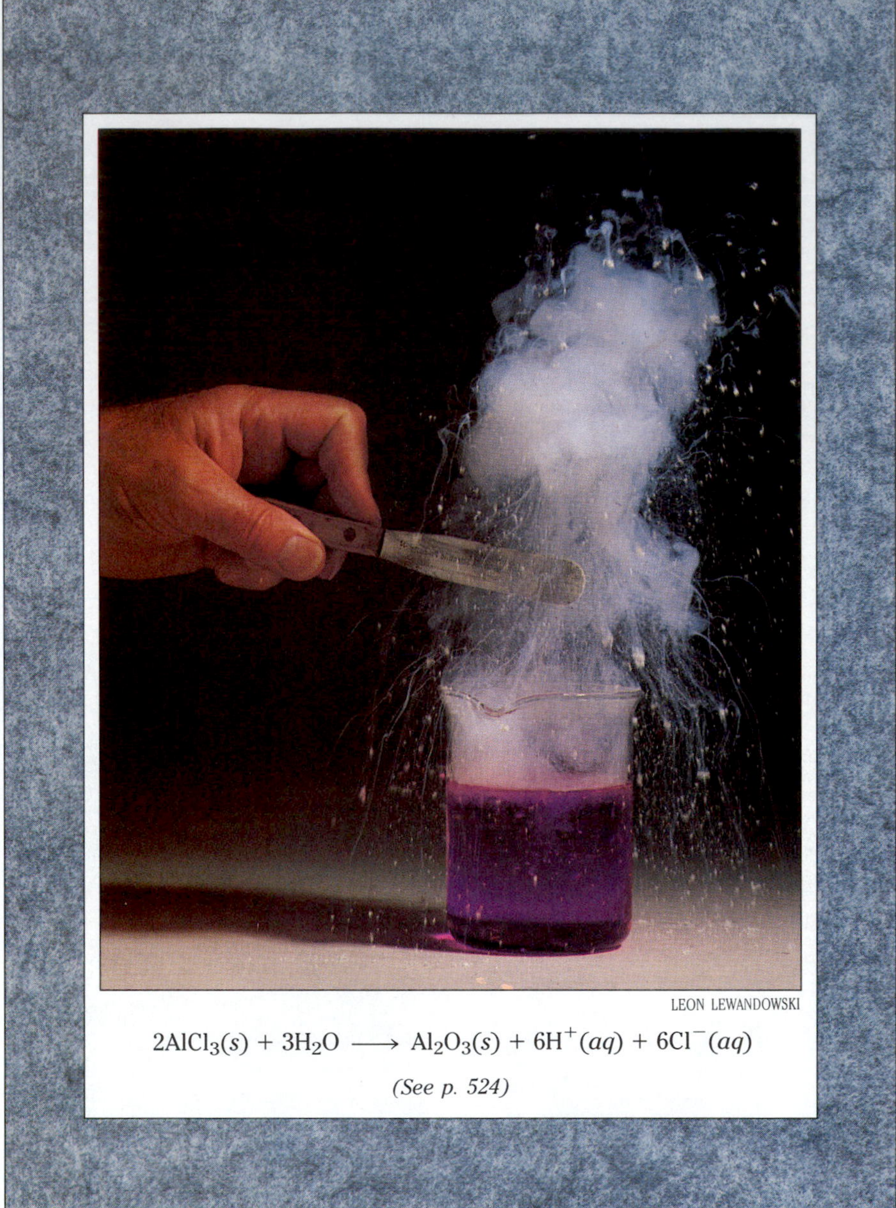

LEON LEWANDOWSKI

$$2AlCl_3(s) + 3H_2O \longrightarrow Al_2O_3(s) + 6H^+(aq) + 6Cl^-(aq)$$

(See p. 524)

CHAPTER OUTLINE

Organic chemistry deals with the compounds of carbon, of which there are literally millions. More than 90% of all known compounds contain carbon atoms. There is a simple explanation for this remarkable fact. Carbon atoms bond to one another to a far greater extent than do atoms of any other element. Carbon atoms may link together to form chains or rings.

$$-\overset{|}{\underset{|}{C}}-\overset{|}{\underset{|}{C}}- \qquad \rangle C{=}C\langle \qquad -C{\equiv}C- \qquad -\overset{|}{\underset{|}{C}}-\overset{|}{\underset{|}{C}}-\overset{|}{\underset{|}{C}}-\overset{|}{\underset{|}{C}}-$$

The bonds may be single (one electron pair), double (two electron pairs), or triple (three electron pairs).

There are a wide variety of different organic compounds that have quite different structures and properties. However, all these substances have certain features in common:

1. *Organic compounds are molecular rather than ionic.* Most of the compounds we discuss consist of small, discrete molecules. Many of them are gases or liquids at room temperature.

2. *Each carbon atom forms a total of four covalent bonds.* This is illustrated by the structures written above. A particular carbon atom may form four single bonds, two single bonds and a double bond, two double bonds, or one single bond and a triple bond. One way or another, though, the bonds add up to four.

Carbon always obeys the octet rule

3. *Carbon atoms may be bonded to each other or to other nonmetal atoms, most often hydrogen, a halogen, oxygen, or nitrogen.* In most organic compounds

— a hydrogen or halogen atom (F, Cl, Br, I) forms one covalent bond, —H, —X
— an oxygen atom forms two covalent bonds, —O— or =O
— a nitrogen atom forms three covalent bonds, $-\overset{|}{N}-$, =N—, or ≡N

These rules hold only for organic compounds

In this chapter, we consider

— the simplest type of organic compound, called a **hydrocarbon**, which contains only two kinds of atoms, hydrogen and carbon. Hydrocarbons can be classified as alkanes (Section 21.1), alkenes and alkynes (Section 21.2), and aromatics (Section 21.3).
— organic compounds containing oxygen atoms in addition to carbon and hydrogen (Section 21.4)
— organic compounds of very high molar mass, known as polymers (Section 21.5)

21.1 SATURATED HYDROCARBONS: ALKANES

One large and structurally simple class of hydrocarbons includes those substances in which all the carbon–carbon bonds are single bonds. These are called *saturated* hydrocarbons or **alkanes**. In the alkanes the carbon atoms are bonded to each other in chains, which may be long or short, straight or branched.

Figure 21.1
Ball-and-stick models of methane, ethane, and propane. The bond angles in all of these compounds are 109.5°, the tetrahedral angle. (Charles Steele)

Methane

Ethane

Propane

The simplest alkanes are methane (CH_4), ethane (C_2H_6), and propane (C_3H_8):

$$\begin{array}{ccc} & H & \\ & | & \\ H- & C & -H \\ & | & \\ & H & \end{array} \qquad \begin{array}{cccc} & H & H & \\ & | & | & \\ H- & C- & C & -H \\ & | & | & \\ & H & H & \end{array} \qquad \begin{array}{ccccc} & H & H & H & \\ & | & | & | & \\ H- & C- & C- & C & -H \\ & | & | & | & \\ & H & H & H & \end{array}$$

methane ethane propane

The two CH_3 groups in C_2H_6 can rotate to give different *conformations*

Around the carbon atoms in these molecules and indeed in any saturated hydrocarbon, there are four single bonds involving sp^3 hybrid orbitals. As would be expected from the VSEPR model, these bonds are directed toward the corners of a regular tetrahedron. The bond angles are 109.5°, the tetrahedral angle. This means that in propane (C_3H_8) and in the higher alkanes, the carbon atoms are arranged in a "zigzag" pattern. (Figure 21.1).

Two different alkanes are known with the molecular formula C_4H_{10}. In one of these, called butane, the four carbon atoms are linked in a "straight chain." In the other, called 2-methylpropane, there is a "branched chain." The longest continuous chain in the molecule contains three carbon atoms; there is a CH_3 branch from the central carbon atom. The geometries of these molecules are shown in Figure 21.2. The structural formulas are

$$\begin{array}{cccccc} & H & H & H & H & \\ & | & | & | & | & \\ H- & C- & C- & C- & C & -H \\ & | & | & | & | & \\ & H & H & H & H & \end{array} \qquad \begin{array}{ccccc} & H & CH_3 & H & \\ & | & | & | & \\ H- & C- & C- & C & -H \\ & | & | & | & \\ & H & H & H & \end{array}$$

butane 2-methylpropane

Butane

2-methylpropane

Figure 21.2
Ball-and-stick models of butane and 2-methylpropane, the isomers of C_4H_{10}. (Charles Steele)

Compounds having the same molecular formula but different molecular structures are called **structural isomers**. Butane and 2-methylpropane are referred to as structural isomers of C_4H_{10}. They are two distinct compounds with their own characteristic physical and chemical properties.

In structural isomers, the atoms are bonded in different patterns

---

Example 21.1 Draw structures for the isomers of C_5H_{12}.

Strategy Start with the straight-chain structure, stringing all five carbon atoms one after the other. Then work with structures containing four carbon atoms in a chain with one branch; find all the nonequivalent structures of this type. Continue this process using a three-carbon chain, which is the shortest one that can be drawn for C_5H_{12}.

Solution For simplicity, we show only the carbon atoms; there is an H atom attached to each bond extending from a C atom.

five-C chain:
```
  |  |  |  |  |
 —C—C—C—C—C—
  |  |  |  |  |
```
isomer (I)

four-C chain:
```
  |  |  |  |
 —C—C—C—C—
  |  |  |  |
   —C—
     |
```
isomer (II)

three-C chain:
```
      |
    —C—
  |  |  |
 —C—C—C—
  |  |  |
    —C—
      |
```
isomer (III)

Figure 21.3
The barbecue grill is fueled by "bottled gas," a mixture of liquid propane (C_3H_8) and liquid butane (C_4H_{10}). (David R. Frazier Photolibrary)

Working with only pencil and paper, you might be tempted to draw other structures, such as

```
  |  |  |  |            |  |  |  |
 —C—C—C—C—           —C—C—C—C—
  |  |  |  |            |  |  |  |
 —C—                       —C—
  |                          |
```

However, a few moments' reflection (or access to a molecular model kit) should convince you that these are in fact equivalent to structures written previously. In particular, the first one, like I, has a five-carbon chain in which no carbon atom is attached to more than two other carbons. The second structure, like II, has a four-carbon chain with one carbon atom bonded to three other carbons. Structures I, II, and III represent the three possible isomers of C_5H_{12}; there are no others.

Remember, these molecules are 3-dimensional

Natural gas, transmitted around the United States and Canada by pipeline, consists largely of methane (80–90%) with smaller amounts of C_2H_6, C_3H_8, and C_4H_{10}. Cylinders of "bottled gas" used with campstoves, barbecue grills, and the like, contain liquid propane (C_3H_8) and butane (C_4H_{10}) (Fig. 21.3). The

Figure 21.4
Petroleum refinery. (Standard Oil)

pressure remains constant as long as any liquid is present, then drops abruptly to zero, indicating that it's time for a recharge.

The higher alkanes are most often obtained from petroleum, a dark brown, viscous liquid dispersed through porous rock deposits. Distillation of petroleum gives a series of fractions of different boiling points (Fig. 21.4). The most important of these is gasoline; distillation of a liter of petroleum gives about 250 mL of "straight-run" gasoline. It is possible to double the yield of gasoline by converting higher or lower boiling fractions to hydrocarbons in the gasoline range (C_5–C_{12}).

Nomenclature

As organic chemistry developed, it became apparent that some systematic way of naming compounds was needed. About 50 years ago, the International Union of Pure and Applied Chemistry (IUPAC) devised a system that could be used for all organic compounds. To illustrate this system, we will show how it works with alkanes.

For straight-chain alkanes such as

$$\underset{\text{propane}}{CH_3{-}CH_2{-}CH_3} \qquad \underset{\text{butane}}{CH_3{-}CH_2{-}CH_2{-}CH_3}$$

the IUPAC name consists of a single word. These names, for up to eight carbon atoms, are listed in Table 21.1.

With alkanes containing a **branched chain**, such as

$$\begin{array}{c} H \\ | \\ CH_3{-}C{-}CH_3 \\ | \\ CH_3 \end{array}$$
2-methylpropane

the name is more complex. A branched-chain alkane such as 2-methylpropane can be considered to be derived from a straight-chain alkane by replacing one or more hydrogen atoms by alkyl groups. The name consists of two parts:

— *a suffix that identifies the parent straight-chain alkane.* To find the suffix, count the number of carbon atoms in the longest continuous chain. For a three-carbon chain, the suffix is *propane;* for a four-carbon chain it is *butane,* and so on.

— *a prefix that identifies the branching alkyl group (Table 21.1) and indicates by a number the carbon atom where branching occurs.* In 2-methylpropane, referred to above, the methyl group is located at the second carbon from the end of the chain:

$$\begin{array}{ccccc} 1 & & 2 & & 3 \\ C & - & C & - & C \\ & & | & & \end{array}$$

Structural isomers have different names; if they don't, they're not isomers

Following this system, the IUPAC names of the isomers of pentane are

$$\underset{\text{pentane}}{CH_3{-}CH_2{-}CH_2{-}CH_2{-}CH_3} \qquad \underset{\text{2-methylbutane}}{\begin{array}{c} H \\ | \\ CH_3{-}C{-}CH_2{-}CH_3 \\ | \\ CH_3 \end{array}} \qquad \underset{\text{2,2-dimethylpropane}}{\begin{array}{c} CH_3 \\ | \\ CH_3{-}C{-}CH_3 \\ | \\ CH_3 \end{array}}$$

Table 21.1 Nomenclature of Alkanes

Straight-Chain Alkanes		Alkyl Groups	
Methane	CH_4	Methyl	$CH_3—$
Ethane	CH_3CH_3	Ethyl	$CH_3—CH_2—$
Propane	$CH_3CH_2CH_3$	Propyl	$CH_3—CH_2—CH_2—$
Butane	$CH_3(CH_2)_2CH_3$		
Pentane	$CH_3(CH_2)_3CH_3$	Isopropyl	$CH_3—CH(CH_3)—$ (H above C, CH_3 below C)
Hexane	$CH_3(CH_2)_4CH_3$		
Heptane	$CH_3(CH_2)_5CH_3$	Butyl	$CH_3—CH_2—CH_2—CH_2—$
Octane	$CH_3(CH_2)_6CH_3$		

Notice that

— *if the same alkyl group is at two branches, the prefix di- is used (2,2-dimethylpropane).* If there were three methyl branches, we would write trimethyl, and so on.

— *the number in the name is made as small as possible.* Thus, we refer to 2-methylbutane, numbering the chain from the left, rather than from the right.

```
1  2  3  4
C—C—C—C
   |
```

Example 21.2 Assign IUPAC names to the following

```
        CH3                                H
        |                                  |
(a) CH3—C—CH2—CH3     (b) CH3—CH2—C—CH2—CH3
        |                                  |
        CH3                                CH2
                                           |
                                           CH3
```

Strategy Find the longest chain and use the proper suffix to identify it. Then find the branching alkyl group; locate the carbon atom where branching occurs. Number the carbon atoms so as to give the lower number for the prefix identifying the alkyl group.

Solution

(a) The longest chain contains four carbon atoms (butane). There are two CH_3 (methyl) groups branching at the second carbon from the end of the chain (2).

The correct name is 2,2-dimethylbutane.

(b) The longest chain, however you count it, contains five carbon atoms. There is a $CH_3—CH_2$ branch at the number three carbon, whichever end of the chain you start from. The IUPAC name is 3-ethylpentane.

21.2 UNSATURATED HYDROCARBONS: ALKENES AND ALKYNES

In an unsaturated hydrocarbon, at least one of the carbon–carbon bonds in the molecule is a multiple bond. There are many types of unsaturated hydrocarbons, only two of which are discussed here:

— **alkenes**, in which there is one carbon–carbon double bond in the molecule

$$\begin{array}{ccc} \diagdown & & \diagup \\ & C{=}C & \\ \diagup & & \diagdown \end{array}$$

— **alkynes**, in which there is one carbon–carbon triple bond in the molecule

$$-C{\equiv}C-$$

Alkenes

The simplest alkene is ethene, C_2H_4 (common name, ethylene).

$$\begin{array}{ccc} H & & H \\ & C{=}C & \\ H & & H \end{array}$$

ethene

Ethene is produced in larger amounts than any other organic compound; about 1.7×10^7 metric tons of C_2H_4 are manufactured in the United States each year. Ethene is used primarily as a starting material for the preparation of other organic substances, including ethyl alcohol (p. 591), ethylene glycol (p. 592), and, in particular, polyethylene (p. 596). Smaller amounts of ethene are used to ripen fruit, picked while still unripe to avoid spoilage.

That's how we get red tomatoes that taste like sawdust

You may recall that we discussed the bonding in ethene in Chapter 7. The double bond in ethene and other alkenes consists of a sigma bond and a pi bond. The ethene molecule is planar. There is no rotation about the double bond, since that would require "breaking" the pi bond. The bond angle in ethene is 120°, corresponding to sp^2 hybridization about each carbon atom. The geometries of ethene and the next member of the alkene series, C_3H_6, are shown in Figure 21.5.

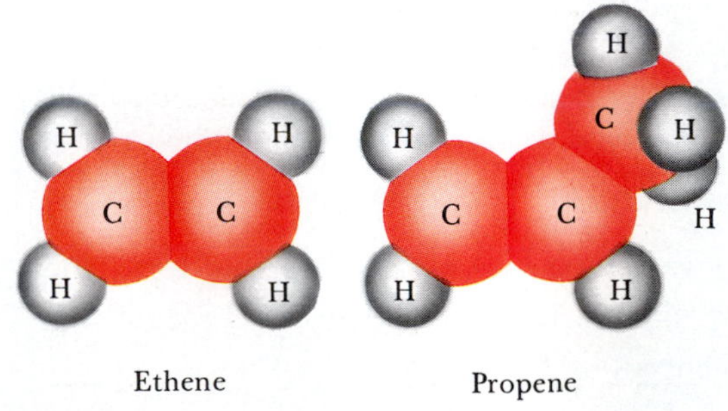

Figure 21.5
Space-filling models of ethene and propene. The ethene molecule is planar. Propene contains three carbon atoms, two of which are joined by a double bond.

The names of alkenes are derived from those of the corresponding alkanes with the same number of carbon atoms per molecule. There are two modifications.

— the ending *-ane* is replaced by *-ene:*

$CH_3—CH_3$ (ethane) $CH_2=CH_2$ (ethene)

— where necessary, a number is used to designate the double bonded carbon; the number is made as small as possible.

$\overset{1}{C}H_2=\overset{2}{C}H—\overset{3}{C}H_2—\overset{4}{C}H_3$ (1-butene) $\overset{1}{C}H_3—\overset{2}{C}H=\overset{3}{C}H—\overset{4}{C}H_3$ (2-butene)

$CH_2=C(CH_3)—CH_2—CH_3$ (2-methyl-1-butene) $CH_3—C(H_3C)=C(H)—CH_3$ (2-methyl-2-butene)

You may be surprised to learn that there are actually two different 2-butenes, differing from each other in molecular geometry.

cis-2-butene (CH_3 and CH_3 on the same side, H and H on the same side), bp = 4°C

trans-2-butene (CH_3 and H on one carbon, H and CH_3 on the other, on opposite sides), bp = 1°C

These compounds are **geometric isomers**. In the *cis*- isomer, the two CH_3 groups (or the two H atoms) are as close to one another as possible. In the *trans*- isomer, the two identical groups are farther apart. The two forms exist because there is no free rotation about the carbon–carbon double bond. Here, as with complex ions (Chap. 15), geometry is responsible for *cis-trans* isomerism; the atoms are bonded to each other in the same way.

If there were free rotation, there would be no geometric isomers

Geometric isomerism is common among alkenes. Indeed, it occurs with all alkenes *except those in which two identical atoms or groups are attached to one of the double-bonded carbons*. Thus, although 2-butene has *cis*- and *trans*- isomers, 1-butene does not because two H atoms are bonded to carbon-1.

Example 21.3 Draw all the isomers of the molecule $C_2H_2Cl_2$ in which two of the H atoms of ethylene are replaced by Cl atoms.

Strategy Draw the C=C skeleton. Attach both chlorine atoms to the same carbon atom; there is only one such compound. Then attach a chlorine atom to each of the carbon atoms. In this case, there are *cis*- and *trans*- isomers.

Solution The isomers of the molecule $C_2H_2Cl_2$ are

1,1-dichloroethene ($Cl_2C=CH_2$)

cis-1,2-dichloroethene (Cl and Cl on the same side)

trans-1,2-dichloroethene (H and Cl on one carbon, Cl and H on the other, Cl atoms on opposite sides)

Cis- and *trans-* isomers differ from one another in their physical, and, to a lesser extent, their chemical properties. They may also differ in their physiological behavior. For example, the compound *cis*-9-tricosene

$$\begin{array}{ccc} CH_3-(CH_2)_7 & & (CH_2)_{12}-CH_3 \\ & C=C & \\ H & & H \end{array}$$

is a sex attractant secreted by the female housefly. The *trans-* isomer is totally ineffective in this area.

Name the compound: $\begin{array}{ccc} H & & Pa \\ & C=C & \\ Ma & & H \end{array}$

Ans: transparent

Alkynes

The IUPAC names of alkynes are derived from those of the corresponding alkenes by replacing the suffix *-ene* with *-yne*. Thus

$H-C\equiv C-H$ (ethyne) $\quad H-C\equiv C-CH_3$ (propyne)

$H-C\equiv C-CH_2-CH_3$ (1-butyne) $\quad CH_3-C\equiv C-CH_3$ (2-butyne)

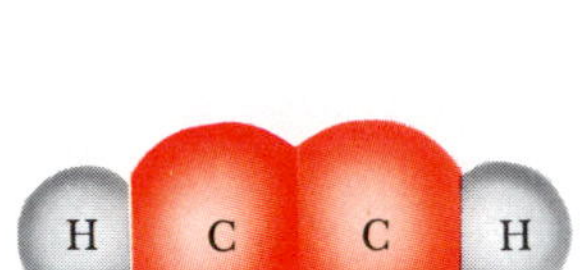

Ethyne (Acetylene)

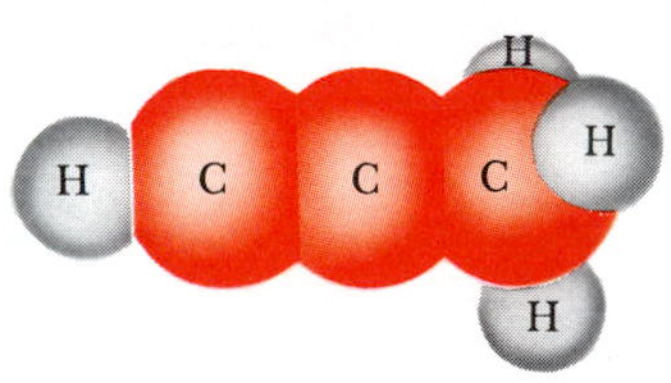

Propyne (Methylacetylene)

Figure 21.6
In acetylene and methylacetylene, two carbon atoms are linked by a triple bond. Both molecules contain four atoms on a straight line.

The most common alkyne by far is the first member of the series, commonly called acetylene. Recall from Chapter 7 that the C_2H_2 molecule is linear, with 180° bond angles. The triple bond consists of a sigma bond and two pi bonds; each carbon atom is sp-hybridized. The geometries of acetylene and the next member of the series, C_3H_4, are shown in Figure 21.6.

Acetylene can be made in the laboratory by allowing water to drop on calcium carbide.

$$CaC_2(s) + H_2O(l) \longrightarrow C_2H_2(g) + CaO(s)$$

The gas produced in this way has a garlic-like odor due to traces of phosphine, PH_3, formed from Ca_3P_2 in the impure calcium carbide. Commercially, acetylene is made by heating methane to about 1500°C in the absence of air:

$$2CH_4(g) \longrightarrow C_2H_2(g) + 3H_2(g)$$

Thermodynamically, acetylene is unstable with respect to decomposition to the elements

$$C_2H_2(g) \longrightarrow 2C(s) + H_2(g) \qquad \Delta G° = -209.2 \text{ kJ at } 25°C$$

At high pressures, this reaction can occur explosively. For that reason, cylinders of acetylene do not contain the pure gas. Instead the cylinder is packed with an inert, porous material that holds a solution of acetylene gas in acetone.

You are probably most familiar with acetylene as a gaseous fuel used in welding and cutting metals (Fig. 21.7). When mixed with pure oxygen in a torch, acetylene burns at temperatures above 2000°C. The heat comes from the reaction

$$C_2H_2(g) + \tfrac{5}{2}O_2(g) \longrightarrow 2CO_2(g) + H_2O(l) \qquad \Delta H = -1300 \text{ kJ}$$

The reaction gives off a brilliant white light, which served as a source of illumination in the headlights of early automobiles.

Also miner's headlamps

Figure 21.7
Acetylene (ethyne) is used in welding. (Courtesy of Bethlehem Steel)

21.3 AROMATIC HYDROCARBONS AND THEIR DERIVATIVES

Aromatic hydrocarbons, sometimes referred to as *arenes*, can be considered to be derived from benzene, C_6H_6. Benzene is a transparent, volatile liquid (bp = 80°C) which was discovered by Michael Faraday in 1825. Its formula, C_6H_6, suggests a high degree of unsaturation, yet its properties are quite different from those of alkenes or alkynes.

The atomic orbital (valence bond) approach regards benzene as a resonance hybrid of the two structures

```
        H                      H
        |                      |
        C                      C
      //  \                  /  \\
  H—C      C—H          H—C      C—H
     |      ||     ⟷       ||      |
  H—C      C—H          H—C      C—H
      \\  /                  \  //
        C                      C
        |                      |
        H                      H
```

This model is consistent with many of the properties of benzene. The molecule is a planar hexagon with bond angles of 120°. The hybridization of each carbon is sp^2. However, this structure is misleading in one respect. Chemically, benzene does not behave as if there were double bonds present.

It is much less reactive than most alkenes

A more satisfactory model of the electron distribution in benzene, based upon molecular orbital theory, assumes that

— each carbon atom forms three sigma bonds, one to a hydrogen atom and two to adjacent carbon atoms
— the three electron pairs remaining are spread symmetrically over the entire molecule to form "delocalized" pi bonds

This model is often represented by the structure

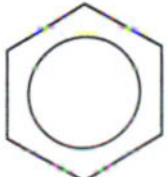

where it is understood that

There are three σ bonds and three π bonds in benzene

— there is a carbon atom at each corner of the hexagon
— there is an H atom bonded to each carbon atom
— the circle in the center of the molecule represents the six delocalized electrons

At one time, benzene was widely used as a solvent, both commercially and in research and teaching laboratories. Its use for that purpose has been largely abandoned because of its toxicity. Chronic exposure to benzene vapor leads to various blood disorders and, in extreme cases, aplastic anemia and leukemia. It appears that the culprits here are oxidation products of the aromatic ring, formed in an attempt to solubilize benzene and thus eliminate it from the body.

Derivatives of Benzene

Monosubstituted benzenes are ordinarily named as derivatives of benzene.

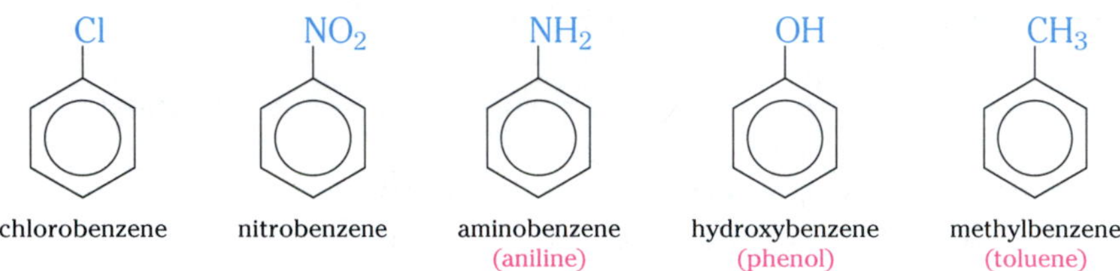

The last three compounds listed are always referred to by their common names, shown in red. Phenol was the first commercial antiseptic; its introduction into hospitals in the 1870s led to a dramatic decrease in deaths from postoperative infections. Its use for this purpose has long since been abandoned because phenol burns exposed tissue, but many modern antiseptics are phenol derivatives. Toluene has largely replaced benzene as a solvent because it is much less toxic. Oxidation of toluene in the body gives benzoic acid, C_6H_5COOH, which is readily eliminated and has none of the toxic properties of oxidation products of benzene.

When there are two groups attached to the benzene ring, three isomers are possible. These are designated by the prefixes *ortho-*, *meta-*, and *para-*, often abbreviated as *o-*, *m-*, and *p-*.

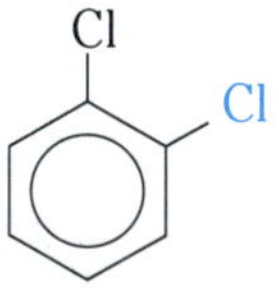

Cl Cl Cl Cl Cl Cl

o-dichlorobenzene 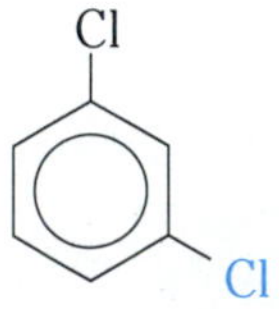 *m*-dichlorobenzene 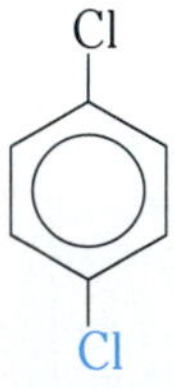 *p*-dichlorobenzene

Numbers can also be used; these three compounds may be referred to as 1,2-dichlorobenzene, 1,3-dichlorobenzene, and 1,4-dichlorobenzene, respectively. As always, the numbers used are as small as possible.

Condensed Ring Structures

In another type of aromatic hydrocarbon, two or more benzene rings are fused together. Naphthalene, the solid that gives mothballs their aromatic odor, is the simplest compound of this type. Fusion of three benzene rings gives two different isomers, anthracene and phenanthrene.

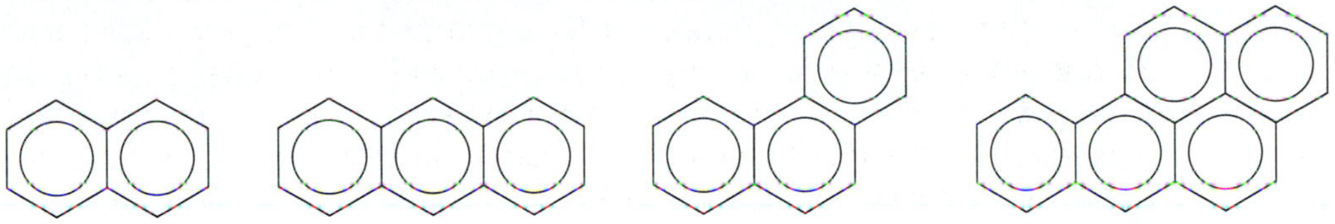

naphthalene $C_{10}H_8$

anthracene $C_{14}H_{10}$

phenanthrene $C_{14}H_{10}$

3,4-benzpyrene $C_{20}H_{12}$

It is important to realize that in these structures there are no hydrogen atoms bonded to the carbons at the juncture of two rings. The eight hydrogen atoms in naphthalene are located as shown below:

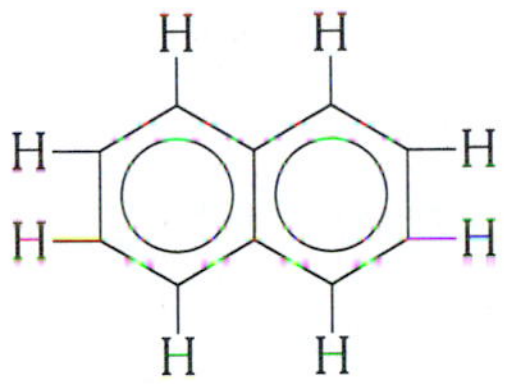

H H
H— —H
H— —H
H H

Certain compounds of this type are potent carcinogens. One of the most dangerous is 3,4-benzpyrene, which has been detected in cigarette smoke. It is believed to be a cause of lung cancer, to which heavy smokers are susceptible.

Several derivatives of naphthalene are carcinogenic

Many natural products contain fused rings. One of the most important compounds of this type is cholesterol, which has the structure

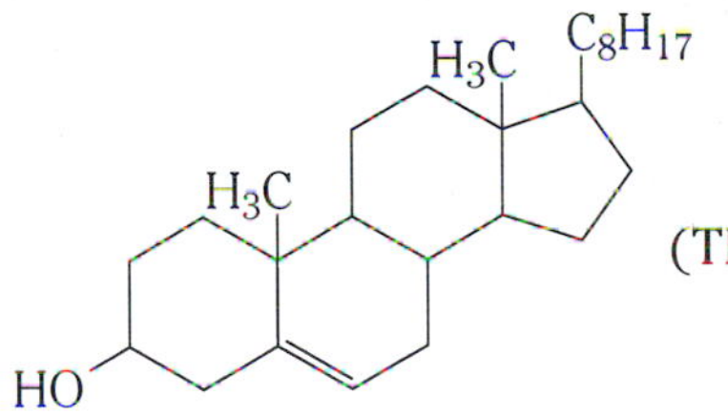

H_3C C_8H_{17}
H_3C
HO

(This molecule has only one *pi* bond.)

cholesterol

Your body contains approximately 150 g of cholesterol. Every day, about one gram of cholesterol is synthesized in the liver; another half gram is taken in food. Cholesterol is the starting material used by the body to synthesize many important steroids, including the sex hormones testosterone (male) and estradiol (female). If there is too high a concentration of cholesterol in blood serum, it can precipitate in plaquelike deposits on the inner walls of arteries. The resultant reduction in the diameter of blood vessels can lead to a heart attack, stroke, or kidney failure.

Table 21.2 Common Functional Groups

Group	Class	Example	Name*
—F, —Cl, —Br, —I	halides	C_2H_5Cl	chloroethane (ethyl chloride)
—OH	**alcohols**	C_2H_5OH	ethanol (ethyl alcohol)
—O—	ethers	$CH_3—O—CH_3$	dimethyl ether
$-\overset{\overset{\large O}{\|}}{C}-H$	aldehydes	$CH_3-\overset{\overset{\large O}{\|}}{C}-H$	ethanal (acetaldehyde)
$-\overset{\overset{\large O}{\|}}{C}-$	ketones	$CH_3-\overset{\overset{\large O}{\|}}{C}-CH_3$	propanone (acetone)
$\mathbf{-\overset{\overset{\large O}{\|}}{C}-OH}$	**carboxylic acids**	$CH_3-\overset{\overset{\large O}{\|}}{C}-OH$	ethanoic acid (acetic acid)
$\mathbf{-\overset{\overset{\large O}{\|}}{C}-O-}$	**esters**	$CH_3-\overset{\overset{\large O}{\|}}{C}-OCH_3$	methyl acetate
$-\overset{\|}{N}-$	amines	CH_3NH_2	aminomethane (methyl amine)
$-\overset{\overset{\large O}{\|}}{C}-\overset{\overset{\large H}{\|}}{N}-$	amides	$CH_3-\overset{\overset{\large O}{\|}}{C}-NH_2$	ethanamide (acetamide)

* Common names are shown in red.

Taking a breathalyzer test for alcohol. The legal limit is 0.1% alcohol. (E.R. Degginger)

21.4 FUNCTIONAL GROUPS

Many organic molecules can be considered to be derived from hydrocarbons by substituting a **functional group** for a hydrogen atom. The functional group can be a nonmetal atom or small group of atoms that is bonded to carbon. Table 21.2 lists the types of functional groups commonly found in organic compounds. One of these, the NH_2 group found in amines, was discussed in Chapter 4. Here, we consider three other classes of compounds: alcohols, carboxylic acids, and esters.

Alcohols

An alcohol is derived from a hydrocarbon by replacing one or more H atoms by —OH groups. Alcohols are named by substituting the suffix *-ol* for the *-ane* suffix of the corresponding alkane. The first four members of the series are (common names shown in red):

CH_3OH	$CH_3—CH_2OH$	$CH_3—CH_2—CH_2OH$	$CH_3-\underset{\underset{\large OH}{\|}}{CH}-CH_3$
methanol (methyl alcohol)	ethanol (ethyl alcohol)	1-propanol (propyl alcohol)	2-propanol (isopropyl alcohol)

Notice that 1-propanol and 2-propanol are structural isomers. In 1-propanol, the —OH group is bonded to a terminal carbon atom; in 2-propanol, it is bonded to the central carbon atom.

About 3×10^9 kg of *methanol* are produced annually in the United States from water gas, a mixture of carbon monoxide and hydrogen:

$$CO(g) + 2H_2(g) \xrightarrow[250\ \text{atm},\ 350°C]{ZnO,\ Cr_2O_3} CH_3OH(g)$$

Methanol in "moonshine" whiskey can cause blindness

Methanol is also formed as a by-product when charcoal is made by heating wood in the absence of air. For this reason, methanol is sometimes called wood alcohol. Methanol is used in jet fuels and as a solvent, gasoline additive, and starting material for several industrial syntheses. Methanol is a deadly poison; ingestion of as little as 25 mL can be fatal. The antidote for methanol poisoning is a solution of sodium hydrogen carbonate, $NaHCO_3$.

Ethanol, the most common alcohol, can be prepared by the fermentation of grains or sugar (Fig. 21.8). It is the active ingredient of alcoholic beverages. There it is present in various concentrations (4–8% in beer, 12–15% in wine, and 40% or more in distilled spirits). The "proof" of an alcoholic beverage is twice the volume percent of ethanol. Thus, an 86 proof bourbon whiskey contains 43% ethanol. The taste of alcoholic beverages is due to impurities; ethanol itself is tasteless and colorless.

Industrial ethanol is made by the reaction of ethene with water:

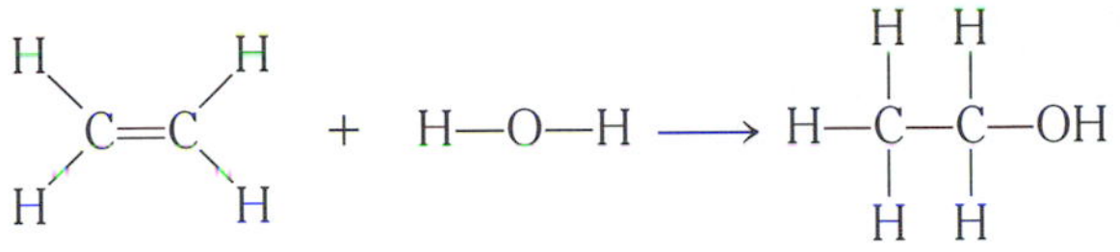

$$H_2C{=}CH_2 + H{-}O{-}H \longrightarrow H{-}\overset{H}{\underset{H}{C}}{-}\overset{H}{\underset{H}{C}}{-}OH$$

Gasohol contains 10% of methanol or ethanol

Frequently, ethanol is "denatured" by adding small quantities of methanol or benzene. This avoids the high federal tax on beverage alcohol; denatured alcohol is poisonous.

Figure 21.8
Wine *(far right)* is produced from the glucose in grape juice *(left)* by fermentation. The reaction is: $C_6H_{12}O_6(aq) \rightarrow 2C_2H_5OH(aq) + 2CO_2(g)$. The purpose of the bubble chamber in the fermentation jug *(center)* is to allow the carbon dioxide to escape but prevent oxygen from entering and oxidizing ethanol to acetic acid. (Courtesy of Prof. James M. Bobbitt, Twenty Mile Vineyard)

Certain alcohols contain two or more —OH groups per molecule. Perhaps the most familiar compounds of this type are ethylene glycol and glycerol:

$$\begin{array}{cc} \text{H} & \text{H} \\ | & | \\ \text{H—C} & \text{—C—H} \\ | & | \\ \text{OH} & \text{OH} \end{array} \qquad \begin{array}{ccc} \text{H} & \text{H} & \text{H} \\ | & | & | \\ \text{H—C} & \text{—C—} & \text{C—H} \\ | & | & | \\ \text{OH} & \text{OH} & \text{OH} \end{array}$$

ethylene glycol glycerol

Ethylene glycol is widely used as an antifreeze. Glycerol is formed as a by-product in making soaps and detergents. It is a viscous, sweet-tasting liquid, used in making drugs, antibiotics, plastics, and explosives (nitroglycerin).

Carboxylic Acids

Carboxylic acids are derived from hydrocarbons by replacing one or more H atoms by a carboxyl group, $\text{—}\underset{\underset{\text{O}}{\|}}{\text{C}}\text{—OH}$, often abbreviated —COOH. The systematic names of these compounds are obtained by adding the suffix *-oic* to the stem of the name of the corresponding alkanes. In practice, these names are seldom used for the first two members of the series, which are commonly referred to as formic acid and acetic acid.

$$\text{H—}\underset{\underset{\text{O}}{\|}}{\text{C}}\text{—OH} \qquad CH_3\text{—}\underset{\underset{\text{O}}{\|}}{\text{C}}\text{—OH}$$

methanoic acid formic acid ethanoic acid acetic acid

Acetic acid is the active ingredient of vinegar, responsible for its sour taste. "White" vinegar is made by adding pure acetic acid to water, forming a 5% solution. "Brown" or "apple cider" vinegar is made from apple juice; the ethanol formed by fermentation is oxidized to acetic acid:

Homemade wine often contains some vinegar

$$C_2H_5OH(aq) + O_2(g) \longrightarrow CH_3COOH(aq) + H_2O$$

The most important chemical property of carboxylic acids is implied by their name; they act as weak acids in water solution.

$$RCOOH(aq) \rightleftharpoons H^+(aq) + RCOO^-(aq)$$

Carboxylic acids vary considerably in acid strength. Acetic acid has a dissociation constant of 1.8×10^{-5}; that of formic acid is about ten times as great, 1.9×10^{-4}. Perhaps the strongest acid of this type is trichloroacetic acid (K_a = 0.20); a 0.10 *M* solution of $Cl_3C\text{—}COOH$ is about 73% ionized. Trichloroacetic acid is an ingredient of over-the-counter preparations used to treat canker sores and remove warts.

Treatment of a carboxylic acid with the strong base NaOH forms the sodium salt of the acid. With acetic acid, the acid-base reaction is

$$CH_3\text{—}\underset{\underset{\text{O}}{\|}}{\text{C}}\text{—OH}(aq) + OH^-(aq) \longrightarrow CH_3\text{—}\underset{\underset{\text{O}}{\|}}{\text{C}}\text{—O}^-(aq) + H_2O$$

Evaporation gives the salt sodium acetate, which contains Na^+ and CH_3COO^- ions. Many of the salts of carboxylic acids have important uses. Sodium and calcium propionate (Na^+ or Ca^{2+} ions, $CH_3CH_2COO^-$ ions) are added to

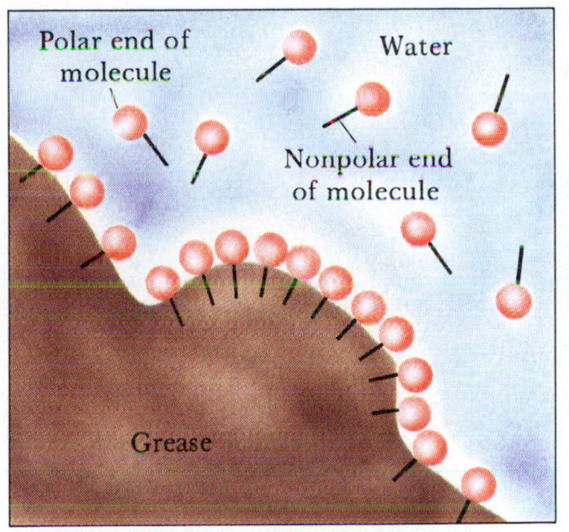

Soap molecules on surface of grease

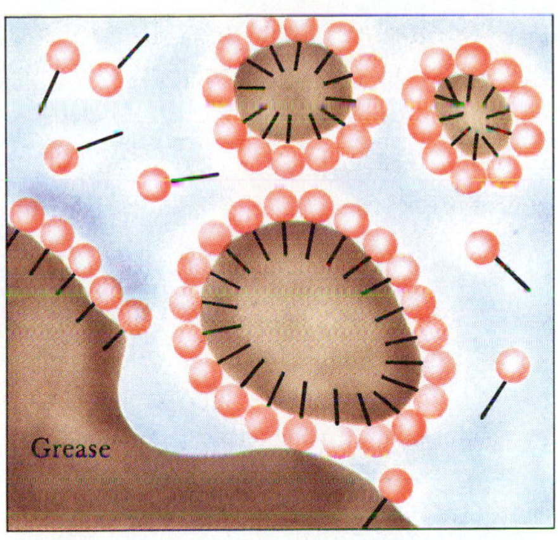

Grease droplets dispersed in wash water

Figure 21.9
The long hydrocarbon chain of a soap ion is a good solvent for grease. The polar end of the ion is soluble in water. In the washing process, the grease is dispersed into the water as droplets surrounded by soap ions.

bread, cake, and cheese to inhibit the growth of mold. *Soaps* are sodium salts of long-chain carboxylic acids such as stearic acid:

$$CH_3(CH_2)_{16}\!-\!\underset{\underset{\displaystyle O}{\|}}{C}\!-\!OH \qquad Na^+,\ CH_3(CH_2)_{16}\!-\!\underset{\underset{\displaystyle O}{\|}}{C}\!-\!O^-$$

stearic acid — sodium stearate, a soap

The cleaning action of soap reflects the nature of the long-chain carboxylate anion (Fig. 21.9). The long hydrocarbon group is a good solvent for greases and oils, while the ionic COO^- group gives high water solubility.

Esters

The reaction between a carboxylic acid and an alcohol forms an **ester**, containing the functional group $-\underset{\underset{\displaystyle O}{\|}}{C}-O-$, often abbreviated —COO—. The reaction between acetic acid and methyl alcohol is typical:

$$CH_3\!-\!\underset{\underset{\displaystyle O}{\|}}{C}\!-\!OH(aq) + HO\!-\!CH_3(aq) \xrightarrow{H^+} CH_3\!-\!\underset{\underset{\displaystyle O}{\|}}{C}\!-\!O\!-\!CH_3(aq) + H_2O$$

acetic acid — methyl alcohol — methyl acetate

The name of an ester consists of two words. The first word (methyl, ethyl, . . .) is the name of the alkyl group of the alcohol. The second word (formate, acetate, . . .) is the name of the acid with the *-ic* suffix replaced by *-ate*. Thus ethyl butyrate (Table 21.3) is made from ethyl alcohol and butyric acid.

$$CH_3CH_2OH \qquad CH_3CH_2CH_2COOH$$

ethyl alcohol — butyric acid

Example 21.4 Show the structure of

(a) the three-carbon alcohol with an —OH group at the end of the chain.
(b) the three-carbon carboxylic acid.
(c) the ester formed when these two compounds react.

Table 21.3 Properties of Esters

Ester	Structure	Odor, Flavor
Ethyl formate	$CH_3CH_2-O-\underset{\underset{O}{\parallel}}{C}-H$	rum
Isobutyl formate	$(CH_3)_2-CHCH_2-O-\underset{\underset{O}{\parallel}}{C}-H$	raspberry
Methyl butyrate	$CH_3-O-\underset{\underset{O}{\parallel}}{C}-(CH_2)_2CH_3$	apple
Ethyl butyrate	$CH_3CH_2-O-\underset{\underset{O}{\parallel}}{C}-(CH_2)_2CH_3$	pineapple
Isopentyl acetate	$(CH_3)_2-CH-(CH_2)_2-O-\underset{\underset{O}{\parallel}}{C}-CH_3$	banana
Octyl acetate	$CH_3-(CH_2)_7-O-\underset{\underset{O}{\parallel}}{C}-CH_3$	orange
Pentyl propionate	$CH_3-(CH_2)_4-O-\underset{\underset{O}{\parallel}}{C}-CH_2CH_3$	apricot

Most esters have pleasant odors

Strategy To work (a) and (b), start by drawing the parent carbon chain. Put the functional group in the specified position; fill out the structural formula with hydrogen atoms. In (c), remember that an —OH group comes from the acid, an H atom from the alcohol.

Solution

(a) $CH_3CH_2CH_2-OH$ (b) $CH_3CH_2-\underset{\underset{O}{\parallel}}{C}-OH$

(c) $CH_3CH_2-\underset{\underset{O}{\parallel}}{C}-O-CH_2CH_2CH_3$

Animal fats and vegetable oils are esters of long-chain carboxylic acids with glycerol. A typical fat molecule might have the structure

Fats are also called triglycerides

$$\begin{array}{r}
CH_3(CH_2)_{14}-COO-CH_2 \\
| \;\; \\
CH_3(CH_2)_7CH{=}CH(CH_2)_7-COO-CH \;\; \\
| \;\; \\
CH_3(CH_2)_{16}-COO-CH_2
\end{array}$$

The three carboxylic acids from which this fat is derived are

— palmitic acid, $CH_3(CH_2)_{14}COOH$

— oleic acid, $CH_3(CH_2)_7CH{=}CH(CH_2)_7COOH$

— stearic acid, $CH_3(CH_2)_{16}COOH$

These acids are typical of those found in fats. Some "fatty acids" are *saturated,* such as palmitic and stearic acid; the hydrocarbon chain contains no multiple bonds. Others, such as oleic acid, are *unsaturated;* there are one or more carbon–carbon multiple bonds in the molecule.

So-called saturated fats contain relatively few carbon–carbon multiple mal products such as lard and butter. Unsaturated fats contain a higher proportion of multiple bonds. They are liquids at room temperature and are found in such vegetable products as corn oil and cottonseed oil. Recently, there has been a trend toward the use of unsaturated as opposed to saturated fats. It appears that saturated fats raise the level of cholesterol in the blood, perhaps contributing to the risk of heart attacks and other circulatory disorders.

Example 21.5 Classify the following as alcohols, carboxylic acids, or esters. More than one functional group may be present in the molecule.

(a) HO—CH₂—CH₂—C(=O)—OH, written as $HO-CH_2-CH_2-C(=O)-OH$ (b) benzene ring bearing O—C(=O)—CH_3 and OH (aspirin)

Strategy Look for the following functional groups:

—OH	—C(=O)—OH	—C(=O)—O—
alcohol	acid	ester

Solution (a) alcohol, carboxylic acid (b) alcohol, ester

Esters in common household items. (Richard Megna/FUNDAMENTAL PHOTOGRAPHS, New York)

21.5 SYNTHETIC ORGANIC POLYMERS

Since about 1930, a wide variety of polymers have been synthesized by chemists. A **polymer** is made up of a large number of small molecular units called *monomers*, combined together chemically. There are two different kinds of synthetic polymers:

- **addition polymers**, in which monomer units add directly to one another; typically only one kind of monomer is involved.
- **condensation polymers**, in which monomer units combine by splitting out (condensing) a simple molecule such as H_2O. Typically, two different monomers react with one another to form a condensation polymer.

In addition polymers, all the reactants end up in the polymer

Addition Polymers

Table 21.4 lists some of the more familiar synthetic addition polymers. Notice that each of these is derived from a monomer containing a carbon—carbon double bond. Upon polymerization, the double bond is converted to a single bond

$$>C=C< \longrightarrow -\overset{|}{\underset{|}{C}}-\overset{|}{\underset{|}{C}}-$$

and successive monomer units add to one another.

More industrial chemists work with polymers than any other class of materials

Table 21.4 Some Common Addition Polymers

Monomer	Name	Polymer	Uses
$\begin{matrix} H & & H \\ & C{=}C & \\ H & & H \end{matrix}$	ethylene	polyethylene	bags, coatings, toys
$\begin{matrix} H & & H \\ & C{=}C & \\ H & & CH_3 \end{matrix}$	propylene	polypropylene	beakers, milk cartons
$\begin{matrix} H & & H \\ & C{=}C & \\ H & & Cl \end{matrix}$	vinyl chloride	polyvinyl chloride, PVC	floor tile, raincoats, pipe, phonograph records
$\begin{matrix} H & & H \\ & C{=}C & \\ H & & CN \end{matrix}$	acrylonitrile	polyacrylonitrile, PAN	rugs; Orlon and Acrilan are copolymers with other monomers.
$\begin{matrix} H & & H \\ & C{=}C & \\ H & & C_6H_5 \end{matrix}$	styrene	polystyrene	cast articles using a transparent plastic
$\begin{matrix} H & & CH_3 \\ & C{=}C & \\ H & & \underset{\Vert \atop O}{C}{-}OCH_3 \end{matrix}$	methyl methacrylate	Plexiglas, Lucite, acrylic resins	high-quality transparent objects, latex paints
$\begin{matrix} F & & F \\ & C{=}C & \\ F & & F \end{matrix}$	tetrafluoroethylene	Teflon	gaskets, insulation, bearings, pan coatings

Figure 21.10
The polyethylene bottle at the left is made of pliable, branched polyethylene. The one at the right is made of semirigid, linear polyethylene.
(Marna G. Clarke)

Perhaps the most familiar addition polymer is **polyethylene**, a solid derived from the monomer ethene (common name ethylene). We might represent the polymerization process as

$$n\left(\begin{matrix} H & & H \\ & C{=}C & \\ H & & H \end{matrix}\right) \longrightarrow \left(\begin{matrix} & H & H & \\ & | & | & \\ - & C & -C & - \\ & | & | & \\ & H & H & \end{matrix}\right)_n$$

ethylene → polyethylene

where n is a very large number, of the order of 2000.

Depending upon the conditions of polymerization, the product may be

— *branched* polyethylene, which has a structure of the type

$$\begin{matrix} & H & H & H & H & H & H & H & H & H & H & H & H & \\ & | & | & | & | & | & | & | & | & | & | & | & | & \\ - & C- & C- & C- & C- & C- & C- & C- & C- & C- & C- & C- & C & - \\ & | & | & | & | & | & | & | & | & | & | & | & | & \\ & H & H & H & CH_2 & H & H & H & H & H & H & CH_2 & H & \\ & & & & | & & & & & & & | & & \\ & & & & CH_3 & & & & & & & CH_3 & & \end{matrix}$$

Here, neighboring chains are arranged in a somewhat random fashion, producing a soft, flexible solid (Fig. 21.10). The plastic bags at the vegetable counters of supermarkets are made of this material.

— *linear* polyethylene, which consists almost entirely of unbranched chains:

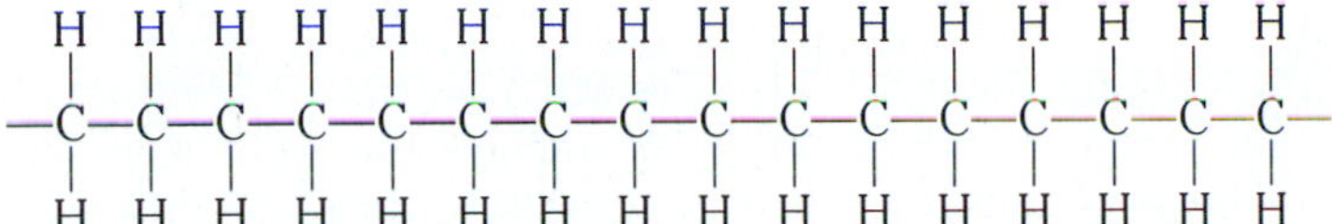

Neighboring chains in linear polyethylene line up nearly parallel to each other. This gives a polymer that approaches a crystalline material. It is used for bottles, toys, and other semirigid objects.

In **polyvinyl chloride**, there is another factor that can complicate the polymer structure. Vinyl chloride, in contrast to ethylene, is unsymmetrical. We might refer to the CH_2 group in vinyl chloride as the "head" of the molecule and the CHCl group as the "tail":

$$\underset{\text{head}}{H_2C}{=}\underset{\text{tail}}{CHCl} \qquad \text{vinyl chloride}$$

In principle at least, vinyl chloride molecules can add to one another to give three different types of polymers:

1. A *head-to-tail* polymer, in which there is a Cl atom on every other C atom in the chain:

$$\begin{array}{cccccccccc} H & H & H & H & H & H & H & H & H & H \\ | & | & | & | & | & | & | & | & | & | \\ -C- & C- & C- & C- & C- & C- & C- & C- & C- & C- \\ | & | & | & | & | & | & | & | & | & | \\ H & Cl & H & Cl & H & Cl & H & Cl & H & Cl \end{array}$$

2. A *head-to-head, tail-to-tail* polymer, in which Cl atoms occur in pairs on adjacent carbon atoms in the chain:

$$\begin{array}{cccccccccc} H & H & H & H & H & H & H & H & H & H \\ | & | & | & | & | & | & | & | & | & | \\ -C- & C- & C- & C- & C- & C- & C- & C- & C- & C- \\ | & | & | & | & | & | & | & | & | & | \\ H & Cl & Cl & H & H & Cl & Cl & H & H & Cl \end{array}$$

The Only Zipper Bag with a Color Change Seal

High-density (left) and low-density (right) polyethylene. (Richard Megna/FUNDAMENTAL PHOTOGRAPHS, New York)

3. A *random* polymer:

```
  H  H  H  H  H  H  H  H  H  H
  |  |  |  |  |  |  |  |  |  |
—C—C—C—C—C—C—C—C—C—C—
  |  |  |  |  |  |  |  |  |  |
  H  Cl H  Cl Cl H  H  Cl H  Cl
```

PVC is a stiff, rugged, cheap polymer

Similar arrangements are possible with polymers made from other asymmetrical monomers (Example 21.6). In practice, the addition polymer is usually of the head-to-tail type. This is the case with polyvinyl chloride and polypropylene.

Example 21.6 Sketch a polymer derived from propene:

```
H       H
 \     /
  C=C
 /     \
H       CH3
```

assuming it to be a

(a) head-to-tail polymer. (b) head-to-head, tail-to-tail polymer.

Strategy In the head-to-tail polymer, the two different groups (H and CH_3 in this case) alternate along the carbon chain. In the other type of polymer, different groups alternate in pairs (two H atoms, then two CH_3 groups, . . .).

Solution

```
     H  H     H  H     H  H     H  H     H  H
     |  |     |  |     |  |     |  |     |  |
(a) —C—C——C—C——C—C——C—C——C—C—
     |  |     |  |     |  |     |  |     |  |
     H  CH3   H  CH3   H  CH3   H  CH3   H  CH3

     H  H     H     H  H  H     H     H  H  H
     |  |     |     |  |  |     |     |  |  |
(b) —C—C——C——C—C—C——C——C—C—C—
     |  |     |     |  |  |     |     |  |  |
     H  CH3   CH3   H  H  CH3   CH3   H  H  CH3
```

Condensation Polymers

In order to produce a condensation polymer, *the molecules involved must have functional groups at both ends of the molecule.* When an alcohol with two —OH groups, HO—R—OH, reacts with a dicarboxylic acid, HOOC—R′—COOH, a **polyester** is formed. The first step in the process is the formation of a simple ester which has a reactive group at both ends of the molecule.

```
HO—R—OH + HO—C—R′—C—OH ⟶
             ‖     ‖
             O     O
dihydroxy      dicarboxylic
alcohol        acid

                 HO—R—O—C—R′—C—OH + H2O
                        ‖     ‖
                        O     O
                 ester with active
                 end groups
```

Thread made of polyester. (Diane Schiumo/FUNDAMENTAL PHOTOGRAPHS, New York)

Figure 21.11
The *Gossamer Albatross*, the first human-powered craft to fly across the English Channel (1979). Lightweight, durable polyesters cover the wings and body.

The COOH group at one end of the ester molecule can react with another alcohol molecule. The OH group at the other end can react with an acid molecule. This process can continue, leading eventually to a long-chain polymer containing 500 or more ester groups. The general structure of the polyester can be represented as

$$-\underset{\substack{\|\\ O}}{C}-R'-\underset{\substack{\|\\ O}}{C}-O-R-O-\underset{\substack{\|\\ O}}{C}-R'-\underset{\substack{\|\\ O}}{C}-O-R-O-$$

section of a polyester molecule

A thin polyester film was used to cover the wings and pilot compartment of the *Gossamer Albatross*, the first human-powered aircraft to cross the English Channel (Fig. 21.11).

One of the most familiar polyesters is Dacron, in which the monomers are ethylene glycol and terephthalic acid:

$$HO-CH_2-CH_2-OH \qquad HO-\underset{\substack{\|\\ O}}{C}-C_6H_4-\underset{\substack{\|\\ O}}{C}-OH$$

ethylene glycol — terephthalic acid

Example 21.7

(a) Show the structure of the ester formed when one molecule of ethylene glycol reacts with one molecule of terephthalic acid.
(b) Draw the structure of a section of the Dacron (Mylar) polymer.

Strategy In (a), split out H_2O between the alcohol and acid molecules. In (b), continue the esterification process at both ends of the molecule obtained in (a).

Solution

(a) $$HO-CH_2-CH_2-O-\underset{\substack{\|\\ O}}{C}-C_6H_4-\underset{\substack{\|\\ O}}{C}-OH$$

(b) $$-\underset{\substack{\|\\ O}}{C}-C_6H_4-\underset{\substack{\|\\ O}}{C}-O-CH_2-CH_2-O-\underset{\substack{\|\\ O}}{C}-C_6H_4-\underset{\substack{\|\\ O}}{C}-O-CH_2-CH_2-O-$$

Figure 21.12
Nylon can be made by the reaction between hexamethylenediamine, $H_2N{-}(CH_2)_6{-}NH_2$ and adipic acid, $HOOC{-}(CH_2)_4{-}COOH$. It forms at the interface between the two reagents. (Charles D. Winters)

Another type of condensation polymer is a **polyamide** formed when a diamine reacts with a dicarboxylic acid:

$$NH_2{-}R{-}\underset{\underset{\textstyle H}{|}}{N}{-}H + HO{-}\underset{\underset{\textstyle O}{\|}}{C}{-}R'{-}\underset{\underset{\textstyle O}{\|}}{C}{-}OH \longrightarrow$$

$$NH_2{-}R{-}\underset{\underset{\textstyle H}{|}}{N}{-}\underset{\underset{\textstyle O}{\|}}{C}{-}R'{-}\underset{\underset{\textstyle O}{\|}}{C}{-}OH + H_2O$$

Condensation can continue to form a long-chain polymer such as nylon-66, which has the structure

$${-}\underset{\underset{\textstyle O}{\|}}{C}{-}(CH_2)_4{-}\underset{\underset{\textstyle O}{\|}}{C}{-}\underset{\underset{\textstyle H}{|}}{N}{-}(CH_2)_6{-}\underset{\underset{\textstyle H}{|}}{N}{-}\underset{\underset{\textstyle O}{\|}}{C}{-}(CH_2)_4{-}\underset{\underset{\textstyle O}{\|}}{C}{-}\underset{\underset{\textstyle H}{|}}{N}{-}(CH_2)_6{-}\underset{\underset{\textstyle H}{|}}{N}{-}$$

In nylon, there are hydrogen bonds between adjacent polymer chains

This polymer was first made in 1935 by Wallace Carothers at Du Pont; it is readily prepared in the laboratory by bringing the two monomers into contact with one another (Fig. 21.12).

PERSPECTIVE • Optical Isomerism

Organic compounds (and certain coordination complexes) show a different type of isomerism from those discussed in this chapter. This is **optical isomerism**; the isomers differ in the effect that they have on plane-polarized light (Fig. 21.A).

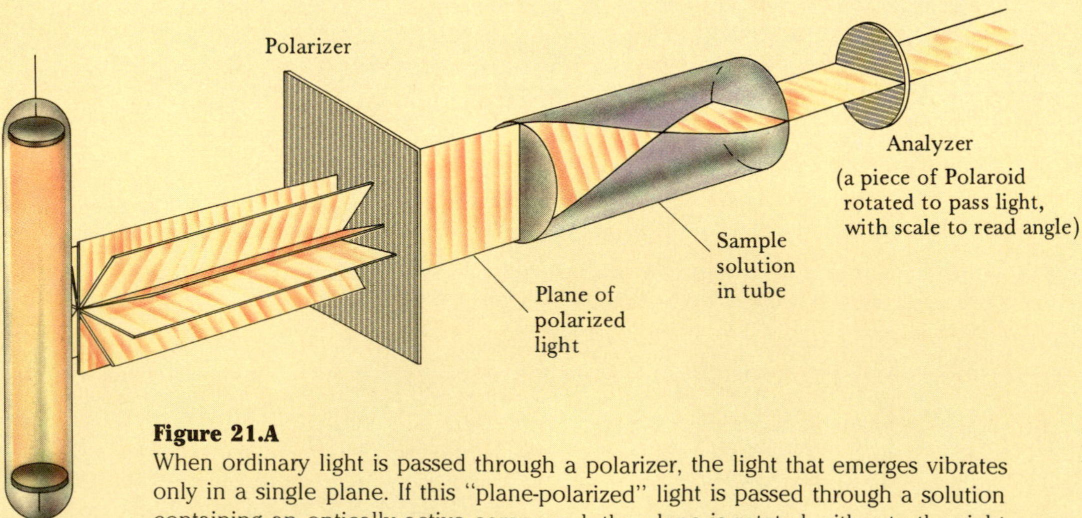

Figure 21.A
When ordinary light is passed through a polarizer, the light that emerges vibrates only in a single plane. If this "plane-polarized" light is passed through a solution containing an optically active compound, the plane is rotated, either to the right (clockwise) or to the left (counterclockwise).

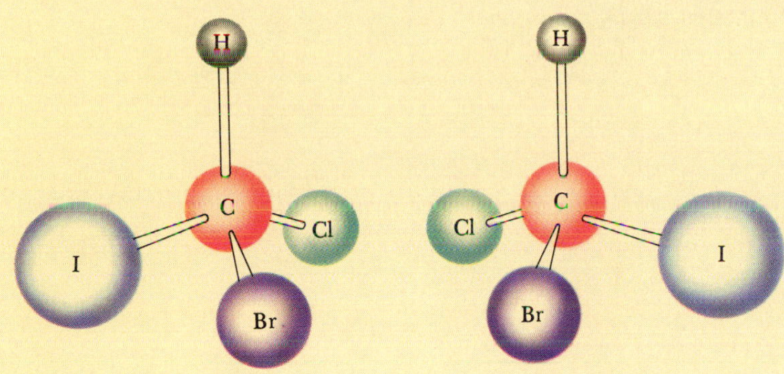

Figure 21.B
The two optical isomers of CHClBrI are mirror images of each other.

Optical isomerism arises because of the tetrahedral nature of the bonding around a carbon atom. It occurs when at least one carbon atom in a molecule is bonded to four different atoms or groups. Consider, for example, the methane derivative CHClBrI. As you can see from Figure 21.B, there are two different forms of this molecule, which are mirror images of one another. The mirror images are not superimposable; that is, you cannot place one molecule over the other so that identical groups are touching. In this sense, the two isomers resemble right- and left-hand gloves.

A carbon atom with four different atoms or groups attached to it is referred to as a **chiral center**. A molecule containing such a carbon atom shows optical isomerism. It exists in two different forms that are nonsuperimposable mirror images. These forms are referred to as optical isomers or *enantiomers*. Molecules may contain more than one chiral center, in which case there can be more than one pair of enantiomers.

Enantiomers ordinarily resemble each other closely in their physical and chemical properties. For example, the two forms of lactic acid have the same melting point (52°C), density (1.25 g/cm^3), and acid dissociation constant ($K_a = 1.4 \times 10^{-4}$):

```
          COOH                COOH
           |                   |
(I) HO—C—H        (II) H—C—OH
           |                   |
          CH3                 CH3
```

enantiomers of lactic acid

On the other hand, enantiomers frequently differ in their physiological activity. This was discovered in 1848 by Louis Pasteur, the father of modern biochemistry. Working with a mixture of the optical isomers of lactic acid, he found that mold growth occurred only with enantiomer II. Apparently, the mold was unable to metabolize enantiomer I. A more modern example of this type involves amphetamine, often used illicitly as an "upper" or "pep pill." Amphetamine consists of two enantiomers:

(I) $C_6H_5-CH_2-\underset{CH_3}{\overset{H}{C}}-NH_2$ (II) $NH_2-\underset{CH_3}{\overset{H}{C}}-CH_2-C_6H_5$

enantiomers of amphetamine

Enantiomer I, called Dexedrine, is by far the stronger stimulant. It is from two to four times as active as Benzedrine, the racemic mixture of the two isomers.

CHAPTER HIGHLIGHTS

KEY CONCEPTS

1. *Draw*
   —structural isomers
   (Example 21.1; Problems 15–24, 57)
   —geometric isomers
   (Example 21.3; Problems 25–28)
2. *Name hydrocarbons*
   (Example 21.2; Problems 1–8)
3. *Write structural formulas for alcohols, carboxylic acids, and esters*
   (Examples 21.4, 21.5; Problems 11–14, 29–32)
4. *Draw structures for*
   —addition polymers
   (Example 21.6; Problems 33–38)
   —condensation polymers
   (Example 21.7; Problems 39–42, 59)

KEY TERMS

alcohol	geometric isomer	polymer
alkane	—*cis-*	—addition
alkene	—*trans-*	—condensation
alkyne	*o-*, *m-*, *p-*	—head-to-head
aromatic	polyamide	—head-to-tail
carboxylic acid	polyester	structural isomer
ester		

SUMMARY PROBLEM

Consider the alkane which has the structure

$$H_3C-\underset{H}{\overset{H}{C}}-\underset{H}{\overset{CH_3}{C}}-CH_3$$

a. Name this alkane.
b. Draw the structures of all the alkanes isomeric with it.
c. Draw the structure of the alkene obtained by removing the two hydrogen atoms shown in red. What is the name of this alkene?

d. Does the alkene referred to in (c) show structural isomerism? geometric isomerism?
e. Sketch a portion of the head-to-tail addition polymer derived from the alkene in (c).
f. Draw the structures of the two alcohols formed by replacing an H atom shown in red with an —OH group.
g. Give the molecular formula of the ester formed when one of the alcohols in (f) reacts with formic acid.

Answers

a. 2-methylbutane

b. 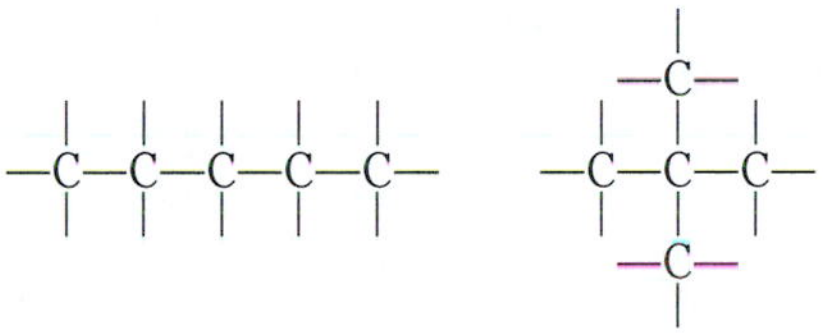
```
 |  |  |  |  |              |
—C—C—C—C—C—                —C—
 |  |  |  |  |           |  |  |
                        —C—C—C—
                         |  |  |
                           —C—
                            |
```

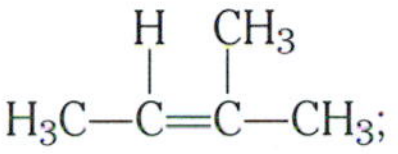

c.
```
      H  CH3
      |  |
H3C—C═C—CH3;
```
2-methyl-2-butene   **d.** yes; no

e.
```
  H    CH3   H    CH3
  |    |     |    |
—C—— C—— C—— C—
  |    |     |    |
 CH3  CH3   CH3  CH3
```

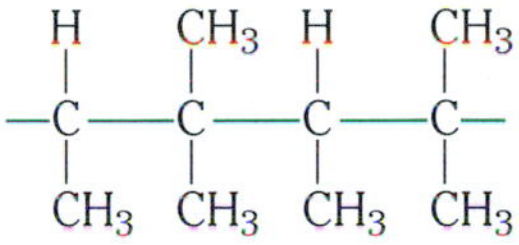

f. 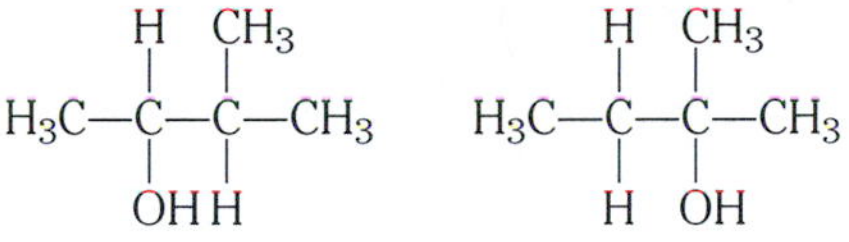
```
      H  CH3               H  CH3
      |  |                 |  |
H3C—C—C—CH3          H3C—C—C—CH3
      |  |                 |  |
     OH  H                 H  OH
```
g. $C_6H_{12}O_2$

QUESTIONS & PROBLEMS

Nomenclature

1. Name the following alkanes.
 a. $CH_3—CH_2—CH(CH_3)—CH_3$
 b. $CH_3—CH_2—CH(CH_2CH_3)—CH_3$
 c. $CH_3—CH(CH_3)—CH(CH_2CH_3)—CH_3$
 d. $CH_3—CH(CH_3)—CH_2—CH(C(H)(CH_3)CH_3)—CH_3$
```
a. CH3—CH2—CH—CH3
              |
             CH3

b. CH3—CH2—CH—CH3
              |
             CH2
              |
             CH3

c. CH3—CH—CH—CH3
        |   |
       CH3 CH2
            |
           CH3

d. CH3—CH—CH2—CH—CH3
        |       |
       CH3   H—C—CH3
                |
               CH3
```
2. Name the following alkanes.
```
a. CH3—(CH2)5—CH—CH3
               |
              CH3
```
 b. $(CH_3)_4C$
```
c. CH3—CH—CH2—C(CH3)3
        |
       CH3

          H
          |
d. CH3—C—(CH2)2—CH—CH3
          |         |
         CH2       CH3
          |
         CH3
```
3. Write structures for the following alkanes.
 a. 3-ethylpentane **b.** 2,2-dimethylbutane
 c. 2-methyl-3-ethylheptane **d.** 2,3-dimethylpentane
4. Write structures for the following alkanes.
 a. 2,2,4-trimethylpentane **b.** 2,2-dimethylpropane
 c. 4-isopropyloctane **d.** 2,3,4-trimethylheptane
5. The following names are incorrect; draw a reasonable structure for the alkane and give the proper IUPAC name.
 a. 5-isopropyloctane **b.** 2-ethylpropane
 c. 1,2-dimethylpropane
6. Follow the directions of Problem 5 for the following names.
 a. 2-dimethylbutane **b.** 4-methylpentane
 c. 2-ethylpropane

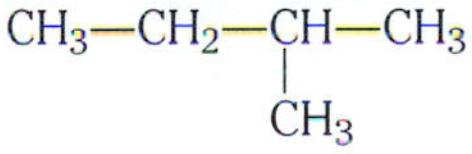

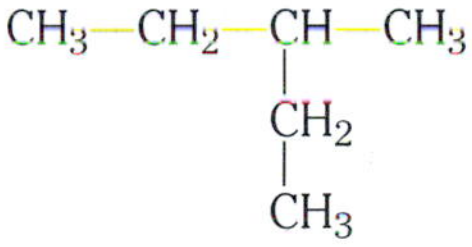

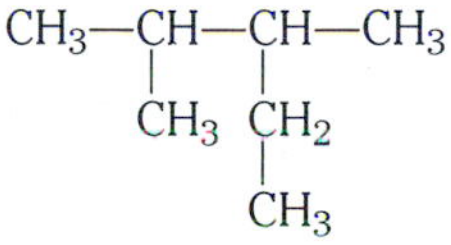

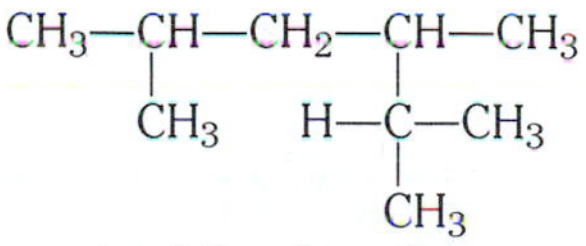

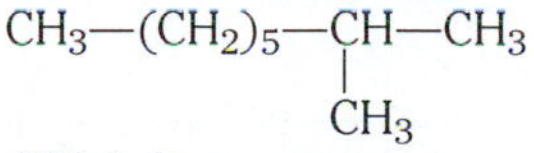

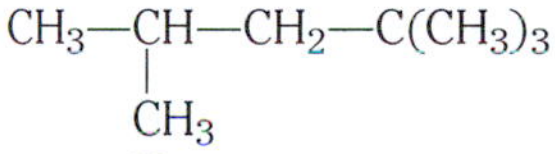

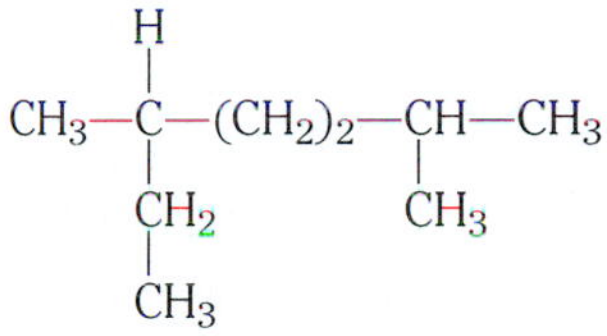

7. Name the following alkenes.
a. $CH_2{=}C{-}(CH_3)_2$
b. $(CH_3)_2{-}C{=}C{-}(CH_3)_2$
c. $CH_3{-}CH{=}CH{-}CH_2CH_3$
d. $CH_3{-}C(CH_2CH_3){=}CH_2$

8. Write structures for the following alkynes.
a. 2-pentyne **b.** 4-methyl-2-pentyne
c. 2-methyl-3-hexyne **d.** 3,3-dimethyl-1-butyne

9. Name the following compounds, using the *ortho-*, *meta-*, *para-* notation.

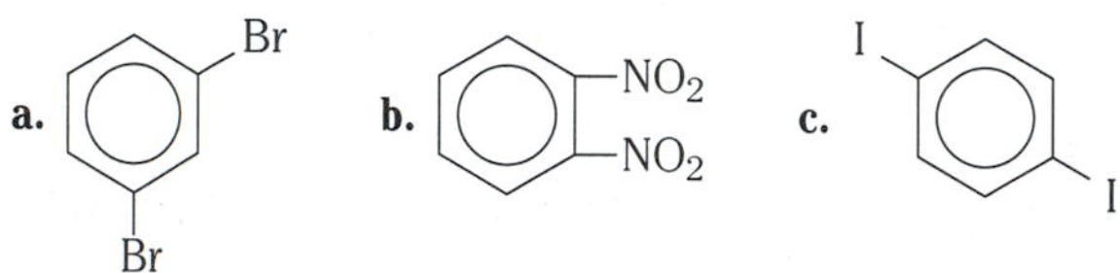

10. Name the following compounds using numbers (1, 2, . . .).

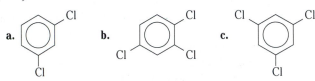

a. (Cl, Cl) b. (Cl, Cl, Cl, Cl) c. (Cl, Cl, Cl)

11. Give the IUPAC and common names of
a. CH_3OH **b.** CH_3COOH **c.** $CH_3CH_2CH_2OH$

12. Give the IUPAC and common names of
a. HCOOH. **b.** CH_3CH_2OH.
c. $CH_3{-}CH(OH){-}CH_3$.

13. Show the structure of
a. methyl formate. **b.** ethyl propionate.
c. propyl acetate. **d.** isopropyl acetate.

14. Name the following esters.
a. $CH_3{-}O{-}C({=}O){-}CH_3$
b. $H{-}C({=}O){-}O{-}CH_2CH_3$
c. $CH_3{-}O{-}C({=}O){-}CH_2CH_3$

Structural Isomerism

15. Draw the structural isomers of the alkane C_6H_{14}.

16. Draw the structural isomers of the alkene C_4H_8.

17. Draw the structural isomers of C_4H_9Cl in which one hydrogen atom of a C_4H_{10} molecule has been replaced by chlorine.

18. Draw the structural isomers of $C_3H_6Cl_2$ in which two of the hydrogen atoms of C_3H_8 have been replaced by chlorine atoms.

19. There are three compounds with the formula C_6H_4ClBr in which two of the hydrogen atoms of the benzene molecule have been replaced by halogen atoms. Draw structures for these compounds.

20. There are three compounds with the formula $C_6H_3Cl_3$ in which three of the hydrogen atoms of the benzene molecule have been replaced by chlorine atoms. Draw structures for these compounds.

21. Write structures for all the structural isomers of double-bonded compounds with the molecular formula C_5H_{10}.

22. Write structures for all the structural isomers of compounds with the molecular formula C_4H_6ClBr in which Cl and Br are bonded to a double-bonded carbon.

23. Draw structures for all the alcohols with molecular formula $C_5H_{12}O$.

24. Draw structures for all the saturated carboxylic acids with four carbon atoms per molecule.

Geometric Isomerism

25. Of the compounds in Problem 21, which ones show geometric isomerism? Draw the *cis-* and *trans-* isomers.

26. Of the compounds in Problem 22, which ones show geometric isomerism? Draw the *cis-* and *trans-* isomers.

27. Maleic acid and fumaric acid are the *cis-* and *trans-* isomers, respectively, of $C_2H_2(COOH)_2$, a dicarboxylic acid. Draw and label their structures.

28. For which of the following is geometric isomerism possible?
a. $(CH_3)_2C{=}CCl_2$ **b.** $CH_3ClC{=}CCH_3Cl$
c. $CH_3BrC{=}CCH_3Cl$

Functional Groups

29. Classify each of the following as a carboxylic acid, ester, and/or alcohol.
a. $HO{-}CH_2{-}CH_2{-}CH_2{-}OH$
b. 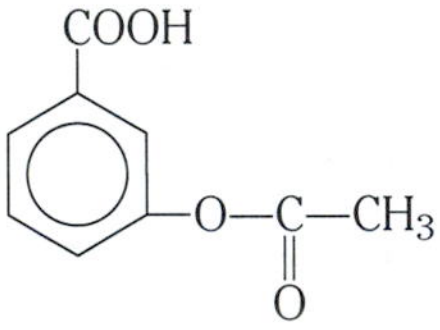 benzene ring bearing COOH and $-O{-}C({=}O){-}CH_3$
c. $CH_3{-}CH(OH){-}COOH$ (H above, OH below the central C)

30. Classify each of the following as a carboxylic acid, ester, and/or alcohol.
a. $CH_3{-}(CH_2)_3{-}OH$
b. $CH_3{-}CH_2{-}C({=}O){-}O{-}CH_2{-}CH_3$
c. $CH_3{-}CH_2{-}O{-}C({=}O){-}(CH_2)_6{-}COOH$

31. Give the structure of
a. a four-carbon straight-chain alcohol with an —OH group not at the end of the chain.
b. a five-carbon straight-chain carboxylic acid.
c. the ester formed when these two compounds react.

32. Give the structure of
a. a three-carbon alcohol with the —OH group on the center carbon.

b. a four-carbon branched-chain carboxylic acid.
c. the ester formed when these two compounds react.

Addition Polymers

33. Consider a polymer made from tetrachloroethylene, C_2Cl_4.
a. Draw a portion of the polymer chain.
b. What is the molar mass of the polymer if it contains 3.2×10^3 tetrachloroethylene molecules?
c. What are the mass percents of C and Cl in the polymer?

34. Consider Teflon, the polymer made from tetrafluoroethylene.
a. Draw a portion of the Teflon molecule.
b. Calculate the molar mass of a Teflon molecule that contains 5.0×10^4 CF_2 units.
c. What are the mass percents of C and F in Teflon?

35. Sketch a portion of the acrylonitrile polymer (Table 21.4), assuming it is a
a. head-to-tail polymer.
b. head-to-head, tail-to-tail polymer.

36. Styrene, $H_2C{=}C(H){-}C_6H_5$, forms a head-to-tail addition polymer. Sketch a portion of a polystyrene molecule.

37. The polymer whose structure is shown below is made from two different monomers. Identify the monomers.

$$-CH_2-CH(Cl)-CH_2-CH(C_6H_5)-CH_2-CH(Cl)-CH_2-CH(C_6H_5)-$$

38. Show the structure of the monomer used to make the following addition polymers.

a. $-CH_2-CH(F)-CH(F)-CH_2-$ **b.** $-CH(CH_3)-CH(CH_3)-CH(CH_3)-CH(CH_3)-$

Condensation Polymers

39. A rather simple polymer can be made from ethylene glycol, $HO-CH_2-CH_2-OH$, and oxalic acid, $HO-C(=O)-C(=O)-OH$. Sketch a portion of the polymer chain obtained from these monomers.

40. Lexan is a very rugged polyester in which the monomers can be taken to be carbonic acid, $HO-C(=O)-OH$, and

$$HO-C_6H_4-C(CH_3)_2-C_6H_4-OH$$

Sketch a section of the Lexan chain.

41. The following condensation polymer is made from a single monomer. Identify the monomer.

$$-N(H)-CH_2-C(=O)-N(H)-CH_2-C(=O)-$$

42. Identify the monomers from which the following condensation polymers are made.

a. $-N(H)-CH_2-CH_2-N(H)-C(=O)-CH_2-C(=O)-$

b. $-C(=O)-C_6H_4-C(=O)-O-CH_2-C(H)(CH_3)-O-$

Optical Isomerism

43. Which of the following can show optical isomerism?
a. 2-bromo-2-chlorobutane
b. 2-methylpropane
c. 2,2-dimethyl-1-butanol
d. 2,2,4-trimethylpentane

44. Which of the following compounds can show optical isomerism?
a. dichloromethane
b. 1,2-dichloroethane
c. bromochlorofluoromethane
d. 1-bromoethanol

45. Locate the chiral carbon(s), if any, in the following molecules.

a. $HO-C(=O)-C(H)(OH)-C(H)(OH)-C(=O)-H$ **b.** $CH_3-C(=O)-C(=O)-OH$

c. $CH_3-CH_2-C(H)(NH_2)-COOH$

46. Locate the chiral carbon(s), if any, in the following molecules.

a. $CH_3-C(H)(OH)-C(H)(OH)-H$ **b.** $H-C(H){=}C(H)-CH_2-OH$

c. $CH_3-C(Cl)(Cl)-C(F)(H)-Cl$

Unclassified

47. Which of the following are expected to show bond angles of 109.5°? 120°? 180°?

a. H_3C-CH_3 **b.** $H_3C-C(H){=}CH_2$
c. $H_3C-C{\equiv}C-CH_3$

48. State at least one use for each of the following hydrocarbons.

a. C_3H_8 **b.** C_2H_4 **c.** C_2H_2 **d.** toluene

49. What does the "circle" represent in the structural formula of benzene?

50. Correct each of the following statements to make it true.

a. Gasoline contains mostly alkenes and alkynes.
b. Benzene is used in chemotherapy.
c. Calcium carbide reacts with water to form ethene.
d. Alkanes show geometric isomerism.
e. Cholesterol is a condensed-ring hydrocarbon.

51. Explain how the following compounds are prepared.

a. methanol **b.** ethanol **c.** acetic acid
d. ethyl acetate

52. Calculate $[H^+]$ and the pH of a 0.10 *M* solution of chloroacetic acid ($K_a = 1.5 \times 10^{-3}$); use successive approximations.

53. Explain what is meant by

a. a saturated fat. **b.** a soap. **c.** the "proof" of an alcoholic beverage. **d.** denatured alcohol.

54. Which of the following monomers could form an addition polymer? a condensation polymer?

a. C_2H_6 **b.** C_2H_4
c. $HO—CH_2—CH_2—OH$ **d.** $HO—CH_2—CH_3$

55. Draw the structures of the monomers that could be used to make the following polymers

a.
```
    Br H  Br H  Br H
    |  |  |  |  |  |
  —C—C—C—C—C—C—
    |  |  |  |  |  |
    H  H  H  H  H  H
```

b.
```
        H           H
        |           |
  —N—C—C—N—C—C—
    |  |  ‖  |  |  ‖
    H  H  O  H  H  O
```

c.
```
    O              O  H               H
    ‖              ‖  |               |
  —C—(CH2)4—C—N—(CH2)6—N—
```

Challenge Problems

56. The structure of cholesterol is given in the text. What is its molecular formula?

57. Draw structures for all the alcohols with molecular formula $C_6H_{14}O$.

58. What mass of propane must be burned to bring one quart of water in a saucepan at 25°C to the boiling point? Use data in Appendix 1 and elsewhere and make any reasonable assumptions.

59. Glycerol, $C_3H_5(OH)_3$, and orthophthalic acid,

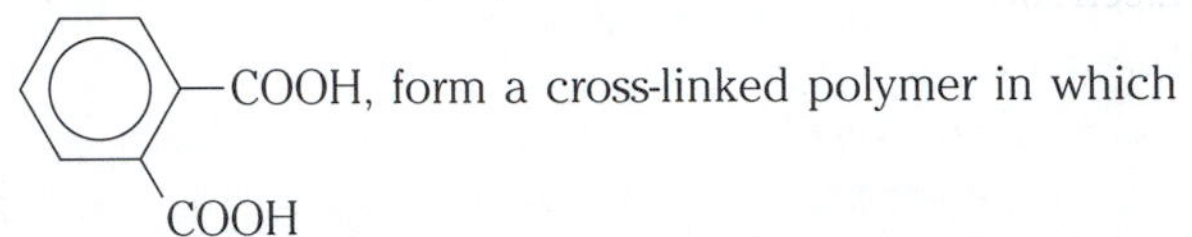

form a cross-linked polymer in which adjacent polymer chains are linked together; this polymer is used in floor coverings and dentures.

a. Write the structural formula for a portion of the polymer chain.
b. Use your answer in (a) to show how cross-linking can occur between the polymer chains to form a water-insoluble, network covalent solid.

Appendix 1 Constants and Reference Data

FUNDAMENTAL CONSTANTS

Acceleration of gravity (standard)	$9.8066\ m/s^2$
Atomic mass unit (amu)	1.6605×10^{-24} g
Avogadro constant	6.0221×10^{23}/mol
Electronic charge	1.6022×10^{-19} C
Electronic mass	9.1094×10^{-28} g
Faraday constant	9.6485×10^{4} C/mol
Gas constant	$0.082058\ L \cdot atm/mol \cdot K$
	$8.3145\ J/mol \cdot K$
Planck constant	$6.6261 \times 10^{-34}\ J \cdot s$
Velocity of light	2.9979×10^{8} m/s
π	3.1416
e	2.7183
$\ln x$	$2.3026 \log_{10}x$

VAPOR PRESSURE OF WATER (mm Hg)

T(°C)	vp	T(°C)	vp	T(°C)	vp	T(°C)	vp
0	4.58	21	18.65	35	42.2	92	567.0
5	6.54	22	19.83	40	55.3	94	610.9
10	9.21	23	21.07	45	71.9	96	657.6
12	10.52	24	22.38	50	92.5	98	707.3
14	11.99	25	23.76	55	118.0	100	760.0
16	13.63	26	25.21	60	149.4	102	815.9
17	14.53	27	26.74	65	187.5	104	875.1
18	15.48	28	28.35	70	233.7	106	937.9
19	16.48	29	30.04	80	355.1	108	1004.4
20	17.54	30	31.82	90	525.8	110	1074.6

THERMODYNAMIC DATA

	ΔH_f° (kJ/mol)	S° (kJ/K · mol)	ΔG_f° (kJ/mol) at 25°C		ΔH_f° (kJ/mol)	S° (kJ/K · mol)	ΔG_f° (kJ/mol) at 25°C
$Ag(s)$	0.0	+0.0426	0.0	$Cd(s)$	0.0	+0.0518	0.0
$Ag^+(aq)$	+105.6	+0.0727	+77.1	$Cd^{2+}(aq)$	−75.9	−0.0732	−77.6
$AgBr(s)$	−100.4	+0.1071	−96.9	$CdCl_2(s)$	−391.5	+0.1153	−344.0
$AgCl(s)$	−127.1	+0.0962	−109.8	$CdO(s)$	−258.2	+0.0548	−228.4
$AgI(s)$	−61.8	+0.1155	−66.2	$Cl_2(g)$	0.0	+0.2230	0.0
$AgNO_3(s)$	−124.4	+0.1409	−33.4	$Cl^-(aq)$	−167.2	+0.0565	−131.2
$Ag_2O(s)$	−31.0	+0.1213	−11.2	$ClO_3^-(aq)$	−104.0	+0.1623	−8.0
$Al(s)$	0.0	+0.0283	0.0	$ClO_4^-(aq)$	−129.3	+0.1820	−8.5
$Al^{3+}(aq)$	−531.0	−0.3217	−485.0	$Cr(s)$	0.0	+0.0238	0.0
$Al_2O_3(s)$	−1675.7	+0.0509	−1582.3	$CrO_4^{2-}(aq)$	−881.2	+0.0502	−727.8
$Ba(s)$	0.0	+0.0628	0.0	$Cr_2O_3(s)$	−1139.7	+0.0812	−1058.1
$Ba^{2+}(aq)$	−537.6	+0.0096	−560.8	$Cr_2O_7^{2-}(aq)$	−1490.3	+0.2619	−1301.1
$BaCl_2(s)$	−858.6	+0.1237	−810.4	$Cu(s)$	0.0	+0.0332	0.0
$BaCO_3(s)$	−1216.3	+0.1121	−1137.6	$Cu^+(aq)$	+71.7	+0.0406	+50.0
$BaO(s)$	−553.5	+0.0704	−525.1	$Cu^{2+}(aq)$	+64.8	−0.0996	+65.5
$BaSO_4(s)$	−1473.2	+0.1322	−1362.3	$CuO(s)$	−157.3	+0.0426	−129.7
$Br_2(l)$	0.0	+0.1522	0.0	$Cu_2O(s)$	−168.6	+0.0931	−146.0
$Br^-(aq)$	−121.6	+0.0824	−104.0	$CuS(s)$	−53.1	+0.0665	−53.6
$C(s)$	0.0	+0.0057	0.0	$Cu_2S(s)$	−79.5	+0.1209	−86.2
$CCl_4(l)$	−135.4	+0.2164	−65.3	$CuSO_4(s)$	−771.4	+0.1076	−661.9
$CHCl_3(l)$	−134.5	+0.2017	−73.7	$F_2(g)$	0.0	+0.2027	0.0
$CH_4(g)$	−74.8	+0.1862	−50.7	$F^-(aq)$	−332.6	−0.0138	−278.8
$C_2H_2(g)$	+226.7	+0.2008	+217.1	$Fe(s)$	0.0	+0.0273	0.0
$C_2H_4(g)$	+52.3	+0.2195	+68.1	$Fe^{2+}(aq)$	−89.1	−0.1377	−78.9
$C_2H_6(g)$	−84.7	+0.2295	−32.9	$Fe^{3+}(aq)$	−48.5	−0.3159	−4.7
$C_3H_8(g)$	−103.8	+0.2699	−23.5	$Fe(OH)_3(s)$	−823.0	+0.1067	−696.6
$CH_3OH(l)$	−238.7	+0.1268	−166.3	$Fe_2O_3(s)$	−824.2	+0.0874	−742.2
$C_2H_5OH(l)$	−277.7	+0.1607	−174.9	$Fe_3O_4(s)$	−1118.4	+0.1464	−1015.5
$CO(g)$	−110.5	+0.1976	−137.2	$H_2(g)$	0.0	+0.1306	0.0
$CO_2(g)$	−393.5	+0.2136	−394.4	$H^+(aq)$	0.0	0.0000	0.0
$CO_3^{2-}(aq)$	−677.1	−0.0569	−527.8	$HBr(g)$	−36.4	+0.1986	−53.4
$Ca(s)$	0.0	+0.0414	0.0	$HCl(g)$	−92.3	+0.1868	−95.3
$Ca^{2+}(aq)$	−542.8	−0.0531	−553.6	$HCO_3^-(aq)$	−692.0	+0.0912	−586.8
$CaCl_2(s)$	−795.8	+0.1046	−748.1	$HF(g)$	−271.1	+0.1737	−273.2
$CaCO_3(s)$	−1206.9	+0.0929	−1128.8	$HI(g)$	+26.5	+0.2065	+1.7
$CaO(s)$	−635.1	+0.0398	−604.0	$HNO_3(l)$	−174.1	+0.1556	−80.8
$Ca(OH)_2(s)$	−986.1	+0.0834	−898.5	$H_2O(g)$	−241.8	+0.1887	−228.6
$CaSO_4(s)$	−1434.1	+0.1067	−1321.8	$H_2O(l)$	−285.8	+0.0699	−237.2

	ΔH_f° (kJ/mol)	S° (kJ/K · mol)	ΔG_f° (kJ/mol) at 25°C		ΔH_f° (kJ/mol)	S° (kJ/K · mol)	ΔG_f° (kJ/mol) at 25°C
$H_2O_2(l)$	−187.8	+0.1096	−120.4	$NO_2^-(aq)$	−104.6	+0.1230	−32.2
$H_2PO_4^-(aq)$	−1296.3	+0.0904	−1130.3	$NO_3^-(aq)$	−205.0	+0.1464	−108.7
$HPO_4^{2-}(aq)$	−1292.1	−0.0335	−1089.2	$N_2O_4(g)$	+9.2	+0.3042	+97.9
$H_2S(g)$	−20.6	+0.2057	−33.6	$Na(s)$	0.0	+0.0512	0.0
$H_2SO_4(l)$	−814.0	+0.1569	−690.1	$Na^+(aq)$	−240.1	+0.0590	−261.9
$HSO_4^-(aq)$	−887.3	+0.1318	−755.9	$NaCl(s)$	−411.2	+0.0721	−384.2
$Hg(l)$	0.0	+0.0760	0.0	$NaF(s)$	−573.6	+0.0515	−543.5
$Hg^{2+}(aq)$	+171.1	−0.0322	+164.4	$NaOH(s)$	−425.6	+0.0645	−379.5
$HgO(s)$	−90.8	+0.0703	−58.6	$Ni(s)$	0.0	+0.0299	0.0
$I_2(s)$	0.0	+0.1161	0.0	$Ni^{2+}(aq)$	−54.0	−0.1289	−45.6
$I^-(aq)$	−55.2	+0.1113	−51.6	$NiO(s)$	−239.7	+0.0380	−211.7
$K(s)$	0.0	+0.0642	0.0	$O_2(g)$	0.0	+0.2050	0.0
$K^+(aq)$	−252.4	+0.1025	−283.3	$OH^-(aq)$	−230.0	−0.0108	−157.2
$KBr(s)$	−393.8	+0.0959	−380.7	$P_4(s)$	0.0	+0.1644	0.0
$KCl(s)$	−436.7	+0.0826	−409.1	$PCl_3(g)$	−287.0	+0.3117	−267.8
$KClO_3(s)$	−397.7	+0.1431	−296.3	$PCl_5(g)$	−374.9	+0.3645	−305.0
$KClO_4(s)$	−432.8	+0.1510	−303.2	$PO_4^{3-}(aq)$	−1277.4	−0.222	−1018.7
$KNO_3(s)$	−369.8	+0.1330	−394.9	$Pb(s)$	0.0	+0.0648	0.0
$Mg(s)$	0.0	+0.0327	0.0	$Pb^{2+}(aq)$	−1.7	+0.0105	−24.4
$Mg^{2+}(aq)$	−466.8	−0.1381	−454.8	$PbBr_2(s)$	−278.7	+0.1615	−261.9
$MgCl_2(s)$	−641.3	+0.0896	−591.8	$PbCl_2(s)$	−359.4	+0.1360	−314.1
$MgCO_3(s)$	−1095.8	+0.0657	−1012.1	$PbO(s)$	−219.0	+0.0665	−188.9
$MgO(s)$	−601.7	+0.0269	−569.4	$PbO_2(s)$	−277.4	+0.0686	−217.4
$Mg(OH)_2(s)$	−924.5	+0.0632	−833.6	$S(s)$	0.0	+0.0318	0.0
$MgSO_4(s)$	−1284.9	+0.0916	−1170.7	$S^{2-}(aq)$	+33.1	−0.0146	+85.8
$Mn(s)$	0.0	+0.0320	0.0	$SO_2(g)$	−296.8	+0.2481	−300.2
$Mn^{2+}(aq)$	−220.8	−0.0736	−228.1	$SO_3(g)$	−395.7	+0.2567	−371.1
$MnO(s)$	−385.2	+0.0597	−362.9	$SO_4^{2-}(aq)$	−909.3	+0.0201	−744.5
$MnO_2(s)$	−520.0	+0.0530	−465.2	$Si(s)$	0.0	+0.0188	0.0
$MnO_4^-(aq)$	−541.4	+0.1912	−447.2	$SiO_2(s)$	−910.9	+0.0418	−856.7
$N_2(g)$	0.0	+0.1915	0.0	$Sn(s)$	0.0	+0.0516	0.0
$NH_3(g)$	−46.1	+0.1923	−16.5	$Sn^{2+}(aq)$	−8.8	−0.0174	−27.2
$NH_4^+(aq)$	−132.5	+0.1134	−79.3	$SnO_2(s)$	−580.7	+0.0523	−519.6
$NH_4Cl(s)$	−314.4	+0.0946	−203.0	$Zn(s)$	0.0	+0.0416	0.0
$NH_4NO_3(s)$	−365.6	+0.1511	−184.0	$Zn^{2+}(aq)$	−153.9	−0.1121	−147.1
$N_2H_4(l)$	+50.6	+0.1212	+149.2	$ZnI_2(s)$	−208.0	+0.1611	−209.0
$NO(g)$	+90.2	+0.2107	+86.6	$ZnO(s)$	−348.3	+0.0436	−318.3
$NO_2(g)$	+33.2	+0.2400	+51.3	$ZnS(s)$	−206.0	+0.0577	−201.3

EQUILIBRIUM CONSTANTS

Dissociation Constants, Weak Acids, K_a

H_3AsO_4	5.7×10^{-3}	HNO_2	6.0×10^{-4}	$N_2H_5^+$	1.0×10^{-8}
$H_2AsO_4^-$	1.8×10^{-7}	H_3PO_4	7.1×10^{-3}	$Al(H_2O)_6^{3+}$	1.2×10^{-5}
$HAsO_4^{2-}$	2.5×10^{-12}	$H_2PO_4^-$	6.2×10^{-8}	$Ag(H_2O)_2^+$	1.2×10^{-12}
$HBrO$	2.6×10^{-9}	HPO_4^{2-}	4.5×10^{-13}	$Ca(H_2O)_6^{2+}$	2.2×10^{-13}
$HCHO_2$	1.9×10^{-4}	H_2S	1.0×10^{-7}	$Cd(H_2O)_4^{2+}$	4.0×10^{-10}
$HC_2H_3O_2$	1.8×10^{-5}	HS^-	1×10^{-13}	$Fe(H_2O)_6^{3+}$	6.7×10^{-3}
HCN	5.8×10^{-10}	H_2SO_3	1.7×10^{-2}	$Fe(H_2O)_6^{2+}$	1.7×10^{-7}
H_2CO_3	4.4×10^{-7}	HSO_3^-	6.0×10^{-8}	$Mg(H_2O)_6^{2+}$	3.7×10^{-12}
HCO_3^-	4.7×10^{-11}	HSO_4^-	1.0×10^{-2}	$Mn(H_2O)_6^{2+}$	2.8×10^{-11}
$HClO_2$	1.0×10^{-2}	H_2Se	1.5×10^{-4}	$Ni(H_2O)_6^{2+}$	2.2×10^{-10}
$HClO$	2.8×10^{-8}	H_2SeO_3	2.7×10^{-3}	$Pb(H_2O)_6^{2+}$	6.7×10^{-7}
HF	6.9×10^{-4}	$HSeO_3^-$	5.0×10^{-8}	$Sc(H_2O)_6^{3+}$	1.1×10^{-4}
HIO	2.4×10^{-11}	$CH_3NH_3^+$	2.4×10^{-11}	$Zn(H_2O)_4^{2+}$	3.3×10^{-10}
HN_3	2.4×10^{-5}	NH_4^+	5.6×10^{-10}		

Dissociation Constants, Weak Bases, K_b

AsO_4^{3-}	4.0×10^{-3}	N_3^-	4.2×10^{-10}	$HSeO_3^-$	3.7×10^{-12}
$HAsO_4^{2-}$	5.6×10^{-8}	NH_3	1.8×10^{-5}	$AlOH^{2+}$	8.3×10^{-10}
$H_2AsO_4^-$	1.8×10^{-12}	N_2H_4	1.0×10^{-6}	$AgOH$	8.3×10^{-3}
BrO^-	3.8×10^{-6}	NO_2^-	1.7×10^{-11}	$CaOH^+$	4.5×10^{-2}
CH_3NH_2	4.2×10^{-4}	PO_4^{3-}	2.2×10^{-2}	$CdOH^+$	2.5×10^{-5}
CHO_2^-	5.3×10^{-11}	HPO_4^{2-}	1.6×10^{-7}	$FeOH^{2+}$	1.5×10^{-12}
$C_2H_3O_2^-$	5.6×10^{-10}	$H_2PO_4^-$	1.4×10^{-12}	$FeOH^+$	5.9×10^{-8}
CN^-	1.7×10^{-5}	S^{2-}	1×10^{-1}	$MgOH^+$	2.7×10^{-3}
CO_3^{2-}	2.1×10^{-4}	HS^-	1.0×10^{-7}	$MnOH^+$	3.6×10^{-4}
HCO_3^-	2.3×10^{-8}	SO_3^{2-}	1.7×10^{-7}	$NiOH^+$	4.5×10^{-5}
ClO_2^-	1.0×10^{-12}	HSO_3^-	5.9×10^{-13}	$PbOH^+$	1.5×10^{-8}
ClO^-	3.6×10^{-7}	SO_4^{2-}	1.0×10^{-12}	$ScOH^{2+}$	9.1×10^{-11}
F^-	1.4×10^{-11}	HSe^-	6.7×10^{-11}	$ZnOH^+$	3.0×10^{-5}
IO^-	4.2×10^{-4}	SeO_3^{2-}	2.0×10^{-7}		

Solubility Product Constants, K_{sp}

Compound	K_{sp}	Compound	K_{sp}	Compound	K_{sp}
AgBr	5×10^{-13}	$Co_3(PO_4)_2$	1×10^{-35}	$Ni_3(PO_4)_2$	1×10^{-32}
$AgC_2H_3O_2$	1.9×10^{-3}	CuCl	1.7×10^{-7}	NiS	1×10^{-21}
Ag_2CO_3	8×10^{-12}	CuBr	6.3×10^{-9}	$PbBr_2$	6.6×10^{-6}
AgCl	1.8×10^{-10}	CuI	1.2×10^{-12}	$PbCO_3$	1×10^{-13}
Ag_2CrO_4	1×10^{-12}	$Cu_3(PO_4)_2$	1×10^{-37}	$PbCl_2$	1.7×10^{-5}
AgI	1×10^{-16}	CuS	1×10^{-36}	$PbCrO_4$	2×10^{-14}
Ag_3PO_4	1×10^{-16}	Cu_2S	1×10^{-48}	PbF_2	7.1×10^{-7}
Ag_2S	1×10^{-49}	$FeCO_3$	3.1×10^{-11}	PbI_2	8.4×10^{-9}
AgSCN	1.0×10^{-12}	$Fe(OH)_2$	5×10^{-17}	$Pb(OH)_2$	1×10^{-20}
AlF_3	1×10^{-18}	$Fe(OH)_3$	3×10^{-39}	PbS	1×10^{-28}
$Al(OH)_3$	2×10^{-31}	FeS	2×10^{-19}	$Pb(SCN)_2$	2.1×10^{-5}
$AlPO_4$	1×10^{-20}	GaF_3	2×10^{-16}	$PbSO_4$	1.8×10^{-8}
$BaCO_3$	2.6×10^{-9}	$Ga(OH)_3$	1×10^{-35}	$Sc(OH)_3$	2×10^{-31}
$BaCrO_4$	1.2×10^{-10}	$GaPO_4$	1×10^{-21}	SnS	3×10^{-28}
BaF_2	1.8×10^{-7}	Hg_2Br_2	6×10^{-23}	$SrCO_3$	5.6×10^{-10}
$BaSO_4$	1.1×10^{-10}	Hg_2Cl_2	1×10^{-18}	$SrCrO_4$	3.6×10^{-5}
Bi_2S_3	1×10^{-99}	Hg_2I_2	5×10^{-29}	SrF_2	4.3×10^{-9}
$CaCO_3$	4.9×10^{-9}	HgS	1×10^{-52}	$SrSO_4$	3.4×10^{-7}
CaF_2	1.5×10^{-10}	Li_2CO_3	8.2×10^{-4}	TlCl	1.9×10^{-4}
$Ca_3(PO_4)_2$	1×10^{-33}	$MgCO_3$	6.8×10^{-6}	TlBr	3.7×10^{-6}
$CaSO_4$	7.1×10^{-5}	MgF_2	7×10^{-11}	TlI	5.6×10^{-8}
$CdCO_3$	6×10^{-12}	$Mg(OH)_2$	6×10^{-12}	$Tl(OH)_3$	2×10^{-44}
$Cd(OH)_2$	5×10^{-15}	$Mg_3(PO_4)_2$	1×10^{-24}	Tl_2S	1×10^{-20}
$Cd_3(PO_4)_2$	1×10^{-33}	$Mn(OH)_2$	2×10^{-13}	$ZnCO_3$	1.1×10^{-10}
CdS	1×10^{-29}	MnS	5×10^{-14}	$Zn(OH)_2$	4×10^{-17}
$Co(OH)_2$	2×10^{-16}	$NiCO_3$	1.4×10^{-7}		

Appendix 2 Properties of the Elements

Element	At. No.	mp(°C)	bp(°C)	E.N.	Ion. Ener. (kJ/mol)	At. Rad. (nm)	Ion. Rad. (nm)
H	1	−259	−253	2.2	1312	0.037	(−1)0.208
He	2	−272	−269		2372	0.05	
Li	3	186	1326	1.0	520	0.152	(+1)0.060
Be	4	1283	2970	1.6	900	0.111	(+2)0.031
B	5	2300	2550	2.0	801	0.088	
C	6	3570	subl.	2.5	1086	0.077	
N	7	−210	−196	3.0	1402	0.070	
O	8	−218	−183	3.5	1314	0.066	(−2)0.140
F	9	−220	−188	4.0	1681	0.064	(−1)0.136
Ne	10	−249	−246		2081	0.070	
Na	11	98	889	0.9	496	0.186	(+1)0.095
Mg	12	650	1120	1.3	738	0.160	(+2)0.065
Al	13	660	2327	1.6	578	0.143	(+3)0.050
Si	14	1414	2355	1.9	786	0.117	
P	15	44	280	2.2	1012	0.110	
S	16	119	444	2.6	1000	0.104	(−2)0.184
Cl	17	−101	−34	3.2	1251	0.099	(−1)0.181
Ar	18	−189	−186		1520	0.094	
K	19	64	774	0.8	419	0.231	(+1)0.133
Ca	20	845	1420	1.0	590	0.197	(+2)0.099
Sc	21	1541	2831	1.4	631	0.160	(+3)0.081
Ti	22	1660	3287	1.5	658	0.146	
V	23	1890	3380	1.6	650	0.131	
Cr	24	1857	2672	1.6	653	0.125	(+3)0.064
Mn	25	1244	1962	1.5	717	0.129	(+2)0.080
Fe	26	1535	2750	1.8	759	0.126	(+2)0.075
Co	27	1495	2870	1.9	758	0.125	(+2)0.072
Ni	28	1453	2732	1.9	737	0.124	(+2)0.069
Cu	29	1083	2567	1.9	746	0.128	(+1)0.096
Zn	30	420	907	1.6	906	0.133	(+2)0.074
Ga	31	30	2403	1.6	579	0.122	(+3)0.062
Ge	32	937	2830	2.0	762	0.122	
As	33	814	subl.	2.2	944	0.121	
Se	34	217	685	2.5	941	0.117	(−2)0.198
Br	35	−7	59	3.0	1140	0.114	(−1)0.195
Kr	36	−157	−152		1351	0.109	
Rb	37	39	688	0.8	403	0.244	(+1)0.148

Element	At. No.	mp(°C)	bp(°C)	E.N.	Ion. Ener. (kJ/mol)	At. Rad. (nm)	Ion. Rad. (nm)
Sr	38	770	1380	0.9	550	0.215	(+2)0.113
Y	39	1509	2930	1.2	616	0.180	(+3)0.093
Zr	40	1852	3580	1.4	660	0.157	
Nb	41	2468	5127	1.6	664	0.143	
Mo	42	2610	5560	1.8	685	0.136	
Tc	43	2200	4700	1.9	702	0.136	
Ru	44	2430	3700	2.2	711	0.133	
Rh	45	1966	3700	2.2	720	0.134	
Pd	46	1550	3170	2.2	805	0.138	
Ag	47	961	2210	1.9	731	0.144	(+1)0.126
Cd	48	321	767	1.7	868	0.149	(+2)0.097
In	49	157	2000	1.7	558	0.162	(+3)0.081
Sn	50	232	2270	1.9	709	0.140	
Sb	51	631	1380	2.0	832	0.141	
Te	52	450	990	2.1	869	0.137	(−2)0.221
I	53	114	184	2.7	1009	0.133	(−1)0.216
Xe	54	−112	−107		1170	0.130	
Cs	55	28	690	0.8	376	0.262	(+1)0.169
Ba	56	725	1640	0.9	503	0.217	(+2)0.135
La	57	920	3469	1.1	538	0.187	(+3)0.115
Ce	58	795	3468	1.1	528	0.182	(+3)0.101
Pr	59	935	3127	1.1	523	0.182	(+3)0.100
Nd	60	1024	3027	1.1	530	0.182	(+3)0.099
Pm	61	1027	2727	1.1	536	0.181	
Sm	62	1072	1900	1.1	543	0.180	
Eu	63	826	1439	1.1	547	0.204	(+2)0.097
Gd	64	1312	3000	1.1	592	0.179	(+3)0.096
Tb	65	1356	2800	1.1	564	0.177	(+3)0.095
Dy	66	1407	2600	1.1	572	0.177	(+3)0.094
Ho	67	1461	2600	1.1	581	0.176	(+3)0.093
Er	68	1497	2900	1.1	589	0.175	(+3)0.092
Tm	69	1356	2800	1.1	597	0.174	(+3)0.091
Yb	70	824	1427	1.1	603	0.193	(+3)0.089
Lu	71	1652	3327	1.1	524	0.174	(+3)0.089
Hf	72	2225	5200	1.3	654	0.157	
Ta	73	2980	5425	1.5	761	0.143	
W	74	3410	5930	1.7	770	0.137	
Re	75	3180	5885	1.9	760	0.137	
Os	76	2727	4100	2.2	840	0.134	
Ir	77	2448	4500	2.2	880	0.135	
Pt	78	1769	4530	2.2	870	0.138	
Au	79	1063	2966	2.4	890	0.144	(+1)0.137
Hg	80	−39	357	1.9	1007	0.155	(+2)0.110

(continued)

Element	At. No.	mp(°C)	bp(°C)	E.N.	Ion. Ener. (kJ/mol)	At. Rad. (nm)	Ion. Rad. (nm)
Tl	81	304	1457	1.8	589	0.171	(+3)0.095
Pb	82	328	1750	1.9	716	0.175	
Bi	83	271	1560	1.9	703	0.146	
Po	84	254	962	2.0	812	0.165	
At	85	302	334	2.2			
Rn	86	−71	−62		1037	0.14	
Fr	87	27	677	0.7			
Ra	88	700	1140	0.9	509	0.220	
Ac	89	1050	3200	1.1	490	0.20	
Th	90	1750	4790	1.3	590	0.180	
Pa	91	1600	4200	1.4	570		
U	92	1132	3818	1.4	590	0.14	

Appendix 3 Review of Mathematics

The mathematics you will use in general chemistry is relatively simple. You will, however, be expected to

— make calculations involving exponential numbers, such as 6.022×10^{23} or 1.6×10^{-10}.
— work with logarithms or inverse logarithms, particularly in problems involving pH:

$$pH = -\log_{10} (\text{conc. } H^+)$$

This appendix reviews each of these topics. It will be helpful to start with electronic calculators, which are very useful for all kinds of calculations in general chemistry.

ELECTRONIC CALCULATORS

A simple "scientific calculator" selling for $20 or less is entirely adequate for general chemistry. Like any calculator, it will allow you to carry out such simple operations as addition, subtraction, multiplication, and division. Beyond that, make sure that the scientific calculator you buy can be used to

— enter and perform operations on numbers expressed in exponential (scientific) notation.
— find a base 10 or natural logarithm or inverse logarithm (number corresponding to a given logarithm).
— raise a number to any power, *n*, or extract the *n*th root of a number.

The first thing you should do after buying a calculator is to learn how to use it. Read the instruction manual and work with the calculator until you become familiar with it. To get started, try carrying out the following operations:

a. $2.2 \times 6.1 = 13.42$
b. $8.1/2.7 = 3$
c. $(64)^{1/2} = 8$
d. $(27)^{1/3} = 3$
e. $3^4 = 81$

(In d and e, you will need to use the $\boxed{y^x}$ or $\boxed{x^y}$ key. Refer to your instruction manual for the sequence of operations, which differs depending upon the brand of calculator.)

f. $\dfrac{16 \times 9}{3 \times 8} = 6$

(This is carried out as a single operation; you do *not* solve for intermediate answers.)

In working with a calculator, you should be aware of one of its limitations. It does not indicate the number of significant figures in the answer. Consider, for example, the operations in a and b on p. 615. Assume that 2.2, 6.1, 8.1, and 2.7 represent experimentally measured quantities. Following the rules for significant figures (Chap. 1), the answers should be 13 and 3.0, in that order, *not* 13.42 and 3, the numbers appearing on the calculator. In another case, if you are asked to obtain the reciprocal of 3.68, your answer should be

$$1/3.68 = 0.272$$

not 0.2717391 . . . , or whatever other number appears on your calculator.

EXPONENTIAL NOTATION

Chemists deal frequently with very large or very small numbers. In one gram of the element carbon there are

$$50{,}140{,}000{,}000{,}000{,}000{,}000{,}000 \text{ atoms of carbon}$$

At the opposite extreme, the mass of a single atom is

$$0.00000000000000000000001994 \text{ g}$$

Numbers such as these are very awkward to work with. For example, neither of the numbers just written could be entered directly on a calculator. Operations involving very large or very small numbers can be simplified by using **exponential (scientific)** notation. To express a number in exponential notation, write it in the form

$$C \times 10^n$$

where C is a number between 1 and 10 (for example, 1, 2.62, 5.8) and n is a positive or negative integer such as 1, −1, −3. To find n, count the number of places that the decimal point must be moved to give the coefficient, C. If the decimal point must be moved to the *left*, n is a *positive* integer; if it must be moved to the *right*, n is a *negative* integer. Thus

$26.23 = 2.623 \times 10^1$ (decimal point moved 1 place to left)

$5609 = 5.609 \times 10^3$ (decimal point moved 3 places to left)

$0.0918 = 9.18 \times 10^{-2}$ (decimal point moved 2 places to right)

Multiplication and Division

A major advantage of exponential notation is that it simplifies the processes of multiplication and division. To *multiply, add exponents:*

$$10^1 \times 10^2 = 10^{1+2} = 10^3 \qquad 10^6 \times 10^{-4} = 10^{6+(-4)} = 10^2$$

To *divide, subtract* exponents:

$$10^3/10^2 = 10^{3-2} = 10^1 \qquad 10^{-3}/10^6 = 10^{-3-6} = 10^{-9}$$

It often happens that multiplication or division yields an answer that is not in standard exponential notation. For example

$$(5.0 \times 10^4) \times (6.0 \times 10^3) = (5.0 \times 6.0) \times 10^4 \times 10^3 = 30 \times 10^7$$

The product is not in standard exponential notation since the coefficient, 30, does not lie between 1 and 10. To correct this situation, rewrite the coefficient as 3.0×10^1 and then add exponents:

$$30 \times 10^7 = (3.0 \times 10^1) \times 10^7 = 3.0 \times 10^8$$

In another case,

$$0.526 \times 10^3 = (5.26 \times 10^{-1}) \times 10^3 = 5.26 \times 10^2$$

Exponential Notation on the Calculator

On all scientific calculators, it is possible to enter numbers in exponential notation. The method used depends upon the brand of calculator. Most often, it involves using a key labeled **EXP**, **EE**, or **EEX**. Check your instruction manual for the procedure to be followed. To make sure you understand it, try entering the following numbers:

$$2.4 \times 10^6 \qquad 3.16 \times 10^{-8} \qquad 6.2 \times 10^{-16}$$

Multiplication, division, and many other operations can be carried out directly on your calculator. Try the following exercises, using your calculator:

a. $(6.0 \times 10^2) \times (4.2 \times 10^{-4}) = ?$
b. $\dfrac{6.0 \times 10^2}{4.2 \times 10^{-4}} = ?$
c. $(2.50 \times 10^{-9})^{1/2} = ?$
d. $3.6 \times 10^{-4} + 4 \times 10^{-5} = ?$

The answers, expressed in exponential notation and following the rules of significant figures, are as follows: a. 2.5×10^{-1} b. 1.4×10^6 c. 5.00×10^{-5} d. 4.0×10^{-4}.

LOGARITHMS AND INVERSE LOGARITHMS

The logarithm of a number ***n*** to the base ***m*** is defined as the power to which ***m*** must be raised to give the number ***n***. Thus:

$$\text{if } m^x = n, \text{ then } \log_m n = x$$

In general chemistry, you will encounter two kinds of logarithms.

1. *Common logarithms,* where the base is 10, ordinarily denoted as $\log_{10}$. If $10^x = n$, then $\log_{10} n = x$. Examples include the following:

$$\log_{10} 100 = 2.000 \qquad (\text{since } 10^2 = 100)$$
$$\log_{10} 1 = 0.000 \qquad (\text{since } 10^0 = 1)$$
$$\log_{10} 0.001 = -3.000 \qquad (\text{since } 10^{-3} = 0.001)$$

2. *Natural logarithms,* where the base is the quantity $e = 2.718. \ldots$ Many of the equations used in general chemistry are expressed most simply in terms of natural logarithms, denoted as ln. If $e^x = n$, then $\ln n = x$.

$$\ln 100 = 4.606 \qquad (\text{i.e., } 100 = e^{4.606})$$

$$\ln 1 = 0 \qquad (\text{i.e., } 1 = e^{0})$$

$$\ln 0.001 = -6.908 \qquad (\text{i.e., } 0.001 = e^{-6.908})$$

Notice that

a. $\mathbf{\log_{10} 1 = \ln 1 = 0}$. The logarithm of 1 to any base is zero, since any number raised to the zero power is 1. That is

$$10^0 = e^0 = 2^0 = \cdots = n^0 = 1$$

b. Numbers larger than 1 have a positive logarithm; numbers smaller than 1 have a negative logarithm. For example:

$$\log_{10} 100 = 2 \qquad \ln 100 = 4.606$$

$$\log_{10} 10 = 1 \qquad \ln 10 = 2.303$$

$$\log_{10} 0.1 = -1 \qquad \ln 0.1 = -2.303$$

$$\log_{10} 0.01 = -2 \qquad \ln 0.01 = -4.606$$

c. The common and natural logarithms of a number are related by the expression:

$$\mathbf{\ln n = 2.303 \log_{10} n}$$

That is, the natural logarithm of a number is 2.303 . . . times its base 10 logarithm.

An *inverse logarithm* is, quite simply, the number corresponding to a given logarithm. In general

$$\text{if } m^x = n, \text{ then inverse } \log_m x = n$$

For example:

$$10^2 = 100 \qquad \text{inverse } \log_{10} 2 = 100$$

$$10^{-3} = 0.001 \qquad \text{inverse } \log_{10} (-3) = 0.001$$

In other words, the numbers whose base 10 logarithms are 2 and −3 are 100 and 0.001, respectively. The same reasoning applies to natural logarithms.

$$e^{4.606} = 100 \qquad \text{inverse } \ln 4.606 = 100$$

$$e^{-6.908} = 0.001 \qquad \text{inverse } \ln -6.908 = 0.001$$

The numbers whose natural logarithms are 4.606 and −6.908 are 100 and 0.001, respectively. Notice that if the inverse logarithm is positive, the corresponding number is larger than 1; if it is negative, the number is smaller than 1.

Finding Logarithms on a Calculator

To obtain a base 10 logarithm using a calculator, all you need do is enter the number and press the **LOG** key. This way you should find that

$$\log_{10} 2.00 = 0.301 \ldots$$

$$\log_{10} 0.526 = -0.279 \ldots$$

Similarly, to find a natural logarithm, you enter the number and press the **LN X** key.

$$\ln 2.00 = 0.693\ldots$$

$$\ln 0.526 = -0.642\ldots$$

To find the logarithm of an exponential number, you simply enter the number in exponential form and take the logarithm in the usual way. This way you should find that:

$$\log_{10} 2.00 \times 10^3 = 3.301\ldots \qquad \ln 2.00 \times 10^3 = 7.601\ldots$$

$$\log_{10} 5.3 \times 10^{-12} = -11.28\ldots \qquad \ln 5.3 \times 10^{-12} = -25.96\ldots$$

The base 10 logarithm of an exponential number can be found in a somewhat different way by applying the relation:

$$\log_{10}(C \times 10^n) = n + \log_{10} C$$

Thus

$$\log_{10} 2.00 \times 10^3 = 3 + \log_{10} 2.00 = 3 + 0.301 = 3.301$$

$$\log_{10} 5.3 \times 10^{-12} = -12 + \log_{10} 5.3 = -12 + 0.72 = -11.28$$

Finding Inverse Logarithms on a Calculator

The method used to find inverse logarithms depends upon the type of calculator. On certain calculators, you enter the number and then press, in succession, the **INV** and either **LOG** or **LN X** keys. With other calculators, you press the **10^x** or **e^x** key. Either way, you should find that

$$\text{inverse } \log_{10} 1.632 = 42.8\ldots \qquad \text{inverse } \ln 1.632 = 5.11\ldots$$

$$\text{inverse } \log_{10} -8.82 = 1.5 \times 10^{-9} \qquad \text{inverse } \ln -8.82 = 1.5 \times 10^{-4}$$

Significant Figures in Logarithms and Inverse Logarithms

For base 10 logarithms, the rules governing significant figures are quite simple:

1. In taking the logarithm of a number, retain after the decimal point in the log as many digits as there are significant figures in the number. (This part of the logarithm is often referred to as the *mantissa;* digits that precede the decimal point comprise the *characteristic* of the logarithm.) To illustrate this rule, consider the following:

$$\log_{10} 2.00 = 0.301 \qquad \log_{10}(2.00 \times 10^3) = 3.301$$

$$\log_{10} 2.0 = 0.30 \qquad \log_{10}(2.0 \times 10^1) = 1.30$$

$$\log_{10} 2 = 0.3 \qquad \log_{10}(2 \times 10^{-3}) = 0.3 - 3 = -2.7$$

2. In taking the inverse logarithm of a number, retain as many significant figures as there are after the decimal point in the number. Thus,

$$\text{inverse } \log_{10} 0.301 = 2.00 \qquad \text{inverse } \log_{10} 3.301 = 2.00 \times 10^3$$

$$\text{inverse } \log_{10} 0.30 = 2.0 \qquad \text{inverse } \log_{10} 1.30 = 2.0 \times 10^1$$

$$\text{inverse } \log_{10} 0.3 = 2 \qquad \text{inverse } \log_{10}(-2.7) = 2 \times 10^{-3}$$

These rules take into account the fact that, as mentioned earlier:

$$\log_{10}(C \times 10^n) = n + \log_{10} C$$

The digits that appear before (to the left of) the decimal point specify the value of n, i.e., the power of 10 involved in the expression. In that sense, they are not experimentally significant. In contrast, the digits that appear after (to the right of) the decimal point specify the value of the logarithm of C; the number of such digits reflects the uncertainty in C. Thus:

$$\log_{10} 209 = 2.320 \qquad \text{(3 sig. fig.)}$$

$$\log_{10} 209.0 = 2.3201 \qquad \text{(4 sig. fig.)}$$

The rules for significant figures involving natural logarithms and inverse logarithms are somewhat more complex than those for base 10 logs. However, for simplicity, we will assume that the rules listed above apply here as well. Thus:

$$\ln 209 = 5.342 \qquad \text{(3 sig. fig.)}$$

$$\ln 209.0 = 5.3423 \qquad \text{(4 sig. fig.)}$$

Operations Involving Logarithms

Since logarithms are exponents, the rules governing the use of exponents apply here as well. The rules that follow are valid for all types of logarithms, regardless of the base. We illustrate the rules with natural logarithms, since that is where you are most likely to use them in working with this text.

Multiplication: $\ln(xy) = \ln x + \ln y$

Example: $\ln(2.50 \times 1.25) = \ln 2.50 + \ln 1.25 = 0.916 + 0.223 = 1.139$

Division: $\ln(x/y) = \ln x - \ln y$

Example: $\ln(2.50/1.25) = 0.916 - 0.223 = 0.693$

Raising to a Power: $\ln(x^n) = n \ln x$

Example: $\ln(2.00)^4 = 4 \ln 2.00 = 4(0.693) = 2.772$

Extracting a Root: $\ln(x^{1/n}) = \frac{1}{n} \ln x$

Example: $\ln(2.00)^{1/3} = \frac{\ln 2.00}{3} = \frac{0.693}{3} = 0.231$

Taking a Reciprocal: $\ln(1/x) = -\ln x$

Example: $\ln(1/2.00) = -\ln 2.00 = -0.693$

Appendix 4 Answers to Questions & Problems

Chapter 1

2. **(a)** length **(b)** mass **(c)** energy **(d)** volume **(e)** density **(f)** pressure **(g)** energy
4. **(a)** 27.12 g **(b)** 35 cm^3 **(c)** 2.87 g/L **(d)** 525 mm
6. **(a)** (4.0×10^1)°C **(b)** 313 K
8. **(a)** −320.4°F **(b)** 77.4 K
10. **(a)** 10^{-3} m^3 **(b)** 1 kg · m/s^2 **(c)** 10^3 kg · m/s^2
12. **(a)** 0.260 atm **(b)** 1.52×10^6 J
14. **(a)** 5 **(b)** 4 **(c)** 3 **(d)** 5 **(e)** 6 **(f)** 2 or 3
16. 0.0270 nm^3 **18.** **(a)** 4 **(b)** 3 **(c)** 3 **(d)** 4
20. **(a)** 3.6 L **(b)** 8.4×10^3 mm **(c)** 72.5 g **(d)** 0.3926 J
22. **(a)** 1.390×10^{-6} mi^2 **(b)** 3.598×10^{-6} km^2 **(c)** 3.598 m^2
24. 1.7×10^9 **26.** 2.3×10^2 min
28. **(a)** 1157 drachms **(b)** 0.260 lb; 3.12 oz
30. 117 g **32.** 8.769 g Al, 13.00 g O, 0.819 g H, 5.761 g Cl
34. **(a)** P **(b)** P **(c)** C **(d)** P
36. 13.6 g/cm^3 **38.** 1.6 g/cm^3
40. 7.0 lb **42.** 737 g
44. **(a)** 7.24×10^{-3} g **(b)** 3.45×10^3 g
46. see text p. 20
48. 0.126% As; highly likely
50. **(a)** compound contains 2 or more elements
(b) an element is a pure substance
(c) solution has uniform composition
(d) distillation requires vaporization
52. a, d **54.** 4.2×10^5 g **56.** \$5.39 $\times 10^5$
58. −24.6°C **59.** 2.0 km^2 **60.** 1.71×10^2 cm
61. 8.1×10^{-3} g

Chapter 2

2. **(a)** green solid **(b)** red solid
4. **(a)** $CuSO_4 \cdot 5H_2O$—blue solid
(b) $CoCl_2 \cdot 4H_2O$—purple solid
6. Lavoisier; see text p. 27
8. **(a)** law of conservation of mass
(b) law of constant composition
(c) none of the laws
10. $\%Hg = \frac{48.43}{52.30} \times 100 = 92.60\%$

$\%Hg = \frac{15.68}{16.93} \times 100 = 92.62\%$

12. E. Rutherford; see text p. 29 **14.** $^{222}_{86}Rn$
16. Differ in neutrons; $^{63}_{29}Cu$ and $^{65}_{29}Cu$
18. **(a)** 53 p^+ **(b)** 78 n **(c)** 53 e^- **(d)** 78 n, 53 p^+
20.

$^{79}_{35}Br$	0	35	44	35
$^{14}_{7}N^{3-}$	−3	7	7	10
$^{75}_{33}As^{5+}$	+5	33	42	28
$^{90}_{40}Zr$	0	40	50	40

22. **(a)** 60 p^+, 60 e^- **(b)** 100 p^+, 100 e^- **(c)** 15 p^+, 18 e^- **(d)** 15 p^+, 10e^-
24. 33, 28; 46, 46; 53, 54; 10, 10
26. **(a)** manganese **(b)** sodium **(c)** arsenic **(d)** tungsten **(e)** phosphorus
28. metals: Mn, Na, W; nonmetal: P; metalloid: As
30. **(a)** 2 **(b)** 8 **(c)** 8 **(d)** 18 **(e)** 18
32. hydrazine—N_2H_4, NH_2; hydrogen peroxide—H_2O_2, HO; xenon tetrafluoride—XeF_4, XeF_4; tetrasulfur tetranitride—S_4N_4, SN; nitrogen trifluoride—NF_3, NF_3; carbon tetrachloride—CCl_4, CCl_4
34. **(a)** Li_2S, Li_3N **(b)** FeO, Fe_2O_3
36. **(a)** Cu_2Se **(b)** $MnCO_3$ **(c)** H_2SO_4 **(d)** $NaHCO_3$ **(e)** $Ni(ClO_4)_2$
38. **(a)** SnS **(b)** $AuCl_3$ **(c)** HNO_3 **(d)** $Ca(ClO_4)_2$ **(e)** KH_2PO_4
40. cesium hydroxide, $NaMnO_4$, $Li_2Cr_2O_7$, ammonium cyanide, $Al_2(SO_4)_3$, barium nitrate
42. **(a)** perchlorate **(b)** chlorite **(c)** hypoiodite **(d)** nitrate
44. **(a)** $HClO_2$ **(b)** HNO_2 **(c)** H_2SO_3 **(d)** HClO
46. see text p. 43
48. $1.0 \times 10^{21} Co^{2+}$ ions, 6.0×10^{21} water molecules
50. **(a)** never true **(b)** never true **(c)** usually true **(d)** usually true. NH_4Cl is ionic.
52. **(a)** CBr_4—carbon tetrabromide
(b) $Ca_3(PO_4)_2$—calcium phosphate
(c) $Ba(NO_2)_2$—barium nitrite
53. g H/g C in ethane = 0.252; g H/g C in ethene = 0.168

$\frac{0.252}{0.168} = \frac{3}{2}$ Formulas could be CH_3 and CH_2; C_2H_6 and C_2H_4

54. 3.71 g/cm^3—lots of empty space
55. 1.4963×10^{-23} g

56. **(a)** 2.5×10^{24} molecules of air **(b)** 2.3×10^{-20}
(c) 2.9×10^{2} molecules

Chapter 3

2. **(a)** 3.083 **(b)** 0.6312 **(c)** 0.3735
4. 151.960 **6.** 0.36% **8.** 4.3% Cr-50, 83.8% Cr-52
10. **(a)** 2 **(b)** 36 and 38 **(c)** peaks at 36 and 38
12. **(a)** 6.7×10^{-7} g **(b)** 3.002×10^{12} atoms
14. **(a)** 3.441×10^{-10} g **(b)** 8.240×10^{22} atoms
16. 7.579×10^{24}
18. **(a)** 10 **(b)** 6.022×10^{24} **(c)** 1.206×10^{26}
(d) 5.974×10^{24}
20. **(a)** 17.03 g/mol **(b)** 84.01 g/mol **(c)** 190.2 g/mol
22. **(a)** 0.0877 mol **(b)** 0.0105 mol **(c)** 187 mol
24. **(a)** 427 g **(b)** 1.07×10^{3} g **(c)** 314 g
26.

0.2500	0.004304	2.592×10^{21}	1.555×10^{22}
11.62	0.2000	1.204×10^{23}	7.226×10^{23}
2.4×10^{-12}	4.2×10^{-14}	2.5×10^{10}	1.5×10^{11}
1.3×10^{-7}	2.2×10^{-9}	1.3×10^{15}	8.0×10^{15}

28. 30.93% Al, 45.86% O, 2.889% H, 20.32% Cl
30. 193.9 kg **32.** 40.0%
34. 38.40% C, 1.50% H, 52.28% Cl, 7.82% O
36. Cr_2O_3
38. **(a)** C_3H_8O **(b)** $C_5H_8O_4NNa$ **(c)** $C_2H_3Cl_3O_2$
40. $C_8H_8O_3$ **42.** $C_2H_6N_2O$ **44.** C_3H_8N, $C_6H_{16}N_2$
46. CuO **48.** 52.68%; 16
50. **(a)** $UO_2(s) + 4HF(l) \longrightarrow UF_4(s) + 2H_2O(l)$
(b) $4PH_3(g) + 8\,O_2(g) \longrightarrow P_4O_{10}(s) + 6H_2O(g)$
(c) $2C_2H_3Cl(l) + 5\,O_2(g) \longrightarrow 4CO_2(g) + 2H_2O(g) + 2HCl(g)$
52. **(a)** $2Al(s) + 3Cl_2(g) \longrightarrow 2AlCl_3(s)$
(b) $Sr(s) + Cl_2(g) \longrightarrow SrCl_2(s)$
(c) $2Li(s) + Cl_2(g) \longrightarrow 2LiCl(s)$
(d) $2Cr(s) + 3Cl_2(g) \longrightarrow 2CrCl_3(s)$
(e) $2Ag(s) + Cl_2(g) \longrightarrow 2AgCl(s)$
54. **(a)** $NH_4NO_3(s) \longrightarrow N_2O(g) + 2H_2O(g)$
(b) $4NH_3(g) + 3\,O_2(g) \longrightarrow 2N_2(g) + 6H_2O(g)$
(c) $2KClO_3(s) \longrightarrow 2KCl(s) + 3\,O_2(g)$
(d) $B_2O_3(s) + 3C(s) + 3Cl_2(g) \longrightarrow 2BCl_3(g) + 3CO(g)$
(e) $2C_6H_6(l) + 15\,O_2(g) \longrightarrow 12CO_2(g) + 6H_2O(l)$
56. **(a)** 51.7 **(b)** 9.29 **(c)** 0.0541 **(d)** 14.95
58. **(a)** 333 g **(b)** 0.02499 mol **(c)** 4.01 g
(d) 0.8148 g
60. **(a)** $NCl_3(g) + 3H_2O(l) \longrightarrow NH_3(g) + 3HClO(aq)$
(b) 5.09 mol **(c)** 105 g
62. **(a)** 294 cm^3 **(b)** 584 g
64. 9.2×10^{3} L
66. **(a)** $H_2(g) + Cl_2(g) \longrightarrow 2HCl(g)$
(b) H_2 **(c)** 15.0 mol **(d)** 1.50 mol
68. 193 g; 49.0% **70.** 0.65 kg
72. 1.51×10^{3} g; 277 cm^3 **74.** 1.79
76. see p. 67 **78.** 2.7×10^{20}
80. 127 g **81.** 893 g/mol **82.** 6.01×10^{23}
83. 3.657 g; 2.973 g **84.** 34.7%
85. **(a)** V_2O_5; V_2O_3 **(b)** 2.271 g **86.** 28%

Chapter 4

2. **(a)** K_2CO_3 (soluble) **(b)** $Mg(OH)_2$ (insoluble)
(c) $Ce(NO_3)_3$ (soluble) **(d)** $NiSO_4$ (soluble)
4. **(a)** Add NaOH solution, filter off $Cu(OH)_2$
(b) Add $SrCl_2$ solution, filter off $SrSO_4$
(c) Add HCl solution, filter off Hg_2Cl_2
6. **(a)** $Fe^{3+}(aq) + 3\,OH^-(aq) \longrightarrow Fe(OH)_3(s)$
(b) $Cd^{2+}(aq) + S^{2-}(aq) \longrightarrow CdS(s)$
$Sr^{2+}(aq) + SO_4{}^{2-}(aq) \longrightarrow SrSO_4(s)$
8. **(a)** $Cu^{2+}(aq) + 2\,OH^-(aq) \longrightarrow Cu(OH)_2(s)$
(b) $Cu^{2+}(aq) + S^{2-}(aq) \longrightarrow CuS(s)$
(c) $Hg_2{}^{2+}(aq) + 2Cl^-(aq) \longrightarrow Hg_2Cl_2(s)$
(d) $Ba^{2+}(aq) + SO_4{}^{2-}(aq) \longrightarrow BaSO_4(s)$
$Fe^{3+}(aq) + 3\,OH^-(aq) \longrightarrow Fe(OH)_3(s)$
(e) $Ni^{2+}(aq) + CO_3{}^{2-}(aq) \longrightarrow NiCO_3(s)$
10. **(a)** no reaction
(b) $Ca^{2+}(aq) + CO_3{}^{2-}(aq) \longrightarrow CaCO_3(s)$
(c) $Pb^{2+}(aq) + S^{2-}(aq) \longrightarrow PbS(s)$
(d) $Fe^{3+}(aq) + 3\,OH^-(aq) \longrightarrow Fe(OH)_3(s)$
12. **(a)** HNO_2 **(b)** H^+ **(c)** $HC_2H_3O_2$ **(d)** H^+
(e) $HC_3H_5O_3$
14. **(a)** OH^- **(b)** $(CH_3)_2NH$ **(c)** OH^- **(d)** C_5H_5N
16. **(a)** weak acid **(b)** strong base **(c)** weak base
(d) strong acid
18. **(a)** $H^+(aq) + OH^-(aq) \longrightarrow H_2O$
(b) $HNO_2(aq) + OH^-(aq) \longrightarrow H_2O + NO_2{}^-(aq)$
(c) $H^+(aq) + C_6H_5NH_2(aq) \longrightarrow C_6H_5NH_3{}^+(aq)$
20. **(a)** $H^+(aq) + C_2H_5NH_2(aq) \longrightarrow C_2H_5NH_3{}^+(aq)$
(b) correct
(c) $OH^-(aq) + HC_2H_3O_2(aq) \longrightarrow C_2H_3O_2{}^-(aq) + H_2O$
(d) correct
(e) correct
22. **(a)** N = +3, O = −2 **(b)** Te = +6, F = −1
(c) N = +3, O = −2 **(d)** S = +2, O = −2
(e) Cl = +7, O = −2
24. **(a)** Ca = +2, C = +3, O = −2
(b) H = +1, S = +6, O = −2
(c) Na = +1, Fe = +3, O = −2
(d) N = +3, O = −2, F = −1
(e) N = −2, H = +1
26. **(a)** R **(b)** O **(c)** R **(d)** O
28. **(a)** FeS oxidized, $NO_3{}^-$ reduced; FeS is reducing agent, $NO_3{}^-$ oxidizing agent
(b) C_2H_4 oxidized, O_2 reduced; C_2H_4 is reducing agent, O_2 oxidizing agent
30. **(a)** $3FeS(s) + 8NO_3{}^-(aq) + 8H^+(aq) \longrightarrow 8NO(g) + 3SO_4{}^{2-}(aq) + 3Fe^{2+}(aq) + 4H_2O$
(b) $C_2H_4(g) + 3\,O_2(g) \longrightarrow 2CO_2(g) + 2H_2O(l)$
32. **(a)** $P_4(s) + 6H_2O \longrightarrow 2PH_3(g) + 2HPO_3{}^{2-}(aq) + 4H^+(aq)$

(b) $3H_3AsO_3(aq) + BrO_3^-(aq) \longrightarrow 3H_3AsO_4(aq) + Br^-(aq)$

(c) $2MnO_4^-(aq) + 5HSO_3^-(aq) + H^+(aq) \longrightarrow 2Mn^{2+}(aq) + 5SO_4^{2-}(aq) + 3H_2O$

(d) $2Sn^{2+}(aq) + O_2(g) + 4H^+(aq) \longrightarrow 2Sn^{4+}(aq) + 2H_2O$

(e) $3Pt(s) + 4NO_3^-(aq) + 18Cl^-(aq) + 16H^+(aq) \longrightarrow 3PtCl_6^{2-}(aq) + 4NO(g) + 8H_2O$

34. **(a)** $Ag(s) + NO_3^-(aq) + 2H^+(aq) \longrightarrow Ag^+(aq) + NO_2(g) + H_2O$

(b) $3CuS(s) + 2NO_3^-(aq) + 8H^+(aq) \longrightarrow 3Cu^{2+}(aq) + 2NO(g) + 3S(s) + 4H_2O$

(c) $4Sn^{2+}(aq) + IO_4^-(aq) + 8H^+(aq) \longrightarrow 4Sn^{4+}(aq) + I^-(aq) + 4H_2O$

36. **(a)** $S_2O_3^{2-}(aq) + 4I_2(s) + 10\,OH^-(aq) \longrightarrow 2SO_4^{2-}(aq) + 8I^-(aq) + 5H_2O$

(b) $3CN^-(aq) + 2MnO_4^-(aq) + H_2O \longrightarrow 3CNO^-(aq) + 2MnO_2(s) + 2\,OH^-(aq)$

(c) $2Cr(OH)_3(s) + ClO_3^-(aq) + 4\,OH^-(aq) \longrightarrow 2CrO_4^{2-}(aq) + Cl^-(aq) + 5H_2O$

38. **(a)** Dissolve 35.1 g KOH to form 0.500 L solution
(b) Dissolve 419 g $CuSO_4$ to form 0.750 L solution

40. **(a)** 0.256 g NaCl; 12.6 g $HC_2H_3O_2$
(b) 0.683 L NaCl; 0.0142 L $HC_2H_3O_2$

42. **(a)** 1.01 *M* **(b)** 1.58 g **(c)** 0.0173 L **(d)** 2.57 g

44. **(a)** 0.99 mol **(b)** 0.99 mol **(c)** 0.99 mol
(d) 1.65 mol

46. **(a)** 0.00831 *M* **(b)** 0.0563 g

48. **(a)** 35.0 mL **(b)** 107 mL **(c)** 249 mL

50. 16.6 mL

52. **(a)** 9.78 mL **(b)** 21.9 mL **(c)** 512 mL

54. **(a)** $2MnO_4^-(aq) + 5C_2O_4^{2-}(aq) + 16H^+(aq) \longrightarrow 2Mn^{2+}(aq) + 10CO_2(g) + 8H_2O$
(b) 0.571 *M*

56. **(a)** $4Fe(OH)_2(s) + O_2(g) + 2H_2O \longrightarrow 4Fe(OH)_3(s)$
(b) 4.76 g

58. **(a)** 53.0 mL **(b)** 0.989 g

60. 0.02428 *M* **62.** 85.9% **64.** 1

66. 41.8% Fe, 59.7% Fe_2O_3 **68.** 0.206 *M*

70. $(CH_3)_2$—N—$CH_2CH_3(aq) + H^+(aq) \longrightarrow (CH_3)_2$—$NH^+$—$CH_2CH_3(aq)$

72. 52.9 mL

74. **(a)** $Au(s) + 4Cl^-(aq) + NO_3^-(aq) + 4H^+(aq) \longrightarrow AuCl_4^-(aq) + NO(g) + 2H_2O$
(b) 4HCl : 1HNO$_3$ **(c)** 17 mL HCl, 3.2 mL HNO_3

76. 93.9% **77.** 0.794 g; Yes **78.** 0.30 L

79. 6.5×10^7 g **80.** 0.0980 *M* Fe^{2+}; 0.0364 *M* Fe^{3+}

Chapter 5

2. 11.35 L, 14.8 g, 300 K

4. 0.208 atm, 21.1 kPa; 604 mm Hg, 80.5 kPa; 9.60×10^2 mm Hg, 1.26 atm

6. **(a)** 414 mm Hg **(b)** 2.69×10^3 mm Hg

8. 969 K = 696°C **10.** 140°F

12. 529 mL **14.** 1.06 atm

16. 0.10 m^3 **18.** 33 atm **20.** 0.00516 mol

22.

735 mm Hg	5.01 L	15°C	0.205 mol	18.9 g
15.8 atm	489 mL	38°C	0.302 mol	27.8 g
1.49 atm	0.885 L	45°C	0.0505 mol	4.65 g
239 kPa	2.75 L	−111°C	0.489 mol	45.0 g

24. **(a)** 0.959 g/L **(b)** 1.37 g/L **(c)** 5.56 g/L

26. 55 g/L vs. 1.80 g/L **28.** C_2H_3NO

30. **(a)** 28.6 g/mol **(b)** 1.17 g/L vs. 1.18 g/L

32. Cl

34. **(a)** $2NF_3(g) + 3H_2O(g) \longrightarrow 6HF(g) + NO(g) + NO_2(g)$
(b) 2.00 L

36. 10.6 mL **38.** **(a)** 2.53 g **(b)** 1.80 L

40. 0.225 atm; CH_4 **42.** **(a)** 736 mm Hg **(b)** 0.160 g

44. 0.083atm **46.** 0.2821 **48.** 46.8 s **50.** 186 m/s

52. 2.3 K **54.** **(a)** more ideally **(b)** less ideally

56. **(a)** 179 atm **(b)** 1.5×10^2 atm

58. 1.4×10^3 L **60.** see text p. 123

62.

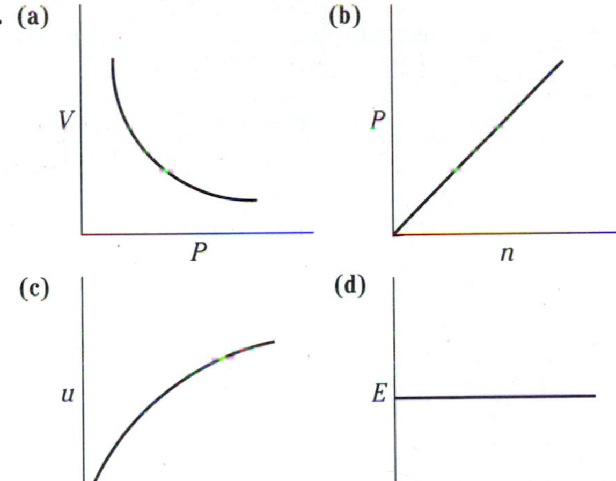

64. **(a)** same **(b)** Cl_2 **(c)** Cl_2 **(d)** Ne

66. **(a)** 4.54×10^{-3} mol **(b)** 0.038 g **68.** 46 g

70. 0.572 g/L; 28.0 g/mol **72.** 1.8 ft

73. 0.0456 L · atm/mol-° R **74.** 24.1% Al

75. 6.72 m **76.** 0.821 atm

77. $\frac{V_A}{V} = \frac{n_A}{n}$; volume fraction = mole fraction

Chapter 6

2. higher energy in excited state

4. **(a)** 5.25×10^4 nm
(b) 3.78×10^{-21} J/particle **(c)** 2.28 kJ/mol

6. **(a)** 1.498×10^4 nm **(b)** infrared
(c) 7.986 kJ/mol

8. **(a)** 4.87×10^{-8} m **(b)** 48.7 nm
(c) 6.16×10^{15} s^{-1}

10. **(a)** $3.82 \times 10^{14}\ s^{-1}$ **(b)** 2.53×10^{-19} J

12. energy absorbed: (1), (2); energy emitted: (3), (4)

14.

n	1	2	3	4
E	−1312	−328	−146	−82

Lyman: **n** = 2, 3, 4 to **n** = 1

Balmer: **n** = 3, 4 to **n** = 2

16. 1005 nm **18.** 7456 nm

20. The probability of finding the hydrogen electron decreases as the distance from the nucleus increases.

22. **(a)** 0 **(b)** 3, 2, 1, 0, −1, −2, −3
(c) $\ell = 2$, $m_\ell = 2, 1, 0, -1, -2$
$\ell = 1$, $m_\ell = 1, 0, -1$; $\ell = 0$, $m_\ell = 0$

24. **(a)** 3s **(b)** 4p **(c)** 1s **(d)** 4d

26. **(a)** p **(b)** d **(c)** s

28. **(a)** 32 **(b)** 10 **(c)** 2

30. **(a)** 3 **(b)** d **(c)** 7 **(d)** 3

32. $b < d < e < c = f < a$

34. **(a)** $1s^22s^22p^63s^2$ **(b)** $1s^22s^22p^63s^23p^64s^23d^7$
(c) $1s^22s^22p^63s^23p^64s^1$
(d) $1s^22s^22p^63s^23p^64s^23d^{10}4p^65s^24d^{10}5p^1$
(e) $1s^22s^22p^63s^23p^64s^23d^{10}4p^65s^24d^{10}5p^66s^24f^{14}5d^{10}6p^2$

36. **(a)** $[He]2s^22p^4$ **(b)** $[Xe]6s^2$ **(c)** $[Ar]4s^23d^8$
(d) $[Kr]5s^24d^{10}5p^4$ **(e)** $[Xe]6s^24f^{14}5d^{10}6p^3$

38. **(a)** Cu **(b)** Nb **(c)** Cl **(d)** H

40. **(a)** 12/20 **(b)** 24/55 **(c)** 18/44

42. **(a)** ground **(b)** excited **(c)** ground
(d) impossible **(e)** impossible **(f)** excited

44. **(a)**

1s	2s	2p	3s	3p
(↑↓)	(↑↓)	(↑↓)(↑↓)(↑↓)	(↑↓)	(↑)(↑)( )

b–d: 1s, 2s, 2p, 3s, 3p, 4s, 3d, 4p, filled plus:

	5s	4d	5p
(b)	(↑↓)	(↑)(↑)(↑)( )( )	
(c)	(↑↓)		
(d)	(↑↓)	(↑↓)(↑↓)(↑↓)(↑↓)(↑↓)	(↑)(↑)(↑)

46. **(a)** Mg **(b)** P **(c)** O

48. **(a)** Cr, Mg **(b)** K, Sc, Cu, Ga **(c)** Kr **(d)** Sb

50. **(a)** 2 **(b)** 5 **(c)** 1

52. **(a)** 1, 2, 13, 14, 15 **(b)** 16 **(c)** 17 **(d)** 18

54. **(a)** $1s^22s^22p^3$; $1s^22s^22p^6$
(b) $1s^22s^22p^63s^23p^64s^1$; $1s^22s^22p^63s^23p^6$
(c) $1s^22s^22p^63s^23p^64s^23d^1$; $1s^22s^22p^63s^23p^6$
(d) $1s^22s^22p^63s^23p^63d^6$; $1s^22s^22p^63s^23p^63d^5$

56. **(a)** 2 **(b)** 0 **(c)** 0 **(d)** 0 **(e)** 0

58. **(a)** $Cl < S < Mg$ **(b)** $Mg < S < Cl$
(c) $Mg < S < Cl$

60. **(a)** Rb **(b)** Br **(c)** Rb

62. **(a)** N^{3-} **(b)** Ba **(c)** Cr **(d)** Cu^+

64. **(a)** $Co > Co^{2+} > Co^{3+}$ **(b)** $Br^- > Cl^- > Cl$

66. **(a)** $LaCl_3$ **(b)** La_2O_3 **(c)** Y_2S_3
They behave like Sc which forms Sc^{3+}

68. $Es < Cf < Bk < Pa$ **70.** 5.34×10^{19} photons/pulse

72. **(a)** no 2 electrons can have the same set of quantum numbers
(b) maximum number of unpaired electrons
(c) wavelength of light emitted
(d) **n** = 1, 2, 3, . . .

74. **(a)** directly proportional
(b) inversely proportional to $\mathbf{n}^2$
(c) 4s fills before 3d

76. **(a)** added electrons repel
(b) has 3 more electrons than Ar
(c) effective nuclear charge increases

78. **(a)** phosphorus, P **(b)** astatine, At
(c) titanium, Ti **(d)** francium, Fr
(e) argon, Ar

79. 2954 kJ/mol

80. $$E = h\nu = \frac{hc}{\lambda} = R_H\left(\frac{1}{4} - \frac{1}{\mathbf{n}^2}\right) = R_H\left(\frac{\mathbf{n}^2 - 4}{4\mathbf{n}^2}\right)$$

$$\lambda = \frac{4hc\mathbf{n}^2}{R_H(\mathbf{n}^2 - 4)} = \frac{364.5\ \mathbf{n}^2}{\mathbf{n}^2 - 4}$$

81. $1s^41p^4$

82. **(a)** 3, 9, 15 **(b)** 27 **(c)** $1s^32s^32p^2$; $1s^32s^32p^93s^2$

83. **(a)** 3.42×10^{-19} **(b)** 581 nm

Chapter 7

2. **(a)**
```
   ..
H—P—H
  |
  H
```
(b)
```
      ..
     :F:
  ..  |  ..
 :F—Si—F:
  ..  |  ..
     :F:
      ..
```

(c)
```
   ..  ..  ..
H—O—N=O
   ..      ..
```
(d)
```
        H
        |
[H—N—H]+
        |
        H
```

4. **(a)**
```
   ..  ..  ..
(:O—I—O:)−
   ..  |   ..
      :O:
       ..
```
(b)
```
   ..  ..  ..
(:F—Cl—F:)+
   ..  ..  ..
```

(c)
```
   ..  ..  ..
(:O—S—O:)2−
   ..  |   ..
      :O:
       ..
```
(d)
```
   ..  ..  ..
(:O—As—O:)3−
   ..  |   ..
      :O:
       ..
```

6. **(a)**
```
 ..  . .  ..
:F— Br —F:
 ..   |   ..
     :F:
      ..
```
(b)
```
        ..
       :O:
  ..    | .   ..
 :F—Xe —O:
  ..    |     ..
       :F:
        ..
```

(c)
```
   .. ...  ..
(:I—I—I:)−
   ..  ..  ..
```
(d)
```
  ..     ..
 :F ..  F:
  ..\ /..
     I
  ../|\..
 :F  |  F:
  .. |  ..
    :F:
     ..
```

8. **(a)**
```
   ..      ..
(H—O—C—O:)−
   ..  ||  ..
      :O:
```
(b)
```
    H
    |        ..
H—C—C—O—H
    |  ||    ..
    H :O:
```

(c)
```
   ..  ..  ..
H—O—S—O—H
   ..  |   ..
      :O:
       ..
```

10. 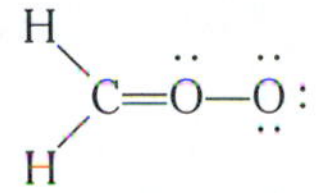 $H_2C{=}\ddot{O}{-}\ddot{O}:$ (other structures are possible)

12. 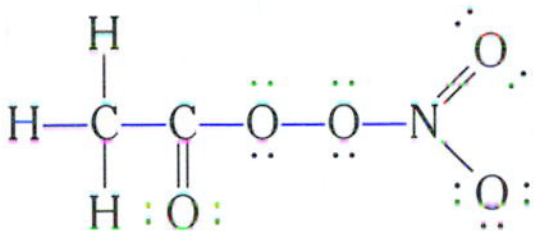 $H{-}CH_2{-}C(=O){-}O{-}O{-}NO_2$

14. 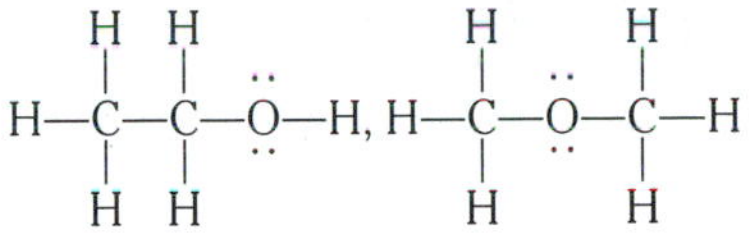 $H{-}CH_2{-}CH_2{-}\ddot{O}{-}H$, $H{-}CH_2{-}\ddot{O}{-}CH_2{-}H$

16. **(a)** Cl_2 **(b)** H_2SO_4 **(c)** CH_4 **(d)** CCl_4

18. **(a)** 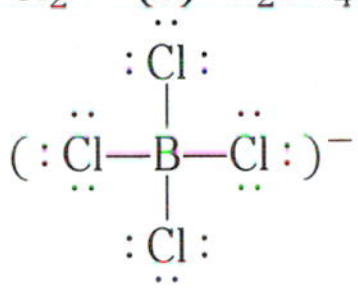 $(BCl_4)^-$ (Cl—B—Cl, with Cl above and below B) **(b)** $(S_2O_3)^{2-}$ (O—S—S, with O above and below the central S)

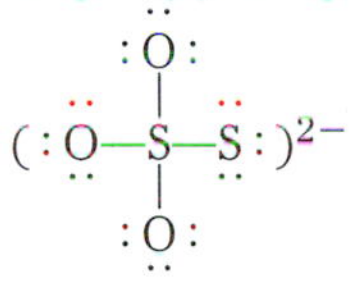

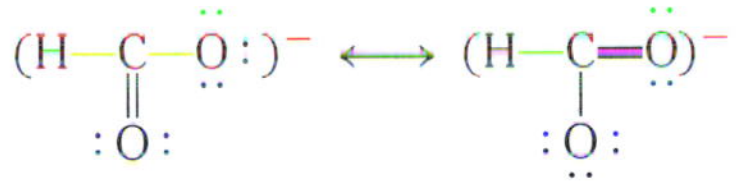

(c) 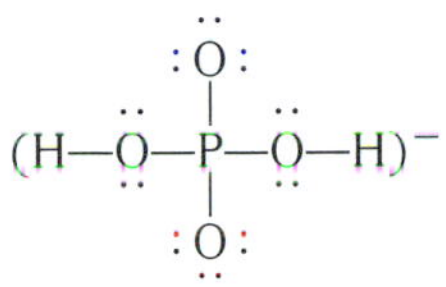 $(H{-}O{-}PO_2{-}O{-}H)^-$ **(d)** 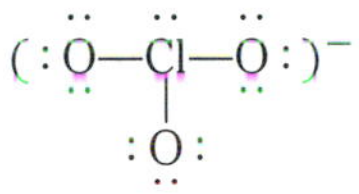 $(O{-}ClO{-}O)^-$

20. **(a)** 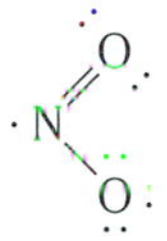 $\cdot NO_2$ **(b)** H—Be—H

(c) 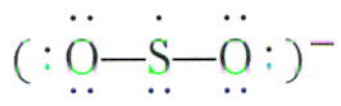 $(:\ddot{O}{-}\dot{S}{-}\ddot{O}:)^-$ **(d)** 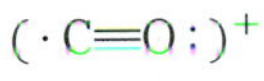 $(\cdot C{\equiv}O:)^+$

22. **(a)** 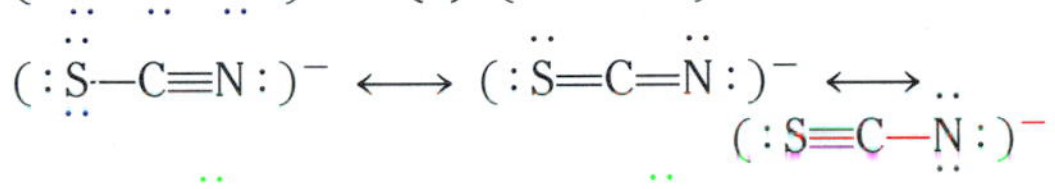 $(:\ddot{S}{-}C{\equiv}N:)^- \longleftrightarrow (:\ddot{S}{=}C{=}\ddot{N}:)^- \longleftrightarrow (:S{\equiv}C{-}\ddot{N}:)^-$

(b) $(H{-}C(=O){-}\ddot{O}:)^- \longleftrightarrow (H{-}C(-\ddot{O}:){=}O)^-$

(c) $H{-}O{-}N(=O){-}O: \longleftrightarrow H{-}O{=}N(-O:){-}O: \longleftrightarrow H{-}O{-}N(-O:){=}O$

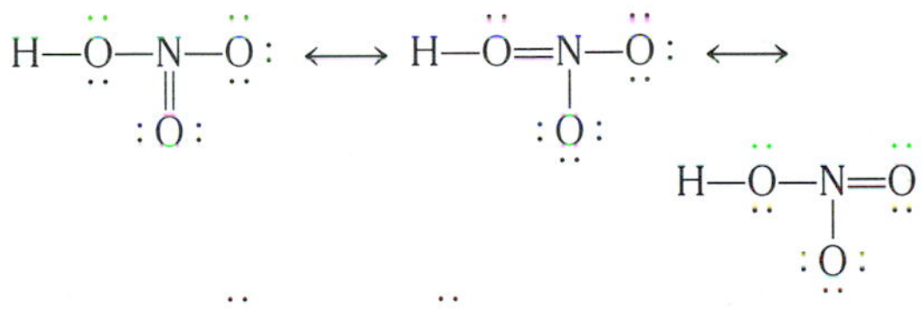

24. **(a)** $H{-}N{\equiv}N{-}\ddot{N}:$ $H{-}\ddot{N}{-}N{\equiv}N:$

(b) no; different skeletons

26. 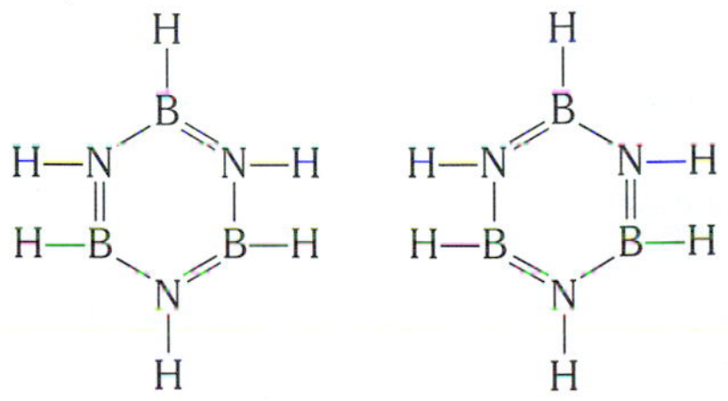 two resonance structures of the six-membered ring $B_3N_3H_6$ (alternating B and N atoms, each bearing one H, with B=N double bonds in alternate positions)

28. **(a)** linear **(b)** bent **(c)** bent **(d)** tetrahedral

30. **(a)** tetrahedral **(b)** bent **(c)** triangular planar **(d)** triangular pyramid

32. **(a)** see-saw **(b)** octahedral **(c)** linear **(d)** T-shaped

34. **(a)** 120° **(b)** 120° **(c)** 120°, 109.5°

36. **(a)** 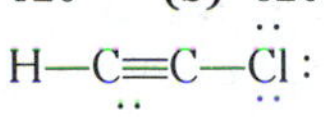 $H{-}C{\equiv}C{-}\ddot{Cl}:$ linear

(b) 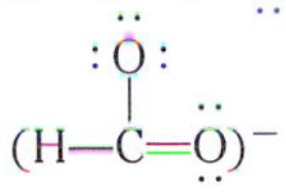 $(H{-}C(-O:){=}O)^-$ triangular planar

(c) 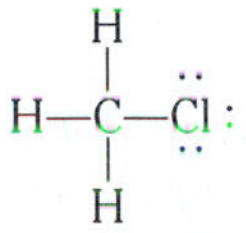 $H{-}CH_2{-}\ddot{Cl}:$ tetrahedral

(d) 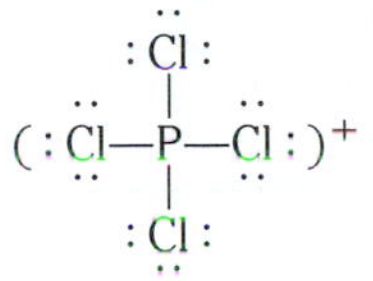 $(PCl_4)^+$ tetrahedral

38. **(a)** square planar **(b)** square pyramid **(c)** triangular bipyramid

40. **(a)** 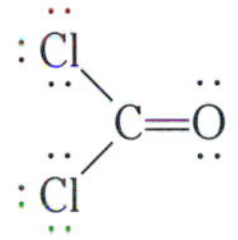 $Cl_2C{=}O$ triangular planar

(b) 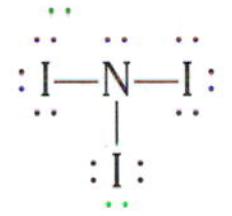 $:\ddot{I}{-}\ddot{N}(-\ddot{I}:){-}\ddot{I}:$ triangular pyramid

(c) 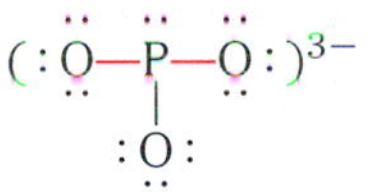 $(:\ddot{O}{-}\ddot{P}(-O:){-}\ddot{O}:)^{3-}$ triangular pyramid

(d) 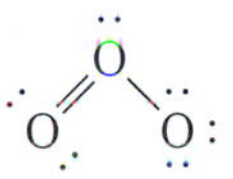 $O{=}\ddot{O}{-}O$ (ozone) bent

42. **(a)** 0, 4, 120° **(b)** 2, 2, 109.5° **(c)** 0, 3, 120° **(d)** 1, 3, 109.5°

44. **(a)** 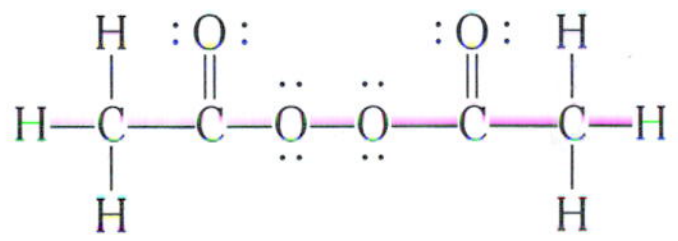 $H{-}CH_2{-}C(=O){-}\ddot{O}{-}\ddot{O}{-}C(=O){-}CH_2{-}H$

(b) 109.5° in CH_3 groups, around central O atoms; 120° around C=O

46. **(a)** 109.5° **(b)** 109.5°, 180° **(c)** 109.5°, 120°

48. PH_3 and H_2S; unshared electron pair present

50. b, c, d

52. d

54. first molecule is polar

56. **(a)** sp **(b)** sp^2 **(c)** sp^2 **(d)** sp^3

58. **(a)** sp^3 **(b)** sp^3 **(c)** sp^2 **(d)** sp^3

60. **(a)** 5, sp^3d **(b)** 6, sp^3d^2 **(c)** 5, sp^3d **(d)** 5, sp^3d

62. **(a)** 6, sp^3d^2 **(b)** 6, sp^3d^2 **(c)** 5, sp^3d

64. all carbons: sp^2; nitrogen—sp^2; one oxygen—sp^2, the other sp^3

66. **(a)** sp^3 **(b)** sp^2 **(c)** sp **(d)** sp^2

68. **(a)** sp^3 **(b)** sp^2 **(c)** sp^3

70. **(a)** 4 sigma **(b)** 3 sigma, 1 pi **(c)** 2 sigma, 2 pi **(d)** 3 sigma, 1 pi

72. **(a)** BCl_3 **(b)** CO_3^{2-} **(c)** NH_4^+ **(d)** CO_3^{2-}

74. **(a)** XeF_2 :F—Xe—F: XeF_4 :F—Xe—F: (with :F: above and below Xe)

(b) linear, nonpolar, sp^3d, 2 sigma, no pi bonds
square planar, nonpolar, sp^3d^2, 4 sigma, no pi bonds

76. 2.94 atm

78. Be = −2; Cl = +1; Structure with single bonds has formal charge of 0 on both Cl and Be.

80. $(:O—Cr—O—Cr—O:)^{2-}$ (each Cr bonded to :O: above and :O: below)

81. b, d

82. **(a)** 5 pi, 21 sigma **(b)** 109.5°, 120°, 109.5°
(c) sp^2, sp^3, sp^2

83.

AX_2E_2	2	2	bent	sp^3	polar
AX_3	3	0	triangular planar	sp^2	nonpolar
AX_4E_2	4	2	square planar	sp^3d^2	nonpolar
AX_5	5	0	trigonal bipyramid	sp^3d	nonpolar

84. $x = 3$; T-shaped, polar, 90° and 180°, sp^3d, 3 sigma bonds

85. H—N—N—H (each N bonded to H below), 109.5° bond angle, yes

86. 6, sp^3d^2, octahedral

87. **(a)** $(:O—S—O:)^{2-}$ (S bonded to :O: above and :O: below) $(:O—S—O:)^{2-}$ (S double-bonded to :O: above and below)

(b) both tetrahedral **(c)** sp^3, sp^3
(d) S = 2; O = −1, −1, −1, −1
S = 0; O = −1, −1, 0, 0

88. **(a)** :Cl—P—Cl: (P bonded to :Cl: above and :O: below) P = 1; Cl = 0, 0, 0; O = −1

(b) :Cl—P—Cl: (P double-bonded to :O: above, bonded to :Cl: below) P = 0; Cl = 0, 0, 0; O = 0

Chapter 8

2. **(a)** $Hg(l) \longrightarrow Hg(s)$; $\Delta H = -2.33$ kJ
(b) $C_{10}H_8(l) \longrightarrow C_{10}H_8(g)$; $\Delta H = 43.3$ kJ
(c) $C_6H_6(s) \longrightarrow C_6H_6(l)$; $\Delta H = 9.84$ kJ

4. **(a)** $2K(s) + \frac{1}{2}O_2(g) \longrightarrow K_2O(s)$
(b) $N_2(g) + 2H_2(g) \longrightarrow N_2H_4(l)$
(c) $N_2(g) + 4H_2(g) + C(s) + \frac{3}{2}O_2(g) \longrightarrow (NH_4)_2CO_3(s)$
(d) $N_2(g) + 2\,O_2(g) \longrightarrow N_2O_4(g)$

6. 25.4°C **8.** 67.8 g

10. **(a)** $KBr(s) \longrightarrow K^+(aq) + Br^-(aq)$
(b) endothermic **(c)** +223 J **(d)** +19.8 kJ

12. -1.38×10^3 kJ **14.** 4302 J/°C **16.** 21.81°C

18. **(a)** 3.87 kJ/°C **(b)** −50.0 kJ **(c)** -1.40×10^3 kJ

20. **(a)** $Ni(CO)_4(g) \longrightarrow Ni(s) + 4CO(g)$
(b) endothermic **(c)** products above reactants
(d) 0.9413 kJ **(e)** 10.62 g

22. **(a)** −84.4 kJ **(b)** −0.449 kJ

24. **(a)** $2Mg(s) + CO_2(g) \longrightarrow C(s) + 2MgO(s)$; $\Delta H = -812$ kJ
(b) 8.77 kJ

26. 1.548×10^4 kJ; 3.700×10^6 cal; 3.700×10^3 kcal

28. condensing $C_6H_6(g)$ **30.** 12.2 kJ

32. $SiO_2(s) + 2Mg(s) + 2Cl_2(g) + 2C(s) \longrightarrow Si(s) + 2MgCl_2(s) + 2CO(g)$
−592.7 kJ; exothermic

34. −226.7 kJ

36. **(a)** −1675.7 kJ/mol **(b)** −205.4 kJ

38. 3.67 kJ

40. **(a)** +354.8 kJ **(b)** −7.1 kJ **(c)** +264.4 kJ

42. **(a)** −633.2 kJ **(b)** −1530.4 kJ

44. **(a)** $C_2H_4(g) + \frac{1}{2}O_2(g) + 2HCl(g) \longrightarrow C_2H_4Cl_2(l) + H_2O(l)$; $\Delta H = -318.7$ kJ
(b) −165.2 kJ/mol

46. −919.9 kJ/mol **48.** −2.91 kJ

50. **(a)** −941 kJ **(b)** −1656 kJ **(c)** −1167 kJ

52. +98 kJ **54.** −78 kJ **56.** +102 kJ

58. **(a)** 77 J **(b)** 22 J

60. **(a)** 40.7 kJ **(b)** 37.6 kJ **(c)** −3.1 kJ

62. **(a)** $C_3H_8(g) + 5\,O_2(g) \longrightarrow 3CO_2(g) + 4H_2O(l)$; $\Delta H = -2219.9$ kJ
(b) 50.2 kJ

64. 4.4×10^3 kcal **66.** see text p. 222

68. 8.452×10^4 kJ **70.** 23.1°C **72.** 0.559

74. 5.7×10^3 kJ **75.** 1.72×10^5 J; 515 g

76. 1 part ice, 3 parts water

77. **(a)** −851.5 kJ **(b)** 6.6×10^3°C **(c)** yes

Chapter 9

2. a > d > c > b

4. **(a)** dispersion **(b)** dispersion and dipole
(c) dispersion and dipole
(d) dispersion and dipole

6. b, d

8. **(a)** molecular vs. ionic **(b)** hydrogen bonding
(c) hydrogen bonding
(d) dispersion forces increase with molar mass

10. a, c, d

12. **(a)** PH_3 (weaker dispersion forces)
(b) C_6H_6 (weaker dispersion forces)
(c) PH_3 (no H-bonds)
(d) C_3H_8 (molecular vs. ionic)

14. **(a)** H-bonds **(b)** dispersion **(c)** dispersion
(d) dispersion

16. **(a)** network covalent or ionic **(b)** molecular **(c)** ionic
18. **(a)** molecular, network covalent, metallic **(b)** network covalent, ionic **(c)** metallic
20. **(a)** metallic **(b)** molecular **(c)** network covalent **(d)** ionic **(e)** molecular
22. **(a)** $C_{12}H_{22}O_{11}$ **(b)** Na_2CO_3 **(c)** SiC **(d)** steel
24. **(a)** C atoms **(b)** Si and C atoms **(c)** Fe^{2+} and Cl^- ions **(d)** C_2H_2 molecules
26. 0.219 nm **28.** 0.0793 nm^3
30. **(a)** 0.151 nm **(b)** no
32. $s(3)^{1/2} = 2(r_{cation} + r_{anion})$
34. **(a)** 4 **(b)** 8 for Cl^- at corner, 2 for Cl^- in center of face
36. **(a)** 6.26 mg **(b)** 0.15 mm Hg **(c)** 0.466 mm Hg
38. **(a)** yes **(b)** 19.83 mm Hg
40. **(a)** 40.6 kJ **(b)** 17 mm Hg
42. 3.8×10^2 mm Hg **44.** 120°C
46. 34 kJ **48.** b, c
50. **(a)** liquid, vapor **(b)** vapor **(c)** liquid
52. **(a)** sublimes **(b)** melts **(c)** condenses **(d)** condenses
54. **(a, b)**

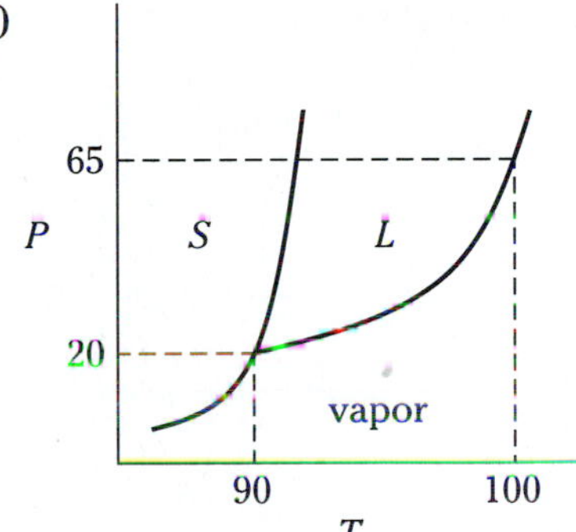

(c) liquid vaporizes
56. fibrous silicates: 1 Si:[2 + 2($\frac{1}{2}$)] O = 1 Si:3 O
layer silicates: 1 Si:[1 + 3($\frac{1}{2}$)] O = 1 Si:$\frac{5}{2}$ O
58. $Ca^{2+}(aq) + 2NaZ(s) \longrightarrow CaZ_2(s) + 2Na^+(aq)$
Water softened by sodium zeolite contains high conc. of Na^+
60. a
62. **(a)** intramolecular forces = covalent bonds **(b)** see Figure 9.5 **(c)** hydrogen bond is special type of dipole force **(d)** see Figure 9.5
64. **(a)** H bonds in ethyl alcohol **(b)** H bonds in HF **(c)** LiF is ionic
66. 0.0436 atm
68. **(a)** liquid and vapor **(b)** 26.7 mm Hg
69. yes
70. 80 atm; 0.60°C; not likely; heat conduction
71. 0.414
72. vapor pressure (C_3H_8) is more sensitive to *T* than gas pressure (N_2)

Chapter 10

2. **(a)** $C_6H_{12}O_6$ **(b)** H_2SO_4 **(c)** K_2CrO_4
4. **(a)** $KClO_4(s) \longrightarrow K^+(aq) + ClO_4^-(aq)$
(b) $Sc_2(SO_4)_3(s) \longrightarrow 2Sc^{3+}(aq) + 3SO_4^{2-}(aq)$
(c) $CaBr_2(s) \longrightarrow Ca^{2+}(aq) + 2Br^-(aq)$
(d) $Ni(ClO_3)_2(s) \longrightarrow Ni^{2+}(aq) + 2ClO_3^-(aq)$
6. **(a)** $Br_2(l) \longrightarrow Br_2(aq)$
(b) $KMnO_4(s) \longrightarrow K^+(aq) + MnO_4^-(aq)$
(c) $(NH_4)_3PO_4(s) \longrightarrow 3NH_4^+(aq) + PO_4^{3-}(aq)$
(d) $C_6H_{12}O_6(s) \longrightarrow C_6H_{12}O_6(aq)$
8. **(a)** 11.2% **(b)** 87.72 **(c)** 0.0251
10. 0.707 m **12.** 0.0027 mol
14. **(a)** 0.139 mol, 0.304 *M* **(b)** 3.38 g, 0.272 L **(c)** 78.1 g, 0.868 mol
16. **(a)** 0.8543, 1.200×10^5, 0.01516
(b) 0.00367, 0.0586, 6.62×10^{-5}
(c) 5.85, 5.85×10^4, 0.00696
(d) 3.13, 33.3, 3.33×10^5
18. **(a)** Dilute 1.67 L of 0.750 *M* acid to 5.00 L with water
(b) Dilute 3.57 L of 0.350 *M* acid to 5.00 L with water
20. **(a)** 0.0383 *M*, 0.115 *M*, 0.0383 *M* **(b)** 0.690 *M*, 0.230 *M*
22. 14.6 *M*, 57.8 m, $X = 0.510$
24. **(a)** 0.940 m, 11.0% **(b)** 2.26 *M*, 2.66 m **(c)** 2.71 *M*, 29.1%
26. **(a)** NaF (ionic) **(b)** NH_3 (H bonds) **(c)** CO_2 (molecular) **(d)** CH_3OH (H bonds)
28. **(a)** greater **(b)** less **(c)** greater **(d)** less
30. **(a)** 0.23 *M* **(b)** 1.1×10^{-5} *M*
32. **(a)** 5.26 mm Hg **(b)** 52.6 mm Hg **(c)** 63.1 mm Hg
34. 345 g **36.** 1.50×10^{-4} atm
38. **(a)** 14.5, 100.42°C **(b)** 15.5 g, 100.42°C
40. −33.5°C; yes **42.** 1.70°C/m **44.** $C_{12}H_{26}O$
46. 6.14% **48.** 0.16 *M* **50.** 6.50×10^4 g/mol
52. **(a)** −0.56°C **(b)** −0.37°C **(c)** −0.74°C
54. nonionized
56. The solubility of CO_2 increases with decreasing temperature. Sap is forced out by CO_2 bubbles.
58. −11°C
60. Add a small crystal of KNO_3. If nothing happens, solution is saturated. If crystal dissolves, solution is unsaturated. If precipitate forms, solution is supersaturated.
62. **(a)** $CaCl_2$ has more ions.
(b) Solution of gas in water is an exothermic process.
(c) HCl in water forms ions.
(d) 1 L of solution contains about one kg of water.
64. **(a)** Seawater has dissolved ions, which lower the freezing point.
(b) It depends on the number of solute particles rather than on the type of solute.
(c) Excess solute crystallizes.
(d) Excess air dissolved at high *P* comes out of solution.

66. **(a)** Test conductivity.
(b) CO_2 becomes less soluble when T increases.
(c) Number of kg of solvent is ordinarily less than the number of moles of solution.
(d) Vapor pressure is lowered.

68. **(a)** 2.2 M **(b)** 2.4 m **(c)** 2.2×10^2 atm **(d)** 105.0°C

70. Add about 1030 g water.

71. $m = \dfrac{n \text{ solute}}{\text{kg solvent}}$; in 1 L of solution, n solute $= M$

$$\text{kg solvent} = \frac{\text{mass solution (g)} - \text{mass solute (g)}}{1000}$$

$$= \frac{(1000 \times d) - M(\mathcal{M})}{1000} = d - \frac{M(\mathcal{M})}{1000}$$

$$m = \frac{M}{d - \dfrac{M(\mathcal{M})}{1000}}$$

In dilute solution, $m \longrightarrow M/d$;
for water, $d = 1.00$ g/mL

72. 49% **73.** 0.0018 g/cm^3

74. **(a)** 2.08 M **(b)** 1.872 mol **(c)** 47.4 L

75. $V_{gas} = \dfrac{n_{gas} \times RT}{P_{gas}}$; $n_{gas} = k \times P_{gas}$; $V_{gas} = kRT$

Chapter 11

2. **(a)** $\Delta[H_2]/\Delta t$ **(b)** $-\frac{1}{2}[\Delta[HI]/\Delta t]$

4. **(a)** 0.039 mol/L · s **(b)** 0.065 mol/L · s **(c)** 0.078 mol/L · s

6. **(a)** $-\frac{1}{2}[\Delta[N_2O_5]/\Delta t]$ **(b)** 0.0384 mol/L · min

8. 0.014 mol/L · min

10. **(a)** 3, 3 **(b)** 1, 1, 2 **(c)** 1, 2, 3 **(d)** 1, 1

12. **(a)** L^2/mol^2 · s **(b)** L/mol · s **(c)** L^2/mol^2 · s **(d)** s^{-1}

14. 0.090 mol/L · s; 1.1 M; 0.14 L/mol · s

16. **(a)** rate = k **(b)** 2.5×10^{-4} mol/L · min **(c)** all concentrations

18. **(a)** 0.0163 L/mol · s **(b)** 1.23 M **(c)** 15.5 M

20. a **22.** 2nd order

24. **(a)** 1st order **(b)** 1st order **(c)** 2.5×10^{-5} mol/L · s

26. **(a)** 2nd order in A, zero order in B, 1st order in C **(b)** rate = $k[A]^2[C]$ **(c)** 0.250 L^2/mol^2 · s **(d)** 0.162 mol/L · s

28. **(a)** rate = $k[Cr(H_2O)_6{}^{3+}][SCN^-]$ **(b)** 0.0072 L/mol · s **(c)** 3.9×10^{-8} mol/L · s

30. 2nd order **32.** 10.0 s; 5.0 s; 2.5 s; zero order

34. **(a)** linear plot obtained for $\ln[C_2H_5Cl]$ vs. t **(b)** 0.017 min^{-1} **(c)** 82 min

36. **(a)** 2.03×10^{-3} min^{-1} **(b)** 341 min **(c)** 0.0450 M

38. **(a)** 1.91 h^{-1} **(b)** 0.852 **(c)** 0.0851 h

40. **(a)** 0.078 L/mol · s **(b)** 85 s

42. 2.9 L/mol · min **44.** 8.0 min; 32 min

46. **(a)** System 3 **(b)** 30 kJ, 105 kJ, 42 kJ

48.

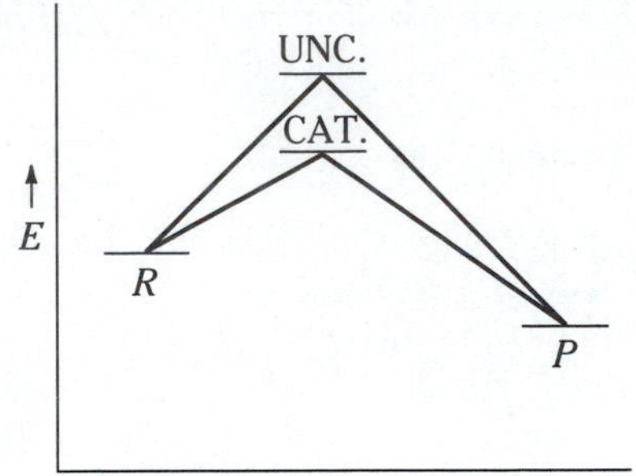

50. 1.9×10^2 kJ **52.** 75 kJ **54.** 57%

56. **(a)** 519°C **(b)** 1.6×10^3 L/mol · s **58.** 1.8 s

60. **(a)** rate = $k[K][HCl]$ **(b)** rate = $k[NO_3][CO]$ **(c)** rate = $k[NO_2]^2$

62. rate = $k_2[N_2O_2][H_2]$; $[N_2O_2] = \dfrac{k_1}{k_{-1}}[NO]^2$

$$\text{rate} = \frac{k_2 k_1}{k_{-1}}[NO]^2[H_2]$$

64. c, d

66. 4.7×10^{-15} mol/L · s; 2.3×10^{-19} mol/L · s

68. 5×10^{-9} mol/L formed per second; very rapid, $\frac{1}{4}$ of NO is gone in 1 s!

70. **(a)** usually decreases **(b)** increases **(c)** increases

72. 1.9 kJ absorbed **74.** 8.5 times faster

76. $\dfrac{-d[A]}{a\,dt} = k[A]$; $\dfrac{-d[A]}{[A]} = ka\,dt$; $\ln\dfrac{[A_0]}{[A]} = kat$

77. **(a)** $\dfrac{-d[A]}{dt} = k[A]^2$; $\dfrac{-d[A]}{[A]^2} = k\,dt$;

$$\frac{1}{[A]} - \frac{1}{[A_0]} = kt$$

(b) $\dfrac{-d[A]}{dt} = k[A]^3$; $\dfrac{-d[A]}{[A]^3} = k\,dt$;

$$\frac{1}{2[A]^2} - \frac{1}{2[A_0]^2} = kt$$

78. rate = $[A]^2[B][C]$ **79.** 0.90 g; 0.055 g

Chapter 12

2. **(a)** 0.350 atm **(b)** forward reaction faster at 45 s; rates about the same at 90s

4. **(a)** $3A(g) \rightleftharpoons B(g)$ **(b)** no, partial pressures still changing

6. **(a)** $K = \dfrac{(P_{I_2})^{1/2}(P_{F_2})^{1/2}}{P_{IF}}$ **(b)** $K = \dfrac{P_{C_{10}H_{12}}}{(P_{C_5H_6})^2}$

(c) $K = \dfrac{(P_{POCl_3})^{10}}{(P_{P_4O_{10}})(P_{PCl_5})^6}$

8. **(a)** $K = \dfrac{P_{Ni(CO)_4}}{(P_{CO})^4}$ **(b)** $K = P_{O_2}$ **(c)** $K = \dfrac{(P_{O_2})^3}{(P_{H_2O})^2}$

10. **(a)** $N_2(g) + Na_2CO_3(s) + 4C(s) \rightleftharpoons 3CO(g) + 2NaCN(s); K = \frac{(P_{CO})^3}{P_{N_2}}$

(b) $Mg_3N_2(s) + 6H_2O(g) \rightleftharpoons 3Mg(OH)_2(s) + 2NH_3(g); K = \frac{(P_{NH_3})^2}{(P_{H_2O})^6}$

(c) $BaCO_3(s) \rightleftharpoons BaO(s) + CO_2(g); K = P_{CO_2}$

12. **(a)** $2HBr(g) \rightleftharpoons H_2(g) + Br_2(g)$
(b) $CO(g) + 2H_2(g) \rightleftharpoons CH_3OH(g)$
(c) $C_2H_6(g) \rightleftharpoons C_2H_4(g) + H_2(g)$
(d) $SO_2(g) + \frac{1}{2}O_2(g) \rightleftharpoons SO_3(g)$
(e) $CO(g) + H_2O(l) \rightleftharpoons CO_2(g) + H_2(g)$

14. **(a)** $K = 1.3$ **(b)** $K = 0.58$

16. 2.0×10^{-10} **18.** 1.1×10^3

20. **(a)** $CO(g) + 2H_2(g) \rightleftharpoons CH_3OH(g)$ **(b)** $K = 6.22 \times 10^{-3}$

22. 2.97×10^{-4} **24.** 1.05×10^{-2}

26. **(a)** no, $Q > K$ **(b)** ←

28. **(a)** → **(b)** ←

30. some of both **32.** 0.18 atm

34. 0.16 atm **36.** 0.063 atm

38. $P_{CO} = P_{H_2O} = 0.234$ atm; $P_{H_2} = P_{CO_2} = 0.266$ atm

40. $P_{CO} = 0.067$ atm; $P_{Br_2} = 0.047$ atm; $P_{COBr_2} = 0.017$ atm

42. **(a)** 2.6 **(b)** $P_{CO} = P_{H_2} = 0.77$ atm; $P_{H_2O} = 0.23$ atm

44. 0.50 atm

46. **(a)** increase **(b)** increase **(c)** no effect **(d)** increase **(e)** increase

48. **(a)** ← **(b)** ← **(c)** no effect

50. (c) or (d) **52.** 1.6×10^{-23}

54. **(a)** 0.0276 **(b)** ≈260°C

56. 5.9×10^{-3} **58.** see text p. 349 **60.** 8.1 g

62. **(a)** upside down **(b)** multiplied by coefficients **(c)** added instead of multiplied **(d)** divided P by volume

64. chemical equation and T

66. $$K = \frac{(P_C)^c(P_D)^d}{(P_A)^a(P_B)^b} = \frac{([C]RT)^c([D]RT)^d}{([A]RT)^a([B]RT)^b} = \frac{[C]^c[D]^d}{[A]^a[B]^b}(RT)^{(c+d)-(a+b)}$$

$\Delta n_g = (c + d) - (a + b)$ $K = K_c(RT)^{\Delta n_g}$

67. 0.0442 **68.** 0.52 atm **69.** $K = 25$; 0.65 atm

70. 1.1 **71.** 0.22 atm; 0.4 g

Chapter 13

2. **(a)** acids: H_2O, HCN; bases: CN^-, OH^-; pairs: HCN/CN^- and H_2O/OH^-
(b) acids: H_3O^+, H_2CO_3; bases: HCO_3^-, H_2O; pairs: H_3O^+/H_2O and H_2CO_3/HCO_3^-
(c) acids: $HC_2H_3O_2$, H_2S; bases: HS^-, $C_2H_3O_2^-$; pairs: $HC_2H_3O_2/C_2H_3O_2^-$ and H_2S/HS^-

4. **(a)** acid **(b)** acid **(c)** base

6. **(a)** $C_2H_3O_2^-$ **(b)** $Fe(H_2O)_4(OH)_2^+$ **(c)** SO_3^{2-} **(d)** CH_3NH_2 **(e)** HS^-

8. **(a)** 1.0; acidic **(b)** −1.0; acidic **(c)** 2.15; acidic **(d)** 8.09; basic

10. **(a)** 1×10^{-5} *M*; 1×10^{-9} *M*
(b) 1.6×10^{-11} *M*; 6.3×10^{-4} *M*
(c) 9.1×10^{-16} *M*; 11 *M*
(d) 2.9×10^{-7} *M*; 3.5×10^{-8} *M*

12. Solution 1; Solution 1 **14.** 10^2; 10^{-2}

16. 3×10^2

18. **(a)** 0.60 *M*; 0.22 **(b)** 0.60 *M*; 0.22 **20.** −0.194

22. **(a)** 0.80 *M*, 1.2×10^{-14} *M*, 0.10
(b) 0.10 *M*, 1.0×10^{-13} *M*, 1.00

24. 12.92

26. **(a)** $Zn(H_2O)_3(OH)^+(aq) + H_2O \rightleftharpoons H_3O^+(aq) + Zn(H_2O)_2(OH)_2(aq)$
(b) $HSO_4^-(aq) + H_2O \rightleftharpoons H_3O^+(aq) + SO_4^{2-}(aq)$
(c) $HNO_2(aq) + H_2O \rightleftharpoons H_3O^+(aq) + NO_2^-(aq)$
(d) $Fe(H_2O)_6^{2+}(aq) + H_2O \rightleftharpoons H_3O^+(aq) + Fe(H_2O)_5(OH)^+(aq)$
(e) $Mn(H_2O)_6^{2+}(aq) + H_2O \rightleftharpoons H_3O^+(aq) + Mn(H_2O)_5(OH)^+(aq)$
(f) $HC_2H_3O_2(aq) + H_2O \rightleftharpoons H_3O^+(aq) + C_2H_3O_2^-(aq)$

28. **(a)** $PH_4^+(aq) \rightleftharpoons H^+(aq) + PH_3(aq)$ $K_a = \frac{[H^+][PH_3]}{[PH_4^+]}$
(b) $HS^-(aq) \rightleftharpoons H^+(aq) + S^{2-}(aq)$ $K_a = \frac{[H^+][S^{2-}]}{[HS^-]}$
(c) $HC_2O_4^-(aq) \rightleftharpoons H^+(aq) + C_2O_4^{2-}(aq)$ $K_a = \frac{[H^+][C_2O_4^{2-}]}{[HC_2O_4^-]}$

30. **(a)** 1×10^{-3} **(b)** 2×10^{-6} **(c)** 2.1×10^{-10}

32. **(a)** B > D > A > C **(b)** C

34. 1.3×10^{-5} **36.** 1.3×10^{-10}

38. **(a)** 7.2×10^{-5} *M* **(b)** 2.9×10^{-5} *M*

40. **(a)** 2.6×10^{-5} *M* **(b)** 3.8×10^{-10} *M* **(c)** 4.59 **(d)** 2.1×10^{-3}%

42. 1.58 **44.** 2.03

46. $[H_2C_5H_5O_7^-] = 0.0093$ *M*
$[HC_2H_5O_7^{2-}] = 1.8 \times 10^{-5}$ *M*

48. **(a)** $(CH_3)_3N(aq) + H_2O \rightleftharpoons OH^-(aq) + (CH_3)_3NH^+(aq)$
(b) $PO_4^{3-}(aq) + H_2O \rightleftharpoons OH^-(aq) + HPO_4^{2-}(aq)$
(c) $HPO_4^{2-}(aq) + H_2O \rightleftharpoons OH^-(aq) + H_2PO_4^-(aq)$
(d) $H_2PO_4^-(aq) + H_2O \rightleftharpoons OH^-(aq) + H_3PO_4(aq)$
(e) $HS^-(aq) + H_2O \rightleftharpoons OH^-(aq) + H_2S(aq)$
(f) $C_2H_5NH_2(aq) + H_2O \rightleftharpoons OH^-(aq) + C_2H_5NH_3^+(aq)$

50. b > d > c > a **52.** b, c

54. **(a)** 2.6×10^{-10} **(b)** 2.3×10^{-11}

56. 11.47 **58.** 2.5×10^{-6}

60. **(a)** acidic **(b)** acidic **(c)** basic **(d)** acidic **(e)** basic **(f)** acidic

62. **(a)** $Al(H_2O)_6^{3+}(aq) + H_2O \rightleftharpoons Al(H_2O)_5(OH)^{2+}(aq) + H_3O^+(aq)$

(b) $NO_2^-(aq) + H_2O \rightleftharpoons OH^-(aq) + HNO_2(aq)$
$NH_4^+(aq) + H_2O \rightleftharpoons H_3O^+(aq) + NH_3(aq)$
(c) $ClO^-(aq) + H_2O \rightleftharpoons OH^-(aq) + HClO(aq)$
(d) $NH_4^+(aq) + H_2O \rightleftharpoons H_3O^+(aq) + NH_3(aq)$
(e) $CO_3^{2-}(aq) + H_2O \rightleftharpoons OH^-(aq) + HCO_3^-(aq)$
(f) $HSO_4^-(aq) + H_2O \rightleftharpoons H_3O^+(aq) + SO_4^{2-}(aq)$

64. $Ba(OH)_2 > LiCN > K_2SO_4 > NH_4Br > HClO_4$
66. (a) LiF, Li_2CO_3, Li_2S, $LiNO_2$
(b) LiCl, $LiNO_3$, LiBr, LiI
(c) Na_2SO_4, K_2SO_4, Li_2SO_4, $CaSO_4$
(d) $CuSO_4$, $Al_2(SO_4)_3$, $ZnSO_4$, $(NH_4)_2SO_4$
68. 0.66 **70.** 2.35
72. (a) neutral (b) acidic (c) basic (d) acidic
74. neutral **76.** 2.1×10^{-17}
77. Test pH of $AgNO_3$ solution; if acidic, AgOH is a weak base.
78. (a) −5.58 kJ (b) −6.83 kJ
79. $\% \text{ diss} = \frac{[H^+]}{[HB]} \times 100; K_a \approx \frac{[H^+]^2}{[HB]}$

$\% \text{ diss} \approx \frac{K_a^{1/2}}{[HB]^{1/2}} \times 100$

% dissociation is inversely proportional to $[HB]^{1/2}$
80. −1.64°C

Chapter 14

2. (a) $HCN(aq) + OH^-(aq) \rightleftharpoons H_2O(aq) + CN^-(aq)$
(b) $ClO^-(aq) + H^+(aq) \rightleftharpoons HClO(aq)$
(c) $NH_4^+(aq) + OH^-(aq) \rightleftharpoons NH_3(aq) + H_2O$
(d) $CH_3NH_2(aq) + HClO(aq) \rightleftharpoons CH_3NH_3^+(aq) + ClO^-(aq)$
4. (a) $OH^-(aq) + H^+(aq) \rightleftharpoons H_2O$
(b) $CN^-(aq) + H^+(aq) \rightleftharpoons HCN(aq)$
(c) $C_6H_5NH_2(aq) + H^+(aq) \rightleftharpoons C_6H_5NH_3^+(aq)$
6. (a) 5.9×10^4 (b) 3.6×10^7 (c) 5.6×10^4
(d) 1.2×10^3
8. (a) 1.0×10^{14} (b) 1.7×10^9 (c) 3.8×10^4
10. (a) any (b) methyl orange (c) phenolphthalein
(d) methyl orange
12. (a) 1.1001 (b) 1.4494 (c) 7.00
14. (a) NH_4^+, Cl^-, H_2O (b) acidic
16. (a) 2.87 (b) 3.00×10^{-3} mol
(c) $HC_2H_3O_2(aq) + OH^-(aq) \rightleftharpoons H_2O + C_2H_3O_2^-(aq)$
(d) 8.79
18. (a) 7.1×10^{-7} *M*, 7.85 (b) 3.6×10^{-7} *M*, 7.55
(c) 7.1×10^{-8} *M*, 6.85 (d) 1.4×10^{-8} *M*, 6.15
20. 4.24
22. (a) lactic acid/lactate (b) lactic acid/lactate
(c) NH_4^+/NH_3
24. (a) 2.1 (b) 2.1 L
26. 8.51 **28.** 3.1 g **30.** 5.48
32. (a) 4.92 (b) 4.85 (c) 5.07
34. (a) 1.8 (b) 9.23 (c) 8.82
36. a, b **38.** 3.25
40. (a) $Cu_2P_2O_7(s) \rightleftharpoons 2Cu^{2+}(aq) + P_2O_7^{4-}(aq)$; $K_{sp} = [Cu^{2+}]^2[P_2O_7^{4-}]$
(b) $Ni_3(AsO_4)_2(s) \rightleftharpoons 3Ni^{2+}(aq) + 2AsO_4^{3-}(aq)$; $K_{sp} = [Ni^{2+}]^3[AsO_4^{3-}]^2$
(c) $Fe(OH)_3(s) \rightleftharpoons Fe^{3+}(aq) + 3\,OH^-(aq)$; $K_{sp} = [Fe^{3+}][OH^-]^3$
(d) $Mg(NbO_3)_2(s) \rightleftharpoons Mg^{2+}(aq) + 2NbO_3^-(aq)$; $K_{sp} = [Mg^{2+}][NbO_3^-]^2$
42. (a) $Ca(IO_3)_2(s) \rightleftharpoons Ca^{2+}(aq) + 2IO_3^-(aq)$
(b) $Ag_2SO_3(s) \rightleftharpoons 2Ag^+(aq) + 2SO_3^{2-}(aq)$
(c) $Mn_3(AsO_4)_2(s) \rightleftharpoons 3Mn^{2+}(aq) + 2AsO_4^{3-}(aq)$
(d) $PbC_2O_4(s) \rightleftharpoons Pb^{2+}(aq) + C_2O_4^{2-}(aq)$
44. (a) 8×10^{-6} *M* (b) 4×10^{-5} *M* (c) 3×10^{-5} *M*
(d) 9×10^{-4} *M*
46. (a) 1×10^{-42} *M* (b) 1×10^{-31} *M*
(c) 3×10^{-22} *M*
48. (a) 9.2×10^{-4} *M* (b) 8.4×10^{-3} *M*; 84%
50. no **52.** yes **54.** 7.8×10^{-10}
56. (a) 1×10^{-16} g/L (b) 1×10^{-32} g/L
(c) 5×10^{-33} g/L
58. symptoms: muscle cramps and convulsions
causes: hyperventilation resulting from fever, infection, or the action of certain drugs
60. (a) 0.64 (b) 37% (c) 14%
62. (a) false; less than (b) true
(c) false; 3.2×10^{-10}
64. no **66.** 12.40
67. (a) 2.38 (b) 4.74 (c) 7.44 (d) 9.23
(e) 11.00 (f) 13.52
68. $K = 0.02$; solubility is 0.06 *M* vs. 1×10^{-4} *M*
69. about 30 mL; phenolphthalein
70. 1.8×10^{-4} *M* **71.** 1×10^{-8} *M*
72. (a) 1×10^{-5} *M* (b) yes, Al^{3+} and Fe^{3+}
(c) almost all (d) 1.6 g
73. $-\log[H^+] = -\log K_a + \left(-\log\frac{[HB]}{[B^-]}\right)$;

$pH = pK_a - \log\frac{[HB]}{[B^-]} = pK_a + \log\frac{[B^-]}{[HB]}$

Chapter 15

2. (a) $4NH_3$ molecules, $2Cl^-$ ions (b) +3
(c) $Co(C_2O_4)_2Cl_2^{3-}$
4. (a) $Cr(C_2O_4)_2(H_2O)_2^-$ (b) $Cr(NH_3)_5SO_4^+$
(c) $Cr(en)(NH_3)_2I_2^+$
6. (a) 6 (b) 4 (c) 2 (d) 6
8. (a) $Ag(en)^+$ (b) $Fe(H_2O)_6^{2+}$ (c) $Zn(CN)_4^{2-}$
(d) $Pt(en)_3^{4+}$
10. 22.61%
12. (a) $Co(NH_3)_4(H_2O)Cl^{2+}$ (b) $Co(NH_3)_5SO_4^+$
(c) $Ni(CN)_4^{2-}$ (d) $Co(NH_3)_5NO_3^{2+}$
14. (a) tetrahydroxoaluminate(III)
(b) diaquadioxalatocobaltate(III)
(c) triamminetrichloroiridium(III)
(d) diamminedibromoethylenediaminechromium(III)

16. **(a)** hexaquairon(III)
(b) amminetribromoplatinate(II)
(c) diamminesilver(I)
(d) chlorobis(ethylenediamine)thiocyanatocobalt(III)

18. 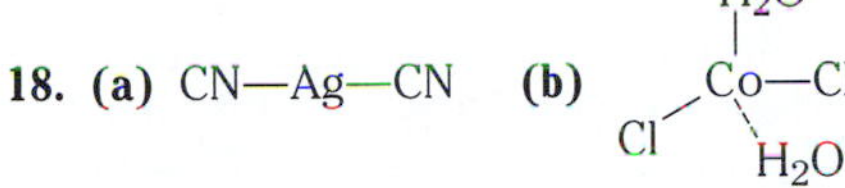

(c) Cl, H_2O, Ni, Cl, H_2O **(d)** Br, en, Pt, Br, en

(e) Cl, H_2O, H_2O, Cr, H_2O, H_2O, Cl

20. H_2O, NH_2, Pd, H_2O, NH_2

22. **(b)** Cl, en, Co, en, Br and Cl, en, Co, Br, en

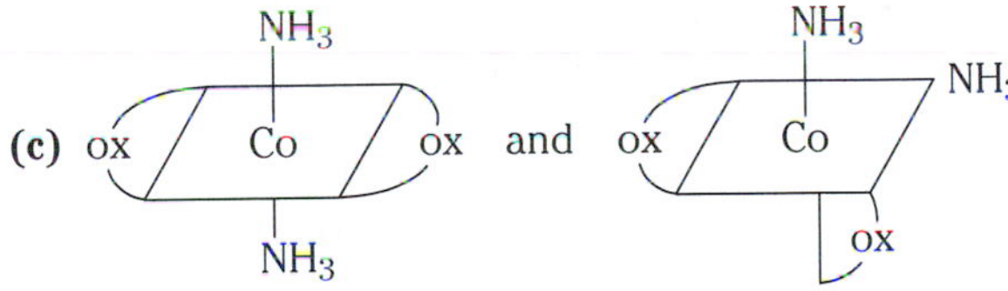

24. 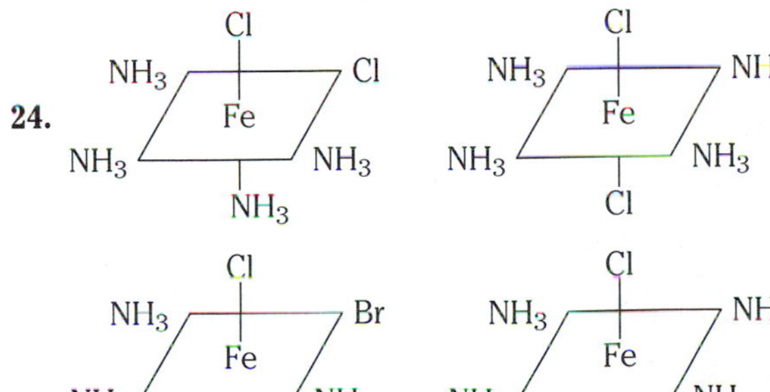

26. **(a)** $[Ar]3d^6$ **(b)** $[Ar]3d^3$ **(c)** $[Ar]3d^{10}$
(d) $[Ar]3d^{10}$ **(e)** $[Kr]4d^3$

28. 3d
(a) (↑↓)(↑)(↑)(↑)(↑) 4 unpaired e^-
(b) (↑)(↑)(↑)( )( ) 3 unpaired e^-
(c) (↑↓)(↑↓)(↑↓)(↑↓)(↑↓) 0 unpaired e^-
(d) (↑↓)(↑↓)(↑↓)(↑↓)(↑↓) 0 unpaired e^-
4d
(e) (↑)(↑)(↑)( )( ) 3 unpaired e^-

30. **(a)** ( )( ) / (↑↓)(↑↓)(↑↓) and (↑)(↑) / (↑↓)(↑)(↑)
(b) ( )( ) / (↑↓)(↑)(↑) and (↑)( ) / (↑)(↑)(↑)

32. two 3d electrons

34. $Co(NH_3)_6{}^{3+}$: ( )( ) / (↑↓)(↑↓)(↑↓) $CoF_6{}^{3-}$ (↑)(↑) / (↑↓)(↑)(↑)

36. **(a)** 4 **(b)** 4 **(c)** 4 **(d)** 2 **(e)** 4

38. 399 nm

40. lungs saturated with O_2, so shift is to oxyhemoglobin formation; tissues low in O_2, so in arterial blood shift is to hemoglobin formation

42. 0.92 g

44. **(a)** false; 6 **(b)** false; 0 **(c)** false; shorter

46. **(a)** forms complex with Fe^{3+}
(b) accepts electron pair
(c) $Ni(NH_3)_6{}^{2+}$

48. **(a)** $PtN_2H_6Cl_4$

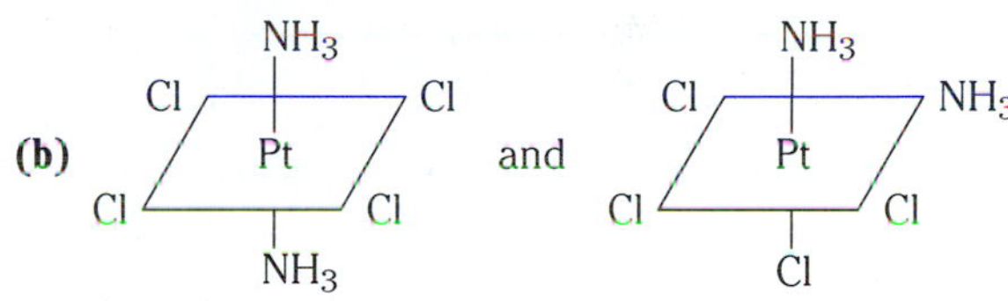

49. $[Pt(NH_3)_4][PtCl_4]$ or $[Pt(NH_3)_3Cl][Pt(NH_3)Cl_3]$

50. **(a)** $CuC_4H_{22}N_6SO_4$

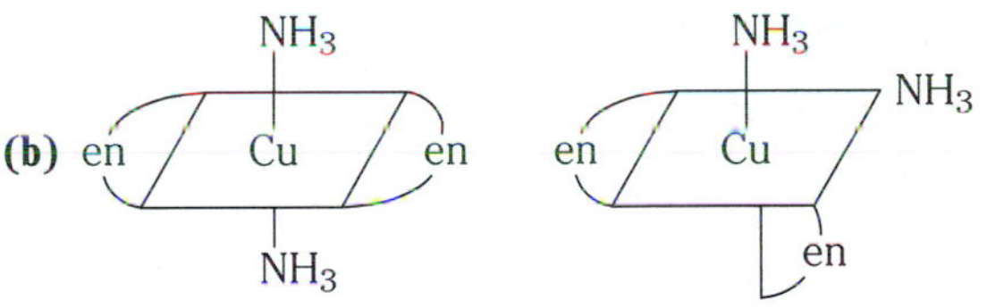

51. **(a)** $Al(OH)_3(s) + OH^- \longrightarrow Al(OH)_4{}^-(aq)$
(b) $Al^{3+}(aq) + 4\,OH^-(aq) \longrightarrow Al(OH)_4{}^-(aq)$
(c) 2×10^2

52. $Cu^{2+}(aq) + 2NH_3(aq) + 2H_2O \longrightarrow Cu(OH)_2(s) + 2NH_4{}^+(aq)$
$Cu(OH)_2(s) + 4NH_3(aq) \longrightarrow Cu(NH_3)_4{}^{2+}(aq) + 2\,OH^-(aq)$
$Cu(NH_3)_4{}^{2+}(aq) + 4H^+(aq) + 4Cl^-(aq) \longrightarrow CuCl_4{}^{2-}(aq) + 4NH_4{}^+(aq)$

Chapter 16

2. b

4. **(a)** no **(b)** no **(c)** yes **(d)** yes

6. **(a)** + **(b)** + **(c)** − **(d)** +

8. **(a)** − **(b)** − **(c)** − **(d)** −

10. **(a)** + **(b)** + **(c)** +

12. **(a)** −332.0 J/K **(b)** +24.9 J/K **(c)** +171.9 J/K
(d) −20.9 J/K

14. **(a)** −15.2 J/K **(b)** −213.9 J/K **(c)** +9.8 J/K

16. **(a)** −131.9 J/K **(b)** −271.2 J/K **(c)** −533.0 J/K
18. **(a)** −49.5 kJ **(b)** −889.2 kJ **(c)** −46 kJ
20. **(a)** +4.6 kJ (nonspontaneous)
(b) +170.4 kJ (nonspontaneous)
(c) +200.5 kJ (nonspontaneous)
(d) −316.4 kJ (spontaneous)
22. **(a)** −29.1 kJ **(b)** +173.2 kJ **(c)** +218.1 kJ
(d) −318.6 kJ
24. **(a)** −519.8 kJ **(b)** −33.5 kJ **(c)** −120.5 kJ
26. no
28. **(a)** +132 J/K; yes, gas produced
(b) +96 J/mol · K **(c)** −510.9 kJ/mol
30. **(a)** +333 J/K; yes **(b)** +284 J/mol · K
(c) −218.8 kJ/mol
32. **(a)** more spontaneous at high T
(b) more spontaneous at low T
(c) nonspontaneous at any T
34. **(a)** 4950 K; at reasonable T, nonspontaneous
(b) 5200 K; at reasonable T, spontaneous
(c) not T dependent
36. 780 K **38.** $\Delta G° = 467.9 - 0.5581\,T$

T	200	400	600	800	1000
$\Delta G°$	356.3	244.7	133.0	21.4	−90.2

(b) 838 K
40. **(a)** $T = 2840$ K **(b)** 841 K **(c)** 903 K
recommend b
42. 282 K **44.** 84.4 J/K
46. **(a)** $\Delta G° = 47.4$ kJ, nonspontaneous
(b) $\Delta G = -66.7$ kJ, spontaneous
48. **(a)** $\Delta G° = -177.6$ kJ **(b)** −118.6 kJ
50. **(a)** −141.8 kJ **(b)** 1×10^{-25} atm
52. −129.7 kJ/mol
54. $C_6H_{12}O_6(aq) + 6\,O_2(g) + 80\,ADP(aq) + 80HPO_4^{2-}(aq) + 160H^+(aq) \longrightarrow 80\,ATP(aq) + 86H_2O + 6CO_2(g)$
56. **(a)** 54.2 kJ **(b)** 2.2×10^{-4}
58. +211 kJ; +107 kJ **60.** 1×10^{-10}
62. arsenic—n-type because it must give up its 5th valence electron (need only 4 covalent bonds) boron—p-type because it has only 3 valence electrons (4 covalent bonds needed)
64. more sunshine, less cloud cover available more days of the year in Mojave desert
66. **(a)** only if $T\Delta S$ can be ignored
(b) at standard conditions
(c) no. moles gas **(d)** only at low temperatures
68. **(a)** 0 K **(b)** 0 **(c)** <0 **(d)** less
70. +2870 kJ + 24(−203.6 kJ) = −2016 kJ
72. 1.6×10^{-5} **73.** 309 K
74. $P_{H_2} = P_{I_2} = 0.032$ atm $P_{HI} = 0.24$ atm
75. **(a)** +6.00 kJ **(b)** 0 **(c)** +22.0 J/K
(d) +0.21 kJ **(e)** −0.23 kJ
76. **(a)** 5.1 kJ **(b)** ≈190 g (if $m \approx 60$ kg)
77. ≈1160°C
78. (1) $\Delta G° = 146.0 - 0.1104\,T$; becomes spontaneous at high T
(2) $\Delta G° = 168.6 - 0.0758\,T$; becomes spontaneous at higher T

Chapter 17

2. **(a)** $3Zn(s) + 2Cr^{3+}(aq) \longrightarrow 3Zn^{2+}(aq) + 2Cr(s)$
(b) $2Sn(s) + O_2(g) + 4H^+(aq) \longrightarrow 2Sn^{2+}(aq) + 2H_2O$
(c) $2Al(s) + 3I_2(s) \longrightarrow 2Al^{3+}(aq) + 6I^-(aq)$
4. **(a)** Zn anode, Cd cathode; e^- move from Zn to Cd. Anions move to Zn, cations to Cd.
(b) Cu anode, Au cathode; e^- move from Cu to Au. Anions move to Cu, cations to Au.
(c) Fe anode, Cu cathode; e^- move from Fe to Cu. Anions move to Fe, cations to Cu.
6. cathode: $Br_2(l) + 2e^- \longrightarrow 2Br^-(aq)$
anode: $2I^-(aq) \longrightarrow I_2(s) + 2e^-$
overall equation: $Br_2(l) + 2I^-(aq) \longrightarrow 2Br^-(aq) + I_2(s)$
cell notation: $Pt|\,I^-|I_2\,||\,Br_2|Br^-\,|Pt$
Electrons move from Pt anode to Pt cathode; anions to anode, cations to cathode.
8. **(a)** Ag **(b)** I^- **(c)** Fe^{2+}
(d) H_2 in basic solution
10. $Al^{3+} < Ni^{2+} < AgBr < ClO_3^- < F_2$
12. oxid. agents: $Cr^{3+} < Sn^{2+} < Br_2$
red. agents: $Cr^{3+} < Hg < Sn^{2+} < H_2$
14. **(a)** Fe^{3+}, Hg_2^{2+}, . . . **(b)** Tl, Pb, . . .
(c) ClO_3^-, Au^{3+}, . . .
16. **(a)** +0.695 V **(b)** +0.383 V **(c)** +0.202 V
18. **(a)** +0.063 V **(b)** +1.178 V **(c)** +0.948 V
20. **(a)** +0.281 V **(b)** +0.625 V **(c)** +0.946 V
22. **(a)** −0.799 V **(b)** +3.668 V **(c)** +1.101 V
24. none
26. **(a)** +0.924 V; yes **(b)** −1.431 V; no
(c) +1.297 V; yes
28. **(a)** no reaction
(b) $MnO_2(s) + 4H^+(aq) + 2Hg(l) \longrightarrow Mn^{2+}(aq) + Hg_2^{2+}(aq) + 2H_2O$
(c) no reaction
30. a, b, c
32. **(a)** no reaction
(b) $2Fe^{2+}(aq) + Br_2(l) \longrightarrow 2Fe^{3+}(aq) + 2Br^-(aq)$
$2I^-(aq) + Br_2(l) \longrightarrow I_2(s) + 2Br^-(aq)$
(c) no reaction
34. **(a)** +579 kJ, 10^{-101} **(b)** −579 kJ, 10^{101}
(c) +1160 kJ, 10^{-203}
36. **(a)** +0.146 V **(b)** −0.146 V
38. +0.216 V, −125 kJ, 8×10^{21}
40. **(a)** −134 kJ **(b)** −73.9 kJ **(c)** −58.5 kJ
42. **(a)** 1.3×10^2 **(b)** 7×10^{39} **(c)** 1×10^{64}
44. **(a)** $E = E° - \frac{0.0257}{2} \ln \frac{[Br^-]^2}{[I^-]^2}$ **(b)** +0.543 V
(c) +0.561 V
46. −1.308 V

48. **(a)** +0.778 V, yes **(b)** −0.436 V, no
50. **(a)** +0.235 V **(b)** $1.1 \times 10^{-4}\,M$ **52.** 4.31
54. **(a)** −0.460 V **(b)** $2 \times 10^{-9}\,M$ **(c)** 2×10^{-10}
56. **(a)** 6.696×10^{25} **(b)** 124.2 A **(c)** 889.5 g
58. **(a)** 0.61 g **(b)** 1.5×10^3 s
60. **(a)** 6.4×10^7 J **(b)** $1.10
62. anodic area: $Fe(s) \longrightarrow Fe^{2+}(aq) + 2e^-$
cathodic area: $\frac{1}{2}O_2(g) + H_2O + 2e^- \longrightarrow 2\,OH^-(aq)$
64. $2Cu(s) + H_2O(g) + SO_2(g) + \frac{3}{2}O_2(g) \longrightarrow Cu(OH)_2 \cdot CuSO_4(s)$
66. **(a)** 3.22×10^4 **(b)** 53.7 A
(c) 48.3 kJ; 0.0134 kWh
68. **(a)** False, $E° < 0$ **(b)** False; to cathode
(c) False, $PbSO_4$ is consumed
70. **(a)** 7.30 L **(b)** 13.78 **72.** 108 amu
74. −0.188 V **75.** 2.007 V
76. **(a)** +0.621 V **(b)** Zn^{2+} increases, Sn^{2+} decreases
(c) 1×10^{21}
(d) $[Zn^{2+}] = 2.0\,M$, $[Sn^{2+}] = 2 \times 10^{-21}\,M$
77. **(a)** +112 kJ; +38.0 kJ; +150 kJ **(b)** −0.389 V
78. 0.414 V

Chapter 18

2. $^{51}_{24}Cr \longrightarrow {}^{0}_{1}e + {}^{51}_{23}V$
4. **(a)** $^{237}_{93}Np \longrightarrow {}^{4}_{2}He + {}^{233}_{91}Pa$
(b) $^{85}_{39}Y \longrightarrow {}^{0}_{1}e + {}^{85}_{38}Sr$
(c) $^{12}_{6}C + {}^{12}_{6}C \longrightarrow {}^{23}_{11}Na + {}^{1}_{1}H$
(d) $^{239}_{94}Pu + {}^{1}_{0}n \longrightarrow {}^{130}_{50}Sn + 4\,{}^{1}_{0}n + {}^{106}_{44}Ru$
6. **(a)** $^{235}_{92}U$ **(b)** $^{231}_{91}Pa$
8. **(a)** $^{238}_{92}U + {}^{16}_{8}O \longrightarrow {}^{249}_{100}Fm + 5\,{}^{1}_{0}n$
(b) $^{26}_{13}Al + {}^{4}_{2}He \longrightarrow {}^{30}_{15}P$
(c) $^{63}_{29}Cu + {}^{1}_{1}H \longrightarrow {}^{63}_{30}Zn + {}^{1}_{0}n$
(d) $^{27}_{13}Al + {}^{2}_{1}H \longrightarrow {}^{4}_{2}He + {}^{25}_{12}Mg$
10. **(a)** $^{1}_{0}n$ **(b)** $^{243}_{97}Bk$ **(c)** $^{9}_{4}Be$ **(d)** $^{1}_{0}n$
12. **(a)** $^{28}_{14}Si$ **(b)** $^{6}_{3}Li$ **(c)** $^{23}_{11}Na$
14. **(a)** Ni **(b)** Se **(c)** Cd
16. 3.20×10^2 g **18.** 0.15
20. 36 h **22.** 2.0×10^{14}
24. $^{210}_{82}Pb \longrightarrow {}^{210}_{83}Bi + {}_{-1}^{0}e$; 4.17×10^4 mCi
26. ≈4500 B.C. **28.** yes **30.** 4.7×10^9 yr
32. **(a)** 2.7×10^9 yr **(b)** 3.4×10^9 yr
34. -2.8×10^{-5} g; -2.5×10^6 kJ **36.** F-19
38. **(a)** -1.76×10^8 kJ **(b)** -2.06×10^8 kJ
40. -3.64×10^{-7} g
42. **(a)** 1.33×10^{-8} g **(b)** 3.34×10^{13} atoms
44. yes; $\Delta m = -0.00322$ g/mol
46. 6.65×10^7 kJ vs 2.06×10^8 kJ
48. Its decay products including Po-218 and Po-214 are intensely radioactive and are readily absorbed in the lung tissue. Ra-222 is a gas and seeps through cracks in concrete from the ground into houses.
50. 3×10^4 atoms/s
52. **(a)** α and β rays have opposite charges
(b) see text p. 498 **(c)** see text p. 509
54. $6.59 \times 10^{-5}\,M$ **56.** 4.84×10^{-11}%
58. 6.5×10^2 mL **60.** 2.13×10^6 L
62. 0.015 L **64.** $5.8 \times 10^{-19}\,M$
65. **(a)** 5.5×10^{-10} g **(b)** 1.2×10^{-3} kJ **(c)** 17 rems
66. **(a)** 1×10^{-13} J **(b)** 6×10^6 m/s
67. **(a)** -5.72×10^8 kJ **(b)** 1.3×10^{28} kJ
(c) 1.8×10^{-11}

Chapter 19

2. $2Al_2O_3(l) \longrightarrow 4Al(l) + 3\,O_2(g)$; 2.91 g
4. $Cu_2S(s) + O_2(g) \longrightarrow 2Cu(s) + SO_2(g)$
6. −211.9 kJ
8. **(a)** $Fe_2O_3(s) + 3CO(g) \longrightarrow 2Fe(l) + 3CO_2(g)$
(b) $C(s) + O_2(g) \longrightarrow CO_2(g)$
10. 2.5×10^3 kWh **12.** 1.7×10^6 L
14. **(a)** potassium nitride, K_3N
(b) potassium iodide, KI
(c) potassium hydroxide, KOH
(d) potassium hydride, KH
(e) potassium sulfide, K_2S
16. **(a)** $Na_2O_2(s) + 2H_2O \longrightarrow 2Na^+(aq) + OH^-(aq) + H_2O_2(aq)$
sodium and hydroxide ions, hydrogen peroxide
(b) $2Ca(s) + O_2(g) \longrightarrow 2CaO(s)$; calcium oxide
(c) $Rb(s) + O_2(g) \longrightarrow RbO_2(s)$;
rubidium superoxide
(d) $SrH_2(s) + 2H_2O \longrightarrow Sr^{2+}(aq) + 2\,OH^-(aq) + 2H_2(g)$
strontium and hydroxide ions, hydrogen gas
18. 0.126 g
20. **(a)** $Co(s) + 2H^+(aq) \longrightarrow Co^{2+}(aq) + H_2(g)$
(b) $3Cu(s) + 2NO_3^-(aq) + 8H^+(aq) \longrightarrow 3Cu^{2+}(aq) + 2NO(g) + 4H_2O$
(c) $Cr_2O_7^{2-}(aq) + 6e^- + 14H^+(aq) \longrightarrow 2Cr^{3+}(aq) + 7H_2O$
22. $3Cd(s) + 12Cl^-(aq) + 2NO_3^-(aq) + 8H^+(aq) \longrightarrow 3CdCl_4^{2-}(aq) + 2NO(g) + 4H_2O$
24. **(a)** $Fe(s) + 3NO_3^-(aq) + 6H^+(aq) \longrightarrow Fe^{3+}(aq) + 3NO_2(g) + 3H_2O$
(b) $4Cr(OH)_3(s) + 3\,O_2(g) + 8\,OH^-(aq) \longrightarrow 4CrO_4^{2-}(aq) + 10\,H_2O$
26. **(a)** Cd ($E° = 1.366$ V) **(b)** Cr ($E° = 1.708$ V)
(c) Co ($E° = 1.246$ V) **(d)** Ag ($E° = 0.165$ V)
28. **(a)** 0.724 V **(b)** 0.942 V
30. **(a)** 9×10^9 **(b)** $2 \times 10^{-4}\,M$
32. Na^+; NH_3; lavender flame
34. **(a)** They form insoluble chlorides.
(b) They form soluble chlorides and sulfides.
36. Both would precipitate as CuS and NiS.
38. **(a)** $Pb^{2+}(aq) + 2Cl^-(aq) \longrightarrow PbCl_2(s)$
(b) $Al^{3+}(aq) + 3\,OH^-(aq) \longrightarrow Al(OH)_3(s)$
(c) $NH_4^+(aq) + OH^-(aq) \longrightarrow NH_3(g) + H_2O$
40. **(a)** At 400°C, only liquid is present. Between 370 and 300°C, solid A is in equilibrium with liquid. Below 300°C, solid B appears.

(b) At 400°C, only liquid is present. Between 360 and 300°C, solid B is in equilibrium with the liquid. At 300°C, solid A appears.

42. **(a)** Wood's metal
(b) Wood's metal, solder, pewter
(c) Duriron, monel, spiegeleisen

44. bones and teeth; hydroxyapatite—$Ca(OH)_2 \cdot 3Ca_3(PO_4)_2$

46. They deactivate enzymes containing —SH groups. They complex with ligands containing sulfur atoms.

48. 208 g **50.** 6.7 **52.** 53.8% Zn, 46.2% Cu

54. $2Ag_2S(s) + 8CN^-(aq) + 3\,O_2(g) + 2H_2O \longrightarrow 4Ag(CN)_2^-(aq) + 2SO_2(g) + 4\,OH^-(aq)$
$2Ag(CN)_2^-(aq) + Zn(s) \longrightarrow Zn(CN)_4^{2-}(aq) + 2Ag(s)$

56. **(a)** Cr^{2+} **(b)** Au^+ **(c)** Co^{2+} **(d)** Mn^{2+}

57. 2% BaO_2

58. **(a)** $Fe(OH)_3(s) + 3H_2C_2O_4(aq) \longrightarrow Fe(C_2O_4)_3^{3-}(aq) + 3H_2O + 3H^+(aq)$
(b) 0.28 L

59. 2.80% **60.** 2.83×10^3 K

61. $Cr_2O_7^{2-}(aq) + 2\,OH^-(aq) \longrightarrow 2CrO_4^{2-}(aq) + H_2O$
$2Ag^+(aq) + CrO_4^{2-}(aq) \longrightarrow Ag_2CrO_4(s)$
$Ag_2CrO_4(s) + 4NH_3(aq) \longrightarrow 2Ag(NH_3)_2^+(aq) + CrO_4^{2-}(aq)$
$2Ag(NH_3)_2^+(aq) + 4H^+(aq) + CrO_4^{2-}(aq) \longrightarrow Ag_2CrO_4(s) + 4NH_4^+(aq)$

Chapter 20

2. **(a)** bromic acid **(b)** potassium hypoiodite
(c) sodium chlorite **(d)** sodium perbromate

4. **(a)** $KBrO_2$ **(b)** $CaBr_2$ **(c)** $NaIO_4$ **(d)** $Mg(ClO)_2$

6. **(a)** NO_3^- **(b)** SO_4^{2-} **(c)** ClO_4^-

8. **(a)** H_2SO_3 **(b)** HClO **(c)** H_3PO_3

10. **(a)** NaN_3 **(b)** H_2SO_3 **(c)** N_2H_4 **(d)** NaH_2PO_4

12. **(a)** H_2S **(b)** N_2H_4 **(c)** PH_3

14. **(a)** NH_3, N_2H_4 **(b)** HNO_3 **(c)** HNO_2
(d) HNO_3

16. **(a)** $2I^-(aq) + SO_4^{2-}(aq) + 4H^+(aq) \longrightarrow I_2(s) + SO_2(g) + 2H_2O$
(b) $2I^-(aq) + Cl_2(g) \longrightarrow I_2(s) + 2Cl^-(aq)$

18. **(a)** $3HClO(aq) \longrightarrow Cl_2(g) + HClO_2(g) + H_2O$
(b) $2ClO_3^-(aq) \longrightarrow ClO_4^-(aq) + ClO_2^-(aq)$

20. **(a)** $Cl_2(g) + 2Br^-(aq) \longrightarrow 2Cl^-(aq) + Br_2(l)$
(b) NR **(c)** NR **(d)** NR

22. **(a)** $Pb(N_3)_2(s) \longrightarrow 3N_2(g) + Pb(s)$
(b) $2\,O_3(g) \longrightarrow 3\,O_2(g)$
(c) $2H_2S(g) + O_2(g) \longrightarrow 2S(s) + 2H_2O$

24. **(a)** $Cd^{2+}(aq) + H_2S(aq) \longrightarrow CdS(s) + 2H^+(aq)$
(b) $H_2S(aq) + OH^-(aq) \longrightarrow H_2O + HS^-(aq)$
(c) $2H_2S(aq) + O_2(g) \longrightarrow 2H_2O + 2S(s)$

26. **(a)** $2H^+(aq) + CaCO_3(s) \longrightarrow CO_2(g) + H_2O + Ca^{2+}(aq)$
(b) $H^+(aq) + OH^- \longrightarrow H_2O$
(c) $Cu(s) + 4H^+(aq) + SO_4^{2-}(aq) \longrightarrow Cu^{2+}(aq) + 2H_2O + SO_2(g)$

28. P

30. **(a)** Pass O_2 through 10^4 V of electric discharge.
(b) Freeze liquid sulfur at 119°C and cool quickly to room temperature.
(c) Heat white phosphorus in the absence of air to about 300°C.

32. **(a)** :Cl—O—Cl: **(b)** :O—N≡N:
(c) P_4 (tetrahedron of four P atoms, each with one lone pair) **(d)** :N≡N:

34. a and b

36. **(a)** $(:O—N{=}O:$ with :O: bonded to N$)^-$ **(b)** $(H—O—S—O:$ with :O: above and :O: below S$)^-$
(c) $(H—O—P—O—H$ with :O: above and :O: below P$)^-$

38. **(a)** :O—N—O—N—O: (each N double-bonded to :O:) **(b)** H—O—N=O: (with :O: bonded to N)
(c) $(:O—S—O:$ with :O: above and :O: below S$)^{2-}$

40. 14.3 L **42.** 0.204 *M* **44.** 7.7×10^4 L; 1.5×10^5 g

46. 12 L; 3.53 **48.** 4.28, 0.10 *M* **50.** 1.9×10^{-3}

52. 0.012 g/100 mL **54.** yes; 0 K

56. **(a)** yes **(b)** 2×10^{16}

58. 510 K **60.** 2.26×10^5 g **62.** 27.4 kg

64. b, c, d **66.** 4.6

68. thyroid gland; synthesis of I-containing hormones: goiter

70. P: found in bones and teeth; small amounts in DNA
S: mostly found in skin, hair, and nails
Cl: found as Cl^- in body fluids; HCl secreted during digestion

72. **(a)** dispersion **(b)** dispersion, dipole
(c) dispersion, H— bonds
(d) dispersion, H— bonds
(e) no intermolecular forces; not a molecule

74. **(a)** +3 **(b)** +4 **(c)** +5 **(d)** −3

76. **(a)** HClO **(b)** S, $KClO_3$
(c) NH_3, NaClO **(d)** HF

78. **(a)** see text p. 571 **(b)** has an unpaired electron **(c)** H^+ is a reactant. **(d)** C is a product.

79. density of sulfur; depth of deposit, purity of S

80. no **81.** 1.30%

82. Assume reaction is: $NaN_3(s) \longrightarrow Na(s) + \frac{3}{2}N_2(g)$
Assume 25°C, 1 atm pressure
mass of NaN_3 is 35 g

Chapter 21

2. **(a)** 2-methyloctane **(b)** 2,2-dimethylpropane **(c)** 2,2,4-trimethylpentane **(d)** 2,5-dimethylheptane

4. **(a)** $CH_3-C(CH_3)_2-CH_2-CH(CH_3)-CH_3$ **(b)** $CH_3-C(CH_3)_2-CH_3$

(c) $CH_3-(CH_2)_2-CH[CH(CH_3)-CH_3]-(CH_2)_3-CH_3$

(d) $CH_3-CH(CH_3)-CH(CH_3)-CH(CH_3)-(CH_2)_2-CH_3$

6. **(a)** $CH_3-C(CH_3)_2-CH_2-CH_3$
2,2-dimethylbutane

(b) $CH_3-(CH_2)_2-CH(CH_3)-CH_3$
2-methylpentane

(c) $CH_3-CH(CH_2-CH_3)-CH_3$
2-methylbutane

8. **(a)** $CH_3-C\equiv C-CH_2-CH_3$

(b) $CH_3-C\equiv C-CH(CH_3)-CH_3$

(c) $CH_3-CH(CH_3)-C\equiv C-CH_2-CH_3$

(d) $H-C\equiv C-C(CH_3)_2-CH_3$

10. **(a)** 1,3-dichlorobenzene **(b)** 1,2,4-trichlorobenzene **(c)** 1,3,5-trichlorobenzene

12. **(a)** methanoic acid; formic acid **(b)** ethanol; ethyl alcohol **(c)** 2-propanol; isopropyl alcohol

14. **(a)** methyl acetate **(b)** ethyl formate **(c)** methyl propionate

16. $>C=C-C-C-$, $-C-C=C-C-$, $-C-C(=C<)-C-$

18. $-C-C-C(Cl)-Cl$, $-C-C(Cl)(Cl)-C-$, $-C(Cl)-C-C(Cl)-$, $-C-C(Cl)-C(Cl)-$

20. Cl Cl Cl (1,2,3); Cl Cl Cl (1,2,4); Cl Cl Cl (1,3,5)

22. $Br-C(Cl)=C-C-C-$, $Br-C=C(Cl)-C-C-$,
$Cl-C=C(Br)-C-C-$, $-C-C(Br)=C(Cl)-C-$,
$-C-C(-C-)=C(Cl)-Br$

24. $-C-C-C-C(=O)-OH$, $-C-C(HO-C=O)-C-$

26. All of the compounds in Problem 22 show cis-trans isomerism except

$-C-C(-C-)=C(Cl)-Br$

where there are two $-CH_3$ groups attached to the same carbon

28. b, c **30.** **(a)** alcohol **(b)** ester **(c)** ester, acid

32.

```
         H                          H  O
         |                          |  ‖
(a) CH3—C—CH3        (b) CH3—C—C—OH
         |                          |
         OH                         CH3

         CH3        H
         |          |
(c) CH3—C—C—O—C—CH3
         |  ‖       |
         H  O       CH3
```

34. (a)

```
   F  F  F  F
   |  |  |  |
 —C—C—C—C—
   |  |  |  |
   F  F  F  F
```

(b) 2.5×10^6 g/mol

(c) 24.02% C, 75.98% F

36.

```
  H  H  H  H
  |  |  |  |
—C—C—C—C—
  |  |  |  |
  H  C6H5 H  C6H5
```

38. (a) $H_2C{=}CHF$ **(b)**

```
H3C—C=C—CH3
     |  |
     H  H
```

40.

```
                       CH3
                       |
—O—C—O—C6H4—C—C6H4—O—
     ‖                 |
     O                 CH3
```

42. (a) $H_2N{-}CH_2{-}CH_2{-}NH_2$ and $HOOC{-}CH_2{-}COOH$

(b)

```
HOOC—C6H4—COOH

                     H
                     |
and HO—CH2—C—OH
                     |
                     CH3
```

44. c, d

46. (a) center carbon **(b)** carbon atom at right

48. (a) present in "bottled gas."
(b) starting material for organic compounds like ethyl alcohol.
(c) fuel used in welding.
(d) nontoxic solvent; benzene replacement

50. (a) Gasoline contains mostly alkanes.
(b) Benzene is a carcinogen (cancer-causing).
(c) CaC_2 reacts with H_2O to form acetylene.
(d) Alkenes show geometric isomerism.
(e) Cholesterol is a fused-ring compound

52. 1.94

54. (a) neither **(b)** addition
(c) condensation **(d)** neither

56. $C_{27}H_{46}O$

57.

```
C—C—C—C—C—C              C—C—C—C—C—C
               |                      |
               OH                     OH

C—C—C—C—C—C              C—C—C—C—C—OH
         |                      |
         OH                     C

         OH                 OH
         |                  |
C—C—C—C—C              C—C—C—C—C
         |                     |
         C                     C

   OH
   |
C—C—C—C—C              HO—C—C—C—C—C
         |                          |
         C                          C

      OH
      |
C—C—C—C—C              C—C—C—C—C
      |                      |
      C                      C—OH

C—C—C—C—C              C—C—C—C—C—OH
      |  |                   |
      C  OH                  C

      OH
      |
C—C—C—C                C—C—C—C—OH
   |  |                   |  |
   C  C                   C  C

   C                      C
   |                      |
C—C—C—C                C—C—C—C
   |                      |  |
   C—OH                   C  OH

   C
   |
C—C—C—C—OH
   |
   C
```

58. approximately 6 g (Assume $H_2O(l)$ is a product; neglect heat capacity of pan.)

59. (a)

```
       C6H4 (ortho)
      /         \                 H
    —C           C—O—CH2—C—CH2—O—
     ‖           ‖                |
     O           O                OH
```

(b) orthophthalic acid condenses with OH groups in 2 adjacent chains.

Appendix 5 Net Ionic Equations

Most of the reactions discussed in this book take place in water solution. All such reactions are represented by net ionic equations. These equations involve ions as reactants and/or products. They contain other species as well, including molecules in solution and solids. As with all chemical equations, net ionic equations contain only those species that take part in the reaction. So-called "spectator ions" are omitted.

In this appendix, we will consider how balanced net ionic equations are written for a variety of reactions in water solution. Many of these equations have appeared before in the text, along with a discussion of the reactions they represent. However, this material was spread over several chapters, interspersed with stoichiometry and equilibrium principles. It should be helpful to summarize in one place all that we have said about net ionic equations.

FORMATION OF A SOLUTION (Chapters 4, 10)

When an ionic solid dissolves in water, the ions separate from one another. This process is described by a very simple type of net ionic equation. The formula of the (soluble) solid appears on the left side of the equation. The right side of the equation shows the cation and the anion, formed when the solid dissolves.

To illustrate this process, consider what happens when

—sodium nitrate ($NaNO_3$) dissolves in water. The products are Na^+ and NO_3^- ions in water solution. The net ionic equation is:

$$NaNO_3(s) \longrightarrow Na^+(aq) + NO_3^-(aq)$$

—sodium sulfate (Na_2SO_4) dissolves in water, forming Na^+ and SO_4^{2-} ions:

$$Na_2SO_4(s) \longrightarrow 2Na^+(aq) + SO_4^{2-}(aq)$$

There are two Na^+ ions in the equation because one mole of Na_2SO_4 contains two moles of Na^+ ions. In general, *the coefficient of the ion in the equation is equal to its subscript in the formula of the solid.*

Example 1 Write balanced net ionic equations for the formation of a water solution of the following ionic solids.

(a) $Al_2(SO_4)_3$ (b) iron(III) nitrate

Strategy First, decide what ions are present. The cation and anion appear as products, followed by the symbol (aq). The coefficient of each ion is equal to its subscript in the formula of the ionic compound. The reactant is the ionic compound, followed by the symbol (s).

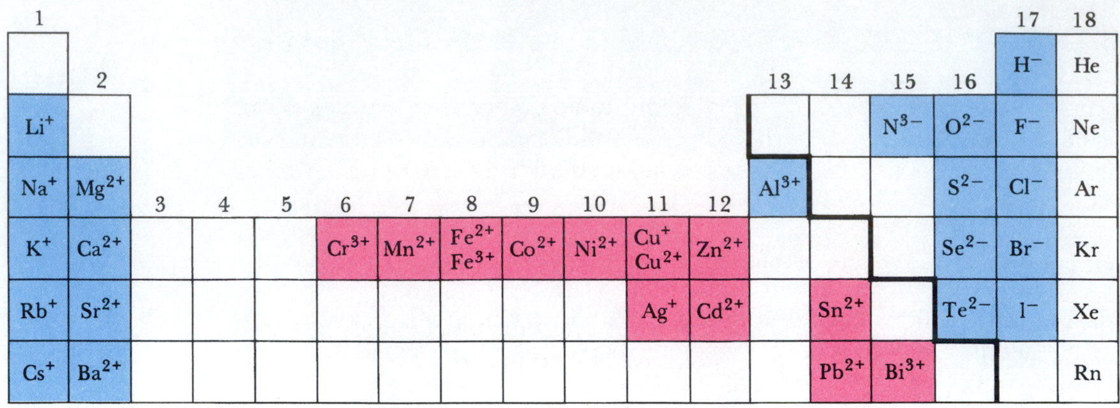

Figure 2.8
Charges of ions found in solid ionic compounds. The step-like diagonal line separates metals from nonmetals and cations from anions. Ions shown in blue have the same number of electrons as the neighboring noble gas atom. For example, S^{2-}, Cl^-, K^+, and Ca^{2+} all have 18 electrons, as does the Ar atom. The cations shown in red do not have a noble gas structure.

Solution

(a) Al^{3+}, $SO_4{}^{2-}$ ions

$$Al_2(SO_4)_3(s) \longrightarrow 2Al^{3+}(aq) + 3SO_4{}^{2-}(aq)$$

(b) Fe^{3+}, $NO_3{}^-$ ions; the formula of the solid must be $Fe(NO_3)_3$.

$$Fe(NO_3)_3(s) \longrightarrow Fe^{3+}(aq) + 3NO_3{}^-(aq)$$

You will notice that in order to write the equation referred to in Example 1, you had to know the charges and/or names of common cations and anions. This is indeed a requirement for writing just about any type of net ionic equation. The charges of monatomic ions were given in Figure 2.8, p. 37. The charges and names of polyatomic ions were listed in Table 2.3, p. 38.

Table 2.3 Some Common Polyatomic Ions

+1	−1	−2	−3
$NH_4{}^+$ (ammonium)	OH^- (hydroxide)	$CO_3{}^{2-}$ (carbonate)	$PO_4{}^{3-}$ (phosphate)
	$NO_3{}^-$ (nitrate)	$SO_4{}^{2-}$ (sulfate)	
	$ClO_3{}^-$ (chlorate)	$CrO_4{}^{2-}$ (chromate)	
	$ClO_4{}^-$ (perchlorate)	$Cr_2O_7{}^{2-}$ (dichromate)	
	CN^- (cyanide)	$HPO_4{}^{2-}$ (hydrogen phosphate)	
	$C_2H_3O_2{}^-$ (acetate)		
	$MnO_4{}^-$ (permanganate)		
	$HCO_3{}^-$ (hydrogen carbonate)		
	$H_2PO_4{}^-$ (dihydrogen phosphate)		

FORMATION OF PRECIPITATE: CATION + ANION (Chapter 4)

Sometimes, when solutions of two different ionic compounds are mixed, an insoluble solid (precipitate) is formed. This solid is an ionic compound, containing the cation from one solution combined with the anion of the other solution.

To write a net ionic equation for a reaction of this type, you follow what amounts to a four-step path. To illustrate that path, let us apply it to answer the following exercise:

Write a balanced net ionic equation for any precipitation reaction that occurs when solutions of $Pb(NO_3)_2$ and NaOH are mixed.

1. *Decide what ions are present in the two solutions.* A solution of $Pb(NO_3)_2$ contains Pb^{2+} cations and NO_3^- anions:

$$Pb(NO_3)_2(s) \longrightarrow Pb^{2+}(aq) + 2NO_3^-(aq)$$

A solution of NaOH contains Na^+ and OH^- ions:

$$NaOH(s) \longrightarrow Na^+(aq) + OH^-(aq)$$

2. *Identify the two possible precipitates, formed by combining a cation from one solution with the anion from the other.* Here, the possibilities are:

— $Pb(OH)_2$, formed by the reaction between Pb^{2+} cations and OH^- anions
— $NaNO_3$, formed by the reaction between Na^+ cations and NO_3^- anions

3. *Using the solubility rules (Table 4.1, p. 76), decide whether either or both of the possible solids will precipitate.* In this case

— $NaNO_3$ does not precipitate (applying the solubility rules for nitrates)
— $Pb(OH)_2$ does precipitate (applying the solubility rules for hydroxides)

4. *Write a balanced net ionic equation for any precipitation reaction that occurs.* The product is the insoluble solid $Pb(OH)_2$. The reactants are the aqueous cation and anion that combined to form the precipitate. In this case, the equation is:

$$Pb^{2+}(aq) + 2\,OH^-(aq) \longrightarrow Pb(OH)_2(s)$$

Table 4.1 Solubility Rules*

NO_3^-	All nitrates are soluble.
Cl^-	All chlorides are soluble except AgCl, Hg_2Cl_2, and $PbCl_2$.
SO_4^{2-}	Most sulfates are soluble; exceptions include $SrSO_4$, $BaSO_4$, and $PbSO_4$.
CO_3^{2-}	All carbonates are insoluble except those of the Group 1 elements and NH_4^+.
OH^-	All hydroxides are insoluble except those of the Group 1 elements, $Sr(OH)_2$, and $Ba(OH)_2$, ($Ca(OH)_2$ is slightly soluble.)
S^{2-}	All sulfides except those of Group 1 and 2 elements and NH_4^+ are insoluble.

* Insoluble compounds are those that precipitate when we mix equal volumes of ~0.1 mol/L solutions of the corresponding ions.

Example 2 Write balanced net ionic equations for any precipitation reaction that occurs when

(a) solutions of $BaCl_2$ and Ag_2SO_4 are mixed.
(b) solutions of nickel(II) nitrate and potassium sulfate are mixed.

Strategy Follow the four-step path referred to above.

Solution

(a) 1. Ions present: Ba^{2+} and Cl^-; Ag^+ and SO_4^{2-}
2. Possible precipitates: $BaSO_4$, AgCl
3. By the solubility rules, both $BaSO_4$ and AgCl are insoluble.
4. There are two reactions, so you write two equations:

$$Ag^+(aq) + Cl^-(aq) \longrightarrow AgCl(s)$$
$$Ba^{2+}(aq) + SO_4^{2-}(aq) \longrightarrow BaSO_4(s)$$

(b) 1. Ni^{2+} and NO_3^-; K^+ and SO_4^{2-}
2. $NiSO_4$, KNO_3
3. Both solids are soluble
4. Since there is no reaction, there is no equation

FORMATION OF PRECIPITATE: CATION + H_2S or NH_3 (Chapter 20)

In principle at least, the simplest way to precipitate a metal sulfide is to mix two solutions, one containing the metal cation, the other the sulfide anion. In practice, however, metal sulfides are most often precipitated by using hydrogen sulfide molecules (H_2S) rather than sulfide anions (S^{2-}). This is, for example, the procedure used in qualitative analysis (Chapter 19). The product is the same, the insoluble metal sulfide. The reactants are the metal cation and an H_2S molecule. The net ionic equation for the reaction is readily written if you keep in mind a simple principle: every H_2S molecule produces one sulfide ion. The reaction that occurs when hydrogen sulfide is added to a water solution containing Cu^{2+} is:

$$Cu^{2+}(aq) + H_2S(aq) \longrightarrow CuS(s) + 2H^+(aq)$$

Notice that:

— the reactants are the cation and an H_2S molecule
— the products comprise the sulfide precipitate and the $2H^+$ ions left over from the H_2S molecule

An analogous situation applies to the precipitation of metal hydroxides such as aluminum hydroxide, $Al(OH)_3$. The simplest way, at least on paper, to form such a precipitate is to mix a solution containing the Al^{3+} cation (e.g., $AlCl_3$) with a solution of a strong base (e.g., NaOH). Alternatively, aluminum hydroxide can be precipitated by adding a weak base such as ammonia to a solution containing the Al^{3+} cation. The OH^- ions required for the precipitation of $Al(OH)_3$ come from the reaction:

$$NH_3(aq) + H_2O \rightleftharpoons OH^-(aq) + NH_4^+(aq)$$

The overall equation for the reaction that occurs when ammonia is added to a solution of $AlCl_3$ is:

$$Al^{3+}(aq) + 3NH_3(aq) + 3H_2O \longrightarrow Al(OH)_3(s) + 3NH_4^+(aq)$$

Notice that, for every OH^- ion required to form the precipitate, one NH_3 molecule and one H_2O molecule appear as reactants; one NH_4^+ ion is formed as a byproduct.

Example 3 Write a balanced net ionic equation for the precipitation reaction that occurs when

(a) H_2S is added to a solution containing Bi^{3+} ions.
(b) NH_3 is added to a solution containing Pb^{2+} ions.

Strategy First decide upon the identity and formula of the precipitate. Then follow the suggestions above for writing the net ionic equation. Remember that:

— one H_2S molecule supplies one S^{2-} ion and forms two H^+ ions as a byproduct.
— one NH_3 molecule (and one H_2O molecule) are required to supply one OH^- ion; one NH_4^+ ion is formed as a byproduct.

Solution

(a) The product is bismuth(III) sulfide. Applying the principle of electroneutrality, the formula of the solid must be Bi_2S_3. Since one mole of Bi_2S_3 contains three moles of S^{2-} ions, three moles of H_2S are required and six moles of H^+ ions are formed.

$$2Bi^{3+}(aq) + 3H_2S(aq) \longrightarrow Bi_2S_3(s) + 6H^+(aq)$$

(b) The product is lead(II) hydroxide, $Pb(OH)_2$. Two OH^- ions are required, so two NH_3 molecules appear as reactants, along with two H_2O molecules; two NH_4^+ ions are formed.

$$Pb^{2+}(aq) + 2NH_3(aq) + 2H_2O \longrightarrow Pb(OH)_2(s) + 2NH_4^+(aq)$$

FORMATION OF AN ACIDIC SOLUTION (Chapters 4, 13)

An acidic water solution is one that contains an excess of H^+ ions*. The species producing these ions is referred to as an acid. It may be either a (neutral) molecule or a (positively charged) cation.

Molecular acids can be divided into two categories.

1. Strong acids (Table 4.2, p. 80), which are completely dissociated in water solution. A simple example is hydrochloric acid, a water solution of HCl. The net ionic equation for the dissociation of HCl in water is written most simply as

$$HCl(aq) \longrightarrow H^+(aq) + Cl^-(aq)$$

* More properly, hydrated H^+ ions, e.g., H_3O^+ (see the discussion of the Brønsted-Lowry model in Chapter 13). To keep the equations simple, we will use H^+ ions rather than H_3O^+ ions.

Table 4.2 Common Strong Acids and Bases

Acid	Name of Acid	Base	Name of Base
HCl	hydrochloric acid	$LiOH$	lithium hydroxide
HBr	hydrobromic acid	$NaOH$	sodium hydroxide
HI	hydriodic acid	KOH	potassium hydroxide
HNO_3	nitric acid	$Ca(OH)_2$	calcium hydroxide
$HClO_4$	perchloric acid	$Sr(OH)_2$	strontium hydroxide
H_2SO_4	sulfuric acid	$Ba(OH)_2$	barium hydroxide

The reactant is the acid molecule; the products include the H^+ ion and the anion (Cl^-) derived from the acid. The corresponding equation for the dissociation of sulfuric acid is:

$$H_2SO_4(aq) \longrightarrow H^+(aq) + HSO_4^-(aq)$$

2. Weak acids, which are only partially dissociated in water. Any molecular acid not listed in Table 4.2 can be assumed to be weak. The net ionic equation for its dissociation is almost identical to those written above except that a double arrow is used to indicate that the reaction does not go to completion. Thus for HF ($K_a = 6.9 \times 10^{-4}$) and H_2CO_3 ($K_a = 4.4 \times 10^{-7}$), we have:

$$HF(aq) \rightleftharpoons H^+(aq) + F^-(aq)$$

$$H_2CO_3(aq) \rightleftharpoons H^+(aq) + HCO_3^-(aq)$$

A large number of cations produce acidic water solutions. All of these cations are weak acids, incompletely dissociated in water solution. The simplest example is the ammonium ion; the net ionic equation for its dissociation is:

$$NH_4^+(aq) \rightleftharpoons H^+(aq) + NH_3(aq)$$

The byproduct here is the NH_3 molecule, the conjugate base of the NH_4^+ ion.

Virtually all transition metal cations act as weak acids in water solution. To understand this behavior, you have to realize that these ions are hydrated in water solution. Typically, the metal cation is bonded to four or six H_2O molecules to give a complex ion.

$$\text{zinc(II): } Zn(H_2O)_4^{2+}$$

$$\text{chromium(III): } Cr(H_2O)_6^{3+}$$

The H^+ ion which makes a solution of zinc(II) or chromium(III) salts acidic comes from the dissociation of one of the water molecules bonded to the metal cation, i.e.

$$Zn(H_2O)_4^{2+}(aq) \rightleftharpoons H^+(aq) + Zn(H_2O)_3(OH)^+(aq)$$

$$Cr(H_2O)_6^{3+}(aq) \rightleftharpoons H^+(aq) + Cr(H_2O)_5(OH)^{2+}(aq)$$

Notice that the byproduct is a complex ion in which one of the H_2O molecules of hydration has been replaced by an OH^- ion. It has a charge one unit less than that of the hydrated cation from which it is formed (e.g., +1 as opposed to +2, +2 as opposed to +3).

Certain main-group cations of high charge and small size, notably aluminum(III), behave very much like transition metal cations as far as acidic properties are concerned. For example, a solution of $Al(NO_3)_3$ is about as acidic as one of $Cr(NO_3)_3$. The net ionic equations written to explain the acidic behavior are very similar:

$$Al(H_2O)_6{}^{3+}(aq) \rightleftharpoons H^+(aq) + Al(H_2O)_5(OH)^{2+}(aq)$$

All metal cations except those derived from strong bases (Li^+, Na^+, K^+; Ca^{2+}, Sr^{2+}, Ba^{2+}) behave this way. When you write a net ionic equation to explain the acidity of a metal cation, remember that

— the reactant is a complex ion in which the bare cation (e.g., Zn^{2+}, Cr^{3+}, Al^{3+}) is bonded to 4 or 6 water molecules
— the H^+ ion is one product
— the other product is the conjugate base of the hydrated cation in which an H_2O molecule has been replaced by an OH^- ion.

Example 4 Write net ionic equations to explain the acidity of water solutions of the following compounds.

(a) H_3PO_4 (b) NH_4Cl (c) $ZnCl_2$

Strategy First identify the species which is responsible for the acidity of the solution. This may be a molecule (H_3PO_4), a simple cation ($NH_4{}^+$), or a hydrated cation ($Zn(H_2O)_4{}^{2+}$). Ignore spectator ions such as Cl^-. Since all of the acids are weak (none of them are listed in Table 4.2), the reactions are reversible. One of the products is an H^+ ion; the other is the conjugate base formed by loss of an H^+ ion from the weak acid.

Solution

(a) The weak acid is the H_3PO_4 molecule; its conjugate base is the $H_2PO_4{}^-$ ion. The net ionic equation is:

$$H_3PO_4(aq) \rightleftharpoons H^+(aq) + H_2PO_4{}^-(aq)$$

(b) Weak acid = $NH_4{}^+$ cation; conjugate base = NH_3 molecule.

$$NH_4{}^+(aq) \rightleftharpoons H^+(aq) + NH_3(aq)$$

(c) Weak acid = $Zn(H_2O)_4{}^{2+}(aq)$ complex cation; conjugate base = $Zn(H_2O)_3(OH)^+(aq)$.

$$Zn(H_2O)_4{}^{2+}(aq) \rightleftharpoons H^+(aq) + Zn(H_2O)_3(OH)^+(aq)$$

FORMATION OF A BASIC SOLUTION (Chapters 4 and 13)

A basic solution is one which has an excess of OH^- ions. As with acids, we can distinguish between strong and weak bases. The strong bases (Table 4.2) are the hydroxides of the alkali and alkaline earth metals, which are completely ionized in water solution, forming cations (e.g., Na^+, Ca^{2+}) and OH^- ions. The net ionic equation here is simply that for the solution process:

$$NaOH(s) \longrightarrow Na^+(aq) + OH^-(aq)$$

$$Ca(OH)_2(s) \longrightarrow Ca^{2+}(aq) + 2\ OH^-(aq)$$

A weak base operates in a quite different way. It reacts with an H_2O molecule to set free an OH^- ion, which makes the solution basic. The H^+ ion formed at the same time bonds to the weak base, converting it to its conjugate weak acid.

A weak base may be a (neutral) molecule such as NH_3. The net ionic equation for the formation of a basic solution when ammonia is added to water is:

$$NH_3(aq) + H_2O \rightleftharpoons OH^-(aq) + NH_4^+(aq)$$

The reaction is reversible ($K_b NH_3 = 1.8 \times 10^{-5}$). The products are an OH^- ion and the NH_4^+ ion, which is the conjugate acid of NH_3. The reaction with amines such as CH_3NH_2 is entirely analogous:

$$CH_3NH_2(aq) + H_2O \rightleftharpoons OH^-(aq) + CH_3NH_3^+(aq)$$

Here again, the byproduct is the species formed when the weak base acquires a proton.

Anions derived from weak acids behave as weak bases. Species in this category include the F^- ion and the CO_3^{2-} ion. The net ionic equations for their reactions with water are:

$$F^-(aq) + H_2O \rightleftharpoons OH^-(aq) + HF(aq)$$

$$CO_3^{2-}(aq) + H_2O \rightleftharpoons OH^-(aq) + HCO_3^-(aq)$$

Notice that regardless of whether the weak base is a molecule or anion:

— the reactants are the weak base (NH_3, CH_3NH_2, F^-, CO_3^{2-}) and an H_2O molecule
— the reaction is reversible($\rightleftharpoons$)
— the products are an OH^- ion and the conjugate acid of the weak base (NH_4^+, $CH_3NH_3^+$, HF, HCO_3^-).

Example 5 Write balanced net ionic equations to explain why solutions of the following compounds are basic.

(a) $C_2H_5NH_2$ (b) Na_3PO_4

Strategy First identify the weak base (molecule or anion). That species is one reactant; the other is an H_2O molecule. The products are an OH^- ion and the conjugate acid of the weak base.

Solution

(a) The weak base is the $C_2H_5NH_2$ molecule; its conjugate acid is the $C_2H_5NH_3^+$ ion. The net ionic equation is:

$$C_2H_5NH_2(aq) + H_2O \rightleftharpoons OH^-(aq) + C_2H_5NH_3^+(aq)$$

(b) The weak base is the PO_4^{3-} anion; the Na^+ cations are spectators. Since the conjugate acid of PO_4^{3-} is HPO_4^{2-}, the net ionic equation is:

$$PO_4^{3-}(aq) + H_2O \rightleftharpoons OH^-(aq) + HPO_4^{2-}(aq)$$

In order to write a net ionic equation to explain the acidity or basicity of a water solution, you must be able to distinguish between

— strong and weak acids and bases (Table 4.2, p. 80)
— conjugate acids and bases (Chapter 13)
— acidic, basic, and spectator ions (Table 13.7, p. 376)

Table 13.7 Acid-Base Properties of Some Common Ions in Water Solution

	Spectator		Basic		Acidic
Anion	Cl^- Br^- I^-	NO_3^- ClO_4^- SO_4^{2-}	$C_2H_3O_2^-$ F^- CO_3^{2-} S^{2-} PO_4^{3-}	CN^- NO_2^-	
Cation	Li^+ Na^+ K^+	Ca^{2+} Ba^{2+}			Mg^{2+} Al^{3+} NH_4^+ transition metal ions

ACID-BASE REACTIONS; STRONG ACID–STRONG BASE (Chapters 4, 14)

When an acidic water solution is mixed with a basic solution, a chemical reaction takes place. The nature of that reaction and of the corresponding net ionic equation depends upon whether the acid and base involved are strong or weak. When both the acid and base are strong, the reaction is between the H^+ ion of the acidic solution and the OH^- ion of the basic solution. The product of this reaction, referred to as **neutralization,** is an H_2O molecule:

$$H^+(aq) + OH^-(aq) \longrightarrow H_2O$$

This simple equation applies to the reaction of any strong acid with any strong base, e.g.,

$$HCl + NaOH \qquad H^+(aq) + OH^-(aq) \longrightarrow H_2O$$

$$HNO_3 + Ca(OH)_2 \qquad H^+(aq) + OH^-(aq) \longrightarrow H_2O$$

The other ions present in these solutions (Cl^-, Na^+, – –) are spectators; they take no part in the reaction and are not included in the equation.

A reaction quite similar to neutralization takes place when a strong acid is added to a water-insoluble metal hydroxide. The H^+ ions of the acid react with the OH^- ions of the solid, converting them to H_2O molecules. The reaction with iron(III) hydroxide is typical:

$$Fe(OH)_3(s) + 3H^+(aq) \longrightarrow Fe^{3+}(aq) + 3H_2O$$

Notice that for every OH^- ion in the solid, one H^+ ion is consumed and one H_2O molecule forms.

Example 6 Write balanced net ionic equations for the reactions of sulfuric acid with

(a) a solution of $Ca(OH)_2$ (b) $Mg(OH)_2(s)$

Strategy In both cases, the H^+ ion found in a solution of the strong acid H_2SO_4 is a reactant. The other reactant is the OH^- ion. In (a), the OH^- ion is in solution; in (b), it is present in the solid. Water molecules are produced in both cases.

Solution

(a) $H^+(aq) + OH^-(aq) \longrightarrow H_2O$
(b) $Mg(OH)_2(s) + 2H^+(aq) \longrightarrow Mg^{2+}(aq) + 2H_2O$

ACID–BASE REACTIONS; WEAK ACID–STRONG BASE (Chapters 4, 14)

The reaction that takes place in water solution between a weak acid and a strong base resembles neutralization. However, the net ionic equations are somewhat different. This happens because the principal species present in the solution of the weak acid is the weak acid molecule rather than an H^+ ion. Specifically, for the reaction between hydrofluoric acid and a strong base, the net ionic equation is

$$HF(aq) + OH^-(aq) \longrightarrow H_2O + F^-(aq)$$

Comparing the equation just written to that for a strong acid with a strong base

$$H^+(aq) + OH^-(aq) \longrightarrow H_2O$$

you can see that

— in each case, the OH^- ion is a reactant and H_2O is a product
— in the weak acid case, the weak acid molecule (HF) is the reactant rather than the H^+ ion; the conjugate base (F^-) is a byproduct.

Example 7 Write balanced net ionic equations for the reaction of a solution of sodium hydroxide with a solution of

(a) HNO_2 (b) NH_4Cl

Strategy First, identify the species that acts as a weak acid, the HNO_2 molecule in (a) and the $NH_4{}^+$ ion in (b). This is one reactant; the other is the OH^- ion. There are two products, the H_2O molecule and the conjugate base of the weak acid.

Solution

(a) $HNO_2(aq) + OH^-(aq) \longrightarrow H_2O + NO_2{}^-(aq)$
(b) $NH_4{}^+(ag) + OH^-(aq) \longrightarrow H_2O + NH_3(aq)$

ACID-BASE REACTIONS; STRONG ACID–WEAK BASE (Chapters 4, 14)

When a strong acid reacts with a weak base, there are two reactants:

— the H^+ ion of the strong acid
— the weak base, which may be a molecule (NH_3, – –) or an anion ($HCO_3{}^-$, $CO_3{}^{2-}$, – –).

There is a single product, the conjugate acid of the weak base. The net ionic equation for the reaction is a relatively simple one:

$$H^+(aq) + NH_3(aq) \longrightarrow NH_4^+(aq)$$

$$H^+(aq) + HCO_3^-(aq) \longrightarrow H_2CO_3(aq) \longrightarrow CO_2(g) + H_2O$$

(Carbonic acid, H_2CO_3, is relatively unstable, decomposing to carbon dioxide and water.)

When a strong acid is added to a solution containing the carbonate ion, CO_3^{2-}, a two-step reaction occurs. The first step is the formation of the HCO_3^- ion, the conjugate acid of CO_3^{2-}. This reacts with excess H^+ ion by the equation written above; the final products are CO_2 and H_2O. The overall equation is

$$2H^+(aq) + CO_3^{2-}(aq) \longrightarrow CO_2(aq) + H_2O$$

A very similar reaction occurs when a strong acid is added to an insoluble metal carbonate such as calcium carbonate

$$2H^+(aq) + CaCO_3(s) \longrightarrow CO_2(g) + H_2O + Ca^{2+}(aq)$$

The effect is to bring the solid carbonate into solution and evolve bubbles of carbon dioxide gas.

Example 8 Write balanced net ionic equations for the reaction of hydrochloric acid with

(a) CH_3NH_2 (b) NO_2^- (c) Ag_2CO_3

Strategy Perhaps the simplest approach here is to reason by analogy with the equations written above for NH_3, HCO_3^-, and $CaCO_3$. There are always two reactants, the H^+ ion and the weak base molecule or anion. Remember that the CO_3^{2-} ion is converted to CO_2 and H_2O.

Solution

(a) $H^+(aq) + CH_3NH_2(aq) \longrightarrow CH_3NH_3^+(aq)$
(b) $H^+(aq) + NO_2^-(aq) \longrightarrow HNO_2(aq)$
(c) $2H^+(aq) + Ag_2CO_3(s) \longrightarrow CO_2(g) + H_2O + 2Ag^+(aq)$

Clearly, in order to write a net ionic equation for an acid-base reaction, you must first decide whether the acid and base are strong or weak. That determines the nature of the species that appear in the equation, both as reactants and products (Table 4.3, p. 82).

Table 4.3 Types of Acid-Base Reactions

Reactants	Reacting Species	Net Ionic Equation
Strong acid–Strong base	H^+, OH^-	$H^+(aq) + OH^-(aq) \longrightarrow H_2O$
Weak acid–Strong base	HB, OH^-	$HB(aq) + OH^-(aq) \longrightarrow H_2O + B^-(aq)$
Strong acid–Weak base	H^+, B	$H^+(aq) + B(aq) \longrightarrow BH^+(aq)$

REDOX REACTIONS; SPONTANEITY (Chapter 17)

In one important class of reactions in aqueous solution, known as oxidation-reduction (redox) reactions, two different elements change oxidation number. One element increases in oxidation number and is said to be oxidized. At the same time another element decreases in oxidation number; it is reduced. Typical reactions of this type are

$$Zn(s) + 2H^+(aq) \longrightarrow Zn^{2+}(aq) + H_2(g)$$

$$Cl_2(g) + 2Br^-(aq) \longrightarrow 2Cl^-(aq) + Br_2(l)$$

In these reactions, zinc and bromine are oxidized (oxid. no. Zn: $0 \rightarrow +2$; oxid. no. Br: $-1 \rightarrow 0$). At the same time, hydrogen (oxid. no. $+1 \rightarrow 0$) and chlorine (oxid. no. $0 \rightarrow -1$) are reduced.

To decide whether or not a redox reaction will take place when two species are mixed, you apply a simple principle. Generally speaking, if the standard voltage, $E°$, is positive, the reaction will take place. The standard voltage is calculated from the relation

$$E° = E°_{ox} + E°_{red}$$

where $E°_{ox}$ and $E°_{red}$ are the standard voltages for the half-reactions of oxidation and reduction. These quantities can be obtained from Table 17.1 p. 468. Values of $E°_{red}$ are listed directly in the table. The quantity $E°_{ox}$ is obtained by changing the sign of the standard potential listed in the table.

Applying these principles to the two reactions referred to above:

(1)
$$\begin{array}{ll} Zn(s) \longrightarrow Zn^{2+}(aq) + 2e^- & E°_{ox} = +0.762\text{ V} \\ \underline{2H^+(aq) + 2e^- \longrightarrow H_2(g)} & \underline{E°_{red} = 0.000\text{ V}} \\ Zn(s) + 2H^+(aq) \longrightarrow Zn^{2+}(aq) + H_2(g) & E° = +0.762\text{ V} \end{array}$$

(2)
$$\begin{array}{ll} Cl_2(g) + 2e^- \longrightarrow 2Cl^-(aq) & E°_{red} = +1.360\text{ V} \\ \underline{2Br^-(aq) \longrightarrow Br_2(l) + 2e^-} & \underline{E°_{ox} = -1.077\text{ V}} \\ Cl_2(g) + 2Br^-(aq) \longrightarrow 2Cl^-(aq) + Br_2(l) & E° = +0.283\text{ V} \end{array}$$

Since the calculated $E°$ values are positive, both of these reactions should, and do, occur spontaneously.

If you are asked to predict whether a redox reaction will take place when two species are mixed, an additional step is involved. Before you can calculate $E°$, you have to decide upon the nature of the oxidation and reduction half reactions. Table 17.1 is helpful here. Species listed in the left column (Li^+, K^+, – – –) can be reduced; species in the right column (Li, K, – – –) can be oxidized. To see how this works out, consider the question:

Will a redox reaction occur when Sn^{2+} and Fe^{2+} are mixed?

Scanning the left column, it appears that there are two possible reductions:

$$Sn^{2+}(aq) + 2e^- \longrightarrow Sn(s) \qquad E°_{red} = -0.141\text{ V}$$

$$Fe^{2+}(aq) + 2e^- \longrightarrow Fe(s) \qquad E°_{red} = -0.409\text{ V}$$

Examination of the right column shows two possible oxidations:

$$Sn^{2+}(aq) \longrightarrow Sn^{4+}(aq) + 2e^- \qquad E°_{ox} = -0.154\text{ V}$$

$$Fe^{2+}(aq) \longrightarrow Fe^{3+}(aq) + e^- \qquad E°_{ox} = -0.769\text{ V}$$

Table 17.1 Standard Potentials in Water Solution at 25°C

Oxidizing Agent	Reducing Agent	E°_{red}(V)
	Acidic Solution	
$Li^+(aq) + e^-$	$\rightarrow Li(s)$	−3.040
$K^+(aq) + e^-$	$\rightarrow K(s)$	−2.936
$Ba^{2+}(aq) + 2e^-$	$\rightarrow Ba(s)$	−2.906
$Ca^{2+}(aq) + 2e^-$	$\rightarrow Ca(s)$	−2.869
$Na^+(aq) + e^-$	$\rightarrow Na(s)$	−2.714
$Mg^{2+}(aq) + 2e^-$	$\rightarrow Mg(s)$	−2.357
$Al^{3+}(aq) + 3e^-$	$\rightarrow Al(s)$	−1.68
$Mn^{2+}(aq) + 2e^-$	$\rightarrow Mn(s)$	−1.182
$Zn^{2+}(aq) + 2e^-$	$\rightarrow Zn(s)$	−0.762
$Cr^{3+}(aq) + 3e^-$	$\rightarrow Cr(s)$	−0.744
$Fe^{2+}(aq) + 2e^-$	$\rightarrow Fe(s)$	−0.409
$Cr^{3+}(aq) + e^-$	$\rightarrow Cr^{2+}(aq)$	−0.408
$Cd^{2+}(aq) + 2e^-$	$\rightarrow Cd(s)$	−0.402
$PbSO_4(s) + 2e^-$	$\rightarrow Pb(s) + SO_4^{2-}(aq)$	−0.356
$Tl^+(aq) + e^-$	$\rightarrow Tl(s)$	−0.336
$Co^{2+}(aq) + 2e^-$	$\rightarrow Co(s)$	−0.282
$Ni^{2+}(aq) + 2e^-$	$\rightarrow Ni(s)$	−0.236
$AgI(s) + e^-$	$\rightarrow Ag(s) + I^-(aq)$	−0.152
$Sn^{2+}(aq) + 2e^-$	$\rightarrow Sn(s)$	−0.141
$Pb^{2+}(aq) + 2e^-$	$\rightarrow Pb(s)$	−0.127
$2H^+(aq) + 2e^-$	$\rightarrow H_2(g)$	0.000
$AgBr(s) + e^-$	$\rightarrow Ag(s) + Br^-(aq)$	0.073
$S(s) + 2H^+(aq) + 2e^-$	$\rightarrow H_2S(aq)$	0.144
$Sn^{4+}(aq) + 2e^-$	$\rightarrow Sn^{2+}(aq)$	0.154
$SO_4^{2-}(aq) + 4H^+(aq) + 2e^-$	$\rightarrow SO_2(g) + 2H_2O$	0.155
$Cu^{2+}(aq) + e^-$	$\rightarrow Cu^+(aq)$	0.161
$Cu^{2+}(aq) + 2e^-$	$\rightarrow Cu(s)$	0.339
$Cu^+(aq) + e^-$	$\rightarrow Cu(s)$	0.518
$I_2(s) + 2e^-$	$\rightarrow 2I^-(aq)$	0.534
$Fe^{3+}(aq) + e^-$	$\rightarrow Fe^{2+}(aq)$	0.769
$Hg_2^{2+}(aq) + 2e^-$	$\rightarrow 2Hg(l)$	0.796
$Ag^+(aq) + e^-$	$\rightarrow Ag(s)$	0.799
$2Hg^{2+}(aq) + 2e^-$	$\rightarrow Hg_2^{2+}(aq)$	0.908
$NO_3^-(aq) + 4H^+(aq) + 3e^-$	$\rightarrow NO(g) + 2H_2O$	0.964
$AuCl_4^-(aq) + 3e^-$	$\rightarrow Au(s) + 4Cl^-(aq)$	1.001
$Br_2(l) + 2e^-$	$\rightarrow 2Br^-(aq)$	1.077
$O_2(g) + 4H^+(aq) + 4e^-$	$\rightarrow 2H_2O$	1.229
$MnO_2(s) + 4H^+(aq) + 2e^-$	$\rightarrow Mn^{2+}(aq) + 2H_2O$	1.229
$Cr_2O_7^{2-}(aq) + 14H^+(aq) + 6e^-$	$\rightarrow 2Cr^{3+}(aq) + 7H_2O$	1.33
$Cl_2(g) + 2e^-$	$\rightarrow 2Cl^-(aq)$	1.360
$ClO_3^-(aq) + 6H^+(aq) + 5e^-$	$\rightarrow \frac{1}{2}Cl_2(g) + 3H_2O$	1.458

Table 17.1 (*Continued*)

Oxidizing Agent	Reducing Agent	E°_{red}(V)
	Acidic Solution	
$Au^{3+}(aq) + 3e^-$	$\rightarrow Au(s)$	1.498
$MnO_4^-(aq) + 8H^+(aq) + 5e^-$	$\rightarrow Mn^{2+}(aq) + 4H_2O$	1.512
$PbO_2(s) + SO_4^{2-}(aq) + 4H^+(aq) + 2e^-$	$\rightarrow PbSO_4(s) + 2H_2O$	1.687
$H_2O_2(aq) + 2H^+(aq) + 2e^-$	$\rightarrow 2H_2O$	1.763
$Co^{3+}(aq) + e^-$	$\rightarrow Co^{2+}(aq)$	1.953
$F_2(g) + 2e^-$	$\rightarrow 2F^-(aq)$	2.889
	Basic Solution	
$Fe(OH)_2(s) + 2e^-$	$\rightarrow Fe(s) + 2\ OH^-(aq)$	−0.891
$2H_2O + 2e^-$	$\rightarrow H_2(g) + 2\ OH^-(aq)$	−0.828
$Fe(OH)_3(s) + e^-$	$\rightarrow Fe(OH)_2(s) + OH^-(aq)$	−0.547
$S(s) + 2e^-$	$\rightarrow S^{2-}(aq)$	−0.445
$NO_3^-(aq) + 2H_2O + 3e^-$	$\rightarrow NO(g) + 4\ OH^-(aq)$	−0.140
$NO_3^-(aq) + H_2O + 2e^-$	$\rightarrow NO_2^-(aq) + 2\ OH^-(aq)$	0.004
$ClO_4^-(aq) + H_2O + 2e^-$	$\rightarrow ClO_3^-(aq) + 2\ OH^-(aq)$	0.398
$O_2(g) + 2H_2O + 4e^-$	$\rightarrow 4\ OH^-(aq)$	0.401
$ClO_3^-(aq) + 3H_2O + 6e^-$	$\rightarrow Cl^-(aq) + 6\ OH^-(aq)$	0.614
$ClO^-(aq) + H_2O + 2e^-$	$\rightarrow Cl^-(aq) + 2\ OH^-(aq)$	0.890

Clearly, since all of the values of E°_{red} and E°_{ox} are negative, it is impossible to obtain a positive value of E°. You would predict, correctly, that no reaction takes place when Sn^{2+} and Fe^{2+} ions are mixed.

Example 9 What, if anything, happens when nitric acid is added to a solution of NaI?

Strategy First, decide what ions are present in the two solutions. Then tabulate the possible oxidations and reductions, using Table 17.1. Finally, see if any combination of values of E°_{ox} with E°_{red} gives a positive E°.

Solution

(1) Ions present: Na^+, I^-; H^+, NO_3^-

(2) Possible reductions

$Na^+(aq) + e^- \longrightarrow Na(s)$ $\quad E^\circ_{red} = -2.714$ V

$2H^+(aq) + 2e^- \longrightarrow H_2(g)$ $\quad E^\circ_{red} = 0.000$ V

$NO_3^-(aq) + 4H^+(aq) + 3e^- \longrightarrow NO(g) + 2H_2O$ $\quad E^\circ_{red} = +0.964$ V

Possible oxidations

$2I^-(aq) \longrightarrow I_2(s) + 2e^-$ $\quad E^\circ_{ox} = -0.534$ V

(3) Combining E°_{red} for NO_3^- and E°_{ox} for I^- gives

$$E^\circ = +0.964\text{ V} - 0.534\text{ V} = +0.430\text{ V}$$

This reaction should and does occur. When nitric acid is added to a solution of sodium iodide, elementary iodine is formed, giving the solution a reddish color.

REDOX REACTIONS; BALANCING EQUATIONS (Chapter 4)

For most reactions, the chemical equation can be balanced by the "trial and error" method described in Chapter 2. However, for certain redox reactions that approach is tedious at best and in some cases virtually impossible. Consider, for example, the reaction referred to in Example 9:

$$I^-(aq) + NO_3^-(aq) \longrightarrow I_2(s) + NO(g) \qquad \text{(acidic solution)}$$

It is by no means obvious how this equation is to be balanced.

To balance redox equations such as this, a systematic procedure is followed. It consists essentially of three major steps.

1. *Divide the equation into two half-equations, oxidation and reduction.* It is helpful here to assign oxidation numbers. For the reaction between I^- and NO_3^-:

$$\text{oxidation: } I^-(aq) \longrightarrow I_2(s) \qquad \text{oxid. no. I: } -1 \longrightarrow 0$$

$$\text{reduction: } NO_3^-(aq) \longrightarrow NO(g) \qquad \text{oxid. no. N: } +5 \longrightarrow +2$$

2. *Balance each half-equation, first with respect to atoms and then with respect to charge.* Proceed as follows:

(a) Balance the number of atoms of the element oxidized or reduced.
(b) Balance oxygen by adding H_2O molecules.
(c) Balance hydrogen by adding H^+ ions.
(d) Balance charge by adding electrons.

In this case, the oxidation half-equation is readily balanced. First (a) balance iodine by putting two I^- ions on the left:

$$2I^-(aq) \longrightarrow I_2(s)$$

Then (d) balance charge by adding two electrons to the right

$$2I^-(aq) \longrightarrow I_2(s) + 2e^-$$

The reduction half-equation is a bit more complex. Proceeding systematically:

(a) nitrogen is already balanced; 1 N atom on both sides
(b) to balance oxygen, add $2H_2O$ molecules to the right

$$NO_3^-(aq) \longrightarrow NO(g) + 2H_2O$$

(c) to balance hydrogen, add $4H^+$ ions to the left

$$NO_3^-(aq) + 4H^+(aq) \longrightarrow NO(g) + 2H_2O$$

(d) to balance charge, add $3e^-$ to the left

$$NO_3^-(aq) + 4H^+(aq) + 3e^- \longrightarrow NO(g) + 2H_2O$$

3. *Combine the two balanced half-equations in such a way as to eliminate electrons.* Notice that there are $2e^-$ on the right in the oxidation half-equation, $3e^-$ on the left in the reduction half-equation. To eliminate electrons, multiply the first half-equation by three, the second by two, and add

$$3[\ 2I^-(aq) \longrightarrow I_2(s) + 2e^-]$$
$$2[\ NO_3^-(aq) + 4H^+(aq) + 3e^- \longrightarrow NO(g) + 2H_2O]$$
$$6I^-(aq) + 2NO_3^-(aq) + 8H^+(aq) \longrightarrow 3I_2(s) + 2NO(g) + 4H_2O$$

This is the final balanced equation for the redox reaction referred to in Example 9.

If a redox reaction takes place in basic solution, an additional step is required.

4. *Eliminate H^+ ions by adding an equal number of OH^- ions to both sides of the equation.* To illustrate, suppose you are asked to write a balanced equation for the reaction between iodide and nitrate ions in basic solution

$$I^-(aq) + NO_3^-(aq) \longrightarrow I_2(s) + NO(g) \qquad \text{(basic solution)}$$

Starting with the equation in acidic solution derived above, you add 8 OH^- ions to both sides

$$\begin{array}{l} 6I^-(aq) + 2NO_3^-(aq) + 8H^+(aq) \longrightarrow 3I_2(s) + 2NO(g) + 4H_2O \\ \qquad + 8\ OH^-(aq) \qquad\qquad\qquad\qquad + 8\ OH^-(aq) \\ \hline 6I^-(aq) + 2NO_3^-(aq) + 8H_2O \longrightarrow 3I_2(s) + 2NO(g) + 4H_2O + 8\ OH^-(aq) \end{array}$$

Simplify by subtracting $4H_2O$ molecules from both sides to obtain the final balanced equation in basic solution

$$6I^-(aq) + 2NO_3^-(aq) + 4H_2O \longrightarrow 3I_2(s) + 2NO(g) + 8\ OH^-(aq)$$

Example 10 Balance the following redox equation, first in acidic and then in basic solution

$$AsO_3^{3-}(aq) + ClO_3^-(aq) \longrightarrow AsO_4^{3-}(aq) + Cl_2(g)$$

Strategy In acidic solution, follow Steps 1–3 described above. To modify the equation for basic solution, add Step 4.

Solution

(1) oxidation: $AsO_3^{3-}(aq) \longrightarrow AsO_4^{3-}(aq)$
reduction: $ClO_3^-(aq) \longrightarrow Cl_2(g)$

(2) oxidation (b) $AsO_3^{3-}(aq) + H_2O \longrightarrow AsO_4^{3-}(aq)$
(c) $AsO_3^{3-}(aq) + H_2O \longrightarrow AsO_4^{3-}(aq) + 2H^+(aq)$
(d) $AsO_3^{3-}(aq) + H_2O \longrightarrow AsO_4^{3-}(aq) + 2H^+(aq) + 2e^-$
reduction (a) $2ClO_3^-(aq) \longrightarrow Cl_2(g)$
(b) $2ClO_3^-(aq) \longrightarrow Cl_2(g) + 6H_2O$
(c) $2ClO_3^-(aq) + 12H^+(aq) \longrightarrow Cl_2(g) + 6H_2O$
(d) $2ClO_3^-(aq) + 12H^+(aq) + 10e^- \longrightarrow Cl_2(g) + 6H_2O$

(3) $5[AsO_3^{3-}(aq) + H_2O \longrightarrow AsO_4^{3-}(aq) + 2H^+(aq) + 2e^-]$
$2ClO_3^-(aq) + 12H^+(aq) + 10e^- \longrightarrow Cl_2(g) + 6H_2O$

$5AsO_3^{3-}(aq) + 2ClO_3^-(aq) + 2H^+(aq) \longrightarrow 5AsO_4^{3-}(aq) + Cl_2(g) + H_2O$
This is the balanced equation in acidic solution.

(4) Add 2 OH^- ions to each side obtaining, after simplification

$$5AsO_3^{3-}(aq) + 2ClO_3^-(aq) + H_2O \longrightarrow 5AsO_4^{3-}(aq) + Cl_2(g) + 2\ OH^-(aq)$$

This is the balanced equation in basic solution.

Problems

1. Write balanced net ionic equations for the dissolving in water of

a. $(NH_4)_2SO_4$ **b.** K_3PO_4 **c.** $Al(NO_3)_3$
d. $Sc_2(SO_4)_3$ **e.** $Ca(ClO_4)_2$

2. Write balanced net ionic equations for any precipitation reaction that occurs when solutions of the following compounds are mixed.

a. $NiCl_2$ and $AgNO_3$ **b.** $Ba(NO_3)_2$ and H_2SO_4
c. NaOH and $BaCl_2$ **d.** $CuSO_4$ and BaS
e. $Al_2(SO_4)_3$ and KOH

3. Write balanced net ionic equations for the precipitation reactions that occur when ammonia or hydrogen sulfide is added to a solution of

a. $MnCl_2$ **b.** $SbCl_3$

4. Write balanced net ionic equations to explain why solutions of the following compounds are acidic.

a. HCN **b.** NH_4Br **c.** H_3PO_4
d. $Al(NO_3)_3$ **e.** $CoSO_4$

5. Write balanced net ionic equations to explain why solutions of the following compounds are basic.

a. $C_2H_5NH_2$ **b.** NaF **c.** K_2CO_3
d. $Ca(NO_2)_2$ **e.** $(CH_3)_2NH_2$

6. Write balanced net ionic equations for the reaction of hydrochloric acid with

a. a solution of $Ca(OH)_2$ **b.** $Al(OH)_3(s)$
c. a solution of KOH

7. Write balanced net ionic equations for the reaction of a solution of sodium hydroxide with a solution of

a. HClO **b.** NH_4NO_3 **c.** H_3PO_4

8. Write balanced net ionic equations for the reaction of a solution of hydrochloric acid with

a. $NH_3(aq)$ **b.** $NO_2^-(aq)$ **c.** $CO_3^{2-}(aq)$
d. $CuCO_3(s)$

9. Describe the redox reaction, if any, that occurs when nitric acid is added to a solution of

a. $SnCl_2$ **b.** $Zn(NO_3)_2$ **c.** SO_2

10. Write balanced net ionic equations for any reactions taking place in Problem 9.

Answers

1. a. $(NH_4)_2SO_4(s) \longrightarrow 2NH_4^+(aq) + SO_4^{2-}(aq)$
b. $K_3PO_4(s) \longrightarrow 3K^+(aq) + PO_4^{3-}(aq)$
c. $Al(NO_3)_3(s) \longrightarrow Al^{3+}(aq) + 3NO_3^-(aq)$
d. $Sc_2(SO_4)_3(s) \longrightarrow 2Sc^{3+}(aq) + 3SO_4^{2-}(aq)$
e. $Ca(ClO_4)_2(s) \longrightarrow Ca^{2+}(aq) + 2ClO_4^-(aq)$

2. a. $Ag^+(aq) + Cl^-(aq) \longrightarrow AgCl(s)$
b. $Ba^{2+}(aq) + SO_4^{2-}(aq) \longrightarrow BaSO_4(s)$
d. $Cu^{2+}(aq) + S^{2-}(aq) \longrightarrow CuS(s)$
$Ba^{2+}(aq) + SO_4^{2-}(aq) \longrightarrow BaSO_4(s)$
e. $Al^{3+}(aq) + 3\ OH^-(aq) \longrightarrow Al(OH)_3(s)$

3. a. $Mn^{2+}(aq) + 2NH_3(aq) + 2H_2O \longrightarrow Mn(OH)_2(s) + 2NH_4^+(aq)$
$Mn^{2+}(aq) + H_2S(aq) \longrightarrow MnS(s) + 2H^+(aq)$
b. $Sb^{3+}(aq) + 3NH_3(aq) + 3H_2O \longrightarrow Sb(OH)_3(s) + 3NH_4^+(aq)$
$2Sb^{3+}(aq) + 3H_2S(aq) \longrightarrow Sb_2S_3(s) + 6H^+(aq)$

4. a. $HCN(aq) \rightleftharpoons H^+(aq) + CN^-(aq)$
b. $NH_4^+(aq) \rightleftharpoons H^+(aq) + NH_3(aq)$
c. $H_3PO_4(aq) \rightleftharpoons H^+(aq) + H_2PO_4^-(aq)$
d. $Al(H_2O)_6^{3+}(aq) \rightleftharpoons H^+(aq) + Al(H_2O)_5(OH)^{2+}(aq)$
e. $Co(H_2O)_6^{2+}(aq) \rightleftharpoons H^+(aq) + Co(H_2O)_5(OH)^+(aq)$

5. a. $C_2H_5NH_2(aq) + H_2O \rightleftharpoons OH^-(aq) + C_2H_5NH_3^+(aq)$
b. $F^-(aq) + H_2O \rightleftharpoons OH^-(aq) + HF(aq)$
c. $CO_3^{2-}(aq) + H_2O \rightleftharpoons OH^-(aq) + HCO_3^-(aq)$
d. $NO_2^-(aq) + H_2O \rightleftharpoons OH^-(aq) + HNO_2(aq)$
e. $(CH_3)_2NH(aq) + H_2O \rightleftharpoons OH^-(aq) + (CH_3)_2NH_2^+(aq)$

6. a. $H^+(aq) + OH^-(aq) \longrightarrow H_2O$
b. $Al(OH)_3(s) + 3H^+(aq) \longrightarrow Al^{3+}(aq) + 3H_2O$
c. $H^+(aq) + OH^-(aq) \longrightarrow H_2O$

7. a. $HClO(aq) + OH^-(aq) \longrightarrow H_2O + ClO^-(aq)$
b. $NH_4^+(aq) + OH^-(aq) \longrightarrow H_2O + NH_3(aq)$
c. $H_3PO_4(aq) + OH^-(aq) \longrightarrow H_2O + H_2PO_4^-(aq)$

8. a. $H^+(aq) + NH_3(aq) \longrightarrow NH_4^+(aq)$
b. $H^+(aq) + NO_2^-(aq) \longrightarrow HNO_2(aq)$
c. $2H^+(aq) + CO_3^{2-}(aq) \longrightarrow CO_2(g) + H_2O$
d. $2H^+(aq) + CuCO_3(s) \longrightarrow CO_2(g) + H_2O + Cu^{2+}(aq)$

9. a. $Sn^{2+}(aq) + NO_3^-(aq) \longrightarrow Sn^{4+}(aq) + NO(g)$
c. $SO_2(g) + NO_3^-(aq) \longrightarrow SO_4^{2-}(aq) + NO(g)$

10. a. $3Sn^{2+}(aq) + 2NO_3^-(aq) + 8H^+(aq) \longrightarrow 3Sn^{4+}(aq) + 2NO(g) + 4H_2O$
c. $3SO_2(g) + 2NO_3^-(aq) + 2H_2O \longrightarrow 3SO_4^{2-}(aq) + 2NO(g) + 4H^+(aq)$

Index/Glossary

Note: Boldface terms are defined in the context used in the text. Italic page numbers indicate figures; t indicates table; q indicates an end-of-chapter question.